Support within Reach Puts Success w…

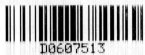

If you were to look into a crystal ball and see your future, you … solving will be a large part of both your professional and pers… questions such as *What should I consider when buying a car? Should I rent or buy a home?* In addition to helping you learn the math you'll need, this text will help you develop problem-solving skills that you'll use every day.

Problem-solving strategies are introduced in Chapter 1, and these techniques and principles form a common theme that is woven throughout the book.

Problem Solving boxes throughout the text recap the specific problem-solving strategies or principles—first introduced in Chapter 1—that are appropriate for the example at hand.

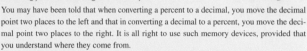

PROBLEM SOLVING

Strategy: The Analogies Principle

You may have been told that when converting a percent to a decimal, you move the decimal point two places to the left and that in converting a decimal to a percent, you move the decimal point two places to the right. It is all right to use such memory devices, provided that you understand where they come from.

Always remember that percent means hundredths. If you understand how to rewrite 0.29 as 29%, then you also know how to rewrite 1.29 as a percent. Similarly, if you know that 19% equals 0.19, then you also know how to rewrite 19.32% as a decimal.

If you want, you can use the following rules:

To convert from a percent to a decimal, drop the percent sign and divide by 100.

To convert from a decimal to a percent, multiply by 100 and add a percent sign.

EXAMPLE 8 **Negotiating a Basketball Contract**

LeBron is negotiating a new contract with the Miami Heat. In order to reduce current taxes, he has agreed to defer a \$15 million bonus to be paid as \$18 million 4 years from now. If the Heat invests the \$15 million now, what is the minimum annual interest rate they need to earn to guarantee that the \$18 million is available in 4 years? Assume that the annual interest rate is compounded monthly.

SOLUTION: Review the Splitting-Hairs Principle on page 12.

To solve this compound interest problem, we will use the formula $A = P(1 + \frac{r}{m})^n$ with $A = 18$, $P = 15$, $m = 12$, and $n = 12 \times 4 = 48$. We will solve for r. Substituting for A, P, m, and n, we solve as follows:

$$18 = 15\left(1 + \frac{r}{12}\right)^{48}$$

$$1.2 = \left(1 + \frac{r}{12}\right)^{48} \quad \text{Divide both sides by 15}$$

$$(1.2)^{\frac{1}{48}} = \left(\left(1 + \frac{r}{12}\right)^{48}\right)^{\frac{1}{48}} = 1 + \frac{r}{12}.^* \quad \text{Raise both sides to the } \tfrac{1}{48}\text{th power}$$

Subtracting 1 from both sides gives us

$$\frac{r}{12} = (1.2)^{\frac{1}{48}} - 1 = 1.003805589 - 1 = 0.003805589.$$

So, $r = 12 \times 0.003805589 \approx 0.046$. Thus the Heat needs to find an investment that pays about 4.6% annual interest compounded monthly.

Now try Exercises 43 and 44.

NEW! Problem-solving voice balloons offer tips for choosing the right problem-solving strategy. These appear in many examples, emphasizing the importance of setting up the problem correctly.

NEW! MyMathLab includes a new exercise type that asks students to choose an appropriate problem-solving strategy before working through a multi-part solution.

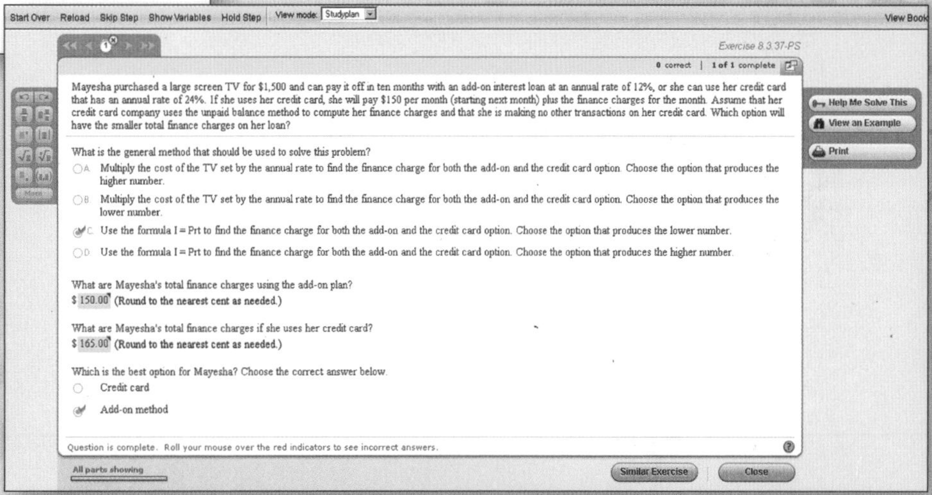

MATHEMATICS
All Around

5TH EDITION

THOMAS L. PIRNOT
Kutztown University of Pennsylvania

PEARSON

Boston Columbus Indianapolis New York San Francisco Upper Saddle River
Amsterdam Cape Town Dubai London Madrid Milan Munich Paris Montréal Toronto
Delhi Mexico City São Paulo Sydney Hong Kong Seoul Singapore Taipei Tokyo

Editorial Director:	Chris Hoag
Editor in Chief:	Anne Kelly
Senior Acquisitions Editor:	Marnie Greenhut
Assistant Editor:	Elle Driska
Senior Managing Editor:	Karen Wernholm
Associate Managing Editor:	Tamela Ambush
Senior Production Project Manager:	Sheila Spinney
Art Director:	Beth Paquin
Digital Assets Manager:	Marianne Groth
Media Producer:	Nicholas Sweeny
Software Development:	Eileen Moore and Mary Durnwald
Executive Marketing Manager:	Roxanne McCarley
Marketing Assistant:	Caitlin Crain
Procurement Manager:	Evelyn Beaton
Procurement Specialist:	Linda Cox
Media Procurement Specialist:	Ginny Michaud
Cover Design:	StudioWink
Text Design:	Infiniti
Production Coordination, Composition, and Illustrations:	Integra Software Services
Cover photo:	Rubik's Cube® used by permission of Seven Towns Ltd. www.rubiks.com/Photo by Pearson Education, Inc.

Library of Congress Cataloging-in-Publication Data

Pirnot, Thomas L.
Mathematics all around / Thomas L. Pirnot, Kutztown University of Pennsylvania.—Fifth edition.
pages cm
Includes bibliographical references and index.
ISBN 10: 0-321-83699-5 (student edition)
ISBN 13: 978-0-321-83699-1 (student edition)
1. Mathematics—Textbooks. I. Title.
QA39.3.P57 2014
510—dc23

2012022670

3 4 5—V011—14 13

PEARSON

www.pearsonhighered.com

ISBN 10: 0-321-83699-5
ISBN 13: 978-0-321-83699-1

DEDICATION

With gratitude to my wife, Ann, whose love and support
are a continuing source of encouragement to me

Contents

Preface ix

1 Problem Solving
Strategies and Principles 1

1.1 Problem Solving 2
1.2 Inductive and Deductive Reasoning 18
1.3 Estimation 28
Chapter Summary 38
Chapter Review Exercises 38
Chapter Test 39

2 Set Theory
Using Mathematics to Classify Objects 42

2.1 The Language of Sets 43
2.2 Comparing Sets 50
2.3 Set Operations 57
2.4 Survey Problems 67
2.5 **Looking Deeper:** Infinite Sets 76
Chapter Summary 81
Chapter Review Exercises 81
Chapter Test 83

3 Logic
The Study of What's True or False or Somewhere in Between 85

3.1 Statements, Connectives, and Quantifiers 86
3.2 Truth Tables 95
3.3 The Conditional and Biconditional 106
3.4 Verifying Arguments 115
3.5 Using Euler Diagrams to Verify Syllogisms 123
3.6 **Looking Deeper:** Fuzzy Logic 130
Chapter Summary 137
Chapter Review Exercises 138
Chapter Test 139

4 Graph Theory (Networks)
The Mathematics of Relationships 142

4.1 Graphs, Puzzles, and Map Coloring 143
4.2 The Traveling Salesperson Problem 157
4.3 Directed Graphs 168
4.4 **Looking Deeper:** Scheduling Projects Using PERT 175
Chapter Summary 184
Chapter Review Exercises 185
Chapter Test 186

5 Numeration Systems
Does It Matter How We Name Numbers? 190

5.1 The Evolution of Numeration Systems 191
5.2 Place Value Systems 200
5.3 Calculating in Other Bases 209
5.4 **Looking Deeper:** Modular Systems 221
Chapter Summary 229
Chapter Review Exercises 230
Chapter Test 230

6 Number Theory and the Real Number System
Understanding the Numbers All Around Us 232

6.1 Number Theory 233
6.2 The Integers 245
6.3 The Rational Numbers 253
6.4 The Real Number System 265
6.5 Exponents and Scientific Notation 275
6.6 **Looking Deeper:** Sequences 285
Chapter Summary 295
Chapter Review Exercises 297
Chapter Test 298

7 Algebraic Models
How Do We Approximate Reality? 300

7.1 Linear Equations 301
7.2 Modeling with Linear Equations 313
7.3 Modeling with Quadratic Equations 321
7.4 Exponential Equations and Growth 330
7.5 Proportions and Variation 340
7.6 Modeling with Systems of Linear Equations and Inequalities 348
7.7 **Looking Deeper:** Dynamical Systems 364
Chapter Summary 371
Chapter Review Exercises 372
Chapter Test 374

8 Consumer Mathematics
The Mathematics of Everyday Life 378

8.1 Percents, Taxes, and Inflation 379
8.2 Interest 388
8.3 Consumer Loans 399
8.4 Annuities 408
8.5 Amortization 416
8.6 **Looking Deeper:** Annual Percentage Rate 424
Chapter Summary 431
Chapter Review Exercises 433
Chapter Test 434

9 Geometry
Ancient and Modern Mathematics Embrace 436

9.1 Lines, Angles, and Circles 437
9.2 Polygons 446
9.3 Perimeter and Area 456
9.4 Volume and Surface Area 468
9.5 The Metric System and Dimensional Analysis 477
9.6 Geometric Symmetry and Tessellations 488
9.7 **Looking Deeper:** Fractals 500
Chapter Summary 509
Chapter Review Exercises 511
Chapter Test 512

10 Apportionment
How Do We Measure Fairness? 516

10.1 Understanding Apportionment 517
10.2 The Huntington–Hill Apportionment Principle 525
10.3 Other Paradoxes and Apportionment Methods 534
10.4 **Looking Deeper:** Fair Division 548
Chapter Summary 556
Chapter Review Exercises 557
Chapter Test 558

11 Voting
Using Mathematics to Make Choices 561

11.1 Voting Methods 562
11.2 Defects in Voting Methods 572
11.3 Weighted Voting Systems 583
11.4 **Looking Deeper:** The Shapley-Shubik Index 592
Chapter Summary 600
Chapter Review Exercises 601
Chapter Test 602

12 Counting
Just How Many Are There? 605

12.1 Introduction to Counting Methods 606
12.2 The Fundamental Counting Principle 614
12.3 Permutations and Combinations 622
12.4 **Looking Deeper:** Counting and Gambling 635
Chapter Summary 640
Chapter Review Exercises 641
Chapter Test 641

13 Probability
What Are the Chances? 643

13.1 The Basics of Probability Theory 644
13.2 Complements and Unions of Events 659
13.3 Conditional Probability and Intersections of Events 668

13.4 Expected Value 682
13.5 **Looking Deeper**: Binomial Experiments 691
Chapter Summary 697
Chapter Review Exercises 698
Chapter Test 699

14 **Descriptive Statistics**
What a Data Set Tells Us 701

14.1 Organizing and Visualizing Data 702
14.2 Measures of Central Tendency 714
14.3 Measures of Dispersion 727
14.4 The Normal Distribution 738
14.5 **Looking Deeper**: Linear Correlation 751
Chapter Summary 759
Chapter Review Exercises 760
Chapter Test 761

Appendix A 764

Answers to Quiz Yourself Problems 767

Answers to Exercises 775

Credits C1

Index I1

Preface

The fifth edition of *Mathematics All Around* builds on the success of earlier editions, each of which has been described by instructors and students as a readable, interesting, and relevant textbook for *today's* liberal arts mathematics students.

Because students live in an increasingly complex world where both information and, unfortunately, misinformation are growing rapidly, it is important that they learn to think critically and become comfortable in the numerical world all around them. This text stresses problem-solving strategies and principles along with the philosophy that students are most successful when they rely on understanding rather than memorization.

The text emphasizes intuitive thinking and visualization to help students understand and remember material more easily. Based on user and reviewer suggestions, I have refined explanations, diagrams, and annotations. As a result, students will find the many interesting topics in the text understandable and easier to master. Many examples have been revised and new examples written to make the overall flow of ideas clearer and more student-friendly. Instructors who have used *Mathematics All Around* have found that students respond enthusiastically to the realistic, modern applications presented throughout the text.

As with previous editions, this new edition provides students majoring in the liberal arts, the social sciences, education, business, and other nonscientific areas an understanding and appreciation of mathematics, the contributions made to mathematics by diverse cultures, and its many fascinating applications. *Mathematics All Around* is particularly appropriate for students needing to satisfy a one- or two-course requirement in mathematics for graduation or to transfer to another institution.

Content Changes

In every chapter I have added and revised examples, clarified explanations, simplified computations, and increased annotations. I have spent numerous hours researching, updating, revising, and adding hundreds of new exercises. Many chapter and section openers have been rewritten to make them more current and relevant to students' lives. Here is a sampling of the exciting content changes in this fifth edition

Chapter 1 New and lively examples illustrate the **problem-solving** strategies and principles that are central to this text. Visualization in problem solving is emphasized. New examples such as course scheduling, the dangers of investing in economic bubbles, and the importance of estimation in understanding current affairs accentuate the role of problem solving in students' lives.

Chapter 2 **Set theory** is illustrated with fresh topics such as social media, Oscar-winning movies, classifying blood types, the price of gasoline, immigration, and Olympic competition. The role of set theory in organizing corporate databases and protecting against terrorism and identity theft is stressed.

Chapter 3 Many solutions to examples in the **logic** chapter have been simplified to improve clarity and reduce calculations, particularly in the section on verifying arguments. New examples and exercises involve analyzing the mistakes candidates make during job interviews, understanding the fine print of credit card agreements, and some delightful new Harry Potter variations on the classic "knight and knaves" puzzles.

Chapter 4 The section on PERT scheduling in **graph theory** has been consolidated to improve its flow. Examples and exercises investigate the mathematics of social media, planning job interviews, scheduling auditions for *America's Got Talent*, and ranking *CrossFit* competitors. Directed graphs are shown using the model of how a virus spreads and how to use technology to solve large-scale instances of this problem.

Chapter 5 New examples and exercises trace **numeration** from ancient times to the exotic mathematics of today's numeration systems—systems that are the basis not only for Bart Simpson's arithmetic homework but also for students' smart phones, Kindles, and tablet computers. The new *Between the Numbers* highlight and exercises explain the role modular arithmetic plays in preventing identity theft.

Chapter 6 The flow of discussion on the properties of the **real number system** has been streamlined. New examples and exercises on scientific notation discuss topics such as the relationship among defense and Medicare spending and the gross domestic product, the individual's share of the national debt, the number of computations required to produce *Toy Story 3*, and the danger of firing a gun in the air to celebrate at midnight on New Year's Eve.

Chapter 7 Using current data and relevant **algebraic models,** new material encourages students to think critically about claims they see in popular media. Chapter 7 now includes material on systems of equations and inequalities. Examples and exercises discuss the increase in digital music sales, the soundness of the Social Security system, the growth of college faculty and administrative salaries, and the revenues of companies such as Apple and Nike.

Chapter 8 Many examples in **consumer mathematics** have been simplified. The computations on refinancing loans have been greatly streamlined, new material on inflation has been included, and the use of current technology is encouraged. New examples explain important financial topics like rising tuition rates, the inflation of cookie prices, the difficulty of paying off excessive credit card debt, refinancing loans, and the importance of saving early to achieve long-term goals. Financial traps such as excessive student debt, payday loans, and adjustable rate mortgages are analyzed. Consumer mathematics examples continue to use notation consistent with TI-83 Plus and TI-84 Plus calculators, and Using Technology exercises now show students how to use TI's **TVM Solver.**

Chapter 9 Examples have been updated and simplified. They include many new and fascinating applications of **geometry** such as why bees build their hives as they do, how to use geometry to create hidden inflation, and the total area of all pizzas delivered on Super Bowl Sunday. Interesting discussions of non-Euclidean and fractal geometry and their applications are included.

Chapters 10 and 11 The discussion of **apportionment** has been streamlined to improve its clarity. Examples and exercises have been updated in both chapters. The power of spreadsheet software like Microsoft Excel® and Apple Numbers® shows students how to simplify tedious apportionment and **voting** computations. The new *Between the Numbers* highlights and corresponding exercises in Chapter 11 encourage students to investigate viable alternatives to our current plurality voting method.

Chapter 12 Several **counting** examples have been rewritten and simplified to improve clarity. Numerous new exercises have been added and the *Between the Numbers* highlight and corresponding exercises show students the folly of relying on winning a lottery to secure their financial futures.

Chapter 13 Many examples of **probability** have been revised or replaced with more current content. New examples and exercises discuss a wide range of applications such as the odds of surviving an airplane crash, the prospect of passing a multiple-choice quiz by guessing, the likelihood that a person testing positive on a test for illegal drugs is actually innocent, and an exploration of how casino owners and insurance companies acquire their vast wealth.

Chapter 14 The **statistics** chapter has been revised using extensive research. Numerous new examples and exercises use real, current data. There is now a greater emphasis on interpreting data graphically. Highlights and end-of-chapter exercises demonstrate how to use technology when doing statistical analysis. The Using Technology exercises in Chapter 14 illustrate Excel's built-in statistical functions as well as its **Analysis ToolPak** to simplify computations.

New and Continuing Pedagogical Features

ENHANCED!

Format

The format makes this text easy to use and makes learning mathematics easier for the student. I introduce ideas conversationally and then illustrate them with clear, simple examples. I have included numerous **visual explanations** to help students "see" the mathematics that they are learning. Carefully annotated diagrams and the effective use of color help reinforce the written explanations of numerous topics in this edition. After reinforcing highlighted points with more examples, students test their understanding by working through the **Quiz Yourself** problems. More detailed examples occur later in a section after the ideas have been clearly discussed. Summary boxes help students locate information when doing exercises.

UPDATED!

Motivation with Emphasis on Applications

Applications throughout the text motivate the discussion of mathematics and increase students' interest in the material. Naturally, I have updated and revised the applications to include more current and fresh ideas so they are of interest to today's students. Chapter openers give an overview of the chapter and also introduce realistic situations relevant to students' lives that motivate the mathematical tools developed in the chapter. I later return with examples of these situations, reinforcing the value of the concepts studied in the chapter. Many section openers have also been rewritten to introduce the concepts in a way students find more interesting.

Objectives

Objectives appear at the beginning of each section to give students a clear idea of what they are going to learn in each section. In addition, reviewing these objectives after reading the section is an excellent way of checking comprehension of the key ideas just covered.

ENHANCED!

Problem Solving

Problem solving continues to be one of the main themes of this text. Section 1.1 discusses strategies and principles that help the student understand and attack problems more effectively. These strategies and principles are strongly reinforced throughout the text with frequent **problem-solving reminders** in the example solutions and also in **Problem-Solving boxes** that review the methods from Chapter 1.

Some Good Advice

In addition to the problem-solving boxes, there are numerous boxes titled **Some Good Advice,** which point out common mistakes, provide further advice on problem solving, and make connections between different areas of mathematics. Both instructors and students appreciate these boxes.

Math in Your Life

Each chapter contains one or more **Math in Your Life** boxes that connect the mathematics to the various majors of students in this course, as well as to outside the classroom. These connections make the mathematics more interesting to learn and easier to understand contextually, as well as highlight the usefulness of mathematics in everyday life.

Quiz Yourself

Each section contains numerous short quizzes called **Quiz Yourself,** which students can use to check their understanding of the material immediately preceding the quiz. These quizzes can be used as a break in the flow of the lecture material and to encourage active learning. A special symbol, 10, indicating when students should try a **Quiz Yourself**

exercise, has been included at the end of appropriate examples to provide students with further assistance while studying the material.

UPDATED!

Highlights

Each chapter contains **Highlight** boxes that discuss the history and practical applications of the topics being presented. These highlights help students understand the material that they are learning in a broader context. A listing of the historical figures mentioned in the text is located at the end of the book and includes individuals such as Aristotle, George Polya, Sophie Germain, and Georg Cantor. Also in the **Highlight** boxes, I encourage students to take advantage of available technology by explaining how to use tools such as graphing calculators and spreadsheets effectively to solve problems.

NEW!

Between the Numbers highlights encourage students to think critically about interpreting numerical information that is relevant to their lives and have been tightly integrated into the text. Examples of these include the dangers of adjustable rate mortgages and excessive credit card debt, the reliability of crowd estimates at a rally, the soundness of the Social Security system, and understanding what graphs do (or do not) tell you. **Between the Numbers Exercises** reinforce ideas developed in the highlight.

Looking Deeper

The last section of each chapter is titled **Looking Deeper,** which discusses an enrichment topic that goes beyond the main body of material in the chapter. This material is included in the Chapter Summary, and tested in the Chapter Review Exercises and the Chapter Test. Topics include fuzzy logic, linear programming, fractals, and counting and gambling.

UPDATED!

Exercises

Exercise sets have undergone a major revision to reduce repetition, enhance variety, and emphasize current applications. Over 800 exercises are either new or updated.

- Students can use **Sharpening Your Skills** exercises to hone their newly acquired skills.
- For **Applying What You've Learned** exercises, I have researched hundreds of sources to vary, update, and enrich these application-oriented exercises with real, current data.
- **Communicating Mathematics** asks the student to write about the mathematics that he or she is learning, and have been enhanced based on reviewers' recommendations.
- To encourage students to read the *Between the Numbers* boxes, **Between the Numbers exercises** have been included in their own category so instructors can assign them.
- More rigorous exercises are grouped in the **Challenge Yourself** category.
- **Using Technology** exercises have been moved to the end of the chapter and are more detailed and comprehensive, showing students how to use both **TI calculators** and **Microsoft's Excel** spreadsheets to solve problems. Mention is also made of how to use mobile technology apps and **Apple's Numbers**® spreadsheet. Technology resources are available through **MyMathLab**®.
- The end-of-chapter **Group Projects** have been upgraded and are now more comprehensive.

Also, more examples are cross-referenced with exercises to help students in doing their homework.

UPDATED!

Summary, Chapter Review Exercises, and Chapter Test

Each chapter has a summary followed by **Chapter Review Exercises** and a **Chapter Test**. I have tightly coordinated the **Chapter Summary** with the chapter objectives and included cross-references in the Chapter Summary to help students locate relevant examples for review more quickly.

The **Chapter Review Exercises** are also cross-referenced to the sections that cover the topics being tested. This enables a student who is not comfortable with a particular concept to return to the section covering that material. The **Chapter Test** then checks students' knowledge of the chapter material by providing a variety of questions that are not referenced to specific sections. Chapter Test problems are presented in a random order to better test students' understanding and ability to do these problems. Several group projects follow each Chapter Test.

Supplements

Supplements for the Instructor

Annotated Instructor's Edition

ISBN 0-321-83727-4 / 978-0-321-83727-1

The Annotated Instructor's Edition contains answers to most exercises on the page where they occur. Answers that do not fit on the page are in the back of the book.

The following supplements are ONLINE ONLY and are available for download in the Pearson Higher Education catalog or inside your MyMathLab course at www.pearsonhighered.com/irc.

Instructor's Solutions Manual

James Lapp

This manual contains detailed, worked-out solutions to all the exercises in the text.

Instructor's Testing Manual

The Testing Manual includes two tests per chapter. These items may be used as actual tests or as references for creating tests.

Insider's Guide

Developed by instructors, for instructors, to aid faculty with course preparation. Included are helpful teaching tips correlated to each section of the text, as well as additional resources for classroom enrichment.

PowerPoint® Lecture Slides

These fully editable lecture slides include definitions, key concepts, and examples for use in a lecture setting and are available for each section of the text.

TestGen®

TestGen (**www.pearsonhighered.com/testgen**) enables instructors to build, edit, print, and administer tests using a computerized bank of questions developed to cover all the objectives of the text. TestGen is algorithmically based, allowing instructors to create multiple but equivalent versions of the same question or test with the click of a button. Instructors also can modify test bank questions or add new questions. Tests can be printed or administered online. The software and test bank are available for download from Pearson Education's online catalog.

Supplements for the Student

Student Edition

ISBN 0-321-83699-5 / 978-0-321-83699-1

Student's Solutions Manual

James Lapp

ISBN 0-321-83737-1 / 978-0-321-83737-0

This manual contains detailed, worked-out solutions to all the odd-numbered section exercises and to all Chapter Review and Chapter Test exercises.

Video Lectures Online

Video lectures in MyMathLab make it easy and convenient for students to watch the videos from anywhere and are ideal for distance learning or supplemental instruction. Videos include optional subtitles.

Media Supplements

MyMathLab Online Course (access code required)

MyMathLab delivers **proven results** in helping individual students succeed.

- MyMathLab has a consistently positive impact on the quality of learning in higher education math instruction. MyMathLab can be successfully implemented in any environment—lab based, hybrid, fully online, traditional—and demonstrates the quantifiable difference that integrated usage has on student retention, subsequent success, and overall achievement.
- MyMathLab's comprehensive online gradebook automatically tracks your students' results on tests, quizzes, homework, and in the study plan. You can use the gradebook to quickly intervene if your students have trouble, or to provide positive feedback on a job well done. The data within MyMathLab is easily exported to a variety of spreadsheet programs, such as Microsoft Excel. You can determine which points of data you want to export, and then analyze the results to determine success.

MyMathLab provides **engaging experiences** that personalize, stimulate, and measure learning for each student.

- **Exercises:** The homework and practice exercises in MyMathLab are correlated to the exercises in the textbook, and they regenerate algorithmically to give students unlimited opportunity for practice and mastery. The software offers immediate, helpful feedback when students enter incorrect answers.
- **Multimedia Learning Aids:** Exercises include guided solution, sample problems, animations, videos, and eText clips for extra help at point-of-use.
- **Expert Tutoring:** Although many students describe the whole of MyMathLab as "like having your own personal tutor," students using MyMathLab do have access to live tutoring from Pearson, from qualified math and statistics instructors.

And, MyMathLab comes from a **trusted partner** with educational expertise and an eye on the future.

- Knowing that you are using a Pearson product means knowing that you are using quality content. That means that our eTexts are accurate and our assessment tools work. Whether you are just getting started with MyMathLab, or have a question along the way, we're here to help you learn about our technologies and how to incorporate them into your course.

To learn more about how MyMathLab combines proven learning applications with powerful assessment, visit **www.mymathlab.com** or contact your Pearson representative.

New to the MyMathLab course

- Problem-solving reinforced with stepped problems asking students to indentify the best problem-solving method first before working the problem.
- New vocabulary quizzes
- Assignable Quiz Yourself problems from the text
- Lecture video assessment questions

MyMathLab Ready to Go Course (access code required)

These new Ready to Go courses provide students with all the same great MyMathLab features that you're used to, but make it easier for instructors to get started. Each course includes pre-assigned homeworks and quizzes to make creating your course even simpler. Ask your Pearson representative about the details for this particular course or to see a copy of this course.

MathXL® Online Course (access code required)

MathXL is the homework and assessment engine that runs MyMathLab. (MyMathLab is MathXL plus a learning management system.)

With MathXL, instructors can

- Create, edit, and assign online homework and tests using algorithmically generated exercises correlated at the objective level to the textbook.
- Create and assign their own online exercises and import TestGen tests for added flexibility.
- Maintain records of all student work tracked in MathXL's online gradebook.

With MathXL, students can

- Take chapter tests and receive personalized study plans and/or personalized homework assignments based on their test results.
- Use the study plan and/or the homework to link directly to tutorial exercises for the objectives they need to study.
- Access supplemental animations and video clips directly from selected exercises.

MathXL is available to qualified adopters. For more information, visit our website at **www.mathxl.com,** or contact your Pearson representative.

Acknowledgments

It has been a rewarding experience to work with the many wonderfully talented people who have helped me in so many ways in writing this edition. In particular, I thank Marnie Greenhut, senior acquisitions editor at Pearson, who oversaw the project and provided the guidance to reshape this edition, and my assistant editor, Elle Driska, who gave wonderful advice and encouragement to bring this edition to fruition while also handling the myriad of administrative details. I appreciate the work of Sheila Spinney, who supervised production of the text, and Beth Paquin, art director who has produced a beautiful and elegant design. Thanks also to Roxanne McCarley for continuing to spread the message of *Mathematics All Around*. Nicholas Sweeny has done a wonderful job producing and managing all of the fine media supplements for the text, including MyMathLab.

I also greatly appreciate Greg Tobin's continuing confidence in me and support for my books over the years.

I would like to thank James Lapp, who wrote the solutions manuals and corrected my errors for this edition. I greatly appreciate the careful work of Joanna Eichenwald and John Samons in checking for accuracy in the text and exercise answers.

My wife, Ann, and my children, Matt, Tony, Joanna, and Mike, continue to be a great source of encouragement and support for me in my writing.

Finally, last, but certainly not least, I thank all my reviewers and users of previous editions for helping me to continue improving *Mathematics All Around*, and reshaping it into a better text. I always listen carefully to your generous advice and appreciate your many constructive suggestions. The following is a list of reviewers of all editions (reviewers of the fifth edition are marked with an "*").

Gisela Acosta, *Valencia Community College*
Randall Allbritton, *Daytona Beach Community College*
Seth Armstrong, *Southern Utah University*
James Arnold, *University of Wisconsin*
Carmen Q. Artino, *The College of Saint Rose*
Bruce Atkinson, *Samford University*
John Atkinson, *Missouri Western State College*
Wayne Barber, *Chemetka Community College*
René Barrientos, *Miami Dade Community College–Kendall*
Thomas Bartlow, *Villanova University*
Linda Barton, *Ball State University*
Kathleen Bavelas, *Manchester Community Technical College*
Kari Beaty, *Midlands Technical College–Airport*
*Molly Beauchman, *Yavapai College*
David Behrman, *Somertset Community College*
Norma Biscula, *University of Maine–Augusta*
Terence R. Blows, *Northern Arizona University*
*Phyllis Bolin, *Lillere Cheonan University*
Warren S. Butler, *Daytona Beach Community College*
Marc Campbell, *Daytona Beach Community College*
*Deborah Casson, *University of New Mexico*
*Matthew Cathey, *Wofford College*
David Cochener, *Austin Peay State University*
Linda F. Crabtree, *Metropolitan Community College*
Daniel Cronin, *New Hampshire Technical College*
Debra Curtis, *Bloomfield College*
Walter Czarnec, *Framingham State College*
Greg Dietrich, *Florida Community College at Jacksonville*
Margaret M. Donlan, *University of Delaware*
Cameron B. Douthitt, *University of Colorado at Boulder*
Gina Poore Dunn, *Lander University*
John Emert, *Ball State University*
*Dan Endres, *University of Central Oklahoma*
Catherine Ferrer, *Valencia Community College*
Tom Foley, *Glendale Community College*
Olivia Garcia, *University of Brownsville*
John Gaudio, *Waubonsee Community College*
John Gimbel, *University of Alaska–Fairbanks*
Gail Gonyo, *Adirondack Community College*
Donald R. Goral, *Northern Virginia Community College*
Glen Granzow, *Idaho State University*
Tom Greenwood, *Bakersfield College*
Joshua Guest, *University of Mississippi*
Philip Gustafson, *Mesa State College*
Cliff A. Harris, *University of New Mexico*
Ward Heilman, *Bridgewater State College*
Nyeita Irish-Schult, *St. Petersburg Junior College*
Stacey Jones, *Benedict College*
Anne Jowsey, *Niagara County Community College*

Vicky Kauffman, *University of New Mexico*
*Dan Kleinfelter, *College of Desert*
Danny T. Lau, *Gainesville State College*
Kathryn Lavelle, *Westchester Community College*
Ana Leon, *Lexington Community College*
Maita Levine, *University of Cincinnati*
Rhonda MacLeod, *Florida State University*
Mahdi Majidi-Zolbanin, *University of Illinois at Springfield*
Rafael Marino, *Nassau Community College*
Anne Martincic, *McHenry County College*
Karen L. McLaren, *University of Maryland–College Park*
Kenneth Myers, *Bloomfield College*
Bill Naegele, *South Suburban College*
Nancy H. Olson, *Johnson County Community College*
John L. Orr, *University of Nebraska–Lincoln*
Joanne Peeples, *El Paso Community College*
Scott Reed, *College of Lake County*
Nancy Ressler, *Oakton Community College*
Nolan Rice, *College of Southern Idaho*
*Lucille Cook Roth, *Wake Technical Community College*
*Lisa Rombes, *Washtenaw Community College*
Len Ruth, *Sinclair Community College*
Gene Schlereth, *University of Tennessee–Chattanooga*
Linda Schultz, *McHenry County College*
Daniel Seabold, *Hofstra University*
Mary Lee Seitz, *Erie Community College–City Campus*
Kara Shavo, *Mercer County Community College*
Mark Sherrin, *Flagler College*
Marguerite Smith, *Merced College*
Jamal Tartir, *Youngstown State University*
Beverly Taylor, *Valencia Community College*
William Thistleton, *State University of New York–Utica/Rome*
Mark Tom, *College of the Sequoias*
Robert R. Urbanski, *Middlesex County College*
Rebecca Walls, *West Texas A & M University*
Sheela L. Whelan, *Westchester Community College*
Carol S. White, *Flagler College*
John L. Wisthoff, *Anne Arundel Community College*
Robert Woods, *Broome Community College*
Rob Wylie, *Carl Albert State University*
Marvin Zeman, *Southern Illinois University*

High School Reviewers

Courtney Carpenter, *Albemarle High School*
David Emanuel, *Prunell Swett High School*
Cathy Hicks, *Wakefield High School*
Scott Maxwell, *Athens Drive High School*

I welcome comments and suggestions from users of this book and if you wish to contact me, my e-mail address is tompirnot@gmail.com.

Tom Pirnot,
Kutztown, PA

To the Student

How to Succeed at Mathematics

I would like to suggest to you some things that you can do before, during, and after class to become a more successful mathematics student.

Prepare ahead.

Although it takes discipline to do so, you will get much more out of your mathematics class if you are able to read the material (even briefly) before it is covered in class. Reading ahead will make lectures more meaningful and questions might occur to you that you would not have thought to ask had you not read the section in advance. You should use the Key Points that are highlighted in the margins at the beginning of each subsection to get an overview of the topics covered before trying to read the section more carefully.

Read slowly and carefully.

Also remember that reading a math book is very different from reading a textbook in other subjects, such as history, sociology, or music. Because mathematical language is very compressed, a few symbols or words usually contain a lot of information. By paying careful attention to the exact meaning of the definitions and notation being used, you will increase your understanding of mathematics. As you read your book, always try to be actively involved with the material. After reading an example, cover the solution and then try to rework the example yourself. If you cannot work all of the way through, restudy the example to see what points you are missing and try again.

Try to understand the big picture.

I cannot emphasize strongly enough that mathematics is much more than a list of facts, formulas, and equations to be memorized. The more that you can see the big picture of how the material is developed, the better your overall understanding of mathematics will be. If your instructor makes comments in class about the reasons for doing a particular example, or how one topic relates to another, be sure to get those comments in your notes.

Remember how to remember.

Although it is important to remember the basic material for a test, you will find that memorizing without understanding is not very useful. Throughout this text, we will use many diagrams, analogies, and examples to help you understand mathematical terminology, formulas, equations, and solution methods. If you work at understanding the intuitive meaning behind mathematical concepts, those ideas will become more permanent in your memory. We call this process "Remembering How to Remember," and you will find this method of studying an effective way to retain information and hone your math skills.

Do homework intelligently.

When doing exercises, first work on those labeled Sharpening Your Skills, which begin each exercise set. These exercises closely follow the order of the topics presented in the section and will reinforce your understanding of the overall plan of the section.

Students sometimes tell me that although they understand the lectures and do the homework, they still are not successful on exams. I believe that this occurs because they are too book-dependent; that is, they can only do a certain type of problem when they know where it is placed in the book. I have helped many such students improve their grades dramatically by encouraging them to make 3×5 flash cards. To make these cards, choose typical problems from each section. Put a question on one side of a card and its solution on the back. Then when studying, shuffle these cards and give yourself a test so

that the questions are coming randomly. This is a great help in finding out whether you can do problems on your own or can only do them when you are looking at a particular section of the text.

Study with a friend.

It is a great study aid if you can team up with other members of your class to study before quizzes and exams. Other students in your class are hearing the same lectures and will understand your instructor's point of view on the material. You can quiz each other by using the flash cards and the Sharpening Your Skills exercises. I have seen many students' grades improve once they start working in study groups.

I sincerely hope that you enjoy reading this book as much as I have enjoyed writing it. Good luck in your mathematical studies!

1

Problem Solving

Strategies and Principles

We are currently preparing students for jobs that don't exist, using technologies that haven't been invented, in order to solve problems that we don't even know are problems yet.*

If you were to look into a crystal ball and see your future, you would see that problem solving will be a large part of both your professional and personal life. However, the problems you encounter will be quite different from the problems that you learn to solve in this mathematics course or any other mathematics course.

As a student I took dozens of mathematics courses, but the real-life problems that I had to solve were very much unlike the ones appearing in my textbooks. Friends, family, and colleagues have asked me for help regarding safely taking body-building supplements to avoid being poisoned, determining whether an oil delivery person was cheating them, organizing the allocation of graduate students on my campus, building a large wooden *(continued)*

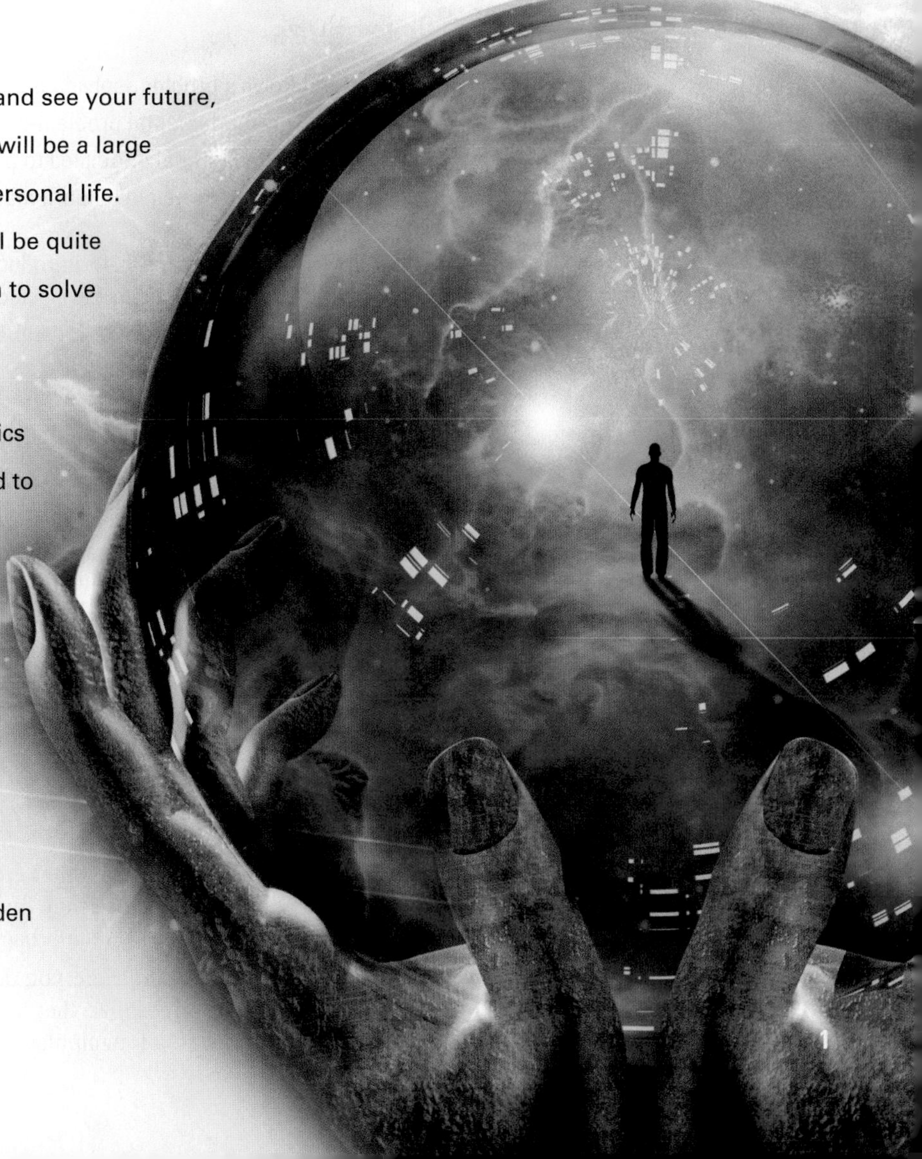

*"Did You Know," YouTube Video

star on top of a church, planning for retirement, ordering concrete for an oddly shaped basketball court, negotiating a labor contract, and so on—none of which were anything like the problems I had seen.

Right now you may have some problems to solve in your own life. Should you live on campus next semester, or in an apartment, or at home? If you rent an apartment how are you going to furnish it? How about selecting a career? Does it make more sense to buy or rent your textbooks? After you graduate, what kind of health insurance (if any) do you want to buy? What factors should you consider in buying a car?

To solve your problems you first have to organize them. Later in this chapter we will show you how you can organize a problem by giving you a simple and systematic technique for constructing a course schedule.

Although it is unlikely that you will see the exact same problems that you are required to solve in this course, understanding the techniques and principles that you learn in this chapter will improve your ability to analyze and solve the real problems that do appear in your life.

1.1 Problem Solving

Objectives

1. Understand Polya's problem-solving method.
2. State and apply fundamental problem-solving strategies.
3. Apply basic mathematical principles to problem solving.
4. Use the Three-Way Principle to learn mathematical ideas.

My goal in writing this section is to introduce you to some practical techniques and principles that will help you to solve many personal and professional problems in your life—such as whether to buy or lease a car, borrow money for graduate school, or organize a large class project. You will find that although real-life problems are often more complex than those found in this text, by mastering the techniques presented here, you will increase your ability to solve problems throughout your life. It is important to remember that you cannot rush becoming a good problem solver. Like anything else in life, the more you practice problem solving, the better you become at it.

Much of the advice presented in this section is based on a problem-solving process developed by the eminent Hungarian mathematician George Polya (see the historical highlight at the end of this section). We will now outline Polya's method.

George Polya's Problem-Solving Method

KEY POINT

George Polya developed a four-step problem-solving method.

Step 1: Understand the problem. It would seem unnecessary to state this obvious advice, but yet in my years of teaching, I have seen many students try to solve a problem before they completely understand it. The techniques that we will explain shortly will help you to avoid this critical mistake.

Step 2: Devise a plan. Your plan may be to set up an algebraic equation, draw a geometric figure, or use some other area of mathematics that you will learn in this text. The plan you choose may involve a little creativity because not all problems succumb to the same approach.

Step 3: Carry out your plan. Here you do what many often think of as "doing mathematics." However, realize that steps 1 and 2 are at least as important as the mechanical process of manipulating numbers and symbols to get an answer.

Math in Your Life

Problem Solving and Your Career

It may seem to you that some of the problems that we ask you to solve in this book are artificial and of no practical use to you. However, consider the following question that was asked of a prospective employee during a job interview. "How many quarters—placed one on top of the other—would it take to reach the top of the Empire State Building?" Although this question may strike you as strange and unrelated to your qualifications for getting a job, according to an article at monster.com, such puzzle-type questions can play a critical role in determining who is hired for an attractive, well-paying job and who is not. What is important about this question is not the answer, but rather the applicant's ability to display creative problem-solving techniques in a pressure situation.

If you are interested in learning more about how your ability to think creatively in solving puzzles can affect your future job prospects, see the book *How Would You Move Mount Fuji? Microsoft's Cult of the Puzzle—How the World's Smartest Company Selects the Most Creative Thinkers* by William Poundstone.

Step 4: Check your answer. Once you think that you have solved a problem, go back and determine if your answer fits the conditions originally stated in the problem. For example, if you are to find the number of snowboarders who participate in the Winter X Games, $19\frac{1}{2}$ people is not an acceptable answer. Or, in an investment problem, it is highly unlikely that your deposit of \$1,000 would earn \$334 interest in a bank account. If your solution is not reasonable, then look for the source of your error. Maybe you have misunderstood one of the conditions of the problem, or perhaps you made a simple computational or algebraic mistake.

KEY POINT

Problem solving relies on several basic strategies.

Problem-Solving Strategies

Problem solving is more of an art than a science. We will now suggest some useful strategies; however, just as we cannot list a set of rules describing how to write a novel, we cannot specify a series of steps that will enable you to solve every problem. Artists, composers, and writers make creative decisions as to how to use their tools, and so you also must be creative in using your mathematical tools.

Mathematics is not as rigid as you may believe from your past experiences. It is important to use the strategies in this section to keep your focus on understanding concepts rather then memorizing formulas. If you do this, you may be surprised to find that a given problem can be solved in several different ways.

PROBLEM SOLVING

Strategy: Draw Pictures

Problems usually contain several conditions that must be satisfied. You will find it useful to draw pictures to understand these conditions before trying to solve the problem.

EXAMPLE 1 Visualizing a Condition in a Word Problem

Four architects are meeting for lunch to discuss preliminary plans for a new performing arts center on your campus. Each will shake hands with all of the others. Draw a picture to illustrate this condition, and determine the number of handshakes.

SOLUTION:

We will use points labeled A, B, C, and D, respectively, to represent the people, and join these points with lines representing the handshakes, as in Figure 1.1.

QUIZ YOURSELF 1

How might your diagram change in Example 1 if we were counting the ways the architects could send text messages to each other? Realize now that a message sent from A to B is not the same as a message sent from B to A.
Hint: Consider putting arrowheads on the edges.

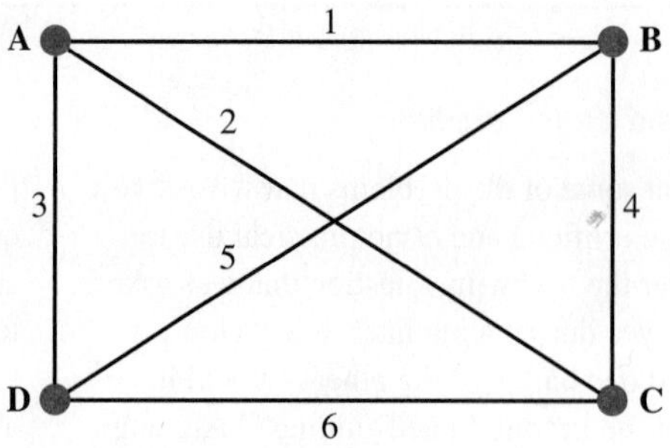

FIGURE 1.1 Visualizing handshakes.

If we represent the handshake between A and B by AB, then we see that there are six handshakes; namely, AB, AC, AD, BC, BD, and CD.

Now try Exercises 1 to 4. 1

In later chapters, you will often be interested in determining all the possibilities that can occur when a series of things are occurring. The next example illustrates an effective way to visualize such situations.

EXAMPLE 2 Drawing a Tree Diagram

Draw a diagram to illustrate the different ways that you can flip three coins.

SOLUTION:

In order to better understand what each coin is doing, we will assume that the coins are a penny, a nickel, and a dime. It is often useful in situations such as this to assume that things happen in distinct stages. For example, we can assume that the penny is flipped first, then the nickel, and finally the dime. We can visualize the possibilities for the three stages in the following diagram (Figure 1.2), which is called a **tree diagram.** The third branch of the tree (shown in red) illustrates that one possibility is that the penny shows a head, the nickel shows a tail, and the dime shows a head. By tracing through this diagram, you can see that there are eight different ways that the three coins can be flipped.

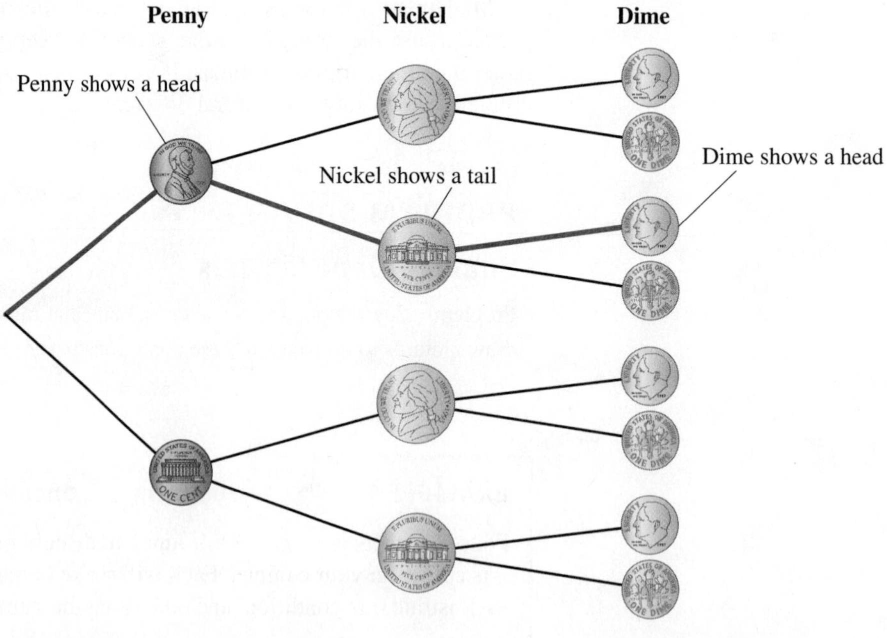

FIGURE 1.2 A tree diagram shows the eight ways to flip three coins.

Now try Exercises 11 and 12.

PROBLEM SOLVING

Strategy: Choose Good Names for Unknowns

It is a good practice to name the objects in a problem so you can remember their meaning easily.

Example 3 combines good naming with the drawing strategy mentioned earlier.

EXAMPLE 3 Combining the Naming Strategy and the Drawing Strategy

Assume that one group of students who are interested in physical fitness is taking Zumba classes and another is taking Pilates. Choose good names for these groups and represent this situation with a diagram.

SOLUTION:

In Figure 1.3, the region labeled Z represents the students taking Zumba and the region labeled P represents the students taking Pilates.

QUIZ YOURSELF 2

Describe the students who are represented by region r_3. What about r_4?

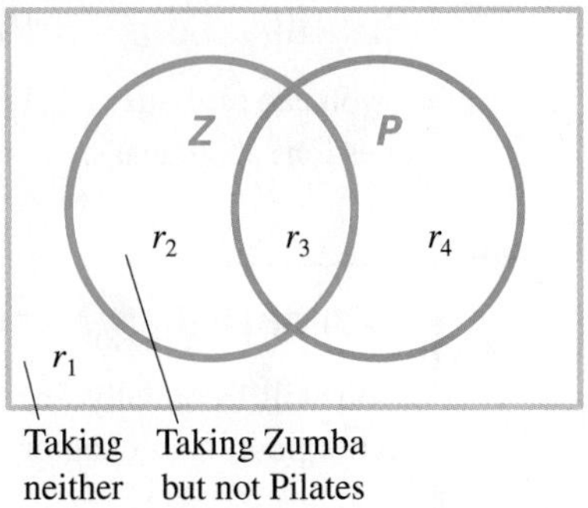

FIGURE 1.3 Representing several groups of people in a diagram.

As you can see in Figure 1.3, the region marked r_2 indicates students who are taking Zumba but not Pilates. Region r_1 represents students who are taking neither.

Now try Exercises 17 and 18. 2

PROBLEM SOLVING

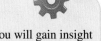

Strategy: Be Systematic

If you approach a situation in an organized, systematic way, frequently you will gain insight into the problem.

EXAMPLE 4 Systematically Listing Options

Javier is buying an iPhone and is considering which optional features to include with his purchase. He has narrowed it down to three choices: extended-life battery, deluxe ear buds, and 64 gigabytes of memory. Depending on price, he will decide how many of these options he can afford. In how many ways can he make his decision?

SOLUTION:

In making his decision, we see that there are four cases.

Choose none *or* one *or* two *or* all three of the options.

We organize these possibilities in Table 1.1.

	Battery	Ear Buds	64 Gigabytes
choose none	No	No	No
choose one	Yes	No	No
	No	Yes	No
	?	?	?
choose two	Yes	Yes	No
	Yes	No	Yes
	?	?	?
choose three	Yes	Yes	Yes

TABLE 1.1 Systematic listing of choices of iPhone options.

QUIZ YOURSELF 3

Complete Table 1.1.

We see that there are eight ways that Javier can make his decision.

Now try Exercises 9, 10, 13, and 14. 3

PROBLEM SOLVING

Strategy: Look for Patterns

If you can recognize a pattern in a situation you are studying, you can often use it to answer questions about that situation.

EXAMPLE 5 Finding Patterns in Pascal's Triangle

You will frequently encounter the pattern that we show in Figure 1.4, called **Pascal's triangle** in later chapters.

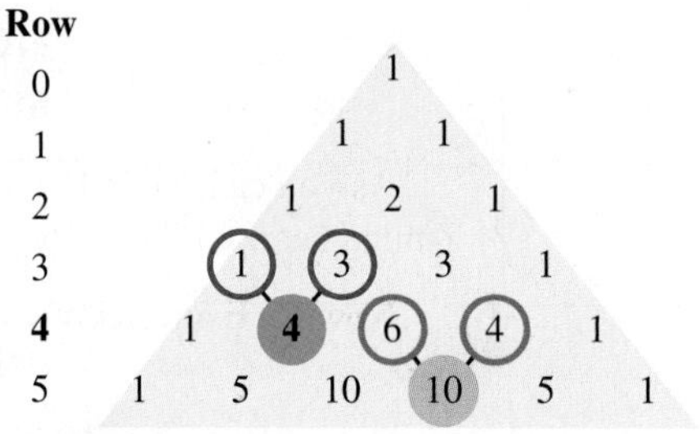

FIGURE 1.4 Pascal's triangle.

Notice how each number is the sum of the two numbers immediately above it, which are a little to the right and a little to the left. Suppose we want to find the total of all the numbers that will be in the ninth row of this diagram.

SOLUTION:

(It will be convenient when we discuss this diagram in later chapters to begin numbering rows with 0 instead of 1.)

Notice that in the zeroth row, the total is 1; in the first row, the total is 2; in the second row, the total is 4; and in the third row, it is 8. We continue this pattern in Table 1.2.

Row	0	1	2	3	4	5	6	7	8	9
Total	1	2	4	8	16	32	64	128	256	512

TABLE 1.2 The sum of the numbers in each row of Pascal's triangle.

We now easily see that the desired total is 512. 4

QUIZ YOURSELF 4

a) List the numbers in the sixth and seventh rows of Pascal's triangle.
b) What are the first two numbers in the 100th row of Pascal's triangle?

If you look closely at Example 5, you may notice that the 1, 3, 3, 1 pattern in row 3 of Pascal's triangle appeared in a slightly disguised form in Example 4. Notice that there was one way to choose no options, three ways to choose one option, three ways to choose two options, and one way to choose all three options. It is interesting how the same pattern in mathematics occurs in seemingly different situations.

PROBLEM SOLVING

Strategy: Try a Simpler Version of the Problem

You can begin to understand a complex problem by solving some scaled-down versions of the problem. Once you recognize a pattern in the way you are solving the simpler problems, then you can carry over this insight to attack the full-blown problem.

In these days of identity theft, it is important when you send personal information, such as your Social Security number or bank account numbers, to another party that the information cannot be intercepted and your identity stolen. In Example 6, we consider a problem similar to the handshake problem you saw earlier.

The situation in Example 6 is very similar to the situation you saw in Example 1, except now we will be counting communication links instead of handshakes. However, it is not practical to draw a diagram with twelve objects connected in all possible ways, so we will solve the problem by looking at simpler versions of the problem until we see a pattern that leads us to a solution.

EXAMPLE 6 Secure Communication Links

Suppose that 12 branches of Bank of America need secure communication links among them so that financial transfers can be made safely. How many links are necessary?

SOLUTION:

Instead of considering all 12 branches, we will look at much smaller numbers of bank branches, count the links, and see if we recognize a pattern. Let's call the branches A, B, C, D, E, F, G, H, I, J, K, and L. In Table 1.3, we will write AB for the link between A and B, AC for the link between A and C, etc.

From looking at these smaller examples, we see an emerging pattern. We notice that as we add new branches, the number of links increases first by 1, then by 2, then by 3, and then by 4, etc.

You can easily see why this is the case if you imagine establishing the branch offices of the banks one at a time. First A is established, so no links are required.

Number of Branches	Branches	Links	Number of Links	
1	A	None	0	
2	A, B	AB	1	⟵ add 1 link
3	A, B, C	AB, AC, BC	3	⟵ add 2 links
4	A, B, C, D	AB, AC, AD, BC, BD, CD	6	⟵ add 3 links
5	A, B, C, D, E	AB, AC, AD, AE, BC, BD, BE, CD, CE, DE	10	⟵ add 4 links

TABLE 1.3 Looking for a pattern in the links between Bank of America branches.

Then when B is built, one link is required between A and B. When C is built, two additional links are needed, namely, AC and BC. When branch D is added, we will need three additional links—AD, BD, and CD. If we continue this pattern as in Table 1.4, we can solve the original problem.

QUIZ YOURSELF 5

Determine the number of ways e-mail can be sent between the various branches. Notice now that e-mail from A to B, written AB, is not the same as e-mail from B to A, written BA.

Number of Branches	1	2	3	4	5	6	7	8	9	10	11	12
Number of Links	0	1	3	6	10	15	21	28	36	45	55	66

TABLE 1.4 Counting the links.

We now see that for 12 branches, there will be 66 links.

Now try Exercises 25 to 30. 5

PROBLEM SOLVING

Strategy: Guessing Is OK

One of the difficulties in solving word problems is that you can be afraid to say something that may be wrong and consequently sit staring at a problem, writing nothing until you have the full-blown solution. Making guesses, even incorrect guesses, is not a bad way to begin. It may give you some understanding of the problem. Once you make a guess, evaluate it to see how close you are to meeting all the conditions of the problem.

Imagine that in a few years, when you have graduated, you decide to purchase a brand-new house. At first you are happy with your decision, but then one day when you open your mail, you are shocked to find a school real estate tax bill for $5,200. You are not alone, because in some areas of the country, both young homeowners and senior citizens are struggling with excessive property taxes. As a result, taxpayer groups have urged politicians to consider other ways of funding public education. We consider a hypothetical case in our next example.

EXAMPLE 7 Solving a Word Problem by Guessing

Suppose that you own a house in a school district that has a yearly budget of $100 million. In order to reduce the portion of the budget borne by property owners, your local taxpayers' organization has negotiated the following political agreement:

1. The amount funded by the state income tax will be three times the amount funded by property taxes.
2. The amount funded by the state sales tax will be $15 million more than the amount funded by property taxes.

How much of the budget will be funded by property taxes?

SOLUTION:

In a later chapter, we will discuss how to solve problems algebraically, but for now, all we want to do is make several educated guesses and then keep adjusting them until we get an acceptable answer. Let us call the amount of the budget due to property taxes p, the amount due to income taxes i, and the amount due to sales taxes s.

QUIZ YOURSELF 6

The Field Museum in Chicago, which houses Sue, the largest preserved *T. Rex*, has acquired three more mechanical dinosaurs for a new exhibit. The combined weight of the three new dinosaurs is 50 tons. If the weight of the Apatosaurus is seven times the weight of the Duckbill, and the Diplodocus is 14 tons more than the Duckbill, what are the weights of each dinosaur?

We will begin by assuming that $p = 20$, $i = 20$, and $s = 60$. We do not agonize about making this starting guess. The important thing is to get the process started and then evaluate and refine our guesses to solve the problem. We will organize our guesses in the following chart.

Guesses for p, i, s (in millions of $)	Evaluation of Guess	
	Good Points	Weak Points
20, 20, 60	Total is 100.	Amounts are not different.
20, 30, 60	Amounts are different.	Total is not 100. The amount i is not three times p. The amount s is not 15 more than p.
20, 60, 35	The amount i is three times p. The amount s is 15 more than p.	The total is greater than 100, so we have to reduce the amount p.
18, 54, 33	The amount i is three times p. The amount s is 15 more than p.	The total is still more than 100.
17, 51, 32	We have a solution. The amount from property taxes is $17 million.	

Now try Exercises 53 to 62. 6

You may not believe that the guessing approach to problem solving that we suggested in Example 7 is doing mathematics. However, if you are making intelligent guesses and systematically refining them, you probably have some solid, intuitive, underlying logical reasons for what you are doing. And, as a result of this thinking, you are doing legitimate mathematics. The problem with guessing is that it can be inefficient and, if the answer is complex, you probably will not be able to find it without doing some algebra.

PROBLEM SOLVING

Strategy: Relate a New Problem to an Older One

An effective technique in solving a new problem is to try to connect it with a problem you have solved earlier. It is sometimes possible to rewrite a condition so that the problem becomes exactly like one you have seen before.

We will now show you the example that we mentioned in the chapter opener regarding how to organize course selection for a semester.

EXAMPLE 8

Anthony is selecting his courses for next semester. Because he works afternoons and also commutes, he must schedule classes that meet only on Monday, Wednesday, and Friday mornings and has decided on the following possibilities: Math at 9, 11, or 12; English at 9 or 12; Sociology at 10, 11, or 12; Art History at 9, 10, or 11. Determine the possible ways that Anthony can schedule these classes.

SOLUTION: Review the Draw Pictures strategy on page 3.

If you think carefully about this problem, you will recognize that Anthony is making a series of decisions: first select a (M)ath class, then select an (E)nglish class, next a

(S)ociology class, and finally an (A)rt History class. *This pattern is somewhat similar to flipping three coins in order as we did in Example 2*, so we can organize our thinking by constructing a tree diagram (Figure 1.5), which is similar to the tree diagram we drew in Example 2.

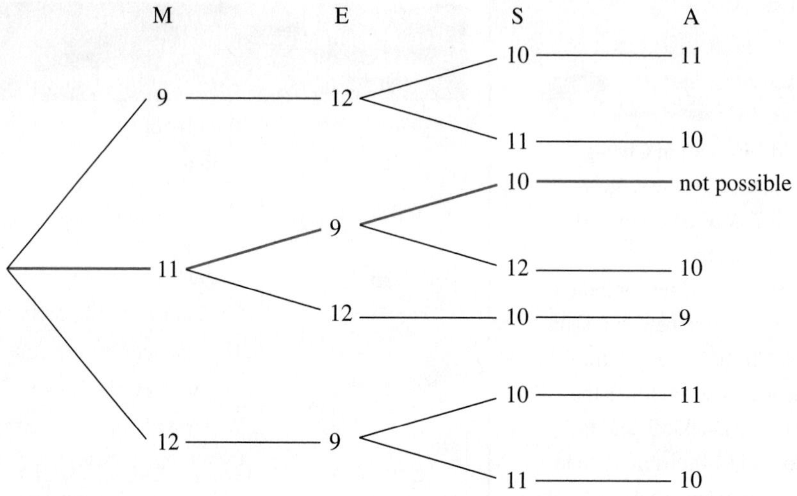

FIGURE 1.5 A tree diagram showing all possible ways to construct Anthony's schedule.

QUIZ YOURSELF 7

Assume that Math and Sociology are taught in buildings on opposite sides of the campus and the distance is such that Anthony cannot take them back-to-back. Use Figure 1.5 to determine what schedules are now satisfactory.

The three leftmost branches of the tree show the possibilities for choosing the math class; the next set of branches shows how to choose the English classes *after* the math class has been selected, and so on.

The top branch of the tree shows that one possibility for Anthony is Math at 9, English at 12 (notice that if Math is at 9, then English cannot also be taken at 9), Sociology at 10, and Art History at 11. The third branch of the tree (in red) shows that if Anthony chooses Math at 11, English at 9, and Sociology at 10, then it is impossible to schedule Art History. So, from the diagram, we see that there are six satisfactory schedules possible.

Now try Exercises 65 to 68. 7

Some Mathematical Principles

KEY POINT

Remembering several fundamental principles helps us with problem solving.

We will now discuss some basic mathematical principles that we will refer to frequently throughout this text.

PROBLEM SOLVING

Strategy: The Always Principle

When we say a statement is true in mathematics, we are saying that the statement is true *100% of the time.* One of the great strengths of mathematics is that we do not deal with statements that are "sometimes true" or "usually true."

EXAMPLE 9 An Algebraic Statement That Is Always True

Find an example to illustrate that the statement

$$(x + y)^2 = x^2 + 2xy + y^2$$

is true.*

*For a review of the order of operations in arithmetic and algebra, see Appendix A.

SOLUTION:

In algebra, we prove that the statement $(x + y)^2 = x^2 + 2xy + y^2$ is true for all numbers x and y. This means that any number we substitute for x and y should make the statement true.

Suppose that $x = 3$ and $y = 4$. Then, substituting in the given statement we get

$$\underbrace{(3+4)^2}_{(x+y)^2} = \underbrace{3^2}_{x^2} + \underbrace{2\cdot3\cdot4}_{2\cdot x\cdot y} + \underbrace{4^2}_{y^2} \quad \text{or} \quad 49 = 9 + 24 + 16.$$

8

QUIZ YOURSELF 8

Find another example to show that

$(x + y)^2 = x^2 + 2xy + y^2$

In a similar way, we accept a mathematical argument only if it holds up under every set of conditions. If we can think of even a single situation in which the argument fails, then the argument is not acceptable.

PROBLEM SOLVING

Strategy: The Counterexample Principle

An example that shows that a mathematical statement fails to be true is called a **counterexample.** Keep in mind that if you want to use a mathematical property and someone can find a counterexample, then the property you are trying to use is not allowable. A hundred examples in which a statement is true do not prove it to be *always* true, yet a single example in which a statement fails makes it a false statement. Be careful to understand that when we say a statement is false, we are not saying that it is always false. We are only saying that the statement is *not always true.* That is, we can find at least one instance in which it is false.

EXAMPLE 10 Counterexamples to False Algebra Statements

Although the following two statements are false, those who are not secure in working with algebra often believe them to be true.

a) $(x + y)^2 = x^2 + y^2$ b) $\sqrt{x + y} = \sqrt{x} + \sqrt{y}$

Provide a counterexample for each statement.

SOLUTION:

a) Assume that $x = 3$ and $y = 4$. Then,

$$(x + y)^2 = (3 + 4)^2 = 7^2 = 49,$$

which is not equal to

$$x^2 + y^2 = 3^2 + 4^2 = 9 + 16 = 25.$$

b) Let $x = 16$ and $y = 9$. Then,

$$\sqrt{x + y} = \sqrt{16 + 9} = \sqrt{25} = 5.$$

However,

$$\sqrt{x} + \sqrt{y} = \sqrt{16} + \sqrt{9} = 4 + 3 = 7.$$

Now try Exercises 31 to 38. 9

QUIZ YOURSELF 9

Find a counterexample to the following statement:

$$\frac{a + b}{a + c} = \frac{b}{c}.$$

PROBLEM SOLVING

Strategy: The Order Principle

When you read mathematical notation, pay careful attention to the *order* in which the operations must be performed. The order in which we do things in mathematics is as important as it is in everyday life. When getting dressed in the morning, it makes a difference whether you first put on your socks and then your shoes, or first put on your shoes and then your socks. Although the difference may not seem as dramatic, reversing the order of mathematical operations can also give unacceptable results. Note that we are not saying that it is *always* wrong to reverse the order of mathematical operations; we are saying that if you reverse the order of operations, you *may* accidentally change the meaning of your calculations.

QUIZ YOURSELF 10

Consider the equation

$$\sqrt{x+y} = \sqrt{x} + \sqrt{y}.$$

a) On the left side of the equation we are performing two operations. What are these operations and what is their order?
b) What is the order of the operations performed on the right side of the equation?
c) Summarize what is wrong with this equation, as we did in Example 11(a).

EXAMPLE 11 Reversing the Order of Operations Can Change the Result of Mathematical Computations

Explain how Example 10 illustrates the order principle.

SOLUTION:

In Example 10, the problem with statements a) and b) is that in each case we carelessly changed the order in which we performed the operations.

a) In the equation $(x + y)^2 = x^2 + y^2$, the left side of the equation tells us to *first* add x and y and *then* square the resulting sum. The right side of the equation tells us to *first* square x and y separately and *then* add these two squares. To say this more succinctly, we can say that to "first add and then square" is not the same as to "first square and then add."
b) The problem here is similar. Test your understanding of the order principle by solving Quiz Yourself 10.

Now try Exercises 39 to 42. 10

Understanding the order principle will be important when you work with set theory in Chapter 2 and also when you study logic in Chapter 3.

PROBLEM SOLVING

Strategy: The Splitting-Hairs Principle

You should "split hairs" when reading mathematical terminology. If two terms are similar but sound slightly different, they usually do not mean exactly the same thing. In everyday English, we may use the words *equal* and *equivalent* interchangeably; however, in mathematics they do not mean the same thing. The same is true for notation. When you encounter different-looking notation or terminology, work hard to get a clear idea of exactly what the difference is. Representing your ideas precisely is part of good problem solving.

EXAMPLE 12 Identifying Differences in Notation

Notice the difference in the following pairs of symbols. It is not important that you know what these symbols mean at this time; all we want you to do is recognize that there are slight differences in the notations.

QUIZ YOURSELF 11

In your own words, explain the differences you see in each pair of symbols. Do not be concerned if you do not know what these symbols mean.

a) $\supset$ and $\supseteq$
b) $\{\emptyset\}$ and $\emptyset$
c) $\cup$ and $\vee$
d) $\{0\}$ and 0

a) $<$ and $\leq$. There is an extra line under the $<$ symbol on the right.
b) $\cap$ and $\wedge$. The symbol on the left is rounded; the symbol on the right is pointed.
c) $\in$ and $\subset$. The symbol on the left has an extra line.
d) $\emptyset$ and 0. The symbol on the right is the number 0; the symbol on the left is something else, not a number. 11

PROBLEM SOLVING

Strategy: The Analogies Principle

Much of the formal terminology that we use in mathematics sounds like words that we use in everyday life. This is not a coincidence. Whenever you can associate ideas from real life with mathematical concepts, you will better understand the meaning behind the mathematics you are learning.

EXAMPLE 13 Relating Mathematical Terms to Everyday Language

In Table 1.5, the left column contains formal mathematical terminology. The right column contains ordinary English words that will help you remember the mathematical concepts.

Mathematical Concept	Related English Ideas
Union	Labor union, marriage union, united
Complement	Complete
Equivalent	Equivalent (in some way the same)
Slope	Ski slope, slope of a roof

TABLE 1.5 Mathematical terms related to words in everyday language.

Now try Exercises 49 to 52. 12

QUIZ YOURSELF 12

Which English words are you reminded of by the following mathematical terminology?

a) intersection
b) simultaneous (as in simultaneous equations)

PROBLEM SOLVING

Strategy: The Three-Way Principle

We conclude this section with a method for approaching mathematical concepts that we illustrate in Figure 1.6.

Whether you are learning a new concept or trying to gain insight into a problem, it is helpful to use the ideas we have discussed in this chapter to approach mathematical situations in three ways.

- *Verbally*—Make analogies. State the problem in your own words. Compare it with situations you have seen in other areas of mathematics.
- *Graphically*—Draw a graph. Draw a diagram.
- *By example*—Make numerical or other kinds of examples to illustrate the situation.

Not every one of these three approaches fits every situation. However, if you get in the habit of using a verbal-graphical-example approach to doing mathematics, you will find that mathematics is more meaningful and less dependent on rote memorization. If you practice approaching mathematics using the strategies and principles that we have discussed, you will find eventually that you are more comfortable and more successful in your mathematical studies.

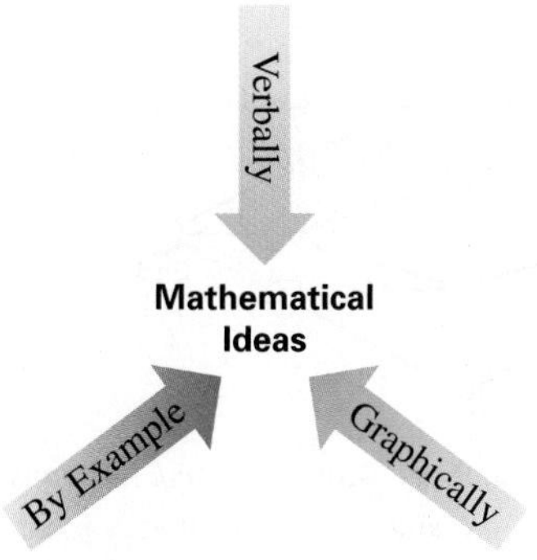

FIGURE 1.6 The Three-Way Principle.

Historical Highlight

George Polya

We could argue that George Polya (1887–1985) is the father of problem solving as we teach it in so many of today's mathematics textbooks (including this one). As a youth, Polya decided to study law but because of his dislike for memorization, he found it tedious and eventually chose a career in mathematics. While working as a mathematics tutor, he began to develop a problem-solving method that led him to write a book called *How to Solve It*. This book, which has sold more than 1 million copies, discusses many of the approaches to problem solving that you learned in this section.

Polya loved to solve problems, and it seemed that he could find them everywhere. One day while he and his wife-to-be, Stella, were walking in a garden in Switzerland, they encountered another young couple six times. Polya wondered what the likelihood was that they would meet the same couple so many times on the same walk. His attempts to answer that question eventually led him to publish research papers on the topic of the random walk problem.

In the 1940s, due to their concern about the increasing influence of Nazism in Europe, George and Stella emigrated to the United States. Polya eventually accepted a position at Stanford University in California, where he conducted research and worked on problem solving until he was well into his nineties.

EXERCISES 1.1

Sharpening Your Skills

In Exercises 1–4, draw a picture to illustrate each situation. You are not being asked to solve any problems—just draw a picture.

1. Five liters of a 10% sugar solution are mixed with pure water to get a 5% solution.
2. Five people clink each other's glasses in a toast.
3. Regan, Chris, and Ava are sending tweets to each other regarding their favorites in the *America's Got Talent* finals.
4. The Democratic, Republican, and Green political parties have no members in common.

In Exercises 5–8, choose names that would be meaningful for each item. Again, you are not being asked to solve the problem—just make a recommendation for some good names.

5. In order to reduce dependence on foreign oil, a presidential commission is considering increasing funding for research in the development of hybrid automobiles, windmill turbines, and solar energy.
6. Sheldon, Leonard, Penny, and Howard are planning a surprise party for Raj.
7. A person makes two investments—one in stocks, the other in bonds.
8. Michael Phelps wants to include calcium and protein in his diet.

In Exercises 9 and 10, list the items mentioned. Try to organize your list in a systematic way.

9. List the different combinations of heads and tails that can occur when a penny and a nickel are flipped.
10. Using the numbers 1, 2, and 3, form as many ordered pairs of numbers as you can. For example, (2, 3) is one such pair; (3, 2) is a different one. You are allowed to repeat numbers, so (1, 1) is an allowable pair.
11. If you draw a tree, as we did in Example 2, to show the number of ways to flip five coins, how many possibilities would there be?
12. If you draw a tree, as we did in Example 2, to show the number of ways to roll two dice (one red and one green), how many possibilities would there be? (*Hint:* Draw the branches for the roll of the first die and then attach more branches to correspond to the roll of the second die.)
13. Choose two recording artists to host the Grammy Awards. Your choices are Beyonce, Rihanna, Usher, Eminem, and Lady Gaga. (Represent the pairs of artists by a pair of letters. For example, Usher and Lady Gaga could be represented by UL. Note that UL is the same as LU.)
14. Follow the arrows in the given diagram to travel from "Begin" to "End."

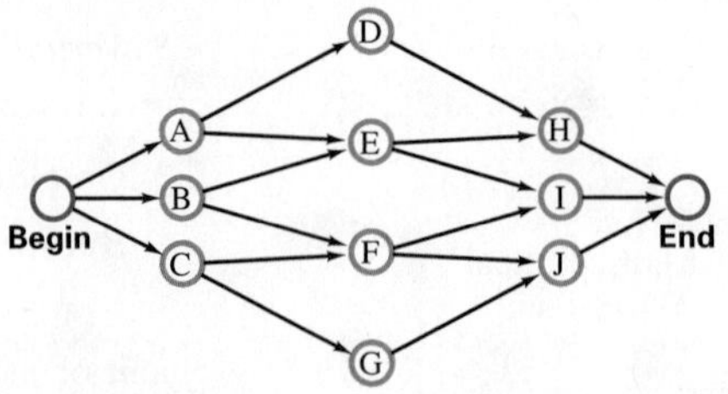

15. Some role playing games, such as *Dungeons and Dragons*, have dice with other than six sides. Assume that you are

rolling two four-sided dice—with faces numbered 1, 2, 3, and 4. Draw a tree diagram and then list all of the possible ordered pairs of numbers that can be obtained when the two dice are rolled.

16. Repeat the previous problem, but now only list pairs of numbers that are different.

17. Consider the following diagram, which represents various groups of people. G is the group of people who are good singers, and A is the group of people who have appeared as contestants on *American Idol.* Describe the people who are in regions r_3 and r_4.

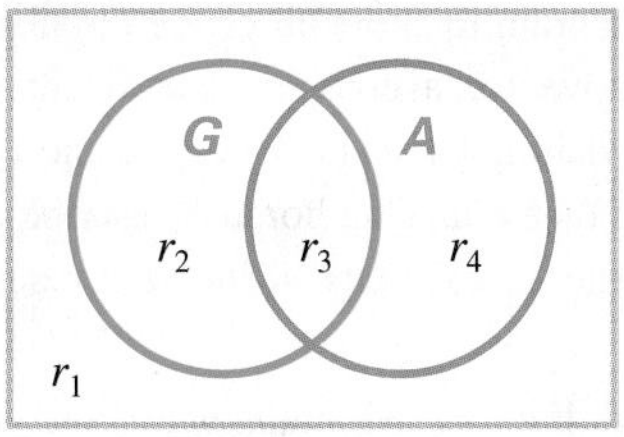

18. Redraw the diagram in Exercise 17, except now label the two groups of people as W, those who are working to reduce global warming, and H, those who are striving to reduce world hunger. Describe the people who are in regions r_2 and r_4.

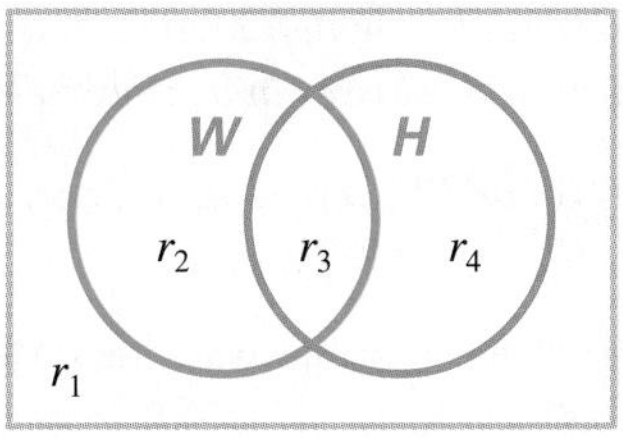

In Exercises 19–24, continue the pattern for five more items in the list. (There may be more than one way to continue the pattern. We will provide only one solution.)

19. 7, 14, 21, 28, . . .

20. 1, 3, 9, 27, . . .

21. *ab*, *ac*, *ad*, *ae*, *bc*, *bd*, *be*, . . .

22. (1, 1), (1, 2), (1, 3), (1, 4), (1, 5), (1, 6), (2, 1), (2, 2), (2, 3), . . .

23. 1, 1, 2, 3, 5, 8, 13, . . .

24. 2, 3, 5, 7, 11, 13, 17, . . .

In Exercises 25–30, we state a problem. Instead of trying to solve that problem, state a simpler problem and solve it instead.

25. Ten people are being honored for their work in reducing pollution. In how many ways can we line up these people for a picture?

26. If you guess at 10 true–false questions, how many different ways can you fill in the 10 answers?

27. Using all the letters of the alphabet, how many two-letter codes can we form if we are allowed to use the same letter twice? For example, *ah*, *yy*, *yo*, and *bg* are all allowable pairs. (*Note: bg* is different from *gb*.)

28. A family has seven children. If we list the genders of the children (for example, *bbggbgb*, where *b* is boy and *g* is girl), how many different lists are possible?

29. An electric-blue Ferrari comes with seven options: run flat tires, front heated seats, polished rims, front parking sensor, carbon interior trim, fender shields, and a custom tailored cover. You may buy the car with any combination of these options (including none). How many different choices do you have?

30. In printing your résumé, you have 10 different colors of paper to use and 20 different font styles. How many different ways can you print your résumé?

In Exercises 31–38, determine whether each statement is true or false. If a statement is true, give two examples to illustrate it. If it is false, give a single counterexample.

31. Months of the year have 31 days.

32. All past presidents of the United States are deceased.

33. $\frac{a}{b} + \frac{c}{d} = \frac{a + c}{b + d}$

34. If $a < b$, then $a^2 < b^2$.

35. If A is the father of B and B is the father of C, then A is the father of C.

36. If X is acquainted with Y and Y is acquainted with Z, then X is acquainted with Z.

37. If the price on a Blu-ray player is increased by 10% and then later reduced by 10%, the price will be the same as the original price.

38. Assume that due to budget concerns, your work-study hourly rate was cut by 20% last summer. This summer you are given a 20% pay raise, are you earning the same wage as you were before the pay cut?

In Exercises 39–42, decide whether the two sequences of operations will give the same results.

39. Squaring a number, then adding 5 to it; adding 5 to a number and then squaring that sum.

40. Squaring two numbers x and y, then subtracting the results; subtracting x and y, then squaring the difference.

41. Adding two numbers x and y, then dividing the result by 3; dividing x and y by 3, then adding the results.

42. Multiplying two numbers x and y by 5, then adding the results; adding x and y, then multiplying the result by 5.

In Exercises 43–48, explain the differences you see in each pair of symbols.

43. 5 and $\{5\}$

44. A and A'

45. U and u

46. $\{1, 2\}$ and $(1, 2)$

47. (2, 3) and (3, 2)

48. ∅ and 0

In later chapters, we will use the following mathematical terminology. Of what common English usage do these terms remind you? Explain.

49. path, circuit, bridge, directed

50. dimension, reflection, translation, transform

51. intercept, best fit, compounding

52. median, deviation, central tendency, correlation

Applying What You've Learned

In Exercises 53–62, do not try to solve each problem algebraically. Instead, make a guess that satisfies one or more conditions of the problem and then evaluate your guess, as we did in Example 7. Keep adjusting your guesses until you have a solution that fits all the conditions of the problem.

53. The local historical society wants to preserve two buildings. The total age of the buildings is 321 years. If one building is twice as old as the other, what are the ages of the two buildings?

54. To celebrate Dunder Mifflin's 40th anniversary, Dwight has made a 40-foot-long sandwich. The sandwich is cut into three unequal pieces. The longest piece is three times as long as the middle piece, and the shortest piece is 5 feet shorter than the middle piece. What are the lengths of the three pieces?

55. Janine worked 15 hours last week. One job as a clerk in a sporting goods store paid her \$7.25 per hour, while her job giving piano lessons paid \$12 per hour. If she earned \$137.25 between the two jobs, how many hours did she work at each job?

56. Vince worked 18 hours last week. Part of the time he worked in a fast-food restaurant and part of the time he worked tutoring a high school student in mathematics. He was paid \$7.25 per hour in the restaurant and \$15 per hour tutoring. If he earned \$161.50 total, how many hours did he work at each job?

57. In a recent National Football League season, Tom Brady, Phillip Rivers, and Aaron Rodgers threw a total of 94 touchdown passes. If Rivers had two more touchdowns than Rodgers and six less than Brady, how many did Brady have?

58. In the home run derby competition prior to baseball's All Star Game, Robinson Cano, Adrian Gonzalez, and Prince Fielder hit a total of 72 home runs. If Cano hit one more than Gonzalez and 23 more than Fielder, how many did Cano hit?

59. Heather has divided \$8,000 between two investments—one paying 8%, the other 6%. If the return on her investment is \$550, how much does she have in each investment?

60. Carlos has \$9,000 in two mutual funds. One fund pays 11% interest and the other fund pays 8%. If his income from the two funds last year was \$936, how much did he invest in each fund?

61. The administration at Center City Community College has formed a 26-person planning committee. There are five times as many administrators as there are students and five more faculty members than students. How many students are there on the committee?

62. Last week Minxia made 55 phone calls to gather support for the reelection of her local state representative. She contacted three times as many senior citizens as she did young adults. The number of middle-age adults was one-half the number of senior citizens. How many senior citizens did she contact?

63. Draw a diagram to show all of the possible routes that a sales representative for a company could take by starting at Los Angeles, visiting Chicago, Houston, and Philadelphia in some order, and then returning home. (Describe each route by listing the first letter of each city visited. For example, LHPCL is one route.)

64. Reconsider Exercise 63, but now assume you have 10 different cities, including Los Angeles. By looking at simpler examples, find a pattern to determine the total number of routes that would be possible. Assume that you are always starting at Los Angeles.

In Exercises 65–68, assume that Menaka has determined the following possibilities for classes in her next semester's schedule:

> *Math at 9, 10, or 12; English at 9, 11, or 12; Sociology at 10 or 12; Art History at 9, 10, or 11.*

65. Determine all the possible ways that Menaka can schedule these classes.

66. Assume now that Menaka cannot take her English and Math classes back-to-back. Now what are her possible schedules?

67. Now assume that Menaka decides that her course load would be too heavy and decides not to schedule Sociology. What are her possible schedules for Math, English, and Art History?

68. In selecting her classes, Menaka finds that the Sociology class at 10 is full, but a new Sociology course at 11 has just been opened. Now determine her possible schedules.

69. Carmelo has been commissioned to create a decorative wall for the 21st Annual X Games consisting of a square array of square tiles in a pattern forming a large X. The following example shows a pattern with 5 rows and 5 columns. If the wall will have a similar pattern with 21 rows and 21 columns, how many of the colored tiles will be needed?

70. If the colored tiles in the figure in Exercise 69 formed a diamond pattern as shown here, how many colored tiles would Carmelo need for a 25 by 25 wall?

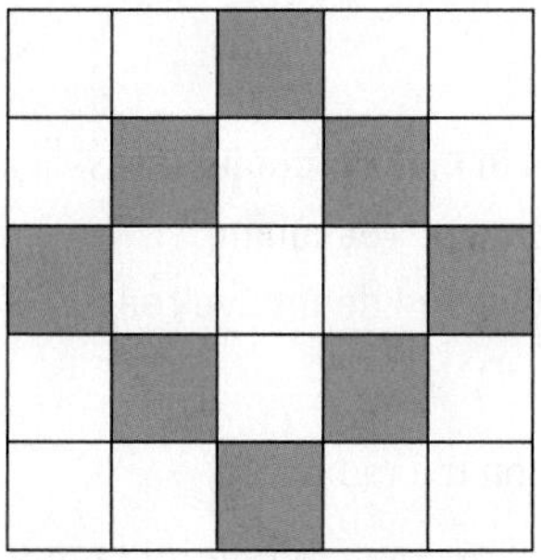

71. The board of trustees for your school is considering a tuition increase. One faction in the board supports raising the tuition 8% this year and 5% the next. Others want the 5% raise this year and the 8% next year. Some say it doesn't matter because in either case, you will be paying the same in the second year. Does it make any difference as to which plan is implemented?

72. Assume that the board of trustees for your school is bound by state law to increase the tuition by no more than 10% over the next three years. If tuition is raised by 2%, then 3%, and then 5%, has the board followed its mandate? Explain.

Communicating Mathematics

73. Students often ignore the first and last step of Polya's problem-solving method. Why do you think that this is so? What are the dangers of ignoring these steps?

74. Discuss the relationship in problem solving between considering simpler examples, listing items systematically, and looking for patterns. How do these techniques relate to George Polya's problem-solving method?

75. Is the Three-Way Principle in learning mathematics different from the way you have approached learning in the past? Identify at least three ways that we have used the Three-Way Principle in solving problems in this section.

76. Think of a real problem-solving situation in your personal life (it may involve some purchasing decision, time management, scheduling, a building project, etc.) or involving a national or global problem (replacing gasoline-powered cars with electric vehicles, reducing the property tax burden on homeowners, funding Social Security, etc.). Describe the ways that you can apply the problem-solving techniques you learned in this section to better understand your selected problem.

Challenge Yourself

77. How many ordered triples are possible if we roll three 6-sided dice?

78. How many ordered triples are possible if we roll three 12-sided dice?

79. Continue the following sequence of pairs of numbers by listing the next two pairs in the sequence: (3, 5), (5, 7), (11, 13), (17, 19), (29, 31), ...

80. Continue the following sequence of pairs of numbers by listing the next two pairs in the sequence: (5, 11), (7, 13), (11, 17), (13, 19), (17, 23), (19, 29), ...

81. We will call the accompanying figure a 5×5 (5-by-5) square. In this figure, we have highlighted a 1×1 square in red and a 3×3 square in blue. Find the number of all possible squares in the figure. Try to be systematic in solving this problem by considering all 5×5 squares, then all 4×4 squares, etc.

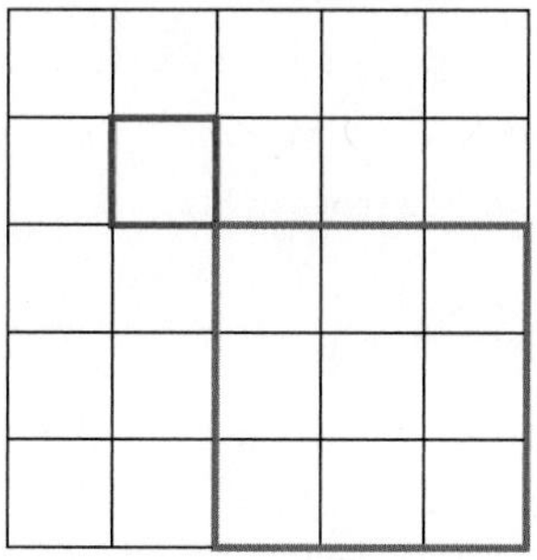

82. This exercise is similar to Exercise 81, except now we are considering all rectangles (including squares). In the given figure, we have shown a 1×4 rectangle in red (one row, four columns) and a 3×2 rectangle in blue (three rows and two columns).

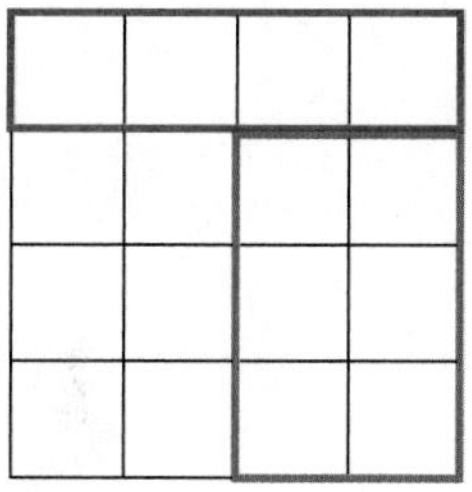

a) How many 2×1 rectangles are there?

b) How many 3×2 rectangles are there?

c) How many rectangles of various dimensions are there?

83. Consider in the following map how many different ways there are to go directly from the Hard Rock Cafe to The Cheesecake Factory.

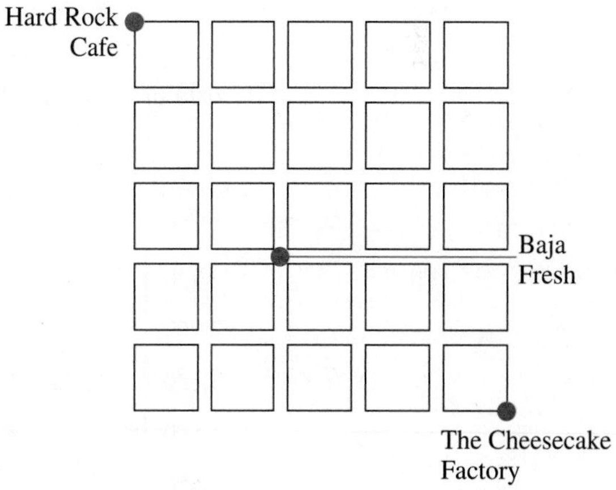

84. Suppose in Exercise 83 that after leaving The Hard Rock Cafe, you want to stop at Baja Fresh before going to The Cheesecake Factory. Now how many direct routes are there?

1.2

Inductive and Deductive Reasoning

Objectives

1. Understand how inductive reasoning leads to making conjectures.
2. Give examples of correct and incorrect inductive reasoning.
3. Understand the difference between inductive and deductive reasoning.

Driving to your morning math class, you hear on the radio:

With rising gold prices, now is the time to invest in gold!

Because the money in your bank account has been earning very little interest, and, as Figure 1.7(a) shows, recently there has been a huge upward trend in gold prices, maybe you should invest in gold.

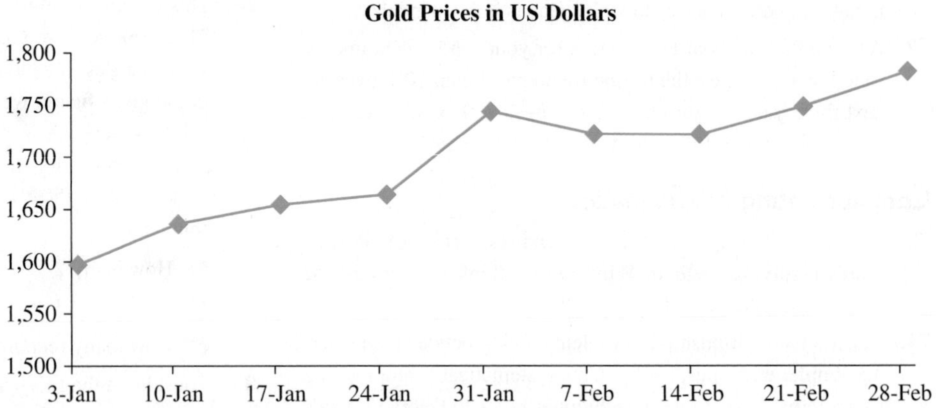

FIGURE 1.7(a) Increase in gold price per ounce over two months in the winter of 2012.

But before you jump on the gold bandwagon, let me show you a few more weeks of data in Figure 1.7(b).

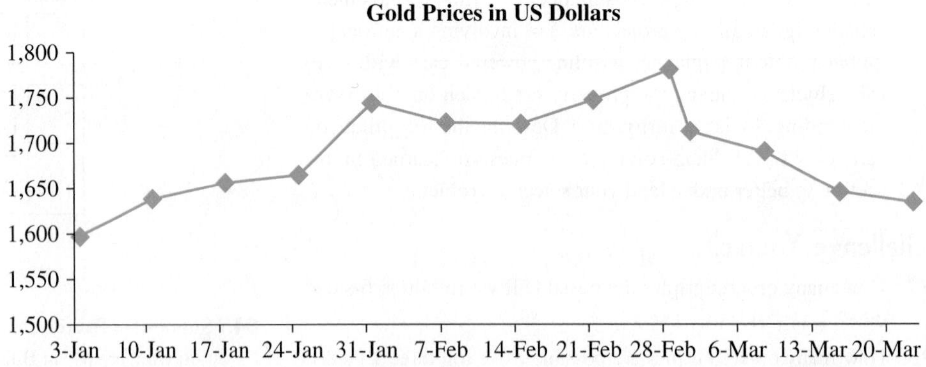

FIGURE 1.7(b) Change in gold prices over three months in the winter of 2012.

You now see that there is risk involved—what goes up can also go down! So if you buy gold now, it is possible to lose money and not have your tuition money available when you need it.

Your struggle with whether or not to invest in gold is an example of **inductive reasoning,** which involves drawing conclusions based on studying previous examples. Inductive reasoning is one type of reasoning that we use in mathematics, science, and everyday life.

Inductive Reasoning

KEY POINT

In inductive reasoning, we begin with specific examples.

> DEFINITION **Inductive reasoning** is the process of drawing a general conclusion by observing a pattern in specific instances. This conclusion is called a **hypothesis** or **conjecture.**

Mathematicians and scientists often make conjectures based on their observations. Mathematicians try to prove that conjectures are true by using the laws of mathematics. Example 1 explains a famous conjecture that was made hundreds of years ago, but mathematicians have not yet been able to prove it.

EXAMPLE 1 Goldbach's Conjecture

In 1742, Christian Goldbach made the conjecture that we could write every even integer greater than 2 as the sum of two prime (not necessarily different) numbers. A prime number is a number such as 3, 5, and 7 that can be divided evenly only by itself and 1. Illustrate Goldbach's conjecture, by expressing the even integers 20, 48, and 100 as the sum of two prime numbers.

SOLUTION:

Write $20 = 13 + 7$, $48 = 11 + 37$, and $100 = 41 + 59$.

Now try Exercises 23 to 26. 13

QUIZ YOURSELF 13

Verify Goldbach's conjecture for a) 38 and b) 46.

Although no one has yet been able to prove Goldbach's conjecture, many believe it to be true based on inductive reasoning.

In Example 2, we ask you to make a conjecture that mathematicians have been able to prove is true.

Historical Highlight

IQ Tests and Inductive Reasoning

Has it ever occurred to you how much of your future is dependent on inductive reasoning? Many intelligence tests, such as college entrance exams and placement tests in the armed services, contain many questions that involve pattern recognition and inductive reasoning. Why should that one ability play such a large role in determining your future? Maybe it shouldn't!

One way of measuring intelligence is the IQ, or intelligence quotient, which is credited to German psychologist Wilhelm Stern. In 1912, Stern defined a child's IQ to be his mental age divided by his chronological age times 100. For example, a 10-year-old child with a mental age of 11 would have an IQ of $11/10 = 1.10 = 110\%$, or 110.

In more recent years, as IQ tests have begun to fall out of favor, many have accepted Howard Gardner's alternative notion of multiple intelligences such as logical-mathematical, musical, bodily kinesthetic, and interpersonal. Others say that we should not test at all and only measure achievement itself.

EXAMPLE 2 A Divisibility Test for 9

Consider the numbers a) 72, b) 491, c) 963, d) 19,856, e) 45,307, and f) 7,538,463. Verify that the numbers a), c), and f) are evenly divisible by 9, but b), d), and e) are not. Add the digits of each number. Do you see any pattern? Make a conjecture.

SOLUTION:

a) $7+2=9$ b) $4+9+1=14$

c) $9+6+3=18$ d) $1+9+8+5+6=29$

e) $4+5+3+0+7=19$ f) $7+5+3+8+4+6+3=36$

Notice that for the numbers a), c), and f) that were divisible by 9, the sum of their digits is evenly divisible by 9 and for the numbers b), d), and e) that were not divisible by 9, the sum of their digits is not divisible. Our conjecture is that in order for 9 to divide evenly into a number, 9 must divide the sum of the digits of the number. 14

QUIZ YOURSELF 14

Test this conjecture: If the sum of an integer's digits is divisible by 8 then the integer is divisible by 8.

PROBLEM SOLVING

Strategy: The Always Principle

Keep in mind that in doing inductive reasoning, you are only making an educated guess. You cannot be sure that your conclusion is true. Recall our discussion of the Always Principle and the Counterexample Principle in Section 1.1.

In Chapter 4, we will discuss an important problem called the Traveling Salesperson Problem (historically, called the Traveling Salesman Problem) that has many real-world applications, such as routing calls through a telephone network, scheduling airline flights, and sending e-mails and text messages through a network of computers. This problem, simply stated, is: What is the best way to have a salesperson travel through a collection of cities with minimal cost? We will not get into the complete solution here as we do in Chapter 4, but in the next example, we will use inductive reasoning to determine how many different routes are possible.

EXAMPLE 3 Determining the Number of Routes for a Salesperson

Suppose that Sharifa is going to visit branch offices of her medical supply company based in Atlanta with branch offices in Boston, New York, Cincinnati, Miami, Detroit, Portland, Los Angeles, Houston, and Kansas City. How many different ways can she begin in Atlanta, visit all branches, and return home?

SOLUTION: Review the Be Systematic Strategy on page 5.

In solving this problem, we will use several of our problem-solving techniques that we discussed in Section 1.1—we will consider simpler examples, draw diagrams, list our examples systematically, and look for a pattern. We will represent the cities by the first letter of their names. Suppose there was only the one office in Atlanta. Then Sharifa would make no visits. If there were one other city—say, Boston—then she would visit Boston and return home. We will represent her one possible trip by a tree diagram as in Figure 1.8(a). The diagram shows that there is only one possible trip: she would travel from A to B. Now suppose that there are two branch offices, Boston and New York. The

diagram would now look like Figure 1.8(b). Here we see that there are two possible trips: ABN and ANB, and then return home.

With three branch offices—Boston, New York, and Cincinnati—the diagram would look like Figure 1.8(c).

We begin to see a pattern that we explain in Table 1.6. We will indicate a route by listing the first letter of the cities in the order that they are visited.

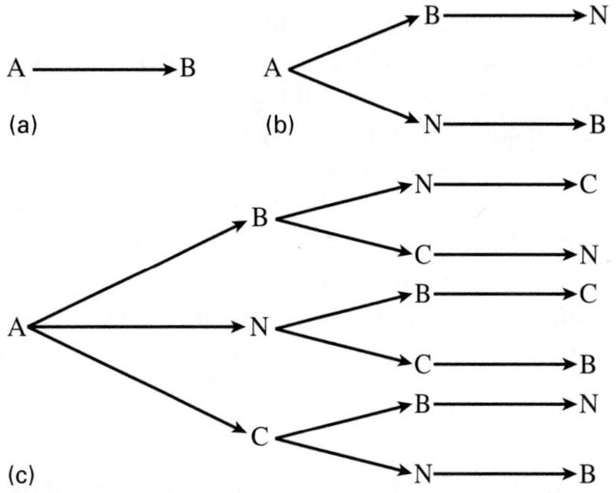

FIGURE 1.8 (a) One branch office, Boston. (b) Two branch offices, Boston and New York. (c) Three branch offices—Boston, New York, and Cincinnati.

QUIZ YOURSELF 15

In Example 3, conjecture how many routes there will be if we have 12 branch offices.

Number of Branch Offices	Branches	Routes	Number of Routes	
0	None	None	0	
1	B	AB	1	
2	B, N	ABN ANB	$2 = 2 \times 1$	⟵ Twice as many routes. She can first go to B or N and then has only one choice for the last city.
3	B, N, C	ABNC ABCN ANBC ANCB ACBN ACNB	$6 = 3 \times 2 \times 1$	⟵ Three times as many routes. Starting at A, she can first go to either B or N or C, then she can visit the remaining two cities in two ways.
4	B, N, C, M	ABNCM ABNMC ABMNC ABMCN ANBCM ANBMC Etc.	$24 = 4 \times 3 \times 2 \times 1$	⟵ Four times as many routes. Now she can start at one of four cities, and then visit the remaining three cities in six ways.

TABLE 1.6 Systematically listing routes as more cities are added.

If there were five cities, Sharifa could first visit one of the five cities and then travel to the other four cities in $4 \times 3 \times 2 \times 1$ different ways. Continuing this pattern for nine cities, the number of possible routes would be an amazing $9 \times 8 \times 7 \times 6 \times 5 \times 4 \times 3 \times 2 \times 1 = 362{,}880$ ways. 15

Between the Numbers

What Will You Do If the Bubble Bursts?

A bubble seems like a harmless thing, but in the business world a bubble can be a prelude to economic catastrophe. A bubble occurs when investors continue to buy a commodity that is rising in price, believing that the price will continue to rise and so enable them to make a profit later. As the bubble grows, investors become willing to pay much more for a commodity than it is actually worth. In effect a bubble is an example of inductive reasoning run amok.

A famous example of an economic bubble occurred in the 17th century in Holland when investors began to invest irrationally in rare tulip bulbs.* Tulip mania, as it was called, drove up the price of exotic tulip bulbs to dizzying heights—for example, an offer was made on a rare bulb called Semper Augustus at an amount equal to the annual income of a wealthy merchant. As people traded their real fortunes for imaginary ones, the frenzy grew until investors began to refuse to pay the inflated prices for the bulbs. At this point panic ensued as investors frantically tried to unload what they now recognized as bad investments. Within weeks, the mania was over.

In the early part of this century, heavy investment in Internet companies caused a "dot-com" bubble, which eventually burst, costing investors billions of dollars. More recently, the global financial crisis of 2007–2010 is believed to be the result of an unsustainable housing bubble.

Incorrect Inductive Reasoning

Sometimes inductive reasoning can mislead us into thinking that something is true that is not, as we see in Example 4.

EXAMPLE 4 False Inductive Reasoning

We want to divide a circle into regions by selecting points on its circumference and drawing line segments from each point to each other point. Figure 1.9 shows the greatest number of regions that we get if we have one point (no line segment is possible for this case), two, three, and four points.

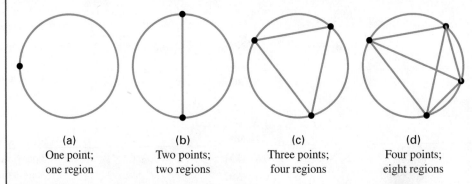

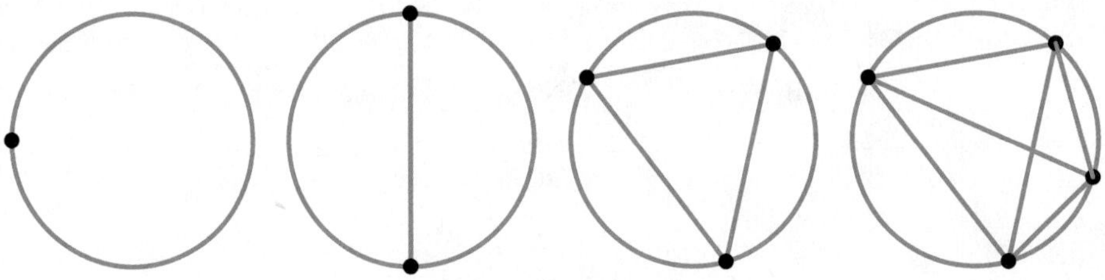

(a) One point; one region — (b) Two points; two regions — (c) Three points; four regions — (d) Four points; eight regions

FIGURE 1.9 Dividing a circle into regions.

Use inductive reasoning to find the greatest number of regions we would get if we had six points on the edge of the circle.

SOLUTION: Review the Draw Pictures Strategy on page 3.

This problem seems somewhat similar to what we did in Example 3. Notice that it appears that each time we add another point, we double the number of regions. It is natural to conjecture that if we have five points, there would be 16 regions, and with six points, we would get 32 regions. However, this is not true. You can try it for yourself by drawing a large circle and picking six points in different ways on the circle. The largest number of regions that you will find is 31, not 32.

*For more on this intriguing story, see the book *Tulipomania* by Mike Dash.

Deductive Reasoning

KEY POINT

Deductive reasoning begins with accepted facts and principles.

In inductive reasoning, we draw a general conclusion by considering a number of specific examples. In a sense, deductive reasoning is a reverse way of reasoning from inductive reasoning.

> DEFINITION In **deductive reasoning,** we use accepted facts and general principles to arrive at a specific conclusion.

In mathematics, we often use general mathematical principles to prove some specific fact. In Example 5, we will show you how to explain what at first seems to be a mystifying number trick.

EXAMPLE 5 Explaining a Number Trick by Using Deductive Reasoning

Consider the following number trick:

1. How many days a week do you eat out?
2. Multiply this number by 2.
3. Add 5 to the number you got in step 2.
4. Multiply the number you obtained in step 3 by 50.
5. If you have already had your birthday this year, add 1,764; if you haven't, add 1,763.
6. Subtract the four-digit year that you were born.
7. You should now have a three-digit number. The first digit is the number of times that you eat out each week. (There is more to this trick that we will investigate in Exercises 67 and 68.)

Use deductive reasoning to explain why this trick works.

SOLUTION:

We will use general algebra principles and deductive reasoning to explain this trick. Let's begin by calling the number of days that you eat out n and go through the trick step-by-step.

Step 2: Multiplying the number by 2 will give us $2n$.
Step 3: Adding 5, we get $2n + 5$.
Step 4: Multiplying by 50 gives us the expression $(2n + 5)50$.
Step 5: Let's assume that you haven't had your birthday yet, so we will add 1,763 to get $(2n + 5)50 + 1{,}763$.
Step 6: Let's assume that you were born in 1995, so we will subtract 1,995 to get $(2n + 5)50 + 1{,}763 - 1{,}995$.

When we simplify the expression in step 6, we get

$$\begin{aligned}(2n + 5)50 + 1{,}763 - 1{,}995 &= 100n + 250 + 1{,}763 - 1{,}995\\ &= 100n + 2{,}013 - 1{,}995\\ &= 100n + 18.\end{aligned}$$

Notice that when we look at this expression in this form, the hundred's digit is n, which is the number of days that you eat out each week. Notice that the difference $2{,}013 - 1{,}995$ has no effect on the hundred's digit. We will have more to say about this difference in the Challenge Yourself Exercises.

The next example is another good illustration of deductive reasoning.

EXAMPLE 6 Using Deductive Reasoning to Solve a Puzzle

Four students, Alex, Carmella, Noah, and Winnie, participate in different college activities (debate team, basketball, orchestra, or theater). Use the following clues to determine the activity of each student.

1. Winnie lives in the same apartment complex as the musician and theater participant.
2. The musician and Noah were friends in high school.
3. Carmella has a heavier course load than the basketball player, but fewer credits than the theater participant.
4. Noah, who has the fewest credits, is not on the debate team.

SOLUTION: Review the Be Systematic Strategy on page 5.

To organize our thinking, we will list all possibilities in the table below. From clue 1, we see that Winnie is not the musician and does not participate in theater, which allows us to place two X's in the table to show that we have eliminated two possibilities.

	Alex	Carmella	Noah	Winnie
Debate				
Basketball				
Orchestra				X-Clue 1
Theater				X-Clue 1

Clue 2 tells us that Noah does not play in the orchestra; from clue 3 we deduce that Carmella is not the basketball player and does not participate in theater. We can place three more X's in the table.

	Alex	Carmella	Noah	Winnie
Debate				
Basketball		X-Clue 3		
Orchestra			X-Clue 2	X-Clue 1
Theater		X-Clue 3		X-Clue 1

Clue 4 tells us that Noah is not on the debate team, but also because Noah has the fewest credits, by clue 3 he cannot be participating in theater. Our table now looks like this.

	Alex	Carmella	Noah	Winnie
Debate			X-Clue 4	
Basketball		X-Clue 3	✓	
Orchestra			X-Clue 2	X-Clue 1
Theater		X-Clue 3	X-Clue 4	X-Clue 1

This forces Noah to be the basketball player, which means neither Winnie nor Alex plays basketball. Now it is clear that Winnie is on the debate team, therefore we can rule out Alex and Carmella as debaters. This forces us to conclude that Alex is involved with theater.

	Alex	Carmella	Noah	Winnie
Debate	X	X	X-Clue 4	✓
Basketball	X-Clue 4	X-Clue 3	✓	X-Clue 4
Orchestra	X	✓	X-Clue 2	X-Clue 1
Theater	✓	X-Clue 3	X-Clue 4	X-Clue 1

So, Alex is participating in theater, Carmella is in the orchestra, Noah is the basketball player, and Winnie is doing debate. Now try Exercises 39 and 40.

EXERCISES 1.2

Sharpening Your Skills

Is each of the following situations an example of inductive or deductive reasoning?

1. It has rained the past three weekends, canceling your softball game. You expect that next Saturday it will rain again.
2. Olivia is calculating what grade she will need in her International Studies class to make the dean's list.
3. As you read the mystery thriller *The Girl with the Dragon Tattoo*, you are keeping track of the author's clues to solve the mystery.
4. You tell your friend Jay to be ready 15 minutes before you actually intend to pick him up because Jay is always late for his appointments.
5. Luis has noticed that the stock market has gone up on the Friday before each of the last three holidays. He plans to buy stock on the Friday before Labor Day to cash in on this trend.
6. Marianne is using the rules of algebra to solve a word problem on a quiz.
7. Brett is calculating his expenses for next year to determine how large his student loan should be.
8. Latisha noticed that on every true–false quiz so far this semester, her instructor has given twice as many false questions as true questions. On the next quiz, if she is not sure of an answer, she will guess "false."
9. The American Conference team has won the Super Bowl the past four times. You expect that the American Conference team will win again this year.

10. Emily estimates that if she can average 50 miles per hour, she will reach San Diego in $5\frac{1}{2}$ hours.

In Exercises 11–16, use inductive reasoning to predict the next term in the sequence of numbers.

11. 1, 4, 7, 10, 13, ?
12. 2, 8, 14, 20, 26, ?
13. 3, 6, 12, 24, 48, ?
14. 5, 15, 45, 135, 405, ?
15. 1, 1, 2, 3, 5, 8, 13, ?
16. 0.1, 0.10, 0.101, 0.1010, 0.10101, ?

In Exercises 17–20, use inductive reasoning to draw the next figure in the pattern. There may be several correct answers.

17.

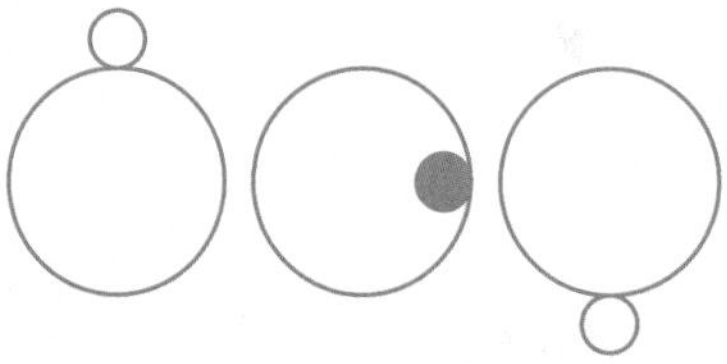

18.

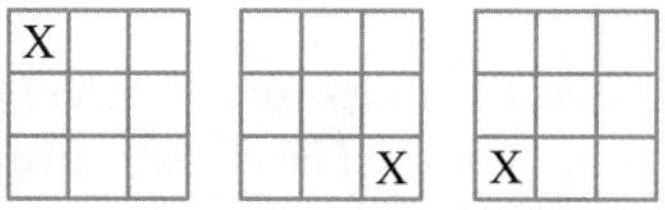

19.

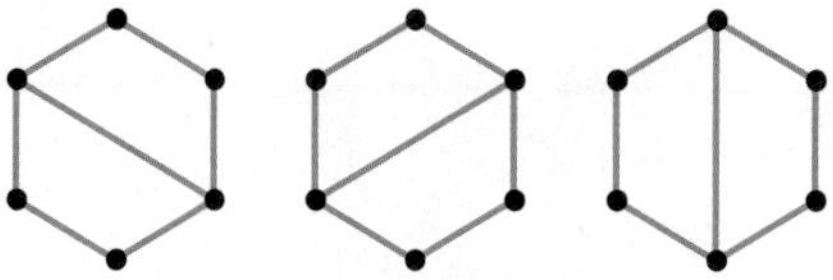

20.

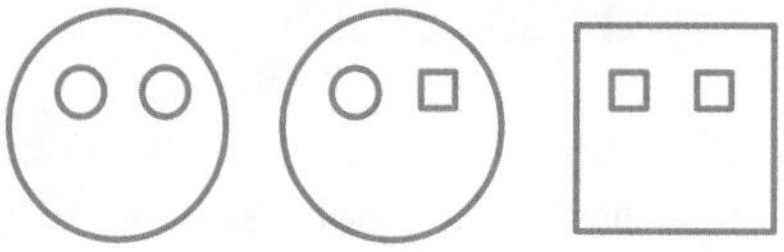

In Exercises 21 and 22, place the next X in a box to continue the pattern.

21.

	X	X		X		X				X				

22.

X	X	X		X			X					X		

Illustrate Goldbach's conjecture for each of the following numbers.

23. 16 **24.** 18

25. 20 **26.** 26

Applying What You've Learned

In each of the next four exercises, we give you a series of numerical equations. Make a conjecture as to what the next two equations in the pattern are. Check your conjecture with a calculator.

27. **a.** $1 + 2 = \dfrac{2 \times 3}{2}$

b. $1 + 2 + 3 = \dfrac{3 \times 4}{2}$

c. $1 + 2 + 3 + 4 = \dfrac{4 \times 5}{2}$

28. **a.** $2 + 4 = 2 \times 3$

b. $2 + 4 + 6 = 3 \times 4$

c. $2 + 4 + 6 + 8 = 4 \times 5$

29. **a.** $1 + 3 = 4$

b. $1 + 3 + 5 = 9$

c. $1 + 3 + 5 + 7 = 16$

30. **a.** $\dfrac{1}{1 \times 2} + \dfrac{1}{2 \times 3} = \dfrac{2}{3}$

b. $\dfrac{1}{1 \times 2} + \dfrac{1}{2 \times 3} + \dfrac{1}{3 \times 4} = \dfrac{3}{4}$

c. $\dfrac{1}{1 \times 2} + \dfrac{1}{2 \times 3} + \dfrac{1}{3 \times 4} + \dfrac{1}{4 \times 5} = \dfrac{4}{5}$

31. In preparation for his induction into the baseball hall of fame, Cal Ripken, Jr. autographed a stack of baseballs to give away to his fans. How many baseballs are in the stack?

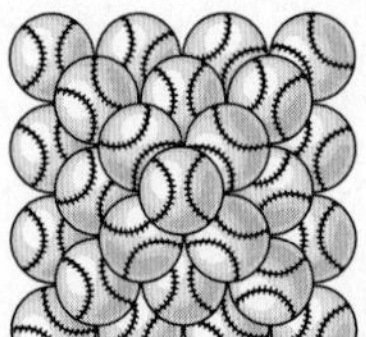

32. If a stack of baseballs similar to those shown in Exercise 31 was six layers high, how many baseballs would be in the stack?

A magic square is a square arrangement of numbers such that if you add the numbers in any row, column, or diagonal, you always get the same sum. In the magic square shown below, called a 3-by-3 square, the sum of any row, column, or diagonal is always 15.

8	1	6
3	5	7
4	9	2

In Exercises 33 and 34, the magic squares use each of the integers from 1 to 16 exactly once. Use deductive reasoning to

a) Determine the total of all the numbers in the square.

b) Determine the total for each row, column, and diagonal.

c) Complete the magic squares.

Explain your reasoning.

33.

7			9
			8
13		2	
4	1		14

34.

16	3		13
5			8
	6		12
4		14	

In Exercises 35–38, follow the instructions for each "trick" starting with several different numbers of your choice. Make a conjecture as to what result you get in each case. Use algebra and deductive reasoning, as we did in Example 5, to explain why your conjecture is correct.

35. **a.** Choose any natural number.

b. Multiply the number by 3.

c. Add 9 to the product you just found.

d. Divide the results of part (c) by 3.

e. Subtract the number that you started with.

36. **a.** Choose any natural number.

b. Multiply the number by 5.

c. Add 20 to the product you just found.

d. Divide the results of part (c) by 5.

e. Subtract 4.

37. **a.** Choose any natural number.

b. Multiply the number by 8.

c. Add 12 to the product you just found.

d. Divide the results of part (c) by 4.

e. Subtract 3.

38. **a.** Choose any natural number.
b. Multiply the number by 15.
c. Add 20 to the product you just found.
d. Divide the results of part (c) by 5.
e. Subtract 4.

39. Four students, Adriana, Caleb, Ethan, and Julia, are traveling to a conference on *Greening the World* to give presentations on recycling, solar power, water conservation, and political issues. Use the given clues to determine who is giving which speech.

1. Julia will be speaking before the speech on water conservation but after the speech on political issues.
2. The speaker on solar power helped Ethan and Julia make their PowerPoint presentations.
3. Ethan, who is speaking last, is not interested in political issues.
4. Caleb will be speaking after Julia.

40. Jessica, Serena, Andre, and Emily ran for president in the recent Student Government Council election. Determine who won the election using the following clues. Who was last in the election?

1. Emily has been a member of the council longer than the people who placed third and fourth in the voting.
2. Andre finished right behind Jessica and has the same major as the third-place finisher.
3. Although Serena did not win, she was happy that she did not finish last.
4. Jessica had 37 fewer votes than Emily.

Exercises 41 to 44 are actual questions taken from an IQ test. Use inductive reasoning to solve them. Explain your thinking as to how you found a solution.

41. GGAGLLGA is to 46336466 as LLGAAGGL is to what number?

42. Assume that NASA scientists have analyzed alien signals from outer space and have deduced the following:

"Foofrug Merduc Lilit" means: Where is your leader?

"Niurus Tuume Gazist" means: Our planet is distant.

"Foofrug Merduc Gazist" means: Where is your planet?

What is the best translation for the word "Lilit"?

43. Committee is to Etimoc as 367768899 is to what number?

44. What is the next number in the following series: $8 - 14 - 12 - 18 - 16 - 22$?

45. Explain why the number of routes for Sharifa in Example 3 increased as it did. For example, if you found that the number of routes for five cities is 120, how would the total number of routes increase for six cities? What about for seven?

46. Show that the conjecture we made in Example 4 is true for five points. Draw a figure that has 31 regions if we are using six points. How was Example 4 an example of false inductive reasoning?

In Exercises 47 and 48, draw the next figure in the sequence.

47.

(a) step 0 (b) step 1 (c) step 2

48.

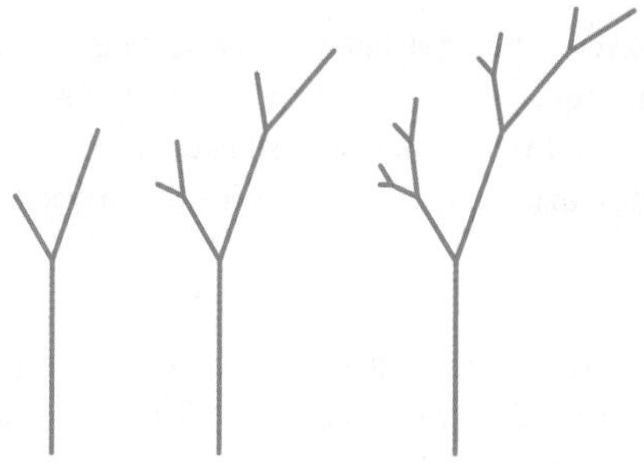

Communicating Mathematics

49. What is the role of inductive reasoning in mathematics? How is inductive reasoning related to statements such as Goldbach's conjecture? Search on the Internet for some examples of mathematical conjectures that have not been proved. If you search for "unsolved problems in mathematics," you should be able to find some that are understandable to you.

50. Write a paragraph relating the concepts of proof, conjecture, theorem, deductive reasoning, and inductive reasoning.

51. Give several examples of how you have used inductive reasoning in your life. Try to give at least one example in which you think your reasoning led you to a correct conclusion and at least one in which you were led to a wrong conclusion.

52. Find an example from the media that illustrates inductive reasoning. It might be from a political speech, a newspaper editorial, or perhaps an advertisement.

Between the Numbers

53. Research the topic of tulip mania that we discussed on page 22. Give examples of the perceived value of tulip bulbs at the height of the mania and then later, after the bubble burst. Why do you think this mania occurred?

54. Research bubbles in U.S. economic history. Try to find an estimate of the amount lost to investors by these bubbles. How is tulip mania similar or dissimilar to our own recent economic bubbles? How might you protect yourself from being harmed by an economic bubble?

Challenge Yourself

In Exercises 55–58, find the next three terms in the given sequences. (Hint: In each case, from the second or third term on, the term is some combination of previous terms.)

55. 2, 3, 7, 13, 27, ...

56. 5, 21, 85, 341, 1365, ...

57. 3, 4, 7, 11, 18, 29, ...

58. 3, 4, 10, 18, 38, 74, ...

59. We will call the rectangle to the right, which has three rows and four columns, a 3 × 4 rectangle. How many squares of various sizes can you find in this rectangle?

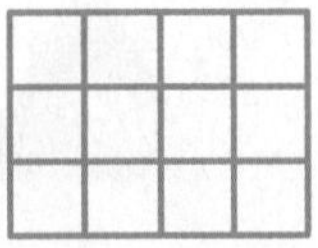

60. Repeat Exercise 59 for a 6 × 4 rectangle.

61. a) Repeat Exercise 59, but now count rectangles of all types, including squares. b) Repeat Exercise 60, now counting rectangles of all types, including squares. It is important when doing this that you are systematic. Count rectangles of sizes 1 × 1, 1 × 2, 1 × 3, 1 × 4, 2 × 1, 2 × 2, and so on.

62. Can you find some general pattern that would enable you to count rectangles of all various sizes in a 10 × 6 rectangle without actually drawing the rectangle and counting? Explain your thinking.

63. Stacking baseballs. If a stack of baseballs (similar to the one shown in Exercise 31) has a rectangular base that is six balls long and four balls wide, and is stacked with as many balls as possible, how many balls will be in the stack? Explain your reasoning. We can only place a ball on the stack if it is resting on four other balls.

64. Stacking baseballs. Redo Exercise 63, but now assume the base is seven balls long and five balls wide.

65. Make up a 3 × 3 magic square of your own using the natural numbers from 1 to 9. Explain how you constructed the magic square.

66. Make up a 4 × 4 magic square of your own using the natural numbers from 1 to 16. Explain how you constructed the magic square.

67. Explaining a trick. In Example 5, we mentioned that there is more to the trick than what we explained. The rest of the trick states that the last two digits of the three-digit number you obtain is your age. However, this part of the trick will only work in the year 2014. Explain why this is so.

68. Explaining a trick. How would you adjust the trick in Exercise 67 so that it will work in the year 2015? Explain how you came up with your adjustment.

1.3 Estimation

Objectives

1. Use rounding to make estimates.
2. Approximate calculations by using compatible numbers.
3. Make estimates based on information presented in graphs, maps, and photographs.

"That can't be right!" Do you ever find yourself saying that to a clerk in a convenience store? On occasion, I do. As I collect my purchases, I often do a rough mental estimate of what I expect to pay when the bill is presented to me. For another example, if you want to leave a tip of roughly 15% for a $37.86 bill in a restaurant, what should you leave? Chances are that you do not have a calculator in your pocket, so I guess that you have to wing it. What's a quick way to decide on the tip?

Unfortunately, in spite of the many forms of technology that we have available to do our calculations for us, chances are there will be times in real life when we have to make a quick, rough estimate, which is the topic of this section. Estimation is also an important part of effective problem solving in mathematics. To check your work, often you will find it useful to be able to make a reliable estimate to decide if your answer is reasonable.

Rounding to Make Estimates

Our first example shows how rounding can give us a quick estimate. Recall that when rounding, if the digit to the right of the digit being rounded is 5 or greater, we round up, otherwise, we round down.

EXAMPLE 1 Estimating a Grocery Bill by Rounding

On your way home from work, you stopped at the market to pick up the items listed in Table 1.7. You also would like to buy a half gallon of ice cream for $3.59, but you remember that you have only $20 in your wallet and you don't want to be caught in the checkout line without enough money. Use rounding to the nearest 10 cents to

a) Estimate the total cost of your purchases.
b) Decide if it is safe to put the ice cream in your cart.

Digit to the right is greater than 5, so we round up. Digit to the right is less than 5 so, we round down.

Item	Cost ($)	Cost Rounded to Nearest 10 Cents
Cereal	4.29	4.30
Milk	2.41	2.40
Bread	1.89	1.90
Lunch meat	3.36	3.40
Pickles	2.37	2.40
Dishwashing liquid	2.87	2.90
	Total: $17.19	Total: $17.30

TABLE 1.7 Estimating the cost of grocery items.

SOLUTION:

a) You can add the rounded prices mentally to get $17.30.
b) You decide that the half gallon is probably too expensive and decide to buy a quart of ice cream instead.

Notice in Example 1 that if you had rounded to the nearest dollar, your estimate would have been $16, and you would have taken the half gallon, which would have put you over $20.

SOME GOOD ADVICE

In Example 1, you do not need a calculator to add the estimated prices. For example, to add 4.30 and 2.40, first add the dollars to get $2 + 4 = 6$. Then, add the cents to get $30 + 40 = 70$ cents, so the total is 6.70. You can easily add in the other numbers one at a time to get the total. Doing this kind of mental arithmetic will make you a stronger mathematics student.

Using Compatible Numbers

KEY POINT

Using compatible numbers simplifies approximate calculations.

Another way to make a quick estimate that is slightly different than rounding is to use compatible numbers. In using compatible numbers, instead of using the given numbers in a problem, we substitute other numbers that are easier to work with. For example, instead of dividing 298 by 14, we might divide 300 by 15 to get 20. Or, instead of multiplying 11 times 73, we multiply 10 times 73 to get 730.

EXAMPLE 2 Using Compatible Numbers to Estimate a Bill

Marcia is buying 19 Blu-ray Discs at $12.89 each. Use compatible numbers to estimate the cost of the Blu-ray Discs.

SOLUTION:

We can replace $12.89 by $13 and 19 by 20 to get $20 \times 13 = 260$. Because we have replaced the $12.89 and 19 by two larger numbers, we know that Marcia's actual bill will be somewhat less than $260.

Now try Exercises 13 to 28.

Estimating Graphical Data

Often, you can use estimation to summarize information that is given to you graphically.

KEY POINT

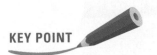

Estimation can help interpret graphical data.

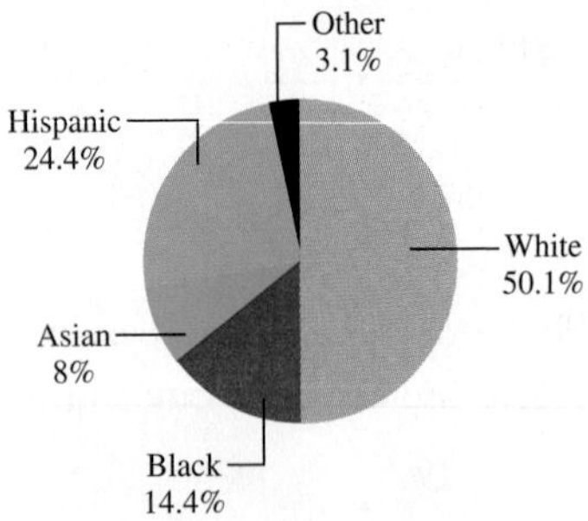

FIGURE 1.10 The U.S. population in the year 2050.

EXAMPLE 3 Using Compatible Numbers to Estimate a Population

According to the U.S. Bureau of the Census, by the year 2050, the population of the United States will be 419,854,000. Use the graph in Figure 1.10 to estimate the number of Hispanics who will be living in the United States in 2050.

SOLUTION:

We can replace the percentage* of Hispanics in 2050, which is estimated to be 24.4%, by the simpler number 25%. Recall that 25% is equal to the fraction $\frac{1}{4}$. Also, we can replace the total U.S. population of 419,854,000 by 400,000,000. Then, multiplying we get

$$\text{Number of Hispanics} = \frac{1}{4} \times 400{,}000{,}000$$
$$= 100{,}000{,}000.$$

Now try Exercises 41 to 48. 16

QUIZ YOURSELF 16

Use compatible numbers and Figure 1.10 to estimate the number of blacks there will be in the United States in 2050.

EXAMPLE 4 Who's Doing the Housework?

A sociology researcher, studying the division of household labor between married couples who are both working, found the data given in the graph in Figure 1.11.

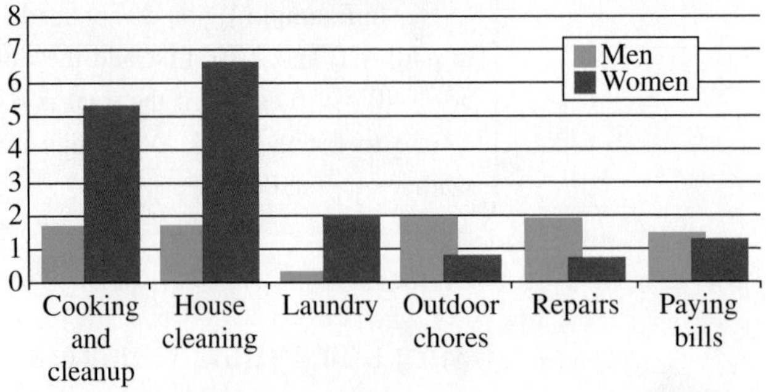

FIGURE 1.11 Household labor done by married couples.

Use this information to estimate answers to the following questions.

a) How many more hours per week do women spend on cooking and cleaning up than men do?

*If you need to review how to work with percents, see Appendix A.

b) How many more hours per week do men spend on outdoor chores, repairs, and paying bills than women do?

c) How many more total hours per week do women spend on household chores than men do?

d) If this weekly discrepancy is constant throughout one year, how many more 8-hour days of work per year is the typical wife in this study putting in at housework than her husband does?

SOLUTION:

We will estimate the amounts shown in the graph and summarize our estimates in Table 1.8.

	Men	Women
Cooking and cleanup	1.8	5.2
House cleaning	1.8	6.8
Laundry	0.2	2.0
Outdoor chores	2.0	0.8
Repairs	2.0	0.6
Paying bills	1.5	1.3
Total	9.3	16.7

Women spend 3.4 more hours here.

Men spend 2.8 more hours here.

TABLE 1.8 Estimates of household labor done by married couples.

a) It appears that women spend $5.2 - 1.8 = 3.4$ hours more per week on cooking and cleanup.

b) From our estimates, the men spend

$$(2.0 + 2.0 + 1.5) - (0.8 + 0.6 + 1.3) = 5.5 - 2.7 = 2.8$$

more hours per week on the three chores mentioned.

c) For total chores, from the graph, it appears that women spend $16.7 - 9.3 = 7.4$ hours more per week on these chores than men do.

d) If we multiply 7.4 by 52 weeks in a year, we get $7.4 \times 52 \approx 385$ more hours per year. Dividing this by 8 hours in a typical work day, we get $\frac{385}{8} = 48.1$ more days of work for women per year.

Now try Exercises 29 to 32 and 37 to 40.

In Examples 5 and 6, we show you some other ways to make estimates.

Historical Highlight

An Amazing Estimate

The ancient Greek mathematician Eratosthenes (276–194 BC) devised a clever way to estimate Earth's circumference. He believed that the cities Alexandria and Syene (today called Aswan) in Egypt were on the same great circle of Earth. He also knew the distance between the cities because it had been measured by a surveyor called a *bematistes,* who was trained to walk in equal steps. The bematistes, using a unit of measurement called a *stadium,* which was 516.73 feet, found the cities to be 5,000 stadia apart.

Using simple but clever geometry calculations, which we explain in Example 3 of Section 9.1, Eratosthenes determined the distance from Alexandria to Syene to be equal to $\frac{1}{50}$ of Earth's circumference. He then used this information to estimate Earth's circumference to be 24,662 miles, which is just 245 miles less than its true value.

EXAMPLE 5 Estimating Distances from a Map

In the Great Alaska Challenge, boats will race from the starting point at Prudhoe Bay (A) to the finish line at Kodiak (F) (see Figure 1.12).

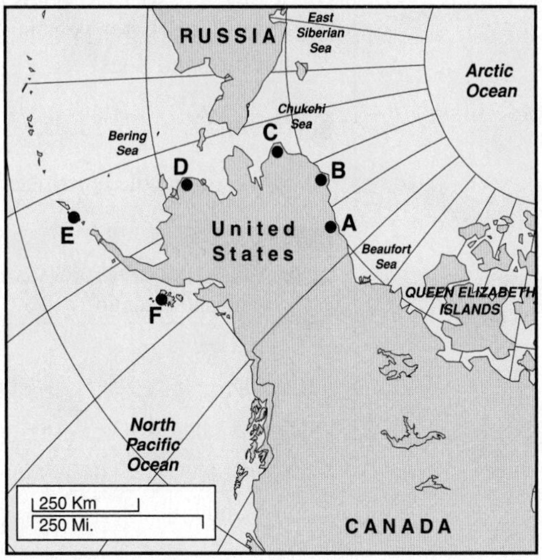

FIGURE 1.12 Map of Alaska.

a) Estimate the length of the race course. Assume that at any point in the race, boats will take the shortest path.

b) If the boat *The Inyuk* travels at 35 miles per hour, how long will it take to complete the course?

SOLUTION:

a) We have marked points A, B, C, and so on, on the map to approximate the total length of the race. If we measure the total length of these distances on the map, we get about 2 inches. The scale shows that $\frac{3}{4}$ inch $=$ 250 miles, so 1 inch $= \frac{4}{3} \times 250 \approx 333$ miles; thus the total length of the race course is approximately 666 miles.

b) If *The Inyuk* travels at 35 miles per hour, then the race will take $\frac{666}{35} \approx 19$.

Now try Exercises 49 and 50.

SOME GOOD ADVICE

When making an estimate, it is a good idea to "estimate your estimate." By that, we mean you should have a rough idea of whether your estimate is too large or too small.

Usually after a large political rally takes place, the news media report an estimate of the size of the crowd. Of course, if the crowd is in the hundreds of thousands, no one has actually counted the people. Example 6 explains a technique that can be used for making this type of estimate. However, the Between the Numbers that follows cautions us to be careful in accepting the reported numbers.

Between the Numbers

Whom Can You Believe?

War Protestors Estimated to Be Somewhere Between 400,000 and Two Million People

If you were to come across this headline on an article on your favorite news Web site, you might think that the people making the estimate really didn't know the size of the crowd. Of course, a news source probably would not give you both estimates. The organizers of the rally would report the crowd to be very large, while those who oppose the rally may want it to seem small. Sometimes, however, a poor crowd estimate is not due to politics, but rather to a lack of proper methods and perhaps using the wrong equipment.

Such was the case in the fall of 1995, when the Million Man March, organized by Afro-American political leader and activist Louis Farrakhan, was reported by U.S. Park Police to be 400,000. Dismayed by the low estimate of crowd size, Farrakhan, who thought the number to be between 1.5 and 2 million, threatened to sue the National Park Service.

The lawsuit was avoided, however, when three days after the march, Dr. Farouk El-Baz, director of The Center for Remote Sensing at Boston University, used enlarged aerial photographs to obtain a more accurate estimate of 870,000, which was acceptable to both Farrakhan and the Park Service. Because of the controversy generated by this bad estimate, the Park Police have since been barred by Congress from issuing crowd estimates.

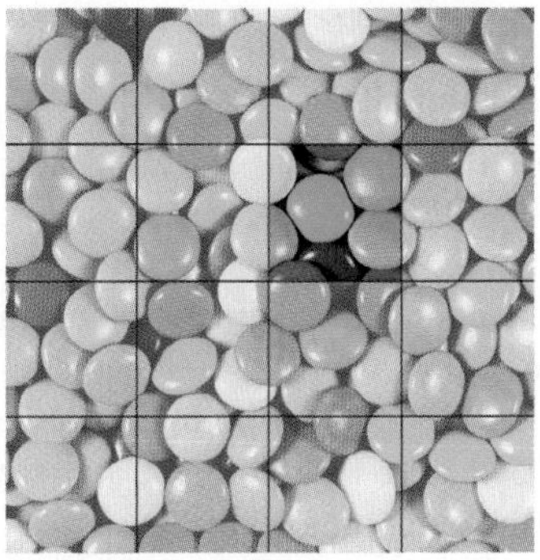

FIGURE 1.13 Estimating number of M&Ms.

EXAMPLE 6 Estimating the Size of a Large Number of Objects

Estimate the number of M&Ms that are visible in Figure 1.13.

SOLUTION:

In Figure 1.13, we have divided the photograph into 16 equally sized rectangles. It appears that each rectangle has approximately the same number of candies. We see that the rectangle we have highlighted contains 10 candies (we will count any candy that is partially visible). Therefore, it is reasonable to estimate that there are $10 \times 16 = 160$ candies in the photograph.

Although there are many formulas in geometry for finding the area of standard geometric figures such as squares, triangles, circles, etc., sometimes we need to estimate the area of an irregularly shaped figure as we do Example 7.

EXAMPLE 7 Estimating an Irregular Area

Suppose that an environmental control officer is spraying for the West Nile virus in a park and needs an estimate of the area of the following region. Assume that each square in the grid in Figure 1.14 has dimension one unit by one unit and represents one acre.

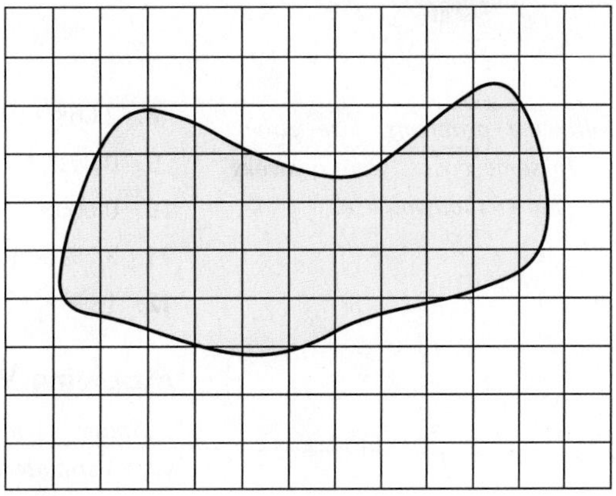

FIGURE 1.14(a) Finding the area of an irregularly shaped region.

SOLUTION: Review the Draw Pictures Strategy on page 3.

We will make our estimate in two stages. First, we find what we call the **inner area** of the figure consisting of all the orange squares completely *contained inside* the shaded region in Figure 1.14(b). If you count the orange squares, you will find that the inner area consists of 24 acres.

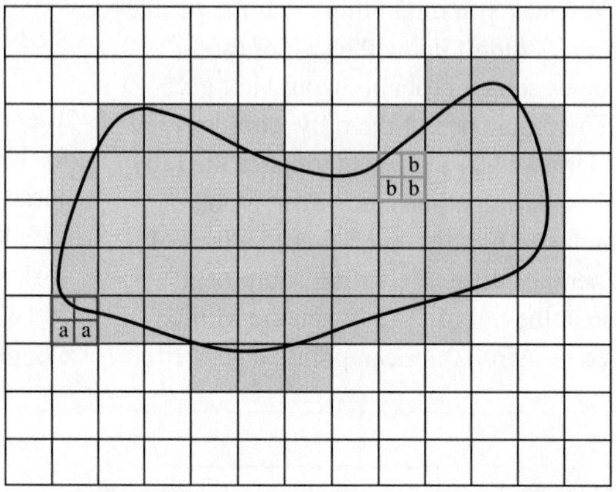

FIGURE 1.14(b) The inner area consists of all the orange squares. The outer area consists of the area covered by either orange or green squares.

Next we will find the **outer area** by adding to the inner area all the green squares in Figure 1.14(b) that *intersect any additional parts* of the region. (For the moment, consider the squares with red and blue borders as green squares.) Counting both the orange and green squares, we see that the outer area consists of 55 acres. Therefore, the area we want is somewhere between 24 (the inner area) and 55 (the outer area) acres.

To finalize our estimate, we will average these to get $\frac{24 + 55}{2} = 39.5$ acres. You can improve this estimate by dividing all of the squares into smaller squares having areas of $\frac{1}{4}$ acre each. The red-bordered square in Figure 1.14(b) shows that by using smaller squares, we would reduce the outer area by not including the two small red squares marked with "a." Similarly, the blue-bordered square shows that by using smaller squares, we will increase the inner area by including the three small blue-bordered squares marked with "b." Using this finer grid will lead to a closer estimate of the true area.

EXERCISES 1.3

Sharpening Your Skills

Estimate the answers to the following problems. Use either rounding or compatible numbers. In some cases, your answers may differ from ours depending on what method you use.

1. $21.6 + 38.93 + 191 + 42.5$
2. $17.18 + 27.2 + 10.31 + 87.6$
3. $34.6 - 15.3$
4. $107.28 - 68.49$
5. 4.75×16.3
6. $1{,}028 \times 14.8$
7. $17.4/3.31$
8. $2{,}068/72.4$
9. 0.091×785
10. $0.0008 \times 4{,}026.3$
11. $8.7\% \times 1{,}024$
12. $18\% \times 683$

Applying What You've Learned

Estimate each of the following answers. Explain how you made your estimate and, where possible, state whether your estimate is larger or smaller than the exact answer. We will provide one

possible answer, and yours may differ from ours depending on how you make your estimate.

13. Training for cross country. Mike is training for the cross country team. He will run 3.7 miles per day, five days a week. How many miles will he run during the next six weeks?

14. Purchasing supplies. Julia is purchasing 18 packs of pencils at $0.94 per pack. What is her bill?

15. Travel time. Olive's father is averaging 47.5 miles per hour on the Hoover's trip to Redondo Beach, CA, for a beauty pageant. If he still has 325 miles to travel and it is now 1:00 PM, about what time should they arrive?

16. The price of gasoline. At a gas station on the turnpike, Raphael spent $57.60 for 14.8 gallons of gasoline. Estimate the price he paid per gallon.

17. Calculating a tip. The Pritchetts went out to dinner and their bill was $118.45. If they want to leave a 15% tip, how much should they leave?

18. Sharing apartment expenses. Ted, Lily, and Marshall share an apartment. Their total utilities bill for last month was $76.38. What is each person's share of the bill?

19. Buying plants. Emily is buying plants for spring planting. She bought three packs of petunias for $2.95 per pack, four packs of vincas for $1.39 per flat, and a package of potting soil for $2.79. What is her total bill?

20. Buying a computer. Shandra bought a new computer for $1,389. If the sales tax is 6%, what is her total bill?

21. Capacity of an elevator. An elevator has a capacity of 2,300 pounds. Alicia and her fifth-grade class of 21 students want to crowd into the elevator. Do you think it is safe? Explain.

22. Capacity of an elevator. Would it be safe for eight linemen for the NFL Washington Redskins to crowd into the elevator mentioned in Exercise 21? Explain.

23. Calculating taxes. Dwight's taxable income is $37,840. If the state income tax rate is 2.4% and the county wage tax is 1.1%, what is the total that he will pay for those two taxes?

24. Estimating a tire warranty. Chuck bought new tires for the Nerd Herd company car that are guaranteed for 42,000 miles. If he drives 11.5 miles each way to work five days a week and then an additional 14 miles each weekend, how many weeks will it be before his warranty expires?

25. Population density. According to the U.S. Bureau of the Census, New Jersey is the most densely populated state with 1,196 inhabitants per square mile and Alaska is the least populated with a density of 1.2 inhabitants per square mile. How many times more densely populated is New Jersey than Alaska?

26. Population density. The United States' average population density is 87.4 inhabitants per square mile while Florida's is 350.6. Roughly how many times greater is Florida's population density than the United States' in general?

27. Calculating tax deductions. Mary Rose works at home as a financial consultant. She wants to claim some of her yearly expenses as tax deductions. Her average monthly basic phone bill is $13.75, electricity is $68.45 per month, and water and sewer is $12.80 per month. If she can claim 1/7 of these expenses as deductions, how much will her annual deduction be?

28. Calculating tax deductions. Ben bought a new car for $19,880 and paid a 2.5% sales tax. He can deduct this amount on his federal income tax return. If he is in the 21% bracket, how much will this deduction save him in taxes?

This graph shows the gender pay gap in yearly salary between men and women according to education in 2010. (Note that college graduates include those that have higher degrees, including professional degrees.)*

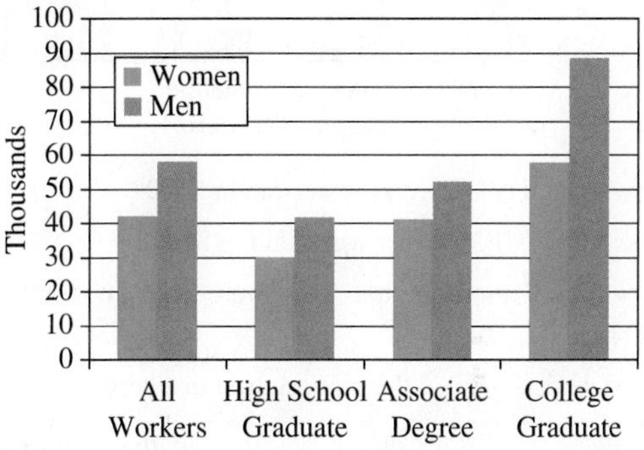

29. Estimate the average yearly salary for male high school graduates. For females with associate degrees.

30. Estimate for all workers how much more men make than women per year.

31. In which category are women farthest from men in salaries? Closest?

32. On average, how much more do male college graduates make than male high school graduates?

The Harris Organization polled 2,309 adult viewers of YouTube to determine whether they would change their viewing habits if short commercials were included before every clip. Use the following pie chart, which summarizes the results of the poll, to solve Exercises 33–36.

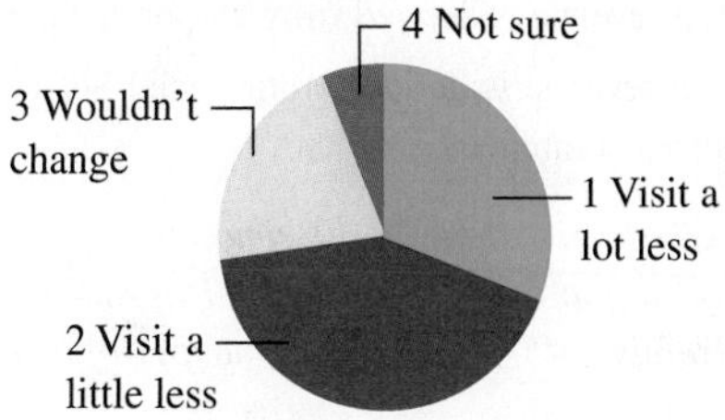

33. Which of the four categories contains the most viewers?

34. Which is the best estimate of the number of viewers that were polled that gave response 3: 500, 300, 1,000, or 1,200?

*Graph from *Essentials of Sociology: A Down-to-Earth Approach,* 9th Edition, by James M. Henslin. Copyright © 2011 James M. Henslin. Printed and electronically produced by permission of Pearson Education Inc., Upper Saddle River, New Jersey.

35. Was response 1 or 2 given more frequently?

36. Estimate how many viewers gave response 4.

The following graph shows the change in the sales of electronic devices from 2004 to 2009 (The New York Times Almanac, 2011). *Use this graph to estimate the answers for Exercises 37–40. We will give the exact answers in the back of the text.*

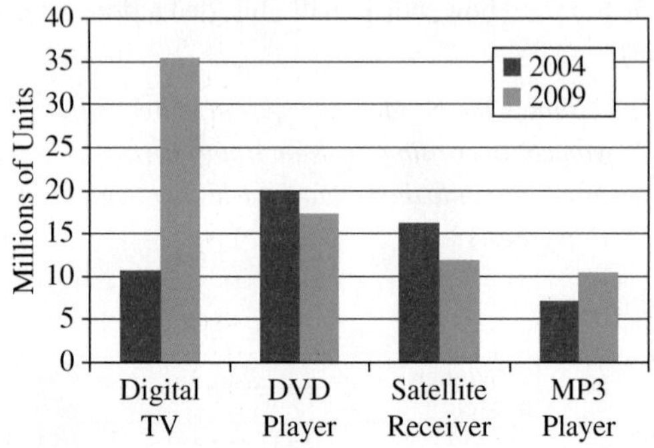

37. How many DVD players were sold in 2009?

38. How many MP3 players were sold in 2004?

39. How many more digital TVs were sold in 2009 than in 2004?

40. What device showed the biggest drop in sales?

The following pie chart shows revenues of the federal government in billions of dollars for the fiscal year 2010. Use this graph to estimate answers to Exercises 41–44. The total revenue is \$2,165 billion.

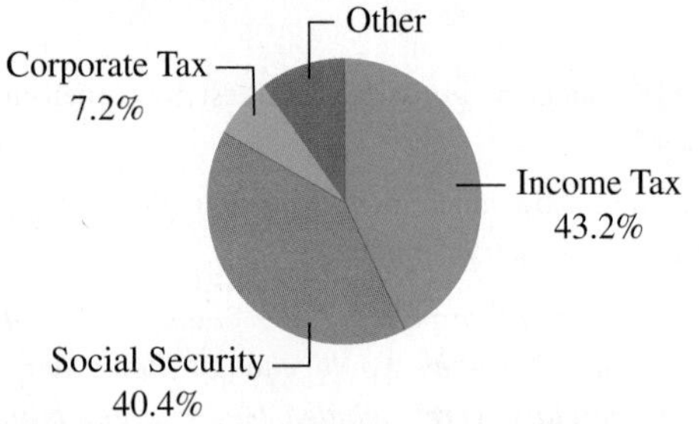

41. How much revenue is derived from Social Security taxes?

42. How much revenue is derived from income taxes?

43. How much revenue is derived from corporate taxes?

44. How much revenue is derived from Social Security taxes and income taxes combined?

The following pie chart shows a distribution of immigration into the United States in a recent year. Use this graph to estimate the answers to Exercises 45–48. The total number of immigrants was 705,361.

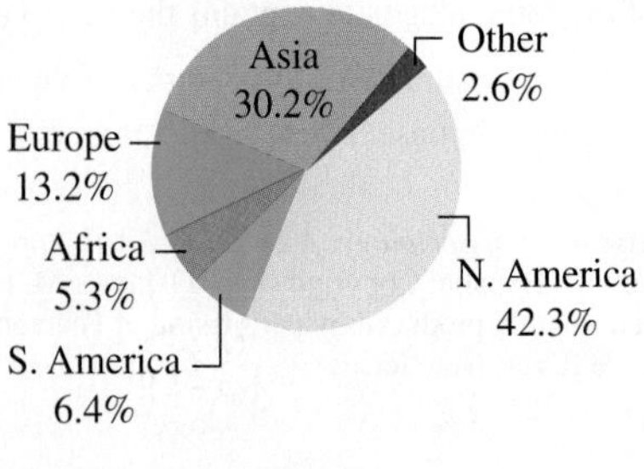

45. How many people immigrated to the United States from Asia?

46. How many people immigrated to the United States from Europe?

47. If 130,661 people immigrated from Mexico, what percentage was that of the total number of immigrants?

48. If 41,034 people immigrated from China, what percentage was that of the total number of immigrants?

Use the given map to estimate the distances (in miles) a sailboat would travel in Exercises 49 and 50.

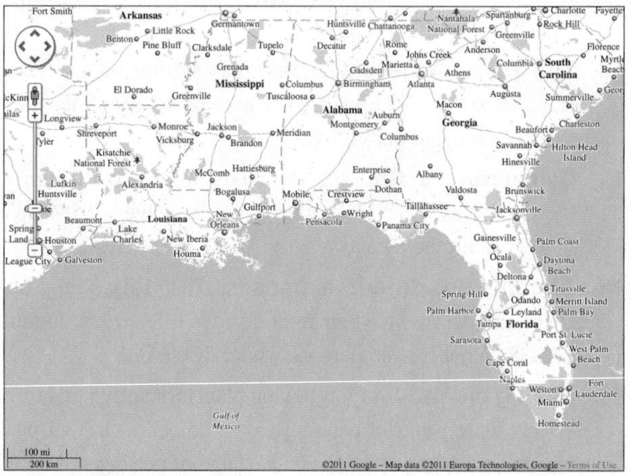

49. From Daytona Beach to Sarasota.

50. From Miami to New Orleans.

Communicating Mathematics

51. Consider the following issues with regard to estimation in mathematics:

a) What do you see as the relationship between estimation and problem solving?

b) How is estimation related to being able to do mental computations?

c) What dangers can result in doing estimation? Give specific examples of the danger of overestimation and underestimation.

52. Ask an acquaintance who runs a household what he or she thinks estimation has to do with each of the following situations (or others that she or he may want to discuss): budgets, cooking, planning a party, paying taxes, planning for college, planning a wedding or other important event, repaying student loans. Compare your results with other students and report on your findings.

Between the Numbers

53. Do online research regarding crowd estimation for various events at the National Mall in Washington, DC, and discuss the controversies that have arisen regarding the discrepancies in the crowd estimates. For example, in addition to the Million Man March example mentioned in this section, you might consider events such as President Obama's inauguration ceremony in 2009, and Glenn Beck's "Restoring Honor" rally in 2010.

54. Research the topic "Estimating Crowds Methodology." Include in your discussion the organizations and methods used in estimating crowds for specific events such as those mentioned in Exercise 53.

Challenge Yourself

55. Buying fertilizer. The Martinezes' yard is 96 feet by 169 feet, and a diagram of their yard is given in Figure 1.15.

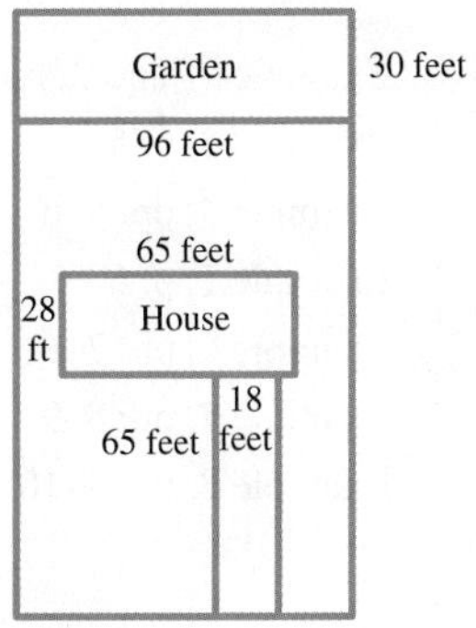

FIGURE 1.15 The Martinezes' yard.

The dimensions of the house, driveway, and garden are also given. The rest of the yard is grass. Mr. Martinez wants to fertilize his lawn. A bag of fertilizer covers 5,000 square feet. Estimate how many bags of fertilizer he will need. Explain how you made your estimate.

56. Purchasing paint. Heidi and Spencer are painting their living room, which is 18.5 feet long and 11 feet wide. The room is 7.75 feet high. If they want to put two coats of paint on the walls and a gallon of paint will cover 200 square feet, how many gallons of paint should they buy? Explain how you arrived at your estimate.

57. Estimating Earth's circumference. Use a map of Egypt to estimate the distance between Alexandria and Aswan. Then multiply your estimate by 50. When Eratosthenes made his estimate of Earth's circumference, he also added an additional 2,000 stadia to his estimate. Do this also. You need to know that there are 5,280 feet in a mile to find the extra miles in your estimate. How does your estimate compare with Eratosthenes' estimate?

58. Assume that the state funding in millions of dollars for a state educational system over the five-year period from 2010 to 2014 is 123.4, 125.2, 126.1, 128.2, 129.3. The graph below shows these data.

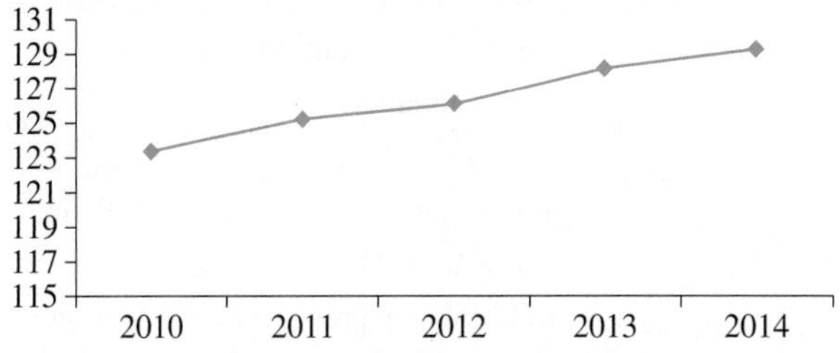

Redraw this graph by changing the vertical scale to emphasize different points to different audiences.

a) You are speaking to parents of college students and you want to emphasize that there has been strong funding support for the system.

b) You are speaking to an anti-tax group and you want to show that government has been trying to keep spending down.

Use the method of Example 7 to estimate the areas of the figures described in Exercises 59 and 60.

59. A circle with radius 4. (The true area is about 50.27 square units.)

60. A square with side 5 that has been rotated 45 degrees. (The true area is 25 square units.)

CHAPTER SUMMARY

Section	Summary	Example
SECTION 1.1	The four steps in Polya's problem-solving method are: 1. Understand the problem. 2. Devise a plan. 3. Carry out your plan. 4. Check your answer.	Discussion, pp. 2–3
	The following strategies are helpful in solving problems:	
	Draw pictures.	Examples 1 and 2, pp. 3, 4
	Choose good names for unknowns.	Example 3, p. 5
	Be systematic.	Example 4, pp. 5–6
	Look for patterns.	Example 5, p. 6
	Try a simpler version of the problem.	Example 6, pp. 7–8
	Guessing is OK.	Example 7, pp. 8–9
	Relate a new problem to an older one.	Example 8, pp. 9–10
	The following principles are useful in problem solving:	
	The **Always** Principle	Example 9, pp. 10–11
	The **Counterexample** Principle	Example 10, p. 11
	The **Order** Principle	Example 11, p. 12
	The **Splitting-Hairs** Principle	Example 12, pp. 12–13
	The **Analogies** Principle	Example 13, p. 13
	The **Three-Way Principle** tells us that we can understand mathematical ideas verbally, graphically, and by making examples.	Discussion, p. 13
SECTION 1.2	**Inductive reasoning** is the process of drawing a general conclusion by observing a pattern in specific instances.	Examples 1–3, pp. 19–21
	Inductive reasoning can lead to a false conclusion.	Example 4, p. 22
	In **deductive reasoning**, we use accepted facts and general principles to arrive at a specific conclusion.	Examples 5–6, pp. 23–25
SECTION 1.3	In **rounding**, we consider the digit to the right of the digit being rounded. If that digit to the right is 5 or more, we round up; otherwise, we round down.	Example 1, p. 29
	In using **compatible numbers**, instead of using the numbers given in a problem, we substitute other numbers that are easier to work with.	Examples 2 and 3, p. 30
	Estimation can help us understand information that is presented to us in maps, graphs, and photographs.	Examples 4–7, pp. 30–34

CHAPTER REVIEW EXERCISES

Section 1.1

1. List the four steps in Polya's problem-solving method.

2. What is a counterexample?

3. Dr. House's Fellowship applicants, Remy, Lawrence, Chris, Amber, and Travis, are working on diagnosing a patient. In how many different ways can we choose two of these students to present their results? Realize that the order in which we choose the students is not important.

4. At a T.G.I. Friday's, you have 8 appetizers, 20 entrées, and 10 desserts. How many different meals can you choose if you select one appetizer, one entrée, and one dessert? Do not solve this problem, but state a simpler version and solve it instead.

5. Picaboo worked 20 hours last week. Part of the time she worked as a stockperson for \$5.65 per hour. The rest of the time she worked as a ski instructor for \$8 per hour. If she earned \$141.20, how many hours did she work at each job? Solve this problem by making and evaluating guesses until you find the answer.

6. Is the following statement true or false?

$$\frac{a}{b} + \frac{c}{d} = \frac{a + c}{b + d}$$

If it is true, give two examples. If it is false, give a counterexample.

7. Explain the Three-Way Principle.

Section 1.2

8. Explain the difference between inductive and deductive reasoning.

9. Do the following situations illustrate inductive or deductive reasoning?
   - **a.** You are following a list of clues in the movie *Source Code* as Colter Stevens works to prevent a bomber from destroying the train.
   - **b.** J.K. Rowling's last four books have sold over 10 million copies. If she writes another book, it will sell over 10 million copies.

10. Use inductive reasoning to predict the next term in the following sequences: a) 2, 7, 12, 17, 22,... b) 3, 4, 7, 11, 18, 29,....

11. Use inductive reasoning to draw the next figure in the pattern.

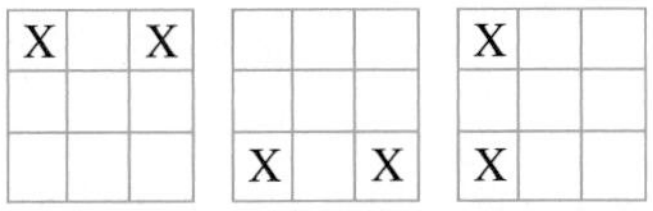

12. Illustrate Goldbach's conjecture for the number 48.

13. Follow the instructions for this "trick" starting with several different numbers of your choice. Make a conjecture as to what result you get in each case. Use algebra and deductive reasoning to explain why your conjecture is correct.
   - **a.** Choose any natural number.
   - **b.** Multiply the number by 8.
   - **c.** Add 12 to the product you just found.
   - **d.** Divide the results of part (c) by 4.
   - **e.** Subtract 3.

Section 1.3

14. Round each of the following numbers to the nearest thousand.
   - **a.** 46,358
   - **b.** 27,541

15. Use compatible numbers to estimate the answers to the following problems. Your answers may differ from ours.
   - **a.** $209.35 - 61.19$
   - **b.** 5.85×15.64

16. Juana is averaging 52.4 miles per hour on her trip to Miami. If she still has 156 miles to travel and it is now 4:00 PM, about what time should she arrive?

17. The given graph compares the average cost of attending college (both four-year and two-year combined) between private and public colleges for several selected years.

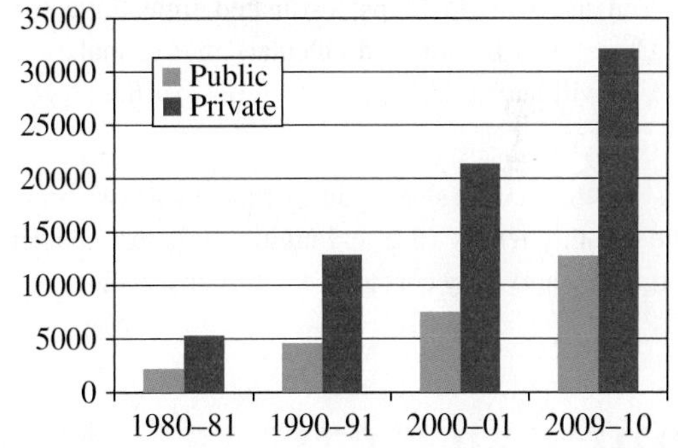

   - **a.** What was the average cost (to the nearest thousand) of attending a private college in the academic year 2000–01? A public college?
   - **b.** Which of the following are true for the time period 1980–81 to 1990–91: The cost of attending a private college i) increased by 50 percent; ii) more than doubled; iii) increased by more than \$10,000; iv) tripled.
   - **c.** Between which two of the given years did the cost of attending a public college increase the most in terms of the percentage increase?

CHAPTER TEST

1. List three problem-solving techniques that we discussed in Section 1.1.

2. Identify which of the following statements is false and give a counterexample.

   a. $\dfrac{a + b}{c} = \dfrac{a}{c} + \dfrac{b}{c}$ **b.** $\dfrac{a}{b + c} = \dfrac{a}{b} + \dfrac{a}{c}$

3. Solve the following problem by making a series of guesses. As of September 2011, the three top-selling video games of all time, Wii Sports, Super Mario Brothers, and Pokémon Red/Green/Blue, sold a total of 118 million copies. If Super Mario Brothers sold 9 million more than Pokémon and 7 million less than Wii Sports, how many copies did Pokémon sell?

4. According to *USA Today*, NASA is tracking 12,000 objects the size of a grapefruit or larger that are orbiting around Earth. The graph on the following page shows the distribution of owners of this space debris.
   - **a.** Estimate the number of objects owned by Russia.
   - **b.** Estimate the number of objects owned by China.

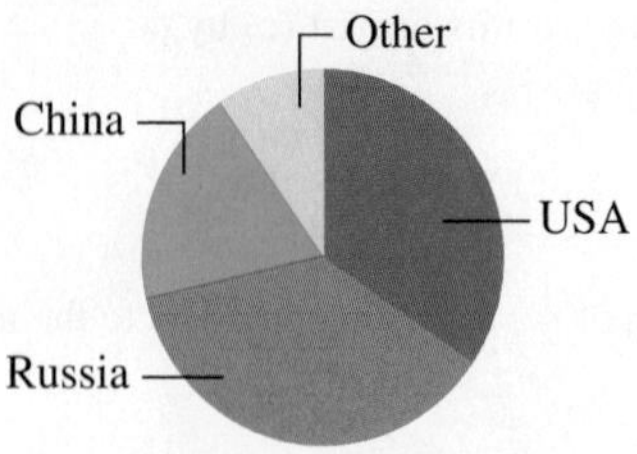

Figure for Exercise 4

5. Round 36,478 a) to the nearest thousand and b) to the nearest hundred.
6. What is The Splitting-Hairs Principle?
7. Explain the difference between inductive and deductive reasoning.
8. What type of reasoning do the following situations illustrate?
   a. Earlier you called your friend Carla and left her a message to call you back. Since Carla is faithful in returning your calls, and has not called back yet, you assume that she has not returned home yet.
   b. You use Google Maps' estimated time for a trip from Chicago to Boston and calculate that to make that time you will have to drive over 63 miles per hour.
9. State the Three-Way Principle.
10. Assume that you are sharing an apartment with two roommates. The monthly rent is \$625 and utilities will cost another \$180. Estimate your yearly cost to live in this apartment.
11. What is the next likely term in the following sequence of numbers?

    1, 2, 6, 15, 31, 56, 92, . . .
12. Continue the following pattern for three more items in the list:

    abc, abd, abe, bcd, bce, bcf, . . .
13. What is the likely next figure in the following sequence?

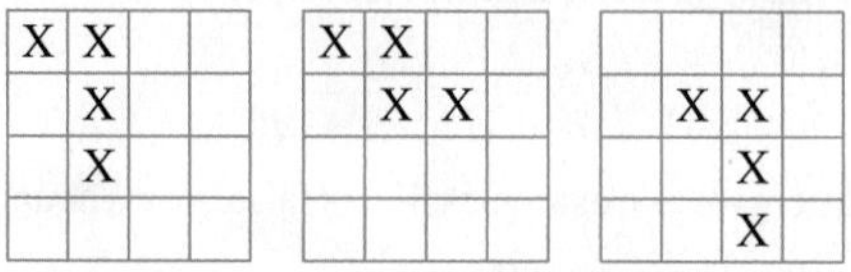

14. Illustrate Goldbach's Conjecture for 60.
15. Determine if the following statement is true or false. If it is false, give a counterexample. If the price of a laptop computer is decreased by 10% and then increased by 10%, the price will be the same as the original.
16. Follow the instructions for the following "trick" by starting with several different values for n. Make a conjecture as to what you will get for an arbitrary n. Prove your conjecture using algebra.
    a. Choose any natural number.
    b. Multiply the number by 4.
    c. Add 40 to the product you just found.
    d. Divide by 2.
    e. Subtract 20.

GROUP PROJECTS

1. **Figurate numbers.** The ancient Greeks were intrigued with certain numbers that they called figurate numbers because they could illustrate the numbers with geometric figures, as we show in Figure 1.16. We represent the first triangular number by T_1, the second by T_2, the third by T_3, etc. Similarly, we represent square numbers by $S_1, S_2, S_3 \ldots$, and the pentagonal numbers by $P_1, P_2, P_3 \ldots$.

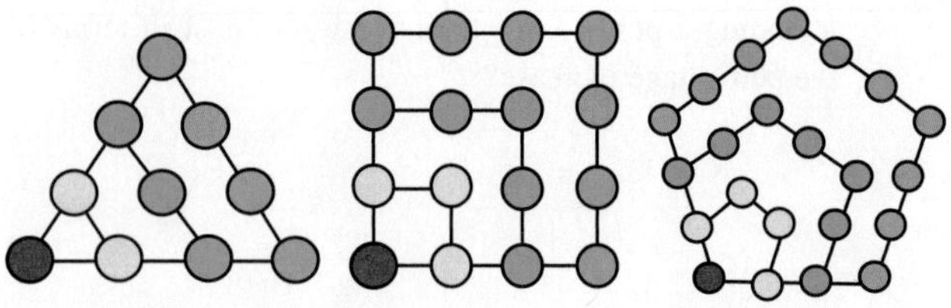

(a) (b) (c)

FIGURE 1.16 (a) triangular (b) square (c) pentagonal

   a) Figure 1.16(a) shows the first four triangular numbers to be 1, 3, 6, and 10. Calculate the next three triangular numbers.
   b) It is known that the nth triangular number is given by the equation $T_n = \frac{n(n+1)}{2}$. Verify this equation for $n = 1, 2, 3, 4, 5, 6, 7$.
   c) Find T_{10} and T_{20}.
   d) The first four square numbers are 1, 4, 9, and 16. Find the next three square numbers.
   e) Consider the following diagram and explain how it shows a relationship between a square number and two triangular numbers. Fill in the question marks in the equation $T_? + T_? = S_?$. Consider other square numbers, look for similar patterns, and express them as similar equations.

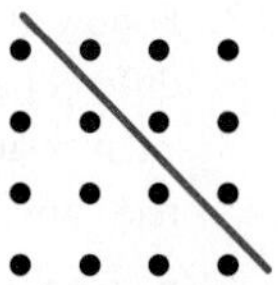

   f) Complete the following equation: $T_? + T_? = S_n$
   g) Search the Internet for a formula for P_n and compute the next three pentagonal numbers.
2. **Estimating crowds.** Get a photo of a crowd scene. Have each member of the group make an estimate of the size of the crowd independently from the rest of the group. Compare your estimates. Average your estimates and compare your average with the averages of the other groups in the class.
3. **Estimating distance on a map.** Get a map and have each member of the group estimate the distance between two cities. Compare your estimates. Average your estimates and compare your average with the averages of the other groups in the class.

USING TECHNOLOGY

1. **Finding a number pattern.**

   a) Use a calculator to find the last digit in the number 2^{100}. If you try to do this on your calculator, you will get an answer that looks something like 1.2676506 E 30, which is the number written in what is called scientific notation (we discuss this in a later chapter). That is not what we are asking you to do. Instead, consider the sequence of computations $2, 2 \times 2, 2 \times 2 \times 2, 2 \times 2 \times 2 \times 2$, etc. until you see a pattern that will allow you to predict the last digit in 2^{100}.

   b) Now find the last digit in the number 7^{100}.

2. **Verifying a conjecture.** In each part in the next column, we give you a series of numerical equations for $n = 1, 2,$ and 3. Make a conjecture as to what the equations will be for $n = 4$ and 5. Finally, use a calculator to check your calculation for $n = 10$.

   a) $1 = \frac{1 \times 2}{2} \quad 1 + 2 = \frac{2 \times 3}{2} \quad 1 + 2 + 3 = \frac{3 \times 4}{2}$

   b) $1 + 3 = 4 \quad 1 + 3 + 5 = 9 \quad 1 + 3 + 5 + 7 = 16$

   c) $\frac{1}{2} = \frac{1}{2} \quad \frac{1}{1 \times 2} + \frac{1}{2 \times 3} = \frac{2}{3} \quad \frac{1}{1 \times 2} + \frac{1}{2 \times 3} + \frac{1}{3 \times 4} = \frac{3}{4}$

3. **Comparing presidential approval ratings.** As I was writing this edition, there was a great deal of political commentary in the media regarding President Obama's approval rating. In researching this topic on the Internet, I found the following fascinating Web site—http://www.gallup.com/poll/124922/Presidential-Approval-Center.aspx—that allowed me to compare President Obama's approval rating with the approval ratings of presidents of the past and thus gave me a historical perspective on the commentary that I was hearing. Go to this site and compare the approval ratings of various presidents that interest you and give a report on what you find.

2

Set Theory

Using Mathematics to Classify Objects

A number of years ago my wife and I were walking on my campus with our young son, who was just learning to talk, when he pointed to a tall man walking towards us and said, "Dada." My wife and I smiled and told him that although he was "a man," he was not "daddy." We explained that his daddy was a man, but not every man was "daddy." At his early age, my son was grappling with a problem we all face throughout our lives—classifying objects into categories, or sets.

Set theory often pops up when you shop online. Concerned about your father's fitness, you might buy *Russian Strength Training Secrets for Every American*, as a gift for Father's Day. The next time you visit that site, you are asked if you would like *The Spartan Warrior Workout: Get Action Movie Ripped in 30 Days*. It is weird that the Web site seems to read your mind—suggesting items that you want before you even ask.

Many stores have "Preferred Customer Cards" that gather information to construct a database of information about your shopping habits, while search engines, such as Google, use set theory to retrieve

huge amounts of information in a matter of seconds. As you study this chapter, you will understand these applications and use set theory to solve problems by organizing data. In Section 2.4, we will show an interesting and important example of how to use set theory to classify blood types, which can be of life-saving importance for someone needing a blood transfusion.

2.1 The Language of Sets

Objectives

1. Specify sets using both listing and set-builder notation.
2. Understand when sets are well defined.
3. Use the element symbol properly.
4. Find the cardinal number of sets.

Did you Google it? If you are savvy about using the Internet, you know that I am asking you whether you used the popular search engine Google to look for information. As I began revising this section, I did a search for "Japanese luxury automobiles" by first specifying each of the individual words separately. When I searched for "Japanese," I received a list of 78,400,000 Web sites, including the Japanese language, cancer research in Japan, Japanese museums, Japanese gardens, as well as many, many other topics. The search for "automobiles" returned 9,080,000 sites, whereas searching for "luxury" returned 34,000,000 sites, including luxury real estate, clothes, and hotels. Finally, searching for "luxury Japanese automobiles," I received a considerably shorter list of "only" 53,000 sites.

In the first three searches, Google was searching through its huge collection of Web addresses to identify a subcollection that contained the word I specified, such as "Japanese." In the last search, we found the sites that had the words "Japanese," "luxury," and "automobiles" in common.

Mathematicians frequently use this notion of grouping together objects with common characteristics so they can treat that collection as a single mathematical object.

In mathematical terms, a collection of objects is called a **set** and the individual objects in this collection are called the **elements** or **members** of the set.

We generally use capital letters to name sets and lowercase letters to denote elements in a set. For example, we may label the set of Olympic gold medal winners by the letter G and denote elements of that set of people by x, a, p, or d. It is important to choose a name for a set to help remember what it is. We could call a group of Republicans X and a group of Democrats Y, but it is easier to work with these sets if we name the set of Republicans R and the Democrats D.

KEY POINT

We represent sets by listing elements or by using set-builder notation.

Representing Sets

We often write a set by **listing** its elements within braces. For example, consider the set of seasons of the year to be the set S. Then, we may write $S = \{\text{spring, summer, fall, winter}\}$. Although it may be possible to list all the elements of a set, it is sometimes inconvenient to do so. If B were the set of all natural numbers* from 1 to 1,000 inclusive, listing all its elements would be cumbersome. Instead, we could write $B = \{1, 2, 3, \ldots, 1{,}000\}$. The first few elements of B are written to establish a pattern. The dots, called an ellipsis, indicate that the list continues in the same manner up to the last number in the set, which is

*We describe some common sets of numbers on page 44.

Highlight

Sets of Numbers Commonly Used in Mathematics

$N = \{x : x \text{ is a natural number}\} = \{1, 2, 3, \ldots\}$

$W = \{x : x \text{ is a whole number}\} = \{0, 1, 2, 3, \ldots\}$

$I = \{x : x \text{ is an integer}\} = \{\ldots, -2, -1, 0, 1, 2, 3, \ldots\}$

$Q = \{x : x \text{ is a rational number}\} = \{x : x \text{ is of the form } \frac{a}{b}, \text{ where } a, b \text{ are integers and } b \neq 0\}$

$R = \{x : x \text{ is a real number}\} = \{x : x \text{ has a decimal expansion}\}$

The set N is also sometimes called the set of counting numbers and denoted by C.

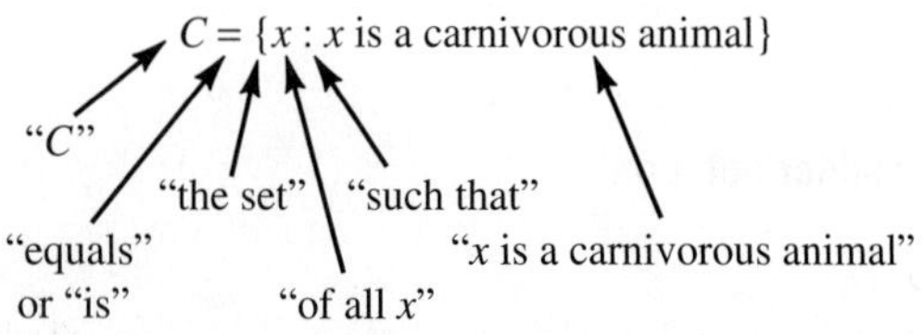

FIGURE 2.1 Reading set-builder notation.

1,000. We would write the set W, consisting of all whole numbers, in a similar fashion as $W = \{0, 1, 2, 3, \ldots\}$. Because we do not put a number after the ellipsis, this means that the list does not end.

If the elements of a set all share some common characteristics that are satisfied by no other object, then we can use **set-builder notation** to represent the set. For example, suppose C is the set of all carnivorous (meat-eating) animals. Using set-builder notation, we write

$$C = \{x : x \text{ is a carnivorous animal}\}.$$

As shown in Figure 2.1, we read this equation as

"C is the set of all x such that x is a carnivorous animal."

Clearly, *lion* is one of the elements of C, but *lamb* is not.

We often write a set represented by the listing method by using set-builder notation, and vice versa.

SOME GOOD ADVICE

It is important to recognize that when you read the set equation in Figure 2.1 out loud, you get a grammatically correct English sentence.

QUIZ YOURSELF 1

Express each set using an alternative method.

a) $A = \{y : y \text{ is a day of the week}\}$

b) $B = \{1, 2, 3, \ldots, 60\}$

EXAMPLE 1 Using Set Notation

Use an alternative method to write each set.

a) $T = \{A^-, A^+, B^-, B^+, AB^-, AB^+, O^-, O^+\}$

b) $B = \{y : y \text{ is a color of the American flag}\}$

c) $A = \{a : a \text{ is a counting number less than 20 and is evenly divisible by 3}\}$

SOLUTION:

a) We can use set-builder notation to write

$$T = \{x : x \text{ is a blood type}\}.$$

b) B can be written using the listing method as $B = \{\text{red, white, blue}\}$.

c) We can write $A = \{3, 6, 9, 12, 15, 18\}$.

Now try Exercises 13 to 22. 1

Well-Defined Sets

KEY POINT

Sets must be well defined.

A set is **well defined** if we are able to tell whether any particular object is an element of that set. Example 2 illustrates this idea.

EXAMPLE 2 Determining Whether a Set Is Well Defined

Which sets are well defined?

a) $A = \{x : x \text{ is a winner of an Academy Award}\}$

b) $T = \{x : x \text{ is tall}\}$

SOLUTION:

a) This set is well defined because we can always determine whether a person belongs to A. Morgan Freeman, Kate Winslet, and Sean Penn would be members of A. Hillary Clinton, Harry Potter, and Albert Pujols would not be members of A because they have never won an Oscar.

b) Whether or not a person belongs to this set is a matter of how we interpret *tall*; therefore, T is not well defined. Can you think of one situation in which a person who is 6 feet tall would be considered tall and a different situation in which that same person would be considered short?

Now try Exercises 23 to 30. 2

QUIZ YOURSELF 2

Which sets are well defined?

a) $\{x : x \text{ is a mountain over } 10{,}000 \text{ feet high}\}$

b) $\{y : y \text{ is a scary movie}\}$

In using set-builder notation, it is possible to have a condition that no object satisfies. For example, the set

$$M = \{m : m \text{ is in your math class and also plays in the National Football League}\}$$

has no elements.

DEFINITION The set that contains no elements is called the **empty set** or **null set.** This set is labeled by the symbol $\varnothing$. Another notation for the empty set is $\{\ \}$.

The set M that we just described is the empty set. We can write this as $M = \varnothing$.

QUIZ YOURSELF 3

Do $\{\ \}$ and $\varnothing$ have the same meaning?

EXAMPLE 3 Using Similar Notations Precisely

a) Does $\{\varnothing\}$ have the same meaning as $\varnothing$? b) Do $\{\varnothing\}$ and $\{0\}$ mean the same thing?

SOLUTION: Review the Splitting-Hairs Principle on page 12.

a) Note that $\{\varnothing\}$ is not the same as $\varnothing$. To make this more clear, you might think of a set as a paper bag that you might get at a supermarket. Then, the empty set $\varnothing$ corresponds to an empty bag 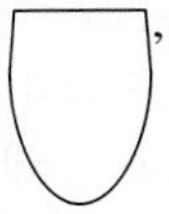, whereas the set $\{\varnothing\}$ could be visualized as one bag containing a second bag, which is empty.

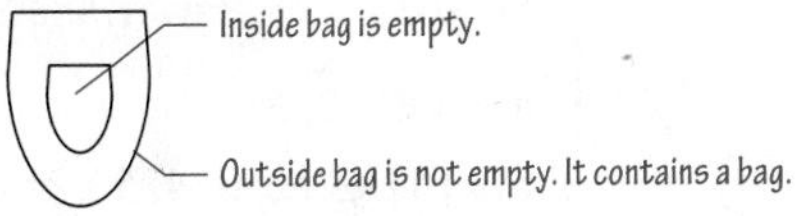

b) Similarly, $\{0\}$ is not the same as $\{\varnothing\}$. If we make bag drawings, then we see that $\{\varnothing\}$ corresponds to a bag containing an empty bag, whereas $\{0\}$ corresponds to a bag containing the number zero.

Now try Exercises 49 to 52. 3

Another set that we frequently use is called the universal set.

Math in Your Life

How Do Babies Learn?

From the time when you were an infant, you needed to make sense out of a world full of new objects, people, sights, sounds, and emotions. How did you ever learn it all? Psychologists who study cognitive development are interested in answering exactly that question. One of the methods by which scientists investigate learning is by presenting subjects with a collection of different objects and trying to understand how the subjects learn to identify the characteristics that are common to various sets of the objects.

For example, in one experiment a researcher was interested in teaching a chimp named Sarah to recognize the similarities and differences in a set of objects. The scientist had plastic symbols for actions such as give and take, other symbols for colors, fruits, etc. One day the scientist wanted Sarah to give him a particular fruit for which he had no plastic symbol. When he showed her the symbols for "give," "apple," and the color "orange," Sarah responded by handing him an orange. As you read this chapter, you will recognize that Sarah was dealing with some of the same basic set theory ideas that you are learning in this chapter.*

DEFINITION The **universal set** is the set of all elements under consideration in a given discussion. We often denote the universal set by the capital letter U.

For example, in a certain problem we may want to use only the counting numbers from 1 to 10. For this discussion, the universal set would be $U = \{1, 2, 3, \ldots, 10\}$. In another situation we might consider only female consumers living in the United States. In this case, the universal set would be $U = \{x : x \text{ is a female consumer living in the United States}\}$.

KEY POINT

The element symbol expresses that an object is a member of a set.

The Element Symbol

We use the symbol $\in$ to stand for the phrase *is an element of.* Although $\in$ looks somewhat like the letter e, these two symbols are *not the same* and should not be confused. The notation "$3 \in A$" expresses that 3 is an element of the set A. If 3 is not an element of set A, we write "$3 \notin A$."

QUIZ YOURSELF 4

Decide whether each statement is true or false.

a) $3 \in \{x : x \text{ is an odd counting number}\}$
b) $2 \notin \varnothing$
c) chocolate $\in \{x : x \text{ is a vitamin}\}$

EXAMPLE 4 Using Set Element Notation

Replace the symbol # in each statement by either $\in$ or $\notin$.

a) 3 # $\{2, 3, 4, 5\}$
b) $\{5\}$ # $\{2, 3, 4, 5\}$
c) Bill Gates # $\{x : x \text{ is a billionaire}\}$
d) jogging # $\{y : y \text{ is an aerobic exercise}\}$
e) the ace of hearts # $\{f : f \text{ is a face card in a standard 52-card deck}\}$

SOLUTION: Review the Splitting-Hairs Principle on page 12.

a) $3 \in \{2, 3, 4, 5\}$
b) $\{5\} \notin \{2, 3, 4, 5\}$. Notice that 5 is a number and is not the same as $\{5\}$, which is a set.
c) Bill Gates $\in \{x : x \text{ is a billionaire}\}$
d) jogging $\in \{y : y \text{ is an aerobic exercise}\}$
e) the ace of hearts $\notin \{f : f \text{ is a face card in a standard 52-card deck}\}$

Now try Exercises 31 to 42. 4

*For an interesting Web site discussing current research in cognitive psychology by Dr. Liz Brannon at Duke University, see http://www.duke.edu/web/mind/level2/faculty/liz/cdlab.htm.

Between the Numbers

Why Should You Care Whether or Not Pluto Is a Planet?

Several years ago, headlines proclaimed that our beloved Pluto was no longer a planet. Following a vigorous debate, the International Astronomical Union declared that according to the new definition of a planet, Pluto had been demoted to being only a dwarf planet. Of course, what we are dealing with here is the importance of the set of planets being well defined. Depending on the definition, Pluto either is one—or it isn't. Alan Stern, the leader of NASA's mission to Pluto, said, "The definition stinks . . . ," and although some have tried to get the definition overturned, today Pluto remains a non-planet.

The notion of something being well defined is very important in your personal life. When you buy insurance on your belongings, your car, or your health, the terms of the policy are carefully defined by insurance company lawyers and, unfortunately, you may not know what you have bought until you make a claim. Then not understanding the definitions in your policy can cost you thousands of dollars.

Cardinal Number

KEY POINT

The cardinal number of a set indicates its size.

In solving set theory problems, we often need to know the number of elements in a set.

DEFINITIONS The number of elements in set A is called the **cardinal number** of set A and is denoted $n(A)$. A set is **finite** if its cardinal number is a whole number. An **infinite** set is one that is not finite.

The following diagram will help you remember the meaning of the notation $n(A)$.

$n(A)$

The n reminds us of the word "number." — Capital A reminds us that we are dealing with a set.

QUIZ YOURSELF 5

Find the cardinal number of each set.

a) $\{2, 4, \ldots, 20\}$

b) $\{\{1, 2\}, \{1, 3, 4\}\}$

c) $\{s : s \text{ is one of the United States}\}$

EXAMPLE 5 Finding the Cardinal Number of a Set

State whether each set is finite or infinite. If it is finite, state its cardinal number using $n(A)$ notation.

a) $P = \{x : x \text{ is a planet in our solar system}\}$

b) $N = \{1, 2, 3, \ldots\}$

c) $A = \{y : y \text{ is a person living in the United States who is not a citizen}\}$

d) $\varnothing$

e) $X = \{\{1, 2, 3\}, \{1, 4, 5\}, \{3\}\}$

SOLUTION:

a) There are eight planets; therefore, this is a finite set. $n(P) = 8$.

b) The set of counting numbers is an infinite set.

c) There are a finite number of people living in the United States who are not citizens; however, we probably do not know $n(A)$.

d) The empty set has no elements, so it is a finite set. Thus, $n(\varnothing) = 0$.

e) If you consider the given bag diagram, you can easily see that the set X contains three objects: the set $\{1, 2, 3\}$, the set $\{1, 4, 5\}$, and the set $\{3\}$. Therefore, $n(X) = 3$.

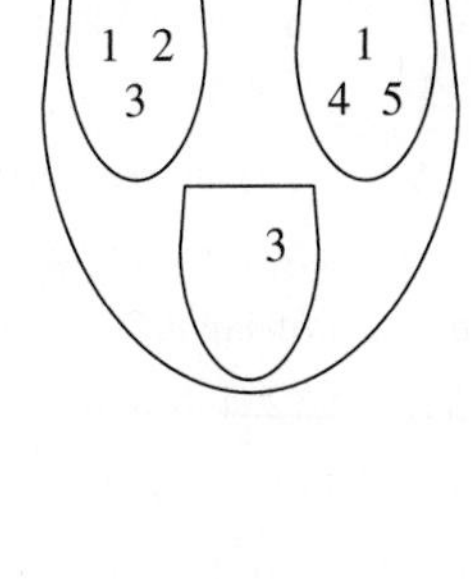

Now try Exercises 43 to 48. 5

EXERCISES 2.1

Sharpening Your Skills

In Exercises 1–12, use set notation to list all the elements of each set.

1. The natural numbers from 10 to 15 inclusive

2. The letters of the alphabet that follow *e* and come before *k*

3. {17, 18, . . ., 25}

4. The integers between −5 and 5, not inclusive

5. The natural numbers less than 30 that are divisible by 4

6. {*y* : *y* is an odd natural number between 6 and 20}

7. The days of the week

8. {*x* : *x* is one of the 50 U.S. states and begins with the word "New"}

9. {*y* : *y* is a natural number less than 0}

10. {11, 13, 15, . . ., 25}

11. The set of female U.S. presidents who served before Barak Obama.

12. {*x* : *x* is a U.S. state that borders no other state}

In Exercises 13–22, use an alternative method to express each set. There may be several acceptable answers.

13. {3, 6, 9, 12}

14. {red, white, blue}

15. {*n* : *n* is a possibility for the number of days in a month of the year}

16. {*z* : *z* is a negative integer}

17. {*y* : *y* is a month of the year}

18. {Aries, Taurus, Gemini, Cancer, . . ., Aquarius, Pisces}

19. {*y* : *y* is a natural number greater than 100}

20. {2, 4, 6, 8, 10, . . .}

21. {2, 4, 6, 8, 10, . . ., 100}

22. {*x* : *x* is a natural number that is evenly divisible by 3}

In Exercises 23–30, determine whether each set is well defined.

23. {*x* : *x* lives in Michigan}

24. {2, 4, 6, 8, 10, . . .}

25. {*y* : *y* has an interesting job}

26. {*t* : *t* has traveled much}

27. {*x* : *x* is a ferocious animal}

28. {*y* : *y* is a mammal}

29. {1, −3, 5, −7, 9, −11, . . .}

30. {*y* : *y* is an easy cell phone number to remember}

In Exercises 31–42, replace each # with either ∈ or ∉ to express a true statement.

31. 3 # {2, 4, 6, 8}

32. 3 # {*x* : *x* is a whole number}

33. George W. Bush # {*x* : *x* is a past president of the United States}

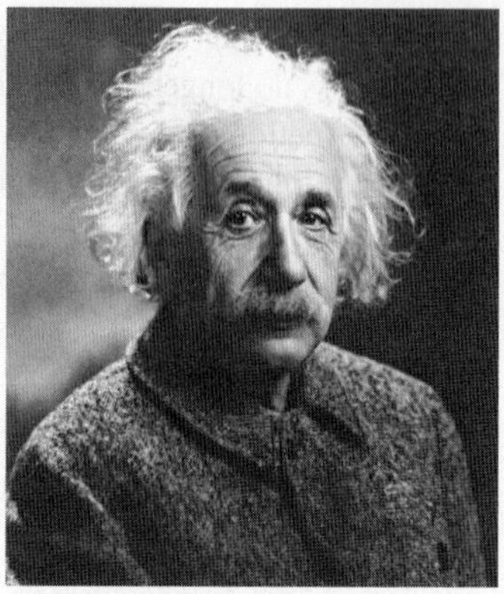

34. Albert Einstein # {*y* : *y* is a a living American poet}

35. SONY # {*w* : *w* is a manufacturer of computers}

36. Oprah Winfrey # {*a* : *a* is a professional ice skater}

37. 5 # {*x* : *x* is a rational number}

38. −5 # {*y* : *y* is a real number}

39. 0 # ∅

40. ∅ # {0}

41. Florida # {*x* : *x* is a state south of Pennsylvania}

42. {Florida} # {*x* : *x* is a state east of the Mississippi}

Find n(A) for each of the following sets A.

43. {1, 3, 5, 7, . . ., 11}

44. {3, 4, 5, . . ., 13}

45. {*x* : *x* is a living American president born before 1900}

46. {*x* : *x* is one of the continental United States}

47. {*x* : *x* is a letter in the word Mississippi}

48. {*x* : *x* is a vowel in the alphabet}

In Exercises 49–52, draw a "bag diagram" similar to what we did in Example 5 on page 47 to illustrate each set, A, before finding its cardinal number.

49. {{1, 2}, {1, 2, 3}}

50. {{1}, ∅, 0, {0}}

51. {{{∅}}}

52. {{1}, {2}, {3}, {1, 2, 3}}

Describe each of the following sets as either finite or infinite.

53. {*x* : *x* is a word written by Shakespeare}

54. {*y* : *y* is the number of people who have walked on the moon}

55. {*y* : *y* is a real number between 4 and 5}

56. {*x* : *x* is an element of the empty set}

In Exercises 57–64, find an element of set A that is not an element of set B. There are many correct answers.

57. A = {*y* : *y* is a number between 4 and 10}
B = {*y* : *y* is a natural number between 4 and 10}

58. A = {*y* : *y* is a member of the human race}
B = {*y* : *y* is a citizen of the United States}

59. A = {*y* : *y* is a manufacturer of electronic products}
B = {*y* : *y* is a company based in the United States}

60. $A = \{y : y \text{ is an animal}\}$
$B = \{y : y \text{ is covered with fur}\}$

61. $A = \{y : y \text{ is a world political leader}\}$
$B = \{y : y \text{ is American}\}$

62. $A = \{y : y \text{ is a car manufacturer}\}$
$B = \{y : y \text{ is a company based in the United States}\}$

63. $A = \{y : y \text{ is a day of the week}\}$
$B = \{y : y \text{ is a weekday}\}$

64. $A = \{y : y$ is a state whose name begins with the letter "A," "B," or "C"$\}$
$B = \{y : y$ is a state whose name begins with the letter "A" or "B"$\}$

Applying What You've Learned

In Exercises 65–68, use the table of general education electives to describe each set in an alternative way.

	Humanities	Writing	World Culture	Cultural Diversity
History012	Yes	Yes	Yes	No
History223	Yes	Yes	Yes	Yes
English010	Yes	Yes	No	No
English220	Yes	Yes	No	No
Psychology200	No	Yes	No	No
Geography115	No	No	Yes	Yes
Anthropology111	Yes	No	Yes	Yes

65. {History012, History223, English010, English220, Anthropology111}

66. {English010, English220, Psychology200}

67. $\{x : x \text{ satisfies a world culture requirement}\}$

68. $\{x : x \text{ does not satisfy a cultural diversity requirement}\}$

In October of 2011, the average price of regular gasoline was \$3.56, or 356 cents, per gallon. Use this information and the given graph to do Exercises 69–72.

Price Per Gallon of Regular Gasoline (October 2011)

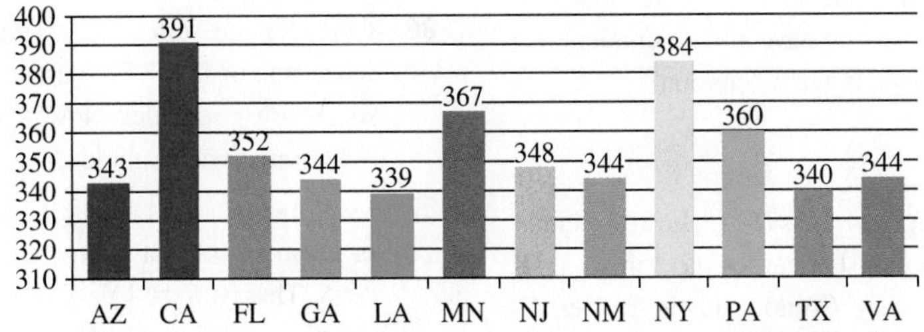

69. Use the listing method to list the set of states L that had regular gasoline at a price less than the national average.

70. Use the listing method to list the set of states G that had regular gasoline at a price greater than the national average.

71. Use set-builder notation to describe the following set of states: {CA, NY}.

72. Use set-builder notation to describe the following set of states: {AZ, GA, LA, NJ, NM, TX, VA}.

Use the given graph, which shows the top 10 parent Internet companies according to the 2011 New York Times Almanac, to do Exercises 73–76.

Audience in Millions

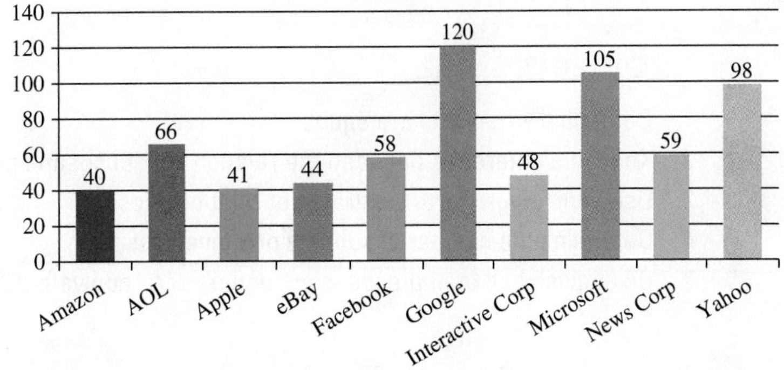

73. Use the listing method to list the set of sites L that had an audience of less than 60 million.

74. Use the listing method to list the set of sites M that had a larger audience than AOL.

75. Use set-builder notation to describe the following set of sites:

$$\{\text{Amazon, Apple, eBay, Interactive Corp}\}.$$

76. Use set-builder notation to describe the following set of sites:

$$\{\text{AOL, Google, Microsoft, Yahoo}\}.$$

Communicating Mathematics

77. The Analogies Principle tells us to look for similarities in terminology and notation with ideas we are already familiar with. How do the notations $=$, $\neq$, $\leq$, and $<$ help you to remember the meaning of corresponding set theory notation?

78. When would it be appropriate to use set-builder notation instead of the listing method to describe a set?

79. Give a careful explanation of the difference in the meanings of $\{\emptyset\}$ and $\emptyset$.

80. How do you "remember how to remember" the meaning of the notation $n(A)$?

Between the Numbers

81. What was the point that we were making in Between the Numbers on page 47? Obtain a copy of an insurance policy and look for several precise definitions in the policy. Describe a situation in which your understanding of the definition could be of critical importance.

82. Describe a situation other than understanding an insurance policy where it is important that you have an exact understanding of the definitions being used in the discussion.

Challenge Yourself

83. Sets of well-known people. Let $U = \{$LeBron James, Babe Ruth, Beethoven, Bach, Leonardo da Vinci, J. K. Rowling, Bart Simpson, Hillary Clinton, Justin Bieber, Eli Manning, Rembrandt, Winston Churchill, Julius Caesar, Shakespeare, Bill Clinton, Beyonce, Harry Potter, Lady Gaga$\}$.

From U we can choose a set of elements that all share some common characteristic and can give this set using both the listing method and set-builder notation—for example, $\{$Bart Simpson, Harry Potter$\} = \{x : x \text{ is a fictional character}\}$. Find as many such sets as you can and give them using both the listing method and set-builder notation.

84. Sets of common objects. Consider the universal set

$U = \{$apple, flat-screen TV, hat, Sirius radio, fish, sofa, washing machine, shoe, dog, automobile, potato chip, toenail clipper, bread, banana, vacuum cleaner, hammer, bed, pizza$\}$.

From U we can choose a set of elements that all share some common characteristic and can give this set using both the listing method and set-builder notation—for example, $\{$flat-screen TV, Sirius radio, washing machine, vacuum cleaner$\}$ $= \{x : x \text{ is an electrical appliance}\}$. Find as many such sets as you can and give them using both the listing method and set-builder notation.

We will define a paradox as a statement that contradicts itself, or a statement that can be proved to be both true and false at the same time. Use this definition in thinking about Exercises 85–87.

85. A paradox. A small town has only one barber, who is male. Furthermore, this barber shaves those men, and only those men, who do not shave themselves. Who shaves the barber? One of two things can happen—either the barber shaves himself or he does not. First assume that the barber shaves himself. What must you conclude? Now assume that the barber does not shave himself. What must you conclude?

86. Consider the statement, "This sentence is false." Is this sentence true or false? First assume that the sentence is true. What must you conclude? Now assume that the sentence is false. What must you conclude?

87. Let A be the set $\{1, 2, 3\}$. Clearly, $A \notin A$. Now think of the set of all such sets that are not elements of themselves and call that set S. That is, $S = \{X : X \text{ is a set and } X \notin X\}$. Now we ask the question, "Is $S \in S$?" First assume that $S \notin S$. What must you conclude? Now assume that $S \notin S$. What must you conclude?

2.2 Comparing Sets

Objectives

1. Determine when sets are equal.
2. Know the difference between the relations of subset and proper subset.
3. Use Venn diagrams to illustrate set relationships.
4. Determine the number of subsets of a given set.
5. Distinguish between the ideas of "equal" and "equivalent" sets.

From the time you were very young, you have been making comparisons like the following: "I can run faster than you." "She's smarter than her brother." "These are my most comfortable shoes." We also often make comparisons in mathematics.

The way we will compare sets may remind you of how you have compared numbers and algebraic expressions, and it will greatly help your understanding if you try to recognize both the similarities and the differences in the ways we make these comparisons.

Set Equality

KEY POINT

Equal sets contain the same members.

One of the most fundamental things we need to know about two sets is when they are considered to be the same.

> **DEFINITION** Two sets A and B are **equal** if they have exactly the same members. In this case, we write $A = B$. If A and B are not equal, we write $A \neq B$.

This definition says that for sets A and B to be equal, every element of A must also be a member of B and every element of B must also be in A.

> **SOME GOOD ADVICE**
>
> When reading a definition, be careful not to assume conditions that are not specifically stated. The set equality definition says nothing about the order in which elements are listed in the sets or whether elements are repeated.

QUIZ YOURSELF 6

Determine whether each statement is true or false.

a) {Socrates, Shakespeare, Beethoven} = {Shakespeare, Beethoven, Socrates}

b) {tiger, gray whale, giant panda} $= \{y : y$ is an endangered species$\}$

EXAMPLE 1 Set Equality

Which pairs of sets are equal?

a) $\{A^-, A^+, B^-, B^+, AB^-, AB^+, O^-, O^+\}$ and $\{A^-, B^-, AB^-, O^-, A^+, B^+, AB^+, O^+\}$

b) $A = \{x : x$ is a citizen of the United States$\}$ and $B = \{y : y$ was born in the United States

SOLUTION:

a) Notice that the left-hand set and the right-hand set contain exactly the same elements. The order of the elements is not important; therefore, the two sets are equal.

b) Because Arnold Schwarzenegger is an element of set A, but is not an element of set B, the sets are not equal. Can you think of any other elements of A that are not in B?

Now try Exercises 1 to 8. 6

Subsets

KEY POINT

One set is a subset of another if all its elements are found in the other set.

Another way we compare sets is to determine whether one set is part of another set.

> **DEFINITION** The set A is a **subset** of the set B if every element of A is also an element of B. We indicate this relationship by writing $A \subseteq B$. If A is not a subset of B, then we write $A \not\subseteq B$.

In order to show that $A \subseteq B$, we must show that every element of A also occurs as an element of B. To show that A is not a subset of B, all we have to do is find one element of A that is not in B.

PROBLEM SOLVING

Strategy: The Analogies Principle*

Similarities in notation and terminology often reflect corresponding similarities in the ideas being represented. Because the notation for "is a subset of," $\subseteq$, reminds us of the notation for "less than or equal to," $\leq$, we might expect that both ideas share similar properties.

EXAMPLE 2 Identifying Subsets

Determine whether either set is a subset of the other.

a) $A = \{1, 2, 3\}$ and $B = \{1, 2, 3, 4\}$

b) $A = \{$Jim Parsons, Edie Falco, Neil Patrick Harris, Melissa McCarthy$\}$ and $E = \{x : x$ has won an Emmy since 2010 $\}$

SOLUTION:

a) Every member of A is also in B, so we can say $A \subseteq B$. Because there is an element of B that is not in A, we write $B \nsubseteq A$.

b) Because every person in set A has won an Emmy since 2010, we can say that $A \subseteq E$. However, Jon Stewart won an Emmy for *The Daily Show* in 2011 but is not a member of A, so E is not a subset of A.

KEY POINT

Venn diagrams represent set relationships graphically.

If A is any set, then $A \subseteq A$ because clearly each element of A is an element of A. Also, the empty set is a subset of every set. For example, $\emptyset \subseteq \{1, 2, 3\}$. Even though this sounds strange, it is true that every element of the empty set is also an element of $\{1, 2, 3\}$. According to the Counterexample Principle from Section 1.1, to show that this statement is false, *you must find a counterexample.* That is, you must find an element of $\emptyset$ that is not in $\{1, 2, 3\}$. Because this is impossible, our statement is true.

Venn Diagrams

An effective problem-solving technique for working with sets is to draw pictures called **Venn diagrams.** Figure 2.2 shows a Venn diagram that illustrates A is a subset of B.

We represent the universal set by the rectangular region, labeled U. The region labeled A is completely contained in region B, indicating that all A's elements are also in B.

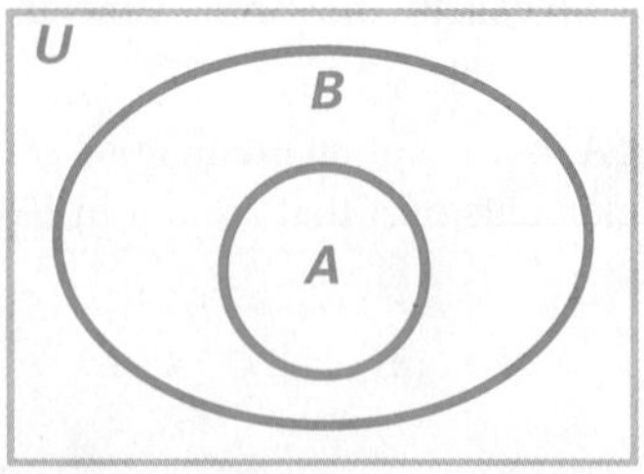

FIGURE 2.2 A Venn diagram illustrating that A is a subset of B.

DEFINITION The set A is a **proper subset** of the set B if $A \subseteq B$ but $A \neq B$. We write this as $A \subset B$. If A is not a proper subset of B, then we write $A \not\subset B$.

From this definition, we see that $\{2, 4, 6, \ldots\} \subseteq \{1, 2, 3, \ldots\}$. Also, $\{2, 4, 6, \ldots\} \subset \{1, 2, 3, \ldots\}$ because $\{1, 2, 3, \ldots\}$ contains elements that are not members of $\{2, 4, 6, \ldots\}$. Also, in Example 2(b), the set $A = \{$Jim Parsons, Edie Falco, Neil Patrick Harris, Melissa McCarthy$\}$ is a proper subset of E because, again, Jon Stewart is not a member of A.

SOME GOOD ADVICE

Distinguish between the notation for subset and proper subset. The lower half of the subset symbol in the notation "$A \subseteq B$" looks somewhat like an equal sign because it is *possible* that sets A and B are equal. Of course, the sets do not have to be equal. When we write $A \subset B$, the lower line is missing to remind us that the sets *cannot* be equal.

*See Section 1.1.

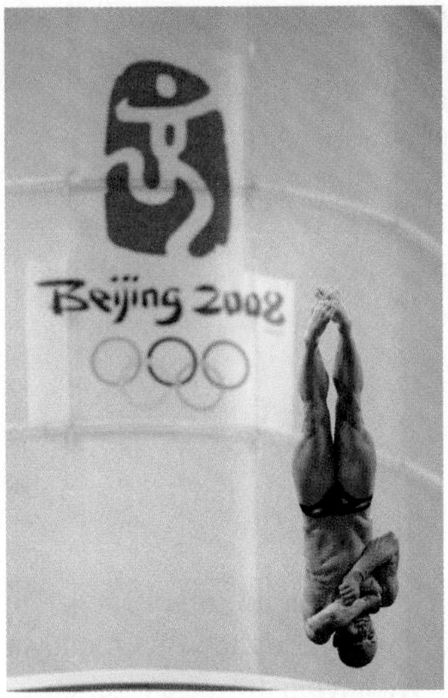

EXAMPLE 3 Identifying Subsets

Consider the following table regarding medalists in the 2008 Summer Olympic Games in Beijing, China.

	Event	Medal	Country
He Wenna	Trampoline	Gold	China
Wu Minxia	Platform diving	Bronze	China
Olha Korobka	Weightlifting (75 kg)	Silver	Ukraine
Ekaterina Khilko	Trampoline	Bronze	Uzbekistan
Nastia Liukin	All-around (Gymnastics)	Gold	United States
Tomasz Majewski	Shot put	Gold	Poland
Kirsty Coventry	100-meter backstroke	Silver	Zimbabwe
Sandra Izbasa	Floor exercise	Gold	Romania
Oksana Chusovitina	Vault	Silver	Germany
Aaron Peirsol	100-meter backstroke	Gold	United States

Assume that this set is the universal set, and define the following sets:

T = the set of trampoline medalists
A = the set of U.S. athletes
G = the set of gold medal winners

Which statements are true?

a) $A \subseteq G$ b) $A \subset G$ c) $T \subset G$

SOLUTION: Review the Analogies Principle on page 13.

a) This is true because every element of A, which equals the set {Liukin, Peirsol}, is also an element of the set G = {Wenna, Liukin, Majewski, Izbasa, Peirsol}.

b) This is also true. We already know that $A \subseteq G$. Because G contains an element that is not in A—for example, Majewski—this means that A is a proper subset of G.

c) This is false. Set T has an element, Khilko, that is not an element of G.

Now try Exercises 9 to 14. 7

QUIZ YOURSELF 7

Decide whether each statement is true or false.

a) $\{2, 4, 6, 8, \ldots\} \subseteq \{1, 2, 3, 4, \ldots\}$
b) In the table in Example 3, $T \subseteq A$.

When one of our cars was totaled in an accident, my wife and I decided to buy a new vehicle. Before we signed on the dotted line, the salesperson asked us if we wanted to consider adding Wi-Fi, OnStar navigation system, SiriusXM radio, or extra paint protection. Suppose you were faced with such a situation. How might you consider the number of ways to make this decision? You might begin by randomly listing different sets of options. We will use w for Wi-Fi, o for OnStar, x for SiriusXM radio, and p for the paint protection. Then some of your choices might be $\{w, o\}$, $\{x, p\}$, $\{o\}$, $\{w, o, p\}$, $\{w, x, p\}$, and so on.

The trouble with this random approach is that after identifying about 11 or 12 subsets, it becomes harder and harder to think of new ones, and you might stop without creating a complete list. As we pointed out in Section 1.1, it is important to attack a problem systematically. We illustrate an organized solution to a similar type of problem in Example 4.

EXAMPLE 4 Finding All Subsets of a Set Systematically

Find all subsets of the set $\{1, 2, 3, 4\}$.

SOLUTION: Review the Be Systematic Strategy on page 5.

We can organize this problem by considering subsets according to their size, going from 0 to 4. This method is illustrated in the following table.

QUIZ YOURSELF 8

Find all subsets of the set {1, 2, 3}.

Size of Subset	Subsets of This Size	Number of Subsets of This Size
0	∅	1
1	{1}, {2}, {3}, {4}	4
2	{1, 2}, {1, 3}, {1, 4}, {2, 3}, {2, 4}, {3, 4}	6
3	{1, 2, 3}, {1, 2, 4}, {1, 3, 4}, {2, 3, 4}	4
4	{1, 2, 3, 4}	1
		Total = 16

Now try Exercises 25 to 28. 8

If we can generalize the solution to one problem, we can often use this knowledge to solve other, related problems. Let us try to generalize the solution that we found in Example 4. To discover a pattern, we look at further examples having different size sets *S*.

S	All Subsets of *S*	Number of Subsets of *S*
∅	∅	1
{1}	∅, {1}	2
{1, 2}	∅, {1}, {2}, {1, 2}	4
{1, 2, 3}	See Quiz Yourself 8	8
{1, 2, 3, 4}	See Example 4	16

QUIZ YOURSELF 9

a) If a set has five elements, how many subsets will it have?

b) How many subsets can be formed using letters of the alphabet?

We see that each time we add an element to *S*, the number of subsets of *S* doubles. The pattern $1 = 2^0, 2 = 2^1, 4 = 2^2, 8 = 2^3$, and $16 = 2^4$ gives us the general relationship that we seek. 9

THE NUMBER OF SUBSETS OF A SET A set that has *k* elements has 2^k subsets.

To get a better idea of why this formula for counting subsets holds, read the discussion of tree diagrams in Exercises 51 and 52.

Equivalent Sets

KEY POINT

Equivalent sets have the same number of elements.

Another relationship that can occur between two sets is that the elements of one set may match up with the elements in the other set.

DEFINITION Sets *A* and *B* are **equivalent,** or in **one-to-one correspondence,** if $n(A) = n(B)$. Another way of saying this is that two sets are equivalent if they have the same number of elements.*

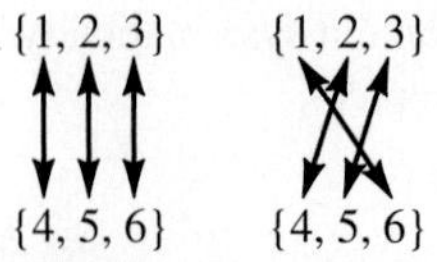

FIGURE 2.3 Two ways to match *A* and *B* in a one-to-one correspondence.

The sets {1, 2, 3} and {4, 5, 6} are equivalent because they have the same number of elements. One way of thinking about equivalent sets is that each element of one set can be matched with exactly one element of the other in a one-to-one correspondence. Figure 2.3 shows two ways to match the elements of {1, 2, 3} and {4, 5, 6}. Can you think of any others? Exercise 69 considers the question of exactly how many one-to-one correspondences there are between two sets.

*We discuss one-to-one correspondences between infinite sets in Section 2.5.

In Example 3, the set of gold medal winners was {Wenna, Liukin, Majewski, Izbasa, Peirsol}, which is not equivalent to the set of U.S. athletes {Liukin, Peirsol}.

SOME GOOD ADVICE

Keep in mind that although the words *equal* and *equivalent* sound similar, *they do not have the same meaning* and cannot be used interchangeably.

EXERCISES 2.2

Sharpening Your Skills

In Exercises 1–8, decide whether each pair of sets is equal. Justify your answer.

1. {1, 3, 5, 7, 9} and {1, 5, 9, 3, 7}
2. {*a*, *e*, *i*, *o*, *u*} and {*a*, *b*, . . . , *u*}
3. {x : x is a counting number between 5 and 19 inclusive} and {y : y is a rational number between 5 and 19 inclusive}
4. {x : x is a counting number between 3 and 10 inclusive} and {y : y is a whole number between 3 and 10 inclusive}
5. {1, 3, 5, . . . , 99} and {x : x is an odd counting number between 0 and 100}
6. {3, 6, 9, 12, 15} and {x : x is a counting number that is a multiple of 3}
7. ∅ and {x : x is a living American born before 1800}
8. ∅ and {∅}

In Exercises 9–14, decide whether each statement is true or false. Justify your answer.

9. {rose, daisy, sunflower, orchid} ⊂ {rose, orchid, sunflower, daisy, carnation}
10. {collie, poodle, beagle, chihuahua, bulldog} ⊆ {bulldog, chihuahua, beagle, poodle, collie}
11. {x : x is a letter in the word *happy*} ⊆ {y : y is a letter in the word *happiness*}
12. {t : t is a letter in the word *Ruth*} ⊂ {z : z is a letter in the word *truth*}
13. ∅ ⊆ {1, 3, 5}
14. ∅ ⊂ ∅

In Exercises 15–24, decide whether each pair of sets is equivalent. Justify your answer.

15. {1, 2, 3, 4, 5} and {*a*, *e*, *i*, *o*, *u*}
16. {2, 4, 6, 8, 10, 12} and {2, 3, 4, . . . , 12}
17. {x : x is a letter in the word *song*} and {x : x is a letter in the word *songs*}
18. {x : x is a letter in the word *tenacity*} and {x : x is a letter in the word *resolve*}
19. ∅ and {∅}
20. {∅} and {0}
21. {1, 3, 5, 7, . . . , 15} and {4, 6, 8, 10, . . . , 18}
22. {a, b, c, d, e, . . . , z} and {3, 4, 5, 6, . . . , 26}
23. {x : x is a day in the year 2012} and {y : y is a day in the year 2011}
24. {x : x was in the starting lineup of the New England Patriots in the 2012 Super Bowl} and {x : x was in the starting lineup of the New York Giants in the 2012 Super Bowl}
25. List all the two-element subsets of the set {1, 2, 3}.
26. List all the two-element subsets of the set {1, 2, 3, 4}.
27. List all the three-element subsets of the set {1, 2, 3, 4}.
28. List all the three-element subsets of the set {1, 2, 3, 4, 5}.
29. If set A has five elements, how many subsets does A have? How many of these subsets are proper?
30. If set A has seven elements, how many subsets does A have? How many of these subsets are proper?

Applying What You've Learned

Use the following table to answer Exercises 31–34.

	Major	Class Rank	GPA	Activities
Allen	Music	Freshman	1.9	Drama
Belinda	Art	Senior	2.8	Newspaper
Carmen	English	Freshman	3.1	Baseball
Dana	History	Freshman	2.9	Drama
Elston	Art	Senior	2.8	Band
Frank	Sociology	Sophomore	3.1	Football
Gina	Chemistry	Junior	2.6	Newspaper
Hector	Physics	Freshman	2.2	Band
Ivana	English	Junior	3.5	Basketball
James	English	Sophomore	2.9	Newspaper

In Exercises 31–34, consider the following sets: U (upperclassmen), L (lowerclassmen), S (science majors), V (GPA above 3.0), A (art majors), T (athletes), and D (involved in drama).

31. Find a set that is equal to V.
32. Find a set that is equivalent to S, but not equal to S.
33. Find a set whose cardinal number is the largest of all the sets.

34. Find a set whose cardinal number is the smallest of all the sets.

Use the table given prior to Exercise 31 to find the number of subsets of each of the following sets.

35. The set of students who are either freshmen or athletes, or both.

36. The set of students who are sophomores and English majors.

37. The set of students who are neither upperclassmen (juniors or seniors) nor have a GPA below 2.5.

38. The set of students who are upperclassmen and have a GPA above 2.5.

39. Domino's Pizza advertises that you can order your pizza plain, with extra cheese, or with any combination of peppers, pepperoni, onion, sausage, anchovies, and olives. In how many different ways can you order your pizza?

40. If Domino's Pizza wants to advertise that there are over 500 ways to order your pizza, how many different toppings must they have available?

41. Burger King advertises that you can "Have it your way." If their burger can be purchased with from none to eight toppings, such as pickle, onion, tomato, etc., in how many different ways can you order your burger?

42. Burger King wishes to outdo Domino's Pizza in Exercise 40 and wants to advertise that you have over a thousand different ways to order your burger. What is the minimum number of different toppings that they must have available?

43. The owners of the Phoenix Flames football team own different amounts of stock in the franchise, so when they vote on an issue they have different amounts of voting power. Assume that Alvarez's and Cianci's votes each will count twice, Belardo's will be counted three times, Devlin's four times, and Espinoza's once. In order for a motion to be passed, the vote must be a total voting weight of at least nine. How many different subsets of this group of owners have a voting weight of at least nine?

44. Five Internet companies are merging so that they can compete with the larger companies. The companies own different stock and so their voting power is weighted accordingly. ComCore's vote will count as four votes, AvantNet's and NanoWeb's will each count as three, eNet's will count as two, and MicroNet's will count only once. For any policy to be passed, a total voting weight of 10 or more is needed. How many different subsets of these companies have a total weight of 10 or more?

Assume that you have in your pocket a penny(S), nickel(P), dime(P), dime(S), quarter(S), quarter(D), and half dollar(D) produced at mints in (S)an Francisco, (P)hiladelphia, and (D)enver. Use the listing method to find the sets described in Exercises 45 to 48. Use 5P to represent a nickel from Philadelphia, 10S to represent a dime from San Francisco, etc.

45. The set totals less than 60 cents and is equivalent to the set of coins minted in San Francisco, but contains no coins minted in San Francisco.

46. The set totals 76 cents, contains a proper subset of the set of coins minted in Denver, and contains fewer than five coins.

47. The set totals 40 cents and contains the set of coins minted in Philadelphia.

48. The set totals 65 cents, contains a proper subset of the coins minted in San Francisco, and contains fewer than four coins.

Communicating Mathematics

49. Which of the following statements are incorrect? Explain the mistake that is being made.

a. $\{1\} \in \{1, 2, 3\}$ **b.** $3 \in \{1, 2, 3\}$

c. $\{2\} \subseteq \{1, 2, 3\}$ **d.** $3 \subseteq \{1, 2, 3\}$

50. What does the Splitting-Hairs Principle that we explained on page 12 have to do with understanding the notations $\in$, $\subseteq$, and $\subset$?

51. How does the following diagram explain why the set $A = \{1, 2\}$ has four subsets?

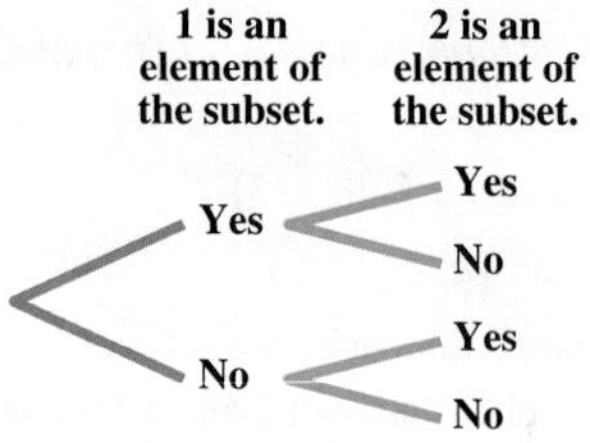

52. Draw a diagram as in Exercise 51 for $B = \{1, 2, 3\}$ and explain why B has eight subsets.

53. On an exam, Pete said that the set B has 25 subsets.

a) Explain why this is not a possible answer.

b) What mistake do you think that Pete made?

c) How many subsets do you think that B had? Why?

54. We have seen that a set with n elements has 2^n subsets. What does this pattern have to do with flipping coins? Taking a true–false test?

Challenge Yourself

When mathematicians find a solution to a problem, they are interested in the practicality of being able to actually compute the solution. You have seen in the box following Example 4 that a set having k elements will have 2^k subsets. We will pursue this idea in Exercises 55–58.

55. Find the largest value for k on your calculator that when you try to compute 2^k, you do not get an overflow. (On the calculator that I was using, the largest k for which I could get an answer was $k = 332$.)

56. If a set has 30 elements, then there will be $2^{30} = 1{,}073{,}741{,}824$ subsets. If you could write one subset every second, how long would it take you to list all 2^{30} subsets? (*Hint:* There are $365 \times 24 \times 60 \times 60$ seconds in a year—we are ignoring leap years.)

57. Find by trial and error the largest k that you could use to list 2^k subsets in 100 years. (Ignore leap years.) Again, you are listing one subset per second.

58. If you were using a computer that could calculate one billion subsets per second, approximately how many years would it take to list all of the subsets in a 100-element set? (*Note:* Your calculator will not give you an exact answer. The answer will be in scientific notation—see Section 6.5.)

In Exercises 59–62, recall that in Section 1.1 we introduced the following arrangement of numbers, which is called Pascal's triangle.

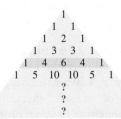

1
1 1
1 2 1
1 3 3 1
1 4 6 4 1
1 5 10 10 5 1
?
?
?

Notice that the fourth line of this triangle contains the numbers 1, 4, 6, 4, 1, which, as we saw in Example 4, are precisely the counts of the number of subsets of a four-element set with 0, 1, 2, 3, and 4 elements, respectively.*

59. With this observation in mind, how do you interpret the fifth line of Pascal's triangle?

60. How do you interpret the sixth line of Pascal's triangle?

61. Tyra Banks is choosing three of the remaining nine contestants in the running to be America's Next Top Model to appear in *Cover Girl Magazine.* In how many ways can this be done? (*Hint:* Use Pascal's triangle as you did in Exercises 59 and 60.)

62. Donald Trump has 10 contestants left on *The Apprentice* and wishes to consider four of them as potential project managers for the next task. In how many ways can this be done? (*Hint:* Use Pascal's triangle as you did in Exercises 59 and 60.)

63. Add the numbers across each row of Pascal's triangle. Do the results look familiar?

64. Notice that the arrangement of numbers in each row is symmetric. See how the 4s are balanced in the fourth row, the 5s are balanced in the fifth row, and so on. Can you give a set theory explanation to account for this?

*We start counting these lines with 0, not 1.

65. Assume the law firm of Dewey, Cheatum, and Howe* has five senior partners and four associates. In order for a policy to be changed, *at least* three of the senior partners must vote for the change. In how many ways can three or more senior partners be chosen?

66. In how many ways can a committee consisting of exactly three senior partners and two associates be formed from the law firm in Exercise 65? Think of forming this committee in two stages:

a. First, think of the number of ways to choose the senior partners and imagine a tree with that many branches.

b. Second, think of the number of ways of choosing the associates.

Imagine attaching the number of branches you determined in part (b) to each of the branches in part (a) to get your final total.

We mentioned that the subset notation, $\subseteq$, and the notation for "less than or equal to," $\leq$, appear to be similar. In Exercises 67–68, for each property of $\leq$, state the corresponding property of $\subseteq$. Next, convince yourself that the newly stated property is indeed a valid set theory property.

67. If $a \leq b$ and $b \leq c$, then $a \leq c$.

68. If $a \leq b$ and $b \leq a$, then $a = b$.

69. By considering examples, find a general pattern that will tell you how many different one-to-one correspondences you can list for a set having n elements.

70. Discuss why it would be impossible with finite sets to have $A \subset B$ and also have a one-to-one correspondence between A and B. Give an example of how this is possible with infinite sets. (*Hint:* See Section 2.5)

*Dewey, Cheatum, and Howe is a fictional law firm that is sometimes mentioned in The Three Stooges movies.

2.3 Set Operations

Objectives

1. Perform the set operations of union, intersection, complement, and difference.
2. Understand the order in which to perform set operations.
3. Know how to apply DeMorgan's laws in set theory.
4. Use Venn diagrams to prove or disprove set theory statements.
5. Use the Inclusion-Exclusion Principle to calculate the cardinal number of the union of two sets.

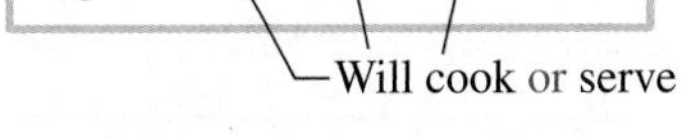

FIGURE 2.4 A Venn diagram illustrating the volunteers at the soup kitchen.

Assume a community service club on your campus is asking for volunteers to cook and serve food, as well as to perform various other tasks at a nearby soup kitchen. In the Venn diagram in Figure 2.4, V is the set of all the volunteers, C is the set of people who cook, and S is the set of people who serve.

There are various subsets of V in this diagram. For example, if we join sets C and S together to form a larger set, then we have the set of people who will either cook or serve the food. We can represent this set by regions r_2, r_3, and r_4.

The set of people who will both cook and serve food is shown in Figure 2.4 as region r_3. You can think of this set as the place where sets C and S overlap.

Finally, we can find the set of people who will only cook, shown as r_2, by starting with set C and removing the elements of set S. 10

You could think of what we were doing in Figure 2.4 as treating the various sets as mathematical objects, and then performing informal operations on them such as "joining," "overlapping," and "removing" to form new sets. We will now give these informal operations the more precise mathematical names of "union," "intersection," and "set difference," and then show you how to do computations with these set operations.

Union of Sets

KEY POINT

We form the union of sets by joining sets together.

Many technical words that we use in mathematics sound exactly like words we use in English. Such is the case with the set theory operation of "union." If you think of how we use *union* in everyday life, you might think of a labor union, the European Union, the United States, a marriage union, etc. In each case, you see the idea of things coming together to make something larger. In forming a set union (see below), we are joining sets together to form a larger set. For example, $\{1, 3, 4, 5\} \cup \{2, 4, 6\} = \{1, 2, 3, 4, 5, 6\}$. Notice that although 4 is an element of both sets, it is not necessary to list it twice.

QUIZ YOURSELF 10

a) Describe the set of people in Figure 2.4 if we remove regions r_2, r_3, and r_4.

b) What regions do you get if you remove all of C from V in Figure 2.4? What set of people do these regions represent?

DEFINITION The **union** of sets A and B, written $A \cup B$, is the set of elements that are members of either A or B (or both). Using set-builder notation,

$$A \cup B = \{x : x \text{ is a member of } A \text{ or } x \text{ is a member of } B\}.$$

The union of more than two sets is the set of all elements belonging to at least one of the sets.

We will now introduce a situation that we will use in several of the examples in this section.

Before deciding on an activity for an exercise campaign, you might consider how convenient it would be to perform the activity and also whether it would require any cost. With this in mind, the given table compares the features of several fitness activities.

Between the Numbers

Getting to Know You

"In Washington, D.C., in the year 2054, murder has been eliminated. The future is seen and the guilty punished before the murder is committed."* Is this only a farfetched plot from the futuristic thriller *Minority Report*, or is it an increasingly possible reality?

In the summer of 2011, Google bought the facial recognition software company PittPatt, acquiring software that identifies people according to photos and videos. PittPatt was developed at Carnegie Mellon University with the financial support of the Defense Department after the terrorist attack on 9/11. If we add this capability to the burgeoning technology of iris scans, voiceprints, smart metering, and GPS tracking, it is clear that the government, insurance companies, and others are getting to know more and more about us.

Some defend these technologies, promising that they will protect us from terrorists and eliminate identity theft. However, others argue that as our personal information is cataloged in huge databases (sets), we are trading security for our privacy and eventually our freedom.

*This quote is taken from a summary of the movie *Minority Report* from the Internet Movie Database.

Activity	Requires Special Location	Requires Special Equipment
Hot room yoga (yo)	Yes	No
Resistance training (rt)	No	Yes
Bicycling (bi)	No	Yes
Calisthenics (ca)	No	No
Hiking (hi)	No	No
Jogging (jo)	No	No
Elliptical machine (el)	Yes	Yes
Tennis (te)	Yes	Yes

These activities form the universal set

$$U = \{\text{yo, rt, bi, ca, hi, jo, el, te}\},$$

where we represent each element by a two-letter abbreviation. Let's define the following subsets of U that we will use in Examples 1 and 2:

E = the set of activities that need special equipment = {rt, bi, el, te}
L = the set of activities that must be done in a special location = {yo, el, te}

EXAMPLE 1 Finding the Union of Sets

Find the union of the following pairs of sets.

a) $A = \{1, 3, 5, 6, 8\}, B = \{2, 3, 6, 7, 9\}$
b) The sets of fitness activities E and L described previously.

SOLUTION: Review the Analogies Principle on page 13.

a) $A \cup B = \{1, 3, 5, 6, 8\} \cup \{2, 3, 6, 7, 9\} =$ the set of elements that are in A or B or both $= \{1, 3, 5, 6, 8, 2, 3, 6, 7, 9\} = \{1, 2, 3, 5, 6, 7, 8, 9\}$.

Notice in our final answer how we listed the elements *in order* and *did not list duplicate elements* because doing so does not affect set equality.

b) $E \cup L = \{\text{rt, bi, el, te}\} \cup \{\text{yo, el, te}\} =$ the set of elements in E or L or both $= \{\text{yo, rt, bi, el, te}\}$

It is good to practice reading the statements in Example 1 aloud to become comfortable with set notation and terminology.

Intersection of Sets

KEY POINT

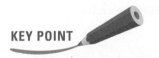

The intersection of sets is the set of elements they have in common.

Another set operation is intersection (see below), which corresponds to our earlier informal notion of overlapping sets. Again, if you think of how we use the word *intersection* in other situations, such as the intersection of streets or the intersection of lines in geometry, you see the essential idea of intersection is a region that is common to both sets. Generally, the intersection of two sets is smaller than either set whereas the union is generally larger.

DEFINITIONS The **intersection** of sets A and B, written $A \cap B$, is the set of elements common to both A and B. Using set-builder notation,

$$A \cap B = \{x : x \text{ is a member of } A \text{ and } x \text{ is a member of } B\}.$$

The intersection of more than two sets is the set of elements that belong to each of the sets. If $A \cap B = \emptyset$, then we say that A and B are **disjoint**.

QUIZ YOURSELF 11

Let $M = \{x : x \text{ is a letter in the word } mathematics\}$ and let $B = \{y : y \text{ is a letter in the word } beauty\}$. Find $M \cup B$ and $M \cap B$.

EXAMPLE 2 Finding the Intersection of Sets

We will find the intersection of the sets discussed in Example 1.

a) $A \cap B = \{1, 3, 5, 6, 8\} \cap \{2, 3, 6, 7, 9\}$ = the set of elements that are in both A and $B = \{3, 6\}$

b) $E \cap L = \{\text{rt, bi, el, te}\} \cap \{\text{yo, el, te}\}$ = the set of elements in E and $L = \{\text{el, te}\}$.

Now try Exercises 1 to 18, which involve only union and intersection. 11

Often you will understand a set theory problem better if you represent it graphically. Figure 2.5 shows how to visualize the union and intersection of two sets.

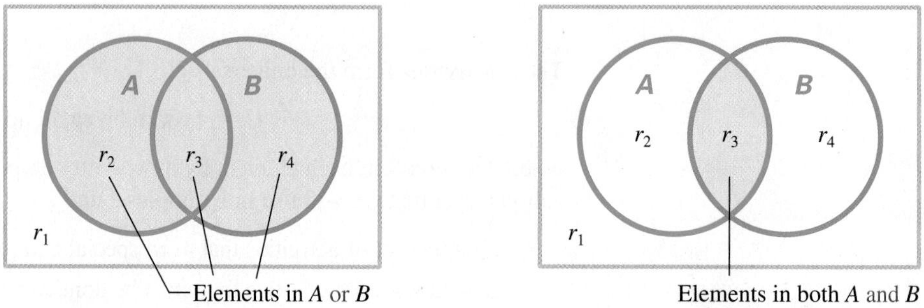

(a) (b)

FIGURE 2.5 (a) Venn diagram of $A \cup B$. (b) Venn diagram of $A \cap B$.

If sets A and B are disjoint, we draw the Venn diagram so that A and B do not overlap.

Set Complement

KEY POINT

The elements not in a set form its complement.

The third set operation that we consider is complementation. When thinking of the complement of a set, think of what elements you need to "complete the set" to get the whole universal set.

DEFINITION If A is a subset of the universal set U, the **complement** of A is the set of elements of U that are *not* elements of A. This set is denoted by A'. Using set-builder notation,

$$A' = \{x : x \in U \text{ but } x \notin A\}.$$

EXAMPLE 3 Finding the Complement of Sets

Find the complement of each set. We have stated a universal set for each set.

a) $U = \{1, 2, 3, \ldots, 10\}$ and $A = \{1, 3, 5, 7, 9\}$.

b) U is the set of people living in the United States, and F is the set of people who have pages on Facebook.

c) U is the set of cards in a standard 52-card deck, and F is the set of face cards.

SOLUTION: Review the Analogies Principle on page 13.

a) A' is the set of elements in U that are not in A, so $A' = \{2, 4, 6, 8, 10\}$.

b) F' is the set of people living in the United States who do not have pages on Facebook.

c) F' is the set of nonface cards.

The complement of A is the shaded region in Figure 2.6.

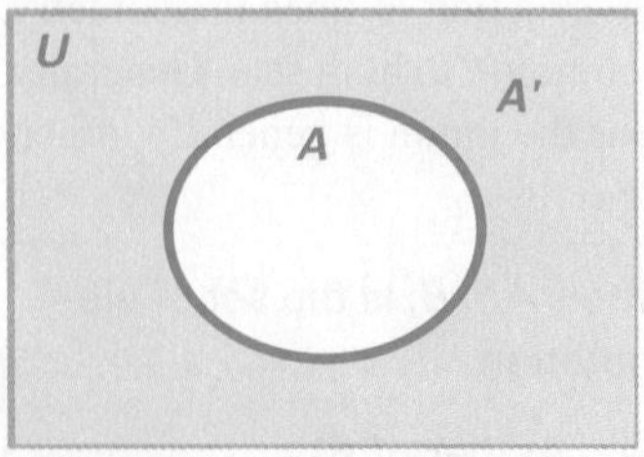

FIGURE 2.6 Venn diagram of the complement of A.

Set Difference

KEY POINT

To form a set difference, begin with one set and remove all elements that appear in a second set.

The last set operation that we discuss in this section is set difference.

DEFINITION The **difference** of sets B and A is the set of elements that are in B but not in A. This set is denoted by $B - A$, Using set-builder notation,

$$B - A = \{x : x \text{ is a member of } B \text{ and } x \text{ is not a member of } A\}.$$

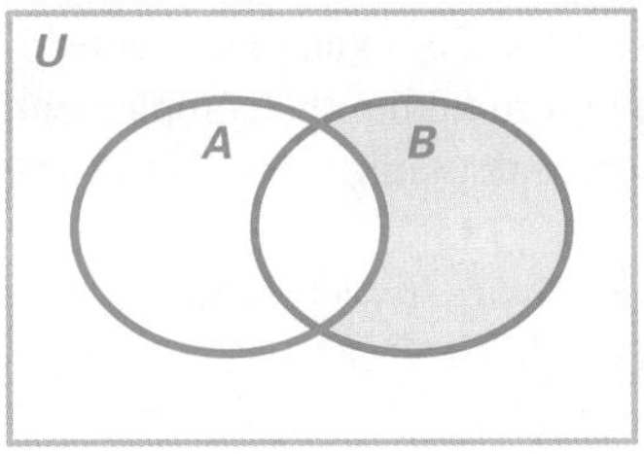

FIGURE 2.7 Venn diagram of $B - A$.

The set difference symbol may remind you of the symbol we use for subtracting numbers—and there is some similarity. When I ask my students how they first learned about subtraction, they usually mention "taking away objects." For example, if you start with five apples and take away two, the result is three. So, a good way to remember how to calculate $B - A$ is to start by listing the elements of set B and then "taking away" the elements that are in set A. The difference of sets B and A is shaded in Figure 2.7.

EXAMPLE 4 Finding the Difference of Sets

a) Find $\{3, 6, 9, 12\} - \{x : x \text{ is an odd integer}\}$.

b) Recall the sets of physical activities $E = \{\text{rt, bi, el, te}\}$ and $L = \{\text{yo, el, te}\}$ from Examples 1 and 2. Find $E - L$ and $L - E$.

SOLUTION: Review the Order Principle on page 12.

a) We start with $\{3, 6, 9, 12\}$ and remove all the odd integers to get $\{6, 12\}$.

Remove elements that are odd.

b) The set $E - L = \{\text{rt, bi, el, te}\} - \{\text{yo, el, te}\} = \{\text{rt, bi}\}$.

Remove elements that are found in L.

Notice that the set $L - E = \{\text{yo, el, te}\} - \{\text{rt, bi, el, te}\} = \{\text{yo}\}$. The fact that $E - L \neq L - E$ is not surprising because we know that if we change the order in which we do a calculation, it often changes the final result. 12

QUIZ YOURSELF 12

a) Let $U = \{a, c, e, g, i, k, m, o, q\}$ and $C = \{x : x \text{ is a letter in the word } game\}$. Find C'.

b) Let $A = \{1, 3, 4, 5, 6\}$ and $B = \{2, 3, 4, 6, 7, 8\}$. Find $B - A$ and $A - B$.

PROBLEM SOLVING

Strategy: The Analogies Principle

It is important to remember the analogies that we have mentioned between set operations and ordinary English words. For *union*, think of joining sets together to get a larger set. *Intersection* reminds you of overlapping streets—you generally get a smaller set. In computing a *complement*, think about finding the elements needed to complete the set to obtain the universal set. In finding a set difference, remember the *take away* model of subtraction. However, it is important to keep in mind that *these informal analogies are not the definitions*—if your instructor asks you for a definition on an exam, use the formal definitions given in the definition boxes.

Order of Set Operations

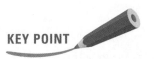

KEY POINT

Set operations must be performed in the correct order.

Just as we must perform arithmetic operations in a certain order, set notation specifies the order in which we perform set operations.

EXAMPLE 5 Order of Set Operations

Let $U = \{1, 2, 3, \ldots, 10\}$, $E = \{x : x \text{ is even}\}$, $B = \{1, 3, 4, 5, 8\}$, and $A = \{1, 2, 4, 7, 8\}$. Find $(A \cup B)' \cap (E' \cup A)$.

SOLUTION:

You should not expect to calculate this set all at once; rather, recognize that it is formed by combining several simpler sets. *The most important thing that you can do now is decide on the order in which to do your calculations.* The notation indicates that some of the operations in this expression must be done before others. For example, in considering $(A \cup B)'$, the parentheses tell us to form the union before taking the complement. In the expression $(E' \cup A)$, we first must find E' before calculating the union. The following is one way to do these calculations.

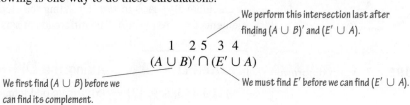

We perform this intersection last after finding $(A \cup B)'$ and $(E' \cup A)$.

1 2 5 3 4

$(A \cup B)' \cap (E' \cup A)$

We first find $(A \cup B)$ before we can find its complement.

We must find E' before we can find $(E' \cup A)$.

We follow these steps in order:

1. $(A \cup B) = \{1, 2, 4, 7, 8\} \cup \{1, 3, 4, 5, 8\} = \{1, 2, 3, 4, 5, 7, 8\}$ Elements in A or B, or both
2. $(A \cup B)' = \{6, 9, 10\}$ Elements not in A union B
3. $E' = \{1, 3, 5, 7, 9\}$ Elements not in E—that is, they are not even
4. $(E' \cup A) = \{1, 2, 3, 4, 5, 7, 8, 9\}$ Elements in E′ or A, or both
5. $(A \cup B)' \cap (E' \cup A) = \{6, 9, 10\} \cap \{1, 2, 3, 4, 5, 7, 8, 9\} = \{9\}$ Elements common to the sets found in steps 2 and 4

Now try Exercises 1 to 12.

Sometimes notation leads us into thinking that something is true when it is not. For example, in algebra we know that $(x + y)^2 \neq x^2 + y^2$, even though it looks like it might be so. Similarly, in set theory if we do not think carefully about the statement $(A \cup B)' = A' \cup B'$, it appears as though it could be true. Example 6 shows that you cannot change the order in which you calculate unions and complements; that is, $(A \cup B)' \neq A' \cup B'$.

QUIZ YOURSELF 13

Let $A = \{1, 2, 5, 7, 8, 9\}$ and $B = \{2, 3, 5, 6, 7\}$ be subsets of the universal set $U = \{1, 2, 3, \ldots, 10\}$. Find the following.

a) $(A \cap B)'$ b) A'
c) B' d) $A' \cup B'$

EXAMPLE 6 Order of Calculating Unions and Complements

Let $U = \{1, 2, 3, 4, 5\}$, $A = \{1, 3, 5\}$, and $B = \{1, 2, 3\}$.

a) Find $(A \cup B)'$. b) Find $A' \cup B'$.

SOLUTION: Review the Order Principle on page 12.

a) As in algebra, the parentheses in the expression $(A \cup B)'$ tell you to work inside the parentheses first before calculating the complement. Hence,

$$(A \cup B)' = (\{1, 3, 5\} \cup \{1, 2, 3\})' = \{1, 2, 3, 5\}' = \{4\}.$$

b) Here, we first take the complements and then find the union, which gives us

$$A' \cup B' = \{1, 3, 5\}' \cup \{1, 2, 3\}' = \{2, 4\} \cup \{4, 5\} = \{2, 4, 5\}.$$

Thus we see that $(A \cup B)' \neq A' \cup B'$. 13

If you look carefully at Example 6, you will see that $(A \cup B)' = A' \cap B'$. This is an example of one of DeMorgan's laws in set theory. Quiz Yourself 13 illustrates the second of DeMorgan's laws.

Historical Highlight

DeMorgan and Boole

Just as biologists classify animals and flowers according to their various characteristics, mathematicians classify mathematical systems according to the properties present in the systems. Set theory is an example of a system called a Boolean algebra, named after the British mathematician George Boole (1815–1864). The son of a shopkeeper, Boole was not able to attend the schools of the more privileged students and had to teach himself Latin, Greek, and mathematics. His book *Investigation of the Laws of Thought*, published in 1854, is one of the classics in the history of mathematics.

Boole, together with Augustus DeMorgan (1806–1871) and others, developed a method of performing logical computations similar to the way in which we do algebra. Their work formed the mathematical basis for all the digital devices that are so prevalent today, such as MP3 players, TiVo, and digital cameras.

DEMORGAN'S LAWS FOR SET THEORY If A and B are sets, then $(A \cup B)' = A' \cap B'$ and $(A \cap B)' = A' \cup B'$.

Viewing set theory as a mathematical system, we can ask whether set operations satisfy some of the same properties that are present in number systems. For example, we know that multiplication is distributive over addition—for all integers a, b, and c, $a \times (b + c) = (a \times b) + (a \times c)$. Example 7 investigates whether intersection is distributive over union.

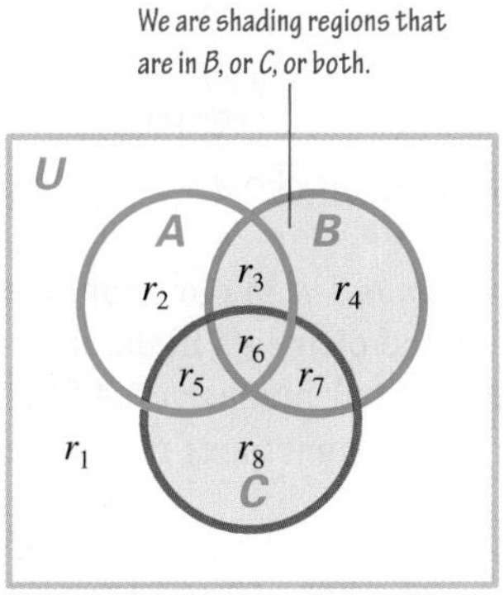

FIGURE 2.8 Venn diagram of $B \cup C$.

EXAMPLE 7 Intersection Distributes over Union

Let A, B, and C be sets in a universal set U. Does intersection distribute over union? That is, does

$$A \cap (B \cup C) = (A \cap B) \cup (A \cap C)?$$

SOLUTION: Review the Draw Pictures Strategy on page 3.

We will answer this question by drawing a Venn diagram for the set on the left-hand side of the equation and another for the set on the right-hand side.

In shading the left-hand set, the parentheses tell us that we must first take the union before considering the intersection. In Figure 2.8, we have shaded $B \cup C$, which consists of regions r_3, r_4, r_5, r_6, r_7, and r_8.

When we intersect $B \cup C$ with set A, we see that of the shaded regions in Figure 2.8, only regions r_3, r_6, and r_5 are also in A. This gives us the Venn diagram of $A \cap (B \cup C)$ shown in Figure 2.9.

Next, we look at the right-hand side of our equation, $(A \cap B) \cup (A \cap C)$. The parentheses tell us to first consider $(A \cap B)$ and $(A \cap C)$ and then take their union. We shade these sets in Figure 2.10.

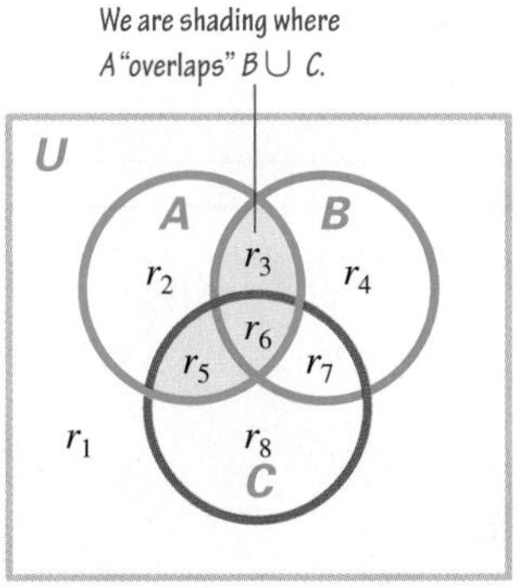

FIGURE 2.9 Venn diagram of $A \cap (B \cup C)$.

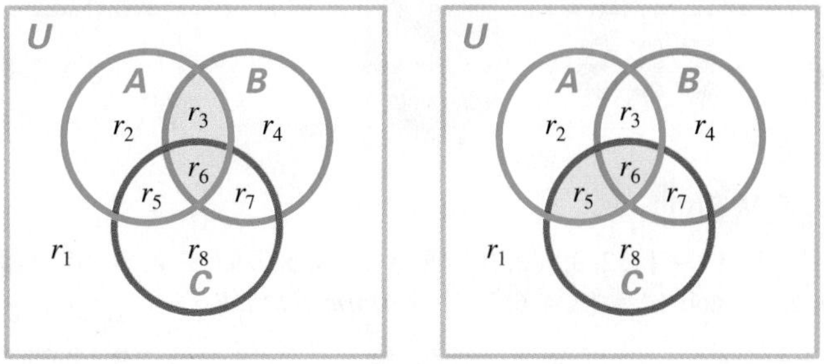

FIGURE 2.10 Venn diagrams of $A \cap B$ and $A \cap C$.

Therefore, $(A \cap B) \cup (A \cap C)$ is made up of regions r_3, r_6, and r_5, which is the same set we shaded in Figure 2.9. Because these two Venn diagrams are the same, this shows that $A \cap (B \cup C) = (A \cap B) \cup (A \cap C)$.

Now try Exercises 23 to 38.

Prior to 1900, physicians were puzzled as to why some patients were saved and others died after receiving blood transfusions. In 1901, Dr. Karl Landsteiner, researcher at the University of Vienna, solved the mystery with his discovery of blood types, for which he later received the Nobel Prize. The next example is based on Landsteiner's classification of blood types, which involves two antigens (proteins that stimulate the bodies immune system) labeled A and B. There are four possibilities: A person can have only A, only B, both A and B, or neither of these antigens. The numbers in the next example are based on Red Cross data.

EXAMPLE 8 Incorrectly Counting the Union of Blood Types

Suppose we test a group of 100 people and find that 40 people have the A antigen and 11 have the B antigen. If we designate the first set of people by A and the second set by B, what is wrong with the following equation: $n(A \cup B) = n(A) + n(B) = 40 + 11 = 51$?

SOLUTION:

In counting the people, we have ignored the fact that some people might possess both the A and B antigens, and so were counted twice. According to the Red Cross, in our group of 100, we can assume that there are 4 people who have both the A and B antigens. So, we have to subtract 4 from our total. That is,

$$n(A \cup B) = n(A) + n(B) - n(A \cap B) = 40 + 11 - 4 = 47.$$

4 people are in both A and B and have been counted twice.

(We will pursue the example of blood typing in greater depth in Section 2.4.)

In counting the elements of $A \cup B$ correctly in Example 8, we *included* in our count the elements of both sets A and B and because of our overcounting the elements in the intersection twice, we subtracted or *excluded* from the count the cardinal number of $A \cap B$. This formula for counting the elements in the union of two sets is sometimes called the **Inclusion-Exclusion Principle.**

THE INCLUSION-EXCLUSION PRINCIPLE If A and B are sets, then
$n(A \cup B) = n(A) + n(B) - n(A \cap B)$.

We must subtract $n(A \cap B)$ so we do not count these elements twice.

EXERCISES 2.3

Sharpening Your Skills

In Exercises 1–12, let $U = \{1, 2, 3, \ldots, 10\}$, $A = \{1, 3, 5, 7, 9\}$, $B = \{1, 2, 3, 4, 5, 6\}$, and $C = \{2, 4, 6, 7, 8\}$. *Perform the indicated operations.*

1. $A \cap B$
2. $A \cup B$
3. $B \cup C$
4. $B \cap C$
5. $A \cup \varnothing$
6. $A \cap \varnothing$
7. $A \cup U$
8. $A \cap U$
9. $A \cap (B \cup C)$
10. $A' \cap (B \cup C')$
11. $(A - B) \cap (A - C)$
12. $A - (B \cup C)$

In Exercises 13–18, consider the universal set U = {*apple, flat-screen TV, hat, satellite radio, fish, sofa, hybrid automobile, potato chip, bread, banana, hammer, pizza*}.

Let M = {x : x *is human-made*}, E = {y : y *is edible*}, G = {t : t *grows on a plant*}. *Find each set.*

13. $M \cap E$

14. $M - E$

15. $E - M$

16. E'

17. $M' \cap G'$

18. $G \cap (M' \cap E)$

Consider the following large and small colored geometric figures. In Exercises 19–22, indicate the number of the region where the specified figure belongs in the given Venn diagram.

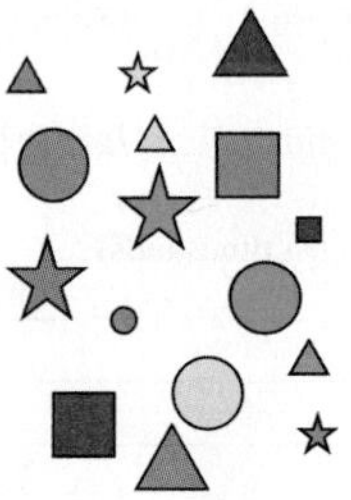

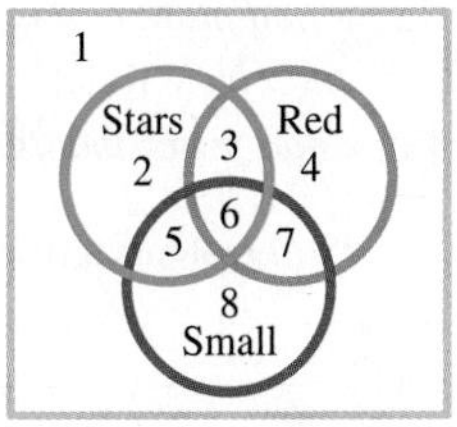

19. ■ **20.** ★ **21.** ☆ **22.** ▲

In Exercises 23–30, represent each set using a Venn diagram.

23. $A - (B \cup C)$

24. $A \cap (B - C)$

25. $(A \cap B) - C$

26. $(A \cup B) - C$

27. $A \cup (B - C)$

28. $A \cup (B \cup C)$

29. $(A \cup (B \cup C))'$

30. $(A \cap (B \cap C))'$

In Exercises 31–38, describe the shaded region using set theory notation.

31.

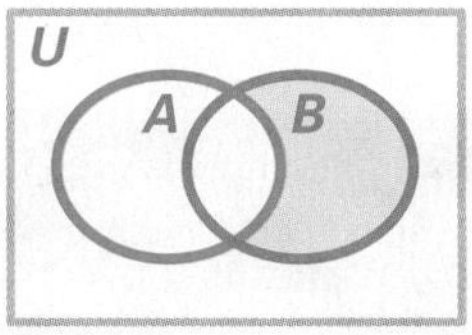

32.

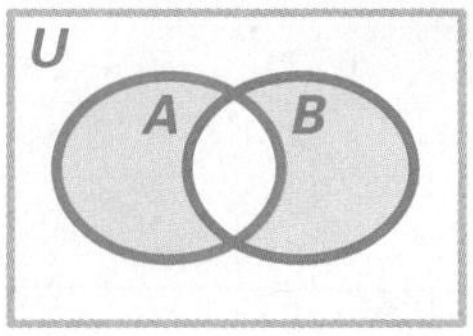

33.

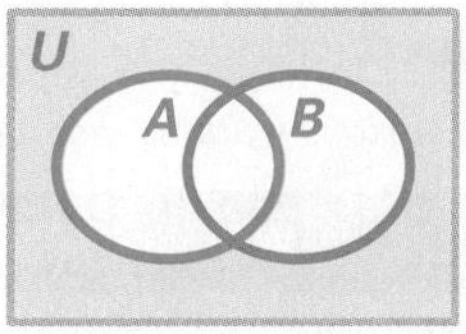

34.

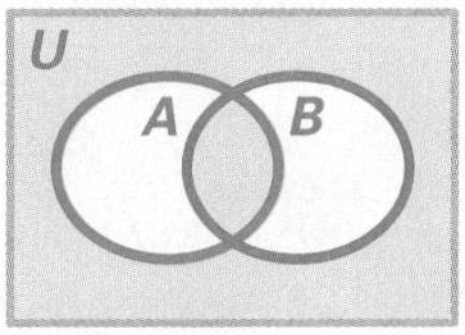

35.

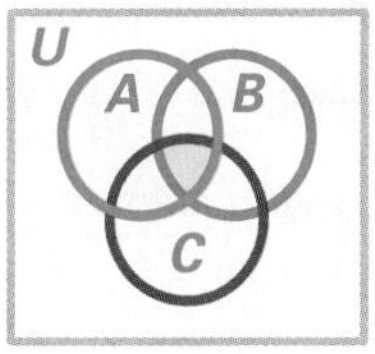

36.

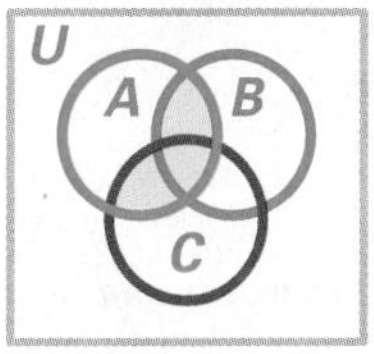

37.

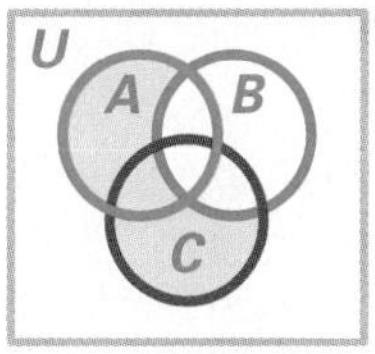

38.

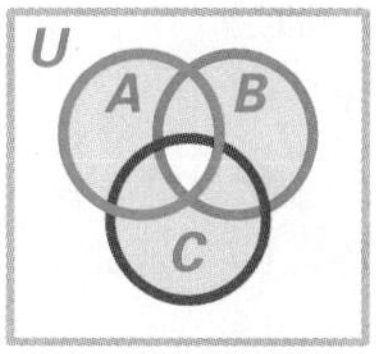

In Exercises 39 and 40, use Venn diagrams to determine if the given pairs of sets are equal.

39. $(A \cup B')'$ and $A' \cap B$

40. $(A' \cap B)'$ and $A \cap B'$

We have indicated the number of elements in each region of the Venn diagram to the right. In Exercises 41–48, find the following.

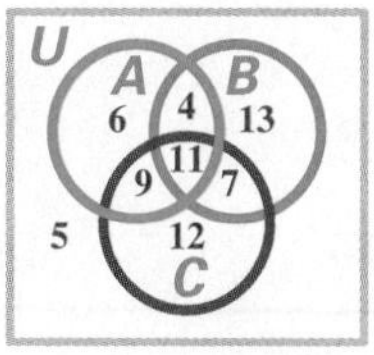

41. $n(A)$

42. $n(A \cup B)$

43. $n(C')$

44. $n(A - C)$

45. $n(A \cap C)$

46. $n(A \cap B \cap C)$

47. $n((A \cup B) \cap C)$

48. $n((A \cup C) - (B \cup C))$

Applying What You've Learned

In the following table, we have listed various features that a person might consider when buying a new car.

	Cost	Size	Warranty (years)	Safety Rating	Antitheft Package
a	$17,800	Subcompact	3	Good	Yes
b	$22,500	Midrange	2	Good	No
c	$19,300	Subcompact	3	Moderate	Yes
d	$21,500	Compact	3	Poor	No
e	$17,200	Subcompact	2	Poor	No
f	$20,700	Compact	3	Poor	No
g	$21,000	Compact	4	Moderate	Yes
h	$23,700	Midrange	4	Good	No

In the universal set U = {a, b, c, . . . , h}, *define subsets of cars with the following characteristics:*

P = *price is above* $20,000
C = *is compact*
G = *has good safety rating*
A = *has antitheft package*
W = *warranty is at least three years*

In Exercises 49–56, first describe each set in words and then find the set.

49. $P \cap C$ **50.** $A \cup G$

51. $W \cap G'$ **52.** $G - A$

53. $P \cap (G \cup W)$ **54.** $G' \cap C'$

55. $P - (G \cup A)$ **56.** $P' - (G \cup C)$

In the following table, we have given nutritional information for the following items available at a fast-food restaurant: Monster (m), *Monster with Cheese* (mc), *Bacon Cheeseburger* (bc), *Hamburger* (h), *Cheeseburger* (c), *Fish Sandwich* (fs), *and Ham Cheesy* (hc).

	% of Minimum Daily Allowance (MDA)			
	Protein	Vitamin A	Vitamin B1	Calcium
m	42	14	25	8
mc	49	21	25	21
bc	49	8	20	17
h	23	3	16	4
c	27	7	16	10
fs	29	—	18	5
hc	37	15	58	19

Let P = {x : x *is an item that provides at least 30% of the MDA for protein*}, C = {x : x *is an item that provides at least 10% of the MDA for calcium*}, B = {x : x *is an item that provides at least 25% of the MDA for vitamin B1*}, A = {x : x *is an item that provides at least 15% of the MDA for vitamin A*}.

In Exercises 57–60, find each set.

57. $P \cap (B \cup A)$ **58.** $(P \cup C) \cap (B \cup A)$

59. $P \cup C \cup B$ **60.** $P \cap C \cap B$

Use the information shown in the following graph regarding the number of immigrants to the given states in 2009 (World Almanac and Book of Facts, *2011*) *to find the sets indicated in Exercises 61–64. Let the universal set be the set of standard state abbreviations, {AZ, CA, FL, IL, MA, NJ, NY, PA, TX, VA}. Also, let*

O = {x : *x had more than 50 thousand immigrants*}
M = {x : *x had more immigrants than Massachusetts, but less than New York*}
F = {x : *x had fewer immigrants than New Jersey*}

Legal Immigrants 2009 (in thousands)

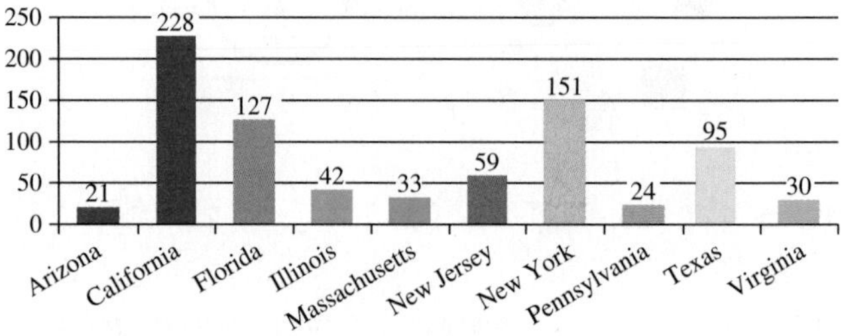

61. $O \cap M$ **62.** $O' \cap F'$

63. $O - M$ **64.** $O \cap (M \cup F)$

According to the 2011 New York Times Almanac, *when adjusted for inflation, the ten top grossing movies of all time are listed in the table below. Define the following sets:*

E = *the set of movies that earned more than $900 million*
O = *the set of movies that won an Oscar*
B = *made before 1970*

Movie	Year	Inflation-Adjusted Gross ($ millions)	Won Oscar
a. Gone With the Wind	1939	1,390	Yes
b. Star Wars	1977	1,225	No
c. The Sound of Music	1965	980	Yes
d. E.T. The Extra Terrestrial	1982	976	No
e. The Ten Commandments	1956	901	No

(Continued)

Movie	Year	Inflation-Adjusted Gross ($ millions)	Won Oscar
f. Titanic	1997	883	Yes
g. Jaws	1975	881	No
h. Doctor Zhivago	1965	854	No
i. The Exorcist	1973	761	No
j. Snow White and the Seven Dwarfs	1937	750	No

Describe each of the following sets in words and then list its elements as a set. For example $B = \{a, c, e, h, j\}$.

65. $E \cap B$ **66.** E' **67.** $O \cap B'$

68. $B - O$ **69.** $E \cap B \cap O$ **70.** $(E \cup B)'$

Communicating Mathematics

71. How does the usage of the terms *union* and *intersection* in the English language help you "remember how to remember" their meaning in set theory?

72. What does the notion of set difference have to do with the idea of removing objects?

73. Students often misstate one of DeMorgan's laws as $(A \cup B)' = A' \cup B'$? Why do you think this is so?

74. Give some examples in this section showing how the Three-Way Principle from Section 1.1 can help you understand set theory concepts.

Between the Numbers

75. Do you have any concerns about government or corporations creating databases of information about you? Give a brief argument either for or against this practice.

76. How do you think the fact that much information about you is available in databases can affect your career? What about the effect in your personal life?

Challenge Yourself

In Exercises 77–80, decide whether each statement is always true. Explain your answer by considering appropriate examples or by drawing a Venn diagram. If you think that a statement is not always true, provide a counterexample. Assume that all sets are finite.

77. If $A \subseteq B$, then $n(A) < n(B)$.

78. If $A - B = \emptyset$, then $A \subseteq B$.

79. If $A \cup B = A \cap B$, then $A = B$.

80. $n(X - Y) = n(X) - n(Y)$

In Exercises 81–84, assume $A \subseteq B$. *Express each set in a simpler way.*

81. $A \cap B$

82. $A - B$

83. $A \cup B$

84. $A' \cap B'$

85. Notice that in comparing number systems with set theory, union is somewhat similar to addition and intersection is somewhat similar to multiplication. For each of the following number system properties, state the corresponding set theory property and determine whether or not it is true.

a. $a \cdot b = b \cdot a$

b. $a + (b + c) = (a + b) + c$

86. Example 7 shows that in set theory, intersection distributes over union. Does union distribute over intersection? Explain your answer.

2.4 Survey Problems

Objectives

1. Label sets in Venn diagrams with various names.
2. Use Venn diagrams to solve survey problems.
3. Understand how to handle contradictory information in survey problems.

Soon after purchasing a new iPad, you might receive an online questionnaire asking, "What is your age? Will you use this product for work? Will you download books, or videos? Play games? Share photos?" Your responses are put into a database, and then set theory is used to extract information when designing and advertising new products.

Government agencies also create huge databases of information to be used in designing and evaluating programs such as Head Start for young children, Stafford loans for college students, the StartRight program for teen moms, and Social Security and Medicare for the elderly.

In this section, you will get a glimpse of how to use set theory to deal with sets of data, but first we need to sharpen our skills in working with Venn diagrams.

Naming Venn Diagrams

KEY POINT

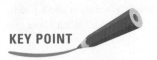

Regions of a Venn diagram can have many different names.

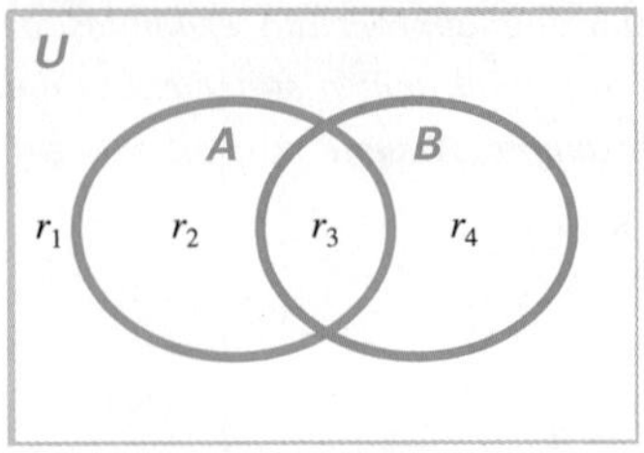

FIGURE 2.11 A Venn diagram with two sets.

In everyday life, it is common for a person to be known by many names. For example, a woman might be known at work as Ms. Smith, at home by Mom or Honey, by her sister as Sis, by her Internet provider as ASmith123, and by the IRS as 178-34-7886. Similarly, the same set can have many different names. For example, consider the four regions shown in the Venn diagram in Figure 2.11.

We can use set notation to name these regions. For example, region r_4 is $B - A$, and region r_1 is $(A \cup B)'$. These names are not unique. Using one of DeMorgan's laws, region r_1 could also be called $A' \cap B'$.

As you can see, we can represent the same set in different ways, just as we often express numbers in different ways. Depending on what calculations we are doing, we may write 6 one time as $3 + 3$ and another time as $5 + 1$. These ideas carry over to Venn diagrams with more than two sets.

EXAMPLE 1 Naming Regions of a Three-Set Venn Diagram

Use Figure 2.12(a) below to answer each question.

a) Name the set we get by combining regions r_6 and r_7.

b) What is a set name for region r_2?

c) Express $A - B$ as the union of two sets.

SOLUTION:

a) Regions r_6 and r_7 are precisely the regions common to B and C; therefore, this set is $B \cap C$. See Figure 2.12(b) below.

b) Region r_2 is clearly *part* of set A; however, elements in region r_2 are not elements of B, nor are they elements of C. Therefore, one name for this set is $A \cap B' \cap C'$. See Figure 2.12(c) below.

Notice that another way you can view this set is to start with set A (regions r_2, r_3, r_5, r_6) and remove those regions that are in $B \cup C$, namely, r_3, r_5, and r_6. Thus, we also can name region r_2 as $A - (B \cup C)$.

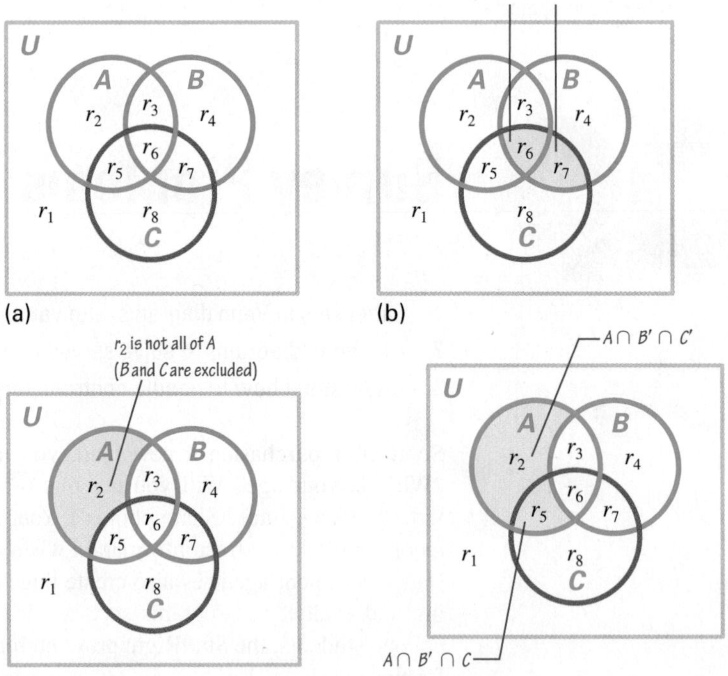

FIGURE 2.12 (a) A Venn diagram with three sets. (b) Venn diagram of $B \cap C$. (c) Venn diagram of $A \cap B' \cap C'$. (d) Venn diagram of $A - B$.

QUIZ YOURSELF 14

Use Figure 2.12 to answer each question.

a) What is a set name for region r_8?

b) Express $B - C$ as the union of two sets.

c) $A - B$ consists of regions r_2 and r_5, which have set names $A \cap B' \cap C'$ and $A \cap B' \cap C$. We can write $A - B = (A \cap B' \cap C') \cup (A \cap B' \cap C)$. See Figure 2.12(d).

Now try Exercises 1 to 10. 14

Survey Problems

KEY POINT

Venn diagrams can organize information in a survey problem.

When we collect data and organize it into sets, we then usually want to analyze the information to answer questions about those sets. These types of problems are often called **survey problems.** Example 2 is an example of a survey problem.

EXAMPLE 2 A Survey Problem Involving Solutions for Global Warming

The National Resource Defense Council, which was instrumental in drafting California's Global Warming Solutions Act, believes that by using technology properly, we can cut U.S. global warming pollution by half. Three of the solutions proposed by the NRDC are using energy-efficient appliances, driving energy-efficient cars, and using renewable energy sources. Assume that you surveyed 100 members of Congress to determine which solutions they favored funding and obtained the following results:

a) 12 favored funding the increased use of renewable energy sources only.

b) 20 recommended funding both energy-efficient appliances and renewable energy sources.

c) 22 favored funding both energy-efficient cars and increased use of renewable energy sources.

d) 14 want to fund all three areas.

From this information, determine the total number who favored increased funding for renewable energy.

SOLUTION: Review the Choose Good Names Strategy on page 5.

We will represent this information by defining the following set:

$$A = \{x : x \text{ favors funding energy-efficient appliances}\}$$
$$C = \{x : x \text{ favors funding energy-efficient cars}\}$$
$$R = \{x : x \text{ favors funding renewable energy sources}\}$$

Then conditions (a) – (d) can be rewritten using set notation as:

a) $n(A' \cap C' \cap R) = 12$ b) $n(A \cap R) = 20$ c) $n(C \cap R) = 22$ d) $n(A \cap C \cap R) = 14$

favor R, not A, not C; favor A and R; favor C and R; favor A, C, and R

We will display this information in a Venn diagram. For example, you can see in Figure 2.13(a) the fact that $n(A \cap C \cap R) = 14$ and $n(A' \cap C' \cap R) = 12$.

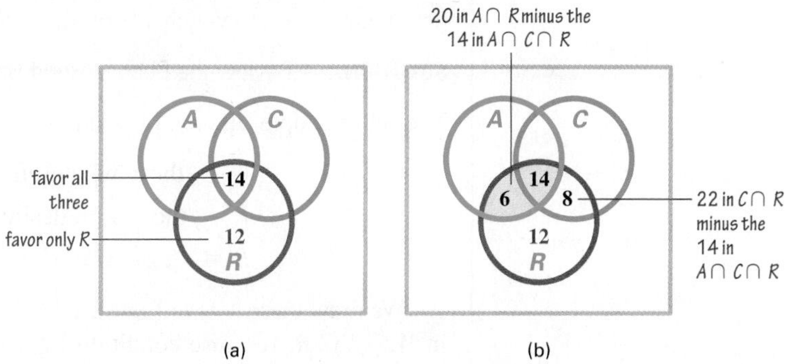

FIGURE 2.13 (a) Venn diagram showing conditions a) and d). (b) Venn diagram showing conditions b) and c).

We have shaded $A \cap R$ in Figure 2.13(b). Because we have already accounted for 14 in $A \cap C \cap R$, there are 6 remaining in $(A \cap R) - C$. Similarly, we can use the fact that $n(C \cap R) = 22$ to get 8 in $(C \cap R) - A$. Adding these numbers, we see that R contains $6 + 14 + 8 + 12 = 40$ members of Congress.

Next we will analyze survey problems that require Venn diagrams having three sets.

In Section 2.3, we discussed Dr. Landsteiner's discovery of the antigens A and B in human blood. Accordingly, a person's blood is classified as A, B, AB (has both), or O (has neither). Blood having the Rh factor is classified as Rh positive (+) and if lacking the factor, Rh negative (−). For example, if a person has the A antigen, lacks the B antigen, and has the Rh factor, her blood would be classified as A^+. Example 3 illustrates the problem we mentioned in the chapter opener.

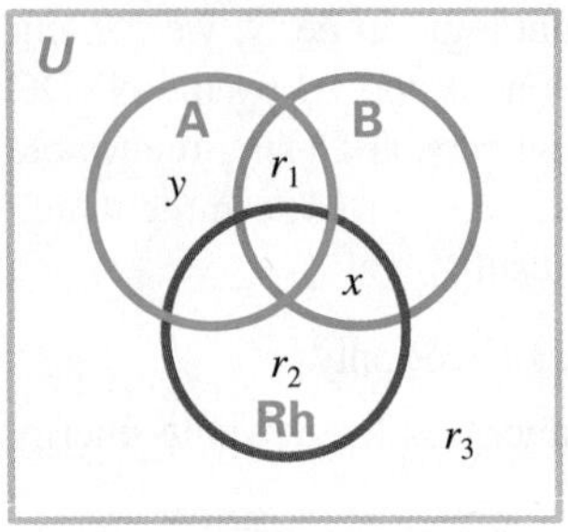

FIGURE 2.14 A Venn diagram illustrating eight blood classifications.

QUIZ YOURSELF 15

State the blood classifications for regions x and y in Figure 2.14.

EXAMPLE 3 Classifying Blood Types

Figure 2.14 shows eight possible blood types. State the classification for blood in regions r_1 to r_3 (for the moment, ignore regions x and y).

SOLUTION:

Region r_1 is contained in both sets A and B, so the blood is type AB. Because r_1 is also outside set Rh, the blood does not contain the Rh factor, so the blood is type AB^-. Region r_2 is in set Rh, but not in either A or B, hence the blood is type O^+. Because r_3 is in none of the sets, r_3 is type O^-.

Now try Exercises 43 to 46. 15

EXAMPLE 4 A Survey of TV Viewer Preferences

A television network conducted a market survey to determine the evening viewing preferences of people in the 18–25 age bracket. The following information was obtained:

a) 3 prefer a reality show early on weekdays.
b) 14 want to watch TV early on weekdays.
c) 21 want to see reality shows early.
d) 8 want reality shows on weekdays.
e) 31 want to watch TV on weekdays.
f) 36 want to watch TV early.
g) 40 want to see reality shows.
h) 13 prefer late, weekend shows that are not reality shows.

From this information, determine how many people do not want to see reality shows and how many prefer to watch TV on the weekend.

SOLUTION: Review the Choose Good Names Strategy on page 5.

The set of people surveyed is a universal set containing three subsets.

$W =$ those who prefer to watch TV on weekdays
$E =$ those who desire early programming
$R =$ those who want to see reality shows

We draw these sets in Figure 2.15. By condition a), we know that there are 3 people in $W \cap E \cap R$. Because condition b) states there are 14 people in $W \cap E$, and we have counted 3 of them, there must be 11 more in this region. We have recorded this information in Figure 2.15(a).

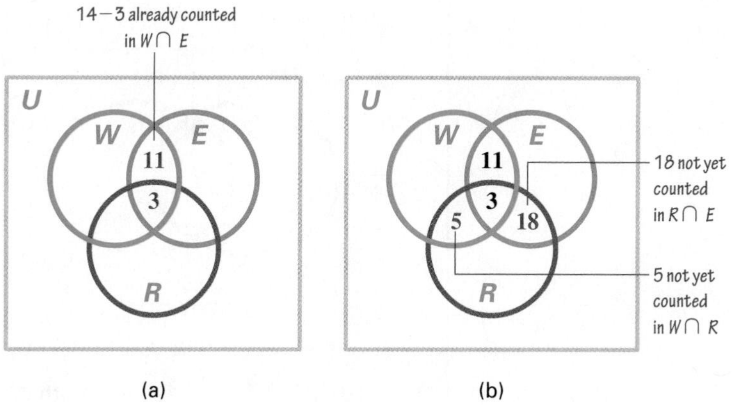

FIGURE 2.15 Counting elements in $W \cap E$, $W \cap R$, and $R \cap E$.

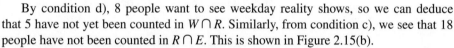

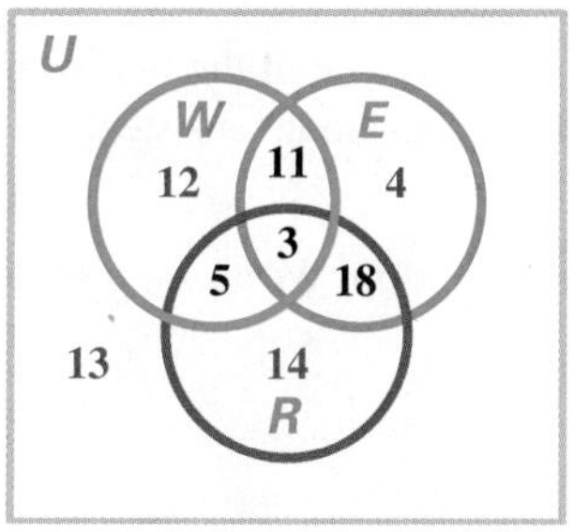

FIGURE 2.16 The number of elements in each region of the Venn diagram.

By condition d), 8 people want to see weekday reality shows, so we can deduce that 5 have not yet been counted in $W \cap R$. Similarly, from condition c), we see that 18 people have not been counted in $R \cap E$. This is shown in Figure 2.15(b).

Condition e) states that 31 people desire weekday programming, but we have already counted 19. Therefore, there are 12 more in set *W*. Similarly, there are 4 more people to be counted in set *E* and 14 more in set *R*. Condition h) tells us that there are 13 people outside sets *W*, *E*, and *R*. We summarize this information in Figure 2.16.

With this information, we can answer the original questions. The people wanting to see nonreality shows fall outside *R*, so they total $12 + 11 + 4 + 13 = 40$. Those who prefer weekend programming are outside *W*. This count is $14 + 18 + 4 + 13 = 49$.

Now try Exercises 27 to 38.

Contradictions in Survey Problems

KEY POINT

A survey problem may contain contradictory information.

It is possible to list a set of conditions on sets that cannot all hold at the same time.

EXAMPLE 5 Inconsistent Survey Data

Suppose that an Internet blog states the following information concerning which Web browsers its readers use:

a) 316 use WebMagic.
b) 478 use iBrowse.
c) 104 use both WebMagic and iBrowse.
d) 567 use only one of the Web browsers.

Find the inconsistency in these data.

SOLUTION: Review the Draw Pictures Strategy on page 3.

We begin by drawing a Venn diagram in Figure 2.17(a). Let *W* be the set of WebMagic users and let *I* be the set of those who use iBrowse.

Figure 2.17(a) shows the 104 people who use both Web browsers. Because *W* contains 316 elements, this means that $316 - 104 = 212$ elements are in $W - I$. Similarly, $478 - 104 = 374$ people use only iBrowse, so $I - W$ contains 374 elements. We show this new information in Figure 2.17(b).

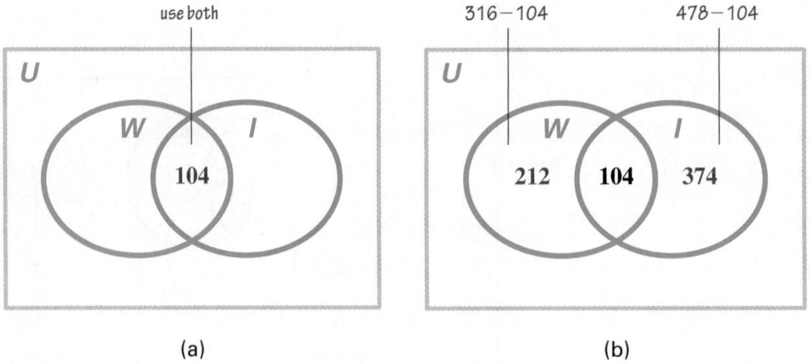

FIGURE 2.17 (a) 104 use both WebMagic and iBrowse. (b) 212 use only WebMagic; 374 use only iBrowse.

According to Figure 2.17(b), 212 + 374 = 586 people use only one of the two Web browsers. This directly contradicts condition d). Therefore, these data are inconsistent.

Information may be organized in a computer's database in table form. We analyze such a situation in Example 6.

EXAMPLE 6 Surveying Media Users

A media company that produces Pilates videos surveyed its customers to determine whether they would prefer to purchase the videos by buying an access code to allow them to download from a Web site, purchase hard copies on a DVD, or access the videos via a podcast. The company has extracted the following table of information from its customer database:

	Access Code (*A*)	Podcast (*P*)	DVD (*D*)	Total
Under 41 (*Y*)oung	20	15	9	44
41 to 55 (*M*)iddle-aged	44	**34**	8	86
Over 55 (*S*)enior	31	14	5	50
Total	95	63	22	180

QUIZ YOURSELF 16

Use the table in Example 6 to find $n(Y \cup D)$.

a) Find the number of elements in $M \cup D$.
b) Find the number of elements in A'.

SOLUTION:

a) The number of elements in $M \cup D$ is the number of people who are middle-aged or prefer DVDs. Thus,

$$n(M \cup D) = n(M) + n(D) - \underbrace{n(M \cap D)}_{\text{counted twice}} = 86 + 22 - 8 = 100.$$

b) A' is the set of people who do not want to buy an access code, so

$$n(A') = 63 + 22 = 85.$$

Now try Exercises 39 to 42. 16

Historical Highlight

Georg Cantor and Set Theory

It is unlikely that Georg Cantor, the nineteenth-century German mathematician who invented many of the set theory ideas that you have been studying, would have foreseen how his theory would have so many applications in today's society.

As you will see in the next section, Cantor discovered some remarkable ideas regarding infinite sets—ideas that were not accepted by his contemporaries. His former teacher Leopold Kronecker called Cantor's methods "a dangerous type of mathematical insanity" and the distinguished Frenchman Henri Poincaré proclaimed, "Later generations will regard [Cantor's set theory] as a disease from which one has recovered."

Today, however, set theory now provides a foundation for virtually all of mathematics. The eminent twentieth-century mathematician David Hilbert said that set theory is "…one of the highest achievements of man's intellectual processes."

EXERCISES 2.4

Sharpening Your Skills

In Exercises 1–4, determine which numbered regions make up the indicated set.

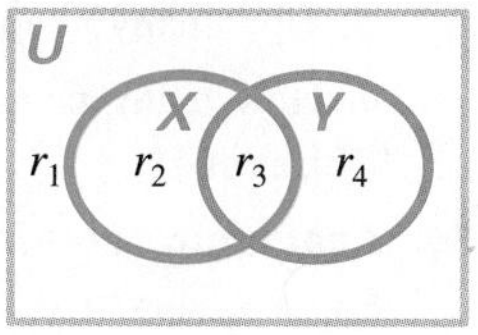

1. X

2. $X \cap Y$

3. $X \cup Y$

4. $Y - X$

In Exercises 5–10, describe each set by referring to the numbered regions, as in Example 1.

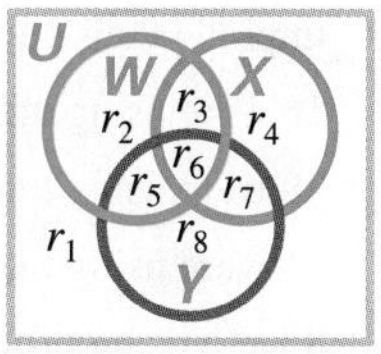

5. W **6.** $W \cap Y$

7. $X - W$ **8.** $W - (X \cup Y)$

9. $X \cap Y \cap W'$ **10.** $X \cap Y \cap W$

The numbers in the regions of the given Venn diagram indicate the number of elements in each region. Use this diagram to solve Exercises 11–16.

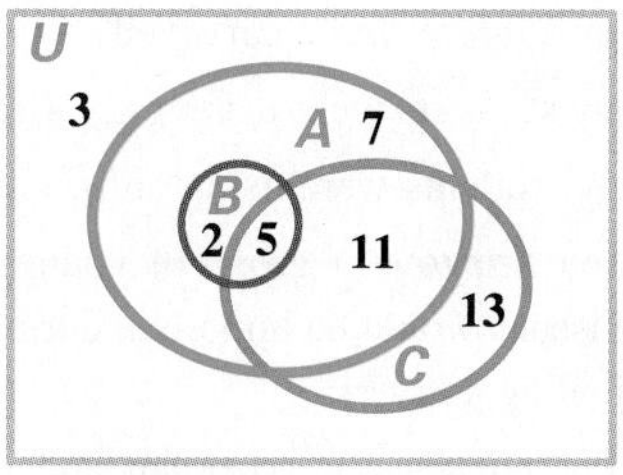

11. $n(A)$ **12.** $n(C')$

13. $n(A - B)$ **14.** $n(B \cup C)$

15. $n(A \cap C)$ **16.** $n(B - A)$

The numbers in the regions of the given Venn diagram indicate the number of elements in each region. Use this diagram to solve Exercises 17–20.

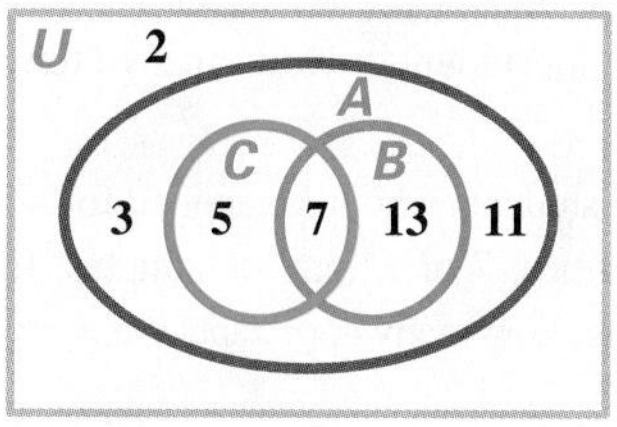

17. $n(B)$ **18.** $n(C')$

19. $n(A - B)$ **20.** $(B \cup C)'$

In Exercises 21–26, find, if possible, the number of elements in sets A, B, and C using the given information. If there is an inconsistency in the information, state where it occurs.

21. $A \cap B = \varnothing$, $n(A \cap C) = 5$, $n(B \cap C) = 3$, $n(C - A) = 7$, $n(A - C) = 2$, $n(U) = 14$.

22. $n(U) = 22$, $n(A - B) = 9$, $n(A \cap B \cap C) = 2$, $n(A - C) = 4$, $n(B \cap C) = 5$, $n(A \cap B) = 3$, $n(A \cup B \cup C)' = 7$.

23. $n(A \cap B) = 5$, $n(A \cap B \cap C) = 2$, $n(B \cap C) = 6$, $n(B - A) = 10$, $n(B \cup C) = 23$, $n(A \cap C) = 7$, $n(A \cup B \cup C) = 31$.

24. $n(B - C) = 4$, $n(C - B) = 9$, $n(A \cap B \cap C) = 3$, $n(B \cup C) = 22$, $n(A - C) = 7$, $n(A \cap B) = 7$, $n(A \cap C) = 5$.

25. $A \cap C = \varnothing$, $n(A \cap B) = 3$, $n(C - B) = 2$, $n(B - C) = 7$, $n(B \cup C) = 16$, $n(A \cup B) = 16$.

26. $A \subset B$, $A \cap C = \varnothing$, $n(C - A) = 8$, $n(A \cup C) = 12$, $n(C - B) = 3$, $n(B \cup C) = 17$.

Applying What You've Learned

27. Automobile accidents. An investigation of a number of automobile accidents revealed the following information:

18 accidents involved alcohol and excessive speed.

26 accidents involved alcohol.

12 accidents involved excessive speed but not alcohol.

21 of the accidents involved neither alcohol nor excessive speed.

How many accidents were investigated?

28. Perceptions of students. The political science department at a university surveyed a number of state politicians regarding their perceptions of college students and gathered the following information:

15 saw students as hardworking and economically advantaged.

25 saw students as hardworking.

11 believed students were economically advantaged but not hardworking.

16 believed students were neither hardworking nor economically advantaged.

How many surveyed believed students were economically advantaged?

29. There are 82 people protesting the destruction of the rain forests. There are 34 males, 23 of whom are students, and 20 female nonstudents. How many female students are protesting?

30. There are 95 students who have applied for a scholarship. There are 18 minorities, 7 of whom are athletes. If there are 20 athletes applying, how many applicants are neither an athlete nor a minority?

31. Survey of vacationers. A survey is taken of 100 people who vacationed at a dude ranch. The following information was obtained:

17 took horseback-riding lessons, attended the Saturday night barbecue, and purchased a tour guide.

28 attended the Saturday night barbecue and purchased a tour guide.

24 took horseback-riding lessons and purchased a tour guide.

42 took horseback-riding lessons but didn't attend the barbecue.

86 took horseback-riding lessons or purchased a tour guide.

14 purchased only a tour guide.

14 did none of these three things.

How many attended the barbecue? How many purchased a tour guide?

32. Social media survey. In a survey of 100 college students, the following information was obtained regarding their use of several social media sites:

21 use Facebook, LinkedIn, and Twitter.

32 use both LinkedIn and Twitter.

44 use Facebook and Twitter.

31 use Facebook but not LinkedIn.

78 use Facebook or Twitter.

12 use only Twitter.

5 use none of these three sites.

How many of those surveyed use Facebook? How many do not use Twitter?

33. Fitness survey. *Personal Fitness Magazine* surveyed a group of young adults regarding their exercise programs and the following results were obtained:

3 were using resistance training, Tae Bo, and Pilates to improve their fitness.

5 were using resistance training and Tae Bo.

12 were using Tae Bo and Pilates.

8 were using resistance training and Pilates.

15 were using resistance training only.

30 were using Tae Bo.

17 were using Pilates but not Tae Bo.

14 were using something other than these three types of workouts.

a. How many people were surveyed?

b. How many people were using only Tae Bo?

c. How many people were using Pilates but not resistance training?

34. Academic services survey. The Dean of Academic Services surveyed a group of students about which support services they were using to help them improve their academic performance and found the following results:

5 were using office hours, tutoring, and online study groups to improve their grades.

16 were using office hours and tutoring.

28 were using tutoring.

14 were using tutoring and online study groups.

8 were using office hours and online study groups but not tutoring.

23 were using office hours but not tutoring.

18 were using only online study groups.

37 were using none of these services.

a. How many students were surveyed?

b. How many students were using only office hours?

c. How many students were using online study groups?

35. World issues survey. A group of young adults who were asked which issues would be important during the next decade gave the following answers:

13 believed that nuclear war, terrorism, and environmental concerns would be important.

43 believed that nuclear war would be important.

17 believed that nuclear war and terrorism would be important.

23 believed that nuclear war or terrorism but not the environment would be important.

28 believed that nuclear war and the environment would be important.

18 believed that terrorism but not nuclear war would be important.

7 believed that only the environment would be a serious issue.

6 believed that none of these issues would be important.

a. How many people were surveyed?

b. How many people believed the environment would be an important issue?

c. How many people believed only terrorism would be an important issue?

36. News sources survey. A survey of young adults was taken to determine which of the various sources they use to obtain news. Of the 36 people who use the Internet, 13 use the Internet only to learn the news. Of the 48 people who use the newspaper, 11 use the newspaper only for news coverage. There are 23 people who use newspapers, the Internet, and television for news coverage. From this information, determine how many use both the Internet and television to learn the news.

37. Mass transit survey. A survey was made of 200 city residents to study their use of mass transit facilities. According to the survey:

83 did not use mass transit.

68 used the bus.

44 used only the subway.

28 used both the bus and subway.

59 used the train.

Explain how you can use this information to deduce that some residents must use both the bus and train.

38. Online music survey. Pandora.com surveyed a group of subscribers regarding which online music channels they use on a regular basis. The following information summarizes their answers:

7 listened to rap, heavy metal, and alternative rock.

10 listened to rap and heavy metal.

13 listened to heavy metal and alternative rock.

12 listened to rap and alternative rock.

17 listened to rap.

24 listened to heavy metal.

22 listened to alternative rock.

9 listened to none of these three channels.

a. How many people were surveyed?

b. How many people listened to either rap or alternative rock?

c. How many listened to heavy metal only?

In Exercises 39–42, use the following table from Example 6 to find the cardinal number of each set.

	Access Code (*A*)	Podcast (*P*)	DVD (*D*)	Total
Under 41 (*Y*)oung	20	15	9	44
41 to 55 (*M*)iddle-aged	44	34	8	86
Over 55 (*S*)enior	31	14	5	50
Total	95	63	22	180

39. $A \cup P$

40. $S \cap D$

41. $A - (M \cup S)$

42. $Y \cap (A \cup D)$

Use the Venn diagram in Example 3 to specify the blood type in the regions listed in Exercises 43–46.

43. $A - (B \cup Rh)$

44. $(A \cup B \cup Rh)'$

45. $(A \cup B) - Rh$

46. $(B \cap Rh) - A$

A person can safely receive a transfusion from another person provided the receiver's blood contains all of the A, B, and Rh factors of the donor. So, a person with B^+ *blood can receive blood from persons with types* B^+, B^-, O^+, *and* O^-, *but not with types* A^+, A^-, AB^+, AB^-. *Use this information to do Exercises 47–50.*

47. Describe using set notation the set of blood types of persons that can receive blood from anyone (universal receivers).

48. Describe using set notation the set of blood types of persons that can donate blood to anyone (universal donors).

49. Which blood types can people having type B^- blood receive?

50. Which blood types can people having type A^+ blood receive?

Communicating Mathematics

51. In Figure 2.11, students often call region r_2 by the name A. What is wrong with this thinking?

52. In Figure 2.12(a), students often call region r_3 by the name $A \cap B$. What is wrong with this thinking?

Challenge Yourself

53. As you saw in Section 2.3, a common mistake in using the Inclusion-Exclusion Principle is to say $n(A \cup B) = n(A) + n(B)$ and forget that elements in $A \cap B$ have been counted twice. Show by giving counterexamples that the following two formulas for counting the elements of three sets are incorrect:

a. $n(A \cup B \cup C) = n(A) + n(B) + n(C)$

b. $n(A \cup B \cup C) = n(A) + n(B) + n(C) - n(A \cap B) - n(A \cap C) - n(B \cap C)$

54. Adjust formula (b) in Exercise 53 to state a proper formula for counting the elements of three sets. (*Hint:* In formula (b) we have subtracted too much.)

55. Notice that if a Venn diagram has one set A, then it divides the universal set into two regions, A and A'. If the diagram has two sets A and B, then there will be four regions: $A \cap B$, $A \cap B'$, $A' \cap B$, and $A' \cap B'$. What is the maximum number of regions for a three-set Venn diagram? Name the regions in a manner similar to what we have just done to one- and two-set Venn diagrams.

56. Thinking along the lines of Exercise 55, what do you expect to be the maximum number of regions possible for a Venn diagram with four sets? Name them as in Exercise 55. Which of these regions are missing in the given Venn diagram with four sets?

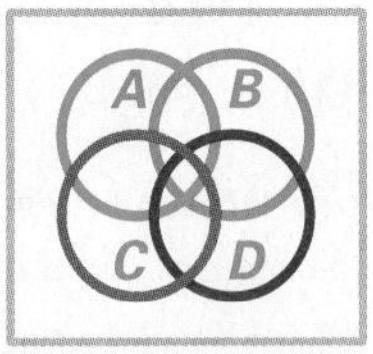

57. Make up a survey question whose conditions will satisfy the given Venn diagram.

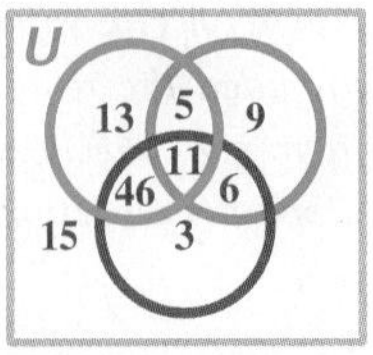

58. Name the regions r_1, r_2, and r_3 in the given four-set Venn diagram. For example, the region labeled r_4 is named $A \cap B \cap C \cap D'$.

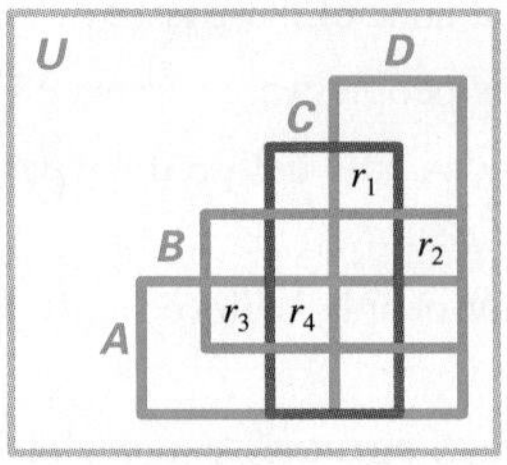

Looking Deeper

2.5

Infinite Sets

Objectives

1. Understand the definition of infinite sets.
2. Show that the set of natural numbers and the set of integers are countable.
3. Recognize that the set of rational numbers is countable.
4. Recognize that not all infinite cardinal numbers are the same size.

To infinity . . . and beyond!
—Buzz Lightyear

Like our lovable space ranger, most people do not have a clear understanding of the word "infinity." In this section we will not talk about infinity—a rather vague idea—but instead we will discuss infinite sets. Over one hundred years ago, the great German mathematician Georg Cantor* proved the surprising fact that infinite sets can have different sizes.

To understand infinite sets better, we will use the notion of one-to-one correspondence to make the notion of infinite cardinal numbers more precise.

> **DEFINITION** Two sets *A* and *B* are in **one-to-one correspondence** if we can pair the elements of the two sets so that every element of *A* is paired with exactly one element of *B* and every element of *B* is paired with exactly one element of *A*.

For example, the set $T =$ {Carreras, Domingo, Pavarotti}, a set of famous tenors, is in one-to-one correspondence with $F =$ {Beyonce, Shakira, Gaga}, a set of popular female vocalists, because we can pair each tenor in *T* with exactly one singer in *F* as follows:

Carreras	Domingo	Pavarotti
↕	↕	↕
Beyonce	Shakira	Gaga

*See Historical Highlight on page 73.

QUIZ YOURSELF 17

Show another one-to-one correspondence between T and F.

Of course, there are several other ways of pairing the elements of T with the elements of F in a one-to-one correspondence, and we will investigate this further in the exercises. 17

We can now say that a set A has cardinal number n, or is finite, if we can place A in a one-to-one correspondence with the set $\{1, 2, 3, \ldots, n\}$. For example, if $A = \{x : x$ is a letter of the alphabet$\}$, then $n(A) = 26$ because we have the following one-to-one correspondence between A and $\{1, 2, 3, \ldots, 26\}$:

$$\begin{array}{ccccc} a & b & c & \cdots & z \\ \updownarrow & \updownarrow & \updownarrow & & \updownarrow \\ 1 & 2 & 3 & \cdots & 26 \end{array}$$

Infinite Sets

Notice that we cannot place a finite set in a one-to-one correspondence with one of its proper subsets. Cantor used this simple observation to define infinite sets.

> **DEFINITION** A set is **infinite** if we can place it in a one-to-one correspondence with a proper subset of itself.

EXAMPLE 1 The Set of Natural Numbers Is Infinite

Show that the set of natural numbers is infinite.

QUIZ YOURSELF 18

Find another one-to-one correspondence between N and one of its proper subsets. Be sure to include the general pattern for the correspondence.

SOLUTION:

Let $N = \{1, 2, 3, \ldots\}$ and $E = \{2, 4, 6, \ldots\}$, then the following is a one-to-one correspondence between N and its proper subset E:

$$\begin{array}{cccccc} 1 & 2 & 3 & \cdots & n & \cdots \\ \updownarrow & \updownarrow & \updownarrow & & \updownarrow & \\ 2 & 4 & 6 & \cdots & 2n & \cdots \end{array}$$

18

Countable Sets

Instead of simply saying that the set of natural numbers is infinite, we can now be more precise.

> **DEFINITION** Cantor called the cardinal number for the set of natural numbers $\aleph_0$. The symbol $\aleph$, aleph, is the first letter of the Hebrew alphabet. We pronounce $\aleph_0$ as "Al-luf null" or "Al-luf naught." Therefore, we can write $n(N) = \aleph_0$. Any set that is either finite or that is in a one-to-one correspondence with N, and therefore has cardinal number $\aleph_0$, is called a **countable set.**

EXAMPLE 2 The Set of Integers Is Countable

Show that I, the set of integers, can be put in a one-to-one correspondence with the set of natural numbers.

SOLUTION:

The following matching shows the correspondence. (In the correspondence, we are assuming n represents a positive integer.)

$$0 \quad 1 \quad -1 \quad 2 \quad -2 \quad 3 \quad -3 \quad \cdots \quad n \quad \quad -n \quad \cdots$$
$$\updownarrow \quad \updownarrow \quad \updownarrow \quad \updownarrow \quad \updownarrow \quad \updownarrow \quad \updownarrow \quad \quad \updownarrow \quad \quad \updownarrow$$
$$1 \quad 2 \quad 3 \quad 4 \quad 5 \quad 6 \quad 7 \quad \cdots \quad 2n \quad 2n+1 \quad \cdots$$

Therefore, $n(I) = \aleph_0$, and I is a countable set.

When we work with infinite sets, we find some surprising results. It would seem that there are so many more rational numbers than there are integers that we might expect that the cardinal number of the rational numbers is larger than \aleph_0. As Example 3 shows, this is not the case.

EXAMPLE 3 ## The Set of Rational Numbers Is Countable

Show that the set of rational numbers is countable.

SOLUTION: Review the Be Systematic Strategy on page 5.

First consider the positive rational numbers as we list them in Figure 2.18. The first row of the arrangement has all positive rational numbers with denominator 1. The second row has all denominators 2, the third row has denominators of 3, and so on. We then trace through the arrangement following the red line in Figure 2.18, skipping over numbers that we have encountered earlier.

We will follow the path through the rational numbers in Figure 2.18 and list those numbers in a straight line, matching them with the natural numbers as follows:

$$1/1, \ 2/1, \ 1/2, \ 1/3, \ 2/2, \ 3/1, \ 4/1, \ 3/2, \ 2/3, \ 1/4, \ 1/5, \ 2/4, \ 3/3, \ 4/2, \ 5/1 \ \cdots$$
$$\updownarrow \quad \updownarrow \quad \updownarrow \quad \updownarrow \quad \text{skip} \quad \updownarrow \quad \updownarrow \quad \updownarrow \quad \updownarrow \quad \updownarrow \quad \updownarrow \quad \quad \text{skip} \quad \quad \updownarrow$$
$$1, \quad 2, \quad 3, \quad 4, \quad \quad 5, \quad 6, \quad 7, \quad 8, \quad 9, \quad 10, \quad \quad \quad \quad 11 \ \cdots$$

Notice that in our matching, we have skipped over rational numbers that we have encountered earlier, such as 2/2, 2/4, 3/3, and 4/2.

When we look at the positive rational numbers in this way, we see that there is a first number, then a second, then a third, and so on, and we are listing each number exactly once. Therefore, there is a one-to-one correspondence between N and the positive rational numbers.

To show that the entire set of rational numbers is countable, we would argue as we did in Example 2 to account for zero and the negative rationals; however, we will not go into the details of how to do it. Therefore, the set of rational numbers has the same cardinal number as the natural numbers, which is \aleph_0. **19**

1/1 2/1 3/1 4/1 5/1 ...
1/2 2/2 3/2 4/2 5/2 ...
1/3 2/3 3/3 4/3 5/3 ...
1/4 2/4 3/4 4/4 5/4 ...
1/5 2/5 3/5 4/5 5/5 ...

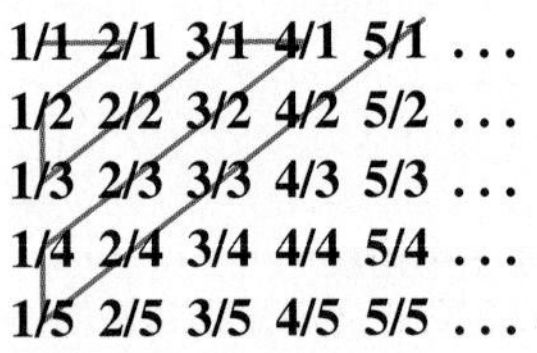

FIGURE 2.18 Listing the positive rational numbers.

QUIZ YOURSELF **19**

a) List the next three numbers after 5 that we would encounter if we were to continue the listing of the positive rational numbers that we began in Example 3.

b) What is the next number that we would skip after 4/2?

The Cardinal Number c

So far, the infinite sets that you have studied have all been countable. We will next show you a set that has a cardinal number greater than \aleph_0. Because the argument that we will show you is a little sophisticated, we will look at a simpler version of our reasoning before going on to the real thing.

When I introduce infinite sets to my students, I often have a class of about 35 students in a room with about 42 chairs. I ask them to tell me as quickly as they can whether there are more students or more chairs in the room. Instantly, a student responds "more chairs." I ask the student how she counted so quickly and she tells me, "I didn't." I ask, "Then how did you know that there are more chairs?" She explains that every student is seated and yet there are empty chairs left over. You will see this same type of reasoning in Example 4.

EXAMPLE 4 A Cardinal Number Greater Than $\aleph_0$

We will reproduce Cantor's argument that the set of real numbers between 0 and 1 has cardinal number greater than $\aleph_0$. We begin by assuming that we *can* put the set of numbers between 0 and 1 in a one-to-one correspondence with the natural numbers and show that no matter how hard we try, there will always be some number that we could not have listed.

Although we would not actually know what the listing would be, for the sake of argument, let us assume that we had listed all the numbers between 0 and 1 as follows:

$$\begin{aligned}
1 &\leftrightarrow 0.6348291347\ldots \\
2 &\leftrightarrow 0.2373261008\ldots \\
3 &\leftrightarrow 0.4821063391\ldots \\
4 &\leftrightarrow 0.6824537128\ldots \\
5 &\leftrightarrow 0.4657189233\ldots \\
&\ \ \vdots \qquad\qquad \vdots
\end{aligned}$$

Although we have assumed that all numbers between 0 and 1 are listed, we will now show you how to construct a number x between 0 and 1 that is not on this list. We want x to be different from the first number on the list, so we will begin the decimal expansion of x with a digit other than a 6 in the tenths place, say $x = 0.5\ldots.$ Because we don't want x to equal the second number on the list, we make the hundredths place not equal to 3, say 4; so far, $x = 0.54\ldots.$ (For this argument to work, we will never switch a number to 0 or 9.)

Continuing this pattern, we make sure that x is different from the third number in the third decimal place, say 3, and different from the fourth number in the fourth decimal place, make it 5, and so on. At this point, $x = 0.5435\ldots.$ By constructing x in this fashion, it cannot be the first number on the list, or the second, or the third, and so on. In fact, x will differ from every number on the list in at least one decimal place, so it cannot be any of the numbers on the list.

This means that our assumption that we were able to match the numbers between 0 and 1 with the natural numbers is wrong. So the cardinal number of this set is not $\aleph_0$. Cantor used the letter c, for the word *continuum,* for this cardinal number. Because there is a number between 0 and 1 that cannot be matched with a natural number, we can argue as we did in our discussion of the seats and the students that the cardinal number c is greater than the cardinal number $\aleph_0$.

Historical Highlight

The Continuum Hypothesis

Among Cantor's remarkable discoveries was that there is an infinite number of infinite cardinal numbers. That is, there is an infinite number of sizes of infinity. The smallest infinite cardinal number he called $\aleph_0$, which as you have seen is the cardinal number of the natural numbers. He proved that the set of all subsets of the natural numbers has another cardinal number $\aleph_1$ that is greater than $\aleph_0$. Cantor believed that perhaps $\aleph_1$ was equal to c. Although he labored for years to prove this conjecture, called the *continuum hypothesis*, Cantor was unable to prove it.

In 1900, the noted German mathematician David Hilbert said that the continuum hypothesis was one of the great unanswered questions in mathematics, and for the first half of the twentieth century, mathematicians could not answer this question. However, in 1963, a mathematician at Princeton University named Paul Cohen made the remarkable discovery that *it is impossible to either prove or disprove the continuum hypothesis.* That is, you can either accept it or reject it. If you accept it, then you are doing Cantorian set theory. If you reject it, then you are doing non-Cantorian set theory. Both are equally consistent but different areas of mathematics.

EXERCISES 2.5

Sharpening Your Skills

In Exercises 1–8, show that each set has cardinal number $\aleph_0$ by showing a one-to-one correspondence between the natural numbers and the given set. Be sure to indicate the general correspondence.

1. $\{4, 8, 12, 16, 20, \ldots\}$

2. $\{5, 10, 15, 20, 25, \ldots\}$

3. $\{8, 11, 14, 17, 20, \ldots\}$

4. $\{7, 11, 15, 19, 23, \ldots\}$

5. $\{2, 4, 8, 16, 32, \ldots\}$

6. $\{3, 9, 27, 81, 243, \ldots\}$

7. $\{1, 1/2, 1/3, 1/4, 1/5, \ldots\}$

8. $\{1/2, 2/3, 3/4, 4/5, 5/6, \ldots\}$

In Exercises 9–12, we give an expression describing the number that corresponds to the natural number n. Use this expression to describe a one-to-one correspondence between the natural numbers and one of its subsets. For example, if we gave you the expression 2n, you would write the following correspondence that we gave in Example 1:

$$\begin{array}{ccccccc} 1 & 2 & 3 & \cdots & n & \cdots \\ \updownarrow & \updownarrow & \updownarrow & & \updownarrow & \\ 2 & 4 & 6 & \cdots & 2n & \cdots \end{array}$$

9. $3n$

10. $2n + 3$

11. $3n - 2$

12. $4n + 5$

In Exercises 13–22, describe a one-to-one correspondence between the given set and one of its proper subsets. For example, if we gave you the set $\{3, 5, 7, 9, 11, \ldots\}$, the nth term is $2n + 1$. You could then write the correspondence by matching the elements of $\{3, 5, 7, 9, 11, \ldots\}$ with the elements of the subset $\{5, 7, 9, 11, 13, \ldots\}$. The general correspondence would match $2n + 1$ with $2n + 3$.

13. $\{2, 4, 6, 8, 10, \ldots\}$

14. $\{5, 10, 15, 20, 25, \ldots\}$

15. $\{7, 10, 13, 16, 19, \ldots\}$

16. $\{6, 9, 12, 15, 18, \ldots\}$

17. $\{2, 4, 8, 16, 32, \ldots\}$

18. $\{3, 9, 27, 81, 243, \ldots\}$

19. $\{1, 1/2, 1/3, 1/4, 1/5, \ldots\}$

20. $\{1/2, 2/3, 3/4, 4/5, 5/6, \ldots\}$

21. $\{1/2, 1/4, 1/6, 1/8, 1/10, \ldots\}$

22. $\{1/2, 1/4, 1/8, 1/16, 1/32, \ldots\}$

In Example 3, we showed you how to match the natural numbers with the positive rational numbers in a one-to-one correspondence. Use that matching to answer the following questions.

23. What rational number corresponds to the natural number 12?

24. What rational number corresponds to the natural number 15?

25. What natural number is matched with the rational number 4/5?

26. What natural number is matched with the rational number 1/6?

Communicating Mathematics

27. In Example 3, what did we do to show that the set of positive rational numbers is countable?

28. What was the essence of the argument in Example 4 that showed the cardinal number of the set of numbers between 0 and 1 is greater than $\aleph_0$?

29. How would you convince someone that the set $\{1, 2, 3, 4, 5\}$ is not infinite?

30. How would you convince someone that the set $\{2, 4, 6, 8, \ldots\}$ is infinite?

31. In Example 3, why did we ignore rational numbers written as 2/2, 2/4, 3/3, and 4/2?

32. In constructing the number x in Example 4, how would you decide what to put in the 99th place?

Challenge Yourself

33. How many one-to-one correspondences are there between the set $\{1, 2, 3\}$ and the set $\{4, 5, 6\}$?

34. How many one-to-one correspondences are there between two four-element sets?

35. The arithmetic of infinite cardinal numbers has some peculiar properties. For example, it is a fact that $1 + \aleph_0 = \aleph_0$. Show this by forming the union of a set whose cardinal number is 1 with a disjoint set whose cardinal number is $\aleph_0$ and then showing that the union of the sets is in one-to-one correspondence with the natural numbers.

36. Another strange infinite arithmetic fact is that $\aleph_0 + \aleph_0 = \aleph_0$. Show why this is true by considering the union of two disjoint sets having cardinal number $\aleph_0$.

37. Imagine that we bend a line segment representing the set of real numbers strictly between 0 and 1 into a semicircle with endpoints missing, as shown in the diagram below. Explain how this diagram shows that the real numbers between 0 and 1 can be put in a one-to-one correspondence with the set of real numbers.

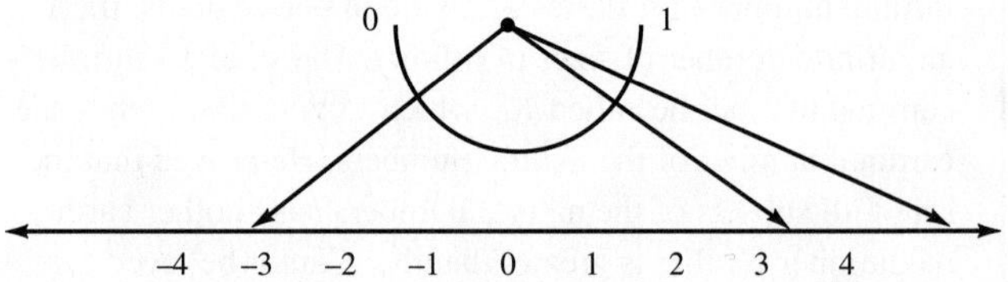

38. Use an argument similar to that of Exercise 37 to show that the points on a square can be put in a one-to-one correspondence with the points on a triangle. (*Hint:* Draw the triangle inside of the square.)

CHAPTER SUMMARY

Section	Summary	Example
SECTION 2.1	We can represent sets by **listing** elements or by using **set-builder notation.**	Example 1, p. 44
	Sets must be **well defined.**	Example 2, p. 45
	The **element symbol** expresses that an object is a member of a set.	Example 4, p. 46
	The **cardinal number** of a set indicates its size.	Example 5, p. 47
SECTION 2.2	Two sets are **equal** if they have exactly the same members.	Example 1, p. 51
	The set A is a **subset** of set B, written $A \subseteq B$, if every element of A is also found in B. If $A \neq B$, then we say that A is a **proper subset** of B and write $A \subset B$.	Example 2, p. 52
	We can use **Venn diagrams** to illustrate subset relationships.	Discussion, p. 52
	A set that has k elements has 2^k subsets.	Example 4, pp. 53–54
	Equivalent sets have the same number of elements.	Discussion, p. 54
SECTION 2.3	a) The **union** of sets A and B, written $A \cup B$, is the set of elements that are elements of either A or B, or both.	Example 1, p. 59
	b) The **intersection** of sets A and B, written $A \cap B$, is the set of elements common to both A and B.	Example 2, p. 60
	c) The **complement** of set A, written A', is the set of elements in the universal set that are not elements of A.	Example 3, p. 60
	d) The **difference** of sets B and A, written $B - A$, is the set of elements that are in B but not in A.	Example 4, p. 61
	We must perform set operations in a proper order.	Example 5, p. 62
	DeMorgan's laws state that $(A \cup B)' = A' \cap B'$ and $(A \cap B)' = A' \cup B'$.	Example 6, p. 62
	We can use **Venn diagrams** to prove or disprove set theory statements.	Example 7, pp. 63–64
	If A and B are sets, then $n(A \cup B) = n(A) + n(B) - n(A \cup B)$. This is often called the **Inclusion-Exclusion Principle.**	Example 8, p. 64
SECTION 2.4	Sets in Venn diagrams can have various names.	Example 1, pp. 68–69
	We can use Venn diagrams to solve survey problems.	Examples 2–4, pp. 69–71
	A survey problem can have contradictory information.	Example 5, pp. 71–72
SECTION 2.5	An **infinite set** can be put in a one-to-one correspondence with a proper subset of itself.	Example 1, p. 77
	The natural numbers and the set of integers are countable.	Examples 1 and 2, pp. 77–78
	The set of rational numbers is countable.	Example 3, p. 78
	The set of real numbers between 0 and 1 is not countable.	Example 4, p. 79

CHAPTER REVIEW EXERCISES

Section 2.1

1. Use an alternative method to express each set.

a. $\{2, 4, 6, 8, \ldots, 18\}$

b. {January, February, . . . , December}

c. $\{x : x$ is a state beginning with the word "New"$\}$

d. $\{y : y$ is a member of the Dallas Cowboys and also a former Miss America$\}$

2. Explain why $\emptyset \neq \{\emptyset\}$.

3. Make up a "bag diagram" to illustrate the set $\{3, \{\emptyset\}, \{\{1, 2\}, \{1, 2, 3\}\}\}$.

4. Find the cardinal number of each of these sets.

a. $\{3, 6, 9, \ldots, 18, 21\}$

b. $\emptyset$

c. $\{\{1, 2, 3\}, \{4, 5\}, \{6\}, \{\emptyset\}\}$

Section 2.2

5. Decide whether each pair of sets is equal. Justify your answer.

a. $\{1, 3, 5, 7, 9\}$ and $\{9, 7, 5, 3, 1\}$

b. $\{2, 3, 4, 5, 4, 6, 4, 7, 8, 9, 10, 9, 11, 12\}$ and $\{2, 3, 4, \ldots, 12\}$

c. $\{2, 4, 6, 8, \ldots\}$ and $\{2, 4, 6, \ldots, 1{,}000\}$

6. Decide whether each statement is true or false. Justify your answer.

a. {daffodil, daisy, rose, carnation} $\subset$ $\{f : f$ is a flower$\}$

b. {daffodil, daisy, rose, carnation} $\subseteq$ $\{f : f$ is a flower$\}$

c. $\{x : x$ is a former president of the United States$\}$ $\subset$ $\{x : x$ is a man$\}$

d. $\emptyset \subseteq \{1, 2, 3\}$

e. $\emptyset \subset \emptyset$

7. Which pairs of sets are equivalent?

a. {baseball, football, basketball, field hockey, volleyball} and {b, f, b, f, v}

b. $\{1, 3, 5, 7, 9, \ldots, 99\}$ and $\{2, 4, 6, 8, 10, \ldots, 100\}$

c. $\{x : x$ is a letter in the word *tulips*$\}$ and $\{x : x$ is a letter in the word *flower*$\}$

d. $\{\emptyset\}$ and $\{0\}$

8. **a.** List all of the subsets of the set $\{a, b, c\}$

b. How many subsets does the set $\{3, 5, 8, 9, 12, 15, 17\}$ have?

Section 2.3

9. Let $U = \{1, 2, 3, \ldots, 10\}$ and let $A = \{2, 5, 7, 8, 9\}$, $B = \{3, 4, 5, 7, 9, 10\}$, and $C = \{5, 3, 8, 9, 2\}$. Find the following sets:

a. $A \cap B$

b. $B \cup C$

c. C'

d. $A - C$

10. Using the same sets as in Exercise 9, find the following sets:

a. $(A \cup B)'$

b. $(A - C) \cup (A - B)$

c. $A' \cap (B' \cup C)$

11. Represent each set using a Venn diagram.

a. $A \cup B$

b. $B \cap C$

c. $A' \cap B \cap C$

d. $(B \cup C) - A$

12. Use DeMorgan's laws to represent $(A \cup B)'$ in a different way.

13. **a.** List three algebraic properties satisfied by the set operation of union.

b. List three algebraic properties satisfied by the set operation of intersection.

c. What algebraic property relates union and intersection?

14. State the Inclusion-Exclusion Principle. What is a common mistake made in using this equation?

Section 2.4

15. Use the following information to answer the given questions: $n(A \cap B) = 4$, $n(A \cap B \cap C) = 1$, $n(B \cap C) = 8$, $n(B - A) = 9$, $n(B \cup C) = 23$, $n(A \cap C) = 6$, $n(A \cup B \cup C) = 35$, $n(B') = 32$.

a. How many elements are there in $C - B$?

b. How many elements are there in A'?

16. A survey was taken of college freshmen regarding the factors that they considered important in choosing a college.

82 said cost.

15 said cost but not academics.

70 said social life.

48 said all three.

56 said cost and social life.

25 said academics but not social life.

16 said academics but not cost.

How many said academics in the survey? How many said both academics and social life?

17. Use the Venn diagram regarding blood classification that we discussed in Section 2.4 to state the blood types that are in each of the following regions:

a. B − (A ∪ Rh)

b. (A ∪ B)′ − Rh

Section 2.5

18. What is the definition of an infinite set?

19. Show that the set of natural numbers is infinite.

20. In matching the rational numbers with the natural numbers as we did in Example 3 in Section 2.5, what rational number matches with 7?

21. In creating the number x in Example 4 in Section 2.5, where we proved that the set of real numbers between 0 and 1 cannot be put in a one-to-one correspondence with the natural numbers, how did we decide what to put in the third decimal place?

CHAPTER TEST

1. Use an alternative method to express each set.
   a. $\{101, 102, 103, 104, \ldots\}$
   b. $\{x : x$ is a month in the year$\}$
   c. $\{y : y$ is a person in your math class and also more than 100 years old$\}$
2. Decide whether each pair of sets is equal. Justify your answer.
   a. {a, b, c, d, e} and {e, a, b, d, c}
   b. $\{\{1\}, \{2\}, \{3\}\}$ and $\{1, 2, 3\}$
   c. $\{2, 4, 6, 8, \ldots\}$ and $\{x : x$ is an even natural number greater than $1\}$
3. Which pairs of sets are equivalent?
   a. {Beyonce, Oprah, Sting, Dr. Phil} and {rose, daisy, collie, Mars, chocolate}
   b. $\{1, 3, 5, 7, 9, \ldots, 99\}$ and $\{2, 4, 6, 8, 10, \ldots, 100\}$
   c. $\{\varnothing\}$ and $\{0\}$
4. Let $U = \{1, 2, 3, \ldots, 10\}$ and let $A = \{1, 2, 5, 6, 9\}$, $B = \{2, 3, 4, 5\}$, and $C = \{2, 4, 6, 8, 10\}$. Find the following sets:
   a. C'
   b. $B - C$
   c. $(A \cap B)'$
   d. $(A' \cap B') \cup C$
5. Explain why $\varnothing \neq \{\varnothing\}$.
6. Find the cardinal number of each of these sets.
   a. $\{2, 4, 8, 16, 32, 64, 128, 256\}$
   b. $\{\varnothing\}$
7. How many subsets does the set $\{1, 2, 3, \ldots, 8\}$ have?
8. Make up a "bag diagram" to illustrate the set $\{\{2\}, \varnothing, \{\{1, 2\}, \{1, 2, 3\}\}\}$.
9. Decide whether each statement is true or false. Justify your answer.
   a. {boxer, poodle, chihuahua, collie} $\subseteq \{d : d$ is a dog$\}$
   b. $\{x : x$ is a letter in the word *love*$\} \subseteq \{x : x$ is a letter in the word *lovely*$\}$
   c. $\varnothing \in \{1, 2, 3\}$
10. Use DeMorgan's laws to represent $(A \cap B)'$ in a different way.
11. Represent each set using a Venn diagram.
    a. $A \cap C$
    b. $(A \cup C) - B$
12. Use the following information to answer the given questions: $n(A \cap C) = 11$, $n(A \cap B \cap C) = 5$, $n(B \cap C) = 8$, $n(C - A) = 10$, $n(B \cup C) = 34$, $n(A \cap B) = 10$, $n(A \cup B \cup C) = 54$, $n(C') = 64$.
    a. How many elements are there in $C - A$?
    b. How many elements are there in B'?
13. A survey was taken of drivers regarding the factors that they considered important in buying a new car.
    a. 84 said cost.
    b. 15 said cost but not gas mileage.
    c. 72 said safety.
    d. 48 said all three.
    e. 56 said cost and safety.
    f. 25 said gas mileage but not safety.
    g. 20 said gas mileage but not cost.

    How many said gas mileage in the survey? How many said both gas mileage and safety?
14. What is the definition of an infinite set?
15. Show that the set of natural numbers is infinite.
16. In matching the rational numbers with the natural numbers as we did in Example 3 in Section 2.5, what rational number matches with 9?
17. In creating the number x in Example 4 in Section 2.5, where we proved that the set of real numbers between 0 and 1 cannot be put in a one-to-one correspondence with the natural numbers, how did we decide what to put in the fifth decimal place?
18. Using the blood type classifications that we discussed in this chapter, use set notation to describe the following blood types:
    a. AB^-
    b. O^+

GROUP PROJECTS

1. **Venn diagrams.** Try to draw a Venn diagram for five sets having 32 distinct regions. If you cannot do this, then search the Internet for examples of such a diagram and present your results.
2. **Fuzzy sets.** In set theory as you have studied it, an element x is either in a set A or it is not. That is, we can say $x \in A$ or $x \notin A$. In **fuzzy set theory,** membership in sets is not so cut and dried. We will start with a set of people $X =$ {Lily, Marshall, Ted, Robin, Barney} and define the fuzzy set T as the set of tall people as follows:

$$T = \{(\text{Ted}, 0.8), (\text{Marshall}, 0.4), (\text{Lily}, 0.7), (\text{Robin}, 0.3), (\text{Barney}, 0.5)\}.$$

Ted, Marshall, Lily, Robin, and Barney are members of T, but they have different **degrees of membership,** depending on how tall they are. For example, Ted's degree of membership is 0.8,

whereas Robin's is 0.3. This means that Ted has the property of "being tall" to a much higher degree than Robin does. Degrees of membership are always between 0 and 1 inclusive. Two fuzzy sets are **equal** if they have the same members, and the degrees of membership of the elements of the two sets are the same.

To compute the **complement** of T, we replace each set membership degree with 1 minus the degree of membership, as follows:

$$T' = \{(\text{Ted}, 0.2), (\text{Marshall}, 0.6), (\text{Lily}, 0.3), (\text{Robin}, 0.7), (\text{Barney}, 0.5)\}.$$

Let us now define S, the fuzzy set of strong people:

$$S = \{(\text{Ted}, 0.5), (\text{Marshall}, 0.6), (\text{Lily}, 0.4), (\text{Robin}, 0.4), (\text{Barney}, 0.6)\}.$$

We compute the **intersection** of T and S by associating with each member of X the smaller degree of membership, as follows:

$$S \cap T = \{(\text{Ted}, 0.5), (\text{Marshall}, 0.4), (\text{Lily}, 0.4), (\text{Robin}, 0.3), (\text{Barney}, 0.5)\}.$$

The **union** of sets T and S is similar, except now we take the larger of the set memberships for each member of X. So,

$$S \cup T = \{(\text{Ted}, 0.8), (\text{Marshall}, 0.6), (\text{Lily}, 0.7), (\text{Robin}, 0.4), (\text{Barney}, 0.6)\}.$$

a. Define some other fuzzy sets using X above.

b. Do several computations using S, T, and the new fuzzy sets that you have defined in (a).

c. Verify that some of the rules you have learned earlier are satisfied in fuzzy set theory. For example, show by computing examples that DeMorgan's laws hold. Show that intersection distributes over union, etc.

USING TECHNOLOGY

1. Using technology to calculate sets. The logical functions AND, OR, and NOT in an Excel spreadsheet behave very much like intersection, union, and complement in set theory. By placing 1s and 0s in certain rows of columns A and B in the spreadsheet in the first column, we have indicated that $A = \{1, 2, 3, 6, 7, 8\}$ and $B = \{1, 3, 5, 7\}$. The 1s and 0s in columns D and E of the worksheet show that $A \cap B = \{1, 3, 7\}$ and $(A \cap B)' = \{2, 4, 5, 6, 8\}$. The formulas that we have typed in cells D9 and E9 show the type of formulas that we have used to do the set computations.

E4 =IF(NOT(D4),1,0)

	A	B	C	D	E	F
1	1	1		1	0	
2	1	0		0	1	
3	1	1		1	0	
4	0	0		0	1	
5	0	1		0	1	
6	1	0		0	1	
7	1	1		1	0	
8	1	0		0	1	
9				IF(AND(A2, B2),1,0)	IF(NOT(D2),1,0)	
10						

Let $U = \{1, 2, 3, 4, 5, 6, 7, 8, 9, 10\}$, $A = \{1, 3, 4, 6, 8\}$, $B = \{2, 3, 4, 6, 9\}$, and $C = \{2, 5, 6, 7, 8\}$. Use Excel as we have shown to do the following set calculations.

a. $A \cap B$ **b.** $A \cup B$ **c.** $A \cap (B \cup C)$ **d.** $(A \cup B)'$ **e.** $A' \cap B'$

3

Logic

The Study of What's True or False or Somewhere in Between

The computer will see you now.

Farfetched? Not really. Scientists have used the logical principles that you will study in this chapter to develop computer programs called expert systems that interview doctors to acquire the expertise needed to make competent medical decisions. Then, by asking you a series of questions and scanning your x-rays, an expert system can make a remarkably accurate diagnosis of your medical condition.

Instead of having a computer program treat your ailment, perhaps you would be more comfortable with a real doctor at Kyung Hee Unversity in Korea controlling a robot over the Internet operating on you. Or how about an expert system acting as an air traffic controller as you land at a busy airport? By analyzing the way we make decisions, scientists have developed expert systems that have a wide variety of applications in your life.

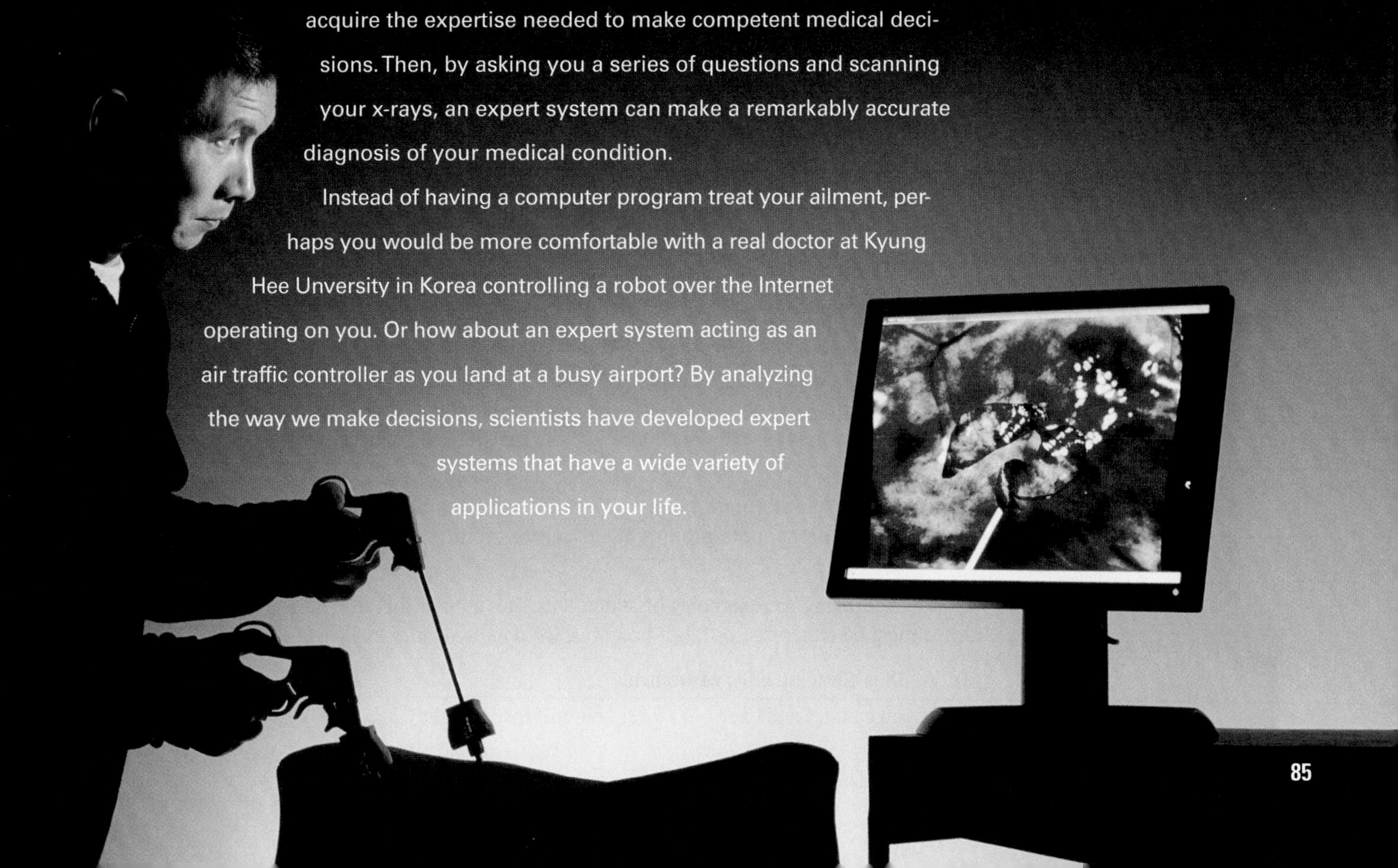

It is quite likely that you have an application of logic in your pocket. Your cell phone and other electronic devices, from digital TVs to iPads, are based on logic.

In Sections 3.4 and 3.5, we will show you how to apply logic to analyze arguments for their correctness and how to defend yourself against some of the false arguments that you encounter in everyday life.

3.1 Statements, Connectives, and Quantifiers

Objectives

1. Identify statements in logic.
2. Represent statements symbolically using five connectives.
3. Understand the difference between the universal and existential quantifiers.
4. Write the negations of quantified statements.

"You just don't understand what I'm saying!" Have you ever blurted that out while arguing with a friend or relative? At such times, wouldn't it be nice if you both could type your arguments into your "logic calculators," hit return, and, in a blink, determine who is correct?

In a sense, that was what the British mathematician George Boole was trying to do when he invented *symbolic logic* in the nineteenth century. Boole believed that he could calculate with logical ideas symbolically—similar to the way we calculate with numerical quantities in algebra. You may recall from algebra that $(x + y)^2 = x^2 + 2xy + y^2$, which is a general statement that is true for any numbers x and y. Similarly, Boole wanted to develop methods for doing routine logical calculations to determine when statements were true or false and whether logical arguments were valid. Although symbolic logic never achieved the success that Boole envisioned, as you will see in the exercise set in Section 3.2, it is important today because it forms the mathematical basis for digital devices that you use every day, such as MP3 players, digital video recorders, satellite radio and TV, and cell phones.

In studying algebra, you learned that to solve problems, the first step is to represent verbal expressions with symbols. For example, instead of using the sentence "The area of a rectangle equals the product of its length times its width," we write the more concise statement "$A = l \times w$." Similarly, in this section, you will learn to represent English statements symbolically; in Section 3.2, we will determine when these statements are true or false.

Statements in Logic

KEY POINT

In symbolic logic, we only care whether statements are true or false—not about their content.

As you begin your study of symbolic logic, keep in mind that *we are only concerned with the truth or falsity of the sentences we analyze and not their content.* As you will see in Section 3.4, an argument can seem ridiculous and yet have a proper logical form. On the other hand, an argument that seems believable may have an unacceptable logical form.

> **DEFINITION** A **statement** in logic is a declarative sentence that is either true or false. We represent statements by lowercase letters such as *p*, *q*, or *r*.

The following are examples of statements. Remember that to be a statement, the sentence must be either true or false; however, *we do not need to know which it is.*

a) AIDS is a leading killer of women.

b) If you eat less and exercise more, you will lose weight.

c) In the last 10 years, we have reduced by 25% the amount of greenhouse gases in the atmosphere that produce global warming.

d) Oprah Winfrey or James Cameron had the highest earnings in the entertainment industry in 2010.

The following are not statements:

e) Come here. (This is not a declarative sentence.)

f) When did dinosaurs become extinct? (This is not a declarative sentence.)

g) This statement is false. (This is a paradox. It cannot be either true or false. If we assume that this sentence is true, then we must conclude that it is false. On the other hand, if we assume it is false, then we conclude that it must be true.)

QUIZ YOURSELF 1

Identify each statement as simple or compound.

a) The 2011 Baltimore Orioles were the best team in the history of baseball.

b) If you break your lease, then you forfeit your deposit.

c) The world's tallest skyscrapers are in Dubai and Chicago.

Notice that statements a)–d) fall into two categories. Statements a) and c) each express a single idea; if we were to remove any part of these sentences, they would no longer make sense. Such sentences are called *simple statements.* In contrast, statements b) and d) clearly contain several ideas connected to make a more complex sentence. These sentences are called *compound statements.* Notice that words such as *if … then* and *or* connect the ideas to form the compound statements. 1

DEFINITIONS A **simple statement** contains a single idea. A **compound statement** contains several ideas combined together. The words used to join the ideas of a compound sentence are called **connectives.**

Connectives

KEY POINT

In logic, we connect ideas using *not, and, or, if... then,* and *if and only if.*

Because the English language is so rich, we can use many words to connect ideas. However, the connectives we use in logic generally fall into five categories: *negation, conjunction, disjunction, conditional,* and *biconditional.* We will now discuss each connective individually.

DEFINITION Negation

Negation is a statement expressing the idea that something is not true. We represent negation by the symbol $\sim$.

EXAMPLE 1 Negating Sentences

Negate each statement:

a) p: The high-definition screen is included in the price of your MacBook.

b) b: The blue whale is the largest living creature.

SOLUTION:

a) The high-definition screen is *not* included in the price of your MacBook. We write this negation symbolically as $\sim p$.

b) The blue whale is *not* the largest living creature. The symbols $\sim b$ represent this statement.

DEFINITION Conjunction

A **conjunction** expresses the idea of *and.* We use the symbol $\wedge$ to represent a conjunction.

The next example illustrates some statements that you might encounter when signing a lease for your new apartment.

EXAMPLE 2 Joining Statements Using *And*

Consider the following statements:

p: The tenant pays utilities. d: A \$150 deposit is required.

a) Express the statement "It is not true that: the tenant pays utilities and a \$150 deposit is required" symbolically.

b) Write the statement $\sim p \wedge \sim d$ in English.

SOLUTION: Review the Order Principle on page 12.

a) This sentence has the form $\sim(p \wedge d)$.

b) The tenant does not pay utilities and a \$150 deposit is not required.

PROBLEM SOLVING

Strategy: The Order Principle

Notice the difference in the form of the statements in Example 2 that we show in this diagram:

a) $\sim(p \wedge d)$ — First use "and." Then negate.

b) $\sim p \wedge \sim d$ — First negate. Then use "and."

These statements sound similar, but do not say the same thing. Keep in mind that changing the order of logical operations can change the meaning of a statement.

DEFINITION Disjunction

A **disjunction** conveys the notion of *or.* We use the symbol $\vee$ to represent a disjunction.

QUIZ YOURSELF 2

Consider the following statements:

d: I will buy a Blu-ray Disc player.

i: I will buy an iPod touch.

Write each statement in symbolic form.

a) I will not buy a Blu-ray Disc player or I will not buy an iPod touch.

b) I will not buy a Blu-ray Disc player and I will buy an iPod touch.

EXAMPLE 3 Joining Ideas Using *Or*

Consider the following statements:

h: We will build more hybrid cars. f: We will use more foreign oil.

a) Write the following statement symbolically:

We will not build more hybrid cars or we will use more foreign oil.

b) Write the symbolic form $\sim(h \vee f)$ in English.

SOLUTION: Review the Order Principle on page 12.

a) The symbolic form for this sentence is $(\sim h) \vee f$.

b) Translating into English, we get:

It is not true that: we will build more hybrid cars or we will use more foreign oil.

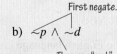

Historical Highlight

The Death of Leibniz's Dream

For hundreds of years, mathematicians pursued Leibniz's dream (see Historical Highlight on page 89) of developing a logical theory that could be used to build all of mathematics. In the 1920s, two British mathematicians, Bertrand Russell and Alfred North Whitehead, tried to develop such a plan and were unsuccessful. All hopes for success were dashed in 1931 when Kurt Gödel, a Princeton mathematician, proved his famous *Incompleteness Theorem*, which showed in a mathematical theory such as elementary arithmetic that there will always exist statements that are "true" but cannot be proved using the theory! Leibniz's dream was finally dead.

QUIZ YOURSELF 7

Prove that $\sim(p \vee q)$ is logically equivalent to $(\sim p) \wedge (\sim q)$.

First compute "and" then "negate."

		2		1	
p	d	$\sim$	$(p$	$\wedge$	$d)$
T	T	**F**	T	T	T
T	F	**T**	T	F	F
F	T	**T**	F	F	T
F	F	**T**	F	F	F

First "negate" then compute "or."

1	3	2
$(\sim p)$	$\vee$	$(\sim d)$
F	**F**	F
F	**T**	T
T	**T**	F
T	**T**	T

Because the final columns (highlighted) in the two truth tables are identical, the two statements are logically equivalent and therefore express exactly the same information.

Now try Exercises 39 to 46. **7**

DeMorgan's Laws

It is easier to remember mathematical facts if we make connections between them and see analogies. Notice that in Example 6, the logical equivalence of $\sim(p \wedge d)$ and $(\sim p) \vee (\sim d)$ looks very similar to one of DeMorgan's laws in set theory, namely $(A \cap B)' = A' \cup B'$. You should expect to see such a parallel because logic and set theory are both Boolean algebras.

DEMORGAN'S LAWS FOR LOGIC If p and q are statements, then

a) $\sim(p \wedge q)$ is logically equivalent to $(\sim p) \vee (\sim q)$.

b) $\sim(p \vee q)$ is logically equivalent to $(\sim p) \wedge (\sim q)$.

By defining logical equivalence, we have taken a large step toward treating logic algebraically. In algebra, when we say $2(x + y) = 2x + 2y$, we are making a general statement that holds for an infinite number of specific cases. For example, we can substitute $x = 3$ and $y = 5$ to get the true statement $2(3 + 5) = 2 \cdot 3 + 2 \cdot 5$. Similarly, when we say that $\sim(p \wedge q)$, is logically equivalent to $(\sim p) \vee (\sim q)$, we are saying that an infinite number of pairs of English language statements are equivalent. For example, if p is the statement "Today is Tuesday" and q is the statement "It is raining," then the following two statements have the same meaning:

"It is not true that: today is Tuesday and it is raining." $\sim(t \wedge r)$

"Today is not Tuesday or it is not raining." $\sim t \vee \sim r$

We can now reason about the meaning of complex statements based on their form rather than their content. Documents such as apartment leases, warranties, car loan applications, college admission forms, and income tax instructions often are written in a legalistic style using the connectives you have been studying.

QUIZ YOURSELF 8

Use one of DeMorgan's laws to rewrite (in English) the following statement taken from a car warranty:

> The car is not more than five years old and has not been driven over 50,000 miles.

EXAMPLE 7 Applying Logic to Legal Documents

Use DeMorgan's laws to rewrite the following statement, which is based on instructions for filing Form 1040-A with the U.S. Internal Revenue Service.

> It is false that: you received interest from a seller-financed mortgage and the buyer used the property as a personal residence.

SOLUTION:

It is easy to rewrite this statement if first we represent it in symbolic form. Let r represent "You received interest from a seller-financed mortgage" and let b represent "The buyer used the property as a personal residence." This statement has the form $\sim(r \wedge b)$.

By DeMorgan's laws this statement is equivalent to $(\sim r) \vee (\sim b)$. We can now rewrite this in English as "You did not receive interest from a seller-financed mortgage or the buyer did not use the property as a personal residence."

Now try Exercises 31–38. 8

SOME GOOD ADVICE

When you use logic to rewrite a statement, the result can sound awkward. You may want to smooth out the grammar so that the sentence sounds better. *This is usually a mistake!* Unless you are quite careful, you can easily change the meaning of a sentence by rewriting it. In logic, *the form of a statement is more important than its literary style.*

KEY POINT

There is an alternative way to construct truth tables.

Some people prefer another method to construct truth tables, which we will now explain.

EXAMPLE 8 Using an Alternative Method to Construct Truth Tables

a) We will reconstruct the truth table for $(\sim p \wedge q) \vee (p \wedge q)$ that you saw in Example 4. Recall that we constructed that table by performing the following steps:

   Step 1: Calculate $\sim p$.

   Step 2: Compute values for $\sim p \wedge q$

   Step 3: Find values for $p \wedge q$.

   Step 4: Calculate values for $(\sim p \wedge q) \vee (p \wedge q)$.

   In the following table, we make separate columns for each of these steps to obtain the same table as we did in Example 4.

		1	2	3	4
p	q	$\sim p$	$\sim p \wedge q$	$p \wedge q$	$(\sim p \wedge q) \vee (p \wedge q)$
T	T	F	F	T	T
T	F	F	F	F	F
F	T	T	T	F	T
F	F	T	F	F	F

b) We will construct a truth table for $(\sim p \wedge q) \vee (\sim r)$ that we explained in Example 5. We constructed that table by performing the following steps:

Step 1: Calculate $\sim p$.

Step 2: Find values for $\sim p \wedge q$.

Step 3: Compute values for $\sim r$.

Step 4: Calculate values for $(\sim p \wedge q) \vee (\sim r)$.

Again, we make separate columns for each step. Notice that the final column in this table is exactly the same as the final column in Example 5.

			1	2	3	4
p	q	r	$\sim p$	$\sim p \wedge q$	$\sim r$	$(\sim p \wedge q) \vee (\sim r)$
T	T	T	F	F	F	F
T	T	F	F	F	T	T
T	F	T	F	F	F	F
T	F	F	F	F	T	T
F	T	T	T	T	F	T
F	T	F	T	T	T	T
F	F	T	T	F	F	F
F	F	F	T	F	T	T

Sometimes notation in logic can be confusing. For example, if we see the expression $\sim p \vee q$, does it mean $(\sim p) \vee q$ or $\sim(p \vee q)$? Generally, I tell my students to assume that the negation sign affects as little as possible, so we would interpret $\sim p \vee q$ as meaning $(\sim p) \vee q$. With the expression $p \wedge q \vee r$, it could mean $(p \wedge q) \vee r$ or $p \wedge (q \vee r)$. There are rules called rules of precedence that cover this situation, but to avoid memorizing these rules we will always use parentheses to make it clear which option we have in mind.

EXERCISES 3.2

Sharpening Your Skills

In Exercises 1–4, use numbers to specify the order in which you would perform the logical operations for each statement.

1. $\sim(p \vee \sim q)$ **2.** $\sim(\sim p \wedge q)$ **3.** $p \wedge \sim(p \vee \sim q)$ **4.** $(p \wedge \sim q) \vee \sim p$

For Exercises 5–12, fill in the missing values in the following truth table.

				2	1	4		3	
p	q	r	$(p$	$\wedge$	$\sim r)$	$\vee$	$(q$	$\vee$	$r)$
T	T	T	T	F	F	**11.** _	T	T	T
T	T	F	T	**5.** _	T	T	T	T	F
T	F	T	T	F	**10.** _	T	F	T	T
T	F	F	T	T	T	T	F	**6.** _	F
F	T	T	F	F	F	T	T	**8.** _	T
F	T	F	F	**9.** _	T	T	T	T	F
F	F	T	F	F	F	T	F	**7.** _	T
F	F	F	F	F	T	**12.** _	F	F	F

State whether the numbers given in Exercises 13–16 are possibilities for the number of the lines in a truth table. If the number is a possibility, state how many different logical variables would be used in constructing the table.

13. 64 **14.** 36 **15.** 72 **16.** 128

In Exercises 17–26, construct a truth table for each statement.

17. $p \wedge \sim q$ **18.** $\sim(\sim p \wedge q)$

19. $\sim(p \vee \sim q)$ **20.** $\sim(p \vee q) \wedge \sim(p \wedge q)$

21. $\sim(p \wedge q) \vee \sim(p \vee q)$ **22.** $(p \vee r) \wedge (p \wedge \sim q)$

23. $(p \wedge r) \vee (p \wedge \sim q)$ **24.** $(p \vee q) \wedge (p \vee r)$

25. $\sim(p \vee \sim q) \wedge r$ **26.** $(p \wedge \sim q) \vee \sim r$

In Exercises 27–30, determine whether we are using the inclusive or *or the* exclusive or *in each sentence.*

27. Pay me now or pay me later.

28. You will earn a tax rebate if your income is less than $23,500 or if you are older than 65.

29. The warranty is void if the appliance has been abused or improperly maintained.

30. The penalty is a $500 fine or 40 hours of community service.

In Exercises 31–38, use DeMorgan's laws to rewrite the negation of each statement.

31. Bill is tall and thin.

32. Yorrel will go to law school or pursue an MBA.

33. Christian will apply for either a loan or work study.

34. Joanna will quit her job and join the Peace Corps.

35. Ken qualifies for a rebate or a reduced interest rate.

36. Pei Li got a larger disk drive and the extended warranty with her computer.

37. The number x is not equal to 5 and s is not odd.

38. The number y divides 10, but it is not even.

In Exercises 39–46, determine whether the pairs of statements are logically equivalent.

39. $\sim(p \wedge \sim q)$, $(\sim p) \vee q$

40. $\sim(p \vee \sim q)$, $(\sim p) \wedge q$

41. $\sim(p \vee \sim q) \wedge \sim(p \vee q)$, $p \vee (p \wedge q)$

42. $\sim(\sim p \vee \sim q)$, $p \vee q$

43. $\sim(p \vee \sim q) \wedge \sim(p \vee q)$, $(\sim p \wedge q) \wedge (\sim p \wedge \sim q)$

44. $(p \wedge \sim q) \vee \sim(p \wedge q)$, $\sim(p \vee \sim q) \vee (\sim p \vee \sim q)$

45. $p \vee (\sim q \wedge r)$, $(p \vee (\sim q)) \wedge (p \vee r)$

46. $p \wedge (\sim q \vee \sim r)$, $(p \wedge (\sim q)) \vee (p \wedge \sim r)$

Applying What You've Learned

Use the following graph that summarizes the status of couples five years after they decide to live together to find the truth values of the statements in Exercises 47–52. Let p *represent the statement "Couples who considered living together as a step towards marriage were most likely to be married after five years." Let* q *represent the statement "Couples who considered living together as a substitute for marriage were most likely to be split up five years later." Let* r *represent "More couples who used living together as a substitute for marriage were married five years later than those who were living together but just dating."*

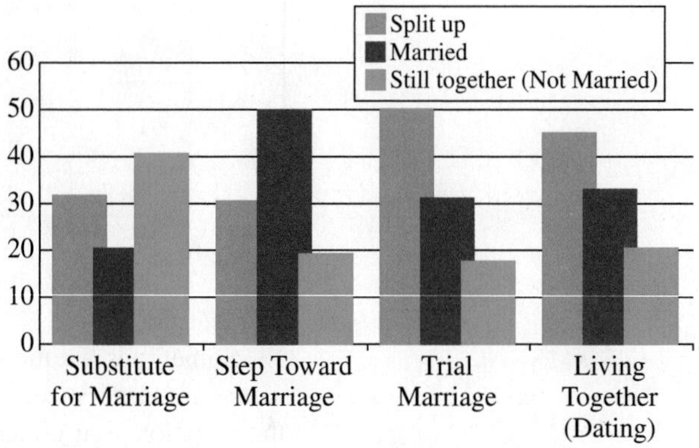

47. $(\sim p) \vee (\sim q)$ **48.** $\sim(p \wedge q)$

49. $(\sim p \vee q) \wedge r$ **50.** $(p \wedge q) \vee \sim r$

51. $p \vee (\sim q \wedge r)$ **52.** $r \wedge \sim(q \vee r)$

The statements in Exercises 53–56 are excerpts from the instructions for filing the Form 1040-A with the Internal Revenue Service. Rewrite them as we did in Example 7.

53. The earned income tax did not reduce the tax you owe or did not give you a refund.

54. You cannot claim yourself or your spouse as a dependent.

55. You are not single and not the head of a household.

56. It is not true that: you are filing a joint return and are covered by a retirement plan at work.

In order to be considered for a Congressional internship, applicants must be majoring in political science, history, international studies, or a related field; be a senior next semester or have graduated within the past year; and have at least a GPA of 3.0. Which of the following applicants described in Exercises 57–60 should be considered for the position?

57. Internship requirements. Marge, a Russian history major, has 24 credits and a GPA of 3.2

58. Internship requirements. Homer, majoring in international politics, will graduate next semester and has a GPA of 3.6.

59. Internship requirements. Bart has been working for a sporting goods store for three months since graduating with a degree in Latin American studies with a GPA of 3.25.

60. Internship requirements. Lisa, a classical dance major with a minor in Asian studies, is entering her senior year and has a GPA of 3.8.

Use the following information provided by the Humane Society regarding pet ownership in the United States to decide if the statements in Exercises 61–64 are true or false.

- *There are approximately 78.2 million owned dogs and 86.4 million owned cats.*
- *Thirty-nine percent of households own at least one dog compared to thirty-three percent that own at least one cat.*
- *Cat owners spend \$219 and dog owners spend \$248 on veterinary visits annually.*
- *Twenty-one percent of dogs and twenty-one percent of cats were adopted from animal shelters.*

61. Pet ownership. There are not fewer cat owners than dog owners, but dog owners spend more than cat owners on veterinary visits.

62. Pet ownership. It is not true that: there are more owned dogs than cats and there are not more households that own cats than dogs.

63. Pet ownership. It is not true that: there are more households with dogs than cats or cat owners spend less than dog owners on veterinary visits annually.

64. Pet ownership. There are not as many dogs as cats adopted from shelters or more households own dogs than cats.

In Section 3.1 (page 94), we showed how to represent computer circuits with logical expressions. We represented a series by the expression $p \wedge q$ *and a parallel circuit by* $p \vee q$.

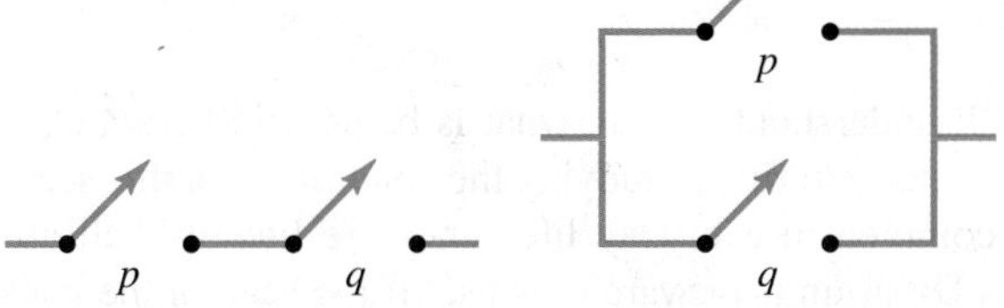

Series circuit Parallel circuit

The logical forms $(p \vee q) \wedge (p \vee r)$ *and* $p \vee (q \wedge r)$ *are equivalent, which means that the following two circuits will behave identically.*

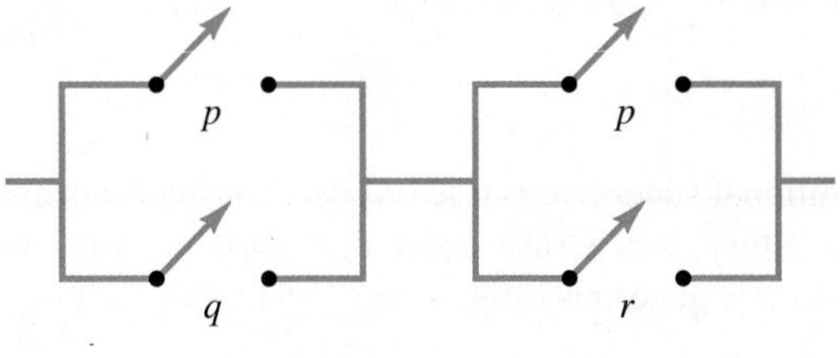

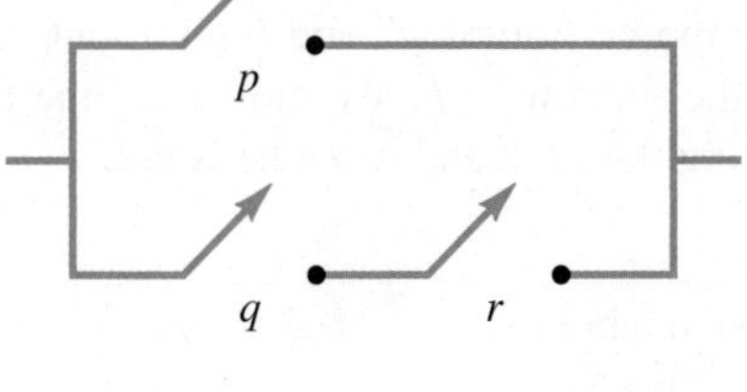

In Exercises 65–68, represent each circuit by a logical form and then rewrite that logical form in an equivalent form. Use truth tables to prove that the second form is equivalent to the first. Draw a circuit that corresponds to the second logical form. If possible, try to choose the second form so that the corresponding circuit has fewer switches than the original circuit.

65.

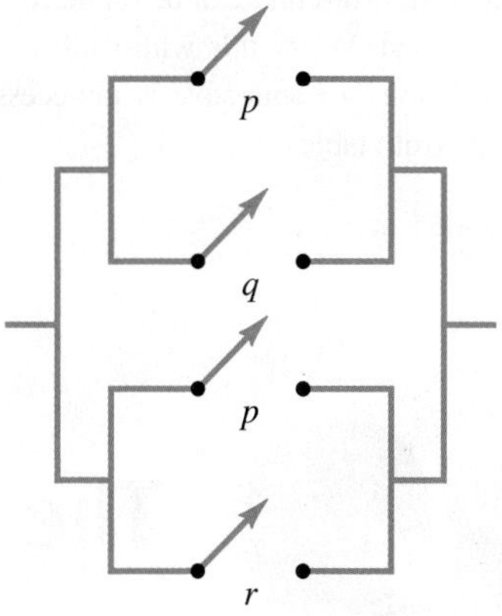

66.

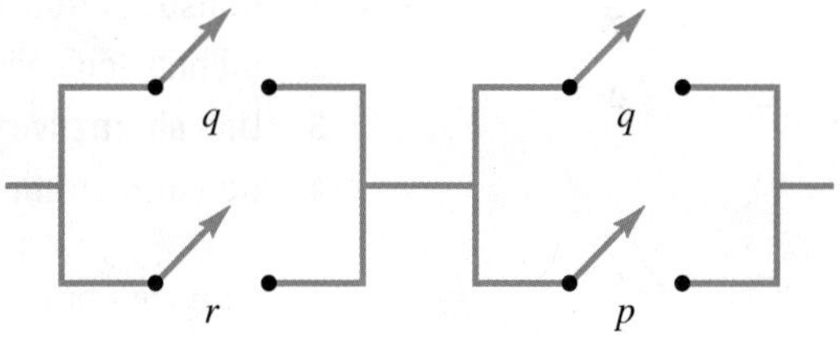

67.

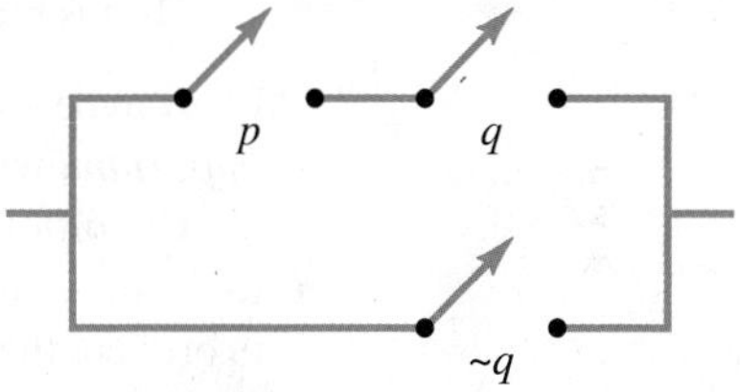

68.

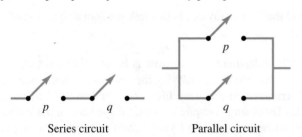

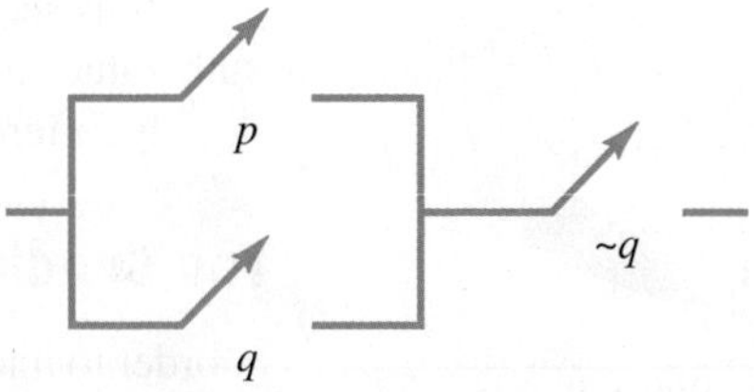

Communicating Mathematics

69. Which *or* do we use in logic—the inclusive or exclusive? How are the truth tables different for these two *ors*? Give an example of a situation in which you would use an inclusive *or* and another situation in which you would use an exclusive *or*.

70. What advantage do you see in using truth tables to determine logical equivalence rather than using our "common sense" and knowledge of the English language?

Challenge Youself

71. In set theory, we showed that a set with k elements has 2^k subsets. If a logical statement has variables p, q, and r, explain how the eight subsets of $\{p, q, r\}$ correspond to the eight lines of the truth table for this statement.

72. The *and* connective is unnecessary in the sense that in a logical expression, the word *and* can be replaced by the equivalent form $\sim(\sim p \vee \sim q)$. Verify this with truth tables. In a similar way, show that the *or* connective is unnecessary. Prove your claim by using truth tables.

The stroke connective has the following truth table:

p	q	$p \mid q$
T	T	F
T	F	T
F	T	T
F	F	T

73. The stroke connective is sometimes called NAND (*not and*). Explain why this is so.

74. Prove that $\sim p$ is logically equivalent to $p \mid p$.

75. Prove that $p \vee q$ is logically equivalent to $(p \mid p) \mid (q \mid q)$.

76. Give a form using only the stroke connective that is equivalent to $p \wedge q$.

3.3 The Conditional and Biconditional

Objectives

1. Construct truth tables for conditional statements.
2. Identify logically equivalent forms of a conditional.
3. Use alternative wording to write conditionals.
4. Construct truth tables for biconditional statements.

"If it ain't broke, don't fix it."

How many times have you heard that sound advice? Is this folk wisdom also telling us,

"If it is broke, then fix it?"

Maybe—or maybe not. You'll understand exactly what is being said here and, *more important, what is not being said* after you finish studying the conditional in this section.

Conditional statements are common in everyday life—they are literally "all around us." The manufacturer of our DVD-burning software tells us, "If the seal on the package is broken, then the software cannot be returned." As a young child at the amusement park, you were no doubt disappointed when you saw a sign warning, "If you are not 48 inches tall, you cannot ride this roller coaster."

In this section, we will use truth tables for conditionals and biconditionals to compute truth values of compound statements. You will notice that sometimes we use conditionals slightly differently in mathematics than in everyday language.

KEY POINT

There is only one way a conditional can be false.

The Conditional

In order to understand when a conditional statement is true or false, consider this example. Mr. Gates, the owner of a small factory, has a rush order that must be filled by next Monday and he approaches you with this generous offer:

If you work for me on Saturday, then I'll give you a $100 bonus.

If we let w represent "You work for me on Saturday" and b represent "I'll give you a $100 bonus," then this statement has the form $w \rightarrow b$. We must examine four cases to determine exactly when Mr. Gates is telling the truth and when he is not.

Case 1 (w is true and b is true.):

You come to work and you receive the bonus.

In this case, Mr. Gates certainly made a truthful statement.

Case 2 (w is true and b is false.):

You come to work and you don't receive the bonus.

Mr. Gates has gone back on his promise, so he has made a false statement.

Case 3 (w is false and b is true.):

You don't come to work, but Mr. Gates gives you the bonus anyway.

Think carefully about this case because it differs from the way we tend to use *if... then* in everyday language. When you listen to Mr. Gates, do not read more into his statement than he actually said. You do not expect to get the bonus if you did not come to work because that is your experience in everyday life. However, Mr. Gates never said that. *You are assuming this condition.* Remember that in logic, a statement is either true or false. Therefore, because Mr. Gates did not say something false, he has told the truth.

Case 4 (w is false and b is false.):

You don't come to work and you don't receive the bonus.

In this case, Mr. Gates is telling the truth for exactly the same reason as in Case 3. Because you did not come to work, Mr. Gates can give you the bonus or not give you the bonus. In either case, he has not told a falsehood and therefore is telling the truth.

p	q	$p \to q$
T	T	T
T	F	F
F	T	T
F	F	T

FIGURE 3.7 A conditional is false *only* when the hypothesis is true and the conclusion is false.

This discussion explains the truth table for the *if... then* connective in Figure 3.7.

DEFINITIONS For a conditional $p \to q$, statement p is called the *hypothesis* and q is called the *conclusion.**

SOME GOOD ADVICE

A good way to remember the truth table for the conditional is that the *only way a conditional can be false* is if it has a true hypothesis and a false conclusion.

We can build truth tables for statements that combine the conditional with the connectives you have studied earlier.

EXAMPLE 1 Constructing a Truth Table for a Conditional Statement

Construct a truth table for the statement $(\sim p \vee q) \to (\sim p \wedge \sim q)$.

SOLUTION: Review the Order Principle on page 12.

Because there are two variables, p and q, the truth table has four lines. After carrying out steps 1 through 5, the values for the hypothesis of the conditional are under the *or* symbol of step 2 (highlighted) and the values for the conclusion are under the *and* symbol of step 5 (highlighted).

In step 6, lines 1 and 3 are the only cases with a true hypothesis and a false conclusion, so we put an F on those lines. The final truth table then looks like this:

		1	2 (hypothesis)		6	3	5 (conclusion)	4
p	q	$(\sim p$	$\vee$	$q)$	$\to$	$(\sim p$	$\wedge$	$\sim q)$
T	T	F	T	T	F	F	F	F
T	F	F	F	F	T	F	F	T
F	T	T	T	T	F	T	F	F
F	F	T	T	F	T	T	T	T

True hypothesis and false conclusion occur on these lines.

Now try Exercises 1 to 16. 9

QUIZ YOURSELF 9

Construct a truth table for the statement $(p \wedge \sim q) \to (\sim p \vee q)$.

*Some call the hypothesis the *antecedent* and the conclusion the *consequent*.

Many formal and informal contracts use the *if…then* connective. We can understand a condition in an agreement by writing it in a symbolic form.

EXAMPLE 2 Analyzing a Refund Policy

Emilio has just purchased a new solar-powered music system for his iPod from overstock.com that has the following return policy:

> *If* you return the system within 30 days in its original packaging, *then* you will be given a full refund on your purchase.

Dissatisfied with his system, Emilio returns it in its original packaging but is denied a refund. What can you deduce from this information?

SOLUTION:

If we let r, p, and f represent the following statements:

r: Emilio returned the music system within 30 days.

p: The system was in its original packaging.

f: He is given a full refund.

we see that the store's policy has the form $(r \wedge p) \rightarrow f$. We know that f is false because Emilio did not get a refund. If we assume that the store is truthful in its policy, then the hypothesis $r \wedge p$ cannot be true because a conditional of the form $T \rightarrow F$ is always false. However, we do know that p is true, because Emilio used the original packaging. If r were also true, that would make $r \wedge p$ true, which we just said cannot be. So we conclude that r is false—the system was not returned within 30 days.

Now try Exercises 75–76.

Derived Forms of a Conditional

There are three typical ways that you might want to rephrase a conditional; however, always remember that when you rewrite a statement in logic, you may also be changing its meaning.

KEY POINT

The converse, inverse, and contrapositive are three derived forms of a conditional.

DEFINITIONS We can derive the following statements from the conditional $p \rightarrow q$:

The converse has the form $q \rightarrow p$.

The inverse has the form $\sim p \rightarrow \sim q$.

The contrapositive has the form $\sim q \rightarrow \sim p$.

Between the Numbers

Always Read the Fine Print!

Have you ever been tempted to buy a large flat-screen TV with the come-on

Buy Now—No Interest Payments for One Year!

Be careful when you see such an ad. Usually the fine print states something like ***if*** you continue to meet certain conditions of the contract, ***then*** the 0% interest rate stays in effect. But, as you saw in the section opener with Mr. Gates, if the hypothesis of this conditional is false—in other words, if you are even one day late in a payment, or perhaps carry a balance past the one-year period for what is called the "tickle rate" of the loan—then the penalties can be huge. You may have to suffer paying late fees, and, now that the agreement has been violated on your part, the interest rates could go as high as 30%. Lenders who make such tempting loans are counting on the fact that you will not be able to pay off your balance by the end of the 0% interest period or that you will slip up in some other way, thus violating the contract.

The following table will help you remember how to construct these derived forms of a conditional:

Name of Derived Form	How It Is Constructed
Conditional	$p \rightarrow q$
Converse	Switch p and q.
Inverse	Negate both p and q. Don't switch.
Contrapositive	Negate both p and q and also switch.

EXAMPLE 3 Rewriting the Converse, Inverse, and Contrapositive of Statements in Words

Write, in words, the converse, inverse, and contrapositive of the statement:

"If marijuana is legalized, then drug abuse will increase."

SOLUTION:

To understand this problem better, we first write this statement symbolically as $m \rightarrow d$, where

m stands for "Marijuana is legalized."

and

d stands for "Drug abuse will increase."

We next write each of the derived forms symbolically.

To form the converse, we *switch* d and m to get the form $d \rightarrow m$. Translating this into words gives us

"If drug abuse will increase, then marijuana is legalized."

To form the inverse, we *negate* both m and d to get the form $\sim m \rightarrow \sim d$, which we can write as

"If marijuana is not legalized, then drug abuse will not increase."

To form the contrapositive, we *negate* both m and d and also *switch* to get the form $\sim d \rightarrow \sim m$. We write this as

"If drug abuse will not increase, then marijuana is not legalized."

Now try Exercises 21 to 28. 10

QUIZ YOURSELF 10

Write, in words, the inverse and contrapositive of the statement:

If the price of downloading videos increases, then people will copy them illegally.

You must be careful when you rephrase a conditional because the derived statement may not be logically equivalent to the original. We now investigate which derived forms of a conditional are logically equivalent.

EXAMPLE 4 Equivalence of the Derived Forms of a Conditional

Which of the statements $p \rightarrow q$, $q \rightarrow p$, $\sim p \rightarrow \sim q$, and $\sim q \rightarrow \sim p$ are logically equivalent?

SOLUTION:

We will answer this question by comparing truth tables for these statements.

logically equivalent (Conditional and Contrapositive)

		Conditional	Converse			Inverse			Contrapositive		
p	q	$p \rightarrow q$	q	$\rightarrow$	p	$\sim p$	$\rightarrow$	$\sim q$	$\sim q$	$\rightarrow$	$\sim p$
T	T	T	T	T	T	F	T	F	F	T	F
T	F	F	F	T	T	F	T	T	T	F	F
F	T	T	T	F	F	T	F	F	F	T	T
F	F	T	F	T	F	T	T	T	T	T	T

logically equivalent (Converse and Inverse)

Because the final columns under the conditional and contrapositive are identical, these statements are logically equivalent. We also see that neither the converse nor the inverse are equivalent to the original conditional; however, they are equivalent to each other.

Example 5 shows how to rephrase the types of statements found in leases, contracts, warranties, and tax forms without changing their meaning.

EXAMPLE 5 Rewriting Legal Statements

Use the contrapositive to rewrite the following statements:

a) If you pay a deposit, then we will hold your concert tickets.

b) If you do not itemize, then you take the standard deduction.

SOLUTION:

The following diagrams show how to write the contrapositives by negating both the hypothesis and conclusion and then interchanging them.

a) Let p represent "You pay your deposit" and h represent "We will hold your concert tickets."

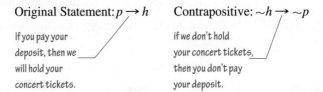

Original Statement: $p \rightarrow h$ — If you pay your deposit, then we will hold your concert tickets.

Contrapositive: $\sim h \rightarrow \sim p$ — if we don't hold your concert tickets, then you don't pay your deposit.

b) Let i represent "You itemize deductions" and s represent "You take the standard deduction."

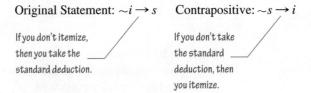

Original Statement: $\sim i \rightarrow s$ — If you don't itemize, then you take the standard deduction.

Contrapositive: $\sim s \rightarrow i$ — If you don't take the standard deduction, then you itemize.

Alternate Wording of Conditionals

There are many ways to express a conditional without using the words *if...then*. Each of the following forms expresses the conditional "if p then q":

q if p	Here the *if* still is associated with p even though it occurs later in the sentence.
p only if q	Recognize that *only if* does not say the same thing as *if*. The *if* condition is the hypothesis; the *only if* condition is the conclusion.
p is sufficient for q.	The sufficient condition is the hypothesis.
q is necessary for p.	The necessary condition is the conclusion.

The following diagram will help you remember the information above:

$$p \longrightarrow q$$

p	$\longrightarrow$	q
↑		↑
if		only if
sufficient		necessary

EXAMPLE 6 Rewriting Statements in *If . . . Then* Form

Write each statement in *if...then* form.

a) Your driver's license will be suspended if you are convicted of driving under the influence of alcohol.

b) You will graduate only if you have a 2.5 grade point average.

c) To hold your reservation, it is sufficient to give us your credit card number.

d) To qualify for a discount on your airline tickets, it is necessary to pay for them two weeks in advance.

SOLUTION: Review the Splitting-Hairs Principle on page 12.

a) Because the *if* goes with the clause "you are convicted of driving...," that clause is the hypothesis. We can rewrite this sentence as

> "If you are convicted of driving under the influence of alcohol, then your driver's license will be suspended."

b) The *only if* goes with the clause "you have a 2.5 grade point average"; therefore, this is the conclusion. We write this sentence as

> "If you graduate, then you have a 2.5 grade point average."

c) The phrase *it is sufficient* identifies the hypothesis. We write the sentence as

> "If you give us your credit card number, then [we will] hold your reservation."

d) The necessary condition is the conclusion. Rewriting this sentence, we get

> "If you qualify for a discount on your airline tickets, then you pay for them two weeks in advance."

Now try Exercises 45 to 52. 11

QUIZ YOURSELF 11

Write each statement using the *if...then* connective.

a) In order to travel on the Amazon, it is necessary to update your immunization.

b) You will increase your cardiovascular fitness only if you exercise three times a week.

KEY POINT

The biconditional means that two statements say the same thing.

The Biconditional

The biconditional (*if and only if*) indicates that two statements mean the same thing. For example, we could say that "Today is Tuesday if and only if tomorrow is Wednesday." In algebra we say "$x + 3 = 7$ if and only if $x = 4$." The notation for the biconditional, which is $p \leftrightarrow q$, suggests that we are saying $p \rightarrow q$ and $q \rightarrow p$ at the same time. We show the truth table for the biconditional in Figure 3.8.

p	q	$p \leftrightarrow q$
T	T	T
T	F	F
F	T	F
F	F	T

FIGURE 3.8 The biconditional is true when *both* p and q have the same value.

Math in Your Life

Thinking About How You Think

You and I think all the time, but do we ever think about how we think? Scientists who study **artificial intelligence** use formal logic to try to understand human thinking in order to build machines and write programs, such as the expert systems that we mentioned in the chapter opener.

How does this apply to you and me? If you were to develop acute abdominal pain while backpacking in a remote area, a local doctor might consult with deDombal's Leeds Abdominal Pain System, an expert system developed at the University of Leeds, to get a handle on your problem. Or I, being in a green frame of mind, might install an intelligent climate control system in my house that uses "common-sense" rules to save energy and improve my comfort. Our drive home from work has become safer, as automobile designers build artificial intelligence into our cars, enabling them to adjust quickly to hazardous driving conditions.

Although researchers still do not understand all the complexities of human thinking, artificial intelligence has become a multibillion-dollar industry that has many important applications.

You should use truth tables to verify that the biconditional $p \leftrightarrow q$ is logically equivalent to the statement $(p \rightarrow q) \wedge (q \rightarrow p)$.

EXAMPLE 7 Computing a Truth Table for a Complex Biconditional

Construct a truth table for the statement $\sim(p \vee q) \leftrightarrow (\sim q \wedge p)$.

SOLUTION: Review the Order Principle on page 12.

		2	1	5	3	4	
p	q	$(\sim$	$(p \vee q))$	$\leftrightarrow$	$(\sim q$	$\wedge$	$p)$
T	T	F	T	T	F	F	T
T	F	F	T	F	T	T	T
F	T	F	T	T	F	F	F
F	F	T	F	F	T	F	F

Another way to verify that two statements are logically equivalent is to join them with a biconditional and see whether this new statement is a tautology. We will investigate this further in the exercises.

EXERCISES 3.3

Sharpening Your Skills

*In Exercises 1–8, assume that p represents a true statement, q a false statement, and r a true statement. Determine the truth value of each statement.**

1. $\sim(p \vee q) \rightarrow \sim p$

2. $(p \wedge \sim q) \rightarrow q$

3. $(p \wedge q) \rightarrow (q \vee r)$

4. $(p \vee \sim q) \rightarrow r$

5. $(\sim p \vee \sim q) \rightarrow r$

6. $r \rightarrow (\sim p \wedge q)$

7. $\sim(\sim p \wedge q) \rightarrow \sim r$

8. $\sim(\sim p \vee r) \rightarrow \sim q$

In Exercises 9–20, construct a truth table for each statement.

9. $p \rightarrow \sim q$

10. $\sim p \rightarrow q$

11. $\sim(p \rightarrow q)$

12. $\sim(q \rightarrow p)$

13. $(p \vee r) \rightarrow (p \wedge \sim q)$

14. $(p \wedge q) \rightarrow (p \wedge \sim r)$

15. $\sim(p \vee r) \rightarrow \sim(p \wedge q)$

16. $\sim(p \wedge r) \rightarrow \sim(p \vee q)$

17. $(p \vee q) \leftrightarrow (p \vee r)$

18. $(p \wedge q) \leftrightarrow (p \wedge r)$

19. $(\sim p \rightarrow q) \leftrightarrow (\sim q \rightarrow p)$

20. $(p \rightarrow \sim q) \leftrightarrow (q \rightarrow \sim p)$

In Exercises 21–28, write in words the converse, inverse, or contrapositive as indicated for each statement.

21. If it rains, it pours. (converse)

22. If this appliance fails within 30 days, then it will be fixed free of charge. (converse)

23. If you buy the all-weather radial tires, then they will last for 80,000 miles. (inverse)

24. If you are older than 18, then you must register with Selective Service. (inverse)

25. If a geometric figure is an equilateral triangle, then its sides are all equal in length. (contrapositive)

26. If a geometric figure is a quadrilateral, then the sum of its interior angles is 180 degrees. (contrapositive)

27. If x evenly divides 6, then x evenly divides 9. (inverse)

28. If x is an even prime number, then x is divisible by 2. (inverse)

Assume that you begin with a statement of the form $p \rightarrow q$. *In Exercises 29–32, describe each of the given forms in a simpler way.*

29. The converse of the inverse

30. The inverse of the converse

31. The contrapositive of the inverse

32. The inverse of the contrapositive

In Exercises 33–36, write the indicated statement in symbolic form.

33. The converse of $(\sim p) \rightarrow q$

34. The inverse of $(\sim p) \rightarrow \sim(q \wedge r)$

*Recall that in interpreting these expressions, we assume that the negation symbol affects as little of the expression as possible. For example, you should interpret $\sim p \wedge q$ as meaning $(\sim p) \wedge q$ rather than $\sim(p \wedge q)$.

35. The contrapositive of $p \rightarrow \sim q$

36. The contrapositive of $\sim(p \vee r) \rightarrow q$

In Exercises 37–40, use the given assumptions to deduce the truth value for q.

37. $p \rightarrow (p \wedge q)$ is false and p is true.

38. $(\sim p) \rightarrow (p \vee q)$ is true and p is false.

39. $(p \wedge \sim q) \rightarrow \sim p$ is false.

40. $(\sim p \vee \sim q) \rightarrow \sim p$ is true and p is true.

In Exercises 41–44, determine which pairs of statements are equivalent. (It is helpful to first write the statements in symbolic form.)

41. If you activate your cell phone before October 1, then you receive 100 free minutes. If you do not receive 100 free minutes, then you do not activate your cell phone before October 1.

42. If you don't register for the LSATs before August 1, then you must pay a $30 late fee. If you register for the LSATs before August 1, then you do not pay a $30 late fee.

43. If it is raining, then use your headlights. If you use your headlights, then it is raining.

44. If Gretchen Bleiles does not score high in the Winter Olympics snowboarding competition, then she will not win a medal. If Gretchen Bleiles qualifies for a medal, then she scores high in the Winter Olympics snowboarding competition.

In Exercises 45–52, rewrite each statement using the words if...then.

45. I'll take a break if I finish my workout.

46. You can return the iPod touch only if you have not opened the package.

47. To qualify for this deduction, it is necessary for you to complete Form 3093.

48. To reserve a campsite, it is sufficient that you pay a small deposit.

49. You will receive a free cell phone only if you sign up before March 1.

50. I'll go to Florida if I can save $850.

51. To get a reduction on your auto insurance, it is sufficient that you remain accident free for three years.

52. To graduate this semester, it is necessary that you complete 18 credits.

Use this graph, which shows the results of a Gallup poll of 18- to 29-year-olds regarding who they think is the greatest U.S. president of all time, to determine if the statements in Exercises 53–56 are true or false.

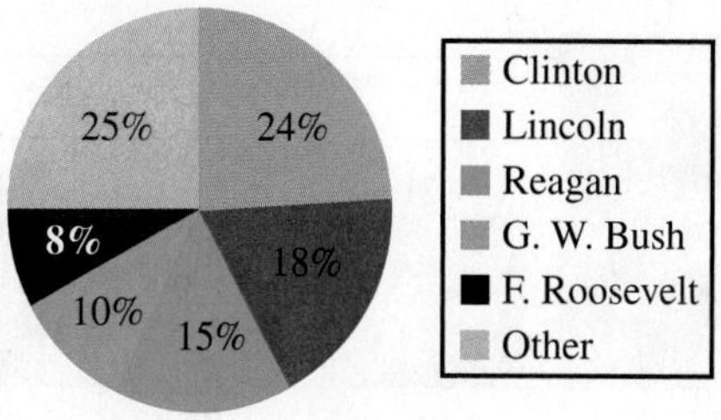

53. If more than half of those surveyed chose Clinton or Lincoln, then more than 10% chose Reagan.

54. If 10% chose Bush, then 10% also chose Roosevelt.

55. If Roosevelt was not ranked higher than Clinton, then Lincoln was not the most preferred.

56. If Bush was not preferred over Lincoln, then Reagan was not preferred by more than 10%.

According to an Accountemps survey appearing in USA Today, the following are mistakes candidates commonly make during a job interview:

- *Little knowledge of the company: 38 percent*
- *Unprepared to discuss skills/experiences: 20 percent*
- *Unprepared to discuss career plans: 14 percent*
- *Lack of eye contact: 10 percent*
- *Lack of enthusiasm: 9 percent*

Use this information to determine if the statements in Exercises 57–60 are true or false.

57. Job interview mistakes. If more than half of the candidates had little knowledge of the company, then another 20 percent lacked enthusiasm and good eye contact.

58. Job interview mistakes. If less than 40 percent were unprepared to discuss their skills or career plans, then more than a quarter lacked enthusiasm.

59. Job interview mistakes. If more than a quarter showed little knowledge of the company, then it was not true that less than 10 percent lacked enthusiasm.

60. Job interview mistakes. If less than 8 percent lacked eye contact and 14 percent were unprepared to discuss their career plans, then half of the candidates had little knowledge of the company.

Perhaps you have heard the term "helicopter parents"—parents who "hover" over their children, protecting them from making mistakes, particularly in school situations. The following graph summarizes some activities in which parents are very involved or even take over for their children when they are applying for college

(Source: collegeboard.com). Use this information to determine if the statements in Exercises 61–64 are true or false.

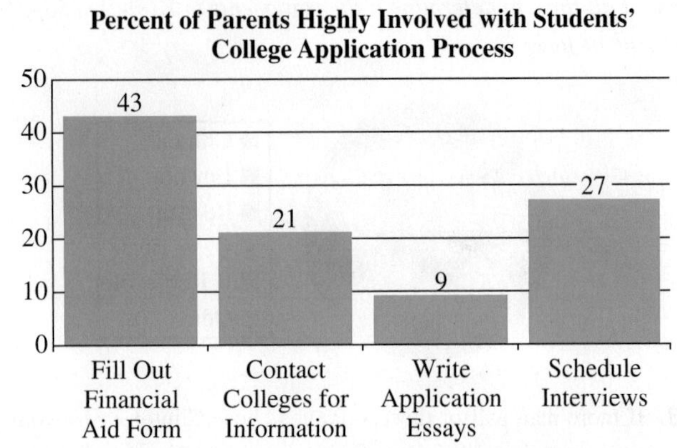

61. If less than 10% of parents help write application essays, then more than 40% fill out financial aid forms.

62. If more than 20% of parents contact colleges for information, then either 27% schedule interviews or less than 40% fill out financial aid forms.

63. If more than 20% are highly involved in scheduling interviews, then more than 10% are involved with writing application essays and contacting colleges for information.

64. If less than 20% schedule interviews, then more than 50% fill out financial aid forms.

Applying What You've Learned

In Exercises 65–68, write the converse, inverse, or contrapositive of each conditional statement based on instructions for filing Form 1040-A with the Internal Revenue Service.

65. Interpreting tax forms. If your gross income is over \$2,250, then you cannot be claimed by someone else as a dependent. (contrapositive)

66. Interpreting tax forms. If you are filing a joint return, then include your spouse's income. (inverse)

67. Interpreting tax forms. If the amount you overpaid is large, then decrease the amount being withheld from your pay. (converse)

68. Interpreting tax forms. If you are a nonresident alien, then you cannot claim an earned income tax credit. (contrapositive)

69. Prove that the biconditional is logically equivalent to the conjunction of two conditionals.

70. Prove that $p \leftrightarrow q$ is logically equivalent to $\sim p \leftrightarrow \sim q$.

Communicating Mathematics

71. Give an example of a true conditional whose inverse is false.

72. Is it possible to have a false conditional whose converse is also false? Explain.

73. Explain why it is reasonable to expect that the forms $p \rightarrow q$ and $\sim p \vee q$ are logically equivalent. Use truth tables to verify that these forms are logically equivalent.

74. Why is it reasonable to expect that the form $p \wedge \sim q$ is logically equivalent to the form $\sim(p \rightarrow q)$? Use truth tables to verify that these forms are logically equivalent.

In Exercises 75 and 76, assume that a credit card company has the following policy:

If you have a platinum credit card with an outstanding balance of more than \$1,000, or you have been a member for at least 10 years, then you qualify for the discount loan rate.

Represent the conditions stated in this policy as follows:

p = *You have a platinum credit card.*

b = *You have a balance of more than \$1,000.*

m = *You have been a member for at least 10 years.*

d = *You qualify for the discount loan rate.*

Based on the partial information given, what can you deduce about each situation? Explain your thinking. (Hint: Recognize that the policy can be written in symbolic form as $((p \wedge b) \vee m) \rightarrow d$.)

75. Jamie is a platinum credit card member and has an outstanding balance of \$750. He does not qualify for the discount loan rate.

76. Carla has a balance of \$1,245 on her platinum credit card and has been a member for 12 years.

Between the Numbers

77. Reading the fine print. Credit card companies usually have the right to raise rates, shorten grace periods, and basically change the rules of the card to work to their advantage at any time. Obtain an agreement for a credit card and carefully explain the rules the customer is agreeing to regarding rates, fees, and penalties.

78. Reading the fine print. Find the contract for your cell phone and explain the rules regarding your agreement. In particular, what penalties do you face if you exceed the limits of your contract?

Challenge Yourself

In Exercises 79 and 80, use your knowledge of logic to rewrite each statement in a simpler, equivalent form using fewer connectives.

79. $(\sim p \vee \sim q) \rightarrow \sim r$

80. $\sim q \rightarrow (\sim r \wedge p)$

In Exercises 81 and 82, construct truth tables for each statement. Study these tables carefully and see if you can recognize how to write these statements in simpler, equivalent forms.

81. $p \vee (r \wedge (p \wedge q))$

82. $p \wedge (r \vee (p \vee q))$

Exercises 83–86 are based on the exercise sets in earlier sections in which we discussed the correspondence between electrical circuits and logical forms. Draw a circuit that corresponds to each form.

83. $p \rightarrow q$ (*Hint:* Consider a form that is equivalent to this conditional that uses other connectives.)

84. $p \leftrightarrow q$

85. $(\sim p) \rightarrow \sim(q \wedge r)$

86. $p \leftrightarrow (q \wedge \sim r)$

3.4 Verifying Arguments

Objectives

1. Use truth tables to show an argument to be valid.
2. Understand how a truth table can show an argument to be invalid.
3. Recognize some common valid argument forms.
4. Recognize some common fallacies.

In a famous Monty Python skit called *The Argument Clinic*,* Michael Palin, playing a character called "Man," arranges to pay to have an argument. After paying his fee for a five-minute argument, he is directed to a room where John Cleese, playing the character of "Mr. Vibrating," awaits. The man knocks on the door and Mr. Vibrating speaks first.

Mr. Vibrating: Come in.
Man: Is this the right room for an argument?
Mr. Vibrating: I've told you *once*.
Man: No you haven't.
Mr. Vibrating: Yes I have.
Man: When?
Mr. Vibrating: Just now!
Man: No you didn't.
Mr. Vibrating: Yes I did!
Man: Didn't.
Mr. Vibrating: Did.
And so the skit continues.

When you hear the word *argument*, you may think of an altercation or verbal fight such as this. However, as you will learn in this section, an argument in logic is a series of statements that have a precise form whose validity we can analyze using truth tables.

KEY POINT

We use truth tables to determine whether arguments are valid.

Verifying Arguments

In this section, we will consider when a collection of statements produces a logical conclusion. Instead of expressing an argument in sentence form, we will represent it symbolically and use truth tables to determine whether it is valid. The following is a simple example of an argument:

1. If Mahala passed the bar exam, then she is qualified to practice law.
2. Mahala passed the bar exam.

---

3. Therefore, Mahala is qualified to practice law.

This argument begins with statements 1 and 2, called *premises*, and ends with the final statement 3, called a *conclusion.* If the two premises are true, then the conclusion must also be true. Another way of saying this is that *the conclusion follows from the premises.*

DEFINITIONS An **argument** is a series of statements called **premises** followed by a single statement called the **conclusion.** An argument is **valid** if whenever all the premises are true, then the conclusion must also be true.

We will use a truth table to determine whether the preceding argument is valid. Let us represent the statement "Mahala passed the bar exam" by p and represent the statement "She is qualified to practice law" by q. Then the argument has the following symbolic form. (The symbol $\therefore$ means *therefore* in the conclusion of an argument.)

$$\left.\begin{array}{l} p \to q \\ p \end{array}\right\} \text{premises}$$
$$\overline{\therefore\ q \ \}\ \text{conclusion}}$$

*Videos of this sketch can be found on YouTube.

We can view this argument as a conditional statement that has the following form:

If the first premise is true
and
the second premise is true
then the conclusion is true.

This argument then has the form $[(p \rightarrow q) \wedge p] \rightarrow q$. A clever way to prove that this argument is valid is by convincing ourselves that the argument cannot be invalid.

Realize that the argument is invalid if the expression $[(p \rightarrow q) \wedge p]$ is true, but at the same time q is false. In other words, the argument is invalid if both premises $(p \rightarrow q)$ and p are true, but the conclusion q is false. We make a truth table with separate columns for each premise and another for the conclusion.

		Premises		Conclusion
p	q	$p \rightarrow q$	p	q
T	T	T	T	T
T	F	F	T	F
F	T	T	F	T
F	F	T	F	F

This is the only line in the table in which both premises are true. Because the conclusion q is also true, the argument cannot be invalid and is therefore valid.

You can ignore these three lines, and, in fact, do not need to compute them.

Because true premises never lead us to a false conclusion, the argument is valid. The form $[(p \rightarrow q) \wedge p] \rightarrow q$ is called the **law of detachment.** Note that we have proved not only that this particular argument is valid but also that *any argument having this form is valid.*

VERIFYING AN ARGUMENT We verify an argument as follows:

1. Make a truth table with separate columns for each premise and the conclusion.
2. Examine *only* the lines in the table in which all of the premises are true.
3. If the conclusion is also true for the lines you examined in step 2, the argument is valid.
4. If the conclusion is false for even one of the lines you examined in step 2, then the argument is false.

EXAMPLE 1 Determining the Validity of an Argument

Determine if the following argument is valid:

If you subscribe to the most popular Netflix plan, then you can watch unlimited movies per month.
You cannot watch unlimited movies per month.

∴ You do not have the most popular Netflix plan.

SOLUTION: Review the Order Principle on page 12.

Let p stand for "You subscribe to the most popular Netflix plan" and let u represent "You can watch unlimited movies per month." Then this argument has the following form:

$$\begin{array}{l} p \rightarrow u \\ \underline{\sim u} \\ \therefore \sim p \end{array}$$

We now consider a truth table for this argument.

		Premises		Conclusion
p	u	$p \to u$	$\sim u$	$\sim p$
T	T	T	F	
T	F	F	T	
F	T	T	F	
F	F	T	T	T

← This is the only line in the table in which both premises are true.

Because the only line in which both premises are true also has a true conclusion, we conclude that this argument is valid.

The argument in Example 1 is an example of the **law of contraposition.**

KEY POINT

If a truth table for an argument has a line in which the premises are all true but the conclusion is false, then the argument is invalid.

Invalid Arguments

EXAMPLE 2 Some Invalid Arguments

Determine whether the following arguments are valid:

a) If I have the new calling plan, then I can text my friends for free.
   I can text my friends for free.
   ___
   Therefore, I have the new calling plan.

b) If the moon is made of green cheese, then Justin Bieber has sold more than 1,000 digital albums.
   Justin Bieber has sold more than 1,000 digital albums.
   ∴ The moon is made of green cheese.

SOLUTION:

a) If we let c represent "I have a new calling plan" and let f represent "I can text my friends for free," then we can write the argument symbolically as

$$\begin{array}{l} c \to f \\ f \\ \hline \therefore c \end{array}$$

We will make a truth table, as we did in Example 1, with columns for both premises and the conclusion.

		Premises		Conclusion
c	f	$c \to f$	f	c
T	T	T	T	
T	F	F	F	
F	T	T	T	F
F	F	T	F	

← This is the only line in the table in which both premises are true. However, the conclusion is false.

Line 3 in the table shows that it is possible for both premises to be true, but the conclusion is false. This means that the argument is invalid.

b) This argument has *exactly the same form* as the argument in part a), and therefore is also invalid. 12

QUIZ YOURSELF 12

Use a truth table to show that argument b) in Example 2 is invalid.

Example 2 illustrates a common invalid argument form called the **fallacy of the converse.**

SOME GOOD ADVICE

Example 2 shows the importance of using truth tables to determine an argument's validity rather than your intuition. It is important to remember that *the form of an argument is more important than the content of the statements we are making.*

Valid Argument Forms and Fallacies

KEY POINT

Arguments are often one of several standard forms.

We list some common forms of valid and invalid arguments.

Valid Arguments

Law of Detachment	Law of Contraposition	Law of Syllogism	Disjunctive Syllogism
$p \rightarrow q$	$p \rightarrow q$	$p \rightarrow q$	$p \vee q$
p	$\sim q$	$q \rightarrow r$	$\sim p$
$\therefore q$	$\therefore \sim p$	$\therefore p \rightarrow r$	$\therefore q$

Invalid Arguments

Fallacy of the Converse	Fallacy of the Inverse
$p \rightarrow q$	$p \rightarrow q$
q	$\sim p$
$\therefore p$	$\therefore \sim q$

EXAMPLE 3 Identifying the Form of an Argument

Identify the form of each argument and state whether it is valid or invalid.

a) If you want to improve your cardiovascular fitness, then take up cross-country skiing.
You take up cross-country skiing.
———
Therefore, you want to improve your cardiovascular fitness.

b) If you sleep through your morning math class, then you will be well rested.
If you are well rested, then you will do well on your math test.
———
Therefore, if you sleep through your morning math class, then you will do well on your math test.

SOLUTION:

a) Let i represent "You want to improve your cardiovascular fitness" and let c represent "You take up cross-country skiing." This argument has the following form:

$$\begin{array}{l} i \rightarrow c \\ c \\ \hline \therefore i \end{array}$$

This is the fallacy of the converse, which is an invalid argument form.

QUIZ YOURSELF 13

Identify the form of the argument and state whether it is valid:

If more is spent on research, then medical science will advance.

More is not spent on research.

Therefore, medical science will not advance.

b) Let s represent "You sleep through your morning math class," let r represent "You will be well rested," and let w represent "You will do well on your math test." The argument then has the following form:

$$\begin{array}{l} s \rightarrow r \\ r \rightarrow w \\ \hline \therefore s \rightarrow w \end{array}$$

This is a syllogism and is therefore valid, even though the reasoning seems to make no sense.

Now try Exercises 1 to 16. 13

You may have been surprised in Example 3(b) that the argument was valid. Keep in mind that *the symbolic form, not the content*, determines the validity of an argument.

We can use logic to analyze more complex arguments than those we have studied so far.

EXAMPLE 4 Analyzing a Complex Argument with Three Variables

Determine whether this argument is valid or invalid:

If we balance the budget or reduce taxes, then there will be more money available to fight pollution.

If we do not balance the budget, then we will not reduce taxes.

We will not reduce taxes.

Therefore, there will be more money to fight pollution.

SOLUTION:

Consider the statements b, "We balance the budget"; r, "We reduce taxes"; and m, "There will be more money available to fight pollution."

This argument has the form:

$$\begin{array}{l} (b \vee r) \rightarrow m \\ \sim b \rightarrow \sim r \\ \sim r \\ \hline \therefore m \end{array}$$

As usual, we will make a truth table for the premises and the conclusion.

			Premises									Conclusion
				1		2		3	5	4	6	
b	r	m	$(b$	$\vee$	$r)$	$\rightarrow$	m	$\sim b$	$\rightarrow$	$\sim r$	$\sim r$	m
T	T	T	T	T	T	T	T	F	T	F	F	
T	T	F	T	T	T	F	F	F	T	F	F	
T	F	T	T	T	F	T	T	F	T	T	T	T
T	F	F	T	T	F	F	F	F	T	T	T	
F	T	T	F	T	T	T	T	T	F	F	F	
F	T	F	F	T	F	F	F	T	F	F	F	
F	F	T	F	F	F	T	T	T	T	T	T	T
F	F	F	F	F	F	(T)	F	T	(T)	T	(T)	(F) ← Invalid

We see that all three premises are true on lines 3, 7, and 8 in the table. However, on line 8 the conclusion is false. Therefore the argument is invalid.

Now try Exercises 17 to 28.

Even though the truth table in Example 4 has only a single failure, by the Always Principle, we must say that the argument is invalid.

Highlight

Logic and National Defense

High Altitude Defense System Intercepts Two Missiles in Test at Pacific Missile Range

In October 2011, the Missile Defense Agency, a subdepartment of the Department of Defense, posted this news release, publicizing a recent success of the Ballistic Missile Defense System, an offspring of the "Star Wars" missile defense system proposed by President Reagan over a quarter of a century ago.

Because the system uses complex computer programs to control radar, lasers, and missiles to destroy incoming nuclear warheads, some concerned scientists believe that flawed computer code may cause a failure in the event of a real nuclear attack.

Verifying program correctness is similar to verifying that a logical argument is valid—but on a much larger scale, because these programs may contain millions of lines of code. Although research on program verification continues, computer scientists have not completely solved the problem of determining when a complex program is completely error free.

EXERCISES 3.4

Sharpening Your Skills

In Exercises 1–16, identify the form of each argument and state whether the argument is valid.

1. If a car has air bags, then it is safe.
This car has air bags.
———
Therefore, this car is safe.

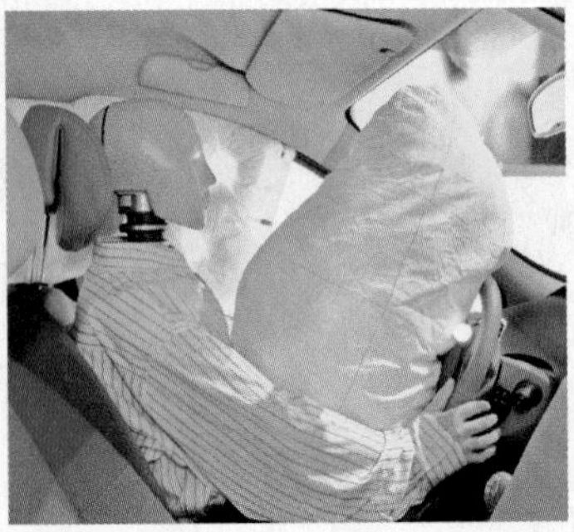

2. If news on inflation is good, then stock prices will increase.
News on inflation is good.
———
Therefore, stock prices will increase.

3. If a movie is exciting, then it will gross a lot of money.
This movie grossed a lot of money.
———
Therefore, it is exciting.

4. If we develop alternative fuels, then we will use less foreign oil.
We are using less foreign oil.
———
Therefore, we are developing alternative fuels.

5. This laptop has the enhanced video card or an optical disc drive.
This laptop does not have the enhanced video card.
———
Therefore, this laptop has an optical disc drive.

6. Either my MP3 player is defective or this download is corrupted.
My player is not defective.
———
Therefore, this download is corrupted.

7. If you pay your tuition late, then you will pay a late penalty.
You do not pay your tuition late.
———
Therefore, you will not pay a late penalty.

8. If you perform maintenance on your PC, then you violate your warranty.

You do not perform maintenance on your PC.

Therefore, you do not violate your warranty.

9. If you watch *The Apprentice*, then you will succeed in business.

If you succeed in business, then you will have a skyscraper named after you.

Therefore, if you watch *The Apprentice*, you will have a skyscraper named after you.

10. If June 1 is Monday, then June 2 is Friday.

If June 2 is Friday, then June 5 is Wednesday.

Therefore, if June 1 is Monday, then June 5 is Wednesday.

11. If you do not buy the sports package, then you will not get the leather seats.

You do get the leather seats.

Therefore, you did buy the sports package.

12. If you love me, then you will do everything I ask.

You do not do everything I ask.

Therefore, you do not love me.

13. If Phillipe joins the basketball team, then he will not be able to work part-time.

Phillipe did not join the basketball team.

Therefore, he will be able to work part-time.

14. If Carrie gets a raise, then she will be able to afford a bigger apartment.

Carrie gets a raise.

Therefore, Carrie will be able to afford a bigger apartment.

15. If January has 28 days, then February has 31.

If February has 31 days, then September also has 31.

Therefore, if January has 28 days, September has 31.

16. Either Tyra Banks or Heidi Klum will be our next vice president.

Tyra Banks will not be our next vice president.

Therefore, Heidi Klum will be our next vice president.

In Exercises 17–28, determine whether each form represents a valid argument.

17. p

$q \rightarrow \sim p$

$\therefore \sim q$

18. p

$\sim q \rightarrow p$

$\therefore \sim p \vee q$

19. $\sim r$

$r \rightarrow q$

$\therefore \sim q \wedge r$

20. p

$\sim q \rightarrow \sim p$

$\therefore q$

21. $\sim q \rightarrow p$

$r \rightarrow \sim q$

$\therefore \sim p \rightarrow r$

22. p

$\sim q \rightarrow \sim p$

$(p \wedge q) \rightarrow r$

$\therefore q \rightarrow r$

23. p

$q \rightarrow \sim p$

$q \rightarrow (r \wedge p)$

$\therefore r$

24. p

$\sim p \rightarrow \sim q$

$q \rightarrow r$

$\therefore r$

25. r

$r \rightarrow \sim q$

$p \vee q$

$\therefore p$

26. r

$r \rightarrow q$

$\sim p \vee \sim q$

$\therefore \sim p$

27. $q \rightarrow \sim p$

$r \rightarrow \sim q$

$\therefore \sim p \rightarrow r$

28. p

$\sim p \rightarrow \sim q$

$(p \wedge q) \rightarrow r$

$\therefore \sim q \rightarrow r$

Applying What You've Learned

For Exercises 29–32, supply a conclusion that will make the argument valid.

29. If Malik has the most expensive Dish TV package, he will get over 300 channels. Malik does not have over 300 channels. Therefore,....

30. If you exercise each day, then you will have more energy. You do not have more energy. Therefore,....

31. Minxia will attend school either in Hawaii or California. She will not go to school in California. Therefore,....

32. Rob will take either the $1,000 rebate or the free financing with his new Nissan Leaf. He decided not to take the rebate. Therefore,....

In Exercises 33–40, determine whether a valid or invalid argument is being described.

33. If the product has a lower price, then the product does not have quality. If the product does not have a lower price or does not have quality, then the product is not reliable. The product has a lower price. Therefore, the product is reliable.

34. If the team wins this game, then it will qualify for the playoffs. The team will play in a tournament or will not qualify for the playoffs. The team wins this game. Therefore, the team will play in a tournament and will qualify for the playoffs.

35. If you buy from a reputable breeder, then your labradoodle will not require shots. If you do not buy from a reputable breeder, then it will cost you extra. It will not cost you extra. Therefore, your labradoodle will not require shots.

36. If Dave is alone tonight, then he will not come to the party. If Dave is not alone tonight or does not come to the party, then he will work on his term paper. Dave will not work on his term paper tonight. Therefore, Dave is not alone tonight.

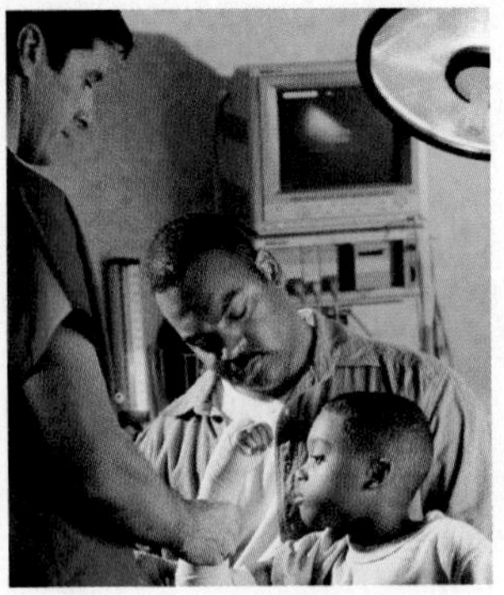

37. If health care is not improved, then the quality of life will not be high. If health care is improved and the quality of life is high, then the incumbents will be reelected. The quality of life is high. Therefore, the incumbents will be reelected.

38. Raul has an IRA. If Raul has an IRA, then he will not withdraw from his CD. Raul will withdraw from his CD or invest in bonds. Therefore, Raul will invest in bonds.

39. Jamie is fluent in Spanish. If Jamie is fluent in Spanish, then she will work in Madrid. She will not visit Mexico or she will not work in Madrid. Therefore, she will visit Mexico.

40. Christian is going to Cancun. If Christian does not go to Cancun, then he will not go on spring break. If Christian goes to Cancun on spring break, then his friends will be envious. Therefore, if Christian goes on spring break, then his friends will be envious.

Communicating Mathematics

It is easy to remember the form of the standard laws and fallacies by thinking about the meaning of their names. For example, we know that $p \rightarrow q$ and $\sim q \rightarrow \sim p$ are equivalent. So in the law of contraposition when we assume premises $p \rightarrow q$ and $\sim q$ to conclude $\sim p$, we are in effect using the equivalence of the conditional with its contrapositive to construct a valid argument. In Exercises 41 and 42, give similar explanations as to how you can remember the stated laws and fallacies.

41. Law of detachment; Law of disjunctive syllogism

42. Fallacy of the converse; Fallacy of the inverse

We have emphasized that the form of a logical argument is more important than its content. Recall in Example 2 that you saw an argument that sounds reasonable, yet has an invalid form. In Exercises 43–46, refer to the tables on page 118 that show forms of some standard valid and invalid arguments. In each exercise, explain your thinking. Try to make your examples interesting by referring to topics currently in the news.

43. Write an argument that sounds reasonable and also has a valid form.

44. Write an argument that sounds reasonable but has an invalid form.

45. Write an argument that sounds unreasonable but has a valid form.

46. Write an argument that sounds unreasonable and also has an invalid form.

Challenge Yourself

Exercises 47–50 are puzzles regarding the residents of a fictional island where all the inhabitants are either knights (who always tell the truth) or knaves (who always lie). In each exercise, use the given information to determine whether the people mentioned are knights or knaves. (Hint: it is useful to consider all possibilities for knaves and knights much as we did in forming truth tables.)*

47. Voldemort says of himself and Dumbledore, "We are both knaves."

48. Gilderoy and Sybil are together. Sybil says, "At least one of us is a knave."

49. Rubeus says, "We are the same kind," and Bellatrix says, "We are of different kinds."

50. Sirius and Alastor were sitting on a bench when a stranger approached Alastor and asked him, "Is either of you a knight?" Alastor responded and from his response, the stranger knew the nature of both Alastor and Sirius.

In a complicated argument with many variables, it is not practical to use truth tables because of their size. We can, however, use valid argument forms to reason without using truth tables. For example, consider the following argument:

$$\begin{array}{l} p \\ p \rightarrow q \\ p \wedge q \rightarrow r \\ \underline{\sim s \rightarrow \sim r} \\ \therefore s \end{array}$$

We assume that all the premises are true, and we reason like this to prove that the argument is valid:

1. *We assumed that p and $p \rightarrow q$ are both true, therefore by the law of detachment, we have that q is true.*
2. *Now both p and q are true, so $p \wedge q$ is also true.*
3. *By the law of detachment again, $p \wedge q$ true and $p \wedge q \rightarrow r$ true force r to be true.*
4. *Because the statement $\sim s \rightarrow \sim r$ is equivalent to its contrapositive, we then know that $r \rightarrow s$ is true.*
5. *Knowing that r and $r \rightarrow s$ are both true, we conclude that s is also true.*

Therefore, by assuming that all the premises are true, we were able to reason that the conclusion s also must be true. This means that the argument is valid. In Exercises 51 and 52, reason similarly to prove that each argument is valid.

51. $$\begin{array}{l} a \wedge b \\ b \rightarrow c \\ \underline{d \rightarrow \sim c} \\ \therefore \sim d \end{array}$$

52. $$\begin{array}{l} p \rightarrow \sim q \\ p \rightarrow r \\ \sim s \rightarrow q \\ \underline{p} \\ \therefore r \wedge s \end{array}$$

*Knights and Knaves puzzles were devised by the renowned logician Raymond Smullyan, and many examples of his puzzles can be found on the Internet

In addition to the argument forms that you studied in this section, there are many situations when a person will use a flawed form of an argument to make his or her point. In Exercises 53–56, we describe some well-known flawed argument forms. In each case, write another flawed argument of the same form.

53. **Circular Reasoning.** In circular reasoning, we support a statement by simply repeating the statement in different or stronger terms. Example: "Arnold Schwarzenegger is a successful governor because he is the best governor our state's ever had."

54. **Slippery Slope.** Slippery slope arguments falsely assume that one thing must lead to some exaggerated other thing. Example: "If we don't fight this tuition increase now, soon nobody will be able to afford college."

55. **Biased Sample.** We draw a conclusion about a population based on a prejudiced sample. Example: You interview 20 students at your library and find that the average student on campus studies more than 25 hours per week.

56. **False Analogy**. Two objects are assumed to be similar and because one object has a certain property, then the other must also. Example: "Running a college is like running a business. In both, the most important thing is the bottom line."

3.5 Using Euler Diagrams to Verify Syllogisms

Objectives

1. Use Euler diagrams to identify a valid syllogism.
2. Use an Euler diagram to identify an invalid syllogism.

We will now analyze arguments called syllogisms—a form of argument that goes back several thousand years to the time of the ancient Greek philosopher Aristotle. A **syllogism** consists of a set of statements called **premises** followed by a statement called a **conclusion.** Syllogisms differ from the arguments that you studied in Section 3.4 in that syllogisms may contain quantifiers such as *all*, *some*, and *none*, whereas the arguments in Section 3.4 did not.

Valid Syllogisms

A syllogism is **valid** if whenever its premises are all true, then the conclusion is also true. If the conclusion of a syllogism can be false even though all the premises are true, then the syllogism is **invalid.** Perhaps the most famous syllogism is

All people are mortal.
Socrates is a person.
___
Therefore, Socrates is mortal.

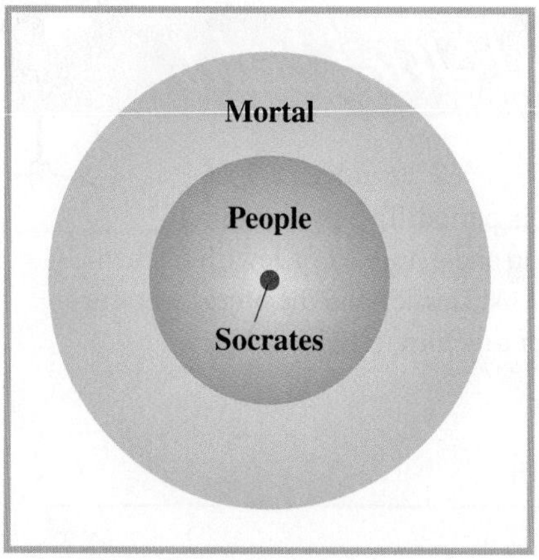

FIGURE 3.9 Euler diagram for the Socrates syllogism.

We can determine whether this syllogism is valid by drawing an **Euler** (pronounced "oiler") **diagram.** In Figure 3.9, we represent the first premise, "All people are mortal," by drawing a small circle labeled "People" inside a larger circle labeled "Mortal." We next place a dot labeled "Socrates" within the circle labeled "People" to indicate "Socrates is a person." Doing this forces the dot "Socrates" to be in the circle "Mortal," so we can see that drawing the two premises in Figure 3.9 forces the conclusion to occur in the diagram. Therefore, the syllogism is valid.

We will often use ∴ for the word *therefore* in the conclusion of a syllogism.

EXAMPLE 1 Determining the Validity of a Syllogism

Use an Euler diagram to determine whether the syllogism is valid.

All poets are good spellers.
Dante is not a good speller.

∴ Dante is not a poet.

SOLUTION: Review the Draw Pictures Strategy on page 3.

Consider the Euler diagrams in Figure 3.10.

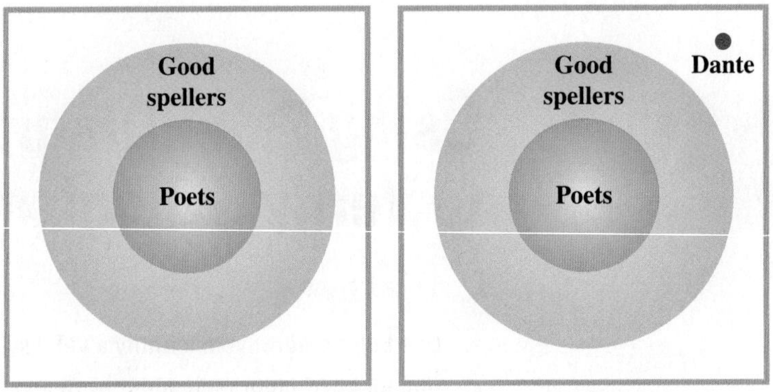

(a) All poets are good spellers. (b) Dante is not a good speller.

FIGURE 3.10

We see from Figure 3.10(b) that if Dante is not a good speller, then he cannot be a poet. Therefore, the conclusion is valid.

Although you may believe that it is possible for a person to be a poet without being a good speller, your feeling about this premise is not relevant to the validity of the argument. Remember: *It is possible for individual premises or the conclusion to be false and yet the syllogism can be valid.* You must rely on Euler diagrams rather than your intuition when deciding whether syllogisms are valid.

Historical Highlight

Polish Logicians

Jan Lukasiewicz, in his papers on Aristotle, made the fundamental ideas of logic available in the Polish language, which led to a school of logic scholars that flourished in Poland in the early part of the twentieth century. Among these was Alfred Tarski, who is considered to be one of the principal logicians of the twentieth century, and Father Salamucha, who formed a group in Krakow to use mathematical logic to modernize Catholic dogma.

Sadly, World War II ended the golden age of Polish logic. During this period, Tarski went to America and Salamucha and a number of Jewish mathematicians were murdered. Lukasiewicz and the others who survived the war chose not to return to Poland.

QUIZ YOURSELF 14

Draw an Euler diagram to determine whether the syllogism is valid:

All credit cards are made of plastic.

This card is not a credit card.

∴ This card is not made of plastic.

Invalid Syllogisms

EXAMPLE 2 Using an Euler Diagram to Show That a Syllogism Is Invalid

Use an Euler diagram to show that the syllogism is invalid.

All tigers are meat eaters.

Simba is a meat eater.

∴ Simba is a tiger.

SOLUTION: Review the Counterexample Principle on page 11.

Figure 3.11 illustrates the two premises. We see in Figure 3.11 that it is possible for Simba to be a meat eater without being a tiger. The argument is therefore invalid because the premises do not force the conclusion to hold.

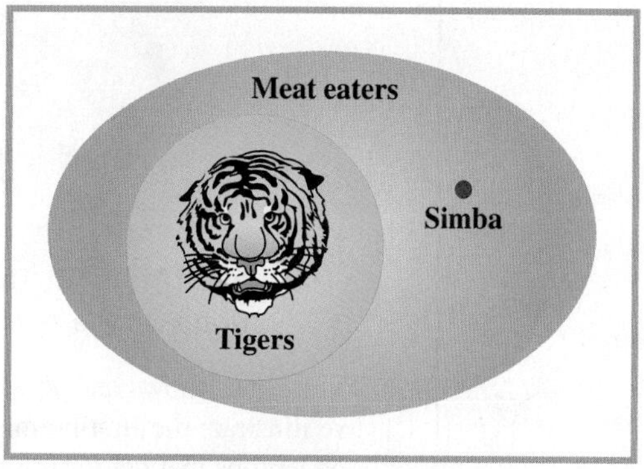

FIGURE 3.11 Simba is a meat eater but not a tiger.

14

Example 3 shows how to handle a premise containing the quantifier *none are* or *no*.

EXAMPLE 3 Analyzing a Syllogism That Contains the Quantifier *No*

Assume that a syllogism begins with the following two premises:

No MacBooks get viruses.

My laptop does not get viruses.

a) Can we conclude that my computer is a MacBook?

b) Can we conclude that my computer is not a MacBook?

SOLUTION: Review the Always Principle on page 10.

a) We need to determine the validity of the syllogism

No MacBooks get viruses.

My laptop does not get viruses.

∴ My laptop is a MacBook.

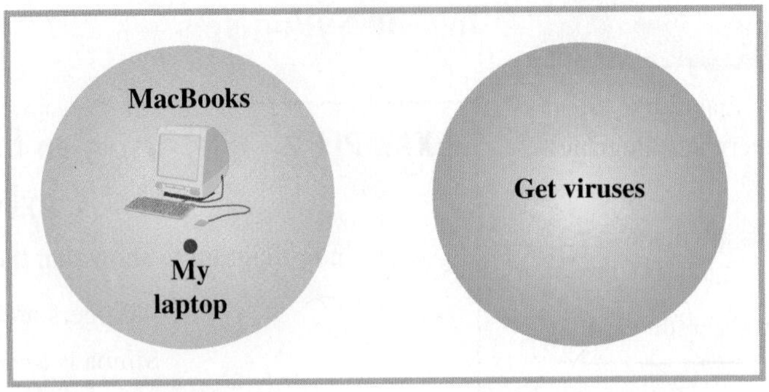

FIGURE 3.12 Because the conclusion holds in this diagram, you may mistakenly think that the syllogism is valid.

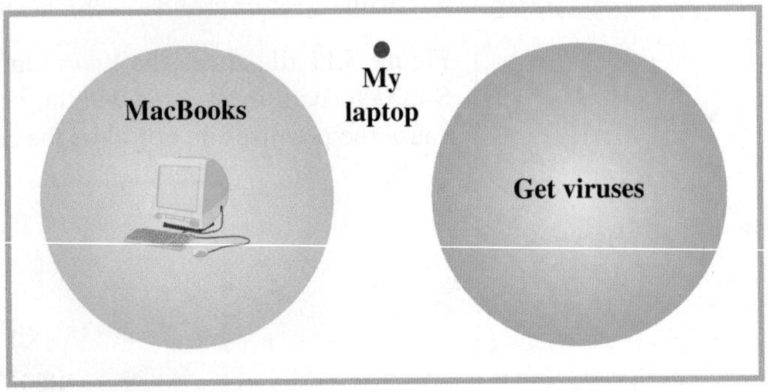

FIGURE 3.13 The conclusion does not hold in this diagram, so the syllogism *cannot* be valid.

Figure 3.12 shows one possible way to draw the first two premises in an Euler diagram. We illustrate the first premise by drawing two disjoint circles to represent MacBooks and laptops that get viruses. The rectangle represents the set of all computers.

From Figure 3.12, you might conclude that my laptop is a MacBook. *That would be a mistake.* As we show in Figure 3.13, there is another way to draw the premise "My laptop does not get viruses."

From Figure 3.13 we see that the syllogism is invalid because the given premises do not force the conclusion to hold. Therefore, you cannot conclude that my laptop is a MacBook.

b) Figure 3.12 shows that the premises can hold and yet we cannot conclude that my laptop is not a MacBook. Therefore, based on the given premises, we cannot conclude that my laptop is not a MacBook.

The quantifier *some* can be misleading in syllogisms because there are several different ways to represent *some* in Euler diagrams. In Figure 3.14, each diagram represents the premise "Some As are Bs."

Notice in Figure 3.14(b) that the fact that "All As are Bs" does not rule out the possibility that "Some As are Bs."

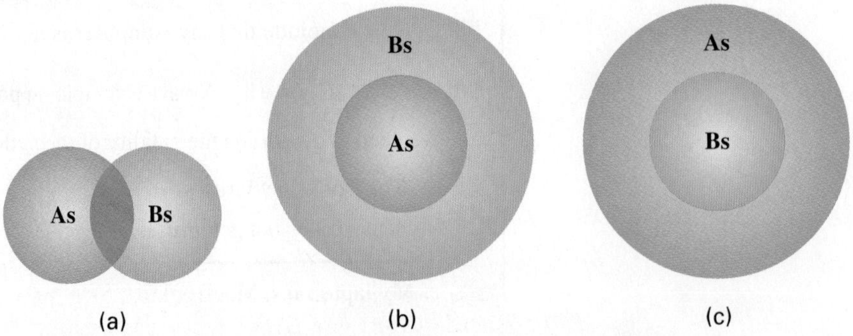

FIGURE 3.14 Euler diagrams illustrating that "Some As are Bs."

QUIZ YOURSELF 15

Determine whether the syllogism is valid:

All firefighters are brave.
Some women are firefighters.

∴ Some women are brave.

EXAMPLE 4 **Analyzing a Syllogism Having the Quantifier *Some***

Is the syllogism valid?

All cars manufactured after 2007 have driver-side air bags.
Some cars with driver-side air bags have passenger-side air bags.

∴ Some cars with passenger-side air bags were manufactured after 2007.

SOLUTION:

Because a single diagram can show a syllogism to be invalid, our strategy will be to find such a diagram. After looking at several diagrams, if we feel that it is impossible to show the argument invalid, then we will say that it is valid. Figure 3.15 shows three ways to illustrate the two premises.

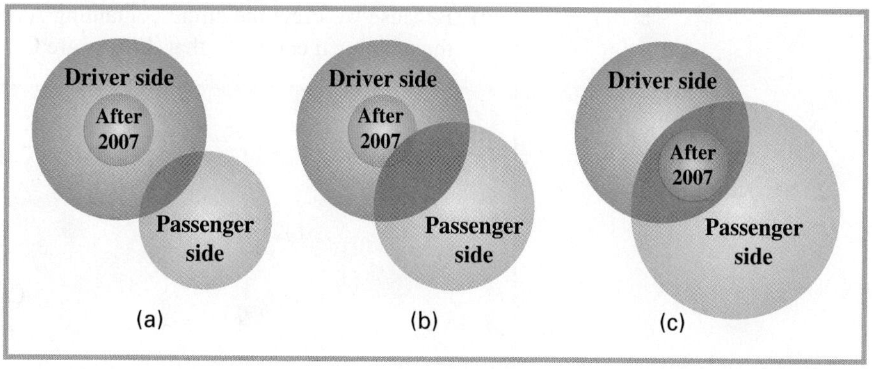

FIGURE 3.15 Three ways to illustrate the premises of the air bag syllogism.

By luck, the first diagram we drew, Figure 3.15(a), shows that *it is possible* that no cars manufactured after 2007 have passenger-side air bags. Therefore, the conclusion does not follow from the premises, so the syllogism is invalid. 15

In Example 4, we cannot conclude that no cars manufactured after 2007 have passenger-side air bags. Clearly the diagrams in Figures 3.15(b) and 3.15(c) show that *it is possible* for some cars manufactured after 2007 to have passenger-side air bags. All we are able to say in Example 4 is that even though the conclusion sounds reasonable, it does not follow from the premises.

PROBLEM SOLVING

Strategy: The Counterexample Principle

From the Counterexample Principle in Section 1.1 you know that one Euler diagram can show a syllogism to be invalid. In testing the validity of a syllogism, it is a good strategy to draw several diagrams, with the idea that one of them may show the syllogism to be invalid. If you are convinced that no diagram can show the syllogism to be invalid, then you can conclude that the syllogism is valid.

The problem with drawing Euler diagrams, particularly for syllogisms using the quantifier *some*, is that whatever diagram you draw will show extra conditions that were not stated as a premise. These extra conditions can mislead you in determining the validity of a syllogism.

QUIZ YOURSELF 16

Find one additional condition in Figure 3.16, other than those given in Example 5, that was not stated as a premise.

EXAMPLE 5 Extra Conditions Are Always Present in an Euler Diagram

Each Euler diagram in Figure 3.16 illustrates the following premises:

All As are Bs.

Some Bs are Cs.

State a condition that is present in each diagram that has not been stated as a premise.

SOLUTION:

There are many possible conditions. We will give one for each diagram.

a) Because the circles containing As and Cs do not intersect, we have drawn the extra condition that "No As are Cs."

b) Because the circles containing As and Cs do intersect, we have drawn the condition that "Some As are Cs."

c) Because we drew the circle containing As inside the circle containing Cs, we have the additional condition that "All As are Cs."

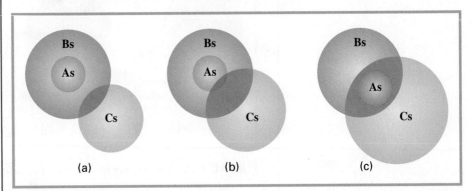

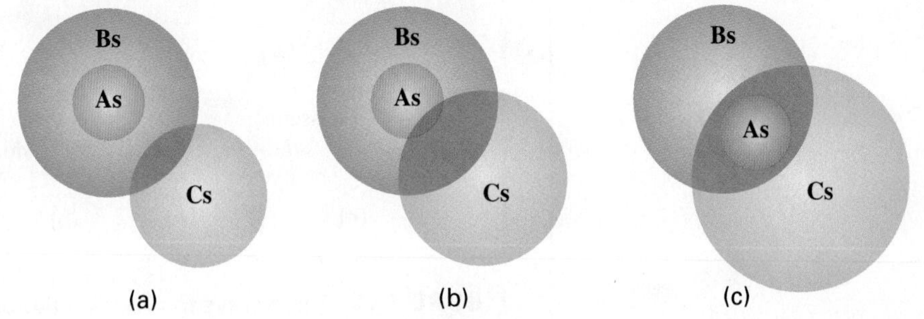

FIGURE 3.16 Euler diagrams illustrating "All As are Bs" and "Some Bs are Cs."

16

In Figure 3.16, several other extra conditions appear in the diagrams that were not stated as premises. It may be hard to show that a syllogism is valid because by the Always Principle, the conclusion must hold in every Euler diagram that illustrates the premises. You must be careful to consider a variety of diagrams before you accept a syllogism as valid.

EXERCISES 3.5

Sharpening Your Skills

In Exercises 1–16, determine whether each syllogism is valid or invalid.

1. All original parts are under warranty.
   This part is an original part.
   ___
   ∴ This part is under warranty.

2. All college students are ambitious.
   Cameron is ambitious.
   ___
   ∴ Cameron is a college student.

3. All imports are subject to a surcharge.
   This jacket is subject to a surcharge.
   ___
   ∴ This jacket is an import.

4. All documentarians are biased.
   Michael Moore is a documentarian.
   ___
   ∴ Michael Moore is biased.

5. All vitamins are healthful.
Milk is healthful.

∴ Milk is a vitamin.

6. All aerobic activities are healthful.
Blogging is an aerobic activity.

∴ Blogging is healthful.

7. All athletes are fit.
Julio is not an athlete.

∴ Julio is not fit.

8. All athletes are fit.
Sheena is not fit.

∴ Sheena is not an athlete.

9. All politicians are honest.
Walt is not honest.

∴ Walt is not a politician.

10. All salespeople are sincere.
Emily is not sincere.

∴ Emily is not a salesperson.

11. Some large corporations are concerned about the environment.
General Motors is not a large corporation.

∴ General Motors is not concerned about the environment.

12. Some children love cookies.
Dani does not love cookies.

∴ Dani is not a child.

13. Some biologists believe the Loch Ness monster exists.
All who believe the Loch Ness monster exists are irrational.
Antawn is not irrational.

∴ Antawn is not a biologist.

14. Some investors are wealthy.
All wealthy people are happy.

∴ Some investors are happy.

15. Some mammals are large.
All large animals are dangerous.

∴ Some mammals are dangerous.

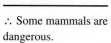

16. Some mammals are large.
Some dangerous animals are large.

∴ Some mammals are dangerous.

Applying What You've Learned

In Exercises 17–24, complete each syllogism so that it is valid and the conclusion is true. There may be several correct answers.

17. Some taxes are unfair.
All unfair taxes should be abolished.

∴

18. All honest politicians should be supported.
Marika should not be supported.

∴

19. No team that plays in a domed stadium has won the Super Bowl.
Some teams that wear red uniforms have won the Super Bowl.

∴

20. Some mathematicians are fine musicians.
All fine musicians are intelligent.

∴

21. All people who like country music have trucks.
All truck owners have dogs.
Some opera singers like country music.

∴

22. All tall people are wealthy.
All wealthy people are admired.
Some basketball players are not admired.

∴

23. All firefighters are courageous.
All courageous people are heroic.
No ballet dancers are courageous.

∴

24. All movie stars are hard working.
All hard-working people are well liked.
No well-liked people are without friends.

∴

Communicating Mathematics

25. Draw an Euler diagram for the statements "All As are Bs" and "Some Bs are Cs."

26. Draw an Euler diagram for the statements "Some As are Bs" and "Some Bs are Cs."

27. Draw an Euler diagram for the statements "No As are Bs" and "All Bs are Cs."

28. In each of your drawings for Exercises 25–27, state a condition in your drawing that is not present in the statement that you are trying to illustrate.

29. Give an example of a valid syllogism that has a false statement for its conclusion.

30. Give an example of an invalid syllogism that has a true statement for its conclusion.

Challenge Yourself

In Exercises 31–34, write a valid syllogism that would be illustrated by each Euler diagram. For example, the Socrates syllogism would be illustrated by a diagram similar to the one in Exercise 31.

31.

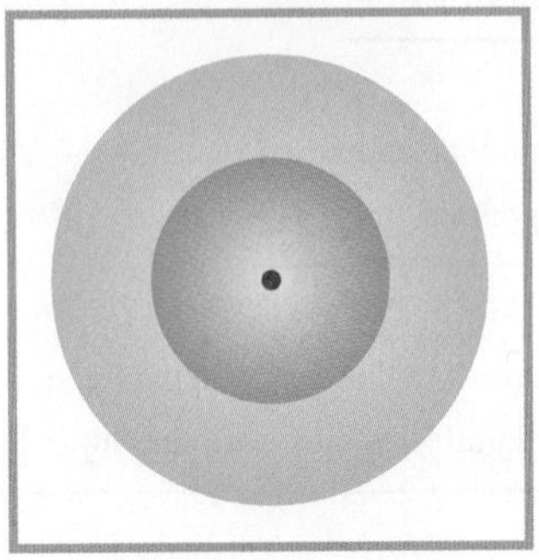

32.

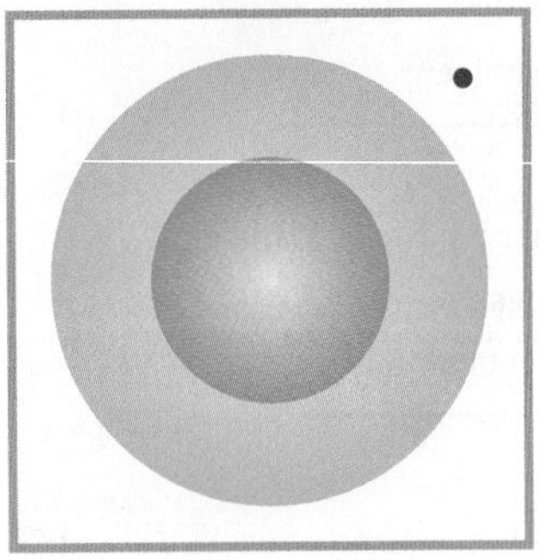

33.

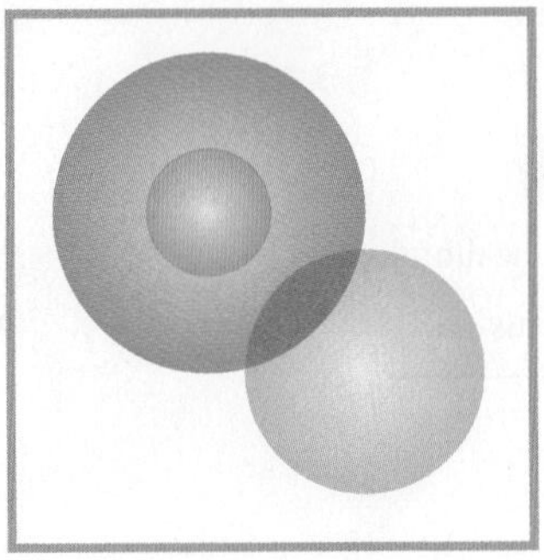

34.

Looking Deeper

3.6 Fuzzy Logic

Objectives

1. Understand the definition of statements in fuzzy logic.
2. Be able to work with connectives in fuzzy logic.
3. Use fuzzy logic to make decisions.

"Aristotle Is Out and Buddha Is In"*

You have been learning Aristotelian logic in this chapter, but increasingly some researchers believe that we would be better off without a right or wrong, true or false approach to problem solving—we should, instead, be using fuzzy thinking to solve our problems.

In fact, by using *fuzzy logic*, researchers have been able to come to your rescue and put an end to some of life's little annoyances. Fuzzy air conditioners make your home more comfortable, with a fuzzy shower controller you can avoid the shock of extreme water temperature changes when a toilet is flushed, and subways operated by fuzzy computers will give you a smoother ride with no need to hang on to a strap during your morning commute.

*Quoted from a book review on fuzzy thinking, by Kirkus Reviews, Kirkus Associates, LP, amazon.com (2007).

In this section, we will teach you how to use fuzzy logic to analyze situations when what is being said is not black or white, but rather some shade of gray.

For example, to say "It is warm today" may not mean the same thing on the first day of March as it does in the middle of August. Although it may be warm today, perhaps it is not as warm as it was two weeks ago. To simply say "It is warm today" ignores exactly how warm it is. The strict condition that statements must be either true or false makes symbolic logic unsuitable for representing many real-life situations.

Fuzzy Statements

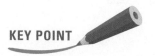

KEY POINT

Statements in fuzzy logic have truth values between 0 and 1.

Mathematicians developed *fuzzy logic* in order to apply the techniques of logic more widely. This is a perfectly acceptable, well-developed area of mathematics. Once we define the meaning of fuzzy statements and explain how fuzzy connectives behave, we can work with them as we did for other (nonfuzzy, or *crisp*) statements.

> DEFINITIONS In **fuzzy logic,** a **statement** is a declarative sentence that has an associated **truth value** between 0 and 1 inclusive.

EXAMPLE 1 The Truth Values of Statements in Fuzzy Logic

Here are some examples of statements in fuzzy logic:

a) "I like peach ice cream," with a truth value of 0.9.

b) "Tiger Woods is a great golfer," with a truth value of 0.92.

c) "Maine is a large state," with a truth value of 0.45.

d) "*Gone with the Wind* is a great movie," with a truth value of 0.73.

e) "You find mathematics interesting," with a truth value of 0.7.

The way we assigned truth values in Example 1 is somewhat arbitrary. You may want to assign a truth value of 0.85 to the statement "*Gone with the Wind* is a great movie" because you believe that it is a greater movie than we do. *This is perfectly OK and is really the point of fuzzy logic.* If we could not differ in our assignment of truth values, we would then both be forced to say that *Gone with the Wind* is a great movie, but we could not express the difference of our opinions of exactly how great the movie is.

Fuzzy Connectives

KEY POINT

Negation, conjunction, and disjunction in fuzzy logic are defined differently than the corresponding connectives in nonfuzzy logic.

We will now consider how connectives work in fuzzy logic. In Example 1, we assigned a truth value of 0.9 to "I like peach ice cream." Because this expresses the degree to which I like peach ice cream, it also says that $1 - 0.9 = 0.1$ indicates the degree to which I *do not like* peach ice cream. The next definition explains how to calculate the truth value of negations in fuzzy logic.

> DEFINITION If p is a statement in fuzzy logic, then the truth value of the **negation** of p, written $\sim p$, is
>
> $$1 - \text{the truth value of } p.$$

QUIZ YOURSELF 17

Calculate the truth value for the negations of statements d) and e) in Example 1.

EXAMPLE 2 Calculating the Truth Values of Negations in Fuzzy Logic

Calculate the truth values for the negations of statements b) and c) in Example 1.

SOLUTION:

b) "Tiger Woods is not a great golfer" has a truth value of $1 - 0.92 = 0.08$.

c) "Maine is not a large state" has a truth value of $1 - 0.45 = 0.55$.

Now try Exercises 9 to 12. **17**

It is easy to compute conjunctions and disjunctions in fuzzy logic.

DEFINITIONS Suppose that p and q are two statements in fuzzy logic.

a) The truth value of the **disjunction** of p and q, written $p \vee q$, is the *maximum* of the truth values for p and q.

b) The truth value of the **conjunction** of p and q, written $p \wedge q$, is the *minimum* of the truth values for p and q.

EXAMPLE 3 Calculating Disjunctions and Conjunctions in Fuzzy Logic

Let i represent the statement "Inflation is driving up the cost of a college education" and let g represent "Less government funding is driving up the cost of a college education." Assume that i has a truth value of 0.65 and g has a truth value of 0.73. Find the truth values of the following statements:

a) "Inflation *and* less government funding are driving up the cost of a college education."

b) "Inflation *or* less government funding is driving up the cost of a college education."

SOLUTION:

a) The statement "Inflation *and* less government funding are driving up the cost of a college education" has the form $i \wedge g$, so its truth value is the minimum of 0.65 and 0.73, which is 0.65.

b) The statement "Inflation *or* less government funding is driving up the cost of a college education" has the form $i \vee g$, so its truth value is the maximum of 0.65 and 0.73, which is 0.73.

Now try Exercises 13 to 16. **18**

QUIZ YOURSELF 18

Let l represent the statement "I like living in a large city" with a truth value of 0.82 and let e represent the statement "I like living on the East Coast" with a truth value of 0.45. Find the truth value of each statement.

a) $l \wedge e$

b) $l \vee e$

We can use our knowledge of fuzzy connectives to analyze more complex statements.

EXAMPLE 4 Computing the Truth Value of a Statement with Three Variables

Let h, s, and c be defined as follows:

h: "Foreign cars have high resale value," with truth value 0.85.

s: "Foreign cars are safer than domestic cars," with truth value 0.73.

c: "Foreign cars cost more than domestic cars," with truth value 0.6.

Find the truth value of the statement

"Foreign-made cars have high resale value, but it is not true that they are safer or cost more than domestic cars."

QUIZ YOURSELF 19

Let p have a truth value of 0.65, let q have a truth value of 0.80, and let r have a truth value of 0.55. Find the truth value of each statement.

a) $\sim(p \wedge q) \vee (\sim r)$
b) $((\sim p) \wedge q) \vee r$

SOLUTION: Review the Order Principle on page 12.

This statement has the following form:

$$\begin{array}{ccc} h \wedge \sim(s & \vee & c) \\ 0.85 \quad\; 0.73 & & 0.6 \end{array}$$

We have written the truth values of h, s, and c to make the calculations easier to follow.

The truth value of $s \vee c$ is the maximum of 0.73 and 0.6, which is 0.73. The truth value of $\sim(s \vee c)$ is then $1 - 0.73 = 0.27$. The truth value of the conjunction of $\sim(s \vee c)$ and h is the minimum of 0.27 and 0.85, namely 0.27. 19

It is awkward to always talk about the minimum of the truth values of h and s, the maximum of the truth values of h and s, and so on, so for simplicity's sake, in calculating the truth value of $h \vee s$, we will write $0.85 \vee 0.73 = 0.85$ and the truth value of $h \wedge s$ as $0.85 \wedge 0.73 = 0.73$. We will handle negation similarly. With this agreement in mind, we can calculate the truth value of $\sim(h \vee \sim(s \wedge \sim h))$:

$$\begin{aligned} \sim(0.85 \vee \sim(0.73 \wedge \sim 0.85)) &= \sim(0.85 \vee \sim(0.73 \wedge 0.15)) \\ &= \sim(0.85 \vee \sim 0.15) \\ &= \sim(0.85 \vee 0.85) \\ &= \sim 0.85 = 0.15 \end{aligned}$$

KEY POINT

We use an equivalent form of the conditional in ordinary logic to define the conditional in fuzzy logic.

It is easy to use truth tables to prove that the conditional $p \rightarrow q$ is logically equivalent to $(\sim p) \vee q$. We use this result in defining how to compute the truth value of conditionals in fuzzy logic.

> **DEFINITION** If p and q are two statements in fuzzy logic, then the truth value of the **conditional,** $p \rightarrow q$, is defined to be the truth value of $(\sim p) \vee q$.

QUIZ YOURSELF 20

Let p have a truth value of 0.65, let q have a truth value of 0.80, and let r have a truth value of 0.55. Find the truth value of each statement.

a) $p \rightarrow q$
b) $(p \wedge q) \rightarrow r$
c) $\sim(p \vee q) \rightarrow \sim r$

EXAMPLE 5 Computing the Truth Values of Conditionals

Let p have a truth value of 0.75 and let q have a truth value of 0.46. Find the truth value of each statement.

a) $p \rightarrow q$ b) $q \rightarrow p$ c) $\sim q \rightarrow \sim p$

SOLUTION:

a) The truth value of $p \rightarrow q$ is the same as the truth value of $(\sim p) \vee q$, which is $(\sim 0.75) \vee 0.46 = 0.25 \vee 0.46 = 0.46$.

b) To calculate the truth value of $q \rightarrow p$, instead we calculate the truth value of $(\sim q) \vee p$, which is $(\sim 0.46) \vee 0.75 = 0.54 \vee 0.75 = 0.75$.

c) The truth value of $\sim q \rightarrow \sim p$ is found by calculating the truth value of $\sim(\sim q) \vee (\sim p)$. We simplify this by replacing $\sim(\sim q)$ with q. The truth value of $q \vee (\sim p)$ is $0.46 \vee (\sim 0.75) = 0.46 \vee 0.25 = 0.46$. It should not surprise you that this is the same truth value that $p \rightarrow q$ had. Why?

Now try Exercises 17 to 24. 20

Fuzzy Decision Making

We will now show how to use fuzzy logic to make decisions. Deciding on what to do when your car needs an expensive repair is certainly a situation in which the factors you consider are not clear-cut. It is a perfect example of a problem that we can analyze using fuzzy logic.

EXAMPLE 6 Making Decisions Using Fuzzy Logic

Assume that your car has broken down and you are trying to decide whether to repair it, buy a newer used car, or buy a brand-new car. In Table 3.1,* we list the characteristics of a car that are important to you and assign a number between 0 and 1 to indicate how important each characteristic is to you. Table 3.2 lists the degree to which each option satisfies these characteristics.

Characteristic	Importance
Cost	0.70
Reliability	0.60
Size (large enough)	0.45
Good gas mileage	0.65
Safety	0.75

TABLE 3.1 Fuzzy truth values describe how important each characteristic is to you.

	Repair	Newer Used Car	New Car
Cost	0.80	0.65	0.30
Reliability	0.40	0.60	0.95
Size (large enough)	0.90	0.75	0.50
Good gas mileage	0.55	0.80	0.85
Safety	0.45	0.60	0.75

TABLE 3.2 Fuzzy truth values describe how well each option satisfies the characteristics that are important to you.

The ideal option for you is the one that scores the highest in the characteristics that are important to you. In other words, we want to consider fuzzy statements of the following type:

"If cost is important to you, then repairing your car satisfies this condition."

"If reliability is important to you, then buying a new car satisfies this condition."

We thus want truth values for fuzzy statements of the form $i \rightarrow c$, where i is a statement that a characteristic is important to you and c is a statement that an option possesses that characteristic. Recall that $i \rightarrow c$ has the same truth value as $(\sim i) \vee c$. Let us look at the truth value for the statement

"If cost is important to you, then repairing your car satisfies this condition."

Here i is the statement "Cost is important to you," which from Table 3.1 has a truth value of 0.70. The statement c is "Repairing your car satisfies this condition (cost)," which from Table 3.2 has a truth value of 0.80. The truth value of $(\sim i) \vee c$ is thus $(\sim 0.70) \vee 0.80 = 0.30 \vee 0.80 = 0.80$.

We will now construct a truth table for all the statements $(\sim i) \vee c$ for each option: repair, buy a used car, and buy a new car.

	i	$\sim i$	Repair c	Repair $(\sim i) \vee c$	Used c	Used $(\sim i) \vee c$	New c	New $(\sim i) \vee c$
Cost	0.70	0.30	0.80	0.80	0.65	0.65	0.30	0.30
Reliability	0.60	0.40	0.40	0.40	0.60	0.60	0.95	0.95
Size	0.45	0.55	0.90	0.90	0.75	0.75	0.50	0.55
Gas mileage	0.65	0.35	0.55	0.55	0.80	0.80	0.85	0.85
Safety	0.75	0.25	0.45	0.45	0.60	0.60	0.75	0.75
Conjunction of all $(\sim i) \vee c$				**0.40**		**0.60**		**0.30**

For an option to be satisfactory, you want to have all the conditions satisfied as well as possible. Therefore, you are interested in cost AND reliability AND size AND gas mileage AND

*In Tables 3.1 and 3.2, the assignment of values between 0 and 1 is somewhat arbitrary. Another person might assign these values differently.

safety. Thus, to finish making your decision, you calculate the *conjunction* of all the expressions $(\sim i) \vee c$ for each option. The conjunction of all the expressions $(\sim i) \vee c$ for repairing your car is the minimum of the truth values in this column, which is 0.40 (highlighted). Similarly we obtain a value of 0.60 (highlighted) for buying a newer used car and 0.30 (highlighted) for buying a new car. Because buying a newer used car has the highest final rating, the best choice for you *according to this rating method* is to buy a newer used car.

Now try Exercises 25 to 28. 21

QUIZ YOURSELF 21

Calculate the truth value of each statement.

a) "If reliability is important to you, then buying a new car satisfies this condition."
b) "If being large enough is important to you, then buying a newer used car satisfies this condition."

The method we used in Example 6 is a real way to use fuzzy logic to make decisions. If you do not agree totally with this process for making the final decision, we invite you to criticize this method and to suggest alternatives in the exercises.

EXERCISES 3.6

Sharpening Your Skills

In Exercises 1–8, assign a truth value between 0 and 1 to each fuzzy statement. Of course, the assignments will vary from student to student.

1. *The Simpsons* is a television show intended for adults.
2. Physicians earn high salaries.
3. Madonna is a good singer.
4. Abraham Lincoln was our best president.
5. The United States should increase its foreign aid.
6. We should spend more money to shelter the homeless.
7. Money brings happiness.
8. Baseball is a difficult sport.

In Exercises 9–12, calculate the truth value of the negation of each fuzzy statement. The truth value of each fuzzy statement is given in parentheses.

9. Picasso was a famous artist. (0.95)
10. Ryan Howard is known mainly for his prowess at pitching. (0.15)
11. Learning a musical instrument can teach self-discipline. (0.75)
12. Space research is vital to national defense. (0.40)

In Exercises 13–16, consider the following fuzzy statements:

p: *Students choose a college because of its location. (0.75)*

q: *Students choose a college because of its cost. (0.85)*

Determine the truth value of each statement.

13. Students choose a college because of its location or cost.
14. Students choose a college because of its location and cost.
15. Students choose a college because of its location and not its cost.
16. It is not true that students choose a college because of its location or its cost.

In Exercises 17–24, assume that p *has a truth value of 0.27,* q *has a truth value of 0.64, and* r *has a truth value of 0.71. Find the truth value of each statement.*

17. $p \wedge \sim q$
18. $\sim(p \vee q)$
19. $\sim(p \vee \sim q)$
20. $(p \vee r) \wedge (p \wedge \sim q)$
21. $\sim(p \vee \sim q) \wedge \sim r$
22. $p \rightarrow r$
23. $q \rightarrow (p \vee r)$
24. $\sim(p \vee q) \rightarrow \sim(q \wedge r)$

Applying What You've Learned

In Exercises 25–28, use the method described in Example 6 to evaluate each situation.

25. You are trying to decide which of two job offers to accept. The first job is a retail sales trainee for a large consumer products company and the second is a marketing data analyst for a communications company. In the first table, we list the characteristics of a job that are important to you; in the second table, we list the degree to which each job satisfies these characteristics. Which job should you take?

Characteristic	Importance
Salary	0.70
Interesting	0.80
Work with people	0.60
Flexible hours	0.75

	Sales Trainee	Data Analyst
Salary	0.60	0.65
Interesting	0.50	0.80
Work with people	0.90	0.30
Flexible hours	0.80	0.60

26. You are trying to decide whether to buy a house or continue to rent an apartment. In the first table, we list the characteristics that are important to you; in the second table, we list the degree to which each situation satisfies these characteristics. What should you do?

Characteristic	Importance
Cost	0.70
Close to job	0.60
Adequate space	0.55
Near city attractions	0.65

	House	Apartment
Cost	0.80	0.65
Close to job	0.40	0.80
Adequate space	0.90	0.75
Near city attractions	0.55	0.90

27. Your younger sister is having a difficult time deciding which of three colleges to attend: Little Small College (LSC), Good Old State (GOS), and Mom's Favorite University* (MFU). In the first table, we list features of each school that are important to your sister and rank these qualities; in the second table, we list the degree to which each school satisfies these characteristics. What should she do?

Characteristic	Importance
Size (not too large)	0.60
Cost	0.80
Academics	0.75
Social life	0.90
Close to home	0.30

	LSC	GOS	MFU
Size (not too large)	0.95	0.60	0.40
Cost	0.40	0.80	0.30
Academics	0.80	0.70	0.65
Social life	0.55	0.75	0.70
Close to home	0.70	0.65	0.95

28. You are trying to decide which of three investments to make: (A)llied InterServe, (B)itLogic, or (C)omQual. In the first table, we list the characteristics of investing that are important to you; in the second table, we list the degree to which each choice satisfies these characteristics. Which investment is your best choice?

Characteristic	Importance
Past performance	0.85
Safety	0.60
Liquidity	0.75
Minimum amount to invest	0.35
Management fee	0.55

	A	B	C
Past performance	0.80	0.65	0.90
Safety	0.45	0.60	0.80
Liquidity	0.60	0.75	0.75
Minimum amount to invest	0.85	0.60	0.65
Management fee	0.45	0.80	0.75

Communicating Mathematics

29. How are the rules for computing the truth tables for connectives that you learned for nonfuzzy logic consistent with the rules for computing truth values of compound statements in fuzzy logic?

30. Discuss some situations in which using fuzzy logic would be more realistic than using two-valued logic.

Challenge Yourself

31. Choose a situation you will face in which you must make a decision, and make the decision using the method in Example 6. Begin by identifying characteristics that are important to you and assign them levels of importance between 0 and 1, inclusive. Next, specify to what degree each choice satisfies each characteristic. Finally, mimic the calculations in Example 6 to make your decision.

32. Do you have any criticisms of the decision-making method that we used in Example 6? Are you comfortable with the way in which decisions are made by this method? Can you suggest any changes? If so, try to implement these changes and then reconsider the examples and exercises of this section to see whether your method results in different decisions being made.

*The reason that this is Mom's favorite university is that Mom would like to see your sister go to a school close to home.

CHAPTER SUMMARY

Section	Summary	Example
SECTION 3.1	A **statement** in logic is a declarative sentence that is either true or false.	Discussion, p. 86
	A **simple statement** contains a single idea. A **compound statement** contains several ideas combined together with words called **connectives.**	Discussion, p. 87
	Negation, represented by $\sim$, expresses the idea that something is *not* true.	Example 1, p. 87
	Conjunction, represented by $\wedge$, expresses the idea of *and.*	Example 2, p. 88
	Disjunction, represented by $\vee$, expresses the idea of *or.*	Example 3, p. 88
	The **conditional,** represented by $\rightarrow$, expresses the notion of *if... then.*	Example 4, p. 89
	The **biconditional,** represented by $\leftrightarrow$, expresses the idea of *if and only if.*	Example 5, p. 89
	Universal quantifiers are words such as *all* and *every* that state that all objects of a certain type satisfy a given property.	Discussion, p. 90
	Existential quantifiers are words such as *some, there exists,* and *there is at least one* that state that there are one or more objects that satisfy a given property.	
	To negate quantified statements we remember that:	Example 6, p. 92
	The phrase *Not all* has the same meaning as *At least one is not.*	
	The phrase *Not some* are has the same meaning as *None are.*	
SECTION 3.2	The **negation** ($\sim$) of a statement reverses its truth value.	Truth table, p. 95 Example 1, pp. 95–96
	A **conjunction** ($\wedge$) is true only when both component parts are true.	Truth table, p. 96 Example 2, p. 96
	A **disjunction** ($\vee$) is false only when both component parts are false.	Truth table, p. 96 Example 3, p. 97
	If a statement has k variables, then its truth table has 2^k lines.	Discussion, p. 99
	Statements are **logically equivalent** if they have the same variables and their truth tables are identical.	Example 6, pp. 100–101
	DeMorgan's laws state:	Example 7, p. 102
	a) $\sim(p \wedge q)$ is logically equivalent to $(\sim p) \vee (\sim q)$.	
	b) $\sim(p \vee q)$ is logically equivalent to $(\sim p) \wedge (\sim q)$.	
SECTION 3.3	A **conditional** ($\rightarrow$) is false only if its hypothesis is true and its conclusion is false.	Truth table, p. 107 Example 1, p. 107
	Derived forms of the **conditional** $p \rightarrow q$ are:	Discussion, pp. 108–109
	Converse: $q \rightarrow p$	Example 3, p. 109
	Inverse: $\sim p \rightarrow \sim q$	
	Contrapositive: $\sim q \rightarrow \sim p$	
	A conditional and its contrapositive are logically equivalent.	Example 4, pp. 109–110
	A **biconditional** ($\leftrightarrow$) is true only when both component parts have the same value (both Ts or both Fs).	Truth table, p. 111 Example 7, p. 112
SECTION 3.4	An **argument is valid** if whenever the premises are true, then the conclusion is true.	Discussion, p. 115 Example 1, pp. 116–117
	An **argument is invalid** if when we compute its truth table, there is even one F.	Example 2, p. 117

(Continued)

Section	Summary	Example
	There are several standard **valid argument** forms:	Example 3, pp. 118–119

Law of Detachment	Law of Contraposition	Law of Syllogism	Disjunctive Syllogism
$p \to q$ p $\therefore q$	$p \to q$ $\sim q$ $\therefore \sim p$	$p \to q$ $q \to r$ $\therefore p \to r$	$p \vee q$ $\sim p$ $\therefore q$

There are several standard **fallacies:** — Example 3, pp. 118–119

Fallacy of the Converse	Fallacy of the Inverse
$p \to q$ q $\therefore p$	$p \to q$ $\sim p$ $\therefore \sim q$

Section	Summary	Example
SECTION 3.5	A **syllogism** consists of a set of statements called **premises** followed by a statement called a **conclusion.** The premises and conclusion of the syllogism may contain quantifiers such as *all*, *some*, and *none.*	Discussion, p. 123
	A syllogism is **valid** if whenever its premises are all true, then the conclusion is also true.	Example 1, p. 124
	If the conclusion of a syllogism can be false even though all the premises are true, then the syllogism is **invalid.**	Example 2, p. 125
SECTION 3.6	**Statements** in fuzzy logic have truth values between 0 and 1 inclusive.	Example 1, p. 131
	In fuzzy logic, the value of the **negation** of p, written $\sim p$, is 1 − the value of p.	Example 2, p. 132
	The truth value of the **disjunction** of p and q, written $p \vee q$, is the *maximum* of the truth values for p and q. The truth value of the **conjunction** of p and q, written $p \wedge q$, is the *minimum* of the truth values for p and q.	Example 3, p. 132
	We can use fuzzy logic to make **decisions.**	Example 6, pp. 134–135

CHAPTER REVIEW EXERCISES

Section 3.1

1. Which of the following are statements? Explain your answers.
   a. Why do these things always happen to me?
   b. The distance from Los Angeles to New York City is 2,000 miles.
   c. Bring me back a pizza.
2. Let v represent the statement "I will buy a new Volt" and let s represent the statement "I will sell my old car." Write each statement in symbolic form.
   a. I will not buy a new Volt or I will sell my old car.
   b. It is not true that: I will buy a new Volt and not sell my old car.
3. Let f represent "Antonio is fluent in Spanish" and let l represent "Antonio has lived in Spain for a semester." Write each statement in English.
   a. $\sim(f \wedge \sim l)$
   b. $\sim f \vee \sim l$
4. Negate each quantified statement and then rewrite it in English in an alternate way.
   a. All writers are passionate.
   b. Some graduates received several job offers.

Section 3.2

5. Let p represent some true statement, q represent some false statement, and r represent some false statement. What is the truth value of the following statements?
   a. $p \wedge (\sim q)$
   b. $r \vee (\sim p \wedge q)$
   c. $\sim(p \vee q) \wedge \sim r$
6. How many rows will be in the truth table for each statement?
   a. $\sim(p \vee q) \wedge \sim(r \vee p)$
   b. $(p \vee q) \wedge (r \vee s) \wedge t$
7. Construct a truth table for each statement.
   a. $\sim(p \vee \sim q)$
   b. $\sim(p \vee \sim q) \wedge \sim r$
8. Negate each statement and then rewrite the negation using DeMorgan's laws.
   a. I will take Pilates or Zumba.
   b. I will not sign the lease or I will not accept the housing agreement.
9. Which pairs of statements are logically equivalent?
   a. $\sim(p \wedge \sim q)$, $(\sim p) \vee q$
   b. $\sim(p \vee \sim q) \wedge \sim(p \vee q)$, $p \vee (p \wedge q)$

Section 3.3

10. Assume that p represents a true statement, q a false statement, and r a true statement. What is the truth value of each statement?

a. $\sim(p \vee q) \rightarrow \sim p$

b. $(p \wedge q) \leftrightarrow (q \vee r)$

c. $((\sim p) \vee (\sim q)) \rightarrow r$

11. Construct a truth table for each statement.

a. $\sim p \rightarrow q$

b. $\sim(p \wedge r) \leftrightarrow \sim(p \vee q)$

12. Write in words the converse, inverse, and contrapositive for the statement "If I go to Starbucks, then I'll have the new Mocha."

13. Rewrite each statement using the words *if . . . then.*

a. The Heat will get to the finals only if they beat the Lakers.

b. To be an astronaut, it is necessary to have a pilot's license.

Section 3.4

14. Identify the form of each argument.

a. If you make lots of money, then you will be happy.
You do make lots of money.
___
∴ You are happy.

b. If Felicia enjoys spicy food, then she will enjoy this Cajun chicken.
Felicia does not enjoy this Cajun chicken.
___
Therefore, Felicia does not enjoy spicy food.

15. Determine whether the form represents a valid argument:

$$\begin{array}{l} \sim p \\ q \rightarrow p \\ (p \vee q) \rightarrow r \\ \hline \therefore r \end{array}$$

16. Use a truth table to determine whether the argument is valid.

If you pay more for your phone plan, then you will have more calling minutes.
If you pay more for your phone plan or have more calling minutes, then you will call your mother more often.
You paid more for your phone plan.
___
Therefore, you will call your mother more often.

Section 3.5

In Exercises 17 and 18, use Euler diagrams to determine whether each syllogism is valid or invalid.

17. Some used cars are expensive.
All expensive cars are safe.
This car is not safe.
___
∴ This is not a used car.

18. All professors are wealthy.
Some professors are absent-minded.
All wealthy people are happy.
___
∴ Some absent-minded people are happy.

Section 3.6

19. Assume that p and q are fuzzy statements having truth values of 0.47 and 0.82, respectively. Compute the truth values for the following statements:

a. $p \wedge \sim q$

b. $\sim(p \vee \sim q)$

c. $p \rightarrow \sim q$

CHAPTER TEST

1. Which of the following are statements?

a. New York City is the largest city in North America.

b. When did the Red Sox last win the pennant?

2. Negate each quantified statement and then rewrite it in English in an alternate way.

a. All rock stars are fine musicians.

b. Some dogs are aggressive.

3. Let p represent the statement "I will pass my lifeguard test" and let f represent "I will have fun this summer." Write each statement in symbolic form.

a. I will pass my lifeguard test or I will not have fun this summer.

b. It is not true that: I will not pass my lifeguard test and have fun this summer.

4. Let t represent "The Tigers will win the series" and let v represent "Verlander will win the Cy Young Award." Write each statement in English.

a. $\sim(t \vee \sim v)$

b. $\sim t \wedge \sim v$

5. How many rows will there be in a truth table for a logical statement having six variables?

6. If p is false and q is true and r is false, what is the truth value of each statement?

a. $\sim(p \vee \sim q)$

b. $\sim(p \vee q) \wedge \sim r$

c. $\sim(\sim r \vee \sim p) \wedge r$

7. Assume that p, q, and r are fuzzy statements having truth values of 0.65, 0.38, and 0.75, respectively. Find the truth values of the following fuzzy statements:

a. $\sim(p \vee r)$

b. $(\sim r) \vee \sim(p \wedge q)$

8. Construct a truth table for each statement.

a. $\sim(p \wedge \sim q)$

b. $(\sim p \vee \sim q) \wedge r$

9. Write each statement using the words *if... then.*
   a. To get enough sources for your research term paper, it is sufficient to go to Wikipedia.
   b. Ticketmaster will mail the concert tickets only if you pay a fee.
10. Negate each statement and then rewrite the negation using DeMorgan's laws.
    a. You can take the final exam or write a term paper.
    b. I will not finish the painting or I will not show it at the gallery.
11. Determine whether the following pairs of statements are logically equivalent.
    a. $\sim(p \vee \sim q), \sim p \wedge q$
    b. $(\sim p \vee \sim q) \wedge (\sim p \vee q), \sim p \wedge q$
12. Write in words the converse, inverse, and contrapositive for the statement "If it glitters, then it is gold."
13. If p is true, q is false, and r is true, what is the truth value of each statement?
    a. $(p \vee \sim q) \rightarrow \sim q$
    b. $(\sim p \wedge q) \rightarrow \sim r$
    c. $(p \vee \sim q) \leftrightarrow \sim(p \wedge q)$
14. Construct a truth table for each statement.
    a. $(\sim p \vee q) \rightarrow \sim(p \wedge q)$
    b. $(\sim p \vee \sim q) \leftrightarrow r$
15. Determine whether the form represents a valid argument.

$$\begin{array}{l} p \\ \sim q \rightarrow \sim r \\ (q \vee r) \rightarrow \sim p \\ \hline \therefore \sim r \end{array}$$

16. Identify the form of each argument.
    a. If it ain't broke, then don't fix it.
       It is broke.
       ______
       Therefore, fix it.
    b. I'll major in music or art history.
       I am not majoring in music.
       ______
       Therefore, I am majoring in art history.
17. In fuzzy logic, we replaced the conditional $p \rightarrow q$ by what logical form?
18. Use a truth table to determine if the argument is valid.
    If you go to eBay, then you will find a bargain or you will waste your money.
    If you do not waste your money, then you will find a bargain.
    You will not go to eBay or you will not find a bargain.
    ______
    Therefore, you will waste your money.
19. Use an Euler diagram to determine whether the syllogism is valid or invalid.
    Some poets are sensitive.
    No sensitive people are selfish.
    Brittany is a poet.
    ______
    Therefore, Brittany is not selfish.

GROUP PROJECTS

1. **Three-valued logic.** In Sections 3.1 to 3.4, you studied a two-valued logic. Statements were either true or false. In the early part of the twentieth century, the Polish logician Jan Lukasiewicz* and others invented three-valued logic. Instead of true or false, we include a third option of "maybe."
   a. Truth tables for this three-valued logic are similar to truth tables for two-valued logic except now, for k variables we would have 3^k lines. We give the truth table for the "and" connective; you construct the truth tables for "not," "or," and "if... then." Recall that in two-valued logic, the statement $p \rightarrow q$ is logically equivalent to $\sim p \vee q$.

p	q	$p \wedge q$
T	T	T
T	M	M
T	F	F
M	T	M
M	M	M
M	F	F
F	T	F
F	M	F
F	F	F

   b. Compute truth tables for some compound statements using two variables p and q, as we did in Section 3.2.
   c. Do DeMorgan's laws hold in three-valued logic? Explain.
   d. Investigate equivalent and nonequivalent derived forms of the conditional as we did in Section 3.3
2. **The stroke connective.** Recall the stroke connective, written $|$, that we introduced in the exercise set in Section 3.2. Try writing some English sentences using only the word *stroke* for connectives. For example, we know that the statements $\sim p$ and $p|p$ are logically equivalent, so in a weather report, instead of saying "It will not rain today," we could say "It will rain today *stroke* it will rain today" and avoid the use of the negation. As another example, consider how a weatherperson would report "If the clouds break, then it will be warm and humid this afternoon," using only the stroke connective.
3. **Analyzing a quote from A. Lincoln.** The following paragraph is an excerpt from Abraham Lincoln's "House Divided" speech given on June 16, 1858.

*See Historical Highlight on page 124.

I believe this government cannot endure permanently half slave and half free. I do not expect the Union to be dissolved—I do not expect the house to fall—but I do expect it will cease to be divided. It will become all one thing, or all the other. Either the opponents of slavery will arrest the further spread of it, and place it where the public mind shall rest in the belief that it is in the course of ultimate extinction; or its advocates will push it forward, till it shall become alike lawful in all the States, old as well as new—North as well as South.

Analyze each sentence with regard to its logical form and then rewrite it in a form that makes the logical structure more clear.

4. **Analyzing a quote from Susan B. Anthony.** The following paragraph is an excerpt from a speech given by Susan B. Anthony after her arrest for casting an illegal vote in the presidential election of 1872.

It was we, the people; not we, the white male citizens; nor yet we, the male citizens; but we, the whole people, who formed the Union. And we formed it, not to give the blessings of liberty, but to secure them; not to the half of ourselves and the half of our posterity, but to the whole people—women as well as men. And it is a downright mockery to talk to women of their enjoyment of the blessings of liberty while they are denied the use of the only means of securing them provided by this democratic-republican government—the ballot.

Analyze each sentence with regard to its logical form and then rewrite it in a form that makes the logical structure more clear.

USING TECHNOLOGY

1. **Using a calculator to do logical computations.** If you have a calculator, look at your manual to see if it has the logical connectives that we have discussed. For example, on my calculator I used the "or" connective to see if the statement $8 > 4$ or $2 = 3$ is true (see accompanying image of calculator screen). The calculator returned the number 1, which stands for "true." If my statement had been false, then the calculator would have returned a 0.

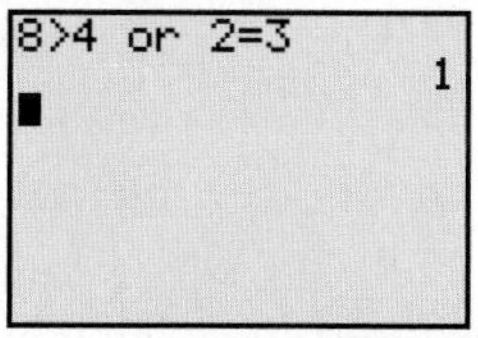

Learn how to use the logical connectives on your calculator and test the following statements to see if they are true or false.

a. $(5 \geq 6/3)$ *and* $1 < 4$ **b.** $6 = 2*3$ *or* $8 = 7*5$

c. Make up some more complicated statements and test them with your calculator to see if they are true or false.

2. **Using Excel to do logical computations.** The Excel spreadsheet has several of the logical functions that you have studied in this chapter. The formulas that we have typed in cells D6 and E6 below show the type of formulas that you can use to do logical computations.

Use Excel to calculate truth tables for the following logical statements. We will use *A*, *B*, and *C* instead of *p*, *q*, and *r*.

a) $(\sim A \wedge B) \vee \sim C$ (Example 5 in Section 3.2) The *if . . . then* statement in Excel has the form IF(AND(B2, C2), 1, 0). If the underlined condition is true, then the whole *if...then* statement is true. Otherwise, the *if...then* statement is false.

b) $A \rightarrow B$

c) $(A \vee B) \rightarrow \sim B$

d) $[(A \rightarrow B) \wedge \sim B] \rightarrow \sim A$ (Example 1 in Section 3.4)

Logic Operations in Excel

	A	B	C	D	E	F
1	TRUE	TRUE		TRUE	FALSE	
2	TRUE	FALSE		FALSE	FALSE	
3	FALSE	TRUE		FALSE	FALSE	
4	FALSE	FALSE		FALSE	TRUE	
5						
6				AND(A1, B1)	NOT(OR(A1,B1))	
7						

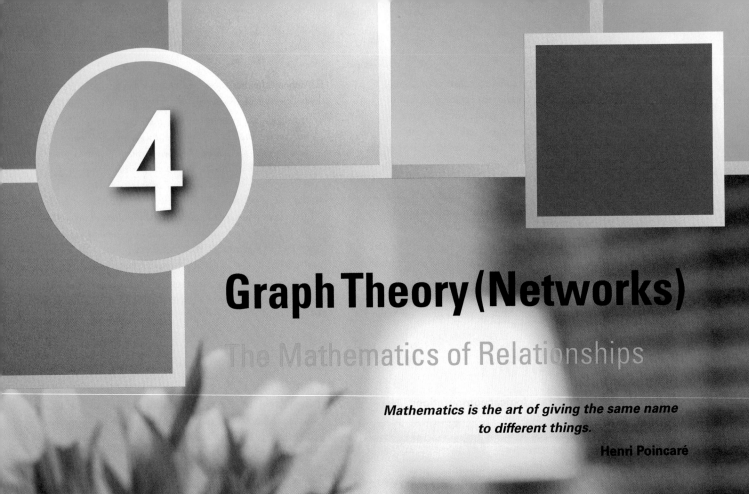

4

Graph Theory (Networks)

The Mathematics of Relationships

Mathematics is the art of giving the same name to different things.

Henri Poincaré

The topic of **graph theory** that you will study in this chapter is a good example of what Poincaré had in mind. We will talk about seemingly unrelated situations such as solving puzzles, coloring maps, taking trips, sending packages, spreading disease, surfing the Internet, seating people at tables, and scheduling projects. Although these situations appear to be quite different, you will see they are all tied together by a common thread—relationships.

Although many believe that mathematicians only work with numbers and algebraic equations, there is much more to mathematics than that. Some mathematicians spend their professional lives investigating the

(continued)

relationships among objects. In this chapter, we will use mathematics to answer questions such as: How are cities related by the cost to travel between them? How are the victims of a fast-spreading disease on campus connected? What is the best way to organize and schedule your semester project?

Researchers in such areas as sociology, biology, political science, and urban studies apply graph theory to solve a variety of problems. In Section 4.3, we will show you how graph theory can help you to schedule interviews for your dream job efficiently. In Section 4.4, you will learn how to organize a plan for a complex project.

4.1 Graphs, Puzzles, and Map Coloring

Objectives

1. Understand graph terminology.
2. Apply Euler's theorem to graph tracing.
3. Understand when to use graphs as models.
4. Use Fleury's theorem to find Euler circuits.
5. Utilize graph coloring to simplify a problem.

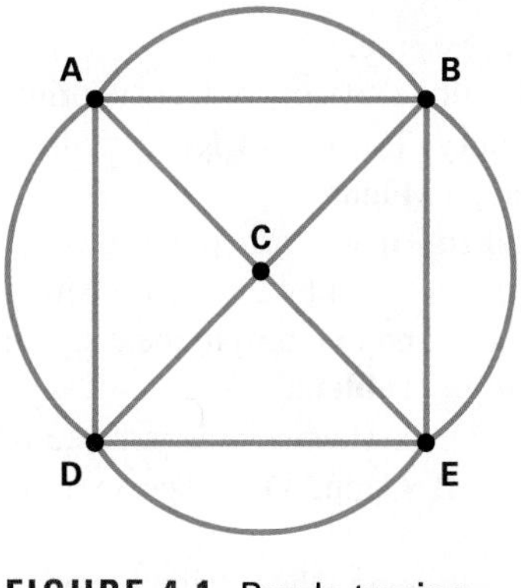

FIGURE 4.1 Puzzle tracing.

When my oldest son was about 5 years old, I gave him the puzzle that you see drawn in Figure 4.1 and that you might want to try. Place your pencil on any dot and trace the figure completely without lifting your pencil and without tracing any part of any line twice. Can you do it?

After a few minutes of struggling, my son came to me and said, "I can't do it.... Nobody can do it." "Why do you say that?" I asked. He replied, "There are too many places to get stuck." Did you also notice that? Several hundred years ago, the brilliant Swiss mathematician Leonhard Euler did and, after learning a little graph theory, you will also see why this problem cannot be done.

The mathematics that you study in this chapter is probably different from the mathematics that you have studied before, and at first you may think that graph theory is only concerned with children's games. However, this field has many practical applications, as you will see later in the chapter.

We will begin by introducing some basic terminology.

Graph Terminology

KEY POINT

Graphs consist of vertices and edges.

DEFINITIONS A **graph** consists of a finite set of points, called **vertices,** and lines, called **edges,** that join pairs of vertices. (Certain types of graphs are called *networks.*)

Figure 4.1 is an example of a graph. Its vertices are the points A, B, C, D, and E, and the 12 connecting lines are the edges. Generally, we use capital letters to designate vertices. If there is only one edge joining a pair of vertices, then we label that edge by the vertices it connects. For example, in Figure 4.1 we can refer to the edge joining vertices A and C as AC or CA; the order is not important. To refer to edge AB would be confusing because there are two edges joining vertices A and B. In this case, we might label one edge e_1 and the other e_2.

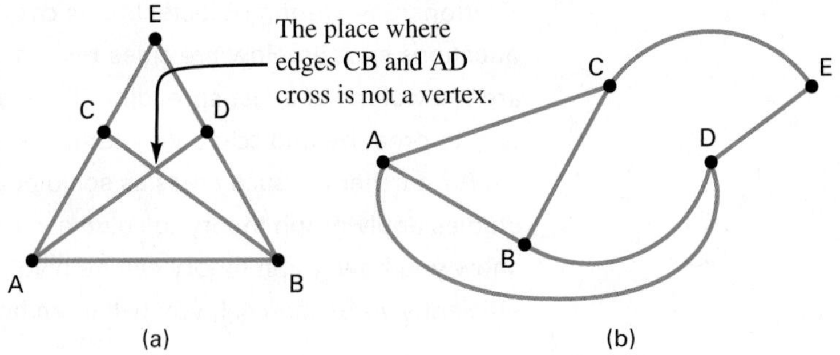

FIGURE 4.2 Two ways to draw the same graph.

The graph in Figure 4.2(a) has five vertices: A, B, C, D, and E. Although edges BC and AD intersect, the point of intersection is not a vertex. We will not consider a point of intersection of a pair of edges to be a vertex unless we place a solid dot there and label it as a vertex.

In drawing a graph, the important information is what vertices we connect by edges. The placement of the vertices and the shape of the edges are unimportant. We could have also drawn Figure 4.2(a) as in Figure 4.2(b), because this graph contains exactly the same vertices and edges. If a graph looks awkward the first time you draw it, you should redraw it to present the information more clearly.

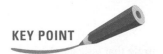

KEY POINT

Graphs represent relationships among objects.

Graph Tracing

We will now introduce several situations that we can model by graphs.

One of the most famous and yet elementary applications of graph theory originated in Koenigsberg, Prussia, during the eighteenth century. The Pregel River divided Koenigsberg into four distinct sections, as shown on the map in Figure 4.3.

Seven bridges connected the four portions of Koenigsberg. It was a popular pastime for the citizens of Koenigsberg to start in one section of the city and take a walk visiting all sections of the city, trying to cross each bridge exactly once and to return to the original starting point. This problem is called the **Koenigsberg bridge problem.**

It might not be apparent that this is a graph theory problem. However, as you see in Figure 4.4, we can model Koenigsberg by using vertices A, B, C, and D to represent the bodies of land and seven edges to represent the bridges.

In 1735, Leonhard Euler discovered a simple way to determine when a graph can be traced.* To **trace** a graph means to begin at some vertex and draw the entire graph without

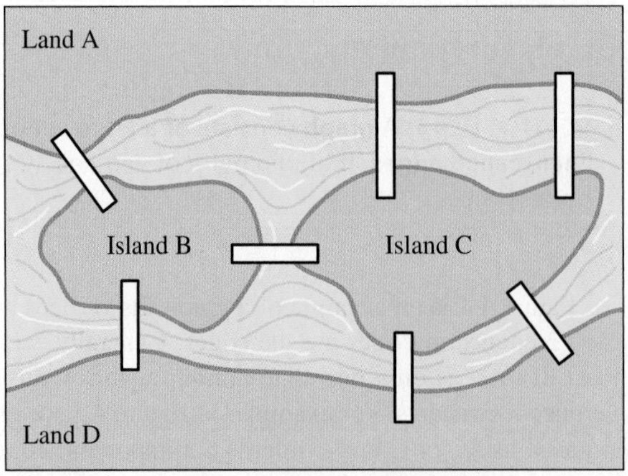

FIGURE 4.3 Map of Koenigsberg.

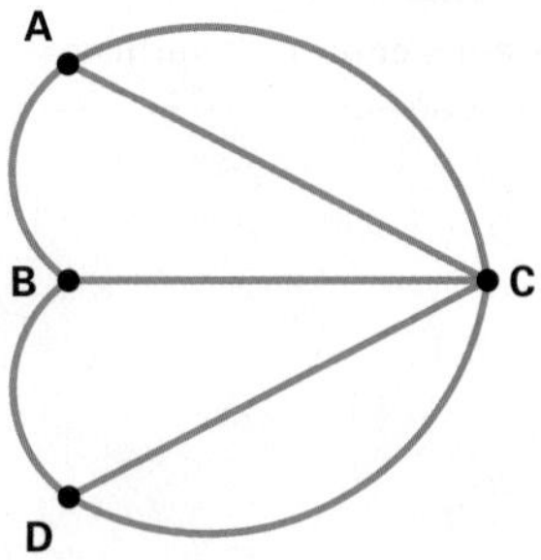

FIGURE 4.4 Graph model of Koenigsberg.

*A very readable translation of Euler's original paper is J. R. Newman, Ed., "Leonhard Euler and Koenigsberg Bridges," *Scientific American* (July 1953), pp. 66–70.

Historical Highlight

Pioneers in Graph Theory

In 1735, the Swiss mathematician Leonhard Euler became the first person to work in graph theory by solving the Koenigsberg bridge problem. Euler was a prolific mathmatician, writing almost 900 papers. He produced mathematics so effortlessly that his biographer Arago said he calculated without effort: "as men breathe, or as eagles sustain themselves in the wind." Incredibly, he did about half of his work during the last 17 years of his life while totally blind. One afternoon, while working on a problem, Euler suffered a stroke and with the words, "I die," he ceased calculating.

Later, in the nineteenth century, the English mathematician Arthur Cayley became interested in the four-color problem (pages 151–152) and subsequently wrote several papers applying graph theory to chemistry.

lifting your pencil and without going over any edge more than once. Before giving the solution to this graph tracing problem, we need a few more definitions.

DEFINITIONS A graph is **connected*** if it is possible to travel from any vertex to any other vertex of the graph by moving along successive edges. A **bridge** in a connected graph is an edge such that if it were removed, the graph would no longer be connected.

Figure 4.5 illustrates these definitions.

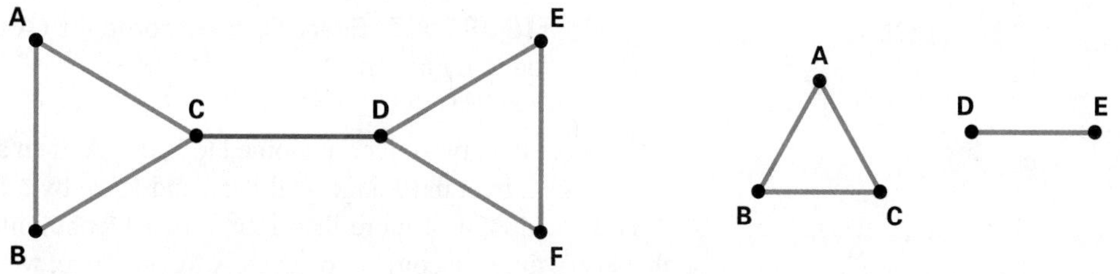

FIGURE 4.5 Examples of a connected graph and a nonconnected graph.

DEFINITIONS A vertex of a graph is **odd** if it is the endpoint of an odd number of edges. Similarly, a vertex is **even** if it is the endpoint of an even number of edges. In general, the **degree** of a vertex is the number of edges joined to that vertex.

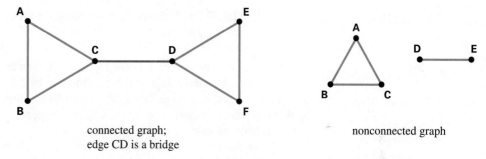

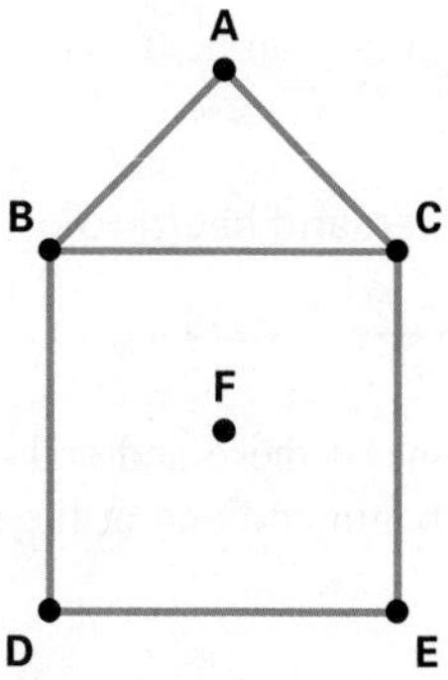

FIGURE 4.6 Odd and even vertices.

EXAMPLE 1 Odd and Even Vertices in a Graph

a) What are the degrees of vertices A, C, and F in the graph in Figure 4.6?

b) Which vertices in the graph are odd? Which are even?

SOLUTION:

a) Vertex A is the endpoint of two edges, so its degree is two. Similarly, the degree of C is three. Because F has no edges joined to it, its degree is zero.

b) Vertices B and C are odd, the others are all even. Note that because zero is an even number, F is an even vertex. 1

*Connected graphs are sometimes called *networks*.

QUIZ YOURSELF 1

Is this graph connected? List its odd and even vertices. What is the degree of G? Are any edges bridges?

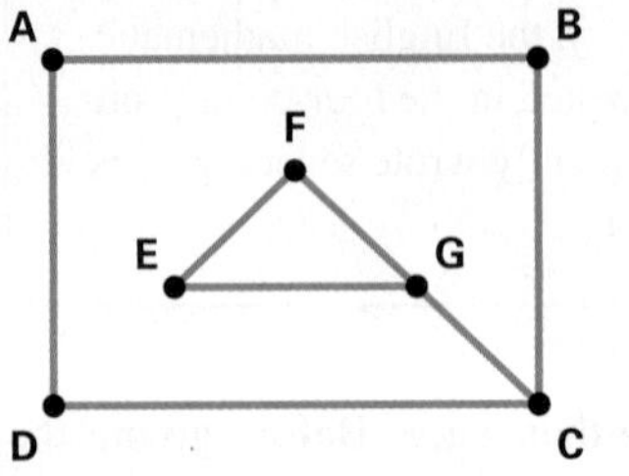

Euler's Theorem

We will now make two observations that tell us when a graph can be traced.

Observation 1: If, while tracing a graph, we neither begin nor end with vertex A, then A must be an even vertex.

This is easy to see. Assume that we are tracing a graph and that vertex A is neither a beginning nor ending vertex. Eventually we must come into A by means of one edge, call it e_1, and then must leave by another edge, call it e_2. See Figure 4.7(a). If no other edges are joined to A, then A is the endpoint of exactly two edges, so it is even.

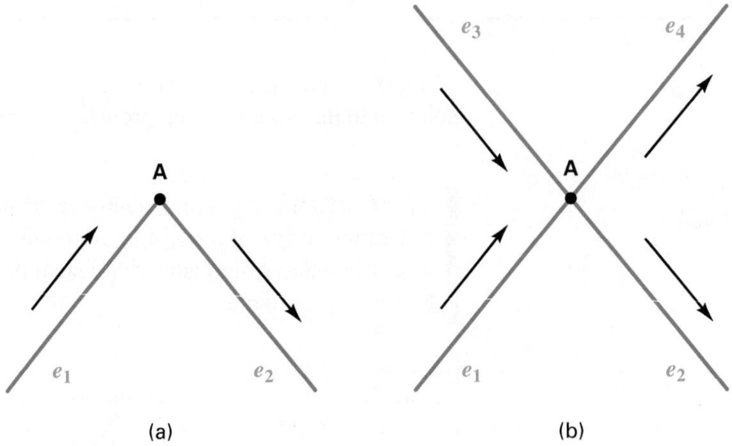

FIGURE 4.7 Every time we come into A by one edge, we must leave by another.

If more than two edges are joined to vertex A, then as we continue tracing, we will come into A again by a third edge, call it e_3, and leave by a fourth edge, e_4. See Figure 4.7(b). If A is the endpoint of more than four vertices, continuing this line of thinking, you can see that every time we come into vertex A by one edge, we must leave by another, so A must be the endpoint of an even number of edges.

From this we conclude that an odd vertex can only be used as either a starting point or ending point when tracing a graph.

Observation 2: If a graph can be traced, then it can have at most two odd vertices.

This follows directly from Observation 1. In tracing the graph, one odd vertex could be the starting vertex and another could be the ending vertex. Because no other vertices could be the starting or ending vertex, all other vertices must be even.

We now paraphrase Euler's theorem, which tells us when a graph can be traced.

KEY POINT

Euler's theorem tells when a graph can be traced.

EULER'S THEOREM A graph can be traced if it is connected and has zero or two odd vertices.

If a graph has two odd vertices, the tracing must begin at one of these and end at the other. If all the vertices are even, then the graph tracing must begin and end at the same vertex. It does not matter at which vertex this occurs.

EXAMPLE 2 Tracing Graphs

Which of the graphs in Figure 4.8 can be traced? Recognize that Figure 4.8(a) is the puzzle graph in Figure 4.1, and Figure 4.8(b) is the graph of Koenigsberg and its bridges.

QUIZ YOURSELF 2

Using Euler's theorem, state which of the following graphs can be traced. For a graph that can be traced, list a sequence of vertices that describes how to trace the graph. For a graph that cannot be traced, state which part of Euler's theorem fails.

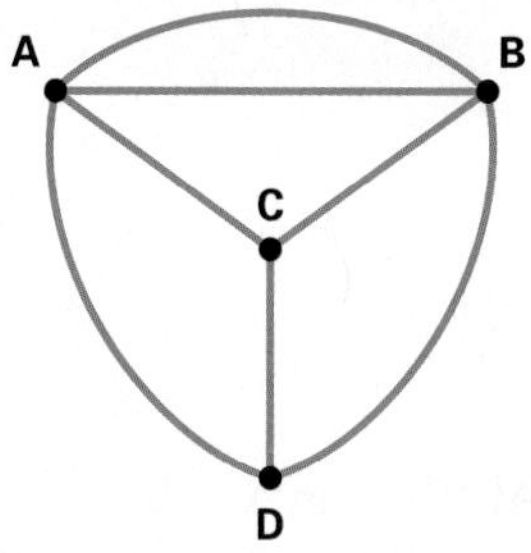

(a)

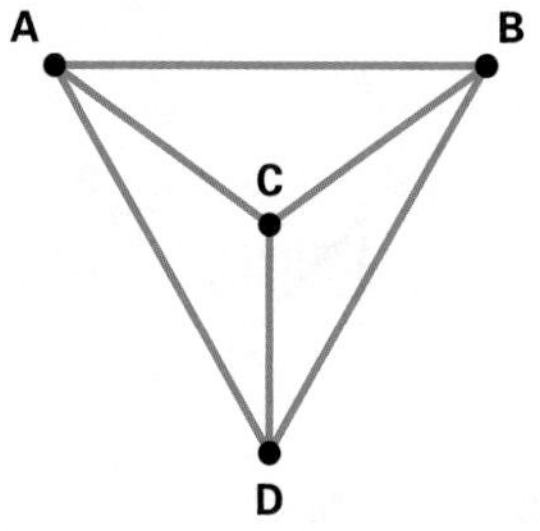

(b)

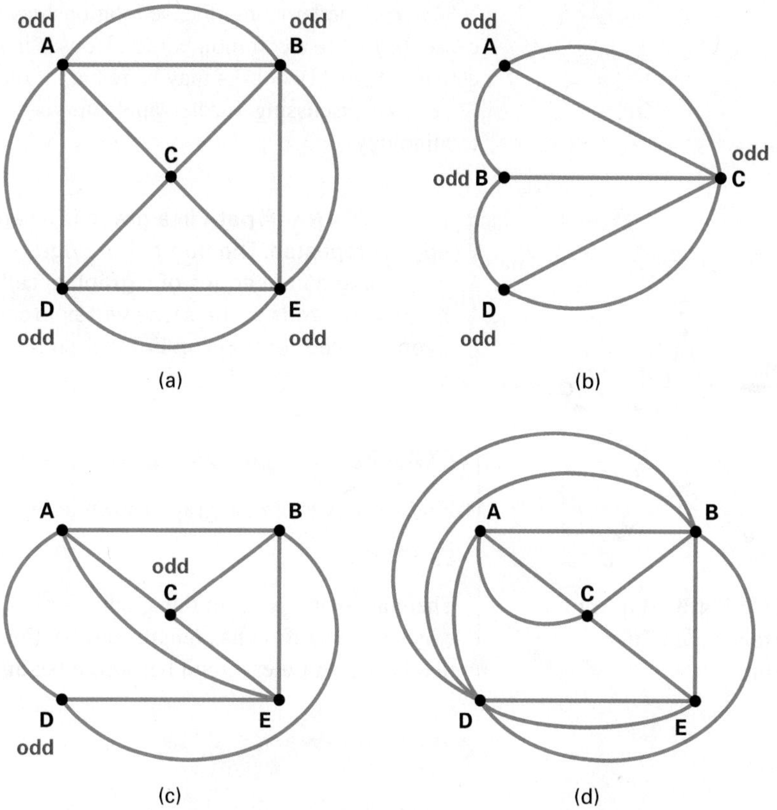

FIGURE 4.8 (a) Puzzle graph. (b) Koenigsberg graph. Only (c) and (d) can be traced by Euler's theorem.

SOLUTION:

a) Notice that vertices A, B, D, and E are all odd; therefore, by Euler's theorem, this graph cannot be traced.

b) All vertices are odd; this means that the Koenigsberg bridge graph cannot be traced.

c) This graph has two odd vertices, C and D. One way to trace this graph is to begin at D and follow this sequence of vertices:

D, A, B, D, E, B, C, A, E, C

d) All vertices in this graph are even. Can you find a sequence of vertices describing how to trace this graph?

Now try Exercises 9 to 16. 2

PROBLEM SOLVING

Strategy: Draw a Picture

The Koenigsberg bridge problem has two features that are present in every graph model:

1. A set of objects—the four bodies of land
2. A relationship among the objects—they were connected by bridges

If these two features are present in a problem, you can model it with a graph as follows:

First, represent the objects by vertices having good names.

Second, join the vertices representing related objects with edges.

We can model many diverse relationships with graphs. Countries might be related because they share a common border. Houses in a town could be related if a fiber-optic cable connects them. Two tasks may be related if one must be done before the other.

Before discussing further applications of Euler's theorem, we will introduce more terminology.

DEFINITIONS A **path** in a graph is a series of consecutive edges in which no edge is repeated. The number of edges in a path is called its **length.** A path containing all the edges of a graph is called an **Euler path.** An Euler path that begins and ends at the same vertex is called an **Euler circuit.** A graph with all even vertices contains an Euler circuit and is called an **Eulerian graph.**

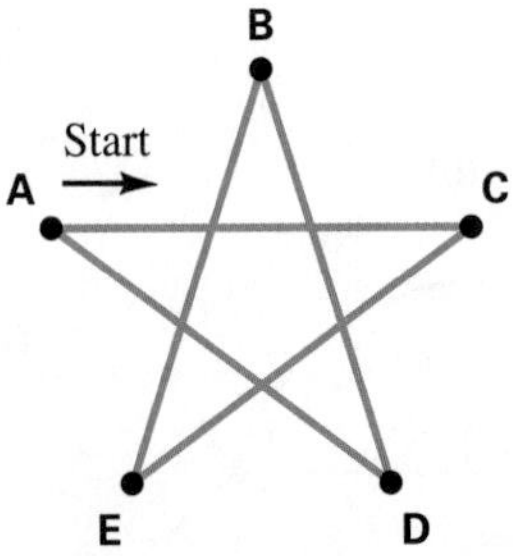

FIGURE 4.9 A graph containing an Euler circuit.

EXAMPLE 3 An Euler Circuit

Find some paths in the graph shown in Figure 4.9.

SOLUTION:

There are many paths in this graph. For example, the path ACEB from A to B has three edges and therefore has length 3. Also, the path ACEBDA is an Euler path of length 5 that is also an Euler circuit because it begins and ends at the same vertex. 3

QUIZ YOURSELF 3

a) Find an Euler path in this graph.
b) Is there an Euler circuit? Explain.

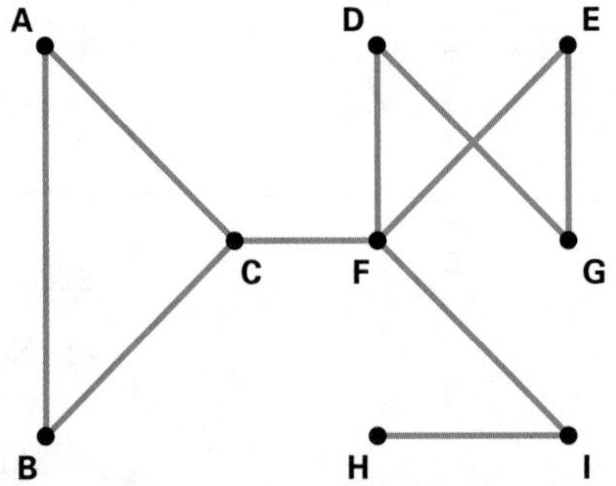

Between the Numbers

It's a Small World After All

The term "six degrees of separation," first mentioned in a 1929 novel by the Hungarian writer Frigyes Karinthy, states that any two people on Earth are connected by a chain of no more than six associations.*

The sociologist Stanley Milgram tested this theory, called "the small-world problem," by randomly selecting people in the Midwest to send packages to a stranger in Massachusetts. They were to send the package to a person they knew who they thought was most likely to know the target. Then that person would do the same, and so on, until the package was delivered. Milgram found that, on the average, the package was delivered after passing through the hands of five to seven people. In similar experiments, other researchers have found that the average length of a path of e-mails between a random sender and receiver was six.

While discussing this with my wife, we realized that because of my appearance on a national television amateur show in my teens, my degree of separation with the Beatles is three and as a result of a graduate course that she took, her degree of separation from Mother Theresa of Calcutta is two.

So what does this have to do with you? In your activities at school—fraternities, clubs, musical organizations, and sports teams—you are creating links in these personal chains that you may be able to use in later life to your benefit. In fact, companies like LinkedIn develop software that will provide you with chains of contacts that can help you in the corporate world.

On the flip side, connectivity can sometimes work against you. Prospective employers are also only a few degrees of separation away from you and can use popular social media such as Facebook to check you out before inviting you for an interview.

*Perhaps you have played the trivia game "Six Degrees of Kevin Bacon."

KEY POINT

We use Fleury's algorithm to find Euler circuits.

Fleury's Algorithm

Although we can use Euler's theorem to find out whether Euler circuits exist for a given graph, it does not tell us how to find them. Until now, we have used trial and error; however, in a large graph, this is not efficient. A method called **Fleury's algorithm** provides a systematic technique for finding Euler circuits. An **algorithm** is a series of steps that we follow to accomplish something. You might think of an algorithm as a recipe to do some mathematical task.

> FLEURY'S ALGORITHM If a connected graph has all even vertices, we can find an Euler circuit for it by beginning at any vertex and traveling over consecutive edges according to these rules:
>
> 1. After you have traveled over an edge, erase it. If all the edges for a particular vertex have been erased, then erase that vertex also.
> 2. Travel over an edge that is a bridge only if there is no alternative.

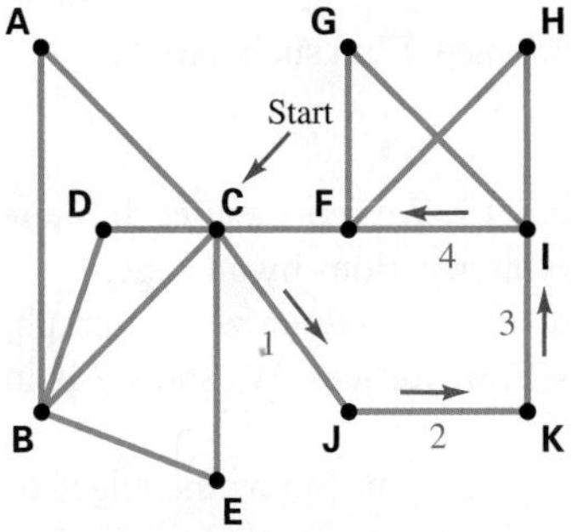

FIGURE 4.10 Graph showing paths in a theme park.

EXAMPLE 4 Using Fleury's Algorithm to Find an Efficient Route

Assume you are doing maintenance work along pathways joining locations A, B, and so on in a theme park, as shown in Figure 4.10. Find an Euler circuit in this graph to make your job efficient by not retracing pathways. Assume that you leave from and return to building C.

SOLUTION: Review the Be Systematic Strategy on page 5.

We will use Fleury's algorithm to find an Euler circuit in this graph.

Step 1: We begin at vertex C and traverse edge CJ, next JK and KI, and then IF. We numbered these edges 1, 2, 3, and 4 in Figure 4.10, indicating the order in which we will travel over them. We erase these edges and also vertices J and K because they no longer have any edges joined to them. This gives us the graph in Figure 4.11.

Step 2: We are now at vertex F. We cannot traverse FC because it is a bridge. So we traverse FG, GI, IH, and HF (marked 5, 6, 7, and 8), erasing these edges and vertices G, I, and H. The graph now looks like Figure 4.12(a).

Step 3: We have no choice now but to traverse FC (edge 9). We follow this with CA and AB (marked 10 and 11). After erasing appropriate edges and vertices, we have the graph in Figure 4.12(b).

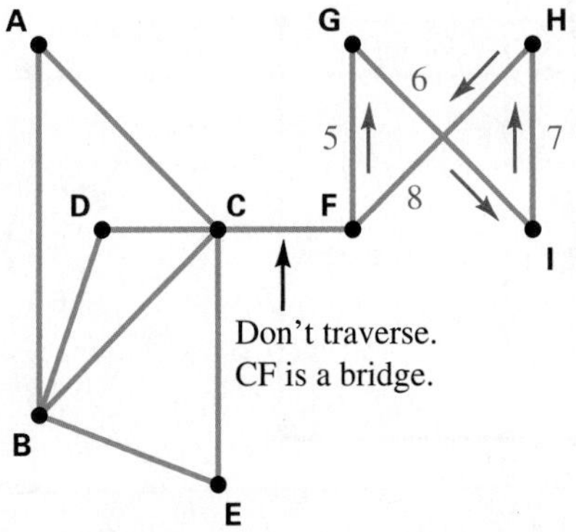

FIGURE 4.11 Step 1.

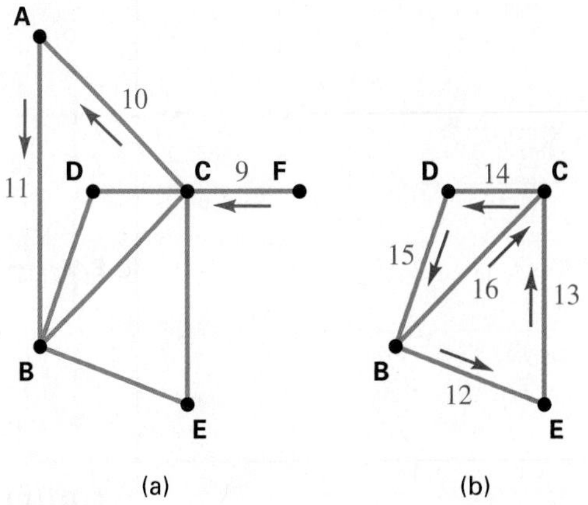

FIGURE 4.12 (a) Step 2. (b) Step 3.

We can now finish the circuit by traveling over BE, EC, CD, DB, and BC (marked 12, 13, 14, 15, and 16). The final circuit is CJKIFGIHFCABECDBC. Notice that we have traversed every edge exactly once and ended at our starting point, vertex C. If you were to follow this circuit, you would cover each path exactly once in performing maintenance on the paths in the park.

Now try Exercises 17 to 20.

Some of the decisions we made in Example 4 were arbitrary, and there are certainly many different ways to construct an Euler circuit for this graph.

Eulerizing a Graph

KEY POINT

We can add edges to convert a non-Eulerian graph to an Eulerian graph.

Have you ever been on a DUCK tour? Cities such as Washington, D.C., Miami, Seattle, Chicago, and many others offer guided tours in authentic, renovated World War II amphibious landing vehicles called DUCKs. On a DUCK tour in Boston, Massachusetts, our tour guide, called a "ConDUCKtor," explained that although we were allowed to "quack" at people on the street, certain areas were zoned by law to be "nonquacking zones," and in these zones, quacking was forbidden. In order to minimize traffic where possible, a tour would not go over the same street twice. In the next example, you will see how to use graph theory to design an optimal DUCK tour.

To solve our problem, we will have to add some edges to a non-Eulerian graph (a graph having odd vertices) so that the new graph is Eulerian. This technique is called **Eulerizing a graph.**

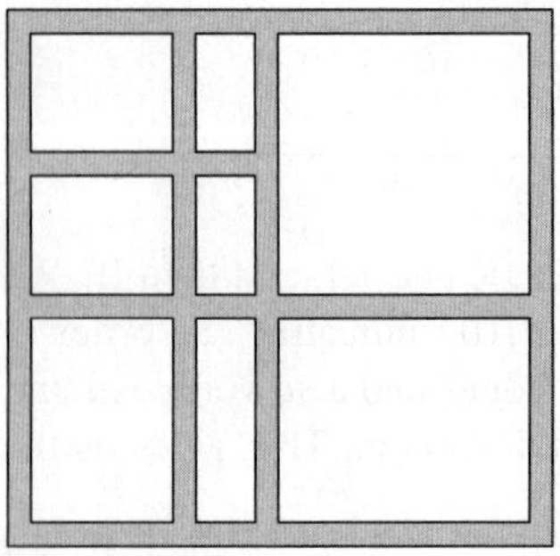

FIGURE 4.13 Map of Boston historic area.

EXAMPLE 5 Designing a DUCK Tour

The Boston DUCK tour company wants to design a DUCK route in a historic area of Boston shown in the map in Figure 4.13. We want to begin and end the tour at the same location and minimize traveling over any street more than once. Find such a route.

SOLUTION:

We can model this map with the graph shown in Figure 4.14. We represent each intersection by a vertex and each section of street joining two intersections by an edge.

We labeled the *odd* vertices in this graph A, B, and so on. To Eulerize this graph, we duplicate some edges so that the new graph has only even vertices. We show this in Figure 4.15.

If we begin our route at the upper right corner of the graph and follow the edges as they are numbered, we will traverse all the edges of the graph and return to our starting point. Notice that the five pairs of duplicate edges—AB, CG, and so on—represent

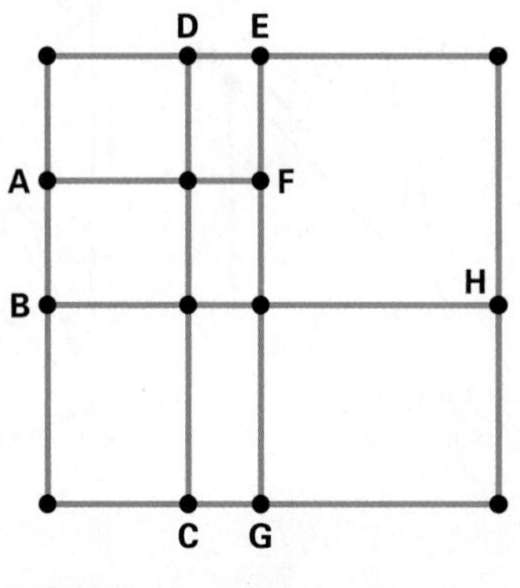

FIGURE 4.14 Graph representing Boston historic area.

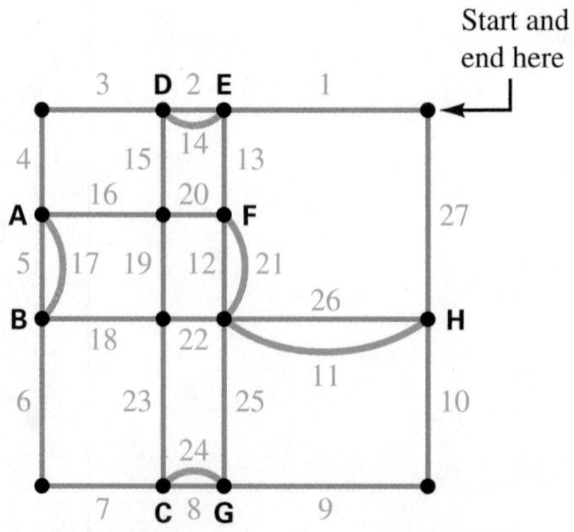

FIGURE 4.15 Route for DUCK tour.

QUIZ YOURSELF 4

Add edges to this graph to make it Eulerian.

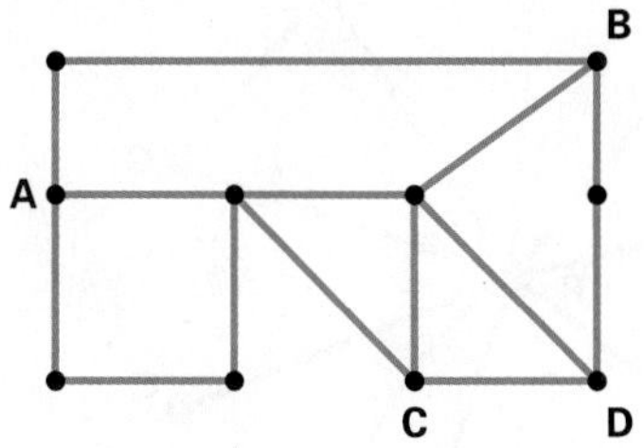

streets that must be traveled twice. Although we will not prove it, there is not a better way to Eulerize the original graph to reduce the number of streets that are traveled twice.

Now try Exercises 33 to 36. 4

When Eulerizing a graph, you can only duplicate edges that are *already present* in the graph. In Example 5, you are not allowed to insert a *single* edge from F to H to Eulerize the graph.

Map Coloring

We now turn our attention to an interesting problem called the **four-color problem.** Although we state this as a puzzle, like the Koenigsberg bridge problem, it has interesting real-life applications. In 1852, Francis Guthrie, a student at University College, London, first posed this famous question to his mathematics professor, Augustus DeMorgan.

KEY POINT

The four-color problem is another application of graph theory.

THE FOUR-COLOR PROBLEM Using at most four colors, is it always possible to color a map so that any two regions sharing a common border receive different colors?

FIGURE 4.16 Map of South America.

Unable to find an answer, DeMorgan communicated the problem to his friend Sir William R. Hamilton at Trinity College, Dublin, Ireland. For more than a hundred years, this problem remained unsolved. However, in 1976, Professors Kenneth Appel and Wolfgang Haken of the University of Illinois announced that they had solved this problem by proving that it is possible to color any map using no more than four colors. However, their proof was treated with some skepticism, because the proof was not done in the traditional method—by hand, where each step could be checked for validity. Instead Appel and Haken programmed a computer to do the proof; the completion of the proof required 1,200 hours of computer time!

For an illustration of the statement of this problem, consider the map of South America in Figure 4.16. Using at most four colors, we want to color this map so that we use a different color for any two countries having a common border. For example, we cannot use the same color for Colombia and Peru; however, we can use the same color for Paraguay and Uruguay.

EXAMPLE 6 Solving the Four-Color Problem for South America

Model the map of South America by a graph and use this graph to color the map using at most four colors.

SOLUTION: Review the Draw Pictures Strategy on page 3.

In this problem, we have a set of countries, some of which are related in that they share a common border. Therefore, we can model this situation by a graph.

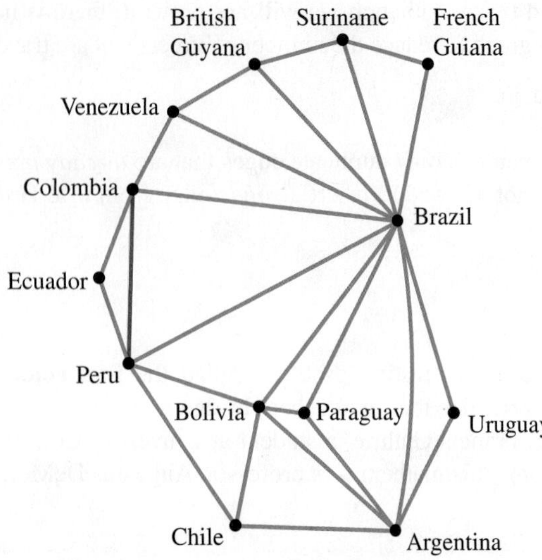

FIGURE 4.17 Graph model of map of South America.

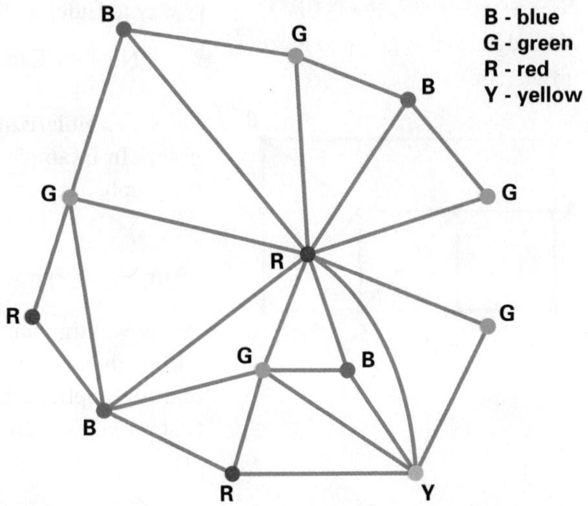

FIGURE 4.18 Coloring of graph of South America.

We will represent each country by a vertex; if two countries share a common border, we draw an edge between the corresponding vertices. This graph appears in Figure 4.17.

Note that we connect the vertices representing Peru and Colombia with an edge because they share a common boundary. We do not connect the vertices representing Argentina and Peru, because they have no boundary in common.

We can rephrase the map-coloring question now as follows: Using four or fewer colors, can we color the vertices of a graph so that no two vertices of the same edge receive the same color? It is easier to think about coloring a graph than it is to think about coloring the original map.

There are several ways to color the graph in Figure 4.17. However, unlike tracing graphs, there is no particular procedure for accomplishing this coloring except trial and error. One possible coloring for the graph appears in Figure 4.18. You may want to color this graph in a different way; however, notice that you cannot do it using fewer than four colors.

Now try Exercises 37 to 44. **5**

Example 7 shows how we can use map coloring for a practical purpose.

QUIZ YOURSELF 5

Color the following "map" by first modeling it with a graph and then coloring the graph, as we did in Example 6.

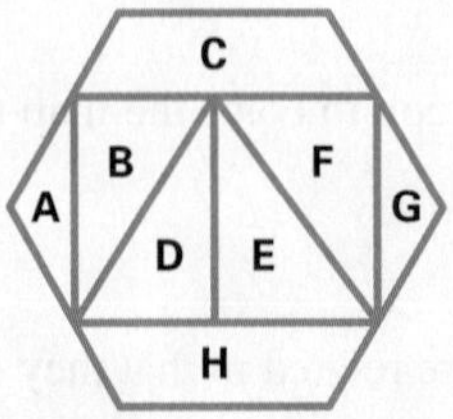

EXAMPLE 7 Using Graph Theory to Schedule Committees

Each member of a city council usually serves on several committees to oversee the operation of various aspects of city government. Assume that council members serve on the following committees: police, parks, sanitation, finance, development, streets, fire department, and public relations.

Use Table 4.1, which lists committees having common members, to determine a conflict-free schedule for the meetings. We do not duplicate information in Table 4.1. That is, because police conflicts with fire department, we do not also list that fire department conflicts with police.

Committee	Has Members in Common With
Police	Public relations, fire department
Parks	Streets, development
Sanitation	Fire department, parks
Finance	Police, public relations
Development	Streets
Streets	Fire department, public relations
Fire department	Finance

TABLE 4.1 Committees that have common members.

SOLUTION: Review the Draw Pictures Strategy on page 3.

We first will model the information in this table with the graph in Figure 4.19. We join two committees by an edge provided they have a conflict. This problem is similar to the map-coloring problem. If we color this graph, then all vertices having the same color represent committees that can meet at the same time. We show one possible coloring of the graph in Figure 4.19.

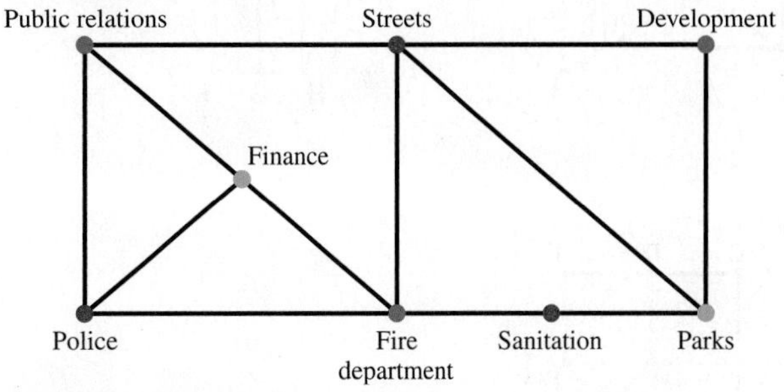

FIGURE 4.19 Graph model of committees with common members.

From Figure 4.19, we see that the police, streets, and sanitation committees have no common members and therefore can meet at the same time. Public relations, development, and the fire department can meet at a second time. Finance and parks can meet at a third time.

Now try Exercises 51 to 54.

EXERCISES 4.1

Sharpening Your Skills

In Exercises 1–8, determine whether the graph is connected. Which vertices are odd? Which vertices are even?

1.

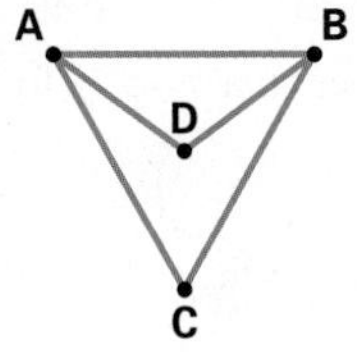

2.

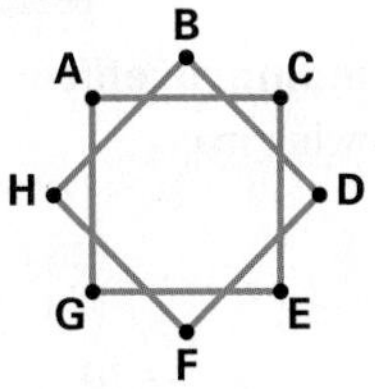

3.

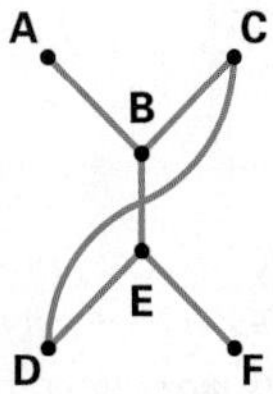

4.

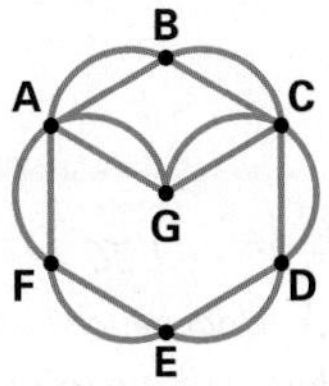

5.

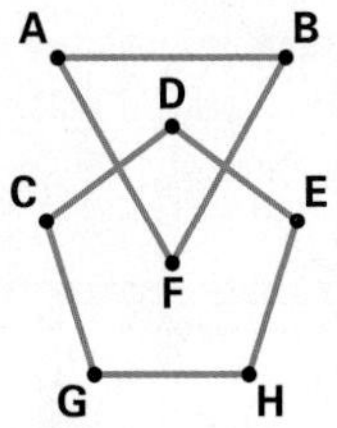

6.

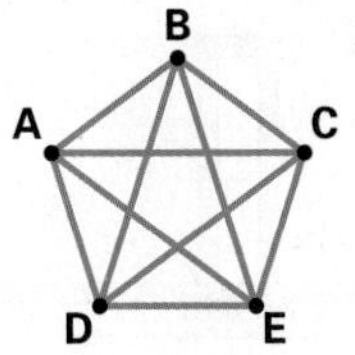

7.

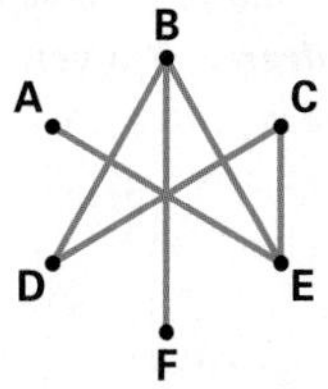

8.

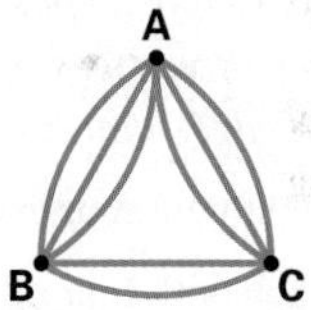

In Exercises 9–16, use Euler's theorem to decide whether the specified graph can be traced. If the graph cannot be traced, tell which condition of the theorem fails.

9. The graph in Exercise 1

10. The graph in Exercise 2

11. The graph in Exercise 3

12. The graph in Exercise 4

13. The graph in Exercise 5

14. The graph in Exercise 6

15. The graph in Exercise 7

16. The graph in Exercise 8

In Exercises 17–20, if the given graph is Eulerian, find an Euler circuit in it. If the graph is not Eulerian, first Eulerize it and then find an Euler circuit. Write your answer as a sequence of vertices, as we did in Example 4. There are many possible correct answers to these exercises. We will provide only one in the answer key.

17.

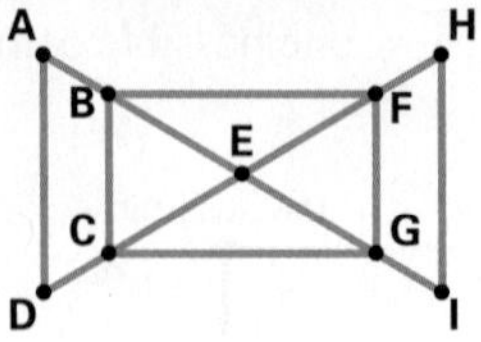

18.

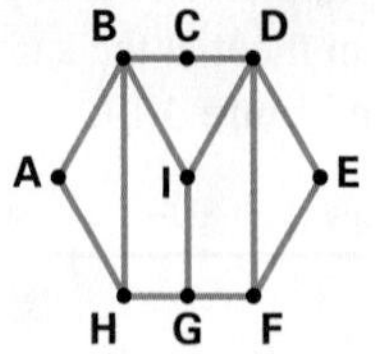

19.

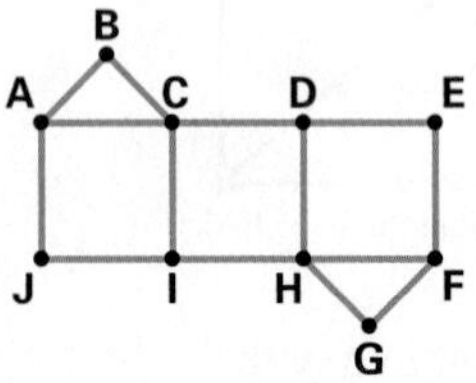

20.

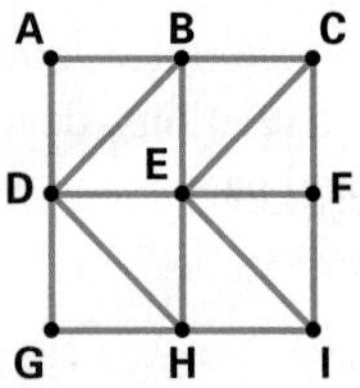

In Exercises 21–28, try to give an example of each graph that we describe. If, after several tries, you cannot find the graph that we have requested, explain why you think that it may be impossible to find that example. (In the answer key, we will give examples that are as simple as possible in the sense that they have the fewest number of vertices and edges.) The degree of a vertex is the number of edges that are joined to that vertex.

21. A graph with four even vertices

22. A graph with four odd vertices

23. A graph with three odd vertices

24. A graph with four vertices of degree two and two vertices of degree three

25. A connected graph with one even vertex and four odd vertices

26. A graph with one odd vertex

27. A graph with six vertices of degree three

28. A graph with five vertices and the largest number of edges so that no edges are repeated. That is, there will be only one edge joining A and B, only one edge joining A and C, and so on.

In Exercises 29–32, remove one edge to make the graph Eulerian.

29.

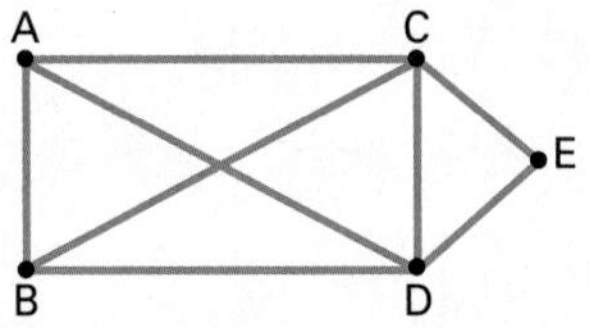

30.

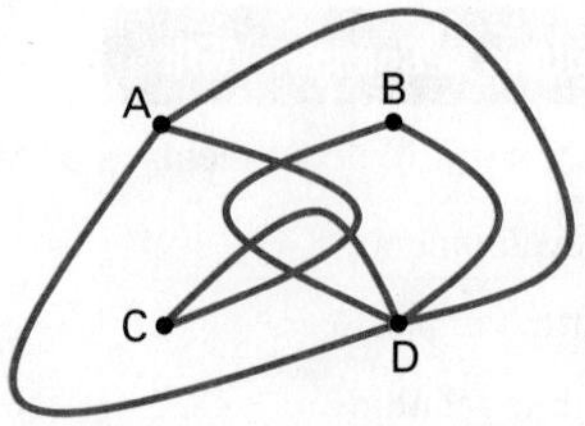

31.

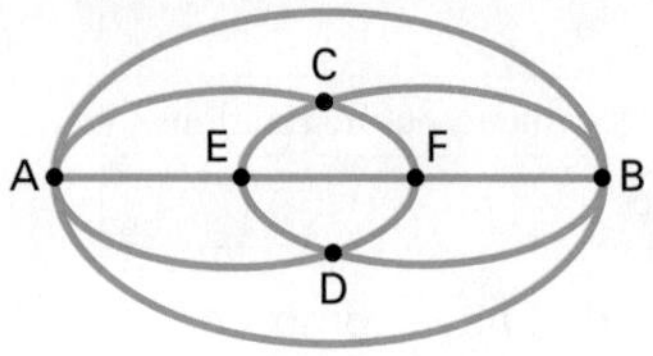

32.

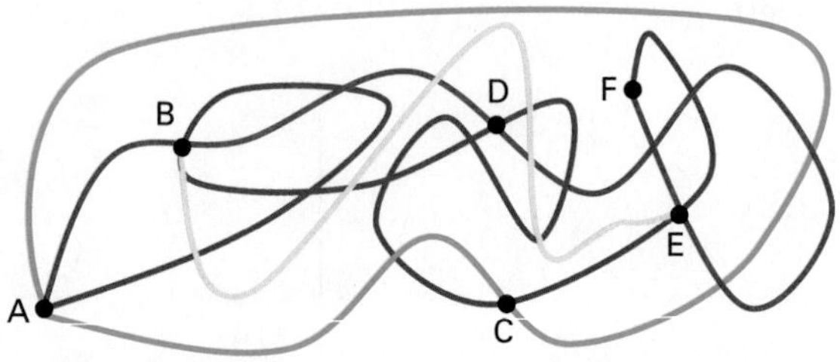

Applying What You've Learned

33. Finding an efficient route. A taxi driver wants to travel over each of the streets indicated in the following map, but does not want to travel over any part of the route more than once. Can this be done? Explain your answer.

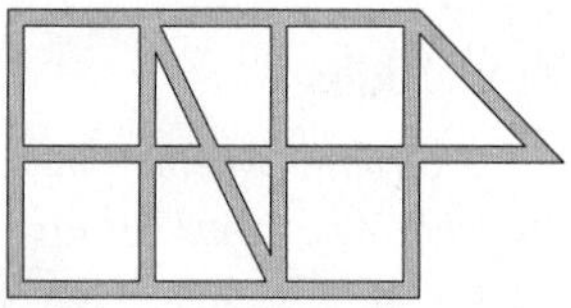

34. Finding an efficient route. Repeat Exercise 33 for the following map.

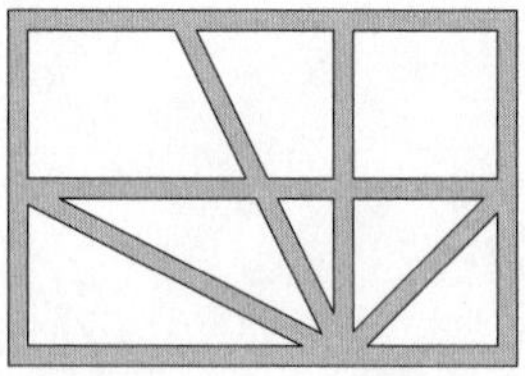

Exercises 35 and 36 are similar to the DUCK tour problem in Example 5. Model each map by a graph and then Eulerize it to design a route that uses a minimal number of streets more than once.

35.

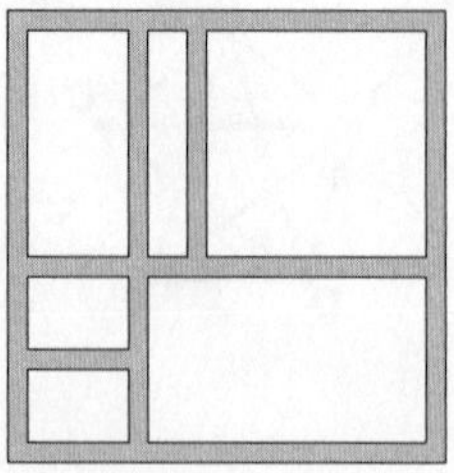

36.

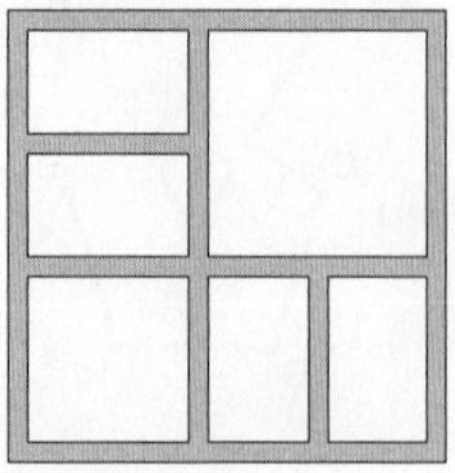

Represent the maps given in Exercises 37–40 by graphs as we did in Example 6. Recall that we join two vertices by an edge if and only if the states that they represent share a stretch of common border.

37.

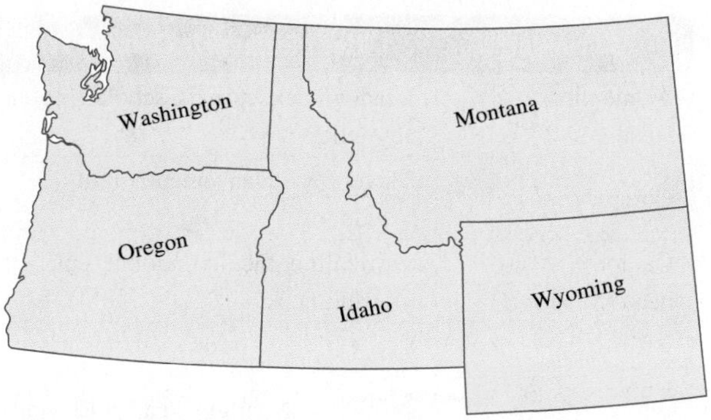

38.

39.

40.

In Exercises 41–44, find the smallest number of colors that can be used to color the specified map so that any two states sharing a stretch of common border are not colored with the same color.

41. The map in Exercise 37

42. The map in Exercise 38

43. The map in Exercise 39

44. The map in Exercise 40

In Exercises 45–48, we give you a group of states. Is it possible to begin in one of the states in the group and travel through all the states without ever crossing the same boundary between two states twice? (Hint: Think of the graphs that you drew in Exercises 37–40.)

45. Use the states that we listed in Exercise 37.

46. Use the states that we listed in Exercise 38.

47. Use the states that we listed in Exercise 39.

48. Use the states that we listed in Exercise 40.

49. Finding an efficient route. Because of Michael's escape from Fox River State Penitentiary, security procedures are being reexamined. In the following floor plan of a section of the prison, if all the doors are open, is it possible for a guard to enter this section from the hallway, pass through each door locking it behind him, and then exit without ever having to open a door that has been previously locked? (*Hint:* Model this with a graph where you consider the set of objects to be the rooms and the hallway and that any two rooms are related if they are connected by an open door.)

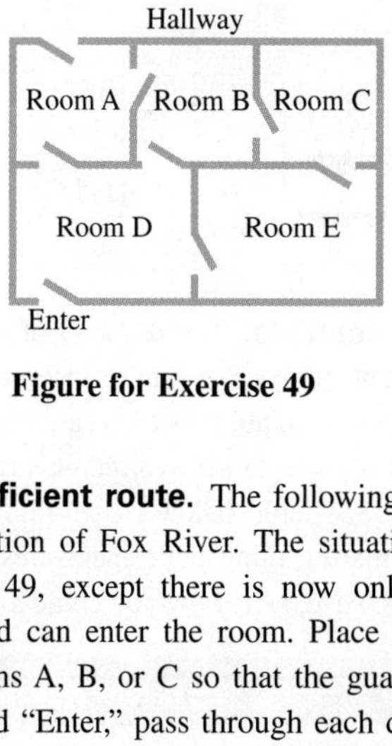

Figure for Exercise 49

50. Finding an efficient route. The following is a floor plan for another section of Fox River. The situation is the same as in Exercise 49, except there is now only one door by which the guard can enter the room. Place an exit door in one of the rooms A, B, or C so that the guard can enter by the door marked "Enter," pass through each door and lock it behind him, and then exit by the door you've placed in the plan. Explain why this is the only possible place to locate the door.

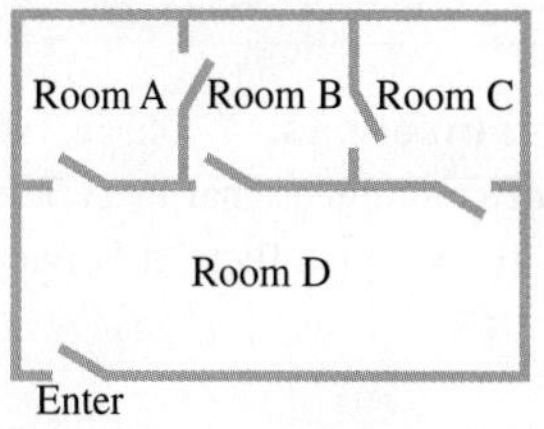

Use the technique presented in Example 7 to answer Exercises 51–54. We do not list duplicates in the tables of information.

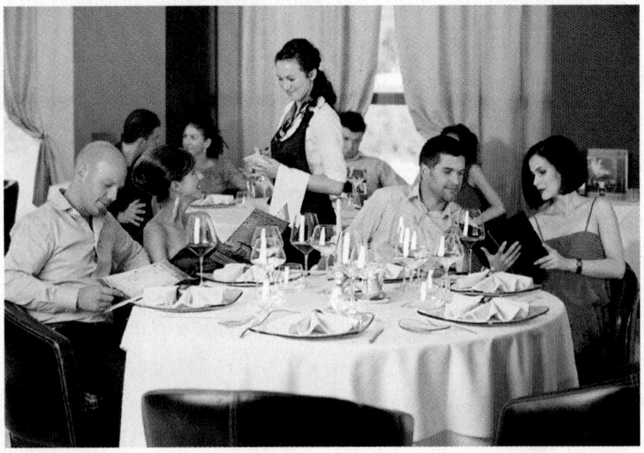

51. **Avoiding Conflicts.** The Griffins are looking forward to a "Family Guy" wedding, but there is concern about the impending rehearsal dinner because certain people invited to the dinner just don't get along with each other. Therefore, it is important that people who are not friendly be seated at different tables. Use the information in the table to determine a satisfactory seating arrangement for the dinner using as few tables as possible.

Dinner Guest	Is Not Friendly With
Peter	Carter
Carter	Glenn, Chris
Glenn	Tom
Lois	Steve
Barbara	Tom, Dianne
Tom	Dianne, Meg
Dianne	Cleveland
Meg	Cleveland

52. **Avoiding conflicts.** The designer of the new Jungle World entertainment complex wants to include several large enclosures in which wild animals can roam freely. If one animal can harm another, then those two animals cannot share the same enclosure. (See table below.) Use this information to determine the smallest number of enclosures necessary to contain the animals. Also, state how you could assign the animals.

Animal	Cannot Be Placed With
Tiger	Zebra, leopard, rhinoceros, giraffe, antelope, ostrich
Leopard	Zebra, boar, antelope, giraffe, ostrich
Crocodile	Ostrich, heron
Boar	Tiger, crocodile, zebra

53. **Scheduling meetings.** A college's student government has a number of committees that meet Tuesdays between 11:00 and 12:00. To avoid conflicts, it is important not to schedule two committee meetings at the same time if the two committees have students in common. Use the following table, which lists possible conflicts, to determine an acceptable schedule for the meetings.

Committee	Has Members in Common With
Academic standards	Academic exceptions, scholarship, faculty union
Computer use	University advancement, event scheduling
Campus beautification	Curriculum, faculty union, event scheduling
Affirmative action	Academic exceptions, scholarship
University advancement	Parking, curriculum, academic standards
Parking	Academic standards, affirmative action
Faculty union	Computer use, event scheduling
Scholarship	Campus beautification

54. **Scheduling meetings.** Repeat Exercise 53 using the following table.

Committee	Has Members in Common With
Academic standards	Scholarship, university advancement
Computer use	University advancement, affirmative action
Campus beautification	Curriculum, academic festival, faculty union, event scheduling, scholarship
Affirmative action	Scholarship
University advancement	Curriculum, academic festival, faculty union
Faculty union	Computer use, event scheduling, academic festival
Long-range planning	Campus beautification, computer use

Communicating Mathematics

55. If in tracing a graph, we neither begin nor end at vertex A, why must A be even? How does that imply that a graph that can be traced cannot have more than two odd vertices?

56. Examine a number of the graphs that we have drawn in this section. For each graph find the sum of the degrees of all the vertices. Do you notice any pattern? Explain what the pattern is and why it must be so.

57. Can an Eulerian graph have a bridge? In order to answer this question, you could examine several of the Eulerian graphs that we have drawn in this section. If you think you have an answer, explain why it must be so.

58. Consider any graph that has some odd and also some even vertices. Add one edge to this graph. Does the number of odd and even vertices change? Why or why not?

Between the Numbers

59. Discuss specific examples of ways that connectivity, as we discussed in Between the Numbers on page 148, has influenced things such as the way we conduct business, national and international affairs, and society in general.

60. Give specific examples of the way connectivity has been used in a positive or negative way in your life.

Challenge Yourself

61. Draw a graph that can be colored with only two colors.

62. Draw a graph that cannot be colored with two colors but can be colored with three. Can you state what configuration of vertices will force you to use at least three colors in coloring a graph?

63. Draw a graph that cannot be colored with three colors but can be colored with four.

64. Can you state what configuration of vertices will force you to use at least four colors in coloring a graph?

65. Different notes on a trumpet are obtained by moving its three valves up and down. The given table indicates the eight possible positions for these three valves and a note that would sound if the valves were in the indicated position. Is it possible to play one of these notes and then, by changing only one valve at a time, to play all the other notes without repeating a note twice? (*Hint:* Consider the notes as the objects of a set and that two notes are related if one can be obtained from the other by moving only one valve.)

Valve 1	Valve 2	Valve 3	Note
Up	Up	Up	C
Up	Up	Down	A
Up	Down	Up	B
Up	Down	Down	E♭
Down	Up	Up	F
Down	Up	Down	D
Down	Down	Up	E
Down	Down	Down	C#

66. If an instrument has four valves, there are sixteen possible positions for the valves. Repeat Exercise 65 for this situation.

67. Assume that the registrar at your school is building a final exam schedule and because the following courses are known to be particularly difficult, the registrar does not want any student to take finals for two of them on the same day. The courses are Anth215, Bio325, Chem264, Mat311, Phy212, Fin323, and ComD265. The asterisks in the given table indicate which courses have students in common and therefore should have their finals on different days. Assume that two courses are related if they have any students in common. Model this set of courses with a graph and use it to design a final exam schedule.

68. Make up a scheduling problem that would be of some interest to you similar to what we did in Exercise 67. Use graph theory to develop a solution.

	Anth215	Bio325	Chem264	Mat311	Phy212	Fin323	ComD265
Anth215		*	*	*		*	*
Bio325	*		*			*	
Chem264	*	*		*	*		
Mat311	*		*		*	*	*
Phy212			*	*			
Fin323	*	*		*			
ComD265	*			*			

Table for Exercise 67

4.2 The Traveling Salesperson Problem

Objectives

1. Understand how to solve the traveling salesperson problem using Hamilton circuits.
2. Determine all Hamilton circuits in a complete graph.
3. Solve the traveling salesperson problem using the brute force algorithm.
4. Solve the traveling salesperson problem using the nearest neighbor algorithm.
5. Solve the traveling salesperson problem using the best edge algorithm.

Some problems in mathematics are so simple to state that a child can understand them, yet the greatest mathematicians in the world cannot solve them. This is the case with a famous and difficult problem in graph theory called **the traveling salesperson problem (TSP).**

The TSP gets its name from the problem of determining the most efficient way for a salesperson to schedule a trip to a series of cities and then return home.

For a good example of a TSP, suppose that you are going to graduate soon from your school in Philadelphia and have been invited for job interviews in New York City, Cleveland, Atlanta, and Memphis (see Figure 4.20). In order to save time and money, you plan to visit all of the cities in one trip and then return to Philadelphia.

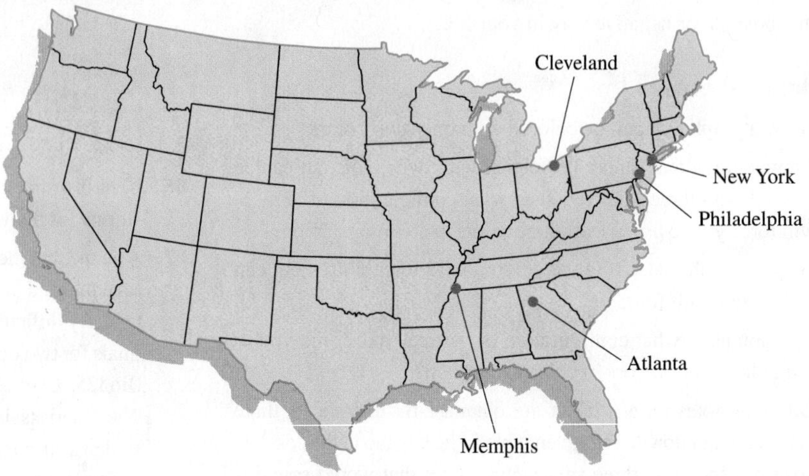

FIGURE 4.20 Cities you must visit.

You have obtained the prices for the flights between each pair of cities,* and want to determine the cheapest way to schedule your trip. Beginning in Philadelphia, you could fly first to New York, then on to Cleveland, Atlanta, and finally to Memphis before returning to Philadelphia. Or, you could first fly to Cleveland, then Atlanta, Memphis, New York, and then home. In order to find the cheapest trip, you would have to consider all possible ways to visit the four cities and return home.

In keeping with the spirit of this chapter, we model this situation with a graph. We represent each city by a vertex and join two vertices by an edge if you might fly from one to the other. In this graph, every pair of vertices will be joined by an edge. For clarity, we will represent each city by a letter—P for Philadelphia, N for New York City, and so on. This gives us the graph in Figure 4.21.

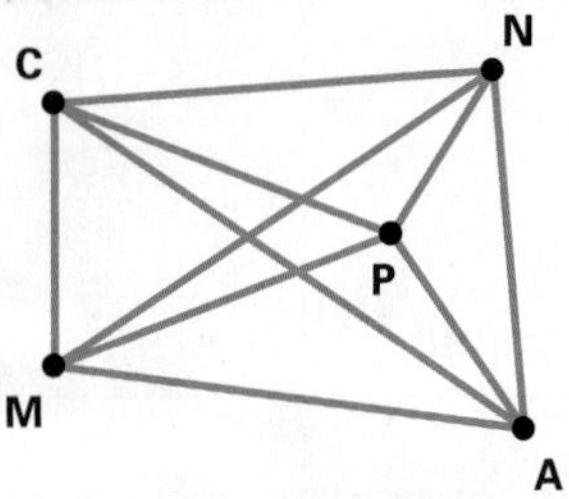

FIGURE 4.21 Graph representing cities you will visit.

Hamilton Paths

We now model your possible trips by paths in this graph. For instance, we can represent the trip (Philadelphia, New York, Cleveland, Atlanta, Memphis, home) by the path PNCAMP, and the trip (Philadelphia, Cleveland, Atlanta, Memphis, New York, home) by the path PCAMNP. If we knew the prices of all the possible flights, we could then solve your problem. But first we need to develop some more graph theory.

DEFINITIONS A path that passes through all the vertices of a graph exactly once is called a **Hamilton path.** If a Hamilton path begins and ends at the same vertex, then it is called a **Hamilton circuit.** If a graph has a Hamilton circuit we will say it is **Hamiltonian.**

PROBLEM SOLVING

Strategy: The Splitting-Hairs Principle

Recall the Splitting-Hairs Principle in Section 1.1. Although the definitions of *Hamilton path* and *Euler path* sound similar, they are not the same. In producing a Hamilton path, you do not have to trace every edge, as you do with an Euler path.

*We will assume that direction is not important here and that a one-way flight between two cities costs the same in either direction.

EXAMPLE 1 Hamilton Paths and Circuits

Find a Hamilton path in each graph shown in Figure 4.22.

SOLUTION:

a) In Figure 4.22(a), the path ABCFDGE (in red) is a Hamilton path. If we return to vertex A, then ABCFDGEA is a Hamilton circuit.

b) The graph in Figure 4.22(b) has no Hamilton paths or circuits. If you start at any vertex, you will see that you cannot follow edges to visit every vertex *exactly once*. For example, if you start at vertex A and then go to vertices B and C, you are in trouble. If you go to D, then you cannot get to E without passing through C a second time. Similarly, if you go to E, you cannot get to D. You may start your path at other vertices, but you will find that no matter how you try you cannot find a Hamilton path.

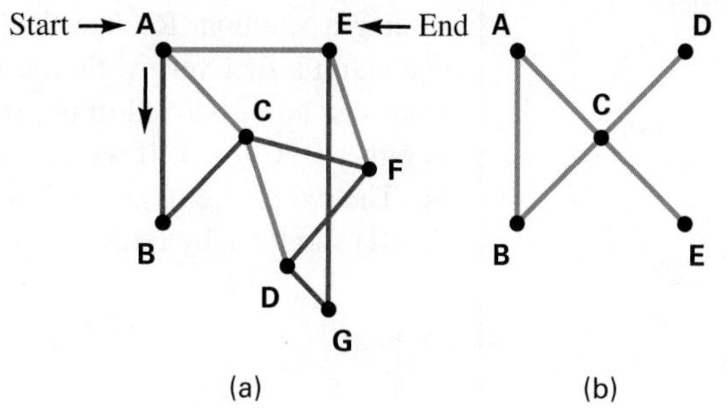

FIGURE 4.22 (a) ABCFDGE is a Hamilton path. (b) No Hamilton path exists.

Now try Exercises 1 to 4.

Unlike the situation with Euler's theorem for graph tracing, we will not give a rule for determining when a graph has a Hamilton path.

Finding Hamilton Circuits

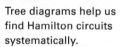

KEY POINT

Tree diagrams help us find Hamilton circuits systematically.

If you think about it, the reason we could not find a Hamilton path in the graph in Example 1(b) was that there were not enough edges. When we went to vertex D, we were stuck. If we had more edges to use, we might have avoided going back through C a second time to get to vertex E. Often we will be dealing with graphs that have every possible edge.

DEFINITION A **complete graph** is one in which every pair of vertices is joined by an edge. A complete graph with n vertices is denoted by K_n.*

EXAMPLE 2 Complete Graphs

Each graph shown in Figure 4.23 is complete.

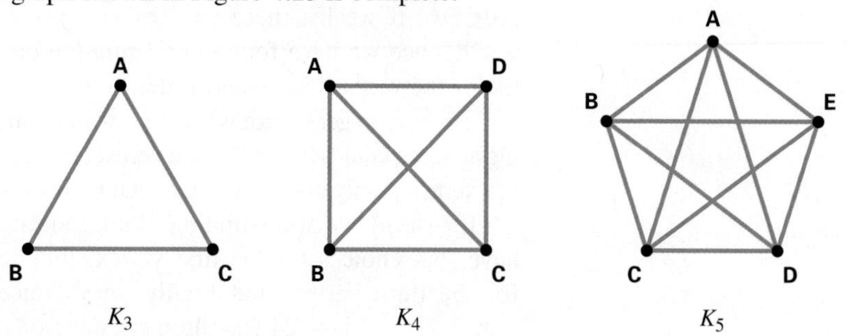

FIGURE 4.23 Complete graphs.

*The notation K for complete graphs is chosen to honor the twentieth-century Polish mathematician Kasimir Kuratowski, who discovered several important theorems in graph theory.

Often we need to find all the Hamilton circuits in a graph. It is easy to do this in complete graphs.

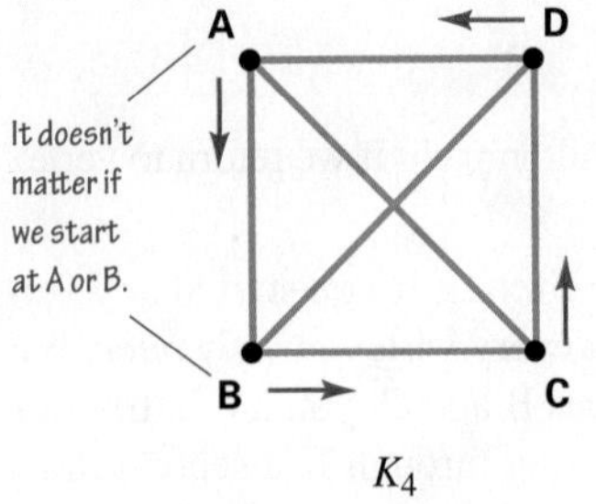

FIGURE 4.24 Circuit ABCDA is the same as BCDAB.

EXAMPLE 3 Finding Hamilton Circuits in K_4.

Find all Hamilton circuits in K_4.

SOLUTION: Review the Be Systematic Strategy on page 5.

Path ABCDA shown in Figure 4.24 is one of the Hamilton circuits in K_4. We will consider path BCDAB to be the same path because it passes through the same vertices in the same order and the only difference is that we are beginning and ending at vertex B rather than vertex A. In light of this remark, we will assume that all Hamilton circuits in this example begin at vertex A.

A good drawing can put us on the right track in understanding this problem and finding a solution. Realize that in constructing a Hamilton circuit in K_4, you have to decide on a first vertex, then a second, then a third, and so on. This sequence of decisions can be visualized in the **tree diagram** in Figure 4.25. Begin at vertex A, then go to either B, C, or D. If we decide to go to B, we can then choose *either* C *or* D, and so on. The red path, highlighted in Figure 4.25, shows how to construct the Hamilton path ACBD and then, by returning to A, get the Hamilton circuit ACBDA.

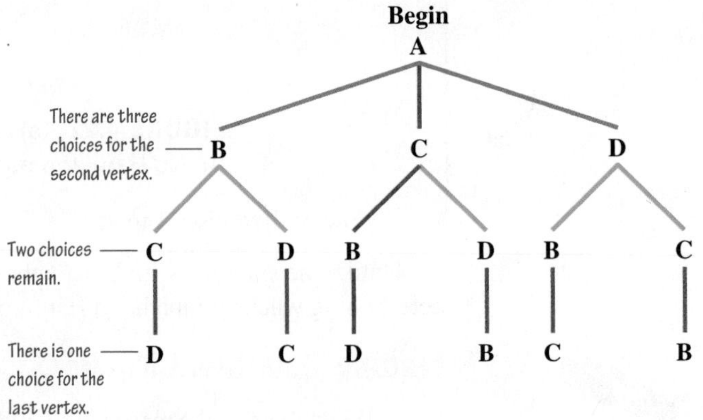

FIGURE 4.25 Finding Hamilton circuits in K_4.

Tracing through the six branches of the tree, we see that K_4 has six Hamilton circuits:

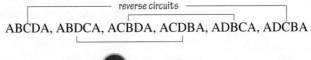

reverse circuits

ABCDA, ABDCA, ACBDA, ACDBA, ADBCA, ADCBA

Now try Exercises 5 and 6. 6

QUIZ YOURSELF 6

a) Write the circuit ACBDA, but begin with vertex B.
b) What is the reversal of ACBDA?

Notice in Example 3 that the Hamilton circuits occur in pairs. Consider the circuit ABCDA. If we list these vertices in reverse order, we get the circuit ADCBA. In any graph, once we have found one Hamilton circuit, we automatically have another one by listing the vertices in reverse order.

It is easy to generalize what we saw in Example 3. In finding Hamilton circuits, after starting at A, we could choose the first vertex (after A) in three ways, the second in two ways, and the last vertex in only one way, so the total number of Hamilton circuits in K_4 is $3 \times 2 \times 1 = 6$.

If instead we were finding Hamilton circuits in K_5, again starting at A, we would have four choices for the first vertex, three choices for the second vertex, two choices for the third vertex, and finally, one choice for the last vertex. So, there would be $4 \times 3 \times 2 \times 1 = 24$ Hamilton circuits in K_5.

If we were finding Hamilton circuits in K_n, after starting at A, we would have $n - 1$ choices for the first vertex, $n - 2$ for the second, and so on. Continuing in this fashion, we would get a tree with $(n - 1) \times (n - 2) \times (n - 3) \times (n - 4) \cdots 3 \times 2 \times 1$ branches.

THE NUMBER OF HAMILTON CIRCUITS IN K_n K_n has $(n - 1) \times (n - 2) \times (n - 3) \times (n - 4) \cdots 3 \times 2 \times 1$ Hamilton circuits. This number is written $(n - 1)!$ and is called $(n - 1)$ *factorial.*

Table 4.2 shows the number of Hamilton circuits in K_n for various values of n.

n	Number of Hamilton Circuits in K_n
3	$2! = 2 \cdot 1 = 2$
4	$3! = 3 \cdot 2 \cdot 1 = 6$
5	$4! = 4 \cdot 3 \cdot 2 \cdot 1 = 24$
10	$9! = 362{,}880$
15	$14! = 87{,}178{,}291{,}200$
20	$19! = 121{,}645{,}100{,}408{,}832{,}000$

TABLE 4.2 Number of Hamilton circuits in K_n.

QUIZ YOURSELF 7

How many Hamilton circuits are there in K_9?

As you can see, as n increases, the number of Hamilton circuits in K_n grows at an incredible rate. 7

EXAMPLE 4 Scheduling Auditions

Sharon and Howie are going to visit St. Louis; Washington, D.C.; Tampa Bay; Los Angeles; Austin; and Charlotte to attend auditions for *America's Got Talent* for the upcoming season. Assume that they are going to start and end their trip in New York City and visit every other city only once. In how many ways can they make their trip?

SOLUTION: Review the Look for Patterns Strategy on page 6.

Realize that what we are asking for is the number of Hamilton circuits that begin at New York and pass through the other cities. Including New York, there are seven cities, so there are $(7 - 1)! = 6! = 6 \times 5 \times 4 \times 3 \times 2 \times 1 = 720$ possible trips.

Now try Exercises 7 and 8.

Historical Highlight

William Rowan Hamilton

Hamilton circuits are named after perhaps the greatest Irish mathematician of all time, William Rowan Hamilton, who was born in Dublin, Ireland, in 1805. In his early teens, he took part in an arithmetic contest with an American youth named Zerah Colburn, who was known as a "lightning calculator." When Hamilton came in second best, he dedicated himself fiercely to studying mathematics.

Hamilton was very skilled at learning languages and could read the works of Euclid in the original Greek, Isaac Newton in Latin, and the eminent mathematician Pierre Simon Laplace in French. At age 17, he sent an error he found in Laplace's work to the Royal Irish Academy, whereby the president of the academy declared him to be a first-rate mathematician.

Hamilton attended Trinity College in Dublin, where he distinguished himself in both mathematics and poetry—saying that although he was a mathematician by profession, he was a poet by heart. His work in optics was so impressive that he was appointed a professor of astronomy while still an undergraduate student!

Although his work seemed theoretical at the time, Hamilton made contributions to many areas of mathematics that today have numerous practical applications.

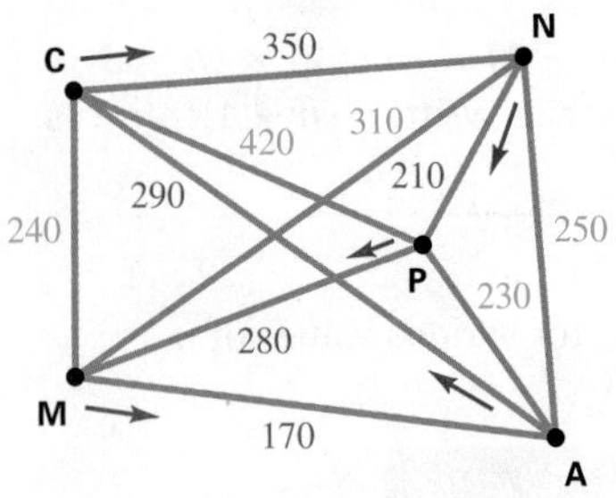

FIGURE 4.26 Graph model of your possible trips. Weights on edges denote the cost of travel between cities.

Solving the TSP by Brute Force

We will now use Hamilton circuits to solve the problem of finding the cheapest way to travel to your interviews that we posed at the beginning of this section. The cost (in dollars) of traveling between the cities is shown on each edge in Figure 4.26.

> **DEFINITIONS** When we assign numbers to the edges of a graph, the graph is called a **weighted graph** and the numbers on the edges are called **weights.** The **weight of a path** in a weighted graph is the sum of the weights of the edges of the path.

In your travel-cost problem, the weights represent money; in other problems, the weights might represent distance or time.

EXAMPLE 5 Solving a TSP Using Brute Force

Use Figure 4.26 to find the sequence of cities for you to visit that will minimize your total travel cost.

SOLUTION:

Another way to state this problem is that we want to find the Hamilton circuit that has the smallest weight. For example, if you choose the circuit PMACNP, the cost of your trip will be

280	+	170	+	290	+	350	+	210	=	$1,300.
Philadelphia to Memphis		Memphis to Atlanta		Atlanta to Cleveland		Cleveland to New York		New York to Philadelphia		Total

Our plan for a solution is simple but tedious. We will calculate the weight of each Hamilton circuit in the graph. The circuit with the smallest weight will be our solution. Because this graph has five vertices there will be $4! = 24$ Hamilton circuits. However, in Table 4.3 we have listed only 12 of these circuits because clearly each circuit has the same weight as its reverse.

From Table 4.3, we see that the smallest weight is $1,200, which corresponds to the circuit PAMCNP. Thus, you should start at Philadelphia and then travel to Atlanta, Memphis, Cleveland, and New York City in that order and then return home.

Now try Exercises 17 to 20.

Hamilton Circuit	Weight ($)
PACMNP	1,280
PACNMP	1,460
PAMCNP	1,200
PAMNCP	1,480
PANCMP	1,350
PANMCP	1,450
PCAMNP	1,400
PCANMP	1,550
PCMANP	1,290
PCNAMP	1,470
PMACNP	1,300
PMCANP	1,270

TABLE 4.3 Hamilton circuits and their weights.

In Example 5, our solution was not clever or sophisticated. We simply considered every possible Hamilton circuit and chose the best one. This approach to the TSP is called **brute force.** Realize how impractical this method would have been if you visited 15 cities. In that case, from Table 4.2 we see that we would have had to consider $14! = 87{,}178{,}291{,}200$ Hamilton circuits.

Although this problem is very large, you might think that a fast computer could solve it fairly quickly. Let's suppose that we used a computer that could examine 100 circuits per second. There are 31,536,000 seconds in a year, so the computer could examine $31{,}536{,}000 \times 100 = 3{,}153{,}600{,}000$ Hamilton circuits per year. Thus, it would still take our computer

$$\frac{14!}{3{,}153{,}600{,}000} = \frac{87{,}178{,}291{,}200}{3{,}153{,}600{,}000} \approx 28 \text{ years}$$

to complete the solution!

Recall that an algorithm is a series of steps that finds a solution for a problem. We will formalize what we did in Example 5.

QUIZ YOURSELF 8

Use the brute force algorithm to find a Hamilton circuit with minimum weight in the graph.

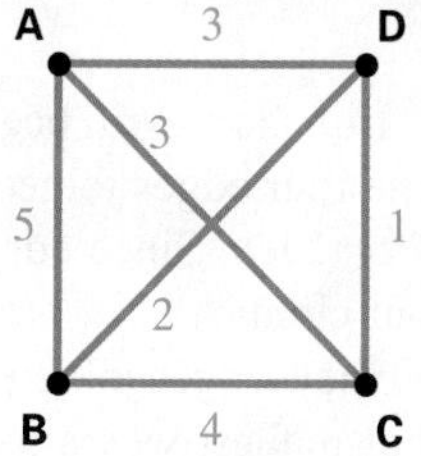

THE BRUTE FORCE ALGORITHM FOR SOLVING THE TSP

Step 1: List all Hamilton circuits in the graph.

Step 2: Find the weight of each circuit found in step 1.

Step 3: The circuits with the smallest weights tell us the solution to the TSP.

The Nearest Neighbor Algorithm

It would be nice if there were an algorithm that would solve a TSP much faster than the brute force algorithm. Unfortunately, at present there is no such algorithm and, in fact, mathematicians do not know whether it is even possible to find such an algorithm.

In a situation such as this, mathematicians ask a different question: "Is there an algorithm that, although it may not find the best solution, does find a pretty good approximation to the best solution?" In fact, a number of algorithms will do this. One rather intuitively simple approach is, when constructing a Hamilton circuit, always choose the next edge that has the smallest weight. This method is called the **nearest neighbor algorithm.**

KEY POINT

There are algorithms requiring less work than the brute force algorithm that give good approximations to solutions to the TSP.

THE NEAREST NEIGHBOR ALGORITHM FOR SOLVING THE TSP

Step 1: Start at any vertex X.

Step 2: Of all the edges connected to X, choose any one that has the smallest weight. (There may be several with smallest weight.) Select the vertex at the other end of this edge. This vertex is called the *nearest neighbor* of X.

Step 3: Choose subsequent *new* vertices as you did in step 2. When choosing the next vertex in the circuit, choose one whose edge with the current vertex has the smallest weight.

Step 4: After all vertices have been chosen, close the circuit by returning to the starting vertex.

Math in Your Life

Can Ants Do Mathematics?

Yes, believe it or not, ants can do mathematics! Of course, they do not use tiny pencils and paper or minicalculators, but in their own way, when you see ants scurrying around their nest, they are actually working on finding the shortest routes to food sources in a way that is similar to the TSP problem discussed in this section.

In more recent years, scientists have applied mathematics to study how ants, bees, and other social insects cooperate using "swarm intelligence" to solve problems. This research has paid off in applications such as finding improved telephone routes in congested lines and helping robots work cooperatively. Eric Bonabeau, a researcher in ant algorithms, foresees a world in which computer "chips are embedded in every object, from envelopes to trashcans to heads of lettuce" and believes that ant algorithms will be necessary to allow these chips to communicate.

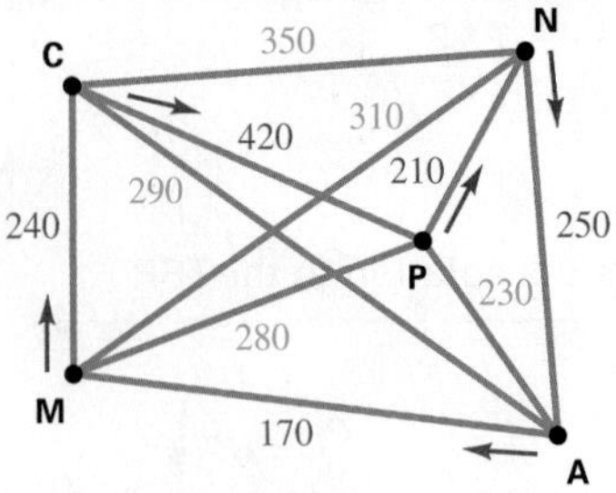

FIGURE 4.27 Graph model of your possible trips.

EXAMPLE 6 Solving a TSP Using the Nearest Neighbor Algorithm

Use the nearest neighbor algorithm to schedule your interview trip.

SOLUTION: Review the Be Systematic Strategy on page 5.

The graph in Figure 4.27 shows the possibilities for your trip. You will begin in Philadelphia, so we start the Hamilton circuit with vertex P. The four edges joined to P (with their weights) are PA (230), PC (420), PM (280), and PN (210). Since edge PN has the smallest weight, we choose N as the second vertex in our circuit.

Because we do not want to return to P, we consider the three new edges joined to N, namely, NA (250), NC (350), and NM (310). Edge NA has the smallest weight, so we add vertex A to our circuit.

We are now at vertex A and must choose either M or C for the fourth vertex. Edge AM (170) has a smaller weight than AC (290), so the fourth vertex in the circuit is M. This means that we complete the circuit by adding vertex C and then returning to P. The Hamilton path we have constructed is therefore PNAMCP, whose weight is

210	+	250	+	170	+	240	+	420	=	$1,290.
Philadelphia to New York		New York to Atlanta		Atlanta to Memphis		Memphis to Cleveland		Cleveland to Philadelphia		Total

Now try Exercises 21 to 24. 9

QUIZ YOURSELF 9

Redo Example 6, but this time assume that you are based in Memphis and must visit each of the other four cities. Again, use the nearest neighbor algorithm to find a Hamilton circuit and state its weight.

Notice from Examples 5 and 6 that although the nearest neighbor algorithm did not find the best solution to your problem, it did find a reasonably inexpensive way to schedule your trip. In Quiz Yourself 9, the nearest neighbor algorithm did find the best solution. Of course, we cannot predict when the nearest neighbor algorithm will find the best solution to the TSP.

The Best Edge Algorithm

We will investigate another way to find an approximate solution to the TSP. In looking at Figure 4.27, it seems a mistake not to try to use edges like AM (weight 170) and PN (weight 210). If we focus our attention on always choosing the "best edge," rather than trying to construct the circuit vertex by vertex, we might be able to find a pretty good approximation to the best solution to the TSP.

THE BEST EDGE ALGORITHM FOR SOLVING THE TSP

Step 1: Begin by choosing any edge with the smallest weight.

Step 2: Choose any remaining edge in the graph with the smallest weight.

Step 3: Keep repeating step 2; however, do not allow a circuit to form until all vertices have been used. Also, because the final Hamilton circuit cannot have three edges joined to the same vertex, never allow this to happen during the construction of the circuit.

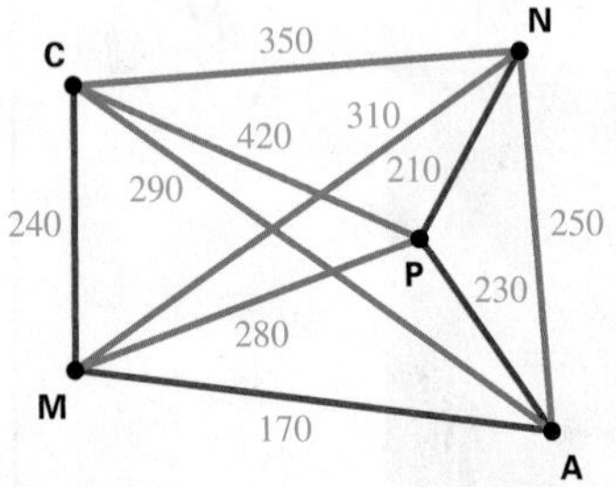

FIGURE 4.28 First four edges selected for your trip using the best edge algorithm.

EXAMPLE 7 Solving a TSP Using the Best Edge Algorithm

Use the best edge algorithm to schedule your interview trip.

SOLUTION:

We begin by selecting edge AM, which has the smallest weight, 170. Next choose edges NP (weight 210), AP (weight 230), and CM (weight 240). We have highlighted the edges we have already selected in Figure 4.28.

The edge with the next smallest weight is AN; however, selecting it forms a circuit, so we do not choose it. We do not choose MP, AC, or MN for the same reason. Selecting the last remaining edge, CN, forms a Hamilton circuit. Notice that it has a weight of 1,200, which also makes it the best solution to your problem.

Now try Exercises 25 to 28.

Although the best edge algorithm found the solution for your TSP in Example 7, there is no guarantee that this will happen for other TSPs.

If you are interested in learning more about this topic, the Web site http://www.tsp.gatech.edu at Georgia Tech is a fascinating site that describes the state of current research on the Traveling Salesperson (Salesman) Problem.

In addition to finding computer algorithms that humans can use to solve the TSP, other scientists such as Lars Chittka at Queen Mary, University of London, are trying to understand how bees, using a brain that is only about the size of a grass seed, can solve the TSP as they gather pollen from hundreds of flowers.

EXERCISES 4.2

Sharpening Your Skills

For many of these exercises, there may be several possible correct answers. We will provide only one in the answer key.

In Exercises 1–4, find a Hamilton circuit in each graph that begins with the specified edge.

1. **a.** AD **b.** EB

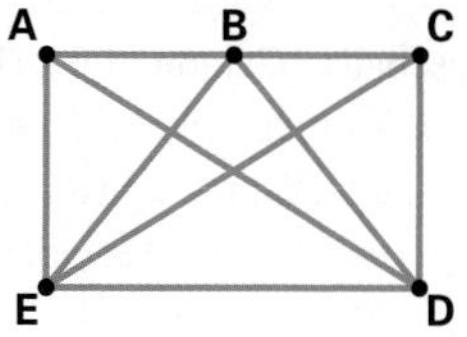

2. **a.** AC **b.** BD

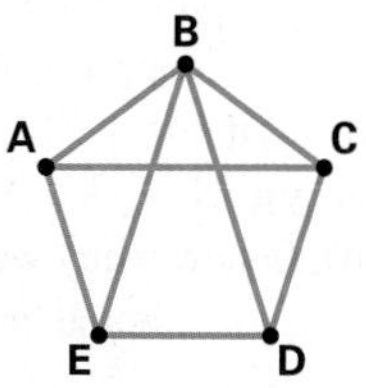

3. **a.** AD **b.** EA

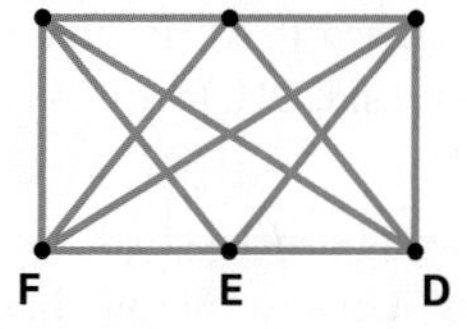

4. **a.** AE **b.** FD

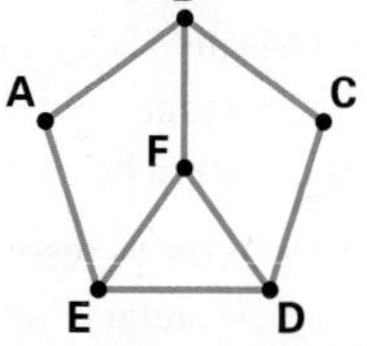

In Exercises 5–8, it helps organize your solution if you construct a tree diagram similar to the one shown in Figure 4.25.

5. Find all the Hamilton circuits that begin with edge AB in the graph shown in Exercise 1.

6. Find all the Hamilton circuits that begin with edge AF in the given graph.

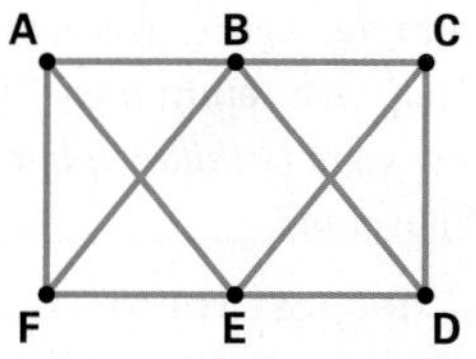

7. How many Hamilton circuits are in K_7?

8. How many Hamilton circuits are in K_8?

9. Draw K_6.

10. Draw K_7.

There are several theorems that will tell you when a graph is Hamiltonian.

***Dirac's theorem:** If a simple* graph has* n *vertices where* $n \geq 3$ *and each vertex has at least* $\frac{n}{2}$ *edges joined to it, then the graph is Hamiltonian.*

***Ore's theorem:** If a graph has* n *vertices where* $n \geq 2$ *and for any two nonadjacent vertices the sum of their degrees is at least* n*, then the graph is Hamiltonian.*

For the graphs in Exercises 11–14, determine: a) Are the conditions of Dirac's theorem satisfied? b) Are the conditions of Ore's theorem satisfied? c) Is the graph Hamiltonian?

11.

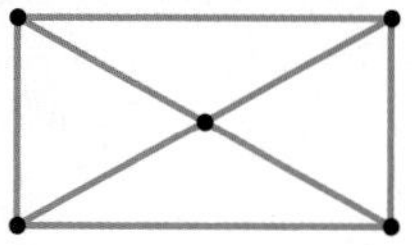

12.

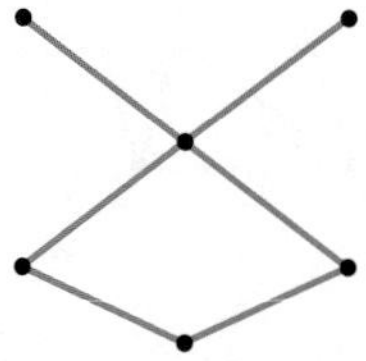

13.

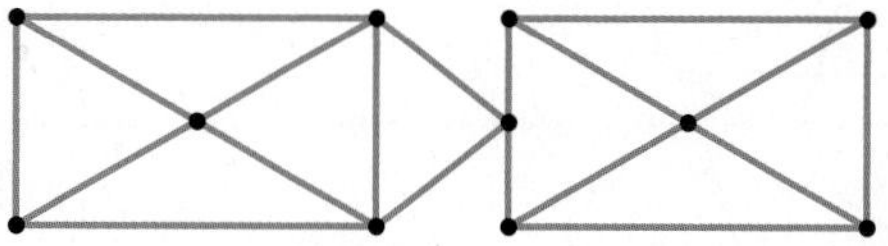

14.

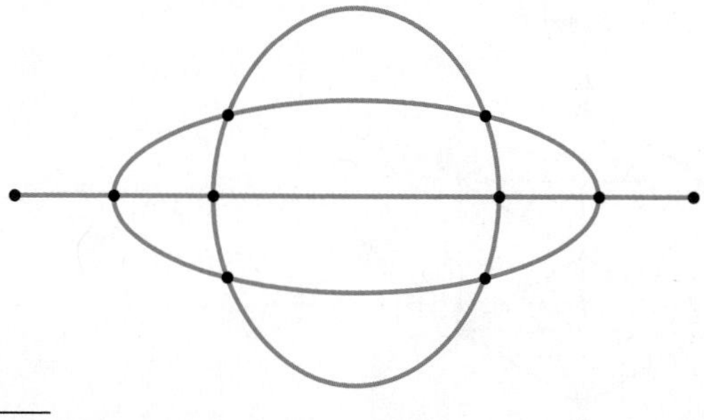

*Simple means that any two vertices in the graph have at most one edge between them and no vertex has an edge to itself.

15. Find the weight of the specified paths in the following graph.

a. AGEDCB **b.** ABCDGEF

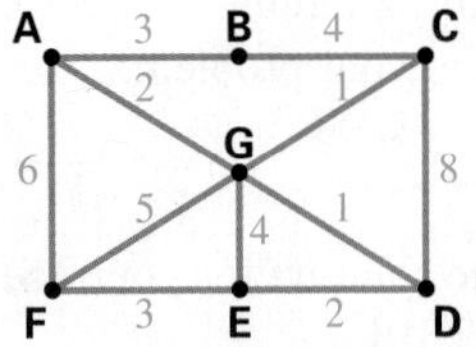

16. Find the weight of the specified paths in the following graph.

a. AGBHCI **b.** FGHIDJ

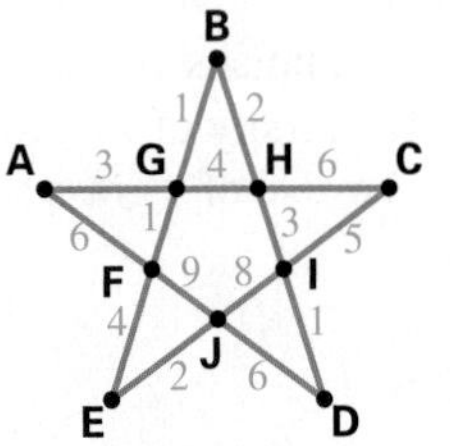

In Exercises 17–20, use the brute force algorithm to find a Hamilton circuit that has minimal weight. Recall that we found all Hamilton circuits in K_4 in Example 3.

17.

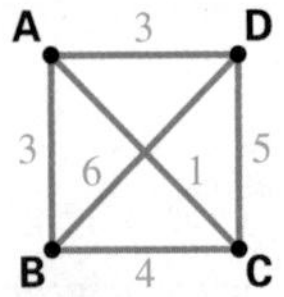

18.

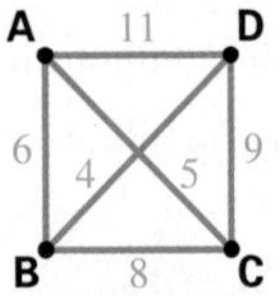

19.

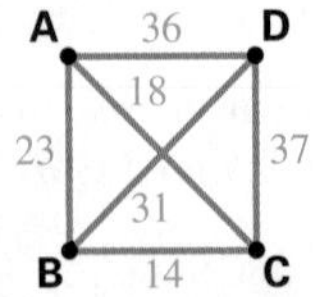

20.

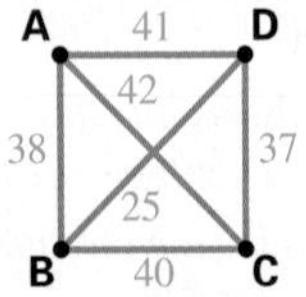

In Exercises 21–24, use the nearest neighbor algorithm to find a Hamilton circuit that begins at vertex A in each graph.

21.

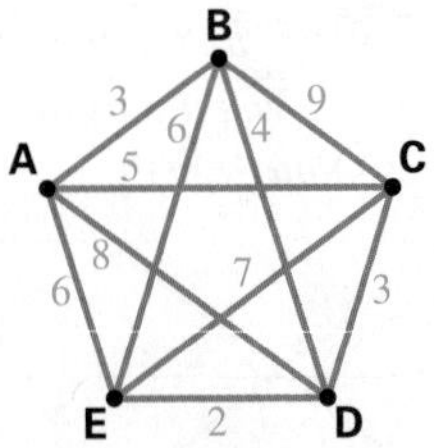

22.

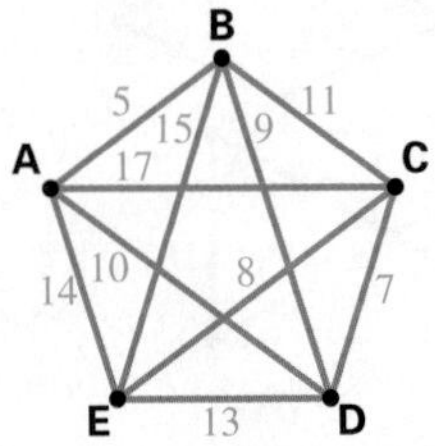

23.

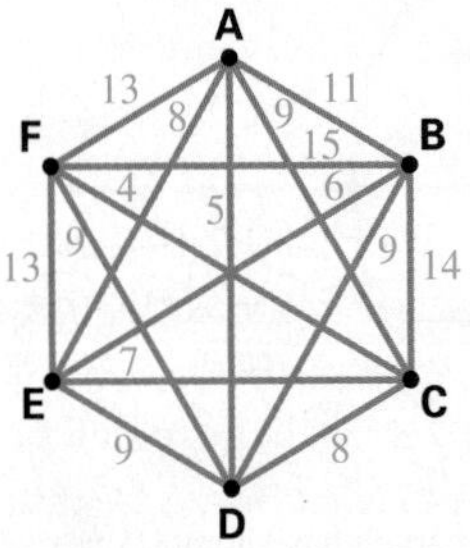

24.

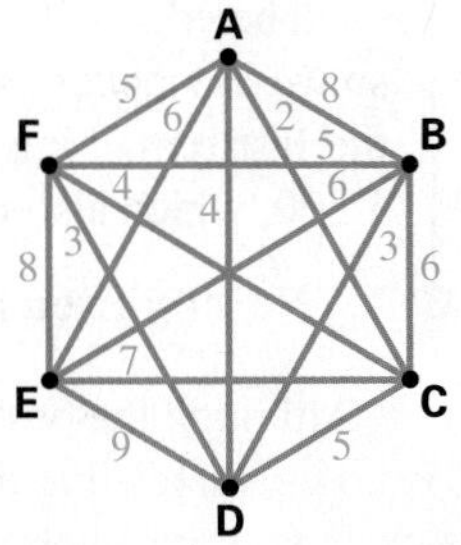

In Exercises 25–28, use the best edge algorithm to find a Hamilton circuit in each graph. List the circuit beginning at vertex A.

25. The graph of Exercise 21

26. The graph of Exercise 22

27. The graph of Exercise 23

28. The graph of Exercise 24

Applying What You've Learned

29. Simon Cowell plans a series of live shows from some of the music capitals of the country—New York, Chicago, New Orleans, Nashville, Memphis, and Los Angeles. His first show will be in New York and he will then travel to the other cities and return to New York. In how many ways can he make this tour?

30. ESPN is planning to do a special series called *Hall of Fame Month* in which they are going to do live broadcasts from different sports halls of fame. Their broadcast team will travel to Cooperstown, NY (baseball); Canton, OH (football); Springfield, MA (basketball); Toronto, Ontario, Canada (hockey); Newport, RI (tennis); and Saint Augustine, FL (golf). In how many ways can they travel to all these cities and return to their home in Bristol, CT?

31. Foodandwine.com states that the best ice cream cities in the United States are Columbus (OH), Scottsdale (AZ), Austin (TX), Cambridge (MA), Honolulu (HI), Minneapolis (MN), New Orleans (LA), and San Francisco (CA). In how many ways can Jamie, who lives in Cleveland, OH, tour these cities and then return home?

32. In order to try to resolve a European financial crisis, the U.S. treasury secretary plans to visit London, Berlin, Madrid, Rome, Paris, Bucharest, Budapest, Warsaw, Vienna, Prague, and Brussels. In how many ways can this trip be done assuming the secretary is leaving from and returning to Washington, D.C.?

In Exercises 33–34, we have added some cities to your interview trip. Instead of drawing a graph, we have listed the cost between pairs of cities in a table. Again, you will start at Philadelphia, visit all the cities, and then return home. Recall that the original cities in this problem were (P)hiladelphia, (N)ew York, (A)tlanta, (M)emphis, and (C)leveland.

33. Finding the cheapest route. Assume that (R)aleigh and (B)oston are to be added to the trip.

Table for Exercise 33

	P	N	A	M	C	R	B
P	0	210	230	280	420	240	430
N	210	0	250	310	350	320	180
A	230	250	0	170	290	90	510
M	280	310	170	0	240	120	500
C	420	350	290	240	0	270	380
R	240	320	90	120	270	0	290
B	430	180	510	500	380	290	0

a. If you were to draw a graph as we did in Example 5 and then use the brute force algorithm, how many Hamilton circuits would you have to consider?

b. If you could examine one Hamilton circuit per minute, how many hours would it take you to solve this TSP using the brute force algorithm?

c. Use the nearest neighbor algorithm to find a Hamilton circuit beginning at Philadelphia.

d. Use the best edge algorithm to find a Hamilton circuit beginning at Philadelphia.

34. Finding the cheapest route. Redo Exercise 33; however, in addition to Raleigh and Boston, assume that we have also added (D)allas to the trip. The cost of travel between Dallas and each of the other cities is given in the following table.

	P	N	A	M	C	R	B
D	510	530	410	370	450	260	560

Use the map to solve Exercises 35 and 36. Assume that all blocks are the same length.

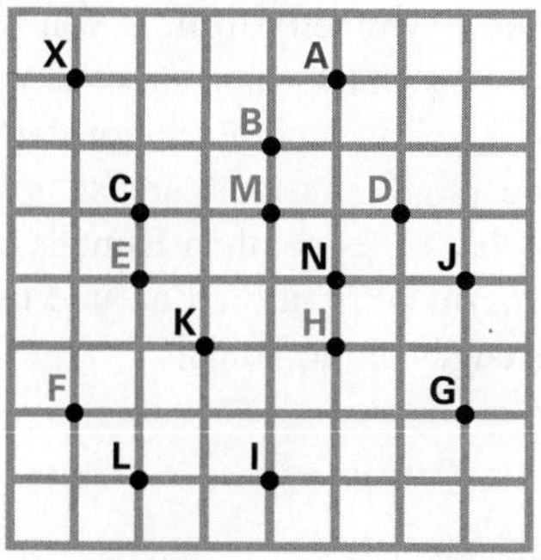

Figure for Exercises 35 and 36

35. Finding the most efficient route. Papa John's Pizza is located at the position marked A on the map. Assume that deliveries must be made to locations B, D, E, H, F, and M, and then the delivery truck must return to the store.

a. Use the nearest neighbor algorithm to devise a delivery route that will minimize the number of blocks traveled. You may assume that all streets are two-way streets. (*Hint:* First make a table of the distances between the various locations, similar to what we did in Exercise 33.)

b. Repeat part (a) using the best edge algorithm.

36. Finding the most efficient route. Repeat Exercise 35, but now assume that the deliveries must be made to B, E, H, I, J, and L.

Use the map for Exercises 37 and 38.

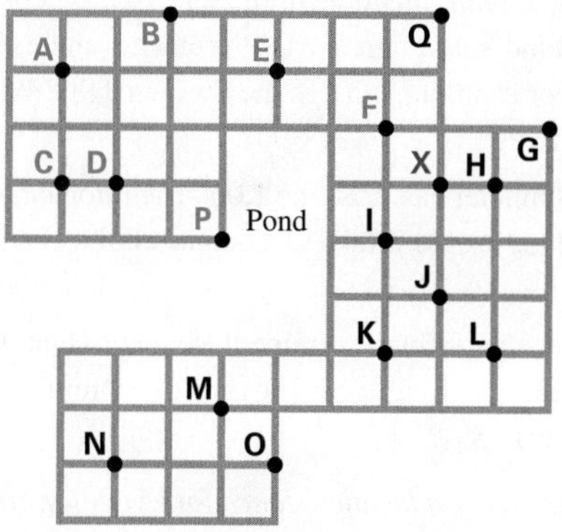

37. Finding the most efficient route. FedEx is located at the position marked X on the map.

a. Use the nearest neighbor algorithm to devise an efficient route to make deliveries to locations A, B, C, D, E, F, and P. The route starts and ends at X.

b. Repeat part (a) using the best edge algorithm.

38. Finding the most efficient route. Repeat Exercise 37, but now deliveries must be made to locations G, H, I, J, K, L, and N.

Communicating Mathematics

39. What is the difference between a Hamilton circuit and an Euler circuit?

40. We discussed three algorithms for solving the TSP—brute force, nearest neighbor, and best edge.

a. Of these three algorithms, which one can be the least feasible in that it can require unreasonable amounts of time to find a solution?

b. For the two algorithms that you did not mention in part (a), although they are faster, they have a disadvantage. What is it?

41. a. Explain why there are 9! Hamilton circuits in K_{10}.

b. Explain why there are $(n - 1)!$ Hamilton circuits in K_n.

c. When asked to find all Hamilton circuits in a complete graph, Michael found 15. How do we know he has made a mistake?

42. Sometimes the popular media ridicule scientific research because on the surface, not knowing what a researcher is trying to discover, the research can sound ridiculous. How would you argue against a newspaper editorial stating that bee brain research is frivolous and should not be funded?

Challenge Yourself

43. People who work with very large mathematical problems are interested in how rapidly the size of a problem grows. In Chapter 2, we showed that a set having n elements will have 2^n subsets. In Chapter 3, we saw that a logical statement having n variables will have 2^n lines in its truth table. Use a calculator or

spreadsheet to compare the size of 2^n with $n!$ for several values of n. What conclusion do you draw?

44. There are a number of wonderful sites devoted to the TSP (traditionally called The Traveling Salesman Problem). The site at Georgia Tech, mentioned on page 165, is a particularly good one. Find some famous TSP problems and explain what they are. For example, what is the Sweden TSP? What is the Mona Lisa TSP?

45. If a computer can examine 1,000 Hamilton circuits per second, how long would it take to examine all the Hamilton circuits in K_{10}?

46. If a computer can examine 1,000,000 Hamilton circuits per second, how long would it take to examine all the Hamilton circuits in K_{20}?

You are arranging a business lunch for (A)guilar, (B)yrne, (C)alvaresi, (D)abrowski, (E)delstein, and (F)oster. They will be seated at a large round table and you want the people seated next to each other to have common interests. Use the information in Exercises 47 and 48 to devise an acceptable seating arrangement. (We will not list information twice, so if A has common interests with C, we will not state that C has common interests with A.) There may be several acceptable answers.

47. A has common interests with C, D, and F; B has common interests with D, E, and F; C has common interests with E and F; D has common interests with F; E has common interests with F.

48. A has common interests with C, D, and E; B has common interests with D, E, and F; C has common interests with E and F; D has common interests with F.

If a weighted graph has vertices A, B, C, and so on, a variation of the nearest neighbor algorithm is to use the algorithm to find a Hamilton circuit starting at A, then find another one starting at B, then find another one starting at C, and so on. Then pick the Hamilton circuit from among these that has the smallest weight. We will call this algorithm the extended nearest neighbor algorithm.

49. Redo Exercise 21 using the extended nearest neighbor algorithm.

50. Redo Exercise 23 using the extended nearest neighbor algorithm.

4.3 Directed Graphs

Objectives

1. Understand how directed graphs model relationships that go in only one direction.
2. Use directed graphs to model influence.
3. Model the spread of a disease by a directed graph.

In real life, you are related to people in many different ways. You are friends with some people, smarter than others, and live on the same street as still others. "Living on the same street" is an example of a relationship that has a symmetry to it. If you live on the same street as Conan O'Brien, then Conan O'Brien lives on the same street as you do.

Other relationships lack this symmetry. "Love" is such a relationship. You may love Chris, but tragically Chris may not love you (sigh). As another example, consider the "mother of" relationship. If Rosalie is the mother of Esme, then Esme is certainly not the mother of Rosalie. Relationships that are nonsymmetric are all around us, and to model them with graphs, we assign directions to the edges of the graphs.

Directed Graphs

KEY POINT

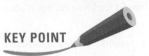

Directed graphs model the flow of information.

When an edge has a direction it is called a **directed edge.** A graph in which all edges are directed is called a **directed graph.** One application of directed graphs is to model the flow of information within a group of people or within an organization.

EXAMPLE 1 Modeling the Spread of Rumors

Assume that we have gathered the following information about how rumors spread in the Pritchett family:

1. If Manny hears a rumor, he will communicate it to Gloria and Jay, but Gloria and Jay will not tell rumors they hear to Manny.
2. Gloria and Jay will tell rumors they hear to each other.

3. Gloria tells rumors she hears to Cameron, but Cameron does not relay rumors he hears to Gloria but he does tell them to Jay.

Model this situation with a directed graph.

SOLUTION: Review the Draw Pictures Strategy on page 3.

Because rumors do not always flow in both directions, we will model this situation with the directed graph in Figure 4.29. The arrows on the edges show the direction in which the rumors travel.

For example, the directed edge from M to G shows that Manny communicates rumors to Gloria. Likewise, the absence of a directed edge from G to M indicates that Gloria does not tell rumors to Manny. 10

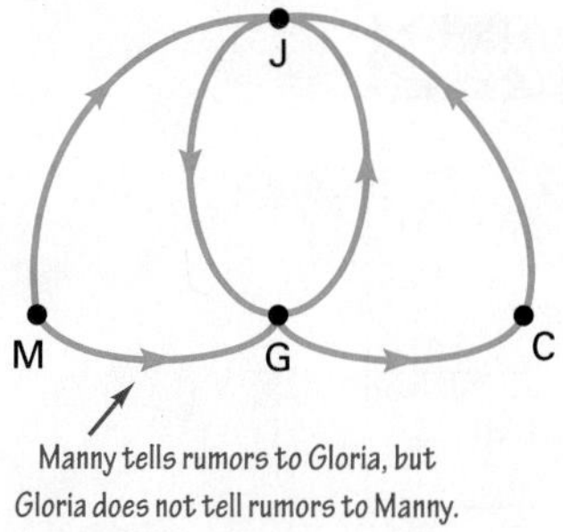

FIGURE 4.29 A directed graph modeling the spread of rumors.

QUIZ YOURSELF 10

If Cameron hears a rumor first, who in Example 1 will not hear the rumor from anyone in the group?

DEFINITIONS Suppose that X and Y are vertices in a directed graph. If it is possible to begin at X, follow a sequence of edges in the directions indicated, and end at Y, then we call the sequence of edges encountered a **directed path from X to Y.** We will denote a directed path by the sequence of vertices encountered along the path. The **length** of a directed path is the number of edges along that path. (Recall that edges cannot be repeated in a path.)

EXAMPLE 2 Finding Paths in a Directed Graph

Consider the directed graph shown in Figure 4.30.

a) What is the length of the directed path ACDE?
b) Is ABCE a directed path?

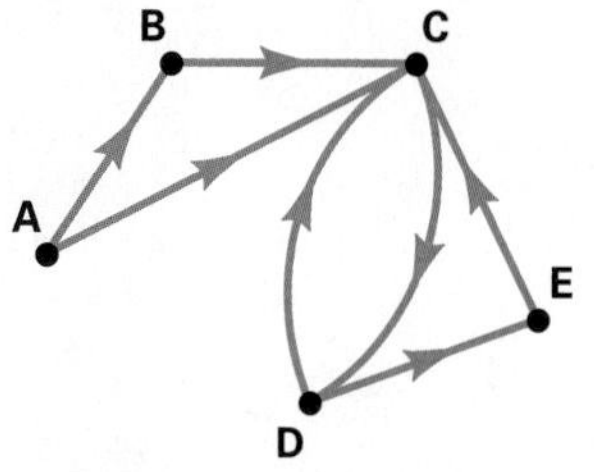

FIGURE 4.30 A directed graph.

SOLUTION:

a) ACDE is a directed path of length 3 from A to E.
b) ABCE is *not* a directed path from A to E because CE has the wrong direction.

Now try Exercises 1 to 4. 11

QUIZ YOURSELF 11

Use the graph to answer the following questions:

a) Can you find two directed paths from A to C?
b) Are there any directed paths from C to A?
c) What is the length of the directed path ABCDB?

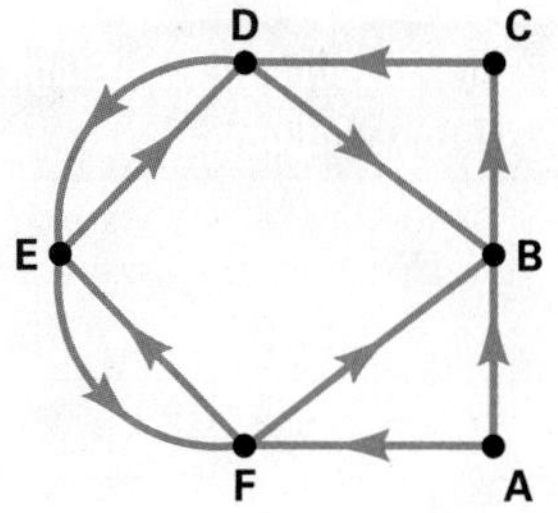

Modeling Influence

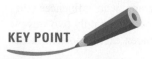

An interesting application of directed graphs is to model how members of a set influence each other. Because influence is usually not exerted equally in both directions, we use directed graphs as models in these situations.

KEY POINT

Directed graphs model how members of a set exert influence.

EXAMPLE 3 Modeling Influence

BigMart wants to build a shopping center and is receiving opposition from the residents of Peacefultown. BigMart's lawyers suggest that the best approach to get approval for the project is to gather the support of the most influential members of the township zoning board. After a careful investigation, BigMart's lawyers determine the patterns of influence listed in Table 4.4 on page 170.

Use Table 4.4 to determine which board member has the most influence.

SOLUTION: Review the Be Systematic Strategy on page 5.

Because influence is not exerted in both directions (Alvarez influences Cohen, but Cohen does not influence Alvarez), we use the directed graph shown in Figure 4.31 to model this relationship among zoning board members.

The directed edges in the graph indicate who influences whom. For instance, the directed edge from A to E reflects that Alvarez has influence over Ellis. Because Alvarez

Council Member	Board Members Influenced by This Person
Alvarez	Cohen, Ellis, Ferraro
Baker	Ferraro
Cohen	Baker, Ellis
Davis	Baker, Cohen, Ellis
Ellis	Baker
Ferraro	Cohen, Ellis

TABLE 4.4 Influence on zoning board members.

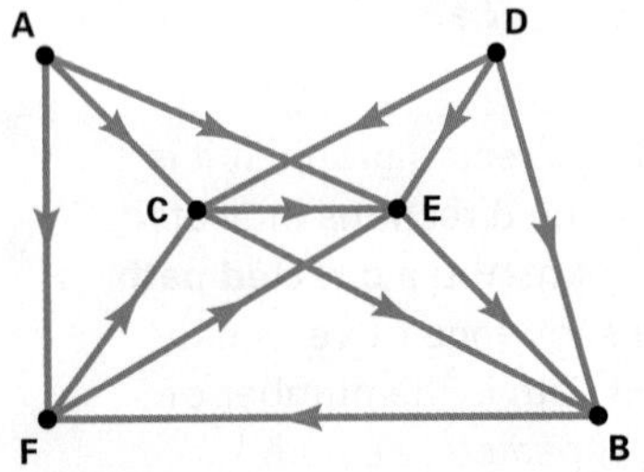

FIGURE 4.31 Directed graph modeling influence on zoning board.

and Davis each exert direct influence over three people, we might say that both are equally influential.

However, let us consider what we will call two-stage influence. We observe that Alvarez influences Cohen, who in turn influences Ellis. Therefore, we will say that Alvarez has two-stage influence over Ellis. In terms of our model, if there is a directed path of length 2 from vertex X to vertex Y, then board member X exerts two-stage influence over Y.

Let us now determine the number of ways that Alvarez can influence Ellis, either directly or in two stages. In Figure 4.31, in addition to the directed edge AE, there are also the directed paths ACE and AFE. Therefore, there are three ways in which Alvarez can influence Ellis in one or two stages.

We will consider paths from A to B, A to C, A to D, A to E, and so on. We summarize this information in Table 4.5, which shows the amount of one- and two-stage influence for each pair of board members. For example, the 3 in row A, column E, of Table 4.5 indicates that there are three paths of length 1 or 2 from A to E. The entry of row C, column B, is 2 because there are two paths of length 1 or 2 from C to B—namely, CB and CEB. In general, the entry in row X, column Y, in the table tells us the number of paths of length 1 or 2 from vertex X to vertex Y in the graph.

two paths of length 1 or 2 from C to B

three paths of length 1 or 2 from A to E

Alvarez's total influence

From \ To	A	B	C	D	E	F	Total Direct and Two-Stage Influence
A	0	2	2	0	3	1	8
B	0	0	1	0	1	1	3
C	0	2	0	0	1	1	4
D	0	3	1	0	2	1	7
E	0	1	0	0	0	1	2
F	0	2	1	0	2	0	5

TABLE 4.5 Total one- and two-stage influence exerted by board members.*

Now that we have determined the amount of one- or two-stage influence, we can rank the board members. Although Alvarez and Davis influence Ellis directly and also by way of Cohen, we see that Alvarez also influences Ferraro, who influences Ellis. There are three ways in which Alvarez exerts influence over Ellis but only two ways in which Davis can. Therefore, it is reasonable to say that Alvarez exerts more influence over Ellis than Davis can. Now try Exercises 5 to 8.

*These calculations can be done very quickly with a calculator, as we show in the Using Technology exercises at the end of this chapter.

Most Influential

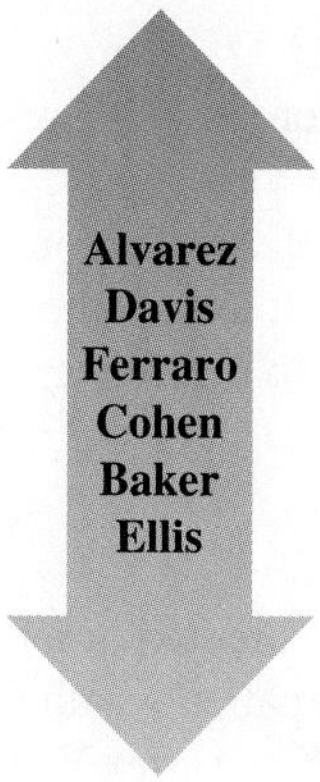

Least Influential

FIGURE 4.32 Ranking of zoning board members according to one- and two-stage influence.

We can now rank board members according to their one- and two-stage influence. Adding the entries in row A, we get

$$0 + 2 + 2 + 0 + 3 + 1 = 8$$

(8 — A's total influence; the terms are, in order: A's influence over A, A's influence over B, A's influence over C, A's influence over D, A's influence over E, A's influence over F)

which means that there are eight ways Alvarez exerts influence over other board members in one or two stages. Adding the entries for each of the other rows, we obtain the last column in Table 4.5. According to the way we are measuring influence, Figure 4.32 shows the ranking of the board members.

Now try Exercises 15 to 20. **12**

QUIZ YOURSELF 12

List in a table the number of directed paths of length 1 or 2 between each pair of vertices in the given graph.

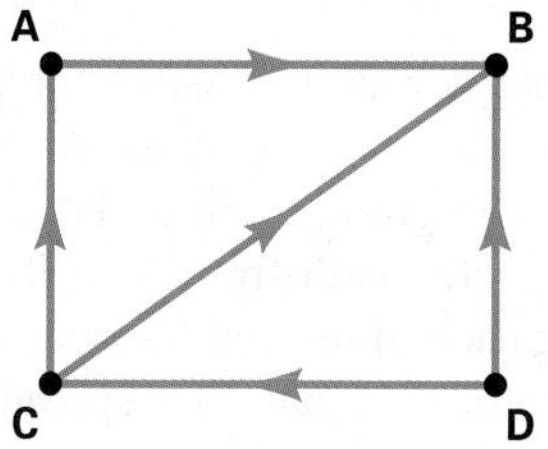

As in Example 3, we often want to rank objects with respect to some property. For example, we could rank football teams by using their number of victories. Assume Ohio State and Nebraska have played the same number of games, but Ohio State has four victories and Nebraska has three. We could then rank Ohio State above Nebraska. Suppose instead that each team has four victories. In order to break a tie in the ranking order, we could consider the notion of two-stage dominance; that is, if Ohio State has beaten Michigan and Michigan has defeated Indiana, then Ohio has demonstrated two-stage dominance over Indiana. If Ohio State has more two-stage dominances over other teams than Nebraska has, then we could rank Ohio State above Nebraska. If the two teams are still tied with regard to the number of two-stage dominances, we might then consider the number of three-stage dominances, and so forth.

Keep this example in mind when solving the exercises at the end of this section in which you are asked to rank objects of a set.

SOME GOOD ADVICE

Realize that when building mathematical models, *you* decide what features you feel are important in the model. You may not agree with the way a model is constructed. For example, should two-stage dominance count the same as one-stage dominance in the last discussion? If not, then you may want to change the method of building the model.

Math in Your Life

You'll Have That Tomorrow

Isn't it marvelous how you can drop off a package and within 24 hours it reaches its destination? I once received a tracking report that went roughly like this: Pickup 2:41 PM—Reading, PA 7:01 PM—Philadelphia, PA 10:52 PM—Chicago, IL 4:10 AM—Delivered 10:27 AM. This report describes how my package traveled over a path in a huge directed graph that models the delivery company's operations.

Scientists working in the area of **operations research** frequently use graph theory to organize the flow of tens of millions of packages each day, coordinate millions of phone calls, and help hoards of airline passengers arrive safely at their destinations.

Modeling Disease

We now use directed graphs to answer a different type of question.

KEY POINT

Directed graphs model disease transmission.

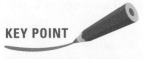

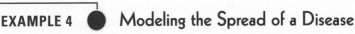

EXAMPLE 4 Modeling the Spread of a Disease

Dr. Erin Mears, a scientist at the Centers for Disease Control and Prevention, has quarantined eight people in Minneapolis who have come down with a deadly bat virus, and she hopes that no others outside this group have contracted the disease. She believes that one person introduced the virus and communicated it to the others in the group. Use the information in Table 4.6 to determine if Dr. Mears' hypothesis is correct.

Patient	Others Within the Group Who Could Have Contracted the Virus from This Patient
Amanda	Dustin, Jackson
Brian	Caterina, Frank, Ina
Caterina	Frank
Dustin	Caterina
Frank	Louisa
Ina	Brian, Frank
Jackson	Amanda, Caterina, Frank
Louisa	Caterina

TABLE 4.6 How a virus might have spread in a town.

SOLUTION: Review the Draw Pictures Strategy on page 3.

We can easily model this situation using the directed graph in Figure 4.33. We represent each of the eight patients with a vertex and draw a directed edge from vertex X to vertex Y, if X could have transmitted the virus to Y.

Because Dustin could have given the virus to Caterina, we draw a directed edge from D to C. Similarly, because Brian could have transmitted the virus to Frank, there is a directed edge drawn from B to F. The directed paths in the graph show how the infection might have spread within the group. For example, the directed path ADCF shows that it was possible for the virus to travel from Amanda to Dustin to Caterina, and finally to Frank. It is important to realize that if the virus spread from X within the group to Y, then there must be a directed path in the graph from vertex X to vertex Y.

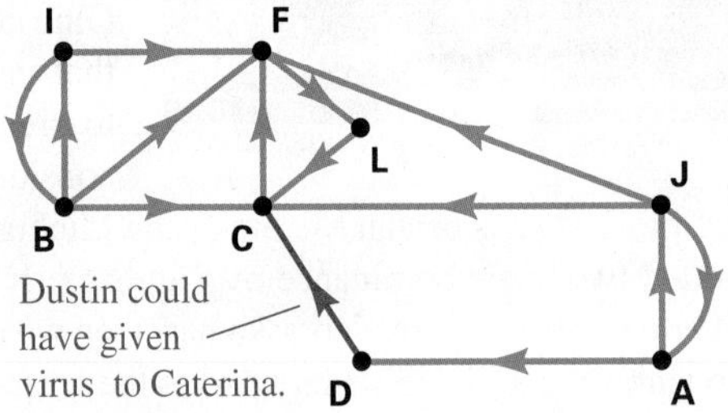

FIGURE 4.33 Directed graph representing spread of a virus.

We see that it was impossible for the virus to have started with Amanda and spread within the group to Brian because there is no directed path from A to B. In fact, checking each of the eight people, we see that it was impossible for the virus to start with any one of them and then spread to all the others. Therefore, if our information and assumption are correct, there is at least one other person in the city who has the virus but who has not yet been identified.

In the movie *Contagion*, the character Dr. Mears is modeled after Rear Admiral Dr. Anne Schuchat, who is the director of CDC's National Center for Immunization and Respiratory Diseases. Her work includes meningitis vaccine studies in West Africa, SARS emergency response in China, and the 2009 H1N1 response. In such cases, when working with large numbers of people, scientists like Dr. Schuchat use large mathematical models to track the infections. We will give you a sense of how this is done in the Group Projects at the end of the chapter.

EXERCISES 4.3

Sharpening Your Skills

For many of these exercises there may be several correct answers. We will provide only one in the answer key.

In Exercises 1–4, use each graph to find the requested items, if it is possible. If it is not possible to find a requested item, explain why not.

1. **a.** Two different directed paths from A to E

b. A directed path from A to C

c. A directed path of length 3 from A to E

d. A directed path of length 2 from A to E

e. A directed path of length 5 from A to A

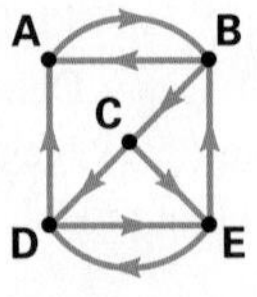

2. **a.** Two different directed paths from A to E

b. A directed path from A to C

c. A directed path of length 3 from A to E

d. A directed path of length 2 from A to E

e. A directed path of length 5 from A to A

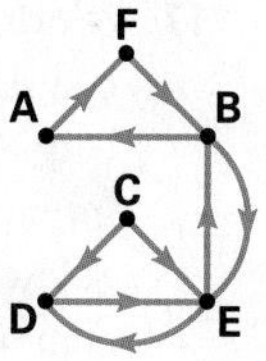

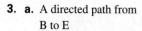

3. a. A directed path from B to E

b. A directed path from A to G

c. A directed path of length 5 from A to E

d. A directed path from F to D

e. A directed path of length 5 from A to C

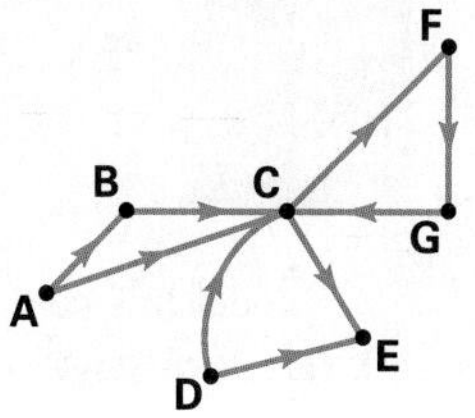

4.

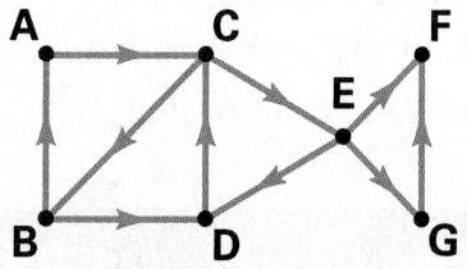

a. Two different directed paths from D to E

b. A directed path from B to F

c. A directed path from F to C

d. A directed path of length 5 from A to E

e. A directed path of length 5 from A to A

5. Construct a table that displays the number of directed paths of length 1 or 2 between each pair of vertices in the graph of Exercise 1.

6. Construct a table that displays the number of directed paths of length 1 or 2 between each pair of vertices in the graph of Exercise 2.

7. Construct a table that displays the number of directed paths of length 1 or 2 between each pair of vertices in the graph of Exercise 3.

8. Construct a table that displays the number of directed paths of length 1 or 2 between each pair of vertices in the graph of Exercise 4.

Applying What You've Learned

9. Modeling the spread of rumors. Ryan, Dwight, Pam, Jim, Angela, and Kevin work in the same office. A rumor has been circulating that their company is being taken over by a Singapore-based firm in a hostile takeover. The given directed graph shows how rumors travel among these six employees.

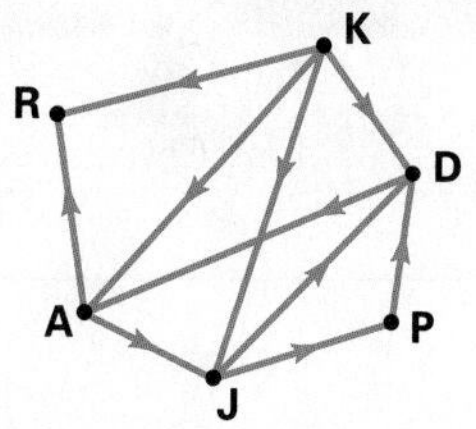

a. Are there any people among these six who could have initiated the rumor?

b. Change the direction of as few edges as possible to make Pam the person who started the rumor.

10. Modeling the spread of classified information. Several news organizations have made public a secret report on how to improve the U.S. diplomatic stature in the Middle East. Based on sources (which we cannot reveal), the given graph indicates how the information could have passed among these organizations.

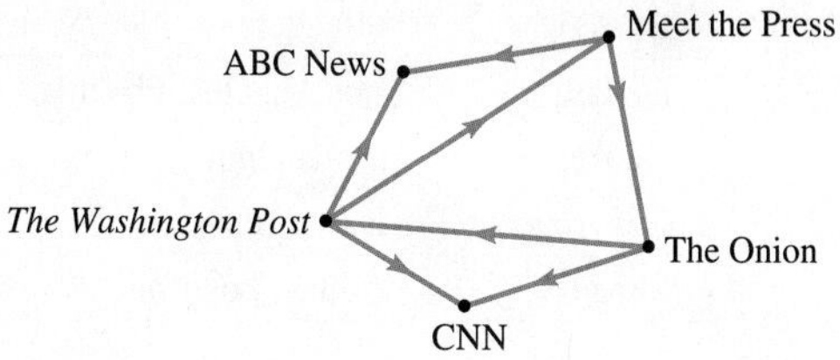

a. Determine which organizations could have first obtained this information.

b. Change the direction of only one edge in the graph so that only the *Washington Post* could have obtained the information first.

11. Modeling the flow of paperwork through a bureaucracy. In a large corporation, five signatures are needed on a form. It is known that certain managers will not give their approval before others. This situation is depicted in the following directed graph. A directed edge from vertex A to vertex B indicates that A must sign the form before B. If a secretary must hand-carry the form from office to office, in which sequence can all the signatures be obtained?

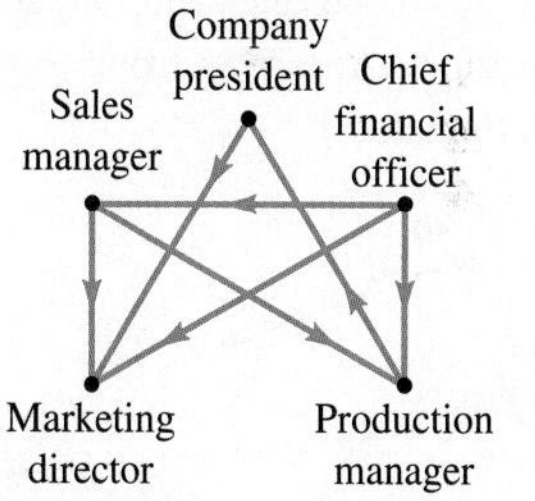

12. Modeling the spread of disease. The following directed graph models the spread of a motovirus among a group of passengers on a cruise ship.

a. Are there any people in this group who could have introduced the virus to the group? Explain.

b. Change the direction of one edge in the graph so that no one could have introduced the virus to the group.

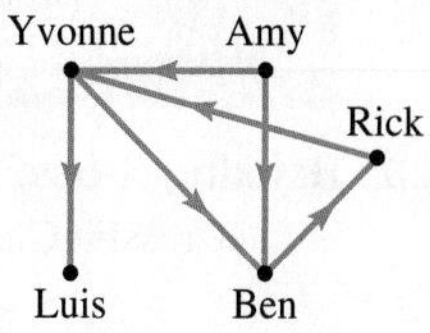

13. Modeling a food chain. The African hare and gazelle are vegetarians that feed primarily on grass. Gazelles and hares are eaten by lions, cheetahs, and humans. Draw a directed graph to model this food chain.

14. Modeling a communications network. An AMBER alert emergency broadcast system has been set to transmit messages among several cities. The following table indicates how this system has been set up. Draw a directed graph modeling this emergency network. Can a message originate in one city and spread throughout the network? Explain.

City	Can Broadcast To
Philadelphia	New York, Detroit
New York	Philadelphia, Boston
Boston	Philadelphia, New York
Dallas	Los Angeles, Phoenix
Detroit	Philadelphia, Dallas
Los Angeles	Dallas, Phoenix
Phoenix	Dallas, Los Angeles

15. Ranking football teams. The given graph shows the results among several football teams in the Southeast Conference. Rank these teams using one- and two-stage dominance.

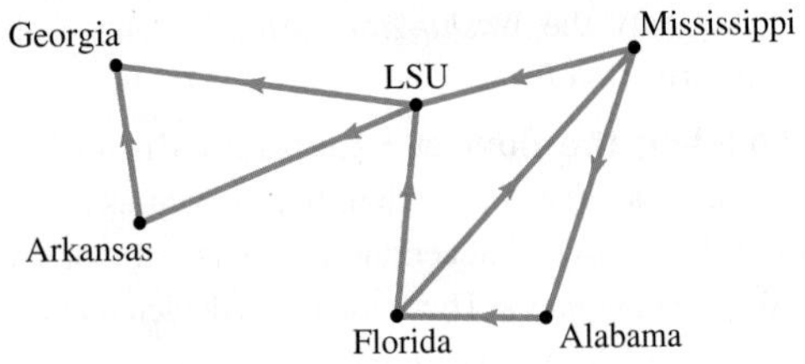

16. Ranking American Gladiators. The graph below shows (hypothetical) partial results after several rounds of head-to-head competition among the American Gladiators in the Hit and Run Contest. Use one- and two-stage dominance to rank these competitors. An arrow from vertex X to vertex Y indicates that X defeated Y.

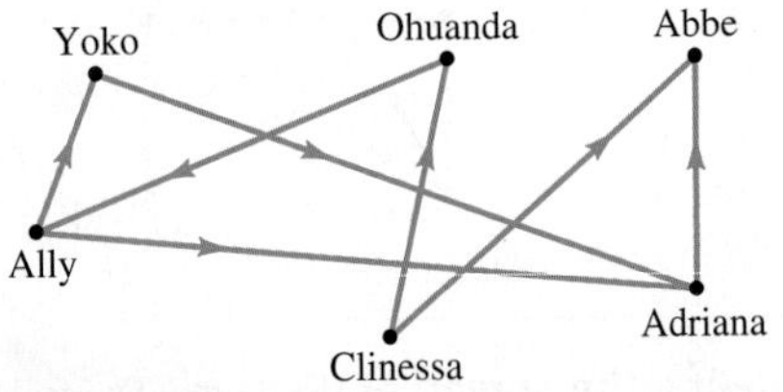

17. Modeling influence. A designer who wishes to obtain the contract for creating a new advertising campaign for Center City Community College wants to determine the most influential members of the selection committee. The table below lists the influences that have been observed among committee members.

Member	Lee	Feinstein	Johnson	Cordaro	Murciano
Influences	No one	Cordaro, Murciano	Lee, Cordaro	Lee	Lee, Johnson

Model this committee with a directed graph and use one- and two-stage influence to determine the most influential committee members.

18. Ranking teams. In a round-robin singles pool tournament, each of six competitors plays each other once. The results of the tournament are shown in the following table:

Player	Defeated
Carla	Rob, Sara, Orlando, Matt
Rob	Tanya, Orlando, Matt
Tanya	Carla, Sara, Matt
Sara	Rob, Matt
Orlando	Tanya, Sara
Matt	Orlando

Use a directed graph to model this situation and determine a ranking order for the players using one- and two-stage dominance.

19. Modeling drink preferences. In a bottled water "paired comparison" test, college students were asked to compare the drinks Propel, Aquafina, Fuji, Dasani, and Vitaminwater. Use the table below to rank the drinks using one- and two-stage preference.

Brand	Propel	Aquafina	Fuji	Dasani	Vitaminwater
Preferred Over	Aquafina, Fuji	Vitaminwater	Aquafina, Dasani, Vitaminwater	Propel, Aquafina, Vitaminwater	Propel

20. Ranking CrossFit competitors. CrossFit is a conditioning program that combines strength, flexibility, and speed. After five events in the CrossFit Games Open, the following results have been obtained:

	Power Lifting	Gymnastics	Kettle Bell	Plyometrics	Sprinting
Competitor	Noah	Tom	Caleb	Ben	Christian
Defeated	Tom, Ben	Christian	Tom, Ben, Christian	Tom, Christian	Noah

Rank the five contenders using one- and two-stage dominance.

*When tracking a real disease involving a large number of people, we represent the data mathematically using a rectangular arrangement of numbers called an **incidence matrix** (plural: **matrices**) to represent the directed graph. For example, we can represent the directed graph*

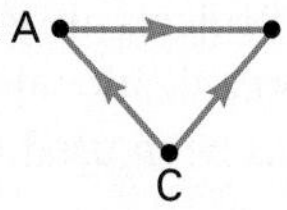

by the matrix $\begin{bmatrix} 0 & 1 & 0 \\ 0 & 0 & 0 \\ 1 & 1 & 0 \end{bmatrix}$.

We think of the rows and columns corresponding to the vertices A, B, and C. The 1 in the top row of the matrix indicates that there is a directed edge from A to B, and the 0s tell us there are no directed edges from A to A or from A to C. The 1s in the bottom row represents the directed edges from C to A and C to B. Represent the directed graphs in Exercises 21 and 22 by incidence matrices.

21.

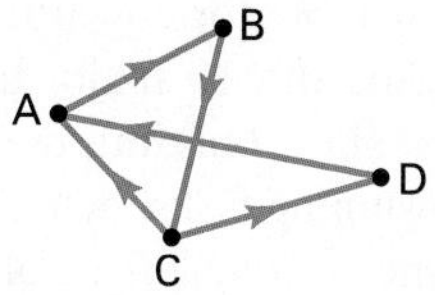

22.

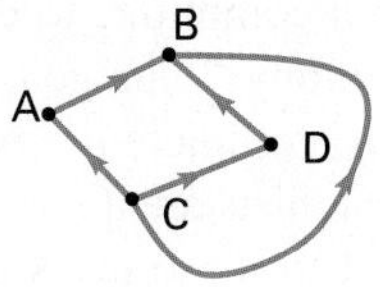

In Exercises 23–26, draw a directed graph corresponding to the given matrix.

23. $\begin{bmatrix} 0 & 1 & 0 & 0 \\ 0 & 0 & 1 & 1 \\ 1 & 0 & 0 & 0 \\ 0 & 1 & 0 & 0 \end{bmatrix}$

24. $\begin{bmatrix} 0 & 1 & 1 & 0 \\ 0 & 0 & 0 & 1 \\ 1 & 0 & 0 & 1 \\ 0 & 1 & 0 & 0 \end{bmatrix}$

25. $\begin{bmatrix} 0 & 1 & 0 & 1 \\ 1 & 0 & 0 & 0 \\ 1 & 1 & 0 & 1 \\ 0 & 0 & 1 & 0 \end{bmatrix}$

26. $\begin{bmatrix} 0 & 1 & 1 & 1 \\ 1 & 0 & 0 & 0 \\ 0 & 0 & 0 & 1 \\ 0 & 0 & 1 & 0 \end{bmatrix}$

Communicating Mathematics

27. When is a directed graph a better model for a relationship than a nondirected graph?

28. In modeling influence, we counted two-stage influence as equal with one-stage influence. Criticize this approach. Can you suggest another method? How would this affect our tables that summarize one- and two-stage influence? Explain.

29. Think of a situation that we have not discussed in this section in which you feel directed graphs could be used as models. Describe the application and explain the reasons why you think it can be represented by a directed graph.

30. In the incidence matrices in Exercises 23 to 26, notice that there are no 1s in the A row–A column, B row–B column, C row–C column, or D row–D column positions. Why is this? Can you think of a real-life application where 1s might occur in these positions?

Challenge Yourself

31. Modeling consumer preferences. When purchasing food products, a consumer would consider ease of preparation, nutritional value, price, and taste. Conduct a consumer survey by asking five people you know to complete a copy of the following ballot.

For each of the following pairs, circle the quality that is more important to you:

Ease of preparation	Nutritional value
Ease of preparation	Price
Ease of preparation	Taste
Nutritional value	Price
Nutritional value	Taste
Price	Taste

On the third line of the ballot, if more people chose taste over ease of preparation, then we can say that taste is preferred over ease of preparation. Use the results of your survey to determine which quality is preferred for each pair, and then summarize your results by means of a directed graph.

32. Modeling car feature preferences. When shopping for an automobile, a buyer may consider each of the following features: economy, safety, comfort, style, manufacturer, and availability of service. Assume you are a car buyer and compare each possible pair of features, indicating which of the two is more important to you. After completing this paired-comparison test, determine a ranking order by importance of the six features.

Looking Deeper

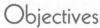

4.4 Scheduling Projects Using PERT

Objectives

1. Understand the language of PERT diagrams.
2. Use PERT diagrams to schedule projects.

Well begun is half done.
—Aristotle

This ancient advice is as valid today as it was 2,000 years ago. When beginning a project—whether planning your senior paper, building your new house, designing a

nuclear submarine, or organizing the Olympic Games—the beginning of a project is the most important part. In particular, good planning is fundamental to completing a project successfully.

Although it is unlikely that you will be in charge of building a submarine or planning the Olympics, you certainly will have to organize many projects in your personal and professional lives. In this section, we will introduce an organizational technique called PERT (Program Evaluation and Review Technique) that has been used for over 50 years in planning large projects.

PERT Diagrams

KEY POINT

PERT diagrams help organize large projects.

In order to understand how PERT helps you organize a project, consider the following hypothetical example.

Imagine that you are serving on a committee to develop a timetable for constructing and populating an orbiting space colony. Your goal is to schedule the 10 major tasks listed in Table 4.7 that are needed to complete the project. Also, since building the life support systems and recruiting the colonists are the most time-consuming tasks, determine if extra resources should be devoted to these tasks to shorten the overall length of the project.

It is tempting to simply add the time required for each task and conclude that the project requires 75 months. However, because some tasks can be done simultaneously, the total time needed will be less than 75 months. The last column of Table 4.7 shows which tasks must precede others.

Should we try to shorten these times? (tasks 3 and 4)

Task	Time Required (months)	Preceding Tasks
1. Train construction workers	6	None
2. Build shell	8	None
3. Build life support systems	14	None
4. Recruit colonists	12	None
5. Assemble shell	10	1, 2
6. Train colonists	10	2, 3, 4
7. Install life support systems	4	1, 2, 3, 5
8. Install solar-energy systems	3	1, 2, 5
9. Test life support and energy systems	4	1, 2, 3, 5, 7, 8
10. Bring colonists to the colony	4	1, 2, 3, 4, 5, 6, 7, 8, 9

TABLE 4.7 Time and precedence of the space colony tasks.

We can display the data in Table 4.7 in a special type of directed graph called a **PERT diagram,** shown in Figure 4.34. We represent each task by a vertex containing the number of months needed for that task and assume that the beginning and end of the project require no time. (*Note:* Sometimes you may need to redraw a PERT diagram several times to make the information stand out.)

If a task X *immediately precedes* a task Y, we draw a directed edge from vertex X to vertex Y in the graph. For example, from Table 4.7 we see that training construction workers *immediately precedes* assembling the shell in space, so we have drawn a directed edge from the vertex "Train construction workers" to the vertex "Assemble shell." Notice that because building the shell on Earth *does not immediately precede* installing the life support systems, we do not draw a directed edge between these two vertices. We will return to the solution of this problem after Quiz Yourself 13. 13

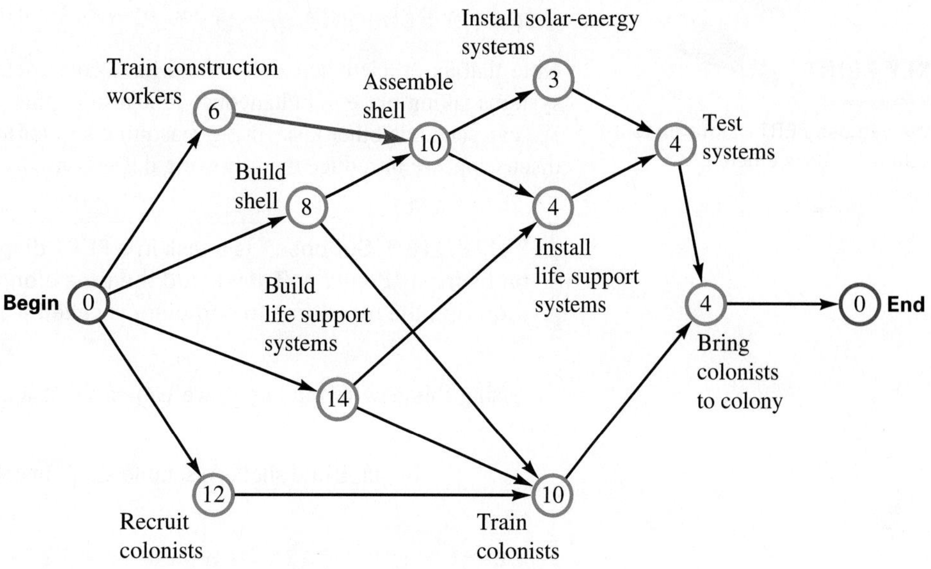

FIGURE 4.34 A PERT diagram for the space colony project.

QUIZ YOURSELF 13

The following table lists the times required to complete the tasks of a particular project and the dependency among these tasks. Draw a PERT diagram for this project.

Task	Days Required	Preceding Tasks
A	3	None
B	4	A
C	6	A
D	2	A, B, C
E	5	A, B, C, D
F	7	A, B, C, D, E

We now return to our space colony scheduling problem.

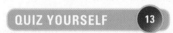

EXAMPLE 1 **Determining When to Install the Life Support Systems in the Space Colony**

Use the PERT diagram in Figure 4.34 to determine the earliest time we could install the life support systems.

SOLUTION: Review the Be Systematic Strategy on page 5.

Figure 4.34 shows that there are three different sequences of tasks that must be done before installing the life support systems:

a) Begin (0 months), Train construction workers (6 months), Assemble shell (10 months); total time is 16 months.

b) Begin (0 months), Build shell (8 months), Assemble shell (10 months); total time is 18 months.

c) Begin (0 months), Build life support systems (14 months); total time is 14 months.

Since sequence b is the most time-consuming sequence of tasks that must precede installing the life support systems, we schedule the installation of the life support systems to begin in the nineteenth month of the project.

KEY POINT

We can use PERT diagrams to schedule tasks efficiently.

Efficient Scheduling

Note that an efficient schedule would have construction of the shell and the life support systems taking place simultaneously because neither of these tasks depends on the other. We can schedule other tasks using reasoning similar to that in Example 1. To simplify our discussion, we introduce the following definition:

> **DEFINITION** Suppose T is a task in a PERT diagram. Let us consider all directed paths from "Begin" to T. If we add the time along each of these paths, any path requiring the most time to complete is called a **critical path** for task T.

Using this new terminology, we can say that a critical path for "Install life support systems" is:

Begin, Build shell, Assemble shell, Install life support systems.

> **SCHEDULING A TASK IN A PERT DIAGRAM** To determine when to schedule a task T in a PERT diagram, do the following:
>
> 1. Find a critical path for task T.
> 2. Add all times along this critical path, with the exception of the time required for task T. This sum gives us the time to be allowed before scheduling T.

EXAMPLE 2 Scheduling the Testing of the Colony's Systems

Use the preceding scheduling procedure to determine when to start the testing of the colony's systems.

SOLUTION: Review the Draw Pictures Strategy on page 3.

From our PERT diagram in Figure 4.34, we see that a critical path for this task is:

Begin, Build shell, Assemble shell, Install life support systems, Test systems.

Examining this path, we see that $0 + 8 + 10 + 4 = 22$ months must be allowed *before* the testing of the colony's systems can begin. Therefore, this task should be scheduled to begin in the twenty-third month.

Table 4.8 lists a schedule for all the tasks of the space colony project.

Task	Month Task Begins
1. Train construction workers	1st
2. Build shell	1st
3. Build life support systems	1st
4. Recruit colonists	1st
5. Assemble shell	9th
6. Train colonists	15th
7. Install life support systems	19th
8. Install solar-energy systems	19th
9. Test life support and energy systems	23rd
10. Bring colonists to the colony	27th

TABLE 4.8 Schedule for the space colony tasks.

EXAMPLE 3 Scheduling the Space Colony Project

Determine the time needed for the entire space colony project.

SOLUTION:

To do this, we must find a critical path for the vertex "End." Such a path is:

Begin, Build shell, Assemble shell, Install life support systems,
Test systems, Bring colonists to the colony, End,

which is highlighted by the red edges in Figure 4.35. From this diagram, we find that we need $0 + 8 + 10 + 4 + 4 + 4 + 0 = 30$ months to complete the project.

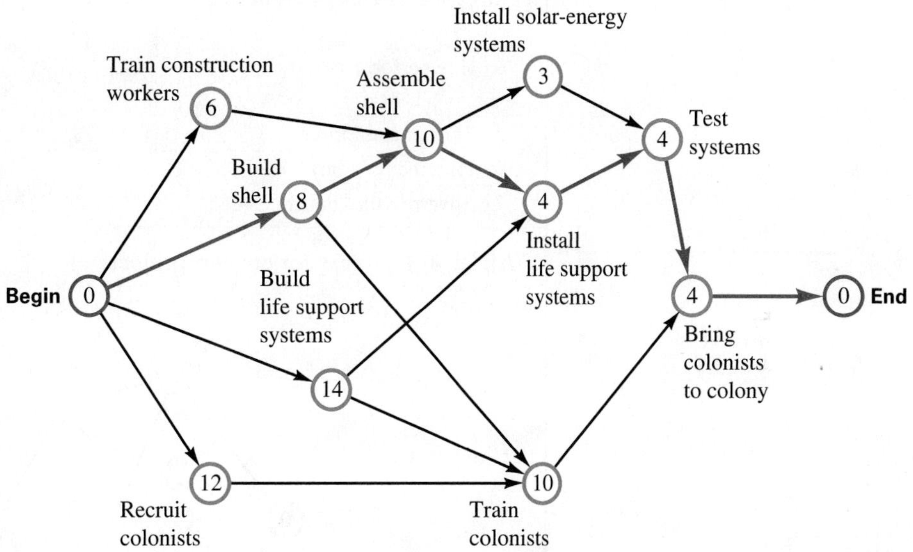

FIGURE 4.35 Critical path for the space colony project. 14

Earlier we were asked to determine if we should invest more resources to shorten the times in building the life support systems (task 3) and recruiting the colonists (task 4). Because neither of these tasks lies on a critical path for the vertex "End," reducing the times, for tasks 3 and 4 will not shorten the time to complete the project.

QUIZ YOURSELF 14

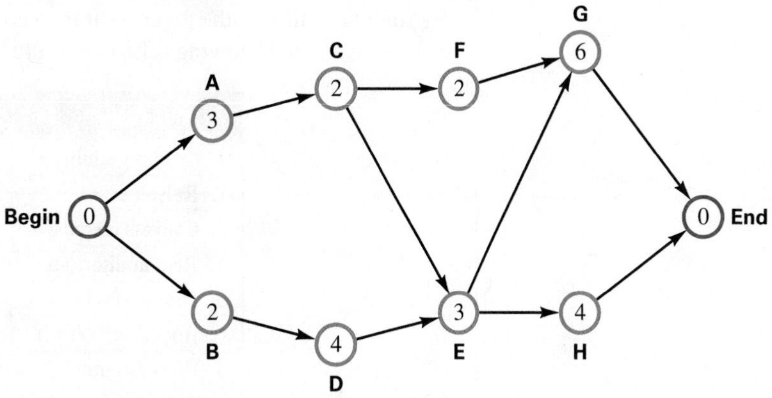

a) Find a critical path for G.
b) What is a critical path for "End"?
c) Assuming that the numbers in the vertices represent days, when should task H be scheduled?
d) How long will the whole project take?

EXAMPLE 4 Using PERT to Organize a Concert

Assume that you are in charge of organizing a concert to raise money to aid victims of a recent earthquake. Your job is to develop a schedule to complete the project in the shortest possible time. The project tasks and their dependencies are given in Table 4.9 and are displayed in the PERT diagram in Figure 4.36.

Task	Time Required (weeks)	Preceding Tasks
1. Obtain permits from the city	2	None
2. Raise funds from local agencies	1	None
3. Canvass local merchants for advertising support	4	None
4. Hire performers	3	Obtain permits, raise funds, canvass merchants
5. Rent an auditorium	2	Obtain permits
6. Print the program	1	All of the above
7. Advertise the concert	2	All of the above except print programs

TABLE 4.9 Tasks for concert project.

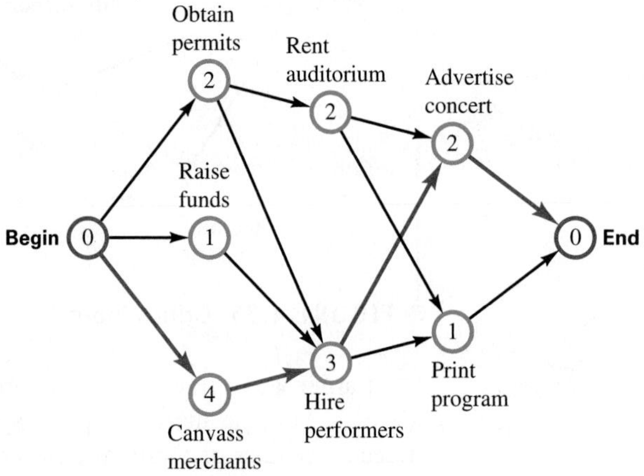

FIGURE 4.36 PERT diagram for concert project.

SOLUTION:

By finding critical paths for each of the vertices in the PERT diagram in Figure 4.36, we easily obtain the following schedule of the tasks:

Task	Week Task Begins
1. Obtain permits	1
2. Raise funds	1
3. Canvass merchants	1
4. Rent auditorium	3
5. Hire performers	5
6. Advertise concert	8
7. Print program	8

We determine the shortest time period to complete this project by finding a critical path for the vertex "End." Such a path (marked in red) is:

Begin, Canvass merchants, Hire performers, Advertise concert, End.

Thus, the entire concert project can be completed in 9 weeks, which is the sum of the times along this path.

EXERCISES 4.4

Sharpening Your Skills

Some of these exercises may have several correct answers. We will give only one in the answer key.

In Exercises 1–4, assume that the time is measured in days.

1. Use the following PERT diagram to answer the following questions.

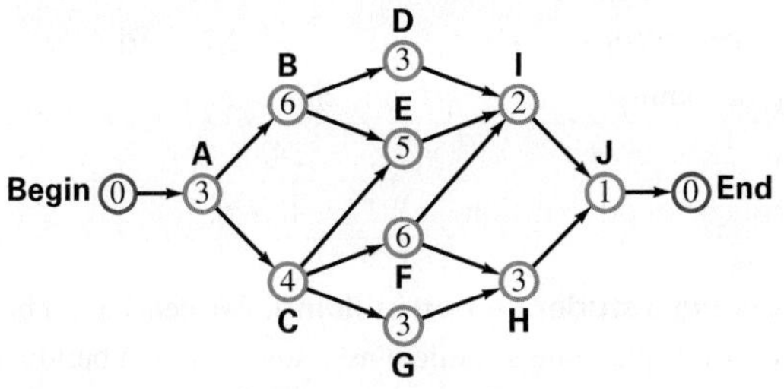

a. Find a critical path for task I.

b. Find a critical path for task E.

c. On what day will task H begin?

d. When will task I be completed?

e. What is the least number of days needed to complete this project?

f. Find a critical path for the vertex "End."

2. Use the following PERT diagram to answer the following questions.

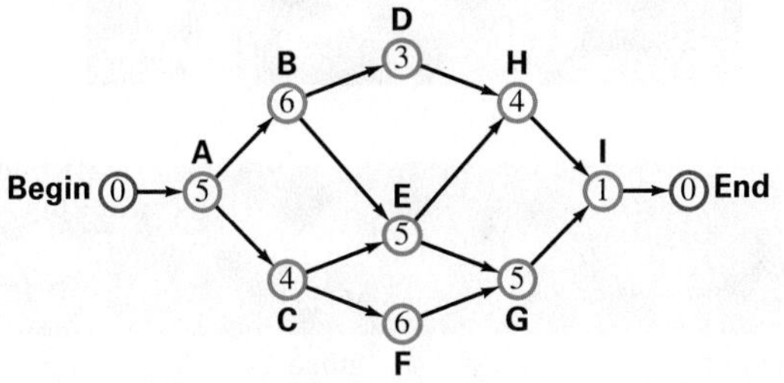

a. Find a critical path for task H.

b. Find a critical path for task G.

c. On what day will task G begin?

d. When will task H be completed?

e. What is the least number of days needed to complete this project?

f. Find a critical path for the vertex "End."

3. Use the following PERT diagram to answer the following questions.

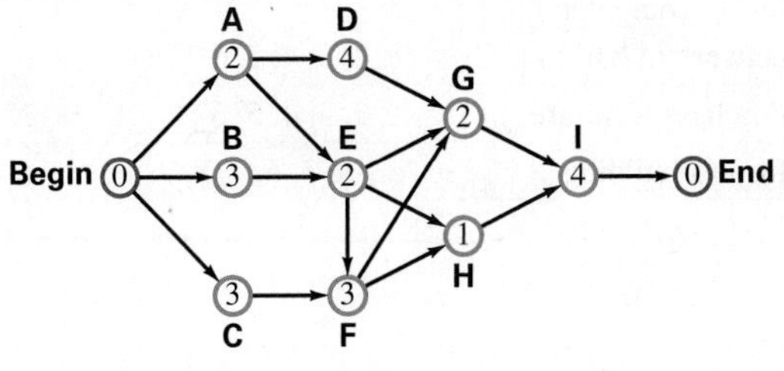

a. Find a critical path for task G.

b. Find a critical path for task H.

c. What is a critical path for "End"?

d. When should task F be scheduled?

e. When should we schedule task G?

f. How many days are required for the whole project?

4. Use the following PERT diagram to answer the following questions.

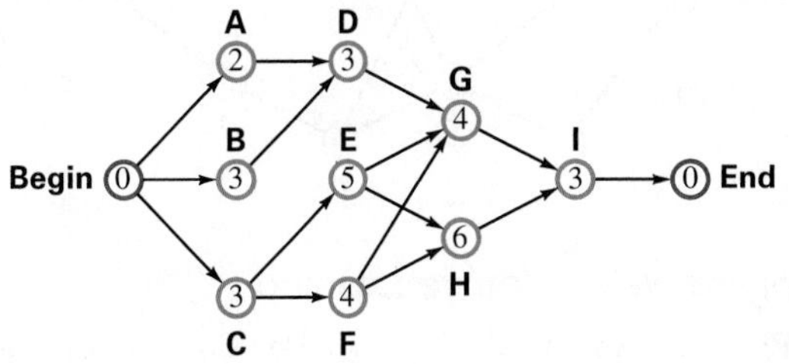

a. Find a critical path for task G.

b. Find a critical path for task H.

c. What is a critical path for "End"?

d. When should task H be scheduled?

e. When should we schedule task G?

f. How many days are required for the whole project?

In Exercises 5–8, use the given PERT diagrams to schedule the tasks so each task is completed in the least possible amount of time. Assume the numbers in the vertices refer to days.

5.

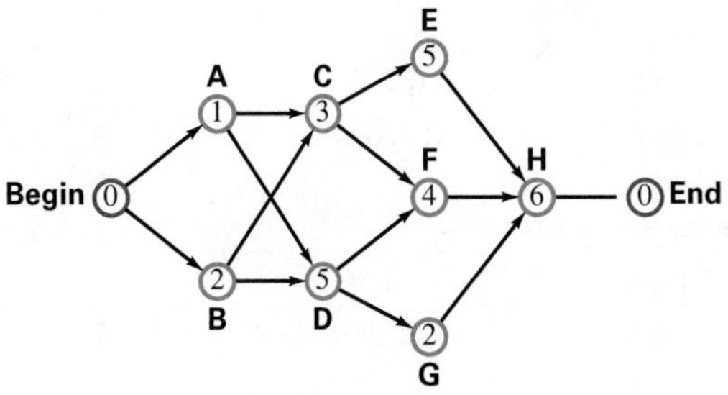

6.

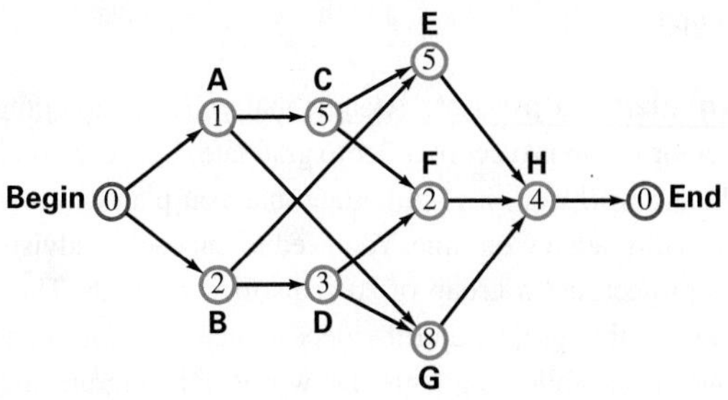

7.

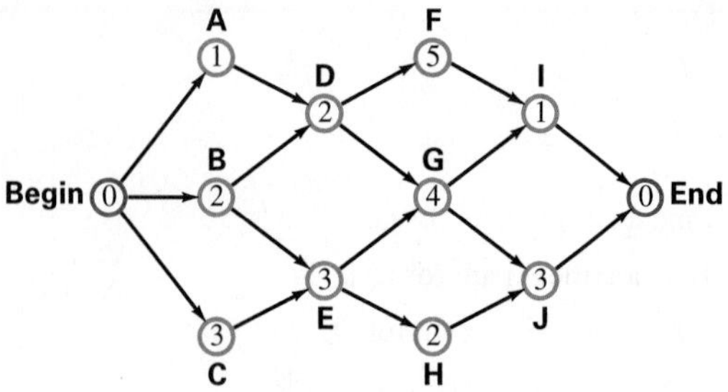

8.

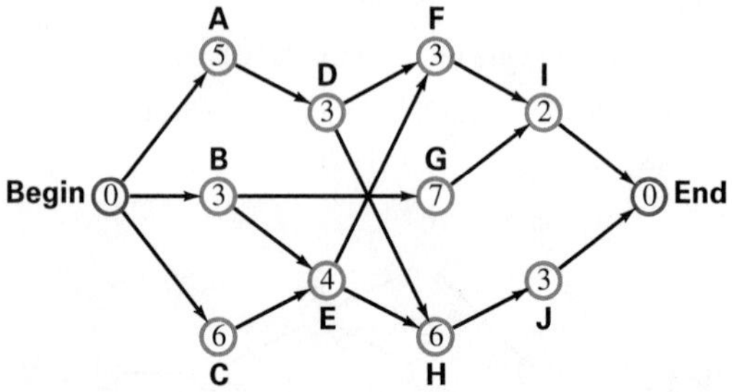

Applying What You've Learned

9. **Planning a festival.** The Earth Day Committee has begun planning for an Earth Day festival. We list the subtasks for this project and the dependencies among them in the following table. Draw a PERT diagram for this project and construct a schedule for the tasks.

Task	Preceding Tasks	Time Required (weeks)
1. Get funding	none	2
2. Get permits	none	1
3. Decide on program	1	2
4. Rent tents	1, 2, 3	2
5. Arrange for speakers, entertainers, etc.	1, 2, 3, 4	4
6. Advertise	1, 3, 5	2
7. Set up tents, booths, etc.	1, 2, 3, 4	1
8. Set up festival	1, 2, 3, 4, 5, 6, 7	1

10. **Organizing a project.** Assume that you need to complete a senior group project in order to graduate. Because you have a heavy workload, it is important that you plan carefully so you will graduate on time. You need to choose an advisor for this project and a group of students to work with. The subtasks for this project and the dependencies among them are listed in the following table. Draw a PERT diagram for this project and give a schedule for the tasks.

Task	Preceding Tasks	Time Required (weeks)
1. Decide on group	none	2
2. Choose advisor	1	1
3. Choose project	1, 2	4
4. Divide responsibilities for project among group	1, 2, 3	2
5. Do research	1, 2, 3	4
6. Develop rough outline of project	1, 2, 3	2
7. Refine outline	1, 2, 3, 4, 5, 6	2
8. Complete project	1, 2, 3, 4, 5, 6, 7	4
9. Arrange for presentation	1, 2, 3, 4, 5, 6, 7, 8	1

11. **Building a student union building.** Waldenville Community College is planning to build a new student union building. The subtasks for this project and their dependencies are listed in the following table. Draw a PERT diagram for this project and give a schedule for the tasks.

Task	Preceding Tasks	Time Required (months)
1. Get funding	none	3
2. Choose contractor	1	2
3. Draw plans	1	4
4. Grade land	1, 2, 3	1
5. Lay underground utilities, sewers, etc.	1, 2, 3, 4	1
6. Build building	1, 2, 3, 4, 5	8
7. Install utilities in building	1, 2, 3, 4, 5, 6	2
8. Install computer network in building	1, 2, 3, 4, 5, 6	1
9. Purchase furniture	1, 2, 3, 4, 5, 6	2
10. Inspect building	1, 2, 3, 4, 5, 6, 7, 8, 9	1

12. Organizing a health program. A developing nation plans to improve available health care for its citizens. Use the following table to draw a PERT diagram for this project and then schedule the tasks so the project can be completed as efficiently as possible.

Task	Preceding Tasks	Time Required (months)
1. Appropriate money	None	3
2. Build health clinics	1	8
3. Construct hospitals	1	18
4. Educate citizens in health practices	1	12
5. Recruit students	1	8
6. Train doctors	1, 3, 5	30
7. Train aides	1, 2, 5	15
8. Inoculate	1, 2, 6, 7	4
9. Conduct follow-up study	All of the above	2

13. Organizing an advertising campaign. The American Legacy Foundation together with MTV is developing an advertising campaign, called *Truth*, to reduce smoking in young adults. A series of newspaper ads, billboards, and TV and radio commercials are to be produced. Use the following table to draw a PERT diagram for this project and then schedule the tasks so the project can be completed efficiently.

Task	Preceding Tasks	Time Required (months)
1. Conduct survey	None	3
2. Develop budget	1	1
3. Hire PR firm	1	6
4. Set up production schedule	1, 2, 3	1
5. Produce ads	1, 2, 3	8
6. Disseminate ads	1, 2, 3, 5	2
7. Evaluate results	1, 2, 3, 4, 5, 6	6

Communicating Mathematics

14. Study Figure 4.35 with the idea of how you would allocate resources to various tasks to shorten the total time for the project. In particular, what happens if you dedicate too many resources to shortening tasks along a critical path? Imagine that you were in charge of budget and could shorten some of the tasks by a total of six months. Where would you shorten tasks to get the shortest time for the project?

15. Write a brief report on a real-life project that can be organized using PERT. Books on operations research and project management often have sections on PERT, which is sometimes called the **critical path method,** or CPM. Find references on the Internet for commercial software that draws PERT diagrams and describe some of these products and their applications.

Challenge Yourself

16. Planning an innovative house. A local electric company wants to build an experimental house that will conserve energy. Company officials believe the project breaks down naturally into the following tasks (the time required for each task is given in parentheses):

1. Draw plans for house (2 months)
2. Design energy systems for house (6 months)
3. Develop new insulation techniques (3 months)
4. Purchase land (4 months)
5. Build conventional shell of house (6 months)
6. Install new energy systems (2 months)
7. Install new insulation (1 month)
8. Finish the remaining conventional parts of the house (3 months)
9. Landscape (1 month)
10. Perform tests to determine energy usage in house (8 months)

Determine a reasonable list of dependencies for these tasks, then draw a PERT diagram and develop a schedule for these tasks.

17. Select a project that interests you. Identify the subtasks and draw a PERT diagram to develop a schedule for the project.

CHAPTER SUMMARY

Section	Summary	Example
SECTION 4.1	A **graph** is a set of points, called **vertices,** and lines, called **edges,** that join pairs of vertices.	Definitions, p. 143
	A graph is **connected** if it is possible to travel from any vertex to any other vertex of the graph by moving along successive edges. A **bridge** in a connected graph is an edge such that if it were removed, the graph would no longer be connected.	Definitions, p. 145 Example 1, p. 145
	A vertex is **odd** if it is an endpoint of an odd number of edges. A vertex is **even** if it is the endpoint of an even number of edges. The **degree** of a vertex is the number of edges joined to that vertex.	
	Euler's theorem states that a graph can be traced if it is connected and has zero or two odd vertices.	Discussion, p. 146 Example 2, pp. 146–147
	A graph can be used to **model** a set of objects that have some relationship among them.	Problem Solving, p. 147
	We can use **Fleury's algorithm** to find Euler circuits.	Example 4, pp. 149–150
	It is always possible to **color a map** with at most four colors so that any two regions sharing a common border will have different colors.	Example 6, pp. 151–152
SECTION 4.2	We use Hamilton circuits to solve the **Traveling Salesperson Problem (TSP),** which asks: What is the most efficient way for a salesperson to schedule a trip to a series of cities and then return home? A **Hamilton path** is a path that passes through all the vertices of a graph exactly once. A **Hamilton circuit** is a Hamilton path that begins and ends at the same vertex.	Discussion, pp. 157–158 Example 1, p. 159
	In a **complete graph,** every pair of vertices is joined by an edge. A complete graph with n vertices is denoted by K_n and has $(n - 1)!$ Hamilton circuits.	Examples 2–3, pp. 159–160
	A **weighted graph** has numbers assigned to its edges. The **weight of a path** is the sum of the weights of the edges of the path.	Discussion, pp. 160–161 Definitions, p. 162
	The **brute force algorithm** for solving the TSP is as follows: Step 1: List all the Hamilton circuits in the graph. Step 2: Find the weight of each circuit found in step 1. Step 3: The circuits with the smallest weights are solutions to the TSP.	Example 5, p. 162
	The **nearest neighbor algorithm** for solving the TSP is as follows: Step 1: Start at any vertex X. Step 2: Choose the edge with the smallest weight that is connected to X. (There may be several edges with the smallest weight.) Select the vertex at the other end of this edge. This vertex is called the *nearest neighbor* of X. Step 3: Choose subsequent *new* vertices as you did in step 2, by considering the weights of the edges connected to the current vertex. Step 4: After all vertices have been chosen, return to the starting vertex.	Example 6, p. 164
	The **best edge algorithm** for solving the TSP is as follows: Step 1: Begin by choosing the edge with the smallest weight. Step 2: Keep choosing *new* edges as in step 1. However, do not let a circuit be formed until all vertices have been used, and never allow three edges to be joined to the same vertex.	Example 7, pp. 164–165
SECTION 4.3	An edge in a graph that is given a direction is called a **directed edge.** A graph with all directed edges is called a **directed graph.**	Discussion, p. 168 Example 1, pp. 168–169
	If we can begin at vertex X in a directed graph and follow a sequence of edges in the direction indicated to reach vertex Y, that path is called a **directed path.**	Example 2, p. 169
	We use directed graphs as **models** when the relationship between objects is not necessarily exerted in both directions. This is the case with influence, dominance, and disease transmission.	Example 3, pp. 169–171 Example 4, p. 172
SECTION 4.4	We can use a directed graph called a **PERT diagram** to schedule the order of tasks in a complex project.	Discussion, pp. 176–177
	A **critical path** for task T is a path from "Begin" to T that requires the most time. The time required before we can begin task T tells us when to schedule T.	Examples 1 and 2, pp. 177, 178

CHAPTER REVIEW EXERCISES

Section 4.1

1.

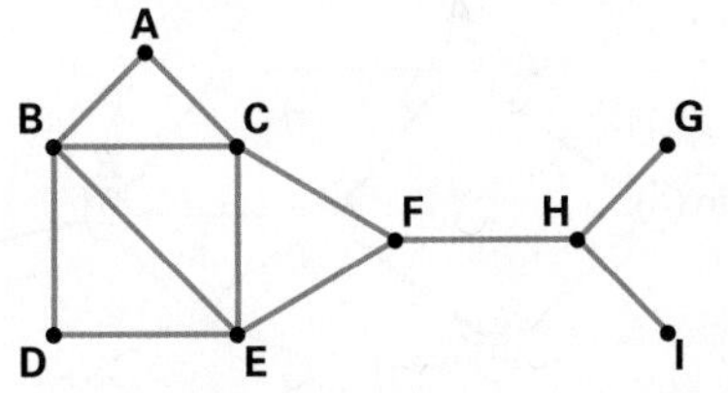

Use the preceding graph to answer the following questions.

a. How many edges does the graph have?

b. Which vertices are odd? Which are even?

c. Is the graph connected?

d. Does the graph have any bridges?

2. Explain how graphs are used to model a collection of objects in which some of the objects are related to each other. Give an example.

3. Which of the following graphs can be traced? Explain your answer by referring to Euler's theorem.

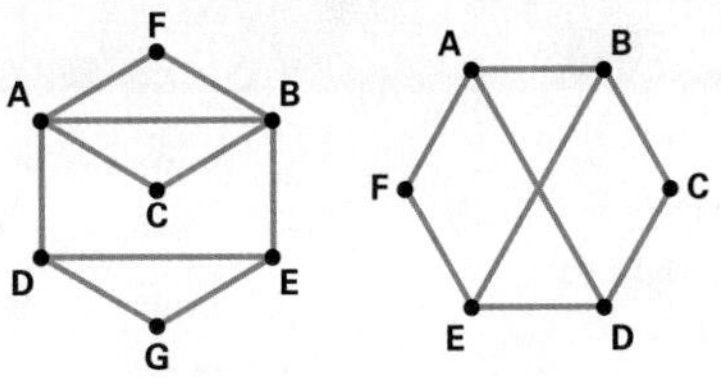

(a) (b)

4. Use Fleury's algorithm to find an Euler circuit in the following graph. Describe the circuit by listing the vertices on the path.

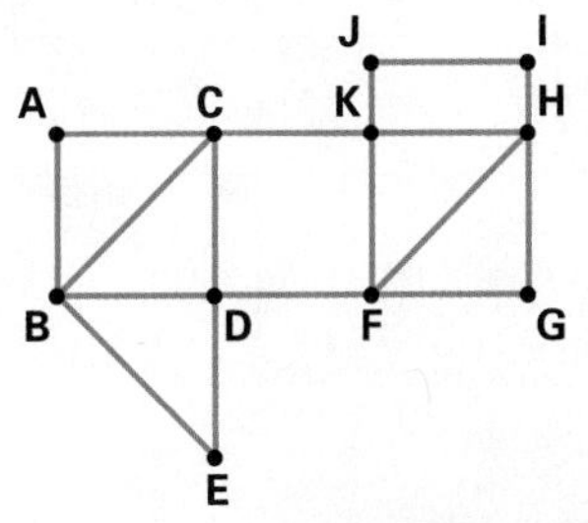

5. Model the following street map with a graph and design an efficient way of traveling through all the streets so a minimal number of streets are traveled more than once.

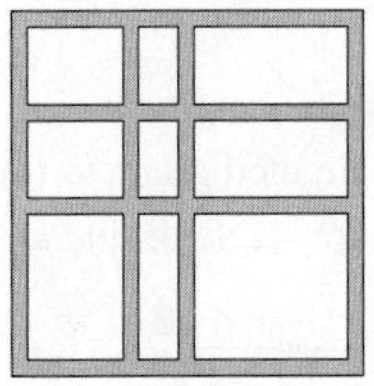

6. Model the following "map" of countries A, B, C,..., M by a graph and then devise a way of coloring the map using a minimal number of colors.

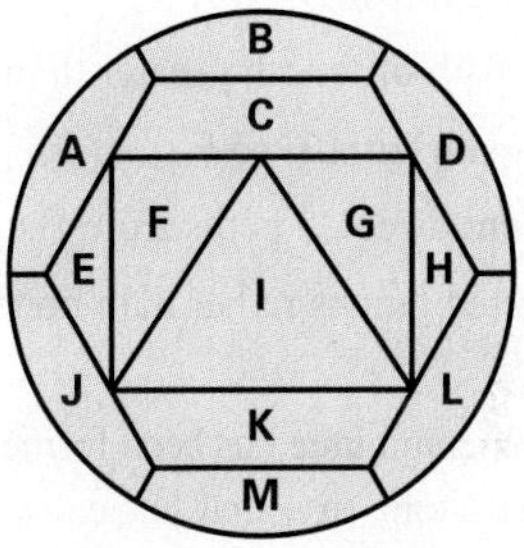

7. Allison, Branden, Colin, Donny, Erica, Jim, Kami, Lance, Marshall, and Nicole are competing on a whitewater rafting trip in the Amazing Race and will be traveling in several canoes. In considering a variety of factors such as weight, skill level, etc., the trip leaders have decided that certain people should not be in the same canoe as certain other people. The table below shows who should not travel together. We will abbreviate using the first letter of each name and will not list duplicate information; for example, if A cannot travel with B, we will not also list that B cannot travel with A. How many canoes are needed? Develop a plan as to who can travel together.

Person	A	B	C	D	E	K	L	M
Cannot Travel With	B, J, D	D, J	K, L, E	J	K, L	D, L	M, N	N

Section 4.2

8. Find all Hamilton circuits that begin at vertex A in K_5 and pass next through vertex B.

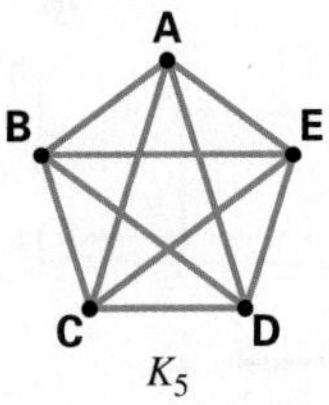

K_5

Use the following weighted graph to answer Exercises 9–11. There may be several possible correct answers. We will provide only one in the answer key.

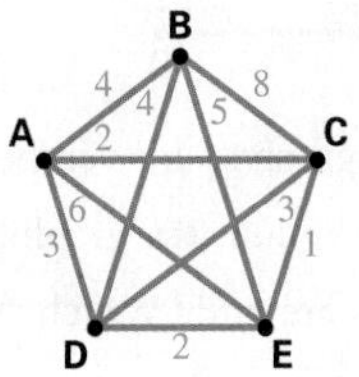

9. Use the brute force algorithm to find a Hamilton circuit that has minimal weight.

10. Use the nearest neighbor algorithm to find a Hamilton circuit that begins at vertex A.

11. Use the best edge algorithm to find a Hamilton circuit that begins at vertex A.

Section 4.3

12. Use the following directed graph to find the requested items, if possible. If it is not possible to find an item, explain why not.

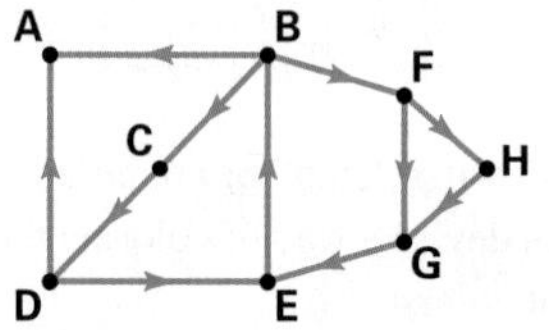

a. A directed path from C to B

b. A directed path of length greater than 5 from C to B

c. A directed path from H to F

d. Two different directed paths from F to B

13. When are directed graphs rather than nondirected graphs used as models?

14. A student action committee has been formed on your campus to lobby local legislators on social justice issues such as poverty, discrimination, environmentalism, taxation, diversity, etc. The influence exerted among committee members is listed in the following table. Use a graph to determine one- and two-stage influence among these people and determine who is the most influential committee member.

Committee Member	Committee Members Influenced by This Person
Pham	Stein, Vaccaro, Alvarez
Vaccaro	Alvarez
Stein	Vaccaro, Bartkowski
Robinson	Stein, Alvarez
Bartkowski	Vaccaro
Alvarez	Stein, Bartkowski

Section 4.4

15. Use the following PERT diagram to answer the following questions.

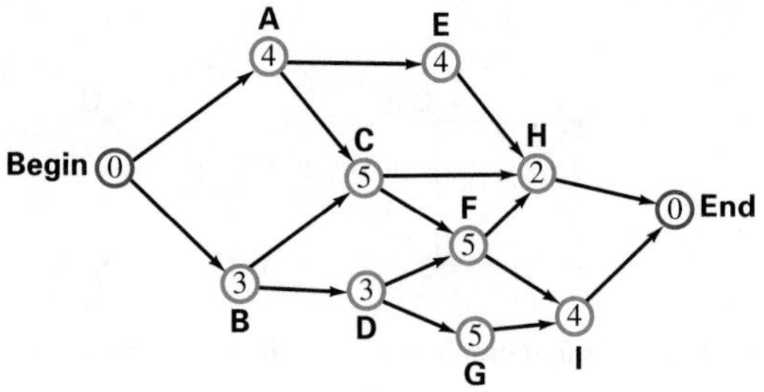

a. Find a critical path for task H.

b. What is a critical path for "End"?

c. When should task F be scheduled?

d. How many days are required for the whole project?

16. Congratulations! You are planning your wedding. The tasks and their dependencies for this happy event are listed in the following table. Develop a schedule for these tasks.

Task	Preceding Tasks	Time in Months
1. Decide who will finance wedding	none	2
2. Decide on date for wedding	none	2
3. Hire wedding planner	1, 2	3
4. Decide on wedding party	1, 2	2
5. Create guest list	1, 2	3
6. Decide on places for ceremony and reception	1, 3	2
7. Decide on music, florist, pictures	1, 3	3
8. Decide on menu	1, 3, 6	2

CHAPTER TEST

1.

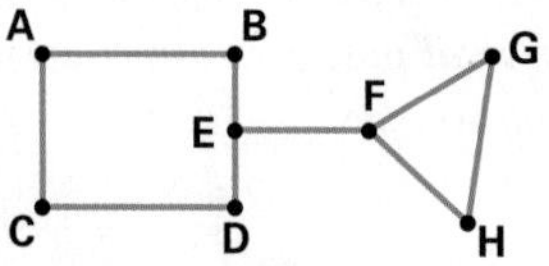

Use the preceding graph to answer the following questions.

a. How many edges does the graph have?

b. Which vertices are odd? Which are even?

c. Is the graph connected?

d. Does the graph have any bridges?

2. Which of the following graphs can be traced? If a graph cannot be traced, explain what part of Euler's theorem fails.

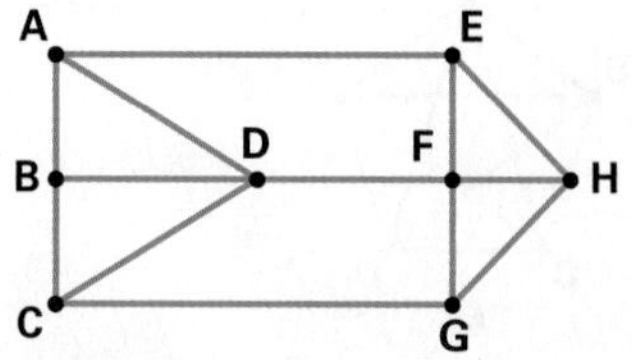

(a)

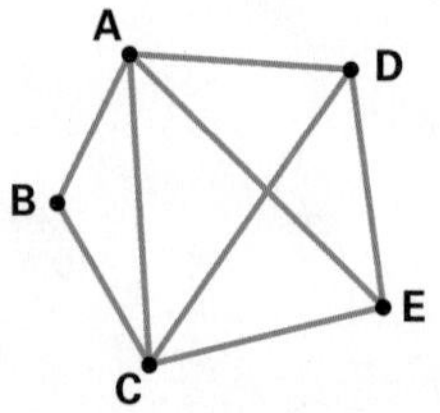

(b)

3. Use Fleury's algorithm to find an Euler circuit in the following graph beginning at vertex A. Describe the circuit by listing its vertices.

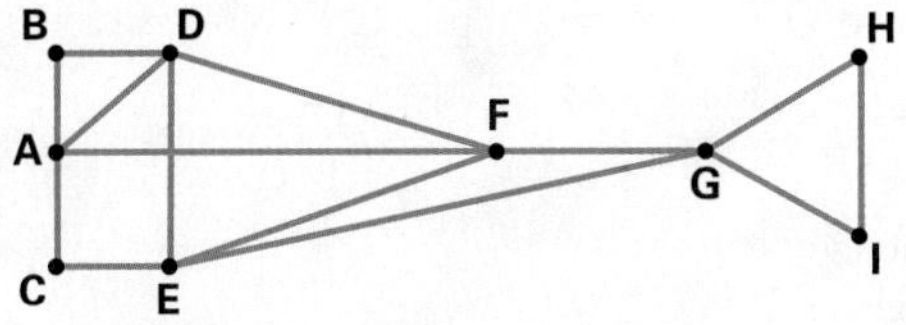

4. Find all Hamilton circuits that begin at vertex A in K_5 and pass next through vertex D.

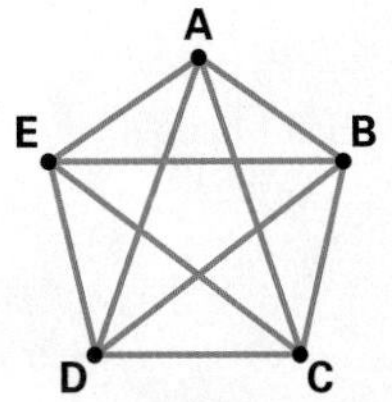

5. Model the following street map with a graph and design an efficient way of traveling through all the streets so that a minimal number of streets are traveled more than once.

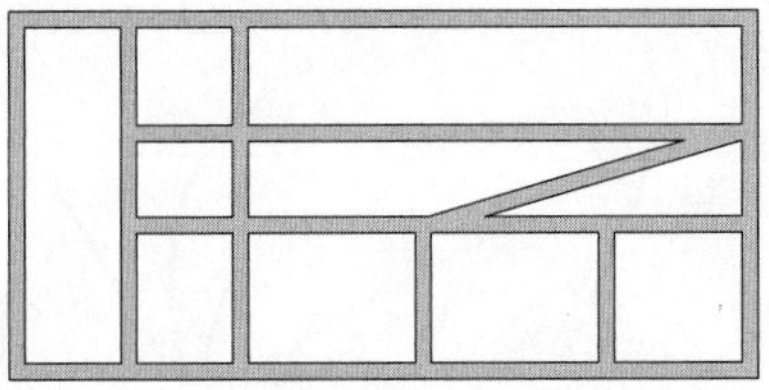

6. Model the following "map" of countries A, B, C,..., H by a graph and then devise a way of coloring the graph using a minimal number of colors.

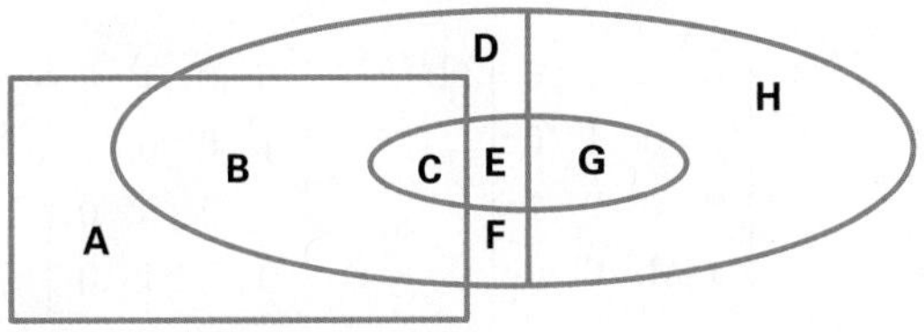

Use the following weighted graph to answer Exercises 7–9. Begin all circuits at vertex A. There may be several possible correct answers. We will provide only one in the answer key.

7. Use the brute force algorithm to find a Hamilton circuit that has minimal weight.

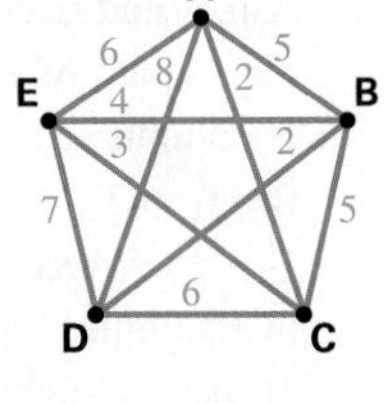

8. Use the nearest neighbor algorithm to find a Hamilton circuit that has minimal weight.

9. Use the best edge algorithm to find a Hamilton circuit that has minimal weight.

10. Use the given directed graph to find the following items. If it is not possible, explain why not.

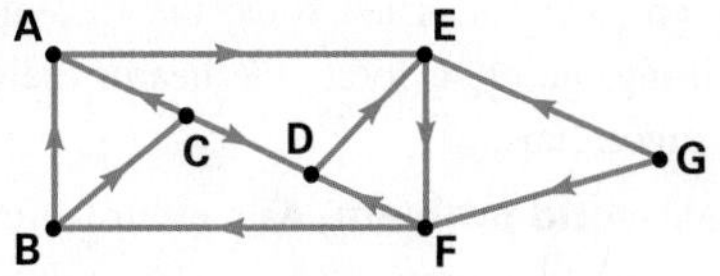

a. A directed path from B to F

b. A directed path from C to G

11. Use the given PERT diagram to answer the following questions.

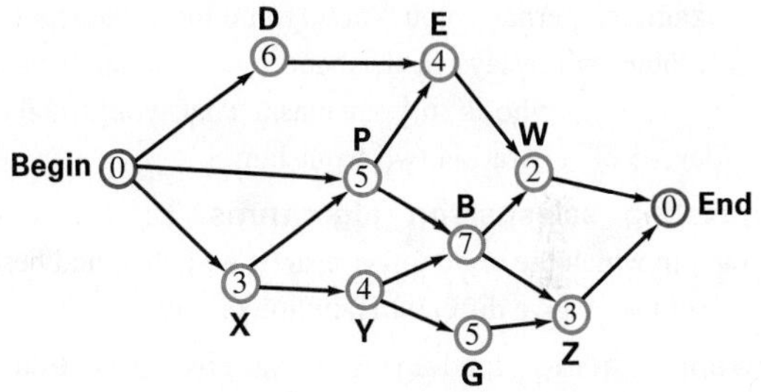

a. Find a critical path for task Z.

b. What is a critical path for "End"?

c. When should task W be scheduled?

d. How many days are required for the whole project?

12. The following graph models one- and two-stage influence among a group of people: A, B, C,..., F. Find the person who has the most influence.

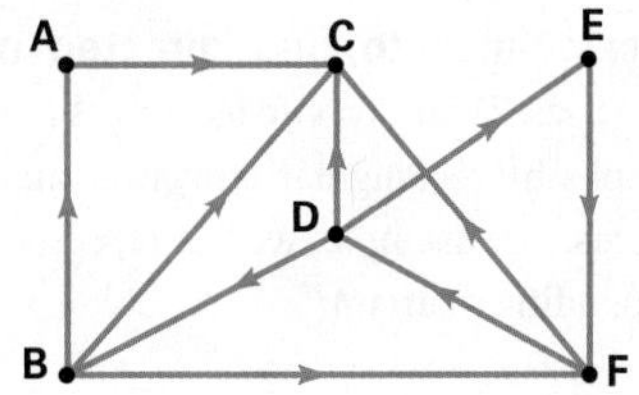

GROUP PROJECTS

1. **The traveling salesperson problem.** As a group, pick five locations either on campus or in town and measure the distance between each pair of locations. If the locations are within walking distance, you can use a pedometer to measure the distance. If the locations are more distant, you can use the odometer on your car. Then solve the traveling salesperson problem using the brute force, the nearest neighbor, and the best edge algorithms.
2. **The small world problem.** As a group, discuss the idea of "six degrees of separation."
   a. Have a contest to see who has the smallest degrees of separation from well-known personalities.
   b. Try to determine the degrees of separation among different members of the class (other than the obvious one, because you all are taking the same class). For example, perhaps you know someone outside of class, Jerome, who plays in the band, and another band member, Jamie, who is in your class. That would make your degree of separation two from Jamie.
3. **Traveling salesperson algorithms.** Draw a weighted graph in which the brute force, nearest neighbor, and best edge algorithms all give different Hamilton circuits.
4. **Graph tracing.** Find maps of various cities around the country and the world and make up problems similar to the Koenigsberg bridge problem. For example, here is a map of the San Francisco area and its corresponding graph.

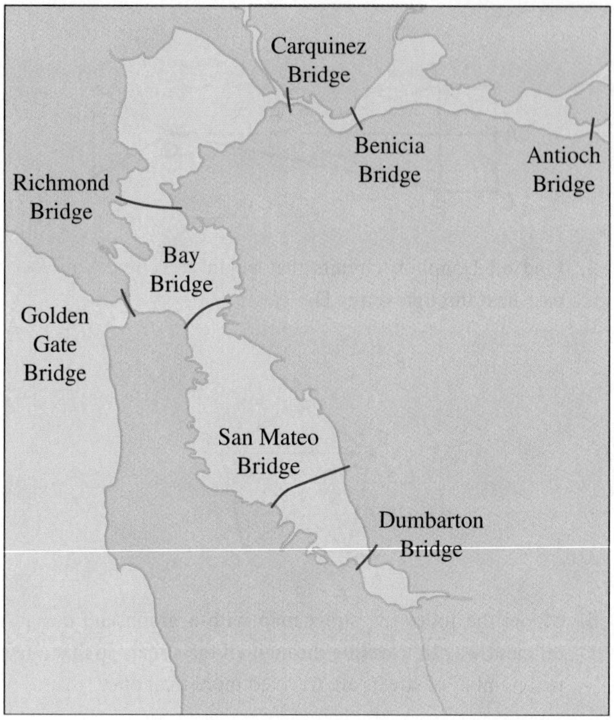

USING TECHNOLOGY

1. **Using a calculator to find directed paths.** We have seen in Exercises 21 to 26 on page 175 that we can represent directed graphs by rectangular arrays of numbers called incidence matrices. For example, we can represent the given graph by a corresponding matrix M.

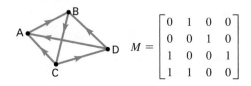

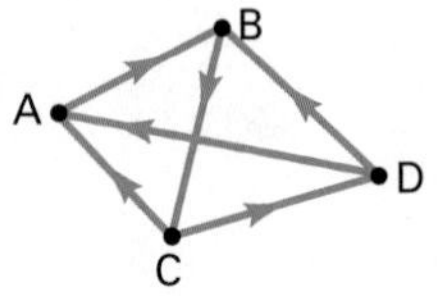

$$M = \begin{bmatrix} 0 & 1 & 0 & 0 \\ 0 & 0 & 1 & 0 \\ 1 & 0 & 0 & 1 \\ 1 & 1 & 0 & 0 \end{bmatrix}$$

Many graphing calculators have the capability to calculate with matrices. We can use this capability to find directed paths in a directed graph. Here is a screen for the matrix M that we entered on a graphing calculator, and another screen for M^2.

$$[M] \begin{bmatrix} 0 & 1 & 0 & 0 \\ 0 & 0 & 1 & 0 \\ 1 & 0 & 0 & 1 \\ 1 & 1 & 0 & 0 \end{bmatrix} \quad [M]\text{^}2 \begin{bmatrix} 0 & 0 & 1 & 0 \\ 1 & 0 & 0 & 1 \\ 1 & 2 & 0 & 0 \\ 0 & 1 & 1 & 0 \end{bmatrix}$$

In each matrix, think of the rows as A, B, C, and D from top to bottom and the columns as A, B, C, and D from left to right. The matrix M^2 counts the number of directed paths of length two in the directed graph. For example, the matrix entry in the A row and C column is 1, which tells us that there is one directed path of length two from A to C. The 2 in the C row and B column tells us that there are two directed paths of length two from C to B, as we can see in the graph.

Now reconsider the disease transmission situation posed in Example 4 on page 172. Model that graph with a matrix M, having eight rows and eight columns. Realize that for the

disease to have spread from one person to all of the others in the group, there must have been directed paths of length seven or less from the vertex representing that person to all of the other vertices. If you consider the sum

$$M + M^2 + M^3 + \cdots + M^7,$$

which counts the number of directed paths of length one, plus the number of directed paths of length two, plus the number of directed paths of length three, and so on, you should be able to see why Dr. Mears' hypothesis was incorrect.

2. **Traveling salesperson applets.** Perform a search of the Internet for "traveling salesman problem applets." You will be able to find sites with interactive programs devoted to the TSP. You may want to search separately or as a group to find a site that you can use easily. Some sites, although they have programs that will give solutions, are difficult to understand. As a group, you may make more sense out of what they are doing than if you work alone. Some sites will allow you to vary the algorithms that are used to solve the problem. Finally, agree on which is the best site and write a brief explanation of what the site does. You may want to present your findings to your class. (You will have better luck searching for "traveling salesman" rather than "traveling salesperson.")

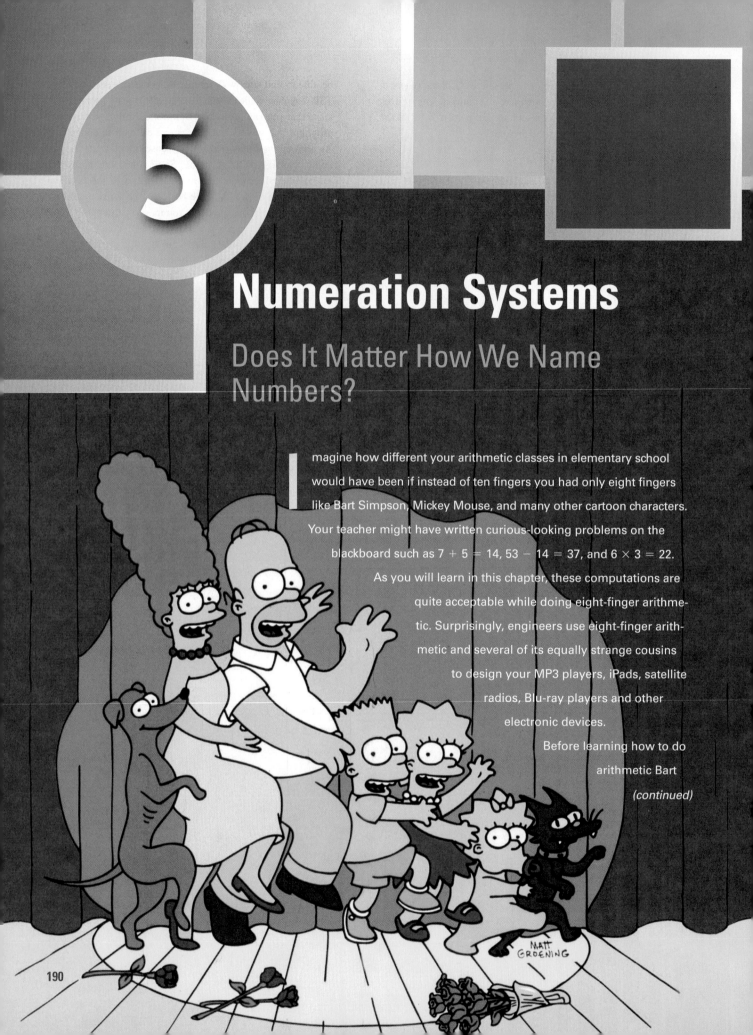

5

Numeration Systems

Does It Matter How We Name Numbers?

Imagine how different your arithmetic classes in elementary school would have been if instead of ten fingers you had only eight fingers like Bart Simpson, Mickey Mouse, and many other cartoon characters. Your teacher might have written curious-looking problems on the blackboard such as $7 + 5 = 14$, $53 - 14 = 37$, and $6 \times 3 = 22$. As you will learn in this chapter, these computations are quite acceptable while doing eight-finger arithmetic. Surprisingly, engineers use eight-finger arithmetic and several of its equally strange cousins to design your MP3 players, iPads, satellite radios, Blu-ray players and other electronic devices.

Before learning how to do arithmetic Bart

(continued)

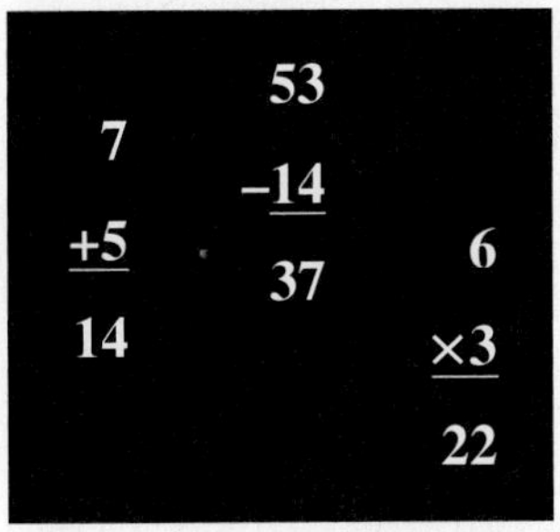

and Mickey's way, you will travel back in time to study Babylonian, Chinese, and Mayan mathematics, where you will see the beginnings of elegant ideas that thousands of years later have blossomed into the Hindu-Arabic system that works so beautifully for us today.

Toward the end of this chapter, we will show you an unusual type of "clock arithmetic" that is widely used in scanning your purchases properly at the grocery store, insuring that your online bank transactions are secure, and keeping your credit card number safe from identity theft.

5.1 The Evolution of Numeration Systems

Objectives

1. Represent numbers and compute in the Egyptian system.
2. Understand the Roman numeration system.
3. Understand the Chinese numeration system.

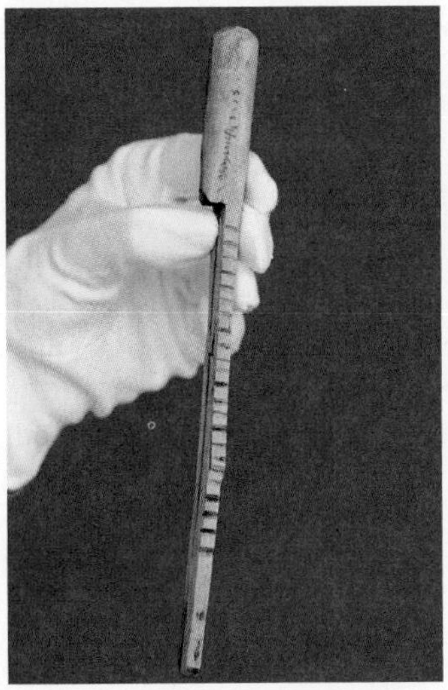

We can easily take a brilliant idea for granted because of its inherent simplicity. Such is the case with our familiar system for representing numbers that is the result of thousands of years of development culminating in the Hindu-Arabic system that we use today. In this section, we will examine some of the early attempts made by various cultures to develop practical numeration systems.

Primitive societies needed only simple counting systems. An early sheepherder might have scratched tally marks in the dirt, tied knots in a vine, cut notches in a tally stick, or matched a pile of small pebbles with his sheep in order to keep track of them. Such early counting methods as these eventually led to the invention of the abstract concept of number. The number 10 could represent 10 sheep, 10 pebbles, or 10 people.

DEFINITIONS A **number** tells us how many objects we are counting; a **numeral** is a symbol that represents a number.

Although the *number* ten would mean the same amount in each culture, the *numeral* for ten (the symbols we use to write the number) varied greatly. For example, Babylonians used the symbol ⟨, Egyptians used ∩, Greeks used the letter iota ι, Romans used X, and we, of course, represent ten by the symbols 10.

We will first discuss the simple grouping systems used by the Egyptians and Romans and then explain the multiplicative grouping system of the Chinese, which has some similarities to our current Hindu-Arabic system that we discuss in Section 5.2.

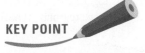

KEY POINT

In a simple grouping system, we represent a number as the sum of the values of its numerals.

The Egyptian Numeration System

The Egyptian hieroglyphic numeration system, which is more than 5,500 years old, is an example of a **simple grouping** system. In the Egyptian system, we form numerals by

Number	Symbol	Name
1	𓏺	Stroke
10	𓎆	Heel bone
100	𓍢	Scroll
1,000	𓆼	Lotus flower
10,000	𓂭	Pointing finger
100,000	𓆐	Fish or tadpole
1,000,000	𓁨	Astonished person

TABLE 5.1 Egyptian numeration symbols.

combining the symbols that represent various powers of 10, as shown in Table 5.1. The value of the numeral is the sum of the values of the numerals. For example, we would write 325 as follows:

𓍢𓍢𓍢𓎆𓎆𓏺𓏺𓏺𓏺𓏺.

3 hundreds, 2 tens, 5 ones

Because the order of the symbols is not important, we could also write 325 as

𓏺𓏺𓍢𓍢𓎆𓍢𓎆𓏺𓏺𓏺

EXAMPLE 1 Converting Between Egyptian and Hindu-Arabic Notation

a) Convert 𓆼𓆼𓆼𓍢𓍢𓍢𓍢𓎆𓎆𓏺𓏺𓏺𓏺𓏺 to Hindu-Arabic notation.

b) Write 1,230,041 in Egyptian notation.

QUIZ YOURSELF 1

a) Write the Egyptian numeral 𓂭𓂭𓆼𓍢𓍢𓍢𓎆𓎆𓎆𓎆𓏺𓏺𓏺𓏺𓏺𓏺 in Hindu-Arabic notation.

b) Write the number 241,536 using Egyptian notation.

SOLUTION:

a) We have 3 thousands, 4 hundreds, 2 tens, and 5 ones, so this Egyptian numeral represents

$$3{,}000 + 400 + 20 + 5 = 3{,}425.$$

b) We will represent 1,000,000 by 𓁨, 200,000 by 𓆐𓆐, 30,000 by 𓂭𓂭𓂭, 40 by 𓎆𓎆𓎆𓎆, and 1 by 𓏺, to get

𓁨𓆐𓆐𓂭𓂭𓂭𓎆𓎆𓎆𓎆𓏺. 1

As you can see, the Egyptian system is not an efficient way to represent numbers. The Egyptian numeral 𓏺𓏺𓏺𓏺𓏺𓏺𓏺𓏺𓏺 requires nine times as many symbols as we use to represent nine using the Hindu-Arabic system. The number 68 requires 14 symbols, 𓎆𓎆𓎆𓎆𓎆𓎆𓏺𓏺𓏺𓏺𓏺𓏺𓏺𓏺. Imagine how awkward it would be to balance your checkbook or total your restaurant bill using this system.

It is straightforward to add and subtract in the Egyptian hieroglyphic system.

EXAMPLE 2 Adding and Subtracting in the Egyptian System

a) Add 𓍢𓍢𓍢𓎆𓎆𓎆𓎆𓎆𓎆𓎆𓏺𓏺 and 𓍢𓍢𓎆𓎆𓎆𓎆𓎆𓏺𓏺𓏺.

b) Subtract 𓎆𓎆𓎆𓎆𓎆𓎆𓏺𓏺 from 𓍢𓍢𓎆𓎆𓎆𓏺𓏺𓏺𓏺 using Egyptian notation.

SOLUTION: Review the Analogies Principle on page 13.

a) We add these two numbers by simply grouping all the symbols together (the addition sign is not part of the Egyptian notation). After obtaining the sum, we regroup 10 heel bones as one scroll.

𓍢𓍢𓍢𓎆𓎆𓎆𓎆𓎆𓎆𓎆𓏺𓏺
\+ 𓍢𓍢𓎆𓎆𓎆𓎆𓎆𓏺𓏺𓏺
𓍢𓍢𓍢𓍢𓍢𓎆𓎆𓎆𓎆𓎆𓎆𓎆𓎆𓎆𓎆𓎆𓎆𓏺𓏺𓏺𓏺𓏺

Rewrite 10 heel bones as one scroll.

We can now rewrite this answer as

𓍢𓍢𓍢𓍢𓍢𓍢𓎆𓎆𓏺𓏺𓏺𓏺𓏺.

Historical Highlight

Solving the Mystery of Egyptian Hieroglyphics

In 1798, the French emperor Napoleon sailed with a large army to conquer Egypt and disrupt the lucrative trade routes to India. Although he was severely defeated, this military disaster turned out to be a scientific triumph for Europe. Napoleon had taken scholars with him to study Egypt's culture, and when they returned, they brought back a wealth of information about this ancient civilization. Unfortunately, much of the material was written in a form of hieroglyphics called **demotic script,** which no one was able to translate.

Fortunately, Napoleon also brought back with him the key to solving this puzzle—a polished stone, called the **Rosetta stone,** that contained writing in Greek, demotic script, and ancient hieroglyphics. Scholars believed that the stone contained three versions of the exact same text and set about using their knowledge of Greek to translate the other two cryptic sections.

After returning from Egypt with Napoleon, the French mathematician Jean-Baptiste Fourier showed some

hieroglyphics to an 11-year-old boy named Jean Francois Champollion. When Fourier stated that no one could read hieroglyphics, the boy replied boldly, "I will do it when I am older," and from then on he dedicated his life to translating hieroglyphics. Legend has it that when Champollion finally solved the mystery of hieroglyphics, he exclaimed, "I've got it," and fainted.

Much of our knowledge of early Egyptian mathematics comes from the Rhind papyrus, named after a Scotsman named A. Henry Rhind, who purchased it in 1858. The scroll was written in 1650 BC by a scribe named Ahmes, who claimed that the document contained "a thorough study of all things, insights into all that exists, [and] knowledge of all obscure secrets." This was an exaggerated promise, because when the papyrus was translated, it was found to contain only a collection of mathematical exercises and rules for doing multiplication and division.

QUIZ YOURSELF 2

Perform the following operations using Egyptian notation:

a) 999∩∩∩∩IIIIII +
 99999999∩∩IIIII

b) 𐦯𐦯𐦯𐦯𐦯∩∩∩∩II −
 𐦯𐦯9999∩∩∩IIIIIIIII

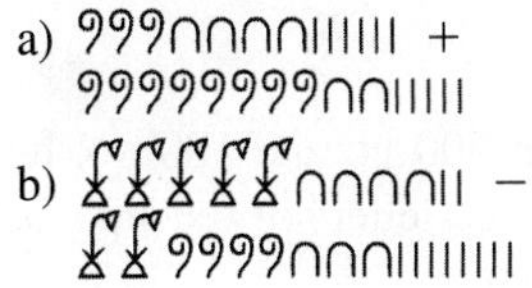

b) We have the reverse problem from part a). Although we can subtract two strokes from four strokes, we cannot subtract six heel bones from three heel bones unless we "borrow." That is, we convert one scroll to 10 heel bones.

Rewrite one scroll as 10 heel bones.

Rewrite 99∩∩∩IIII 9∩∩∩∩∩∩∩∩∩∩∩∩IIII
 − ∩∩∩∩∩∩II as − ∩∩∩∩∩∩II
 9∩∩∩∩∩∩∩II
 └── 13 − 6 heel bones

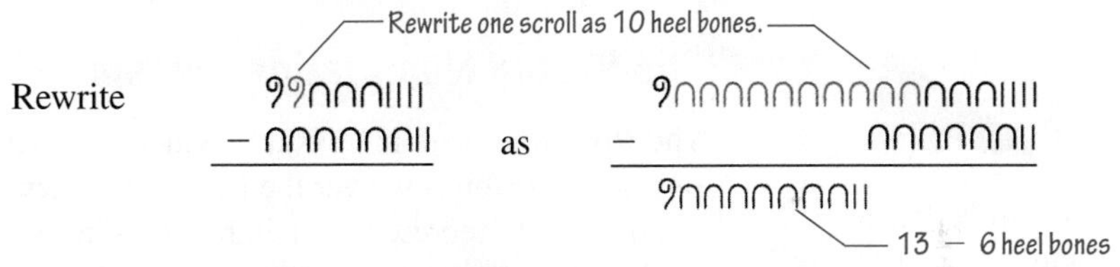

Now try Exercises 13 to 20. 2

Notice in Example 2 how the Egyptian calculations remind us of the way we carry and borrow in performing addition and subtraction using Hindu-Arabic notation.

SOME GOOD ADVICE

You can check your work in performing Egyptian arithmetic by converting all the numbers to Hindu-Arabic notation and then redoing the computations.

Power of 2	Value
2^0	1
2^1	2
2^2	4
2^3	8
2^4	16
2^5	32
2^6	64

TABLE 5.2 Powers of 2.

The Egyptians also had a method for doing multiplication, as we show in Example 3.* This method is based on the fact that any positive integer can be written as the sum of powers of 2 (see Table 5.2). For example, $19 = 1 + 2 + 16$ and $81 = 1 + 16 + 64$.

*In explaining this doubling method, we will use Hindu-Arabic numerals rather than the cumbersome Egyptian numerals.

EXAMPLE 3 Computing Area Using the Egyptian Method of Doubling

A craftsman is restoring a rectangular wall of the Temple of Queen Hatshepsut (the fifth pharaoh of the Eighteenth Dynasty of Ancient Egypt) with tiles. If the wall measures 13 feet by 21 feet, use the Egyptian method of doubling to determine how many square feet must be covered.

SOLUTION: Review the Be Systematic Strategy on page 5.

First we write 13 as 1 + 4 + 8. To find the area, we will calculate 13 × 21 as follows:

$$13 \times 21 = (1 + 4 + 8) \times 21^{*} = 1 \times 21 + 4 \times 21 + 8 \times 21 = 21 + 84 + 168 = 273.$$

We find these products by repeatedly doubling 21 in Table 5.3.

Power of 2	Times 21
1	21
2	21 + 21 = 42
4	42 + 42 = 84
8	84 + 84 = 168
16	168 + 168 = 336

13 = 1 + 4 + 8

13 × 21 = 21 + 84 + 168

TABLE 5.3 Powers of 2 times 21.

So, the craftsman must use 273 square feet of tile for the temple wall.

Now try Exercises 21 to 24. 3

QUIZ YOURSELF 3

a) Represent 25 as the sum of powers of 2.
b) Use the Egyptian method of doubling to calculate 25 × 43.

The Roman Numeration System

KEY POINT

The Roman numeration system is a more sophisticated simple-grouping numeration system.

The Roman numeration system, which was developed between 500 BC and 100 AD, has several improvements over the Egyptian system. The Romans used letters of the alphabet as numerals to represent certain numbers, as we show in Table 5.4.

We show a simple example of translating a Roman numeral to Hindu-Arabic notation in Example 4.

Number	Roman Numeral
1	I
5	V
10	X
50	L
100	C
500	D
1,000	M

TABLE 5.4 Roman numerals.

EXAMPLE 4 A Simple Conversion of a Roman Numeral to Hindu-Arabic Notation

Convert DCLXXVIII to Hindu-Arabic notation.

SOLUTION:

It is important to recognize that in DCLXXVIII, a numeral with a smaller value never occurs to the left of a numeral with a larger value. In this case we simply add the values of the numerals as follows:

$$\begin{array}{ccccccccc} D & C & L & X & X & V & I & I & I \\ \downarrow & \downarrow & \downarrow & \downarrow & \downarrow & \downarrow & \downarrow & \downarrow & \downarrow \end{array}$$
$$500 + 100 + 50 + 10 + 10 + 5 + 1 + 1 + 1 = 678.$$

One big advantage that the Roman system had over the Egyptian system was a **subtraction principle** that allowed the Romans to represent numbers more concisely than the

*We are using the property that multiplication distributes over addition here.

Historical Highlight

The Algorists versus the Abacists*

In 1299, the rulers of Florence, Italy, passed a law forbidding bankers from using Hindu-Arabic numerals (our commonly used numerals 1, 2, 3, . . .). Although the more efficient Hindu-Arabic numerals had been introduced to Europe 500 years earlier, they were still outlawed as late as the fourteenth century.

To understand why such a law might be necessary, we first must understand how computations were done using Roman numerals. Archaeologists have found stone and marble counting boards called abaci† (plural of *abacus*), which were tablets with four grooves cut into them, as we show in Figure 5.1.

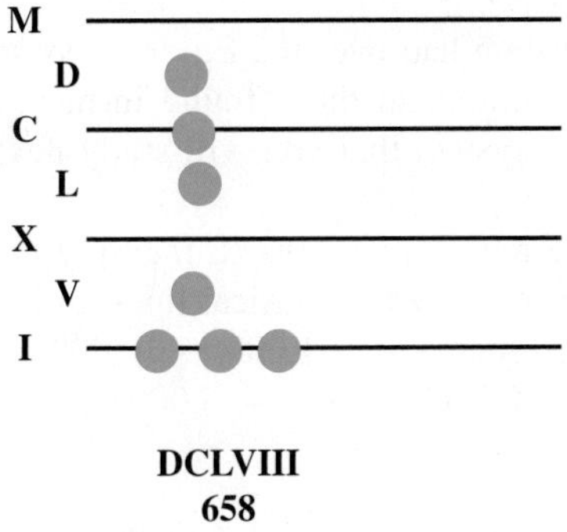

FIGURE 5.1 Roman abacus showing 658.

The grooves corresponded to 1, 10, 100, and 1,000, and the spaces represented the intermediate values, 5, 50, and 500. The **abacist** (person using the abacus) represented Roman numerals by placing small stones in the grooves and spaces as shown in Figure 5.1. Computations were done by moving and regrouping stones. For example, two stones in the 5 space could be replaced by one stone in the 10 groove. It was easy for the ordinary person to understand such calculations.

In contrast, **algorists** did Hindu-Arabic calculations with pen and paper—but people were suspicious of this new method in which one ink mark could represent two, three, or even four counters. Even worse, algorists had a symbol that stood for no counters at all! In medieval times, the notion of zero was confusing to many people and not necessary to those who used Roman numerals. Furthermore, unscrupulous merchants and bankers could cheat uneducated customers by changing a Hindu-Arabic numeral such as a 0 to a 6 or a 1 to a 4.

Eventually, because of the efficiency of Hindu-Arabic notation, the algorists won the battle, which is probably why you do not calculate with counting boards and pebbles today.

Egyptians could. With the subtraction principle, *if the value of a numeral is ever less than the value of the numeral to its right*, then the value of the left numeral is subtracted from the value of the numeral to its right. For example,

$$\begin{aligned} \text{IV represents } 5 - 1 &= 4, \\ \text{IX represents } 10 - 1 &= 9, \\ \text{XL represents } 50 - 10 &= 40, \\ \text{and CM represents } 1{,}000 - 100 &= 900 \end{aligned}$$

There are two restrictions on this subtraction principle:

1. We can only subtract the numerals I, X, C, and M. For example, we cannot use VL to represent 45.
2. We can only subtract numerals from the next two higher numerals. For instance, we can only subtract I from V and X; therefore, we cannot use IC to represent 99.

Notice that the subtraction principle allows us to write multiples of 4 and 9 more efficiently than we could using Egyptian notation. For example, instead of writing 99 as LXXXXVIIII, we can write it as XCIX, saving six symbols.

*For more on this, see "The Algorists vs. the Abacists: An Ancient Controversy on the Use of Calculators," *The College Mathematics Journal*, Vol. 24, No. 3, May 1993.

†Chinese and Japanese abaci were invented much later and are much different from the Roman ones. We will discuss these other abaci in Section 5.2.

QUIZ YOURSELF 4

a) Identify two errors made in writing the Roman numeral LDIL.
b) Convert DXLVIII to Hindu-Arabic notation.
c) Convert |IX| and $\overline{V}$ to Hindu-Arabic notation.

EXAMPLE 5 Translating Between Roman Numerals and Hindu-Arabic Notation

a) Convert MCMXLIII to Hindu-Arabic notation.
b) Write 492 in Roman numerals.

SOLUTION:

a) The following diagram shows how to interpret this numeral. Remember, every time we see a smaller numeral to the left of a larger numeral, we must subtract.

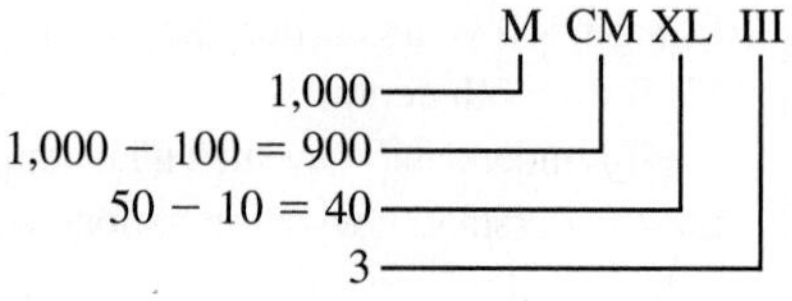

Thus, MCMXLIII represents 1,000 + 900 + 40 + 3 = 1,943.

b) You can write 400 as CD, 90 as XC, and 2 as II to express 492 as CDXCII.

A second advantage the Roman system had over the Egyptian system was that it used a multiplication principle that is similar to those found in more advanced numeration systems, such as the Chinese system that you will study next and also our Hindu-Arabic system.

In the Roman system, a bar above a symbol means to multiply the value of the symbol by 1,000. Also, bracketing a symbol by two vertical lines multiplies the value of the symbol by 100. Thus, $\overline{X}$ represents $10 \times 1{,}000 = 10{,}000$, $|V|$ represents $5 \times 100 = 500$, and $|\overline{L}|$ represents $50 \times 1{,}000 \times 100 = 5{,}000{,}000$.

Now try Exercises 25 to 42. 4

EXAMPLE 6 Calculating a Roman Sum

A Roman art dealer has sold two statues, one for CCCXXVIIII denarii (plural of *denarius*, a Roman coin) and another for CCCXIII. Add these two numbers on a counting board and represent the sum as a Roman numeral.

SOLUTION: Review the Draw Pictures Strategy on page 3.

We will place all of the counters on a single board and then simplify.

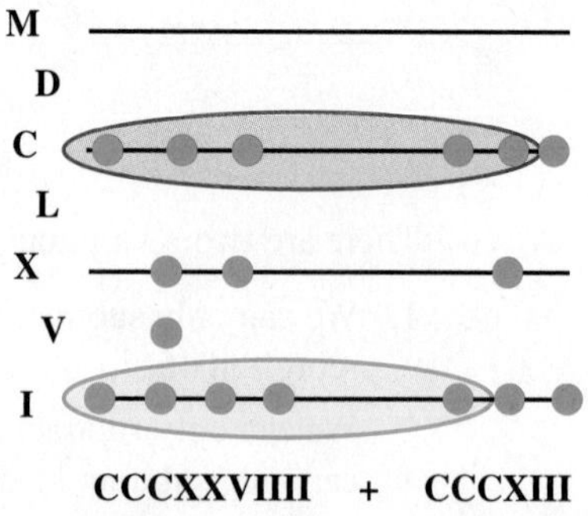

The five ones (circled in blue) can be replaced by a single **V** (five) and the five hundreds (circled in red) can be replaced by a single **D** (five hundred) as shown on the next board. Finally, the two **V** counters (circled in purple) can be replaced by a single **X** counter to get the final sum **DCXXXXII.**

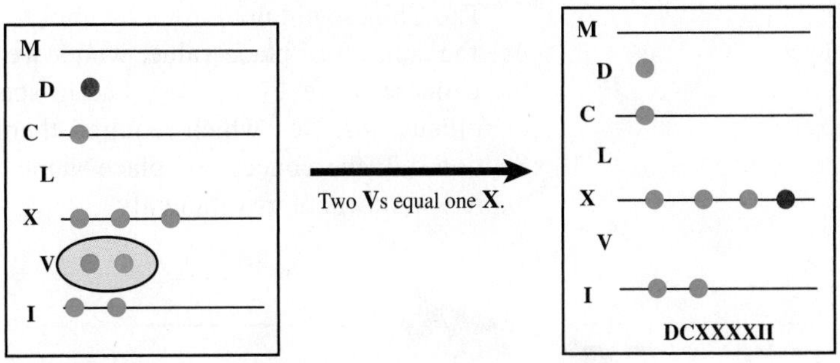

Now try Exercises 67 to 70.

The Chinese Numeration System

KEY POINT

The Chinese numeration system is a multiplicative numeration system.

The traditional Chinese numeration system uses multiples of powers of 10 such as 10, 100, and 1,000 to represent numbers and is based on the notation shown in Table 5.5. These symbols originated during the Han dynasty, which extended from 206 BC to 220 AD. Thus, these symbols have not changed much during the past 2,000 years.*

This system is an example of a **multiplicative system** because we form numerals by writing products of integers between 1 and 9, inclusive, and powers of 10. Traditionally, the ancient Chinese wrote numerals vertically. For example, to write 300, they would write

三 ← 3
百 ← times 100,

and they would write 5,000 as

五 ← 5
千 ← times 1,000.

However, the Chinese now write these symbols horizontally, and we will also follow this modern practice. (There are also other features of modern Chinese notation that we will not go into here but will discuss in the exercises.) Therefore, we will write 300 as 三百 and 5,000 as 五千.

Chinese Numeral	Value
一	1
二	2
三	3
四	4
五	5
六	6
七	7
八	8
九	9
十	10
百	100
千	1,000

TABLE 5.5 Chinese numerals.

EXAMPLE 7 Translating Between Chinese and Hindu-Arabic Notation

a) Express 九百四十二 in Hindu-Arabic notation.

b) Write 3,542 in Chinese notation.

SOLUTION:

a) As you can see in Figure 5.2, the first two symbols represent the product of 9 and 100, or 900. The second two symbols represent 4 times 10, or 40, and the last figure represents 2. Thus, the number we are representing is 942.

Notice how we write the units numeral without multiplying by a power of 10.

九百 → 9×100 四十 → 4×10 二 → 2

FIGURE 5.2 Translating Chinese numerals.

b) To express this number, we need 3 thousands, 5 hundreds, 4 tens, and 2 units. We write this as

三千   五百   四十   二

↑   ↑   ↑   ↑

$$3 \times 1{,}000 \quad + \quad 5 \times 100 \quad + \quad 4 \times 10 \quad + \quad 2.$$

Now try Exercises 43 to 54.

*The Chinese numeration system that we describe here was used for representing numbers (much as we use Roman numerals to number the Super Bowls) but *not* for doing calculations. Another Chinese system, called a "rod system," was actually used as early as the thirteenth or fourteenth centuries to do computations. Some believe that this early rod system was adopted by the Arabs and Indians and led to our current Hindu-Arabic numeration system.

The Chinese did not have a symbol for zero in the traditional notation, nor did they use the concept of place value, which we will discuss in Section 5.2. Therefore, when the Chinese wrote 六 (6), they had to specify whether they meant 6 tens, 6 hundreds, or 6 thousands, etc., which required them to use an extra symbol. As you will see in Section 5.2, the concept of place value allows us to avoid these extra symbols and express numbers more efficiently.

EXERCISES 5.1

Sharpening Your Skills

Write the Egyptian numerals using Hindu-Arabic numerals.

1. 𓍢𓍢𓍢𓎆𓎆𓏺𓏺𓏺𓏺
2. 𓎆𓍢𓍢𓍢𓍢𓎆𓎆𓏺𓏺
3. 𓂭𓂭𓆼𓍢𓍢𓍢𓎆𓎆𓏺𓏺𓏺𓏺
4. 𓆐𓆐𓂭𓂭𓂭𓆼𓆼𓍢𓎆𓎆𓏺𓏺𓏺
5. 𓁨𓁨𓆐𓂭𓂭𓆼𓍢𓍢𓎆
6. 𓁨𓆐𓆐𓂭𓂭𓂭𓍢𓍢𓎆𓏺𓏺𓏺

Write each Hindu-Arabic numeral using Egyptian numerals.

7. 245 **8.** 362 **9.** 3,245
10. 23,416 **11.** 245,310 **12.** 2,036,042

Perform each of the following addition problems using Egyptian notation.

13. 𓍢𓍢𓍢𓎆𓎆𓏺𓏺𓏺𓏺𓏺 + 𓍢𓍢𓍢𓍢𓍢𓍢𓍢𓍢𓎆𓎆𓏺𓏺𓏺𓏺𓏺𓏺𓏺𓏺
14. 𓍢𓎆𓎆𓎆𓎆𓎆𓎆𓎆𓎆𓏺𓏺𓏺𓏺𓏺𓏺 + 𓍢𓍢𓍢𓍢𓍢𓍢𓎆𓎆𓎆𓎆𓏺𓏺𓏺𓏺𓏺𓏺𓏺
15. 𓁨𓂭𓂭𓂭𓂭𓆼𓍢𓍢𓎆 + 𓁨𓆐𓂭𓂭𓂭𓂭𓂭𓂭𓂭𓂭𓍢
16. 𓆼𓆼𓆼𓆼𓆼𓍢𓍢𓍢𓍢𓍢𓍢 + 𓂭𓆼𓆼𓆼𓆼𓆼𓆼𓍢𓍢𓍢𓍢𓍢𓎆

Perform each of the following subtraction problems using Egyptian notation.

17. 𓍢𓍢𓍢𓍢𓍢𓍢𓎆𓎆𓏺𓏺𓏺𓏺𓏺𓏺𓏺𓏺 − 𓍢𓍢𓍢𓎆𓎆𓎆𓎆𓎆𓏺𓏺𓏺𓏺
18. 𓆼𓍢𓍢𓎆𓎆𓎆𓏺𓏺𓏺 − 𓍢𓍢𓍢𓎆𓎆𓏺𓏺𓏺𓏺
19. 𓂭𓂭𓂭𓍢𓍢𓍢𓍢𓍢 𓂭𓂭𓆼𓆼𓆼𓆼𓍢𓍢𓍢𓎆𓎆𓎆𓎆𓎆𓏺𓏺𓏺𓏺
20. 𓁨𓆐𓂭𓂭𓂭 − 𓁨𓂭𓂭𓆼

Use the Egyptian method of doubling to calculate the following products.

21. 14 × 43 **22.** 11 × 57
23. 21 × 126 **24.** 35 × 121

Write each Roman numeral using Hindu-Arabic numerals.

25. DLXIV **26.** CLXIX
27. MCMLXIII **28.** MDCXXXVI
29. $\overline{\text{V}}$MMDXLIV **30.** $\overline{\text{X}}$MMMCDLIV
31. |D|CCLXII **32.** |M|DLVII
33. |$\overline{\text{V}}$|MCDXX **34.** |$\overline{\text{L}}$|MMMDX

Write each numeral in Roman notation. (There may be several correct answers.)

35. 278 **36.** 947
37. 444 **38.** 999
39. 4,795 **40.** 3,247
41. 89,423 **42.** 98,546

Write each Chinese numeral as a Hindu-Arabic numeral. (Recall that we are writing Chinese numerals horizontally rather than in the traditional vertical form.)

43. 四百三十六 **44.** 八千五百二十五
45. 五千六十七 **46.** 四千三百四
47. 九千九百九十九 **48.** 四千九十七

Write each numeral using Chinese numerals.

49. 495 **50.** 726
51. 2,805 **52.** 3,926
53. 9,846 **54.** 8,054

Applying What You've Learned

55. The Great Pyramid at Giza was completed in 𓆼𓆼𓍢𓍢𓍢𓍢𓍢𓍢𓎆𓎆𓎆𓎆𓎆 BC. Write this date using Hindu-Arabic notation.

56. Cheops, the builder of the Great Pyramid at Giza, died in 𓆼𓆼𓍢𓍢𓍢𓍢𓍢𓎆𓎆𓏺𓏺𓏺𓏺𓏺𓏺𓏺𓏺 BC. Write this date using Hindu-Arabic notation.

57. An Egyptian merchant has a warehouse that contains 𓆐𓆐𓂭𓂭𓂭𓆼𓆼𓆼𓆼𓍢𓍢𓍢𓍢𓍢𓎆𓎆𓎆𓎆 square feet of storage. If he purchases another warehouse containing 𓆐𓆼𓆼𓎆𓎆𓎆𓎆𓎆𓎆𓎆𓎆 square feet of storage, what is the total square feet of storage that he now has? Do the calculation using Egyptian notation and double-check your answer using Hindu-Arabic notation.

58. An ancient Egyptian merchant had on hand 𓂭𓆼𓆼𓆼𓆼𓍢𓍢𓍢𓎆𓎆𓎆𓎆𓎆 bushels of wheat and sold 𓂭𓆼𓆼𓍢𓍢𓍢𓍢𓍢𓎆𓎆𓎆 to another merchant. How many bushels of wheat does he have remaining? Do the calculation using Egyptian notation and double-check your answer using Hindu-Arabic notation.

Using Egyptian notation, the number 100, symbolized as 𓍢, would be preceded by 99, which requires nine ∩s and nine |s for a total of 18 symbols. How many symbols would be required to write the Egyptian numeral preceding each of the following Egyptain numerals?

59. 𓂭 **60.** 𓆼 **61.** 𓁨 **62.** 𓆐

Frequently, Roman numerals are used today in movie credits to specify the date that a movie was released. Translate each of the following movie dates into Roman notation.

63. *Gone with the Wind* (1939)

64. *On the Waterfront* (1954)

65. *Forrest Gump* (1994)

66. *The Hunger Games* (2012)

The counting boards in Exercises 67–70 show addition problems similar to the one we did in Example 6. Show each sum using as few counters as possible.

67.

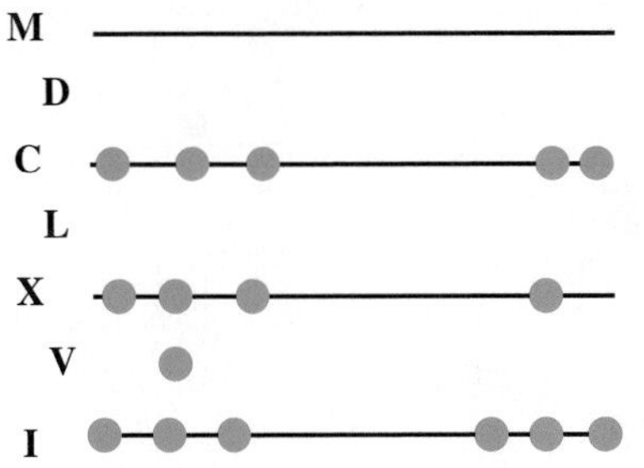

CCCXXXVIII + CCXIII

68.

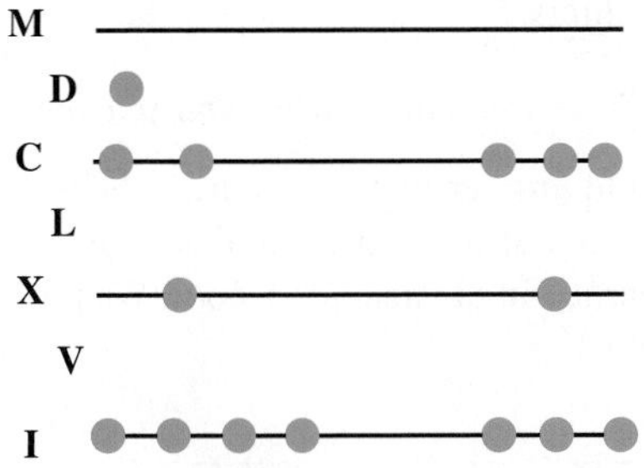

DCCXIIII + CCCXIII

69.

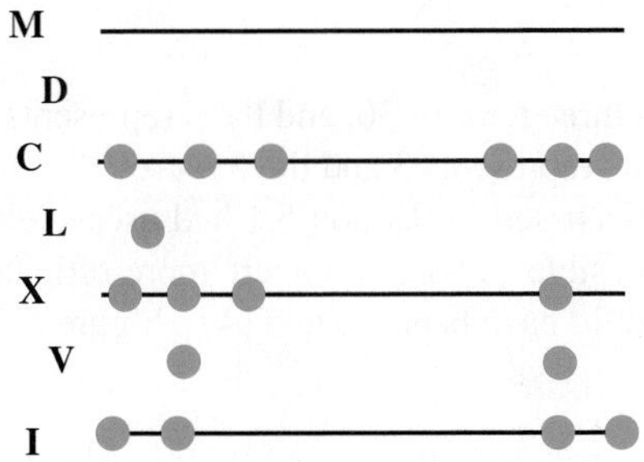

CCCLXXXVII + CCCXVII

70.

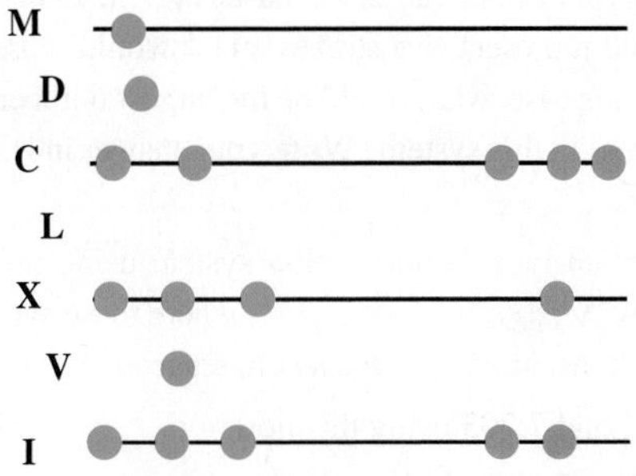

MDCCXXXVIII + CCCXII

71. The oldest discovery of Chinese written numerals is from the Yin dynasty (1523–1027 BC). Express these dates using Chinese numerals.

72. When Marco Polo visited China in 1274, he was impressed by the grandeur of the court of Kublai Khan. Express this date in Chinese numerals.

Communicating Mathematics

73. What is the difference between "number" and "numeral"?

74. Explain two advantages of the Roman numeration system over the Egyptian system.

75. The traditional Chinese numeration system had no symbol for zero. What complication did this cause in writing numerals?

76. Research the Ionic Greek numeration system, which is an example of a ciphered numeration system. Write a paragraph explaining how this system works. What is one advantage and one disadvantage of this system?

Challenge Yourself

77. In the Egyptian numeration system, whenever we have 10 of one symbol we regroup using a single symbol representing the next highest power of 10. For example, we replace 10∩s by one 𓍢. If we follow this rule, what is the largest number that you can express using the Egyptian notation that we discussed in this section? Write your answer using Hindu-Arabic notation.

78. Suppose that Egyptian numeration was based on 5 rather than 10. That is, ∩ would represent five strokes and 9 would represent five ∩s, etc. In this case, what would be the largest number that you could express in this system? Write your answer in Hindu-Arabic notation.

79. Invent an Egyptian type of numeration system using the following symbols: √, ⊗, ∇, ∞, ≈, ↑, ◇, where these symbols represent units, tens, hundreds, and so on, respectively.

a. Write 3,142 and 7,203 using this notation.

b. Find 3,142 + 7,203 similar to the way we added in Example 2.

c. Find 7,203 − 3,142 similar to the way we subtracted in Example 2.

80. Write the number 1,999 in Roman numerals as many ways as you can. (*Hint:* Realize that it is not mandatory to use the subtraction principle in writing Roman numerals.)

Egyptian mathematics had a unique way of writing fractions as sums of unit fractions, that is, fractions of the form $\frac{1}{n}$. *For example, the fraction* $\frac{2}{9}$ *could be written as* $\frac{1}{6}+\frac{1}{18}$ *and also as* $\frac{1}{5}+\frac{1}{45}$. *They would not represent a number like* $\frac{2}{3}$ *as* $\frac{1}{3}+\frac{1}{3}$; *they would use two different unit fractions. Unit fractions were written by placing the symbol* ⬭, *which looked somewhat like an eye, over the numeral. For example,* $\frac{1}{3}$ *could be written as* ⬭ |||.

In Exercises 81–84, write the given fractions as the sum of unit fractions, using our common notation rather than the cumbersome Egyptian notation. (There may be several correct answers but we will give only one.)

81. $\frac{2}{3}$ **82.** $\frac{2}{15}$ **83.** $\frac{2}{7}$ **84.** $\frac{2}{33}$

5.2 Place Value Systems

Objectives

1. Understand the Babylonian place value system.
2. Understand the Mayan numeration system.
3. Understand and compute in the Hindu-Arabic system.
4. Use galleys and Napier's rods to calculate products.

What is the most important invention in the history of mathematics and science?

Before reading further, think about how you would answer that question.... When I ask my classes that question, I get a variety of answers, but never the one that I am looking for—the simple numeration system that we all learned in elementary school. We probably don't notice it because it works so well.

DEFINITION In a **place value system,** also called a **positional system,** the placement of the symbols in a numeral determines the value of the symbols.

For example, in the numeral 35, the 3 represents three tens, or 30, and the 5 represents five ones, or 5. However, in the numeral 53, the 3 now represents 3 and the 5 represents 50.

If the Chinese numeration system that we discussed in Section 5.1 had used the concept of place value, then it would have been possible to write numbers more efficiently. For example, the number 543 in Figure 5.3(a) could have been written as in Figure 5.3(b).

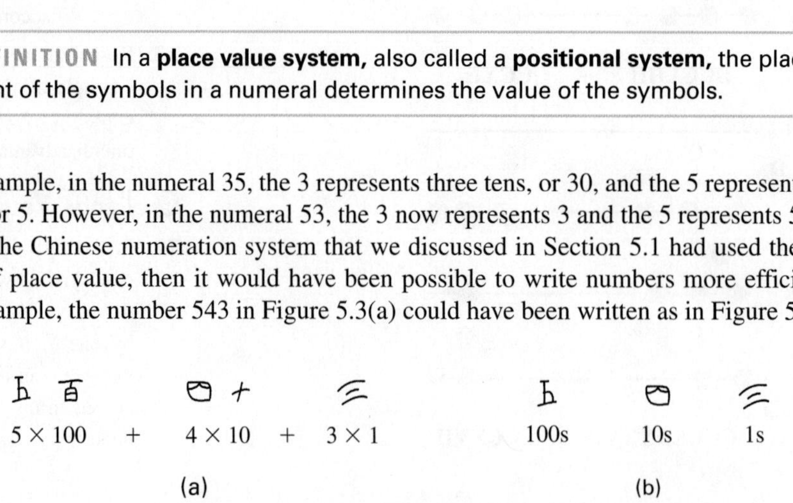

FIGURE 5.3 (a) Chinese notation (b) with the concept of place value.

KEY POINT

The Babylonians developed an early example of a place value system.

With this modification, the rightmost symbol represents units, the symbol to its left represents tens, the next symbol to the left represents hundreds, etc. As you will see later, in order to have a true place value system, it also would have been necessary to invent a symbol for zero.

We will now discuss several place value systems, beginning with the ancient Babylonian system.

The Babylonian Numeration System

The Babylonians* had a primitive place value system that was based on powers of 60 and hence is called a *sexagesimal* system. There were two symbols in the Babylonian system: 𒁹, which represented 1, and 𒌋, which represented 10. These symbols were written in wet clay with wedge-shaped sticks, and when the clay hardened, there remained a permanent record of the calculations. To write small numbers, this system worked much like the Egyptian system. For example, the number 23 could be written as 𒌋𒌋𒁹𒁹𒁹. However, to represent larger numbers, the Babylonians used several groups of these symbols, separated by spaces, and multiplied the value of these groups by increasing powers of 60, as we illustrate in Example 1.

EXAMPLE 1 Converting Babylonian Numerals to Hindu-Arabic Numerals

Convert 𒁹𒁹 𒌋𒌋𒌋𒁹 𒌋𒌋𒁹𒁹𒁹 to Hindu-Arabic notation.

SOLUTION:

We interpret the three groups of numerals as follows: the right group represents units, the middle group represents 60s, and the left group represents 60^2s, as shown in Figure 5.4.

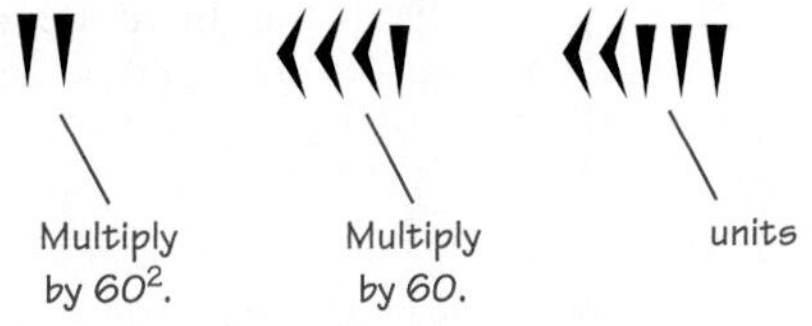

FIGURE 5.4 Babylonian notation.

Thus, this notation represents the number

$$2 \times 60^2 + 31 \times 60 + 23 = 2 \times 3{,}600 + 31 \times 60 + 23 = 9{,}083.$$

Because the early Babylonians did not have a symbol for zero, it could be hard to tell exactly how many spaces were between groups of symbols. For example, it could be difficult to determine whether 𒁹 𒌋𒁹𒁹𒁹𒁹 represents $1 \times 60 + 14$ or $1 \times 60^2 + 14$. (A later Babylonian system had a symbol for zero that avoided this problem.)

The Babylonians used the symbol 𒇲 to indicate subtraction. Thus, the numeral 𒌋𒌋𒇲𒁹𒁹𒁹 represents $20 - 3 = 17$.

To convert Hindu-Arabic numerals to Babylonian numerals, we must divide by powers of 60, similar to the way that we convert seconds to hours and minutes. For example, to convert 7,717 seconds to hours and minutes, we first divide by 3,600 (there are 3,600 seconds in 1 hour) to get the number of whole hours.

$$\begin{array}{r} 2 \text{ hours} \\ 3{,}600\overline{)7{,}717} \\ \underline{7{,}200} \\ 517 \text{ seconds} \end{array}$$

*Ruins of ancient Babylon are located almost 60 miles south of modern-day Baghdad in Iraq. The Babylonian civilization lasted from 2000 BC to about 600 BC.

Thus, we have 2 whole hours with 517 seconds left over. Now, we divide 517 by 60 to find the whole number of minutes.

$$\begin{array}{r} 8 \text{ minutes} \\ 60\overline{)517} \\ \underline{480} \\ 37 \text{ seconds} \end{array}$$

So, we see that 7,717 seconds corresponds to 2 hours, 8 minutes, and 37 seconds.

EXAMPLE 2 Converting from Hindu-Arabic to Babylonian Notation

Write 12,221 as a Babylonian numeral.

SOLUTION:

First divide by $3{,}600 = 60^2$, which is the largest power of 60 that will divide into 12,221.

$$\begin{array}{r} 3 \text{ times } 60^2 \\ 3{,}600\overline{)12{,}221} \\ \underline{10{,}800} \\ 1{,}421 \text{ units left over} \end{array}$$

The quotient 3 tells us how many 60^2s are present in the number. Next, divide 1,421 by 60.

$$\begin{array}{r} 23 \text{ times } 60 \\ 60\overline{)1{,}421} \\ \underline{1{,}380} \\ 41 \text{ units left over} \end{array}$$

The quotient 23 tells us that there are 23 60s in the number, and the remainder 41 tells us that there are 41 units left over. The number 12,221 can now be written as $3 \times 60^2 + 23 \times 60 + 41$. We write this in Babylonian notation as

Now try Exercises 1 to 12. 5

QUIZ YOURSELF 5

a) Convert

to a Hindu-Arabic numeral.

b) Convert 7,573 to Babylonian notation.

You might wonder why the Babylonians chose such a strange base as 60 for their mathematical system. Some speculate that it was due to the fact that they did fraction calculations by combining unit fractions such as $\frac{1}{2}, \frac{1}{3}, \frac{1}{4}, \frac{1}{5}$, and so on. For example, they would write $\frac{7}{12}$ as $\frac{1}{3} + \frac{1}{4}$. The number 60 was a convenient number to use in these situations because it has many different divisors. In computing a sum such as $\frac{1}{12} + \frac{1}{5} + \frac{1}{10} + \frac{1}{15} + \frac{1}{15} = \frac{31}{60}$, the denominator 60 would frequently arise. Thus, using the base 60 made it easier for them to do fractional computations. You can see the influence of a base-60 system today in that we measure an hour as 60 minutes, or $3{,}600 = 60^2$ seconds. Also, in geometry, we have 360 degrees in a circle.

The Mayan Numeration system

The Maya Indians (see Historical Highlight on page 203) had a numeration system based on the number 20. It used patterns of dots and bars to count, as shown in Figure 5.5.

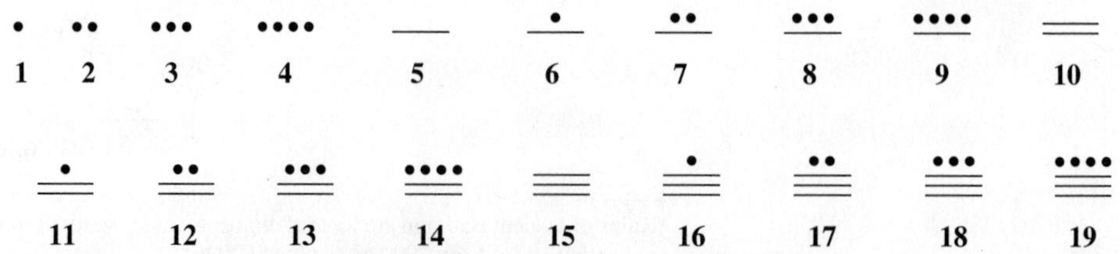

FIGURE 5.5 Counting with Mayan numerals.

Historical Highlight

Mayan Mathematics and Astronomy

The Maya Indians, who lived on the Yucatan Peninsula in Central America from about 200 BC to 1540 AD, made important contributions not only to mathematics, but also to astronomy and art. Although their numeration system was based on the number 20, the product 20 × 18 played a prominent role in their calculations because their calendar consisted of 18 *uinals* ("months") of 20 days each, plus 5 extra days added on at the end of the year. They had an extremely precise estimate of the length of a year—365.242000 days versus our current estimate of 365.242198 days.

Mayan astronomers also described the movements of the sun, moon, and planets and accurately predicted solar eclipses. Their accomplishments were truly remarkable considering that they did not know how to make glass and so, unlike later astronomers, did not have the advantage of telescopes.

QUIZ YOURSELF 6

Convert the given Mayan numeral to Hindu-Arabic notation.

To count beyond 19, the Mayans positioned symbols vertically, with the lowest position representing units, the next higher position representing 20s, the next higher position representing 20 × 18,* the next representing 20 × 18 × 20, and so on, as illustrated in Figure 5.6. The symbol  represents zero. (6)

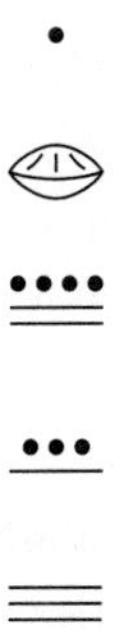

$1 \times 20 \times 18 \times 20^2$	=	144,000
$0 \times 20 \times 18 \times 20$	=	0
$14 \times 20 \times 18$	=	5,040
8×20	=	160
15	=	15
		149,215

FIGURE 5.6 Representing a large number with Mayan notation.

KEY POINT

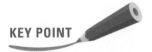

The Hindu-Arabic numeration system is a place value system based on 10.

The Hindu-Arabic Numeration System

The oldest known examples of Hindu-Arabic numerals were found on a stone column in India and are believed to have been written about 250 BC. That Hindu system had no symbol for zero, and scholars are not certain as to when our current Hindu-Arabic system became fully developed. The Persian mathematician al-Khowarizmi became aware of Hindu notation, and in 825, he wrote a book whose title translated into English means "The Book of al-Khowarizmi on Hindu Number." After this, Hindu notation was adopted by the Arabs, and in 1202, the Italian mathematician Leonardo Fibonacci, after studying in the Middle East, wrote a book on arithmetic and algebra titled *Liber Abaci* (A Book on the Abacus), which helped spread Hindu-Arabic numerals throughout Europe.

The Hindu-Arabic system is a place value system based on 10, unlike some of the earlier systems you have studied. One feature of this numeration system is that we can write all numerals using only the digits† 0, 1, 2, . . . , 9. Thus we do not need special symbols for

*The reason they used 20 × 18, rather than 20^2, is explained in Historical Highlight.

†The word *digit* is the Latin word for *finger*. Because we have 10 fingers, it is not surprising that many numeration systems are based on the number 10.

10, 100, 1,000, etc., as did some of the earlier numeration systems. Also, the invention of zero, used as a place holder, makes it easy to distinguish between numbers such as 5,001, 501, and 51. Contrast this approach with that of the early Babylonians, who used spaces between groups of symbols.

We can write Hindu-Arabic numerals in *expanded form* to show explicitly how each digit is multiplied by a power of 10. For example,

$$6{,}582 = 6 \times 10^3 + 5 \times 10^2 + 8 \times 10^1 + 2 \times 10^0.$$

(Recall that $10^0 = 1$, $10^1 = 10$, $10^2 = 10 \times 10 = 100$, $10^3 = 10 \times 10 \times 10 = 1{,}000$, etc.)

EXAMPLE 3 Writing Hindu-Arabic Numbers in Expanded Form

a) Write 53,024 in expanded form.

b) Write $4 \times 10^3 + 0 \times 10^2 + 2 \times 10^1 + 5 \times 10^0$ using Hindu-Arabic notation.

SOLUTION:

a) $53{,}024 = 5 \times 10^4 + 3 \times 10^3 + 0 \times 10^2 + 2 \times 10^1 + 4 \times 10^0$

b) $4 \times 10^3 + 0 \times 10^2 + 2 \times 10^1 + 5 \times 10^0 = 4$ thousands + 0 hundreds + 2 tens + 5 ones
$= 4{,}025$

Now try Exercises 25 to 36.

We can use expanded notation to explain the algorithms to perform the arithmetic operations that you learned in elementary school.

EXAMPLE 4 Using Expanded Notation to Explain the Addition Algorithm

Calculate 4,625 + 814 using expanded notation.

SOLUTION:

Using expanded notation, we can write this problem as

$$\begin{aligned} 4{,}625 &= \quad 4 \times 10^3 + 6 \times 10^2 + 2 \times 10^1 + 5 \times 10^0 \\ +\ 814 &= + \qquad\qquad\quad 8 \times 10^2 + 1 \times 10^1 + 4 \times 10^0. \\ \hline 5{,}439 &= \quad 4 \times 10^3 + 14 \times 10^2 + 3 \times 10^1 + 9 \times 10^0 \end{aligned}$$

We take 10 of the 10^2s and represent them as one 10^3 in the column to the left.

The computations in the 10^0 and 10^1 places are straightforward; however, when we add 6×10^2 and 8×10^2, we get 14×10^2. We cannot express 14 using a single place, so instead we think of 14 as $10 + 4$. Then we can write $14 \times 10^2 = (10 + 4) \times 10^2 = 10 \times 10^2 + 4 \times 10^2$. (We are using the fact that multiplication distributes over addition here.) This gives us 4 ten squareds and an additional ten cubed, making 5 ten cubeds. In expanded notation, we can write this as

$$5 \times 10^3 + 4 \times 10^2 + 3 \times 10^1 + 9 \times 10^0,$$

which equals 5,439.

Now try Exercises 37 to 40.

The way we handled the 14 in the answer in Example 4 explains why you were taught to "put down the 4 and carry the 1 to the next column to the left" in performing this addition.

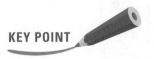

KEY POINT

Hindu-Arabic notation allows us to simplify computations.

EXAMPLE 5 Using Expanded Notation to Explain the Subtraction Algorithm

Calculate 728 − 243 using expanded notation.

SOLUTION:

Using expanded notation, we can write this problem as

$$\begin{array}{rcl} 728 & = & 7 \times 10^2 + 2 \times 10^1 + 8 \times 10^0 \\ \underline{-\ 243} & = & \underline{-\ 2 \times 10^2 + 4 \times 10^1 + 3 \times 10^0.} \end{array}$$

The computations in the 10^2 and 10^0 places would be straightforward, except for the fact that we cannot directly subtract 4×10^1 from 2×10^1.

To remedy this, we express one of the ten squareds as 10×10^1. This gives us 12 tens and 6 ten squareds. We can now rewrite the subtraction as

We took one 10^2 and treated it as ten 10^1s.

$$\begin{array}{rcl} 728 & = & 6 \times 10^2 + 12 \times 10^1 + 8 \times 10^0 \\ \underline{-\ 243} & = & \underline{-\ 2 \times 10^2 + \ \ 4 \times 10^1 + 3 \times 10^0} \\ 485 & = & 4 \times 10^2 + \ \ 8 \times 10^1 + 5 \times 10^0 \end{array}$$

We now can write $4 \times 10^2 + 8 \times 10^1 + 5 \times 10^0 = 485$.

Now try Exercises 41 to 44.

When you learned to do subtraction, you probably were taught to do this same "borrowing," but in a less drawn-out fashion and without using expanded notation.

An important benefit of Hindu-Arabic notation is that we can do basic numerical calculations very simply using pencil and paper rather than an abacus or counting board. In Example 6, we explain the galley method for doing multiplication, which you will see is a forerunner of the multiplication method that we use today. This method was popular in Italy in the fifteenth century; however, because printers found it cumbersome to typeset galleys (particularly when doing division), this method eventually gave way to our more modern way of doing multiplication.

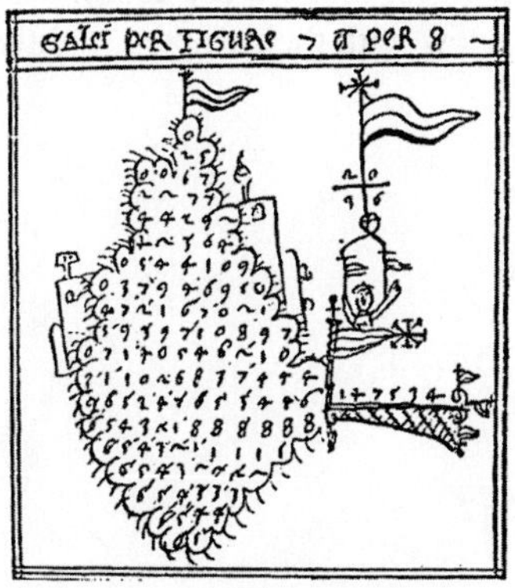

An example of using the galley method to do division.*

EXAMPLE 6 Multiplying Using the Galley Method

Multiply 685 and 49 using the galley method.

SOLUTION: Review the Be Systematic Strategy on page 5.

We begin by constructing the rectangle, divided into triangles, which is called a galley,† in Figure 5.7(a) on the following page. Then we compute the partial products in each box of the galley as shown in Figure 5.7(b). For example, we put the 2 and 4 in the triangles of the upper left-hand box because the product of 6 and 4 is 24.

Next we add the numbers along the diagonals, starting at the bottom right, as shown in Figure 5.8. If the sum along a diagonal is greater than 9, place the units digit at the end of the diagonal and carry the 1 to the channel to the left.

*From Frank J. Swetz (Ed.), *Five Fingers to Infinity* (Chicago: Open Court Pub. Co., 1994).

†It is also possible to do division using the galley method, and when doing so, the pattern of numbers resembles a ship, or galley. This method is also sometimes called the *gelosia* method. *Gelosia* in Italian means "window."

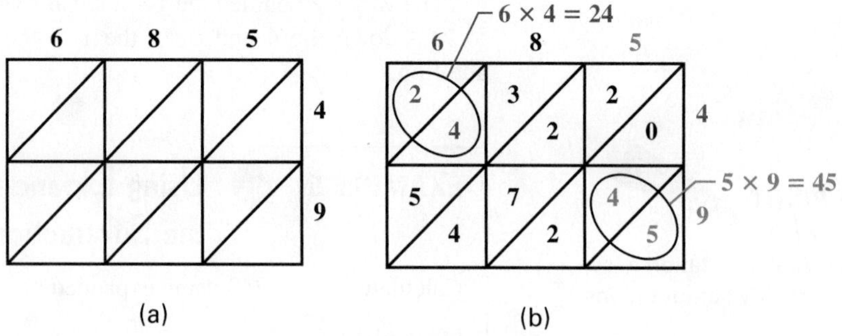

FIGURE 5.7 A galley to multiply 685 and 49.

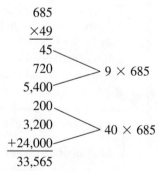

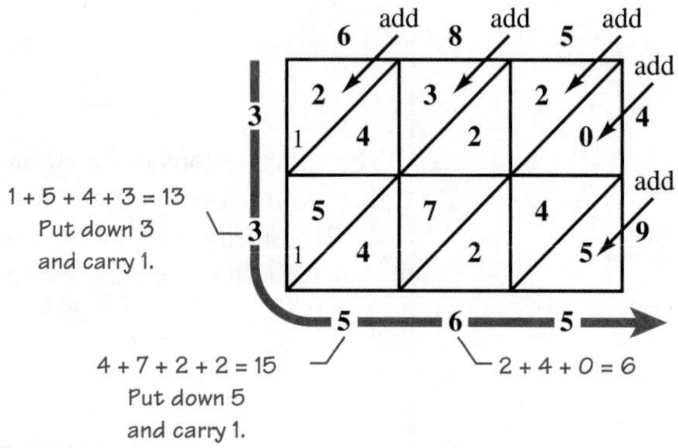

FIGURE 5.8 Computing the product by adding partial products.

Follow the numbers as indicated by the red arrow to get the final product of 33,565.

Now try Exercises 49 to 54. 7

QUIZ YOURSELF 7

Multiply 328 × 39 using the galley method. Show the galley as well as your final answer.

The galley method is much like the multiplication algorithm that we use today. The channels correspond to the columns in the multiplication below with the numbers slightly rearranged.

$$
\begin{array}{rl}
685 & \\
\underline{\times 49} & \\
45 & \\
720 & \Big\} \; 9 \times 685 \\
5{,}400 & \\
200 & \\
3{,}200 & \Big\} \; 40 \times 685 \\
\underline{+24{,}000} & \\
33{,}565 &
\end{array}
$$

KEY POINT

Napier's rods are a variation on the galley method.

In the seventeenth century, the English mathematician John Napier developed a device called **Napier's rods** or **Napier's bones** for doing multiplication. This device consists of a series of strips, each labeled at the top with a digit, 0, 1, 2,…, 9. The remainder of each strip lists all multiples of the label at the top, as shown in Figure 5.9. There is an additional strip called the *index*, which is also shown in Figure 5.9 on the following page.

To calculate a product such as 6 × 325, you select strips for 3, 2, and 5 and place them side by side next to the index, as shown in Figure 5.10. We then use the small galley, formed by the three boxes next to the 6 in the index, to compute the product, 1,950, as we did in Example 6. See Figure 5.11.

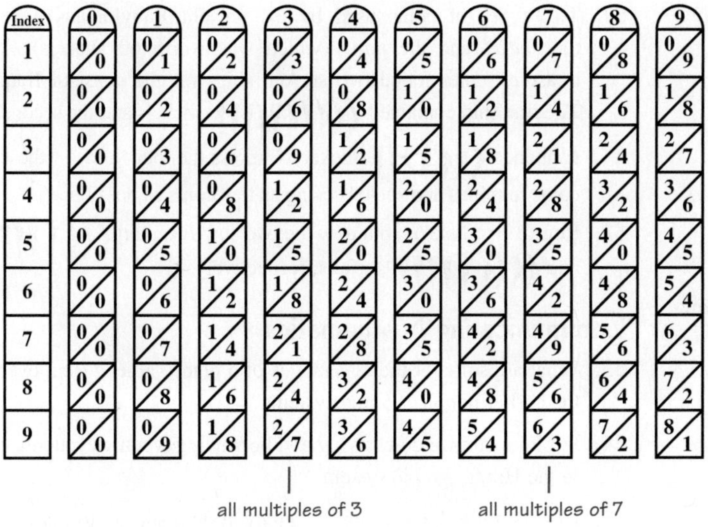

FIGURE 5.9 Napier's rods.

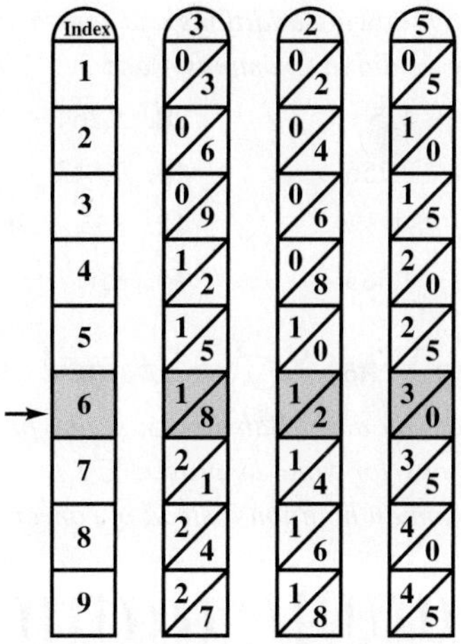

FIGURE 5.10 Using Napier's rods to compute 6 × 325.

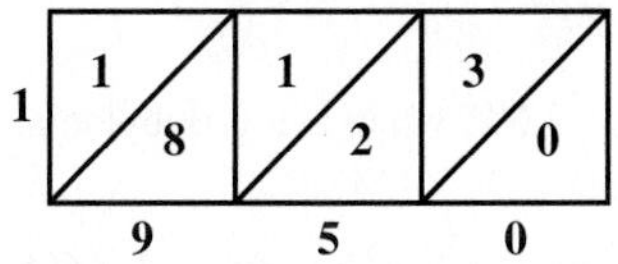

FIGURE 5.11 6 × 325 = 1,950.

We will investigate how to use Napier's rods to do more complex multiplications in Exercises 59 and 60.

EXERCISES 5.2

Sharpening Your Skills

Write the following Babylonian numerals as Hindu-Arabic numerals.

1.  **2.** 

3. 

4. 

Write each number using Babylonian notation.

5. 8,235 **6.** 7,331

7. 18,397 **8.** 26,411

9. 123,485 **10.** 227,597

11. 188,289 **12.** 173,596

Translate each of the following Mayan numerals to Hindu-Arabic notation.

13.  **14.**  **15.**  **16.** 

In Exercises 17–20, write the numeral that precedes the given Mayan numeral.

17. 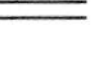 **18.**  **19.**  **20.** 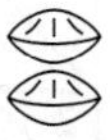

What does 5 represent in each of the following numerals?

21. 37,521 **22.** 53,184

23. 105,000 **24.** 5,023,671

Write each number using expanded notation.

25. 25,389 **26.** 37,248 **27.** 278,063

28. 820,634 **29.** 1,200,045 **30.** 3,002,608

Write each number using standard Hindu-Arabic notation.

31. $5 \times 10^3 + 3 \times 10^2 + 6 \times 10^1 + 8 \times 10^0$

32. $8 \times 10^3 + 2 \times 10^2 + 1 \times 10^1 + 4 \times 10^0$

33. $3 \times 10^5 + 7 \times 10^4 + 0 \times 10^3 + 0 \times 10^2 + 8 \times 10^1 + 2 \times 10^0$

34. $6 \times 10^5 + 0 \times 10^4 + 8 \times 10^3 + 2 \times 10^2 + 0 \times 10^1 + 4 \times 10^0$

Write the expressions in Exercises 35 and 36 using standard expanded notation. For example, $7 \times 10^2 + 5 \times 10^2 + 4 \times 10^2 = 16 \times 10^2 = 1 \times 10^3 + 6 \times 10^2$.

35. $8 \times 10^2 + 3 \times 10^2 + 6 \times 10^2$

36. $8 \times 10^3 + 6 \times 10^3 + 5 \times 10^3$

Perform the following additions and subtractions using expanded notation as we did in Examples 4 and 5.

37. 2,863 + 425 **38.** 5,264 + 583

39. 3,482 + 2,756 **40.** 7,843 + 1,692

41. 926 − 784 **42.** 835 − 362

43. 5,238 − 1,583 **44.** 3,417 − 2,651

Applying What You've Learned

Assume that you are a Babylonian scribe performing the following operations for your employer. Try to do these calculations using Babylonian notation instead of converting to Hindu-Arabic notation.

45. 

46. 

47. 

48. 

In Exercises 49–52, a) Use the galley method to perform the multiplication, and b) rewrite the calculations in a more conventional fashion, as we did in the discussion following Example 6.

49. 23 × 876 **50.** 56 × 371

51. 293 × 465 **52.** 473 × 628

Use the partially completed galleys to determine the numbers that are being multiplied.

53. 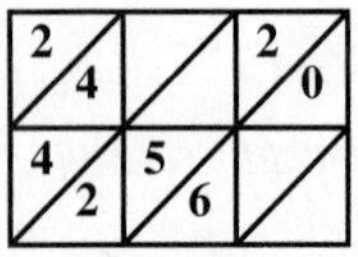

54. 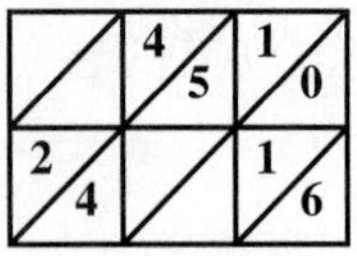

Use Napier's rods to perform each multiplication.

55. 8 × 492 **56.** 5 × 728

57. 6 × 924 **58.** 4 × 834

59. Devise a method for using Napier's rods to multiply numbers having several digits, such as 324 × 615. (*Hint:* Think of 324 as 300 + 20 + 4.) Use this method to multiply 324 × 615.

60. Use the method that you invented in Exercise 59 to multiply 426 × 853.

61. Two types of coins used by the ancient Babylonians were the *shekel* and the *mina* (which was worth 60 shekels). A treasure hunter found two sunken ships off the coast of Iraq. The first ship contained  shekels and the second contained  shekels. Express the total value of these coins in minas and shekels using as few coins as possible.

62. Repeat Exercise 61, but now the two ships contain  and  shekels, respectively.

Communicating Mathematics

63. What are some of the advantages of a place value system over the earlier nonplace value systems?

64. Explain why we do carrying and borrowing when we calculate in the Hindu-Arabic system.

65. How is multiplying using the galley system similar to the way that we do multiplication today? Give an example.

66. How is multiplying using Napier's rods similar to the galley method?

67. Why might we confuse 65 and 3,605 when using Babylonian notation?

68. Give an example of two numbers, other than 65 and 3,605, that might be confused when using Babylonian notation. Why might they be confused?

69. What difficulty does the ancient Chinese numeration system have in representing very large numbers? How does the Hindu-Arabic system resolve this problem?

70. Place the following types of numeration systems in order beginning with the earliest.

a. place value system

b. multiplicative grouping

c. tally marks

d. simple grouping

Give an example of each type of system. Explain why you decided to order the systems as you did.

Challenge Yourself

The modern Chinese numeration system is a base-10 place value system that uses the symbol ○ *for zero. Rewrite each of the following numerals using this modern notation. Realize that now you do not need special symbols for 10, 100, 1,000, etc.*

71. 四百三十六 **72.** 八千五百二十五

73. 五千六十七 **74.** 四千三百四

75. 九千九 **76.** 四千九十

77. Expanded notation uses non-negative powers of ten to write numerals to the left of the decimal point; for example, $5{,}683 = 5 \times 10^3 + 6 \times 10^2 + 8 \times 10^1 + 3 \times 10^0$. Think about how we use negative powers of ten to write an expanded form for numerals written to the right of the decimal point such as 0.375. What is the meaning of 0.3 in terms of a power of ten? What about 0.07? 0.005?

Write the following numbers in expanded notation.

78. 372.4678 **79.** 205.6003 **80.** 418.03006

5.3

Calculating in Other Bases

Objectives

1. Count in non-base-10 systems.
2. Understand the similarity of notation in a base-5 and a base-10 system.
3. Perform arithmetic operations in a non-base-10 system.
4. Understand the relationship between binary, octal, and hexadecimal systems.

Imagine that you could listen in on a romantic conversation between two computers who had fallen in love. As the one computer looks lovingly at the other, he says longingly:

> "1001000 1001111 1010111 1000100 1001111 1001001 1001100 1001111 1010110 1000101 1011001 1001111 1010101–1001100 1000101 1010100 1001101 1000101 1000011 1001111 1010101 1001110 1010100 1010100 1001000 1000101 1010111 1000001 1011001 1010011".*

This string of 0s and 1s is meaningless to us, but, in fact, that is the way the first computer might profess his love to the other in what is called binary, or base-2, notation. (In Exercises 75 to 78, you will learn how to translate this binary code.)

You have already seen the Babylonian numeration system, which is based on 60, and the Mayan system that uses 20, as well as our Hindu-Arabic base-10 system. Now, you will learn how to work not only in a base-2 system but also in several other systems based on 5, 8, and 16. As you will see shortly, we can build numeration systems using any integer greater than 1. These systems are not just mathematical curiosities, because various societies around the world use other bases, such as 2, 3, 4, 5, or 8, for their numeration systems.

Although we will emphasize the base-5 numeration system in this section, you will find that the principles we explain are easy to carry over to systems having other bases.

Non-Base-10 Systems

KEY POINT

The principles of the Hindu-Arabic numeration system carry over to systems that use other bases.

We will now explain how to count using bases other than 10, but first, we must examine carefully how our Hindu-Arabic base-10 system works. We begin counting 1, 2, 3, etc., until we reach 9 and then, instead of writing a single symbol for 10, we write "one zero." It is important to remember that the 1 represents 1 times the base, and the zero represents 0 units, as shown below.

$$\underset{1 \times 10^1}{1}\ \underset{0 \times 10^0}{0}$$

Similarly, in any base, b, we follow the same pattern—the numeral 10 represents 1 times the base plus 0 units, as seen in the following diagram.

$$\underset{1 \times b^1}{1}\ \underset{0 \times b^0}{0}$$

Keeping this in mind, we can now count in base 5, remembering that every time we get a 5 in any position, we "put down a zero and carry a one." We start counting 1, 2, 3, 4, and because the base, 5, is next, we write 10_5. The subscript 5 tells us that the base is 5 instead of 10. Continuing this pattern, when we reach 14_5, the next number would require

*This introduction may sound silly, but at one time or another, you may have accidentally dialed a fax machine, or some other electronic device, and heard a variety of hissing, screeching, and beeping sounds as you listened in on a binary conversation. The translation of this statement is: "HOWDOILOVEYOU—LETMECOUNTTHEWAYS."

us to write 5 in the units place, which we cannot do—so we write 20_5 instead. Therefore, we count in base 5 as follows:

We do not use a single symbol to represent five.

$$1, 2, 3, 4, 10_5, 11_5, 12_5, 13_5, 14_5, 20_5, 21_5, 22_5, \ldots, 44_5.$$

We cannot count any more objects in the 5^0 place or the 5^1 place.

If we add 1 more to 44_5, we would have 5 in the units place, so we write a 0 and carry the 1. However, we now have 5 in the 5's place, so we again write 0 and carry a 1 to the five-squared place. Thus the number following 44_5 is 100_5.

In reading a base-5 numeral such as 23_5, do not say "twenty-three base five" because "twenty" will remind you of base 10 and perhaps confuse you when you are doing computations. Say "two three base five" instead.

In a base-8 system, called an **octal** system, we would count

$$\begin{array}{l} 1, 2, 3, 4, 5, 6, 7, 10_8, 11_8, 12_8, \ldots, 17_8, 20_8, 21_8, \ldots, 77_8, 100_8, \ldots. \\ \text{decimal: } 8 \quad 9 \quad 10 \qquad 15 \quad 16 \quad 17 \qquad 63 \quad 64 \end{array}$$

However, in counting in a base-16 system, called a **hexadecimal** system, we run into a problem. If we start counting in the usual way—1, 2, 3, 4, 5, 6, 7, 8, 9—we cannot write 10_{16} next because 10_{16} represents $1 \times 16 + 0 \times 1$, which is decimal 16—not 10. To solve this problem, it is customary to use the symbols A, B, C, D, E, and F to represent the decimal numbers 10, 11, 12, 13, 14, and 15. Thus, in base 16, we would count

$$\begin{array}{l} 1, 2, 3, \ldots, 9, A, B, C, D, E, F, 10_{16}, 11_{16}, \ldots, 1F_{16}, 20_{16}, 21_{16}, \ldots, FF_{16}, 100_{16}, \ldots. \\ \text{decimal: } 10 \; 11 \; 12 \; 13 \; 14 \; 15 \; 16 \quad 17 \qquad 31 \quad 32 \quad 33 \qquad 255 \quad 256 \end{array}$$

In designing digital devices such as satellite radios and MP3 players, engineers use a base-2, or **binary,** system. This system uses only the digits 0 and 1, so we count as follows:

$$\begin{array}{l} 1, 10_2, 11_2, 100_2, 101_2, 110_2, 111_2, 1000_2, 1001_2, \ldots. \\ \text{decimal: } 1 \quad 2 \quad 3 \quad 4 \quad 5 \quad 6 \quad 7 \quad 8 \quad 9 \end{array}$$

QUIZ YOURSELF 8

Count to the decimal number 10 using a base-3 numeration system.

8

Expanded notation works in non-base-10 systems very much the way it works in the Hindu-Arabic system. For example, we write the numeral $2{,}304_5$ in expanded notation as

$$2 \times 5^3 + 3 \times 5^2 + 0 \times 5^1 + 4 \times 5^0.$$

SOME GOOD ADVICE

It will help you to work in different bases if you always keep two things in mind.

1. In base b, you only use the numerals $0, 1, 2, \ldots, b - 1$. For example, in base 8 you can only use 0, 1, 2, 3, 4, 5, 6, and 7. You cannot use 8 or 9.
2. Just as in base 10, the places in the numeral represent multiples of powers of the base.

We will now use expanded notation to convert numerals from non-base-10 systems to Hindu-Arabic numerals.

EXAMPLE 1 Converting to Decimal Notation

Convert $4{,}302_5$ to decimal notation.

SOLUTION:

To convert to decimal notation, we write $4{,}302_5$ in expanded notation as

$$\begin{aligned} 4{,}302_5 &= 4 \times 5^3 + 3 \times 5^2 + 0 \times 5^1 + 2 \times 5^0 \\ &= 4 \times 125 + 3 \times 25 + 0 \times 5 + 2 \times 1 \\ &= 500 + 75 + 0 + 2 = 577. \end{aligned}$$

Now try Exercises 1 to 22. 9

QUIZ YOURSELF 9

Convert 354_6 to decimal notation.

```
                 4
Ans*5+3
                23
Ans*5+0
               115
Ans*5+2
               577
```

There is an alternate method, called **Horner's Method,** for converting numbers in one base to base ten. This method avoids using powers of the base and can be easily done on some calculators (see TI screen). We illustrate Horner's Method in the following table by converting $4{,}302_5$ from Example 1 to base ten.

Horner's Method

Step	Calculation	Result
1. Call the left digit of the number that you are converting to base ten "result."	We choose 4 from $4{,}302_5$ to be "result."	4
2. Multiply "result" by the base (which in this case is 5), and add to that product the next digit of the number you are converting. Assign that sum to "result."	$4 \times 5 + 3 = 23$ →	23
3. Keep repeating step 2 until you reach the last digit.	$23 \times 5 + 0 = 115$ →	115
⋮		
4. Stop when you have added the rightmost digit (which in this case is 2).	$115 \times 5 + 2 = 577$ →	577 This is your answer.

We will now explain how to convert a decimal numeral to a base-5 numeral. In this process, which we explain in Example 2, we first determine the number of units, next the number of 5s, then the number of 25s, and so on.

Math in Your Life

What Do Numbers Sound Like?*

You probably don't realize it, but when you download a song, it is transmitted as a string of base-2 numbers called bits. How many numbers? I'm glad you asked.

To record music digitally, it is first changed into electrical waves that are then sampled at a rate of 44,100 times per second. Next, each sample is coded using 16 bits. To record in stereo, this coding must be done for each ear.

So, for 1 second of stereo music, you must gather $44{,}100 \times 16 \times 2 = 1{,}411{,}200$ bits of information. If you were to listen to a 3-minute song, you would really be listening to a string of roughly one-fourth of a billion 0s and 1s. Printing out all of these bits would require over 76,000 pages, which, if laid end-to-end, would stretch over 13 miles!

*This example is based on "The Math of Online Music Trading" (February 2002) by Keith Devlin. See www.maa.org/devlin/devlin_02_02.html.

EXAMPLE 2 Converting from Decimal to Base-5 Notation

Write 384 as a base-5 numeral.

SOLUTION:

The following division shows that when dividing 5 into 384, we get a quotient of 76 with 4 units left over.

$$\begin{array}{r} 76 \\ 5\overline{)384} \\ \underline{380} \\ 4 \text{ units left over} \end{array}$$

We next divide 76 by 5. This is equivalent to dividing the original number by 25, which is 5 squared. From this division, we see that after dividing by 25, there is 1 five left over. This means that in the base-5 numeral, there will be a 1 in the fives place.

$$\begin{array}{r} 15 \\ 5\overline{)76} \\ \underline{75} \\ 1 \text{ five left over} \end{array}$$

We continue this process and express the divisions more compactly, as in Figure 5.12.

Remainders

$$
\begin{array}{llll}
5 \mid 384 & 4 & \text{———} & 5^0 = \textbf{units} \\
\quad 5 \mid 76 & 1 & \text{———} & 5^1 = \textbf{fives} \\
\qquad 5 \mid 15 & 0 & \text{———} & 5^2 = \textbf{twenty-fives} \\
\qquad\quad 5 \mid 3 & 3 & \text{———} & 5^3 = \textbf{one-twenty-fives} \\
\qquad\qquad 0
\end{array}
$$

Read remainders in reverse order.

Stop when quotient is zero.

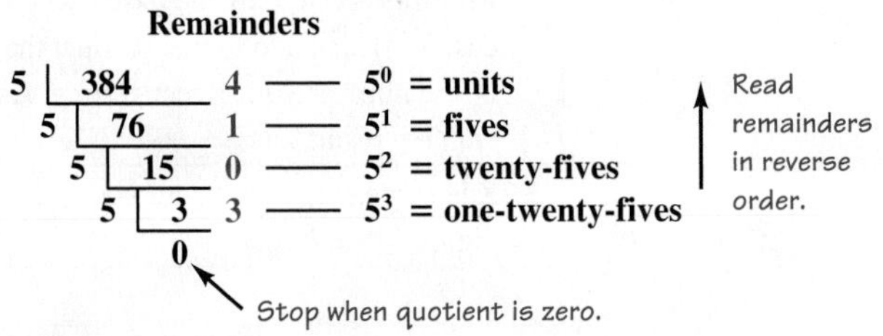

FIGURE 5.12 Repeatedly dividing by 5 to find a base-5 numeral.

In Figure 5.12, we continued dividing each quotient by 5 until we obtained a quotient of 0. Reading the remainders in reverse order will give us the base-5 numeral for 384, which is $3{,}014_5$.

Now try Exercises 23 to 36. **10**

It is important to remember, in Example 2, that we stopped the process when we reached *a quotient of zero*—not a remainder of zero.

QUIZ YOURSELF **10**

Convert the decimal number 113 to base 5.

PROBLEM SOLVING

Strategy: The Analogies Principle

In Section 1.1, we recommended that to understand mathematical ideas, it is useful to make analogies with situations that you have seen before. If you thoroughly understand the concept of place value in our Hindu-Arabic numeration system, you can easily modify the techniques you have used before to understand how to compute in other place value systems.

KEY POINT

Arithmetic operations in other bases are similar to base-10 operations.

Arithmetic in Non-Base-10 Systems

Imagine that you had grown up in a base-5 world. As a small child, perhaps you learned to count by watching *Sesame Street*. Instead of hearing repeatedly the ditty

"one, two, three, four, five, six, seven, eight...nine...ten,"

you may have learned to count

"one, two, three, four, fen, fenone, fentwo, thirfeen, fourfeen, twenfy, twenfy-one," etc.

It is doubtful that you would have used the cumbersome name "one zero base five" for 5 or "one one base five" for 6, etc. We have invented the shorter, fanciful names "fen, fenone, fentwo" for numbers such as 10_5, 11_5, and 12_5. Do not bother to memorize these make-believe names.

Your early days of elementary school would have been much easier because you would have had fewer number facts to memorize. You would not have had to learn that $7 + 8 = 15$ because problems using the digits 7 and 8 would never arise. Also, some number facts would be written differently in a base-5 world. For example, you would write the decimal fact that $4 + 4 = 8$ as $4 + 4 = 13_5$. Using our previous made-up terminology, you would say

"four plus four equals thirfeen."

QUIZ YOURSELF 11

Find $2 + 4$ and $2 + 3$ in Table 5.6.

You would have spent a good deal of time memorizing the addition facts that we list in Table 5.6. For example, in Table 5.6, we see that $3 + 4 = 12_5$. 11

+	0	1	2	3	4
0	0	1	2	3	4
1	1	2	3	4	10_5
2	2	3	4	10_5	11_5
3	3	4	10_5	11_5	12_5 ($3 + 4 = 12_5$)
4	4	10_5	11_5	12_5	13_5

TABLE 5.6 Base-5 addition facts.

We will use the addition facts in Table 5.6 to do the addition in Example 3.

EXAMPLE 3 Adding in Base 5

$$\text{Add } \begin{array}{r} 342_5 \\ +\ 223_5 \\ \hline \end{array}$$

SOLUTION: Review the Analogies Principle on page 13.

We add much like we do in performing decimal addition. First we add the units, $2 + 3 = 5$ decimal, which we express as 10_5. We place the 0 in the units place and carry the 1 to the fives place, as we show in Figure 5.13(a).

$$\begin{array}{r} 1 \\ 342_5 \\ +\ 223_5 \\ \hline 0_5 \end{array} \quad \begin{array}{l} 2 + 3 = 10_5 \\ \text{write 0, carry 1.} \end{array} \qquad\qquad \begin{array}{r} 11 \\ 342_5 \\ +\ 223_5 \\ \hline 20_5 \end{array} \quad \begin{array}{l} 1 + 4 + 2 = 12_5 \\ \text{write 2, carry 1.} \end{array}$$

(a) (b)

FIGURE 5.13 Adding in base 5.

We next add the numerals in the fives place: $1 + 4 + 2 = 12_5$ as we show in Figure 5.13(b). Note how we write down the 2 and carry the 1 to the five-squared place. We finish by adding the numerals in the five-squared place, carrying as necessary.

$$\begin{array}{r} 111 \\ 342_5 \\ +\ \ 223_5 \\ \hline 1120_5 \end{array}$$

$1 + 3 + 2 = 11$, write 1, carry 1.

EXAMPLE 4 Subtracting in Base 5

Subtract $\begin{array}{r} 424_5 \\ -\ 143_5 \\ \hline \end{array}$.

SOLUTION: Review the Analogies Principle on page 13.

Subtraction in base 5 is also much like base-10 subtraction. We begin by subtracting three units from four units to get one unit.

In the fives place, we cannot subtract 4 fives from 2 fives, so we borrow 1 twenty-five and write it as 5 fives. Thus, we now have 7 fives in the fives place, which we have written as 12_5, as shown in Figure 5.14. We complete the subtraction by subtracting 1 twenty-five from the 3 remaining twenty-fives.

$$\begin{array}{r} 3 \\ \not{4}\,{}^{1}2\,4_5 \\ -\ 1\,4\,3_5 \\ \hline 2\,3\,1_5 \end{array}$$

Borrowing 1 twenty-five we now have 12_5, or decimal 7 fives. Subtracting, we get 3 fives.

FIGURE 5.14 Subtracting in base 5 may require borrowing.

Now try Exercises 37 to 54. 12

QUIZ YOURSELF 12

a) Add $315_6 + 524_6$.

b) Subtract $325_6 - 131_6$.

SOME GOOD ADVICE

Although it is useful to check base-5 computations by converting all numbers to base 10, you will increase your skill in working in other bases if you try to do all computations by thinking in base 5 even though it may seem a little difficult at first.

QUIZ YOURSELF 13

Find 2×4 and 4×4 in Table 5.7.

In order to understand how to multiply in other bases, we first need to know our multiplication table. The multiplication facts will look different from what you are used to seeing in base 10. For example, in Table 5.7, we see that $3 \times 4 = 22_5$, which is the base-5 representation of decimal 12. You should verify the other base-5 multiplication facts given in Table 5.7. 13

×	0	1	2	3	4
0	0	0	0	0	0
1	0	1	2	3	4
2	0	2	4	11_5	13_5
3	0	3	11_5	14_5	22_5
4	0	4	13_5	22_5	31_5

$3 \times 4 = 22_5$

TABLE 5.7 Base-5 multiplication facts.

Before giving an example of multiplication, it is useful to review the discussion following Example 6 of Section 5.2 explaining the meaning of galley multiplication in base 10. Recall that we wrote all the partial products in multiplying 685 and 49 as follows:

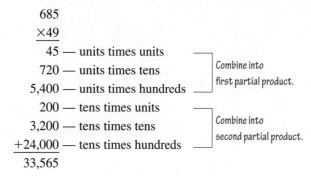

$$\begin{array}{rll} 685 & & \\ \times 49 & & \\ \hline 45 & \text{— units times units} & \\ 720 & \text{— units times tens} & \text{Combine into first partial product.} \\ 5{,}400 & \text{— units times hundreds} & \\ 200 & \text{— tens times units} & \\ 3{,}200 & \text{— tens times tens} & \text{Combine into second partial product.} \\ +24{,}000 & \text{— tens times hundreds} & \\ \hline 33{,}565 & & \end{array}$$

Notice how

$$\begin{aligned} \text{units times units} &= \text{units}, \\ \text{units times tens} &= \text{tens}, \\ \text{tens times tens} &= \text{hundreds, and so on.} \end{aligned}$$

Also, when we do this multiplication in the usual way, we combine the first three products into a single partial product and the second three products into a second partial product. We will follow this pattern in doing base-5 multiplication.

EXAMPLE 5 Multiplying in Base 5

Multiply

$$\begin{array}{r} 134_5 \\ \times 32_5 \\ \hline \end{array}.$$

SOLUTION: Review the Analogies Principle on page 13.

We first multiply 2 units times 4 units to get 8 units in base 10, or 13_5 units. We write down 3 units and carry 1 to the fives place.

$$\begin{array}{rl} 1 & \text{— 5 units expressed as 1 five.} \\ 134_5 & \\ \times 32_5 & \\ \hline 3_5 & \text{— 3 units} \end{array}$$

We next multiply 2 units times 3 fives to get 6 fives, plus the 1 five we carried, gives 7 fives, which we write as 12_5. We write the 2 and carry the 1 to the five-squared column.

$$\begin{array}{rl} 11 & \text{— 5 fives expressed as 1 five squared.} \\ 134_5 & \\ \times 32_5 & \\ \hline 23_5 & \text{— 2 fives} \end{array}$$

We complete the first partial product by multiplying the 2 units times 1 five squared and add the 1 five squared, which was carried, to get 3 five squareds.

$$\begin{array}{r} 134_5 \\ \times 32_5 \\ \hline 323_5 \end{array}$$

We compute the second partial product in a similar way. We begin by multiplying 3 fives times 4 units to get 22_5 fives. We write 2 fives and carry a 2 to the five-squared column.

$$\begin{array}{r} 2 \\ 134_5 \\ \times\ 32_5 \\ \hline 323_5 \\ 2 \\ \hline \end{array}$$

— We indent one place because we are multiplying by 3, which represents 3 × 5.

We complete the second partial product as follows:

$$\begin{array}{r} 22 \\ 134_5 \\ \times\ 32_5 \\ \hline 323_5 \\ 1012 \\ \hline \end{array}$$

We then add the partial products to get the final product.

$$\begin{array}{r} 134_5 \\ \times\ 32_5 \\ \hline 323_5 \\ 1012 \\ \hline 10{,}443_5 \end{array}$$

— Adding in base 5.

QUIZ YOURSELF 14

Multiply:

a) $34_5 \times 42_5$
b) $56_8 \times 47_8$

Now try Exercises 55 to 58. 14

Before doing division, you must have a very good understanding of multiplication.* For example, consider how we do the first step in the following decimal division:

$$\begin{array}{r} 2 \\ 18\overline{)4721} \\ 36 \\ \hline 11 \end{array}$$

— Experience in base 10 tells us that 18 divides into 47 twice

Based on years of experience in working with the decimal system and our ability to make good estimations, we see that 18 divides into 47 twice, but not three times. In dividing,

$$23_5\overline{)41325}_5$$

— We lack experience in base 5 here, so we make a table of multiples of 23_5.

we do not have the same deep background that enables us to find the trial quotient, so we have to use another approach, as we show in Example 6.

*Several years ago, while waiting in the parents' lounge for my son to finish his saxophone lesson, a parent asked me why I thought children in her daughter's grade always had trouble with learning division. She told me the problem was so serious that for the past several years, the teacher in the previous year cut short teaching multiplication to give the children a head start in learning division. I told her, "I think that that's the problem. Children cannot learn to divide unless they are comfortable with multiplication."

EXAMPLE 6 Dividing in Base 5

Perform the division

$$23_5\overline{)4132_5}.$$

SOLUTION:

To be able to estimate trial quotients, we have to know the multiples of 23_5, which we list below. (You should verify these facts.)

$$23_5 \times 0 = 0$$
$$23_5 \times 1 = 23_5$$
$$23_5 \times 2 = 101_5$$
$$23_5 \times 3 = 124_5$$
$$23_5 \times 4 = 202_5$$

From this list, we see that 23_5 divides into 41_5 once but not twice. So we start the division by placing a 1 in the quotient.

$$\begin{array}{r l} 1 & \\ 23_5\overline{)4132_5} & \\ \underline{23} & \text{— Base-5 subtraction} \\ 13 & \end{array}$$

Remember that when subtracting $41_5 - 23_5$, *we are doing a base-5 subtraction.* We next bring down the digit 3 and add another digit to the quotient. From the previous list, we see that 23_5 divides into 133_5 three times with a remainder of 4.

$$\begin{array}{r l} 13 & \\ 23_5\overline{)4132_5} & \\ \underline{23} & \\ 133 & \\ \underline{124} & \text{— Base-5 subtraction} \\ 4 & \end{array}$$

Next, we bring down the 2 and complete the division.

$$\begin{array}{r l} 131_5 & \\ 23_5\overline{)4132_5} & \\ \underline{23} & \\ 133 & \\ \underline{124} & \\ 42 & \\ \underline{23} & \text{— Base-5 subtraction} \\ 14_5 & \end{array}$$

Thus, 23_5 divides into $4{,}132_5$, with a quotient of 131_5 and a remainder of 14_5. Now try Exercises 59 to 62.

SOME GOOD ADVICE

To check to see if we have properly done a division, such as the one we did in Example 6, you can multiply the divisor, 23_5, by the quotient, 131_5, and then add the remainder, 14_5. The result should be the dividend, 4132_5.

Binary, Octal, and Hexadecimal Systems

KEY POINT

Binary, octal, and hexadecimal notation are closely related.

As we have mentioned, digital devices such as computers, MP3 players, and iPhones use the binary system. You might think of the 1s and 0s as representing "on" and "off" commands controlling microscopic switches on electronic chips. The values 0 and 1 are called

Binary		Octal
000	=	0
001	=	1
010	=	2
011	=	3
100	=	4
101	=	5
110	=	6
111	=	7

FIGURE 5.15 Three binary places correspond to one octal place.

bits, which is a contraction of the words **binary digit.** A video chip in your smart phone, for example, may use the 16-bit binary pattern 1011001101011110 to represent red. Because it is tedious to remember such long strings of bits, often we rewrite these commands in a form that is easier for humans to remember.

Figure 5.15 shows how we can represent three binary places efficiently as one octal place. This means that we can express the previous 16-bit command using fewer octal numerals, as we show in Example 7.

EXAMPLE 7 Converting from Binary to Octal and Hexadecimal

a) Write the binary command 1011001101011110 using octal notation.

b) Write the binary command 1011001101011110 using hexadecimal notation.

SOLUTION:

a) We begin by arranging the bits in 1011001101011110 in groups of three *beginning from the right.* We then interpret each group of three bits as an octal number, as shown here. (For the sake of efficiency, we often omit the subscripts that indicate the base.)

Begin grouping from the right.

binary	⟶	1	011	001	101	011	110
octal	⟶	1	3	1	5	3	6

We see that we can represent the binary command 1011001101011110 more succinctly as the octal command 131536. Notice how much easier it is to remember the octal form of the command rather than the binary form.

b) Figure 5.16 shows that we can represent four binary places as one hexadecimal place. Thus, we first group the bits in 1011001101011110 in groups of four, again beginning from the right. We then interpret each group of four bits as a hexadecimal number, as we show here.

binary	⟶	1011	0011	0101	1110
hexadecimal	⟶	B(11)	3	5	E(14)

We can now represent the binary command 1011001101011110 more concisely as the hexadecimal command B35E.

Now try Exercises 63 to 66.

Binary		Hexadecimal
0000	=	0
0001	=	1
0010	=	2
0011	=	3
.		.
.		.
.		.
1110	=	E
1111	=	F

FIGURE 5.16 Four binary places correspond to one hexadecimal place.

Historical Highlight

From the Abacus to the Computer

The earliest mechanical calculating device is the abacus, which dates back to 300 BC. Early abaci, similar to the Roman abacus we described in Historical Highlight on page 195, were widely used in Europe prior to the adoption of written Hindu-Arabic numerals. Some believe that Christians introduced the abacus to China beginning about 1200 AD and later to Japan and Korea.

We show a typical Chinese abacus, or suan-pan, in Figure 5.17. The suan-pan consists of beads that slide on wires that represent powers of 10, as we show in Figure 5.17. Beads below the horizontal bar represent 1; beads above the bar represent 5. Beads next to the bar are active and are

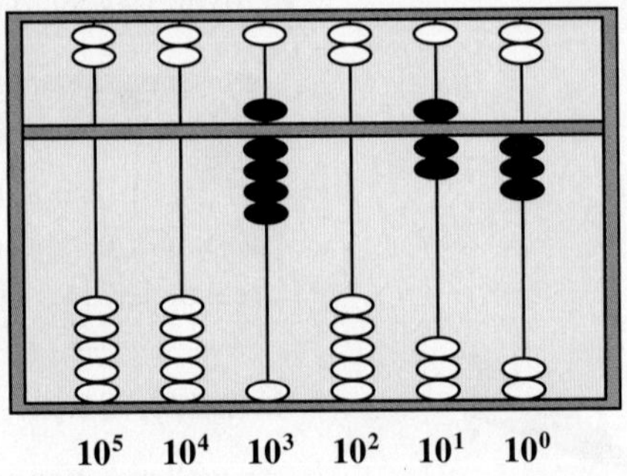

FIGURE 5.17 A Chinese suan-pan representing 9,073.

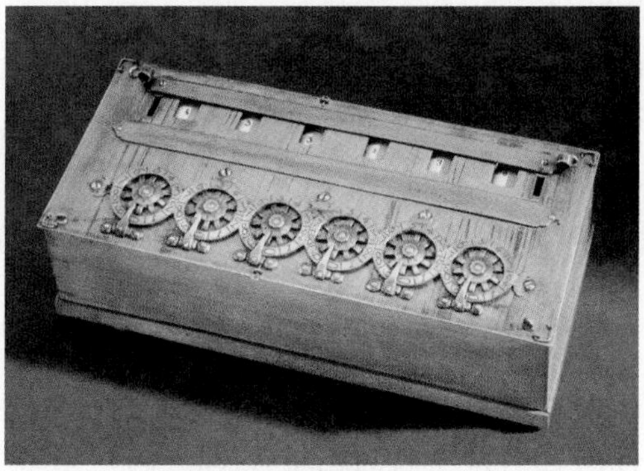

(a)

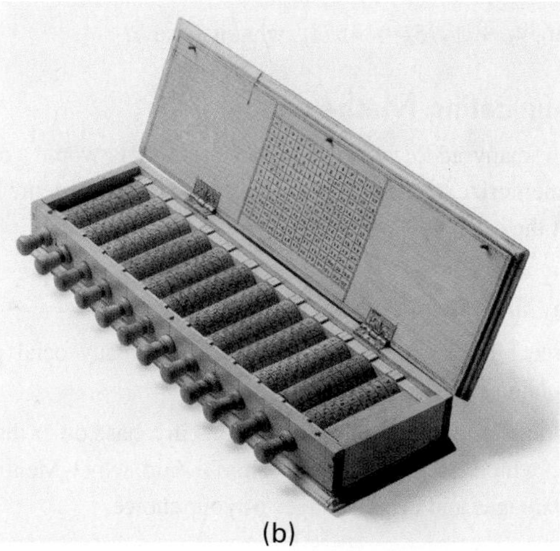

(b)

FIGURE 5.18 (a) Pascal's and (b) Leibniz's calculators.

FIGURE 5.19 Babbage's analytic engine.

used to represent the number. The suan-pan in Figure 5.17 shows how we would represent 9,073.

In 1642, the renowned French mathematician Blaise Pascal invented an adding machine, developing principles that would be used in later calculators (see Figure 5.18(a)).

Later in the century, the German Gottfried Leibniz improved on Pascal's design (see Figure 5.18(b)). The British mathematician Charles Babbage invented a complicated calculating machine called an analytical engine in 1826 (see Figure 5.19). Although Babbage was unable to actually construct his machines, his designs laid the foundation for today's modern computers. IBM later constructed working models of Babbage's machines using his designs, and you can see working models of Babbage's machines made of Legos on the Internet.

The first electronic computer was constructed by J. Presper Eckert and John Mauchly at the University of Pennsylvania in the early 1940s. The computer was so large (more than 30 tons) and so expensive (about a half million dollars) that some of the leading businessmen of the time believed that there would be a worldwide demand for only about a half-dozen of these costly machines.

EXERCISES 5.3

Sharpening Your Skills

List the numbers that precede and follow each of the given numbers in the given base.

1. 24_5

2. 500_6

3. 1011_2

4. 77_8

5. EF_{16}

6. 100_{16}

Write each number as a base-10 numeral.

7. 432_5

8. 243_5

9. 504_6

10. 555_6

11. $100{,}111_2$

12. $100{,}101_2$

13. $1{,}110{,}101_2$

14. $1{,}100{,}111_2$

15. 267_8

16. 137_8

17. 704_8

18. 561_8

19. $2F4_{16}$

20. $18E_{16}$

21. $D08_{16}$

22. $C3B_{16}$

Convert each decimal number to a numeral in the given base.

23. 334 base 5

24. 1,298 base 5

25. 1,838 base 6

26. 3,968 base 6

27. 103 base 2

28. 51 base 2

29. 94 base 2

30. 107 base 2

31. 3,403 base 8

32. 2,297 base 8

33. 2,792 base 16

34. 2,219 base 16

35. 3,562 base 16

36. 3,827 base 16

Perform each addition or subtraction.

37. $3{,}412_5 + 231_5$

38. $3{,}215_6 + 423_6$

39. $2{,}735_9 + 3246_9$

40. $2{,}067_8 + 2{,}443_8$

41. $5415_7 + 2436_7$

42. $563A_{12} + 2B39_{12}$

43. $2A18_{16} + 43B_{16}$

44. $BF2E_{16} + A35_{16}$

45. $11{,}011_2 + 10{,}101_2$

46. $100{,}111_2 + 10{,}111_2$

47. $2{,}412_5 - 321_5$

48. $1{,}325_6 - 453_6$

49. $4263_7 - 2436_7$

50. $653A_{12} + 23B9_{12}$

51. $A83_{16} - 43B_{16}$

52. $6C2E_{16} - A35_{16}$

53. $111{,}011_2 - 10{,}101_2$

54. $100{,}101_2 - 10{,}011_2$

Perform each multiplication or division.

55. $41_5 \times 23_5$

56. $24_5 \times 31_5$

57. $302_5 \times 43_5$

58. $413_5 \times 34_5$

59. $3{,}412_5 \div 24_5$

60. $2{,}143_5 \div 32_5$

61. $4{,}132_5 \div 42_5$

62. $4{,}402_5 \div 14_5$

Write each binary number first as an octal number and then as a hexadecimal number.

63. 1011101101_2

64. 1011110111_2

65. 1111101001_2

66. 1010100101_2

Write each number as a binary number.

67. 246_8

68. 573_8

69. $A3E_{16}$

70. $B8C_{16}$

71. Convert 3524_8 to hexadecimal.

72. Convert 6235_8 to hexadecimal.

73. Convert $3AC_{16}$ to octal.

74. Convert $D7B_{16}$ to octal.

Applying What You've Learned

Many electronic devices store and communicate information using the ASCII coding system, which is an acronym for American Standard Code for Information Interchange. The ASCII coding system uses 7-bit binary codes to represent characters. For example, the binary equivalents of the numbers 65 to 90 represent the capital letters A to Z. Using ASCII coding, we represent A by 1000001, B by 1000010, C by 1000011, and so forth. In Exercises 75–78, translate each binary string into English. (First group the bits into 7-bit groups.)

75. 10000111000001100111010001001011001

76. 10010001000101100110010011001001111

77. 1001100100111110101101000101

78. 10101001010010101010110101001001000

79. Convert the amount $5.43 to quarters, dimes, nickels, and pennies using as few coins as possible.

80. A company sells promotional coffee mugs with discounts as follows: a box of 36 mugs has a cheaper price per mug than a box of 6 and a box of 6 mugs has a cheaper price per mug than buying single mugs. If your company wishes to buy 320 mugs as gifts for its customers, what is the cheapest way to place your order?

81. If $1435_a = A65_b$, which base is larger, a or b?

82. If $7265_b = 5143_c$, which base is larger, b or c?

83. If $2051_b + 1434_b = 3525_b$, what is base b?

84. If $3654_b + 1715_b = 4571_b$, what is base b?

Communicating Mathematics

85. How many addition and multiplication facts would you have to memorize in a base-8 system? In a base-16 system? Why is that the case?

86. Why did we use A in a hexadecimal system to represent 10? Why could we not use 10?

87. Three binary places correspond to how many octal places? Explain.

88. If you were forced to do calculations in a base other than base ten, which base would you choose and why? Mention the advantages and disadvantages of your choice.

Challenge Yourself

Make the following conversions.

89. Convert 201221_3 to base 5.

90. Convert 4523_7 to base 8.

91. Convert $B05_{16}$ to base 9.

92. Convert 365_8 to base 4.

Consider a base-4 system with the symbols , , , *and*  *corresponding to 0, 1, 2, and 3, respectively.*

93. Count to decimal 17 in this system.

94. Translate  to a decimal number.

95. Translate the decimal number 28 to this system.

96. Add  and  in this system.

In Exercises 97–100 find the missing values for x and y.

97.
$$\begin{array}{r} 3x2y_8 \\ +yxx_8 \\ \hline 4171_8 \end{array}$$

98.
$$\begin{array}{r} 2xy1_6 \\ +yx2_6 \\ \hline xx2x_6 \end{array}$$

99.
$$\begin{array}{r} 4x3_{13} \\ +3y_{13} \\ \hline y08_{13} \end{array}$$

100.
$$\begin{array}{r} x2C_{16} \\ +y7_{16} \\ \hline B23_{16} \end{array}$$

Looking Deeper

5.4 Modular Systems

Objectives

1. Understand modular systems.
2. Perform operations in modular systems.
3. Solve congruences in modular systems.
4. Understand some applications of modular systems.

In the late third century, the Chinese mathematician Sun Tzu asked his students:

> *We have things of which we do not know the number; if we count them by threes, the remainder is 2; if we count them by fives, the remainder is 3; if we count them by sevens, the remainder is 2. How many things are there?*

Your first reaction to this question might be, "Who cares?" It does in fact seem to be a curious problem of no apparent use.

However, three hundred years later, the noted Indian mathematician Brahmagupta was also fascinated with this same type of question. And, about 1200 years after that, the great eighteenth-century German mathematician Carl Friedrich Gauss developed a strange clock-like arithmetic that enabled mathematicians to solve such problems. We show you how to use a spreadsheet to solve these problems in the Technology Projects at the end of this chapter.

Strangely enough, Gauss' "clock arithmetic" now is critical in your personal life. It is used in scanning your groceries in a supermarket, protecting your online purchases from theft, and allowing banks to transfer trillions of dollars securely.

Gauss' arithmetic works like your bedroom clock that ticks off the numbers 1, 2, 3, . . . ,12 and then recycles them again. Such a system is called a modulo-m system, which we will study next. These systems have many properties in common with the familiar system of integers. As you will see, in modulo-m systems, we can count, perform arithmetic operations, and solve equations. In addition, modulo-m systems have interesting, real-life applications.

KEY POINT

We use a clock to visualize the operations in a modulo-m system.

Modulo-m Systems

> **DEFINITIONS** If m is an integer greater than 1, then a **modulo-m system*** consists of the numbers 0, 1, 2, . . . , $m - 1$. Counting and arithmetic operations are performed in a manner corresponding to movements on an m-hour clock. The number m is called the **modulus** of the system.

In order to understand a modulo-12 system, we draw a 12-hour clock as shown in Figure 5.20. Notice how we have replaced the 12 on the clock with the number 0—the reason for doing this will become clear shortly. The numbers 0, 1, 2,. . . , 11 now form a modulo-12 system. If we were to begin at 0 and count to 53 in this system, at what number would we stop? Of course, we could begin counting on the clock, 1, 2, 3, 4, 5, 6, 7, 8, 9, 10, 11, 0, 1, 2, and so on until we counted 53 numbers. However, there is an easier way. Notice that if we count on the clock to multiples of 12, such as 12, 24, 36, and

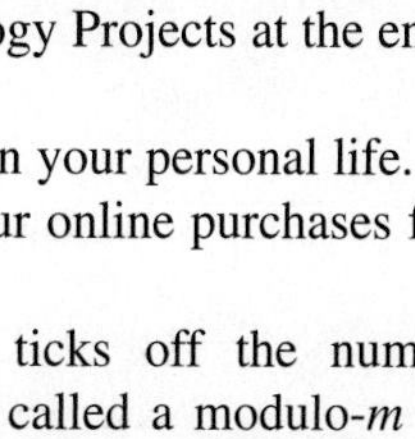

FIGURE 5.20 A 12-hour clock.

*Modulo-m systems are also called **modular arithmetic systems,** or simply **modular systems.**

then 48, we return to the 0 position (see Figure 5.21). If we now count an additional 5, we reach 53, which is position 5 on this 12-hour clock. Thus, counting to 53 in a modulo-12 system will give us 5.

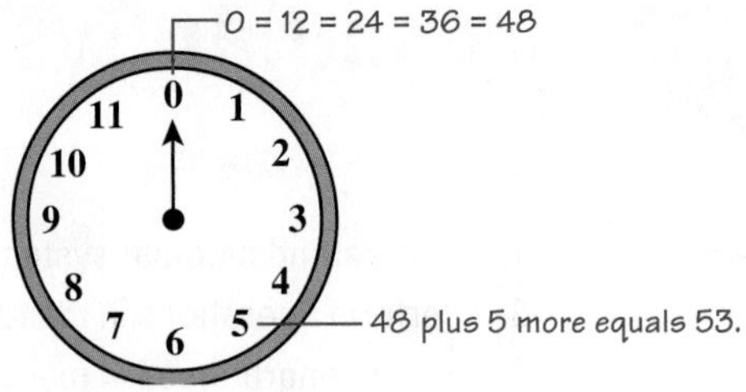

FIGURE 5.21 Counting to 53 on a 12-hour clock.

SOME GOOD ADVICE

Because multiples of 12 are the same as 0 on a 12-hour clock, a quicker way to determine the 5 would have been to divide 53 by 12, and keep the remainder of 5.

We can visualize the days of the week, Monday (1), Tuesday (2), Wednesday (3), Thursday (4), Friday (5), Saturday (6), and Sunday (7) on the 7-hour clock shown in Figure 5.22. Notice that we replace the 7 by 0 on this clock. The numbers 0, 1,..., 6 form a modulo-7 system.

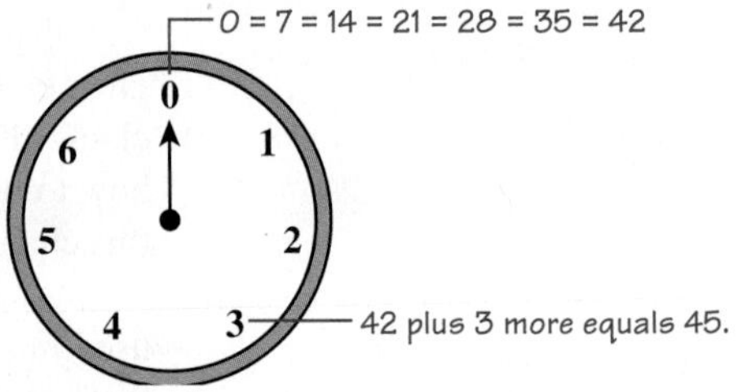

FIGURE 5.22 Counting to 45 on a 7-hour clock.

If we count to 45 on this clock, we see that we reach 0 at each multiple of 7, namely 7, 14, 21, 28, 35, and 42. If we now count 3 more, we reach 45. This means that 45 in a modulo-7 system is the same as 3. Notice again that dividing 45 by 7 and keeping the remainder 3 is a quicker way to solve this problem. **15**

QUIZ YOURSELF 15

a) Count to 75 on a 12-hour clock.
b) Count to 41 on a 7-hour clock.

We can express the fact that 53 occupies the same position as 5 on a 12-hour clock more precisely.

DEFINITION We say that a **is congruent to** b **modulo** m, written $a \equiv b \pmod{m}$, provided that m evenly divides $a - b$. Note, if you find it more convenient to divide m into $b - a$, this is also acceptable.

Because 12 divides $53 - 5 = 48$ evenly, this means that $53 \equiv 5 \pmod{12}$. We read this as, 53 is congruent to 5 modulo 12. Similarly, because $45 - 3 = 42$ is a multiple of 7, we write $45 \equiv 3 \pmod{7}$.

PROBLEM SOLVING

Strategy: The Splitting-Hairs Principle

Remember the Splitting-Hairs Principle from Section 1.1. The three lines $\equiv$ look somewhat like the equal sign but do not mean exactly the same thing. When we write $a \equiv b \pmod{m}$, we are not saying that a and b are equal in the usual sense, but that in terms of an m-hour clock, we can consider them to be the same.

EXAMPLE 1 **Determining When Numbers Are Congruent**

Determine which statements are true.

a) $39 \equiv 15 \pmod{12}$
b) $33 \equiv 19 \pmod{8}$
c) $25 \equiv 48 \pmod{7}$
d) $11 \equiv 35 \pmod{6}$

SOLUTION:

a) $39 - 15 = 24$, which is evenly divisible by 12. Therefore, $39 \equiv 15 \pmod{12}$ is a true statement.
b) $33 - 19 = 14$, which is not evenly divisible by 8. Thus, $33 \equiv 19 \pmod{8}$ is false.
c) Instead of dividing 7 into $25 - 48$, we consider $48 - 25 = 23$. Because 7 does not divide evenly into 23, this statement is false.
d) Because 6 divides evenly into $35 - 11 = 24$, this statement is true.

Now try Exercises 5 to 10. 16

QUIZ YOURSELF 16

Determine which statements are true.

a) $11 \equiv 33 \pmod{12}$
b) $27 \equiv 59 \pmod{8}$

You might wonder why we have not been discussing negative numbers. This is because in a modulo-m system, we do not need negative numbers. As you can see on the 5-hour clock in Figure 5.23, the number -2 corresponds to moving 2 in a counterclockwise direction. If we do this, we stop at 3. This means that -2 is congruent to 3 and so -2 and 3 can be used interchangeably.

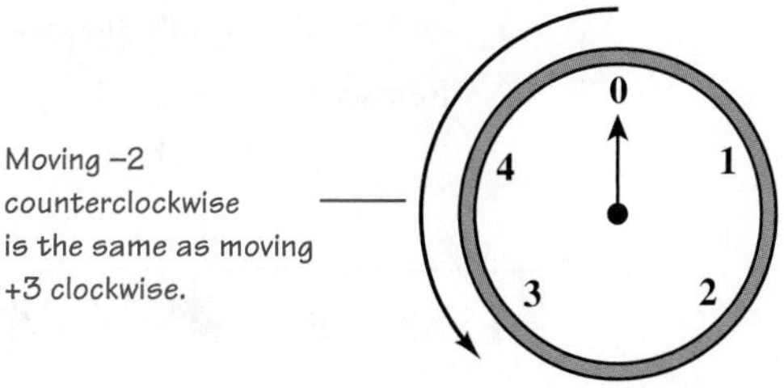
Moving −2
counterclockwise
is the same as moving
+3 clockwise.
0
1
2
3
4

FIGURE 5.23 -2 is congruent to 3 modulo 5.

When doing calculations in a modulo-m system, you can use any two congruent numbers interchangeably. Therefore, if you are calculating in a modulo-5 system and get a result of 19, you can replace it by 4 because $19 \equiv 4 \pmod{5}$. We will use this principle frequently when we are performing arithmetic operations in modular systems.

Between the Numbers

Are You Really Who You Think You Are?

But he that filches from me my good name / Robs me of that which not enriches him / And makes me poor indeed.

—Shakespeare, *Othello,* Act iii, Scene 3

The U.S. Bureau of Justice Statistics states that in 2010, almost 9 million American households were victims of identity theft resulting in the loss of billions of dollars. By "shoulder surfing" (looking over your shoulder at your ATM) or electronically snooping as you use free Wi-Fi at the coffee shop, criminals can in effect become you—buying a new Wii game console with "your" new credit card, withdrawing money from your bank account, and then declaring bankruptcy in your name.

When sensitive information is sent electronically, it is critical to code the information to prevent hacking. One popular method for encrypting messages, based on modular arithmetic, is the **RSA algorithm,** named after its inventors Ronald Rivest, Adi Shamir, and Leonard Adelman, which we will discuss in Exercises 54 to 57.

For more information about identity theft, you and your family might want to visit the following sites:

http://www.ftc.gov/os/2007/11/SynovateFinalReportIDTheft2006.pdf
http://www.ojp.usdoj.gov/programs/identitytheft.htm

KEY POINT

Operations in a modulo-m system are closely related to operations on the integers.

Operations in Modulo-m Systems

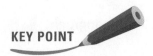

It is easy to add, subtract, and multiply in a modulo-m system. For example, in a modulo-7 system, $6 + 4 = 10$. However, because $10 \equiv 3 \pmod 7$, we can write $6 + 4 \equiv 3 \pmod 7$. It helps to see this computation if you draw the clock in Figure 5.24. If we begin at 0 and count to 6 and then count 4 more, we see that we stop at 3. All modulo-m operations are done in a similar way.

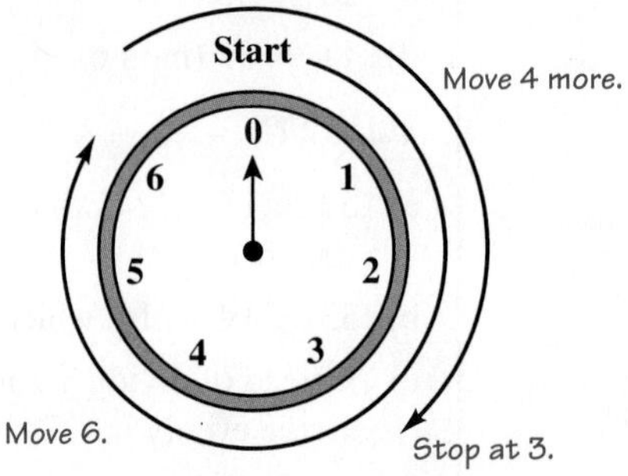

FIGURE 5.24 A 7-hour clock showing $6 + 4 \equiv 3 \pmod 7$.

PERFORMING ARITHMETIC OPERATIONS IN A MODULO-M SYSTEM

To add, subtract, and multiply* in a modulo-m system:

1. Perform the operation as usual.
2. Replace the result in (1) by one of the numbers $0, 1, 2, \ldots, m - 1$, which is congruent to the result in part (1).

EXAMPLE 2 Adding and Subtracting in a Modulo-m System

a) Find $7 + 4 \pmod 8$ b) Find $2 - 5 \pmod{12}$.

SOLUTION:

a) We are being asked to compute $7 + 4$ in a modulo-8 system, or in effect to do this addition on an eight-hour clock. Now $7 + 4 = 11$; however, $11 \equiv 3 \pmod 8$. Therefore, we can write $7 + 4 \equiv 3 \pmod 8$.

b) We begin by calculating $2 - 5 = -3$. However, as we pointed out earlier, we do not need negative numbers in a modulo-m system. On a 12-hour clock, to represent -3, we think of beginning at 0 and moving counterclockwise three numbers, as shown in Figure 5.25. Because we stop at 9, we can say $2 - 5 \equiv 9 \pmod{12}$.

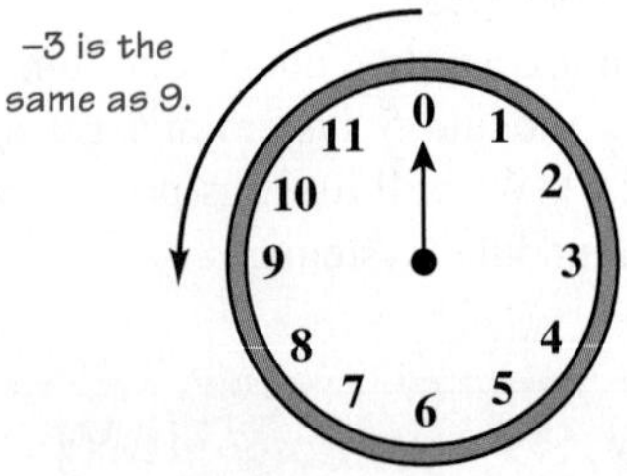

FIGURE 5.25 A 12-hour clock showing $2 - 5 = -3 \equiv 9 \pmod{12}$.

Now try Exercises 11 to 14.

PROBLEM SOLVING

Strategy: Verify Your Answers

To check a subtraction problem in the integers, such as $8 - 5 = 3$, you can perform the addition $3 + 5 = 8$. Similarly, to check subtraction in a modulo-m system, we do a corresponding addition. In Example 2(b), we can check that $2 - 5 \equiv 9 \pmod{12}$ by showing $9 + 5 \equiv 2 \pmod{12}$.

Multiplication is straightforward in a modulo-m system.

*We do division differently. See Exercise 58.

QUIZ YOURSELF 17

Perform the following operations.

a) $4 + 6 \pmod 8$
b) $3 - 8 \pmod{12}$
c) $4 \times 5 \pmod 6$

EXAMPLE 3 Multiplying in a Modulo-*m* System

a) Find $9 \times 7 \pmod{12}$. b) Find $5 \times 8 \pmod 9$.

SOLUTION:

a) We first calculate $9 \times 7 = 63$. Next, we divide 63 by 12 to get a remainder of 3. Therefore, $9 \times 7 \equiv 3 \pmod{12}$.

b) We begin by calculating $5 \times 8 = 40$. Dividing 40 by 9, we get a remainder of 4. Thus, $40 \equiv 4 \pmod 9$, so $5 \times 8 \equiv 4 \pmod 9$.

Now try Exercises 15 to 20. 17

Solving Congruences

KEY POINT

We can use trial and error to solve congruences.

In a modulo-*m* system, we solve congruences instead of equations. As you will see in Example 4, we can solve congruences by trial and error.

EXAMPLE 4 Solving Congruences Using Trial and Error

a) Solve $4 + x \equiv 2 \pmod 5$. b) Solve $2x - 3 \equiv 3 \pmod 6$.
c) Solve $2x \equiv 3 \pmod 6$.

SOLUTION:

a) We will test each of the numbers 0, 1, 2, 3, and 4 to see which ones solve the congruence.

$4 + 0 \equiv 2 \pmod 5$ (False; not a solution.)
$4 + 1 \equiv 2 \pmod 5$ (False; not a solution.)
$4 + 2 \equiv 2 \pmod 5$ (False; not a solution.)
$4 + 3 \equiv 2 \pmod 5$ (True; 3 is a solution.)
$4 + 4 \equiv 2 \pmod 5$ (False; not a solution.)

Thus we find that 3 is the only solution of this congruence.

b) When we test each number from 0 to 5, we find that both 0 and 3 are solutions.

$2 \times 0 - 3 \equiv 3 \pmod 6$ (True; 0 is a solution.)
$2 \times 1 - 3 \equiv 3 \pmod 6$
$2 \times 2 - 3 \equiv 3 \pmod 6$
$2 \times 3 - 3 \equiv 3 \pmod 6$ (True; 3 is a solution.)
$2 \times 4 - 3 \equiv 3 \pmod 6$
$2 \times 5 - 3 \equiv 3 \pmod 6$

c) If you substitute 0, 1, 2, . . . , 5 for x in the congruence, you will find that none of these numbers makes the congruence true. (Verify this.) Therefore, this congruence has no solutions.

Now try Exercises 21 to 30. 18

QUIZ YOURSELF 18

Solve the following congruences.

a) $4 - x \equiv 7 \pmod 8$
b) $3x \equiv 0 \pmod 9$

In Example 5, you must solve a pair of congruences simultaneously, which is similar to the problem Sun Tzu posed to his students.

EXAMPLE 5 Solving a Pair of Congruences

Lauren and Whitney are arranging a collection of designer jeans for display in a Teen Vogue fashion show. If they arrange the jeans in stacks of five, they have three left over. If they arrange the jeans in stacks of four, they have two left over. What is the smallest number of jeans that they can have?

SOLUTION:

Let us say that x is the number of jeans. When Lauren and Whitney stack the jeans five at a time, they have three left over, which tells us that if we divide x by 5, the remainder

QUIZ YOURSELF 19

Find the smallest positive integer that satisfies $x \equiv 5 \pmod 7$ and $x \equiv 6 \pmod 9$.

is 3. In terms of congruences, this means that $x \equiv 3 \pmod 5$. We conclude that x is one of the following numbers:

$$3, 8, 13, \mathbf{18}, 23, 28, 33, 38, \ldots.$$

Similarly, when they stack the jeans four at a time, they have two left over, which tells us that $x \equiv 2 \pmod 4$. This means that x must be in the following list:

$$2, 6, 10, 14, \mathbf{18}, 22, 26, 30, \ldots.$$

The smallest number that is found in both of these lists is 18, so the smallest number of jeans Lauren and Whitney have to display is 18.

Now try Exercises 31 to 34. 19

FIGURE 5.26 Kretschmer Wheat Germ product code.

An interesting application of modular arithmetic appears on many of the products that we use each day. Consider the 12-digit Universal Product Code (UPC) found underneath the bar code on a jar of Kretschmer Wheat Germ, shown in Figure 5.26. The first digit, 0, identifies the product. The next five digits, 30000, identify the manufacturer. The next group of digits, 02270, provides information about the product. The last digit, in this case 2, is called a *check digit*. The check digit provides a way of determining if a UPC is a valid code.

Example 6 explains how to compute the check digit for a UPC.

EXAMPLE 6 Using Modular Arithmetic to Find Check Digits

If the 12-digit UPC is of the form $a_1a_2a_3a_4a_5a_6a_7a_8a_9a_{10}a_{11}a_{12}$,* then we compute the check digit in two steps.

1. First we compute the expression

$$3a_1 + a_2 + 3a_3 + a_4 + 3a_5 + a_6 + 3a_7 + a_8 + 3a_9 + a_{10} + 3a_{11}.$$

That is, we multiply the first digit in the UPC by 3; add the second digit; then add three times the third digit; add the fourth digit, and so on.

2. Choose the check digit, a_{12}, so that the sum in step (1) plus a_{12} is congruent to 0 modulo 10.

a) Verify that the UPC for Kretschmer Wheat Germ in Figure 5.26 is a valid code.

b) Suppose that in the middle of typing this UPC you accidentally typed the pattern 772 instead of 227. Show why this would not be a valid UPC.

SOLUTION:

a) In the Kretschmer Wheat Germ UPC, which is 030000022702, the first digit, a_1, is 0, the second digit, a_2, is 3, the third digit, a_3, is 0, and so on. Therefore,

$$3a_1 + a_2 + 3a_3 + a_4 + 3a_5 + a_6 + 3a_7 + a_8 + 3a_9 + a_{10} + 3a_{11}$$
$$= 3 \times 0 + 3 + 3 \times 0 + 0 + 3 \times 0 + 0 + 3 \times 0 + 2 + 3 \times 2 + 7 + 3 \times 0 = 18.$$

To find the check digit, a_{12}, we must find a number such that $18 + a_{12} \equiv 0 \pmod{10}$. It is easy to see that if $a_{12} = 2$, then $18 + 2 \equiv 0 \pmod{10}$. Therefore, the check digit for this UPC is 2, so the UPC is a valid code.

b) If you had typed 030000077202 for the UPC, then in computing $3a_1 + a_2 + 3a_3 + a_4 + 3a_5 + a_6 + 3a_7 + a_8 + 3a_9 + a_{10} + 3a_{11}$, you would calculate $3 \times 0 + 3 + 3 \times 0 + 0 + 3 \times 0 + 0 + 3 \times 0 + 7 + 3 \times 7 + 2 + 3 \times 0 = 33$. This would mean that to have $33 + a_{12} \equiv 0 \pmod{10}$, the check digit would have to be 7, not 2. This means that you typed an invalid UPC.

Identification numbers occur in many other places, such as on bank checks, U.S. postal money orders, driver's licenses, express shipping labels, airline tickets, and ISBN numbers on books. Although there are many different ways to calculate check digits other than the method that we explained in Example 6, most methods are based on modular arithmetic. We will discuss several of these other methods in the exercises.

*Don't let this notation scare you. All we are saying is that the UPC consists of 12 numbers. The first number is called a_1, the second is called a_2, and so forth.

EXERCISES 5.4

Sharpening Your Skills

Count to the number that is specified on the clock that is indicated.

1. 43 on a 12-hour clock

2. 21 on a 6-hour clock

3. 39 on a 7-hour clock

4. 69 on an 8-hour clock

Determine which of the following statements are true.

5. $35 \equiv 59 \pmod{12}$

6. $31 \equiv 14 \pmod{7}$

7. $11 \equiv 43 \pmod{8}$

8. $29 \equiv 18 \pmod{6}$

9. $46 \equiv 78 \pmod{10}$

10. $53 \equiv 75 \pmod{11}$

Perform the following operations.

11. $3 + 4 \pmod{5}$

12. $6 + 8 \pmod{11}$

13. $4 - 9 \pmod{12}$

14. $5 - 6 \pmod{8}$

15. $4 \times 9 \pmod{12}$

16. $5 \times 6 \pmod{8}$

17. $2 \times 5 + 4 \pmod{8}$

18. $3 \times 11 + 5 \pmod{12}$

19. $6 - 3 \times 5 \pmod{8}$

20. $3 - 4 \times 2 \pmod{12}$

Solve the following congruences.

21. $6 + x \equiv 4 \pmod{8}$

22. $5 + x \equiv 2 \pmod{6}$

23. $3 - x \equiv 4 \pmod{7}$

24. $5 - x \equiv 7 \pmod{10}$

25. $2 - x \equiv 8 \pmod{9}$

26. $3 - x \equiv 4 \pmod{5}$

27. $3x \equiv 5 \pmod{7}$

28. $5x \equiv 3 \pmod{8}$

29. $4x \equiv 2 \pmod{10}$

30. $6x \equiv 4 \pmod{8}$

Find the smallest positive integer that solves all of the given congruences.

31. $x \equiv 3 \pmod{7}$, $x \equiv 4 \pmod{5}$

32. $x \equiv 1 \pmod{6}$, $x \equiv 7 \pmod{8}$

33. $x \equiv 2 \pmod{10}$, $x \equiv 2 \pmod{8}$, $x \equiv 0 \pmod{6}$

34. $x \equiv 5 \pmod{9}$, $x \equiv 1 \pmod{11}$, $x \equiv 7 \pmod{8}$

Applying What You've Learned

35. In a Jules Verne novel, Phileas Fogg took an imaginary trip around the world in 80 days. If he were counting the days of his trip on a 7-hour clock, what would the clock show when he landed?

36. Columbus's first voyage to America took 71 days. If he were counting the days of his voyage on a 7-hour clock, what would the clock show when he landed in America?

37. Packaging candy. A package designer for a chocolate manufacturer is designing boxes to hold a certain number of candies. When arranging the pieces six in a row, there will be four left over. When arranging the pieces five in a row, there will be three left over. What is the smallest number of pieces that the manufacturer might be considering to place in the box?

38. Arranging a marching band. A marching band is considering different configurations for its upcoming half-time show. When the members are arranged eight or 10 in a row, there are two members left over. If they are arranged 12 in a row, there are six left over. What is the smallest number of members that the band can have?

Exercises 39 and 40 concern Sun Tzu's and Brahmagupta's (slightly simplified) problems that we mentioned in the section opener.

39. Find the smallest positive integer such that if we divide by three, the remainder is 2; if we divide by five, the remainder is 3; if we divide by seven, the remainder is 2.

40. Find the smallest positive integer such that if we divide by two, three, and four, the remainder is 1 but seven divides the number evenly.

While taking a class on the culture of China, you have learned about the Chinese zodiac in which people fall into one of 12 categories, depending on the year of their birth. The categories, numbered 0 to 11, correspond to the following animals: (0) monkey, (1) rooster, (2) dog, (3) pig, (4) rat, (5) ox, (6) tiger, (7) rabbit, (8) dragon, (9) snake, (10) horse, and (11) goat. Those who believe in this zodiac think that the year of a person's birth influences both their personality and fortune in life.

To find your zodiac sign, divide the year of your birth by 12. The remainder then determines your sign. When we divide 1995 by 12, the remainder is 3, so a person born in 1995 is a pig. Use this information in Exercises 41 and 42.

41. Estimating a person's age. You want to date Chris, who is in your Chinese culture class. However, before asking for a date, you want to find out Chris's age without asking directly. Chris is at least 18 and seems to be younger than 30 years old. If according to the Chinese zodiac Chris is a dog, then in what year was Chris born?

42. Estimating a person's age. A person who was between 40 and 50 years old in the year 2003, and who according to the Chinese zodiac is a pig, was born in what year?

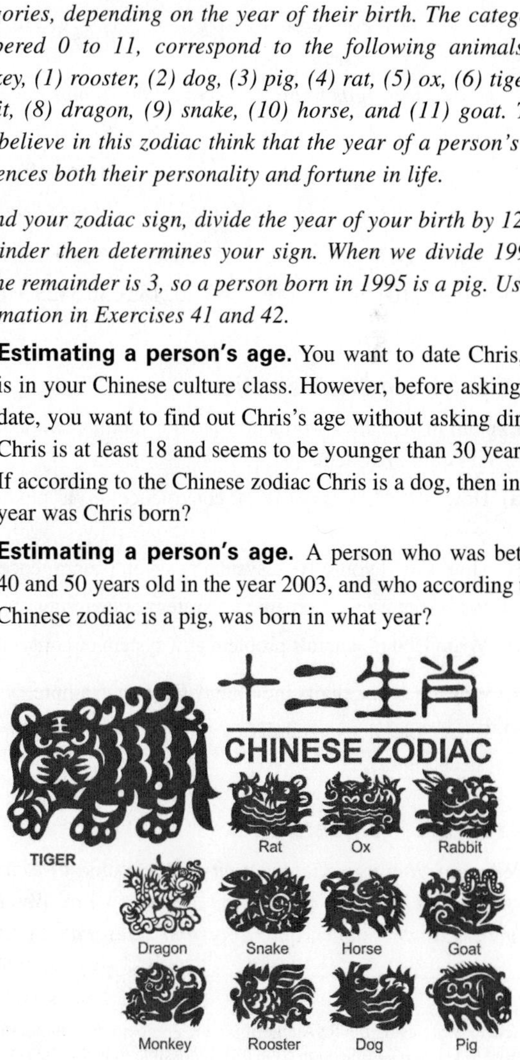

If you look at the back cover of the student edition of this textbook, you will see a 10-digit number called an ISBN (International Standard Book Number). This book's ISBN is 0-321-83699-5 . The first digit, 0, indicates that the book is from an English-speaking country. The digits 321 identify the publisher (in this case, Pearson), and the third group of digits, 83699, identifies the book. The last digit, 5, is a check digit. Finding check digits for ISBNs is more complicated than finding check digits for UPCs. We calculate an ISBN check digit in three steps.

1. *Multiply the first digit by 10, the second digit by 9, the third digit by 8, continuing until you multiply the ninth digit by 2, as in Table 5.8. Then find the total of these products. In this example, the total is 182.*
2. *Divide the total by 11, keep the remainder, which is 6, and call it* r.
3. *Find the check digit* c *such that* r + c = 11. *In this case* c = 5.

10 × 0 =	0
9 × 3 =	27
8 × 2 =	16
7 × 1 =	7
6 × 8 =	48
5 × 3 =	15
4 × 6 =	24
3 × 9 =	27
2 × 9 =	18
Total	182

TABLE 5.8 Table for calculating ISBN numbers.

Find the missing digit d *in each of the following ISBNs.*

43. *A Visit from the Goon Squad*—03075*d*2839

44. *Finding the Next Starbucks*—15918*d*1348

45. *Time and Materials*—0061349*d*07

46. *Tinkers*—193413*d*12X (*Note:* The symbol X represents ten.)

A common method for calculating check digits for credit cards, called the Luhn algorithm, works as follows. Start with a hypothetical credit card number such as 4928 4613 1325 687c. (c is the check digit.)

Step 1: Beginning with the second digit from the right, in this case the 7, replace every other digit (highlighted below) with its double. Notice that 6 was replaced by 12 and 7 was replaced by 14.

	4	9	2	8	4	6	1	3	1	3	2	5	6	8	7 (Begin)	c
Alternate digits were doubled →	8	9	4	8	8	6	2	3	2	3	4	5	12	8	14	

Step 2: Add all the digits in the numbers in the second row of the table to get a sum. In our example the sum is

$$8 + 9 + 4 + 8 + 8 + 6 + 2 + 3 + 2 + 3 + 4 + 5 + 1 + 2 + 8 + 1 + 4 = 78.$$

Step 3: Calculate the sum (mod 10), *which is 78* (mod 10) = *8.*

Step 4: Subtract the result in step 3 from 10 to get the check digit. In this case the check digit is 10 − 8 = 2.

Find the check digit for the credit card numbers in Exercises 47–50.

47. 4563 2625 2104 353*c*

48. 5218 3235 3423 346*c*

49. 3162 4425 3482 291*c*

50. 5104 6143 2371 412*c*

Communicating Mathematics

51. Explain in your own words how to add, subtract, and multiply in a modulo-*m* system.

52. **a)** How would you solve a congruence in a modulo-12 system?

b) How would you solve a system of several congruences?

c) Write Sun Tzu's problem as a system of congruences.

d) Write Brahmagupta's problem as a system of congruences.

53. **a)** Why are check digits important? Give an example.

b) What role does modular arithmetic play in computing check digits?

Between the Numbers

The RSA algorithm uses the equation $c = m^e$ (mod n) *to encode a message. Here,* c *is the coded form of the letter* m. *We choose the numbers* n* *and* e *according to certain rules that we will not explain. To illustrate this method, suppose that we want to encode the letter C, which we will represent by the number 3. Let* n = 35 *and* e = 5. *(Do not be concerned how we picked 35 and 5.) Then to encode 3, we compute* $c = 3^5$(mod 35) = 243 (mod 35) = 33. *So 33 is the coded form of the letter C. Use this method to encode each letter in Exercises 54–57.*

54. L

55. O

56. V

57. E

Challenge Yourself

58. When we do usual division of integers, to say that 12/4 = 3 is the same as saying that 12 = 4 · 3. More generally, if $a/b = x$, then $a = b \cdot x$. Use this idea to devise a way to perform division in a modulo-*m* system. That is, how would you explain how to solve a congruence such as $4/6 \equiv x \pmod 8$? Use the method you devise to find

a. 5/3 (mod 8)

b. 2/6 (mod 8)

59. Solve the congruence $2/x \equiv 5 \pmod 6$.

*In practice, when using the RSA algorithm, the number *n* is a number having more than 500 digits, and unless you know how to factor *n* (see Section 6.1), it is impossible to break the code.

CHAPTER SUMMARY

You will learn the items listed in this summary more thoroughly if you keep the following advice in mind:

1. Focus on "remembering how to remember" the ideas. What pictures, word analogies, and examples help you to remember these ideas?
2. Practice writing each item without looking at the book.
3. Make 3 × 5 flash cards to break your dependence on the book. Use these cards to give yourself practice tests.

Section	Summary	Example
SECTION 5.1	A **number** tells us how many objects we are counting; a numeral is a symbol that represents a **number.**	Discussion, p. 191
	In a **simple grouping system,** such as the Egyptian numeration system, the value of a number is the sum of the values of its numerals.	Example 1, p. 192
	Addition and **subtraction** in the Egyptian system are based on grouping and regrouping its basic symbols.	Example 2, pp. 192–193
	The Egyptians performed multiplication using a "**doubling method,**" which is based on the fact that we can write any positive integer as a sum of powers of 2.	Example 3, p. 194
	The **Roman numeration system** is more sophisticated than the Egyptian's system. It has a *subtraction principle* that we use to write numbers more concisely and a *multiplication principle* that makes it easier to write large numbers.	Example 5, p. 196 pp. 194–196
	The **Chinese numeration system** is a *multiplicative system.* We form Chinese numerals by writing integers between 1 and 9, inclusive, multiplied by powers of 10.	Example 7, p. 198
SECTION 5.2	In a **place value system,** also called a **positional system,** the placement of the symbols in a numeral determines the value of the symbols.	Discussion, p. 200
	The **Babylonian numeration system** was a primitive place value system based on powers of 60 and is called a *sexagesimal system.*	Example 1, p. 201 Example 2, p. 202
	The **Mayans** had a positional numeration system based on the number 20.	Discussion, p. 203
	The **Hindu-Arabic numeration system** is a place value system based on 10. We write Hindu-Arabic numerals in **expanded notation** to show how each digit in the numeral is multiplied by a power of 10.	Discussion, p. 204 Example 3, p. 204
	In the Hindu-Arabic system, expanded notation helps us to understand the algorithms to perform **arithmetic operations.**	Examples 4 and 5, pp. 204–205
	In the **galley method,** we compute partial products in each box of the galley and then add along diagonals to compute the final product.	Example 6, pp. 205–206
	Napier's rods are a variation of the galley method that enables us to compute partial products quickly.	Discussion, pp. 206–207
SECTION 5.3	**Counting** in a non-base-10 numeration system is very similar to counting in a base-10 system.	
	A **base-5 system** uses the digits 0, 1, 2, 3, and 4. In expanded notation, we use powers of 5 rather than powers of 10.	Example 1 and 2, pp. 211 and 212
	Arithmetic operations in a non-base-10 system are done much like the operations in a base-10 system.	Examples 3, 4, and 5, pp. 213–216
	The **binary** system has a base of 2; the **octal** system has a base of 8; the **hexadecimal** system has a base of 16.	Example 7, p. 218
SECTION 5.4	A **modulo-*m* system** consists of the numbers 0, 1, 2, 3,..., $m - 1$. Counting and arithmetic operations are performed in a manner corresponding to movements on an m-hour clock. The number m is called the **modulus** of the system.	Discussion, p. 221
	We say that ***a* is congruent to *b* modulo *m*** if m evenly divides $a - b$.	Example 1, p. 223
	To **add, subtract, or multiply in a modular system,** we perform the operation as usual, divide the result by ***m*,** and keep the remainder.	Examples 2 and 3, pp. 224 and 225
	We **solve congruences** by using trial and error.	Examples 4 and 5, pp. 225–226

CHAPTER REVIEW EXERCISES

Section 5.1

1. Write 𓁨𓆐𓆐𓂭𓂭𓂭𓆼𓆼𓍢𓍢𓎆 using Hindu-Arabic numerals.

2. Find 𓆼𓍢𓍢𓍢𓎆𓎆𓏺𓏺 − 𓍢𓎆𓎆𓎆𓎆𓎆𓏺𓏺𓏺𓏺𓏺 using Egyptian notation.

3. Use the Egyptian method of doubling to calculate 37 × 53.

4. Write 4,795 in Roman notation.

5. Write 七千五百九十三 as a Hindu-Arabic numeral.

6. Do the words *number* and *numeral* mean the same thing? Explain.

7. Explain two advantages of the Roman numeration system over the Egyptian system.

8. Why were people in Europe at first suspicious of Hindu-Arabic numerals?

Section 5.2

9. Write 11,292 using Babylonian notation.

10. Subtract 4,237 − 2,673 using expanded notation.

11. Use the galley method to multiply 46 × 103.

12. Why is zero important in a place value system?

Section 5.3

13. Write the following Mayan numerals in Hindu-Arabic notation:

a. **b.**

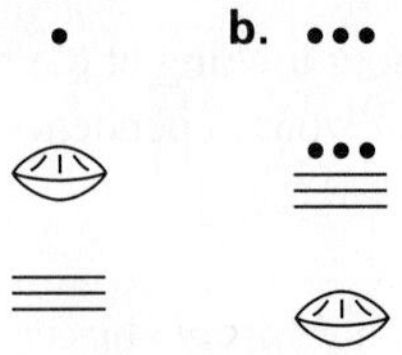

14. Write 342_5 and $B3D_{16}$ as base-10 numerals.

15. Write decimal 3,403 in base-8 notation.

16. Add $10{,}111_2 + 11{,}001_2$.

17. Divide $4{,}312_5 \div 23_5$.

18. Write $1{,}011{,}100{,}010_2$ first as an octal number and then as a hexadecimal number.

19. Convert 463_7 to base-5 notation.

Section 5.4

20. Count to 76 on an 8-hour clock.

21. Determine which of the following statements are true:

a. $54 \equiv 72 \pmod{6}$ **b.** $29 \equiv 75 \pmod{11}$

22. Perform the indicated operations.

a. $8 + 11 \pmod{12}$ **b.** $7 - 9 \pmod{11}$

c. $4 \times 8 \pmod{9}$

23. Solve the following congruences:

a. $3x \equiv 9 \pmod{12}$ **b.** $4x \equiv 3 \pmod{8}$

CHAPTER TEST

1. Write 3,685 in Roman notation.

2. Give two advantages of the Hindu-Arabic numeration system over the Chinese numeration system.

3. Write 264_7 and $A3E_{16}$ as base-10 numerals.

4. Write 𓁨𓁨𓆐𓂭𓂭𓆼𓆼𓆼𓍢𓍢𓍢𓍢𓎆𓎆 using Hindu-Arabic numerals.

5. Write the following Mayan numerals in Hindu-Arabic notation:

a. **b.**

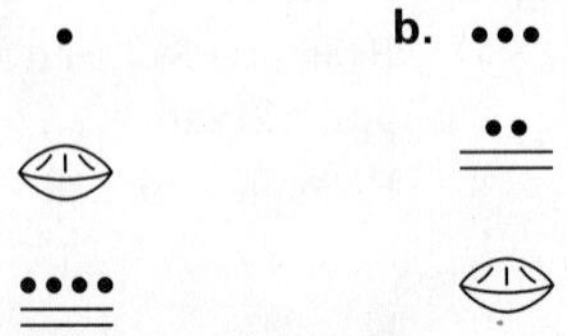

6. Determine which of the following statements are true:

a. $43 \equiv 57 \pmod{8}$ **b.** $16 \equiv 52 \pmod{6}$

7. Find 𓁨𓁨𓆐𓂭𓂭𓆼𓆼𓆼𓍢𓍢𓍢𓍢𓎆𓎆 − 𓁨𓆐𓆐𓂭𓂭𓂭𓆼𓆼𓍢𓍢𓍢𓎆𓎆𓎆 using Egyptian notation.

8. Write 五千三百〇十八 as a Hindu-Arabic numeral.

9. Count to 57 on a 6-hour clock.

10. Give an example of three different numerals that represent the same number.

11. Add $101{,}101_2 + 110{,}101_2$.

12. Write 10,937 using Babylonian notation.

13. Use the galley method to multiply 67 × 238.

14. Write decimal 2,305 as a base-8 numeral.

15. Perform the indicated operations.

a. $9 + 12 \pmod{13}$ **b.** $6 - 10 \pmod{12}$

c. $5 \times 4 \pmod{6}$

16. Add 1,738 + 526 using expanded notation.

17. What problem did the lack of zero cause in the ancient Chinese system?

18. Solve the following congruences:

a. $4x \equiv 8 \pmod{12}$

b. $4x \equiv 7 \pmod{10}$

19. Find $3{,}142_5 \div 24_5$.

20. Use the Egyptian method of doubling to calculate 26 × 59.

21. Write $110{,}101{,}011{,}110_2$ first as an octal number and then as a hexadecimal number.

22. Find $42_5 \times 34_5$.

GROUP PROJECTS

1. **Conversion shortcuts.** In Example 7 in Section 5.3, we discussed a shortcut method to make conversions between binary, octal, and hexadecimal. These conversions worked because 8 and 16 are powers of 2. With this in mind, you would expect to be able to make similar types of conversions between bases where one base is a power of the other—such as between 3 and 9, 2 and 4, and 4 and 16. Use this notion to invent conversion techniques to solve Exercises a)–f).

   a. Convert $201{,}221_3$ to base 9.

   b. Convert $101{,}101{,}011{,}101_2$ to base 4.

   c. Convert A0B5_{16} to base 4.

   d. Convert 768_9 to base 3.

   e. Convert $301{,}233_4$ to base 16.

   f. Convert $312{,}032_4$ to base 16.

2. **Simplifying modular arithmetic computation.** If we wanted to find 78,945,213 (mod 12), we could use long division to divide 12 into 78,945,213 and find the remainder 9. This method is long and tedious, and a quicker way is to use a calculator. Dividing 78,945,213/12, we get 6,578,767.75. If we next subtract 6,578,767 from 6,578,767.75 we get 0.75, and if we then multiply 0.75 by 12, we will get the remainder 9 that we found by doing long division.

   a. Verify this process by finding the following by first doing long division and then using the calculator process: 356 (mod 8); 4,291 (mod 12); 2,543 (mod 16).

   b. Verify the calculation $126 \text{ (mod 8)} \times 419 \text{ (mod 8)} = 126 \times 419 \text{ (mod 8)}$. Make up several more examples like this to show that $a \text{ (mod } n) \times b \text{ (mod } n) = a \times b \text{ (mod } n)$.

   c. Try to find $16{,}253{,}478 \times 3{,}547{,}635$ (mod 12). What happens? Thinking about parts (a) and (b) above, can you find a simpler way to do this calculation?

   d. Find: $645{,}398 \times 423{,}651$ (mod 8); $74{,}562 \times 341{,}576$ (mod 8); $2{,}123{,}426 \times 783{,}542$ (mod 8).

3. **Negative exponents in other bases.** In decimal notation, digits to the right of the decimal point correspond to negative powers of ten. For example,

$$0.326 = \frac{3}{10} + \frac{2}{100} + \frac{6}{1000} = 3 \times 10^{-1} + 2 \times 10^{-2} + 6 \times 10^{-3}.$$

   Interpret what each of the following would represent in the given base and then convert the result to a base-ten number.

   a. 0.423 in base 5

   b. 0.3204 in base 8

   c. 0.01011 in base 2

USING TECHNOLOGY

1. **Java applets and modular arithmetic.** There are a number of excellent Web sites that have Java applets to do modular arithmetic—conversions, addition, subtraction, etc. Search for "applet" + "modular arithmetic" to find such an applet and use it to verify the calculations in the examples of Section 5.3.

2. **Solving systems of congruences with spreadsheets.** You can use an Excel or Numbers spreadsheet to solve systems of congruences. For example, to solve the system $x \equiv 5 \text{ (mod 6)}$, $x \equiv 7 \text{ (mod 8)}$, we want to find a common number in the following two lists:

$$5, 5+6, 5+6+6, 5+6+6+6, \ldots \text{ and}$$
$$7, 7+8, 7+8+8, 7+8+8+8, \ldots.$$

   To the right is an Excel example that shows you how to do this. Cell D2 shows the kind of formulas we are using to generate the lists. In this case we see that 23 appears in both lists and is the solution we seek since it is congruent to 5 modulo 6 and congruent to 7 modulo 8.

D2 | fx =D1+6

	A	B	C	D	E
1	5	7		5	
2	11	15		11	
3	17	23			
4	23	31			
5	29	39			
6	35	47			
7	41	55			
8	47	63			
9					

   a) Write a similar Excel or Numbers spreadsheet to illustrate how to solve Sun Tzu's and Brahmagupta's problems that we mentioned in Section 5.4.

   b) Use an Excel or Numbers spreadsheet to solve the system of congruences $x \equiv 3 \text{ (mod 12)}$ and $x \equiv 5 \text{ (mod 7)}$.

6

Number Theory and the Real Number System

Understanding the Numbers All Around Us

Atoms are very tiny things. In fact, they are so small that it would take about 10 million of them to make a length that is the width of the capital "T" in the word "Theory" in the title of this chapter.

If we were able to line up all the atoms in your body and stretch them out in a straight line, how far do you think they would reach? Would they stretch from New York City to Chicago? Could they stretch around the world? We will answer this question later in this chapter, and the answer will astound you.

Have you ever thought about what it would be like to take a trip around our solar system? If a rocket were to travel at 25,000 miles per hour, how long would it take to travel to a distant planet such as Neptune?

Many people do not have a good understanding of how our national debt has increased in the last 50 years. Since 1960, the

(continued)

U.S. population has almost doubled, but the national debt has grown from about $290 billion to over $15 trillion—that is, it is now more than 50 times larger than it was in 1960.

Later in this chapter, we will discuss how to represent very large and very small numbers and will calculate your share of the national debt.

The author John Paulos wrote a book about a condition that he calls *innumeracy,* which is the mathematical equivalent of illiteracy. We expect that after studying this chapter, you will have a better understanding of the way numbers work and that you will certainly be more "numerate."

6.1 Number Theory

Objectives

1. Determine when numbers are prime.
2. Apply divisibility tests to factor numbers into products of prime factors.
3. Find the GCD and LCM of two natural numbers.
4. Use the GCD and LCM to solve applied problems.

From the time you were a small child, you have known about a set of numbers that has fascinated the greatest minds in the history of mathematics for thousands of years. The numbers that we use for counting, $\{1, 2, 3, \ldots\}$, are called the **natural numbers,** or **counting numbers.** This simple set of numbers has elegant properties and patterns that have intrigued mathematicians from ancient times to the present. The study of the natural numbers and their properties is called **number theory.**

Prime Numbers

KEY POINT

Prime numbers are the building blocks of the integers.

One question that people have studied since the time of the ancient Greeks is exactly which natural numbers can be written as a product of *other* natural numbers and which cannot. For example, $30 = 2 \cdot 3 \cdot 5$, whereas $17 = 17 \cdot 1$. The fact that we cannot write 17 as a product of natural numbers without using 17 greatly interests mathematicians. In order to investigate this question, we have to explain what it means for one natural number to divide another.

DEFINITIONS If a and b are natural numbers, then we say a **divides** b, written $a|b$, provided there is a natural number q such that $b = qa$. Other ways of stating this is that a is a **divisor** of b, a is a **factor** of b, or that b is a **multiple** of a. A way of testing to see if a divides b is to actually divide b by a and see if you get a zero remainder.

For example, we can say that 5 divides 30 because there is a natural number, namely 6, such that $30 = 5 \cdot 6$. We can write this fact more concisely as $5|30$. Also, $17|17$ because $17 = 17 \cdot 1$. When we write a natural number as a product of natural numbers, we say that we have **factored** the number. Some factorizations of 72 are $8 \cdot 9$ and $2 \cdot 2 \cdot 2 \cdot 3 \cdot 3$. Numbers like 17, which have only trivial factorizations, are particularly important in number theory.

The accompanying iPad screen shot shows a calculator app testing for divisibility. Because 5 divides 30, no digits follow the decimal point. Because 7 does not divide 45, we see several digits after the decimal point.

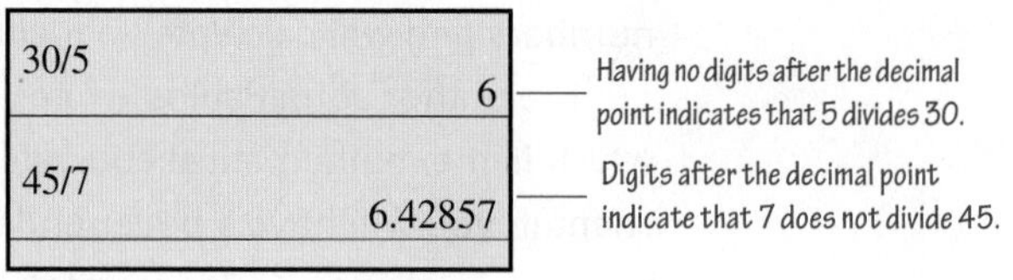
30/5
6
45/7
6.42857
Having no digits after the decimal point indicates that 5 divides 30.
Digits after the decimal point indicate that 7 does not divide 45.

DEFINITIONS A natural number, greater than 1,* that has only itself and 1 as factors is called a **prime number.** A natural number, greater than 1, that is not prime is called **composite.**

Numbers such as 2, 5, 17, and 29 are prime numbers because their only divisors are themselves and 1. On the other hand, numbers such as 4, 33, 87, and 102 are composite because you can find divisors for them that are other than 1 and the number itself.

One very well-known way to generate a list of primes is called the **Sieve of Eratosthenes,** which is named after the Greek mathematician Eratosthenes of Alexandria, whom we discuss in Historical Highlight on page 238. We explain his process in Example 1.

EXAMPLE 1 Finding a List of Prime Numbers

Use the Sieve of Eratosthenes to find all prime numbers less than 50.

SOLUTION:

First we list all natural numbers from 1 to 50 in Figure 6.1 and then systematically cross off all nonprimes according to the following steps:

1. Cross off 1 because it is not prime.
2. Circle 2 because it is prime and then cross off all other multiples of 2, such as 4, 6, 8, . . . , 50.
3. The next number in the list that has not been crossed off is 3, which is prime. Circle it and cross off all remaining multiples of 3, namely 9, 15, 21, . . . , 45.
4. Next circle 5, a prime, and cross off its remaining multiples, 25 and 35.
5. Finally circle 7 and cross off its one remaining multiple, 49.
6. Because the next prime, which is 11, is greater than the square root of 50, we can stop looking for composites and circle all remaining numbers in the list. If you think about it, you will see that it is impossible to have a composite number of the form $a \times b$ where both a and b are 11 or greater and yet the product is less than 50.

It would be a good idea for you to reproduce the above steps† to produce Figure 6.1 from scratch and memorize the primes you found to use in later computations.

Now try Exercises 25 and 26.

*There are technical reasons, which we will not get into, why mathematicians consider 1 to be neither prime nor composite.

†If you wanted to find all primes between 1 and 500, it would take a long time to create the list. You can find interactive Java applets on the Internet that will allow you to use this sieve without the tedium of writing a long list of numbers.

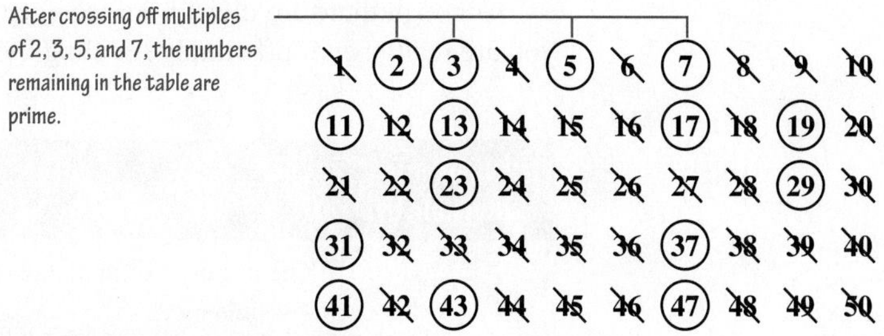

FIGURE 6.1 Using the Sieve of Eratosthenes.

EXAMPLE 2 Identifying Prime Numbers

Determine whether the following numbers are prime or composite.

a) 83 b) 87 c) 397

SOLUTION: Review the Be Systematic Strategy on page 5.

a) To see if 83 has any factors, we start dividing it by primes such as 2, then 3, then 5, and so on. *We don't need to check to see if any composites divide 83, because if a composite like 6 divides 83, then the primes 2 and 3 would have divided 83.* We keep checking for divisibility by primes until we reach the square root of 83. Notice that

$$9 < \sqrt{83} < 10.$$

$9^2 = 81$, so 9 is less than the square root of 83.

$10^2 = 100$, so 10 is greater than the square root of 83.

As in Example 1, if you do not find a prime divisor of 83 by the time you reach the square root of 83, you will not find another divisor until you reach 83 because multiplying two integers greater than 9 will give a product that is at least 100. Because 2, 3, 5, and 7 do not divide 83, we conclude that 83 is prime.

b) The number 87 factors as $3 \cdot 29$. Again, in testing 87, you only need to check primes up to the square root of 87, which is between 9 and 10.

c) Because $\sqrt{397} \approx 19.92$, to determine if 397 is prime, we must check all primes less than or equal to 19 to see if they divide 397. You can do this by hand, or to speed the process, you can use your calculator, as we show in the accompanying TI-84 screen. The nonzero digits after the decimal point show that neither 2 nor 3 divides 397 evenly. If you do the division, you will find that no primes from 2 to 19 divide 397; therefore, 397 is prime.

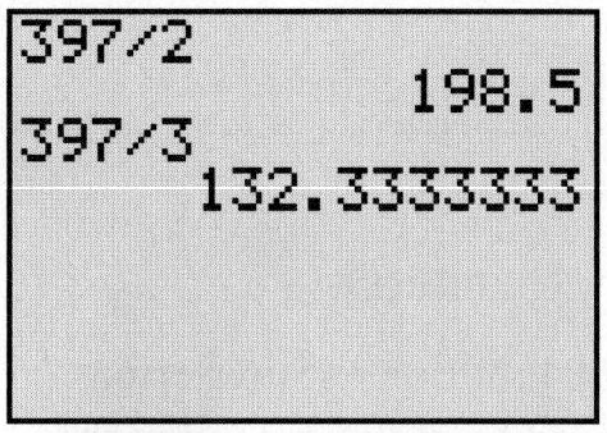

Now try Exercises 13 to 16. 1

QUIZ YOURSELF 1

Identify the following numbers as either prime or composite.

a) 89
b) 187
c) 143
d) 211

Divisibility Tests and Factoring

In checking larger numbers for primality, it is useful to have some quick tests for divisibility so that we don't actually have to do a lengthy division. For example, it would be nice to be able to look at the number 21,021 and have a quick test to see whether 3 divides into it without actually having to do the division. Table 6.1 lists some divisibility tests. We will examine why these tests work in the exercises.

Although there are divisibility tests for 7 and also 11, they are hard to remember, and you are really better off simply dividing by 7 or 11 rather than using the test.

Number Is Divisible by	Test	Example
2	The last digit of the number is divisible by 2.	2 divides 13,578 because 2 divides 8.
3	The sum of the digits is divisible by 3.	3 divides 21,021 because 3 divides $2 + 1 + 0 + 2 + 1 = 6$.
4	The number formed by the last two digits is divisible by 4.	102,736 is divisible by 4 because 4 divides 36.
5	The last digit is 0 or 5.	607,895 is divisible by 5.
6	The number is divisible by both 2 and 3.	802,674 is divisible by both 2 and 3, so it is divisible by 6.
8	The number formed by the last three digits is divisible by 8.	8 divides 230,264 because 8 divides 264.
9	The sum of the digits is divisible by 9.	2,081,763 is divisible by 9 because 9 divides $2 + 0 + 8 + 1 + 7 + 6 + 3 = 27$.
10	The number ends in 0.	12,865,890 is divisible by 10.

TABLE 6.1 Divisibility tests for some small numbers.

EXAMPLE 3 Applying Divisibility Tests

Test the number 11,352 for divisibility by

a) 2 b) 3 c) 4 d) 5
e) 6 f) 8 g) 9 h) 10

SOLUTION:

a) The last digit of 11,352 is divisible by 2, so 2 divides the number.

b) The sum of the digits is $1 + 1 + 3 + 5 + 2 = 12$, which is divisible by 3, so 11,352 is also divisible by 3.

Math in Your Life

What's the Buzz on Prime Numbers?

During the late spring of 2004, billions of mysterious insects called *cicadas* emerged from the ground in the eastern United States to live briefly in the fresh air and sunshine, mate, and then die. You might wonder, "What does this incredible swarm of insects have to do with mathematics?" Well, periodical cicadas, as they are called, emerge every 13 or 17 years—a curious choice of numbers. Why would the cicadas settle on prime numbers for their life cycles?

Paleontologist Stephen Gould proposed that by emerging every 13 or 17 years, cicadas minimize their chance of synchronizing with predators who will destroy them. For example, a 17-year cycle for the cicadas would synchronize with a 5-year predator only every 85 years. Mathematician Glenn Webb, working at Vanderbilt University, has created mathematical models to investigate Gould's theory and has indeed found that by emerging every 13 or 17 years, cicadas do improve their chances of survival. Other scientists, however, are skeptical and believe that it is only a coincidence that the cicada's life cycles happen to be prime numbers. Webb admits, "I don't know if there will ever be a satisfying scientific resolution."

c) The last two digits of 11,352 form the number 52, and 52 is divisible by 4; therefore, 11,352 is also divisible by 4.

d) The last digit is neither 0 nor 5, so 5 does not divide this number.

e) Both 2 and 3 divide 11,352; therefore, 6 also divides this number.

f) The number formed by the last three digits is 352, which is divisible by 8, so 8 divides 11,352.

g) In part b), we saw that the sum of the digits of 11,352 is 12. Because 9 does not divide 12, 9 does not divide 11,352.

h) The number does not end in 0, so 10 does not divide the number.

Now try Exercises 9 to 12. 2

QUIZ YOURSELF 2

Test the number 30,690 for divisibility by

a) 2 b) 3 c) 4
d) 5 e) 6 f) 8
g) 9 h) 10

In chemistry, we form compounds by combining simpler objects called atoms. For example, a molecule of table salt is formed by combining one sodium and one chlorine atom. Similarly, in mathematics, objects are built from simpler objects. For example, $120 = 2 \cdot 2 \cdot 2 \cdot 3 \cdot 5$ and $84 = 2 \cdot 2 \cdot 3 \cdot 7$. The fundamental theorem of arithmetic states that every natural number greater than 1 is "built" by multiplying a unique combination of prime numbers. This theorem appeared in another way in Euclid's *Elements* and was stated in its present form by the great German mathematician Karl Friedrich Gauss.

THE FUNDAMENTAL THEOREM OF ARITHMETIC Every natural number greater than 1 is a unique product of prime numbers, except for the order of the factors. (Product could mean a single prime number.)

Example 4 shows us one way to find the prime factorization of a natural number by using factor trees.

EXAMPLE 4 Factoring a Natural Number

Factor 4,620.

SOLUTION:

To factor 4,620, we first try to think of a way to write it as a product of two smaller numbers. For example, $4{,}620 = 462 \cdot 10$. It really doesn't matter how you write it as a product, *what is important is that you somehow express 4,620 as a product of smaller, simpler numbers.* We can represent this preliminary factorization graphically by the diagram in Figure 6.2.

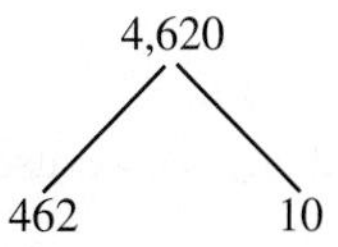

FIGURE 6.2 A first factorization of 4,620.

We call this a **factor tree.** As we add more branches, you will see that this diagram looks like an upside-down tree. Next, we factor 462 and 10 and add new branches to the tree, as in Figure 6.3(a). Using the divisibility test, we see that 3 divides 231.

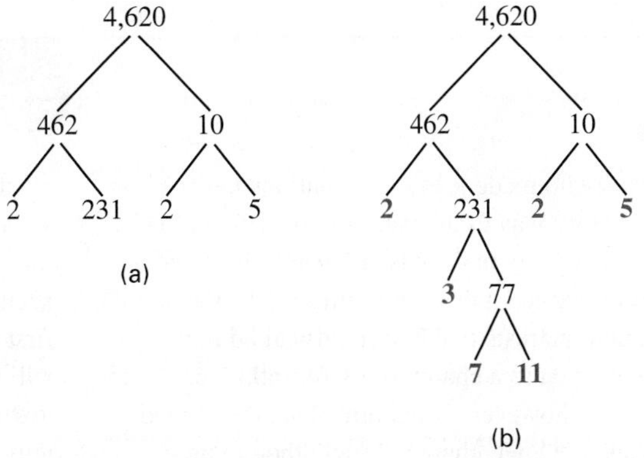

FIGURE 6.3 Completing the factorization of 4,620 using a factor tree.

If we now multiply all the primes at the ends of the branches (shown in red) in Figure 6.3(b), we see that $4{,}620 = 2 \cdot 3 \cdot 7 \cdot 11 \cdot 2 \cdot 5 = 2^2 \cdot 3 \cdot 5 \cdot 7 \cdot 11$.

Now try Exercises 27 to 40. 3

QUIZ YOURSELF 3

a) Draw a factor tree to factor 1,560.
b) Write the factorization.

If you were factoring a number like 2,400 and recognized that it is the product of 24 and 10 and 10, then you could begin drawing your tree with three branches, as in Figure 6.4. As we said, it really doesn't matter how you start your tree, the important thing is to express 2,400 as a product of smaller natural numbers.

It is important to understand that the fundamental theorem of arithmetic tells us that although we may factor a natural number several different ways, we will always end up with the same factorization when we are finished.

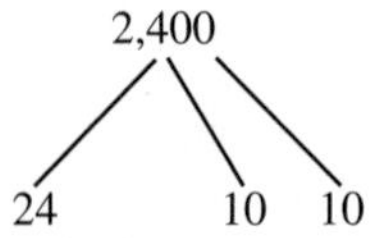

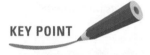

FIGURE 6.4 A possible first step in factoring 2,400.

Greatest Common Divisors and Least Common Multiples

In many applications of number theory, we need to find the greatest common divisor and least common multiple of natural numbers.

> **DEFINITION** The **greatest common divisor** or **GCD** of two natural numbers is the largest natural number that divides both numbers.

KEY POINT

We use prime factorization to find the GCD and LCM of numbers.

For small numbers, you can find the GCD by looking at the numbers and identifying the largest natural number that divides both numbers. For example, to find the GCD of 24 and 40, we list the divisors of both numbers. The common divisors are highlighted in red.

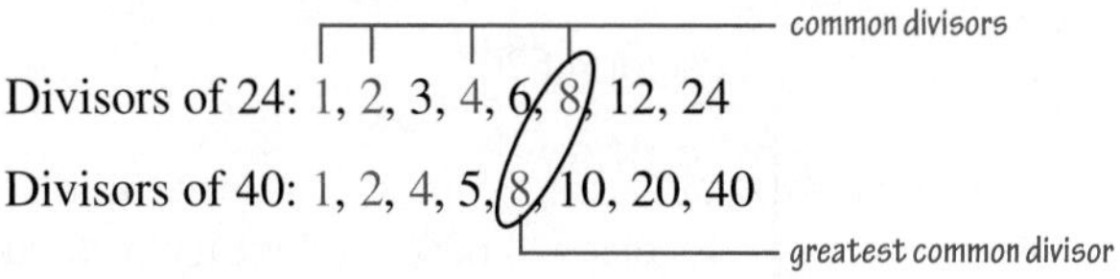

From these lists, we see that the largest number dividing both 24 and 40 is 8.

For larger numbers, it is tedious to list all divisors of both numbers, so we use the approach shown in Example 5, which involves factoring both numbers.

Historical Highlight

Eratosthenes

One of the Greeks who made great contributions to the field of number theory was Eratosthenes (276–194 BC). He studied at Plato's academy in Athens and was considered to be one of the most learned men of antiquity. Eratosthenes had so many varied interests that his friends called him Pentathis, a name given to a champion in five athletic events. His enemies, however, called him Beta, the second letter of the Greek alphabet, charging that although he worked in many fields, he was outstanding in none.

In his work *Geographia,* he argued that the world was round—1,700 years before Columbus made his famous voyage in 1492. He also produced the most accurate map of the known world of his time and was the first to use a grid of lines of longitude and latitude that is still used in maps today. Although writers of mathematics textbooks remember him for his prime sieve, he is probably more famous for devising a method for measuring the circumference of Earth. See Section 9.1.

EXAMPLE 5 Using Prime Factorization to Find the GCD

Find the GCD of 600 and 540.

SOLUTION: Review the Be Systematic Strategy on page 5.

We first factor both numbers using factor trees to get

One 3 divides both. One 5 divides both.

$$600 = 2 \cdot 2 \cdot 2 \cdot 3 \cdot 5 \cdot 5 \text{ and } 540 = 2 \cdot 2 \cdot 3 \cdot 3 \cdot 3 \cdot 5.$$

Two 2s divide both.

QUIZ YOURSELF 4

Find the GCD of 126 and 588.

If we think about what "greatest common divisor" means, we recognize that we need to find as many primes as possible that divide *both* numbers at the same time. We see that two 2s will divide both 600 and 540; however, three 2s will not. Similarly, only one 3 divides both numbers and only one 5 divides both numbers. Therefore, the GCD of 600 and 540 is $2 \cdot 2 \cdot 3 \cdot 5 = 60$. 4

Another way of thinking about what we did in Example 5 is to write $600 = 2^3 \cdot 3^1 \cdot 5^2$ and $540 = 2^2 \cdot 3^3 \cdot 5^1$. Then in forming the GCD, we multiplied the 2^2, the 3^1, and the 5^1, which were the *smallest powers* of the primes that divide both numbers. So the GCD is $2^2 \cdot 3^1 \cdot 5^1$.

There is an alternate way to find the GCD of two numbers that does not involve factoring. We use this method, called the **Euclidean algorithm,** to solve Exercises 49 to 52.

PROBLEM SOLVING

Strategy: The Three-Way Principle

The meaning of "greatest common divisor" tells you how to compute the GCD of two numbers. The word *greatest* tells you to find the *largest* natural number that divides evenly into *both* numbers. That is why, for instance, we used 3^1, rather than 3^3, in constructing the GCD of 600 and 540, because 3^1 divides both numbers; however, 3^3 does not divide 600.

Next we will discuss the least common multiple of two numbers.

DEFINITION The **least common multiple** or **LCM** of two natural numbers is the smallest natural number that is a multiple of both numbers.

To find the LCM of small numbers such as 8 and 6, we could list multiples of both numbers and then choose the smallest common multiple as follows:

common multiples

Multiples of 8: 8, 16, 24, 32, 40, 48, ...

Multiples of 6: 6, 12, 18, 24, 30, 36, 42, 48, ...

least common multiple

However, for larger numbers, we can use a prime factorization approach to find the LCM, similar to what we did in Example 5.

EXAMPLE 6 Using Prime Factorization to Find the LCM

Find the LCM of 600 and 540.

SOLUTION:

As in Example 5, we start by factoring both numbers. Recall that

600 requires three 2s. *540 requires three 3s.*

$$600 = 2 \cdot 2 \cdot 2 \cdot 3 \cdot 5 \cdot 5 \text{ and } 540 = 2 \cdot 2 \cdot 3 \cdot 3 \cdot 3 \cdot 5.$$

600 requires two 5s.

Now, think about the words *least common multiple.* Here we recognize that we need to find the *smallest* natural number that is *divisible by all the prime factors* of 600 and 540. However, we don't want to use any more primes in our number than we absolutely have to. The LCM will have to have three 2s, three 3s, and two 5s. So, the LCM of 600 and 540 is

$$2 \cdot 2 \cdot 2 \cdot 3 \cdot 3 \cdot 3 \cdot 5 \cdot 5 = 5{,}400.$$

Although this number seems large, realize that we cannot omit any of the prime factors. For example, if we omit one of the 3s, then 540 will not divide into the number. If we omit one of the 5s, then 600 will not divide into the number.

Now try Exercises 41 to 48.

Another way of thinking about what we did in Example 6 is to write $600 = 2^3 \cdot 3^1 \cdot 5^2$ and $540 = 2^2 \cdot 3^3 \cdot 5^1$. Then in forming the LCM, we multiplied the 2^3, the 3^3, and the 5^2, which were the *highest powers* of the primes that divide either number. So, the LCM is $2^3 \cdot 3^3 \cdot 5^2$.

FINDING THE GCD AND LCM BY USING FACTORIZATION To find the GCD and LCM of two numbers, do this:

1. Factor both numbers, and write each as a product of powers of primes.
2. To calculate the GCD, multiply the *smallest* powers of any primes that are common to both numbers.
3. To calculate the LCM, multiply the *largest* powers of all primes that occur in either number.

If we look at what happened in Examples 5 and 6 carefully, we see the following pattern:

$$600 = \boxed{2^3}\ \textcircled{3^1}\ \boxed{5^2}$$
$$540 = \textcircled{2^2}\ \boxed{3^3}\ \textcircled{5^1}$$

You see that when we multiply the circled factors we get the GCD, and when we multiply the boxed factors we get the LCM. This means that the GCD times the LCM gives us the product of 600 and 540. Another way of saying this is that once you have found the factors that make up the GCD, the product of the remaining factors is the LCM.

We can state this relationship that we just observed as follows:

RELATIONSHIP BETWEEN GCD AND LCM If a and b are two numbers, then the following relationship holds:

$$\text{GCD}(a, b) \times \text{LCM}(a, b) = a \times b.$$

We illustrate this relationship in Example 7.

EXAMPLE 7 Using the GCD to Find the LCM

Assume that you have found the GCD of 480 and 1320 to be 120. Find the LCM of 480 and 1320 without factoring.

SOLUTION:

By our observation above, we know that

$$\text{GCD}(480, 1320) \times \text{LCM}(480, 1320) = 480 \times 1320.$$

Substituting 120 for GCD(480, 1320), we get

$$120 \times \text{LCM}(480, 1320) = 480 \times 1320.$$

Dividing both sides of this equation by 120 gives us

$$\text{LCM}(480, 1320) = \frac{480 \times 1320}{120} = 4 \times 1320 = 5{,}280.$$

We know that 120 divides 480, so it makes our work simpler if we divide the 120 into 480 to get 4 before multiplying.

Applying the GCD and LCM

We find applications of the GCD in situations where we want to take several larger objects and represent them as a collection of smaller objects of the same size. For example, a clerk in a store may want to arrange a few large stacks of dishes into several smaller, equal-size stacks, or a cabinet maker might want to cut several boards of different lengths to make smaller boards of equal size.

The LCM occurs in situations where we have two types of objects and we want to take enough of each to get a larger object of the same size. For example, one person has every fifth day off and another may have every sixth day off, and we ask when both will be off together.

For over 40 years, Japan has been a leader in developing high-speed trains that can travel at over 300 miles per hour with remarkable reliability and safety records. The next example discusses a problem of scheduling the so-called "bullet trains" leaving the Tokyo station.

EXAMPLE 8

Assume that bullet trains have just departed from Tokyo to Osaka, Niigata, and Akita. If a train to Osaka departs every 90 minutes, a train to Niigata departs every 120 minutes, and a train to Akita departs every 80 minutes, when will all three trains again depart at the same time?

Highlight

There's Big Money in Big Numbers

Mathematicians are fascinated with finding certain types of very large prime numbers. A prime p is called a Sophie Germain prime (see Historical Highlight on page 259) if $2p + 1$ is also prime. For example, 2, 3, 5, and 11 are Sophie Germain primes because $2 \cdot 2 + 1 = 5, 2 \cdot 3 + 1 = 7, 2 \cdot 5 + 1 = 11$, and $2 \cdot 11 + 1 = 23$ are prime numbers. Thirteen is not a Sophie Germain prime because $2 \cdot 13 + 1 = 27$, which is not prime. It is conjectured that there are infinitely many Sophie Germain primes, but this has not been proven. At the time of writing this Highlight, the largest known Sophie Germain prime was $18543637900515 \times 2^{666667} - 1$, which has 200,701 digits. However, this is not the largest known prime. In October of 2009, UCLA was awarded a $100,000 prize for finding a Mersenne prime (a prime of the form $2^p - 1$) that had almost 13 million digits. The Electronic Frontier Foundation will award $250,000 to the first person to find a prime number with at least a billion digits. You can join the search for this number by going to www.mersenne.org.

SOLUTION:

Trains to Osaka will leave in 90, 180, 270, . . . minutes; trains to Niigata leave in 120, 240, 360, . . . minutes; trains to Akita leave in 80, 160, 240, . . . minutes. We want to know the smallest number of minutes until all three trains again leave at the same time. That is, we want to find the LCM of 90, 80, and 120. We can factor these numbers as follows:

greatest powers of 3 and 5 in any of the three numbers

greatest power of 2 in any of the three numbers

$$90 = 2 \times 3^2 \times 5 \qquad 80 = 2^4 \times 5 \qquad 120 = 2^3 \times 3 \times 5$$

Thus, the LCM of these three numbers is $2^4 \times 3^2 \times 5 = 720$ minutes, or, dividing by 60 minutes, we get $\frac{720}{60} = 12$ hours.

Now try Exercises 67 to 73.

EXERCISES 6.1

Sharpening Your Skills

Determine whether each of the following statements is true.

1. 8 divides 56.
2. 21 is a multiple of 2.
3. 27 is a multiple of 6.
4. 5 is a divisor of 35.
5. 9 is a factor of 96.
6. 14 divides 42.
7. 7 is a divisor of 63.
8. 6 is a factor of 76.

Check each of the following numbers for divisibility by each of these numbers: 2, 3, 4, 5, 6, 8, 9, 10. State the numbers that divide the given number.

9. 141,270
10. 18,036
11. 47,385
12. 476,376

In Exercises 13–16, use a calculator to determine if the first number divides the second.

13. 447 and 17,433
14. 453 and 25,825
15. 671 and 32,881
16. 893 and 338,447

Provide a counterexample to show that each of the following statements is false.

17. If 2 and 4 divide a, then 8 divides a.
18. If 3 and 6 divide a, then 18 divides a.
19. If 10 and 4 divide a, then 40 divides a.
20. If 4 and 6 divide a, then 24 divides a.

Estimate the square root of each of the following numbers, n, by finding a natural number a *such that* $a < \sqrt{n} < a + 1$.

21. 95 **22.** 138 **23.** 153 **24.** 229

25. Use the Sieve of Eratosthenes to find all prime numbers between 51 and 100, inclusive.

26. Use the Sieve of Eratosthenes to find all prime numbers between 101 and 120, inclusive.

Factor each of the following natural numbers. If the number is prime, state so.

27. 231 **28.** 89 **29.** 113 **30.** 153
31. 227 **32.** 143 **33.** 119 **34.** 443
35. 980 **36.** 396 **37.** 621
38. 805 **39.** 319 **40.** 403

Find the GCD and LCM of each of the following pairs of natural numbers.

41. 20, 24
42. 60, 72
43. 56, 70
44. 66, 110
45. 216, 288
46. 675, 1,125
47. 147, 567
48. 275, 363

There is another way to find the GCD of two natural numbers called the ***Euclidean algorithm.*** *The Euclidean algorithm is as follows:*

Suppose that we want to find the GCD of two numbers such as 24 and 88.

First we divide the smaller number, 24, into the larger number, 88, as shown in Figure 6.5.

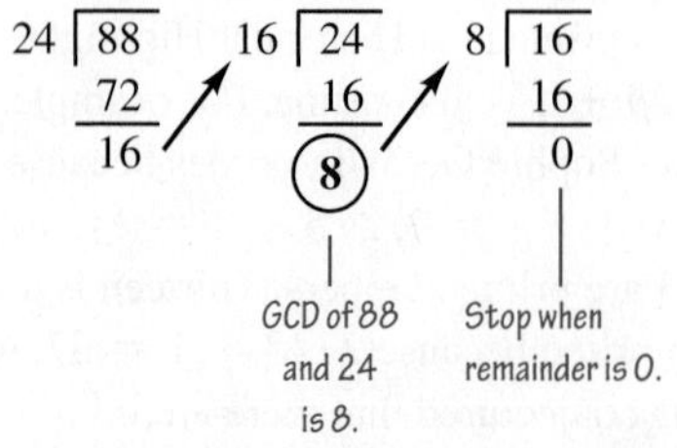

FIGURE 6.5 The Euclidean algorithm.

Next we divide the remainder, 16, into the previous divisor, 24, to get a new remainder, 8. We continue this process, always dividing the remainder into the previous divisor. So, next we divide 8 into 16. At this point, the remainder is 0, so we stop. The previous remainder, 8, is the GCD of 24 and 88.

Use the Euclidean algorithm to find the GCD of each of the following pairs of natural numbers.

49. 280, 588

50. 84, 1,200

51. 495, 1,575

52. 99, 1,155

You have seen that if you multiply the GCD and LCM of two natural numbers a *and* b, *the product is the same as the product* a • b. *In Exercises 53–56, first find the GCD of the two numbers and then divide the GCD into the product of the two numbers to find the LCM. To find the GCD, you can use the Euclidean algorithm.*

53. 12, 27

54. 16, 56

55. 90, 120

56. 17, 178

You can find the GCD of three numbers a, b, *and* c *by first finding GCD(*a, b*), call it* d, *and then finding the GCD of* d *and* c. *The same process works for finding the LCM of three numbers. Use this method to find the GCD and LCM of the triples of numbers in Exercises 57–60.*

57. 120, 90, 84

58. 72, 99, 132

59. 64, 56, 100

60. 99, 165, 143

Applying What You've Learned

61. Stacking iPad covers. A college bookstore manager has 24 plain and 16 leather iPad covers. If she wishes to display the covers with the same number in each stack and only one type of cover in each stack, what is the largest number of covers that can be in a stack? If she is not concerned with the largest number of covers per stack, but wants at least 2 per stack, what are her possibilities?

62. Packaging sports cards. A sports card dealer has 54 baseball cards and 72 football cards. He wishes to sell them in packages of the same size with only one type of card in each package. What is the largest size of the packages that he could have? If the dealer is not concerned with the largest number of cards per package, but wants at least 3 per package, what are his possibilities?

63. Servicing a car. Your new vintage 1965 Corvette convertible requires an oil change every 2,500 miles and replacement of all fluids every 3,000 miles. If these services have just been performed by your dealer, how many miles from now will both be due at the same time?

64. Training for a race. In training for a half-marathon, you and your friend Mike are running around a circular track. Assume that you start out together and it takes you 12 minutes to run completely around the track and it takes Mike only 8 minutes. When will Mike first pass you? If you plan to run for 90 minutes, at what other times will Mike pass you?

65. Cicadas. In addition to avoiding their predators, another advantage of the 13- and 17-year emergence pattern for cicadas is that the two species rarely emerge during the same year to compete for food. Suppose that type A cicada has a 13-year emergence cycle and type B cicada has a 17-year cycle. If both type A and type B cicadas emerged during 2010, when is the next year that both cicadas will emerge during the same year?

66. Cicadas. Continuing the discussion in Exercise 65, suppose that a predator of both type A and type B cicadas emerges every 5 years. If the predator also emerged in 2010, when will both types of cicadas and the predator all emerge together again?

67. Scheduling flights. Assume that Jet Blue flights for Miami leave every 35 minutes and flights for Dallas leave every 20 minutes. If flights to Miami and Dallas have just departed, how many minutes will it be before this will happen again?

68. Scheduling nurses' shifts. Carla and Laverne are nurses who occasionally work in the emergency room. Carla works in the emergency room every 6 weeks and Laverne works in the emergency room every 8 weeks. If Carla and Laverne have just worked together in the emergency room, in how many weeks will they work in the emergency room together again?

69. Storing medical supplies. In a medical supply room, there are 36 packages of type O positive blood and 30 packages of type AB negative blood. It is important not to confuse these two types of blood. If we want to stack piles of each type of blood on a shelf, with only one type of blood in a pile, what is the largest number of packets of blood that we can have in each pile if each pile is of the same size?

70. Displaying store merchandise. A sporting goods store has 20 instructional DVDs on skiing and 12 DVDs on snowboarding. The owner of the store wants to display the DVDs on a shelf with stacks of the same size and each stack consisting of only one type of DVD. What is the most number of DVDs in each stack that will accomplish this?

71. Tiling a museum floor. In the Central America room of the Ancient Civilization Museum, we want to tile the floor with replicas of ancient Aztec tiles. If the floor measures 33 feet by 21 feet, what is the largest-size square tile that we can use to tile the floor without having to cut any tiles?

72. Scheduling lawn service. At the Berkshire Country Club, the lawn service cuts the grass every 8 days and the pest service sprays for insects every 30 days. If both services have just been to the country club, in how many days will both be there again?

73. Monitoring pollution. The Environmental Protection Agency is monitoring the Lackawanna River to determine

the cause of a fish kill. A steel mill releases hot water into the river every 48 hours, and a plastics plant discharges pollutants into the river every 54 hours. If both hot water and pollutants from the plastics plant have just been discharged, in how many hours will both be released into the river again?

74. Movie showings. At the 24/7 Monster MoviePlex, *Harry Potter* shows every 120 minutes, *Hugo* shows every 140 minutes, and *The Muppets* shows every 90 minutes. If all three movies have just begun, in how many minutes will they again all begin at the same time?

Communicating Mathematics

75. If you were using the Sieve of Eratosthenes to find the primes up to 300, what is the largest prime whose multiples you would cross off before you knew that the remaining numbers in the table were prime?

76. Explain why the divisibility test for 4 works by considering the number 36,824. It helps to think of 36,824 as 36,800 + 24.

77. Explain why the divisibility test for 3 works by considering the number 5,712. It helps to think of the following equations:

$$\begin{aligned} 5{,}712 &= 5(1{,}000) + 7(100) + 1(10) + 2 \\ &= 5(999 + 1) + 7(99 + 1) + 1(9 + 1) + 2 \\ &= 5 \cdot 999 + 7 \cdot 99 + 1 \cdot 9 + (5 + 7 + 1 + 2) \end{aligned}$$

78. Do a number of examples of finding the GCD and LCM of two numbers first using the prime factorization method that we showed in Example 5 and then using the Euclidean algorithm. Which do you prefer? Does one method seem to work better with larger numbers? Explain your thinking.

Challenge Yourself

Use your calculator to test the following numbers to see if they are prime. To do this,

1. *Find the square root of the number,* n, *that you are checking.*
2. *Divide* n *by all prime numbers up to and including* $\sqrt{n}$. *If none of these primes divides* n, *then* n *is a prime number.*

79. 493 **80.** 577 **81.** 677 **82.** 713

83. Mersenne primes. A Mersenne prime is a prime of the form $2^p - 1$, where p is a prime number. For example, $7 = 2^3 - 1$ is a Mersenne prime.

a. Find the first five Mersenne primes.

b. Give an example of a prime p such that $2^p - 1$ is not prime.

84. Mersenne primes. As of August 2008, the largest known Mersenne prime was $2^{43112609} - 1$, which we will call x. If we were to write out x, it would have 12,978,189 digits. To put the size of this number in perspective, assume that a typical page in your word processor has 46 lines with 90 characters per line. Do the following calculations:

a. Divide 12,978,189 by the number of characters per page to find the number of pages that would be required to print out x.

b. Take the number of pages that you found in part (a) and multiply it by 11 inches to determine the length of the paper (in inches) required to print x. Next, divide this number by 12 to find the length of the paper in feet.

c. Divide the number that you found in part (b) by 5,280 (the number of feet in 1 mile) to determine the length of the paper in miles.

There are many famous conjectures regarding prime numbers. In Exercises 85 and 86, we investigate two of the most famous conjectures.*

85. A pair of **twin primes** is a pair of numbers that differ by 2 that are both prime. For example, 17 and 19 are a pair of twin primes. Also 41 and 43 are another pair of twin primes. The **twin prime conjecture** states that there are an infinite number of twin primes. Find the next three twin primes that are greater than 43.

86. Goldbach's conjecture states that any even number greater than 2 can be written as the sum of two primes. For example, 16 = 11 + 5, 26 = 13 + 13, and 40 = 3 + 37. Write each of the following as a sum of two primes (there may be several ways to do this):

a. 80 **b.** 100 **c.** 200

87. If both p and q divide the natural number n, then does $p \cdot q$ divide the number n? Either explain why this must be true or else provide a counterexample to this statement.

88. The divisibility test for 6 is that the number must be divisible by both 2 and 3. Explain why the divisibility test for 8 is *not* that the number must be divisible by both 2 and 4.

89. Devise a divisibility test for 15.

90. Happy numbers. Happy numbers are numbers such that if we add the squares of their digits to get a new number, then do this again to that new number and continue to do this process to each new number, we eventually get 1. For example, if we begin with 32, we compute $3^2 + 2^2 = 13$, then we compute $1^2 + 3^2 = 10$. Finally, $1^2 + 0^2 = 1$. Because we eventually got a result of 1, we say that 32 is "happy." If you try this same method with 16, you will see that the results you get will repeat but never reach 1; therefore, 16 is "not happy."

a. Determine if 5, 7, 12, 19, and 68 are happy.

b. List the first five happy primes.

*You can think of a conjecture as an educated guess that has not yet been proved.

6.2 The Integers

Objectives

1. Model the rules for adding and subtracting integers.
2. Understand the rules for multiplying and dividing integers.
3. Perform computations using integers.

Although the natural numbers were known in various forms to many ancient peoples, zero did not appear until many years later. The Hindus are credited with inventing zero, which later became part of our Hindu-Arabic system of notation.

DEFINITION We adjoin zero to the set of natural numbers to form the set {0, 1, 2, 3, . . .}, called the set of **whole numbers.**

Many mathematicians were slow to accept zero, arguing that we could not have a number that represented nothing. The Greek mathematician Diophantus said that to solve the equation $3x + 20 = 8$ and get a solution of -4 was absurd. As late as the sixteenth century, many European mathematicians would use negative numbers in doing calculations but would not accept zero or a negative number as an answer to a problem. Now, of course, once we learn the rules for calculating with negative numbers, we can calculate with them as easily as we do with positive numbers.

DEFINITION The set of **integers** is the set {. . . , −3, −2, −1, 0, 1, 2, 3, . . .}.

In this section, we will investigate the rules for adding, subtracting, multiplying, and dividing integers.

SOME GOOD ADVICE

Keep in mind that if you understand *why* a rule works, that will help you to remember *what* the rule is telling you to do.

Adding and Subtracting Integers

KEY POINT

We can explain addition rules by considering movement on a number line.

A common model that we can use to explain the addition of integers is movement on the number line. In this model, we think of integers as points on the number line, as shown in Figure 6.6.

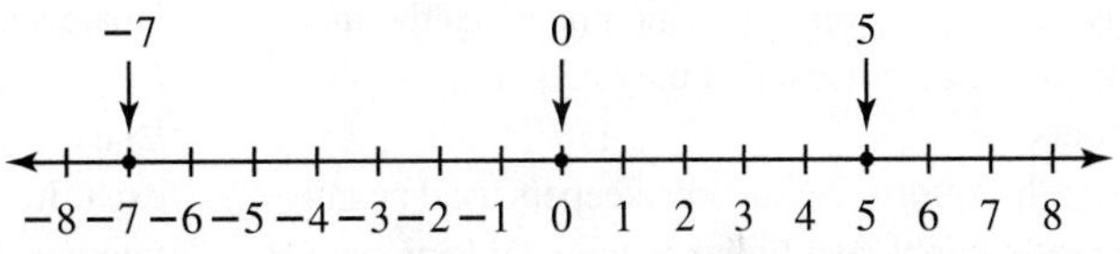

FIGURE 6.6 Representing integers on the number line.

Then, beginning at 0, we interpret positive numbers as movement to the right and negative numbers as representing movement to the left. We illustrate this in Example 1.

EXAMPLE 1 Representing Integer Addition as Movement on the Number Line

Perform the following additions:

a) (+3) + (+5) b) (+9) + (−12) c) (−149) + (137)

SOLUTION:

a) Starting at 0, think of +3 as a movement 3 places to the right. Then think of +5 as an additional movement of 5 places to the right, as in Figure 6.7. Thus, we see that (+3) + (+5) = +8.

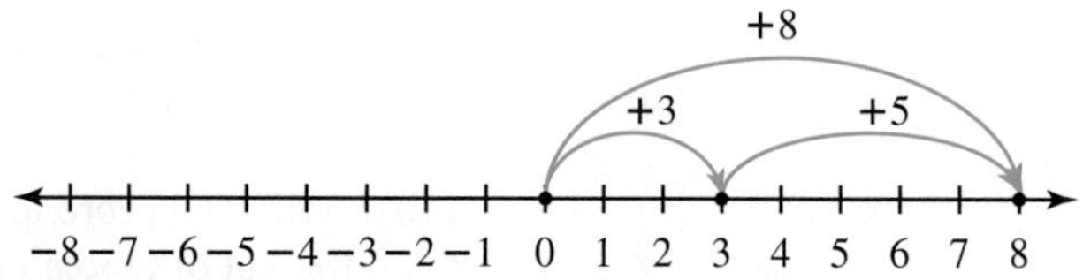

FIGURE 6.7 (+3) + (+5) = +8.

b) To calculate (+9) + (−12), we first move 9 places to the right and then move 12 places to the left, as we show in Figure 6.8.

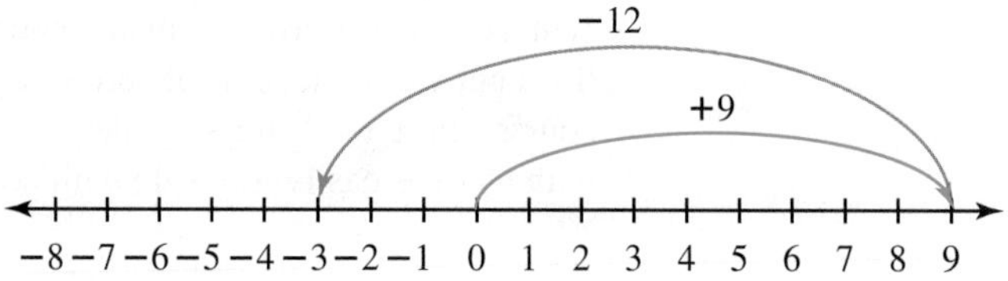

FIGURE 6.8 (+9) + (−12) = −3.

c) It is not practical to draw a number line when calculating with such large numbers; however, you still can imagine movements along the number line. If you visualize moving 149 units to the left and then 137 units to the right, you see that the net effect is that you have moved 149 − 137 = 12 units to the left, so (−149) + (+137) = −12.

Now try Exercises 1 to 6. 5

QUIZ YOURSELF 5

Find:

a) (+4) + (6)
b) (+8) + (−13)
c) (−3) + (−7)
d) (+145) + (−123)

Historical Highlight

Negative Numbers

Historians tell us that there is no evidence of negative numbers in the mathematics of ancient civilizations such as the Babylonians and the Greeks. In fact, Greek mathematics was so heavily based on geometry, which involves the measuring of lengths and distances, that they had no need for negative numbers.

By the seventh century, Indian bookkeepers used negative numbers to represent debt, and Indian mathematicians such as Brahmagupta showed that they had a clear idea of how to use negative numbers in doing scientific calculations.

During the Renaissance, the Italian mathematician Girolamo Cardano used negative numbers in his 1545 work *Ars Magna,* which was a text on algebraic equations. However, a century later, the Frenchman René Descartes called negative numbers "false roots" and his contemporary, Blaise Pascal, believed numbers "less than zero" could not exist. It was not until the eighteenth century that negative numbers were generally accepted.

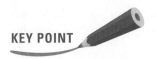

KEY POINT

We define subtraction of integers in terms of addition.

Mathematicians consider the set of integers as an abstract system having only two operations—addition and multiplication. At first, they do not mention subtraction and division, considering these operations to be somewhat secondary operations.

Now we will define subtraction in terms of addition, and later we will define division in terms of multiplication. However, first, we need a definition.

> **DEFINITION** Two integers x and y are **opposites,** or additive inverses, if and only if $x + y = 0$.

For example, -8 is the opposite of $+8$ because $(-8) + (+8) = 0$. Also, because $(-13) + (+13) = 0$, -13 and $+13$ are opposites. Because $0 + 0 = 0$, we say that 0 is its own opposite. As you can see in Figure 6.9, opposites are balanced on opposite sides of 0.

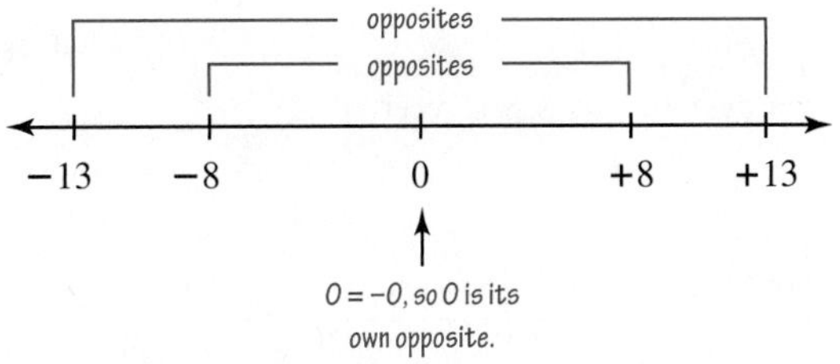

FIGURE 6.9 Opposite numbers.

> **DEFINITION** If a and b are integers, then $a - b = a + (-b)$.

This definition means to compute $3 - 8$, you change the subtraction to an addition, as in the following diagram:

$$3 - 8 = 3 + (-8) = -5$$

Add the opposite of 8. Result is −5.

QUIZ YOURSELF 6

Convert the following subtraction problems to addition problems and find the answers:

a) $(+5) - (+12)$

b) $(-3) - (-9)$

We compute $(-6) - (-19)$ as follows:

$$(-6) - (-19) = (-6) + (+19) = +13$$

Add the opposite of −19. Result is +13.

Now try Exercises 7 to 14. 6

SOME GOOD ADVICE

You may have heard in the past that "two negatives make a positive." This is not a rule in mathematics, but rather a memory device that sometimes works and sometimes does not. For example, $(-6) + (-2)$ does not make a positive. Rather than thinking of "two negatives make a positive," it is better to think of $-(-6)$ as the opposite of -6, or $+6$, and think of $3 - (-5)$ as $3 + (+5)$.

Multiplying and Dividing Integers

KEY POINT

We can use a money model to explain integer multiplication.

You can understand the rules for multiplying integers if you consider the following model. Call today day zero, and assume that you have \$100 in your wallet. Also assume that each day, one of two things happens, either you receive \$5 in the mail from your rich uncle or you spend \$5 for a pizza. We assume that you are making no other changes in your finances.

We will now consider the product of integers a and b. The integer a will represent changes in days—a positive a represents days in the future, a negative a represents days in the past. The number b will represent changes in your finances—a positive b means you receive \$5, a negative b means you spend \$5.

For example, we interpret $(+3)(-5)$ to mean that for the next 3 days, you spend \$5 each day on pizza. The product $(-4)(+5)$ means for the past 4 days, you have been receiving \$5 from your uncle. With this in mind, consider Table 6.2.

Interpretation of $a \cdot b$	Multiplication Rule
$(+3)(+5) = +15$ — You will gain \$15 in 3 days. For the next 3 days, you receive \$5.	A positive times a positive is a positive.
$(+3)(-5) = -15$ — You will lose \$15 in 3 days. For the next 3 days, you spend \$5.	A positive times a negative is a negative.
$(-3)(+5) = -15$ — You had \$15 less 3 days ago. For the past 3 days, you received \$5.	A negative times a positive is a negative.
$(-3)(-5) = +15$ — You had \$15 more 3 days ago. For the past 3 days, you spent \$5.	A negative times a negative is a positive.

TABLE 6.2 Rules for multiplying integers.

We can state these rules for multiplying integers more succinctly.

RULES FOR MULTIPLYING INTEGERS If a and b are integers, then:

a) If a and b have the same sign, then $a \cdot b$ is positive.

b) If a and b have opposite signs, then $a \cdot b$ is negative.

EXAMPLE 2 Applying the Rules for Multiplying Integers to the Stock Market

a) Assume that stock for Itech, an international technology company, has been losing value at \$3 per day for each of the last 8 days. How much more was the stock worth 8 days ago?

b) If the stock continues to lose its value at \$3 per day, how much less will the stock be worth in 4 days?

SOLUTION:

a) To find the worth 8 days ago, we represent "8 days ago" by -8 and "losing value at \$3 per day" by -3. We then calculate the product $(-8)(-3) = +24$. So, the stock was worth \$24 more 8 days ago.

b) The change in value in 4 days will be $(+4)(-3) = -12$, so the stock will be worth \$12 less.

Now try Exercises 15 to 24. 7

QUIZ YOURSELF 7

Find the following products:

a) $(-4)(-6)$

b) $(+8)(-7)$

KEY POINT

We define division in terms of multiplication.

Like subtraction, we consider division to be somewhat of a secondary operation, and we define division in terms of multiplication.

DEFINITION If a, b, and c are integers, then $a/b = c$ means that $a = b \cdot c$.

Try to imagine that you did not know any division facts but were comfortable with doing multiplication. If you knew the definition of division, to compute 8/2 you would consider

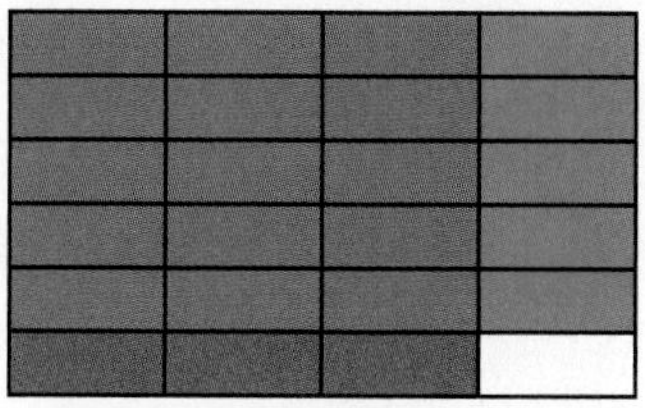

FIGURE 6.14 Representation of $\frac{3}{4} \times \frac{5}{6} = \frac{15}{24}$.

If we overlap these two figures, we get Figure 6.14, which shows that if we take $\frac{5}{6}$ of the $\frac{3}{4}$, we get 15 of the 24 sections, or $\frac{15}{24}$, which equals $\frac{5}{8}$.

Another way of looking at Figure 6.14 is $\frac{5}{6} \cdot \frac{3}{4} = \frac{5 \cdot 3}{6 \cdot 4} = \frac{15}{24} = \frac{5}{8}$. This helps us remember the multiplication rule for rational numbers.

MULTIPLYING RATIONAL NUMBERS

$$\frac{a}{b} \cdot \frac{c}{d} = \frac{a \cdot c}{b \cdot d}$$

EXAMPLE 4 Multiplying Rational Numbers

Find

a) $\frac{4}{3} \cdot \frac{3}{15}$ b) $\frac{18}{25} \cdot \frac{10}{81}$ c) $\left(\frac{9}{4}\right) \cdot \left(-\frac{12}{5}\right)$

SOLUTION:

a) Applying the multiplication rule, we get $\frac{4}{3} \cdot \frac{3}{15} = \frac{4 \cdot 3}{3 \cdot 15} = \frac{12}{45} = \frac{4}{15}$.

b) You can use the multiplication rule to get $\frac{18}{25} \cdot \frac{10}{81} = \frac{180}{2{,}025}$ and then reduce this number; however, this is a lot of unnecessary work.

If you recognize that eventually you are going to cancel common factors, it is easier to do that first before multiplying the numerators and denominators. You can cancel 9 from the first numerator and second denominator, and then cancel 5 from the first denominator and the second numerator. (You can also do the cancellation in one step if you prefer.) Finally, multiply the simpler numbers $\frac{2}{5}$ and $\frac{2}{9}$ to get $\frac{4}{45}$, as we show in Figure 6.15.

$$\frac{\overset{2}{\cancel{18}}}{25} \cdot \frac{10}{\underset{9}{\cancel{81}}} = \frac{2}{\underset{5}{\cancel{25}}} \cdot \frac{\overset{2}{\cancel{10}}}{9} = \frac{2}{5} \cdot \frac{2}{9} = \frac{4}{45}$$

Cancel 9. Cancel 5. Now multiply.

FIGURE 6.15 Canceling before multiplying saves work.

c) *The rules for multiplying signed rational numbers are the same as those for multiplying integers*, so the product of a positive number and a negative number is a negative number. We will first multiply without regard to signs, remembering that the final product is negative.

$$\frac{9}{\underset{1}{\cancel{4}}} \cdot \frac{\overset{3}{\cancel{12}}}{5} = \frac{9}{1} \cdot \frac{3}{5} = \frac{27}{5}$$

So the answer is $-\frac{27}{5}$. Although you may recall that $-\frac{27}{5}$, $\frac{-27}{5}$, and $\frac{27}{-5}$ are all equal, we tend to write the answer as $-\frac{27}{5}$. 12

QUIZ YOURSELF 12

Find:

a) $\frac{24}{9} \cdot \frac{21}{20}$

b) $\left(-\frac{8}{15}\right) \cdot \left(-\frac{25}{12}\right)$

KEY POINT

Division of rational numbers is based on multiplication.

To understand the rule for dividing rational numbers, consider how we might think about dividing $\frac{\frac{3}{4}}{\frac{5}{8}}$. What makes this division complicated is that we have $\frac{5}{8}$ in the denominator. One way to get rid of the $\frac{5}{8}$ is to multiply it by its reciprocal, $\frac{8}{5}$. However, if we multiply the denominator by $\frac{8}{5}$, we must also multiply the numerator by $\frac{8}{5}$. Multiplying by $\frac{8}{5}$ gives us a 1 in the denominator and the product $\frac{3}{4} \cdot \frac{8}{5}$ in the numerator, as we see here.

$$\frac{\frac{3}{4}}{\frac{5}{8}} = \frac{\frac{3}{4} \times \frac{8}{5}}{\frac{5}{8} \times \frac{8}{5}} = \frac{\frac{3}{4} \times \frac{8}{5}}{1} = \frac{3}{\underset{1}{\cancel{4}}} \times \frac{\overset{2}{\cancel{8}}}{5} = \frac{6}{5}$$

Instead of going through this lengthy process each time we divide, we use the following rule, which tells us in effect to "invert the denominator and multiply."

DIVIDING RATIONAL NUMBERS

$$\frac{\frac{a}{b}}{\frac{c}{d}} = \frac{a}{b} \cdot \frac{d}{c}$$

EXAMPLE 5 ● Dividing Rational Numbers

Perform the following divisions:

a) $\dfrac{\frac{25}{12}}{\frac{10}{3}}$ b) $\dfrac{-\frac{11}{6}}{\frac{7}{9}}$

SOLUTION:

a) Applying the division rule,* we invert the denominator and multiply.

$$\frac{\frac{25}{12}}{\frac{10}{3}} = \frac{\overset{5}{\cancel{25}}}{\underset{4}{\cancel{12}}} \cdot \frac{\overset{1}{\cancel{3}}}{\underset{2}{\cancel{10}}} = \frac{5}{4} \cdot \frac{1}{2} = \frac{5}{8}$$

b) *The rules for division of rational numbers are the same as the rules for division of integers*, so a negative number divided by a positive will result in a negative number. We will remember that fact and divide without regard to signs.

Again, we invert and multiply to get $\dfrac{\frac{11}{6}}{\frac{7}{9}} = \frac{11}{\underset{2}{\cancel{6}}} \cdot \frac{\overset{3}{\cancel{9}}}{7} = \frac{11}{2} \cdot \frac{3}{7} = \frac{33}{14}$, so the answer is $-\frac{33}{14}$.

Now try Exercises 27 to 42. **13**

QUIZ YOURSELF **13**

Find:

a) $\dfrac{\frac{5}{28}}{\frac{11}{14}}$ b) $\dfrac{-\frac{33}{14}}{-\frac{11}{8}}$

Note that you have to be careful how you interpret a quotient such as $\dfrac{\frac{4}{8}}{16}$. In this case, the longer horizontal bar tells us that we are dividing $\frac{4}{8}$ by 16, so you can think of $\dfrac{\frac{4}{8}}{16}$ as $\dfrac{\frac{4}{8}}{\frac{16}{1}}$. You might find it enlightening to compute $\dfrac{\frac{4}{8}}{16}$ and $\dfrac{4}{\frac{8}{16}}$. You will see that you do not get the same result.

Example 6 requires us to use several operations on rational numbers at the same time.

EXAMPLE 6 Using the Scale Function in a Graphics Program

Suppose you are drawing a landscaping plan using a graphics package, such as Adobe CS3: Master Collection, and have drawn a rectangle measuring $\frac{8}{3}$ inches long by $\frac{5}{6}$ inch wide representing a rectangular Japanese rock garden.

a) If you scale down the length by a factor of $\frac{3}{4}$ and increase the width by a factor of $\frac{9}{8}$, what are the new dimensions of the sides of the garden in your drawing?

b) How does the area of the new rectangle compare with the area of the original rectangle in your drawing?

*When multiplying and dividing rational numbers with your calculator, you should put parentheses around the operands as follows: $\left(\frac{4}{3}\right) \cdot \left(\frac{3}{15}\right)$ and $\left(\frac{25}{12}\right) \cdot \left(\frac{3}{10}\right)$; otherwise, the calculator results may be incorrect.

SOLUTION:

a) We will multiply the length of the original rectangle by the scaling factor of $\frac{3}{4}$ to get a length of $\frac{8}{3} \cdot \frac{3}{4} = \frac{24}{12} = 2$ inches. We multiply the width of the original rectangle by the scaling factor $\frac{9}{8}$ to get a new width of $\frac{5}{6} \cdot \frac{9}{8} = \frac{45}{48} = \frac{15}{16}$ inch.

b) To answer this question, we will divide the area of the new rectangle by the area of the old rectangle. The area of the new rectangle is $2 \cdot \frac{15}{16} = \frac{30}{16} = \frac{15}{8}$ square inches. The area of the old rectangle is $\frac{8}{3} \cdot \frac{5}{6} = \frac{40}{18} = \frac{20}{9}$ square inches. Dividing the new area by the old area, we get

$$\frac{\frac{15}{8}}{\frac{20}{9}} = \frac{\overset{3}{\cancel{15}}}{8} \cdot \frac{9}{\underset{4}{\cancel{20}}} = \frac{27}{32}.$$

So, the area of the new garden in the drawing is $\frac{27}{32}$ the area of the old garden in the drawing.

Mixed Numbers

KEY POINT

Mixed numbers help us understand the size of rational numbers.

In working with rational numbers, you could get an answer such as $\frac{123}{8}$. A rational number such as this, in which the numerator is greater than the denominator, is called an **improper fraction.**

Although the answer may be correct in its present form, we have to think a little to understand the size of this number. However, if we divide 123 by 8 to get the quotient 15 and remainder 3 (Figure 6.16), then we see that $\frac{123}{8} = 15\frac{3}{8}$, which is not quite 15 and $\frac{1}{2}$.

We can use this example to give a general rule for writing an improper fraction as a mixed number.

$$\begin{array}{r} 15 \\ 8\overline{)123} \\ \underline{120} \\ 3 \end{array}$$

15 — Quotient tells us that 8 divides into 123 fifteen whole times.

3 — Remainder tells us that 3 eighths are left over.

FIGURE 6.16 Converting $\frac{123}{8}$ to $15\frac{3}{8}$.

CONVERTING AN IMPROPER FRACTION TO A MIXED NUMBER To convert the improper fraction $\frac{a}{b}$ to a mixed number, perform the division

$$\begin{array}{r} q \\ b\overline{)a} \\ \cdot \\ \underline{\cdot} \\ r \end{array}$$

Then, write $\frac{a}{b}$ as $q + \frac{r}{b} = q\frac{r}{b}$.

Historical Highlight

Sophie Germain

Sophie Germain was born in France in 1776, at a time when women were discouraged from studying mathematics. Because women were barred from enrolling at the Ecole Polytechnique in Paris, she secretly obtained lecture notes of various professors at this prestigious university. Students were allowed to make written comments on the lectures of their professors and, even though she was not a student, she submitted her commentaries using the pseudonym M. Leblanc. Unaware that his "student" was a woman, Joseph Lagrange, one of the eminent mathematicians of the time, had high praise for the work of M. Leblanc.

When Lagrange discovered Leblanc's true identity, he extolled Germain as one of the promising young mathematicians of the time. In 1816, she was awarded a prize from the French Academy for her paper on the mathematics of elastic surfaces.

EXAMPLE 7 Converting an Improper Fraction to a Mixed Number

Convert the following improper fractions to mixed numbers:

a) $\frac{45}{6}$ b) $\frac{133}{8}$

SOLUTION:

$$\begin{array}{r} 7 \\ 6\overline{)45} \\ \underline{42} \\ 3 \end{array} \quad \begin{array}{l} \text{Quotient} \\ \\ \\ \text{Remainder} \end{array}$$

a) As you can see in the accompanying figure, when we divide 45 by 6, we get a quotient of 7 and a remainder of 3. Thus, $\frac{45}{6} = 7\frac{3}{6} = 7\frac{1}{2}$.

b) Similarly, in b), $\frac{133}{8} = 16\frac{5}{8}$.

We can also convert mixed numbers into improper fractions. For example, we think of $5\frac{3}{4}$ as $5 + \frac{3}{4} = \frac{5 \cdot 4}{4} + \frac{3}{4} = \frac{23}{4}$. This calculation is the basis for the following conversion rule.

Now try Exercises 47 to 54.

CONVERTING A MIXED NUMBER TO AN IMPROPER FRACTION The mixed number $q\frac{r}{b}$ equals the improper fraction $\frac{q \cdot b + r}{b}$.

Don't let this formula intimidate you. All we are saying is you multiply q times b and add r to get the numerator. Then use b for the denominator.

EXAMPLE 8 Converting a Mixed Number to an Improper Fraction

Convert the following mixed numbers to improper fractions:

a) $5\frac{2}{7}$ b) $8\frac{3}{11}$

SOLUTION:

a) $5\frac{2}{7} = \frac{5 \cdot 7 + 2}{7} = \frac{37}{7}$ b) $8\frac{3}{11} = \frac{8 \cdot 11 + 3}{11} = \frac{91}{11}$. 14

QUIZ YOURSELF 14

a) Convert $\frac{23}{4}$ to a mixed number.

b) Convert $2\frac{3}{5}$ to an improper fraction.

In calculating with mixed numbers, you will often find it useful to first convert the numbers to improper fractions before doing the computation. For example, to compute $3\frac{1}{2} \times 2\frac{2}{5}$, rewrite it as $\frac{7}{2} \times \frac{12}{5} = \frac{84}{10} = \frac{42}{5} = 8\frac{2}{5}$.

Repeating Decimals

KEY POINT

Rational numbers have repeating decimal representations.

If you were to divide 3/16 on your calculator, the answer would look something like 0.1875000. We will now discuss how to write rational numbers in decimal form and then how to write a decimal number as the quotient of two integers.

EXAMPLE 9 Writing a Rational Number in Decimal Form

Write each of the following rational numbers as a decimal number:

a) $\frac{5}{8}$ b) $\frac{7}{16}$ c) $\frac{82}{111}$

SOLUTION:

a) If you divide the numerator by the denominator, you will get $\frac{5}{8} = 0.625$.

b) Dividing, we find that $\frac{7}{16} = 0.4375$.

c) Although this seems like a strange choice for an example, as you will see, it is very interesting. If you divide $\frac{82}{111}$ on your calculator, you might get an answer that looks something like 0.7387387387. This is not quite correct.

If you do the division by hand, as in Figure 6.17, you will see that the digits in the quotient will repeat indefinitely. Once you have calculated a quotient of 0.738 and get a remainder of 82, the calculations that you just did will repeat all over again to give you the quotient 0.738738.

$$
\begin{array}{r}
.738 \\
111 \overline{\smash{)}82.000000} \\
777 \\
\hline
430 \\
333 \\
\hline
970 \\
888 \\
\hline
820 \\
\end{array}
$$

We are again dividing 111 into 820, so the digits in the quotient repeat all over again.

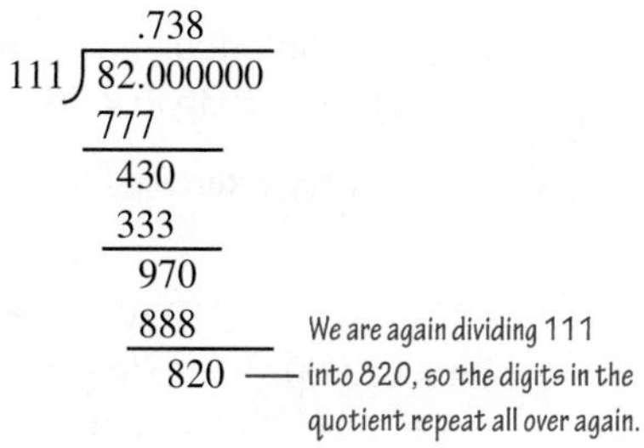

FIGURE 6.17 $\frac{82}{111}$ has a repeating decimal expansion.

The true decimal expansion of $\frac{82}{111}$ is actually the infinite repeating decimal 0.738738738738. . . . We usually write the part of a decimal that repeats with a bar over it, so we would write $\frac{82}{111}$ as $0.\overline{738}$.

Now try Exercises 55 to 62. **15**

QUIZ YOURSELF **15**

Write $\frac{21}{33}$ as a decimal number.

When dividing a rational number to get a decimal expansion, we can have only a finite number of possible remainders, so at some point, the type of repetition that we saw in Example 9(c) must always occur.

> **DECIMAL EXPANSIONS OF RATIONAL NUMBERS** When writing a rational number in decimal form, we always get either a terminating expansion, as in a) and b) in Example 9, or an infinite repeating expansion, as in c). (We could think of a terminating expansion as a repeating expansion in which 0 repeats from some point on.)

We will now explain how to write decimal numbers as quotients of integers. If a number has a terminating expansion such as 0.124, we remember that we read this number as 124 thousandths (see Figure 6.18).

So we can write 0.124 as

$$
\frac{124}{1{,}000} = \frac{31 \cdot \cancel{4}}{250 \cdot \cancel{4}} = \frac{31}{250}.
$$

4 divides both numerator and denominator.

$$0.1\,2\,4$$
tenths ⎤
hundredths ⎤
thousandths ⎤

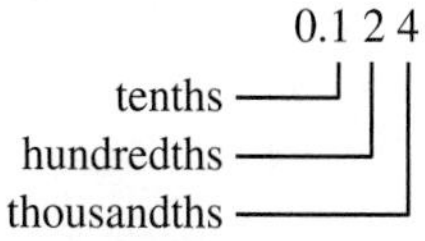

FIGURE 6.18 Reading 0.124.

To write a repeating decimal such as $0.\overline{36}$ as a quotient of integers is a little more complicated, as we explain in Example 10.

EXAMPLE 10 **Writing a Repeating Decimal as a Quotient of Integers**

Write $x = 0.\overline{36}$ as a quotient of integers.

SOLUTION:

The problem in writing 0.36363636363636 . . . as a quotient of integers is that we have to deal with the infinite "tail" of 363636. . . . The technique that we use is to create another number with the same tail that x has. We then subtract the two numbers to get a number without a repeating infinite tail.

QUIZ YOURSELF 16

a) Write 0.2548 as a quotient of integers.

b) Write $x = 0.\overline{54}$ as a decimal number.

Consider the number $100 \cdot x = 36.36363636\ldots$. If we subtract $100 \cdot x - x$, we get

$$\begin{aligned} 100 \cdot x &= 36.36363636\ldots \\ - \quad x &= -0.36363636\ldots \\ \hline 99 \cdot x &= 36 \end{aligned}$$

so the infinite tails have disappeared. Solving $99 \cdot x = 36$, we find that $x = \frac{36}{99} = \frac{4}{11}$. Thus, $0.36363636363636\ldots = \frac{4}{11}$.*

Now try Exercises 63 to 74. 16

If in Example 10 we were dealing with a number such as $x = 0.\overline{634}$, then we would subtract $1{,}000 \cdot x - x$ to get rid of the infinite tail.

A number with a nonrepeating decimal expansion is not a rational number, and we will discuss such numbers in the next section.

EXERCISES 6.3

Sharpening Your Skills

Write all answers in this exercise set in lowest terms.

Which of the following pairs of rational numbers are equal?

1. $\frac{2}{3}, \frac{8}{12}$
2. $\frac{5}{6}, \frac{10}{18}$
3. $\frac{12}{14}, \frac{14}{16}$
4. $\frac{11}{6}, \frac{18}{20}$
5. $\frac{22}{14}, \frac{30}{21}$
6. $\frac{5}{16}, \frac{10}{32}$
7. $\frac{5}{14}, \frac{15}{42}$
8. $\frac{9}{7}, \frac{54}{42}$

Reduce each fraction.

9. $\frac{15}{35}$
10. $-\frac{30}{48}$
11. $\frac{-24}{72}$
12. $\frac{60}{135}$
13. $\frac{225}{350}$
14. $-\frac{132}{96}$
15. $\frac{143}{154}$
16. $\frac{-120}{216}$

Perform the following operations. Express your answer as a positive or negative quotient of two integers in reduced form.

17. $\frac{2}{3} + \frac{1}{2}$
18. $\frac{3}{4} + \frac{1}{6}$
19. $\frac{1}{6} - \frac{1}{2}$
20. $\frac{13}{16} - \frac{5}{8}$
21. $\frac{7}{16} - \frac{1}{3}$
22. $\frac{5}{12} + \frac{3}{14}$
23. $\frac{3}{4} + \frac{5}{6} + \frac{7}{8}$
24. $\frac{1}{3} + \frac{2}{5} + \frac{5}{6}$
25. $\frac{1}{8} - \frac{2}{3} + \frac{1}{2}$
26. $\frac{2}{9} - \frac{2}{27} + \frac{1}{4}$

Perform the following operations. Express your answer as a positive or negative quotient of two integers in reduced form.

27. $\frac{2}{3} \cdot \frac{1}{2}$
28. $\frac{5}{16} \cdot \frac{4}{15}$
29. $\frac{1}{6} \div \frac{1}{2}$
30. $\frac{5}{16} \div \frac{3}{8}$
31. $\frac{7}{8} \div \left(-\frac{5}{24}\right)$
32. $-\frac{7}{32} \cdot \frac{8}{35}$
33. $\left(\frac{7}{18} \cdot \left(-\frac{3}{4}\right)\right) \div \left(\frac{7}{9}\right)$
34. $\left(\frac{14}{25} \div \frac{4}{5}\right) \cdot \left(\frac{10}{3}\right)$
35. $\left(\frac{11}{30} \div \left(-\frac{1}{6}\right)\right) \cdot \left(\frac{15}{4}\right)$
36. $\left(\frac{7}{4} \cdot \frac{8}{21}\right) \div \left(\frac{2}{5}\right)$

Perform the following operations. Express your answer as a positive or negative quotient of two integers in reduced form.

37. $\frac{5}{3} \cdot \left(\frac{8}{15} + \frac{2}{3}\right)$
38. $\frac{8}{9} \cdot \left(\frac{5}{6} + \frac{1}{4}\right)$
39. $\frac{22}{27} \cdot \left(\frac{3}{11} + \frac{2}{3}\right)$
40. $\frac{10}{9} \cdot \left(\frac{2}{5} - \frac{2}{3}\right)$
41. $\frac{7}{30} \div \left(\frac{1}{6} - \frac{3}{14}\right)$
42. $\frac{11}{40} \div \left(\frac{3}{5} - \frac{1}{4}\right)$

*Notice that $0.\overline{36} = \frac{36}{99}$. You would find that $0.\overline{634} = \frac{634}{999}$; in general, you can often make this type of conversion quickly by following this pattern.

Each of the following divisions represents a conversion of an improper fraction to a mixed number. Write that conversion as an equation with the improper fraction on the left and the mixed number on the right.

43. $\begin{array}{r} 15 \\ 5\overline{)79} \\ \underline{75} \\ 4 \end{array}$

44. $\begin{array}{r} 4 \\ 8\overline{)35} \\ \underline{32} \\ 3 \end{array}$

45. $\begin{array}{r} 9 \\ 3\overline{)29} \\ \underline{27} \\ 2 \end{array}$

46. $\begin{array}{r} 17 \\ 8\overline{)143} \\ \underline{136} \\ 7 \end{array}$

Convert each improper fraction to a mixed number.

47. $\frac{27}{4}$ **48.** $\frac{139}{8}$ **49.** $\frac{121}{15}$ **50.** $\frac{214}{12}$

Convert each mixed number to an improper fraction.

51. $2\frac{3}{4}$ **52.** $5\frac{3}{8}$ **53.** $9\frac{1}{6}$ **54.** $4\frac{5}{12}$

Write each rational number as a terminating or repeating decimal.

55. $\frac{3}{4}$ **56.** $\frac{5}{8}$ **57.** $\frac{3}{16}$

58. $\frac{27}{32}$ **59.** $\frac{9}{11}$ **60.** $\frac{16}{33}$

61. $\frac{4}{13}$ **62.** $\frac{4}{7}$

Write each decimal as a quotient of two integers in reduced form.

63. 0.64 **64.** 0.075 **65.** 0.836

66. 0.345 **67.** 12.2 **68.** 4.068

69. $0.\overline{4}$ **70.** $0.3\overline{8}$ **71.** $0.\overline{189}$

72. $0.\overline{21}$ **73.** $0.3\overline{18}$ **74.** $0.\overline{384615}$

Applying What You've Learned

75. Living expenses. Andre spends $\frac{1}{3}$ of his paycheck on rent, $\frac{1}{4}$ on food, and $\frac{1}{6}$ on utilities. What fractional part of his paycheck does he have left for other expenses?

76. Scholarships allocation. At Central State College, $\frac{1}{8}$ of the students have athletic scholarships and $\frac{1}{3}$ of the students have academic scholarships. (Assume that no students have both.) What fractional part of the student body has one or the other of these scholarships?

77. Dividing the purse in a tournament. Assume that in a Texas Hold'em tournament, the winner receives $\frac{1}{3}$ of the purse, the person who comes in second receives $\frac{1}{4}$, and the remainder of the purse is split among four other players. What fractional part of the purse does each of these four players receive?

78. Advertising budget. The Great Downhill Ski Company spends $\frac{1}{3}$ of its advertising budget on print media, $\frac{2}{5}$ of the budget on TV ads, and $\frac{1}{6}$ of the budget on radio ads. What part of the budget remains for other types of advertising?

Use the following graph, which describes the proportion of women officers in four branches of the armed services in 2008, to answer Exercises 79 and 80 (Source: Department of Defense).

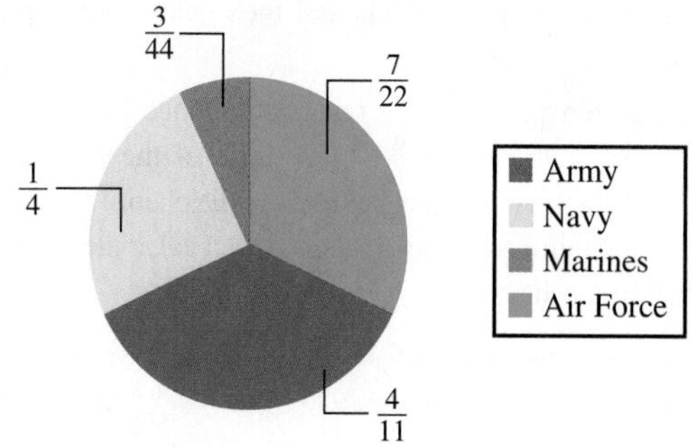

79. What fractional part of women officers were in either the Army or the Navy?

80. What fractional part of women officers were in neither the Army nor the Air Force?

81. Tiling walls of a museum. The director of the New Central American Art Museum is redecorating the walls of a room with replicas of ancient Inca tiles. The square tiles are $8\frac{1}{2}$ inches. If the room is 17 feet high, how many $8\frac{1}{2}$-inch tiles will be needed vertically? Will any of the tiles have to be cut?

82. Buying art supplies. Isabella is an artist who normally purchases her spray fixative in $6\frac{2}{3}$-ounce cans. If the economy size contains $1\frac{2}{5}$ as much spray, how many ounces will it contain?

83. Resurfacing a road. A paving crew is resurfacing three stretches of scenic, country roads, of lengths $2\frac{1}{3}$, $4\frac{3}{8}$, and $3\frac{1}{4}$ miles. What is the total length that will be resurfaced?

84. Dredging a creek. To prevent flooding, two sections of the Brandywine creek will be dredged. The first section is $\frac{3}{8}$ of a mile long and the second is $\frac{2}{3}$ of a mile. If more than a mile is being dredged, a special permit from the Environmental Protection Agency will be required. Will a permit be required?

85. Decorating a band shell. In preparing to honor a local hero, a committee will decorate the four sides of the park's rectangular band shell with patriotic bunting. If the band shell measures $32\frac{2}{3}$ by $18\frac{1}{2}$ feet, how much bunting is required?

86. Building a cabinet. Adrian is building a cabinet for his new Bose audio system. He will use a special West Indian mahogany for two sides and the top. If the sides measure $3\frac{2}{3}$ feet and the top measures $4\frac{3}{4}$ feet, how much mahogany will he need?

87. Modifying a recipe. In *The Fanny Farmer* cookbook, the recipe for Hungarian Goulash, which serves four, calls for (among other ingredients) $1\frac{1}{2}$ tablespoons of lemon juice. If you want to increase this recipe to serve 10, how much lemon juice should you use?

88. Modifying a recipe. In *The Fanny Farmer* cookbook, the recipe for 1 cup of horseradish cream sauce (to accompany roast beef) calls for $\frac{3}{4}$ cup of heavy cream. If you want to make $1\frac{1}{2}$ cups of this sauce, how many cups of heavy cream should you use?

89. Unit pricing.* Milan can buy a $5\frac{1}{4}$-ounce tube of toothpaste for \$2.60 or a $7\frac{1}{8}$-ounce tube of the same toothpaste for \$3.60. Which size is the better buy? Explain. (*Hint:* Convert the mixed numbers to decimals and then calculate the price per ounce for each size.)

90. Unit pricing. Janita can buy a $37\frac{1}{2}$-ounce bottle of orange juice for \$1.75 or a $44\frac{3}{4}$-ounce bottle of the same juice for \$2.70. Which size is the better buy? Explain. (*Hint:* Convert the mixed numbers to decimals and then calculate the price per ounce for each size.)

Use the given graph, which gives a projection for the racial-ethnic makeup of the United States in 2025, to answer Exercises 91 and 92. (Source: Sociology *by James Henson, Allyn & Bacon, 2007)*

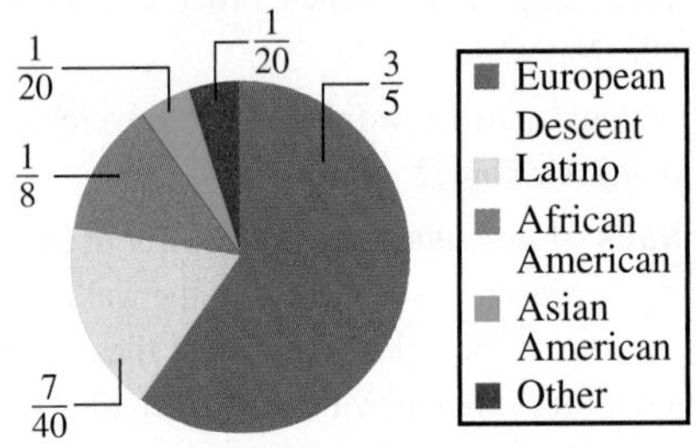

91. Racial-ethnic makeup. What fractional part of the projected U.S. population is expected to be of neither European descent nor Latino?

92. Racial-ethnic makeup. What fractional part of the projected U.S. population is expected to be African American, Latino, or Asian American?

93. Minimizing waste on a remodeling project. Ty Pennington is placing $3\frac{3}{8}$-foot strips of bamboo flooring on the floor of a house for the show *Extreme Makeover: Home Edition* and wants to minimize the waste caused by having leftover pieces that are too short. How many $3\frac{3}{8}$-foot strips can he cut from a 12-foot strip? If Ty buys the strips in 15-foot lengths, will there be more or less waste per 15-foot strip than for a 12-foot strip?

94. Estimating a painting job. Paige is estimating the cost of a painting job. A room measures $10\frac{1}{4}$ feet on two walls and $13\frac{1}{2}$ feet on the other two walls. Also, the room is $8\frac{3}{4}$ feet high. What is the total area of the four walls of the room?

Communicating Mathematics

95. Give a counterexample to show that it is incorrect to add rational numbers as follows:

$$\frac{a}{b} + \frac{c}{d} = \frac{a+c}{b+d}$$

96. Explain why the decimal expansion of a rational number has to repeat.

97. One way to add mixed numbers is to add the integer parts and the fractional parts separately and then simplify. For example,

$$12\frac{3}{4} + 14\frac{2}{3} = 12 + 14 + \frac{3}{4} + \frac{2}{3} = 26 + \frac{9}{12} + \frac{8}{12} = 26 + \frac{17}{12} = 26 + 1 + \frac{5}{12} = 27\frac{5}{12}.$$

Make up several examples of adding mixed numbers using this method and also the method using improper fractions that we discussed following Example 7. Which method do you prefer? Explain.

98. One method for adding rational numbers is to use the following formula

$$\frac{a}{b} + \frac{c}{d} = \frac{ad+bc}{bd}.$$

Perform the additions $\frac{5}{6} + \frac{3}{4}$ and $\frac{29}{36} + \frac{31}{54}$ and express your answers in reduced form. What advantage and what drawback do you see in using this method?

99. Draw a diagram to illustrate the product $\frac{3}{8} \cdot \frac{2}{3}$ similar to the discussion that preceded Example 4. Explain how your diagram illustrates the product.

Perform each of the following divisions using the technique that we used to calculate $\frac{\frac{3}{4}}{\frac{5}{8}}$ *in the discussion prior to Example 5.*

100. $\dfrac{\frac{3}{2}}{\frac{1}{4}}$ **101.** $\dfrac{\frac{5}{6}}{\frac{2}{3}}$ **102.** $\dfrac{\frac{5}{12}}{\frac{3}{8}}$

Challenge Yourself

103. Making a bookshelf. Antonio is making a student bookshelf from boards and cinder blocks. He has a piece of weathered pine that is $10\frac{7}{8}$ feet long and wants to cut it into four pieces of equal length. If we ignore the width of the saw's cut, how long will each shelf be? Express your answer in terms of feet.

104. Making a bookshelf. Redo Exercise 103 taking into account that each saw cut will be $\frac{1}{8}$ of an inch wide.

105. Hanging a mirror. Suppose that you want to hang a mirror that is $40\frac{1}{2}$ inches wide on a wall that is 6 feet (72 inches) wide. You want to center the mirror and put two hooks into the wall to support the mirror at $\frac{1}{3}$ and $\frac{2}{3}$ distance from one side of the mirror to the other side (see Figure 6.19). How far should the hooks be placed from the edges of the wall?

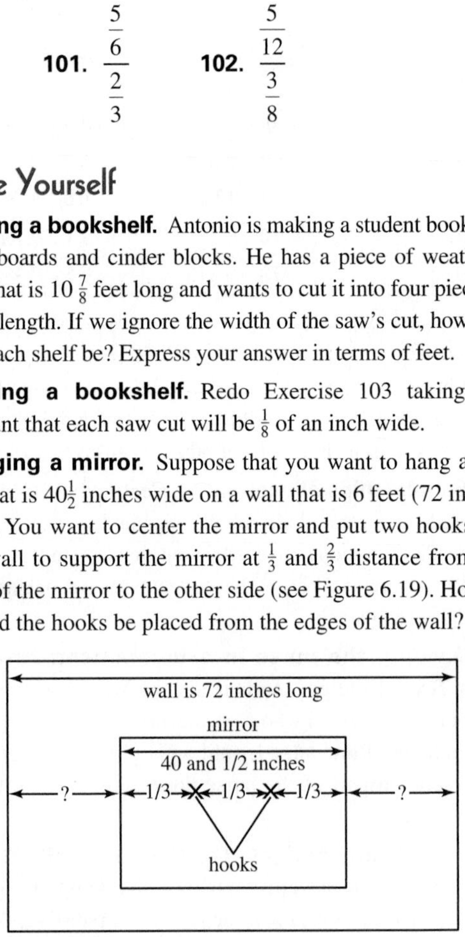

FIGURE 6.19 Hanging a mirror.

*You may find it interesting to go to a market and find examples of products where the smaller size costs less per unit than does the larger "economy" size.

106. Publishing a textbook. A publisher wants to increase the content in a new edition of a textbook without increasing the number of pages. The current printed area on each page is $5\frac{1}{4}$ inches wide and $8\frac{1}{4}$ inches long. If we increase the width of the printed area to $6\frac{1}{8}$ inches and increase the length to $8\frac{1}{2}$ inches, by how many square inches will the printed area increase per page?

107. Decorating project. You need 12 wood strips $3\frac{1}{4}$ inches wide to finish a home decorating project and can buy the strips in 8-, 10-, 12-, or 15-inch widths. What width will give you the least waste?

108. Buying paint. Habitat for Humanity operates ReStore resale outlets selling used furniture, building materials, paint, etc. They sell partial cans of paint. If you need $2\frac{1}{2}$ gallons of latex primer and can buy $\frac{2}{3}$, $\frac{3}{4}$, and $\frac{7}{8}$ gallons of primer, will you have enough for your job?

6.4 The Real Number System

Objectives

1. Understand that irrational numbers have nonrepeating decimal expansions.
2. Compute using radicals.
3. Understand the properties of common number systems.

In the previous section we discussed rational numbers, which might suggest that there exist other numbers that are not rational; that is, irrational numbers. "Irrational" seems like an extreme word to describe numbers. Are such numbers nonsensical? Or crazy? Understanding a little history will help explain why the ancient Greeks would think of some numbers as being irrational.

About 500 BC, a school of ancient Greek mathematicians led by Pythagoras of Samos (see Historical Highlight on page 269) discovered an amazing number, the square root of 2, which we write as $\sqrt{2}$. Their discovery was based on the Pythagorean theorem, which we illustrate in Figure 6.20(a). This theorem states that for a right triangle, the square of the length of the hypotenuse (the longest side) is equal to the sum of the squares of the lengths of the other two sides. Or, $c^2 = a^2 + b^2$. If the right triangle has two sides of length 1, as in Figure 6.20(b), then $c^2 = 1^2 + 1^2 = 2$, or $c = \sqrt{2}$.

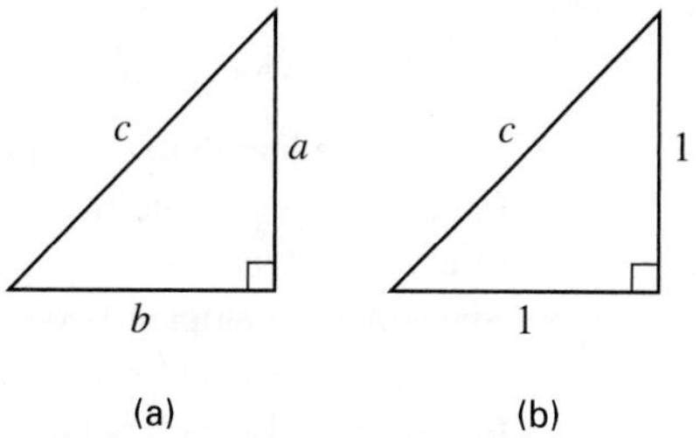

FIGURE 6.20 (a) $c^2 = a^2 + b^2$; (b) $c^2 = 1^2 + 1^2 = 2$.

What is amazing about this discovery is that we cannot write $\sqrt{2}$ as the quotient of two integers, and therefore, it is not a rational number. Prior to that time, the Greeks believed that all numbers were rational. So, when the Greeks discovered that $\sqrt{2}$ was not a quotient of integers, it didn't make sense—it was irrational. We will give Pythagoras' proof that $\sqrt{2}$ is not rational in Exercise 93.

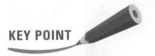

Irrational Numbers

KEY POINT

Irrational numbers have nonrepeating decimal expansions.

You saw in Section 6.3 that the rational numbers are precisely those numbers that have repeating decimal expansions. (Recall that we consider terminating decimals to be repeating decimals with the repeated digit being zero.) This means that any number that is not rational, such as $\sqrt{2}$, must have a nonrepeating decimal expansion.

> DEFINITION An **irrational number** is a number that is not a rational number and therefore has a nonrepeating decimal expansion.

A number such as 5.12112111211112... is an example of an irrational number. Although there is a pattern to the digits in this expansion, there is no single block of numbers that repeats from some point on. Other examples of irrational numbers would be 15.1234567891011121314... and 0.102003000400005000006....

If you use your calculator to find $\sqrt{2}$, you may get an answer such as $\sqrt{2} = 1.414213562$. Because this is a terminating (and hence repeating) decimal, what your calculator is telling you is not quite true. In reality, your calculator gave you a very close rational number approximation, which is a little less than $\sqrt{2}$. *In fact, any number of the form $\sqrt{n}$, where n is not a perfect square, will be an irrational number.*

Although you probably have had experience with many more rational numbers than irrational numbers, you may find it surprising that there are more irrational numbers than there are rational numbers (See Section 2.5 for a discussion of infinite sets.). We will discuss several famous irrational numbers: π (pi), ϕ (phi), and e.

The number $\pi = 3.141592654...$ frequently occurs in geometry and is defined as the ratio of the circumference of a circle to its diameter. For example, a circle with diameter 2 will have a circumference of 2π. See Figure 6.21.

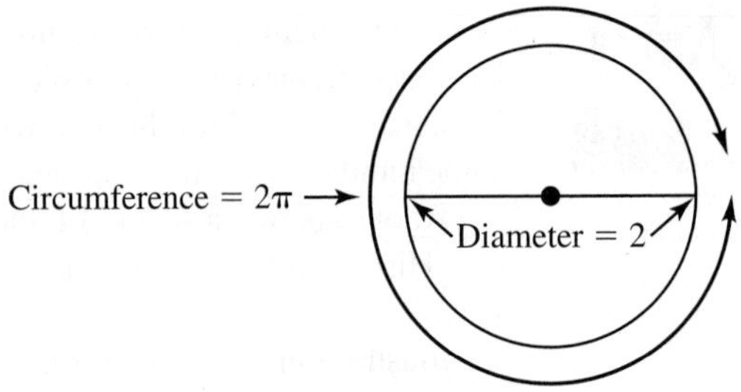

FIGURE 6.21 A circle with diameter 2 has a circumference of 2π.

The Greeks thought that the number $\phi = \dfrac{\sqrt{5}+1}{2} = 1.618033989...$ described the ratio of the length to the width of the most beautiful rectangle. The architecture of both the Greek Parthenon and the United Nations building (see page 292), as well as many works of art, are based on this ratio. The number $e = 2.718281828...$, named after the famous Swiss mathematician Leonhard Euler (pronounced "oiler"), has many important applications in science and engineering. It is probably the only number on your calculator keypad that is named after a person.

EXAMPLE 1 Approximating π, ϕ, and e

Although $\pi = 3.141592654...$, $\phi = 1.618033989...$, and $e = 2.7182818284...$ are irrational numbers and therefore have nonrepeating decimal expansions, there are methods for computing good approximations to these numbers.

a) Approximate π by using the first 20 terms of the following series:

$$4 - \frac{4}{3} + \frac{4}{5} - \frac{4}{7} + \frac{4}{9} - \frac{4}{11} + \frac{4}{13} \cdots$$

b) Use your calculator to approximate ϕ as follows:

Step 1: Enter 1.

Step 2: Press the $\boxed{1/x}$ or $\boxed{x^{-1}}$ key.

Step 3: Add 1 to the result in step 2.

Step 4 and beyond: Keep repeating steps 2 and 3—that is, invert your answer and add 1 repeatedly until you get the first five decimals after the decimal point in the decimal expansion of ϕ.

c) Approximate e by evaluating the expression $(1 + 1/n)^n$ for $n = 10, 100, 1{,}000$, and 10,000.

SOLUTION:

a) If you compute $4 - \frac{4}{3} + \frac{4}{5} - \frac{4}{7} + \frac{4}{9} - \frac{4}{11} + \frac{4}{13} - \cdots - \frac{4}{39}$, you should get the result 3.091623807. As you add more terms to this expression, you will get values closer to π. I did this (using a computer) for 100 terms and got 3.131592904. For 10,000 terms, the approximation was 3.141492654.

b) The following calculator screens show the beginning of the process and then the calculations at a much later stage for evaluating ϕ:

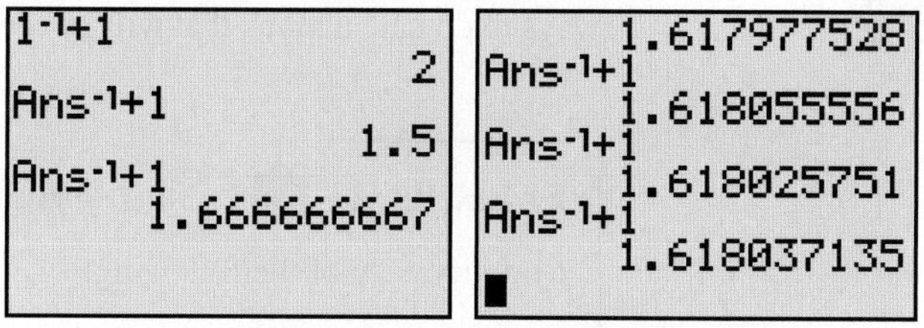

c) Using my calculator, I obtained the following results:

$$(1 + 1/10)^{10} = 2.59374246$$

$$(1 + 1/100)^{100} = 2.704813829$$

$$(1 + 1/1{,}000)^{1{,}000} = 2.716923932$$

$$(1 + 1/10{,}000)^{10{,}000} = 2.718145927$$

Computing with Radicals

In order to do exact calculations with irrational numbers, we often express them using radicals (roots).

DEFINITIONS A number such as $\sqrt{x}$ is called a **radical.** The symbol $\sqrt{\ }$ is called a **radical sign** and x is called the **radicand.**

Because scientific and mathematical formulas often contain radicals, you should understand the following rules, which are essential to working with radicals.

MULTIPLYING AND DIVIDING RADICALS If $a \geq 0$ and $b \geq 0$, then

$$\sqrt{ab} = \sqrt{a}\sqrt{b}, \text{ and if } b \neq 0, \text{ then } \sqrt{\frac{a}{b}} = \frac{\sqrt{a}}{\sqrt{b}}.$$

EXAMPLE 2 Multiplying and Dividing Radicals

Use the rules for multiplying and dividing radicals to rewrite each of the following. Use a calculator to verify your answers. (Remember that your calculator is only giving you approximations instead of exact results.)

a) $\sqrt{6 \cdot 5}$ b) $\dfrac{\sqrt{7}}{\sqrt{15}}$

SOLUTION:

a) $\sqrt{6 \cdot 5} = \sqrt{6}\,\sqrt{5}$. Using a calculator, $\sqrt{6 \cdot 5} = \sqrt{30} = 5.477225575$, and $\sqrt{6}\,\sqrt{5} = 5.477225575$. Note when calculating $\sqrt{6}\,\sqrt{5}$, calculate this product by first calculating $\sqrt{6}$; do not clear the calculator. Then press the multiplication key. Next calculate $\sqrt{5}$ and then press "Enter" ("=" key). (You should practice duplicating our calculations until you are confident in working with your calculator.)

b) $\dfrac{\sqrt{7}}{\sqrt{15}} = \sqrt{\dfrac{7}{15}}$. Calculating both sides, we get 0.6831300511.

We can use the product and quotient rules for radicals to simplify radicals. In simplifying a radical, we want to eliminate all perfect squares (numbers such as $2 \cdot 2 = 4$, $3 \cdot 3 = 9$, $5 \cdot 5 = 25$, and so on) under the radical. There are usually several different ways to do calculations that will give you the same simplification.

EXAMPLE 3 Simplifying Radicals

Simplify each of the following:

a) $\sqrt{15}\,\sqrt{5}$ b) $\sqrt{\dfrac{54}{12}}$

SOLUTION:

a) We notice that 5 is a factor of 15 and 5 also appears in the second radical sign. So, if we combine the two radicals into one radical, then we can simplify the 5 squared in the single radical. Thus,

$$\sqrt{15}\sqrt{5} = \sqrt{15 \cdot 5} = \sqrt{3 \cdot 5 \cdot 5} = \sqrt{3 \cdot 25} = \sqrt{3}\sqrt{25} = \sqrt{3} \cdot 5 = 5 \cdot \sqrt{3}.$$

b) There are several ways to do this. We will begin by using the quotient rule for radicals. Thus,

Identify perfect squares in 54 and 12.

$$\sqrt{\frac{54}{12}} = \frac{\sqrt{54}}{\sqrt{12}} = \frac{\sqrt{6 \cdot 9}}{\sqrt{3 \cdot 4}} = \frac{\sqrt{6} \cdot \sqrt{9}}{\sqrt{3} \cdot \sqrt{4}} = \frac{\sqrt{6} \cdot 3}{\sqrt{3} \cdot 2} = \sqrt{\frac{6}{3}} \cdot \frac{3}{2} = \sqrt{2} \cdot \frac{3}{2} = \frac{3}{2}\sqrt{2}.$$

Now try Exercises 11 to 16 and 23 to 34. 17

QUIZ YOURSELF 17

Simplify the following:

a) $\sqrt{45}$

b) $\sqrt{\dfrac{15}{48}}$

Prior to the invention of calculators, people had to use pencil and paper computations to make rational number estimates of numbers such as $\frac{2}{\sqrt{3}}$. It was easier to do such calculations if quotients did not have radicals in the denominator. Shortly, we will discuss how to rewrite quotients such as $\frac{2}{\sqrt{3}}$ in this preferred form.

SOME GOOD ADVICE

In doing a calculation, you will find it to your advantage to try to work with smaller numbers whenever possible. For example, if you write $\sqrt{72}\,\sqrt{48}$ as $\sqrt{72 \cdot 48} = \sqrt{3{,}456}$, then you have the problem of finding the factors in 3,456.

On the other hand, if you recognize that $\sqrt{72} = \sqrt{2 \cdot 36} = 6\sqrt{2}$ and $\sqrt{48} = \sqrt{3 \cdot 16} = 4\sqrt{3}$, then you can do this calculation quickly as $\sqrt{72}\sqrt{48} = 6\sqrt{2} \cdot 4\sqrt{3} = 24\sqrt{6}$.

KEY POINT

Rationalizing the denominator makes a quotient more understandable.

Sometimes when we calculate with radicals, we get an answer such as $\frac{2}{\sqrt{3}}$. We have no difficulty evaluating this expression with modern calculators; however, without using a calculator, it is a little difficult to estimate the size of this number. If we write the number in another way, without a radical in the denominator, we can get a better intuitive understanding of the number. The process of writing a quotient so that it has no radicals in the denominator is called **rationalizing the denominator.** We illustrate this technique in Example 4.

EXAMPLE 4 Rationalizing Denominators

Rationalize the denominator in the following expression:

$$\frac{7}{\sqrt{12}}$$

SOLUTION:

If we factor 12, we find that $12 = 2 \cdot 2 \cdot 3$. The product $2 \cdot 2$ is already a perfect square, so we need another factor of 3 to have the radical contain a perfect square. Thus, we will multiply the numerator and denominator by $\sqrt{3}$. So we have

$$\frac{7}{\sqrt{12}} = \frac{7 \cdot \sqrt{3}}{\sqrt{12} \cdot \sqrt{3}} = \frac{7 \cdot \sqrt{3}}{\sqrt{36}} = \frac{7 \cdot \sqrt{3}}{6} = \frac{7}{6}\sqrt{3}.$$

Now try Exercises 35 to 42. **18**

QUIZ YOURSELF 18

Rationalize the denominator in $\frac{3}{\sqrt{15}}$.

Historical Highlight

The Pythagoreans

Shortly before 500 BC, Pythagoras founded a school in the Greek settlement of Crotona, Italy, that was to have a profound influence on mathematics for the next 2,500 years. The school, which had about 300 students (including about 30 women), was somewhat like a secret society or fraternity. All students studied the same subjects, or *mathemata*—number theory, music, geometry, and astronomy. In addition, they studied logic, grammar, and rhetoric.

Beginners, called *acoustici*, listened silently as Pythagoras lectured them from behind a curtain. After three years of obedient study, they became *mathematici*, and the secrets of the society were revealed to them. Many of these "secrets" are the mathematical theorems that we still study today.

The Pythagoreans would not eat beans or drink wine. They believed that one's soul could leave the body and live in another person or animal. Accordingly, they would not eat meat or fish so that they would not consume the residence of another's soul.

Around 500 BC, there was a political revolt in Crotona. The school was burned, and Pythagoras was killed. There are conflicting reports of his death. One romantic account tells that as he fled for his life, Pythagoras stopped at the edge of a sacred field of beans, allowing his enemies to kill him rather than trample his sacred plants.

KEY POINT

We add and subtract radicals with the same radicand.

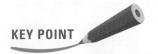

It is a little more complicated to add and subtract radicals. If we add $5\sqrt{3} + 4\sqrt{3}$, we can think "five of something plus four of something equals nine of something." Thus, $5\sqrt{3} + 4\sqrt{3} = 9\sqrt{3}$. However, if we want to add $5\sqrt{2} + 4\sqrt{7}$, we are, so to speak, adding apples and oranges. We cannot add five of one thing to four of another thing. In order to add (or subtract) expressions with radicals, all the radicals must have the same radicand. We illustrate this in Example 5.

EXAMPLE 5 Adding Expressions Containing Radicals

Perform the operations.

a) $5\sqrt{3} + 8\sqrt{12}$ b) $\sqrt{45} + \sqrt{12}$

SOLUTION:

a) Here the radicands are different. However, we see that 12 has a factor of 4, so we can simplify $\sqrt{12}$. Therefore,

$$\begin{aligned} 5\sqrt{3} + 8\sqrt{12} &= 5\sqrt{3} + 8\sqrt{4\cdot 3} = 5\sqrt{3} + 8\sqrt{4}\sqrt{3} \\ &= 5\sqrt{3} + 8\cdot 2\sqrt{3} = 5\sqrt{3} + 16\sqrt{3} = 21\sqrt{3}. \end{aligned}$$

b) We rewrite $\sqrt{45} + \sqrt{12}$ as $3\sqrt{5} + 2\sqrt{3}$ and see that because the radicands are different, we cannot simplify this expression any further.

Now try Exercises 17 to 22. 19

QUIZ YOURSELF 19

Find:

a) $3\sqrt{5} + 8\sqrt{5}$
b) $3\sqrt{20} - 2\sqrt{45}$

Properties of the Real Numbers

KEY POINT

Number systems share many common properties.

Figure 6.22 summarizes the relationships between the number systems that you have studied in this chapter.

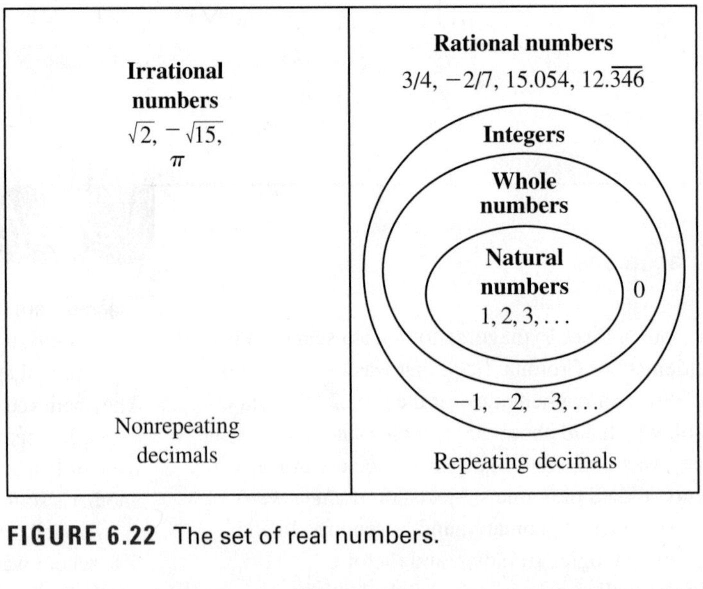

FIGURE 6.22 The set of real numbers.

If we combine the set of rational numbers with the set of irrational numbers, we get the set of real numbers. The set of real numbers, together with the operations of addition and multiplication,* form a mathematical system that has the properties listed in Table 6.3.

*Recall that in Section 6.2 we said subtraction and division are not independent operations, but are based on addition and multiplication.

Property	Definition	Examples
Closure property for addition and multiplication	If a and b are real numbers, then $a + b$ and $a \cdot b$ are also real numbers.	3 and π are both real numbers, so $3 + \pi$ is a real number as well. $\frac{1}{2}$ and $\sqrt{3}$ are real numbers, so $\frac{1}{2} \cdot \sqrt{3}$ is also a real number.
Commutative property for addition and multiplication	If a and b are real numbers, then $a + b = b + a$ and $a \cdot b = b \cdot a$. (The *order* in which we combine the numbers is not important.)	$3 + \sqrt{2} = \sqrt{2} + 3$ and $(-5) \cdot \pi = \pi \cdot (-5)$
Associative property for addition and multiplication	If a, b, and c are real numbers, then $(a + b) + c = a + (b + c)$ and $(a \cdot b) \cdot c = a \cdot (b \cdot c)$. (The way we *group* the numbers is not important.)	$(3 + \sqrt{5}) + \sqrt{7} = 3 + (\sqrt{5} + \sqrt{7})$ and $\left((-6) \cdot \frac{1}{2}\right) \cdot \pi = (-6) \cdot \left(\frac{1}{2} \cdot \pi\right)$
Zero is an **identity element for addition,** and 1 is an **identity element for multiplication.**	If a is any real number, then $0 + a = a + 0 = a$ and $1 \cdot a = a \cdot 1 = a$.	$0 + 5 = 5 + 0 = 5$ and $1 \cdot \sqrt{10} = \sqrt{10} \cdot 1 = \sqrt{10}$
If x is a real number, then $-x$ is an **additive inverse** for x. If $x \neq 0$, then $\frac{1}{x}$ is a **multiplicative inverse** for x.	For any real number x, $x + (-x) = (-x) + x = 0$ and $x \cdot \left(\frac{1}{x}\right) = \left(\frac{1}{x}\right) \cdot x = 1$.	$3 + (-3) = (-3) + 3 = 0$ and $\pi \cdot \frac{1}{\pi} = \frac{1}{\pi} \cdot \pi = 1$
Multiplication **distributes** over addition.	If a, b, and c are real numbers, then $a \cdot (b + c) = (a \cdot b) + (a \cdot c)$.	$\frac{1}{2} \cdot (4 + 6) = \left(\frac{1}{2} \cdot 4\right) + \left(\frac{1}{2} \cdot 6\right) = 5$

TABLE 6.3 Properties of the real numbers.

PROBLEM SOLVING

Strategy: The Always Principle

In checking whether a property such as closure holds for an operation, remember the Always Principle from Section 1.1. If we say that the property holds, then it must hold for every possibility *without even one exception*. For example, the set of rational numbers is not closed under division because we cannot divide by the rational number 0.

Examples 6 and 7 illustrate some properties of the real number system.

EXAMPLE 6 Identifying Properties of the Real Numbers

Identify the property being used in each of the following statements.

a) $3 + 4$ is an integer
b) $\sqrt{2} + \pi = \pi + \sqrt{2}$
c) $0 + 3.6 = 3.6 + 0 = 3.6$
d) $1.24 \times (4.678 \times 3.9) = (1.24 \times 4.678) \times 3.9$
e) $\sqrt{5} + -\sqrt{5} = 0$
f) $\sqrt{2} \times \frac{1}{\sqrt{2}} = 1$
g) $3 \times (\sqrt{2} + 5) = 3 \times \sqrt{2} + 3 \times 5$

Between the Numbers

A Million-Dollar Comma

Just as we use the associative property in mathematics to regroup numbers, we use commas in English to regroup ideas. Although punctuation may strike you as an unimportant and tedious topic, in fact, changing punctuation often changes the meaning of a sentence.*

An article in the *New York Times* business section on October 25, 2006, describes how an extra comma in a contract allowed a Canadian phone company to pull out of a deal earlier than expected, thereby saving the phone company one million dollars. Although the intent of the agreement seemed clear, the regulator ruled that both parties had to live by what was written in the contract, rather than what they thought they had agreed upon.

SOLUTION: Review the Analogies Principle on page 13.

a) The integers have the closure property for addition.
b) Addition is commutative in the real numbers.
c) Zero is an identity element for addition.
d) Multiplication is associative in the real numbers.
e) $\sqrt{5}$ and $-\sqrt{5}$ are additive inverses of each other.
f) Every nonzero real number has a multiplicative inverse.
g) Multiplication distributes over addition in the real numbers.

Often students get the commutative and associative properties confused. It is easy to remember the difference if you think how we use similar words in English. "Commutative" reminds us of the word *commuter*, who is a person who goes back and forth between two places. So the commutative property involves numbers changing places. Similarly, the associative property means that we are *reassociating*, or regrouping, numbers.

EXAMPLE 7 Finding Counterexamples

Provide a counterexample for each of the following statements.

a) Subtraction is commutative in the set of integers.
b) Division is associative in the set of real numbers.
c) Every rational number has a multiplicative inverse.

SOLUTION: Review the Counterexample Principle on page 11.

a) $2 - 5 \neq 5 - 2$, since $2 - 5 = -3$ but $5 - 2 = 3$.
b) $8 \div (4 \div 2) \neq (8 \div 4) \div 2$, because $8 \div (4 \div 2) = 8 \div 2 = 4$, but $(8 \div 4) \div 2 = 2 \div 2 = 1$.
c) Zero has no multiplicative inverse. A multiplicative inverse for 0 would be a number a such that $a \times 0 = 1$, which is impossible since any number multiplied by 0 will result in a product of 0.

Now try Exercises 53 to 58. 20

QUIZ YOURSELF 20

Give examples to illustrate each of the following facts:

a) Addition is associative on the set of real numbers.
b) Multiplication is commutative on the set of integers.
c) Multiplication is not closed in the set of negative integers.

*See Exercises 87 and 88 regarding "A panda eats shoots and leaves" versus "A panda eats, shoots, and leaves."

EXERCISES 6.4

Sharpening Your Skills

Which of the following numbers are rational, and which are irrational?

1. $\frac{3}{8}$ **2.** 5.0136

3. 1.23456789101112. . . **4.** $\sqrt{10}$

5. 3.1416 **6.** 0.10110111

7. 0.101101110 . . . **8.** $\sqrt{81}$

9. Give an example to show that $\sqrt{n}$ can be a rational number.

10. Give an example to show that $\sqrt{n}$ can be an irrational number.

Simplify the given radicals.

11. $\sqrt{18}$ **12.** $\sqrt{27}$ **13.** $\sqrt{75}$

14. $\sqrt{48}$ **15.** $\sqrt{189}$ **16.** $\sqrt{240}$

If possible, combine the radicals into a single radical.

17. $2\sqrt{3} - 4\sqrt{2}$ **18.** $5\sqrt{7} - 3\sqrt{5}$

19. $\sqrt{20} + 6\sqrt{5}$ **20.** $5\sqrt{12} - 4\sqrt{3}$

21. $\sqrt{50} + 2\sqrt{75}$ **22.** $\sqrt{28} + 2\sqrt{63}$

Perform the indicated operation, and simplify if possible.

23. $\sqrt{18}\sqrt{2}$ **24.** $\sqrt{15}\sqrt{5}$ **25.** $\sqrt{12}\sqrt{15}$

26. $\sqrt{72}\sqrt{10}$ **27.** $\sqrt{28}\sqrt{21}$ **28.** $\sqrt{27}\sqrt{33}$

29. $\frac{\sqrt{24}}{\sqrt{6}}$ **30.** $\frac{\sqrt{54}}{\sqrt{6}}$ **31.** $\frac{\sqrt{32}}{\sqrt{18}}$

32. $\frac{\sqrt{45}}{\sqrt{20}}$ **33.** $\frac{\sqrt{96}}{\sqrt{72}}$ **34.** $\frac{\sqrt{63}}{\sqrt{112}}$

Rationalize the denominator and simplify.

35. $\frac{3}{\sqrt{5}}$ **36.** $\frac{4}{\sqrt{3}}$ **37.** $\frac{12}{\sqrt{6}}$

38. $\frac{21}{\sqrt{14}}$ **39.** $\frac{10}{\sqrt{22}}$ **40.** $\frac{8}{\sqrt{10}}$

41. $\frac{3\sqrt{10}}{\sqrt{72}}$ **42.** $\frac{3\sqrt{5}}{\sqrt{135}}$

Find (a) a rational number and (b) an irrational number between the two given numbers. Give your answers in decimal form. There are many correct answers. (Hint: Write all numbers in decimal form before answering the questions.)

43. 0.43 and 0.44 **44.** 1.245 and 1.246

45. $0.\overline{4578}$ and $0.45\overline{78}$ **46.** $0.\overline{123}$ and $0.12\overline{31}$

47. $\frac{4}{7}$ and $\frac{5}{7}$ **48.** $\frac{5}{8}$ and $\frac{3}{4}$

Put the numbers in order from smallest to largest.

49. a. 0.345345 **b.** $0.\overline{345}$ **c.** $0.3\overline{45}$ **d.** $0.345345\overline{34}$

50. a. $0.\overline{261}$ **b.** $0.2\overline{61}$ **c.** $0.\overline{26}$ **d.** 0.2626

51. a. $\frac{4}{9}$ **b.** $\frac{5}{9}$ **c.** $0.4\overline{54}$ **d.** $0.5\overline{54}$

52. a. $\frac{5}{13}$ **b.** $0.3\overline{84}$ **c.** $\frac{4}{9}$ **d.** $0.\overline{38}$

Determine which of the following are true or false. If a statement is false, give a counterexample.

53. The product of two irrational numbers is irrational.

54. The product of an integer and a rational number is a rational number.

55. Every negative real number has a multiplicative inverse.

56. If n is any positive integer, then $\sqrt{n}$ is irrational.

57. The rational numbers are closed under division.

58. If n is any integer, then $\sqrt{n^2}$ is a rational number.

State which property of the real numbers we are illustrating.

59. $3(4 + 5) = 3 \cdot 4 + 3 \cdot 5$

60. $6(8 + 2) = (8 + 2)6$

61. $3 + (6 + 8) = (6 + 8) + 3$

62. $5(3 \cdot 2) = (5 \cdot 3)2$

63. $3 + (6 + 8) = (3 + 6) + 8$

64. $5(3 \cdot 2) = (3 \cdot 2)5$ **65.** $7 + 0 = 7$

66. $8 + (-8) = 0$

67. $4 + (2 + (-1)) = 4 + ((-1) + 2)$

68. $7(8 - 2) = 7 \cdot 8 - 7 \cdot 2$

Applying What You've Learned

In Exercises 69–72, first express your answer as a radical expression and then as a decimal rounded to the nearest tenth.

69. We can use the equation $D = \sqrt{2H}$ to approximate the distance D, in miles, that you can see to the horizon at a height of H feet. How many miles could a forest ranger see to the horizon if his eyes are 6 feet above ground level and he is standing on a fire tower that is 42 feet high?

70. Using the equation in Exercise 69, how far to the horizon could you see from a hot air balloon that is 160 feet above the ground?

71. Police investigating an accident can use the equation $v = 2\sqrt{5L}$ to estimate the speed at which a vehicle was traveling when the driver hit the brakes. Here, v is the speed of the vehicle in miles per hour and L is the length of the skid marks in feet. If a car leaves a skid mark of 120 feet, how fast was the car going when the driver hit the brakes?

72. Repeat Exercise 71, but now assume that the skid mark is 160 feet.

73. The time for a pendulum to swing all the way forward and then back (called its **period**) depends on its length. If a pendulum is L feet long, then we use the formula $T = 2\pi\sqrt{\frac{L}{32}}$ to calculate

T, the time in seconds of its period. If a child is swinging from a 20-foot rope attached to a tree branch hanging over a pond, how long would it take for him to swing all the way out and then back again?

74. Repeat Exercise 73 for a 30-foot rope.

75. Squaring up the base of a shed. If a triangle has sides a and b and c, satisfying the equation $a^2 + b^2 = c^2$, then sides a and b are perpendicular to each other. Carpenters sometimes use this fact in "squaring up" objects. Assume that the base of a shed with sides of lengths 9 and 12 feet is not quite a rectangle (shown below). What should the length of the diagonal c be to make the two sides perpendicular?

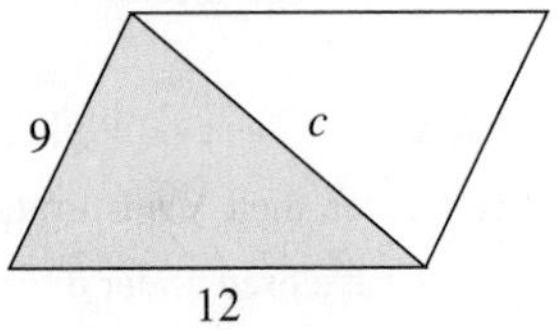

76. Squaring up the base of a shed. Repeat Exercise 75, but now assume that the sides of the base have lengths 8 and 15 feet.

77. Marius Pudzianowski, a frequent champion in the World's Strongest Man competition, is competing in a contest in which forces of 120 and 160 pounds are applied to two ropes that form a right angle, as shown in the figure. Marius must hold a third rope, attached to the vertex of the right angle, for as long as he can. Find the force that Marius is resisting, which corresponds to length of the hypotenuse in the right triangles in the figure.

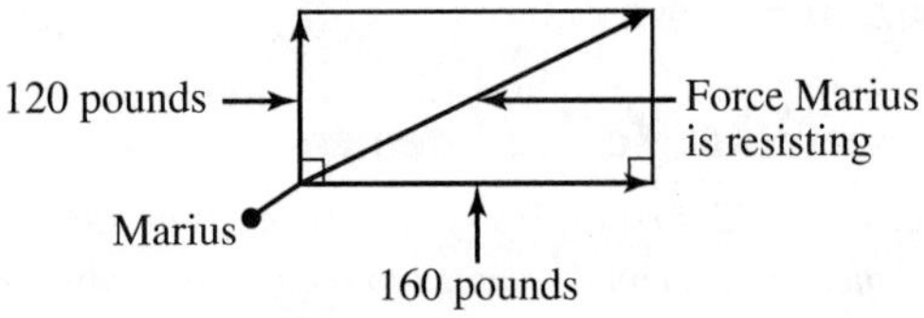

78. Repeat Exercise 77 for weights of 180 and 240 pounds.

79. The equation $t = \frac{\sqrt{d}}{4}$ estimates the time t in seconds that it takes an object to fall a distance of d feet. Assume that you are on an observation deck 1,600 feet above the ground on the Taipei 101 skyscraper in Taiwan, China. If your MP3 player slips out of your hand, how long will it take to hit the ground?

80. Repeat Exercise 79 assuming you are now on an observation deck that is 1,000 feet above the ground on the Empire State Building.

81. A dangerous tradition. In this exercise we will investigate the dangerous practice that some have of firing a gun into the air at midnight on New Year's Eve. Assume that someone fires a gun straight up and that the bullet reaches a height of one-half mile (a mile is 5,280 feet).

a. Use the information given in Exercise 79 to determine how long it will take the bullet to hit the ground. (See disclaimer below.*)

b. Assuming that an object that falls for t seconds will hit the ground at a speed of $32t$ feet per second, find the speed of the bullet in feet per second when it hits the ground.

c. Convert feet per second to feet per hour and then divide by 5,280 to find the speed of the bullet in miles per hour.

82. A dangerous tradition. Repeat Exercise 81 for a bullet that reaches a height of one-quarter mile.

Communicating Mathematics

83. What is incorrect about the following statements?

a. $\sqrt{2} = 1.414213562$ **b.** $\pi = \frac{22}{7}$

84. What is one way to distinguish rational numbers from irrational numbers?

85. Can a real number be both rational and irrational?

86. Can the sum of a rational number and an irrational number be rational? Explain.

Between the Numbers

We mentioned in Between the Numbers on page 272 that changing the placement of commas can change the meaning of a sentence. For example:

> *A panda walks into a cafe and orders a bamboo sandwich. After finishing his meal, he pulls out a gun and shoots the waiter. The police arresting him ask, "Why did you do that?" The panda responds, "I have to because it says in my encyclopedia that a panda eats, shoots, and leaves."*

The panda has misplaced his commas and changed the meaning of the sentence. It should read "... that a panda eats shoots and leaves," meaning that his diet consists of shoots and leaves.

87. Rewrite Sam I Am's statement "I like green, eggs, and ham" to read correctly.

88. My daughter recently said to her husband, "We have cream cheese and bagels in the freezer." What do you think she really said?

Challenge Yourself

89. Continue the calculation of π in Example 1(a) for 20 more terms. How does this approximation compare with the one we obtained in Example 1?

90. Do the calculation of ϕ in Example 1(b) for 20 iterations. What is your result?

*This is what we call a "textbook exercise" and we are ignoring very important factors such as air resistance and spin on the bullet. In a 2006 episode of *MythBusters*, they neither confirmed nor denied that a falling bullet alone could be fatal. Nonetheless, firing weapons into the air does cause fatalities, probably due to the fact that the weapon is usually not fired straight up into the air and hence the bullet maintains much of its original velocity from being fired.

91. There are many interesting Web sites devoted to the irrational number pi. Have a contest to find the site that reports on computing the most digits in the decimal expansion of pi.

92. A mnemonic (a memory device) for memorizing the digits in the decimal expansion of pi is a sentence such as "Can I have a small container of coffee?" If you count the letters in each word of this sentence, you will get 3, 1, 4, 1, 5, 9, 2, and 6, which are the first eight digits in the infinite decimal expansion of pi. Search the Internet for other mnemonics, and then have a contest to see who can write the longest mnemonic for memorizing pi. (When I did this search, I found mnemonics in Bulgarian, Dutch, German, French, and Polish, as well as in many other languages.)

93. We will now give a proof that $\sqrt{2}$ is irrational. Explain why each of the steps in the proof is valid.

a. Assume that $\sqrt{2}$ can be written as a quotient of integers $\frac{a}{b}$, where the quotient is in reduced form. So, $\sqrt{2} = \frac{a}{b}$.

b. $2 = \dfrac{a^2}{b^2}$ **c.** $2 \cdot b^2 = a^2$

d. a^2 is even, that is, divisible by 2.

e. a cannot be odd, so it must be even.

f. $a = 2 \cdot k$ for some integer k.

g. $2 \cdot b^2 = (2k)^2$. **h.** $2 \cdot b^2 = 4k^2$.

i. $b^2 = 2k^2$. **j.** b^2 is even.

k. b is even.

l. We have a contradiction, so our assumption that $\sqrt{2}$ can be written as a quotient of integers $\frac{a}{b}$ is false. Therefore, $\sqrt{2}$ is not rational.

94. Do a similar proof to show that $\sqrt{3}$ is irrational.

95. Find the length of line segment AB in the given figure.

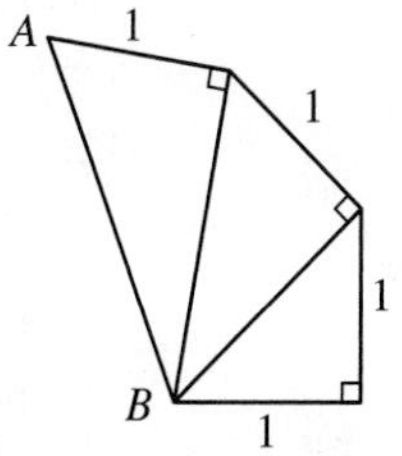

96. The triple of natural numbers 3, 4, 5 is called a Pythagorean triple. Why do you think that this is so? (*Hint:* Consider right triangles.)

97. Find another Pythagorean triple different from the ones found in Exercises 75 and 76.

98. If a, b, and c is a Pythagorean triple, what about the triple $3a$, $3b$, and $3c$? What about any triple of the form na, nb, and nc, where n is a natural number? Explain your answer.

6.5 Exponents and Scientific Notation

Objectives

1. Understand and apply the basic rules of exponents.
2. Perform operations using scientific notation.
3. Solve problems using scientific notation.

We hear about really, really small and really, really large numbers all the time. For example, biologists tell us that the HIV virus is about one micron in size, or about 0.000004 inch. On the other hand, some astronomers estimate the distance to the edge of the universe to be about 46.6 billion light-years. It is not only difficult to understand such extreme numbers but it can be even more difficult to calculate with them.

In this section you will learn how to calculate efficiently with such numbers to find a good estimate of your share of the national debt. And then, we will give an amazing answer to the question posed in the chapter opener regarding how far the atoms in a person's body would stretch if arranged in a straight line. But first we will need to discuss some rules for working with exponents.

KEY POINT

All exponent rules are based on the definition of exponents.

Exponents

In order to write such very large and very small numbers efficiently, we will use exponential notation. Some examples of exponential notation are $10^3 = 10 \cdot 10 \cdot 10 = 1{,}000$ and $2^5 = 2 \cdot 2 \cdot 2 \cdot 2 \cdot 2 = 32$. In general, we have the following definition.

DEFINITION If a is any real number and n is a counting number, then

$$a^n = \underbrace{a \cdot a \cdot a \cdot \ldots \cdot a}_{\text{product of } n\ a\text{'s}}.$$

The number a is called the **base,** and the number n is called the **exponent.**

This simple definition is the basis for all the rules for exponents that we will discuss in this section.

EXAMPLE 1 Evaluating Expressions with Exponents

Write each of the following expressions in another way using the definition of exponents, then evaluate the expression.

a) $3 \cdot 3 \cdot 3 \cdot 3 \cdot 3$ b) 2^4 c) 0^6 d) $(-2)^4$ e) -2^4 f) 5^1

SOLUTION: Review the Order Principle on page 12.

a) $3 \cdot 3 \cdot 3 \cdot 3 \cdot 3 = 3^5 = 243$
b) $2^4 = 2 \cdot 2 \cdot 2 \cdot 2 = 16$
c) $0^6 = 0 \cdot 0 \cdot 0 \cdot 0 \cdot 0 \cdot 0 = 0$
d) $(-2)^4 = (-2) \cdot (-2) \cdot (-2) \cdot (-2) = +16$
e) This is not the same as d). Notice that here the exponent, 4, takes precedence* over the negative sign. So, first we multiply the four 2s and then insert the negative sign. Thus,

First raise 2 to the fourth power.

$$-2^4 = -(2 \cdot 2 \cdot 2 \cdot 2) = -16$$

f) $5^1 = 5$

Suppose in a physics class you encountered the expression $x^3 \cdot x^4$. You could simplify this expression by using only the definition of exponents as follows:

$$x^3 \cdot x^4 = \underbrace{(x \cdot x \cdot x)}_{3\ x\text{'s}} \cdot \underbrace{(x \cdot x \cdot x \cdot x)}_{4\ x\text{'s}} = \underbrace{x \cdot x \cdot x \cdot x \cdot x \cdot x \cdot x}_{7\ x\text{'s}} = x^7.$$

Thus, $x^3 \cdot x^4 = x^7$.

By thinking of a simple example such as this, you can remember the following rule for rewriting certain products:

PRODUCT RULE FOR EXPONENTS If x is a real number and m and n are natural numbers, then

$$x^m x^n = x^{m+n}.$$

Notice in this rule that all the bases are the same. This rule would not apply to an expression such as x^3y^5 because the bases are different.

EXAMPLE 2 Applying the Product Rule for Exponents

Use the product rule for exponents to rewrite each of the following expressions, if possible:

a) $2^5 \cdot 2^9$ b) $3^2 \cdot 5^4$

*For a review of the order of operations, see Appendix A.

QUIZ YOURSELF 21

Use the product rule to rewrite $3^4 \cdot 3^2$.

SOLUTION:

a) $2^5 \cdot 2^9 = 2^{5+9} = 2^{14}$

b) We cannot apply the product rule here. The product $3^2 \cdot 5^4$ contains two 3s and four 5s as factors and we cannot combine these six factors into a single expression. We could, however, rewrite this as $3 \cdot 3 \cdot 5 \cdot 5 \cdot 5 \cdot 5 = 5{,}625$. 21

Suppose that you did not recall the rule for simplifying the expression $(y^3)^4$. You could think of $(y^3)^4$ as (y^3) multiplied by itself repeatedly as follows:

$$(y^3)^4 = (y^3)(y^3)(y^3)(y^3) = (y \cdot y \cdot y)(y \cdot y \cdot y)(y \cdot y \cdot y)(y \cdot y \cdot y) = y^{12}.$$

You see from this example that we simplified the expression by multiplying the exponents, which leads us to the following rule:*

POWER RULE FOR EXPONENTS If x is a real number and m and n are natural numbers, then

$$(x^m)^n = x^{m \cdot n}.$$

EXAMPLE 3 Applying the Power Rule for Exponents

Use the power rule for exponents to simplify each of the following:

a) $(2^3)^2$ b) $((-2)^3)^4$

QUIZ YOURSELF 22

Use the power rule for exponents to simplify each of the following:

a) $(5^4)^3$

b) $((-2)^2)^4$

SOLUTION: Review the Order Principle on page 12.

a) $(2^3)^2 = 2^{3 \cdot 2} = 2^6 = 64$

b) $((-2)^3)^4 = (-2)^{3 \cdot 4} = (-2)^{12} = 4{,}096$ 22

PROBLEM SOLVING

Strategy: The Three-Way Principle

The Three-Way Principle in Section 1.1 suggests that a good way to remember what exponent rule to use in a particular situation is to think of simple examples. Then use the definition of exponents to recall how we derived the rule. For example, to remember whether you add or multiply exponents in simplifying the expression $(a^8)^7$, think what you would do with $(x^3)^2$. If you write this expression as $x^3 \cdot x^3 = x \cdot x \cdot x \cdot x \cdot x \cdot x = x^6$, then you recall that you are supposed to multiply exponents in this situation.

We often have to divide expressions containing exponents. For example, suppose that we want to simplify the expression $\frac{x^7}{x^3}$. Using the definition of exponents, we can rewrite the numerator and denominator as follows:

$$\frac{x^7}{x^3} = \frac{x \cdot x \cdot x \cdot x \cdot \cancel{x} \cdot \cancel{x} \cdot \cancel{x}}{\cancel{x} \cdot \cancel{x} \cdot \cancel{x}} = \frac{x \cdot x \cdot x \cdot x}{1} = x \cdot x \cdot x \cdot x = x^4.$$

This example leads us to the following quotient rule for exponents:

QUOTIENT RULE FOR EXPONENTS If x is a nonzero real number and both m and n are natural numbers, then

$$\frac{x^m}{x^n} = x^{m-n}.$$

*We will explain some other power rules for exponents in the exercises.

This rule is fine provided m is greater than n. However, if m equals n, then $m - n = 0$ and if m is less than n, then $m - n < 0$. To understand what to do in these cases, let's look at some examples. Suppose that $m = n = 3$, then $\frac{x^m}{x^n} = \frac{x^3}{x^3} = \frac{\cancel{x}\cdot\cancel{x}\cdot\cancel{x}}{\cancel{x}\cdot\cancel{x}\cdot\cancel{x}} = 1$. Also, if $m = 3$ and $n = 5$, we get $\frac{x^m}{x^n} = \frac{x^3}{x^5} = \frac{\cancel{x}\cdot\cancel{x}\cdot\cancel{x}}{\cancel{x}\cdot\cancel{x}\cdot\cancel{x}\cdot x\cdot x} = \frac{1}{x\cdot x} = \frac{1}{x^2}$. These examples will help you to remember the following definitions:

DEFINITIONS If $a \neq 0$, then $a^0 = 1$, and if n is a natural number, then $a^{-n} = \frac{1}{a^n}$.

EXAMPLE 4 Applying the Quotient Rule for Exponents

Use the quotient rule to simplify the following expressions and write your answer as a single number:

a) $\frac{3^7}{3^5}$ b) $\frac{7^5}{7^8}$ c) $\frac{17^9}{17^9}$

SOLUTION:

a) $\frac{3^7}{3^5} = 3^{7-5} = 3^2 = 9$ b) $\frac{7^5}{7^8} = 7^{5-8} = 7^{-3} = \frac{1}{7^3} = \frac{1}{343}$

c) $\frac{17^9}{17^9} = 17^0 = 1$ 23

QUIZ YOURSELF 23

Simplify and evaluate these expressions.

a) $\frac{6^8}{6^{10}}$ b) $\frac{(-8)^{11}}{(-8)^9}$

c) $\frac{11^0}{11^0}$

We explained the rules for working with exponents by using very simple examples in which the exponents m and n were both natural numbers. Without proving it, we will simply state that the *rules don't change* when you are working with other types of exponents (see Figure 6.23).

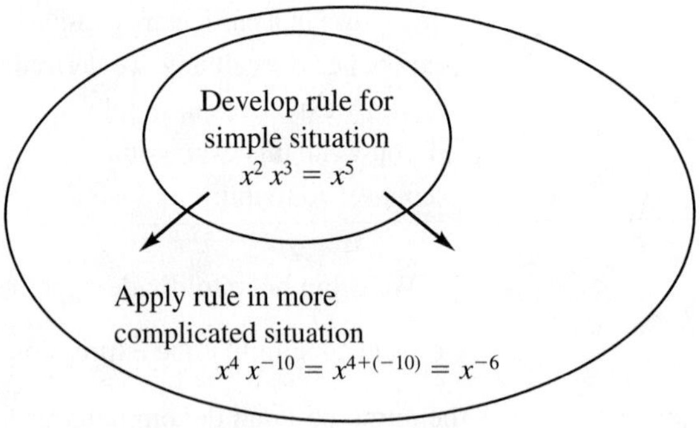
Develop rule for
simple situation
$x^2 x^3 = x^5$
Apply rule in more
complicated situation
$x^4 x^{-10} = x^{4+(-10)} = x^{-6}$

FIGURE 6.23 The rules don't change as we work in more complicated situations.

For example, if you are dealing with the expression $3^{-4}\cdot 3^6$, you can rewrite it as $3^{-4}\cdot 3^6 = 3^{-4+6} = 3^2 = 9$. In other mathematics courses, you may see rational exponents, or even irrational exponents! However, if you remember that the rules are the same when dealing with these more complicated examples, you will be able to do your computations confidently.

KEY POINT

We can extend the exponent rules to all integer exponents.

EXAMPLE 5 Using Exponent Rules When the Exponents Are Not Positive

Assume that the exponent rules extend to these expressions and then simplify as in Example 4.

a) $5^{-4} \cdot 5^{7}$ b) $(2^{-3})^{2}$ c) $\dfrac{(-3)^{-2}}{(-3)^{-7}}$ d) $\dfrac{12^{-3}}{12^{-3}}$

SOLUTION:

(We are showing you only one way to apply the rules of exponents to get these solutions. There are other valid ways to solve these problems.)

a) We apply the product rule to get $5^{-4} \cdot 5^{7} = 5^{-4+7} = 5^{3} = 125$.

b) Here, we apply the power rule:

$$(2^{-3})^{2} = 2^{-3 \cdot 2} = 2^{-6} = \frac{1}{2^{6}} = \frac{1}{64}.$$

c) Now, we use the quotient rule to get

$$\frac{(-3)^{-2}}{(-3)^{-7}} = (-3)^{-2-(-7)} = (-3)^{-2+7} = (-3)^{5} = -243.$$

d) We'll use the definition of a zero exponent to simplify

$$\frac{12^{-3}}{12^{-3}} = 12^{-3-(-3)} = 12^{-3+3} = 12^{0} = 1.$$

Now try Exercises 1 to 28. 24

QUIZ YOURSELF 24

Simplify:

a) $3^{8} \cdot 3^{-6}$ b) $(2^{-2})^{-1}$

Scientific Notation

KEY POINT

We use scientific notation to represent very large and very small numbers.

As we mentioned at the beginning of this section, scientists often have to deal with very large and very small numbers. In order to be able to understand and calculate with such extreme quantities, mathematicians have developed scientific notation. Before giving the formal definition, we will illustrate scientific notation with several examples.

Astronomers often measure distances in light-years, which is the distance that light travels in one year. A light-year is approximately 5,865,696,000,000 miles. To convert such a very large number to scientific notation, we keep dividing the number by 10 until the number is between 1 and 10. We do this by moving the decimal point to the left and keeping track of how many places we have moved the decimal point. In this case, we need to move the decimal point 12 places to the left (Figure 6.24(a)). Because moving the decimal point to the left makes the number smaller, we must multiply by 10^{12} to restore the size of the number (Figure 6.24(b)).

In physics, the electric charge on an electron is 0.0000000000000000016 coulomb. To write this number in scientific notation, we move the decimal point 19 places to the right to get 1.6, which, in effect, multiplies the number by 10^{19}. To restore the number to its small size, we must divide by 10^{19} or, what is equivalent, multiply by 10^{-19}. Thus, $0.00000000000000000016 = 1.6 \times 10^{-19}$.

5,865,696,000,000.

Moving the decimal point 12 places to the left makes the number smaller.

(a)

5.865696×10^{12}

Multiplying by 10^{12} restores the size of the number.

(b)

FIGURE 6.24 (a) Moving the decimal point to the left. (b) Multiplying by 10^{12}.

DEFINITION A number is written in **scientific notation** if it is of the form $a \times 10^{n}$, where $1 \le a < 10$ and n is any integer.

Remember that for a number of the form $a \times 10^n$ to be in scientific notation, the a must be equal to or greater than 1 and less than 10. The two numbers in the following diagram are not in scientific notation:

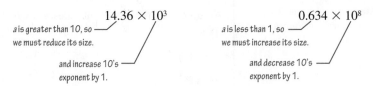

14.36×10^3 — *a* is greater than 10, so we must reduce its size, and increase 10's exponent by 1.

0.634×10^8 — *a* is less than 1, so we must increase its size, and decrease 10's exponent by 1.

We can write these numbers in scientific notation as 1.436×10^4 and 6.34×10^7.

If your calculator has the capability to express numbers in scientific notation, it probably has one of the following keys:

[EE] [EEx] [EXP]

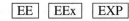

To enter 6.34×10^7 on your calculator, you might enter

[6] [.] [3] [4] [EE] [7]

Calculators also display scientific notation in different ways. One popular way to indicate 6.34×10^7 is to display 6.34 E 7. The E 7 tells us that the exponent of 10 is 7 in scientific notation. Another popular method uses a space instead of the E to display 6.34×10^7 as 6.34 7.

RULES FOR CONVERTING A DECIMAL NUMBER TO SCIENTIFIC NOTATION

If the number x is very large,

a) Make the number smaller by moving the decimal point to the left until the number is equal to or greater than 1 and less than 10.

b) Restore the size of x by multiplying the number you obtained in part a) by 10 raised to the power, which is the number of places that you moved the decimal point to the left. (If you move the decimal point three places to the left, multiply by 10^3.)

If x is very small,

c) Make the number larger by moving the decimal point to the right until the number is equal to or greater than 1 and less than 10.

d) Restore the size of x by multiplying the number you obtained in part c) by 10 raised to the power, which is the *negative* of the number of places that you moved the decimal point to the right. (If you move the decimal point eight places to the right, multiply by 10^{-8}.)

Between the Numbers

What's in a Number?

We are bombarded with numbers. We might hear that unemployment is at 8.9%, the national debt is almost $15 trillion, there is one foreclosure per 350 households, college loan debt is almost $1 trillion, the gross domestic product (the value of all goods and services produced within a country) is over $14.5 trillion, and so on. It is important when dealing with such information to try to put it in context.

For example, the amount allocated in 2010 for defense in the United States budget was $719 billion. By itself it is hard to understand what this number means. However, if we compare it to the total budget of $3.7 trillion, we see that it is slightly over 19%. So we might ask, "How does that compare with other years?" In 1970, $81 billion was spent for defense in a budget of $195 billion, or about 41.5%.

However, in this comparison we have ignored discretionary budget items, which could involve military spending. Always realize that in any discussion of this type, the presenter may be slanting the information for his or her purposes.

EXAMPLE 6 **Converting from Standard Notation to Scientific Notation**

Rewrite the numbers in the following statements in scientific notation:

a) The length of Manhattan Island is approximately 708,660 inches.

b) The MacBook Air takes 0.0000000034 of a second to perform an operation.

SOLUTION:

a) Move the decimal point in 708,660. to the left five places to get 7.08660. Now multiply this number by 10^5 to get 7.08660×10^5.

b) Move the decimal point in 0.0000000034 to the right nine places to get 3.4. Now multiply this number by 10^{-9} to get 3.4×10^{-9}.

EXAMPLE 7 **Converting from Scientific Notation to Standard Notation**

Rewrite the numbers in the following statements in standard notation:

a) The precise length of a year is 3.1556925511×10^7 seconds.

b) The diameter of an atom is about 2.0×10^{-10} meter.

SOLUTION:

a) We move the decimal point in 3.1556925511 to the right seven places to get 31,556,925.511 seconds.

b) Move the decimal point in 2.0 ten places to the left to get 0.0000000002 meter.

Now try Exercises 29 to 44. 25

QUIZ YOURSELF 25

a) Convert 53,728.41 to scientific notation.

b) Convert 2.45×10^{-3} to standard notation.

KEY POINT

We use rules of exponents to multiply and divide numbers in scientific notation.

It is easy to multiply and divide numbers written in scientific notation. For example,

$$
\begin{aligned}
&(3 \times 10^2)\cdot(4 \times 10^5) & \\
&= (3 \times 4)\cdot(10^2 \times 10^5) & \text{Regroup factors.}\\
&= 12 \times 10^7. & \text{Add exponents.}
\end{aligned}
$$

Also,

$$
\begin{aligned}
&\frac{8 \times 10^6}{2 \times 10^4} & \\
&= \frac{8}{2} \times \frac{10^6}{10^4} & \text{Multiplication of rational numbers.}\\
&= 4 \times 10^{6-4} & \text{Subtract exponents when dividing.}\\
&= 4 \times 10^2. & \text{Simplify exponent.}
\end{aligned}
$$

PROBLEM SOLVING

Strategy: The Analogies Principle

Notice that in multiplying and dividing numbers written in scientific notation, you are not doing anything new. You are simply applying properties such as associativity and commutativity, as well as the rules for working with rational numbers, that you have learned earlier.

Applications of Scientific Notation

We will now use scientific notation to solve some applied problems.

EXAMPLE 8 Calculating What You Owe on the National Debt

In 2011, the population of the United States was approximately 307 million and the national debt was about $15.097 trillion.

a) Represent both of these numbers in scientific notation.

b) Use the representations you found in part a) to calculate your share of the national debt. Then express your answer in standard notation.

SOLUTION:

a) 307 million is equal to 3.07×10^8 and 15.097 trillion is 15,097,000,000,000, which equals 1.5097×10^{13}.

b) Dividing the debt by the population, we get

$$\frac{1.5097 \times 10^{13}}{3.07 \times 10^8} = \frac{1.5097}{3.07} \times 10^{13-8} = \frac{1.5097}{3.07} \times 10^5$$
$$\approx (4.92 \times 10^{-1}) \times 10^5 = 4.92 \times 10^4.$$

We can rewrite this as $4.92 \times 10^4 = 4.92 \times 10{,}000 = 49{,}200$. So your share of the debt is about $49,200!

EXAMPLE 9 Finding the Total Length of All the Atoms in a Person's Body

If all the atoms in a person weighing 175 pounds were lined up in a straight line, how long would that line be? Compare this length with the distance to the nearest star, Proxima Centauri. Use the following information* to answer the question:

The person's body contains 3.4×10^{27} atoms.

The diameter of an atom is 2×10^{-10} meter (a meter is slightly more than 39 inches).

The distance to the nearest star, Proxima Centauri, is 4.6 light-years.

A light-year is 9.46×10^{15} meters.

SOLUTION:

We first multiply the number of atoms in the person's body by the diameter of a single atom.

$$(3.4 \times 10^{27})(2 \times 10^{-10}) = (3.4 \times 2)(10^{27} \times 10^{-10})$$
$$= 6.8 \times 10^{27+(-10)} = 6.8 \times 10^{17}$$

Thus, we find that total length of the atoms in the body is 6.8×10^{17} meters.

Next, we find the distance, in meters, to Proxima Centauri. To do this, we multiply $4.6 \times 9.46 \times 10^{15}$. This gives us

a must be between 1 and 10.

$$4.6 \times 9.46 \times 10^{15} = 43.516 \times 10^{15} = 4.3516 \times 10^{16}.$$

*This information is taken from Cesare Emiliani, *The Scientific Companion* (New York: Wiley, 1988), pp. 1–10.

c) Finally, we divide the length we found in part a) by the distance we found in part b) to get

$$\frac{6.8 \times 10^{17}}{4.3516 \times 10^{16}} = \frac{6.8}{4.3516} \times \frac{10^{17}}{10^{16}} \approx 1.56 \times 10 = 15.6$$

Thus, we have found that the total length of the atoms in the person's body is roughly 16 times the distance to the nearest star!

Generally, we will not be giving you exercises as complicated as Example 9; however, we just wanted to show you the power of using the rules of exponents and scientific notation.

EXERCISES 6.5

Sharpening Your Skills

Evaluate each expression.

1. $2 \cdot 2 \cdot 2 \cdot 2 \cdot 2$ **2.** 5^3 **3.** -2^4

4. 0^3 **5.** -3^2 **6.** 5^{-2}

7. 9^1 **8.** 3^{-4} **9.** 3°

10. $(-5)^2$

Use the rules for exponents to first rewrite each of the following expressions and then evaluate the new expression.

11. $3^2 \cdot 3^4$ **12.** $(-2)^3 \cdot (-2)^5$

13. $(7^2)^3$ **14.** $8^0 \cdot 8^2$

15. $5^4 \cdot 5^{-6}$ **16.** $4^2 \cdot 4^3$

17. $(3^2)^{-3}$ **18.** $2^{-4} \cdot 2^{-3}$

19. $(-3)^{-2}(-3)^3$ **20.** $(3^2)^4$

21. $\frac{5^9}{5^7}$ **22.** $(7^{-1})^{-3}$

23. $\frac{6^{-2}}{6^{-4}}$ **24.** $\frac{(-3)^6}{(-3)^9}$

25. $(3^{-4})^0$ **26.** $(7)^{-2} \cdot (7)^6$

27. $\frac{2^9}{4^3}$ (*Hint:* Rewrite 4.)

28. $9^3 \cdot 27^{-2}$ (*Hint:* Rewrite 9 and 27.)

Rewrite each of the following numbers in scientific notation.

29. 4,356,000 **30.** 3,200,000,000

31. 783 **32.** 0.000258

33. 0.0024 **34.** 28

35. 0.008 **36.** 8,056

Rewrite each of the following numbers in standard notation.

37. 3.25×10^4 **38.** 4.7×10^8

39. 1.78×10^{-3} **40.** 7.41×10^{-8}

41. 6.3×10^1 **42.** 9.7×10^1

43. 4.5×10^{-7} **44.** 8×10^{-7}

Each of the following numbers is not written in scientific notation. Rewrite them so that they are in scientific notation.

45. 23.81×10^6 **46.** 426.5×10^5

47. 0.84×10^3 **48.** 0.03×10^8

Use scientific notation to perform the following operations. Leave your answer in scientific notation form.

49. $(3 \times 10^6)(2 \times 10)^5$

50. $(4 \times 10^3)(2 \times 10^4)$

51. $(1.2 \times 10^{-3})(3 \times 10^5)$

52. $(4.2 \times 10^{-2})(1.83 \times 10^{-4})$

53. $(8 \times 10^{-2}) \div (2 \times 10^3)$

54. $(4.8 \times 10^4) \div (1.6 \times 10^{-3})$

55. $\frac{(5.44 \times 10^8)(2.1 \times 10^{-3})}{(3.4 \times 10^6)}$

56. $\frac{(1.4752 \times 10^{-2})(5.7 \times 10^4)}{(4.61 \times 10^{-3})}$

57. $\frac{(9.6368 \times 10^3)(4.15 \times 10^{-6})}{(1.52 \times 10^4)}$

58. $\frac{(3.5445 \times 10^{-3})(2.8 \times 10^{-5})}{(8.34 \times 10^6)}$

Rewrite each number using scientific notation before performing the operation. Leave your answer in scientific notation form.

59. $67,300,000 \times 1,200$

60. $83,600 \times 4,200,000$

61. $6,800,000 \times 2,300,000$

62. $1,750,000 \times 3,400,000$

63. 0.00016×0.0025

64. 0.000325×0.000008

Applying What You've Learned

In these exercises, use the fact that one million is 1,000,000 = 10^6, one billion is 1,000,000,000 = 10^9, and one trillion is 1,000,000, 000,000 = 10^{12}.

Express each very large or very small quantity mentioned in Exercises 65–72 in scientific notation.

65. The Smithsonian Institution estimates that at any moment there are 10,000,000,000,000,000,000 insects on Earth.

66. According to the U.S. Energy Information Administration, one kilogram (2.2 pounds) of uranium-235 releases 56,000,000,000,000 BTUs of energy.

67. The U.S. federal budget for 2010 was $3,720,000,000,000.

68. Your new MP3 player has 420,000,000,000 bytes of memory.

69. The size of one HIV virus is 0.0000001 meter (a meter is slightly longer than a yard).

70. The *E. coli* bacterium is 0.000002 meter long.

71. The diameter of the nucleus of a gold atom is 0.000000000000014 meter.

72. The diameter of a molecule of water is 0.0000000001 meter.

Use the following graphs to do Exercises 73–78 and express your answers in scientific notation.

U.S. Population Growth

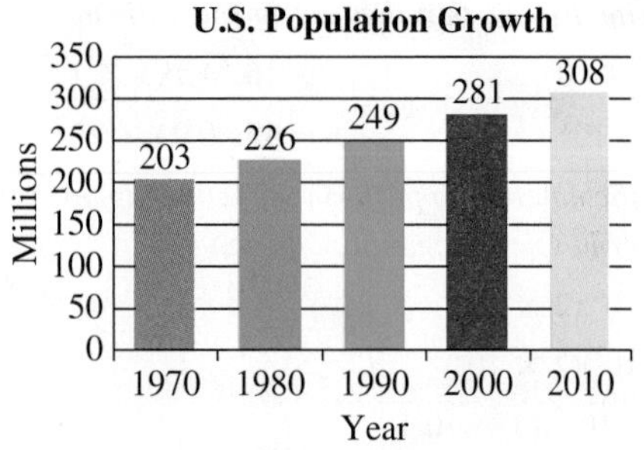

U.S. National Debt

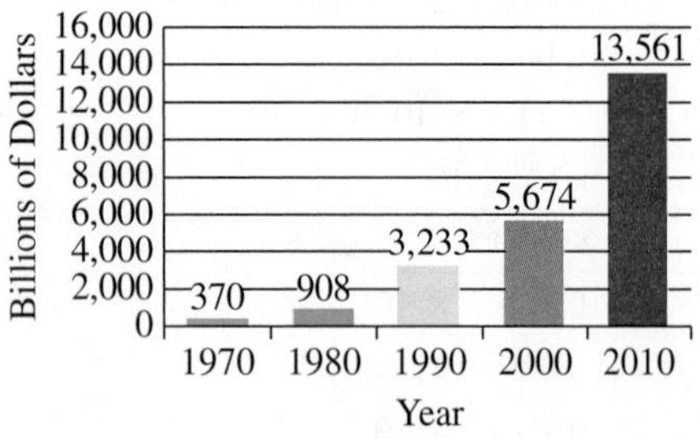

73. National debt. Calculate the amount of national debt owed by each person in the United States in 1990.

74. National debt. Repeat Exercise 73 for the year 2010.

75. Increase in U.S. population. Represent the increase in U.S. population from 1970 to 2010.

76. Increase in U.S. debt. Represent the increase in U.S. debt from 1970 to 2010.

77. Comparing populations. In 2010, the world's population was about 6,850,000,000. How many times larger was the world's population than the U.S. population?

78. Comparing populations. In 2010, China's population was 1.33 billion. How many times larger was China's population than the U.S. population?

79. Comparing weights. How many times heavier is a 130-pound person than a mosquito? Assume that there are 454 grams to a pound and that a mosquito weighs 1×10^{-3} gram.

a. First find the number of grams that the person weighs.

b. Divide the number that you found in part (a) by 1×10^{-3}.

c. Interpret your answer in part (b).

80. Comparing astronomical distances. The distance from Earth to the sun (on average) is about 390.76 times the distance from Earth to the moon, which is 238 thousand miles. What is the average distance from Earth to the sun expressed in scientific notation?

81. Land per person in the United States. By 2005, the U.S. population had grown to approximately 296 million, and the land area of the United States was about 3.5×10^6 square miles. How many people per square mile were there in the United States in 2005?

82. Comparing population density. Alaska has a population of 698 thousand and a land area of 570,665 square miles and New Jersey has a population of 8,707 thousand with a land area of 7,354 square miles (*Statistical Abstract of the United States*). How many more people per square mile are there in New Jersey than Alaska? Express your answer in scientific notation.

83. In 2007, the U.S. population was 301 million and the country's defense budget was $439.3 billion. How much was spent on defense for each person in the United States?

84. Convert 1 million seconds to days. (*Hint:* There are $60 \times 60 \times 24$ seconds in a day.)

85. Convert 1 billion seconds to years.

86. On July 4, 2006, how many seconds old was the United States? (Ignore leap years.)

87. Health care spending. By 2016, it is estimated that health care spending will be $4.1 trillion, or $12,800 per resident. Use this information to calculate the estimated U.S. population in 2016.

88. Space flight.* The sun is about 93,000,000 miles away. If a rocket ship travels from Earth to the sun at approximately 25,000 miles per hour, how many hours will the flight take?

*In dealing with astronomical distances, we have simplified these problems in the sense that we are assuming the distances are straight-line distances. In reality, a spaceship would not fly in a straight line through space. However, to take these factors into account would complicate the problem.

89. Space flight. In 1977, scientists sent the spaceship Voyager II to Neptune, which is about 2.8 billion miles away. If the spacecraft averaged about 25000 miles per hour, how long did the journey take?

90. Comparing astronomical distances. Earth is 93 million miles from the sun and Pluto is 3.6 billion miles from the sun. How many more times is Pluto's distance from the sun than Earth's?

91. Calculations for an animated movie. Computer-generated animations require tremendous amounts of computing power. A complex scene can require thousands of processors to spend hours of computer time just to produce $\frac{1}{24}$ of a second of an animated movie. Suppose that a complex frame requires 8.424×10^{11} calculations. If a bank of 10,000 processors are each doing 2.6 million calculations per second, how long will it take to generate that frame?

92. Calculations for an animated movie. Assuming the information in Exercise 91, how many calculations would be required to produce *Toy Story 3,* which runs for 103 minutes?

93. Calculating a light-year. Assuming that light travels 186,000 miles per second and there are 365 days in a year, how many miles will light travel in a year?

94. Astronomical distances. The space shuttle orbits Earth at about 18,000 miles per hour. How many years would it take the space shuttle to travel one light-year (the distance light travels in a year)?

Communicating Mathematics

95. Explain why you must rewrite the following numbers if they are to be in scientific notation. Then, rewrite them properly in scientific notation.

a. 125.436×10^3 **b.** 0.537×10^8

96. Explain the difference in evaluating -2^4 and $(-2)^4$. How does that affect your answer when calculating these expressions?

97. Provide a counterexample to show that $(x + y)^2 = x^2 + y^2$ is not true. How should you evaluate the left side of the equation versus the right side of the equation?

98. What are the advantages of using scientific notation?

Between the Numbers

99. Medicare spending. In 1970, the federal budget was \$195.6 billion, of which \$6.2 billion was spent on Medicare. By 2010, these numbers had risen to \$3.72 trillion and \$451.6 billion, respectively. Express the increase in Medicare spending from 1970 to 2010 in scientific notation.

100. Medicare spending. Continuing Exercise 99, what percent of the federal budget was spent on Medicare in 1970 and in 2010?

Challenge Yourself

101. Does $\left(x^m\right)^n = x^{(m^n)}$? Consider simple examples.

102. Look at simple examples to complete the following rule of exponents. $\left(\frac{a}{b}\right)^n =$

103. The length of one billion dollars. The one-dollar bill is 6.14 inches long. If a billion dollar bills were placed end-to-end, how many miles long would that string of bills be?

104. In 2010 the national debt was \$13.786 trillion. If 13.786 trillion one-dollar bills were placed end-to-end, could that string of bills stretch from Earth to the moon? From Earth to the sun? (You will need to research these distances.)

105. Comparing computer memory. In the late 1970s the Radio Shack TRS80 personal computer had 48K—48,000—bytes of memory and cost \$600. In the fall of 2011, you could buy an iPhone 4S with 64 gigabytes—64 billion—bytes of memory for \$400. Ignoring the rest of the hardware and using these figures, calculate how many times greater was computer memory's cost in 1970 than in 2011.

106. Calculating energy usage. The British thermal unit is the amount of energy needed to raise the temperature of one pound of water by one degree Fahrenheit. If the United States uses one quadrillion BTUs of energy every 3.7 days, how many BTUs of energy does the United States use in one 365-day year?

Looking Deeper

6.6 Sequences

Objectives

1. Understand arithmetic sequences.
2. Understand the difference between geometric and arithmetic sequences.
3. Recognize applications of the Fibonacci sequence.

Congratulations. When you signed your contract for your new job, you agreed to a monthly salary of \$2,750 per month; however, today your supervisor gave you a pleasant

surprise. During your first year, you will receive a salary increase of \$50 each month. Because your salary will be different every month, you wonder what your total income will be for the next year. You could get out your calculator and add the amounts

$$2750, 2800, 2850, 2900, \ldots,$$

but, as you will soon see, there is a faster way to do this.

In spite of your good news, you don't seem to be feeling so well. In fact, several days ago, you picked up a single virus that has been doubling ever since. The following list of numbers describes the growth of the virus in your body:

$$1, 2, 4, 8, 16, 32, 64, \ldots.$$

You have just had a good experience and a bad experience with lists of numbers called sequences. We will study the patterns found in several different types of sequences in this section.

> **DEFINITION** A **sequence** is a list of numbers that follows some rule or pattern. The numbers in the list are called the **terms** of the sequence. We often name the terms $a_1, a_2, a_3, a_4, a_5, \ldots$. We read this list "*a* sub one, *a* sub two, *a* sub three," and so on.

In the sequence of monthly salaries, we get each successive term by adding 50 to the previous number. In this sequence

$$a_1 = 2{,}750,$$
$$a_2 = a_1 + 50 = 2{,}750 + 50 = 2{,}800,$$
$$a_3 = a_2 + 50 = 2{,}800 + 50 = 2{,}850, \text{ and so forth.}$$

In the sequence describing the virus growth, we get successive terms by doubling the previous terms. Here,

$$a_1 = 1,$$
$$a_2 = 2 \cdot a_1 = 2 \cdot 1 = 2,$$
$$a_3 = 2 \cdot a_2 = 2 \cdot 2 = 4,$$
$$a_4 = 2 \cdot a_3 = 2 \cdot 4 = 8, \text{ and so on.}$$

Arithmetic Sequences

The salary sequence is an example of an arithmetic sequence.

KEY POINT

Terms in an arithmetic sequence differ by a fixed constant.

> **DEFINITION** An **arithmetic sequence** is a sequence in which each term after the first term differs from the preceding term by a fixed constant amount called the **common difference** of the sequence.

EXAMPLE 1 Writing Terms of Arithmetic Sequences

Find the common difference for the following arithmetic sequences. Then write the next three terms of the sequence.

a) 3, 7, 11, 15, ... b) 5, 3, 1, −1, ...

SOLUTION:

a) We can find the common difference of an arithmetic sequence by subtracting the first term from the second term, so the common difference is $7 - 3 = 4$. The next three terms of the sequence are $15 + 4 = 19$, $19 + 4 = 23$, and $23 + 4 = 27$.

b) The common difference is $3 - 5 = -2$. Therefore, the next three terms of the sequence are -3, -5, and -7.

In order to get more insight as to how arithmetic sequences behave, we will rewrite sequence a) in Example 1 as follows:

1st	2nd	3rd	4th	n^{th}
3,	7,	11,	15, . . .	
$3 + 0 \cdot 4$,	$3 + 1 \cdot 4$,	$3 + 2 \cdot 4$,	$3 + 3 \cdot 4, \ldots$	$3 + (n - 1) \cdot 4$

From this we see that in the first term, we have added 4 *zero* times; in the second term, we have added 4 *one* time; in the third term, we have added 4 *two* times; and so on. In general, in each term we have added one less 4 than the number of the term. From this pattern, you can see that the sixteenth term would have fifteen 4s added, or would equal $3 + 15 \cdot 4 = 63$. This leads us to the following formula for the nth term of an arithmetic sequence:

FORMULA 1 The nth Term of an Arithmetic Sequence. The nth term of an arithmetic sequence with first term a_1 and common difference d is

$$a_n = a_1 + (n - 1)d.$$

EXAMPLE 2 Finding the nth Term of Arithmetic Sequences

a) Find the 12th term of the arithmetic sequence whose first term is 2,750 and whose common difference is 50.

b) Find the 20th term of the arithmetic sequence whose first term is 5 and whose common difference is -2.

SOLUTION:

a) Here, $n = 12$, $a_1 = 2{,}750$, and $d = 50$. Using Formula 1, we have

$$a_{12} = a_1 + (n - 1)d = 2{,}750 + (12 - 1)50 = 2{,}750 + 11 \cdot 50 = 3{,}300.$$

b) Now, $n = 20$, $a_1 = 5$ and $d = -2$. Using Formula 1, we have

$$a_{20} = a_1 + (n - 1)d = 5 + (20 - 1)(-2)$$
$$= 5 + 19(-2) = 5 - 38 = -33.$$

26

QUIZ YOURSELF 26

Find the 16th term in the arithmetic sequence 5, 8, 11, 14, ...

As in the salary example, we sometimes want to add the first n terms in an arithmetic sequence. If we look carefully at a concrete example, we will see a pattern that leads us to a general formula for finding such sums.

Suppose that we wish to find the sum of the first six terms of the arithmetic sequence 2, 5, 8, 11, 14, 17, ... Rather than finding this sum directly, we will look at it in a slightly different way. Consider the sum $2 + 5 + 8 + 11 + 14 + 17$ and the same sum written in reverse order as $17 + 14 + 11 + 8 + 5 + 2$. If we place one sum under the other and add, we get

$$\begin{array}{r} 2 + 5 + 8 + 11 + 14 + 17 \\ +17 + 14 + 11 + 8 + 5 + 2. \\ \hline 19 + 19 + 19 + 19 + 19 + 19 \end{array}$$

What you are seeing is that we got the sum, 19, which is $a_1 + a_6$, six times. Therefore, because we have added every term twice, $6(a_1 + a_6)$ is exactly twice the sum that we originally wanted. Therefore, $\frac{6(a_1 + a_6)}{2}$ is the sum we want. This example leads us to the following formula:

FORMULA 2 The Sum of the First n Terms of an Arithmetic Sequence. The sum of the first n terms of an arithmetic sequence is given by

$$\frac{n(a_1 + a_n)}{2}.$$

EXAMPLE 3 Summing Terms in an Arithmetic Sequence

Find the sum of the first 20 terms in the arithmetic sequence 3, 7, 11, 15, ...

SOLUTION:

We will use Formula 2, $\frac{n(a_1 + a_n)}{2}$, but to do so, we first need to find a_{20}. We know that $a_1 = 3$ and $d = 4$, so using Formula 1 for finding the nth term of a sequence, we get

$$a_{20} = a_1 + (20 - 1)d = 3 + 19 \cdot 4 = 3 + 76 = 79.$$

Now we can now apply Formula 2. We know that $a_1 = 3$, $a_{20} = 79$, and $n = 20$. Therefore, the sum is

$$\frac{20(3 + 79)}{2} = \frac{20(82)}{2} = 820.$$

Now try Exercises 13 to 20.

We can now answer the question we posed at the beginning of this section.

EXAMPLE 4 Finding the Yearly Total of an Arithmetic Sequence of Monthly Salaries

Assume that you receive \$2,750 as a salary this month, and for the next 11 months, you receive a monthly raise of \$50. What is the total salary that you receive over the 12 months?

SOLUTION:

The monthly salaries form an arithmetic sequence with the first term 2,750 and the monthly difference of $d = 50$. In Example 2, part a), we found that for this sequence, $a_{12} = 3{,}300$. So, using Formula 2, the sum of your salaries over 12 months is

$$\frac{n(a_1 + a_2)}{2} = \frac{12(\overset{a_1}{2{,}750} + \overset{a_{12}}{3{,}300})}{2} = \frac{\overset{6}{\cancel{12}}(2{,}750 + 3{,}300)}{\cancel{2}} = 6 \cdot 6{,}050 = 36{,}300.$$

27

QUIZ YOURSELF 27

Find the sum of the first 16 terms in the sequence 5, 8, 11, 14, ...

Geometric Sequences

KEY POINT

We multiply by a fixed constant to get new terms in a geometric sequence.

You have seen that we generate each new term in an arithmetic sequence by adding the same fixed constant to the previous term. We generate each new term in a geometric sequence by multiplying the previous term by the same fixed constant. The virus sequence 1, 2, 4, 8, 16, 32, . . . that we discussed earlier is an example of a geometric sequence because to get each new term, we multiply the previous term by 2.

> DEFINITION A **geometric sequence** is a sequence in which each term, after the first term, is a nonzero constant multiple of the preceding term. The constant by which we are multiplying is called the **common ratio.**

EXAMPLE 5 Writing Terms in a Geometric Sequence

Find the common ratio of each of the following geometric sequences. Then write the next three terms of the sequence.

a) 2, 6, 18, 54, . . . b) 3, −6, 12, −24, . . .

SOLUTION:

a) We can find the common ratio of a geometric sequence by dividing the second term of the sequence by the first term. So, the common ratio of this sequence is $\frac{6}{2} = 3$. The next three terms of the sequence are

$$3 \cdot 54 = 162, 3 \cdot 162 = 486, \text{ and } 3 \cdot 486 = 1{,}458.$$

b) The common ratio is $-\frac{6}{3} = -2$. The next three terms of the sequence are

$$(-2) \cdot (-24) = 48, (-2) \cdot 48 = -96, \text{ and } (-2) \cdot (-96) = 192.$$

To understand how geometric sequences behave, we will rewrite sequence a) in Example 5 as follows:

1st	2nd	3rd	4th	n^{th}
2,	6,	18,	54, . . .	
$2 \cdot 3^0$,	$2 \cdot 3^1$,	$2 \cdot 3^2$,	$2 \cdot 3^3$, . . .	$2 \cdot 3^{(n-1)}$

From this we see that in the first term, we have multiplied 2 by 3^0; in the second term, we have multiplied 2 by 3^1; in the third term, we have multiplied 2 by 3^2; and so on. In general, in each term we have multiplied 2 by one less 3 than the number of the term. In general, in the nth term there are $n - 1$ factors that are 3. This gives us the following formula:

FORMULA 3 The *n*th Term of a Geometric Sequence. The nth term of a geometric sequence with common ratio r is

$$a_n = a_1 \cdot r^{n-1}.$$

We will use this formula in Example 6.

EXAMPLE 6 Finding the *n*th Term of Geometric Sequences

a) Find the eighth term of the geometric sequence whose first term is 5 and whose common ratio is 3.

b) Find the sixth term of the geometric sequence whose first term is 3 and whose common ratio is -2.

c) Find the number of viruses in your body after 18 hours if initially you have 1 and the virus doubles every hour.

SOLUTION:

a) Here $n = 8$, $a_1 = 5$, and $r = 3$. Using Formula 3, we have

$$a_8 = a_1 \cdot r^{8-1} = 5 \cdot 3^7 = 10{,}935.$$

b) Now, $n = 6$, $a_1 = 3$, and $r = -2$. Again using Formula 3, we have

$$a_6 = a_1 \cdot r^{6-1} = 3 \cdot (-2)^5 = 3 \cdot (-32) = -96.$$

c) In this case, $n = 18$, $a_1 = 1$, and $r = 2$, so

$$a_{18} = a_1 \cdot r^{18-1} = 1 \cdot (2)^{17} = 1 \cdot 131{,}072 = 131{,}072.$$

So, there are 131,072 copies of the virus in your body after 18 hours.

Now try Exercises 21 to 26. 28

QUIZ YOURSELF 28

Find the twelfth term in the sequence 2, 6, 18, 54, ...

Math in Your Life

If It Sounds Too Good to Be True . . .

You have perhaps received a letter assuring you of riches that goes something like this:

> *This letter is not a joke. It really works! Follow these instructions carefully and do not break the chain. Send one dollar to the person who is first on the list, cross off that person's name and add your own name to the bottom of the list. Then send the new list to five people asking them to follow these same instructions. If you do not break the chain, after several weeks, you will receive over 1 million dollars in the mail.*

At first, this plan seems like it might work, provided that no one breaks the chain. In step 1, you send five letters, then each of these five people sends five more letters for 25 or 5^2 letters. At the next stage, these 25 people send five letters each to get $125 = 5^3$ letters. If the list has 15 names, it would seem that by the time your name gets to the top of the list, you would be on easy street.

The problem, however, is that 5, 25, 125, 625, . . . is a geometric sequence with $a_1 = 5$ and $r = 5$. If the original list has 15 names, by the time your name reaches the top of the list and you start cashing in, $a_{15} = 5(5^{14}) = 5^{15}$. However, 5^{15} is roughly 30 billion, which is about five times the size of the whole world's population! Even if no one breaks the chain, you will probably never get to see any money because we run out of people before your name ever gets to the top of the list.

Schemes such as chain letters that rely on geometric growth, or "pyramid," cannot work and are usually illegal.

Example 7 shows the dramatic difference between arithmetic and geometric growth.

EXAMPLE 7 Comparing Arithmetic and Geometric Growth

Suppose that you have won a lottery and can choose one of the following options:

a) You can receive \$20,000 this month, \$30,000 next month, \$40,000 the following month, and so forth for 30 months. Calculate the total payments.

b) The lottery agency will not give you any money now, but will set aside 1 cent in an account and double it to 2 cents next month, double it again to 4 cents the following month, and continue doubling it for 30 months and then give you the amount in the account at that time. Calculate the final amount.

c) Which option is the better deal? (Before doing the calculations, decide which option you would choose.)

SOLUTION:

a) With this option, we are summing the first 30 terms of an arithmetic sequence in which $a_1 = 20{,}000$, $d = 10{,}000$, and $n = 30$. To use Formula 2, we need to know the 30th term, a_{30}. Using Formula 1, we get

$$a_{30} = a_1 + (n-1)d = 20{,}000 + (30-1)10{,}000$$
$$= 20{,}000 + 290{,}000 = 310{,}000.$$

Now, we use Formula 2 to get the sum of the 30 monthly payments as follows:

$$\frac{n(a_1 + a_n)}{2} = \frac{30(20{,}000 + 310{,}000)}{2} = \frac{30 \times 330{,}000}{2} = 4{,}950{,}000.$$

So, with option a) you would receive a total of \$4,950,000.

b) In this option, we are finding the *n*th term of a geometric sequence with $a_1 = 1$, $r = 2$, and $n = 30$. We use Formula 3 to get $a_{30} = 1 \cdot 2^{30-1} = 2^{29} = 536{,}870{,}912$ cents, which is \$5,368,709.12.

c) You will get more than \$400,000 extra by choosing option b).

Example 7 shows how remarkably fast quantities can grow if we are using a geometric sequence versus an arithmetic sequence.

The Fibonacci Sequence

KEY POINT

We add consecutive pairs of terms to generate terms in a Fibonacci sequence.

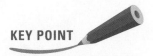

In 1202, Leonardo of Pisa, also known as Fibonacci, wrote a book, *Liber Abaci*, in which he introduced the Hindu-Arabic numeration system to Europe. His intention was to explain this new system in order to replace the more tedious Roman system of numeration for doing calculations. He included in his text the following problem, which has given rise to perhaps the most famous sequence in the history of mathematics.

EXAMPLE 8 The Fibonacci Sequence

Assume that every pair of baby rabbits will mature during their second month and produce a pair of baby rabbits during their third month. If we place a single pair of adult rabbits in a pen, how many rabbits (both babies and adults) will there be in the pen in the eighth month? We are assuming that once adults start producing babies, they do so every month from then on.

SOLUTION:

Figure 6.25 will help you understand the problem.

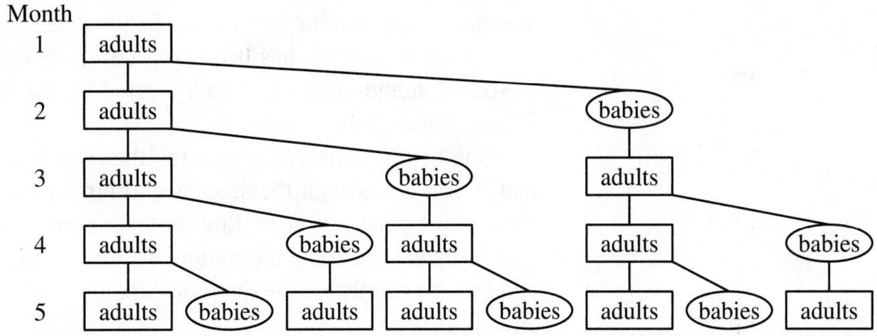

FIGURE 6.25 The growth of rabbits leads to the Fibonacci sequence.

QUIZ YOURSELF 29

Write the next two terms that follow the term 89 in the Fibonacci sequence.

If we count the pairs of rabbits each month, we get the following sequence: 1, 2, 3, 5, 8, ... The pattern that is beginning to emerge is that from the third term on, each term in the sequence is the sum of the previous two terms. Study Figure 6.25 carefully to see why this must be so. If we continue this pattern, there will be $5 + 8 = 13$ pairs in the sixth month, $8 + 13 = 21$ pairs in the seventh month, and $13 + 21 = 34$ pairs in the eighth month. 29

The pattern that we saw in Example 8 is essentially the Fibonacci sequence defined below.

DEFINITION The **Fibonacci sequence** is the sequence that begins

$$1, 1, 2, 3, 5, 8, 13, 21, 34, 55, 89, \ldots,$$

where each term after the second is the sum of the previous two terms. We often label the terms in the Fibonacci sequence by $F_1, F_2, F_3, F_4, \ldots$ rather than $a_1, a_2, a_3, a_4, \ldots$

The Fibonacci sequence occurs in nature in many different ways. The seeds of some flowers, for example, daisies and sunflowers, are arranged in spirals going in two different directions. If you count the number of spirals going in one direction and then count the number of spirals going in the other direction, you will often find that the numbers you get are a pair of consecutive numbers in the Fibonacci sequence, such as 21 and 34, or 34 and 55 (see Figure 6.26(a)). The hexagonal "bumps" on the skin of a pineapple and the woody leaf-like structures on a pine cone also occur in two types of spirals (see Figure 6.26(b)).

(a)

(b)

FIGURE 6.26 The Fibonacci sequence appears in (a) the spirals in a sunflower and (b) the patterns on a pineapple.

$\frac{1}{1} = 1$
$\frac{2}{1} = 2$
$\frac{3}{2} = 1.5$
$\frac{5}{3} = 1.667$
$\frac{8}{5} = 1.6$
$\frac{13}{8} = 1.625$
$\frac{21}{13} = 1.615$
$\frac{34}{21} = 1.619$
$\frac{55}{34} = 1.618$
$\frac{89}{55} = 1.618$
$\frac{144}{89} = 1.618$
$\frac{233}{144} = 1.618$

TABLE 6.4 Quotients obtained by dividing consecutive Fibonacci numbers.

Again, if you count the spirals, you will get a pair of Fibonacci numbers. Later in this section, we will show you how the Fibonacci sequence describes the pattern in the shell of the chambered nautilus (see Figure 6.29(b)).

So much research has been done on the Fibonacci sequence that there is a Fibonacci Association and even a research journal called *The Fibonacci Quarterly* devoted to publishing research on its properties.

One of the remarkable patterns that was discovered about the Fibonacci sequence is that if you go through the sequence, dividing each term by the term preceding it, you get the numbers that we see in Table 6.4.* It seems as though these numbers are settling down and going toward some fixed number. And, in fact, they are. It was proved about 200 years ago that the number these quotients are approaching is $\frac{\sqrt{5}+1}{2}$, which we will call ϕ (phi), and approximate it by 1.618.

The number ϕ, which we discussed briefly in Section 6.4, is often called the **golden ratio** and was known to the ancient Greeks, although they obtained it in a different way. They believed that any rectangle whose one side was $\frac{\sqrt{5}+1}{2}$ times as long as its other side was perfectly proportioned, and they used this principle in art and architecture. Such a rectangle is called a **golden rectangle.** For example, they used golden rectangles in designing the Parthenon (Figure 6.27(a)). In modern times, architects used golden rectangles to determine the rectangular shapes in the sides of the United Nations building (Figure 6.27(b)). The pleasing shape of the golden rectangle is also found in art and in sculpture from ancient to modern times.

(a)

(b)

FIGURE 6.27 The architectures of (a) the Parthenon and (b) the United Nations building are based on the golden ratio.

*We used a spreadsheet to generate these numbers, and in many cases we are approximating the true quotient, which has an infinite, repeating expansion.

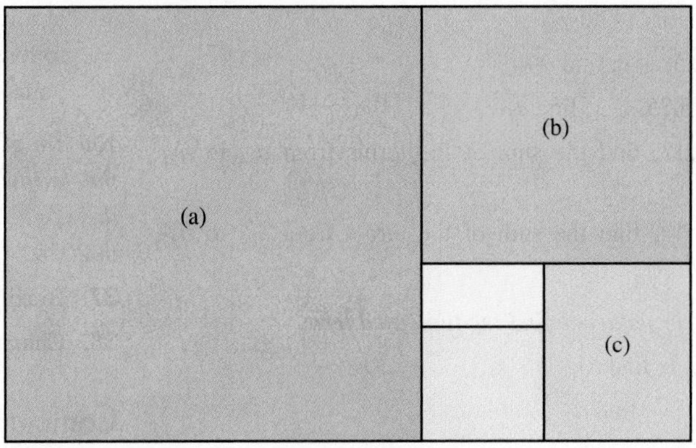

FIGURE 6.28 Cutting off squares from a golden rectangle gives other golden rectangles.

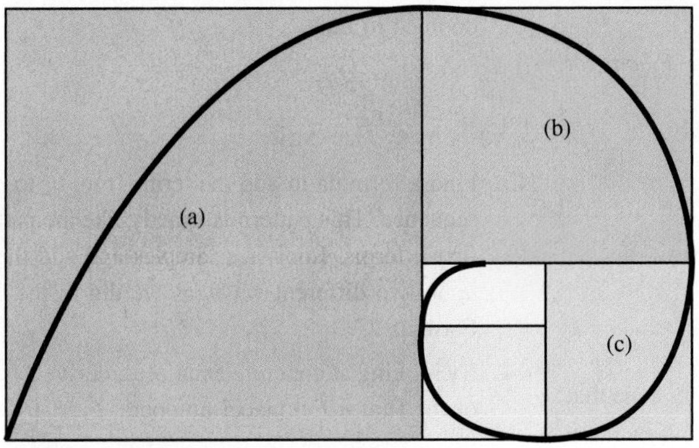

(a)

(b)

FIGURE 6.29 (a) Spiral generated from golden rectangles. (b) Chambered nautilus shell.

An interesting property of a golden rectangle is that if you cut off square (a) on one end of the rectangle, as we show in Figure 6.28, the smaller rectangle that remains is again a golden rectangle. If you then cut off square (b), again the part that remains is a golden rectangle. If you continue doing this, and connect the opposite diagonals of the squares that you are cutting off with a smooth curve, you generate a spiral that is much like the shape of a chambered nautilus shell (see Figure 6.29).

EXERCISES 6.6

Sharpening Your Skills

In Exercises 1–12, identify the sequence as either arithmetic or geometric. List the next two terms of each sequence.

1. 5, 8, 11, 14, ...

2. 11, 7, 3, −1, ...

3. 8, 24, 72, 216, ...

4. 4, 12, 20, 28, ...

5. $1, \frac{1}{2}, \frac{1}{4}, \frac{1}{8}, \ldots$

6. 0.1, 0.01, 0.001, 0.0001, ...

7. 10, 5, 0, −5, ...

8. 4, 17, 30, 43, ...

9. 2, 4, 8, 16, ...

10. $1, \frac{1}{3}, \frac{1}{9}, \frac{1}{27}, \ldots$

11. 1.5, 2.0, 2.5, 3.0, ...

12. 1, −1, 1, −1, ...

For each arithmetic sequence find the specified term a_n and then find the sum of the terms from a_1 to a_n, inclusive.

13. 5, 8, 11, 14, ... ; find a_{11}.

14. 11, 17, 23, 29, ... ; find a_9.

15. 2, 8, 14, 20, ... ; find a_{15}.

16. $-6, -2, 2, 6, \ldots$; find a_{22}.

17. $1, 1.5, 2.0, 2.5, \ldots$; find a_{20}.

18. $3, 3.25, 3.5, 3.75, \ldots$; find a_{11}.

19. In Exercise 13, find the sum of the terms from a_{14} to a_{21}, inclusive.

20. In Exercise 14, find the sum of the terms from a_{21} to a_{37}, inclusive.

For each geometric sequence find the indicated term.

21. $1, 3, 9, 27, \ldots$; find a_{11}.

22. $3, 6, 12, 24, \ldots$; find a_9.

23. $1, \frac{1}{2}, \frac{1}{4}, \frac{1}{8}, \ldots$; find a_7.

24. $2, -4, 8, -16, \ldots$; find a_{10}

25. $2, 0.2, 0.02, 0.002, \ldots$; find a_6.

26. $5, 50, 500, 5{,}000, \ldots$; find a_9.

In Exercises 27–30, we give you two terms in the Fibonacci sequence, and you must find the specified term.

27. $F_{11} = 89$ and $F_{13} = 233$. Find F_{12}.

28. $F_{22} = 17{,}711$ and $F_{23} = 28{,}657$. Find F_{21}.

29. $F_{13} = 233$ and $F_{15} = 610$. Find F_{14}.

30. $F_{23} = 28{,}657$ and $F_{24} = 46{,}368$. Find F_{25}.

Applying What You've Learned

31. **Finding total salary.** Redo Example 4, but now assume that your starting salary is $2,875 and you are getting an increase of $35 per month. Find the amount that you will earn in a year.

32. **Displaying products.** There is a pyramid of cans against the wall of a supermarket. There are 11 cans on the first row, 10 on the second row, 9 on the third row, and so on. What is the total number of cans on the stack?

33. **Growth in a bank account.** You have $1,200 in your ING direct account that is paying an interest rate of 3.5% (0.035) per year. How much money will you have in that account in 6 years? (*Note:* $r = 1.035$)

34. **Chain letters.** Reconsider the discussion of chain letters in Math in Your Life on page 290. Assume that you receive a chain letter and at each stage, the person who receives the letter must send the letter to eight people. If you send your letters and the chain is unbroken, at what stage will the number of letters sent at that stage exceed the population of the world, which in 2008 was about 6.7 billion people?

35. **Height of a bouncing ball.** A ball is dropped from a height of 8 feet. The height of the bounce is always 7/8 of the distance the ball has dropped. What will be the height of the fifth bounce?

36. **Gambling strategy.** A surefire way to never lose when betting at a casino game is to use the following strategy: If you lose, then always double your bet and play again until you win. Assume that you have $2,000. If you are gambling at a roulette wheel and you begin with a $3 bet and continue to lose, at what stage will you be unable to continue doubling your bet?

You are given the population and the growth rate as of 2011 for the following countries. Assume that the growth rate remains the same from year to year. Determine the size of the population in 2020.

37. Brazil: population = 195 million; growth rate = 1.134%

38. China: population = 1,338 million; growth rate = 0.493%

Communicating Mathematics

39. Make up an example to verify Formula 1.

40. Make up an example to verify Formula 2.

41. Make up an example to verify Formula 3.

42. Explain, by referring to sequences, why chain letters are doomed to fail.

Challenge Yourself

43. Find a formula to add the terms from a_k to a_n in an arithmetic sequence. This pattern is exactly like the pattern for adding the first n terms. Look at examples and add the terms from a_k to a_n in two different ways, as we did in the discussion prior to Formula 2.

44. By looking at concrete examples, derive a formula for the sum of the first n Fibonacci numbers, $F_1 + F_2 + F_3 + \cdots + F_n$.

45. By looking at concrete examples, derive a formula for the sum of the squares for two consecutive Fibonacci numbers, F_n and F_{n+1}.

In Exercises 46 and 47, express the requested term of a Fibonacci-like sequence in terms of the first two terms. For example, the seventh term of the sequence 3, 4, 7, 11, 18, 29, 47, 76, 123, ... can be written as $47 = 5 \times 3 + 8 \times 4$.

46. $6, 7, 13, 20, 33, 53, \ldots$; 8th term

47. $3, 8, 11, 19, 30, 49, 79, \ldots$; 10th term

48. If $a, b, a + b, \ldots$ is a Fibonacci-like sequence, express the nth term in terms of a and b.

The Binet form of Fibonacci numbers uses the formula

$$F_n \approx \frac{\left(\frac{1+\sqrt{5}}{2}\right)^n + \left(\frac{1-\sqrt{5}}{2}\right)^n}{\sqrt{5}}.$$

For example, to calculate F_8, *substitute* 8 *for* n *in the previous formula. Using a calculator, due to roundoff errors, the author got* 21.01904 *for* F_8, *whose value is actually* 21. *Use this formula and a calculator to approximate the Fibonacci numbers in Exercises 49–52.*

49. F_5

50. F_7

51. F_9

52. F_{11}

CHAPTER SUMMARY

Section	Summary	Example
SECTION 6.1	The numbers that we use for counting, 1, 2, 3, . . . , are called the **natural numbers,** or **counting numbers.** The study of the natural numbers is called **number theory.** If a and b are natural numbers, then we say a **divides** b and we write $a\|b$, if there is a natural number q such that $b = q \cdot a$. We also say that a is a **divisor** of b, a is a **factor** of b, or b is a **multiple** of a.	Discussion, p. 233
	A natural number, greater than 1, that has only itself and 1 as factors is called a **prime number.** A natural number greater than 1 that is not prime is called **composite.**	Definition, p. 234
	We can use the **Sieve of Eratosthenes** to find prime numbers.	Example 1, p. 234
	There are simple ways to test a number for **divisibility** by 2, 3, 4, 5, 8, 9, and 10. We can use divisibility tests and **factor trees** to find **the prime factorization** of numbers. The **fundamental theorem of arithmetic** tells us that, except for order, this factorization will be unique.	Table 6.1, p. 236 Example 3, pp. 236–237 Example 4, pp. 237–238
	The **greatest common divisor** (**GCD**) of two natural numbers is the largest natural number that divides both numbers. The **least common multiple** (**LCM**) of two natural numbers is the smallest natural number that is a multiple of both numbers. We use prime factorization to find the GCD and LCM of two numbers.	Example 5, p. 239 Example 6, p. 240
	We use the GCD and LCM to solve applied problems.	Example 8, pp. 241–242
SECTION 6.2	The set of **whole numbers** is the set $\{0, 1, 2, 3, \ldots\}$. The set of **integers** is the set $\{\ldots, -3, -2, -1, 0, 1, 2, 3, \ldots\}$.	Definition, p. 245
	We explain **addition rules** for integers by considering movement on a number line.	Example 1, p. 246
	Subtraction is a subsidiary operation to addition. If a and b are integers, then we define $a - b$ as $a + (-b)$.	Definition, p. 247
	We use a **money model** to explain integer multiplication. If a and b have the *same* sign, then $a \cdot b$ is *positive.* If a and b have *opposite* signs, then $a \cdot b$ is *negative.*	Definition, pp. 247–248 Example 2, p. 248
	Division is a subsidiary operation to multiplication. If a, b, and c are integers, then $\frac{a}{b} = c$ means that $a = b \cdot c$.	Definition, p. 248
	If a and b have the *same* sign, then $\frac{a}{b}$ is *positive.* If a and b have *opposite* signs, then $\frac{a}{b}$ is *negative.*	Example 3, p. 249
SECTION 6.3	The set of **rational numbers, Q,** is the set of all numbers that can be written in the form $\frac{a}{b}$, where a and b are both integers and $b \neq 0$. The top number a is called the **numerator** and the bottom number b is called the **denominator.**	Discussion, p. 253
	Two rational numbers $\frac{a}{b}$ and $\frac{c}{d}$ are **equal** if and only if $a \cdot d = b \cdot c$. We can also show two rational numbers to be equal by **canceling common factors** from the numerator and the denominator, that is, $\frac{a \cdot c}{b \cdot c} = \frac{a}{b}$. When all common factors have been canceled, the rational number is in **lowest terms.**	Example 1, p. 254 Example 2, p. 255
	We **add** and **subtract** rational numbers as follows: $$\frac{a}{b} + \frac{c}{d} = \frac{a \cdot d + b \cdot c}{b \cdot d} \quad \text{and} \quad \frac{a}{b} - \frac{c}{d} = \frac{a \cdot d - b \cdot c}{b \cdot d}$$	Example 3, p. 256
	We **multiply** and **divide** rational numbers as follows: $$\frac{a}{b} \cdot \frac{c}{d} = \frac{a \cdot c}{b \cdot d} \quad \text{and} \quad \frac{\frac{a}{b}}{\frac{c}{d}} = \frac{a}{b} \cdot \frac{d}{c}$$	Example 4, p. 257 Example 5, p. 258

(*Continued*)

Section	Summary	Example
	To convert an **improper fraction** $\frac{a}{b}$ to a **mixed number,** divide a by b to get the quotient q and remainder r. Then, $$\frac{a}{b} = q\frac{r}{b}.$$	Example 7, p. 260
	The mixed number $q\frac{r}{b}$ equals the improper fraction $\frac{q \cdot b + r}{b}$.	Example 8, p. 260
	Rational numbers have **repeating decimal expansions.**	Example 9, pp. 260–261
SECTION 6.4	A number that is not a rational number is called an **irrational number** and has a *nonrepeating* decimal expansion.	Discussion, p. 266
	If $a \geq 0$ and $b \geq 0$, then	Discussion, p. 267
	$$\sqrt{ab} = \sqrt{a}\sqrt{b} \text{ and } \sqrt{\frac{a}{b}} = \frac{\sqrt{a}}{\sqrt{b}}.$$	Example 2, p. 268
	The real numbers have the following properties. For any real numbers a, b, and c:	Discussion, p. 270
	Closure: $a + b$ and $a \cdot b$ are real numbers.	Table 6.3, p. 271
	Commutativity: $a + b = b + a$ and $a \cdot b = b \cdot a$.	Examples 6 and 7,
	Associativity: $(a + b) + c = a + (b + c)$ and $(a \cdot b) \cdot c = a \cdot (b \cdot c)$.	pp. 271–272
	Identity element: $a + 0 = 0 + a$ and $a \cdot 1 = 1 \cdot a = a$.	
	Inverse property: $a + (-a) = (-a) + a = 0$ and if $a \neq 0$, then $a \cdot \frac{1}{a} = \frac{1}{a} \cdot a = 1$.	
	Distributivity: $a \cdot (b + c) = a \cdot b + a \cdot c$.	
SECTION 6.5	If a is any real number and n is a counting number, then	Definition, p. 276
	$$a^n = \underbrace{a \cdot a \cdot a \cdot \cdots \cdot a}_{\text{product of } n\ a\text{'s}}$$	Example 1, p. 276
	If x is a real number and m and n are natural numbers, then	
	$$x^m x^n = x^{m+n}$$	Example 2, pp. 276–277
	$$\left(x^m\right)^n = x^{m \cdot n}$$	Example 3, p. 277
	$$\frac{x^m}{x^n} = x^{m-n}, x \neq 0$$	Example 4, p. 278
	$$x^0 = 1, \text{ and if } x \neq 0, \text{ then } x^{-n} = \frac{1}{x^n}$$	
	We write numbers in **scientific notation** in the form $a \times 10^n$, where $1 \leq a < 10$. We use the rules of exponents to calculate with numbers written in scientific notation.	Definition, p. 279 Examples 6 and 7, p. 281
	We can use scientific notation to solve applied problems involving **very large** or **very small** numbers.	Examples 8 and 9, pp. 282–283
SECTION 6.6	A **sequence** is a list of numbers, called **terms,** that follow some pattern. We often name the terms $a_1, a_2, a_3, a_4, \ldots$.	Definition, p. 286
	An **arithmetic sequence** is a sequence in which any two consecutive terms differ by a fixed constant called the **common difference.**	Definition, p. 286 Example 1, p. 286
	The ***n*th term** of an arithmetic sequence with first term a_1 and common difference d is $a_n = a_1 + (n - 1)d$.	Example 2, p. 287
	The **sum of the first *n* terms** of an arithmetic sequence is $\frac{n(a_1 + a_n)}{2}$.	Examples 3 and 4, p. 288
	A **geometric sequence** is a sequence in which each term (except the first) is a nonzero constant multiple of the preceding term. The constant that we multiply by is called the **common ratio.**	Definition, p. 288 Example 5, pp. 288–289

Section	Summary	Example
	The **nth term** of a geometric sequence with first term a_1 and common ratio r is $a_n = a_1 \cdot r^{(n-1)}$.	Example 6, p. 289
	The **Fibonacci sequence** is the sequence that begins $1, 1, 2, 3, 5, 8, 13, 21, 34, 55, 89, \ldots,$ where each term after the second is the sum of the two previous terms. We label the terms of a Fibonacci sequence by $F_1, F_2, F_3, F_4, \ldots$.	Definition, p. 291
	As n gets larger, the quotient $\frac{F_{n+1}}{F_n}$ approaches ϕ, the **golden ratio,** which is approximately 1.618.	Discussion, p. 292

CHAPTER REVIEW EXERCISES

Section 6.1

1. Use the Sieve of Eratosthenes to find all prime numbers between 70 and 90.
2. Estimate the square root of 180 by finding a natural number a such that $a < \sqrt{180} < a + 1$.
3. Are 191 and 441 prime or composite? If the number is composite, factor it into a product of primes.
4. Check 1,080,036 for divisibility by 3, 4, 5, 6, 8, 9, and 10.
5. Find the LCM and GCD of 1,584 and 1,320.
6. Explain how you use prime factorization to get the GCD and the LCM of two natural numbers.

Section 6.2

7. Is 2,772 divisible by 9? 4? 8?
8. How can you use a calculator to determine if 11 divides 2,585?
9. Find the following:

   a. $-5 + 14$ **b.** $-13 - (-24)$

   c. $(-12)(-6)$ **d.** $\frac{48}{-3}$
10. Use a money example to illustrate how to compute the product $(-4)(+3)$.
11. Use the definition of division in terms of multiplication to compute $\frac{-24}{-8}$.
12. If the high temperature in northern Alaska is 17 degrees Fahrenheit at the beginning of October, and at the beginning of November it is -3 degrees Fahrenheit, what is the difference in these two temperatures?
13. Use the definition of division in terms of multiplication to explain why we cannot divide 5 by 0.
14. Simplify $\left(\frac{-3 + 9}{-4 - 2}\right) \cdot \left(\frac{12 - (-4)}{-7 - 1}\right)$.
15. If stock for Pineapple computers currently sells for \$43 a share but has lost \$6 for each of the past 4 days, what was the value of the stock 4 days ago?

Section 6.3

16. Are the numbers $\frac{28}{65}$ and $\frac{14}{35}$ equal? Explain your answer.
17. Perform the following computations:

    a. $\frac{4}{9} \cdot \left(\frac{3}{4} - \frac{1}{3}\right)$ **b.** $\frac{3}{7} \div \left(\frac{2}{3} + \frac{3}{14}\right)$
18. Convert $\frac{53}{8}$ to a mixed number.
19. Find $\left(3\frac{1}{2}\right)\left(4\frac{1}{7}\right)$.
20. Write the following numbers as quotients of integers:

    a. 0.375 **b.** $0.\overline{63}$
21. You have purchased a $1\frac{1}{2}$-pound Vanilla Bean cheesecake from the Cheesecake Factory. If the cake is to be divided among 12 people, how much will each person's piece weigh?
22. If a recipe for a hot chili sauce that serves 16 requires $2\frac{1}{2}$ cups of hot peppers and $3\frac{1}{4}$ cups of tomatoes, how much peppers and tomatoes would be required to make an amount of sauce to serve 24 people?
23. Explain by giving a specific example to show why we invert and multiply when dividing rational numbers.

Section 6.4

24. How is the decimal expansion of an irrational number different from that of a rational number?
25. Simplify the following:

    a. $\sqrt{108}$

    b. $\frac{\sqrt{15}}{\sqrt{5}}$ (Rationalize the denominator, if necessary.)
26. Does every real number have a multiplicative inverse? Explain.
27. Give an example to show that division of rational numbers is not associative.

Section 6.5

28. Evaluate each expression:

a. $3^6 \cdot 3^{-2}$ **b.** $(2^4)^{-2}$ **c.** $\frac{6^8}{6^5}$ **d.** $\frac{8^{-6}}{8^{-4}}$

29. Explain the difference between evaluating -2^4 and $(-2)^4$.

30. Write 0.000456 and 1,230,000 in scientific notation.

31. Rewrite 1.325×10^6 and 8.63×10^{-5} in standard notation.

32. Simplify $\frac{(3.6 \times 10^3)(2.8 \times 10^{-5})}{(4.2 \times 10^4)}$ and write the answer in scientific notation.

33. In a recent year, the world's population was 6.6 billion and Mexico's population was 109 million. How many times larger was the world's population than Mexico's population? Express your answer in scientific notation.

Section 6.6

34. Identify the following sequences as arithmetic or geometric and give the next two terms of the sequence:

a. 6, −12, 24, −48, 96, . . .

b. 11, 16, 21, 26, 31, . . .

35. Consider the arithmetic sequence 4, 7, 10, 13, 16, ...

a. What is the 30th term of this sequence?

b. What is the sum of the first 30 terms of this sequence?

36. Consider the geometric sequence 4, 12, 36, 108, ... What is the 10th term of this sequence?

37. If $F_{12} = 144$ and $F_{13} = 233$ are two terms in the Fibonacci sequence, what are F_{10} and F_{15}?

CHAPTER TEST

1. Use the Sieve of Eratosthenes to find all prime numbers between 100 and 120.

2. Estimate the square root of 200 by finding a natural number a such that $a < \sqrt{200} < a + 1$.

3. Are the numbers 241 and 539 prime or composite? If the number is composite, factor it into a product of primes.

4. Check 2,542,128 for divisibility by 3, 4, 5, 6, 8, and 9.

5. Find the GCD and LCM of 1,716 and 936.

6. If $a = 2^3 3^9 5^3 7^2$ and $b = 2^8 3^6 5^2 7^9$, what is the GCD of a and b? The LCM?

7. Find the following:

a. $-15 - (-9)$ **b.** $-8 + 11$

c. $(-18)(+3)$ **d.** $\frac{-56}{-7}$.

8. Identify the following sequences as arithmetic or geometric and give the next two terms of the sequence:

a. 12, 15, 18, 21, 24, . . .

b. 6, 18, 54, 162, . . .

9. Use a money example to illustrate how to compute the product $(+7)(-5)$.

10. Use the definition of division in terms of multiplication to compute $\frac{-21}{3}$.

11. If the average daily temperature in Alaska is 50 degrees Fahrenheit for July and −19 degrees Fahrenheit for December, what is the difference in these two temperatures?

12. Use the definition of division in terms of multiplication to explain why we cannot divide 8 by 0.

13. If $F_{16} = 987$ and $F_{17} = 1{,}597$ are two terms in the Fibonacci sequence, what are F_{14} and F_{20}?

14. Are the numbers $\frac{198}{213}$ and $\frac{68}{72}$ equal? Explain your answer.

15. Perform the following computations:

a. $\frac{3}{5} \cdot \left(\frac{7}{9} - \frac{3}{4}\right)$ **b.** $\frac{4}{5} \div \left(\frac{2}{5} - \frac{1}{4}\right)$

16. Convert $\frac{47}{3}$ to a mixed number.

17. Find $4\frac{1}{2} \div 3\frac{3}{8}$

18. Write the following numbers as a quotient of integers:

a. 0.573 **b.** $0.\overline{57}$

19. Consider the arithmetic sequence 11, 20, 29, 38, 47, ...

a. What is the 20th term of this sequence?

b. What is the sum of the first 20 terms of this sequence?

20. If a digital photo that measures $4\frac{1}{2}$ inches by $6\frac{1}{3}$ inches is being increased to make a poster by enlarging it $5\frac{1}{4}$ times, what will be the dimensions of the poster?

21. Explain by giving a specific example to show why we invert and multiply when dividing rational numbers.

22. What is the difference between the decimal expansions of a rational number and an irrational number?

23. Simplify the following:

a. $\sqrt{180}$ **b.** $\frac{3}{\sqrt{15}}$

24. Give an example to show that the division of rational numbers is not associative.

25. Evaluate each expression:

a. $2^7 \cdot 2^{-4}$ **b.** $(3^2)^{-2}$ **c.** $\frac{5^7}{5^4}$ **d.** $\frac{3^{-2}}{3^{-5}}$

26. Consider the geometric sequence −2, 6, −18, 54, ... What is the 10th term of this sequence?

27. Explain the difference between evaluating $(-3)^2$ and -3^2.

28. Write 15,460,000 and 0.00000000623 in scientific notation.

29. In a recent year, the population of China was approximately 1,322,000,000, whereas the population of the United States was roughly 304,000,000. How many times larger was China's population than the population of the United States?

GROUP PROJECTS

1. **An abstract system.** The following table defines an abstract operation represented by the symbol $*$. This operation has no real meaning, but behaves as follows: If you choose any row, say the A row, and any column, say the B column, then the result of the operation is found in the intersection of that row and column. So we see in the table that $A * B = D$. Similarly, $C * P = A$. Use this table to answer the following questions:

*	D	I	A	C	B	P
D	D	I	A	C	B	P
I	I	D	P	B	C	A
A	A	P	B	I	D	I
C	C	B	I	D	P	A
B	B	C	D	P	A	C
P	P	A	I	B	C	D

   a. What is $C * I$? $D * B$?

   b. Is $A * B = B * A$? Is $C * D = D * C$?

   c. Is $*$ commutative? (*Hint:* Recall the Always Principle.)

   d. What is the identity element for $*$?

   e. What is B's inverse?

2. **Investigating data trends.**

   a. Do investigations similar to Exercises 99 and 100 in Section 6.5. Group members should consult various sources such as almanacs, the *Statistical Abstract of the United States*, and Internet Web sites. It is interesting to compare spending for defense versus education, or health and human services versus international assistance. What about trends in transportation spending or science research? Different sources will make different emphases depending on their point of view.

   b. Gather the information in part (a) over a period of time to recognize trends and present your information graphically, perhaps using the chart-making capability in products such as Microsoft Word.

USING TECHNOLOGY

1. **Generating sequences in Excel.** Below is an example of an Excel spreadsheet that generates the terms and the sum of the terms for the sequence 3, 7, 11, 15, 19, ... Write similar spreadsheets to generate the terms and the sum of the terms for the following sequences:

   a. 3, 9, 15, 21, 27, 33, . . .   **b.** 2, 6, 18, 54, 162, 486, . . .   **c.** 1, 1, 2, 3, 5, 8, 13, . . .

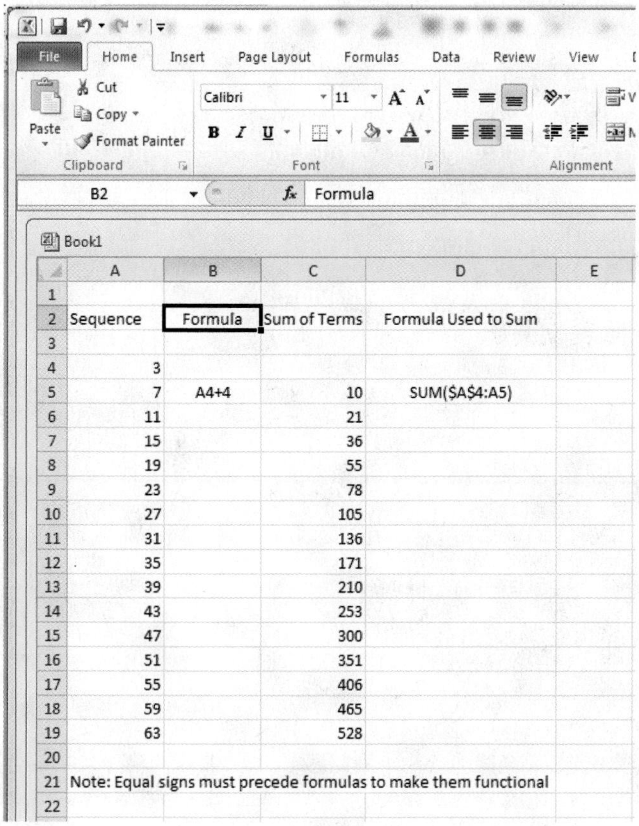

2. The Fibonacci sequence is related to a number of different concepts in mathematics. Do research on the Internet to find connections between the Fibonacci sequence and each of the following and give an example illustrating the connection:

   a. Pascal's triangle

   b. Lucas numbers

3. **Integer sequences.** Many interesting sequences can be found at http://oeis.org/selfsimilar.html, which is the site for the Online Encyclopedia of Integer Sequences. Find some sequences that interest you and present them to your class.

7

Algebraic Models

How Do We Approximate Reality?

We've all seen it. A car speeds down a track, then violently smashes into a concrete wall. Windows shatter, air bags puff out, and crash dummies lurch forward while sensors record data to be used in designing safer cars to protect you and me in real accidents. Recently, however, TV commercials have featured human-shaped clouds of lights—virtual crash dummies—riding in invisible cars, being whipped about in simulated crashes. These mathematical models of the human body—with 2,000,000 data points—can gather almost 17,000 times the information collected by using traditional crash dummies.

Scientists build mathematical models to design not only your car, but also the roads and bridges you drive over. Traffic engineers model the timing of the traffic lights on your way to class, and insurance companies' financial models help determine your auto insurance rates. In fact, before most products and services are ever created, researchers build models to predict their behavior.

(continued)

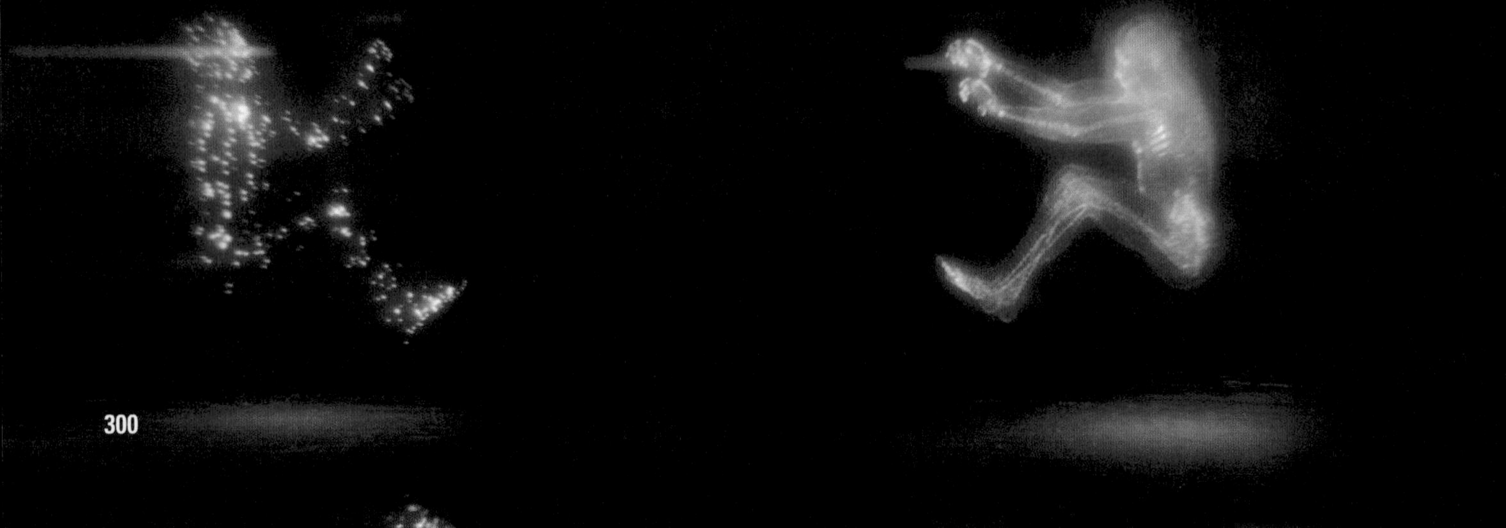

In this chapter, we will model various real-world problems with equations such as the exponential models in Section 7.4 that explain how hard it is to pay off your student loan or credit card debt.

By the end of this chapter, you will have developed an appreciation of how models are used and when they are appropriate and when they are not.

7.1

Linear Equations

Objectives

1. Solve linear equations.
2. Use intercepts to graph linear equations.
3. Apply the slope–intercept form of a line in solving problems.

A state legislator investigating the rise in college tuition might wish to compare data on increasing costs at four-year versus two-year colleges. Or, as a consumer, you may want to compare the costs of plans offered by several cell phone providers. In this section you will study linear equations, which, although quite simple, can effectively model such situations.

Solving Linear Equations

KEY POINT

Linear equations have a constant rate of change between the variables.

In spite of its simplicity, a linear equation often provides us with a reasonably good model for a complex set of data.

DEFINITION A **linear equation** in two variables is an equation that can be written in the form

$$Ax + By = C,$$

where A, B, and C are real numbers and A and B are not both zero. When a linear equation is written in this form, it is in **standard form.**

The equations $-3x + y = 6$ and $5p = 6q - 4$ are examples of linear equations in two variables. In a linear equation in two variables, there is a constant rate of change between the two variables. To understand what we mean by this, let's rewrite the first equation in the form $y = 6 + 3x$. Notice that if we increase x by a certain amount, then y increases by three times as much. For example, if we increase x from 10 to 15, which is an increase of 5, then y increases from 36 to 51, which is an increase of 15. If we increase x from 1,000 to 1,007, then y increases from 3,006 to 3,027, which is an increase of 21.

A **solution** for an equation is a number or numbers such that if we substitute them for the variables in the equation, the resulting statement is true. Solutions of linear equations are ordered pairs of numbers. For example, the **ordered pair** (3, 15)* is a solution for the equation $-3x + y = 6$ because if we substitute 3 for x and 15 for y, we get the true statement $-3(3) + 15 = 6$. Two equations are **equivalent** if they have the same solutions. In solving an equation, we often use the following rules to rewrite it in a simpler, equivalent form. These rules apply to all types of equations.

---

*In the pair (3, 15), the number 3 is called the *first coordinate* and the number 15 is called the *second coordinate.* These coordinates are also called the x-coordinate and y-coordinate, respectively.

REWRITING EQUATIONS IN AN EQUIVALENT FORM

1. You can add or subtract the same expression from both sides of an equation to get an equivalent equation.
2. You can multiply or divide both sides of an equation by the same *nonzero* expression to get an equivalent equation.

We will apply some of these rules in Example 1 to solve for one variable in terms of another—a technique that we often use in solving applied problems.

EXAMPLE 1 Rewriting Equations in Equivalent Forms

a) Solve the equation $6x + 4y = 12$ for y.

b) The equation $F = \frac{9}{5}C + 32$ converts the temperature in degrees Celsius, C, to the temperature in degrees Fahrenheit, F. Solve this equation for C.

SOLUTION:

a) "Solve the equation for y" means that we want the equation in the form $y = \ldots$, so we perform the following steps:

$$6x + 4y = 12 \quad \text{Original equation}$$

$$4y = 12 - 6x \quad \text{Subtract } 6x \text{ from both sides.}$$

$$y = \frac{12 - 6x}{4} \quad \text{Divide both sides by 4.}$$

$$y = \frac{12}{4} - \frac{6}{4}x \quad \text{Rewrite using distributivity.}$$

$$y = 3 - \frac{3}{2}x. \quad \text{Simplify.}$$

b) We can solve for C by rewriting this equation in the following series of steps:

$$F = \frac{9}{5}C + 32 \quad \text{Original equation}$$

$$F - 32 = \frac{9}{5}C \quad \text{Subtract 32 from both sides.}$$

$$\frac{5}{9}(F - 32) = \frac{5}{9}\cdot\frac{9}{5}C \quad \text{Multiply both sides by } \frac{5}{9}.$$

$$\frac{5}{9}(F - 32) = C \quad \text{Simplify.}$$

We can now use this equation to convert degrees Fahrenheit to degrees Celsius. For example, a nice, warm 95-degree-Fahrenheit day at the beach converts to a chilly sounding $C = \frac{5}{9}(95 - 32) = \frac{5}{9}(63) = 35$ degrees Celsius.

Now try Exercises 9 to 20. 1

QUIZ YOURSELF 1

a) Solve the equation $6x + 4y = 12$ for x.
b) Solve the equation $P = 2L + 2W$ for W.

Historical Highlight

René Descartes and Analytical Geometry

René Descartes was born in France in 1596. Because of his poor health, his early teachers allowed him the luxury of remaining in bed as late as he pleased. He attributed his later success to this habit because it gave him time to think and reflect on philosophy and mathematics. In his famous work *Discourse on Method*, Descartes developed analytical geometry, in which we represent points, lines, circles, and planes as numbers and equations.* The advantage that analytical geometry has over traditional geometry is that we can use algebraic methods such as solving equations to solve geometric problems.

Descartes also contributed to many other areas of science. His work in biology was so important that some have called him the "father of modern biology." He made significant discoveries in several areas of physics, and he is credited with being the first person to satisfactorily explain the rainbow.

Unfortunately, this genius met with an untimely end while tutoring Queen Christine of Sweden, who insisted on rising at 5 o'clock to study philosophy with him. The early hour and the severe Swedish winter were too much for Descartes—he caught pneumonia and died on February 11, 1650.

SOME GOOD ADVICE

When solving equations, you may have been told that "when a quantity is taken from one side of an equation to the other, plus signs change to minus signs, and vice versa." It is unwise to rely on memory devices such as this. Our advice for working with equations can be traced back to the early Greeks, who had rules such as "equals added to equals give equals." Remember when rewriting equations, you are *adding*, *subtracting*, *multiplying*, or *dividing* both sides of the equation by the same quantity.

If you were to plot several solutions for the linear equation $-3x + y = 6$, such as (3, 15), (−2, 0), (1, 9), and (2, 12), you would find that they all seem to lie on a line, as we see in the graph below.

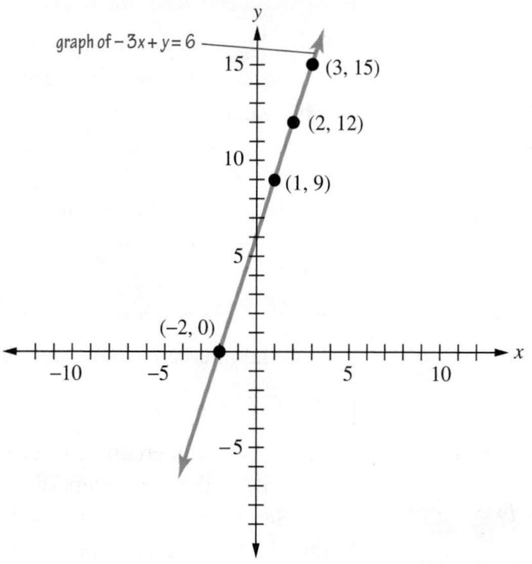

What you see here is no coincidence; in general, we have the following result.

THE GRAPH OF A LINEAR EQUATION The graph of a linear equation is always a straight line.

*As he lay in bed watching a fly walk on the ceiling, Descartes realized that he could describe the fly's position by the pair of numbers telling its distance from two perpendicular walls.

Intercepts

KEY POINT

We use intercepts to graph a linear equation.

The graph of a linear equation in two variables is a straight line, so to graph a linear equation, all you have to do is plot two points and then draw the line through those points. The graph of a linear equation gives us a visual way of understanding the relationship between the two variables in a model. Often, the two simplest points that you can use to graph a linear equation are called the x- and y-intercepts.

> **DEFINITIONS** The ***x*-intercept** of the graph of a linear equation is the point where the graph crosses the x-axis. The ***y*-intercept** is the point where the graph crosses the y-axis. (See Figure 7.1.)

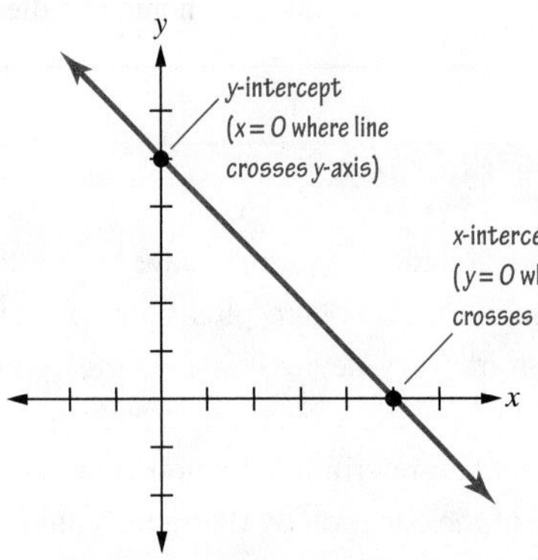

FIGURE 7.1 Plotting x- and y-intercepts.

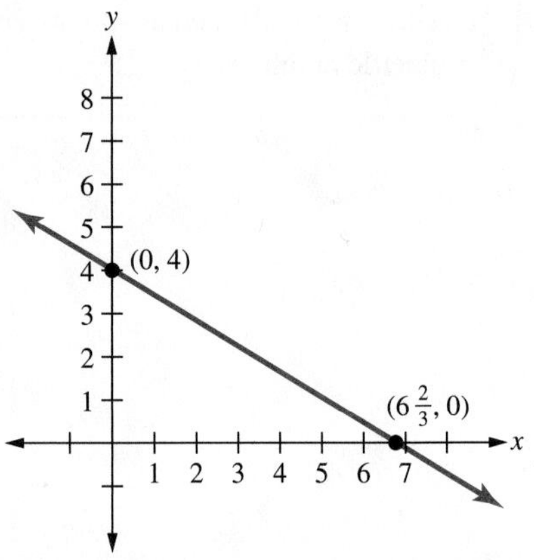

FIGURE 7.2 Using intercepts to graph $3x + 5y = 20$.

SOME GOOD ADVICE

You may mistakenly try to calculate the x-intercept by setting $x = 0$. Keep in mind that the term "x-intercept" means the point where the line "intercepts" or crosses the x-axis, so the y-coordinate of the intersection point is 0.

In Example 2, we use simple algebra to find intercepts.

EXAMPLE 2 Using Intercepts to Graph a Linear Equation

Find the intercepts and graph the equation $3x + 5y = 20$.

SOLUTION:

To find the x-intercept, we set $y = 0$. This gives us $3x + 5(0) = 20$, or $3x = 20$. Thus, $x = 6\frac{2}{3}$, so the x-intercept of the line is the point $\left(6\frac{2}{3}, 0\right)$.

Setting $x = 0$, we get $3(0) + 5y = 20$. This simplifies to $5y = 20$, which means that the y-intercept is the point (0, 4). We graph this line in Figure 7.2.

Now try Exercises 21 to 32. **2**

QUIZ YOURSELF 2

Find the intercepts and graph the equation $4x + 3y = 18$.

EXAMPLE 3 Interpreting Intercepts in a Financial Situation

Suppose you have saved \$2,205 to use for living expenses next semester that you expect to be \$315 per month. Express this information in an equation using the variables r (for remaining cash) and m (for the number of months you are spending the money). Interpret the meaning of the intercepts for the graph of this equation.

SOLUTION: Review the Choose Good Names Strategy on page 5.

You are starting with \$2,205, and each month you will reduce this amount by \$315. Therefore, the equation

$$r = \underbrace{2{,}205}_{\text{original amount}} - \underbrace{315m}_{\text{monthly expense}}$$

describes your financial state at the end of m months. The r-intercept corresponds to the time when $m = 0$; that is,

$$r = 2{,}205 - 315(\underbrace{0}_{\text{no months have elapsed}}) = 2{,}205$$

At this point, you still have all your money.

To find the m intercept, we set $r = 0$. This is the time at which you run out of savings. To find this time, we will solve the equation

$$\underbrace{0}_{\text{no remaining cash}} = 2{,}205 - 315m.$$

Adding $315m$ to each side of the equation, we get $315m = 2{,}205$, and then, dividing both sides by 315, we obtain $m = 7$. Therefore, you can expect to deplete your savings at the end of 7 months.

When using linear equations as models, we often want to know the steepness of an equation's graph. An economist may want to know how rapidly inflation is affecting the cost of a college education or, perhaps during a drought, government officials would want to know how rapidly a town's water supply is being depleted. We measure the steepness of a line by its slope.

DEFINITION If (x_1, y_1) and (x_2, y_2) are two points on a line and $x_1 \neq x_2$, then the **slope** of the line is defined as

$$m = \frac{\text{rise}}{\text{run}} = \frac{\text{change in } y}{\text{change in } x} = \frac{y_2 - y_1}{x_2 - x_1}.$$

Highlight

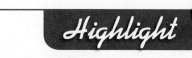

Using Technology to Draw Graphs

Many equations are much too complicated for you to get an accurate graph by merely plotting a few points. Fortunately, technology such as graphing calculators can make your job easier. We used a graphing calculator to get the graph of $y = \dfrac{8x}{\sqrt{x^4 + 10}}$ shown in Figure 7.3.

If you had been graphing this equation by hand, you might have accidentally chosen to plot the three points that we have selected in Figure 7.3 and been completely fooled as to what the graph looked like. You might ask then, "How many points should I plot?" That is not an easy question to answer and the truth is that mathematicians who work with such equations do not plot points, but use areas of mathematics beyond algebra to understand their complexity.

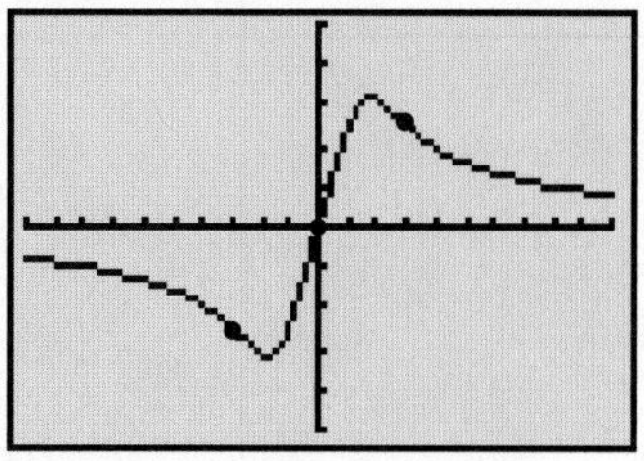

FIGURE 7.3 Graph of a complicated equation drawn with a graphing calculator.

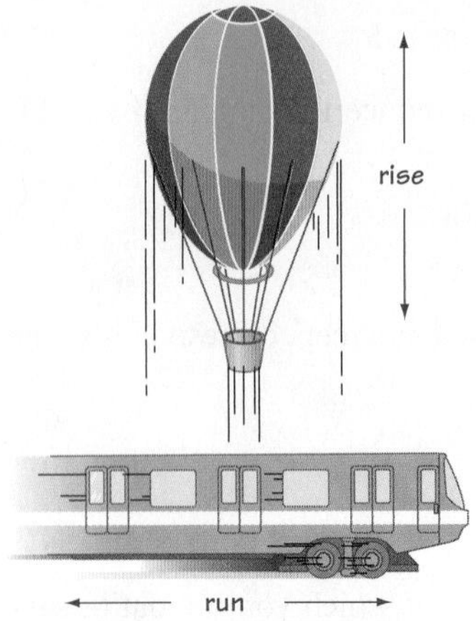

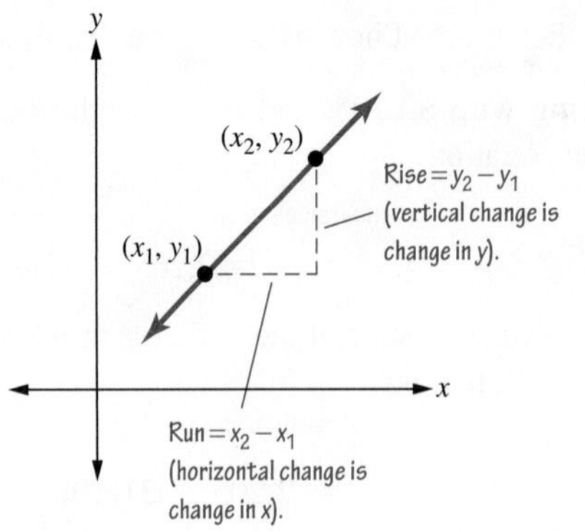

FIGURE 7.4 The rise and run determine a line's slope.

You can remember the meaning of the words *rise* and *run* by thinking about their use in everyday language. We watch the sun rise and see a cake or a hot air balloon rising. In each case something is going up, so *rise* corresponds to a vertical change. For the word *run*, think of a person running across a field or a train running down a track. In these cases, *run* means a horizontal change. Figure 7.4 illustrates the relationship between slope, rise, and run.

EXAMPLE 4 Finding the Slope of a Line

Calculate the slope of the line containing the points (1, 5) and (6, 45).

SOLUTION:

The slope of the line is

$$\text{slope} = \frac{\text{rise}}{\text{run}} = \frac{\text{change in } y}{\text{change in } x} = \frac{45 - 5}{6 - 1} = \frac{40}{5} = 8.$$

Now try Exercises 37 to 42. 3

QUIZ YOURSELF 3

Find the slope of the line through the points (7, 5) and (11, 10).

Some are worried that there may not be enough young workers to fund programs such as Social Security in the future. Example 5 illustrates how we can use slopes to understand this concern.

EXAMPLE 5 Using Slope to Compare Population Data

This graph shows some changes in the U.S. population from 2000 to 2009. Find the slope of the line segment representing the change in

a) the number of 20- to 29-year-olds.
b) the number of 60- to 69-year-olds.

What do these data suggest?

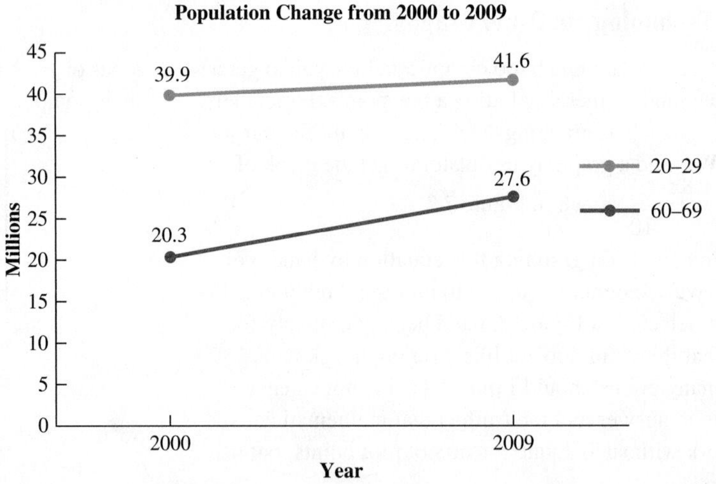

SOLUTION:

a) The line segment representing the 20- to 29-year-olds contains the points (2000, 39.9) and (2009, 41.6), so its slope is

$$\frac{\text{rise}}{\text{run}} = \frac{41.6 - 39.9}{2009 - 2000} = \frac{1.7}{9} \approx 0.19.$$

b) The segment for the 60- to 69-year-olds contains the points (2000, 20.3) and (2009, 27.6), so its slope is

$$\frac{\text{rise}}{\text{run}} = \frac{27.6 - 20.3}{2009 - 2000} = \frac{7.3}{9} \approx 0.81.$$

Perhaps concern about funding programs such as Social Security is justified, because comparing these slopes we see that the rate of increase of the 60- to 69-year-olds is over four times the rate of increase in the 20- to 29-year-olds over this nine-year period.

Now try Exercises 57 to 64.

SOME GOOD ADVICE

In the definition of slope, if the run is equal to 1, then the slope is simply $y_2 - y_1$. This means that we can interpret the slope of a line as the amount of change in y corresponding to a change of 1 in x.

By estimating the rise and the run (including whether it is positive or negative), we can get a good idea of the slope of the line without doing any calculations. The line in Figure 7.5(a) has a positive slope because both rise and run are positive. The line in Figure 7.5(b) has negative slope because the run is positive but the rise is negative.

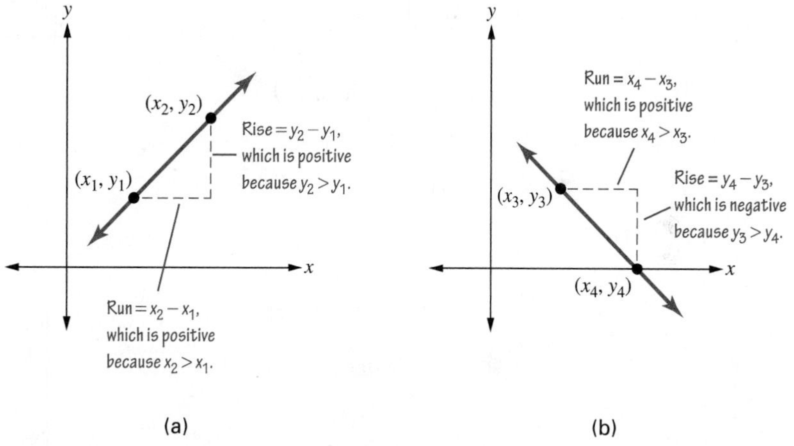

FIGURE 7.5 (a) A line with a positive slope rises from left to right. (b) A line with a negative slope falls from left to right.

Notice that if a line is horizontal, then its rise is zero, so slope $= \frac{\text{rise}}{\text{run}} = \frac{0}{\text{run}}$, which is equal to zero. Similarly, if a line is vertical, then slope $= \frac{\text{rise}}{\text{run}} = \frac{\text{rise}}{0}$, which is undefined because we cannot divide by zero.

The Slope–Intercept Form of a Line

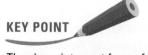

KEY POINT

The slope–intercept form of a linear equation tells us geometric information about its graph.

DEFINITION A linear equation is in **slope–intercept form** if it is written in the form $y = mx + b$. The number m is the slope of the line that is the graph of the equation and $(0, b)$ is the y-intercept.

The equation $y = 2x + 3$ is in slope–intercept form. The slope is 2, and the y-intercept is (0, 3). Examples 6 and 7 show that we can recognize useful geometric information about a line if we write its equation in slope–intercept form.

EXAMPLE 6 Graphing a Linear Equation in Slope–Intercept Form

Graph the linear equation $y = 3x + 4$.

SOLUTION:

We can read off the y-intercept of the graph immediately; because

$$y = 3x + \underbrace{4}_{y\text{-intercept}},$$

we see that the y-intercept is the point (0, 4). We can easily find another point to graph the line if we let $x = 2$. (Although you can choose any value for x that you want, you will get a better graph if you separate the points a little bit.) This gives us

$$y = 3(\overset{x}{2}) + 4 = 6 + 4 = 10,$$

so (2, 10) is a second point on the graph. We then plot the points (0, 4) and (2, 10) to draw the graph as in Figure 7.6.

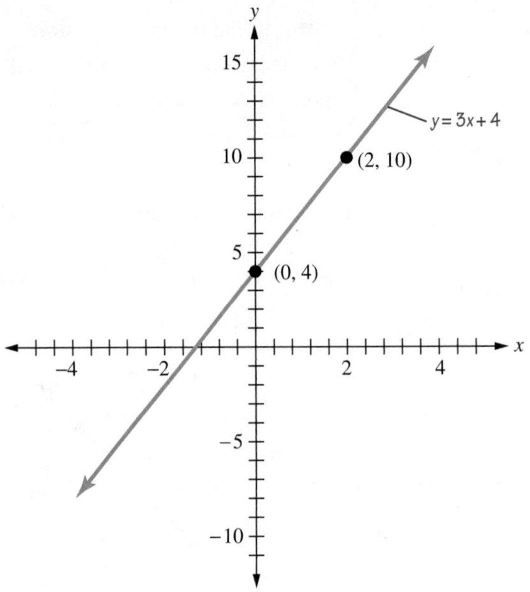

FIGURE 7.6 Graph of $y = 3x + 4$.

Now try Exercises 49 to 52.

EXAMPLE 7 Comparing Jet-Ski Rental Plans

A store that rents jet skis has two rental plans. In plan A, the customer pays a base fee of \$35.00 plus \$12.50 per hour. In plan B, the customer pays a base fee of \$22.50 plus \$15 per hour. Model each of these plans by a linear equation in slope–intercept form. Graph the equations and estimate at what point rental plan A is preferable to plan B.

SOLUTION:

The cost of renting a jet ski for x hours using plan A is the base fee of \$35 plus x times \$12.50. If we let y represent the total rental cost of a jet ski, we can model plan A by the equation

$$y = \underbrace{(12.50)}_{\text{hourly rate = slope}}x + \underbrace{35}_{\text{base fee = } y\text{-intercept}}. \quad \text{(Plan A)}$$

Similarly, we model plan B by

hourly rate = slope — base fee = y-intercept

$$y = (15)x + 22.50. \quad \text{(Plan B)}$$

Notice that although plan B is cheaper initially, because the slope of B's graph is greater than the slope of A's, eventually plan B will cost more.

In Figure 7.7, letting $x = 0$ and $x = 1$, we found the points (0, 35) and (1, 47.50) to graph A's equation and the points (0, 22.50) and (1, 37.50) to graph B's equation.

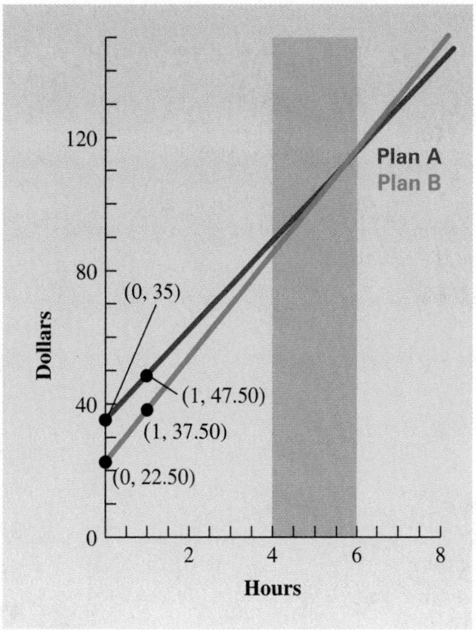

FIGURE 7.7 Graph of plan A versus plan B.

We can see from Figure 7.7 that for at least 4 hours, B's graph lies below A's, so plan B is cheaper, but by the time we reach 6 hours, B has become more expensive. So, it appears from the graphs that around 5 hours, B becomes the more expensive plan.

Now try Exercises 81 and 82.

We also can use algebra to solve Example 7. The point of intersection of the two lines in Figure 7.7 tells us where the two costs will be equal. This gives us the equation

$$12.50x + 35 = 15x + 22.50.$$

If you were to solve this equation, you would find that $x = 5$ is a solution.

The fact that we got the same solution both algebraically and geometrically demonstrates Descartes's reasoning in developing analytical geometry—what we can represent geometrically we can represent algebraically, and vice versa.

It is important when selecting a model that the model is appropriate for the situation you are modeling. When you write a linear equation in slope–intercept form as $y = mx + b$, realize that an increase of 1 for x causes an increase of m for y, an increase of 2 for x causes an increase of $2m$ for y, an increase of 5 for x causes an increase of $5m$ for y, and so on. The point is that a change in x causes an increase of m times that change for y.

If the situation you are modeling does not satisfy this property, then a linear equation is not a good choice for a model. In subsequent sections of this chapter, we discuss several real-life situations that do not satisfy this property; in those cases, we use nonlinear equations to model them.

EXERCISES 7.1

Sharpening Your Skills

Solve each equation by using the rules for rewriting equations. It is easier to solve an equation that has fractions or decimals if you multiply each side of the equation by a suitable constant to make the coefficients easier to work with.

1. $3x + 4 = 5x - 6$ **2.** $4 - 2x = 9x + 13$

3. $4 - 2y = 8 + 3y$ **4.** $5y - 6 = 14y + 12$

5. $\frac{1}{2}x - 6 = \frac{1}{5}x + 3$ **6.** $\frac{1}{3}y + 4 = \frac{1}{4}y + 3$

7. $0.3y + 2 = 0.5y - 3$ **8.** $0.4y - 0.2 = 0.6y + 3$

Solve each equation for the stated variable.

9. $P = 2l + 2w$; solve for w.

10. $m = \frac{a + b}{2}$; solve for a.

11. $z = \frac{x - \mu}{\sigma}$;† solve for μ.

12. $z = \frac{x - \mu}{\sigma}$; solve for σ.

13. $A = P(1 + rt)$; solve for r.

14. $A = \frac{1}{2}h(b + B)$; solve for b.

15. $2x + 3y = 6$; solve for x.

16. $4x - 5y = 3$; solve for y

17. $V = lwh$; solve for l.

18. $A = \frac{1}{2}hb$; solve for b.

19. $S = 2\pi rh + 2\pi r^2$; solve for h.

20. $A = 2lw + 2lh + 2hw$; solve for w.

Graph each equation by first finding the x- and y-intercepts.

21. $3x + 2y = 12$ **22.** $x - 5y = 10$

23. $4x - 3y = 16$ **24.** $5x + 4y = -20$

25. $\frac{1}{3}x + \frac{1}{2}y = 3$ **26.** $x - \frac{1}{5}y = 2$

27. $\frac{1}{6}x - 2y = \frac{3}{4}$ **28.** $x - \frac{1}{4}y = 2$

29. $0.2x = 4y + 1.6$

30. $0.3y = 1.2x + 0.6$

31. $0.4x - 0.3y = 1.2$

32. $0.2x + 0.5y = 2$

In the given graph call the x-intercept of each line (p, 0) and the y-intercept (0, q). Identify the line(s) having the given properties in Exercises 33–36.

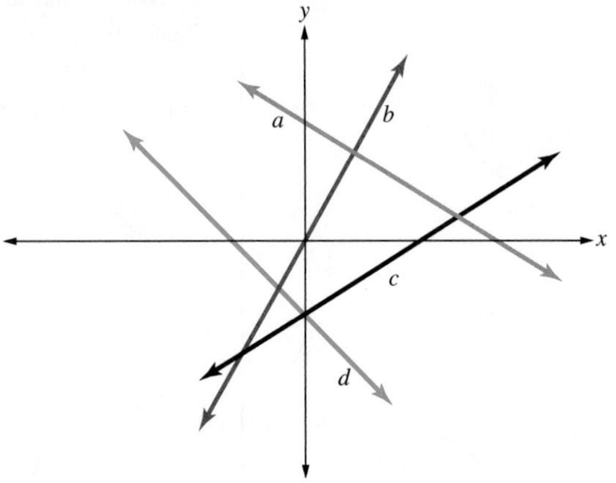

33. p and q are both positive.

34. p equals q.

35. $p \times q$ is negative.

36. p and q are both negative.

Find the slope of the line passing through the two given points.

37. (2, 5) and (6, 8) **38.** (4, 1) and (7, 3)

39. (3, 6) and (8, 2) **40.** (9, 1) and (6, 4)

41. (3, −4) and (5, 1) **42.** (8, −5) and (9, 2)

In Exercises 43–46, list all the lines in the given figure that satisfy the stated condition.

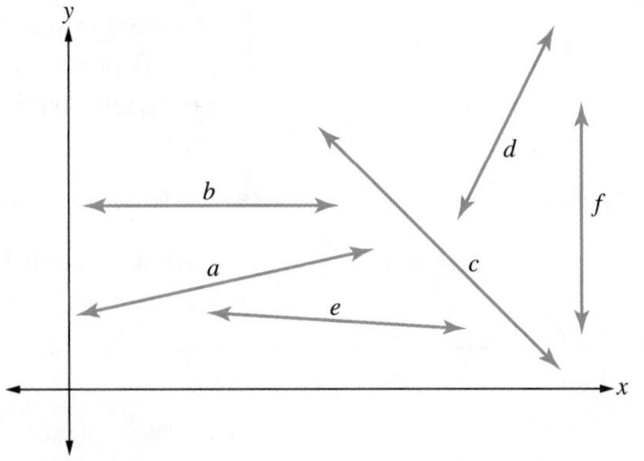

43. Slope is positive.

44. Slope is negative.

45. Slope does not exist.

46. Slope is zero.

47. Explain why the slope of a horizontal line is 0.

48. Explain why a vertical line has no slope.

In Exercises 49–52, state the y-intercept and slope of the graph of each equation.

49. $y = 4x - 3$ **50.** $y = 3x + 5$

51. $y = -5x - 3$ **52.** $y = -2x + 5$

† μ and σ are the lowercase Greek letters mu and sigma.

Applying What You've Learned

To solve Exercises 53–56, first write an equation that describes each situation. Use meaningful names for the variables.

53. Computing health club charges. A health club charges a yearly membership fee of $95, and members must pay $2.50 per hour to use its facilities. How many hours did Jillian use the club last year if her bill was $515?

54. Computing movie club charges. Netflix is now selling its used DVDs at $12 apiece. There is a $5 membership fee to join. If Roeper had a bill of $233 dollars, how many DVDs did he buy?

55. Computing charges for word processing. Thiep does word processing and charges $16 for documents up to 10 pages and $1.30 per page extra for pages beyond 10. If he billed a customer $44.60, how long was the customer's document?

56. Computing overtime. Darryl works at a warehouse where he is paid $9 per hour up to 40 hours. If he works over 40 hours, he is paid 1.5 times his usual wage. If he earned $468 last week, how many hours did he work?

According to the U.S. Bureau of the Census, there were 80 people per square mile (this is called the population density) in the United States in 2000. By 2010, the number of people per square mile had grown to 87.4 (Source: Bloomberg.com). Use this information to develop a linear equation in slope–intercept form to solve Exercises 57–60. In developing your equation, think of 2000 as year zero.

57. Write the equation that you are using to model this information.

58. Estimate the population density in 1960.

59. What do you expect the population density to be in 2020?

60. When will the population density be 100?

The Higher Education Price Index graphs below show the change in college faculty and administrative salaries between 2003 and 2009 using a base year of 1983. For example, an index of 240.7 for faculty salaries in 2005 would mean that college faculty salaries in 2005 were 240.7 percent of what college faculty salaries were in 1983. Use these graphs to do Exercises 61–64.

Higher Education Price Index for Faculty and Administrative Salaries

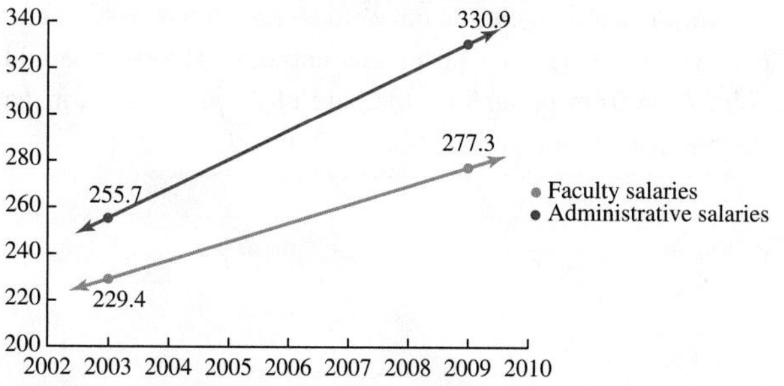

Source: Statistical Abstract of the United States

61. a. Find the slope of the blue line representing the change in the index for faculty salaries.

b. Treating 2003 as year 0, write the equation of the line representing the change in the faculty salaries index in slope–intercept form.

62. Repeat Exercise 61 for the red line representing the change in the index for administrative salaries.

63. Use your equation from Exercise 61 to predict when the index for faculty salaries will be 300.

64. a. Write an equation describing the difference between the faculty and administrative indices.

b. Use this equation to predict in what year the difference will exceed 100 points.

65. In 2000, the revenue for Starbucks was $2.17 billion and increasing at a rate of $0.47 billion per year (*Source:* Wikinvest).

a. Write a linear equation of a line in slope–intercept form that models this information. (Treat 2000 as year 0.)

b. Use your model from part (a) to predict Starbucks revenues in 2020.

66. Repeat Exercise 65 for Nike, using the fact that in 2000, Nike's revenue was $9 billion and increasing at a rate of $0.48 billion per year (*Source:* Wikinvest).

67. The *Wall Street Journal* reports that students graduating in 2011 (year 0) will have an average debt of $22,900 and increasing at a rate of 8% (0.08) per year. Write a linear equation in slope–intercept form to model this information and use it to predict the average student loan debt for students graduating in 2017. (*Source:* usnews.com/money)

68. Use your model in Exercise 67 to estimate what the average student debt will be in 2025.

In Exercises 69–74, write an equation that models each situation.

69. Calculating earnings. Juno has two part-time jobs. Her job in a diner pays $5.60 per hour and her work in a pet store pays $7.35 per hour. She earns $133 in a given week.

70. Determining rental payments. Jared furnished his new apartment by renting a living room set and a TV from a local "rent to own" store. He has l more payments on the living room set of $22 each, and t payments remaining on the TV of $13 each. The total he still owes is $341.

71. Investing in stock. Tamyra has invested $14,500 in the stock market. Some of the stock is in Time-Warner, which costs $35 a share, and the rest is in CDW, which costs $55 a share.

72. Grading. RJ's psychology professor is grading the course based on a contract system. A project is worth 40 points, and a report on a journal article is worth 10 points. RJ earned 250 points toward his final grade.

73. Manufacturing furniture. Aidan owns a small woodworking company. He knows that it takes 9 board feet to construct an end table and 15 to construct a coffee table. He has 342 board feet available.

74. Advertising a business. Bobby Flay is deciding how to advertise a new gourmet restaurant that he is opening. An online ad costs $300 and a TV ad costs $700. He has $7,500 to spend on the ads.

75. Comparing commuting costs. You travel a toll road that requires an 85-cent token. If you buy a special car-pool sticker for \$9.00, then tokens cost only 70 cents and you can use the express lane. At what point is the car-pool plan cheaper?

76. Comparing food plans. College students can purchase points that can be used in the food service areas instead of cash. If you initially pay the basic food service fee of \$35, then points can be purchased for 23 cents each; otherwise, points cost 30 cents each. How many points must you use for it to be cheaper to pay the basic food service fee?

Communicating Mathematics

77. Students often confuse the x-intercept of a line with the y-intercept. Why do you think this happens? What is a good way to remember what the x- and y-intercepts are?

78. Why is the slope of a horizontal line zero? Why does a vertical line have no slope?

79. Explain when it is appropriate to use a linear equation as a model. Give an example of a situation we have not covered in this section that you believe fits a linear equation as a model.

80. Give an example for which you feel it would not be appropriate to use a linear equation as a model. Explain why you feel this way.

Challenge Yourself

We can describe each situation in Exercises 81 and 82 by a pair of linear equations as we did in Example 7. Use the algebraic method that we described after Example 7 to determine when both deals are the same.

81. Chuck can work at the Buy More for a base pay of \$225 per week plus a \$45 commission for each computer system that he sells. At Circuit Town, his base pay would be \$400 with a \$20 commission for each computer system that he sells. Interpret what the solution to this system tells you. How should Chuck decide which position to take?

82. Cassandra is comparing two cell phone calling plans. BT&T charges \$12.75 per month and 7 cents per minute. Cingleton charges \$14.15 per month and 5 cents per minute. Interpret what the solution to this system tells you. How should Cassandra decide which plan to take?

83. Comparing investments. You want to invest \$5,000 that a wealthy relative has given to you. There is a certificate of deposit that pays 3.8% interest. You are in the 14% federal income tax bracket, however, and will have to pay federal income tax on interest earned by the CD. There are several bonds that you could also purchase that are exempt from federal income tax; however, they pay less interest. What interest do you need to earn on these bonds to equal the return that you would get from the CD?

84. Comparing investments. Redo Exercise 83, except now you are considering investing in bonds that are exempt from both the federal tax of 14% and the 2.1% state tax.

The table shows the average number of currency units per dollar in a recent year according to the Federal Reserve Bank.

Country	One Dollar Equals
Japan	105 yen
France	0.68 euro
India	39 rupees
Colombia	2,380 pesos
Mexico	11 pesos

Use this information to make the conversions in Exercises 85–88.

85. Converting currency.
a. dollars to yen **b.** yen to dollars

86. Converting currency.
a. dollars to rupees **b.** rupees to dollars

87. Converting currency. euros to Colombian pesos

88. Converting currency. yen to Mexican pesos

89. The points (3, 11), (7, 19), (11, 27), and (15, 35) all lie on a straight line. Choose several different pairs of points and compute the slope of the line. What do you notice?

90. If (x_1, y_1) and (x_2, y_2) are different points, we know that they determine a unique line. Suppose we compute the slope as $\frac{y_2 - y_1}{x_2 - x_1}$. Does it make a difference if we compute $\frac{y_1 - y_2}{x_1 - x_2}$ instead? Make up several examples and explain why it does or does not make a difference.

Two lines are parallel if they have the same slope. In Exercises 91–94, give equations in slope–intercept form that satisfy the conditions given for line l.

91. l is parallel to the line whose equation is $y = 3x + 5$; l passes through (2, 1).

92. l is parallel to the line whose equation is $y = -2x + 4$; l passes through (5, 3).

93. l is parallel to the line through (2, 4) and (4, 14); l passes through (2, 3).

94. l is parallel to the line through (3, 5) and (6, 11); l passes through (4, 8).

95. Wheelchair accessibility. To comply with the Americans with Disabilities Act, the town library must be made wheelchair accessible. Consider the following diagram of the entrance, which has enough room to construct a ramp as shown. Presently, there are three steps, each 6 inches high and 6 inches deep, leading up to this entrance. How far should point P be from point B (at the base of the first step) for the ramp to have a slope of 8% (that is, 0.08)?

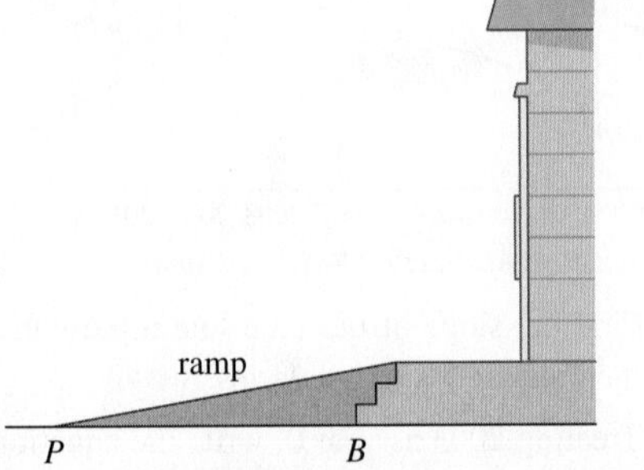

96. Wheelchair accessibility. Repeat Exercise 95, but now assume that there are six steps instead of three.

7.2

Modeling with Linear Equations

Objectives

1. Build a linear model using a point and the slope.
2. Use two points to build a linear model.
3. Understand how to use the line of best fit to model real data.

If you were to search the Internet to find examples of linear equations being used to model real data (see Exercise 35), you might be astounded by the number of applications of this idea that you would find. Realize, however, that a model is simply that—a model. You have no doubt canceled a trip to the beach or the ballpark because the meteorologist's prediction (using a model of the weather) was off-target. Parents who invest in college savings accounts and others who retire early, based on a slick-talking salesman's financial model, often have to make an adjustment years later when the model fails to accurately predict their future.

Engineers who build a physical model of an automobile to study gas mileage might ignore items that do not affect air resistance, such as the sound system. Similarly, we often omit features of mathematical models to make them easier to work with.

The following table shows there are different ways to provide information that determines the same linear equation as a model.

Method of Specifying Equation	Information Provided
Write the equation in standard form.	$3x + 2y = 6$
State the x- and y-intercepts of the graph of the equation.	x-intercept is (2, 0). y-intercept is (0, 3).
Specify the slope and y-intercept of the graph of the equation.	Slope is $-\frac{3}{2}$ and y-intercept is 3. Slope–intercept form of the equation is $y = -\frac{3}{2}x + 3$.

Building a Model with a Point and the Slope

KEY POINT

A point and the slope determine a line.

We can find a linear equation if we know the slope of its graph and a point on its graph.

EXAMPLE 1 Using the Slope and a Point to Determine a Linear Equation

Find a linear equation of the line with slope 3 passing through the point (4, 5).

SOLUTION:

We will assume that we can write the equation of the line in slope–intercept form, $y = mx + b$. The slope of the graph of the equation is $m = 3$, so we can rewrite this equation as

$$y = \overset{m}{3}x + b.$$

We need to find b. Because (4, 5) lies on the line, we substitute 4 for x and 5 for y, to get

$$\overset{y}{5} = 3(\overset{x}{4}) + b.$$

Subtracting 12 from both sides gives us $-7 = b$. The equation we want is therefore

$$y = 3x - 7.$$

Now try Exercises 1 to 8. 4

QUIZ YOURSELF 4

Find an equation of the line passing through (2, −3) having slope −4. Write your answer in slope–intercept form.

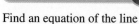

EXAMPLE 2 Building a Model Using a Point and the Slope

As part of a settlement of lawsuits against tobacco companies, the U.S. government required cigarette manufacturers to pay for antismoking education programs. Assume that, as a result of such programs, the smoking rate among 18- to 25-year-olds is dropping at a rate of 0.6% per year and that after 3 years, the smoking rate is 38.8%. If we model this reduction by a linear equation, what would we predict the smoking rate among 18- to 25-year-olds to be in 10 more years?

SOLUTION: Review the Choose Good Names Strategy on page 5.

We will model this decrease in the smoking rate by a linear equation in slope–intercept form written as

$$r = mt + b,$$

where t represents the time in years from the beginning of the campaign, and r represents the rate of smoking in this age group. To determine the relationship between the time t and the smoking rate r, we must first find m and b. (*Note:* In this example, t plays the role of x, and r plays the role of y.)

The change in the smoking rate per year is -0.6% per year. This means that the slope of the graph is $m = -0.6$. Substituting this value for m, we get

$$r = \underbrace{-0.6}_{\text{slope}}t + b.$$

We now use our rules for rewriting equations to find b.

$$38.8 = -0.6(3) + b \quad \text{Original equation}$$
$$38.8 = -1.8 + b \quad \text{Simplify product.}$$
$$38.8 + 1.8 = b \quad \text{Add 1.8 to both sides.}$$
$$40.6 = b \quad \text{Simplify.}$$

So our equation describing the reduction in smoking is therefore

$$r = -0.6t + 40.6.$$

We now use this equation to predict the smoking rate 10 years from now when $t = 13$. Substituting the value $t = 13$, we get

$$r = (-0.6)13 + 40.6 = -7.8 + 40.6 = 32.8.$$

Therefore, if the decline in smoking continues at the same rate, in 10 more years the smoking rate among 18- to 25-year-olds will be 32.8%.

Now try Exercises 17 to 22.

Although the model in Example 2 was easy to construct, you may feel that it has drawbacks that outweigh its simplicity. Perhaps you feel that a linear equation is inappropriate because a change in time does not produce a corresponding change in the smoking rate. Or you could argue that there are factors besides advertising that should be considered. It also may seem unreasonable to use the model over such a long period of time. As a result of such concerns, a researcher studying this problem might develop a more complex model and modify it frequently to make it more accurate.

KEY POINT

We can use two points to find a linear equation.

Building a Model with Two Points

In modeling data, we often use two data points to write a linear equation. Example 3 shows how to find a linear equation if we know two points on its graph.

EXAMPLE 3 Using Two Points to Find a Linear Equation

Find an equation of the line passing through the points (4, 2) and (8, 5).

SOLUTION: Review the Relate a New Problem to an Older One Strategy on page 9.

We will write the equation in slope–intercept form as $y = mx + b$, so we need to find m and b. Using points (4, 2) and (8, 5), we find the slope:

$$m = \frac{\text{rise}}{\text{run}} = \frac{5-2}{8-4} = \frac{3}{4}.$$

Substituting for m, we get

$$y = \overbrace{\frac{3}{4}}^{m}x + b.$$

Because (4, 2) lies on the graph,* we substitute 4 for x and 2 for y to get

$$\underbrace{2}_{y} = \frac{3}{4}\underbrace{(4)}_{x} + b = 3 + b.$$

Subtracting 3 from both sides, we find that $b = -1$. The equation we want is therefore

$$y = \frac{3}{4}x - 1.$$

Now try Exercises 9 to 16. 5

QUIZ YOURSELF 5

Find an equation of the line passing through $(-3, 5)$ and $(6, 0)$. Write the equation in slope–intercept form.

PROBLEM SOLVING

Strategy: Relate a New Problem to an Older One

In Example 3, we used the *Relate a New Problem to an Older One Strategy* that we discussed in Section 1.1. We did not approach Example 3 as a brand-new problem. Rather, we recognized that once we had the slope, we could solve the problem as we had in Example 1. A good problem solver often modifies a technique that is used in one situation and applies it in another.

EXAMPLE 4 Finding a Linear Equation for a Model Based on Two Data Points

Rachel sells gourmet cupcakes on the Internet. In the fourth month of operation, she sold 480 dozen cupcakes, and in the seventh month she sold 792 dozen. Assume that we can model the increase in her business by a linear equation. She estimates that with her current equipment, she can bake a maximum of 1,500 dozen cupcakes per month. If she wants her business to keep growing, during what month will she exceed her current capacity to produce baked goods?

SOLUTION: Review the Relate a New Problem to an Older One Strategy on page 9.

We will model Rachel's situation by a linear equation of the form

$$d = mt + b,$$

*We could have also used (8, 5); however, we generally use the point with the simpler coordinates.

where t is the time in months that her business has been in operation and d is the number of dozens of cupcakes she can sell. (In this problem, t plays the role of x, and d plays the role of y.)

We will use the points (4, 480) and (7, 792) to find the slope of the graph of the equation as follows:

$$m = \frac{\text{rise}}{\text{run}} = \frac{792 - 480}{7 - 4} = \frac{312}{3} = 104.$$

Substituting for m in the original equation, we get

$$d = \overset{\text{slope}}{104}t + b.$$

Now we can use either (4, 480) or (7, 792) to find b. As a rule of thumb, whenever possible, we use smaller numbers rather than larger ones, so we will use (4, 480). Substituting 4 for t gives us the following equation, which we then solve for b:

$$\begin{aligned} \overset{d}{480} &= 104(\overset{t}{4}) + b && \text{Original equation} \\ 480 &= 416 + b && \text{Simplify.} \\ 480 - 416 &= b && \text{Subtract 416 from both sides.} \\ 64 &= b. && \text{Simplify} \end{aligned}$$

Thus, a linear equation describing the growth in Rachel's sales is

$$d = 104t + 64.$$

We now want to find when Rachel will be able to sell 1,500 dozen, so we set $d = 1{,}500$ and solve the equation $1{,}500 = 104t + 64$ in the usual way.

$$\begin{aligned} 1{,}500 &= 104t + 64 && \\ 1{,}436 &= 104t && \text{Subtract 64.} \\ \frac{1{,}436}{104} &= t && \text{Divide by 104.} \end{aligned}$$

Therefore, $t = \frac{1{,}436}{104} \approx 13.8$, which means that she will exceed her capacity to produce cupcakes during the fourteenth month.

Now try Exercises 23 to 28.

The Line of Best Fit

KEY POINT

The line of best fit gives the best linear approximation of data.

The next example illustrates model building with real data. Real data are often messier to work with than the example data in problems because the data points usually do not fall on a straight line. Instead, they may have a somewhat linear pattern and we must find a line that best fits the data.

EXAMPLE 5 **Modeling Digital Music Sales**

Figure 7.8 shows a graph of digital music and music video sales for the years 2005 to 2008 in billions of dollars as reported by the Recording Industry Association of America (RIAA). Although these points do not quite lie on a straight line, we still may wish to model these data with a linear equation.

Historical Highlight

Solving Equations

Although the ancient Babylonian, Greek, and Hindu mathematicians had algebraic methods to solve quadratic equations, it was the work of a ninth-century Arabic mathematician named Mohammed ibn-Musal al-Khowarizmi who made the term *algebra* a household word. In his book *Al-jabr wa'l muqabalah*, he introduced Europeans to the methods of earlier mathematicians for solving quadratic equations. Some believe his title refers to balancing and completing algebraic expressions, which are processes we commonly use in algebra.

Having mastered the quadratic, it was natural to next ask if there was a method to solve *cubic equations*—those that have an x^3 term in them. In 1545, the Italian Girolamo Cardano published his famous book *Ars Magna* in which he showed not only how to solve cubics but also *quartic equations* (equations with an x^4 term). At this point, it seemed mathematicians would invent methods to solve equations having any power of x. This turned out not to be the case.

In the nineteenth century, Niels Henrik Abel from Norway and Evariste Galois of France proved the remarkable result that there is no mechanical formula similar to the quadratic formula for solving *quintic equations* (equations with an x^5 term).

Modeling with Quadratic Equations

POPCORN

We will now apply our knowledge of quadratic equations to model building. After the release of a new book, movie, or DVD, there are four periods in its life cycle.

- Stage 1: Sales increase rapidly. With a new DVD, for example, the number of purchasers grows rapidly soon after the DVD is introduced.
- Stage 2: The sales are still growing, but the increase from week to week is not as great as in the early phase.
- Stage 3: The product is still selling, but now each week's sales are a little lower than the week before.
- Stage 4: The market is saturated, and sales are now dropping rapidly.

We can represent this situation by a parabola opening downward, as in Figure 7.13.

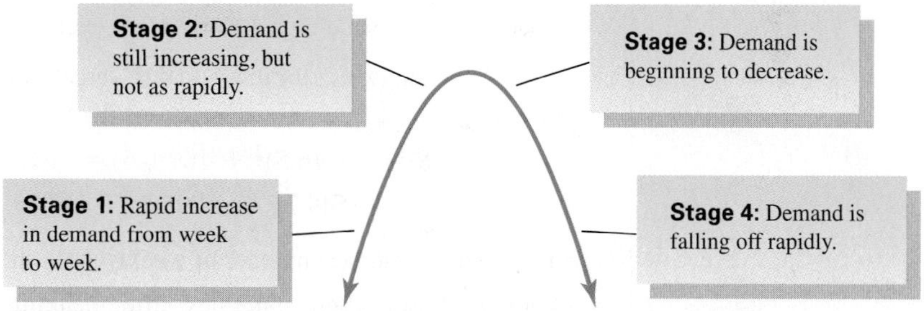
Stage 2: Demand is still increasing, but not as rapidly.
Stage 3: Demand is beginning to decrease.
Stage 1: Rapid increase in demand from week to week.
Stage 4: Demand is falling off rapidly.

FIGURE 7.13 A parabola models the product life cycle for a new product.

We use this pattern to model sales of a particular DVD in Example 5.

EXAMPLE 5 Modeling DVD Sales with a Quadratic Equation

Assume that a producer knows that the demand for a live concert by a popular artist on DVD can be modeled by the equation $S = -13n^2 + 169n$. Here, n is the number of weeks since the introduction of the DVD, and S is the number of thousands of the DVDs sold during week n.

a) When do we expect the sales for the DVD to peak?

b) After sales have peaked, when does this model predict that the sales will sink below 100,000 copies per week?

SOLUTION: Review the Draw Pictures Strategy on page 3.

a) The equation is quadratic with a negative n^2 coefficient, so its graph is a parabola opening downward, as shown in Figure 7.14.

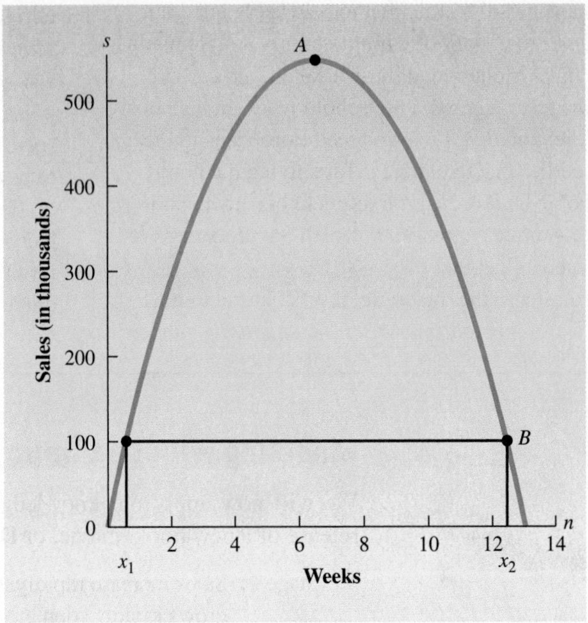

FIGURE 7.14 Parabola modeling DVD sales.

Point A, the vertex of the parabola, shows where the weekly sales are the greatest. Because $S = -13n^2 + 169n$, $a = -13$, $b = 169$, and $c = 0$. Therefore, the first coordinate of the vertex is

$$n = \frac{-b}{2a} = \frac{-169}{2(-13)} = \frac{-169}{-26} = 6.5.$$

Therefore, sales should peak during the seventh week.

Substituting 6.5 for n, we find the greatest number of weekly sales to be

Substitute n.

$$S = -13(6.5)^2 + 169(6.5) = -13(42.25) + 169(6.5)$$
$$= -549.25 + 1{,}098.5 = 549.25.$$

Therefore, the highest number of weekly sales should be about 549,250.

b) Point B shows where the sales, after peaking, drop to 100,000 per week. From Figure 7.14, it appears that there are two values for x that will give sales of 100,000. The first, x_1, appears to be between 0 and 2, and the other, x_2, is slightly past 12. To find these values exactly, we will solve the quadratic equation

$$-13n^2 + 169n = 100.$$

Subtracting 100 from both sides of this equation, we get

$$-13n^2 + 169n - 100 = 0.$$

Using the quadratic formula, we find

$$n = \frac{-169 \pm \sqrt{169^2 - 4(-13)(-100)}}{2(-13)}$$

$$= \frac{-169 \pm \sqrt{28{,}561 - 5{,}200}}{-26} = \frac{-169 \pm \sqrt{23{,}361}}{-26}.$$

Math in Your Life

Who's Watching You Watch TV?

As you recall from our *Star Trek* example at the beginning of this section, although we have correct data, using an inappropriate model may result in an unreliable prediction. The opposite can happen. The model might be valid, but the data are wrong.

In the TV ratings wars, there is a fierce competition that influences millions of advertising dollars for networks. However, because of the intermixing of technologies—radio is on the Web, TV is on cell phones, and the Web is on TV—it is becoming more difficult for ratings companies to gather reliable data. One company, Arbitron, has an innovative way to find out what you watch. They ask people to wear a device called the portable people meter, or PPM, which is about the size of a pager, to record exactly how much radio and TV programming you are exposed to each day. The PPM records everything about your viewing—including your flipping habits—and then transmits this information to be analyzed while you sleep. (*Source:* New York Times)

Thus, $n = 0.62$ and $n = 12.38$. Because we are interested in the point after sales have peaked, we ignore $n = 0.62$ and use $n = 12.38$ weeks as our solution. According to this model, by the 13th week, sales will have dropped below 100,000.

The points graphed below show the number of text messages sent in the United States from 2004 to 2008. The graph has a pattern similar to an upward-opening parabola, suggesting that we might use **quadratic regression** to model these data. Using the data points (2004, 4.7) to (2008, 110.4), we had Excel graph the line and the parabola of best fit. The higher correlation coefficient r for the parabola suggests that the parabola is a better model for these data than the line.

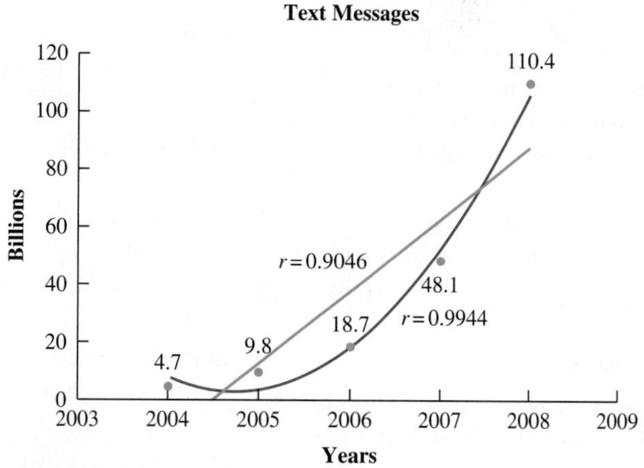

Source: CTIA-The Wireless Association

EXERCISES 7.3

Sharpening Your Skills

Solve each quadratic equation.

1. $x^2 - 10x + 16 = 0$

2. $x^2 - 7x + 12 = 0$

3. $2x^2 - 5x + 3 = 0$

4. $6x^2 - 11x + 4 = 0$

5. $3x^2 + 7x - 6 = 0$

6. $2x^2 + x - 3 = 0$

7. $5x^2 - 17x - 12 = 0$

8. $4x^2 + 12x + 9 = 0$

Answer the following questions for each quadratic equation. Then draw the graph.

Is the equation's graph opening up or down?
What is the vertex of the graph?
What are the x*-intercepts?*
What is the y*-intercept?*

9. $y = -x^2 + 6x - 8$

10. $y = x^2 - 4x - 5$

11. $y = -4x^2 + 8x + 5$

12. $y = x^2 + 5x + 4$

13. $y = 4x^2 - 4x - 2$ **14.** $y = -2x^2 - 10x - 8$

15. $y = 3x^2 + 7x - 6$ **16.** $y = 2x^2 + x - 3$

17. $y = -x^2 + 7x - 12$ **18.** $y = -4x^2 - 12x - 9$

Applying What You've Learned

19. Movie attendance. Dreamworks movie studio tries to release a "blockbuster" movie each summer. Assume that these statistics describe ticket sales for such a movie:

Week 2: 5 million tickets sold

Week 4: 7 million tickets sold

Week 6: 8 million tickets sold

We did quadratic regression to find that $A = -0.125x^2 + 1.75x + 2$ is an equation of the parabola of best fit for these data.

a. Show that the graph of this equation passes through the three data points regarding attendance.

b. According to this model, what is the maximum value that A can attain?

20. Business profits. New business owners typically experience a loss for a certain amount of time before the losses bottom out. Eventually the businesses show a profit. This pattern of profit and loss suggests a situation that could be modeled by a quadratic equation whose graph is a parabola opening up. Assume that you have begun a part-time business photographing weddings and that over the first three months of business you sustain losses of \$200, \$130, and \$70. (We can rephrase this by saying your cumulative profit at the end of month 1 was −\$200, at the end of month 2 was −\$330, and at the end of month 3 was −\$400.) Using quadratic regression, we can show that $P = 30x^2 - 220x - 10$ is the best quadratic equation that fits these data.

a. Verify that this equation fits the given data.

b. Using this model, at the end of what month do you expect to sustain your greatest cumulative loss? What is this loss?

c. At the end of what month do you expect to show your first cumulative profit? (By that, we mean that your total profits from the time the business started are greater than your total losses.)

d. What will your cumulative profit be by the end of month 10?

e. How much do you expect to earn during month 10?

21. DVD sales. Assume that the sales for the DVD in Example 5 are modeled by the equation $S = -4n^2 + 16n - 12$, where S is the number of sales (in millions) in week n.

a. When do we expect the sales for the DVD to peak?

b. According to this model, what is the largest value for S?

c. When do we expect the sales to drop to zero?

22. Presidential popularity ratings. After a U.S. president makes a politically unpopular decision, his approval rating usually drops. After some time, the approval rating rises again. Assume that before the president signs an unpopular tax bill, the approval rating is at 48%. One week after the president signs the bill, the rating is 41%, at two weeks it is 39%, and at three weeks it is 42%. Using quadratic regression we can show that $A = 2.5x^2 - 9.5x + 48$ is the best quadratic equation that fits the data.

Using this model, how many weeks after the signing of the tax bill will the approval rating be back to what it was before the signing of the tax bill?

We model many physical relationships using quadratic equations.

23. Falling bodies. A plane is dropping emergency food supplies to relieve famine in a less-developed country. Crates are dropped, and the height of the crate above the ground at time t is given by the equation $H = 160 - 16t^2$.

a. Graph this equation.

b. Are there any values for t that it would not make sense to use for this equation because of the physical characteristics of a falling crate?

c. Find the time at which the crate will strike the ground.

24. Rocket flight. Assume that if a model rocket is fired directly upward (with a certain velocity), its distance above the ground at time t is given by the equation $d = 100t - 16t^2$.

a. Graph this equation.

b. Are there any values for t that it would not make sense to use for this equation because of the physical characteristics of a rocket?

c. Find the time at which the rocket reaches its highest point.

d. Find the time at which the rocket returns to the ground.

25. The graph shows the U.S. prison population for 1995, 2000, 2005, 2008, and 2009. We used these data to find equations of the line of best fit, $y = 19.95x + 518$, and the parabola of best fit, $y = -0.76x^2 + 30.79x + 501.73$.

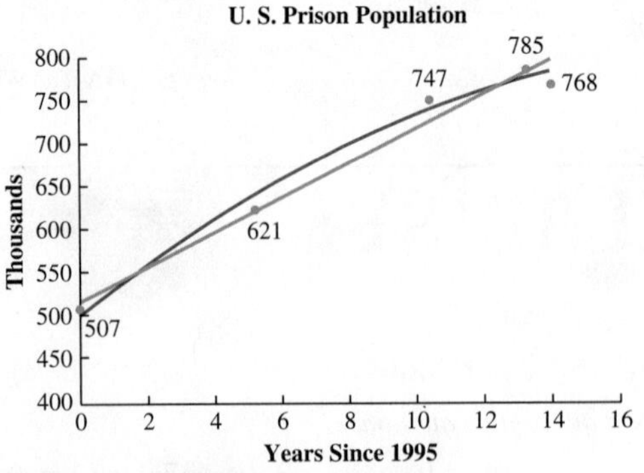

Source: U.S. Bureau of Justice Statistics

Which of these models predicts the prison population better in:

a. 1995 (year 0)?

b. 2000 (year 5)?

c. 2009 (year 14)?

26. The graph shows Apple Inc. revenues in 2004, 2006, 2008, and 2010. The equations for the line and parabola of best fit for these data are $y = 9.2x + 3.73$ and $y = 1.36x^2 + 1.06x + 9.15$, respectively. In 2011, Apple's revenue was 108.25 billion dollars.

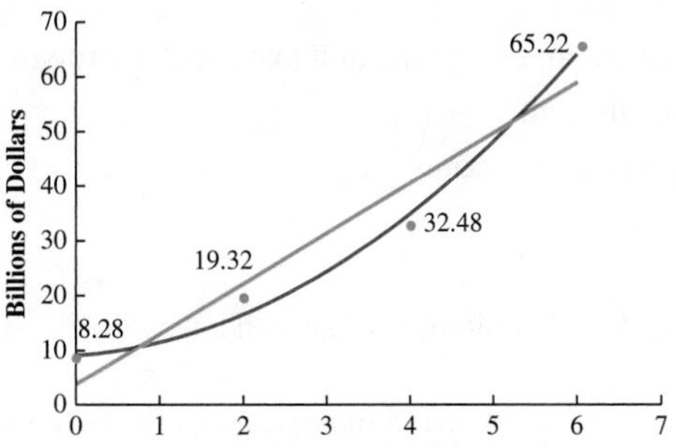

Source: Wikinvest.com

a. What do the line of best fit and the parabola of best fit predict Apple's revenue to be in 2011 (year 7)?

b. Which model appears to be the better predictor overall? Use that model to predict Apple's revenue in 2020.

27. What is the difference between linear regression and quadratic regression?

28. What does the correlation coefficient tell us about the line and parabola of best fit?

Communicating Mathematics

29. Returning to Exercise 23, explain why it is reasonable for the graph representing the falling crate to be shaped as it was. How does this conform to your intuition? Would a linear equation have been an acceptable model?

30. Returning to Exercise 24, explain why it is reasonable for the graph representing the rocket flight to be shaped as it was. How does this conform to your intuition? Would a linear equation have been an acceptable model?

31. **Running a race.** The equation $d = 0.15t^2 + 8t$ describes the distance Usain Bolt, who is running a 100-yard dash, has traveled in t seconds.

a. Graph this equation.

b. Are there any values for t that it would not make sense to use in this equation due to the physical characteristics of this problem?

c. From the graph, determine where the runner is running more slowly and where the runner is running faster.

d. How long does it take for him to finish the race?

32. **Running a race.** Would it make sense to use the model in Exercise 31 for Bolt running a mile race? Explain.

Challenge Yourself

33. The accompanying figure shows a time-lapse photo of a bouncing golf ball. The photographer captured all the images of the golf ball by using a strobe light that flashed once every 0.03 second. It is known from basic physics that the model for the path of a bouncing ball is a quadratic equation. Discuss how you would obtain the numerical data that then could be used to do quadratic regression to obtain the equation.

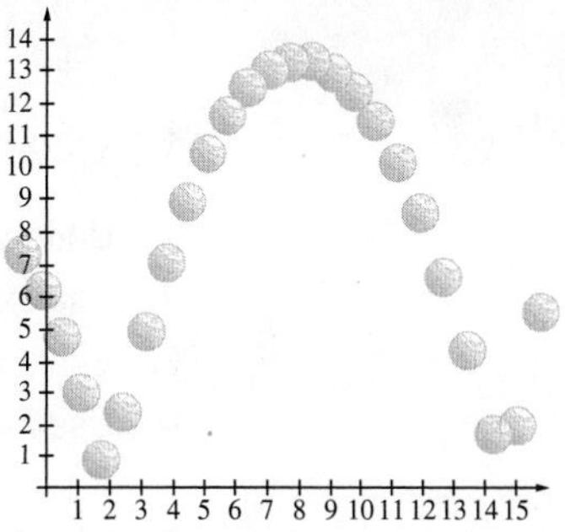

34. **Modeling labor force data.** In Exercise 35 from Section 7.2, we modeled the size of the U.S. labor force with the equation $n = 1.81t + 145.66$, which represented the line of best fit for the data points (0, 146), (1, 147), (2, 149.2), and (3, 151.3). The parabola of best fit for these same four data points is given by $n = 0.275t^2 + 0.985t + 145.935$.

a. Add a line to the table that you began in Exercise 35 from Section 7.2 as follows:

	2003 (year 0)	2004 (year 1)	2005 (year 2)	2006 (year 3)	2009 (year 6)
Actual Data	146	147	149.2	151.3	154.1
Values predicted by your model	146	147.77	149.54	151.31	156.62
Values predicted by the line of best fit	145.66	147.47	149.28	151.09	156.52
Values predicted by parabola of best fit	145.935				

b. Does the line of best fit or the parabola of best fit seem to model the data better? Explain your answer.

35. **Modeling foreign travel expenditures.** In Exercise 36 from Section 7.2, we modeled the amount spent on foreign travel by U.S. residents with the equation $n = 6.56t + 78.66$, which represented the line of best fit for the data points (0, 81.8), (1, 80.5), (2, 91.8), and (3, 99.9). The parabola of best fit for these same four data points is given by $n = 2.35t^2 - 0.49t + 81.01$.

a. Add a line to the table that you began in Exercise 36 from Section 7.2 as follows:

	2002 (year 0)	2003 (year 1)	2004 (year 2)	2005 (year 3)	2009 (year 7)
Actual Data	81.8	80.5	91.8	99.9	105.4
Values predicted by your model	81.8	87.83	93.86	99.89	124.01
Values predicted by the line of best fit	78.66	85.22	91.78	98.34	124.58
Values predicted by parabola of best fit	81.01				

b. Does the line of best fit or the parabola of best fit seem to model the data better? Explain your answer.

7.4 Exponential Equations and Growth

Objectives

1. Understand the difference between linear, quadratic, and exponential growth.
2. Use exponential equations to model growth.
3. Solve exponential equations using the log function.
4. Use logistic models to describe growth.

It seems that every few years we are bombarded by the media with a new crisis—the Avian flu, the eventual collapse of the Social Security system, global warming, famine in developing countries, the mortgage crisis, the spread of AIDS in Africa—the list seems to go on forever. In this section, we will introduce you to a new type of equation that can be used to model many of these situations. You can begin to understand the relationships that we model in this section by recalling the way money grows in a bank account.

KEY POINT

Money compounded in a bank account grows exponentially.

Exponential Growth

Suppose that you deposit \$1,000 (called the *principal*) in an account paying 8% interest per year. To keep our calculations simple, we will assume that you deposit the money at the beginning of the year and the interest is paid all at once at the end of the year.

At the end of the first year, you have the original \$1,000 plus the interest it has earned in 1 year. That is, the amount at the end of the first year equals

$$\$1{,}000 + \underset{\text{Interest}}{\underline{8\% \times \$1{,}000}} = \$1{,}000 + \underset{\text{Interest}}{\underline{0.08 \times \$1{,}000}}$$

$$= \$1{,}000 + \underset{\text{Interest}}{\underline{\$80}} = \$1{,}080. \text{ — amount at end of first year}$$

So, at the beginning of the second year, you have \$1,080 earning 8% interest. Therefore, the amount in your account at the end of the second year is

$$\$1{,}080 + \underset{\text{Interest}}{\underline{8\% \times \$1{,}080}} = \$1{,}080 + \underset{\text{Interest}}{\underline{0.08 \times \$1{,}080}}$$

$$= \$1{,}080 + \underset{\text{Interest}}{\underline{\$86.40}} = \$1{,}166.40. \text{ — amount at end of second year}$$

To compute the amount for successive years, we perform similar calculations. That is, we leave the interest in the account and then calculate the new interest based on the total of the principal and interest previously earned. The process of earning interest on interest is called **compounding.**

Table 7.1 shows how compounding works and shows the pattern in the increased value of the account.

Year	Beginning Balance	+	Interest for Current Year	=	Balance at End of Year
1	1,000	+	$0.08 \times 1{,}000$	=	$1{,}000(1.08) = 1{,}080$
2	1,080	+	$0.08 \times 1{,}080$	=	$1{,}080(1.08) = 1{,}000(1.08)(1.08)$
3	1,166.40	+	$0.08 \times 1{,}166.40$	=	$1{,}166.40(1.08) = 1{,}000(1.08)(1.08)(1.08)$

TABLE 7.1 Calculating compound interest.

Notice in Table 7.1 that the amount in your account

at the end of the *first* year is: (original deposit) $\times$ $(1.08)^1$,
after the *second* year is: (original deposit) $\times$ $(1.08)^2$,
and, after the *third* year is: (original deposit) $\times$ $(1.08)^3$.

In general, we have the following formula:

> **THE COMPOUND INTEREST FORMULA** If we invest the amount P, called the *principal*, in an account earning a yearly interest rate r and we compound the interest for n years, then the amount in the account, A, is
>
> $$A = P(1 + r)^n.*$$

We apply this formula in Example 1.

QUIZ YOURSELF 8

a) Redo Example 1 using an interest rate of 1.25%.
b) How much less do you earn with the lower interest rate?

EXAMPLE 1 Using the Compound Interest Formula

Suppose that you deposit \$1,000 in an account that is compounded annually at a rate of 8%. Find the amount in this account after 30 years.

SOLUTION:

The principal P is \$1,000, the number of years n is 30, and the rate r is 0.08. Thus, the amount in the account at the end of 30 years will be

$$A = P(1 + r)^n = 1{,}000(1 + 0.08)^{30} = 1{,}000(1.08)^{30} = \$10{,}062.66.$$

Now try Exercises 5 to 12. 8

Year	Balance at End of Year
1	$1{,}000(1.08)^1 = 1{,}080$
2	$1{,}000(1.08)^2 = 1{,}166.40$
3	$1{,}000(1.08)^3 = 1{,}259.71$
4	$1{,}000(1.08)^4 = 1{,}360.48$
5	$1{,}000(1.08)^5 = 1{,}469.32$

TABLE 7.2 Growth of an account over 5 years.

In Example 1, we compounded the interest once a year. Many financial institutions compound interest more frequently. To learn how to do this, see Section 8.2.

Let's plot the amount in the account described in Example 1 over a period of years to see better how the value of the account is growing. In Table 7.2, we first calculate the amount in this account for each of the first 5 years and plot these values in Figure 7.15, which shows the relationship between time and the amount in the account.

Between the Numbers

Did Anything Bother You in the Last Example?

Over the course of our marriage, my wife and I have often been asked by eager young investment counselors for a few minutes of our time. After introducing their investment plan, they begin talking about the details with something like this:

Assuming a rate of return of 8%, in 30 years, your investment will be worth... let me punch it in the computer here... ah, there it is...

and up pops some fantastic amount on the screen.

The critical words here are "***assuming a rate of return of 8%***." Between 2008 and 2010 the Dow Jones stock market index fell over 3,000 points, and by 2011 many investments were paying less than 1% return. At the same time, people were losing their homes due to increased interest on adjustable rate mortgages and others were drowning in insurmountable credit card and student loan debt.

When the exponential model is working for you, it can be a wonderful thing—when it is working against you, it can be a financial tragedy.

*To raise a quantity to a power on your calculator, you must use the y^x key (or the ^ key). For example, to compute $(1 + 0.08)^{12}$, you first enter 1.08, then press the y^x key (or the ^ key), then press 12, and then press the = key or Enter.

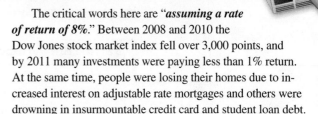

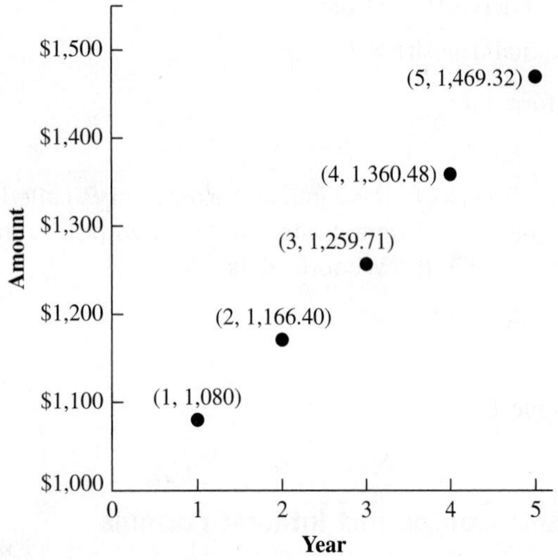

FIGURE 7.15 The balance in a bank account over 5 years.

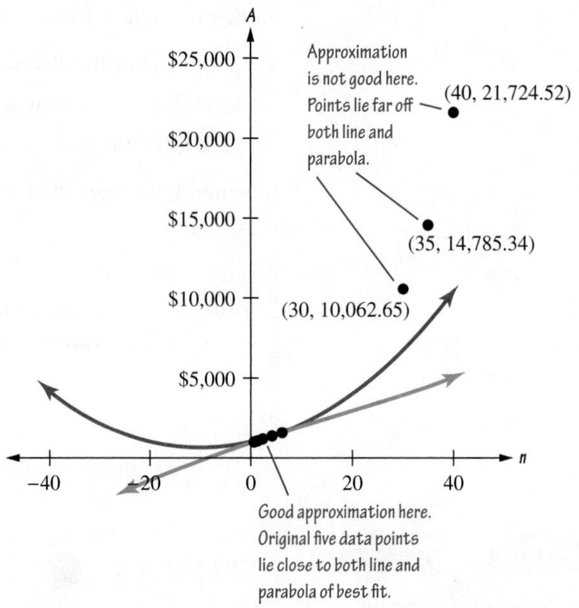

FIGURE 7.16 Graph comparing the balance in a bank account and the line and parabola of best fit.

Because the points in Figure 7.15 appear to be close to lying on a straight line, you might want to use the line of best fit to approximate these data. However, if you look very closely, you will see a slight upward curving, so you may decide instead to find the parabola of best fit.

We used a graphing calculator to find the linear equation that best models the data to be

$$A = 97.272n + 975.366,$$

and the best quadratic model is given by

$$A = 3.74n^2 + 74.84n + 1{,}001.54,$$

where A is the amount in the account after n years.

We graphed the line and parabola of best fit in Figure 7.16. In addition, we graphed the original five data points and also the points (30, 10,062.65), (35, 14,785.34), and (40, 21,724.52), which correspond to the value of the account after 30, 35, and 40 years, respectively.

As time increases, the points representing the amounts in the bank account are farther and farther from both the line and parabola of best fit. The equation describing these data is neither linear nor quadratic.

The type of equation we used to calculate compound interest is also useful for modeling other important phenomena, so we will give it a name.

DEFINITION An **exponential equation** is an equation of the form

$$y = a \cdot b^x.$$

The compound interest formula, $A = P(1 + r)^n$, is one example of an exponential equation.

SOME GOOD ADVICE

We use an exponential equation to model situations in which *the rate of change of a quantity is proportional to its size.* For example, the more money you have in a bank account, the more interest you earn; the larger the population of a country, the more children will be born to increase the population.

Exponential Models

KEY POINT

An exponential equation is a simple model of population growth.

We can use an exponential equation to model population growth. To do this, we use an initial population instead of an initial bank balance, and a yearly growth rate instead of an interest rate.

EXAMPLE 2 Modeling Population Growth with an Exponential Function

According to the U.S. Bureau of the Census, in 2011 the U.S. population was approximately 307 million, with an annual growth rate of 0.7%. If this growth rate continues at that rate until 2053, what will the population of the United States be then?

SOLUTION:

We will use an exponential equation with

$$P = 307 \text{ million}, \quad r = 0.007, \quad \text{and } n = 2053 - 2011 = 42.$$

Therefore,

$$P(1 + r)^n = 307(1 + 0.007)^{42} = 307(1.007)^{42} = 411.5 \text{ million}.$$

According to this exponential model, we expect the U.S. population to be about 411.5 million in 2053.

You may think it strange that I chose the year 2053, but as I was writing this example I compared it with the same example from an earlier edition of this text, using data from 2000 with a population of 281 million and an annual growth rate of 1.2%. In that example, the projected population for 2053 was about 529 million! The point is that even with a larger base population in 2011, the seemingly slight difference in the growth rate (1.2% − 0.7% = 0.5%) yielded a projection of almost 117 million less by 2053.

Now try Exercises 13 to 18. 9

QUIZ YOURSELF 9

In 2008, Mexico had a population of roughly 108 million and a yearly growth rate of 1.153%. Assuming that this growth rate continues for the next 10 years, what will Mexico's population be at that time?

When using a mathematical model as in Example 2, it is always important to consider the appropriateness of the model. For example, you should ask whether it is reasonable to assume that the growth rate in the United States will stay at 0.7% for 42 years.

Often in the media, researchers make dire predictions based on mathematical models regarding such things as population growth, global warming, and the bankruptcy of the Social Security system. These predictions may or may not be valid, depending on the assumptions made by the researcher. As an educated person, you should have an understanding of what a mathematical model can tell us and what it cannot. If the assumptions underlying a model are not sound, then its predictions are worthless.

In predicting future bank balances, population, or similar quantities that are growing exponentially, we often want to know, "How long will it take for this quantity to double?"

KEY POINT

We use the log function to find the time it takes a quantity that is growing exponentially to double.

Before we can answer this question, we need another tool. Many calculators have a key labeled either "log" or "log x," which stands for the *common logarithmic function.* Pressing this key has the effect of reversing the operation of raising 10 to a power. For example, suppose that you compute $10^5 = 100{,}000$ on your calculator. If you next press the "log" key, the display will show "5." If you enter 1,000, which is 10 raised to the third power, and press the "log" key, the display will show "3." Practice finding the log of powers of 10, such as 100 and 1,000,000.

The log function has an important property that will help us solve equations of the form $a = b^x$ for x. For example, we may want to find a value of x such that $5 = 3^x$. Although it may seem strange to you, it is possible to find a number, although not an integer, such that if we raise 3 to that power, we will get 5. In order to do that, we need the following property of the log function:

EXPONENT PROPERTY OF THE LOG FUNCTION

$$\log y^x = x \log y$$

To understand this property, use your calculator to verify the following:

$$\log 3^5 = 5 \log 3 \quad \text{and} \quad \log 8^2 = 2 \log 8$$

EXAMPLE 3 Using the Log Function to Solve an Equation

Solve the equation $5 = 3^x$.

SOLUTION:

We take the log of both sides of the equation to get

$$\log 5 = \log 3^x.$$

We then use the exponent property of the log function to rewrite the equation as

$$\log 5 = x \log 3.$$

If we divide both sides of the equation by log 3, we get

$$\frac{\log 5}{\log 3} = x.$$

We now find $x = \dfrac{\log 5}{\log 3} \approx \dfrac{0.69897}{0.47712} \approx 1.46.$

If you use your calculator, you will find that $3^{1.46}$ is approximately 5. (Because of the way we rounded numbers, you will not get exactly 5 as a result.)

Now try Exercises 19 to 24.

We now can find how long it takes for a quantity to double.

EXAMPLE 4 Doubling a Population

In 2011, India's population was 1.2 billion with an annual growth rate of 1.58% (*Source:* indiaonlinepages.com). Assuming that the growth rate will remain the same, in what year will India's population double?

SOLUTION:

We want to know when the population will be 2.4 billion. We will use the growth model $A = P(1 + r)^n$, $P = 1.2$, $r = 0.0158$, and A is the future population of 2.4 billion. So we want to solve the equation

$$\overset{A}{2.4} = \overset{P}{1.2}(1.\overset{r}{0158})^n$$

for n.

First, we divide both sides of the equation by 1.2 to get

$$2 = (1.0158)^n.$$

Next, we take the log of both sides, giving us

$$\log 2 = \log (1.0158)^n.$$

Then, using the exponent property for the log, we get

$$\log 2 = n \log (1.0158).$$

Dividing both sides of the equation by log 1.0158, we find that

$$n = \frac{\log 2}{\log 1.0158} \approx 44.2.$$

So, we expect the population of India to double by 2056.

Now try Exercises 39 to 42. 10

QUIZ YOURSELF 10

Redo Example 4 but now assume that the growth rate is 2.1%.

We will now show you how to build exponential models using two data points. Recall that an exponential model has the form $y = a \cdot b^x$.

EXAMPLE 5 Building an Exponential Model for Inflation

Just as money grows exponentially in a bank account, the price of a product increases over the years exponentially due to inflation. Suppose that in 2006, a pair of Ugg® boots cost \$130 and in 2011, due to inflation, the same pair cost \$150.

a) Develop an exponential model to describe the inflation in this price.

b) Use the model in part a) to estimate the cost of this pair of boots in 2020.

SOLUTION:

a) We will assume the model that we want to find has the form $y = a \cdot b^x$, so we must find a and b. We will call 2006 year 0 and year 2011 year 5.

First we will find a. If we let $x = 0$ and $y = 130$, then our model becomes $130 = a \cdot b^0$. Recall that any number except 0 raised to the 0 power is equal to 1, so $b^0 = 1$. Therefore, $130 = a \cdot b^0 = a \cdot 1 = a$. We can now write our model as $y = 130 \cdot b^x$.

Next, we have to find b. We know that when $x = 5$, $y = 150$. Substituting these values for x and y in our model gives us the following equation, which we solve for b:

$$150 = 130 \cdot b^5 \quad \text{Substitute 130 for } a.$$

$$b^5 = \left(\frac{150}{130}\right) \approx 1.154 \quad \text{Divide both sides by 130.}$$

$$b = \sqrt[5]{1.154} \approx 1.03^* \quad \text{Take the fifth root of both sides.}$$

Our model is now $y = 130 \cdot (1.03)^x$.

b) We want to find the value for y when $x = 2020 - 2006 = 14$. So, the estimated cost of the boots will be $130 \cdot (1.03)^{14} = \$196.64$.

Now try Exercises 55 to 60. 11

QUIZ YOURSELF 11

Assume that the price of a leather coat was \$180 in 2006 and the same coat cost \$210 in 2010.

a) Find an exponential equation to model this inflation.

b) Use your model in part a) to estimate the cost of the coat in 2015.

Logistic Models

KEY POINT

Limited resources may cause a population to grow according to a logistic model.

In the 1980s, many were alarmed by the rapid growth of AIDS. Figure 7.17 contains data from a report by the Global Aids Policy Commission showing the increase of AIDS cases in North America from 1985 through 1991.

Looking at the graph in Figure 7.17, it might appear that the growth of AIDS was following an exponential pattern. However, as we see in the following table, unlike our earlier bank account example, *the growth rate is not the same from year to year.*

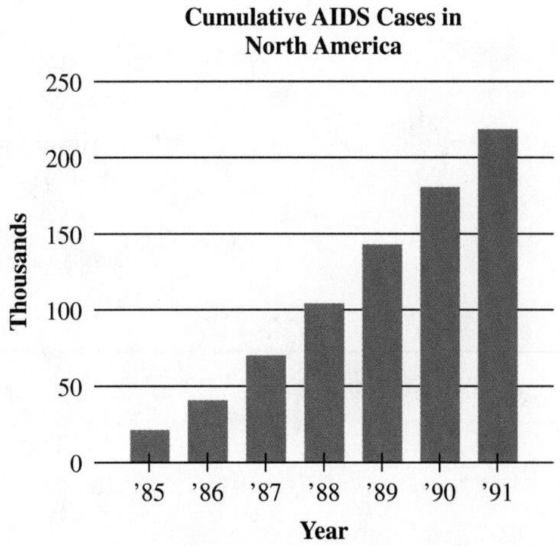

FIGURE 7.17 The growth of AIDS cases in North America, 1985–1991.

*You can take the fifth root of 1.154 by using the y^x key or calculating y^x, with $y = 1.154$ and $x = \frac{1}{5}$.

Years	Cumulative AIDS Cases in North America (in thousands)	Percent Increase from Previous Year
1985	22	
1986	41	86
1987	70	71
1988	104	49
1989	143	38
1990	180	26
1991	219	22

Note decrease in rate of growth from year to year.

Thus, an exponential model is not appropriate to model the data, so we need to introduce another type of model.

KEY POINT

A logistic model takes into account limits on population growth.

In studying the growth of populations, demographers often use a logistic model. A *logistic model* takes into account the fact that as populations grow, there are limits in terms of space, food, and so on that prevent the population from following a true exponential growth pattern.

Recall that in computing compound interest, we used equations of the type

$$\text{balance at end of year} = (1 + \text{rate})(\text{balance at end of previous year}).$$

Translating this into population growth, we would write

$$\text{population at end of year} = (1 + \text{growth rate})(\text{population at end of previous year}).$$

In both cases, we assumed that the money in the bank account and the population would grow at the same rate from year to year. However, with restrictions on the amount of food, water, space, and so on, any environment can sustain only a limited population. As the population grows, the percentage of this capacity that has been used up influences the rate of growth. If a population has a growth rate of 3%, in the beginning it is reasonable to multiply the existing size of the population by (1 + 0.03) to estimate the population the next year. However, as the capacity to sustain further population diminishes, we must reduce the growth rate accordingly.

In a logistic model, it is common to represent the maximum capacity that an environment can support by 1, or 100%. For our model, we will define a quantity P_n, read "P sub n," that represents the percentage of the maximum capacity that has been attained by the population in year n.

For example, to say $P_5 = 0.40$ means that at the end of year 5, the population we are studying has attained 40% of its maximum capacity. If this same population had an original growth rate of 0.03, to calculate P_6 we would reduce this growth rate by the factor $1 - P_5 = 0.60$ to get a growth rate of 0.018, as shown in the following diagram.

original growth rate

40% of the capacity to sustain a population has been used up.

$$0.03 \times (1 - P_5) = 0.03 \times 0.6 = 0.018$$

Only 60% of the capacity to sustain the population remains.

During year 6, we have only 60% of the original growth rate.

The idea is that because 40% of the capacity for growth already has been used up by the population, future growth rate can only be 60% of what it was originally.

We are now ready to give a precise definition of a logistic growth model.

DEFINITIONS Logistic Growth Model

Assume that a population is growing originally at rate r. We let P_n denote the percentage of the maximum capacity that the population has attained in year n. Moreover, P_n satisfies the following equation:

$$P_{n+1} = [1 + r(1 - P_n)]P_n.$$

This collection of equations for $n = 0, 1, 2, \ldots$ is called a **logistic model.**

To calculate P_{n+1}, we reduce the growth rate by multiplying it by $1 - P_n$. We will refer to this quantity as the *rate reduction factor.* So the logistic growth equation can be written as

$$P_{n+1} = [1 + r(\text{rate reduction factor})]P_n.$$

It is useful to recompute the rate reduction factor each time we calculate a new value for P_{n+1}. We illustrate how to use the logistic growth model in Example 6.

EXAMPLE 6 Using the Logistic Growth Model to Predict Population Growth

Assume that a population is growing initially at a rate of 3% per year. Also assume that at the end of the fifth year, the population is at 40% of its maximum size. At what percentage of its maximum size will the population be 1 year later?

SOLUTION:

We are told that $P_5 = 0.40$, $r = 0.03$, and we want to find P_6. Figure 7.18 shows how to use the logistic growth model.

$$P_{5+1} = [1 + 0.03(1 - P_5)]P_5$$

P_{5+1}: percentage of capacity at end of year 6
0.03: original rate of growth
$(1 - P_5)$: rate reduction factor
P_5: percentage of capacity at end of year 5

FIGURE 7.18 Using P_5 to calculate P_6.

At the end of year 5, the rate reduction factor will be $1 - 0.4 = 0.6$; thus,

$$\begin{aligned} P_6 &= [1 + (0.03)(1 - P_5)]P_5 \\ &= [1 + (0.03)(0.6)](0.4) \\ &= (1 + 0.018)(0.4) \\ &= (1.018)(0.4) \\ &= 0.4072 \end{aligned}$$

So we see that at the end of the sixth year, the population is at 0.4072, or 40.72% of its maximum capacity.

Now try Exercises 29 to 32.

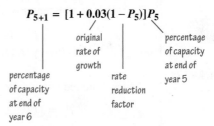

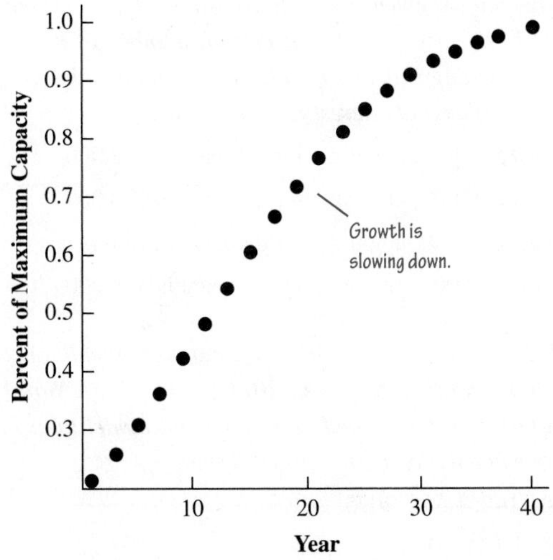

FIGURE 7.19 Graph illustrating logistic growth.

Notice that in Example 6, we did not compute P_6 by substituting 6 into some equation, as we did in previous models. Instead, we described P_6 in terms of P_5. If we had not been given P_5 explicitly, then we would have to calculate it by finding P_4. But to find P_4 would require knowing P_3, and so on. If we keep going, eventually we would need to know P_0. In our population examples, we will always assume that P_0 is the percentage of the capacity we have before the first year begins. An equation that is defined so that to find a given value requires knowing previous values is called a **recursive equation.**

In Figure 7.19, we have plotted the population growth for a hypothetical population over a period of 40 years based on a logistic growth model. It is easy to see that at first the population grows rapidly in a way similar to exponential growth, but as we reach year 15, because the population is at about 60% of its maximum capacity, the growth slows considerably.

EXAMPLE 7 Using the Logistic Model to Predict Future Population

Biologists have chosen a nearly extinct type of lemur to populate a small island. They have introduced a number of lemurs that constitute 20% of the population that the island is capable of supporting. If the growth rate of the lemur population is 10%, find what percentage of the maximum lemur population will be on the island in 3 years.

SOLUTION:

To find the percentage of the maximum population at the end of 3 years, we will find P_1, P_2, and finally P_3.

P_0 is the initial percentage of the maximum population of lemurs, which is 0.20. We will round off calculations to three decimal places to keep them from becoming too unwieldy. The first rate reduction factor is $1 - P_0 = 1 - 0.20 = 0.80$.

$$\begin{aligned} P_1 &= [1 + (0.10)(\text{rate reduction factor})]P_0 \\ &= [1 + (0.10)(0.8)](0.20) \\ &= (1 + 0.08)(0.20) = (1.08)(0.20) = 0.216 \end{aligned}$$

The rate reduction factor is now $1 - P_1 = 1 - 0.216 = 0.784$.

$$\begin{aligned} P_2 &= [1 + (0.10)(0.784)](0.216) \\ &= (1 + 0.0784)(0.216) = (1.0784)(0.216) = 0.233 \end{aligned}$$

The new rate reduction factor is $1 - P_2 = 1 - 0.233 = 0.767$.

$$\begin{aligned} P_3 &= [1 + (0.10)(0.767)](0.233) \\ &= (1 + 0.0767)(0.233) = (1.0767)(0.233) = 0.251 \end{aligned}$$

At the end of 3 years, the island will have slightly more than 25% of its maximum lemur population.

Now try Exercises 33 to 38. 12

QUIZ YOURSELF 12

Assume that a population is growing initially at a rate of 4% per year. Also assume that at the end of the third year, the population is at 60% of its maximum size. At what percentage of its maximum size will the population be 1 year later?

EXERCISES 7.4

Sharpening Your Skills

In Exercises 1–4, you are given the principal in a bank account at the beginning of a year and a rate of interest that is compounded annually. Calculate the amount in the account at the end of the year.

1. \$1,000; 5%

2. \$3,000; 6%

3. \$4,000; 2.5%

4. \$3,000; 6.5%

In Exercises 5–8, you are given the principal in a bank account, a yearly interest rate, and the time the money is in the account. Assuming that no withdrawals are made, use the compound interest formula to compute the amount in the account after the specified time period. Assume compounding is done annually.

5. \$5,000; 5%; 5 years

6. \$7,500; 7%; 6 years

7. \$4,000; 8%; 2 years

8. \$8,000; 4%; 3 years

In Exercises 9–12, you are given an initial deposit in a bank account and the amount in the account after a certain number of years. Assume that compounding is done annually, and that no withdrawals are taken. Find the annual interest rate on the investment.

9. Initial deposit \$15,000; amount after 12 years is \$22,000

10. Initial deposit \$20,000; amount after 20 years is \$50,000

11. Initial deposit \$10,000; amount after 8 years is \$13,000

12. Initial deposit \$12,000; amount after 10 years is \$18,000

In Exercises 13–18, we give you a country's population (in millions) and the population growth rate (%) in 2011 (Source: CIA World Factbook). Use this information and an exponential model to estimate the country's population in the specified year.

13. Bangladesh: 142; 1.566: 2025

14. Brazil: 192; 1.134: 2020

15. Japan: 128; −0.278: 2022

16. Russia: 143; −0.47: 2025

17. Indonesia: 238; 1.069: 2030

18. Pakistan: 178; 1.573: 2030

In Exercises 19–24, use the log function to solve the given equation for x.

19. $5^x = 20$

20. $2^x = 15$

21. $10^x = 3.2$

22. $8^x = 4.65$

23. $(3.4)^x = 6.85$

24. $(15.7)^x = 155.5$

In Exercises 25–28, you are given the population (in millions) of a country in 2000 and 2011 (Source: United Nations). Use the exponential model $A = P(1 + r)^n$ to find the growth rate r for the specified country over this period.

25. Mexico: (2000) 97; (2011) 112

26. Vietnam: (2000) 70; (2011) 86

27. Russia: (2000) 147; (2011) 143

28. Ethiopia: (2000) 77; (2011) 82

In Exercises 29–32, assume that a population is growing initially at the specified rate. You are given a value for P_n and are to use the logistic growth model to compute the value of P_{n+1}. Explain what your calculations tell you.

29. rate = 3%; $P_4 = 0.36$

30. rate = 5%; $P_7 = 0.48$

31. rate = 4.5%; $P_8 = 0.72$

32. rate = 5.5%; $P_3 = 0.51$

In Exercises 33–38, redo the calculations of Example 7 regarding the growth of the lemur population. Use the specified initial growth rate and the given value for P_0 to find P_3. Your answers may vary from ours due to roundoff error.

33. rate = 8%; $P_0 = 0.30$

34. rate = 12%; $P_0 = 0.40$

35. rate = 10%; $P_0 = 0.25$

36. rate = 15%; $P_0 = 0.35$

37. rate = 4.5%; $P_0 = 0.60$

38. rate = 4.25%; $P_0 = 0.20$

Applying What You've Learned

39. Compound interest. If your parents have $10,000 in a college savings account that is paying an interest rate of 5%, which is being compounded annually, how many years will it take to double if the interest rate stays the same?

40. Compound interest. For the account in Exercise 39, how many years will it take to triple?

41. Population growth. In 2011, the population of the United States was 307 million and the growth rate was 0.7%. If the growth rate remains the same, in what year will the population be double what it was in 2011?

42. Population growth. In 2011, the population of Brazil was 192 million and the growth rate was 1.134%. If the growth rate remains the same, in what year will the population be double what it was in 2011?

If a population has an annual growth rate of r percent, then $\frac{70}{r}$ is a good estimate of the population's doubling time. Recall in Example 4 that India's annual growth rate was 1.58%. We can estimate India's population's doubling time to be $\frac{70}{r} = \frac{70}{1.58} \approx 44.3$, which is approximately the same result that we obtained in Example 4. Use this method to estimate the population's doubling time for the countries in Exercises 43–46 (Source: CIA World Factbook).

43. Australia: 2011 growth rate 1.148%

44. Canada: 2011 growth rate 0.794%

45. Democratic Republic of the Congo: 2011 growth rate 2.614%

46. North Korea: 2011 growth rate 0.538%

In Exercises 47–50, the growth rate is negative, which is called exponential decay instead of exponential growth.

47. In 2011, Bulgaria had a population of 7.4 million and a growth rate of −0.781%. Assuming that this rate remains constant, estimate the population of Bulgaria in 2025.

48. In 2011, South Africa had a population of 50.6 million and a growth rate of −0.38%. Assuming that this rate remains constant, estimate the population of South Africa in 2025.

49. Radioactive decay. Assume that a radioactive material decays with an annual growth rate of −0.35%. How many years will it take 100 pounds of the material to decay to 50 pounds?

50. Radioactive decay. Repeat Exercise 49, but now assume that the rate of growth is −0.14%.

When a drug such as a pain killer or an antibiotic is introduced into the body, the kidneys work to eliminate the drug. We will assume that after an hour, 15% of any drug in the body is eliminated. After another hour, 15% of the remaining drug is eliminated, and so on.

51. Modeling drug concentration. If you receive an injection of 500 mg of Novocain, how much of the drug will remain in your body after 3 hours?

52. Modeling drug concentration. If you take a 500-mg dose of erythromycin, how much of the drug will be in your body after 4 hours?

53. Modeling drug concentration. If a patient experiences some numbness as long as 150 mg of Novocain remain in the body, after how many complete hours should a patient stop feeling numbness if initially injected with 500 mg of Novocain?

54. Modeling drug concentration. If your physician wants to keep the level of erythromycin in your body above 200 mg and you are given an initial dose of 500 mg, at what time (to the nearest hour) should you take your next dose?

55. Modeling inflation. A pair of women's Saucony running shoes cost $120 in 2006 and the same model shoes cost $135 in 2010 due to inflation.

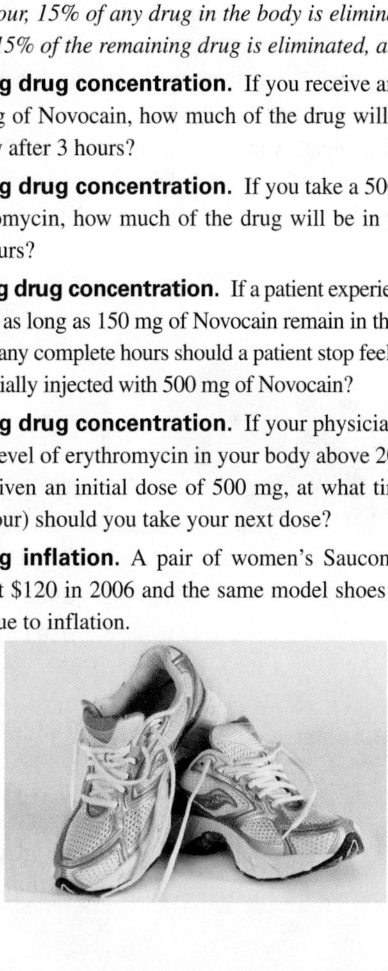

a. Develop an exponential model to describe the rate of inflation over this time period.

b. If inflation continues at the same rate, use your model in part (a) to estimate the price of the shoes in 2016.

56. Modeling inflation. Repeat Exercise 55 for a Nikon digital recorder that cost \$320 in 2005 and \$385 in 2010.

57. Modeling inflation. In the 1990s, Albania suffered from a period of runaway inflation. The inflation was so bad that if it had continued, a fast-food meal that cost \$4.65 would have cost \$1,712 10 years later. Develop an exponential model to determine the yearly rate of inflation over this time period.

58. Modeling inflation. In the 1990s, Hungary also suffered from severe inflation. If that inflation had continued at the same rate, a pair of athletic shoes that cost \$80 would have cost \$945 10 years later. Develop an exponential model to determine the yearly rate of inflation over this time period.

59. Finding an interest rate. If \$1,000 grows to \$1,302 over 6 years, determine the interest rate during that period.

60. Interest rate reduction. If a \$10,000 investment decreases to \$8,850 over a four-year period, determine the (negative) interest rate over that time.

Communicating Mathematics

61. When is an exponential model appropriate? Give examples.

62. How does a logistic model differ from an exponential model?

Between the Numbers

63. Suppose you graduate with a student loan of \$20,000 with an annual interest rate of 9%. If you were to pay only the interest on this loan, but nothing on the principal, how much would you have to pay each month?

64. If you graduate with a student loan of \$25,000 with an annual interest rate of 8%, and did not make any payments for 5 years, how much would you owe at that time?

Challenge Yourself

65. Imagine that you are investing \$1,000 for 20 years at 4% interest, compounded annually. If you were given the choice of either doubling the amount you are investing or doubling the interest rate, which would you choose?

66. Repeat Exercise 65, but now your choice is to either triple the amount you are investing or double the interest rate. Would you make a different decision if you were investing for 30 years?

Use the logistic growth model to solve Exercises 67 and 68. You may use a trial-and-error method to find your answer.

67. Assume that Carl dammed a stream on his land and intends to use the pond that has formed for bass fishing. He stocks the pond with 200 adult bass and believes that at maximum capacity, the pond could support 800 adult bass. Because he wants the bass population to grow, he will not fish from the pond until it has reached a population of 300 fish. Assuming a yearly growth rate of 18%, during which year will he be able to begin fishing?

68. Suppose that the pond in Exercise 67 can support 1,000 fish, the growth rate is 25%, and Carl wants to have 300 fish in the pond by the beginning of the fourth year. Assume that he must purchase fish for stocking in multiples of 50. What is the smallest amount that he can purchase to achieve his goal?

69. Suppose that you could begin with one cent and have that amount doubled for 30 days or you could start with one dollar and have that amount doubled every other day for the 30-day period. Which would you choose?

70. Begin with the choices from Exercise 69, but now add a third choice. You can choose to start with \$100 to be doubled every other day, starting with the second day. For how many days is starting with the \$100 the better deal?

7.5 Proportions and Variation

Objectives

1. Solve problems using ratios and proportions.
2. Understand the capture–recapture method for estimating populations.
3. Use direct and inverse variation to solve problems.

Do you know how many tigers there are in the world? What about Florida manatees? Or wild yaks? How would you go about counting such animals? Is it possible to search all the oceans, bays, and rivers of the world and count the number of leatherback turtles? Yet, biologists and ecologists need to do this. In Example 2, we will show you a surprisingly simple method to estimate such populations, which is based on the notions of ratio and proportion that we will introduce next.

Ratio and Proportion

We encounter the notion of ratio in many situations. For example, if you drive 86.4 miles and use 2.7 gallons of gasoline, then the quotient $\frac{86.4}{2.7} = 32$ miles per gallon is an example of a ratio.

> **DEFINITIONS** A **ratio** is a quotient of two numbers. We write the ratio of the numbers a to b as $a{:}b$ or $\frac{a}{b}$. A **proportion** is a statement that two ratios are equal.

The statement $\frac{3}{4} = \frac{6}{8}$ is an example of a proportion. Notice in the proportion $\frac{3}{4} = \frac{6}{8}$ if you cross multiply, $\frac{3}{4} \times \frac{6}{8}$, you get the products $3 \cdot 8 = 4 \cdot 6$. This illustrates a general principle that you will use in working with proportions.

> **CROSS-MULTIPLICATION PRINCIPLE** If $\frac{a}{b} = \frac{c}{d}$, then $a \cdot d = b \cdot c$. The quantities $a \cdot d$ and $b \cdot c$ are called **cross products.**

Sometimes one of the quantities in a proportion is unknown and we have to solve for it, as we see in Example 1.

Several years ago I had the opportunity to visit for several weeks in Colombia, South America. Colombia's currency is called the *peso*, and as we traveled around the country, I frequently had to make the type of conversion that we see in our first example.

EXAMPLE 1 Estimating Prices in a Foreign Country

Suppose you are shopping for a mummified piranha from the Amazon River as a gift for your cousin and see a price tag of 20,000 pesos in a gift shop. If earlier that morning, you exchanged 120 American dollars for 237,960 pesos, what is the cost in dollars for this piranha?

SOLUTION:

We can solve this problem easily by setting up the following proportion, which compares the ratio of dollars to pesos this morning with the ratio of dollars to pesos in the gift shop:

$$\frac{120}{237{,}960} = \frac{x}{20{,}000}.$$

(dollars in the numerators, pesos in the denominators)

If we now cross multiply, we get the equation

$$(120)(20{,}000) = 237{,}960 \cdot x, \text{ or } 2{,}400{,}000 = 237{,}960 \cdot x.$$

Dividing both sides of this equation by 237,960, we get

$$x = \frac{2{,}400{,}000}{237{,}960} \approx 10.09.$$

So the piranha will cost slightly over 10 American dollars.*

Now try Exercises 1 to 6. 13

QUIZ YOURSELF 13

If you can type 30 pages of your term paper in 2.5 hours, how long will it take you to type the entire paper, which is 54 pages long?

*Actually, when I really had to do this type of conversion I remembered that there were 1,983 pesos in 1 dollar, so I estimated that there were roughly 2,000 pesos to a dollar. For a 20,000-peso purchase, I would knock off three zeros and divide by 2 to get a reasonable estimate of the cost in dollars. Doing this with the piranha would give me an estimate of $10.

SOME GOOD ADVICE

In Example 1, we set up the proportion

$$\frac{\text{dollars this morning}}{\text{pesos this morning}} = \frac{\text{dollars in the gift shop}}{\text{pesos in the gift shop}}.$$

It would not have mattered if we had started with

$$\frac{\text{pesos}}{\text{dollars}} = \frac{\text{pesos}}{\text{dollars}},$$

because we would have obtained the same answer.

However, it is important to be consistent. Whatever comparison you make on the left-hand side of the equation, you must be sure to make the comparison in the same way on the right-hand side.

The Capture–Recapture Method

As we alluded to in the introduction of this section, an interesting application of proportions occurs in estimating the size of wildlife populations. If, for example, we wanted to survey the population of leatherback turtles, we could capture a number of turtles, tag them, and release them back to the wild. At a later time, after the tagged turtles have had sufficient time to mix thoroughly with the population, we would capture a new sample of turtles. The proportion of tagged turtles in the new sample allows us to estimate the total turtle population. We illustrate this method in Example 2.

EXAMPLE 2 The Capture–Recapture Method for Estimating Population Size

Marine biologists around the world capture 832 female breeding leatherback turtles, tag them, and release them back into the wild. Several months later, the biologists take a second sample of 900 female breeding leatherback turtles, of which 7 have been tagged. Use these data to estimate the population of female leatherback breeding turtles.

SOLUTION:

We assume that the ratio of all tagged turtles to the total number of all turtles in the population is equal to the ratio of tagged turtles in the second sample to the total number in the second sample. In other words, we can say that

$$\frac{\text{number of tagged turtles in population}}{\text{number of turtles in population}} = \frac{\text{number of tagged turtles in sample}}{\text{number of turtles in sample}}.$$

Let n be the number of turtles in the population. Also, there are 832 tagged turtles in the population and 7 tagged turtles in the sample of 900 turtles. Thus, the previous equation becomes

tagged in population —— / number in population —— $$\frac{832}{n} = \frac{7}{900}.$$ —— tagged in sample / —— number in sample

If we cross multiply, we get $7n = 832 \cdot 900 = 748{,}800$. Dividing both sides by 7 gives us that $n \approx 106{,}971$. Thus, from this survey, the number of female leatherback breeding turtles is almost 107,000.

Now try Exercises 29 to 32. 14

QUIZ YOURSELF 14

What would the number of turtles in the population have been in Example 2 if the second sample contained 1,000 turtles, of which 8 had been tagged?

Math in Your Life

What's a Country to Do?

The U.S. Constitution mandates that a census be conducted every 10 years to count each and every person residing in the United States on the night of April 1 of the census year. From 1790, with marshals riding from town to town on horseback, to the present, we have tried our best to do what is increasingly an impossible task. What's a country to do?

There have been political debates, lawsuits, and court decisions as to whether or not the capture–recapture method could be used to estimate certain portions of the U.S. population (such as the homeless) rather than doing an actual count as the Constitution specifies. In 1996, a blue-ribbon panel of the American Statistical Association found sampling to be a scientifically sound method for conducting the census. However, in 1997, Congress tried to prevent the U.S. Bureau of the Census from using any sampling method. Eventually the U.S. Supreme Court made a compromise ruling that said that although sampling could be used to count portions of the population, it could not be used in apportioning representatives to the House of Representatives.

Currently, statisticians and the U.S. Bureau of the Census are working to improve methods for counting the population before the next census in 2020.

KEY POINT

Direct variation relates quantities that increase or decrease in the same way.

Variation

Do you have a text messaging plan? If you do, then you are familiar with the mathematical notion of direct variation. For example, if your plan charges you 5 cents per text message, then the equation $c = 0.5t$ models the relationship between the number of text messages, t, and the cost, c. This is an example of *direct variation*; as the number of text messages increases or decreases, your cost varies in exactly the same way.

In another type of variation, called *inverse variation*, as one quantity increases, the other quantity decreases. The number of gallons of heating oil used per month and the outside average monthly temperature might be related in this way because the higher the temperature, the fewer gallons of heating oil you would use, and vice versa. We will now make these notions more precise.

DEFINITIONS We say that y **varies directly** as x, or that y is **directly proportional** to x, if $y = kx$, where k is a nonzero constant. The constant k is called the **constant of variation** or the **constant of proportionality**.

QUIZ YOURSELF 15

Suppose that p varies directly as q and that $p = 154$ when $q = 22$. Find p when $q = 19$.

EXAMPLE 3 Solving a Direct Variation Problem

Suppose that y varies directly as x and that $y = 50$ when $x = 15$. Find y when $x = 24$.

SOLUTION:

Often, the first step in solving a variation problem is to use the information that you are given about x and y to calculate the constant of variation. Because we are told that y varies directly as x, we begin with the equation $y = kx$ and substitute 50 for y and 15 for x to get

$$50 = k(15).$$

Next, divide both sides of this equation by 15 to get

$$k = \frac{50}{15} = \frac{10}{3}.$$

The second step is to use our knowledge that $k = \frac{10}{3}$ to rewrite the equation $y = kx$ as $y = \frac{10}{3}x$. Then, substituting 24 for x gives us $y = \frac{10}{3} \cdot 24 = \frac{240}{3} = 80$.

Now try Exercises 7 to 22. 15

Sometimes in dealing with direct variation, there may be more than two related quantities. This more complex relationship is called *joint variation.* However, as you will see in Example 4, the two-step solution process that we used in Example 3 is essentially the same.

EXAMPLE 4 Calculating the Strength of a Beam

A beam in a viewing booth for officials at the Daytona 500 auto race is being replaced. The current beam is 3.5 inches wide and 6 inches deep and will support a weight of 1,200 pounds. If a replacement beam with the same length is 3 inches wide and 7 inches deep, how much weight can it support? Use the fact that the strength of a beam is directly proportional to its width and the square of its depth.

SOLUTION: Review the Relate a New Problem to an Older One Strategy on page 9.

Step 1: We begin with the equation $s = kwd^2$ and substitute the values given to us for w, d, and s to find the constant of variation k. That is,

$$1{,}200 = k(\underset{w}{3.5})(\underset{d}{6})^2, \text{ or } 1{,}200 = k \cdot 126.$$

Dividing both sides of this equation by 126, we get

$$k = \frac{1{,}200}{126} = \frac{200}{21}.$$

Step 2: We can now use this value for k to rewrite the equation $s = kwd^2$ as $s = \frac{200}{21}wd^2$. If we substitute the values 3 for w and 7 for d, we find that

$$s = \overset{k}{\frac{200}{21}}wd^2 = \frac{200}{21} \cdot 3 \cdot (7^2) = 1{,}400 \text{ pounds.}$$

If we say that two quantities vary inversely, this means that as one increases, the other decreases.

KEY POINT

Inverse variation relates quantities that increase or decrease in opposite ways.

DEFINITIONS We say that y **varies inversely** as x, or that y is **inversely proportional** to x, if $y = \frac{k}{x}$, where k is a nonzero constant.

A good example of inverse variation is the relationship between speed and time to travel a certain distance—the faster you go, the shorter the time it takes to travel a certain distance.

EXAMPLE 5 Time Saved by Speeding

Suppose that the speed limit is 65 miles per hour and a person (not you, of course—your friend) likes to drive just a little over the speed limit. If your friend takes a trip and it takes him $1\frac{1}{2}$ hours going at 65 miles per hour, how much time will he save by going 70 miles per hour? Use the fact that time is inversely proportional to speed.

SOLUTION:

We will use the equation $t = \frac{k}{s}$, where t is time and s is speed. We will think of $1\frac{1}{2}$ hours as 90 minutes. So the previous equation becomes $90 = \frac{k}{65}$. Multiplying both sides of this equation by 65 gives us $90 \cdot 65 = 1 \cdot k = k$. Multiplying this out, we find that $k = 90 \cdot 65 = 5{,}850$.

Next, we rewrite our time–speed equation as $t = \frac{5{,}850}{s}$. Substituting the value 70 for s gives us $t = \frac{5{,}850}{70} \approx 83.6$ minutes, or 1 hour and 23.6 minutes. So all your friend has saved by speeding for over an hour is $90 - 83.6 = 6.4$ minutes.

Now try Exercises 33 and 34. 16

QUIZ YOURSELF 16

Repeat Example 5 using a speed of 75 miles per hour.

It is possible to have a combination of both direct variation and inverse variation present in a relationship among several quantities. We will refine the model of the strength of a beam in Example 4 to illustrate this *combined variation.*

EXAMPLE 6 Calculating the Strength of a Beam

As we stated in Example 4, the strength of a beam is directly proportional to its width and the square of its depth. However, as you probably know from practical experience, the longer a beam is, the less weight it will support. That is, the strength of a beam varies inversely as its length. Suppose a wooden beam that is used to support lighting for an outdoor Shakespearean theater is 10 feet long, 3 inches wide, and 4 inches deep and will support a load of 600 pounds.

a) If the length of the beam is increased to 15 feet, how many pounds will the beam support?

b) If we want the 15-foot beam to support the same weight of 600 pounds, how wide should the beam be to give us the same strength?

SOLUTION: Review the Relate a New Problem to an Older One Strategy on page 9.

a) We will model this situation with the equation

$$s = k\frac{w \cdot d^2}{l}.$$

s varies directly as *w* and d^2.
s varies inversely as *l*.

We find the constant of variation, *k*, by substituting for the variables as follows:

$$600 = k\frac{3 \cdot 4^2}{10}.$$

(*s*, *w*, d^2, *l*) We substituted 3 for *w*, 4 for *d*, 10 for *l*, and 600 for *s*.

Therefore,

$$6{,}000 = k(48).$$

We multiplied both sides by 10 and simplified $3 \cdot 4^2$.

Dividing both sides by 48, we get

$$k = \frac{6{,}000}{48} = 125.$$

We can then rewrite the strength equation as

$$s = 125\frac{w \cdot d^2}{l}.$$

If we now keep the values 3 for *w* and 4 for *d*, but substitute 15 for *l*, we get

$$s = 125\frac{3 \cdot 4^2}{15} = \frac{125 \cdot 3 \cdot 16}{15} = 400.$$

(*k*, *w*, d^2)

Thus, the longer beam will support only 400 pounds.

b) To solve this problem, we will use the values 4 for *d*, 15 for *l*, and 600 for *s*, and we must solve for the value of *w*. Substituting, we get the equation

$$600 = 125\frac{w \cdot 4^2}{15} = \frac{125 \cdot w \cdot 16}{15}$$

or

$$600 \cdot 15 = 125 \cdot 16 \cdot w.$$

Therefore, $9{,}000 = 2{,}000 \cdot w$, or $w = \frac{9{,}000}{2{,}000} = \frac{9}{2} = 4.5$. So, the longer beam should be 4.5 inches wide in order to support 600 pounds.

Now try Exercises 35 to 38.

EXERCISES 7.5

Sharpening Your Skills

Solve for x in the following proportions.

1. $24:x = 18:3$
2. $35:4 = x:2$
3. $\frac{50}{4} = \frac{x}{5}$
4. $\frac{x}{8} = \frac{14}{4}$
5. $\frac{30}{40} = \frac{27}{x}$
6. $\frac{150}{x} = \frac{60}{40}$

Solve each of the following variation problems by first writing the variation as an equation and finding the constant of variation. Then answer the question that we ask.

7. Assume that y varies directly as x. If $y = 37.5$ when $x = 7.5$, what is the value for y when $x = 13$?
8. Assume that y varies inversely as x. If $y = 10$ when $x = 4$, what is the value for y when $x = 6$?
9. Assume that r varies inversely as s. If $r = 12$ when $s = \frac{2}{3}$, what is the value for r when $s = 8$?
10. Assume that d varies directly as the square of t. If $d = 24$ when $t = 4$, what is the value for d when $t = 10$?
11. Assume that a varies directly as the square of b. If $a = 16$ when $b = 6$, what is the value for a when $b = 15$?
12. Assume that y varies jointly as x and z. If $y = 60$ when $x = 4$ and $z = 5$, what is the value for z when $x = 6$ and $y = 45$?
13. Assume that D varies inversely as C. If $D = \frac{3}{4}$ when $C = 2$, what is the value for D when $C = 24$?
14. Assume that A varies directly as the square of r. If $A = 314$ when $r = 10$, what is the value for A when $r = 6$?
15. Assume that r varies jointly as x and y. If $r = 12.5$ when $x = 2$ and $y = 5$, what is the value for r when $x = 8$ and $y = 2.5$?
16. Assume that m varies inversely as n. If $m = 6$ when $n = \frac{2}{3}$, what is the value for m when $n = 15$?
17. Assume that y varies jointly as w and x^2. If $y = 504$ when $w = 4$ and $x = 6$, what is the value of x when $w = 10$ and $y = 6{,}860$?
18. Assume that r varies jointly as s and t^2. If $r = 5{,}600$ when $s = 14$ and $t = 8$, what is the value of s when $t = 22$ and $r = 18{,}150$?
19. Assume that y varies directly as w and inversely as x. If $y = 4$ when $x = 10$ and $w = 6$, what is the value of y when $x = 15$ and $w = 3$?
20. Assume that p varies directly as q and inversely as r. If $p = 6$ when $q = 8$ and $r = 5$, what is the value of r when $p = 6$ and $q = 4$?
21. Assume that y varies jointly as x^2 and w and inversely as z. If $y = 15$ when $x = 2.5$, $w = 8$, and $z = 20$, what is the value of y when $x = 8$, $w = 7$, and $z = 14$?
22. Assume that d varies jointly as a^2 and b and inversely as c. If $d = 288$ when $a = 6$, $b = 10$, and $c = 4$, what is the value of d when $a = 20$, $b = 11$, and $c = 4$?

Applying What You've Learned

In Exercises 23–28, set up a proportion to solve the given problem.

23. **Calculating drug dosage.** The dosage of a particular drug is proportional to the patient's body weight. If the dosage for a 150-pound woman is 6 milligrams, what would the dosage be for her daughter Maria, who weighs 65 pounds?
24. **Calculating a speed limit.** While volunteering in Africa, Brad and Angelina rented a car whose speedometer is calibrated in both miles and kilometers per hour. The marking on the speedometer shows that 30 miles per hour corresponds to 48 kilometers per hour. If the speed limit in Africa is 100 kilometers per hour, what is the speed limit in miles per hour?
25. **Estimating a distance.** While driving from Kenya to Tanzania, Brad sees a road sign that says "Kilimanjaro 56 kilometers." How many miles does he still have to drive? (See Exercise 24.)
26. **Buying fertilizer.** Garth's front lawn is a rectangle measuring 120 feet by 40 feet. If a 25-pound bag of "Weed 'N' Feed" will treat 2,000 square feet, how many bags must Garth buy to treat his lawn? (Assume that he must buy whole bags.)
27. **Mowing grass.** Caroline has a part-time job on campus mowing grass. If it takes her $1\frac{1}{2}$ hours to mow a 60,000-square-foot lawn, how long will it take her to mow the rectangular lawn in front of the student center that is 200 feet wide and 650 feet long?
28. **Cost of carpet.** Jose is a hotel manager who paid \$864 to have a carpet installed in a conference room that measures 18 feet by 27 feet. How much would he have to pay to have the same carpet installed in a room that measures 24 by 33 feet?

Use the capture–recapture method in Exercises 29–32. Use the word equation stated in Example 2.

29. **Estimating a wildlife population.** Biologists capture, tag, and release 400 bald eagles. Several months later, of 240 bald eagles that are captured, 8 are tagged. Estimate the population of bald eagles.
30. **Estimating a wildlife population.** Marine biologists capture, tag, and release 100 Florida manatees. Several months later, of 90 captured manatees, 5 have tags. Estimate the population of Florida manatees.
31. **Estimating a wildlife population.** It is estimated that there are 1,000 grizzly bears living in a certain region. Assume that biologists capture, tag, and release 55 bears. Several months later, they capture a sample of 95 bears. How many would you expect to find tagged?

32. **Estimating a wildlife population.** A lake contains 1,530 largemouth bass. Employees of the state fish commission catch, tag, and release 60 of the bass. If two months later they recapture 106 largemouth bass, how many would you expect to find tagged?

Exercises 33 and 34 are based on Example 5.

33. **Speeding.** If the speed limit is 60 miles per hour and your trip takes 2 hours, how much time do you save by traveling at 65 miles per hour?

34. **Speeding.** If the speed limit is 65 miles per hour and your trip takes 2 hours, how much time do you save by traveling at 75 miles per hour?

Exercises 35–38 are based on the strength of a beam model in Example 6.

35. **Strength of a beam.** If a beam that is 6 inches wide, 8 inches deep, and 4 feet long can support a weight of 672 pounds, how much weight could the same type of beam that is 4 inches wide, 6 inches deep, and 8 feet long support?

36. **Strength of a beam.** If a beam that is 3 inches wide, 10 inches deep, and 10 feet long can support a weight of 480 pounds, how much weight could the same type of beam that is 4 inches wide, 6 inches deep, and 8 feet long support?

37. **Strength of a beam.** A beam that is 4 inches wide, 8 inches deep, and 12 feet long can support a weight of 1,280 pounds. If the same type of beam that is 3 inches wide and 6 inches deep can support a weight of 540 pounds, how long is it?

38. **Strength of a beam.** A beam that is 5 inches wide, 10 inches deep, and 16 feet long can support a weight of 875 pounds. If the same type of beam with the same length is 4 inches wide and can support a weight of 343 pounds, how deep is it?

39. **Converting dollars to euros.** If one euro = 1.2937 U.S. dollars, how many euros can we get for $500 U.S. dollars?

40. **Converting dollars to pesos.** If 100 Mexican pesos equal $7.16 U.S. dollars, how many pesos would we get for $200 U.S. dollars?

41. **Estimating water usage.** Assume that the amount of water that Mario uses for irrigation at his vineyard is inversely proportional to the amount of rainfall. If he uses 30,000 gallons during a month in which there is 3 inches of rain, how much water would he use in a month that has 5 inches of rain?

42. **Determining the maximum strength of a spring.** Hooke's law states that the length a spring can be stretched is directly proportional to the force applied to the spring. However, if too much force is applied to the spring, the spring can be stretched to that point at which it can no longer return to its original shape. If a force of 8 pounds stretches a spring 6 inches and stretching the spring beyond 10 inches will ruin the spring, what is the maximum force that can be applied to the spring without damaging it?

43. **Finding the distance a skydiver falls.** The distance that a body falls varies directly as the square of the time that it is falling. If Eileen jumps from a plane and falls 144 feet in the first 3 seconds, how many feet will she fall in 5 seconds?

44. **Adjusting a photographer's lighting.** The illumination from a light source is inversely proportional to the square of the distance from the light source. Ansel is taking a portrait with his light source set 4 feet from his subject and finds that the illumination is twice as bright as it should be. If he wants to reduce the illumination to one-half of what it is now, at what distance should he place his light?

45. **Braking distance of a car.** The braking distance of a car varies directly as the square of the speed of the car. Assume that a car traveling at 30 miles per hour (mph) can stop 43 feet after the brakes are applied. How long is the braking distance for that same car traveling at 60 mph?

46. **Braking distance of a car.** Assume that a car traveling at 80 mph has a braking distance of 305 feet. How long is the braking distance for that same car traveling at 50 mph?

47. **Calculating the amount of heating oil needed.** Assume that the amount of heating oil used during a given month is inversely proportional to the outside temperature. If Andrea's pet store uses 504 gallons of oil in a month that has an average temperature of 42°F, how much oil would she expect to use in a month (of the same length) with an average daily temperature of 36°F?

48. **Determining waterpark attendance.** In the summer, the monthly attendance at Six Flag's waterpark varies directly as the temperature and inversely as the number of days of rain during the month. If during a given month the average daily temperature is 88°F and it rains 8 days, the total attendance for the month is 3,200. What attendance should we expect for a month (of the same length) if it rains 12 days and the average daily temperature is 92°F?

49. **Finding gas pressure.** If the temperature is held constant, the pressure of a gas in a container varies inversely as the volume of the container. Assume that the gas pressure in a large piston cylinder (the container) is 4 pounds per square inch when the volume of the cylinder is 120 cubic inches. If the volume of the cylinder is reduced to 75 cubic inches, what is the pressure of the gas?

50. **Finding gas pressure.** Redo Exercise 49. Assume now that the gas pressure in a large cylinder is 12 pounds per square inch when the volume of the cylinder is 6.5 cubic inches. If the pressure of the gas in the cylinder is increased to 48 pounds per square inch, what is the volume of the cylinder?

Communicating Mathematics

51. If y varies directly as x and x increases, what does y do?

52. If y varies inversely as x and x increases, what does y do?

53. If y varies inversely as x, does x vary inversely or directly as y?

54. Assume that last semester at Urbanopolis City College, out of a student body of 5,250 there were 1,470 students who made the dean's list. If this semester the percentage who made the

dean's list is the same and 1,554 made the dean's list, what is the current size of the student body at UCC? In solving this problem, Justin set up the equation

$$\frac{5{,}250}{1{,}470} = \frac{1{,}554}{x}.$$

What answer did Justin get and why is it obviously incorrect? Explain what is wrong with Justin's approach.

As you saw in Example 6, the strength of a beam is directly proportional to its width and the square of its depth and inversely proportional to its length. An equation describing this combined variation is

$$s = k\frac{w \cdot d^2}{l}.$$

Use this information to explain how the strength of a beam would change in each of the following situations. It may help if you make up concrete examples.

55. The width of the beam is doubled.

56. Both the width and the length of the beam are doubled.

57. Both the length and the depth of the beam are tripled.

58. The length, width, and depth of the beam are all doubled.

59. Braking distance of a car. In Exercise 45 we stated that the braking distance of a car varies directly as the square of its speed. When finding the stopping time of a car, in addition to the braking distance, we must also consider the reaction distance (the distance the car travels before the driver applies the brakes), which varies directly as the speed of the car. Assume that the braking distance for a car traveling at 40 mph is 76 feet and the reaction distance is 88 feet. What is the total stopping distance of a car traveling at 60 mph?

60. Braking distance of a car. Redo Exercise 59, but now assume that at 35 mph, the reaction distance of the car is 77 feet and the braking distance is 59 feet. What is the total stopping distance of a car traveling at 75 mph?

61. Suppose that in solving a problem, you encounter the ratio $m{:}n$. What would the ratio $n{:}(m + n)$ represent? Explain your answer.

62. What is $(m{:}n) \times (n{:}m)$? Explain. Give an example.

Challenge Yourself

63. If the ratio of PCs to Macintoshes on East Central State's campus is 7:2, what is the ratio of PCs to all computers on campus? (Assume that there are no other computers except PCs and Macs.) Explain your thinking. You may want to give an example to support your conclusion.

7.6 Modeling with Systems of Linear Equations and Inequalities

Objectives

1. Represent systems of linear equations and inequalities graphically.
2. Solve a system of linear equations using the elimination method.
3. Solve a system of linear inequalities graphically.
4. Use systems of linear equations and inequalities as models.

Often, as one technology grows in popularity, another declines in favor. For example, as more people stream their favorite songs from the cloud, fewer buy their music on CDs.

As another example, consider the following graph. It shows that from 2004 to 2009, as digital TV sales increased, there was a corresponding decline in the sale of DVD players.

This graph models the change in sales of both digital TVs and DVD players with linear equations. If you were a marketing analyst for a big box electronics store, you might be interested in where the red graph crosses the blue graph.

We can model many problems by a collection of linear equations called a **system of linear equations.** Every day scientists use such systems having thousands of equations to describe such complex things as the weather or the U.S. economy. For our purposes, however, we will only be working with simpler systems having two equations and two unknowns.

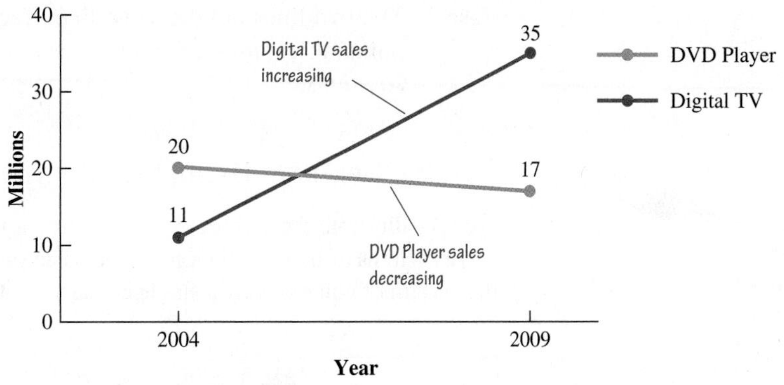

Source: New York Times Almanac, 2011

Systems of Linear Equations

KEY POINT

A system of linear equations represents a pair of lines.

A **solution** of a pair of linear equations is an ordered pair of numbers that satisfies both equations. For example, if you substitute 5 for x and 3 for y in the system

$$2x + 4y = 22$$
$$x - 6y = -13,$$

you will get

$$\overset{x=5}{2 \cdot 5} + 4 \cdot \overset{y=3}{3} = 22$$
$$5 - 6 \cdot 3 = -13.$$

QUIZ YOURSELF 17

Which ordered pair is a solution for the system

$$3x - 5y = 4$$
$$2x + 4y = 10?$$

a) (2, 3)
b) (3, 1)
c) (−3, −1)

Because the ordered pair (5, 3) makes both equations true, it is a solution for the system. Although (7, 2) makes the first equation true in the system, it does not make the second equation true (verify). Therefore, it is not a solution for the system. 17

Recall from Section 7.1 that the graph of a linear equation in two unknowns is a straight line. Therefore, if you graph a system of two linear equations in two unknowns, you will get a *pair* of lines. There are three possible situations, as we show in Figure 7.20 below.

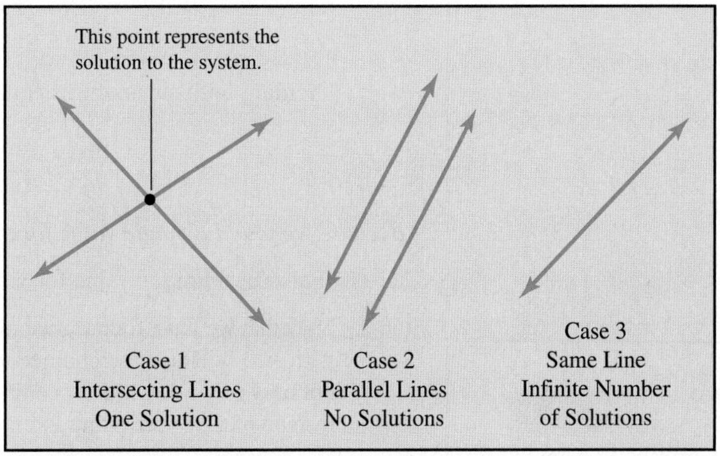

FIGURE 7.20 A geometric representation of the three cases that may arise in solving a system of linear equations.

Case 1: The lines intersect in a single point. The ordered pair that represents this point is the *unique solution* for the system.

Case 2: The lines are distinct parallel lines and therefore don't intersect at all. Because the lines have no common points, this means that the system has *no solutions.*

Case 3: The two lines are the same line. Because the lines have an infinite number of points in common, this means that the system will have *an infinite number of solutions.*

Solving Systems Using the Elimination Method

KEY POINT

We use the elimination method to solve systems of linear equations.

We will illustrate these three cases by using a technique called the **elimination method** to solve systems of linear equations.* The basic strategy of the elimination method is to replace the system of equations by a single equation that is easier to solve, as we see in Example 1.

EXAMPLE 1 Solving a System That Has One Solution

Solve the system

$$\begin{aligned} 3x + 4y &= 10 \\ 5x - 6y &= 4. \end{aligned} \qquad (1)$$

SOLUTION:

To use the elimination method, we will multiply both equations in the system by some constants so that one of the variables appears with *opposite coefficients* in the two equations. If we then add these two equations, the variable with the opposite coefficients will drop out, leaving us with a single equation having one unknown.

Step 1: We have to decide which unknown to eliminate first. We recognize that if we decide to eliminate the y terms, we can avoid multiplying by negative numbers, which will make our work a little bit easier. So, we will multiply the top equation in system (1) by 3 and the bottom equation by 2 to get the system

$$\begin{aligned} 9x + 12y &= 30 \\ 10x - 12y &= 8. \end{aligned} \qquad (2)$$

Now, adding corresponding sides of both equations in (2) causes the y to drop out, giving us

$$\begin{array}{r} 9x + 12y = 30 \\ +10x - 12y = \ \ 8 \\ \hline 19x + \ \ 0 \ \ = 38. \end{array}$$

Dividing both sides of this equation by 19, we get

$$\frac{19x}{19} = \frac{38}{19}, \quad \text{or} \quad x = 2.$$

This tells you that there is one solution to the system.

Because we found a single value for x, we know that we are dealing with Case 1.

There will also be a unique value for y, which we will find next.

Step 2: Now that we have found x, we must find y. Since the value for x is easy to work with, we will use a technique called **back-substitution** to find y. (If x were not easy to work with, we could go back to the original system and solve for y instead of x.) Substituting 2 for x in the first equation of system (1) we get

$x = 2$

$$3(2) + 4y = 10.$$

Subtracting 6 from both sides, this simplifies to $4y = 4$, so $y = 1$. Thus, the solution for this system is (2, 1). You should verify that (2, 1) solves both equations. 18

QUIZ YOURSELF 18

Use the elimination method to solve the following system:

$$\begin{aligned} 5x + 2y &= -10 \\ 2x + 3y &= 7 \end{aligned}$$

*We will explain another method for solving systems, called the **substitution method,** in the exercises.

SOME GOOD ADVICE

If you look at a situation from both a geometric and algebraic point of view, you often gain insight into how to solve a problem or check your solution. If the geometric and algebraic representations of a solution do not agree, then you must go back and find your error.

In Figure 7.21, we graph system (1) from Example 1 on a graphing calculator, where you can see that the lines appear to intersect at the point (2, 1).

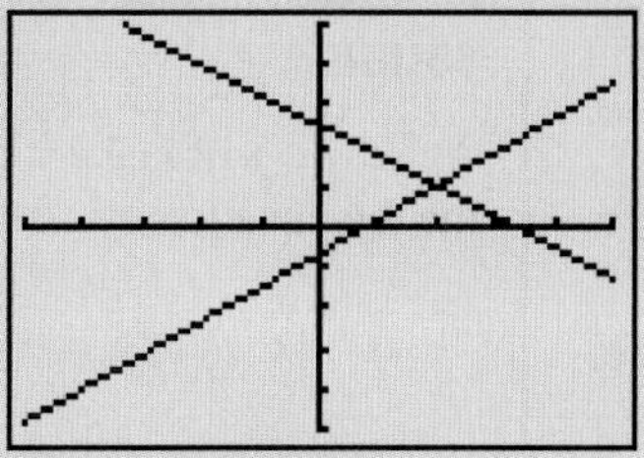

FIGURE 7.21 Graph of system (1).

EXAMPLE 2 Trying to Solve a System That Has No Solutions

Solve the system

$$-\frac{3}{2}x + y = \frac{5}{4}$$
$$3x - 2y = 1.$$

SOLUTION:

It is a good idea to make your work a little easier in a situation such as this by multiplying both sides of the top equation by 4 to clear the fractions.

$$(\overset{2}{\cancel{4}})\left(-\frac{3}{\cancel{2}}\right)x + (4)y = (\cancel{4})\frac{5}{\cancel{4}}$$

So our system becomes the simpler-looking system

$$-6x + 4y = 5$$
$$3x - 2y = 1.$$

If we multiply the bottom equation by 2 and add the equations to eliminate x from the system, we get a surprising result.

$$-6x + 4y = 5$$
$$+6x - 4y = 2$$
$$0 + 0 = 7$$

This tells you that there is no solution to the system.

We have assumed that there is a solution to this system. By doing this, we are now forced to accept that $0 + 0 = 7$. Our only way out of this predicament is to conclude that our assumption is wrong, so a solution for this system does not exist.*

In trying to eliminate one of the variables, we obtained a false statement. We conclude that there are no points common to both lines and therefore we are dealing with Case 2.

A system that has no solutions, such as the one in Example 2, is said to be **inconsistent.**

*If you have studied the logic chapter, you can look at what we have done as: If there is a solution, then $0 + 0 = 7$. The contrapositive states: If $0 + 0 \neq 7$, then there is no solution.

EXAMPLE 3 Solving a System Having Infinitely Many Solutions

Solve the system

$$\begin{aligned} 0.2x - 0.1y &= 0.3 \\ 0.1x - 0.05y &= 0.15. \end{aligned} \qquad (1)$$

SOLUTION:

Again we will make it easy on ourselves by getting rid of the decimals. If we multiply the top equation by 10 and the bottom equation by 100, we can rewrite system (1) as

$$\begin{aligned} 2x - y &= 3 \\ 10x - 5y &= 15. \end{aligned} \qquad (2)$$

If we now multiply the top equation by -5 and add the two equations, we can eliminate x from the system. Again we get a surprising result.

$$\begin{aligned} -10x + 5y &= -15 \\ +10x - 5y &= 15 \\ \hline 0 + 0 &= 0 \end{aligned}$$

This tells you that there are an infinite number of solutions to the system.

You might be tempted to confuse this with Case 2. However, unlike Example 2, the statement $0 + 0 = 0$ is a *true* statement. In trying to solve for x, we obtained a statement that in no way restricts x. This means that we can use any x we want. If we substitute 5 for x in the top equation of system (2), we get $2 \cdot 5 - y = 3$. Solving for y gives us $y = 7$. Thus, (5, 7) is one of the many pairs that are solutions for this system.

If in solving the system we obtain a true statement that does not involve either x *or* y*, this tells us that there are an infinite number of ways to solve the system, so the two lines must be the same.*

Now try Exercises 5 to 18. 19

QUIZ YOURSELF 19

a) Solve the system

$$\begin{aligned} 5x - 3y &= 3 \\ -10x + 6y &= 4. \end{aligned}$$

b) What is your geometric interpretation of the result from part a)?

A system that has an infinite number of solutions, such as the system in Example 3, is said to be **dependent.**

We summarize what you saw happening in Examples 1 to 3.

USING THE ELIMINATION METHOD TO SOLVE SYSTEMS OF LINEAR EQUATIONS

Case 1: We find a single value for both *x* and *y*. The pair (*x*, *y*) corresponds to the point of intersection of the lines represented by the equations in the system.

Case 2: We obtain an obviously false statement. We conclude that there are no solutions to this system, which represents a pair of distinct parallel lines.

Case 3: We obtain a statement that is always true and that does not contain either *x* or *y*. Any value can be used for *x* as the first coordinate of a solution of the system. There are an infinite number of solutions to the system, and the two equations of the system represent the same line.

KEY POINT

A system of equations models a set of relationships between two quantities.

Modeling with Systems of Equations

Now that you know how to solve systems of equations, you can use them to model situations involving several relationships among a set of variables. The next several examples illustrate this.

Math in Your Life

Linear Equations and Angry Birds

In building models, linear equations are easy to work with, so mathematicians often use them instead of more complex equations. Many times a good approximation can give us an acceptable answer to our problem.

For example, if you look carefully at a smart phone's screen, you may be able to see angular edges instead of the smooth face of a character. Curved surfaces take too long to compute, so game designers use tiny planar surfaces and sophisticated shading techniques instead of drawing the true curved surfaces.* Planes are the geometric representation of linear equations in three variables. So, the game designer is replacing nonlinear surfaces in three-dimensional space with linear approximations to speed up the computations. In other words, using linear equations for models allows you to play Angry Birds on your iPhone at real-time speed.

The **law of demand** in economics states that as the price of an item increases, the consumer is less willing to purchase it. According to the **law of supply,** as the price of an item increases, the producer is willing to produce more of the item. If the price is low, consumer demand increases, but because producers are not willing to produce much, there will be a shortage of the product. On the other hand, if the price is high, producers will produce more of the product, but because consumers are not willing to pay the high prices, there is a surplus of the item. Eventually the market for the item adjusts so that there will be a price at which the quantity demanded and the quantity supplied are equal. This point is called an **equilibrium point.** We illustrate this notion of equilibrium in Example 4.

EXAMPLE 4 The Supply and Demand for Hand-Crafted Jewelry

Malik sells hand-crafted turquoise, Native American jewelry at regional craft shows. At a price of \$20 per necklace, he is willing to buy 30 necklaces from his suppliers. However, at this price, his suppliers will only provide him with 9 necklaces. However, if he will pay \$60 per necklace, he can only sell 15 necklaces per show. At this higher price, his suppliers will provide him with 29 necklaces. Assuming that the equations relating price to demand and to supply are both linear, what should the price per necklace be for supply to equal demand?

SOLUTION:

We summarize the supply–demand information in the following graph:

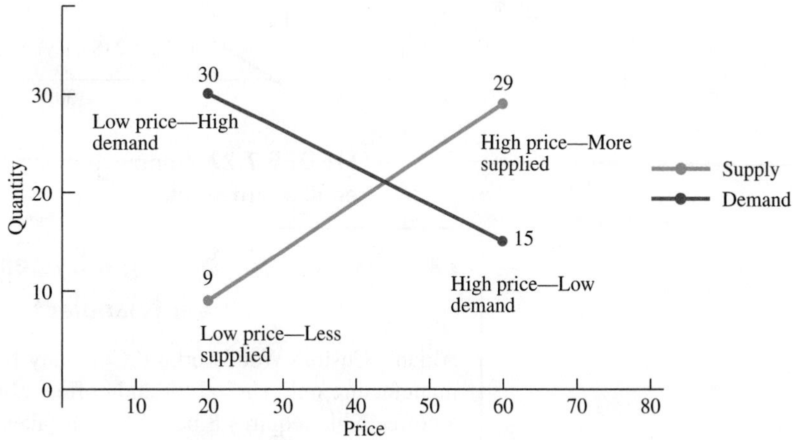

*As video game technology improves, it is harder to see these edges; however, be assured that in creating three-dimensional figures, the common technique is to approximate curved surfaces by flat polygons and then render them by using exotic shading methods. The same technique is used in creating an animated movie such as *Toy Story 3*.

The Supply Equation: Because the supply equation is linear, its graph is a line passing through the points (20, 9) and (60, 29). We can use the method we explained in Example 3 on page 315 to find the supply equation to be $x - 2y = 2$ (verify).

The Demand Equation: Similarly, the graph of the demand equation is a line passing through the points (20, 30) and (60, 15) and again, using the method on page 315, we find the demand equation to be $3x + 8y = 300$ (verify).

Therefore, the following system describes this supply and demand situation:

$$\begin{aligned} x - 2y &= 2 \qquad \text{(supply)} \\ 3x + 8y &= 300 \qquad \text{(demand)} \end{aligned}$$

The Equilibrium Point: We now solve this system to find an equilibrium point.

Multiplying the supply equation by 4 and adding gives us

$$\begin{aligned} 4x - 8y &= 8 \\ +3x + 8y &= 300 \\ \hline 7x \quad\quad &= 308 \end{aligned}$$

Dividing both sides of the equation by 7 results in $x = 44$.

We will now **back-substitute** 44 for x in the demand equation, $3x + 8y = 300$, to get

$$3(44) + 8y = 300, \quad \text{or} \quad 132 + 8y = 300.$$

Subtracting 132 from both sides, we get $8y = 300 - 132 = 168$. So, $y = \frac{168}{8} = 21$.

This means that at $44 per necklace, Malik will be willing to buy 21 necklaces from the supplier, who will be willing to sell 21 necklaces at this price.

Now try Exercises 55 to 58.

In the spirit of Descartes's analytic geometry, it is enlightening to look at this supply–demand situation again geometrically in Figure 7.22.

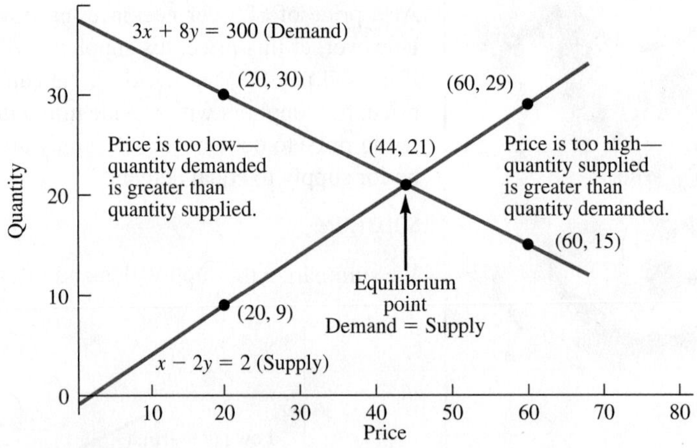

FIGURE 7.22 Supply and demand graphs intersect at the equilibrium point.

EXAMPLE 5 Solving a System of Equations in a Manufacturing Situation

Aidan's Custom Woodworking Company builds cherry furniture. This week, they will manufacture only end tables and coffee tables. An end table requires 6 board feet and a coffee table requires 8 board feet. It takes 2 hours of labor to make an end table and 4 hours of labor to make a coffee table. The company has 1,200 board feet of cherry wood available and also 480 hours of labor. Assume that we want to use all the labor and wood. Describe the conditions on wood and labor as a system of two linear equations in two unknowns. Solve this system and interpret your answer.

SOLUTION: Review the Choose Good Names Strategy on page 5.

It is helpful to organize this information as in Table 7.3, which we will call a **resource table.** We will represent the number of end tables by e and the number of coffee tables by c.

Resources	Needed for an End Table (e)	Needed for a Coffee Table (c)	Available
Wood	6 board feet	8 board feet	1,200 board feet
Labor	2 hours	4 hours	480 hours

TABLE 7.3 Resource table for furniture manufacturing problem.

$6e + 8c = 1{,}200$ — total wood available
- $8c$: wood needed for c coffee tables
- $6e$: wood needed for e end tables

FIGURE 7.23 Equation describing use of wood.

First let's consider the condition on the wood needed to manufacture the tables. Because we use 6 board feet for each end table and 8 board feet for each coffee table and have 1,200 board feet available, we get the equation as seen in Figure 7.23.

Let's now look at the restrictions on labor. We use 2 hours for each end table and 4 for each coffee table. Therefore, the equation $2e + 4c = 480$ describes how to use the 480 available hours to construct e end tables and c coffee tables. We will now solve the system:

$$6e + 8c = 1{,}200 \quad \text{(wood)}$$
$$2e + 4c = 480. \quad \text{(labor)}$$

We multiply the second equation by -2 to get

$$6e + 8c = 1{,}200$$
$$-4e - 8c = -960.$$

When we add these two equations, we get the equation $2e = 240$, whose solution is $e = 120$. If we back-substitute 120 for e in the second equation of the original system, we get $2(120) + 4c = 480$, which means that $c = 60$. Therefore, if the company manufactures 120 end tables and 60 coffee tables, all 1,200 board feet of wood and 480 hours of labor will be used up.

Now try Exercises 45 to 54.

Solving Linear Inequalities

In life you frequently deal with quantities that are not equal. Your parents may expect you to earn *at least* \$2,000 toward next year's tuition, or your doctor might want you to keep your weight *below* 160 pounds.

In building mathematical models, we also often deal with quantities that are not equal. The techniques we use for working with inequalities are similar to those that you have already learned in working with equations.

DEFINITION A **linear inequality in two variables** is a statement we can write in one of the following forms:

$ax + by \geq c$ (read $ax + by$ is greater than or equal to c),

$ax + by > c$ (read $ax + by$ is greater than c),

$ax + by \leq c$ (read $ax + by$ is less than or equal to c),

$ax + by < c$ (read $ax + by$ is less than c),

where a, b, and c are real numbers with both a and b not equal to zero.

Some examples of linear inequalities are $3x + 2y \leq 6$, $x - 5y > 7$, $y < 6$, and $x \geq 2$. We will usually omit reference to the number of variables.

As with linear equations in two variables, solutions of linear inequalities are ordered pairs of numbers. Because ordered pairs of real numbers correspond to points in the plane, we will often use the terminology "ordered pair of numbers" and "point" interchangeably.

KEY POINT

The solution to a linear inequality in two variables is a half-plane.

EXAMPLE 6 The Solution Set of an Inequality

Which of the following pairs are solutions of the inequality $2x + 3y \le 6$?

a) (2, 3) b) (0, 2) c) (−4, 3) d) (5, 2) e) (2, −4)

f) (1, −6) g) (1, 4) h) (−2, −5) i) (4, 3) j) (−1, 2)

SOLUTION: Review the Three-Way Principle (look at examples) on page 13.

We will make a table to check several of these points.

	x	y	$2x + 3y$	Is $2x + 3y \le 6$?
a)	2	3	$2 \cdot 2 + 3 \cdot 3 = 13$	False
b)	0	2	$2 \cdot 0 + 3 \cdot 2 = 6$	True
c)	−4	3	$2 \cdot (-4) + 3 \cdot 3 = 1$	True
d)	5	2	$2 \cdot 5 + 3 \cdot 2 = 16$	False
e)	2	−4	$2 \cdot 2 + 3 \cdot (-4) = -8$	True
f)	1	−6	$2 \cdot 1 + 3 \cdot (-6) = -16$	True

We will leave it to you to check the pairs g) to j) in Quiz Yourself 20. 20

QUIZ YOURSELF 20

Test points g) to j) in Example 6 to determine which ones satisfy the inequality $2x + 3y \le 6$.

We plot the ten points from Example 6 and Quiz Yourself 20 in Figure 7.24, which gives us some insight into the solution set of $2x + 3y \le 6$.

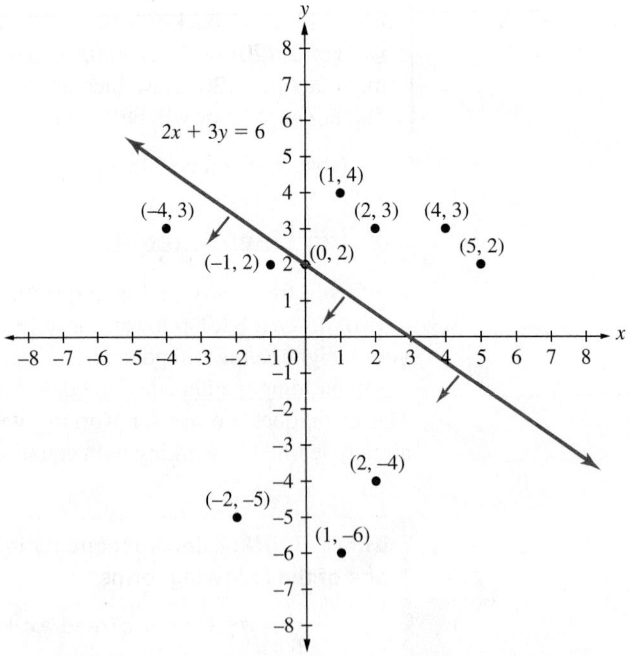

FIGURE 7.24 Points below or on the line $2x + 3y = 6$ are solutions; points above the line are not solutions.

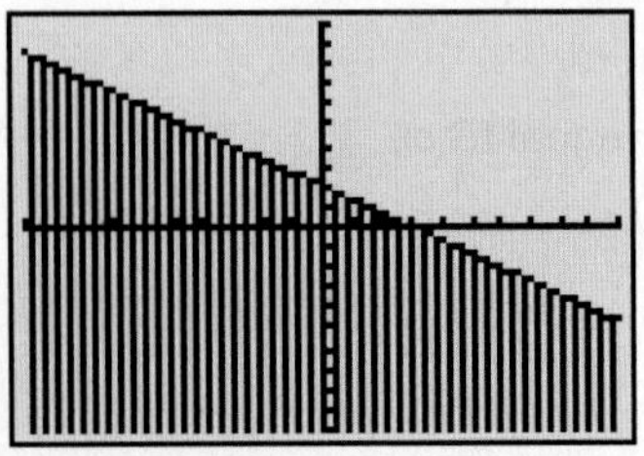

Graph of $2x + 3y \le 6$ drawn with a graphing calculator.

Notice in Figure 7.24 that all of the pairs we tested that satisfy the inequality lie below or on the line $2x + 3y = 6$. Pairs represented by points above the line fail to satisfy the inequality. It can be proved that the graph of the solution set for $2x + 3y \le 6$ is the **half-plane** consisting of the line $2x + 3y = 6$ and all points below it. We drew red arrows in Figure 7.24 to indicate which half-plane is the solution set for $2x + 3y \le 6$. (*Note:* If the inequality were $2x + 3y < 6$, then the line $2x + 3y = 6$ would not be part of the solution set. In this case, we would draw the line dotted rather than solid.)

What happened in Example 6 is true for any linear inequality in two variables.

THE SOLUTION SET FOR A LINEAR INEQUALITY IN TWO VARIABLES The solution set for a linear inequality in two variables is always a half-plane with either a solid or dotted border.

KEY POINT

The one-point test determines the half-plane in solving a linear inequality.

A simple way to determine which side of the line (which half-plane) is the solution set to a linear inequality is called the **one-point test.**

SOLVING LINEAR INEQUALITIES USING THE ONE-POINT TEST
To solve a linear inequality in two variables, follow these steps:

1. Change the inequality to an equation and graph the line. If the inequality contains $\leq$ or $\geq$, draw a solid line. If the inequality contains $<$ or $>$, draw a dotted line.
2. Choose a point that is clearly either above or below the line you drew in step 1. The point (0, 0) is usually the best point to use unless the line passes through or is very close to (0, 0).
3. If the coordinates of the point you chose in step 2 satisfy the inequality, then the side of the line containing that point contains solutions for the inequality. Otherwise, the opposite side of the line contains solutions. We will refer to this test as the **one-point test.**
4. Use arrows or shading to indicate which side of the line contains the solutions to the inequality.

We will illustrate this method in Example 7.

EXAMPLE 7 Using the One-Point Test to Solve an Inequality

Use the one-point test to solve $4x - 3y \geq 9$.

SOLUTION:

Step 1: Graph the line $4x - 3y = 9$. Because the inequality is $\geq$, rather than $>$, we will graph $4x - 3y = 9$ with a solid line in Figure 7.25.

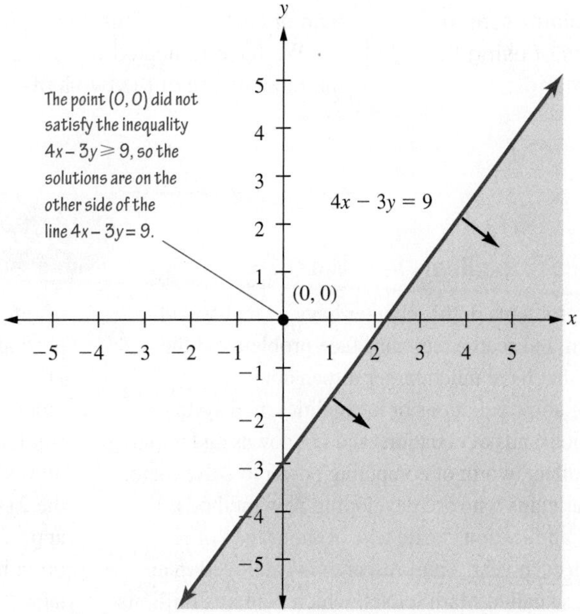

FIGURE 7.25 Using the one-point test to graph $4x - 3y \geq 9$.

Step 2: Use the one-point test to determine which side of the line contains solutions to the inequality. If we test (0, 0), we see that $4(0) - 3(0) \geq 9$ is not true. Therefore, the (0, 0) side of the line does not contain solutions. We choose the other side instead. The solution consists of all points on or below the line, as we show in Figure 7.25.

Now try Exercises 27 to 34. 21

QUIZ YOURSELF 21

Solve $2x - 5y > 10$.

SOME GOOD ADVICE

In working with inequalities, do not look for shortcuts to obtain the graphs more quickly. You cannot assume that if the symbols $\leq$ or $<$ appear in the inequality, the solution to the inequality occurs below the line. Also, don't assume that the symbols $\geq$ or $>$ mean that you should select the upper half-plane. In order to choose the proper half-plane, you should always use the one-point test.

Solving Systems of Inequalities

KEY POINT

The solution of a system of inequalities is the intersection of the solution sets for the individual inequalities.

Just as we can have systems of equations, we can construct systems of two or more inequalities. In solving a system of inequalities, we solve the inequalities separately and then find the intersection of the individual solution sets.

EXAMPLE 8 Solving a System of Inequalities

Solve the system

$$\begin{aligned} 2x - 3y &< -6 \\ x + y &\leq 7. \end{aligned}$$

SOLUTION: Review the Relate a New Problem to an Older One Strategy on page 9.

We first graph $2x - 3y = -6$ using a dotted line and $x + y = 7$ using a solid line, as in Figure 7.26.

Testing (0, 0) with the first inequality, we get $2 \cdot 0 - 3 \cdot 0 = 0$, which is greater than -6, so (0, 0) is not a solution. This means that the region above the line is the desired half-plane to solve $2x - 3y < -6$.

When we test (0, 0) with the second inequality, we get $0 + 0 = 0$, which is less than or equal to 7. Thus, the (0, 0) side of the line $x + y = 7$ is the half-plane we want.

We have indicated these regions with arrows in Figure 7.27 and have also shaded the intersection of the two half-planes to show the solution to the system of inequalities.

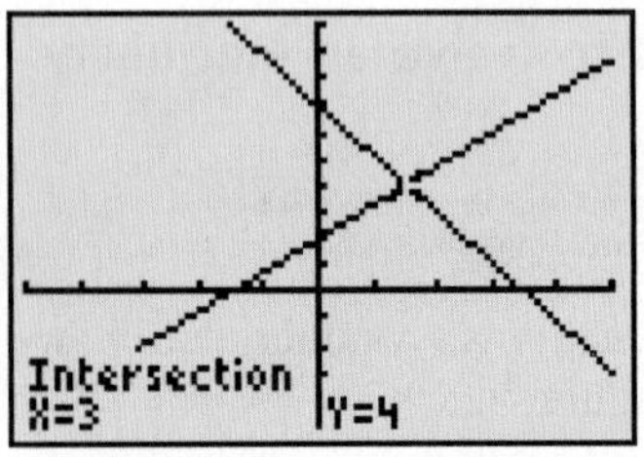

Graphing calculator screen showing results of using the intersect command.

Between the Numbers

Mathematics in the Real "Real World"

Some of the problems in this chapter have a "real-world" flavor to them, but mathematicians face problems in the real "real world" that have much larger dimensions.

Instead of a few equations or inequalities, real systems might have thousands of equations and unknowns and require millions of dollars' worth of computing power to solve them.

Mathematicians who are developing new methods, theories, and applications write tens of thousands of research papers each year. Their research is cataloged in an online database called MathSciNet, which contains millions of references from thousands of mathematics journals. One might question, What is the point of so much research? The answer is, we never know where today's research might be used in the future. For example, Boolean algebra, developed in the 19th century, now has myriad applications in computer science and technology in the 21st century. As an educated citizen, it is important to appreciate that basic research in mathematics and science often has long-range benefits that cannot be seen in the near-term future.

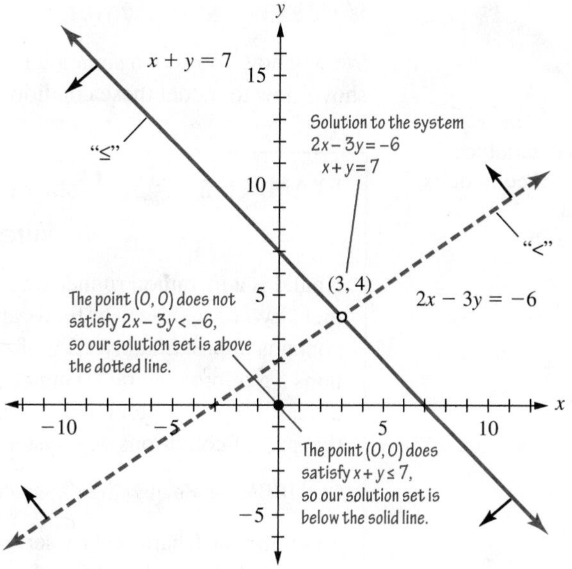

FIGURE 7.26 The dotted line contains points that are not part of the solution of $2x - 3y < -6$.

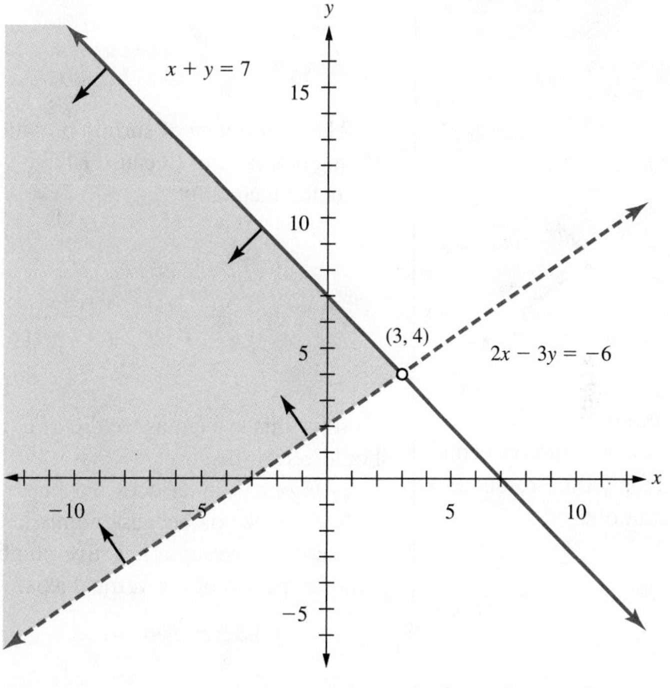

FIGURE 7.27 Graph of the solution of $\begin{matrix} 2x - 3y < -6 \\ x + y \le 7 \end{matrix}$.

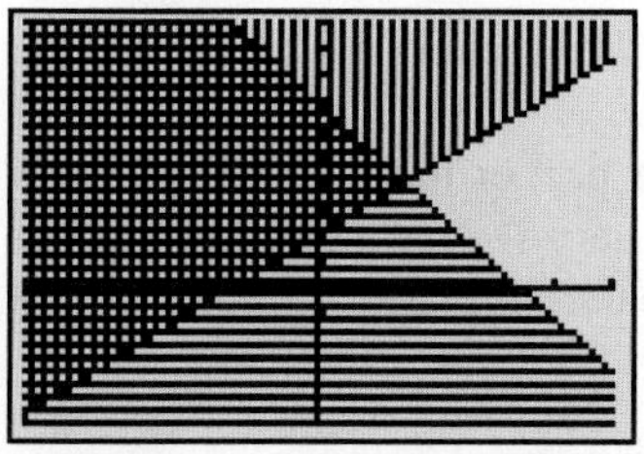

Graphing calculator graph of the solution set of $\begin{matrix} 2x - 3y < -6 \\ x + y \le 7 \end{matrix}$.

Notice that the point (3, 4) is a solution to the system of linear equations $2x - 3y = -6$ and $x + y = 7$. We found this solution using the elimination method as shown earlier. The dot is open rather than solid because (3, 4) lies on the line $2x - 3y = -6$, which is not part of the solution of $2x - 3y < -6$. If both boundary lines had been drawn solid, then we would also have drawn (3, 4) as a solid dot. A point such as (3, 4) that is the intersection of two boundary lines of a solution set is called a **corner point.**

QUIZ YOURSELF 22

Solve the system $\begin{matrix} 3x - 4y \le 8 \\ x + 2y \ge 6 \end{matrix}$.

Now try Exercises 35 to 44. 22

Modeling with Systems of Inequalities

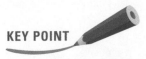

KEY POINT

We use a system of linear inequalities in two variables to represent a set of conditions on two quantities.

We will now consider a situation involving several conditions imposed on a diet. Example 9 shows how to model these conditions with a system of linear inequalities.

EXAMPLE 9 Using Inequalities to Represent Nutritional Requirements

Khalid is a marathon runner who is interested in the amount of protein and calcium in his diet. Two of his favorite foods are fried shrimp and broccoli. A serving of fried shrimp contains approximately 15 g of protein and 60 mg of calcium. A spear of broccoli contains 5 g of protein and 80 mg of calcium. Assume that, as part of his diet, he wants to get at least 60 g of protein and 600 mg of calcium from fried shrimp and broccoli. Express this pair of conditions as a system of inequalities and graph its solution set.

SOLUTION: Review the Choose Good Names Strategy on page 5.

Assume that Khalid eats s servings of shrimp and b spears of broccoli. Because each serving of shrimp has 15 g of protein, if Khalid eats s servings of shrimp, he will consume $15s$ g of protein. Each spear of broccoli has 5 g of protein, so eating b spears of broccoli provides $5b$ g of protein. He wants the amount of protein to be at least 60 g, which we can model with the inequality

$$15s + 5b \geq 60.$$

grams of protein in s servings of shrimp ($15s$); grams of protein in b spears of broccoli ($5b$)

Also, s servings of shrimp provide $60s$ mg of calcium and b spears of broccoli have $80b$ mg of calcium. Because Khalid wants the amount of calcium to be at least 600 mg, we get the inequality

$$60s + 80b \geq 600$$

milligrams of calcium in s servings of shrimp ($60s$); milligrams of calcium in b spears of broccoli ($80b$)

$$15s + 5b \geq 60$$
$$60s + 80b \geq 600.$$

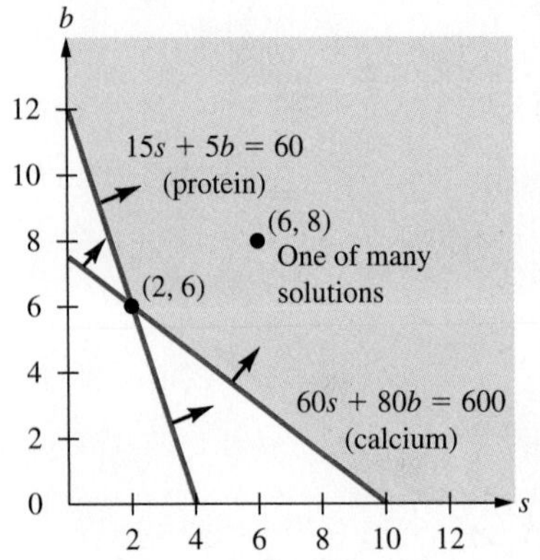

FIGURE 7.28 Shaded region represents the number of servings of shrimp and broccoli needed to satisfy the protein and calcium conditions.

We solve this system as we did in Example 8 and show the solution in Figure 7.28. You should verify this.

In Figure 7.28, it does not make sense to shade the part of the solution set below the horizontal axis because points in that region have a negative second coordinate and correspond to eating a negative number of broccoli spears. Similarly, we do not shade points to the left of the vertical axis.

Now try Exercises 61 to 68.

EXERCISES 7.6

Sharpening Your Skills

The solution for each system is a pair of integers. Determine the solution by graphing each system, and check your answer by substituting values in both equations.

1. $2x + 5y = 18$
$-3x + 4y = -4$

2. $3x + 2y = 9$
$-x + 2y = -3$

3. $2x - 5y = 18$
$3x + y = 10$

4. $-2x + 3y = -5$
$-x + y = -2$

Use the elimination method to solve the following systems of linear equations:

5. $-3x + 2y = 5$
$6x + 4y = 22$

6. $4x - y = 13$
$5x + 8y = -30$

7. $3x - 2y = -14$
$-4x + y = 22$

8. $-5x + 4y = -44$
$2x + y = 15$

9. $8x - 2y = -2$
$-4x + y = 3$

10. $6x + 9y = -4$
$2x + 3y = -2$

11. $x - y = -2$
$-4x + 4y = 8$

12. $10x - 4y = -4$
$-5x + 2y = 2$

13. $6x - 9y = 8$
$-4x + 6y = 10$

14. $12x - 4y = 2$
$-9x + 3y = 11$

15. $12x - 8y = -1$
$-2x + 5y = 2$

16. $12x - 27y = -1$
$5x + 2y = 4$

17. $3x - 9y = -3$
$-4x + 2y = 0$

18. $4x - 5y = -4$
$-8x + 15y = 13$

We will illustrate the ***substitution method,*** *which is another method for solving systems of linear equations, by solving the system*

$$\begin{aligned} 3x - 4y &= 13 \\ x - 2y &= 5 \end{aligned}.$$

Step 1: We solve one of the equations for either x *or* y*, depending on which is easier. In this case, we solve the second equation for* x*, giving us* $x = 2y + 5$.

Step 2: We next back-substitute the expression $2y + 5$ *for* x *in the first equation in the system, giving us* $3(2y + 5) - 4y = 13$. *We rewrite this equation as* $6y + 15 - 4y = 13$*, which simplifies to* $2y = -2$. *Thus,* $y = -1$.

Step 3: Now back-substitute -1 *for* y *in either equation. We choose to substitute* -1 *for* y *in equation 1, which gives us* $3x - 4(-1) = 13$. *Thus* $3x + 4 = 13$*, so* $x = 3$. *The solution to the system is therefore* $(3, -1)$.

Use the substitution method to solve the systems in Exercises 19–24. As with the elimination method, you may discover some systems to be either inconsistent or dependent.

19. $x - 2y = -7$
$3x + 5y = 12$

20. $x + 3y = 11$
$-x + 4y = 10$

21. $-2x + y = 6$
$-2x + 3y = 14$

22. $5x + 4y = 9$
$3x + y = 11$

23. $x + y = 6$
$2x = 8 - 2y$

24. $x + y = 3$
$x - y = -7$

Determine which points satisfy the given inequality.

25. $3x + 4y \geq 2$

a. $(3, 5)$ **b.** $(1, -2)$ **c.** $(0, 0)$ **d.** $(-4, 6)$

26. $2x - 4y \geq 5$

a. $(5, 2)$ **b.** $(6, 0)$ **c.** $(0, -3)$ **d.** $(-2, 3)$

Use the one-point test to solve the inequality. Shade the solution.

27. $3x + 4y \geq 12$

28. $2x - 4y \leq 6$

29. $2x - 4y < 12$

30. $5x + 4y > 10$

31. $x \geq 3y - 9$

32. $4y \leq 10 - 2x$

33. $4x - 8 < 2y$

34. $2x - 6 < 3y$

Solve each system. Indicate all corner points and shade the solution set.

35. $2x - 3y \leq -5$
$x - 2y \leq -8$

36. $2x - 5y \leq 23$
$3x + y \leq 9$

37. $2x - y \geq 3$
$x - y \leq -1$

38. $3y \geq 13 + x$
$9y \leq 23 - x$

39. $-4x + 3y \leq 23$
$3x > 19 - 5y$

40. $5x - 6y \geq 21$
$3y - 6 < 8x$

41. $3x + 5y \leq 32$
$y \geq 4$

42. $5x - 2y \geq 3$
$x \leq 3$

43. $2x + 5y > 26$
$x < 8$

44. $-3x + 2y < 12$
$y \geq 6$

Applying What You've Learned

Use the elimination method to solve each system of two linear equation in two unknowns.

45. College basketball victories. When Tennessee's Lady Vols beat Georgia on February 5, 2009, Pat Summitt became the first Division I college coach (man or woman) to win 1,000 games. The two teams scored a total of 116 points and Tennessee had 30 more points than Georgia. What was the final score?

46. Lopsided victory. In 1916, Georgia Tech, coached by John Heisman (of Heisman trophy fame), defeated Cumberland College by a score of 220 to 0 for the most lopsided victory in the history of college football. Georgia Tech rushed for 1,716 more yards than Cumberland and the total rushing yards by both teams was 1,524 yards. How many yards rushing did each team have?

47. Dietary requirements. An average bagel contains 30 mg of calcium and 2 mg of iron. One ounce of cream cheese contains 25 mg of calcium and 0.4 mg of iron. If Leyla wants to eat a combination of bagels and cream cheese that contains exactly 245 mg of calcium and 10 mg of iron, how much of each should she eat?

48. Dietary requirements. A slice of cheese pizza contains 40 g of carbohydrates and 220 mg of calcium. A 12-oz cola contains 40 g of carbohydrates and 15 mg of calcium. If Karl eats several slices of cheese pizza and drinks cola, how much of each must Karl eat to get a nutritional benefit of exactly 200 g of carbohydrates and 690 mg of calcium?

49. Comparing satellite systems. Menaka is considering two satellite TV systems. WorldCom charges $179 for installation and $17.50 per month. Satellite, Inc. charges $135 for installation and $21.50 per month. Interpret what the solution to this system tells you. How should Menaka decide which system to take?

50. Comparing job offers. Betty can work as freelance editor for *Mode* magazine for a base salary of $20 per hour plus 25 cents per page. At *Blush* magazine, she can earn $22.10 per hour plus 18 cents per page. Interpret what the solution to

this system tells you. How should Betty decide which position to take?

51. **Busiest airports.** In 2011 the two busiest airports in the world were Atlanta's Hartsfield-Jackson International and Beijing's Capital International. If the total number of passengers between the two airports was 98 million and Atlanta had 10 million more than Beijing, how many passengers did each airport handle? (*Source:* Airports Council International)

52. **Tourist travel.** According to the World Tourism Organization, the two European countries that had the most tourists in a recent year were France and Spain. The two countries together had 138 million visitors, and France had 20 million more visitors than Spain. How many tourists visited each country?

The following graph shows the change in motor vehicle production for the United States and China from 2000 to 2010.

Motor Vehicle Production

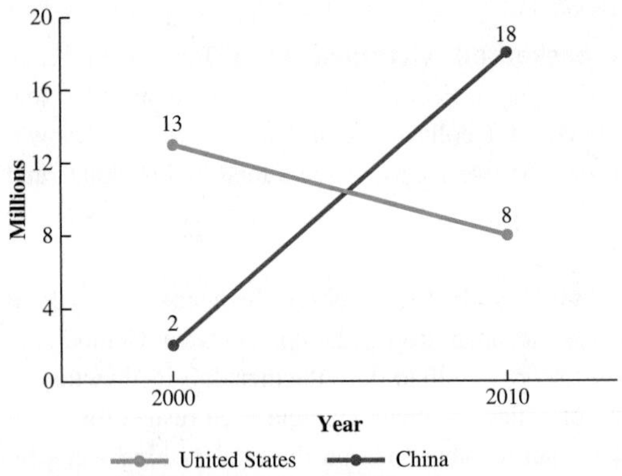

Source: The International Organization of Motor Vehicle Manufacturers

53. **Motor vehicle production.** Write a pair of linear equations in standard form describing the change in automobile production from 2000 to 2010 for the United States and China. (Treat 2000 as year 0.)

54. **Motor vehicle production.** Solve the system in Exercise 53 and interpret your solution.

Use the elimination method in Exercises 55–58 to find an equilibrium point for the described supply–demand situation. You may assume that the supply and demand equations are both linear.

55. **Supply and demand—tutoring.** A tutoring service provides tutoring for a fee. When the tutoring service charges \$8/hour, there is a demand for 30 tutors per week. When the hourly rate rises to \$15/hour, the demand drops to 9 tutors per week. On the other hand, at \$8/hour, the service is able to supply only 9 tutors per week, whereas at the rate of \$15/hour, it is able to provide 37 tutors per week. Find the price per hour at which the number of tutors demanded and the number of tutors supplied will be equal.

56. **Supply and demand—housing.** In doing a low-income housing survey, the local chamber of commerce found that if an apartment rents for \$400 per month, there were only 275 available; however, when the rental price rose to \$550 per month, the supply increased to 350. In contrast, at a rental price of \$400 per month, there was a demand for 450 units, but when the rental price increased to \$550 per month, the demand dropped to only 200 units. Find the monthly rental price at which the number of apartments demanded and the number of apartments supplied will be equal.

57. **Supply and demand—used books.** A bookstore buys and sells used textbooks. For a current sociology text, it has found that if it pays an average price of \$30 for each used book, it will be able to buy 60 texts for resale and will have 95 customers who are willing to buy the texts after markup. If it pays \$42 per used text, it will be able to buy 120 texts but will be able to sell only 50 of them. What price should it offer for the used texts in order to sell all that it buys?

58. **Supply and demand—selling bagels.** Dunkin' Donuts makes fresh bagels each morning. The proprietors will make 4 dozen bagels if the price is \$7/dozen and the demand will be 10 dozen. At a price of \$16/dozen, they will make 16 dozen but will be able to sell only 4 dozen. How should the bagels be priced so that all bagels made will be sold?

If a company produces and sells x units of a certain product, there will be a ***cost,*** *C, of producing the product described by the equation*

$$C = \textit{fixed cost} + (\textit{cost per unit produced}) \cdot x.$$

The ***revenue,*** *R, is the amount of money that comes in from selling x units of the product and is described by the equation*

$$R = (\textit{price per unit sold}) \cdot x.$$

The point where C and R are equal is called the ***break-even point,*** *and it tells us the point at which the company starts to make a profit. Find the break-even point for the situations described in Exercises 59 and 60.*

59. **Manufacturing.** The EarBudz Company is planning to manufacture a new line of high-end noise-reduction ear buds. Fixed costs will be \$126,000 and it will cost \$12 to produce each pair of ear buds. The ear buds will sell for \$40 a pair.

60. **Manufacturing.** The Trekker shoe company is planning to manufacture a new line of workout "toe shoes." Fixed costs will be \$266,000 and it will cost \$30 to produce each pair of shoes. The shoes will sell for \$125 a pair.

Use a system of inequalities to represent the situations described in Exercises 61–68, then solve the system using the methods developed in this section. In drawing the graph of your solution set, you may want to take into consideration that certain quantities cannot be negative and therefore you may want to restrict the solution set appropriately as we did in Example 9.

61. **Manufacturing.** Scott owns a manufacturing company that produces two models of entertainment centers. The Athens requires 4 feet of fancy molding and on average takes 4 hours to manufacture. The Barcelona needs 15 feet of molding and 3 hours to manufacture. In a given week, there are 120 hours of labor available, and the company has 360 feet of molding to use for the centers.

62. **Investing.** Gina is considering investing no more than \$3,000 total in a pharmaceutical company and Facebook stock. She intends to invest at least three times as much money in the pharmaceutical stock as she puts into Facebook stock.

63. Small business. Tami Taylor sells decorated hats and T-shirts at a Dillon Panther's football game. It takes 30 minutes to decorate a hat and 20 minutes to decorate a T-shirt. She has a total of 600 minutes to spend decorating the items.

64. Small business. Jaleel has a part-time business preparing slides for small businesses. It takes him 10 minutes to produce a plain-text slide and 18 minutes to produce a slide with graphics. He intends to spend 300 minutes per week at his business.

65. Diet. As part of her preparation for the World Cup in women's soccer, Cassandra is taking several nutritional supplements. Quantum contains 4 mg of niacin and 80 mg of calcium. NutraPlus contains 2 mg of niacin and 220 mg of calcium. She wants to supplement her diet by taking at least 12 mg of niacin and 960 mg of calcium.

66. Advertising. A newly opened Espresso Bar is deciding on advertising to generate business. The owners have \$2,400 total to spend for ads in the local newspaper and on the local radio station. A radio ad costs \$100, and a newspaper ad costs \$300. They want to have at least three times as many radio ads as they have newspaper ads.

67. Ordering smart phones. A local Better Purchase big box electronics store sells both Apple and Android smart phones. Katy, the manager, wants to order between 30 and 60 phones, inclusive, and wants at least twice as many Apples as Androids.

68. Weaving fair trade items. Magdalene Ngubui Nkambeh, who lives in Cameroon, weaves fair trade baskets and totes to support her family. In a year she can weave no more than 600 items but will easily weave at least 450. She plans to weave at least three times as many totes as baskets.

Communicating Mathematics

69. We can graph a system of two linear equation as a pair of lines.

a. What are the three cases that can arise in graphing these lines?

b. How do you identify which case you are dealing with and what does that tell you about solution(s) to the system?

70. Does an equation representing demand, as in Example 4, have positive or negative slope? Why is this so? What about the equation representing supply?

71. Describe what the solution set of a linear inequality in two variables looks like. What is a quick way to find such a solution set?

72. When solving a system of linear inequalities, we may decide not to shade the region below the x-axis or to the left of the y-axis. Why would we do this? Give an example.

Between the Numbers

73. As we mentioned in Between the Numbers on page 358, textbook problems are often simpler than real-life problems. Describe at least three aspects of Khalid's diet in Example 9 that might be relevant that we have not included. We are not asking for a mathematical discussion here, just an indication of what might be included in a more comprehensive description of the problem.

74. Read an article that gives you mathematical information about some current topic. Mention at least three relevant issues that are not mentioned in the article.

Challenge Yourself

Solve the systems of inequalities in Exercises 75 and 76 and explain why they have no solutions.

75. $2x + 3y \geq 18$
$4x + 6y \leq 16$

76. $5x - 2y < 10$
$-5x + 2y \leq -20$

We can have more than two inequalities in a system. Solve the following systems and find all corner points:

77. $2x + 3y \leq 25$
$5y \geq 20 + x$
$y \leq x + 5$
$x \geq 0, y \geq 0$

78. $3x + 2y \leq 22$
$x + y \leq 8$
$x + 4y \leq 24$
$x \geq 0, y \geq 0$

79. $x - 2y \leq 8$
$2x + y \leq 19$
$x \geq y + 5$
$x \geq 0, y \geq 0$

80. $6y \leq 14 + 5x$
$2y \leq 23 - 2x$
$13y \geq 56 - 2x$

In Exercises 81–84, use the graph to write a system of inequalities whose solution is the indicated region. Ignore the axes in describing these regions.

81. Region 1

82. Region 2

83. Region 3

84. Region 4

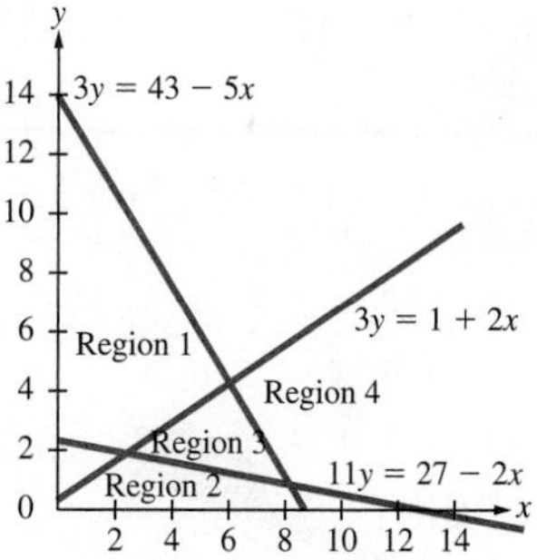

Looking Deeper

7.7 Dynamical Systems

Objectives

1. Understand dynamical systems as models.
2. Identify stable and unstable dynamical systems.

Why are weather predictions sometimes so unreliable? With the vast technology we have to gather and analyze data, it would seem possible to build mathematical models that are more reliable than they are at present. The problem in modeling weather is not only that it is much more complex than the situations we discussed in this chapter but also that many systems are unstable. This means that small changes in what is put into the model may result in very large changes in what comes out of the model.

We can use a coin to illustrate an unstable state. A coin can be in one of three resting states: head, tail, or on edge. If the coin is very close to being in the head state (perhaps the coin is held with one edge touching a table and the opposite side lifted $\frac{1}{10}$ of an inch), when the coin is released, it will go to the head state. There are obviously many states close to the head state such that if the coin is in one of these states, the coin approaches the head state.

Contrast the head state with the edge state. If we place the coin almost on edge and release it, the coin will not move toward the edge state, but rather it will approach either the head state or the tail state. If a system is close to a *stable state*, then the system will move toward that state. In contrast, if a system is close to an *unstable state*, the system's behavior may be very unpredictable.

Dynamical Systems

The systems we study in this section are similar to the bank account and logistic growth models you studied in Section 7.4. In the banking example, knowing the amount in the account at time n allowed us to compute the amount at time $n + 1$. Similarly, in the logistic growth model, knowing the lemur population at time n enabled us to predict the population at time $n + 1$.

In both of these cases, we could think of the model as giving us a sequence of numbers where we can calculate each number in the sequence provided that we know the previous one. For the banking example, we started with a deposit of \$1,000 and then for each successive year, the amount in the account was 1.08 times the amount in the account the previous year, as you can see below.

$$\underbrace{A_0 = 1{,}000}_{\text{initial amount}},\quad \underbrace{A_1 = 1{,}080}_{1.08 \cdot A_0},\quad \underbrace{A_2 = 1{,}166.40}_{1.08 \cdot A_1},\quad \underbrace{A_3 = 1,259.71}_{1.08 \cdot A_2},\ldots,\underbrace{A_{n+1}}_{1.08 \cdot A_n},\ldots$$

For the growth of the lemurs, we calculated the percentage of maximum population attained using the percentage attained the previous year, as follows:

$$\underbrace{P_0 = 0.20}_{\text{initial percentage}},\quad \underbrace{P_1 = 0.216}_{\text{depends on } P_0},\quad \underbrace{P_2 = 0.233}_{\text{depends on } P_1},\quad \underbrace{P_3 = 0.251}_{\text{depends on } P_2},\ldots,\underbrace{P_{n+1}}_{\text{depends on } P_n},\ldots$$

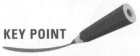

KEY POINT

Each state of a dynamical system is determined by its previous state.

In both examples, the amount that we calculated depended upon the amount present during the previous time period.

We can describe both the banking example and the logistic growth example by a mathematical model called a *dynamical system*.

DEFINITION A **dynamical system*** is a sequence of numbers A_0, A_1, A_2, A_3, and so on such that for each value of n,

$$A_{n+1} = \text{an expression involving } A_n.$$

We now look at Example 1 from Section 7.4 as a dynamical system.

EXAMPLE 1 A Bank Account as a Dynamical System

The amount of $1,000 is placed in an account that is paying 8% interest compounded yearly. Model this situation by a dynamical system.

SOLUTION:

From our earlier discussion, we see that the amount present in the account at the end of each year is 1.08 times as great as the amount present the previous year. We can put this a little more compactly by saying that

$$A_{n+1} = 1.08 \cdot A_n, \quad \text{for years } n = 0, 1, 2, 3, \ldots, \text{ where } A_0 = 1{,}000.$$

Notice that this dynamical system is really an infinite number of equations.

Like the changes in a bank account and changes in the lemur population, if we had enough data and sufficiently complicated equations to describe weather changes, we could view the present weather as an initial state W_0. Then, W_1 would be the weather 1 hour later, W_2 the weather 2 hours later, and so on. If our model were accurate (and our system were stable), then $W_{8,760}$ would be the state of the weather in 8,760 hours, or 1 year from now.

Just as equations can stand alone as mathematical objects without being explained in real-life terms, so can dynamical systems. It is common to identify a dynamical system with a rule that defines it. Here are two examples.

$$(1) \quad A_{n+1} = 5A_n + 3$$
$$(2) \quad A_{n+1} = 2A_n^2 + A_n - 5$$

System (1) is an example of a *linear system*, and system (2) is called *nonlinear*. Can you see why these names would be appropriate? Notice that a dynamical system is really an infinite number of equations. For example, system (1) corresponds to the collection of equations

$$A_1 = 5A_0 + 3,$$
$$A_2 = 5A_1 + 3,$$
$$A_3 = 5A_2 + 3,$$

and so on.

KEY POINT

Dynamical systems model many phenomena.

If you have ever been prescribed antibiotics by a physician, you have had personal experience with a dynamical system.

EXAMPLE 2 Modeling Antibiotics in the Blood by a Dynamical System

Suppose that your doctor has prescribed a 500-mg dose of an antibiotic to be taken three times a day. Assume that the drug enters your bloodstream instantly and that over an 8-hour period, your body eliminates 60% of the drug. Model the amount of drug in your blood by a dynamical system. Use this model to compute the amount of the drug in your system after three doses.

*There are many different categories of dynamical systems, depending on what properties the system satisfies. For example, systems such that A_{n+1} depends only on A_n are sometimes called *first-order systems.* A *second-order system* would be one in which A_{n+1} depends on both A_n and A_{n-1}. We will not deal with such systems for now, and in order to make the discussion clearer, we have slightly simplified the terminology.

SOLUTION:

Because you are taking the drug every 8 hours, each unit of time in our model corresponds to an 8-hour time interval. Let us represent the amount of drug present at time interval n by D_n. At $n = 5$, for example, we see that your body has eliminated 60% of the amount of drug that was present 8 hours earlier, so we see that before you take your next dose, your blood contains

$$D_5 = 0.40(D_4).$$

We are beginning the fifth interval. (points to D_5)

60% of the drug present at the beginning of the fourth time interval has been eliminated. (points to 0.40)

amount present at beginning of fourth time interval (points to D_4)

However, you now take a dose of the drug, which puts 500 mg of the drug into your body. So we get

$$D_5 = 0.40(D_4) + 500.$$

new dose (points to 500)

drug that was remaining in your blood (points to $0.40(D_4)$)

What is true for 5 is true for any time, so we see that

$$D_{n+1} = 0.40(D_n) + 500.$$

The only question that we still must answer is, What should be the value of D_0? If we want your first dose to correspond to time 1, the second to time 2, and so on, then we could say that at time 0, which is 8 hours before you took your first dose, there is no drug in your bloodstream. Therefore,

$$D_{n+1} = 0.40(D_n) + 500, \qquad n = 0, 1, 2, \ldots,$$

with the initial value $D_0 = 0$, describes this dynamical system.

We now calculate D_3, but to do this, we need to first compute D_0, D_1, and D_2.

$$\begin{aligned} D_0 &= 0, \\ D_1 &= 0.40(D_0) + 500 = 0.40(0) + 500 = 500, \\ D_2 &= 0.40(D_1) + 500 = 0.40(500) + 500 = 700, \\ D_3 &= 0.40(D_2) + 500 = 0.40(700) + 500 = 780. \end{aligned}$$

Now try Exercises 1 to 6. **23**

QUIZ YOURSELF 23

Redo Example 2, but now assume that the body eliminates 80% of the drug between doses and that each dose is 400 mg.

KEY POINT

Once a dynamical system reaches an equilibrium value, it stays at that value.

n	D_n (rounded to one decimal place)
1	500
2	700
3	780
4	812
5	824.8
6	829.9
7	832.0
8	832.8
9	833.1
10	833.2
11	833.3
12	833.3

Equilibrium Values and Stability

In Example 2, the amount of drug in your bloodstream is increasing, and we might wonder what the effects are of taking the drug for a long period of time. Is there the possibility that you could become poisoned by the amount of drug in your bloodstream? To look at this question more closely, we generate a table of values for D_n.

We see that as n gets larger, D_n appears to be getting closer to 833.3. In fact, as you will see shortly, if D_n ever equals $\frac{5{,}000}{6} = 833.3333\ldots$, then from that point on, all subsequent values of D_n will also equal $833.333\ldots$.

The value $\frac{5{,}000}{6}$ is an example of an equilibrium value for the dynamical system in Example 2.

DEFINITION An **equilibrium value** for a dynamical system is a number a such that if $A_n = a$, then A_{n+1} will also equal a.

Equilibrium values tell us about the long-term behavior of a system. It is easy to find the equilibrium values for some systems. For example, consider the system

$$A_{n+1} = -3A_n + 5.$$

Assume that at some point, $A_{n+1} = A_n$ and both take on the same value a. Substituting a for both A_n and A_{n+1} in the equation on the previous page, we get

$$a = -3a + 5.$$

Adding $3a$ to both sides of this equation and then dividing by 4 gives us

$$a = \frac{5}{4}.$$

We can check that $\frac{5}{4}$ is an equilibrium point by substituting it for A_n in the equation as follows:

$$A_{n+1} = -3A_n + 5 = -3 \cdot \frac{5}{4} + 5 = \frac{-15}{4} + \frac{20}{4} = \frac{5}{4}.$$

FINDING EQUILIBRIUM VALUES FOR DYNAMICAL SYSTEMS To find an equilibrium value a for a dynamical system, we do the following:

1. Write the equation A_{n+1} = an expression involving A_n.
2. Substitute a for both A_n and A_{n+1} in the equation in step 1.
3. Solve the equation in step 2 for a.

KEY POINT

An equilibrium value may be unstable.

An equilibrium value for a system may or may not be stable. You saw in the drug example that the numbers seemed to get closer and closer to 833.333 . . . and, as you will see in a moment, this is indeed what is happening. In Example 3, we look at another, similar dynamical system, but you will see very different behavior.

EXAMPLE 3 An Unstable Equilibrium Value for a Dynamical System

a) Find an equilibrium value for the dynamical system $A_{n+1} = 4A_n - 5$.

b) Is this equilibrium value stable?

SOLUTION:

a) Substituting a for A_n and A_{n+1} in the equation $A_{n+1} = 4A_n - 5$, we get $a = 4a - 5$. Solving this equation, we find that $a = \frac{5}{3} = 1.6666\ldots$.

b) We now generate a table of values for A using two values for A_0 that are extremely close to $\frac{5}{3}$. The first value for A_0 is 1.66, which is slightly less than $\frac{5}{3}$; the second value for A_0 is1.67, which is just a little bit greater than $\frac{5}{3}$.

n	A_n	A_n	A_n
0	$\frac{5}{3}$	1.66	1.67
1	$\frac{5}{3}$	1.64	1.68
2	$\frac{5}{3}$	1.56	1.72
3	$\frac{5}{3}$	1.24	1.88
4	$\frac{5}{3}$	−.04	2.52
5	$\frac{5}{3}$	−5.16	5.08
6	$\frac{5}{3}$	−25.64	15.32
7	$\frac{5}{3}$	−107.56	56.28
8	$\frac{5}{3}$	−435.24	220.12
9	$\frac{5}{3}$	−1,745.96	875.48
10	$\frac{5}{3}$	−6,988.84	3,496.92

QUIZ YOURSELF 24

Find equilibrium values for the given dynamical systems and determine whether these equilibrium values are stable.

a) $A_{n+1} = 3A_n - 1$

b) $B_{n+1} = 0.5B_n - 3$

Notice that although we take values very close to $\frac{5}{3}$ as initial values A_0 for this system, we see that by the time we compute A_{10}, the values we are getting are very far away from the initial value. Thus, the equilibrium value $\frac{5}{3}$ is *not stable.* Much like the coin standing on edge, if we begin even a little bit away from $\frac{5}{3}$, we wind up with values that are quite far away from $\frac{5}{3}$.

Now try Exercises 7 to 10. 24

The behavior shown by the dynamical system in Example 3, sometimes called the **butterfly effect,** refers to the (hypothetical) fact that a butterfly flapping its wings in South America could cause a slight disturbance in the atmosphere that would grow and eventually lead to a typhoon in China.

There is a theorem, which we will state without proof, that tells us when an equilibrium value is stable and when it is not.

STABILITY OF EQUILIBRIUM VALUES FOR DYNAMICAL SYSTEMS

Suppose that a is an equilibrium value for the system

$$A_{n+1} = mA_n + b.$$

1. If $-1 < m < 1$, then the equilibrium value a is stable.
2. If $m < -1$ or $m > 1$, then a is unstable.
3. If $m = -1$, then the values for A_n will oscillate between two values.

Dynamical systems have many applications. An interesting note about one of the causes of war was given by Lewis F. Richardson in his book *Arms and Insecurity: A Mathematical Study of the Causes and Origins of War.* Example 4 illustrates Richardson's theory.

EXAMPLE 4 Modeling an Arms Race with a Dynamical System

In the early part of the twentieth century, France-Russia and Germany-Austria-Hungary were opposing alliances. According to Richardson, the dynamical system

$$D_{n+1} = \frac{5}{3}D_n - \frac{380}{3}, \quad \text{where } D_0 = 199,$$

models total annual defense spending between the two alliances beginning in 1909 (year 0 in our model).

a) Find an equilibrium value for this system.

b) Is this equilibrium value stable?

c) What does this model predict will happen as n increases?

SOLUTION:

a) We solve the equation $a = \frac{5}{3}a - \frac{380}{3}$ to find an equilibrium value. Multiplying this equation by 3, we get $3a = 5a - 380$. This simplifies to $380 = 2a$, so the equilibrium value is 190.

b) Because the coefficient of D_n is $\frac{5}{3}$, which is greater than 1, the equilibrium value 190 is unstable.

c) The equilibrium value of 190 is not stable. Therefore, if we begin with the value $D_0 = 199$, as n increases, the values for D_n might be quite far away from 190. We calculate some values for D_n in Table 7.4.

Year = n	Amount spent on Defense = D_n
0	199
1	205
2	215
3	231.7
4	259.5
5	305.7
6	382.9
7	511.5
8	725.8

TABLE 7.4 Growth in defense spending, according to Richardson's model of the European arms race.

You can see in Table 7.4 that as n increases, D_n (the amount spent on defense) also increases rapidly. Because it is impossible for both sides to increase defense expenditures indefinitely, eventually one or the other will be in a position where it feels threatened and will declare war.

EXERCISES 7.7

Sharpening Your Skills

For each dynamical system, calculate the value of A_n.

1. $A_{n+1} = 2A_n - 1, A_0 = 3$; find A_1 and A_2.

2. $A_{n+1} = 3A_n + 2, A_0 = 1$; find A_1 and A_2.

3. $A_{n+1} = -3A_n + 4, A_0 = -2$; find A_3.

4. $A_{n+1} = 2.5A_n - 3, A_0 = 2$; find A_3.

5. $A_{n+1} = 1.8A_n - 2, A_0 = 4$; find A_4.

6. $A_{n+1} = -0.8A_n - 2, A_0 = 1.5$; find A_4.

Find the equilibrium value for each dynamical system. Comment on whether the value you find is stable or unstable.

7. $A_{n+1} = 2A_n + 3$

8. $A_{n+1} = 4A_n - 5$

9. $B_{n+1} = 0.25B_n + 4$

10. $B_{n+1} = 0.10B_n - 2$

Applying What You've Learned

Model each situation with a dynamical system. Use your model to answer the question.

11. Compound interest. You deposit $1,000 in a bank account paying 5% yearly interest that is compounded annually. How much will be in your account at the end of 2 years?

12. Compound interest. You deposit $1,500 in a bank account paying 6% yearly interest that is compounded annually. How much will be in your account at the end of 3 years?

13. Wildlife growth. An island is populated with lemurs that have a growth rate of 8%. The island is initially populated with 30% of the island's capacity to sustain these lemurs. What percentage of the lemur population's maximum capacity will be attained at the end of 2 years?

14. Wildlife growth. Repeat Exercise 13, but now assume the growth rate is 12% and the island is initially populated with 20% of its maximum capacity.

15. Antibiotic level. You take a 250-mg dose of an antibiotic every 4 hours. Your body eliminates 40% of the drug in a 4-hour period. How much antibiotic will be in your bloodstream after three doses?

16. Antibiotic level. You take a 1,000-mg dose of an antibiotic every 12 hours. Your body eliminates 75% of the drug in a 12-hour period. How much antibiotic will be in your bloodstream after three doses?

17. Building a college fund. It is January 1 and you have just made a $3,000 deposit into a college fund for your daughter. The fund pays 5% annually and you intend to let your money accumulate interest and add $3,000 to the account each January 1.

a. Model this investment with a dynamical system.

b. How much will be in your account at the end of the fourth year on December 31?

18. Building a college fund. Repeat Exercise 17, but now you are depositing $4,000 each year and the annual interest rate is 3.75%.

a. Model your investments by a dynamical system.

b. Assume that you keep these investments for three years. What will you be earning at that time?

Communicating Mathematics

19. What do we mean when we say that a dynamical system is really an infinite number of equations?

20. What is an equilibrium value for a dynamical system? What is the difference between stable and unstable equilibrium values?

21. If a is an equilibrium value for the dynamical system $A_{n+1} = mA_n + b$, how do you tell whether a is stable?

22. Explain the butterfly effect.

Challenge Yourself

Living plants and animals all contain the chemical element carbon. A certain percentage of that carbon is radioactive, and scientists believe that the percentage has remained constant for thousands of years. Radioactive carbon decays, so that when an animal dies a tiny bit of the radioactive carbon is lost each year. It is known that the amount of radioactive carbon that remains in a fossil at the end of a year is approximately 0.99988 *of the amount that was present at the beginning. Thus, the following dynamical system describes radioactive carbon decay in a fossil:*

$$C_{n+1} = 0.99988 \cdot C_n \quad \text{for } n = 0, 1, 2, 3, \ldots.$$

This system behaves exactly like the compound interest situation (except the amount of radioactive carbon is decreasing), so it is easy to see that after k *years, the amount of radioactive carbon in the fossil will be* $C_k = 0.99988^k \cdot C_0$. *We will always assume that the amount of radioactive carbon at time* 0 *will be* 1; *that is,* $C_0 = 1$. *Use this model to answer Exercises 23 and 24. You can solve these equations using the log function, as you did in Section 7.4.*

23. Carbon dating. A fossilized bone is found that contains 90% of the original radioactive carbon that was present. To the nearest 100 years, how old is the bone?

24. Carbon dating. A leaf of a fossilized plant is found that contains 60% of the original radioactive carbon that was present. To the nearest 100 years, how old is the plant?

Radioactive material decays over time and the amount remaining at time t is given by the equation

$$\text{amount remaining} = \text{initial amount} \times \left(\frac{1}{2}\right)^{\frac{t}{h}},$$

where h is the half-life of the material, which is the time it takes the material to decay to one-half its original amount. Use this equation in Exercises 25 and 26.

25. Assume that we have 50 pounds of radioactive plutonium, which has a half-life of 24,000 years. How much plutonium will still remain after 50,000 years?

26. If we have 100 pounds of radioactive uranium-238, which has a half-life of about 4.7 billion years, how much will still remain after 1 billion years?

Assume that the logistic equation $P_{n+1} = [1 + r(1 - P_n)]P_n$ *models the growth of wild turkeys on a large parcel of state game land. If we want to allow hunting on this land, we adjust this equation by subtracting some number from* P_{n+1} *to account for the turkeys killed by hunting. In the following questions, suppose that the game land is capable of supporting a maximum of 1,000 turkeys.*

27. Managing wildlife. Assume that currently there are 500 turkeys and we want to allow 80 to be harvested per year. The growth rate of the turkeys is 10% per year. Modify the growth equation accordingly, and use it to predict the turkey population at the end of 3 years.

28. Managing wildlife. Assume that currently there are 750 turkeys and we want to allow 100 to be harvested per year. The growth rate of the turkeys is 10% per year. Modify the growth equation accordingly, and use it to predict the turkey population at the end of 3 years.

CHAPTER SUMMARY

Section	Summary	Example
SECTION 7.1	A **linear equation** in two variables, written in **standard form,** has the form $Ax + By = C$. A **solution** for an equation is a number (or numbers) such that if we substitute them for the variables in the equation, the resulting statement is true. Two equations are **equivalent** if they have the same solutions. In **solving equations,** we can add or subtract the same expression from both sides of the equation. We can also multiply or divide both sides of the equation by the same *nonzero* expression.	Definition, p. 301 Discussion, pp. 301–302 Example 1, p. 302
	The ***x*-intercept** (where $y = 0$) of the graph of a linear equation is the point where the graph crosses the x-axis. The ***y*-intercept** (where $x = 0$) of the graph of a linear equation is the point where the graph crosses the y-axis. We can use the x- and y-intercepts to graph a linear equation.	Example 2, p. 304
	If (x_1, y_1) and (x_2, y_2) are two points on a line and $x_1 \neq x_2$, then the **slope** of the line, m, is defined as $m = \frac{\text{rise}}{\text{run}} = \frac{y_2 - y_1}{x_2 - x_1}$. A linear equation is in **slope–intercept** form if it is written in the form $y = mx + b$, where m is the slope and b is the y-intercept of the line.	Example 4, p. 306 Definition, p. 307 Example 6, p. 308
SECTION 7.2	We can find a linear equation if we know the slope of its graph and one point on its graph.	Examples 1 and 2, pp. 313, 314
	We can use two points to find a linear equation by first using the points to find the slope and then applying an earlier technique.	Examples 3 and 4, pp. 315–316
	We can use a graphing calculator to find the **line of best fit** for a set of data points.	Discussion, p. 318 Highlight, p. 318
SECTION 7.3	A **quadratic equation** is an equation in the form $y = ax^2 + bx + c$, where a, b, and c are real numbers and $a \neq 0$. A **solution** of a quadratic equation is a pair of numbers that satisfies the equation.	Discussion, p. 321 Example 1, p. 322
	The graph of a quadratic equation is a **parabola.** The lowest or highest point on a parabola is called its **vertex.** The vertex of the graph of $y = ax^2 + bx + c$ occurs when $x = \frac{-b}{2a}$. The **solution(s)** of the quadratic equation $ax^2 + bx + c = 0$ are $x = \frac{-b \pm \sqrt{b^2 - 4ac}}{2a}$.	Discussion, p. 322 Example 2, p. 322 Example 3, p. 323
	In the quadratic equation $y = ax^2 + bx + c$, if $a > 0$, then the graph of the parabola opens up. If $a < 0$, then the graph of the parabola opens down. We **graph** a quadratic equation by 1. Determining if the parabola opens up or down, 2. Finding its vertex, 3. Finding the x- and y-intercepts of the graph.	Example 4, p. 324
	We can use a quadratic equation to **model** data.	Example 5, pp. 325–327
SECTION 7.4	The **compound interest formula** says that money grows in a bank account according to the formula $A = P(1 + r)^n$, where P is the principal, r is the yearly interest rate, and n is the number of years.	Example 1, p. 331
	We can use an exponential equation to model **population growth.**	Example 2, p. 333
	The **log function** reverses the operation of raising 10 to a power. The log function has the property that $\log y^x = x \log y$.	Examples 3 and 4, p. 334
	A **logistic model** takes into account limits of population growth. The equation $P_{n+1} = [1 + r(1 - P_n)]P_n$ describes logistic growth.	Definition, p. 336 Examples 6 and 7, pp. 337, 338

(Continued)

Section	Summary	Example
SECTION 7.5	A **ratio** is a quotient of two numbers. A **proportion** is a statement that two ratios are equal. The **cross-multiplication principle** says that if $\frac{a}{b} = \frac{c}{d}$, then $a \cdot d = b \cdot c$. The quantities $a \cdot d$ and $b \cdot c$ are called **cross products.**	Discussion, p. 341 Example 1, p. 341
	The **capture–recapture method** is used for estimating the size of populations.	Example 2, pp. 342–343
	We say that y **varies directly** as x, or that y is directly proportional to x, if $y = kx$, where k is a nonzero constant, called the **constant of variation** or the **constant of proportionality.** Joint variation is direct variation in which there are several quantities.	Definitions, p. 343 Example 3, p. 343 Example 4, p. 344
SECTION 7.6	A collection of linear equations is called a **system of linear equations.** A **solution** of the system is an ordered pair of numbers that makes all equations in the system true. We represent a system of two linear equations as a **pair of lines.**	Discussion, pp. 348–349
	A system of two linear equations can have:	
	Case 1: A unique solution	Example 1, p. 350
	Case 2: No solutions	Example 2, p. 351
	Case 3: An infinite number of solutions	Example 3, p. 352
	We can use a system of linear equations to **model** a set of relationships between two quantities.	Example 4, pp. 353–354
	The **laws of supply and demand** state that as price increases, the producer produces more of an item but consumers buy less.	
	An **equilibrium point** is a point at which supply equals demand.	Example 5, pp. 354–355
	A **linear inequality in two variables** is a statement that can be written in one of the following forms: $ax + by \geq c, \quad ax + by > c, \quad ax + by \leq c, \quad ax + by < c,$ where a, b, and c are real numbers with $a \neq 0$ and $b \neq 0$.	Discussion, p. 355
	A **solution** of a linear inequality in two variables is an ordered pair of numbers. The set of all solutions of a linear inequality is a **half-plane.**	Example 6, p. 356
	We use the **one-point test** to determine which half-plane is the solution set of a linear inequality.	Example 7, pp. 357–358
	The **solution of a system of inequalities** is the intersection of the solution sets for the individual inequalities. The intersection of two boundary lines of a solution set is called a **corner point.**	Example 8, pp. 358–359
	We can use a system of inequalities to **model** a set of relationships between two variables.	Example 9, p. 360
SECTION 7.7	A **dynamical system** is a sequence of numbers A_0, A_1, A_2, A_3, and so on such that for each value of n, $A_{n+1} =$ an expression involving A_n.	Definition, p. 365 Examples 1 and 2, pp. 365–366
	An **equilibrium value** for a dynamical system is a number a such that if $A_n = a$, then A_{n+1} will also equal a. An equilibrium value may be **stable** or **unstable.**	Discussion, p. 366 Example 3, pp. 367–368

CHAPTER REVIEW EXERCISES

Section 7.1

1. Solve the following equations:

   a. $\frac{2}{3}x + 2 = \frac{1}{6}x + 4$ **b.** $0.3x - 2 = 3.5x - 0.4$

2. Solve $A = P(1 + rt)$ for r.

3. Darryl works at a warehouse job where he is paid \$5 per hour up to 40 hours. If he works over 40 hours, he is paid twice his usual wage. Assume that he always works at least 40 hours per week.

   a. Model this situation with a linear equation.

   b. How much will he earn if he works 46 hours in a given week?

4. Graph $3x + 5y = 20$ by plotting the intercepts and drawing a line through them.
5. Find the slope of the line passing through (2, 5) and (6, 8).
6. Match the line in the diagram with the given information regarding its slope. Assume that the scale on both axes is the same.
   Line 1: positive slope; less than 1
   Line 2: negative slope; between -1 and 0
   Line 3: slope does not exist
   Line 4: slope greater than 1

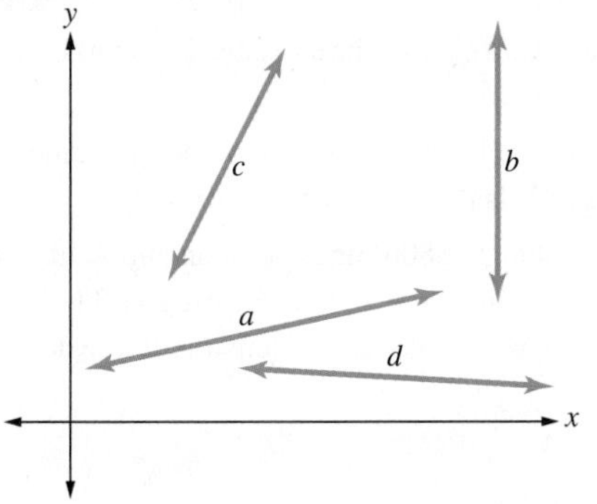

7. Nugyen is considering two satellite TV systems. Global Communications charges \$240 for installation and \$39 per month. World Communications charges \$350 for installation and \$28 per month. After how many months will World Communications be the better buy?

Section 7.2

8. Find an equation of the line with slope -4 that passes through (2, 5).
9. Find a linear equation whose graph passes through (3, 4) and (6, 9).
10. The amount spent by Americans on foreign travel increased from \$81.8 billion in 2002 to \$99.9 billion in 2005.
    a. Model this information with a linear equation.
    b. Use your model to estimate the amount that will be spent by Americans on foreign travel in 2015.
11. What do we mean by the line of best fit?
12. Explain when it is appropriate to use a linear function as a model.

Section 7.3

13. Solve the quadratic equation $2x^2 + 7x = 4$.
14. Answer the following questions for the graph of the equation

$$y = -2x^2 - 10x - 8.$$

    a. Is its graph opening up or down?
    b. What is the vertex of the graph?
    c. What are the x-intercepts?
    d. What is the y-intercept?
    e. Graph the equation.
15. What is quadratic regression?
16. Assume that the equation $A = -0.125x^2 + 2x + 1.125$ models attendance of a recently released movie for x weeks. During what week will the movie attain its highest attendance?

Section 7.4

17. If \$10,000 is placed in an investment paying 4.8% yearly interest compounded annually, how much will be in the account after 5 years?
18. How long will it take for the account in Exercise 17 to double?
19. Assume that a population is growing initially at a 3% rate and that we are using a logistic growth model to describe the population growth. If $P_3 = 0.50$, what is P_4?
20. Assume that a room at a hotel in Disney World cost \$220 a night in 2004 and that same room (due only to inflation) cost \$280 a night in 2008. Use an exponential model to estimate the cost of that room in 2012.
21. What is the difference between an exponential model and a logistic model?

Section 7.5

22. Solve for x in the following proportions:
    a. $25:8 = x:2$
    b. $\frac{30}{4} = \frac{x}{5}$
23. If it requires 3.5 gallons of sealer to coat a 840-square-foot driveway, how much sealer will be needed to coat a 1,500-square-foot driveway?
24. Marine biologists capture, tag, and release 180 penguins. Several months later, of 55 penguins that are captured, 12 are tagged. Estimate the population of penguins.
25. Assume that y varies inversely as x. If $y = 14$ when $x = 3$, what is the value of y when $x = 18$?
26. Assume that d varies jointly as a^2 and b, and inversely as c. If $d = 144$ when $a = 3$, $b = 7$, and $c = 14$, what is the value of d when $a = 8$, $b = 11$, and $c = 16$?
27. The strength of a beam is directly proportional to its width w and the square of its depth d, and inversely proportional to its length l. How would the strength of the beam change if the width and length are doubled?

Section 7.6

In Exercises 28–30, (a) solve the system and (b) explain the geometric significance of your solution.

28. $4x - 3y = 27$
    $x + 2y = -7$
29. $\frac{3}{4}x - \frac{1}{2}y = \frac{5}{4}$
    $-x + \frac{2}{3}y = -\frac{5}{3}$
30. $-4x + 5y = 3$
    $12x = 15y - 10$
31. The U.S. Olympic basketball team won a basketball game by 17 points. If 123 total points were scored, what was the final score?

32. Last month, Carmen had five Big Macs and three McChicken sandwiches for a total of 3,780 calories. If instead Carmen had eaten three Big Macs and five McChicken sandwiches, the total calories would have been 3,420. How many calories are there in each sandwich?

33. West Central State College is making holiday wreaths to raise money for a homeless shelter. From past sales experience, it was learned that if the wreaths are priced at $12 apiece, it will sell 32, but if the price is raised to $24, it will sell only 24. On the other hand, at a price of $12, the college is willing to make only 26 wreaths, but at a price of $24, it will make 30. What price should the college set so that its supply will equal the demand for the wreaths?

34. Use the one-point test to solve $6x + 5y > 20$. Shade the solution.

35. Solve the system $\begin{matrix} 2x + 5y \geq 24 \\ 2y \leq 2x + 18 \end{matrix}$. Find all corner points of the solution set.

36. The art club is selling custom-decorated T-shirts and hats to raise money for a trip. There is a profit of $4 on each T-shirt and $2 on each hat. The number of hats sold is at least 15 more than the number of T-shirts sold and the total profit is no more than $180. Use a system of inequalities to model this situation. Then solve the system using the methods developed in this chapter. In drawing your solution set, take into consideration that certain quantities cannot be negative and restrict the solution set appropriately.

Section 7.7

37. Calculate A_1, A_2, A_3 for the dynamical system $A_{n+1} = 4A_n + 3$, where $A_0 = 6$.

38. Find an equilibrium value for the dynamical system in Exercise 37. Is that value stable or unstable?

39. You are taking an 800-mg dose of an antibiotic every 8 hours. Your body eliminates 50% of the drug in 8 hours. How much of the drug will be in your system after three doses?

CHAPTER TEST

1. Solve the following equations:

a. $\frac{3}{4}x + 5 = \frac{2}{3}x + 4$

b. $0.25x + 4 = 1.5x - 0.2$

2. Solve $X = a(1 + b)$ for b.

3. Find the slope of the line passing through the points (3, 5) and (8, 12).

4. Graph $5x - 4y = 10$ by plotting the intercepts and drawing a line through them.

5. Brandon subscribes to a cell phone plan that charges $30 per month for 1,200 minutes and $0.045 per minute for each minute over 1,200.

a. Model Brandon's plan with a linear equation.

b. If he talked on his cell phone 1,520 minutes, what was his bill?

6. Assume that a Wilson mini-autographed basketball cost $16 in 2002, and in 2008 (due only to inflation) that cost rose to $18.50. Use an exponential model to predict the cost of that same basketball in 2015.

7. Match the terms in the left column with those in the right column in the following table:

1. Positive slope	a. Line is falling from left to right.
2. No slope	b. Line is rising from left to right.
3. Negative slope	c. Line is horizontal.
4. Slope is zero.	d. Line is vertical.

8. Wilfredo is considering two DVD rental plans. Fliks 'R' Us costs $40 to join and charges $1.50 for each DVD rented. DVD Mania costs $22 to join and charges $2 per rental. At what point is Fliks 'R' Us a better deal?

9. When is it appropriate to use a linear equation as a model?

10. Find an equation of the line that passes through the points (2, 5) and (6, 8).

11. Model the following situation with a linear equation, but do not try to solve it. Shanaya has two part-time jobs on campus. Her job at the dining hall pays $6.30 per hour, and her work in the cognitive psychology lab pays $8.25 per hour. She earns $137.70 in a given week.

12. What is linear regression?

13. There were 36.5 million people living in poverty in the United States in 2008, up 1.8 million from 2002. Model this information with a linear equation, and use this model to predict the number living in poverty in 2020.

14. Calculate A_1, A_2, A_3 for the dynamical system $A_{n+1} = 4A_n + 2$, where $A_0 = 5$.

15. Find an equilibrium value for the dynamical system in Exercise 14. Is that value stable or unstable?

16. Answer the following questions for the graph of the equation

$$y = -x^2 + 11x - 24.$$

a. Is the graph opening up or down?

b. What is the vertex of the graph?

c. What are the x-intercepts?

d. What is the y-intercept?

e. Graph the equation.

17. If $5,000 is placed in an investment paying 2.4% yearly interest compounded annually, how much will be in the account after 8 years?

18. How long will it take the account in Exercise 17 to double?

19. What is the formal name for finding the parabola of best fit?

20. Assume that the equation $S = -1.2x^2 + 8x$ models the sales of a recently released music CD, where x is the number of weeks since the release. During what week will sales reach their highest point?

In Exercises 21–23, (a) solve the system and (b) explain the geometric significance of your solution.

21. $6x = 2y - 4$

$8y - 24x = 10$

22. $-3y + 4x = 6$

$8x - 12 = 6y$

23. $3x + 4y = -6$

$2x - 3y = 13$

24. You are taking a 400-mg dose of a pain reliever every 12 hours. Your body eliminates 65% of the drug in 12 hours. How much of the drug will be in your system after four doses?

25. Assume that a population is growing initially at a 2% rate and that we are using a logistic model to describe the population growth. If $P_8 = 0.40$, what is P_9?

26. Census takers are using the capture–recapture method to count the number of homeless people in a large city. They first identify 75 people as being homeless and several weeks later, in another survey of 120 homeless people, 5 are from the first group of homeless that was identified. How many homeless people are there in the city?

27. Shandra works in a fast-food restaurant for \$5.65 per hour and is paid \$8.50 per hour to give swimming lessons. Last week she spent 5 more hours giving swimming lessons than she worked in the restaurant. If she earned \$212.30, how many hours did she work in the restaurant?

28. Krispy Kreme has found that if it prices a box of six deluxe donuts at \$3, it will sell 27 boxes; if it raises the price to \$8, it will sell only 17 boxes. On the other hand, Krispy Kreme will make only 17 boxes if the price is \$3, but will make 32 boxes if the price is \$8. How many boxes should it make if it wants the supply to equal the demand?

29. Use the one-point test to solve $4x - 3y \geq 8$. Shade the solution.

30. If the pitcher Cliff Lee gets 4 hits in his first 35 at bats and continues at that rate, how many hits can we expect if he bats 90 times during the season?

31. Solve the system $\begin{aligned} x + y &\leq 14 \\ 2x - 5y &\leq -14 \\ 5x - 2y &\geq 7 \end{aligned}$. Find all corner points of the solution set.

32. The strength of a beam varies directly as the square of its depth d and inversely as its length l. How would the strength of the beam change if its depth and length were both doubled?

33. Assume that y varies inversely as x. If $y = 8$ when $x = \frac{2}{3}$, what is the value of y when $x = 4$?

34. Assume that s varies jointly as x and y and inversely as t. If $s = 16$ when $x = 2$, $y = 4$, and $t = 6$, what is the value of s when $x = 3$, $y = 5$, and $t = 9$?

35. Mickal sells decorative inlaid wood boxes as souvenirs to visitors to the Grand Canyon. A small box requires 2 square feet of wood and the large box requires 6 square feet of wood. It takes 1 hour to make a small box and 2 hours to make a large box. He has 90 square feet of wood and plans to work no more than 40 hours next week. Model this situation with a system of inequalities, and then solve this system.

GROUP PROJECTS

1. Modeling real data. Gather data from different sources on a current topic such as expenditures on education, national defense, global warming, or energy usage. Have different people gather the same data from different sources: almanacs, the *Statistical Abstract of the United States*, the Bureau of the Census, and various online Web sites.

a. Compare your data for consistency. You will find that different sources will give different figures for the same "factual data."

b. Consider some issue such as "education spending has decreased" and have some in the group present data in favor of that statement and some present data opposed to that statement.

c. You can construct linear models as we did in Example 5 of Section 7.2. By choosing different data points, you can construct different linear models for the same data. For example, if you have four data points A, B, C, and D, a line through A and D will probably be different from a line through B and C. Create different models to make predictions and see who can make the highest and lowest predictions for a given quantity. For example, see who can make the highest and lowest predictions for the cost of attending a four-year public college in 2020. It is also interesting to build a model on data from earlier years such as 2005 to 2010 and compare your predictions with actual data from years such as 2012.

2. Comparing models of real data. Using the regression methods that we discussed in Sections 7.2 and 7.3 (see Highlight on page 318), use data that you have gathered in Exercise 1 to construct linear, quadratic, and exponential models and, by comparing the correlation coefficients, determine which type of model best fits the data.

USING TECHNOLOGY

1. **Modeling data with Excel.** To model data with Excel, as we did in constructing the graph of text messages data on page 327, do the following:

   a. Enter your data in the spreadsheet.

   b. Select (highlight) all the data and click on the *Insert* tab, click on the *Scatter* icon (for a scatterplot), and choose the first option (*Scatter with only Markers*).

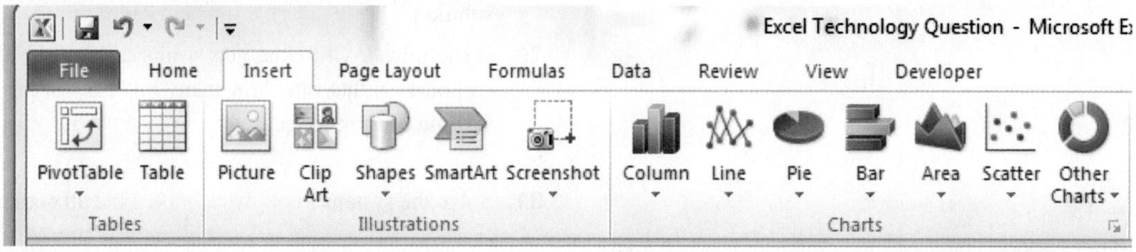

   c. The screen should now look like this:

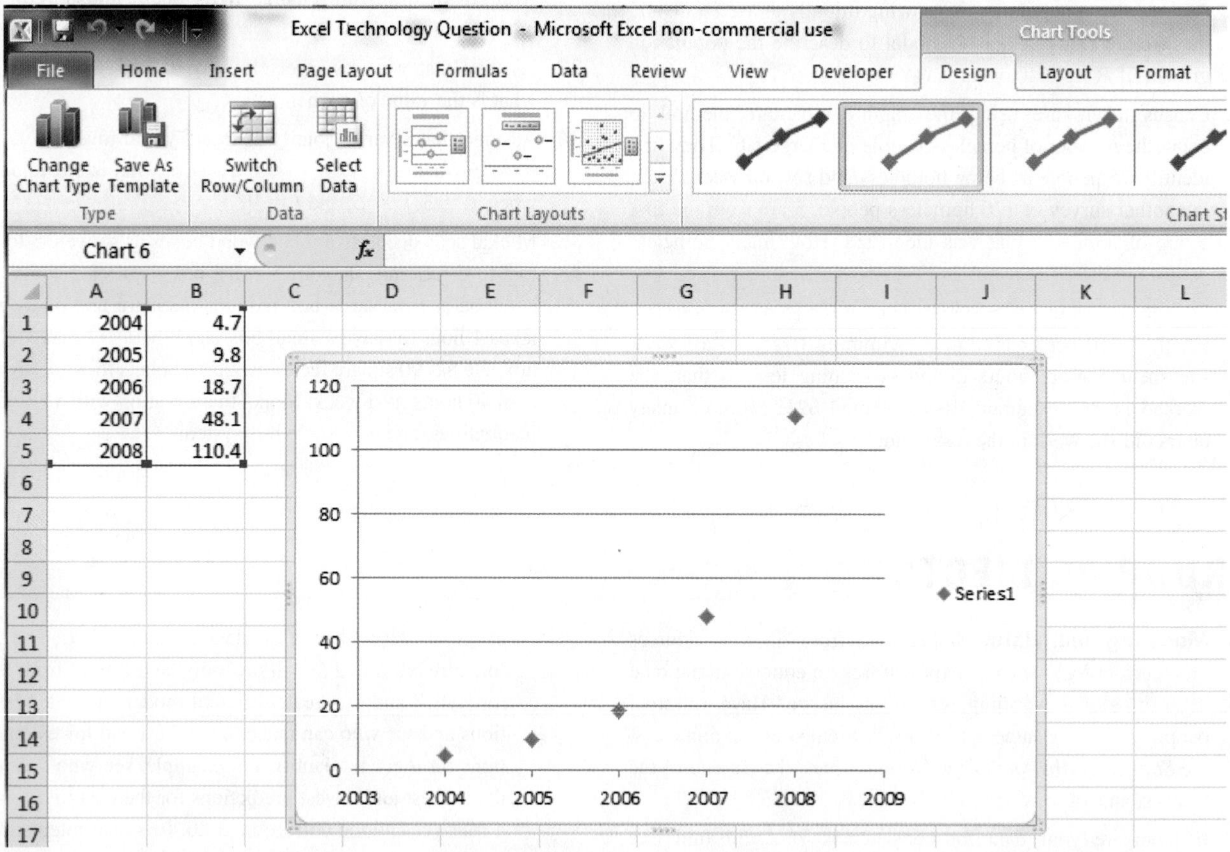

   d. Click on the graph and then select *Layout*, then select *Trendline* and then, if you want the line of best fit, click on *Linear*. If you want to show the equation for the line of best fit or the parabola of best fit, click on *More Trendline Options*. With a little practice, you should be able to construct lines and parabolas of best fit with their equations.

For practice, reproduce the graphs on page 327 and in Exercises 25 and 26 in Section 7.3, and then construct a graph of data of your choosing.

2. **Graphing linear systems.** You can graph systems of linear equations and inequalities with a graphing calculator. We show how to do this with a TI graphing calculator.

   a. Press the [Y =] key and enter equations of lines $Y_1 = 40 - x$ and $Y_2 = 6 + 3*x$, as we show in Screen 1 on the next page.

   b. Press the [GRAPH] key to show the graphs of the two lines, as we show in Screen 2.

   c. To graph inequalities, use the back arrow to position the cursor at the extreme left of the lines defining Y_1 and Y_2 and

keep pressing the [ENTER] key to select whether you want the upper half-plane or the lower half-plane for the inequality, as shown in Screen 3.

d. Then press [GRAPH] to show the graph of the system of inequalities, as in Screen 4.

Use this method to reproduce the graphs of some of the systems of equations and inequalities in Section 7.6.

Screen 1

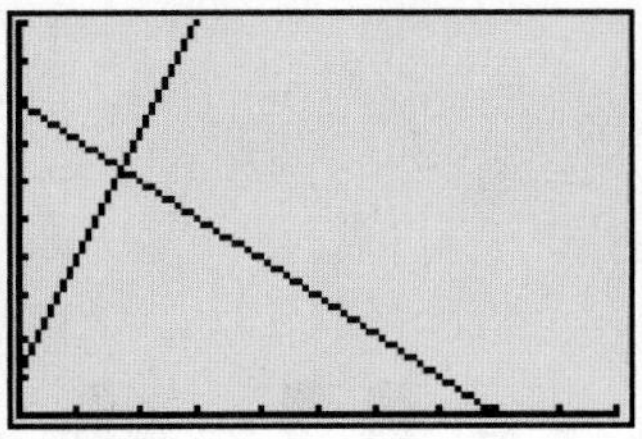

Screen 2

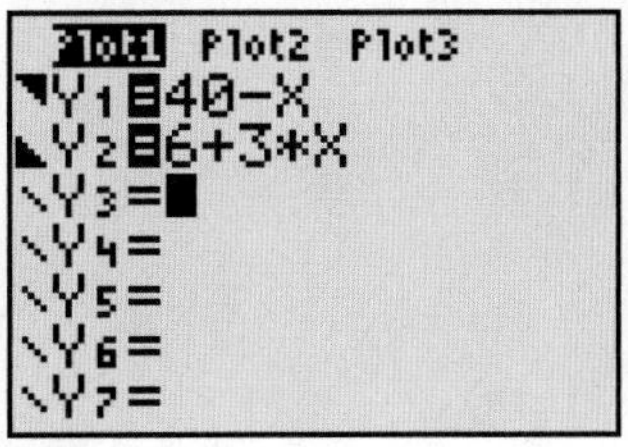

Screen 3

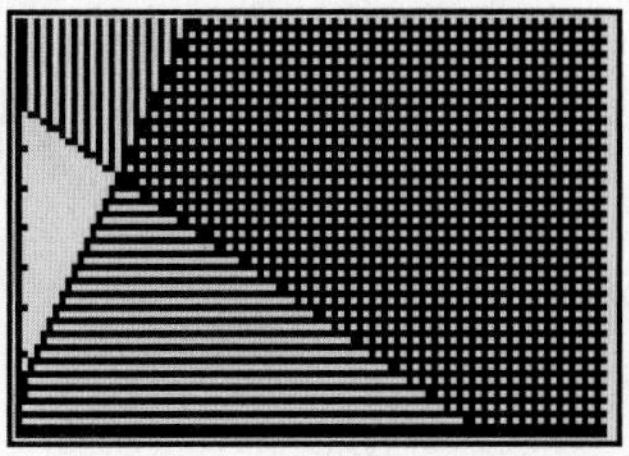

Screen 4

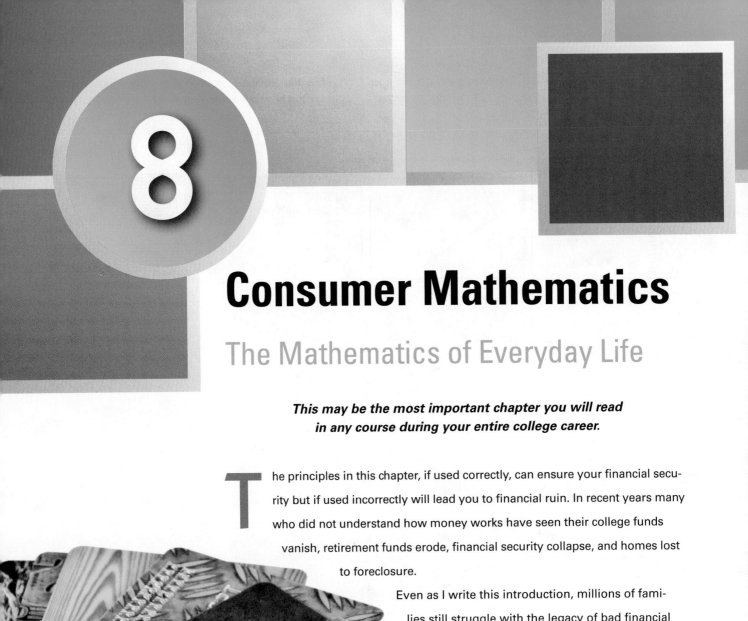

8

Consumer Mathematics

The Mathematics of Everyday Life

This may be the most important chapter you will read in any course during your entire college career.

The principles in this chapter, if used correctly, can ensure your financial security but if used incorrectly will lead you to financial ruin. In recent years many who did not understand how money works have seen their college funds vanish, retirement funds erode, financial security collapse, and homes lost to foreclosure.

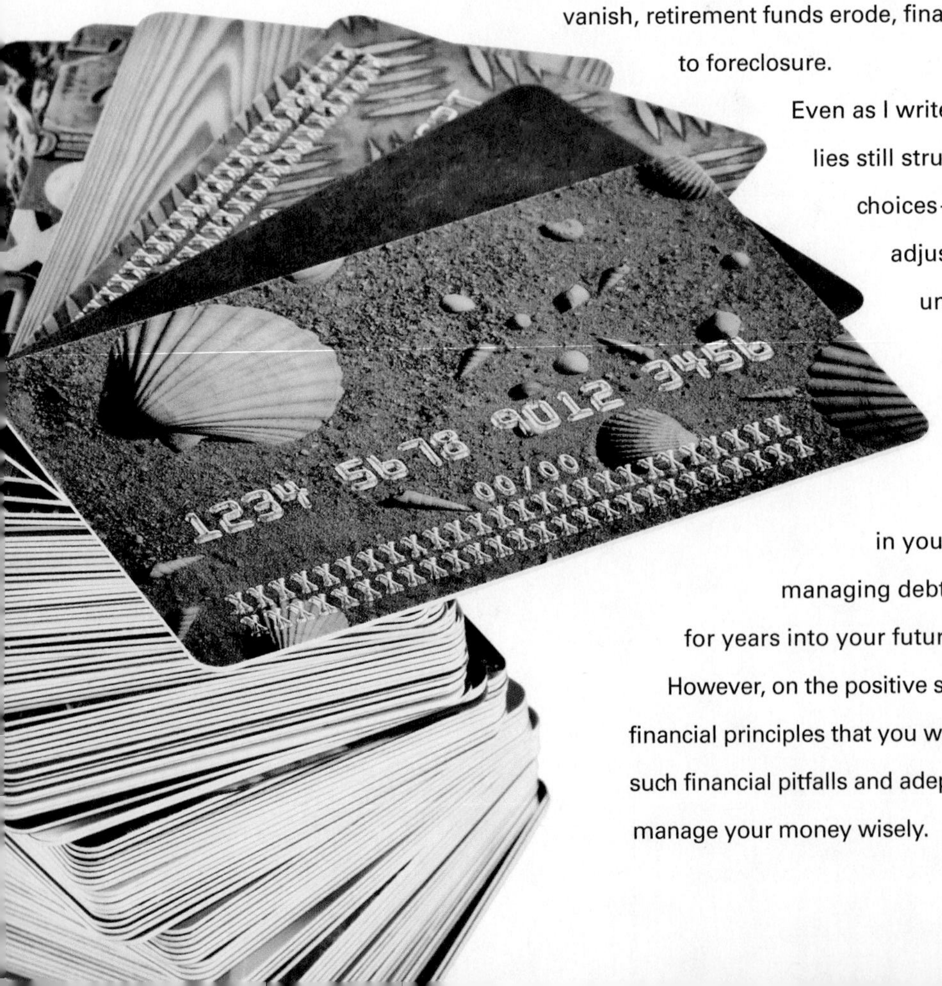

Even as I write this introduction, millions of families still struggle with the legacy of bad financial choices—overwhelming credit card debt, adjustable rate mortgages that are now unpayable, speculative investments gone bad, and the crushing long-term burden of record student loans. In Section 8.2, we will explain the mathematics of how this can happen in your present life—for example, how mismanaging debt can leave you with a burden to carry for years into your future.

However, on the positive side, by understanding and applying the financial principles that you will learn in this chapter, you can avoid such financial pitfalls and adeptly use the power of mathematics to manage your money wisely.

8.1 Percents, Taxes, and Inflation

Objectives

1. Understand how to calculate with percent.
2. Use percents to represent change.
3. Apply the percent equation to solve applied problems.
4. Use percent in calculating income taxes.

Throughout this chapter, we will be discussing various aspects of your future financial life—student loans, credit card borrowing, investments, and mortgages. To understand these ideas and also much of the other information that you encounter daily, you must be comfortable with the notion of percent.

Percent

KEY POINT

Percent means "per hundred."

The word *percent* is derived from the Latin "per centum," which means "per hundred." Therefore, 17% means "seventeen per hundred." We can write 17% as $\frac{17}{100}$ or in decimal form as 0.17. In this chapter, we will usually write percents in decimal form.

EXAMPLE 1 Writing Percents as Decimals

a) Write each of the following percents in decimal form:

36% 19.32%

b) Write each of the following decimals as percents:

0.29 0.354

SOLUTION:

a) Think of 36% as the decimal thirty-six hundredths, or 0.36, as we show in Figure 8.1(a). To write 19.32%, recall the problem-solving advice from Section 1.1 that it is often helpful to solve a simpler problem instead. If we had asked you to write 19% as a decimal, you would write it as 0.19. Now, once you have positioned the 1 and the 9 properly in the decimal, write the 3 and the 2 immediately to their right, as we show in Figure 8.1(b). Thus, 19.32% is equal to 0.1932.

(a)

36 percent
36 hundredths
. 36 (decimal form of 36 hundredths)
hundredths

(b)

At first, ignore 32.
19. 32 percent
19. 32 hundredths
.19 32 (decimal form of 19 hundredths)
.19 32
hundredths

FIGURE 8.1 The key to converting percents to decimals is what you write in the hundredths place.

QUIZ YOURSELF 1

a) Write 17.45% as a decimal.
b) Write 0.05% as a decimal.
c) Write 2.45 as a percent.
d) Write 0.025 as a percent.

b) To translate decimals to percents, we first examine what numbers are in the tenths and hundredths place. To rewrite 0.29, we see that we have 29 hundredths, so 0.29 equals 29%.

To rewrite 0.354, recognize that 0.35 would be 35%, so 0.354 is 35.4%.

Now try Exercises 1 to 16. 1

PROBLEM SOLVING

Strategy: The Analogies Principle

You may have been told that when converting a percent to a decimal, you move the decimal point two places to the left and that in converting a decimal to a percent, you move the decimal point two places to the right. It is all right to use such memory devices, provided that you understand where they come from.

Always remember that percent means hundredths. If you understand how to rewrite 0.29 as 29%, then you also know how to rewrite 1.29 as a percent. Similarly, if you know that 19% equals 0.19, then you also know how to rewrite 19.32% as a decimal.

If you want, you can use the following rules:

To convert from a percent to a decimal, drop the percent sign and divide by 100.

To convert from a decimal to a percent, multiply by 100 and add a percent sign.

We often have to convert a fraction to a percent. Example 2 shows how to do this.

EXAMPLE 2 Converting a Fraction to a Percent

Write $\frac{3}{8}$ as a percent.

SOLUTION: Review the Relate a New Problem to an Older One Strategy on page 9.

Because we already know how to rewrite a decimal as a percent, we first convert $\frac{3}{8}$ to a decimal and then rewrite the decimal as a percent.

If we divide the denominator into the numerator, we get $\frac{3}{8} = 0.375$. Next, we see that 0.375 is equal to 37.5%. Thus, $\frac{3}{8} = 37.5$ percent.

Now try Exercises 17 to 22.

Percent of Change

Example 3 shows how you can use the simple notion of percent to compare changes in data over long periods of time.

EXAMPLE 3 Changes in Defense Spending as a Percent of the Federal Budget over Time

According to the Office of Management and Budget, in 1970 the U.S. government spent \$82 billion for defense at a time when the federal budget was \$196 billion. Forty years later, in 2010, spending for defense was \$692 billion and the budget was \$3,721 billion. What percent of the federal budget was spent for defense in 1970? In 2010?

SOLUTION:

In 1970, \$82 billion out of \$196 billion was spent for defense. We can write this as the fraction $\frac{82}{196} \approx 0.418 = 41.8\%$. In 2010, this fraction was $\frac{692}{3{,}721} \approx 0.1859 \approx 18.6\%$.

Thus, you see, as a percentage of the federal budget, defense spending was considerably less in 2010 than it was in 1970.

The media often use percentages to explain the change in some quantity. For example, you may hear that, on a particularly bleak day on Wall Street, the stock market is down 1.2%. Or on a good day, the evening news tells us that consumer confidence in the economy is up 13.5% over last month. To make such statements, we have to understand several quantities.

The percent of change is always in relationship to a previous or **base amount.** We then compare a **new amount** with the base amount as follows:

$$\text{percent of change} = \frac{\text{new amount} - \text{base amount}}{\text{base amount}}.$$

If the new amount is less than the base amount, then the percent of change will be negative. We illustrate this idea in Example 4.

QUIZ YOURSELF 2

The population of Florida increased from 12.9 million in 1990 to 16 million in 2000. What was the percent of increase of Florida's population over these 10 years?

EXAMPLE 4 Rising Cookie Prices

According to the Bureau of Labor Statistics, during 2011 the average price of a package of chocolate chip cookies rose from \$3.99 to \$4.18. What was the percent of increase in the cost of these cookies?

SOLUTION:

In this example, the base amount is \$3.99 and the new amount is \$4.18. To find the percent of increase, we calculate

$$\frac{\text{new amount} - \text{base amount}}{\text{base amount}} = \frac{4.18 - 3.99}{3.99} = \frac{0.19}{3.99} \approx 0.048 = 4.8\%.$$

Thus, the price of these cookies rose by almost 5% during 2011. 2

Merchants often use percents to describe the deals that they are giving to the public when they have a sale. The increase that a merchant adds to his base price is called **markup.**

EXAMPLE 5 Investigating a Sale Price on a New Car

Monte's Autorama is having an end-of-year clearance in which the TV ads proclaim that all cars are sold at 5% markup over the dealer's cost. Monte has a new Honda CR-V for sale for \$23,295. On the Internet, you find out that this particular model has a dealer cost of \$21,339. Is Monte being honest in his advertising?

SOLUTION: Review the Relate a New Problem to an Older One Strategy on page 9.

To find the percent markup on this particular Honda CR-V, you can calculate the percent of markup, which is the same as the percent of change in the base price (dealer's cost) of the car.

We calculate as in Example 4:

$$\text{percent of markup} = \frac{\overbrace{\text{selling price}}^{\text{new amount}} - \overbrace{\text{dealer cost}}^{\text{base amount}}}{\underbrace{\text{dealer cost}}_{\text{base amount}}}$$

$$= \frac{23{,}295 - 21{,}339}{21{,}339} = \frac{1{,}956}{21{,}339} \approx 0.092 = 9.2\%.$$

Thus, Monte is not being quite truthful here because his markup on this model is 9.2%. which is well above the amount he stated in his TV ads.

Now try Exercises 51 to 54.

KEY POINT

Many percent problems are based on the same equation.

The Percent Equation

We will conclude this section with several examples of percent problems; however, it is important for you to recognize that these problems are all variations of the same equation.

In each case, we will be taking some *percent* of a *base* quantity and setting it equal to an *amount.* We can write this as the equation

$$\text{percent} \times \text{base} = \text{amount}.$$

We will call this equation the **percent equation.**

You saw this pattern in Example 5, where the percent was 9.2% = 0.092, the base (dealer's price) was \$21,339, and the amount (markup) was \$1,956. Notice that 0.092 × 21,339 = 1,956. In the remaining examples, we will give you two of the three quantities percent base, and amount and ask you to find the third.

EXAMPLE 6 Using the Percent Equation

a) What is 35% of 140?

b) 63 is 18% of what number?

c) 288 is what percent of 640?

SOLUTION:

We will illustrate the percent equation graphically to solve each problem.

a) The base is 140 and the percent is 35% = 0.35.

$$\underset{0.35}{\text{percent}} \times \underset{140}{\text{base}} = \text{amount}$$

So the amount is 0.35 × 140 = 49.

b) Using the percent equation again, we get

$$\underset{0.18}{\text{percent}} \times \text{base} = \underset{63}{\text{amount}}.$$

Therefore, we have 0.18 × base = 63, or, dividing both sides of this equation by 0.18, we get

$$\text{base} = \frac{63}{0.18} = 350.$$

c) Here the base is 640 and the amount is 288.

$$\text{percent} \times \underset{640}{\text{base}} = \underset{288}{\text{amount}}.$$

Thus, we have percent × 640 = 288. Dividing both sides of this equation by 640, we get

$$\text{percent} = \frac{288}{640} = 0.45 = 45\%.$$

Now try Exercises 23 to 32. 3

QUIZ YOURSELF 3

a) What is 15% of 60?

b) 18 is 24% of what number?

c) 96 is what percent of 320?

EXAMPLE 7 Calculating Sports Statistics

In the 2010–2011 season, the Chicago Bulls had the best record in the National Basketball Association with 62 wins and 20 losses. What percent of their games did they win?

Between the Numbers

Pay Careful Attention to What They *Don't* Tell You

When negotiating, people often want you to focus on their numbers and distract you from noticing other information that might be useful in making your decision. For example, consider the following situation that occurred several years ago when our faculty union was negotiating with the Commonwealth of Pennsylvania for a new contract.

The state negotiator recommended that we "back-load" the contract over 3 years by accepting raises of 0%, 2%, and 3% over 3 years instead of raises of 3%, 2%, and 0%. He stated that at the end of three years, the percentage of increase would be the same in either case, because

$$1 \cdot (1.02)(1.03) = (1.03)(1.02) \cdot 1 = 1.0506,$$

giving a 5.06% increase. If we focus just on the percent of increase by the third year, the negotiator was telling the truth. So you might ask, "What's the difference?"

In the spirit of the Three-Way Principle, let's look at what effect both types of raises will have on a salary of \$100 in Table 8.1.

With front-loading, we earn \$13.12 − \$7.06 = \$6.06 more money on our initial \$100 than with back-loading. Over three years, a person earning \$50,000 (which is 500 times as large as \$100) would earn 500(\$6.06) = \$3,030 more with front-loading versus back-loading.

	Back-Loading	Front-Loading
Original Amount	\$100	\$100
Amount in 1st year	\$100 + 0% (\$100) = \$100	\$100 + 3% (100) = \$103
Amount in 2nd year	\$100 + 2% (100) = \$102	\$103 + 2% (103) = \$105.06
Amount in 3rd year	\$102 + 3% (102) = \$105.06	\$105.06 + 0% (105.06) = \$105.06
	In 3 years, we gain 2 + 5.06 = \$7.06 more than if we had no raise at all.	In 3 years, we gain 3 + 5.06 + 5.06 = \$13.12 more than if we had no raise at all.

TABLE 8.1 Comparing back-loading versus front-loading.

SOLUTION:

Again, you can use the percent equation; however, you have to be careful. The base is not 62, but rather the total number of games played, which is 62 + 20 = 82. The amount is the number of victories, 62. So we have

$$\text{percent} \times \underset{82}{\text{base}} = \underset{62}{\text{amount}}.$$

Dividing both sides of the equation above, percent × 82 = 62, by 82 gives us percent $= \frac{62}{82} \approx 0.756 = 75.6\%$.

EXAMPLE 8 Increase in Student Loan Debt

According to a report in *The Atlantic*, based on data from the New York Federal Reserve, the amount of outstanding student loans in 2011 was \$550 billion, an increase of 551% over the corresponding amount in 1999. Find the amount of outstanding student loans in 1999.

SOLUTION: Review the Relate a New Problem to an Older One Strategy on page 9.

In this case the base is unknown. In deciding what percent to use in the percent equation, you have to be careful. The 511% is *only the increase*. The amount \$550 billion represents 100% of the debt owed in 1999 plus the 511% increase. Therefore, the percent we will use in the percent equation is

$$100\% + 511\% = 611\% = 6.11.$$

Substituting in the percent equation, we get

$$\overset{\text{percent}}{6.11} \times \text{base} = \overset{\text{amount}}{550}.$$

Dividing both sides of this equation by 6.11 gives us

$$\text{base} = \frac{550}{6.11} \approx 90.$$

So, the amount of outstanding student debt in 1999 was about $90 billion.

Taxes

Calculating various kinds of taxes relies heavily on using percents properly.

EXAMPLE 9 Calculating Your Income Tax

Table 8.2 is taken from the instructions for filling out Form 1040 to compute the federal income tax for a person whose marriage status is single.

	If your taxable income is over—	But not over—	The tax is	Of the amount over—
Line 1	$0	7,550	.... 10%	0$
Line 2	7,550	30,650	$755.00 + 15%	7,550
Line 3	30,650	74,200	$4,220.00 + 25%	30,650
Line 4	74,200	154,800	$15,107.50 + 28%	74,200
Line 5	154,800	336,550	$37,675.50 + 33%	154,800
Line 6	336,550	....	$97,653.00 + 38%	336,550

TABLE 8.2 Federal income taxes due for a single person.

a) If Jaye is unmarried and has a taxable income* of $41,458, what is the amount of federal income tax she owes?

b) How did the IRS arrive at the $4,220 amount in column 3 of line 3?

SOLUTION:

a) In calculating this tax, you first must identify the line of the table that is relevant to Jaye's situation. Because her income is above $30,650 and below $74,200, we will use line 3 (highlighted) from Table 8.2.

So Jaye must pay $4,220 + 25% of the amount of taxable income over $30,650. Therefore, her tax is

$$4{,}220 + (\overset{25\%}{0.25})(\overset{\text{amount over } \$30{,}650}{41{,}458 - 30{,}650}) = 4{,}220 + (0.25)(10{,}808) = 4{,}220 + 2{,}702 = \$6{,}922.$$

b) The table is treating Jaye's income as being divided into two parts. Up to $30,650, she is being taxed according to the instructions on line 2 of the table. The tax on $30,650 would be $755 + 15% of the amount of taxable income over $7,550. So her tax is

$$755 + (0.15)(30{,}650 - 7{,}550) = 755 + (0.15)(23{,}100) = 755 + 3{,}465 = \$4{,}220.$$

Now try Exercises 65 to 68. 4

QUIZ YOURSELF 4

Use Table 8.2 to calculate Aliyah's federal income tax that is due if her taxable income last year was $85,500.

*It is too complicated to get into detail here, but in essence, after you have totaled your wages, tips, interest earned, etc., you then reduce this total by making various kinds of adjustments such as exemptions, deductions for charitable contributions, work-related expenses, and other deductions to calculate what is called your *taxable income*.

Inflation

Inflation is a rise in the level of prices of goods and services over a period of time. One common measure of inflation is the **Consumer Price Index** (CPI), which is maintained by the Bureau of Labor Statistics, U.S. Department of Labor. The base CPI is 100, which corresponds to average prices during the years 1982 to 1984. If you were to look up the CPI for dairy products in 2009, you would find it to be 197.0, which means that, on the average, dairy products in 2009 cost 197% of what they cost in 1982–84.

EXAMPLE 10 How Does the Increase in College Tuition Compare with the CPI?

In 2009, the overall CPI was 214, which meant that, on the average, prices were 214% of what they were in 1982–84. Use the given graph to compare the rise in tuition at public and private colleges with the CPI.

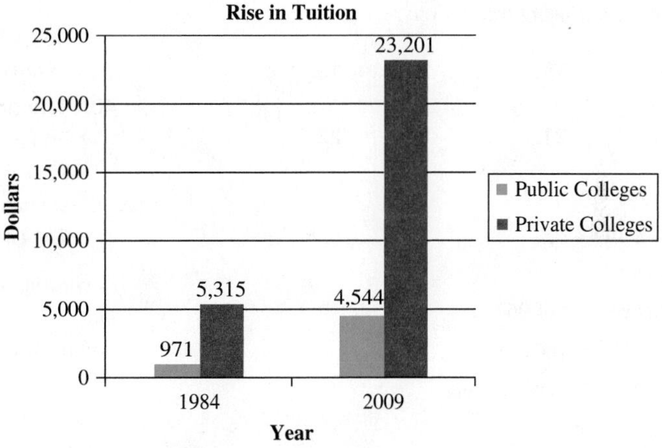

Source: Statistical Abstract of the United States, 2011–2021

SOLUTION:

For public colleges, we will use the percent equation

$$\text{percent} \times \underbrace{\text{base}}_{\text{1984 tuition}} = \underbrace{\text{amount}}_{\text{2009 tuition}}.$$

where the base is 971 and the amount is 4,544.

Thus,

$$\text{percent} \times 971 = 4{,}544.$$

Solving for percent, we get

$$\text{percent} = \frac{4{,}544}{971} \approx 4.68 = 468\%.$$

In the case of private colleges, the base is 5,315 and the amount is 23,201, or

$$\text{percent} = \frac{23{,}201}{5{,}315} \approx 4.37 = 437\%.$$

In both cases, the rise in college tuition is more than twice the CPI of 214.

Now try Exercises 69 to 72.

EXERCISES 8.1

Sharpening Your Skills

Convert each of the following percents to decimals.

1. 78% **2.** 65% **3.** 8% **4.** 3%

5. 27.35% **6.** 83.75%

7. 0.35% **8.** 0.08%

Write each of the following decimals as percents.

9. 0.43 **10.** 0.95 **11.** 0.365

12. 0.875 **13.** 1.45 **14.** 2.25

15. 0.002 **16.** 0.0035

Convert each of the following fractions to percents.

17. $\frac{3}{4}$ **18.** $\frac{7}{8}$ **19.** $\frac{5}{16}$

20. $\frac{9}{25}$ **21.** $\frac{5}{2}$ **22.** $\frac{11}{8}$

23. 12 is what percent of 80?

24. What is 24% of 125?

25. 77 is 22% of what number?

26. 33.6 is what percent of 96?

27. What is 12.25% of 160?

28. 47.74 is 38.5% of what number?

29. 8.4 is what percent of 48?

30. What is 23% of 140?

31. 29.76 is 23.25% of what number?

32. 149.5 is what percent of 130?

Applying What You've Learned

33. Cookie sales. In a certain year, the top-selling cookie in America was Nabisco's Chips Ahoy, with sales of $294.6 million. The total cookie sales for that year were $3,124 million. What percent of the total cookie sales was due to Chips Ahoy? (*Source:* Information Resources, Inc.)

34. Pizza sales. In a certain year, DiGiorno sold $478.3 million worth of frozen pizzas. If total frozen pizza sales were $2,844.8 million, what percent of frozen pizza sales was due to DiGiorno? (*Source:* Information Resources, Inc.)

35. The price of a new home. According to the U.S. Bureau of the Census, from 2000 to 2007 the average price of a new home in the United States increased by 51.5% to $314 thousand. To the nearest thousand, what was the average price of a new home in 2000?

36. The price of a new home. According to the U.S. Bureau of the Census, in 2007 the average price of a new home in the United States was $314 thousand. By 2010, the price had dropped by 13.1%. To the nearest thousand, what was the average price of a new home in 2010?

37. Spanish-speaking radio stations. In 2001 there were 574 Spanish-speaking radio stations in the United States. By 2010, this number had increased by 40.4%. How many Spanish-speaking radio stations were there in 2010? (*Source: World Almanac and Book of Facts 2011*)

38. The price of mail. In 1991, the price of mailing a one-ounce letter was 29 cents. By 2010, the cost was 44 cents. What was the percent of increase?

39. The price of a necklace. In 1969, actor Richard Burton bought a pearl, diamond, and ruby necklace, called La Peregrina, for his wife, Elizabeth Taylor, for $37,000. In 2011 this necklace was auctioned for $11,818,500. The auction price was what percent of the original purchase price? (*Source: The Guardian*)

40. The price of jewelry. Continuing Exercise 39, La Peregrina broke the previous auction record for pearls, which was $7,096,000 for the Baroda pearls. What percent higher was the auction price of La Peregrina than the Baroda pearls? (*Source: The Guardian*)

According to eBizMBA, the most visited Web sites in 2012 (in millions of visitors) are given by the following graph. Use this information to do Exercises 41–44.

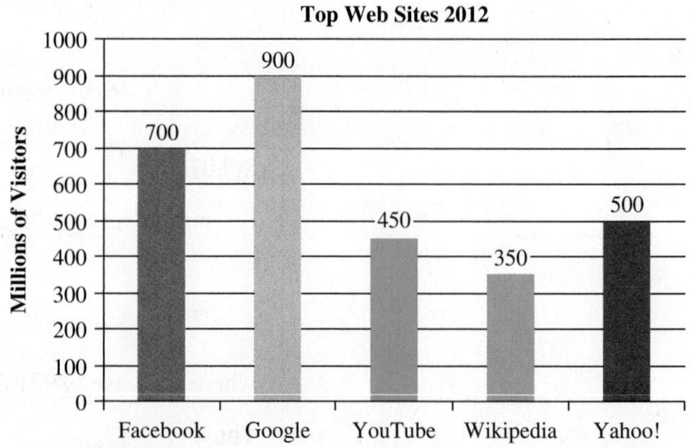

Source: eBizMBA.com

41. Comparing Web site visitors. What percent of the total visitors of these five sites visited Google?

42. Comparing Web site visitors. What percent of the total visitors to these sites was due to Facebook and YouTube combined?

43. Comparing Web site visitors. What percent greater was the number of visitors to Yahoo! than to Wikipedia?

44. Comparing Web site visitors. What percent smaller was the number of visitors to Yahoo! than to Google?

45. Autographs. According to *The Useless Information Society*, only 6% of Beatles' autographs in circulation are estimated to be real. If there are 81 authentic autographs of the Beatles in circulation, then how many nonauthentic autographs are there in circulation?

46. **Happy Meal sales.** In a certain fiscal quarter, 40% of McDonald's profits came from Happy Meals. If Happy Meals profits were $508 million, how much profit came from selling other items?

47. **Population.** According to the U.S. Bureau of the Census, from 2000 to 2009 the population of the United States grew from 281 million to 307 million. What was the percent of increase?

48. **Population.** From 2000 to 2009 California's population grew from 33.9 million to 37 million. What was the percent of increase?

49. **Music downloads.** The Recording Industry Association of America reports that in 2009, downloaded music album sales were 562% of what they were in 2005. If there were 13.6 million albums downloaded in 2005, how many were downloaded in 2009?

50. **Music video sales.** The Recording Industry Association of America found that music video sales decreased by 30.2% from 2005 to 2009. If there were 33.8 million videos sold in 2005, how many were sold in 2009?

51. **Dealer markup on a car.**
    a. If a dealer buys a car from the manufacturer for $19,875 and then sells it for $21,065, what is his markup?
    b. In doing this problem, Angela got an incorrect answer of 5.6%. What mistake did she make?

52. **Dealer markup on a computer.**
    a. A computer retailer buys a multimedia computer for $1,850 and then sells it for $2,081. What is her markup on the computer?
    b. In doing this problem, Raj got an incorrect answer of 11.1%. What mistake did Raj make?

53. **Markup on a boat.** Carlos is a boat dealer who bought a sailboat for $11,400 and then sold it for $12,711. What percent was his markup on the boat?

54. **Markup on an appliance.** Anna, who owns a small appliance store, bought a food processor for $524 and then sold it for $589.50. What percent was her markup on the food processor?

The graph below shows the sales (in thousands) of the five best-selling cars in 2009. Use this information to solve Exercises 55–58.

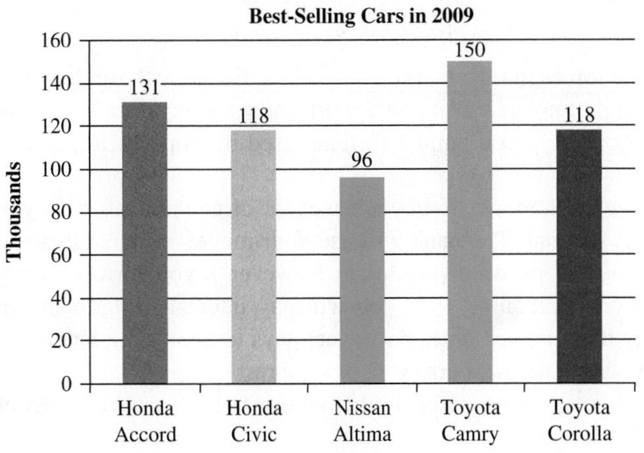

55. **Car sales.** What percent of the sales of these five cars was due to the Camry?

56. **Car sales.** What percent of the sales of these five cars was due to Hondas?

57. **Car sales.** What percent greater were the sales of the Corolla than the Altima?

58. **Car sales.** What percent smaller were the sales of the Accord than the Camry?

59. **Salary increase.** Marcy is working in a 2-year probationary period as a paralegal. She currently earns $28,000 a year and after her probationary period, her salary will increase by 35%. What will her yearly salary be at that time?

60. **Salary decrease.** Jed works for Metal Fabricators, Inc., which just lost a large contract and has asked employees to take a 12% pay cut. If Jed now makes $34,500 per year, what will his yearly pay be after the pay cut?

61. **Buying a grill.** Kirby is buying a new gas barbecue grill that has been reduced for an end-of-summer sale by 15% to $578. What was the original price of the grill?

62. **Increasing workload.** Gisela is an insurance claims adjuster. Last quarter she settled 124 claims and this quarter she settled 155 claims. What is the percent of increase in the claims that she settled this quarter over last quarter?

63. **Change in the stock market.** Due to a slump in the economy, Omarosa's mutual fund has dropped by 12% from last quarter to this quarter. If her fund is now worth $11,264, how much was her fund worth last quarter?

64. **Gas mileage.** Renaldo bought a new Honda Civic hybrid that gets 26.3% more miles per gallon than he got with his old Civic. If his new car gets 48 miles per gallon, what was the gas mileage on his old car?

For Exercises 65–68, use Table 8.2 in Example 9 to calculate the federal income tax due for the given taxable income.

65. $148,000
66. $28,750
67. $47,800
68. $440,000

Recall that the Consumer Price Index (CPI) is a measure of inflation obtained by comparing current prices with base prices in 1982–84. However, it is a mistake to interpret a change in the CPI as the actual percent of inflation. For example, from 2008 to 2009 the CPI of candy and chewing gum rose from 123.2 to 130.2. However, the rate of inflation for these items was only $\frac{130.2 - 123.2}{123.2} \approx 0.057 = 5.7\%$. *For Exercises 69–72, use this information to complete the missing items in the following table.*

Item	CPI for Item in 2000	CPI for Item in 2009	Percent of Increase
Ice cream	164.4	196.6	**69.**___
Gasoline	128.6	**70.**___	57%
Food	**71.**___	218.2	30%
Pets and pet products	144.3	194.9	**72.**___

Communicating Mathematics

Assume that you are studying with a friend for a quiz on percents. Your friend insists on blindly memorizing which way to "move the decimal point" in doing percent problems. Explain how you would help your friend understand how to do the conversions in Exercises 73–76 without relying solely on memorization.

73. You are converting 28.35% to a decimal.

74. You are converting 1.285 to a percent.

75. You are converting 0.0375 to a percent.

76. You are converting 1.375% to a decimal.

Between the Numbers

77. Contract negotiations. Assume that you are negotiating a new three-year contract. Are you better off with yearly raises of 3%, 1%, and then 2% or with yearly raises of 2%, 1%, then 3%? Explain.

78. Contract negotiations. Assume that you are earning $50,000 per year and want a 3.5% raise. A negotiator offers you instead a one-time bonus of $2,000. What are the advantages and disadvantages of her offer?

Challenge Yourself

79. Depreciation on a new car. If a new car costs $18,000 and depreciates at a rate of 12% per year, what will be the value of the car in 4 years?

80. Compounding raises. If you are given a raise of 8% this year and a raise of 5% the following year, what single raise would give you the same yearly salary in the second year?

81. Calculating prices. If a merchant increases the price of a home entertainment system by x% and then later reduces the price by x%, is the price of the system the same as the original price? If not, what is the relationship between the two prices? Explain your answer by using appropriate examples.

82. Calculating prices. If a merchant reduces the price of a luxury speedboat by x% and then later increases the price by x%, is the price of the boat the same as the original price? If not, what is the relationship between the two prices? Explain your answer by using appropriate examples.

83. Inflating prices by reducing size. If a half-gallon (64 ounces) of ice cream costing $4.79 is reduced in size to $1\frac{1}{2}$ quarts (48 ounces) and the price is reduced to $3.89, what is the percent of increase in the price of the ice cream?

84. Inflating prices by reducing size. If a candy bar weighs 6.8 ounces and costs $1.29 and is then reduced in size to 6.4 ounces but the price remains the same, what is the percent of increase in the cost of the candy bar?

8.2 Interest

Objectives

1. Understand the simple interest formula.
2. Use the compound interest formula to find future value.
3. Solve the compound interest formula for different unknowns, such as the present value, length, and interest rate of a loan.

The most powerful force in the universe is....

How would you finish this quote? The world-renowned physicist Albert Einstein said,

. . . compound interest.

Are you surprised that of all the forces that he might have picked, Einstein chose this one? In this section, we will explain how interest can either work for you—or against you. As you will see, used properly, it can help you build a fortune; used improperly, it can lead you to financial ruin.

If you want to accumulate enough money to buy a newer car or go on a vacation, you could deposit money in a bank account. The bank will use your money to make loans to other customers and pay you interest for using your funds. However, if you borrow money from the bank, say to take a college course, then you will pay interest to the bank. In essence, **interest** is the money that one person (a borrower) pays to another (a lender) to use the lender's money. Savers earn interest; borrowers pay interest.

We will discuss simple and compound interest in this section, and discuss the cost of consumer loans in Section 8.3.

Simple Interest

KEY POINT

Simple interest is a straightforward way to compute interest.

The amount you deposit in a bank account is called the **principal.** The bank specifies an **interest rate** for that account as a percentage of your deposit. The rate is usually expressed as an annual rate. For example, a bank may offer an account that has an annual interest rate of 5%. To find the interest that you will earn in such an account, you also need to know how long the deposit will remain in the account. The time is usually stated in years. There is a simple formula that relates principal, interest earned, interest rate, and time. In words,

$$\text{interest earned} = \text{principal} \times \text{interest rate} \times \text{time}.$$

When we compute interest this way, it is called **simple interest.**

FORMULA FOR COMPUTING SIMPLE INTEREST We calculate simple interest using the formula

$$I = Prt,$$

where I is the interest earned, P is the principal, r is the annual interest rate, and t is the time in years.

EXAMPLE 1 Calculating Simple Interest

If you deposit \$500 in a bank account paying 6% annual interest, how much interest will the deposit earn in 4 years if the bank computes the interest using simple interest?

SOLUTION:

In this example:

P is the principal, which is \$500
r is the annual interest rate, which is 6% (written as 0.06)
t is the time, which is 4 (years)

Thus, the interest earned is

$$I = Prt = 500 \times 0.06 \times 4 = 120.$$

In 4 years, this account earns \$120 in interest.

Now try Exercises 1 to 4.

KEY POINT

Future value equals principal plus interest.

To find the amount that will be in your account at some time in the future, called the **future value** or **future amount,** we add the interest earned to the principal. The principal is often called the **present value.** We will represent the future value by A, so we can say

$$A = \text{principal} + \text{interest} = P + I.$$

If we replace I by Prt, we get the formula $A = P + Prt = P(I + rt)$.

COMPUTING FUTURE VALUE USING SIMPLE INTEREST To find the future value of an account that pays simple interest, use the formula

$$A = P(1 + rt),$$

where A is the future value, P is the principal, r is the annual interest rate, and t is the time in years.

EXAMPLE 2 Finding the Future Value and the Present Value of Accounts

a) *Future Value:* Assume that you deposit \$1,000 in an account paying 3% annual interest for 6 years. Use the simple interest formula to calculate the future value of this account.

b) *Present Value:* Assume that you plan to save \$2,500 take a white-water rafting trip on the Rio Grande River in New Mexico in 2 years. Your bank offers a certificate of deposit (CD) that pays 4% annual interest computed using simple interest. How much must you put in this CD now to have the necessary money in 2 years?

SOLUTION: Review the Relate a New Problem to an Older One Strategy on page 9.

a) We see that $P = 1{,}000$, $r = 0.03$, and $t = 6$. Therefore,

$$A = \overset{P}{1{,}000}(1 + (\overset{r}{0.03})(\overset{t}{6})) = 1{,}000(1 + 0.18) = 1{,}000(1.18) = 1{,}180.$$

Thus, your bank account will have \$1,180 at the end of 6 years.

b) Because we want to know the amount that we are to deposit *now,* we are looking for present value. We do not need a new formula to find this; instead, we can use the future value formula $A = P(1 + rt)$, but this time we will solve for P instead of A.

We know that $A = 2{,}500$, $r = 4\% = 0.04$, and $t = 2$. Therefore,

$$2{,}500 = P(1 + (0.04)(2)).$$

We can rewrite this equation as

$$2{,}500 = P(1.08).$$

Dividing both sides of the equation by 1.08, we get

$$P = \frac{2{,}500}{1.08} \approx 2314.814815.$$

We will round this *up* to \$2,314.82 to guarantee that if you put this amount in the CD now, in 2 years you will have the \$2,500 you need for your white-water rafting trip.*

Now try Exercises 5 to 10. 5

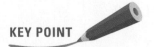

QUIZ YOURSELF 5

Redo Example 2(b), but now assume that you want to save \$2,400 in 4 years and the CD has an annual interest rate of 5%.

SOME GOOD ADVICE

In Example 2(b), we used the earlier formula for computing future value to find the present value rather than stating a new formula to solve this specific problem. You will find it easier to learn a few formulas well and use them, together with simple algebra, to solve new problems rather than trying to memorize separate formulas for every type of problem.

Compound Interest

KEY POINT

Compounding pays interest on previously earned interest.

It seems fair that if money in a bank account has earned interest, the bank should compute the interest due, add it to the principal, and then pay interest on this new, larger amount. This is in fact the way most bank accounts work. Interest that is paid on principal plus previously earned interest is called **compound interest.** If the interest is added yearly, we say that the interest is *compounded annually.* If the interest is added every three months, we say the interest is *compounded quarterly.* Interest also can be compounded monthly and daily.

---

*When calculating a deposit to accumulate a future amount, we will always round up to the next cent.

EXAMPLE 3 Calculating Compound Interest the Long Way

Assume that you want to replace your sailboat with a larger one in 3 years. To save for a down payment for this purchase, you deposit \$2,000 for 3 years in a bank account that pays 10% annual interest,* compounded annually. How much will be in the account at the end of 3 years?

SOLUTION:

We will perform the compound interest calculations one year at a time in the following table. In compounding the interest, we will use the future value from the previous year as the new principal at the beginning of the year. Notice that the quantity $(1 + rt) = (1 + 0.10 \times 1) = (1.10)$ remains the same throughout the computations.

Year	Principal (Beginning of Year) P	Future Value (End of Year) $P(1 + rt) = P(1.10)$
1	\$2,000	\$2,000(1.10) = \$2,200
2	\$2,200	\$2,200(1.10) = \$2,420
3	\$2,420	\$2,420(1.10) = \$2,662

6

QUIZ YOURSELF 6

Continue Example 3 to calculate the amount in your account at the end of the fourth year.

PROBLEM SOLVING

Strategy: Verify Your Answer

You should always check answers to see whether they are reasonable. In Example 3, if we had used simple interest to find the future value, we would have obtained $A = 2{,}000(1 + (0.10)(3)) = 2{,}000(1.30) = 2{,}600$. The interest we found in Example 3 is a *little* larger because as the interest is added to the principal each year, the bank is paying interest on an increasingly larger principal.

If we were to continue the process that we used in Example 3 for a longer period of time, say for 30 years, it would be very tedious. In Figure 8.2 we look at the same computations in a different way, keeping in mind that the amount in the account at the end of each year is 1.10 times the amount in the account at the beginning of the year.

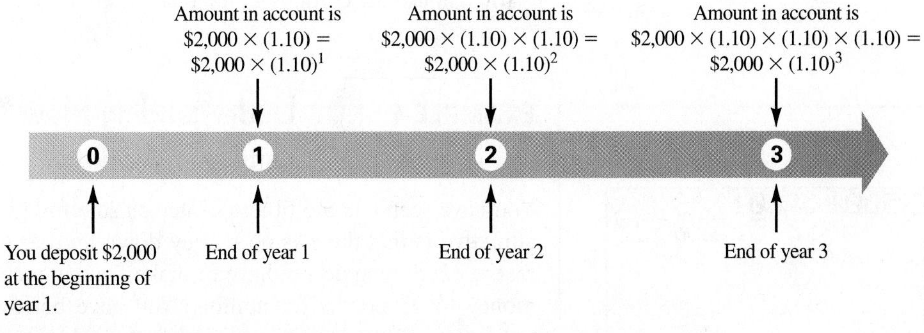

FIGURE 8.2 10% interest being compounded annually.

*An interest rate of 10% would be extraordinarily high. However, we will often choose rates in examples and exercises to keep the computations simple.

If we were to continue the pattern shown in Figure 8.2 to compute the future value of the account at the end of 30 years, we would see that

$$A = 2{,}000(1.10)^{30} \approx 2{,}000(17.44940227) \approx 34{,}898.80.^*$$

This large amount shows how your money can grow if it is compounded over a long period of time.

In general, if we deposit a principal P in an account paying an annual interest rate r for t years, then the future value of the account is given by the formula

money you will have in the future → A; money you have now → P

$$A = P(1 + r)^t.$$

In the example that we just calculated, $P = 2{,}000$, $r = 0.10$, and $t = 30$. It is important to understand that this formula for calculating compound interest only works for the case when r is the *annual interest rate* and t is time being measured in *years*. Do not bother to learn this formula because in just a moment we will give you a similar compounding formula that works for more general situations. 7

QUIZ YOURSELF 7

Calculate the future value of an account containing \$3,000 for which the annual interest rate is 4% compounded annually for 10 years.

Solving for Unknowns in the Compound Interest Formula

KEY POINT

Knowing the principal, the periodic interest rate, and the number of compounding periods, it is easy to determine future value.

All banks and most other financial institutions compound interest more frequently than once a year. For example, many banks send savings account customers a monthly statement showing the balance in their accounts. So far in our discussion of compounding, we have used a yearly interest rate. If compounding takes place more frequently, then the interest rate must be adjusted accordingly. For example, a yearly interest rate of $12\% = 0.12$ corresponds to a monthly interest rate of $\frac{12\%}{12} = \frac{0.12}{12} = 0.01 = 1\%$. If the interest is being compounded quarterly, the quarterly interest rate would then be $\frac{12\%}{4} = \frac{0.12}{4} = 0.03 = 3\%$.

In order to handle situations such as these, we will modify the formula $A = P(1 + r)^t$ slightly.

THE COMPOUND INTEREST FORMULA Assume that an account with principal P is paying an annual interest rate r and compounding is being done m times per year. If the money remains in the account for n time periods, then the future value, A, of the account is given by the formula

$$A = P\left(1 + \frac{r}{m}\right)^n.$$

Notice that in this formula, we have replaced r by $\frac{r}{m}$, which is the annual rate divided by the number of compounding periods per year, and t by n, which is the number of compounding periods.

You can use the compound interest formula for computing compound interest to compare investments.

EXAMPLE 4 Understanding How "No Payments Until..." Works

```
3500*(1.01)^6
     3715.320527
```

You have seen a home fitness center on sale for \$3,500 and what really makes the deal attractive is that there is no money down and no payments due for 6 months. Realize that although you do not have to make any payments, the dealer is not loaning you the money for 6 months for nothing. You have borrowed \$3,500 and, in 6 months, your payments will be based upon that fact. Assuming that your dealer is charging an annual interest rate of 12%, compounded monthly, what interest will accumulate on your purchase over the next 6 months?

---

*To ensure greater accuracy, we often show calculations with eight decimal places. If your calculations do not agree with ours, it may be due to the difference in the way we are rounding our calculations.

SOLUTION:

To determine the interest that has accumulated, we will find the future value of your "loan" (assuming that you make no payments) and subtract \$3,500 from that. We will use the formula for calculating future value with $P = 3{,}500$, $r = 0.12$, $m = 12$, and $n = 6$. Therefore,

$$A = P\left(1 + \frac{r}{m}\right)^n = 3{,}500\left(1 + \underbrace{\frac{0.12}{12}}_{\text{monthly interest rate}}\right)^{\overset{\text{number of months}}{6}} = 3{,}500(1.01)^6 = 3{,}715.33.$$

So the accumulated interest is \$3,715.33 − \$3,500 = \$215.33.

Now try Exercises 11 to 18. **8**

QUIZ YOURSELF 8

Sarah deposits \$1,000 in a CD paying 6% annual interest for 2 years. What is the future value of her account if the interest is compounded quarterly?

Example 5 illustrates a different way to use the compound interest formula.

EXAMPLE 5 Finding the Present Value for a College Tuition Account

Upon the birth of a child, a parent wants to make a deposit into a tax-free account to use later for the child's college education. Assume that the account has an annual interest rate of 4.8% and that the compounding is done quarterly. How much must the parent deposit now so that the child will have \$60,000 at age 18?

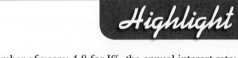

SOLUTION: Review the Relate a New Problem to an Older One Strategy on page 9.

We will use the compound interest formula $A = P\left(1 + \frac{r}{m}\right)^n$ to find P.

Using the values $A = 60{,}000$, *future value we want to save for college*

$r = 0.048$, *annual interest rate of 4.8%*

$m = 4$, *4 compounding periods per year*

and $n = 72$, *18 years × 4 compounding periods per year*

Highlight

Doing Financial Calculations with a Calculator

When doing financial computations, often technology can speed up your work. We will use a calculator to reproduce the solution to Example 5.

On my calculator, if we press the [APPS] key and then choose "1: Finance," Screen 1 comes up. The letters TVM stand for "Time Value of Money." Then by choosing option 1, we get Screen 2. Now we can enter the values 18 for N, the number of years; 4.8 for I%, the annual interest rate; 60,000 for FV, the future value; and 4 for C/Y, the number of compounding periods per year. Next we position the cursor over PV (present value) and press the keys [ALPHA] [ENTER]. The amount −25418.75939 for present value means that we must deposit \$25,418.76 now to have the desired \$60,000 in 18 years (Screen 3).

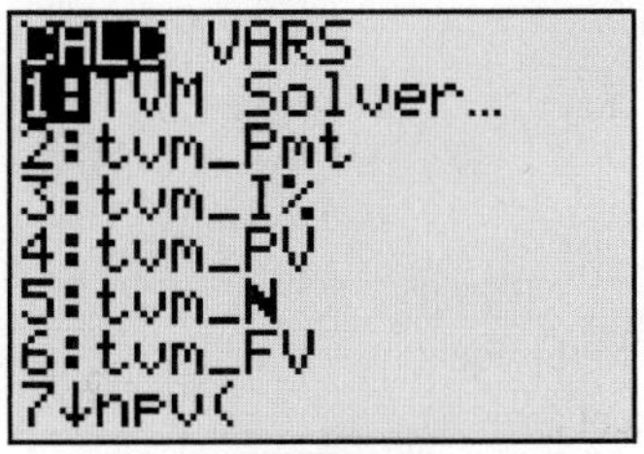

Screen 1

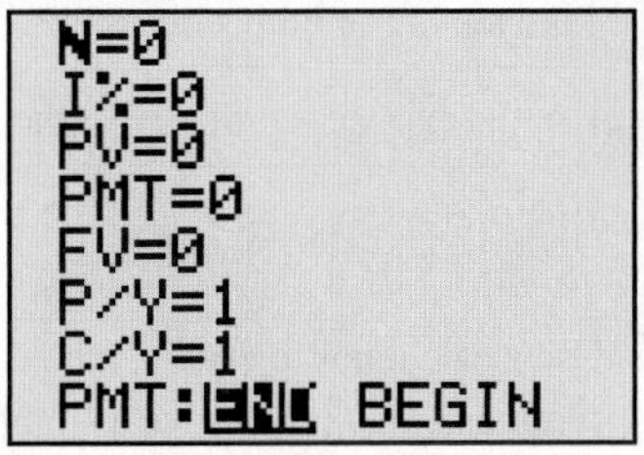

Screen 2

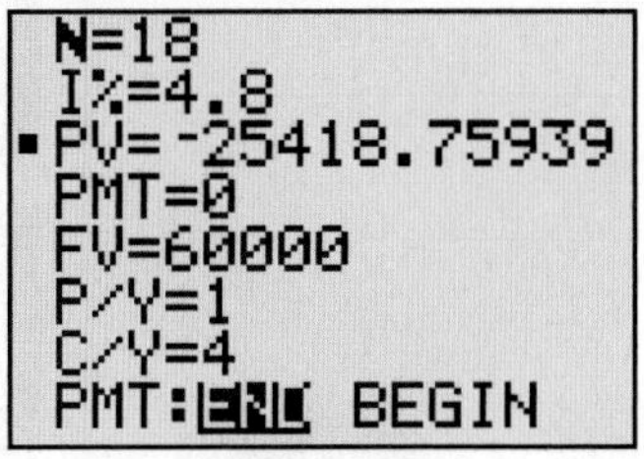

Screen 3

we have

$$60{,}000 = P\left(1 + \frac{0.048}{4}\right)^{72} = P(1 + 0.012)^{72}.$$

Solving for P we get

$$P = \frac{60{,}000}{(1.012)^{72}} = \frac{60{,}000}{2.360461386} \approx 25{,}418.76.$$

A deposit slightly over \$25,400 now will guarantee \$60,000 for college in 18 years.

Now try Exercises 45 and 46.

Although \$60,000 may seem like a lot of money, realize that *inflation*, the increase in the price of goods and services, will also cause the cost of a college education to increase.

So far we have used the formula $A = P\left(1 + \frac{r}{m}\right)^n$ to find A and P. Sometimes we want to find r or n. To do this, we need to introduce some new techniques.

KEY POINT

We use the log function to solve for n in the formula

$$A = P\left(1 + \frac{r}{m}\right)^n.$$

If you want to solve for n in the formula $A = P\left(1 + \frac{r}{m}\right)^n$, you need to be able to solve an equation of the form $a^x = b$, where a and b are fixed numbers. A property of logarithmic functions enables you to solve such equations. Many calculators have a key labeled either "log" or "log x," which stands for the **common logarithmic function.** Pressing this key reverses the operation of raising 10 to a power. For example, suppose that you compute $10^5 = 100{,}000$ on your calculator. If you next press the log key, the display will show 5. If you enter 1,000, which is 10 raised to the third power, and press the log key, the display will show 3. Practice finding the log of powers of 10 such as 100 and 1,000,000. If you enter 23 and then press the log key, the display will show 1.361727836. The interpretation of this result is that $10^{1.361727836} = 23$. (We will not discuss what it means to raise 10 to a power such as 1.361727836.) The log function has an important property that will help us solve equations of the form $a^x = b$.

EXPONENT PROPERTY OF THE LOG FUNCTION

$$\log y^x = x \log y$$

To understand this property, you should use your calculator to verify the following:

$$\log 4^5 = 5 \log 4$$
$$\log 6^3 = 3 \log 6$$

Example 6 illustrates how to use the exponent property to solve equations.

EXAMPLE 6 Solving an Equation Using the Exponent Property of the Log Function

Solve $3^x = 20$.

SOLUTION:

We illustrate the steps required to solve this equation.

Step 1	Take the log of both sides of the equation.	$\log 3^x = \log 20$
Step 2	Use the exponent property of the log function.	$x \log 3 = \log 20$
Step 3	Divide both sides by log 3.	$x = \frac{\log 20}{\log 3}$
Step 4	Use a calculator to evaluate the right side of the equation (your calculator may give a slightly different answer).	$x = 2.726833028$

Now try Exercises 25 to 28. 9

QUIZ YOURSELF 9

Solve $6^x = 15$.

Example 7 illustrates how to use the exponent property of the log function to solve a problem.

EXAMPLE 7 Calculating the Time for an Investment to Double

Collectors often buy classic DVDs hoping to get a good return on their investment. DeDe has invested \$8,500 in old *Stars Wars* DVDs, which are currently increasing in value at 6.5% per year. If this increase continues at the same rate, in how many years will DeDe's investment double?

SOLUTION:

We are looking for a future value of 2 × \$8,500, or \$17,000. So we can use the future value formula $A = P(1 + \frac{r}{m})^n$ with the following values:

$$\begin{aligned} A &= 17{,}000, && \text{Double original amount of \$8,500} \\ P &= 8{,}500, && \text{Initial investment} \\ r &= 0.065, && \text{Annual interest rate} \\ \text{and} \quad m &= 1. && \text{Compounding is done once a year} \end{aligned}$$

Now we will solve the equation $17{,}000 = 8{,}500\left(1 + \frac{0.065}{1}\right)^n$ for n.

$$\begin{aligned} 17{,}000 &= 8{,}500(1.065)^n \\ 2 &= (1.065)^n && \text{Divide both sides by 8,500} \\ \log 2 &= \log(1.065)^n && \text{Take log of both sides} \\ \log 2 &= n \times \log(1.065) && \text{Use exponent property of log} \\ n &= \frac{\log 2}{\log(1.065)} \approx 11 && \text{Divide by log(1.065) and calculate quotient} \end{aligned}$$

Thus DeDe's investment will double in about 11 years.

Now try Exercises 29 to 32. 10

QUIZ YOURSELF 10

Do Example 7 again, but now assume that the interest rate on DeDe's investment is 4%.

There is a quick way to estimate doubling time using the following rule.

THE RULE OF 70 To estimate the **doubling time** of a quantity, divide 70 by the growth rate. For example, in Example 7, the annual growth rate was 6.5% so the doubling time is roughly $\frac{70}{6.5} \approx 10.769$.

The last situation that we will consider is how to solve the compound interest equation $A = P\left(1 + \frac{r}{m}\right)^n$ for r. To do this, we have to be able to solve an equation of the form $x^a = b$, where a and b are fixed numbers. We show how to solve such an equation in Example 8.

EXAMPLE 8 Negotiating a Basketball Contract

LeBron is negotiating a new contract with the Miami Heat. In order to reduce current taxes, he has agreed to defer a \$15 million bonus to be paid as \$18 million 4 years from now. If the Heat invests the \$15 million now, what is the minimum annual interest rate they need to earn to guarantee that the \$18 million is available in 4 years? Assume that the annual interest rate is compounded monthly.

SOLUTION: Review the Splitting-Hairs Principle on page 12.

To solve this compound interest problem, we will use the formula $A = P(1 + \frac{r}{m})^n$ with $A = 18, P = 15, m = 12$, and $n = 12 \times 4 = 48$. We will solve for r. Substituting for A, P, m, and n, we solve as follows:

$$18 = 15\left(1 + \frac{r}{12}\right)^{48}$$

$$1.2 = \left(1 + \frac{r}{12}\right)^{48} \quad \text{Divide both sides by 15}$$

$$(1.2)^{\frac{1}{48}} = \left(\left(1 + \frac{r}{12}\right)^{48}\right)^{\frac{1}{48}} = 1 + \frac{r}{12}.^{*} \quad \text{Raise both sides to the } \tfrac{1}{48}\text{th power}$$

Subtracting 1 from both sides gives us

$$\frac{r}{12} = (1.2)^{\frac{1}{48}} - 1 = 1.003805589 - 1 = 0.003805589.$$

So, $r = 12 \times 0.003805589 \approx 0.046$. Thus the Heat needs to find an investment that pays about 4.6% annual interest compounded monthly.

Now try Exercises 43 and 44.

SOME GOOD ADVICE

Be careful to distinguish between the situations in Examples 7 and 8. In Example 7, we used the log function to solve an equation of the form $a^x = b$. In Example 8, we solved an equation of the form $x^a = b$ by raising both sides of the equation to the $\frac{1}{a}$ power.

EXERCISES 8.2

Sharpening Your Skills

In Exercises 1–4, use the simple interest formula I = Prt *and elementary algebra to find the missing quantities in the table below.*

	I	*P*	*r*	*t*
1.		$1,000	8%	3 years
2.	$196		7%	2 years
3.	$700	$3,500		4 years
4.	$1,920	$8,000	6%	

In Exercises 5–10, use the future value formula A = P(1 + rt) *and elementary algebra to find the missing quantities in the table below.*

	A	*P*	*r*	*t*
5.		$2,500	8%	3 years
6.		$1,600	4%	5 years
7.	$1,770		6%	3 years
8.	$2,332		3%	2 years
9.	$1,400	$1,250		2 years
10.	$966	$840	5%	

In Exercises 11–18, you are given the principal, the annual interest rate, and the compounding period. Use the formula for computing future value using compound interest to determine the value of the account at the end of the specified time period.

11. $5,000, 5%, yearly; 5 years

12. $7,500, 7%, yearly; 6 years

13. $4,000, 8%, quarterly; 2 years

14. $8,000, 4%, quarterly; 3 years

15. $20,000, 8%, monthly; 2 years

---

*In algebra, $(a^x)^y = a^{xy}$. That is why $\left(\left(1 + \frac{r}{12}\right)^{48}\right)^{\frac{1}{48}} = \left(1 + \frac{r}{12}\right)^{48 \times \frac{1}{48}} = \left(1 + \frac{r}{12}\right)^{1} = 1 + \frac{r}{12}$.

16. \$10,000, 6%, monthly; 5 years

17. \$4,000, 10%, daily; 2 years

18. \$6,000, 4%, daily; 3 years

Savings institutions often state a ***nominal rate****, which you can think of as a simple annual interest rate, and the* ***effective interest rate****, which is the actual interest rate earned due to compounding. Given the nominal rate, it is easy to calculate the effective interest rate as follows. Assume that you invest \$1 in an account paying an interest rate of 6% compounded monthly. Using the compound interest formula* $A = P\left(1 + \frac{r}{m}\right)^n$, *with* $P = 1$, $r = 0.06$, $m = 12$, *and* $n = 12$, *we would get* $A = \left(1 + \frac{0.06}{12}\right)^{12} \approx 1.0617$ *So the effective interest rate is* $1.0617 - 1 = 0.0617$, *or 6.17%. Use this method to find the effective interest rate for the investments in Exercises 19–22.*

19. nominal yield, 7.5%; compounded monthly

20. nominal yield, 10%; compounded twice a year

21. nominal yield, 6%; compounded quarterly

22. nominal yield, 8%; compounded daily

In Exercises 23 and 24, you are given an annual interest rate and the compounding period for two investments. Decide which is the better investment.

23. 5% compounded yearly; 4.95% compounded quarterly

24. 4.75% compounded monthly; 4.70% compounded daily

In Exercises 25–32, solve each equation.

25. $3^x = 10$ **26.** $2^x = 12$

27. $(1.05)^x = 2$ **28.** $(1.15)^x = 3$

In Exercises 29–32, you have \$1,500 to invest. Find the time in years that it takes your investment to double with annual compounding (a) using the method of Example 7 and (b) using the Rule of 70.

29. 3.5% **30.** 8.5% **31.** 3% **32.** 5%

In Exercises 33–36, use the compound interest formula $A = P(1 + r)^t$ *and the given information to solve for either* t *or* r.

33. $A = \$2{,}500$, $P = \$2{,}000$, $t = 5$

34. $A = \$400$, $P = \$20$, $t = 35$

35. $A = \$1{,}500$, $P = \$1{,}000$, $r = 4\%$

36. $A = \$2{,}500$, $P = \$1{,}000$, $r = 6\%$

Applying What You've Learned

37. Buying an entertainment system. You have purchased a home entertainment system for \$3,600 and have agreed to pay off the system in 36 monthly payments of \$136 each.

a. What will be the total sum of your payments?

b. What will be the total amount of interest that you have paid?

38. Buying a car. You have purchased a used car for \$6,000 and have agreed to pay off the car in 24 monthly payments of \$325 each.

a. What will be the total sum of your payments?

b. What will be the total amount of interest that you have paid?

Often, through government-supported programs, students may obtain "bargain" interest rates such as 6% or 8% to attend college. Frequently, payments are not due and interest does not accumulate until you stop attending college. In Exercises 39 and 40, calculate the amount of interest due 1 month after you must begin payments.

39. Borrowing for college. You have borrowed \$10,000 at an annual interest rate of 8%.

40. Borrowing for college. You have borrowed \$15,000 at an annual interest rate of 6%.

In Exercises 41–44, we will assume that the lender is using simple interest to compute the interest on the loan.

41. Borrowing for a trip. You plan to take a trip to the Grand Canyon in 2 years. You want to buy a certificate of deposit for \$1,200 that you will cash in for your trip. What annual interest rate must you obtain on the certificate if you need \$1,500 for your trip?

42. Paying interest on late taxes. Jonathan wants to defer payment of his \$4,500 tax bill for 4 months. If he must pay an annual interest rate of 15% for doing this, what will his total payment be?

43. Borrowing from a pawn shop. Sanjay has borrowed \$400 on his father's watch from the Main Street Pawn Shop. He has agreed to pay off the loan with \$425 one month later. What is the annual interest rate that he is being charged?

44. Borrowing from a bail bondsman. If a person accused of a crime does not have sufficient resources, he may have a bail bondsman post bail to be released until a trial is held. Assume that a bondsman charges a \$50 fee plus 8% of the amount of the bail. If a bondsman posts \$20,000 for a trial that takes place in 2 months, what is the interest rate being charged by the bondsman? (Treat the \$50 fee plus the 8% as interest on a \$20,000 loan for two months.)

In Exercises 45 and 46, Ann and Tom want to establish a fund for their grandson's college education. What lump sum must they deposit in each account in order to have $30,000 in the fund at the end of 15 years?

45. Saving for college. 6% annual interest rate, compounded quarterly

46. Saving for college. 7.5% annual interest rate, compounded monthly

The Consumer Price Index (CPI) is a measure of inflation obtained by comparing current prices with base prices in 1982–84. We can find the inflation rate for a given time period by calculating the percent of change in the CPI over that time period. For example, the CPI in 2005 was 195.3 and in 2009 it was 214.5, so the inflation rate was $\frac{214.5 - 195.3}{195.3} \approx 0.098 = 9.8\%$. *Use the given graph to do Exercises 47–50.*

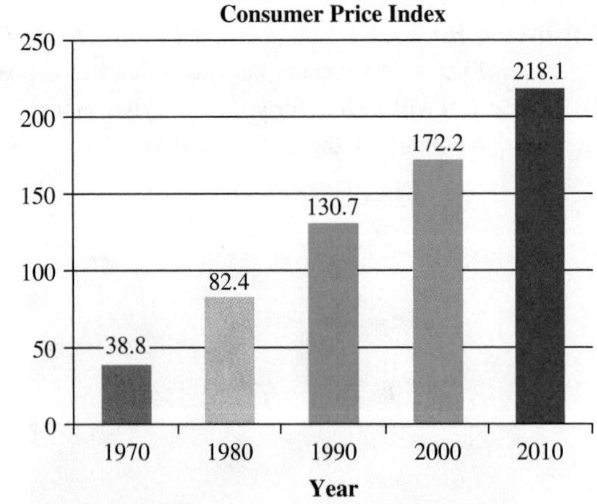

47. Effect of inflation on the cost of sneakers.

a. Calculate the inflation rate from 1970 to 2010.

b. If a pair of sneakers cost $33 in 1970, use the inflation rate from part (a) to estimate the cost of the sneakers in 2010.

48. Effect of inflation on the cost of textbooks.

a. Calculate the inflation rate from 1980 to 2000.

b. If a textbook cost $47 in 1980, use the inflation rate from part (a) to estimate the cost of the book in 2000.

49. Effect of inflation on the cost of cars.

a. Calculate the inflation rate from 1980 to 2010.

b. If a car cost $17,000 in 2010, use the inflation rate from part (a) to estimate what the car would have cost in 1980.

50. Effect of inflation on the cost of jeans.

a. Calculate the inflation rate from 1990 to 2010.

b. If a pair of jeans cost $80 in 2010, use the inflation rate from part (a) to estimate what the cost of the jeans would have been in 1990.

51. Inflation. From 1992 to 1995, Albania experienced a yearly inflation rate of 226%. Determine the price of a $4.65 fast-food meal after 5 years at a 226% inflation rate.

52. Inflation. The inflation rate in Hungary during the mid-1990s was about 28%. Determine the price of a $96 pair of athletic shoes after 10 years at a 28% inflation rate.

53. Comparing investments. Jocelyn purchased 100 shares of Jet Blue stock for $23.75 per share. Eight months later she sold the stock at $24.50 per share.

a. What annual rate, calculated using simple interest, did she earn on this transaction?

b. What annual rate would she have to earn in a savings account compounded monthly to earn the same money on her investment?

54. Comparing investments. Dominick purchased a bond for $2,400 to preserve a wildlife sanctuary and 10 months later he sold it for $2,580.

a. What annual rate, calculated using simple interest, did he earn on this transaction?

b. What annual rate would he have to earn in a savings account compounded monthly to earn the same money on his investment?

55. Investment earnings. Emily purchased a bond valued at $20,000 for highway construction for $9,420. If the bond pays 7.5% annual interest compounded monthly, how long must she hold it until it reaches its full face value?

56. Investment earnings. Lucas purchased a bond with a face value of $10,000 for $4,200 to build a new sports stadium. If the bond pays 6.5% annual interest compounded monthly, how long must he hold it until it reaches its full face value?

57. Saving for retirement. Allison is 30 years old and plans to retire at age 65 with $1,200,000 in her retirement account. What amount would she have to set aside now in an investment paying 6% annual interest if the compounding is done daily (assume 365 days in a year)?

58. Saving for retirement. Anibal wants to have $200,000 to buy a smaller home when he retires in 20 years. How much must he place in an investment now that pays 4.5% annual interest compounded monthly to have that amount?

59. Purchase of Manhattan. There is much folklore regarding Peter Minuit's purchase of Manhattan Island in 1626 from a Native American tribe. Let's assume that his purchase price was, as often mentioned, 24 U.S. dollars. If that amount were placed in an account paying 5% annual interest and allowed to compound annually until 2015, how much would be in the account?

60. Purchase of Manhattan. Some estimate the (2012) current value of Manhattan to be about $800,000,000,000. Assuming this to be true, what would have been a fair price for Peter Minuit to pay for Manhattan in 1626, assuming an interest rate of 5% compounded annually?

61. Devaluing money. Inflation erodes the value of money. Assuming an annual inflation rate of 4%, what would a 2010 dollar be worth in 2020? That is, find an amount A in 2010 that would be worth $1.00 in 2020.

62. Devaluing money. Assuming an inflation rate of 4%, when would a 2010 dollar be worth 50 cents?

Communicating Mathematics

63. What is the difference between simple interest and compound interest?

64. What is the meaning of each variable in the compound interest formula $A = P\left(1 + \frac{r}{m}\right)^n$?

65. Under what circumstances will $A = P(1 + r)^t$ and $A = P(1 + \frac{r}{m})^n$ give you the same answers to a compound interest problem?

66. Looking back on Exercises 59 and 60, we must admit that they are somewhat silly. Mention some factors that we are ignoring that make those problems unrealistic.

Between the Numbers

67. When you use a credit card, the credit card company charges a fee to the merchant, which ranges from 1 percent to 3 percent, on your purchase. Suppose you bought a high-definition TV for $350 and asked if the seller would give you a discount for paying cash. What might be a reasonable discount?

68. What disadvantage do you see that credit card fees have for buyers who don't use credit cards?

Challenge Yourself

Some banks advertise that money in their accounts is compounded continuously. To get an understanding of what this means, apply the compound interest formula using a very large number of compounding periods per year. In Exercises 69 and 70, divide the year into 100,000 compounding periods per year. Apply the compound interest formula for finding future value to approximate what the effective annual yield would be if the compounding were done continuously for the stated nominal yield.

69. nominal yield, 10%

70. nominal yield, 12%

If the principal P *is invested in an account that pays an annual interest rate of* r% *and the compounding is done continuously, then the future value,* A, *that will be in the account after* t *years is given by the formula*

$$A = Pe^{rt}.$$

The number e *is approximately* 2.718281828.

71. Use the formula for continuous compounding to find the effective annual yield if the compounding in Exercise 69 is done continuously.

72. Use the formula for continuous compounding to find the effective annual yield if the compounding in Exercise 70 is done continuously.

According to the National Debt Clock, in 2012 the United States' national debt was $15.2 trillion. Assume this number is your base amount and is staying fixed (which it is not) as you solve Exercises 73–76.

73. If the population of the U.S. at that time was 312 million, what was your share of the national debt?

74. Assuming an interest rate of 5%, how fast was the debt increasing per year in 2012 due to interest?

75. Assuming a 366-day year (2012 is a leap year), how fast was the debt increasing per minute in 2012 due to interest?

76. Assume you finish college in 2018. How much interest would accrue from 2012 to 2018 if no payments were made?

8.3 Consumer Loans

Objectives

1. Determine payments for an add-on loan.
2. Compute finance charges on a credit card using the unpaid balance method.
3. Use the average daily balance method to compute credit card charges.
4. Compare credit card finance charge methods.

Debt is a bottomless sea.

—Thomas Carlyle

In the latter part of the last decade, the United States was hit with an economic firestorm resulting in the Great Recession, which is generally recognized as the worst economic crisis since the Great Depression almost one hundred years ago. Although there were other factors for this financial explosion, one critical factor mentioned by economists was an extremely indebted U.S. economy.

From ancient times to the present, sources such as the Bible, William Shakespeare, Benjamin Franklin, and others have warned about the dangers of unbridled credit. In this section we will explain the mathematics of credit and show you how to use it wisely to avoid drowning in Carlyle's bottomless sea.

Imagine that you have just signed the lease for your first apartment and now all you have to do is furnish it. If you buy living room furniture for \$1,100, which you pay for in payments, you are taking out an installment loan. Loans having a fixed number of payments are called *closed-ended credit* agreements (or **installment loans**). Each payment is called an **installment.** The size of your payments is determined by the amount of your purchase and also by the interest rate that the seller is charging. The interest charged on a loan is often called a **finance charge.**

The Add-On Interest Method

KEY POINT

The add-on interest method is a simple way to compute payments on an installment loan.

We use the simple interest formula from Section 8.2 to calculate the finance charge for an installment loan. To determine the payments for an installment loan, we add the simple interest due on the loan to the loan amount and then divide this sum by the number of monthly payments.

FORMULA FOR DETERMINING THE MONTHLY PAYMENT OF AN INSTALLMENT LOAN

$$\text{monthly payment} = \frac{P + I}{n},$$

where P is the amount of the loan, I is the amount of interest due on the loan, and n is the number of monthly payments.

This method is sometimes called the **add-on interest method** because we are adding on the interest due on the loan before determining the payments.

EXAMPLE 1 Determining Payments for an Add-On Interest Loan

A new pair of Bose speakers for your home theater system costs \$720. If you take out an add-on loan for 2 years at an annual interest rate of 18%, what will be your monthly payments?

SOLUTION:

We first use the simple interest formula to calculate the interest:

$$I = Prt = 720(0.18)2 = 259.20.$$

Next, we add the interest to the purchase price:

$$720 + 259.20 = 979.20.$$

To find the monthly payments, we divide this amount by 24:

$$\frac{979.20}{24} = 40.80.$$

Monthly payments are therefore \$40.80.

Now try Exercises 1 to 8. 11

QUIZ YOURSELF 11

Suppose that you take an installment loan for \$360 for 1 year at an annual interest rate of 21%. What are your monthly payments?

In Example 1, the annual interest rate of 18% is quite misleading. If we think about it, the purchase price was \$720, so it would be fair to say that you are paying off \$720/24 = \$30 of the loan amount each month; the other \$10.80 is interest. When you reach the last month, although you only owe \$30 on the purchase, you are still paying \$10.80 in interest. Simple arithmetic shows that 10.80/30 = 0.36. So in a certain sense, the interest rate for the last month of the loan is actually 36%. Because you are paying 36% interest for one month, this is equivalent to an annual interest rate of 12 × 36% = 432%.

What we want to point out here is that although simple interest is easy to compute, as you pay off the loan amount, the actual interest you are paying on the outstanding balance is higher than the stated interest rate.

When you use your credit card to pay for gas at a gas station, you are using *open-ended credit.* With open-ended credit, the calculation of finance charges can be more complicated than with closed-ended credit. Although you may be making monthly payments on your loan, you may also be increasing the loan by making further purchases.

There are several ways that credit card companies compute finance charges. We will look at two methods and compare them at the end of this section. You will see that if you understand the method being used to compute your finance charges, you can use this information to reduce the cost of borrowing money.

The Unpaid Balance Method

KEY POINT

The unpaid balance method computes finance charges on the balance at the end of the previous month.

The first method that we will discuss for computing finance charges is called the **unpaid balance method.** With this method, the interest is based on the previous month's balance.

> **THE UNPAID BALANCE METHOD FOR COMPUTING THE FINANCE CHARGE ON A CREDIT CARD LOAN** This method also uses the simple interest formula $I = Prt$; however,
>
> P = previous month's balance + finance charge + purchases made − returns − payments.
>
> The variable r is the annual interest rate, and $t = \frac{1}{12}$.

EXAMPLE 2 Using the Unpaid Balance Method for Finding Finance Charges

Assume that the annual interest rate on your credit card is 18% and your unpaid balance at the beginning of last month was \$600. Since then, you purchased ski boots for \$130 and sent in a payment of \$170.

a) Using the unpaid balance method, what is your credit card bill this month?

b) What is your finance charge next month?

SOLUTION:

a) We will list the items that we need to know to compute this month's balance.

Previous month's balance: \$600

Finance charge on last month's balance: $\$600 \times 0.18 \times \left(\frac{1}{12}\right) = \$9$

(0.18: annual interest rate; $\frac{1}{12}$: Time is 1 month.)

Purchases made: \$130

Returns: \$0

Payment: \$170

Therefore, you owe

Previous month's balance + finance charge + purchases made − returns − payments

$$= 600 + 9 + 130 - 0 - 170 = \$569.$$

b) The finance charge for next month will be $\$569 \times 0.18 \times \left(\frac{1}{12}\right) = \$8.54$.

Now try Exercises 21 to 26. 12

QUIZ YOURSELF 12

Assume that the annual interest rate on your credit card is 21%. Your outstanding balance last month was \$300. Since then, you have charged a purchase for \$84 and made a payment of \$100. What is the outstanding balance on your card at the end of this month? What is next month's finance charge on this balance?

Note that you can use the unpaid balance method to your advantage by making a large purchase early in the billing period and then paying it off just before the billing date. This is not to the credit card company's advantage because you can use the credit card company's money for free for almost a whole month.

Because credit is so readily available, it can be tempting to borrow today without thinking of the consequences tomorrow. Often a borrower does not understand the debt that is being assumed and how difficult it will be to pay off. This definitely can be the case with student loans.

EXAMPLE 3 Paying Off a Student Loan

Assume that you graduate from college with \$35,000 in student loans* with a 24% annual simple interest rate. In order to reduce your debt as quickly as possible, beginning next month you are going to pay \$800 per month towards the loan. After your first payment, how much will you owe on your loan?

SOLUTION:

We will first calculate the interest due on your loan. Using the simple interest formula $I = P \times r \times t$, where $P = 35{,}000$, $r = 0.24$, and $t = \frac{1}{12}$, we find that the interest due the first month is $I = 35{,}000 \times 0.24 \times \frac{1}{12} = \$700$.

If you make a payment of \$800, the amount going to reduce the principal will be only \$100. So you still owe $\$35{,}000 - \$100 = \$34{,}900$ on your loan!

Now try Exercises 43 and 44. 13

QUIZ YOURSELF 13

Assume that you have a student loan debt of \$10,000 at a simple interest rate of 21%. If you make a first payment on the loan of \$300, how much will go towards interest? How much will go towards the principal? How much will you still owe?

Example 3 shows how hard it can be to pay off a large debt. The best practice is to pay off as much of your outstanding balance as you can to avoid paying a large amount of interest.

When borrowing, be careful to be aware of the annual interest rate. For example, credit card companies who offer to lend you cash often have a very high interest rate and borrowing from one company to pay off another can actually increase your debt.

The Average Daily Balance Method

A more complicated method for determining the finance charge on a credit card is called the **average daily balance method,** which is one of the most common methods used by credit card companies. With this method, the balance is the average of all daily balances for the previous month.

KEY POINT

The average daily balance method computes finance charges based on the balance in the account for each day of the month.

THE AVERAGE DAILY BALANCE METHOD FOR COMPUTING THE FINANCE CHARGE ON A CREDIT CARD LOAN

1. Add the outstanding balance for your account for each day of the month.
2. Divide the total in step 1 by the number of days in the month to find the average daily balance.
3. To find the finance charge, use the formula $I = Prt$, where P is the average daily balance found in step 2, r is the annual interest rate, and t is the number of days in the month divided by 365.

*The numbers in this example are based on a real case in which a friend asked me what the best way would be to pay off his \$35,000 credit card debt.

Between the Numbers

Will You Be Part of Generation Broke?

You've heard of "Generation X" and "Generation Y" but have you heard of "Generation Broke"? Numerous articles chronicle the worries of a generation of today's college students graduating with a crippling amount of student loan debt. *USA Today* states that outstanding student loan debt reached $1 trillion in 2011 and the Department of Education reports that 8.8% of all borrowers now default on their student loans. Banking laws have been changed so that even declaring bankruptcy will not allow borrowers to avoid paying their student loan debt.

There are many issues to consider when thinking about your loans. With private loans, a lender can charge interest rates and fees that can greatly affect your ability to repay loans. Although standard repayment plans are for 10 years, borrowers can receive extensions for as many as 25 years—a scary thought!

For a good source of information about the pitfalls of student loans, go to projectonstudentdebt.org to read the article "Top 10 Student Loan Tips for Recent Graduates."

EXAMPLE 4 Using the Average Daily Balance Method for Finding Finance Charges

Suppose that you begin the month of September (which has 30 days) with a credit card balance of $240. Assume that your card has an annual interest rate of 18% and that during September the following adjustments are made on your account:

September 11: A payment of $60 is credited to your account.
September 18: You charge $24 for iTunes downloads.
September 23: You charge $12 for gasoline.

Use the average daily balance method to compute the finance charge that will appear on your October credit card statement.

SOLUTION:

To answer this question, we must first find the average daily balance for September. The easiest way to calculate the balance is to keep a day-by-day record of what you owe the credit card company for each day in September, as we do in Table 8.3.

Transaction	Day	Balance	Number of Days × Balance
Balance on Sept. 1	1, 2, 3, 4, 5, 6, 7, 8, 9, 10	$240	$10 \times 240 = \$2{,}400$
Payment of $60 on Sept. 11	11, 12, 13, 14, 15, 16, 17	$180	$7 \times 180 = \$1{,}260$
Charge $24 on Sept. 18	18, 19, 20, 21, 22	$204	$5 \times 204 = \$1{,}020$
Charge $12 on Sept. 23	23, 24, 25, 26, 27, 28, 29, 30	$216	$8 \times 216 = \$1{,}728$
			Total = $6,408

TABLE 8.3 Daily balances for September.

The average daily balance is therefore

$$\frac{(10 \times 240) + (7 \times 180) + (5 \times 204) + (8 \times 216)}{30}$$

$$= \frac{2{,}400 + 1{,}260 + 1{,}020 + 1{,}728}{30} = \frac{6{,}408}{30} = 213.6.$$

We next apply the simple interest formula, where $P = \$213.60$, $r = 0.18$, and $t = \frac{30}{365}$. ($t = \frac{30}{365}$ because we are using the credit card for 30 days out of a 365-day year.) Thus, $I = Prt = 213.6(0.18)\left(\frac{30}{365}\right) = 3.16$. Your finance charge on the October statement will be $3.16.

Now try Exercises 29 to 32. 14

QUIZ YOURSELF 14

Recalculate the average daily balance in Example 4, except now assume you bought the iTunes downloads on September 3 instead of September 18. Make a table similar to Table 8.3.

Historical Highlight

Credit and Interest

Credit cards were not widely used in the United States until the 1950s when cards such as Diners Club, Carte Blanche, and American Express made the use of plastic money more popular. Today, Americans charge about $1 trillion per year on their cards.

Credit is not a modern idea. Surprisingly, there are ancient Sumerian documents dating back to about 3000 BC that show the regular use of credit in borrowing grain and metal. Interest on these loans was often in the 20% to 30% range—similar to the 18% to 21% charged on many of today's credit cards. As the use of credit increased, so did its misuse. Many societies wrote laws to prevent its abuse—particularly the charging of unfairly high interest rates, which is called *usury*.

Credit and interest can appear in many diverse forms. The Ifugao tribe of the Philippines charges 100% on a loan. If rice is borrowed, then at the next harvest, the loan must be paid in double. In Vancouver, Canada, the Kwakiutl have a system of credit based on blankets. The rules of interest state that if five blankets are borrowed, in 6 months they become seven. In Northern Siberia, loans are made in reindeer, usually at a 100% interest rate.

As you will see in Example 5, the amount of finance charges you pay on a loan will vary depending on the method used to compute the charges.

Comparing Financing Methods

EXAMPLE 5 Comparing Methods for Finding Finance Charges

Suppose that you begin the month of May (which has 31 days) with a credit card balance of $500. The annual interest rate is 21%. On May 11, you use your credit card to pay for a $400 car repair, and on May 29, you make a payment of $500. Calculate the finance charge that will appear on the statement for next month using the two methods we have discussed.

SOLUTION:

Method	*P*	*r*	*t*	Finance Charge = *I* = *Prt*
Unpaid balance	last month's balance + finance charge − payment + charge for car repair $= 500 + 8.75 + 400 - 500$ $= 408.75$	21%	$\frac{1}{12}$	$(408.75)(0.21)\left(\frac{1}{12}\right) = \$7.15$
Average daily balance	$\frac{10 \times 500 + 18 \times 900 + 3 \times 400}{31}$ $= \frac{22{,}400}{31}$ $= 722.58$	21%	$\frac{31}{365}$	$(722.58)(0.21)\left(\frac{31}{365}\right) = \$12.89$

With the unpaid balance method, the finance charge is $7.15; with the average daily balance method, the finance charge is $12.89.

Example 5 shows that the exact same charges on two different credit cards having the same annual interest rate can result in very different finance charges. If you understand the method your credit card company is using, you may be able to schedule your purchases and payments to minimize your finance charges.

In deciding how to use credit, you must consider many other issues that we have not discussed in this section. Some credit card companies charge an annual fee; others return part of your interest payments. For some credit cards, there is a grace period.

If you reduce your balance to zero during this grace period, then you pay no finance charges. A credit card may have a low introductory rate that changes to a much higher rate at a later time.

One common enticement is that you can make a purchase and pay no interest payments until several months later. With such deals, however, you must be careful. Often, if you do not pay off the purchase completely at the end of the interest-free period, then all the interest that would have accumulated is added to your balance. It is difficult to understand all the pros and cons of the many different types of credit contracts. However, if you read credit agreements carefully and remember the principles that you learned in this section, you will be an intelligent consumer who will be able to use credit wisely.

EXERCISES 8.3

Sharpening Your Skills

In Exercises 1–4, compute the monthly payments for each add-on interest loan. The amount of the loan, the annual interest rate, and the term of the loan are given.

1. $900; 12%; 2 years

2. $840; 10%; 3 years

3. $1,360; 8%; 4 years

4. $1,710; 9%; 3 years

5. Paying off a computer. Luis took out an add-on interest loan for $1,280 to buy a new laptop computer. The loan will be paid back in 2 years and the annual interest rate is 9.5%. How much interest will he pay? What are his monthly payments?

6. Paying off furniture. Mandy bought furniture costing $1,460 for her new apartment. To pay for it, her bank gave her a 5-year add-on interest loan at an annual interest rate of 10.4%. How much interest will she pay? What are her monthly payments?

7. Financing equipment. Angela's bank gave her a 4-year add-on interest loan for $6,480 to pay for new equipment for her antiques restoration business. The annual interest rate is 11.65%. How much interest will she pay? What are her monthly payments?

8. Paying for a sculpture. Mikeal purchased an antique sculpture from a gallery for $1,320. The gallery offered him a 3-year add-on loan at an annual rate of 9.75%. How much interest will he pay? What are his monthly payments?

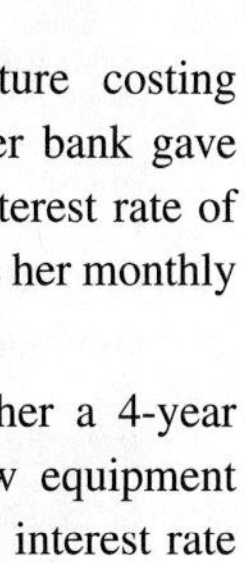

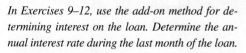

In Exercises 9–12, use the add-on method for determining interest on the loan. Determine the annual interest rate during the last month of the loan.

9. $900; 12%; 2 years

10. $840; 10%; 3 years

11. $1,360; 8%; 4 years

12. $1,710; 9%; 4 years

Applying What You've Learned

13. Financing a boat. Ben is buying a new boat for $11,000. The dealer is charging him an annual interest rate of 9.2% and is using the add-on method to compute his monthly payments.

a. If Ben pays off the boat in 48 months, what are his monthly payments?

b. If he makes a down payment of $2,000, how much will this reduce his monthly payments?

c. If he wants to have monthly payments of $200, how large should his down payment be?

14. Financing a swimming pool. Mr. Phelps is buying a new swimming pool for $14,000. The dealer is charging him an annual interest rate of 8.5% and is using the add-on method to compute his monthly payments.

a. If Mr. Phelps pays off the pool in 48 months, what are his monthly payments?

b. If he makes a down payment of $3,000, how much will this reduce his monthly payments?

c. If he wants to have monthly payments of $250, how large should his down payment be?

15. Financing rare coins. Anna is buying $15,000 worth of rare coins as an investment. The dealer is charging her an annual interest rate of 9.6% and is using the add-on method to compute her monthly payments.

a. If Anna pays off the coins in 36 months, what are her monthly payments?

b. If she makes a down payment of $3,000, how much will this reduce her monthly payments?

c. If she wishes to have monthly payments of $300, how large should her down payment be?

16. Financing a musical instrument. Walt is buying a music synthesizer for his rock band for $6,500. The music store is charging him an annual interest rate of 8.5% and is using the add-on method to compute his monthly payments.

a. If Walt pays off the synthesizer in 24 months, what are his monthly payments?

b. If he makes a down payment of $1,500, how much will this reduce his monthly payments?

c. If he wants to have monthly payments of $150, how large should his down payment be?

17. Financing exercise equipment. Kobe wants to buy some new exercise equipment for his home gym for $132,000 financed at an annual interest rate of 20% using the add-on method. If Kobe wants to pay off the loan in 3 years, what will be his monthly payment?

18. Financing exercise equipment. Continuing Exercise 17, if Kobe decides instead to pay off the loan in 4 years, how much more interest will he pay than if he paid off the loan in 3 years?

19. Financing a new car. Beyonce is buying a new Lamborghini Murcielago for $354,000. The dealer is providing financing at an annual rate of 12% using the add-on method. If Beyonce wants to pay off the car in 3 years, what will her monthly payment be?

20. Financing a new car Continuing Exercise 19, if Beyonce pays off the car in 2 years, how much will she save in interest than if she paid off the car in 3 years?

In Exercises 21–26, use the unpaid balance method to find the finance charge on the credit card account. Last month's balance, the payment, the annual interest rate, and any other transactions are given.

21. Computing a finance charge. Last month's balance, $475; payment, $225; interest rate, 18%; bought ski jacket, $180; returned camera, $145

22. Computing a finance charge. Last month's balance, $510; payment, $360; interest rate, 21%; bought exercise equipment, $470; bought fish tank, $85

23. Computing a finance charge. Last month's balance, $640; payment: $320; interest rate, 16.5%; bought dog, $140; bought pet supplies, $35; paid veterinarian bill, $75

24. Computing a finance charge. Last month's balance, $340; payment, $180; interest rate, 17.5%; bought coat, $210; bought hat, $28; returned boots, $130

25. Computing a finance charge. Last month's balance, $460; payment, $300; interest rate, 18.8%; bought plane ticket, $140; bought luggage, $135; paid hotel bill, $175

26. Computing a finance charge. Last month's balance, $700; payment, $480; interest rate, 21%; bought ring, $210; bought theater tickets, $142; returned vase, $128

A number of years ago, an acquaintance of mine took out a 30-year loan with an annual rate of 8% to put an addition on his house. His banker encouraged him to put other expenses into the loan if he wished, so he increased the loan in order to purchase furniture, a new car, and a computer. Although his loan was probably not an add-on loan, assume that it was for the purpose of the calculations in Exercises 27 and 28.

27. Buying a computer with a long-term loan. Suppose the computer cost $1,200.

a. What was the amount of interest that he paid on the computer part of his loan?

b. What was the total cost of his computer?

28. Buying a car with a long-term loan. Suppose the car cost $11,000

a. What was the amount of interest that he paid on the car part of his loan?

b. What was the total cost of his car?

In Exercises 29–32, use the average daily balance method to compute the finance charge on the credit card account for the previous month. The starting balance and transactions on the account for the month are given. Assume an annual interest rate of 21% in each case.

29. Computing a finance charge. Month: August (31 days); previous month's balance: $280

Date	Transaction
August 5	Made payment of $75
August 15	Charged $135 for hiking boots
August 21	Charged $16 for gasoline
August 24	Charged $26 for restaurant meal

30. Computing a finance charge. Month: October (31 days); previous month's balance: $190

Date	Transaction
October 9	Charged $35 for a book
October 11	Charged $20 for gasoline
October 20	Made payment of $110
October 26	Charged $13 for lunch

31. Computing a finance charge. Month: April (30 days); previous month's balance: $240

Date	Transaction
April 3	Charged $135 for a coat
April 13	Made payment of $150
April 23	Charged $30 for DVDs
April 28	Charged $28 for groceries

32. Computing a finance charge. Month: June (30 days); previous month's balance: $350

Date	Transaction
June 9	Made payment of $200
June 15	Charged $15 for gasoline
June 20	Charged $180 for skis
June 26	Made payment of $130

In Exercises 33–36, redo the specified exercise using the unpaid balance method to calculate the finance charges.

33. Exercise 29

34. Exercise 30

35. Exercise 31

36. Exercise 32

37. Comparing financing methods. Mayesha purchased a large-screen TV for $1,000 and can pay it off in 10 months with an add-on interest loan at an annual rate of 10.5%, or she can use her credit card that has an annual rate of 18%. If she uses her credit card, she will pay $100 per month (beginning next month) plus the finance charges for the month. Assume that Mayesha's credit card company is using the unpaid balance method to compute her finance charges and that she is making no other transactions on her credit card. Which option will have the smaller total finance charges on her loan?

38. Comparing financing methods. Repeat Exercise 37, but now assume that Mayesha purchased an entertainment center for $2,000, the rate for the add-on loan is 9.6%, and she is paying off the loan in 20 months.

39. Accumulated interest. Home Depot advertises 0% financing for 3 months for purchases made before the new year. The fine print in the advertisement states that if the purchase is not paid off within 3 months, the purchaser must pay interest that has accumulated at a monthly rate of 1.75%. Assume that you buy a refrigerator for $1,150 and make no payments during the next 3 months. How much interest has accumulated on your purchase during this time?

40. Accumulated interest. Repeat Exercise 39, but now assume that the purchase is for $1,450 and the annual interest rate is 24%.

Communicating Mathematics

41. Explain what we mean by an add-on loan. What is being added on?

42. Explain the difference between the unpaid balance and the average daily balance methods of calculating your credit card bill. Which method do you think would best work to your advantage when borrowing?

Between the Numbers

43. Government loans. In Example 3 we assumed an interest rate of 24%, which could be possible with a private lender. You might be able to obtain government student loans that have much lower rates. Redo Example 3 assuming you have an interest rate of 8.5%, and again calculate how much you would still owe on your loan after making one payment.

44. Bad financial advice. An online article recommends that you should never borrow more for college than the amount you expect to earn as your yearly salary. The example given is that if you expect to earn $50,000 yearly and have a loan of $50,000, you can pay it off in ten years paying 10%, or $5,000, of your yearly income. Assuming that your loan interest rate is 6.8%, what is terribly wrong with this example? What if the interest rate is 8.5%? (*Source:* http://finances.msn.com/saving-money-advice/6888576)

Challenge Yourself

45. In Example 5, we found that the average daily balance method gave the highest finance charges. Explain why this happened.

46. Make up transactions on a hypothetical credit card so that the unpaid balance method will give you lower finance charges than the average daily balance method does. Assume an annual interest rate of 18%.

47. Make up transactions on a hypothetical credit card so that the average daily balance method will give you lower finance charges than the unpaid balance method does. Assume an annual interest rate of 21%.

48. In our discussions about credit, we ignored the fact that whatever money is not used to pay off a loan can be invested. Assume that you can earn 5% on any money that you do not use for paying off a loan. However, any money you earn as interest is subject to federal, state, and local taxes. Assume that these taxes total 20%. Discuss how that might affect your decision to pay off your credit card debt.

49. Bad financial advice. The same article mentioned in Exercise 44 has an online retirement calculator that assumes your income will increase an average of 3% per year and that your investments will pay a yearly return of 8%. Gather current information regarding rates of return on investments to determine how reliable this calculator would be. (*Source:* http://money.msn.com/retirement/retirement-calculator.aspx)

50. Search online for articles about personal finance. Identify assumptions in these articles that are unrealistic regarding interest rates, projected salaries, inflation, etc.

8.4 Annuities

Objectives

1. Calculate the future value of an ordinary annuity.
2. Perform calculations regarding sinking funds.

He that rises late must trot all day.
—Benjamin Franklin

Although Benjamin Franklin is talking here about the folly of using our time unwisely, the same is true about the way we use our money. If you wish to have something that you are dreaming about for the future—an exotic vacation, a dream house, or a jazzy new car—you have to be wise and begin saving for it today. The longer you wait, the harder it is to achieve your goals.

In this section, we will show you how systematic saving and the power of compound interest will enable your money to grow to achieve long-term goals. Such an account in which you make a series of regular payments to accumulate money for the future is called an *annuity*.

KEY POINT

We make regular payments into an annuity.

Annuities

An **annuity** is an interest-bearing account into which we make a series of payments of the same size. If one payment is made at the *end* of every compounding period, the annuity is called an **ordinary annuity.** The **future value of an annuity** is the amount in the account, including interest, after making all payments.

To illustrate the future value of an annuity, suppose that in January you begin making payments of $100 at the end of each month into an account paying 12% yearly interest compounded monthly. How much money will be in this account for a summer vacation beginning on July 1?

This problem is different from those in Section 8.2. In the earlier problems, we deposited a lump sum that earned a stated interest rate for an entire period. In this problem, we are depositing a *series* of payments, and each payment earns interest for a different number of periods.

The January deposit earns interest for February, March, April, May, and June. Using the formula for compound interest, this deposit will grow to

$$100(1 + 0.01)^5 = \$105.10.$$

However, the May deposit earns interest for only 1 month and therefore grows to only

$$100(1 + 0.01)^1 = \$101.$$

The June deposit earns no interest at all. We illustrate this pattern with the time line in Figure 8.3.

If we compute how much each deposit contributes to the account and sum these amounts, we will have the value of the annuity on July 1.

January	$100(1.01)^5 = \$105.10$
February	$100(1.01)^4 = \$104.06$
March	$100(1.01)^3 = \$103.03$
April	$100(1.01)^2 = \$102.01$
May	$100(1.01)^1 = \$101.00$
June	$100(1.01)^0 = \$100.00$
	Total = $615.20

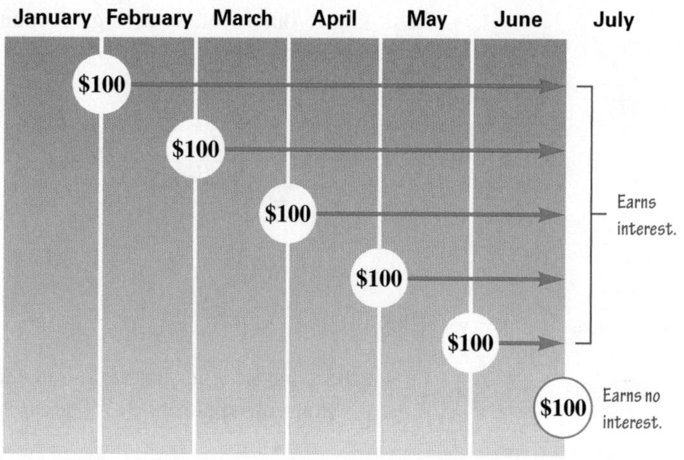

FIGURE 8.3 Time line for ordinary annuity with deposits at end of January,..., June.

We can express the value of this annuity as

$$100(1.01)^5 + 100(1.01)^4 + 100(1.01)^3 + 100(1.01)^2 + 100(1.01)^1 + 100.$$

By factoring out the 100, we can write the value of the annuity in the form

$$100[(1.01)^5 + (1.01)^4 + (1.01)^3 + (1.01)^2 + (1.01)^1 + 1]. \quad (1)$$

Notice that although you deposited \$100 at the end of each month for *six* months, the first deposit earned interest for *five* months, the second deposit earned interest for *four* months, and so on.

Notice also that in equation (1), if we call the number 1.01 by the name x, the expression $(1.01)^5 + (1.01)^4 + (1.01)^3 + (1.01)^2 + (1.01)^1 + 1$ has the form $x^5 + x^4 + x^3 + x^2 + x^1 + 1$, which can be written in a simpler way, as we show in Example 1.

EXAMPLE 1 **Simplifying Annuity Computations**

Show that $x^5 + x^4 + x^3 + x^2 + x^1 + 1 = \dfrac{x^6 - 1}{x - 1}$.

SOLUTION:

To show this relationship, we multiply the polynomials $x^5 + x^4 + x^3 + x^2 + x^1 + 1$ and $x - 1$ in the usual way.

$$\begin{array}{r} x^5 + x^4 + x^3 + x^2 + x^1 + 1 \\ \times\, x - 1 \\ \hline x^6 + x^5 + x^4 + x^3 + x^2 + x^1 \quad\;\; \\ - x^5 - x^4 - x^3 - x^2 - x^1 - 1 \\ \hline x^6 \qquad\qquad\qquad\qquad\qquad\quad - 1 \end{array}$$

This result shows that $(x^5 + x^4 + x^3 + x^2 + x^1 + 1)(x - 1) = x^6 - 1$. Dividing this equation by $x - 1$, we obtain our desired relationship.

Now try Exercises 1 and 2. 15

QUIZ YOURSELF 15

Write $x^3 + x^2 + x^1 + 1$ as a quotient of two polynomials. (Do computations that are similar to those we did in Example 1.)

Following the pattern of Example 1, we can prove in general that if n is a positive integer, then

$$x^n + x^{n-1} + x^{n-2} + \cdots + x^2 + x^1 + 1 = \frac{x^{n+1} - 1}{x - 1}. \quad (2)$$

Returning to our vacation savings account example, we can use equation (2) to simplify our calculations. Because

$$(1.01)^5 + (1.01)^4 + (1.01)^3 + (1.01)^2 + (1.01)^1 + 1 = \frac{(1.01)^6 - 1}{1.01 - 1}$$
$$= \frac{1.061520151 - 1}{1.01 - 1} = \frac{0.061520151}{0.01} \approx 6.1520, \qquad (3)$$

we can write

$$100[(1.01)^5 + (1.01)^4 + (1.01)^3 + (1.01)^2 + (1.01)^1 + 1] \approx 100(6.1520) = \$615.20.$$

This is the same amount that we found earlier.

Let's examine equation (3) more carefully.

$$100((1.01)^5 + (1.01)^4 + (1.01)^3 + (1.01)^2 + (1.01)^1 + 1) = 100\left[\frac{(1.01)^6 - 1}{1.01 - 1}\right].$$

One plus annual rate divided by 12

Number of monthly payments, n

Regular deposit, R

This simplifies to $1.01 - 1 = 0.01 =$ annual rate divided by 12, where 12 is the number of compounding periods per year.

KEY POINT

The future value of an annuity depends on the size of the payment, the interest rate, and the number of payments.

We generalize this pattern in the following formula.

FORMULA FOR FINDING THE FUTURE VALUE OF AN ORDINARY ANNUITY Assume that we are making n regular payments, R, into an ordinary annuity. The interest is being compounded m times a year and deposits are made at the end of each compounding period. The future value (or amount), A, of this annuity at the end of the n periods is given by the equation

$$A = R\frac{\left(1 + \frac{r}{m}\right)^n - 1}{\frac{r}{m}}.$$

To calculate this expression, you should do the following steps:

1st: Find $\frac{r}{m}$ and add 1.

2nd: Raise $1 + \frac{r}{m}$ to the n power and then subtract 1.

3rd: Divide the amount that you found in step 2 by $\frac{r}{m}$.

4th: Multiply the quantity that you found in step 3 by R.

Graphing calculator screen verifies calculations in Example 2. (See highlight on page 393.)

EXAMPLE 2 Finding the Future Value of an Ordinary Annuity

Assume that we make a payment of \$50 at the end of each month into an account paying a 6% annual interest rate, compounded monthly. How much will be in that account after 3 years?

SOLUTION:

This account is an ordinary annuity. The payment R is 50, the monthly rate $\frac{r}{m}$ is $\frac{6\%}{12} = \frac{0.06}{12} = 0.005$, and the number of payments n is $3 \times 12 = 36$. Using the formula for finding the future value of an ordinary annuity, we get

$$A = \underbrace{50}_{R}\left[\frac{\overbrace{(1.005)}^{1+\frac{r}{m}}{}^{\overbrace{36}^{n}} - 1}{\underbrace{0.005}_{\frac{r}{m}}}\right] = 50\left(\frac{1.19668053 - 1}{0.005}\right)$$

$$= 50\left(\frac{0.19668053}{0.005}\right)$$

$$= 50(39.336105) \approx \$1{,}966.81.^{*}$$

Now try Exercises 5 to 14. **16**

QUIZ YOURSELF 16

Redo Example 2, except now assume that you are depositing \$75 per month for 2 years.

If you make regular payments into an annuity for many years, the value of the annuity can become enormous due to the compounding of interest. Figure 8.4 shows that if the annual interest rate is 6.6%, then in roughly 19 years, the amount of interest that the annuity has earned exceeds the amount of the deposits. The fact that the interest curve is rising so rapidly indicates that the future value of your account is also growing rapidly.

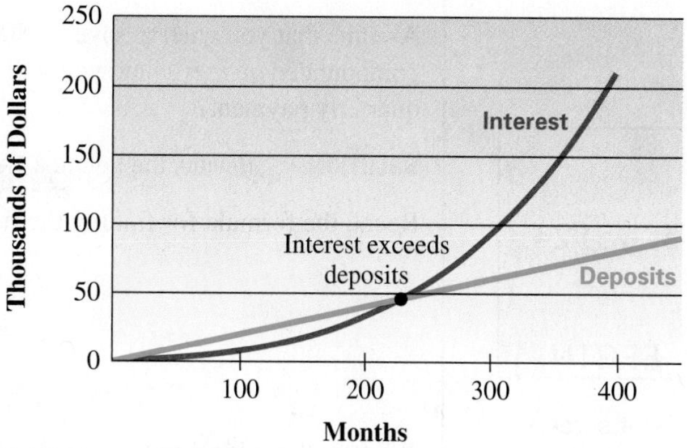

FIGURE 8.4 At an annual rate of 6.6%, the amount of interest earned in an ordinary annuity exceeds the amount of deposits in about 230 months.

Sinking Funds

KEY POINT

With a sinking fund, we make payments to save a specified amount.

You may want to save regularly to have a fixed amount available in the future. For example, you may want to save \$1,800 to travel to Jamaica in 2 years. The question is, how much should you save each month to accomplish this? The account that you establish for your deposits is called a **sinking fund.** You could estimate the amount to save each month by simply dividing 1,800 by 24 months to get $\frac{1{,}800}{24} = \$75$ per month. Because your estimate ignores the interest that your deposits will generate, the actual amount you would need to put aside each month is somewhat less. Knowing exactly how much you need to save each month could be important if you were on a tight budget.

Because a sinking fund is a special type of annuity, it is not necessary to find a new formula to answer this question. We can use the formula for calculating the future value of an ordinary annuity that we have stated earlier. In this case, we know the value of *A* and we want to find *R*.†

*When using your calculator, you should hold off on rounding your answers as long as possible. If you round off too soon, your answers will differ slightly from the answers in this text.

†In making payments into a sinking fund, we will always round the payment *up* to the next cent.

Between the Numbers

Whom Do You "Trust"?

The "trust" that we are referring to is the Social Security trust fund. When you first started working, you may have been dismayed to see a deduction from your first paycheck labeled FICA, which is an acronym for the Federal Insurance Contributions Act. This law mandates that workers must contribute a certain amount of their wages to the Social Security trust fund.

When you contribute to Social Security, you are not actually saving your money for *your retirement*, but your taxes are paying for *someone else's retirement*. The idea is that when you retire, younger workers will then pay for your retirement. However, some see a huge problem with this. Right now, roughly 50 million Americans receive Social Security benefits with contributions supported by approximately 200 million U.S. workers. If we think of this as a ratio, every person on Social Security is supported by 4 workers. In 1950, the ratio was 16 workers for every Social Security beneficiary. It is projected that by 2030, the ratio will be 2 workers for every beneficiary and the fund will be in trouble. So what can you do to protect your retirement?

The government has been encouraging people to make plans for their retirement by establishing tax-deferred annuities to guarantee that when they retire they will have money to supplement Social Security benefits. *Tax deferred* means that the money you set aside in the annuity is not taxed now but at a later date when you start withdrawing from the annuity. As you will see in the exercises, there can be a huge financial benefit to saving this way.

EXAMPLE 3 **Calculating Payments for a Sinking Fund**

Assume that you wish to save \$1,800 in a sinking fund in 2 years. The account pays 6% compounded quarterly and you will also make payments quarterly. What should be your quarterly payment?

```
N=8
I%=6
PV=0
▪ PMT=-213.45124…
FV=1800
P/Y=4
C/Y=4
PMT:END BEGIN
```

Graphing calculator screen verifies calculations in Example 3.

SOLUTION: Review the Relate a New Problem to an Older One Strategy on page 9.

Recall the formula for finding the future value of an ordinary annuity:

$$A = R\frac{\left(1+\frac{r}{m}\right)^n - 1}{\frac{r}{m}}. \qquad (4)$$

We want the value A of the annuity to be \$1,800, the monthly rate $\frac{r}{m}$ is $\frac{6\%}{4} = \frac{0.06}{4} = 0.015$, and the number of payments n is $2 \times 4 = 8$. Substituting these values in equation (4), we get

$$1{,}800 = R\frac{(1+0.015)^8 - 1}{0.015} = R(8.432839106).$$

Dividing both sides of the equation by 8.432839106, we get the quarterly payment

$$R = \frac{1{,}800}{8.432839106} \approx \$213.46.$$

Now try Exercises 15 to 18. **17**

QUIZ YOURSELF 17

What payments must you make to a sinking fund that pays 9% yearly interest compounded monthly if you want to save \$2,500 in 2 years?

SOME GOOD ADVICE

You may be tempted to memorize a new formula to solve sinking fund problems. This is not necessary, because once you have learned to solve annuity problems, you can use the same formula (and a little bit of algebra) to solve sinking fund problems.

KEY POINT

We use the log function to find how long it takes for an annuity to accumulate a specified value.

Sometimes in working with annuities, we want to know how long it will take to save a certain amount. That is, in the annuity formula we want to find n. This problem is a little more complicated than those we have solved so far. To solve such problems, we will use the exponent property for the log function, which we introduced in Section 8.2. We will show you how to use this property in Example 4.

EXAMPLE 4 Finding the Time Required to Accumulate $1,000,000

Suppose you have decided to retire as soon as you have saved $1,000,000. Your plan is to put $200 each month into an ordinary annuity that pays an annual interest rate of 8%. In how many years will you be able to retire?

SOLUTION:

We can use the future value formula for an ordinary annuity to solve this problem:

$$A = R\frac{\left(1+\frac{r}{m}\right)^n - 1}{\frac{r}{m}}$$

Here, A is the future value of $1,000,000, $\frac{r}{m}$ is the monthly interest rate of $\frac{0.08}{12} \approx 0.00667$, and R is 200. We must find n, the number of months for which you will be making deposits. Therefore, we must solve for n in the following equation:

$$\underbrace{1{,}000{,}000}_{A} = \underbrace{200}_{R}\left[\frac{(1 + \overbrace{0.00667}^{r/m})^n - 1}{\underbrace{0.00667}_{r/m}}\right]$$

We solve this equation with the following steps:

$$1{,}000{,}000 = 200\left[\frac{(1 + 0.00667)^n - 1}{0.00667}\right]$$

$$6{,}670 = 200[(1 + 0.00667)^n - 1]$$ Multiply both sides of the equation by 0.00667.

$$34.35 = (1.00667)^n$$ Divide both sides by 200 and add 1 to both sides.

$$\log 34.35 = \log(1.00667)^n$$ Take the log of both sides.

$$\log 34.35 = n\log(1.00667)$$ Use the exponent property of the log function.

$$n = \frac{\log 34.35}{\log(1.00667)} = 531.991532 \approx 532$$ Divide both sides by log(1.00667) and use calculator to find n.

This tells how many months it will take you to save $1,000,000. Dividing 532 by 12 gives us $\frac{532}{12} = 44.33$ years until your retirement.

Now try Exercises 23 to 28. 18

QUIZ YOURSELF 18

If you put $150 per month into an ordinary annuity that pays an annual interest rate of 9%, how long will it take for the annuity to have a value of $100,000?

EXERCISES 8.4

Sharpening Your Skills

In Exercises 1 and 2, simplify each algebraic expression, as in Example 1.

1. $x^7 + x^6 + \cdots + x^2 + x^1 + 1$

2. $x^8 + x^7 + \cdots + x^2 + x^1 + 1$

Exercises 3 and 4 are based on the vacation account example at the beginning of this section. Assume that, beginning in January, you make payments at the end of each month into an account paying the specified yearly interest. Interest is compounded monthly. How much will you have available for your vacation by the specified date?

3. Monthly payment, $100; yearly interest rate, 6%; August 1

4. Monthly payment, $200; yearly interest rate, 3%; May 1

In Exercises 5–14, find the value of each ordinary annuity at the end of the indicated time period. The payment R, *frequency of deposits* m *(which is the same as the frequency of compounding), annual interest rate* r, *and the time* t *are given.*

5. Amount, \$200; monthly; 3%; 8 years

6. Amount, \$450; monthly; 2.4%; 10 years

7. Amount, \$400; monthly; 9%; 4 years

8. Amount, \$350; monthly; 10%; 10 years

9. Amount, \$600; monthly; 9.5%; 8 years

10. Amount, \$500; monthly; 7.5%; 12 years

11. Amount, \$500; quarterly; 8%; 5 years

12. Amount, \$750; quarterly; 9%; 3 years

13. Amount, \$280; quarterly; 3.6%; 6 years

14. Amount, \$250; quarterly; 4.8%; 18 years

In Exercises 15–18, find the monthly payment R *needed to have a sinking fund accumulate the future value* A. *The yearly interest rate* r *and the time* t *is given. Interest is compounded monthly. Round your answer up to the next cent.*

15. $A = \$2{,}000$; $r = 6\%$; $t = 1$

16. $A = \$10{,}000$; $r = 12\%$; $t = 5$

17. $A = \$5{,}000$; $r = 7.5\%$; $t = 2$

18. $A = \$8{,}000$; $r = 4.5\%$; $t = 3$

Solve each equation for x.

19. $3^x = 20$

20. $5^x = 15$

21. $\dfrac{8^x + 2}{5} = 12$

22. $\dfrac{5^x - 8}{10} = 14$

In Exercises 23–28, use the formula for finding the future value of an ordinary annuity,

$$A = R\frac{\left(1 + \frac{r}{m}\right)^n - 1}{\frac{r}{m}},$$

to solve for n. *You are given* A, R, *and* r. *Assume that payments are made monthly and that the interest rate is an annual rate.*

23. $A = \$10{,}000$; $R = 200$; $r = 9\%$

24. $A = \$12{,}000$; $R = 400$; $r = 8\%$

25. $A = \$5{,}000$; $R = 150$; $r = 6\%$

26. $A = \$8{,}000$; $R = 400$; $r = 5\%$

27. $A = \$6{,}000$; $R = 250$; $r = 7.5\%$

28. $A = \$7{,}500$; $R = 100$; $r = 8.5\%$

Applying What You've Learned

In Exercises 29–34, assume that the compounding is being done monthly.

29. Saving for a scooter. Matt is saving to buy a new Vespa. If he deposits \$75 at the end of each month in an ordinary annuity that pays an annual interest rate of 6.5%, how much will he have saved in 30 months?

30. Saving for a trip. Angelina wants to save for an African safari. She is putting \$200 each month in an ordinary annuity that pays an annual interest rate of 9%. If she makes payments for 2 years, how much will she have saved for her trip?

31. Saving for a car. Kristy Joe deposits \$150 each month in an ordinary annuity to save for a new car. If the annuity pays a monthly interest rate of 0.85%, how much will she be able to save in 3 years?

32. Saving for retirement. Cohutta is saving for his retirement in 10 years by putting \$500 each month into an ordinary annuity. If the annuity has an annual interest rate of 9.35%, how much will he have when he retires?

33. Saving for a vacation home. Wendy has set up an ordinary annuity to save for a retirement home in Florida in 15 years. If her monthly payments are \$400 and the annuity has an annual interest rate of 6.5%, what will be the value of the annuity when she retires?

34. Saving for retirement. Thiep has set up an ordinary annuity to save for his retirement in 20 years. If his monthly payments are \$350 and the annuity has an annual interest rate of 7.5%, what will be the value of the annuity when he retires?

The annual returns on investments for most of our examples and exercises are much higher than the actual returns you may encounter. It is very unlikely that you will find safe investments that have annual returns such as 8%, 9.5%, and 12%. In Exercises 35–38, we ask you to redo Example 4 to find the time to accumulate a million dollars using the rates and monthly payments given.

	Annual Rate	Monthly Payment
35.	3.5%	\$200
36.	3.5%	\$500
37.	2.0%	\$800
38.	1.5%	\$1,000

39. Saving for a condominium. Kanye wants to save \$14,000 in 8 years by making monthly payments into an ordinary annuity for a down payment on a condominium at the shore. If the annuity pays 0.7% monthly interest, what will his monthly payment be?

40. Saving to start a business. Victory is making monthly payments into an ordinary annuity that pays 0.8% monthly interest to save enough for a down payment to start her own business. If she wants to save \$10,000 in 5 years, what should her monthly payments be?

41. **Saving for a new range.** Sandra Lee is making monthly payments into an ordinary annuity. She wants to have $600 in the fund to buy a new convection range in 6 months, and the account pays 8.2% annual interest. What are her monthly payments to the account?

42. **Saving for exercise equipment.** Lennox is making monthly payments into an ordinary annuity. He wishes to have $1,150 in 10 months to buy exercise equipment. His account pays 9% annual interest. What are his monthly payments to the account?

Tax-deferred annuities work like this: If, for example, you plan to set aside $400 per month for your retirement in 30 years in a tax-deferred plan, the $400 is not taxed now, so all of the $400 is invested each month. In a nondeferred plan, the $400 is first taxed and then the remainder is invested. So, if your tax bracket is 25%, after you pay taxes, you would have only 75% of the $400 to invest each month. However, in the tax-deferred plan, all of your money is taxed when you withdraw the money. In the nondeferred plan, only the interest that you have earned is taxed.

In Exercises 43–48, we give the amount you are setting aside in an ordinary annuity each month, your current tax rate, the number of years you will contribute to the annuity, and your tax rate when you begin withdrawing from the annuity. Answer the following questions for each situation:

(a) Find the value of the tax-deferred and the nondeferred accounts.

(b) Calculate the interest that was earned in both accounts. This will be the value of the account minus the payments you made.

(c) If you withdraw all money from each account and pay the relevant taxes, which account is better and by how much?

	Monthly Payment	Number of Years	Annual Interest Rate	Current Tax Rate	Future Tax Rate
43.	$300	30	6%	25%	18%
44.	$400	25	4.5%	25%	15%
45.	$400	20	4%	30%	30%
46.	$600	30	4.6%	25%	25%
47.	$500	35	3.4%	25%	30%
48.	$500	30	4.8%	18%	25%

In Exercises 49–52, assume that monthly deposits are being placed in an ordinary annuity and interest is compounded monthly.

49. **Saving for a fire truck.** The Reliance Volunteer Fire Company wants to take advantage of a state program to save money to purchase a new fire truck. The truck will cost $400,000, and members of the finance committee estimate that with community and state contributions, they can save $5,000 per month in an ordinary annuity paying 10.8% annual interest. How long will it take to save for the truck?

50. **Saving for new equipment.** BioCon, a bioengineering company, must replace its water treatment equipment within 2 years. The new equipment will cost $80,000, and the company will transfer $3,800 per month into an ordinary annuity that pays 9.2% interest. How long will it take to save for this new equipment?

51. **Saving for a condominium.** Kirsten wants to save $30,000 for a down payment on a condominium at a ski resort. She feels that she can save $550 per month in an ordinary annuity that has a 7.8% annual interest rate. How long will it take to acquire her down payment?

52. **Saving for a business.** Leo needs to save $25,000 for a down payment to start a photo restoration business. He intends to save $300 per month in an ordinary annuity that pays an annual interest rate of 6%. How long will it take for him to save for the down payment?

53. **Saving for retirement.** At age 21 Julio begins saving $1,000 each year until age 35 (15 payments) in an ordinary annuity paying 6% annual interest compounded yearly and then leaves his money in the account until age 65 (30 years). His friend Max begins at age 41 saving $2,000 per year in the same type of account until age 65 (25 payments). How much does each have in his account at age 65?

54. **Saving for retirement.** At age 31 Julio begins saving $2,000 each year until age 40 (10 payments) in an ordinary annuity paying 8% annual interest compounded yearly and then leaves his money in the account until age 65 (25 years). His friend Max begins at age 41 saving $2,000 per year until age 65 (25 years). How much does each have in his account at age 65?

55. **Saving for retirement.** In Exercise 53, how much would Max have to save per year to have the same amount as Julio at age 65?

56. **Saving for retirement.** In Exercise 54, how much would Max have to save per year to have the same amount as Julio at age 65?

Communicating Mathematics

57. How does using the ordinary annuity formula on page 410 to find the future value of an annuity differ from using it to find the payments to be made into a sinking fund?

58. If you made 10 yearly payments of $3,000 into an ordinary annuity, would you earn more or less interest than if you put a lump sum of $30,000 into that same account? Explain.

Between the Numbers

59. **Social Security.** Research some of the suggested options for making the Social Security program financially sound. Which options do you prefer?

60. **Social Security.** Some recommend that younger workers should be encouraged to have their own retirement accounts instead of having government-backed Social Security. What do you see as the positive and negative aspects of this option?

Challenge Yourself

61. **Buying office equipment.** If you are going to buy a new copier for your small business that costs $10,000, are you better off paying cash for it or making five yearly payments of $2,500 each at the end of each year? Think about the future

value of investing the lump sum of \$10,000 versus the value of investing the \$2,500 each year in an annuity, and consider two cases: one where you earn 3% on your investments and one where you earn 8%.

62. Buying office equipment. What if the copier in Exercise 61 costs \$15,000 and you are considering paying it off with three yearly payments of \$6,000. Again consider interest rates of 3% and 8% in making your decision.

63. Comparing annuities. The difference between an *annuity due* and an ordinary annuity is that with an annuity due, the payment is made at the *beginning* of the month rather than at the end of the month. This means that each payment generates one more month of interest than with an ordinary annuity.

a. How does this change the formula for finding the future value of an annuity?

b. Use this formula to find the value of the annuity in Example 2, assuming that the annuity is an annuity due.

64. Comparing annuities. Continuing Exercise 63, calculate the amount that you will have in an ordinary annuity and an annuity due paying 6% annual interest if you pay \$200 into the annuities each month for 20 years.

8.5 Amortization

Objectives

1. Calculate the payment to pay off an amortized loan.
2. Construct an amortization schedule.
3. Find the present value of an annuity.
4. Calculate the benefit of refinancing a loan.

Congratulations! You just bought a new home—it's lovely—and in a good neighborhood. Only 360 payments and it's all yours. When you make such a large purchase, you usually have to take out a loan that you repay in monthly payments. The process of paying off a loan (plus interest) by making a series of regular, equal payments is called **amortization,** and such a loan is called an **amortized loan.**

If you were to make such a purchase, one of the first questions you might ask is, "What are my monthly payments?" Of course, the lender can answer this question, but you may find it interesting to learn the mathematics involved with paying off a mortgage so that you can answer that question yourself.

Amortization

KEY POINT

Paying off a loan with regular payments is called amortization.

Assume that you have purchased a new car and after your down payment, you borrowed \$10,000 from a bank to pay for the car. Also assume that you have agreed to pay off this loan by making equal monthly payments for 4 years. Let's look at this transaction from two points of view:

Banker's point of view: Instead of thinking about your payments, the banker might think of this transaction as a future value problem in which she is making a \$10,000 loan to you now and compounding the interest monthly for 4 years. At the end of 4 years, she expects to be paid the full amount due. Recall from Section 8.2 that this future value is

$$A = P\left(1 + \frac{r}{m}\right)^n.$$

Your point of view: For the time being, you could also ignore the question of monthly payments and choose to pay the banker in full with one payment at the end of 4 years. In order to have this money available, you could make monthly payments into a sinking fund to have the amount A available in 4 years. As you saw in Section 8.4, the formula for doing this is

$$A = R\frac{\left(1 + \frac{r}{m}\right)^n - 1}{\frac{r}{m}}.$$

Thus, to find your monthly payment, we will set the amount the banker expects to receive equal to the amount that you will save in the sinking fund and then solve for R.

FORMULA FOR FINDING PAYMENTS ON AN AMORTIZED LOAN Assume that you borrow an amount P, which you will repay by taking out an amortized loan. You will make m periodic payments per year for n total payments and the annual interest rate is r. Then, you can find your payment by solving for R in the equation

$$P\left(1+\frac{r}{m}\right)^n = R\left(\frac{\left(1+\frac{r}{m}\right)^n - 1}{\frac{r}{m}}\right).^*$$

Do not let this equation intimidate you. You have done the calculation on the left side many times in Section 8.2 and the computation on the right side in Section 8.4. Once you find these two numbers, you do a simple division to solve for R, as you will see in Example 1.

EXAMPLE 1 Determining the Payments on an Amortized Loan

Assume that you have taken out an amortized loan for \$10,000 to buy a new car. The yearly interest rate is 18% and you have agreed to pay off the loan in 4 years. What is your monthly payment?

Graphing calculator confirms our computations in Example 1.

SOLUTION:

We will use the preceding equation. The values of the variables in this equation are

$$P = 10{,}000$$
$$n = 12 \times 4 = 48$$
$$\frac{r}{m} = \frac{18\%}{12} = 0.015$$

We must solve for R in the equation

monthly interest rate; number of payments; amount of loan

$$10{,}000(1+0.015)^{48} = R\left[\frac{(1+0.015)^{48}-1}{0.015}\right].$$

If we calculate the numerical expressions on both sides of this equation as we did in Sections 8.2 and 8.4, we get

$$20{,}434.78289 = R(69.56521929).$$

Therefore, your monthly payment is

$$R = \frac{20{,}434.78289}{69.56521929} \approx \$293.75.$$

Now try Exercises 1 to 6. 19

QUIZ YOURSELF 19

What would your payments be in Example 1 if you agree to pay off the loan in 5 years?

Instead of doing the calculations in Example 1 by hand, you can use technology such as the TI graphing calculator's TVM Solver, which we explained in Highlight on page 393, to find present and future values, payments, and interest rates. We show you how to use Excel spreadsheets to do financial computations in Highlight on page 420.

*Certainly we could do the necessary algebra to solve this equation for R. Then we could use this new formula for solving problems to find the monthly payments for amortized loans. We chose not to do this because our philosophy is to minimize the number of formulas that you have to memorize to solve the problems in this chapter. We will round payments on a loan *up* to the next cent.

Amortization Schedules

Payments that a borrower makes on an amortized loan partly pay off the principal and partly pay interest on the outstanding principal. As the principal is reduced, each successive payment pays more toward principal and less toward interest. A list showing payment-by-payment how much is going to principal and interest is called an **amortization schedule.** We illustrate such a schedule in Example 2.

EXAMPLE 2 Constructing an Amortization Schedule

The Cullen family wishes to borrow \$200,000 to finance a new summer house. They have obtained a 30-year mortgage at an annual rate of 6%, which has a monthly payment of \$1,199.10. Construct an amortization schedule for the first three payments on this loan.

SOLUTION: Review the Be Systematic Strategy on page 5.

First payment: At the end of the first month, the Cullens have borrowed \$200,000 for 1 month at a monthly interest rate of $\frac{6}{12}$ percent, which we write as 0.005. So, using the simple interest formula, the interest they owe is

$$\underset{P}{\$200{,}000} \times \underset{r}{0.005} \times \underset{t}{1} = \underset{I}{\$1{,}000}.$$

The monthly payment is \$1,199.10, so \$1,199.10 − \$1,000 = \$199.10 is applied towards the principal. The balance is now

$$\$200{,}000 - \$199.10 = \$199{,}800.90.$$

Second payment: The interest for the second month is

$$\$199{,}800.90 \times 0.005 \times 1 = \$999.00,$$

so \$1,199.10 − \$999.00 = \$200.10 goes towards the principal. Table 8.4 shows the schedule for the first three months.

	Payment Number	Monthly Payment	Interest Paid	Paid on Principal	Balance
					\$200,000
Month 1	1	\$1,199.10	\$1,000.00	\$199.10	\$199,800.90
Month 2	2	\$1,199.10	\$999.00	\$200.10	\$199,600.80
Month 3	3	\$1,199.10	\$998.00	\$201.10	\$199,399.70

TABLE 8.4 Making an amortization table for a lengthy mortgage.

Now try Exercises 13 to 16. 20

QUIZ YOURSELF 20

Compute the fourth line of Table 8.4.

When paying off a mortgage, it can be discouraging to see how much of the early payments go towards interest. However, each month the principal decreases, so as time goes on, less is paid for interest and more towards the principal. We used an Excel spreadsheet to calculate Table 8.5 to show what happens to the Cullens' mortgage after 10, 20, and finally, 30 years.

	Payment Number	Monthly Payment	Interest Paid	Paid on Principal	Balance
End of Year 10	119	$1,199.10	$840.45	$358.65	$167,732.06
	120	$1,199.10	$838.66	$360.44	$167,371.62
End of Year 20	239	$1,199.10	$546.58	$652.52	$108,663.44
	240	$1,199.10	$543.32	$655.78	$108,007.66
End of Year 30	358	$1,199.10	$17.81	$1,181.29	$2,381.38
	359	$1,199.10	$11.91	$1,187.19	$1,194.18
	360	$1,199.10	$5.97	$1,193.13	$1.06

TABLE 8.5 The Cullens pay off their mortgage.

KEY POINT

We use the formula for finding the size of monthly payments to determine the present value of an annuity.

Finding the Present Value of an Annuity

When buying a car, your budget determines the size of the monthly payments you can afford, and that determines how much you can pay for the car you buy. Assume that you can afford car payments of $200 per month for 4 years and your bank will grant you a car loan at an annual rate of 12%. We can think of this as a future value of an annuity problem where R is 200, $\frac{r}{m}$ is 1%, and n is 48 months. We know from Section 8.4 that the future value of this annuity is

$$A = 200\left[\frac{(1 + 0.01)^{48} - 1}{0.01}\right] = \$12{,}244.52.$$

This result does not mean that now you can afford a $12,000 car! This amount is the *future value* of your annuity, not what that amount of money would be worth in the *present*.

DEFINITION If we know the monthly payment, the interest rate, and the number of payments, then the amount we can borrow is called the **present value of the annuity.**

Between the Numbers

Can They Really Do That to You?

How would you feel if you took out a $200,000 mortgage for a house, faithfully made all of your payments on time, and at the end of 1 year owed $201,118? Incredibly, this can actually happen if you have an adjustable rate mortgage, or ARM. Some ARMs allow you to make payments that *do not even cover the interest* on the loan, so the amount you owe increases even though you make your payments on time.

ARMs can have other very serious problems for the consumer. With an ARM, it is possible to start with a low interest rate, say 4%, and with yearly increases after several years *your interest rate could be much higher.* Mortgage lenders use an index, often tied to government securities, to determine how much to increase your interest rate. There are many different types of ARMs—some limit the rate increase from year to year, and others limit the maximum rate that can be charged. However, even with these limits, your monthly payments in an ARM could increase from $900 to $1,400 over a 3-year period, causing you great financial distress.

The *Consumer Handbook on Adjustable Rate Mortgages*, available from the Federal Reserve Board, is an excellent guide to ARMs and contains numerous examples, cautions, and a worksheet to help you make sensible decisions regarding mortgages.

Highlight

Using a Spreadsheet to Make an Amortization Schedule

The following Excel spreadsheet shows how to calculate the amortization schedule for the Cullens' mortgage in Example 2. The loan is for \$200,000 with monthly payments of \$1,199.10. We first show the spreadsheet displaying the formulas in the cells of the spreadsheet. (*Note:* You must insert an equal sign to make the formulas operational.)

	A	B	C	D	E	F	G
1			Monthly	Interest			
2			Payment	Rate		Balance	
3			\$1,199.10	0.06		\$200,000	
4							
5		End of	Monthly	Interest	Paid on		
6		Month	Payment	Paid	Principal	Balance	
7		0				F3	
8		1	\$C\$3	(\$D\$3)/12*F7	C8-D8	F3-E8	
9		2	\$C\$3	(\$D\$3)/12*F8	C9-D9	F8-E9	
10		3	\$C\$3	(\$D\$3)/12*F9	C10-D10	F9-E10	

Here is the same spreadsheet when the formulas are evaluated.

	A	B	C	D	E	F	G
1			Monthly	Interest			
2			Payment	Rate		Balance	
3			\$1,199.10	0.06		\$200,000	
4							
5		End of	Monthly	Interest	Paid on		
6		Month	Payment	Paid	Principal	Balance	
7		0				\$200,000	
8		1	\$1,199.10	\$1,000.00	\$199.10	\$199,800.90	
9		2	\$1,199.10	\$999.00	\$200.10	\$199,600.80	
10		3	\$1,199.10	\$998.00	\$201.10	\$199,399.71	*

To generate a new schedule, all we have to do is change the values in cells C3, D3, and F3 and the entire spreadsheet will be recalculated. (A similar Numbers spreadsheet could have been written on an iPad.)

We can find the present value of an annuity by setting the expression for the future value of an account using compound interest equal to the expression for finding the future value of an annuity and solving for the present value P.

FINDING THE PRESENT VALUE OF AN ANNUITY Assume that you are making m periodic payments per year for n total payments into an annuity that pays an annual interest rate of r. Also assume that each of your payments is R. Then to find the present value of your annuity, solve for P in the equation

$$P\left(1+\frac{r}{m}\right)^n = R\left(\frac{\left(1+\frac{r}{m}\right)^n - 1}{\frac{r}{m}}\right).$$

Again, you have done the computations on both sides of this equation many times in Sections 8.2 and 8.4.

*Note discrepancy due to roundoff.

EXAMPLE 3 Determining the Price You Can Afford for a Car

If you can afford to spend \$200 each month on car payments and the bank offers you a 4-year car loan with an annual rate of 12%, what is the present value of this annuity?

SOLUTION:

To solve this problem, we can use the formula for finding payments on an amortized loan:

$$P\left(1+\frac{r}{m}\right)^n = R\left(\frac{\left(1+\frac{r}{m}\right)^n - 1}{\frac{r}{m}}\right). \tag{1}$$

We know $R = 200$, $\frac{r}{m} = 1\% = 0.01$, and $n = 48$ months. If we substitute these for the variables in equation (1), we get

$$P(1+0.01)^{48} = 200\left[\frac{(1+0.01)^{48}-1}{0.01}\right]. \tag{2}$$

Calculating the numerical expressions on both sides of equation (2) gives us

$$P(1.612226078) = 12{,}244.52155.$$

Now dividing both sides of this equation by 1.612226078, we find

$$P = \left(\frac{12.244.52155}{1.612226078}\right) \approx \$7{,}594.79.$$

You may find this answer surprising, but the mathematics of this problem are clear. If you can only afford payments of \$200 per month, then you can only afford to finance a car loan of about \$7,600!

Now try Exercises 27 to 32. 21

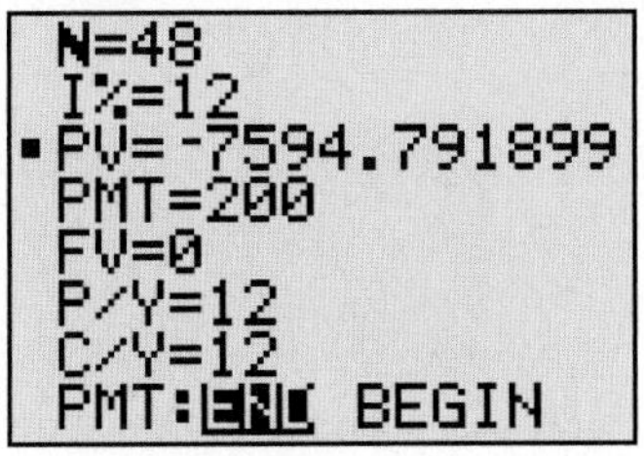

Graphing calculator screen confirms our computations in Example 3.

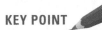

QUIZ YOURSELF 21

Redo Example 3, but now assume that you can afford payments of \$250 per month.

KEY POINT

Refinancing a loan lowers the size of monthly payments and reduces the total interest paid.

Refinancing a Loan

Occasionally you may be forced to borrow money at a high interest rate in order to buy a house or car on credit. If interest rates later decline, it may be wise to consider paying off the remaining debt on the first loan by taking out a second loan at a lower interest rate. This procedure is called **refinancing** the loan.

Before giving you an example of the benefit of refinancing, we will provide Table 8.6, which gives the monthly payments required to pay off a \$1,000 loan given the interest rate and the length of the loan. We will use Table 8.6 in Example 4 and also in several exercises. Technology can do the same calculations for you very quickly.*

	Number of Years for the Loan				
Annual Interest Rate	3	4	10	20	30
4%	\$29.53	\$22.58	\$10.12	\$6.06	\$4.77
5%	29.97	23.03	10.61	6.60	5.37
6%	30.42	23.49	11.10	7.16	6.00
8%	31.34	24.41	12.13	8.36	7.34
10%	32.27	25.36	13.22	9.65	8.78
12%	33.21	26.33	14.35	11.01	10.29

TABLE 8.6 Monthly payments on a \$1,000 loan.

*The TI graphing calculator's TVM Solver was used to calculate this table. You also could do it (of course much more slowly) by following the procedure used in Example 1. I have rounded payments up in this table.

EXAMPLE 4 Refinancing a Loan

You bought a house and took out a 30-year mortgage for \$100,000 at an annual interest rate of 8%, but, after 10 years, you still owe \$87,682. Assume you can now refinance your loan for 20 years at an annual rate of 6%.

a) What are your monthly payments on the original loan?
b) How much will your new monthly payments be?
c) How much interest will you save over the 20 years?

SOLUTION:

a) Using Table 8.6, we see that the monthly payments on a 30-year loan for \$1,000 at an annual interest rate of 8% would be \$7.34, so the payment for a \$100,000 loan would be 100 × \$7.34 = \$734.

b) If you refinance, you are taking out a 20-year loan for the \$87,682 you still owe at an annual rate of 6%. Using Table 8.6 again, we see that the monthly payment on \$1,000 would be \$7.16, so your monthly payment now would be 87.682 × \$7.16 = \$627.80. So you can save over \$100 a month by refinancing.

c) If you continued making payments according to the original loan of \$734 for 20 years (240 payments), you would pay a total of 240 × \$734 = \$176,160. If instead you pay \$627.80 for 20 years, you pay a total of 240 × \$627.80 = \$150,672.

The difference, \$176,160 − \$150,672 = \$25,488, is the amount that you save on interest by refinancing.

Now try Exercises 35 to 38. 22

QUIZ YOURSELF 22

Redo Example 4, assuming that the original mortgage is for \$140,000 at an annual interest rate of 6%. After 10 years you still owe \$117,160. You refinance for 20 years at an annual rate of 4%.

EXERCISES 8.5

Sharpening Your Skills

In Exercises 1–6, you are given a loan amount P, an annual interest rate r, and the length of the loan in years. Represent the number of monthly payments by n. Find the monthly payment R necessary to pay off the loan by doing the following:

a) Calculate $P\left(1+\frac{r}{12}\right)^n$ and call this number A.

b) Calculate $\left(\frac{\left(1+\frac{r}{12}\right)^n-1}{\frac{r}{12}}\right)$ and call this number B.

c) Let $R = A/B$

	Amount	Rate	Time
1.	\$5,000	10%	4 years
2.	\$6,000	12%	3 years
3.	\$8,000	8%	10 years
4.	\$40,000	4%	20 years
5.	\$120,000	6%	30 years
6.	\$150,000	6%	30 years

In Exercises 7–12, use Table 8.6 on page 421 to find the monthly payments on the given loans.

	Amount	Rate	Time
7.	\$4,000	8%	4 years
8.	\$60,000	6%	30 years
9.	\$8,500	4%	3 years
10.	\$40,000	10%	20 years
11.	\$100,000	12%	10 years
12.	\$200,000	6%	30 years

In Exercises 13–16, we give you the annual interest rate on the loan and a line of an amortization schedule for that loan. Complete the next line of the schedule. Assume that payments are made monthly.

	Annual Interest Rate	Payment	Interest Paid	Paid on Principal	Balance
13.	10%	\$126.82	\$35.82	\$91.00	\$4,207.57
14.	8%	\$188.02	\$13.25	\$174.77	\$1,812.99
15.	8.4%	\$246.01	\$32.04	\$213.97	\$4,362.49
16.	6.5%	\$73.07	\$2.71	\$70.36	\$430.25

Applying What You've Learned

17. **Paying off a mortgage.** Assume that you have taken out a 30-year mortgage for $100,000 at an annual rate of 6%.
    a. Use Table 8.6 on page 421 to find the monthly payment for this mortgage.
    b. Construct the first three lines of an amortization schedule for this mortgage.
    c. Assume that you have decided to pay an extra $100 per month to pay off the mortgage more quickly. Find the first three lines of your payment schedule under this assumption.
18. **Paying off a mortgage.** Repeat Exercise 17, but now assume that the mortgage is a 20-year mortgage for $80,000 and the annual rate is 8%.

Use Table 8.6 to find the monthly payment for each amortized loan in Exercises 19–22. Find the total interest paid on the loan. Assume that all interest rates are annual rates.

19. **Paying off a boat.** Wilfredo bought a new boat for $13,500. He paid $2,000 for the down payment and financed the rest for 4-years at an interest rate of 12%.
20. **Paying off a car.** Beatrice bought a new car for $14,800. She received $3,500 as a trade-in on her old car and took out a 4-year loan at 8% to pay the rest.
21. **Paying off a consumer debt.** Franklin's new snowmobile cost $13,500. After his down payment of $2,500, he financed the remainder at 10% for 4 years.
22. **Paying off a consumer debt.** Richard's used motorcycle cost $9,000. He paid $1,100 down and financed the rest at 8% for 10 years.

In Exercises 23 to 26, assume that all mortgages are 30-year adjustable rate mortgages. a) Use Table 8.6 on page 421 to find the monthly mortgage payments. b) Make a rough estimate of your monthly payments for year 3 by recalculating the monthly payments for a 30-year mortgage for the original amount at the new interest rate.

23. $P = \$200{,}000$; beginning interest rate, 4%; rate increases 2% each year
24. $P = \$180{,}000$; beginning interest rate, 5%; rate increases to 10% by year 3
25. $P = \$220{,}000$; beginning interest rate, 5%; rate increases 2% then 1%
26. $P = \$160{,}000$; beginning interest rate, 4%; rate increases 1% then 1%

In Exercises 27–32, find the present value of each annuity. Assume that all rates are annual rates.

27. **The value of a lottery prize.** Marcus has won a $1,000,000 state lottery. He can take his prize as either 20 yearly payments of $50,000 or a lump sum of $425,000. Which is the better option? Assume an interest rate of 10%.
28. **The value of a lottery prize.** Belinda has won a $3,400,000 lottery. She can take her prize as either 20 yearly payments of $170,000 or a lump sum of $1,500,000. Which is the better option? Assume an interest rate of 10%.
29. **Present value of a car.** If Addison can afford car payments of $350 per month for 4 years, what is the price of a car that she can afford now? Assume an interest rate of 10.8%.
30. **Present value of a car.** If Pete can afford car payments of $250 per month for 5 years, what is the price of a car that he can afford now? Assume an interest rate of 9.6%.
31. **Planning for retirement.** Shane has a retirement plan with an insurance company. He can choose to be paid either $350 per month for 20 years, or he can receive a lump sum of $40,000. Which is the better option? Assume an interest rate of 9%.
32. **Planning for retirement.** Nico has a retirement plan with an investment company. She can choose to be paid either $400 per month for 10 years, or she can receive a lump sum of $30,000. Which is the better option? Assume an interest rate of 9%.
33. **Paying off a new car.** You take out a 5-year amortized loan to buy a new car. After making monthly payments of $246.20 for 3 years, you still owe $5,416. If you decided to pay the loan off, how much will you save in interest?
34. **Paying off scuba equipment.** In order to pay for new scuba equipment, you took out a 3-year amortized loan. After 18 monthly payments of $78.57, you still owe $1,298, so you decide to pay off the loan. How much do you save in interest?

In Exercises 35–38, use Table 8.6 to find the monthly payments on the original loan; the monthly payments on the new loan; and the total amount saved on interest by refinancing. All interest rates are annual rates.

35. **Refinancing a vacation home.** In order to buy a vacation home, Neal and Lilly took out a 30-year mortgage for $120,000 at an annual interest rate of 8%. After 10 years, they refinanced the unpaid balance of $105,218 at an annual rate of 6%.
36. **Refinancing a restaurant.** Jamie took out a 10-year loan for $235,000 at an annual interest rate of 12% to modernize his Italian restaurant. After 6 years he decided to refinance the unpaid balance of $127,960 at a rate of 8%.
37. **Paying off tuition.** Rihanna took out a 4-year amortized loan for $18,000 at 10% to pay her tuition at a music conservatory. After 1 year, she refinanced the unpaid balance of $14,404 at a 5% interest rate.
38. **Paying off a business loan.** Sheila took out a 20-year, $140,000 loan at an annual interest rate of 10% to invest in a flower shop. After 10 years, she refinanced the unpaid balance of $102,240 at an annual rate of 8%.

Increasing the down payment on your mortgage reduces both the size of your monthly payments and the total interest paid. In Exercises 39–42, calculate (a) the reduction in your monthly payment by increasing the down payment by the amount specified, and (b) the amount saved on interest over the life of the loan. Assume all mortgages are for 30 years and use Table 8.6 to find the monthly payments.

	Amount of Loan	Interest Rate	Down Payment	Increase in Down Payment
39.	$160,000	10%	$32,000	$8,000
40.	$200,000	6%	$40,000	$10,000
41.	$240,000	4%	$50,000	$25,000
42.	$300,000	8%	$75,000	$20,000

Communicating Mathematics

43. In the discussion preceding Example 3, we mentioned that if you make payments of $200 per month into an account for 4 years, the future value of the account is over $12,000. However, that does not mean that you can afford a $12,000 car. Explain this.

44. What are the benefits of refinancing a loan?

45. When might it not be to your benefit to refinance a loan?

46. If you could make extra payments on a mortgage, is it more beneficial to do it earlier during the loan term or later? Explain.

Between the Numbers

47. What do you see as the benefits and possible pitfalls of adjustable rate mortgages?

48. How can you ensure that you do not wind up in the situation we described in Between the Numbers on page 419?

Challenge Yourself

49. Total cost of a mortgage. Lenders often require home buyers to pay "points." A point is 1% of the purchase price of the home. In addition, the buyer also pays closing costs at the time of purchase. If you are taking out a 30-year mortgage for $140,000, calculate the total costs for option A and option B below; that is,

total cost = points + closing costs + total mortgage payments.

Which option, A or B, has the greater total cost? Assume that points and closing costs are paid separately and are not included in the mortgage. Use Table 8.6 for monthly payments.

A: Annual interest rate 5%; 2 points; closing costs $4,500

B: Annual interest rate 6%; 1 point; closing costs $2,500

50. Total cost of a mortgage. Repeat Exercise 49 for a 20-year, $180,000 mortgage with options A and B.

A: Annual interest rate 4%; 4 points; closing costs $4,500

B: Annual interest rate 5%; 1 point; closing costs $1,500

51. In deciding to refinance at a lower interest rate, how does it affect your payments if you decide to refinance early in the loan period versus later in the loan period? For example, for a 60-month loan, would the new payments be larger, smaller, or the same if you refinance after 12 months instead of 36 months? Make up numerical examples to answer this question. Explain your answer.

52. Some mortgage agreements allow the borrower to make payments that are larger than what is required. Because this extra money goes toward reducing principal, increasing your payments may allow you to pay off the mortgage many years earlier, thus saving a great amount of interest. Assume you take out a 30-year amortized loan at 8% for $100,000 and your monthly payments will be $733.77. Suppose that instead of making the specified payment, you increase it by $100 to $833.77. How much do you save on interest over the life of the loan if you make the larger payment?

Looking Deeper

8.6 Annual Percentage Rate

Objectives

1. Calculate the annual percentage rate from a table.
2. Estimate the annual percentage rate.

A fool and his money are soon parted.
—Dutch Proverb

Have you ever heard of that saying? It certainly applies when you are borrowing money. Because the mathematics of borrowing money is complicated, there are unscrupulous money lenders who will try to take advantage of you. That is why Congress passed a law requiring lenders to inform consumers of the true cost of borrowing money.

The Annual Percentage Rate

To illustrate the problem, assume that you agree to repay a loan for $3,000 (plus the interest) in three yearly payments using an add-on interest rate of 10%. What is your true interest rate? It depends on how you look at this agreement. Using the add-on method described in Section 8.3, we compute the interest using the formula $I = Prt = (3{,}000)(0.10)(3) = \$900$. Thus the amount to be repaid in three equal installments is $3{,}000 + 900 = \$3{,}900$. Each payment is therefore $\frac{3{,}900}{3} = \$1{,}300$, of which $1,000 is being paid on the principal and $300 is interest.

Let's look at this loan systematically. In the following table we solve the simple interest equation $I = P \times r \times t$ for the interest rate r.

Year	Principal	Paid on Principal	Paid on Interest	$I = P \times r \times t$	r
1	$3,000	$1,000	$300	$300 = 3{,}000 \times r \times 1$	0.10 = 10%
2	$2,000	$1,000	$300	$300 = 2{,}000 \times r \times 1$	0.15 = 15%
3	$1,000	$1,000	$300	$300 = 1{,}000 \times r \times 1$	0.30 = 30%

Principal is decreasing each year. ↑ (Principal column)
Amount of interest paid is the same each year. ↑ (Paid on Interest column)
Therefore, the interest rate is rising each year. ↑ (r column)

It seems fair to say that since you only owed $1,000 the third year, but paid $300 for interest, the interest rate is $300/1{,}000 = 0.30 = 30\%$ for the third year.

What is the true interest rate?

The "true" interest rate we are looking for is called the **annual percentage rate,** or APR. Looking at our previous calculations, we see that

interest for the first year + interest for the second year + interest for the third year = $900.

Using the interest formula $I = Prt$, we can rewrite this as

You borrowed $2,000 for one year at *r* percent interest.

$$\underbrace{3{,}000 \times r \times 1}_{\text{You borrowed \$3,000 for one year at } r \text{ percent interest.}} + \overbrace{2{,}000 \times r \times 1} + \underbrace{1{,}000 \times r \times 1}_{\text{You borrowed \$1,000 for one year at } r \text{ percent interest.}} = 900.$$

Collecting like terms, we get $(6{,}000)(r)(1) = 900$. Solving this equation gives us $r = 0.15$. Thus, the annual percentage rate is 15%. You can verify that if you borrow $3,000 for 1 year at 15% and then $2,000 for 1 year at 15% and then $1,000 for 1 year at 15%, the total interest for the 3 years is $900.

> **SOME GOOD ADVICE**
>
> In doing these calculations, it is easy to make a minor computational error that will throw the answer off by a large amount. To detect such errors, always ask yourself, does the answer seem reasonable?

Suppose that we borrowed $6,000 at a simple add-on interest rate of 10% and agreed to repay it by making 60 monthly payments. In calculating the APR, we should remember that we are repaying $100 per month, so we have really borrowed $6,000 for 1 month,

$5,900 for 1 month, $5,800 for 1 month, etc. Instead of having three terms on the left side of the equation as we did previously, we would have 60 terms. To avoid such lengthy computations, lenders use tables similar to Table 8.7 to determine the APR. Because most loans are repaid with monthly payments, for the remainder of this section on APR, to keep the discussion simple, we will only consider payment plans having monthly payments.

	Finance Charge per $100						
	APR						
Number of Payments	10%	11%	12%	13%	14%	15%	16%
6	$2.94	$3.23	$3.53	$3.83	$4.12	$4.42	$4.72
12	$5.50	$6.06	$6.62	$7.18	$7.74	$8.31	$8.88
24	$10.75	$11.86	$12.98	$14.10	$15.23	$16.37	$17.51
36	$16.16	$17.86	$19.57	$21.30	$23.04	$24.80	$26.57
48	$21.74	$24.06	$26.40	$28.77	$31.17	$33.59	$36.03

TABLE 8.7 Finding the annual percentage rate.*

In order to use Table 8.7, you must first know the **finance charge** on a loan, which is the total amount the borrower pays to use the money. This amount may include interest and fees. Then you must find the finance charge per $100 of the amount financed. To find this, divide the finance charge by the amount financed and multiply by 100. For example, if you borrow $780 and pay a finance charge of $148.20, then the finance charge per $100 of the amount financed is

$$\frac{\text{finance charge}}{\text{amount borrowed}} \times 100 = \frac{148.20}{780} \times 100 = 0.19 \times 100 = \$19.$$

USING TABLE 8.7 TO FIND THE APR ON A LOAN

1. Find the finance charge on the loan if it is not already given to you.
2. Determine the finance charge per $100 on the loan.
3. Use the line of Table 8.7 that corresponds to the number of payments to find the number closest to the amount found in step 2.
4. The column heading for the column containing the number found in step 3 is the APR.

EXAMPLE 1 Using the APR Table

Hector has agreed to pay off a $3,500 loan on his home theater by making 24 monthly payments. If the total finance charge on his loan is $460, what is the APR he is being charged?

SOLUTION:

The finance charge per $100 financed is

$$\frac{\text{finance charge}}{\text{amount borrowed}} \times 100 = \frac{460}{3{,}500} \times 100 \approx 0.1314 \times 100 \approx \$13.14.$$

Because Hector is making 24 monthly payments, we use the row in Table 8.7 for 24 payments, as we show in Figure 8.5. Reading across this line, we find that the closest amount to $13.14 is $12.98. The column heading for this column shows the approximate APR for Hector's loan, which is 12%.

*We have kept this table simple to emphasize how it is used. A real table would have more columns for the APR, such as 14.5% and 14.25%.

QUIZ YOURSELF 23

Assume that Jason will repay a loan for \$11,250 by making 36 payments. Assume that the finance charge is \$1,998.

a) What is the finance charge per \$100 financed?

b) What is the APR?

12% — Approximate APR

Number of monthly payments

			\$3.53				
			\$6.62				
24	\$10.75	\$11.86	\$12.98	\$14.10	\$15.23	\$16.37	\$17.51

Closest amount to \$13.14

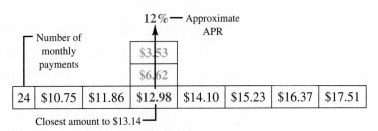

FIGURE 8.5 Using Table 8.7 to find the APR for Hector's loan.

Now try Exercises 3 to 10. 23

We can find the APR if we know the number and size of payments on a loan.

EXAMPLE 2 Finding the APR Using Table 8.7

Jessica is considering buying a used Toyota Venza costing \$11,850. The terms of the sale require a down payment of \$2,000 and the rest to be paid off by making 48 monthly payments of \$250 each. What APR will she be paying on the car financing?

SOLUTION: Review the Relate a New Problem to an Older One Strategy on page 9.

The amount being financed is the purchase price minus the down payment, which is 11,850 − 2,000 = \$9,850. Because her payments amount to 48 × 250 = \$12,000, this makes the finance charge equal to 12,000 − 9,850 = \$2,150. The finance charge per \$100 financed is therefore

$$\frac{\text{finance charge}}{\text{amount borrowed}} \times 100 = \frac{2,150}{9,850} \times 100 \approx 0.2183 \times 100 = \$21.83.$$

We now look at the row for 48 payments in Table 8.7. The number in this row that is closest to \$21.83 is \$21.74. Looking at this column's heading, we find that the APR for this car financing is about 10%.

Now try Exercises 15 to 22. 24

QUIZ YOURSELF 24

In Example 2, assume that Jessica's payments are \$265 instead of \$250. What is the APR that she is being charged?

Since the Consumer Credit Protection Act was passed, it is not as common for lenders to offer add-on interest loans because they must reveal the APR. One way merchants can avoid having to state the APR is by offering the consumer the opportunity to rent rather than purchase the product outright. With a "rent-to-own" contract, because there is no loan, the merchant does not have to reveal the APR. If we consider these rental agreements as loans, we would often find that the APRs are outrageously high.

Knowing the APR enables consumers to determine the best deal when shopping for a loan.

Math in Your Life

Short of Cash? We Can Help

If you are short of cash and were to search the Internet for "payday loans," you could find many sites that will offer you a very short-term loan to tide you over until your next payday. One site that I found will loan you \$100 for 7 days provided you are willing to then pay back \$125. If you take the time to check the page where the company states its annual percentage rates, you would find that you are going to be charged an APR of 1,303.57%.

Another way of looking at this loan is to consider what would happen if you were not to repay this loan for 1 year. The loan P is \$100, the weekly interest rate r is 0.25, and time t is 52 weeks. So, applying the compound interest formula, at the end of the year you would owe

$$P(1 + r)^t = 100(1.25)^{52} \approx \$10,947,644.25.$$

That's right, your \$100 loan would have grown to a debt of almost \$11 million!

EXAMPLE 3 Computing the Cost of "Renting to Own"

Jeff is considering renting a high-definition TV from a nearby rent-to-own store. He can rent the TV, which he saw priced at $479 in a local store, for $19.95 a month. If he rents the TV for 36 months, then the TV is his to keep. Analyze this rental agreement to determine whether Jeff is making a wise decision.

SOLUTION:

There is really not much difference here between what Jeff is considering and purchasing the TV at another store with an agreement to make monthly payments of $19.95. Of course, with the rent-to-own agreement, Jeff can stop renting before 36 months. If Jeff were to rent the TV until he owned it, he would make 36 payments of $19.95, so he would pay a total of $36 \times 19.95 = \$718.20$. His finance charge would be $718.20 - 479 = \$239.20$. The finance charge per $100 financed is therefore

$$\frac{\text{finance charge}}{\text{amount borrowed}} \times 100 = \frac{239.20}{479} \times 100 \approx 0.4994 \times 100 = \$49.94.$$

If we consider his rental as a 36-payment loan, we can try to use Table 8.7. Unfortunately, 49.94 is so large that we cannot find it in Table 8.7, which implies that the interest rate is quite high. Using other methods, which we explain in Highlight on page 429, we find the APR to be about 28.5%. Jeff should carefully consider whether this rental is worth the cost.

Now try Exercises 27 and 28.

Estimating the APR

As you saw in Example 3, it is difficult to calculate an APR without using technology; however, there is a formula that gives a good estimate of an APR for the special case of an add-on interest loan.

FORMULA TO APPROXIMATE THE ANNUAL PERCENTAGE RATE We can approximate the annual percentage rate for an add-on interest loan by using the formula

$$\text{APR} \approx \frac{2nr}{n+1},$$

where r is the annual interest rate and n is the number of payments.

EXAMPLE 4 Estimating an APR

Minxia must borrow $4,000 to pay tuition for her last year in college. Her bank will give her an add-on interest loan at 7.7% for 3 years. Use the formula above to estimate Minxia's APR.

SOLUTION:

In this problem, $n = 3 \times 12 = 36$ and $r = 7.7\% = 0.077$. So Minxia's annual percentage rate is

$$\text{APR} \approx \frac{2nr}{n+1} = \frac{2 \times 36 \times 0.077}{36 + 1} = \frac{5.544}{37} = 0.1499 = 14.99\%.$$

Now try Exercises 11 to 14. 25

QUIZ YOURSELF 25

Use the formula to approximate the APR for an add-on interest loan for $5,500 at an annual interest rate of 9.6% that will be repaid in 48 months.

Highlight

Using a Graphing Calculator to Find an APR

If we consider Jeff's rental in Example 3 as a loan with payments of \$19.95 per month, then we can reason as we did in Section 8.5 when determining payments on a mortgage. Recall that we thought of the banker offering a loan where interest was computed monthly and we thought of the borrower as paying it off by making monthly payments into an annuity. The equation that we used was

$$P\left(1+\frac{r}{m}\right)^n = R\left(\frac{\left(1+\frac{r}{m}\right)^n - 1}{\frac{r}{m}}\right).$$

Knowing P, r, m, and n, it was easy to solve for R. The situation here is different. We know P, m, n, and R and we need to solve for r, which is quite difficult. However, using a calculator such as a TI graphing calculator, we can solve for r as we show in the accompanying screen. Thus, the APR is roughly 28.5%.

```
N=36
▪I%=28.52782389
PV=479
PMT=-19.95
FV=0
P/Y=12
C/Y=12
PMT:END BEGIN
```

EXERCISES 8.6

Sharpening Your Skills

In Exercises 1 and 2, use an approach similar to the discussion preceding Example 1 to find the APR of the loan. Realize that Table 8.7 does not apply to these situations because the payments are not monthly. You are given the amount of the loan, the number and type of payments, and the add-on interest rate.

1. Loan amount, \$6,000; three yearly payments; rate = 8%
2. Loan amount, \$8,000; four yearly payments; rate = 12%

Find the finance charge per \$100 for each loan.

3. Loan, \$1,800; finance charge, \$270
4. Loan, \$3,000; finance charge, \$840
5. Loan, \$2,000; finance charge, \$260
6. Loan, \$5,000; finance charge, \$1,125

Use Table 8.7 to find the APR to the nearest whole percent. Assume that all interest rates are annual rates.

7. **Finding the APR on a loan.** Michael has agreed to pay off a \$3,000 loan by making 24 monthly payments. The total finance charge on his loan is \$420.
8. **Finding the APR on a loan.** Daisy has agreed to pay off a \$4,500 loan by making 24 monthly payments. The total finance charge on her loan is \$600.
9. **Finding the APR on a loan.** Luisa pays a finance charge of \$165 on a 6-month, \$4,000 loan.
10. **Finding the APR on a loan.** Wesley pays a finance charge of \$310 on a 12-month, \$5,000 loan.

In Exercises 11–14, estimate the annual percentage rate for the add-on loan using the given number of payments and annual interest rate. Use the formula on page 428.

11. $n = 36$; $r = 6.4\%$
12. $n = 48$; $r = 4.8\%$
13. $n = 42$; $r = 7\%$
14. $n = 30$; $r = 8\%$

Applying What You've Learned

Use Table 8.7 to find the APR to the nearest whole percent. Assume that all interest rates are annual rates.

15. **Finding the APR on a remodeling loan.** Thiep took out a \$10,000 add-on loan to remodel his house and will repay it by making 24 payments of \$485.
16. **Finding the APR on a car loan.** Diana took out an \$8,000 add-on loan to buy a car and will repay it by making 36 payments of \$270.
17. **Finding the APR on a consumer loan.** Pete took out a \$4,500 add-on loan to buy a sound system for his band and will repay it by making 48 payments of \$116.50.
18. **Finding the APR on a consumer loan.** Amanda took out a \$1,500 add-on loan to buy a new computer and will repay it by making 24 payments of \$71.25.

19. **Finding the APR on a travel loan.** Emily took out a 24-month, $2,000 add-on loan at an interest rate of 8% to visit China.

20. **Finding the APR on a vehicle loan.** John took out a 48-month, $26,000 add-on loan at an interest rate of 7.9% to pay off his truck.

21. **Finding the APR on a loan.** What is the APR of a 36-month add-on loan with an interest rate of 8.2%?

22. **Finding the APR on a loan.** What is the APR of a 48-month add-on loan with an interest rate of 8.75%?

In Exercises 23–26, decide which option, A or B, is the better way to repay a $5,000 loan. Assume that you are making monthly payments. Use Table 8.6 in Section 8.5 for the payments on the amortized loan.

23. A: An amortized loan at 10% for 3 years

    B: An add-on loan at 6% for 3 years

24. A: An amortized loan at 12% for 4 years

    B: 48 payments of $135

25. A: An amortized loan at 8% for 4 years

    B: 48 payments of $120

26. A: An amortized loan at 12% for 3 years

    B: An add-on loan at 6% for 3 years

In Exercises 27 and 28, think of the rent-to-own agreement as though it were an add-on loan. If the consumer rents until the item is paid for, find the finance charge per $100 financed. Although Table 8.7 does not contain enough columns to estimate the APR, guess as to what you think it might be.

27. **Evaluating a rent-to-own agreement.** Marcus rents a TV worth $375 for monthly payments of $18.75. After 2 years, he will own the TV.

28. **Evaluating a rent-to-own agreement.** Maria rents furniture worth $1,375 for monthly payments of $49. After 3 years, she will own the furniture.

Communicating Mathematics

29. **Comparing loan options.** If you have the choice of taking an add-on loan or an amortized loan with the same annual rate and for the same length of time, which do you think is preferable? Explain why you should be able to decide this without making up examples.

30. In what way is a rent-to-own agreement different from an add-on interest loan?

Challenge Yourself

31. **Payday loans.** Assume that you take a payday loan for $100 and have to repay the loan with $110 in 1 month. What is the monthly interest rate? What do you think the lender would claim the APR is? If you left this loan unpaid for one year, what would you owe? What do you think the APR is?

32. **Payday loans.** Repeat Exercise 31, but now you are taking out a payday loan for $500 and must repay $600 in 3 months.

33. We often see advertisements stating that we can consolidate our loans and have more manageable payments. The advertiser may do this for us by extending the length of our loans. Consider an add-on loan for $1,000 at an interest rate of 10% for 3 years and then for 4 years. How does this affect the APR? Is it more financially sound to pay off loans in a short period of time or a longer period of time? (Here we are only thinking about the APR—of course, there may be other considerations.)

34. Is the APR affected by the size of the loan?

CHAPTER SUMMARY

Section	Summary	Example
SECTION 8.1	The word **percent** is derived from the Latin "per centum," which means "per hundred." Examining what numbers are in the tenths and hundredths place in a decimal will allow you to **convert a decimal to a percent** properly. Deciding which numbers to put in the tenths and hundredths place will help you to **convert a percent to a decimal** properly.	Discussion, p. 379 Examples 1–3, pp. 379–380
	Percent of change is given by $$\text{percent of change} = \frac{\text{new amount} - \text{base amount}}{\text{base amount}}.$$	Examples 4 and 5, p. 381
	Many percent problems are based on the equation percent $\times$ base $=$ amount. This equation is used to solve many applied problems.	Example 6, p. 382 Examples 7 and 8, pp. 382–384
	Calculating with percents often occurs when **calculating taxes.**	Example 9, p. 384
	Inflation is a rise in the price of goods and services over a period of time.	Example 10, p. 385
SECTION 8.2	**Interest** is the money that a borrower pays to use a lender's money. The amount borrowed is called the **principal.** The **interest rate** is specified as a percentage of the principal. We use the formula $I = Prt$, where I is the interest, P is the principal, r is the interest rate, and t is the time. The equation $A = P(1 + rt)$ computes the **future value** of an account using simple interest. A is the future value (or amount), P is the principal, r is the interest rate, and t is the time. We find the **present value** of an account by solving the equation $A = P(1 + rt)$ for P.	Discussion, pp. 388–389 Example 1, p. 389 Example 2, p. 390
	To find A, the **future value** of an account using **compound interest,** we use the formula $A = P\left(1 + \frac{r}{m}\right)^n$, where P is the **principal,** r is the annual **interest rate,** m is the number of **compounding periods** per year, and n is the total number of compounding periods.	Discussion, pp. 391–392 Example 4, pp. 392–393
	We find **present value** by solving the compound interest equation for P.	Example 5, pp. 393–394
	The **common logarithmic function** reverses the operation of raising 10 to a power.	Example 6, p. 394
	To solve for n in the compound interest formula, take the log of both sides of the equation and use the exponent property of the log function. To solve for $\frac{r}{m}$, divide both sides of the equation $A = P\left(1 + \frac{r}{m}\right)^n$ by P, raise both sides of the resulting equation to the $\frac{1}{n}$ power, and then subtract 1 from both sides.	Example 7, p. 395 Example 8, pp. 395–396
	The **Rule of 70** states that to estimate the doubling time of a quantity, divide 70 by the annual growth rate.	Discussion, p. 395
SECTION 8.3	When using the **add-on interest method** to compute the monthly payments on a loan, we use the following formula: payment $= \frac{P + I}{n}$, where P is the amount of the loan, I is the interest due on the loan, and n is the number of monthly payments.	Example 1, p. 400
	The **unpaid balance method** for computing finance charges uses the simple interest formula $I = Prt$, where $P =$ pervious month's balance $+$ finance charge $+$ purchases made $-$ returns $-$ payments. The variable r is the annual interest rate and $t = \frac{1}{12}$.	Example 2, p. 401

(Continued)

Section	Summary	Example
	Use the **average daily balance method** to compute a finance charge: 1. Add the outstanding balance for your account for each day of the previous month. 2. Divide this total by the number of days in the previous month. 3. Use the formula $I = Prt$, where P is the average daily balance found in step 2, r is the annual interest rate, and t is the number of days in the previous month divided by 365.	Example 4, p. 403
	The exact same charges on two different credit cards may result in different finance charges.	Example 5, p. 404
SECTION 8.4	When we make a series of regular payments into an interest-bearing account, this account is called an **annuity.** If payments are made at the end of every compounding period, the annuity is called an **ordinary annuity.** The **interest** in an annuity is compounded with the same frequency as the payments. The sum of all the deposits plus all interest is called the **future value** of the account.	Discussion, p. 408
	Assume that we are making a regular payment, R, at the end of each compounding period for an annuity that has an annual interest rate, r, which is being compounded m times per year. Then the value of the annuity after n compounding periods is $$A = R\frac{\left(1 + \frac{r}{m}\right)^n - 1}{\frac{r}{m}}.$$	Example 2, pp. 410–411
	A **sinking fund** is an account into which we make regular payments for the purpose of saving some specified amount in the future. The interest is compounded with the same frequency as the payments. To find the regular **payments** that must be made into a sinking fund to save the amount A, solve for R in the following equation: $$A = R\frac{\left(1 + \frac{r}{m}\right)^n - 1}{\frac{r}{m}}.$$	Discussion, p. 411 Example 3, p. 412
SECTION 8.5	The process of paying off a loan (plus interest) by making a series of regular equal payments is called **amortization,** and such a loan is called an **amortized loan.** We assume that P is the amount **borrowed,** r is the annual **interest rate,** m is the **number** of compounding periods per year, n is the total number of compounding periods, and R is the **payment** that is made regularly.	Discussion, pp. 416–417
	To find the regular **payment** due on an amortized loan, solve for R in the following equation: $$P\left(1 + \frac{r}{m}\right)^n = R\frac{\left(1 + \frac{r}{m}\right)^n - 1}{\frac{r}{m}}.$$	Example 1, p. 417
	A list showing payment-by-payment how much applies to principal and interest on an amortized loan is called an **amortization** schedule.	Example 2, p. 418
	To find the **present value** of your annuity, solve for P in the following equation: $$P\left(1 + \frac{r}{m}\right)^n = R\frac{\left(1 + \frac{r}{m}\right)^n - 1}{\frac{r}{m}}.$$	Example 3, p. 421
	Refinancing is taking out a second loan at a lower interest rate to pay off a first loan.	Example 4, p. 422

Section	Summary	Example
SECTION 8.6	The **annual percentage rate**, or **APR**, is a standardized version of the "true" interest rate on a loan.	Discussion, p. 425
	To use the APR table to calculate an annual percentage rate, we first find the **finance charge per $100** of the amount financed, which equals $\frac{\text{finance charge}}{\text{amount borrowed}} \times 100$. We can then use Table 8.7 to find the APR.	Example 1, pp. 426–427 Example 2, p. 427
	We can **estimate** the annual percentage rate for an add-on loan using the formula APR $\approx \frac{2nr}{n+1}$, where n is the number of payments and r is the annual interest rate.	Example 4, p. 428

CHAPTER REVIEW EXERCISES

Section 8.1

1. Convert 0.1245 to a percent.
2. Convert 1.365 percent to a decimal.
3. Convert $\frac{11}{16}$ to a percent.
4. 2,890 is what percent of 3,400?
5. In 2007, M&M sales were $238.4 million, which was 13.2% of the total chocolate candy sales. What was the total amount spent on chocolate candy in 2007?
6. Use Table 8.2 to calculate the federal income tax that Maribel owes if her taxable income is $56,400.
7. If a pair of running shoes costs $130 in 2014, what will they cost after 5 years of inflation at 4.3%?

Section 8.2

8. Find the future value of an account paying simple interest if $P = \$1{,}500$, $r = 9\%$, and $t = 2$ years.
9. You have agreed to pay off an $8,000 car loan with 24 monthly payments of $400 each. Use the simple interest formula to determine the interest rate that you are being charged.
10. Jacob wants to defer his income taxes of $11,400 for 6 months. If he must pay an 18% penalty (compounded monthly) to do this, what will his tax bill be?
11. Palma wants to establish a fund for her granddaughter's college education. What lump sum must she deposit in an account that pays an annual interest rate of 6%, compounded monthly, if she wants to have $10,000 in 10 years?
12. If you invest $1,000 in an account that pays an annual interest rate of 6.4%, compounded monthly, how long will it take for your money to double?
13. If $A = \$1{,}400$, $P = \$1{,}200$, and $t = 5$, solve $A = P(1 + r)^t$ for r.
14. If you have an investment that is earning 5.5% annually, use the Rule of 70 to estimate how many years it will take for your investment to double.

Section 8.3

15. Bernie purchased a riding lawn mower for $1,320. The store offered him a 3-year add-on loan at an annual rate of 8.25%. How much interest will he pay? What are his monthly payments?
16. Use the unpaid balance method to calculate the finance charge on Joanna's credit account if last month's balance was $1,350, she made a payment of $375, she bought hiking boots for $120, and she returned a jacket for $140. Assume an annual interest rate of 21%.
17. Calculate the finance charges for the following credit card account for August (which has 31 days) using the average daily balance method. July's balance was $275 and the annual interest rate is 18%.

Date	Transaction
August 6	Made payment of $75
August 12	Charged $115 for clothes
August 19	Charged $20 for gasoline
August 24	Charged $16 for lunch

Section 8.4

18. Piers is saving for his retirement by putting $175 each month into an ordinary annuity. If the annuity pays an annual interest rate of 9.35%, how much will he save for his retirement in 10 years?
19. Find the monthly payment needed to have a sinking fund accumulate to $2,000 in 36 months if the annual interest rate is 6%.
20. Solve $\frac{3^x - 4}{2} = 10$ for x.
21. You are making monthly payments of $300 into an ordinary annuity that pays 9% annual interest. How long will it take to accumulate $10,000?

22. Assume that you are saving \$350 a month in a retirement annuity that has an interest rate of 4.2%. Assume that your income taxes for the life of the annuity are 25% and that they drop to 18% when you retire. How much more do you earn in 30 years in a tax-deferred account than a nondeferred one?

Section 8.5

23. Find the monthly payment necessary to pay off a 4-year amortized loan of \$5,000 if the annual interest rate is 10%.
24. Complete the first two lines of an amortization schedule for a 20-year, \$100,000 mortgage if the annual interest rate is 8%.
25. Jesse has won a \$1,000,000 state lottery. She can take her prize as either 20 yearly payments of \$50,000 or a lump sum of \$500,000. Which is the better option? Assume an interest rate of 8%.
26. Assume that you borrow \$180,000 in a 30-year adjustable rate mortgage with an initial interest rate of 4.5%. The interest rate increases over 4 years to 12.5%. What are your monthly payments in the first year? The fifth year?
27. Assume that you take out a 10-year loan for your business for \$235,000 at an annual interest rate of 12% and after 6 years decide to refinance the unpaid balance of \$127,960 at a rate of 8%. Use Table 8.6 on page 421 to find the difference in your monthly payments.

Section 8.6

28. Ann took out a \$1,800 add-on loan for a professional-quality color printer, which she will repay with 12 payments of \$163. Use Table 8.7 to find her APR to the nearest percent.
29. Use the formula on page 428 to estimate the annual percentage rate on a loan that has an annual interest rate of 8% that will be repaid in 20 months.

CHAPTER TEST

1. Convert 0.3624 to a percent.
2. Convert 23.45 percent to a decimal.
3. Convert $\frac{7}{16}$ to a percent.
4. Use the unpaid balance method to calculate the finance charge on Marion's credit account if last month's balance was \$950, she made a payment of \$270, and she bought a ski parka for \$217 and gloves for \$23. Assume an annual interest rate of 24%.
5. Find the future value of an account paying simple interest if $P = \$3{,}400$, $r = 2.5\%$, and $t = 3$ years.
6. 994 is what percent of 2,840?
7. Best Buy is selling an MP3 player that normally sells for \$169.99 for \$149.99. What is the percent of the reduction on the price of this player?
8. Hiro took out a \$1,600 add-on loan for a handmade, split-bamboo, Japanese fly-fishing rod that he will repay with 24 payments of \$75. Use Table 8.7 to find his APR to the nearest percent.
9. You are paying off your large-screen projection TV that cost \$3,000 with 24 monthly payments of \$162.50. What is the annual simple interest rate that you are being charged?
10. If Danica deposits \$4,000 in an account that has an interest rate of 3.6% that is compounded monthly, what is the value of the account in 4 years?
11. If you invest \$1,000 in an account that pays an interest rate of 4.8% compounded monthly, how long will it take for your money to double?
12. Assume that you borrow \$220,000 in a 30-year adjustable rate mortgage with an initial interest rate of 3.2%. The interest rate increases 1.5% per year for the next 4 years. What are your monthly payments in the first year? The fifth year?
13. If you have an investment that is earning 8.5% annually, use the Rule of 70 to estimate how many years it will take for your investment to double.
14. José wants to establish a trust for his nephew who is now 3 years old. He will deposit a lump sum in an account with an annual interest rate of 4.2% compounded monthly. How much must he deposit now if he wants his nephew to have \$15,000 when he turns 21?
15. To pay for scuba diving equipment costing \$1,560, Carmen took out a 2-year add-on interest loan at an annual rate of 10.5%. How much interest will he pay? What are his monthly payments?
16. If $A = \$2{,}400$, $P = \$2{,}100$, and $t = 3$, solve $A = P(1 + r)^t$ for r.
17. From 2005 to 2006, the price of a new home rose from \$250 thousand to \$257 thousand. What was the percent of increase?
18. Calculate the finance charges for the following credit card account for April (which has 30 days) using the average daily balance method. March's balance was \$425, and the annual interest rate is 21%.

Date	Transaction
April 4	Made payment of \$85
April 10	Charged \$25 for gasoline
April 15	Charged \$15 for lunch
April 25	Charged \$80 for concert tickets

19. Grace is saving for her son's college education by putting \$200 each month into an ordinary annuity. If the annuity pays an annual interest rate of 5.15%, how much will she have saved in 8 years?
20. Solve $\dfrac{5^x - 4}{3} = 10$ for x.
21. Use the formula on page 428 to estimate the annual percentage rate on a loan that has an annual interest rate of 8% that will be repaid in 20 months.
22. Find the monthly payment needed to have a sinking fund accumulate to \$1,800 in 36 months if the annual interest rate is 4%.
23. Find the monthly payment necessary to pay off an 8-year amortized loan of \$20,000 if the annual interest rate is 9%.

24. Assume that you are making monthly payments of $450 into an ordinary annuity that pays 3.75% annual interest. How long will it take to accumulate $9,000?

25. Assume that you take out a 10-year loan for $130,000 at an annual interest rate of 8% and after 6 years decide to refinance the unpaid balance of $64,608 at a rate of 6%. Use Table 8.6 to find the difference in your monthly payments.

26. Use Table 8.2 to calculate the federal income tax that Jakob owes if his taxable income is $48,600.

27. Mike has won the $1,000,000 prize on *Deal or No Deal*. He can take his prize as a lump sum of $750,000 or as an ordinary annuity with 10 yearly payments of $100,000. Which is the better option? Assume an annual interest rate of 4%.

28. Assume that you have taken out a 20-year amortized loan of $140,000 at an annual interest rate of 7.5%.
   a. What is the monthly payment on your loan?
   b. Write the first two lines of the amortization table for this loan.

29. If a pair of skis costs $450 in 2014, what will they cost after 5 years of inflation at 3.4%?

30. Estelle has taken an amortized loan at 9.6% for 5 years to pay off her new car, which costs $16,500. What is her monthly payment on the loan?

GROUP PROJECTS

1. **Student loan advice.** Go to http://projectonstudentdebt.org/recent_grads.vp.html, which gives 10 tips regarding student loans. As a group, discuss these tips and agree on three that you think are most important. Write several paragraphs on what you have found.

2. **Adjustable rate mortgages.** Go to http://files.consumerfinance.gov/f/201204_CFPB_ARMs-brochure.pdf to learn more about adjustable rate mortgages and write several paragraphs describing how they work. Explain what is meant by a) the index; b) the margin; c) No-Doc/Low-Doc loans; d) interest rate caps; e) periodic adjustment caps; f) lifetime caps; g) payment caps; and h) types of ARMS. Also discuss several cautions mentioned at this site.

3. Get federal income tax tables for filing single, married filing jointly, and married filing separately and investigate if there are any advantages for a married couple to file one way versus the other.

USING TECHNOLOGY

1. **Creating an amortization schedule.** Implement the spreadsheet that calculates the amortization schedule given in Highlight on page 420.

2. **Using a spreadsheet's financial functions.** Excel has many built-in financial functions such as FV, PV, NPER, and PMT. Investigate what these functions do and illustrate them with examples.

3. **Financial calculations with a graphing calculator.** We will explain how to use a TI graphing calculator, but other brands may have similar capabilities.

   Step 1: Press the APPS key to get to the financial applications screen.

   Step 2: Choose the "1: Finance" option to get to Screen 1.

   Step 3: Select the TVM Solver to do most of the calculations in this chapter. You will see Screen 2.

   Screen 2 shows an amortization problem. In this problem, I wanted to know how much still remained to be paid on a 30-year loan at a 6% interest rate after 20 years of payments; that is, when there were 10 years, or 120 payments, still remaining. I had calculated the monthly payments earlier on another screen. To find FV, I placed the cursor on that line, pressed ALPHA, and then ENTER.

   The variables are as follows:

   `N = 240` is the number of monthly payments made.
   `I% = 6` is the annual interest rate. (Note that we use 6% rather than 0.06.)
   `PV = 140000` is the present value (the original amount of the mortgage).
   `FV = -75605.02074` is the future value. The negative sign means we still owe this amount.
   `P/Y = 12` is the number of payments per year.
   `C/Y = 12` is the number of compounding periods per year.
   `PMT : END BEGIN` lets you choose whether payments are made at the end or the beginning of the month. Because "END" is highlighted, payments are at the end of each month.

   Use TVM Solver (or a similar feature on your calculator) to verify the calculations in the following examples:

   a) Example 3 in Section 8.4
   b) Example 4 in Section 8.4
   c) Several entries in Table 8.6 in Section 8.5

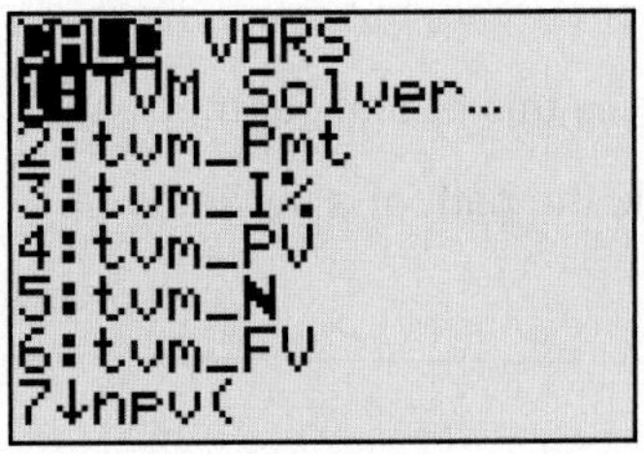

Screen 1

Screen 2

9

Geometry

Ancient and Modern Mathematics Embrace

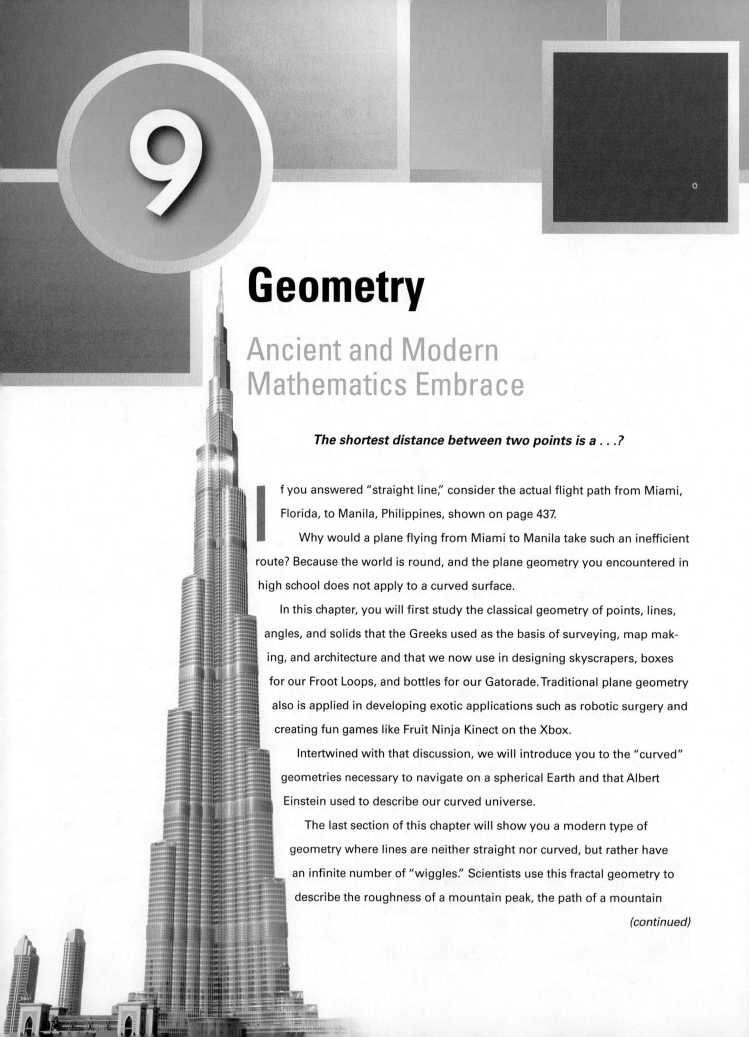

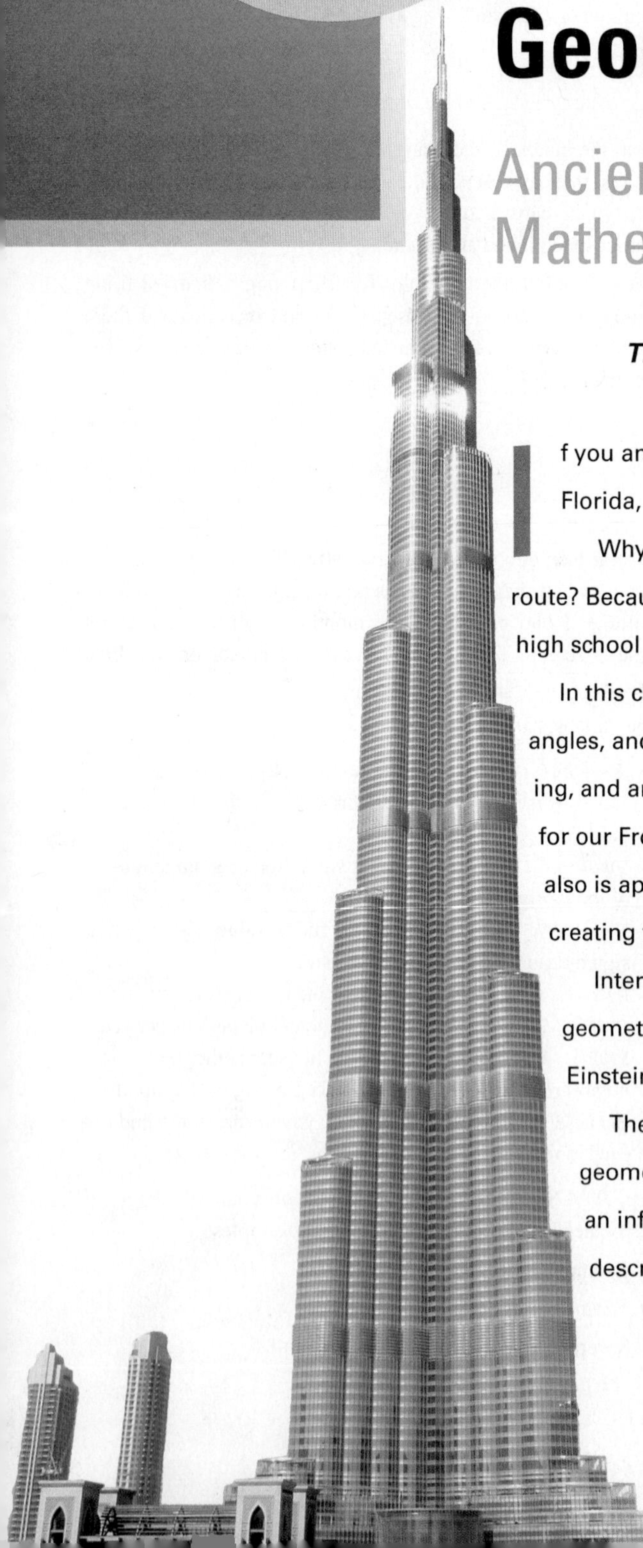

The shortest distance between two points is a . . .?

If you answered "straight line," consider the actual flight path from Miami, Florida, to Manila, Philippines, shown on page 437.

Why would a plane flying from Miami to Manila take such an inefficient route? Because the world is round, and the plane geometry you encountered in high school does not apply to a curved surface.

In this chapter, you will first study the classical geometry of points, lines, angles, and solids that the Greeks used as the basis of surveying, map making, and architecture and that we now use in designing skyscrapers, boxes for our Froot Loops, and bottles for our Gatorade. Traditional plane geometry also is applied in developing exotic applications such as robotic surgery and creating fun games like Fruit Ninja Kinect on the Xbox.

Intertwined with that discussion, we will introduce you to the "curved" geometries necessary to navigate on a spherical Earth and that Albert Einstein used to describe our curved universe.

The last section of this chapter will show you a modern type of geometry where lines are neither straight nor curved, but rather have an infinite number of "wiggles." Scientists use this fractal geometry to describe the roughness of a mountain peak, the path of a mountain

(continued)

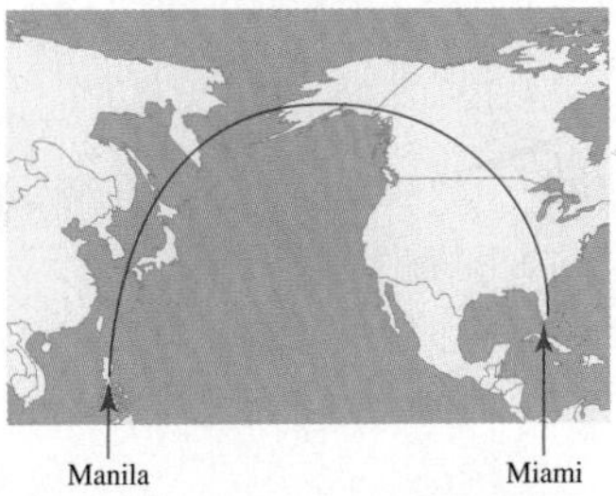

stream as it flows to the sea, the ups and downs of the stock market, and the irregular beating of a healthy heart.

As a consumer bonus, in Section 9.3 we will show how marketers can use geometry against you, devising subtle ways to inflate the price of their products with clever packaging.

9.1 Lines, Angles, and Circles

Objectives

1. Understand the basic properties of geometric objects such as points, lines, and planes.
2. Work with the fundamental properties of angles.
3. Solve problems involving the relationships among angles, arcs, and circles.

If you were to look backwards in the history of mathematics and science to ancient times, you would recognize that perhaps the most important idea invented in the past 2,500 years was the method of deductive reasoning that was put forth in the most famous textbook in the history of the world—*Elements* by Euclid of Alexandria. *Elements*, which was written about 300 BC, is an introduction to elementary mathematics—including number theory and geometry—that gave us a method of reasoning that mathematicians still use today.

Points, Lines, and Planes

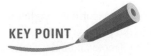

KEY POINT

Two lines in the plane are either parallel or intersecting.

Euclid begins his discussion of geometry with intuitive descriptions of three *undefined* terms, saying that a **point** is "that which has no part," a **line** has "length but no breadth," and a **plane** has "length and breadth only."

To discuss some of the basic properties of lines and angles, we need to introduce some terminology and notation. We will label points with capital letters, such as A, B, and C, and lines with lowercase letters, such as l and m, or we may include subscripts, such as l_1 or l_2.

As we show in Figure 9.1, any point on a line divides the line into three parts—the point and two *half lines*. A **ray** is a half line with its endpoint included. In Figure 9.1(b), the open dot means that the point A is not included in the half line, whereas the solid dot in Figure 9.1(c) means that A is part of the ray. A piece of a line joining two points and including the points is called a **line segment.** 1

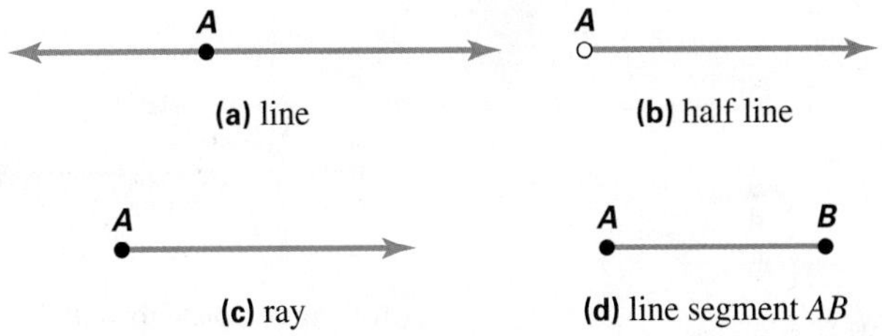

FIGURE 9.1 Some basic terminology regarding lines.

Although we do not try to define precisely what a plane is, you can think of it as being an infinite two-dimensional surface such as an infinitely large flat sheet of paper.

According to Euclid's geometry, lines either intersect or are parallel. **Parallel lines** are lines that lie on the same plane and have no points in common. In Figure 9.3, on the

QUIZ YOURSELF 1

Identify each object shown in Figure 9.2.

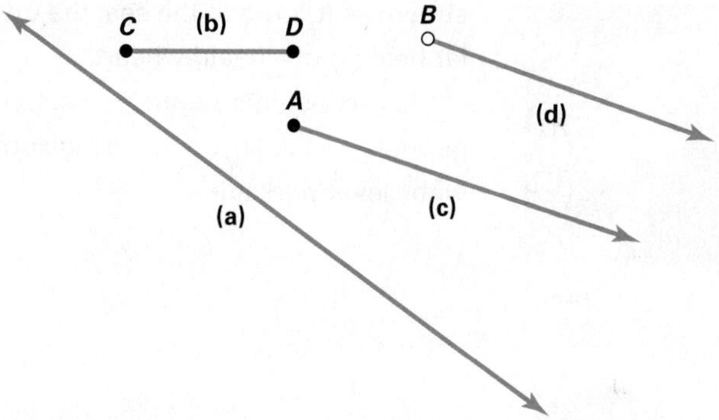

FIGURE 9.2 Line terminology.

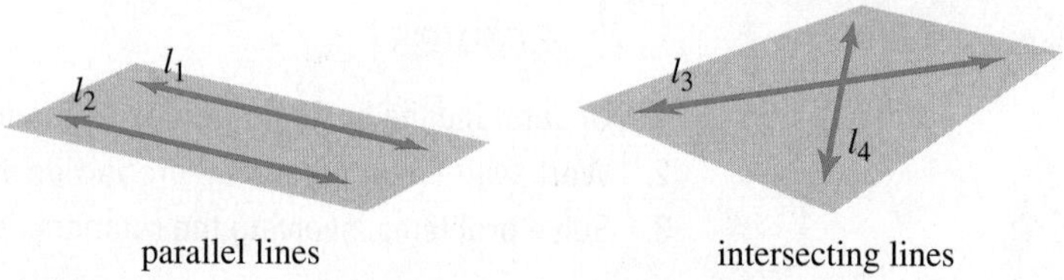

FIGURE 9.3 Parallel lines and intersecting lines.

next page, lines l_1 and l_2 are parallel. We express this as $l_1 \| l_2$. If two different lines lying on the same plane are not parallel, then they have a single point in common and are called **intersecting lines.** In Figure 9.3, lines l_3 and l_4 are intersecting lines.

Angles

Two rays having a common endpoint form an **angle.** In Figure 9.4, we form an angle by rotating ray *AB*, called the **initial side,** about point *B* to finish in the position corresponding to ray *BC*, called the **terminal side.** We use the symbol $\angle$ to denote an angle; therefore, we can call the angle in Figure 9.4 $\angle ABC$, or simply $\angle B$. Point *B* is called the **vertex** of the angle.

KEY POINT

We measure angles in degrees.

We measure angles in units called *degrees.** The symbol ° represents the word *degrees.* If you rotate the initial side of an angle one complete revolution about the vertex so that the terminal side coincides with the initial side, you will have formed a 360° angle, as we show in Figure 9.5(a). If you rotate the initial side only $\frac{1}{360}$ of the way around the vertex to reach the terminal side, that angle has size 1°. A 36° angle is shown in Figure 9.5(b).

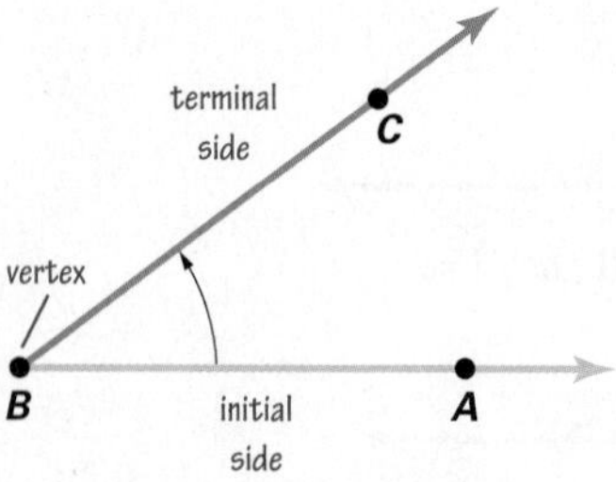

FIGURE 9.4 An angle formed by rotating a ray about point *B*.

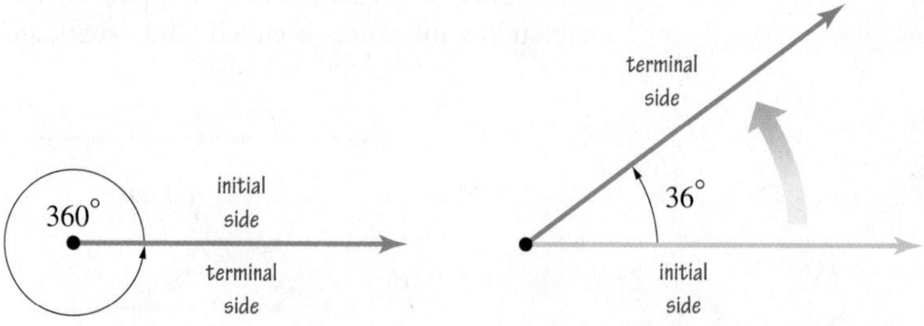

(a) one complete revolution

(b) $\frac{1}{10}$ of a complete revolution

FIGURE 9.5 We form a 36° angle by rotating the initial side $\frac{1}{10}$ of the way around the vertex.

---

*There are other possible units that are used to measure angles, such as radians. We will not discuss these other measures in this book.

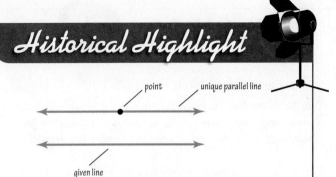

Historical Highlight

Non-Euclidean Geometry (Part 1)

Several years ago, I flew to France to visit my sister. As the jetliner climbed to an altitude of 6 miles, I was surprised to watch on the cabin's monitor that we headed northeast towards Greenland, rather than straight across the ocean towards France. A few hours later, it was clear we were following a curved path to Paris rather than a straight line. The reason for this is that although on a flat surface the shortest distance between two points is a straight line, the same is not true for curved surfaces such as the Earth's.

On a sphere, the shortest distance between two points is an arc of a *great circle.** A great circle, as shown in Figure 9.17, is a circle on a sphere that has the same center as the sphere. Therefore, in spherical geometry, we think of a "line" as a great circle.

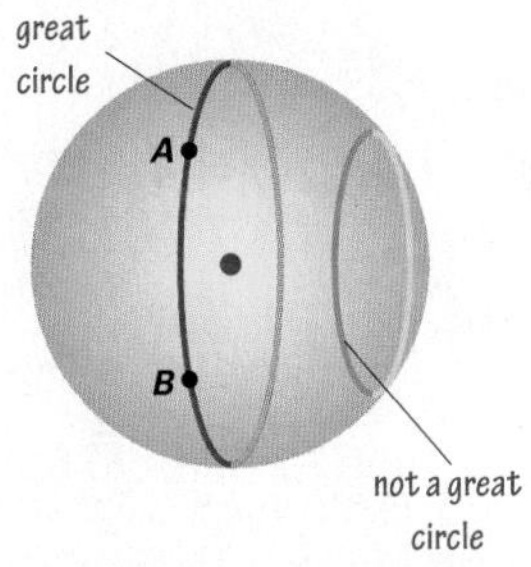

FIGURE 9.17 The distance between points *A* and *B* on a sphere is an arc of a great circle.

There can be no parallel lines in spherical geometry because any two great circles on a sphere intersect in two points. In Figure 9.18, the two great circles intersect at points *P* and *Q*.

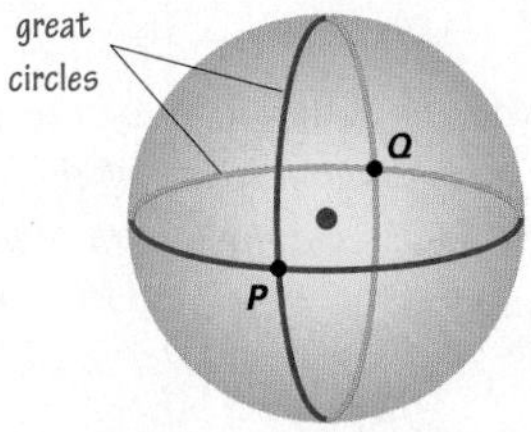

FIGURE 9.18 Two great circles intersect at points *P* and *Q*.

Notice that what we consider to be lines and what properties these lines have depend on the surface on which we are drawing the lines.

Euclid's fifth postulate† states that through a point not on a given line there is *exactly one* line that is parallel to the given line, as we show in Figure 9.19.

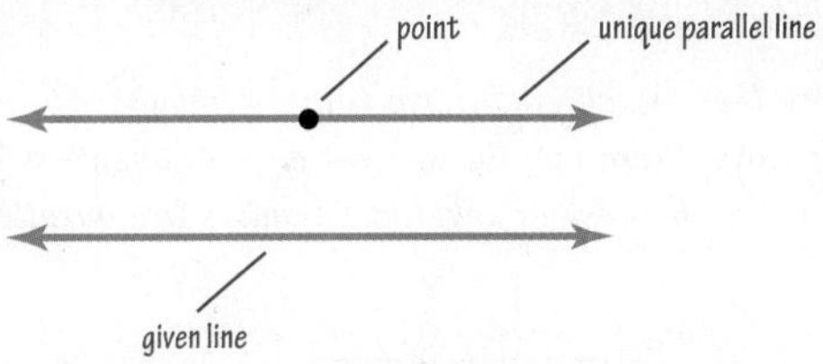

FIGURE 9.19 Euclid's fifth postulate.

However, for centuries mathematicians speculated that Euclid's fifth postulate could be proved from his other postulates. Unable to prove the fifth postulate, mathematicians decided to take another approach. They asked: "What if we assume that the fifth postulate is false? What would our geometry be like?"

There are two ways to deny the fifth postulate:

1) Assume that for a line and a point not on that line, there are no lines parallel to the given line.
2) Assume that for a line and a point not on that line, there are at least two lines parallel to the given line.

The first approach led to geometries such as spherical geometry that we mentioned previously. Around 1850, the great German mathematician Bernhard Riemann invented such a geometry.

In the early part of the nineteenth century, Karl Friedrich Gauss from Germany, Janos Bolyai from Hungary, and Nicolai Lobachevsky from Russia each independently developed a geometry using the second approach. We can visualize this type of geometry on a surface called a *pseudosphere.* To form a pseudosphere, we rotate a curve called a *tractrix* about a line, as shown in Figure 9.20.

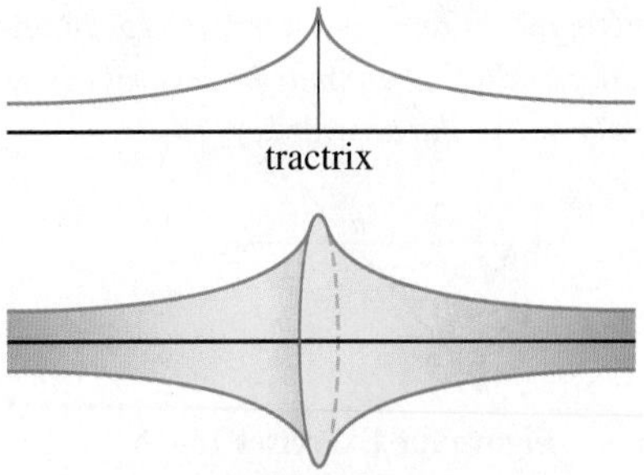

FIGURE 9.20 A pseudosphere is a model of a non-Euclidean geometry.

The surface looks somewhat like the bells of two trumpets joined together. We will discuss these non-Euclidean geometries further in Section 9.2.

---

*Arcs of circles that are not great circles will not give the shortest distance between two points on the surface of a sphere.

†This postulate was an assumption made by Euclid about geometry that could not be proved.

EXERCISES 9.1

Sharpening Your Skills

In Exercises 1–8, match each term with the numbered angles in the given figure. There may be several correct answers. We will state only one in the answer key. Lines l *and* m *are parallel.*

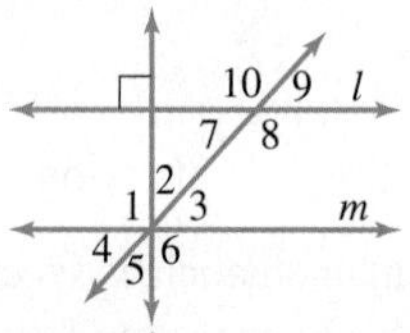

Figure for Exercises 1–8

1. vertical angles
2. complementary angles
3. alternate interior angles
4. right angle
5. obtuse angle
6. corresponding angles
7. supplementary angles
8. acute angle

In Exercises 9–14, determine whether each statement is true or false. Remember by the Always Principle that if a statement is true, it must always be true without exception. If you believe that a statement is false, you should try to find a counterexample.

9. If two angles are complementary, then they must be equal.

10. If two angles (with measure greater than 0°) are complementary, then each must be an acute angle.

11. If two equal angles are supplementary, then each is a right angle.

12. An obtuse angle cannot be complementary to another angle.

13. An angle cannot be the complement of one angle and the supplement of another angle at the same time.

14. The supplement of an acute angle must be an acute angle.

Use the given figure to answer Exercises 15–18. Assume that lines l and m are parallel. There may be several correct answers. We will state only one in the answer key.

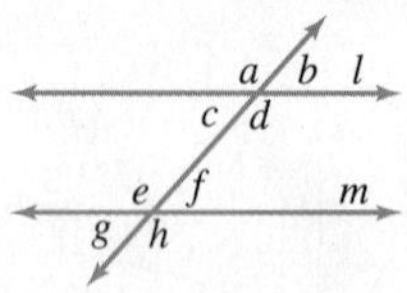

Figure for Exercises 15–18

15. Find a pair of obtuse, alternate interior angles.

16. Find a pair of acute, alternate exterior angles.

17. Find a pair of acute, corresponding angles.

18. Find a pair of obtuse, corresponding angles.

In Exercises 19–24, find the measure of a complementary angle and a supplementary angle for each angle.

19. 30°
20. 108°
21. 120°
22. 45°
23. 51.2°
24. 110.4°

In Exercises 25–30, find the measures of angles a, b, *and* c *in each figure. Lines* l *and* m *are parallel.*

25.

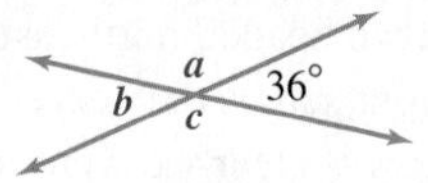

26.

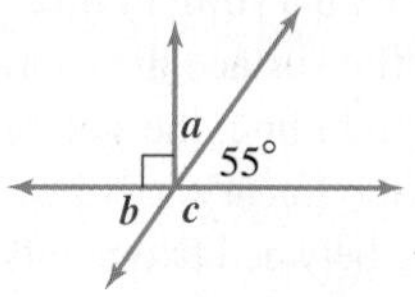

27.

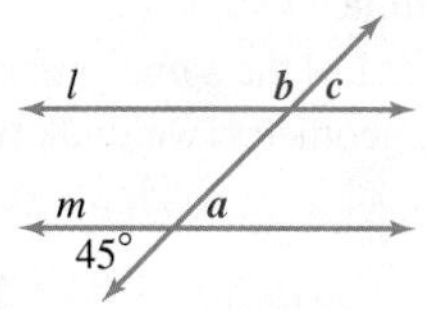

28.

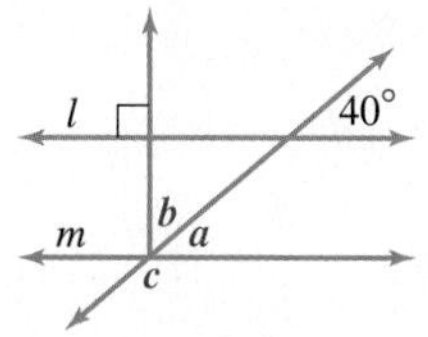

29.

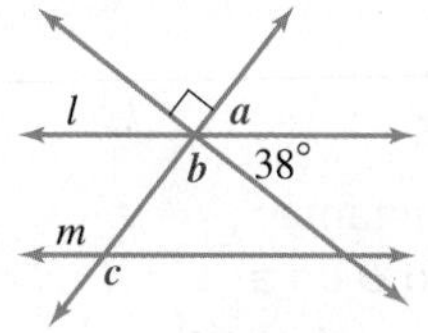

30.

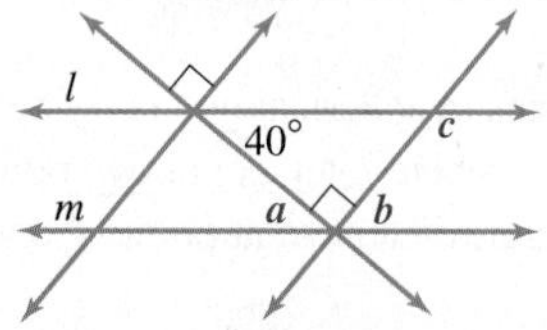

In Exercises 31–36, you are given two of the following three pieces of information: the circumference of the circle, the measure of the central angle ACB, and the length of arc AB. Find the third piece of information.

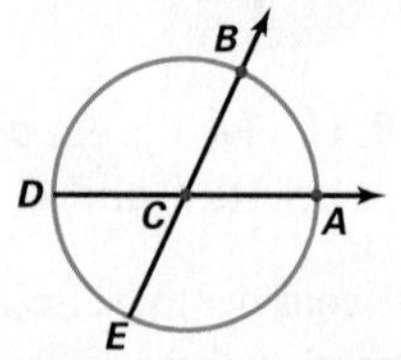

Figure for Exercises 31–40

31. circumference = 24 feet; $m\angle ACB = 90°$

32. circumference = 150 centimeters; $m\angle ACB = 72°$

33. circumference = 12 meters; length of arc AB = 4 meters

34. circumference = 240 inches; length of arc AB = 40 inches

35. $m\angle ACB = 30°$; length of arc AB = 100 millimeters

36. $m\angle ACB = 120°$; length of arc AB = 9 feet

Applying What You've Learned

Continuing the situation from Exercises 31–36, use the given information to answer Exercises 37–40.

37. circumference $= 18$ feet; $m\angle ACB = 60°$. Find the length of arc BD.

38. $m\angle ACB = 30°$; length of arc $AE = 10$ meters. Find the circumference.

39. circumference $= 30$ inches; length of arc $DE = 3$ inches. Find $m\angle BCD$.

40. circumference $= 120$ centimeters; length of arc $AB = 12$ centimeters. Find $m\angle DCE$.

In Exercises 41–44, solve for x. *Assume that lines* l *and* m *are parallel.*

41.

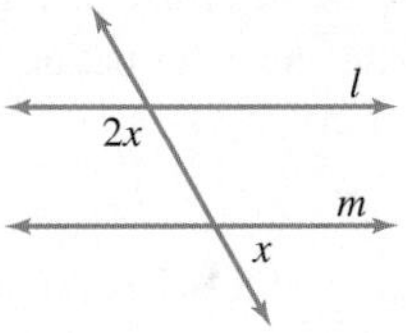

42.

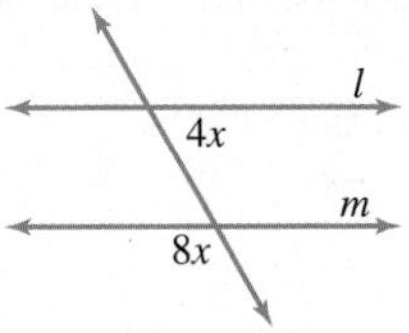

43.

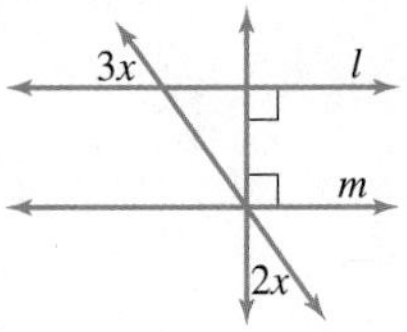

44.

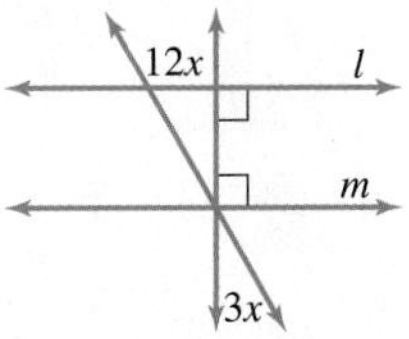

An angle such as $\angle$ABC *is called an* ***inscribed angle*** *and it can be proved that the measure (in degrees) of an inscribed angle is half the measure (in degrees) of the arc it cuts. So if the measure of arc* AC *is* 60°, *then the measure of* $\angle$ABC *would be* 30°. *Recall that* $\angle$ADC *is a central angle.*

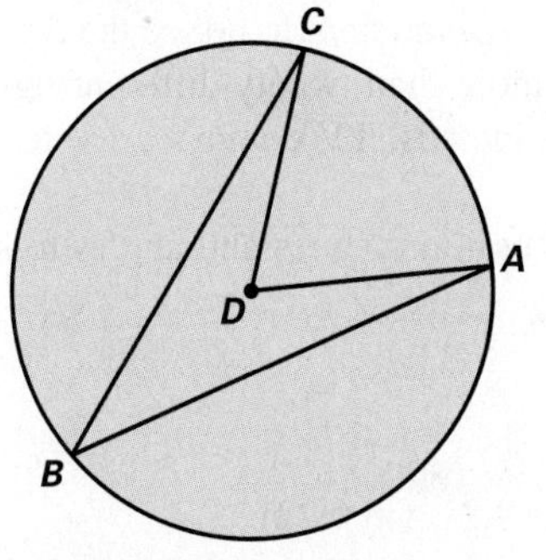

Figure for Exercises 45–48

45. If the circumference of the circle is 30 inches and $m\angle B = 24°$, then what is the length of arc AC?

46. If $m\angle D = 30°$ and the circumference of the circle is 15 inches, then what is the length of arc AC?

47. If the circumference of the circle is 60 inches and the length of arc AC is 20 inches, then what is $m\angle B$?

48. If $m\angle B = 24°$ and the length of arc AC is 2 feet, what is the circumference of the circle?

49. Two lines in a plane can intersect forming four angles (some may have the same measure). What is the greatest number of angles we can form using three lines?

50. Solve Exercise 49 for four lines.

51. In Example 3, your friend has erred in measuring the angle the vertical pole makes with the sun's rays because the correct circumference of Earth is closer to 25,000 miles than 24,000 miles. If we use the circumference of 25,000 miles in Example 3, what should the angle measurement taken by your friend have been?

52. Reconsider Example 3. Assume that you do not know how far your friend is from you. At high noon, your time, your friend tells you that the sun's rays make an 18° angle with the vertical pole he has in the ground. Assume that the true circumference of Earth is 25,000 miles. How far away is your friend?

In Exercises 53 and 54 find the measure of angle x *to make a hole-in-one at the miniature golf course hole. Use the following two facts to find* x:

1) The angle the ball makes as it hits a flat surface has the same measure as the angle the ball makes as it leaves the surface. (For example, angles a *and* b *in Exercise 53 are equal.)*

2) The sum of the measures of the interior (inside) angles of a triangle totals 180°.

53.

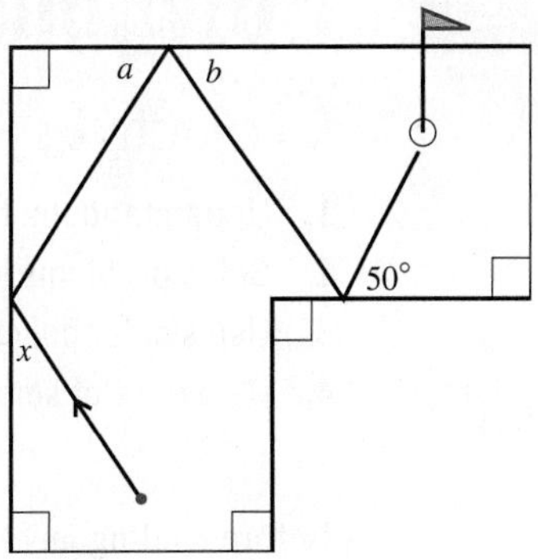

54.

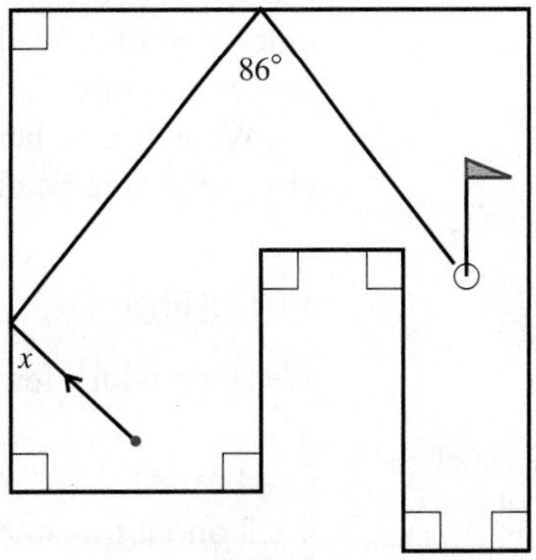

Communicating Mathematics

55. What is the difference between supplementary and complementary angles?

56. When a pair of parallel lines is cut by a transversal, name three types of angles that are pairwise equal.

57. Explain how you remember the meaning of *alternate exterior angles*.

58. Explain how you remember the meaning of *interior angles on the same side of the transversal*.

Challenge Yourself

In Exercises 59–62, use the fact that two intersecting lines form four angles. In each case explain your answer.

59. Can all four angles be equal?

60. What is the largest number of acute angles that could be formed?

61. Could exactly one of the angles be obtuse?

62. If one of the angles is obtuse, what can you say about the other angles?

63. If $\angle A$ is complementary to $\angle B$ and supplementary to $\angle C$, then what can you say about $m\angle C - m\angle B$?

64. If $\angle A$ and $\angle B$ are both supplementary to $\angle C$, what can you deduce?

65. Assume that $m\angle A = 30°$. What is the measure of the supplement of the complement of $\angle A$?

66. In Exercise 65, what is the measure of the complement of the supplement of $\angle A$?

67. Draw a diagram that has four lines and six points and in which each line passes through exactly three of the six points.

68. Draw a diagram that has 5 lines and 10 points and in which each line passes through exactly 4 of the 10 points.

69. Consider your solutions to Exercises 49 and 50. Without drawing a diagram, find the largest number of angles that can be formed by 10 lines. (*Hint:* Consider the number of intersections that can be formed by five lines, six lines, and so on.)

70. A boat lost at sea has a radio that transmits a distress signal. Anyone receiving the signal can determine the direction of the signal, but not the distance. Explain why if one ship receives the signal, it cannot notify a search plane of the exact location of the boat. If two different ships receive the signal, the exact location of the boat can now be determined. Explain how to find the boat's position with the information from both ships.

9.2 Polygons

Objectives

1. Understand the basic terminology and properties of polygons.
2. Solve problems involving angle relationships of polygons.
3. Use similar polygons to solve problems.
4. Be aware of some differences between Euclidean and non-Euclidean geometries.

Before reading any further in this section, take a moment to notice the different geometric shapes around you. When I did this, I saw more than twenty different figures. Surely geometry is all around us. We see it in magazine ads, TV commercials, religious symbols, art, architecture, and product design.

We will now build on our previous discussion of lines and angles to study a familiar class of geometric objects called *polygons*.

Polygons

KEY POINT

Polygons are special types of plane figures made up of line segments.

We begin with a few definitions (see Figures 9.21 and 9.22).

DEFINITIONS A plane figure is **closed** if we can draw it without lifting the pencil and if the starting and ending points are the same. A plane figure is **simple** if we can draw it without lifting the pencil and in drawing it we never pass through the same point twice, with the possible exception of the starting and ending points.

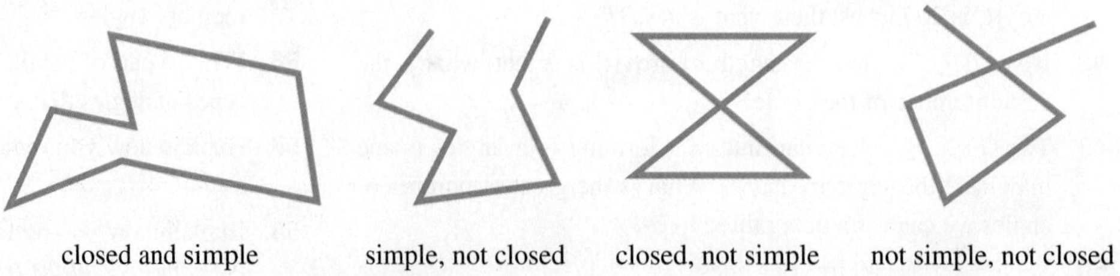

FIGURE 9.21 Closed and simple plane figures.

DEFINITIONS A **polygon** is a simple, closed plane figure consisting only of line segments, called **edges,** such that no two consecutive edges lie on the same line. We call an endpoint of an edge a **vertex** (plural, *vertices*). A polygon is **regular** if all of its edges are the same length and all of its angles have the same measure.

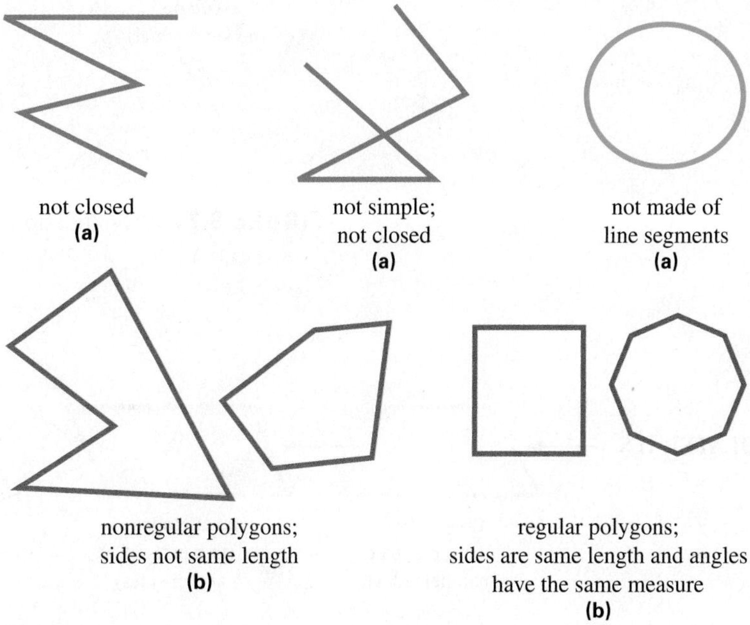

FIGURE 9.22 (a) Nonpolygons. (b) Polygons.

Number of Sides	Name of Polygon
3	Triangle
4	Quadrilateral
5	Pentagon
6	Hexagon
7	Heptagon
8	Octagon
9	Nonagon
10	Decagon

TABLE 9.2 Names of polygons.

We classify polygons according to the number of their sides. Table 9.2 lists the names of polygons having up to 10 sides.

There is another property that a polygon may have regarding its shape.

DEFINITION A polygon is **convex** if for any two points X and Y inside the polygon, the entire line segment XY also lies inside the polygon.

Figure 9.23 shows a convex and a nonconvex polygon.

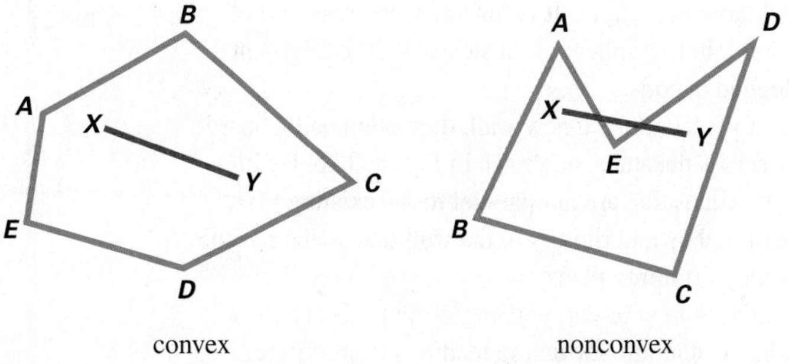

FIGURE 9.23 A convex and a nonconvex polygon.

Figures 9.24 and 9.25 show some special types of triangles and quadrilaterals. In these figures, an object following an arrow has all the properties of the objects preceding the arrow. For example, in Figure 9.24 we see that every equilateral triangle is also an isoceles triangle. Note that an isoceles triangle may not necessarily be an equilateral triangle.

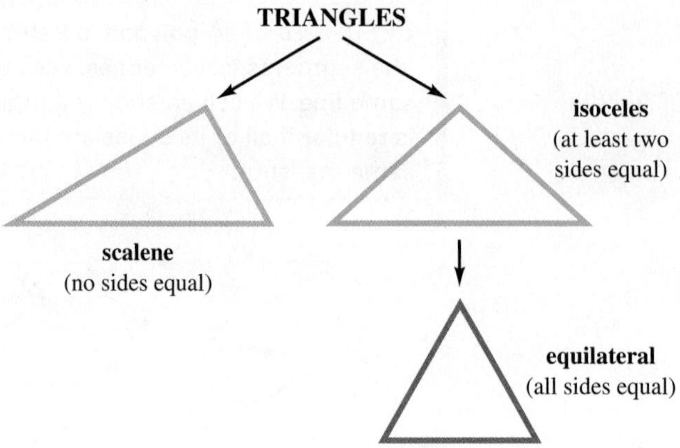

FIGURE 9.24 Classification of triangles.

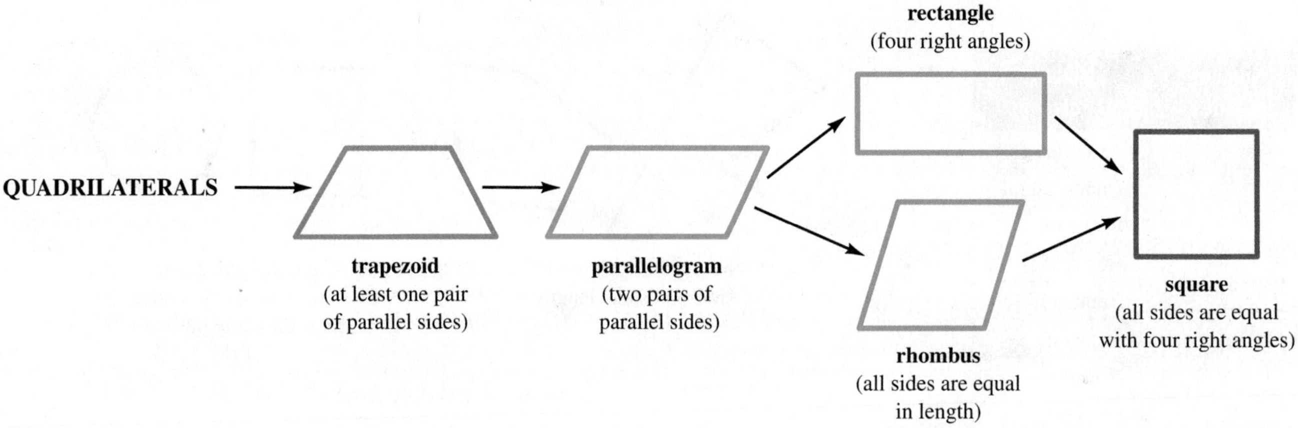

FIGURE 9.25 Classification of quadrilaterals.*

Highlight

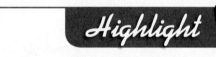

Geometry in Everyday Things

If the floor to your new apartment is not perfectly even, you might consider buying a coffee table with three legs instead of four. Any three points always lie on a plane, so the three-legged table won't wobble and spill your drinks, but the four-legged table just might. It is for this same reason that surveyors and photographers often steady their equipment on three-legged tripods.

When carpenters build a new wall, they often nail a board diagonally across the studs, as shown in Figure 9.26. By introducing new lines that are not parallel to the existing lines in the structure, they add rigidity to the wall that keeps it from shifting as they lift it into place.

While driving in your car, you may want to locate the nearest Italian restaurant. By taking readings from different locations, your car's global positioning system (GPS) determines that your car lies on two nonparallel lines. Because these lines intersect in a unique point, the system can find your exact location and direct you to the nearest restaurant.

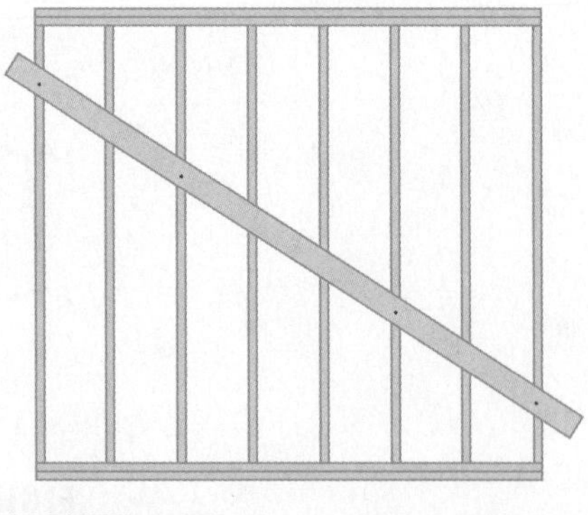

FIGURE 9.26 The diagonal board gives stability to the frame.

*Some define a trapezoid to be a quadrilateral that has *exactly one* pair of parallel sides. According to this alternate definition, a parallelogram would not be a trapezoid.

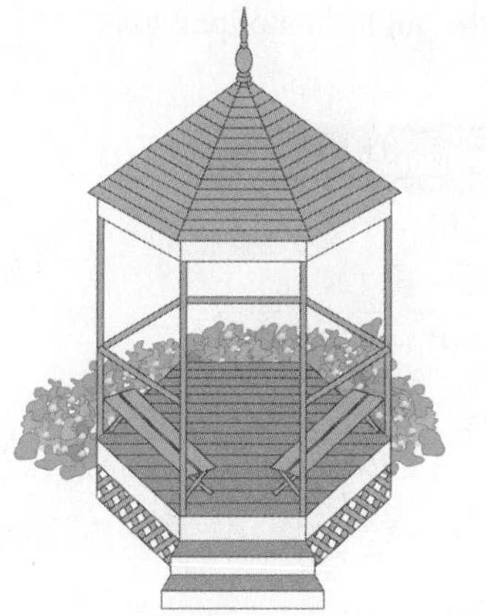

Polygons and Angles

We can use polygons to solve real-life problems such as designing a deck or building a house. However, to do this, we often need to know the sum of measures of the interior angles of the polygons. For example, imagine how difficult it would be to design a hexagon-shaped gazebo without knowing the measure of the interior angles of the floor.

EXAMPLE 1 The Angle Sum of a Triangle Is 180°

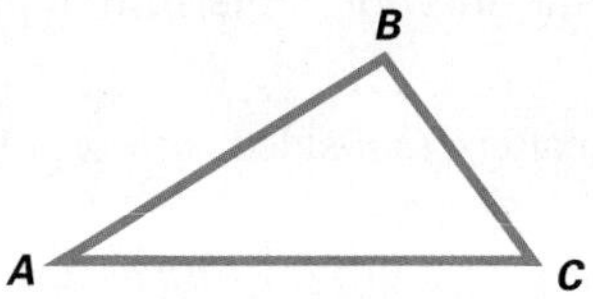

Find the sum of the measures of the interior angles in ΔABC. (The notation ΔABC is read "triangle *ABC*.")

SOLUTION:

We begin by constructing a line *m* that contains line segment *AC* and a second line *l* through point *B* that is parallel to *m*, as shown in Figure 9.27.

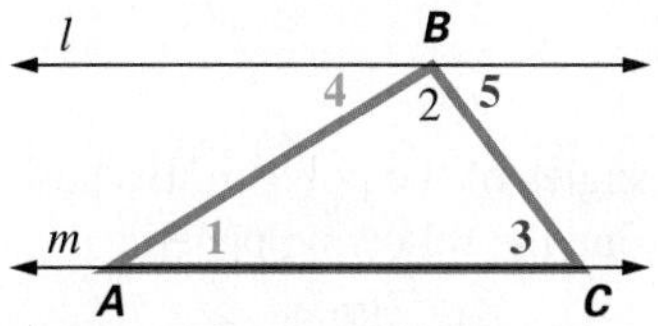

FIGURE 9.27 Lines *l* and *m* are parallel lines cut by transversals that form equal alternate interior angles.

Because angles 1 and 4 are alternate interior angles, they are equal. Similarly, angles 3 and 5 are equal. Therefore, the sum of the measures of angles 1, 2, and 3 equals the sum of the measures of angles 4, 2, and 5, which is 180°.

Now try Exercises 11 to 14.

Now that we know the sum of the angles of a triangle, we can use this information to find the interior angle sum of other polygons.

EXAMPLE 2 The Angle Sum of a Pentagon

Find the interior angle sum of the convex pentagon *ABCDE.*

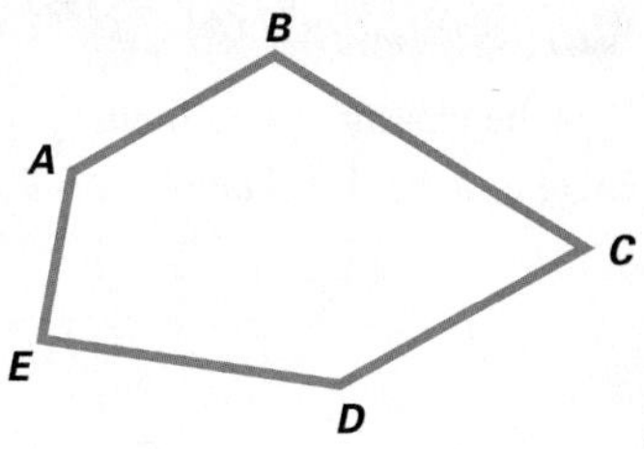

SOLUTION: Review the Relate a New Problem to an Older One Strategy on page 9.

Because we already know the interior angle sum for a triangle, it makes sense to divide the pentagon into a collection of triangles, as shown in Figure 9.28.

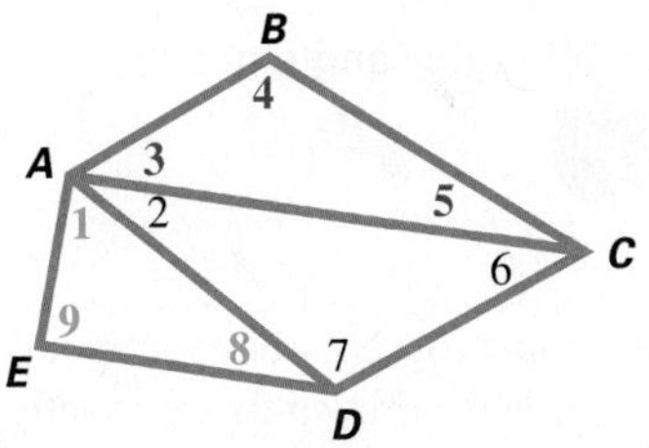

FIGURE 9.28 A pentagon divided into a collection of triangles.

Notice in Figure 9.28 that $\angle A$ is made up of three smaller angles, 1, 2, and 3. Also, $\angle C$ consists of angles 5 and 6 and $\angle D$ is made up of angles 7 and 8.

The sum of the measures of the interior angles of the pentagon is

$$\begin{array}{ccccccccc} m\angle A & + & m\angle B & + & m\angle C & + & m\angle D & + & m\angle E \\ \uparrow & & \uparrow & & \uparrow & & \uparrow & & \uparrow \\ m\angle 1 + m\angle 2 + m\angle 3 & & m\angle 4 & & m\angle 5 + m\angle 6 & & m\angle 7 + m\angle 8 & & m\angle 9 \end{array}$$

Notice that $m\angle A = m\angle 1 + m\angle 2 + m\angle 3$, $m\angle B = m\angle 4$, $m\angle C = m\angle 5 + m\angle 6$, and so on, so the interior angle sum of the pentagon is

$$m\angle 1 + m\angle 2 + m\angle 3 + \cdots + m\angle 9.$$

This is the same as the interior angle sum of the three triangles ΔAED, ΔADC, and ΔACB, which equals $3 \times 180° = 540°$.

Now try Exercises 15 to 20. 6

QUIZ YOURSELF 6

Apply the method used in Example 2 to find the interior angle sum of quadrilateral *ABCD.*

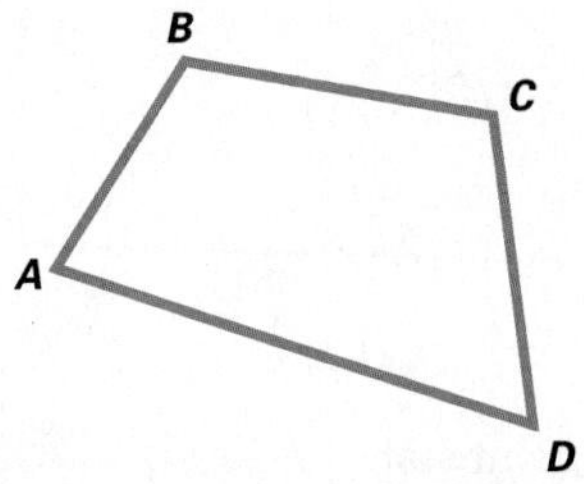

PROBLEM SOLVING

Strategy: Relate a New Problem to an Older One

As we mentioned in Section 1.1, you can often solve a new problem by relating it to one that you have seen before. We solved the problem in Example 2 by relating it to the earlier problem of finding the angle sum of a triangle.

Table 9.3 summarizes what you have seen so far about the angle sum of polygons.

KEY POINT

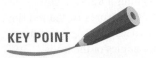

The number of sides of a polygon determines the sum of the measures of its interior angles.

Polygon	Number of Sides	Interior Angle Sum
Triangle	3	$180° = 1 \times 180°$
Quadrilateral	4	$360° = 2 \times 180°$
Pentagon	5	$540° = 3 \times 180°$

TABLE 9.3 Interior angle sum of polygons.

We can generalize this pattern to n-sided convex polygons.

ANGLE SUM OF A POLYGON The sum of the measures of the interior angles of a convex polygon having n sides is $(n - 2) \times 180°$.

If a polygon is regular, we can say a little more about the angles of the polygon. Because each angle of a regular polygon has the same measure, we find the following pattern:

each interior angle of a regular triangle (an equilateral triangle) has $\frac{180°}{3} = 60°$,
each interior angle of a regular quadrilateral (a square) has $\frac{360°}{4} = 90°$, and
each interior angle of a regular pentagon has $\frac{540°}{5} = 108°$.

We can also generalize this pattern.

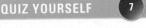

QUIZ YOURSELF 7

Find the measure of each interior angle of a regular octagon.

INTERIOR ANGLES OF A REGULAR POLYGON Each interior angle of a regular polygon with n sides has measure $\frac{(n - 2) \times 180°}{n}$. 7

Although many word-processing and computer illustration programs have graphics capabilities, you often need to understand basic geometry in order to draw a figure exactly the way you want it to look.

EXAMPLE 3 Designing a Logo

Imagine that *Dancing with the Stars* is taking its act on tour to your city and your advertising company has been hired to design a giant star-shaped billboard to advertise this event. Determine the angle measure of each point of the star.

SOLUTION:

Consider the star in Figure 9.29.

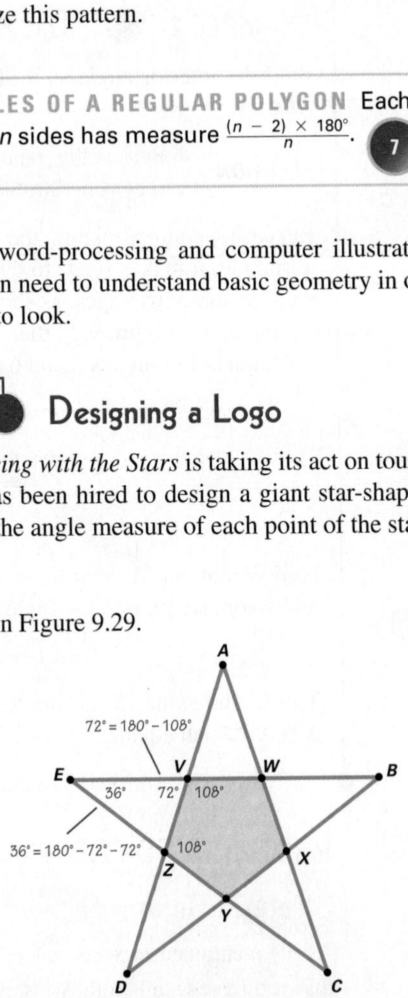

FIGURE 9.29 Constructing the five-point star.

The shaded pentagon $VWXYZ$ is a regular pentagon, so each of its interior angles has measure

$$\frac{(5-2)\times 180^\circ}{5} = 108^\circ.$$

Because a straight angle equals 180°, angles EZV and EVZ each have measure

$$180^\circ - 108^\circ = 72^\circ.$$

This means that angle E has measure $180^\circ - 72^\circ - 72^\circ = 36^\circ$.

KEY POINT

Similar polygons have proportional sides and equal angles.

Similar Polygons

On January 14, 2004, Michael Arad and Peter Walker presented a model of their winning entry,* *Reflecting Absence*, in the World Trade Center Site Memorial Competition. Before beginning work on the final memorial, a larger, partial model was constructed in the Brooklyn Navy Yard to examine the interplay of geometry and light at different times of day and night.

As a graphic artist, you might create a small, preliminary version of an advertisement before enlarging it to billboard size. Similarly, fashion designers often make small sample garments that are later resized after they have sold the designs to stores. In each case, we are working with objects that have the same shape but different sizes. Likewise, in geometry, we often work with similar figures.

DEFINITION Two polygons are **similar** if their corresponding sides are proportional and their corresponding angles are equal.

Polygons A and B in Figure 9.30 are similar. Sometimes it is useful to know if two triangles are similar. It can be proved in geometry that if one triangle has two angles equal to two angles in a second triangle, then the two triangles are similar. We will use this fact in Example 4.

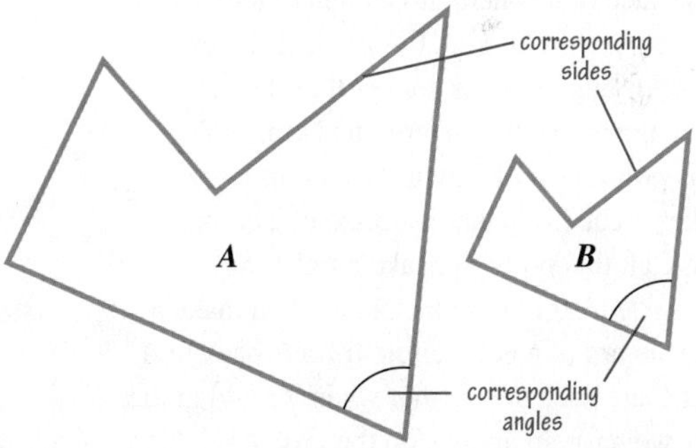

FIGURE 9.30 A and B are similar polygons.

EXAMPLE 4 Using Similar Triangles to Build a Wilderness Bridge

Yau-Man and Rupert, racing against other teams in a wilderness survival contest, encounter a deep gorge whose bridge is missing. They plan to cross by cutting a tree to fall across the gorge to use as a temporary bridge. They have measured the distances in two right triangles, as shown in Figure 9.31. Use this information to find the distance, d, across the gorge.

*For a wonderful discussion and photos of the 5,201 entries in the World Trade Center Site Memorial Competition, go to http://www.wtcsitememorial.org/.

SOLUTION:

Angles BAC and EAD are vertical angles and therefore equal. Also, angles D and C are right angles, so they also are equal. Therefore, triangles ACB and ADE are similar, and so their corresponding sides are proportional. This gives us the ratio

$$\frac{\text{length of } BC}{\text{length of } ED} = \frac{\text{length of } AC}{\text{length of } AD}.$$

Substituting for these four lengths, we get $\frac{d}{6} = \frac{20}{8}$. Cross multiplying gives the equation $d \cdot 8 = 6 \cdot 20 = 120$. Dividing both sides of the equation by 8 gives us $d = \frac{120}{8} = 15$. Therefore, a 15-foot tree will do the job.

Now try Exercises 31 and 32.

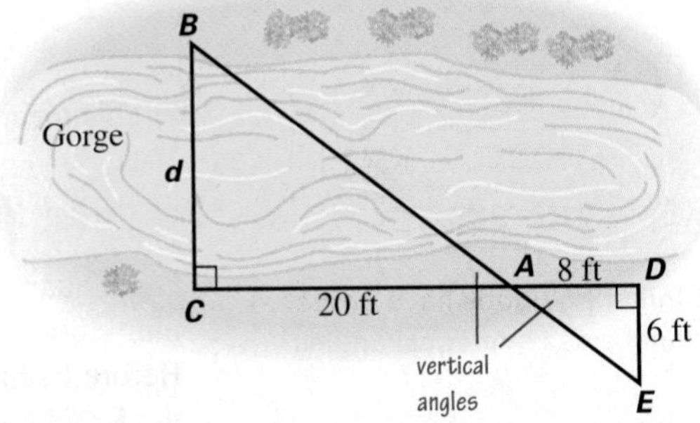

FIGURE 9.31 Building a bridge.

Historical Highlight

Non-Euclidean Geometry (Part 2)

Because the fundamental axioms of non-Euclidean geometry are different from Euclid's axioms, it is not surprising that when we rephrase well-known Euclidean geometry theorems in non-Euclidean terms, they sound somewhat strange. The Euclidean theorem that the interior angle sum of a triangle is 180° in Riemannian geometry becomes

the interior angle sum of a triangle is greater than 180°.

If we consider the surface of a sphere, it is not hard to understand why this theorem should be true. Figure 9.32 shows a triangle whose interior angle sum is greater than 180°. Imagine beginning at the top of the sphere, the North Pole so to speak (call this point A), and drawing an arc of a great circle to meet another great circle that is horizontal, comparable to the equator. Call this point B. Make a right angle, draw an arc along this "equator" to point C, and then make a right angle and draw an arc of a great circle from C back to A. The angles at B and C are each 90° so that when we add in the measure of angle A, we have an angle sum that exceeds 180°.

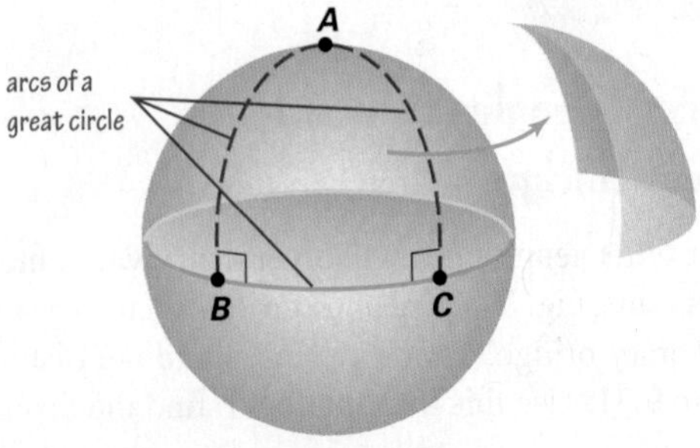

FIGURE 9.32 Triangle with angle sum greater than 180°.

On a pseudosphere, which has a shape like two bells of a trumpet, triangles look something like the curved triangle drawn in Figure 9.33. In this type of non-Euclidean geometry, triangles have an interior angle sum that is less than 180°.

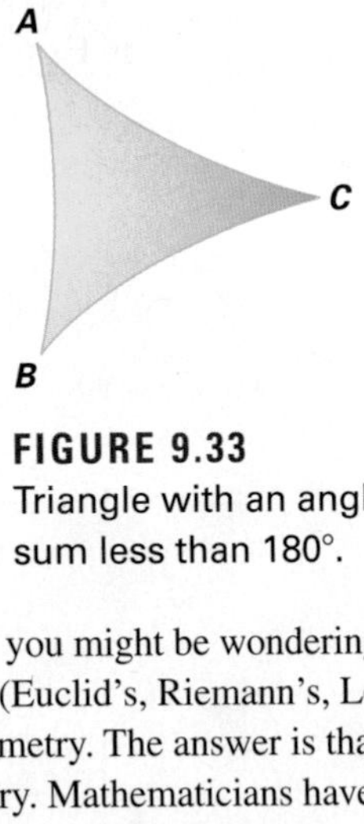

FIGURE 9.33 Triangle with an angle sum less than 180°.

At this point, you might be wondering which of the three geometries (Euclid's, Riemann's, Lobachevsky's) is the "correct" geometry. The answer is that there is no one "correct" geometry. Mathematicians have proved that all three types of geometries are perfectly consistent mathematical systems. Even more remarkable, they have also proved that for either one of these non-Euclidean geometries to be consistent, the other two types must also be consistent. So in this sense, no one of the geometries is better. As to which geometry we use, it all depends on what application we have in mind. For surveying land and building buildings, Euclidean geometry is appropriate; however, Einstein found that to understand the vastness of the universe, non-Euclidean geometry is a better tool.

EXERCISES 9.2

Sharpening Your Skills

In Exercises 1–6, determine whether each statement is true or false. Remember by the Always Principle that if a statement is true, it must always be true without exception.

1. A trapezoid is a parallelogram.
2. A rhombus is a parallelogram.
3. A regular polygon is convex.
4. If two polygons have corresponding angles equal, then the polygons are similar.
5. If two polygons have corresponding sides equal, then the polygons are similar.
6. A convex polygon can have an interior angle sum of 400°.

In Exercises 7–10, state whether each figure is a polygon. For those that are not polygons, state what part of the definition fails.

7. 

8. 

9. 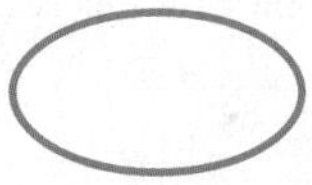

10. 

We have indicated the measures of the angles of the triangles in Exercises 11–14 in terms of x. Find the measure of angle A.

11. 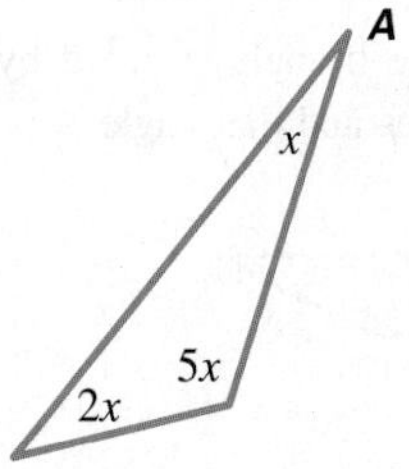

12. 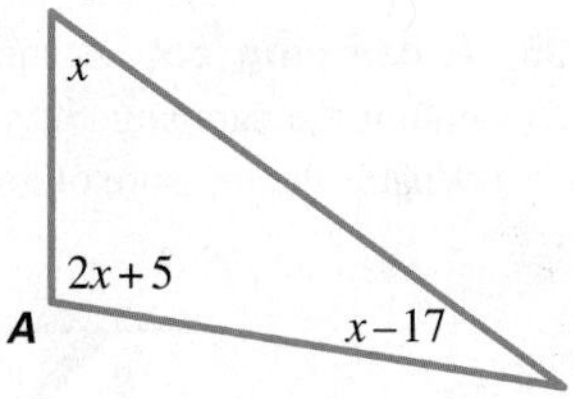

13. 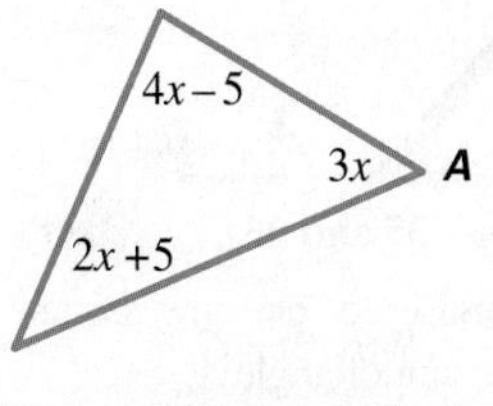

14. 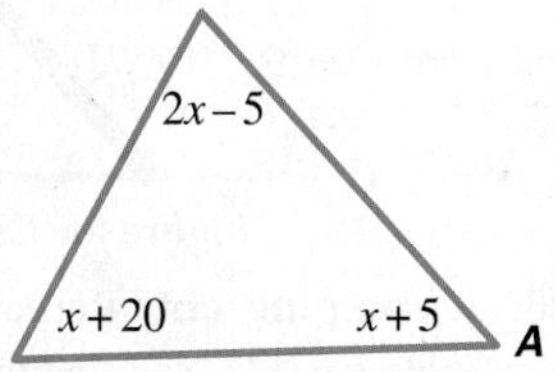

15. If we divide a regular hexagon into triangles as we did in Example 2, how many triangles will there be? What will be the interior angle sum of the hexagon?
16. If we divide a regular octagon into triangles as we did in Example 2, how many triangles will there be? What will be the interior angle sum of the octagon?
17. What is the measure of an interior angle of a regular 20-sided polygon?
18. What is the measure of an interior angle of a regular 12-sided polygon?
19. If each interior angle of a regular polygon measures 160°, how many sides does the polygon have?
20. If each interior angle of a regular polygon measures 135°, how many sides does the polygon have?

In Exercises 21 and 22, find x. *(*Note: *in each pair, the two triangles are similar.)*

21. 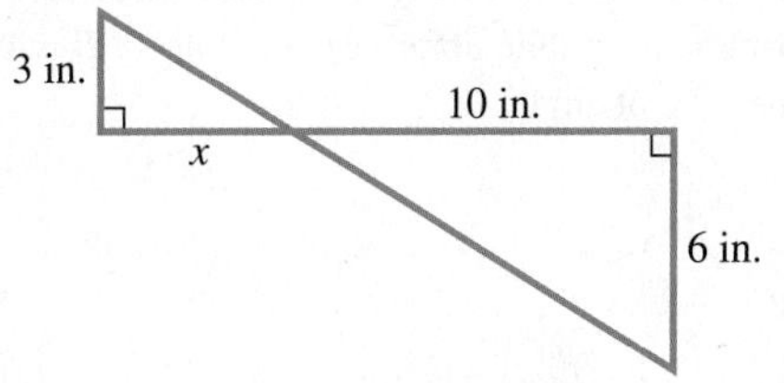

22. 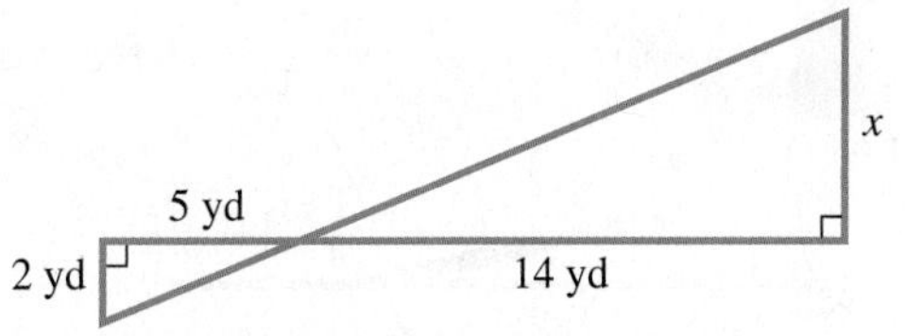

In Exercises 23–26, assume that in each pair the figures are similar. Given the lengths of sides and measures of angles in the left figure, what information do you know about the right figure?

23. 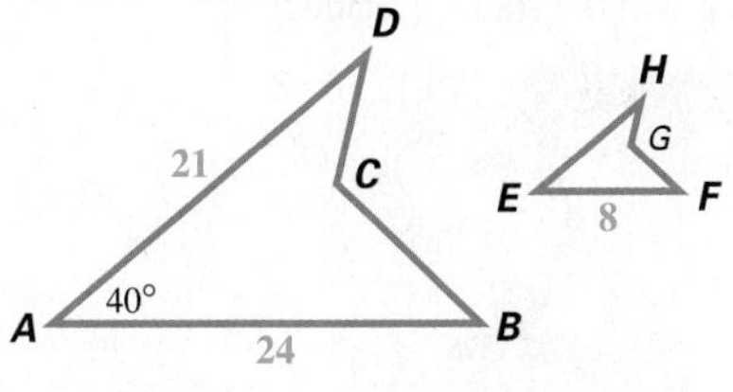

24. 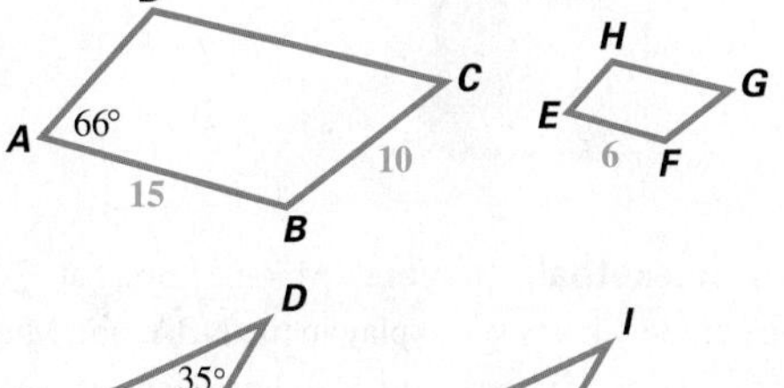

25. 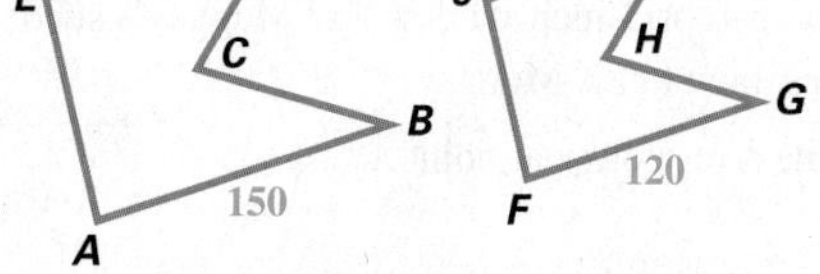

26. 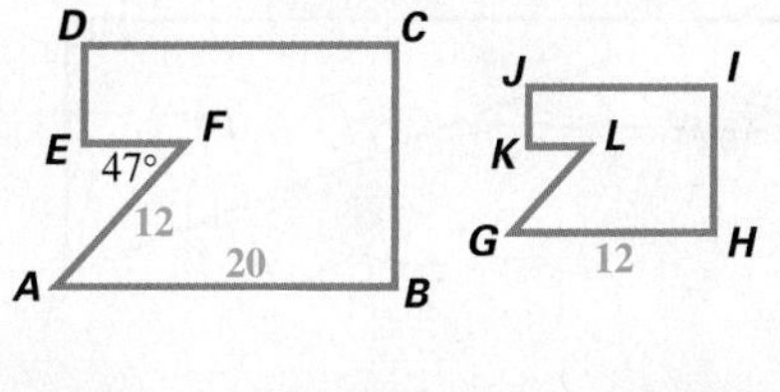

Applying What You've Learned

27. Supporting a windmill. A guy wire supports an energy-producing windmill as shown. If angle A is 138°, what is the measure of angle B?

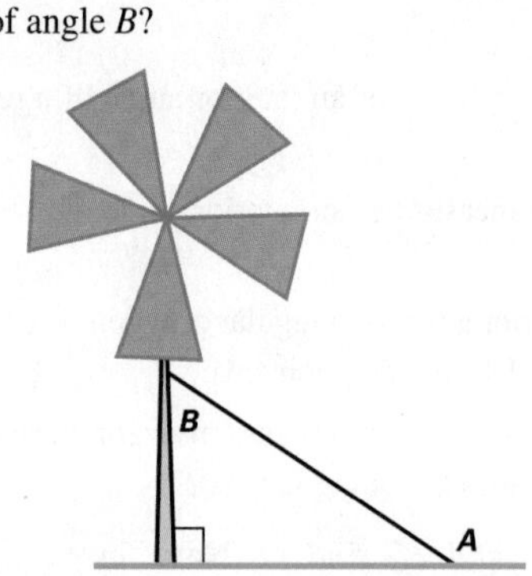

28. An accessibility ramp. A ramp was constructed to make access to Springfield's public library easier. If angle B is 165°, what is the measure of angle A?

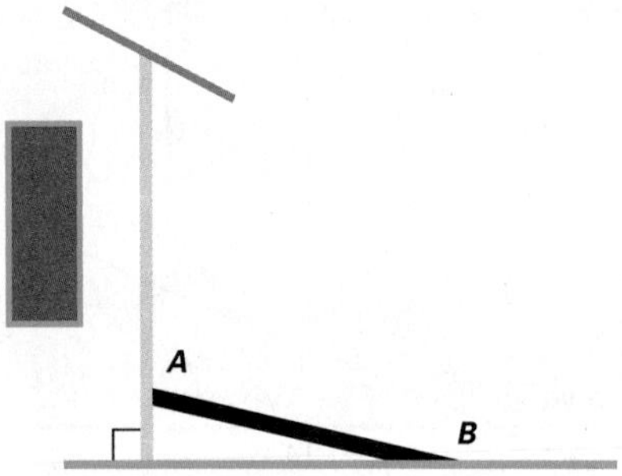

29. The Russians erected the world's largest full-figure statue, *Motherland*, which is 270 feet tall, as a World War II memorial. If at a certain time of day, a 6-foot person casts a 10-foot shadow, how long will the statue's shadow be?

30. Height of basketball players. Manute Bol, at 7 feet 7 inches one of the tallest ever to play in the NBA, and Muggsy Bogues, one of the shortest, are standing side by side. If Manute casts a 15-foot-2-inch shadow and Muggsy's shadow is 10 feet 6 inches, how tall is Muggsy?

31. How wide is the river below at point A?

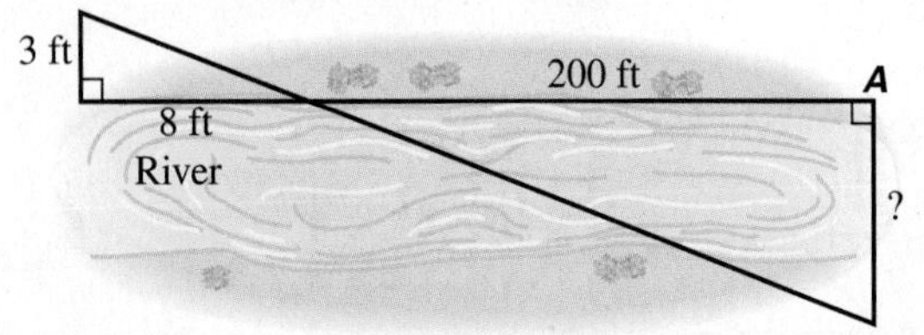

32. How wide is the river below at point A?

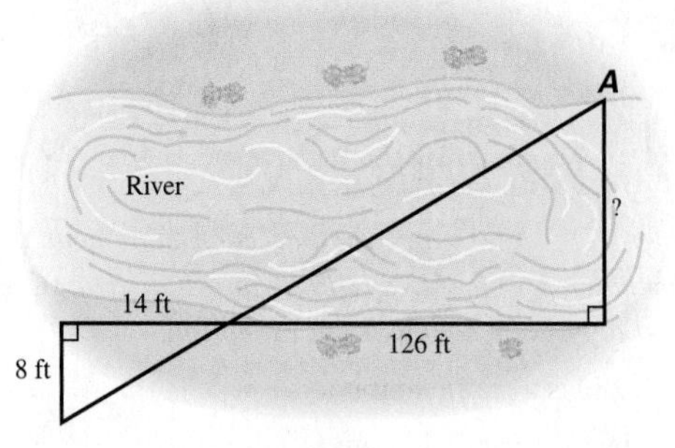

33. If triangles ABC and ADE are similar in this diagram, what is the length of the pond?

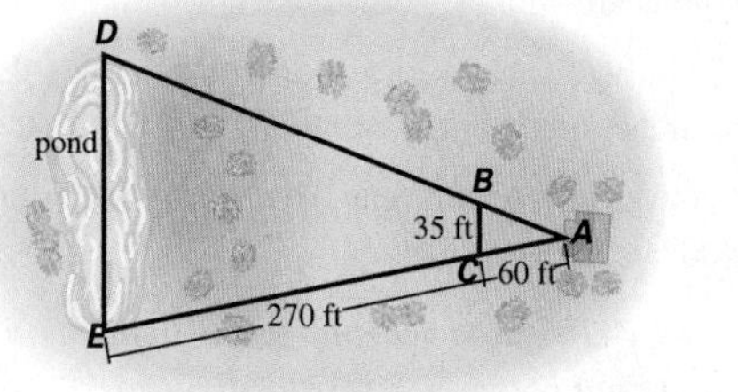

34. Emily wants to place a surveillance camera at location C to observe the Graysons at location X. The camera cannot be seen directly from position X because of a large clump of bushes. The camera will take pictures by aiming at a building, M, with shiny, mirror-like sides, as shown in the diagram. How far should the camera be positioned from point I for this setup to work?

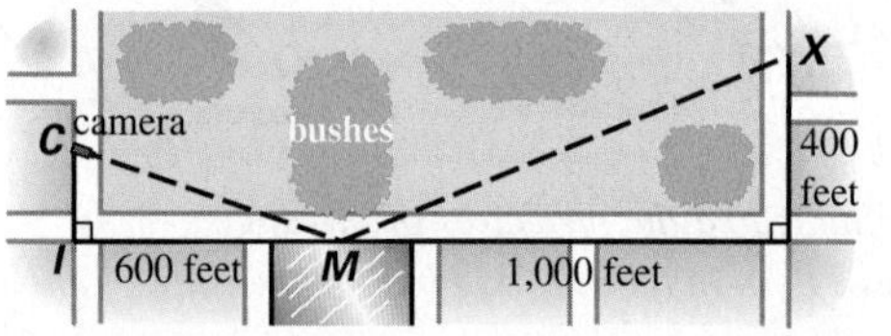

35. A camping cot. Assume that the triangles formed by the end of the camping cot are isosceles and that angle A is 84°. What is the measure of angle B?

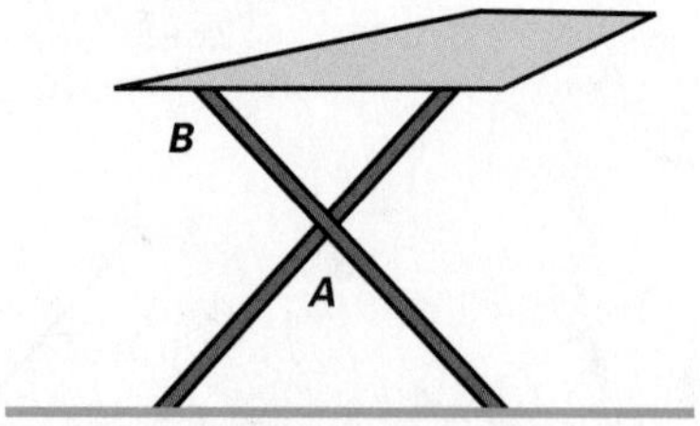

Figure for Exercises 35 and 36

36. A camping cot. Repeat Exercise 35, but now assume that angle B is 140° and find the measure of angle A.

Some Japanese furniture makers construct elegant wooden furniture without using any glue or mechanical fasteners. To assemble the furniture, the pieces of wood must be cut precisely so that they fit together tightly.

37. A furniture maker is constructing a chair using a horizontal beam with a cross section in the shape of a regular pentagon.

All 3 sides are 2 feet
All 4 sides are 1.5 feet
All 6 sides are 1 foot

FIGURE 9.38 An equilateral triangle, a square, and a regular hexagon.

Square: Because the square has sides equal to 1.5 feet, its area is $1.5 \times 1.5 = 2.25$ square feet.

Hexagon: Notice in Figure 9.38 that the hexagon can be divided into six equilateral triangles with sides 1 foot long. Using Heron's formula on the pink triangle, we find

$$s = \frac{1}{2}(a + b + c) = \frac{1}{2}(1 + 1 + 1) = \frac{3}{2}.$$

So, the area of the pink triangle is

$$A = \sqrt{s(s-a)(s-b)(s-c)} = \sqrt{\frac{3}{2}\left(\frac{3}{2} - 1\right)\left(\frac{3}{2} - 1\right)\left(\frac{3}{2} - 1\right)}$$

$$= \sqrt{\frac{3}{2} \times \frac{1}{2} \times \frac{1}{2} \times \frac{1}{2}} = \sqrt{\frac{3}{16}} \approx 0.433 \text{ square feet.}$$

QUIZ YOURSELF 9

Use Heron's formula to find the area of a triangle having sides of length 5, 7, and 8 inches.

The area of the hexagon is six times this, or $6 \times 0.433 = 2.598$ square feet. So with the same 6-foot perimeter, the hexagon has the greatest area. Thus, as Pappus claims, the hexagon is the most efficient of the three forms and it seems that bees *do* know their geometry!

Now try Exercises 21 to 26. 9

We can use the area formula for a triangle to develop area formulas for other figures such as the trapezoid shown in Figure 9.39(a).

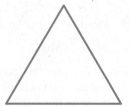

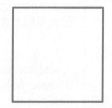

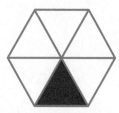

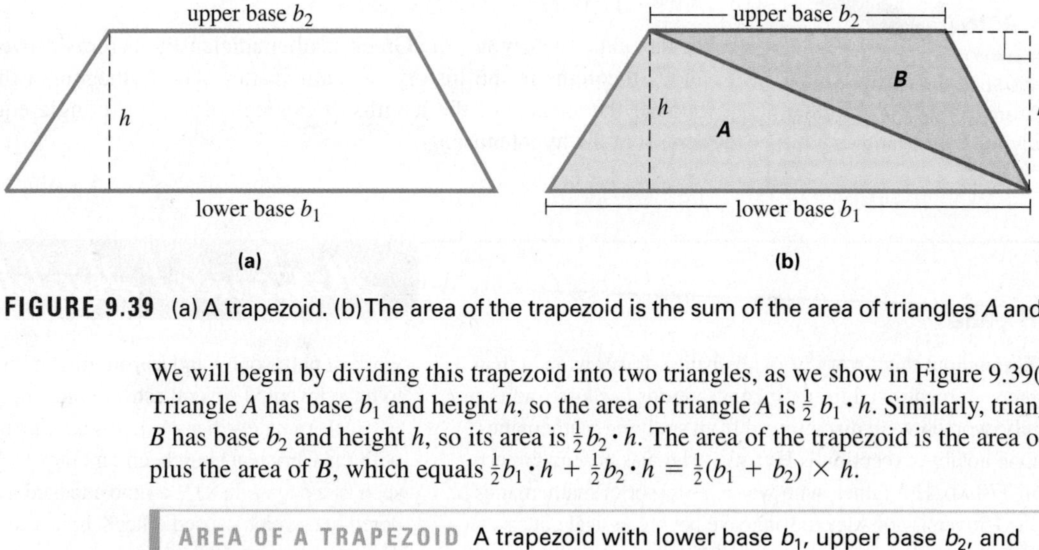
upper base b_2
h
lower base b_1
(a)
upper base b_2
B
A
h
h
lower base b_1
(b)

FIGURE 9.39 (a) A trapezoid. (b) The area of the trapezoid is the sum of the area of triangles *A* and *B*.

We will begin by dividing this trapezoid into two triangles, as we show in Figure 9.39(b). Triangle *A* has base b_1 and height h, so the area of triangle *A* is $\frac{1}{2} b_1 \cdot h$. Similarly, triangle *B* has base b_2 and height h, so its area is $\frac{1}{2} b_2 \cdot h$. The area of the trapezoid is the area of *A* plus the area of *B*, which equals $\frac{1}{2}b_1 \cdot h + \frac{1}{2}b_2 \cdot h = \frac{1}{2}(b_1 + b_2) \times h$.

AREA OF A TRAPEZOID A trapezoid with lower base b_1, upper base b_2, and height h has area

$$A = \frac{1}{2}(b_1 + b_2) \times h.$$

EXAMPLE 3 Finding the Area of Trapezoids in the Base of a Statue

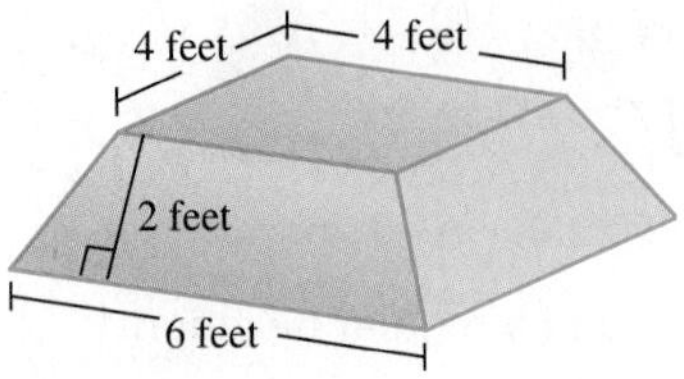

FIGURE 9.40 Base for statue formed by one square and four trapezoids.

A sculptor has created a statue for the lobby of the town's historical society building. Because the building is old and the town engineer wants to reduce the weight of the platform the statue will rest on, he recommends a hollow base instead of a solid one. The platform will be formed by joining four congruent trapezoids and one square, as shown in Figure 9.40. To determine which material will be best for the platform, the sculptor needs to know the surface area of the platform. Use the formulas we have developed so far to determine this surface area.

SOLUTION: Review the Relate a New Problem to an Older One Strategy on page 9.

We know that the area of the top of the platform is $4 \times 4 = 16$ square feet. Each side is a trapezoid with lower base 6 feet, upper base 4 feet, and height 2 feet. Therefore, using the formula for the area of a trapezoid, the area of each face is

$$\frac{1}{2}(b_1 + b_2) \times h = \frac{1}{2}(6 + 4) \times 2 = 10 \text{ square feet.}$$

The total area of the four trapezoidal sides plus the top is $4 \times 10 + 16 = 56$ square feet.

Now try Exercises 5 and 6. 10

QUIZ YOURSELF 10

Find the area of the given trapezoid.

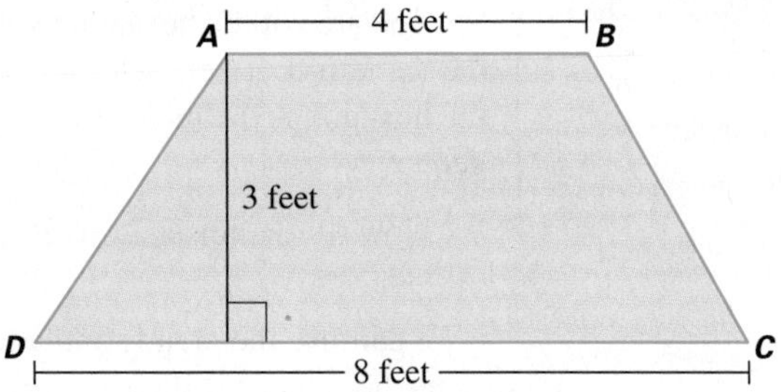

The Pythagorean Theorem

KEY POINT

We can use the Pythagorean theorem to find measurements involving right triangles.

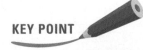

In the sixth century BC, the Greek mathematician Pythagoras proved one of the most famous theorems in the history of mathematics. The Pythagorean theorem states that the sum of the squares of the lengths of the legs of a right triangle equals the square of the length of the hypotenuse.

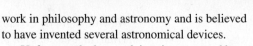

Historical Highlight

Hypatia

Throughout the history of mathematics, fewer women than men are mentioned for their contributions because traditionally, women were discouraged from studying mathematics. One notable exception is Hypatia, who was born in Greece in 370 AD. Her father, who was a professor of mathematics at the University of Alexandria, gave her a classical education, which included mathematics. She lectured on both philosophy and mathematics at Alexandria and wrote major papers on geometry, including the work of Euclid. She also did work in philosophy and astronomy and is believed to have invented several astronomical devices.

Unfortunately, her work in science caused her problems with the Christian church, and moreover, as a Greek she was seen as a pagan. In 415, a mob attacked and brutally murdered her, which caused other scholars to flee Alexandria. This tragic event marked the end of the golden age of Greek mathematics and, some believe, the beginning of the Dark Ages in Europe.

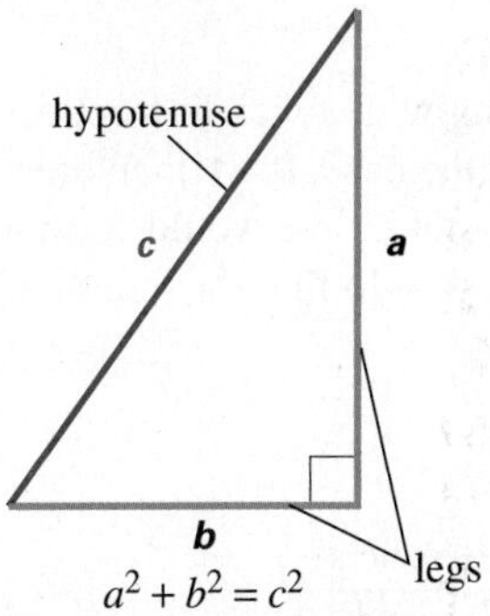

FIGURE 9.41 The Pythagorean theorem.

THE PYTHAGOREAN THEOREM If a and b are the legs of a right triangle and c is the hypotenuse (the side opposite the right angle, shown in red), then $a^2 + b^2 = c^2$ (see Figure 9.41).

If we know the lengths of any two sides of a right triangle, we can use the Pythagorean theorem to find the length of the third side.

EXAMPLE 4 Using the Pythagorean Theorem

Use the lengths of the two given sides to find the length of the third side in the triangles in Figure 9.42.

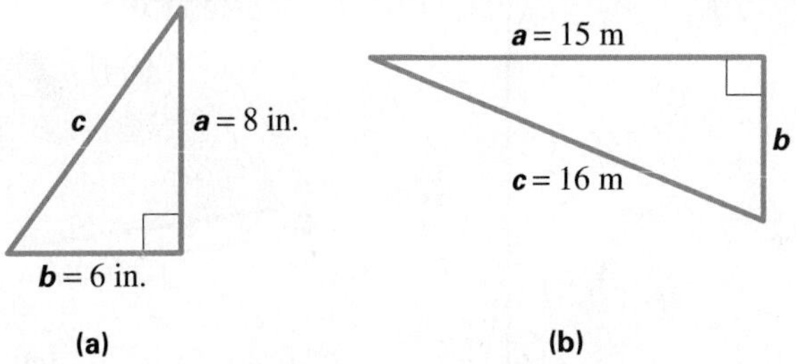

FIGURE 9.42 Using the Pythagorean theorem.

SOLUTION:

a) We are given $a = 8$ and $b = 6$, so we must find c. If we substitute the values for a and b in the Pythagorean theorem, we get

$$c^2 = 8^2 + 6^2 = 64 + 36 = 100.$$

Therefore, $c = \sqrt{100} = 10$ inches.

b) In this case, we are given that $a = 15$ and $c = 16$, and we must find b. Substituting these values in the Pythagorean theorem, we get

$$15^2 + b^2 = 16^2,$$

or

$$225 + b^2 = 256.$$

So,

$$b^2 = 256 - 225 = 31.$$

This means that

$$b = \sqrt{31} \approx 5.57 \text{ meters.}$$

Now try Exercises 27 to 32. 11

QUIZ YOURSELF 11

Assume that the hypotenuse of a right triangle measures 20 inches and one leg measures 10 inches. What is the length of the other leg?

To solve a problem, it may be necessary to use the Pythagorean theorem several times, as we demonstrate in Example 5.

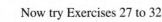

EXAMPLE 5 Using the Pythagorean Theorem to Find the Height of a Pyramid

The Great Pyramid, near Cairo, Egypt, is the tomb of the Egyptian pharaoh Khufu (also called Cheops by the Greeks) and was considered to be one of the seven wonders of the ancient world. An archaeologist wants to determine the height of this pyramid. This pyramid has a square base measuring 230 meters on each side, and she has found the distance from one corner of the base to the tip of the pyramid to be 219 meters. What is the height of the pyramid?

SOLUTION:

If we imagine this pyramid to be hollow, we could drop a string with a weight from the tip of the pyramid, point T, to point M, which is the middle of the base. If we then drew a line segment from M to a corner of the base, calling this point C, we would form a right triangle ΔTMC, as we see in Figure 9.43(a). Our goal now is to find the length of line segment TM.

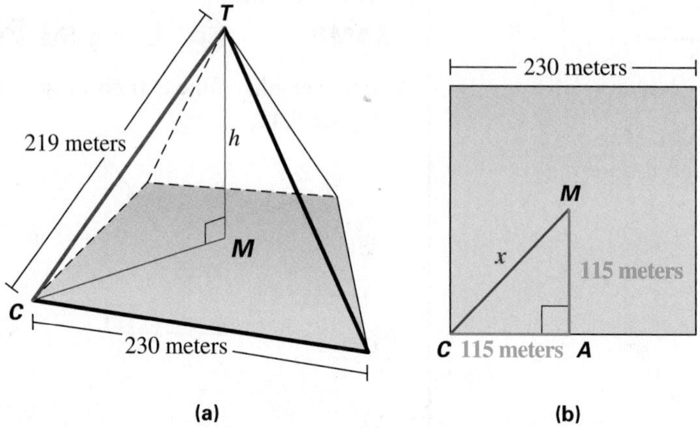

FIGURE 9.43 (a) The Great Pyramid. (b) The base of the pyramid; M is the center of the base, C is the corner of the base, and A is the midpoint of a side of the base.

We will do the solution in two stages.

Stage 1: We will find the length of line segment CM. We draw the base of the pyramid in Figure 9.43(b), showing segment CM. Because each edge of the base is 230 meters long, MA and CA are both 115 meters long. We can then apply the Pythagorean theorem to triangle ΔMAC as follows:

$$(\text{length of } CM)^2 = (\text{length of } CA)^2 + (\text{length of } AM)^2,$$

or

$$x^2 = 115^2 + 115^2 = 13{,}225 + 13{,}225 = 26{,}450.$$

Thus, $x = \sqrt{26{,}450} = 115\sqrt{2} \approx 162.6$.

Stage 2: Now that we have found the length of CM, we can find the length of TM in the right triangle ΔTMC shown in Figure 9.44. Using the Pythagorean theorem again, we get

$$(\text{length of } TC)^2 = (\text{length of } CM)^2 + (\text{length of } TM)^2.$$

Substituting for the various lengths in ΔTMC gives the equation

$$219^2 = (115\sqrt{2})^2 + h^2,$$

or

$$h^2 = 219^2 - (115\sqrt{2})^2 = 47{,}961 - 26{,}450 = 21{,}511.$$

Taking the square root of both sides, we find that $h \approx 146.7$ meters.

Now try Exercises 57 and 58.

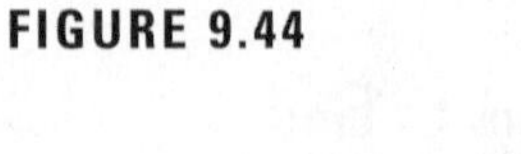

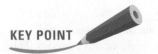

FIGURE 9.44

Circles

KEY POINT

We use simple formulas to compute the circumference and area of circles.

As with polygons, we often need to calculate the circumference and area of circles. The derivation of these formulas is not as intuitive as those for the polygons we have been discussing, so we will state them without proof. The number π, which appears in these formulas, is the Greek letter pi. It stands for the measure of the circumference of a circle divided by the length of its diameter. We will use 3.14 to approximate pi.

CIRCUMFERENCE AND AREA OF A CIRCLE A circle with radius r has circumference $C = 2\pi r$ and area $A = \pi r^2$.

These formulas tell us that a circle with radius 10 feet has a circumference of $C = 2\pi r = 2\pi(10) \approx 62.8$ feet and area $A = \pi r^2 = \pi(10)^2 \approx 314$ square feet.

EXAMPLE 6 Laying Out a Basketball Court

A man is designing a basketball court for his children. The court is in the form of a segment of a circle, as shown in Figure 9.45. There will be a fence at the rounded end of the court, and he will paint the court with a concrete sealer.

a) How much fencing is required?

b) What is the area of the surface of the court?

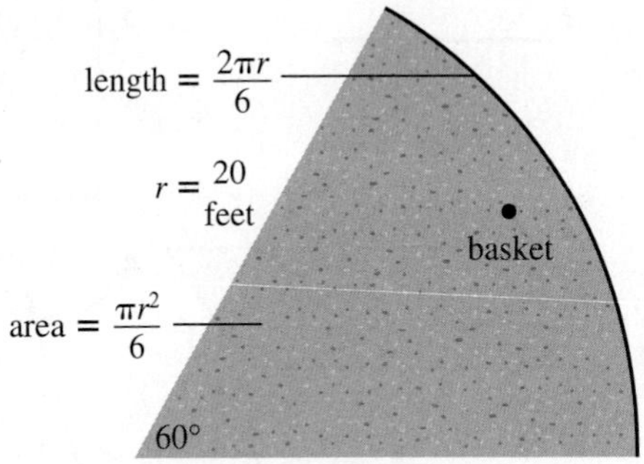

FIGURE 9.45 Basketball court.

SOLUTION:

Since a circle contains 360°, the court is one-sixth of the interior of a circle with a radius of 20 feet.

a) The circumference of a circle with a 20-foot radius is $2\pi r = 2\pi(20) = 40\pi \approx 125.6$ feet. The fencing he needs will be one-sixth of this, or approximately 20.93 feet. The man should buy about 21 feet of fencing.

b) The area of a circle with a 20-foot radius is $\pi(20)^2 = 400\pi \approx 1{,}256$ square feet. One-sixth of this is about 209.3 square feet.

Now try Exercises 59 and 60. 12

QUIZ YOURSELF 12

Find the circumference and area of a circle that has radius 8.

Between the Numbers

The Geometry of Inflation

When manufacturers resize packages, we sometimes wind up with less product for the same price. Recently I noticed that one of my favorite juices, which had been in a round bottle, now came in a hexagonally shaped bottle. To understand the consequences of this, suppose that we inscribe a regular hexagon with side of length 1 inside a circle with radius of length 1 as in Figure 9.46.

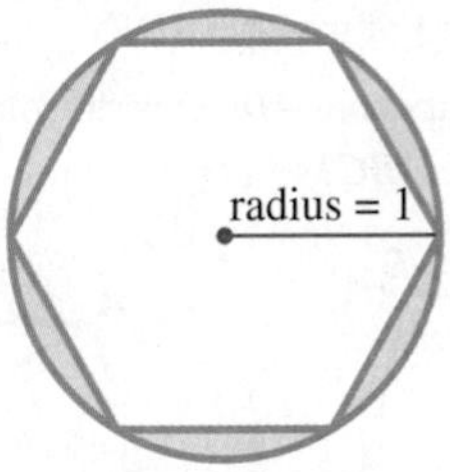

FIGURE 9.46 An inscribed hexagon.

Doing the exact same calculations that we did in Example 2, we find the area of the inscribed hexagon to be 2.598 square units. However, the area of a circle with radius 1 is $\pi r^2 = \pi \times 1^2 = \pi \approx 3.14159$ square units. So the circular cross section has $3.14159 - 2.598 = 0.54359$ more square units of area. Looking at this in terms of percentages, we get

$$\frac{0.54359}{3.14159} \approx 0.1730 = 17.3\%,$$

which means the hexagonal bottle would contain roughly 17.3% less juice for the same price.

EXERCISES 9.3

Sharpening Your Skills

*In Exercises 1–12, find the area of each figure.**

1.

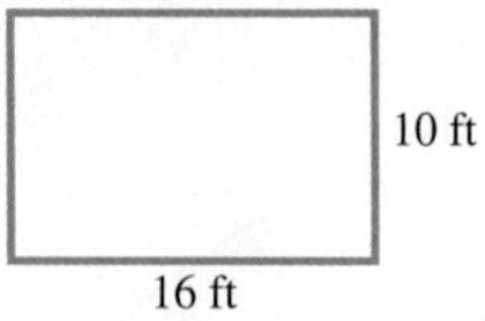

2.

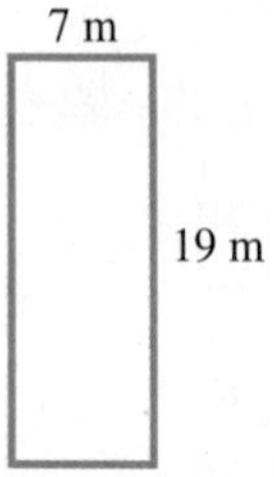

3.

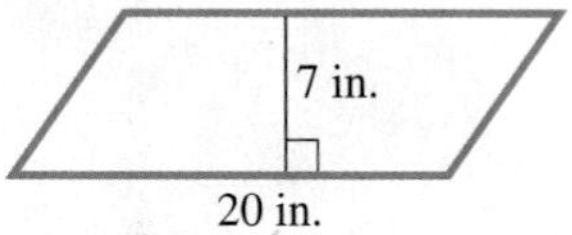

4.

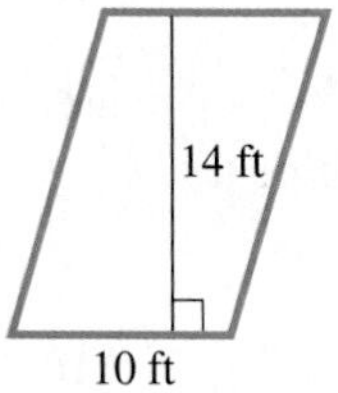

5.

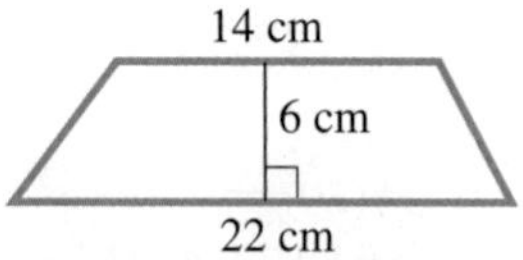

6.

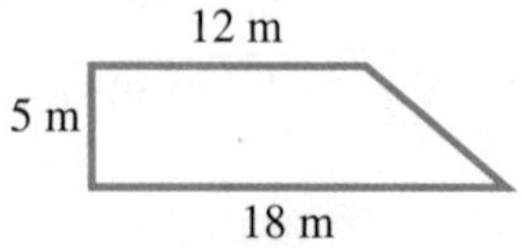

7.

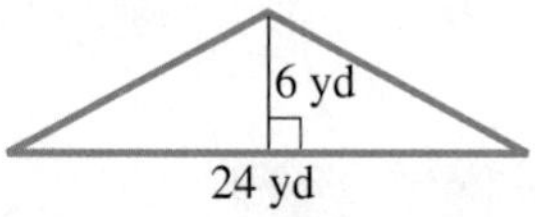

8.

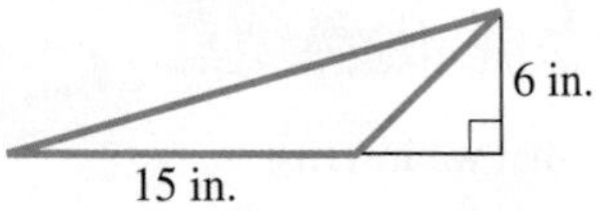

9.

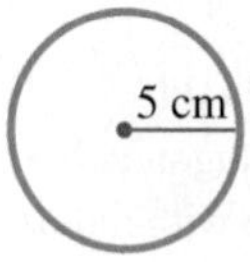

10.

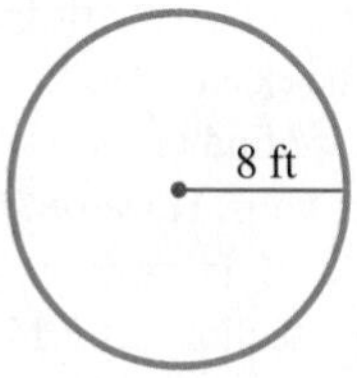

11.

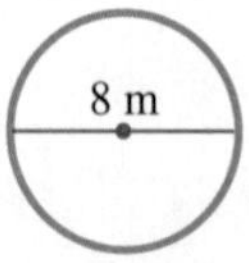

12.

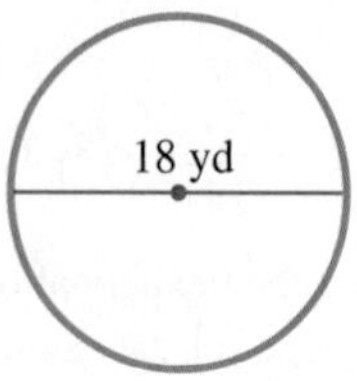

In Exercises 13–18, find the area of the shaded regions. Recall that to solve a new problem, it is helpful to relate it to a problem that you have seen before.

13.

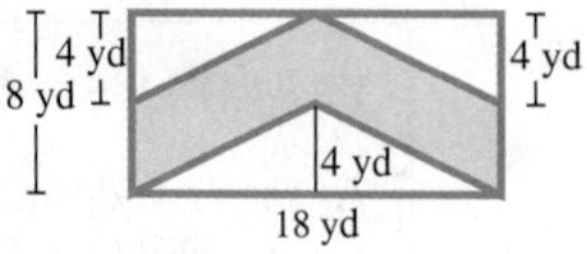

14.

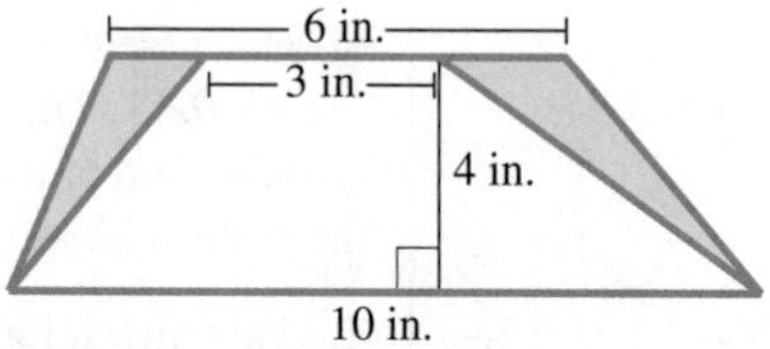

15.

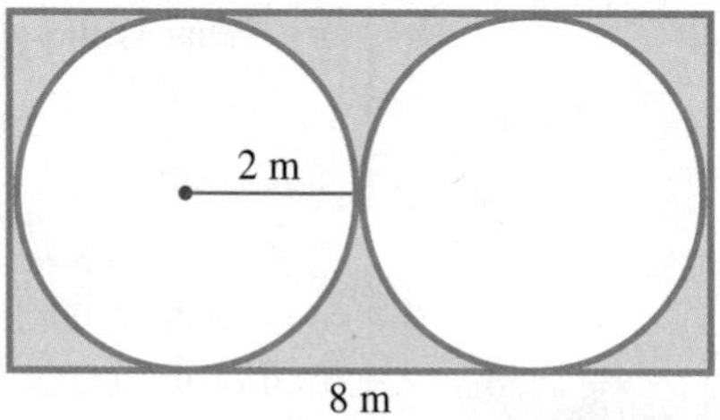

16.

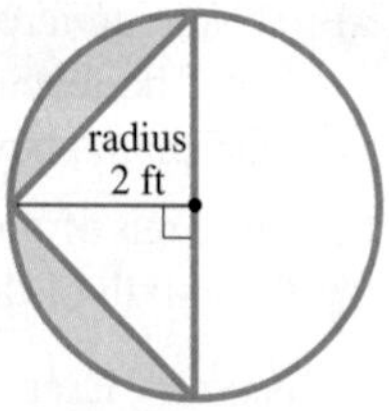

17.

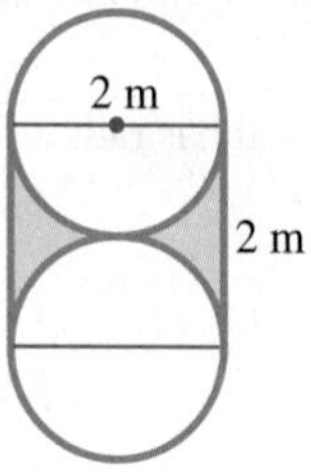

18.

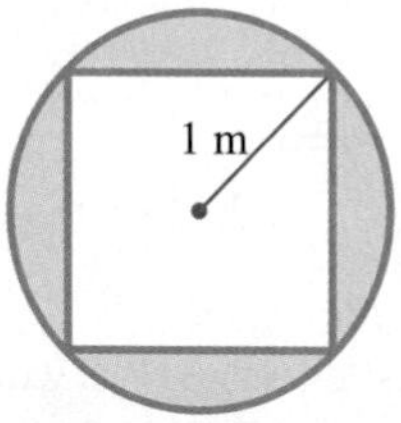

19. The area of trapezoid $ABCD$ is 54 square feet and line segment DE is 6 feet long.

a. Find the height of trapezoid $ABCD$.

b. Find the area of triangle ADE.

20. The area of trapezoid $ABCD$ is 80 square inches. What is the area of triangle ABC?

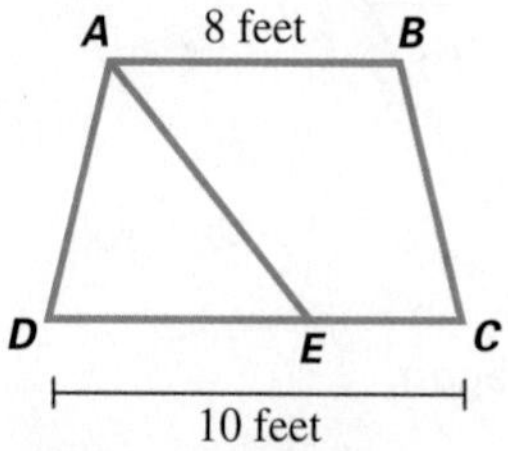

Figure for Exercise 19

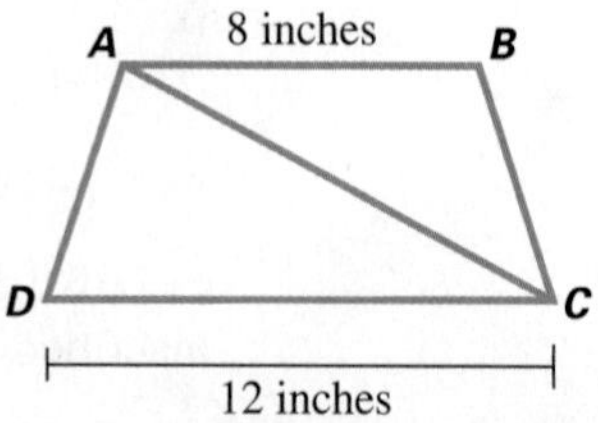

Figure for Exercise 20

*Throughout this exercise set, approximate π by 3.14.

In Exercises 21–24, use Heron's formula to find the area of each triangle.

21.

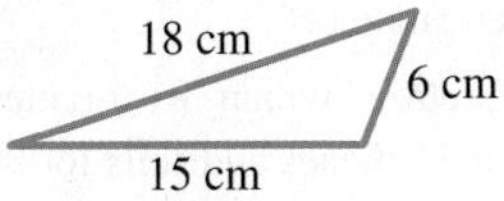

22.

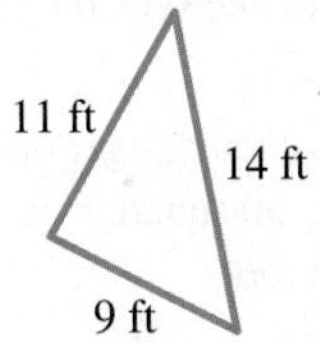

23.

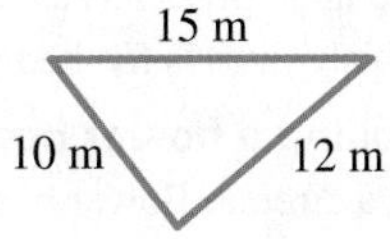

24.

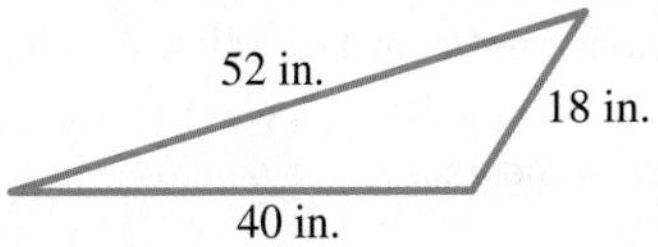

In Exercises 25 and 26, find the height h *for each triangle.*

25.

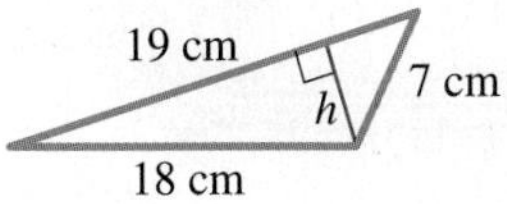

26.

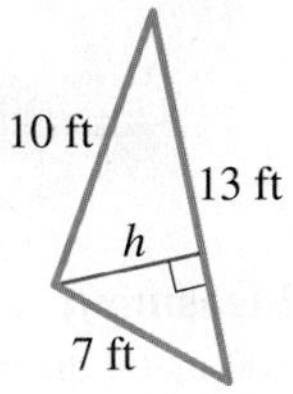

In Exercises 27–30, find the length of side x *for each triangle.*

27.

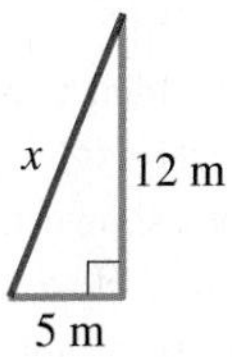

28.

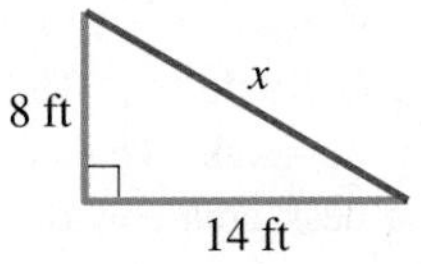

29.

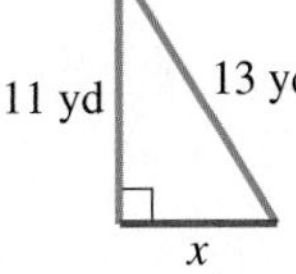

30.

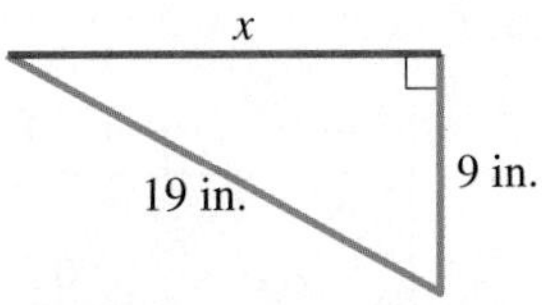

In Exercises 31 and 32, find the area of triangle ABC.

31.

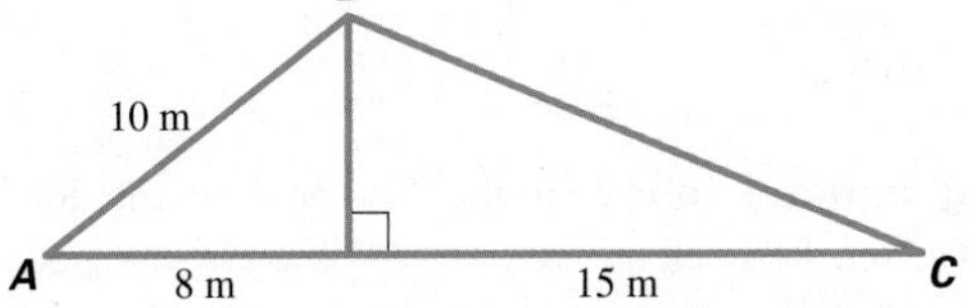

32.

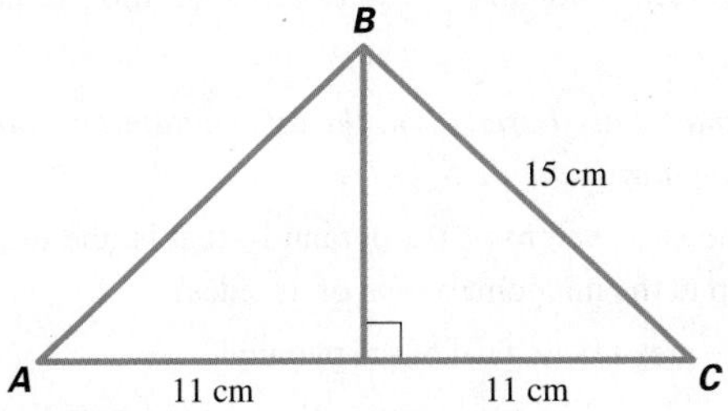

A geoboard is a board with rows of nails spaced 1 inch apart in both the vertical and horizontal directions. In Exercises 33–36, we stretched a rubber band around some of the nails. Find the area of the enclosed figures.

33.

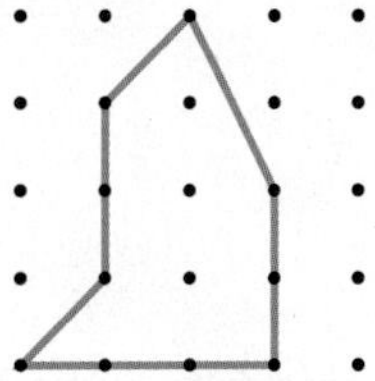

34.

35.

36.

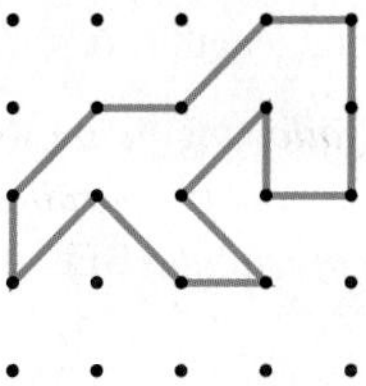

Applying What You've Learned

In Exercises 37–40, state whether perimeter or area would be the more appropriate quantity to measure.

37. You are covering an archeological plot with a tarpaulin.

38. Napoleon wants to buy fencing for a paddock to graze llamas.

39. You are putting a decorative stencil at the top of the walls in your bedroom.

40. You are surfacing your rooftop Japanese garden with slate tiles.

41. Finding distances on a baseball diamond. The bases on a baseball diamond are 90 feet apart. What is the distance from home plate to second base? (All angles in the diamond are right angles.)

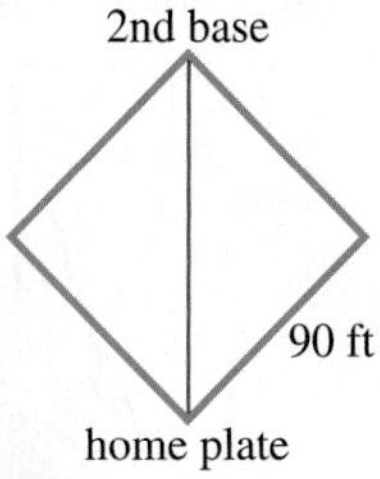

42. Finding the area of home plate. In baseball, home plate is shaped as shown ($m\angle FDE = 90°$).

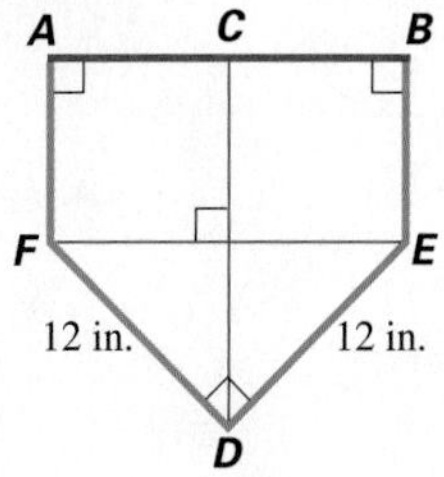

a. What is the length of line segment *AB*?

b. If we assume that line segment *CD* has the same length as segment *AB*, what is the area of home plate?

43. Find the length of line segment AB in the given figure.

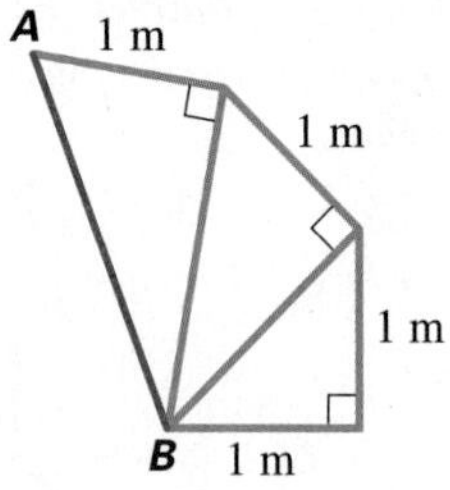

44. Continuing the pattern in Exercise 43, construct a line segment having length $\sqrt{6}$.

Use the following figure to answer Exercises 45 and 46. Assume that the area of the parallelogram ABCD *is 60 square inches and the area of triangle* BEC *is 6 square inches.*

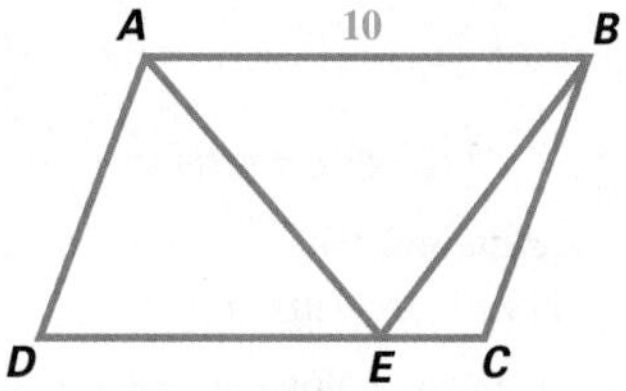

45. What is the area of triangle ADE?

46. What is the area of the trapezoid $ABCE$?

Use the following figure to answer Exercises 47 and 48. Assume that the area of triangle BAE *is 30 square yards, the area of the trapezoid* ABDF *is 66 square yards, and the area of triangle* BDC *is twice the area of triangle* AEF.

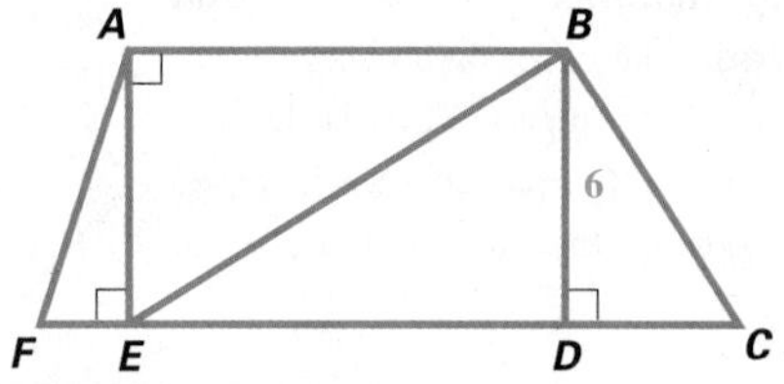

47. Find the area of triangle BDC.

48. Find the area of trapezoid $ABCF$.

49. Making a stained glass window. A stained glass window in a museum is in the shape of a rectangle with a semicircle on top. The height of the rectangular part of the window is twice the base. If the base of the window measures 6 feet, what is the area of the window?

50. Making a stained glass window. Consider a window like the one in Exercise 49. Assume that the semicircular top has a radius of 2 feet. We want the rectangular part of the window to have the same area as the semicircular top. What should the dimensions of the rectangle be?

51. Comparing pizzas. At Cifaretto's Italian Ristorante, the medium pizza has a diameter of 12 inches and sells for \$5.99. The large pizza has a diameter of 16 inches and sells for \$8.99. Which pizza is the better buy? Explain.

52. Making a flower bed. A gardener has enough tiger lilies to fill a circular flower bed having an area of 50 square feet. Find the radius of the flower bed to the nearest foot.

53. Measuring a running track. A running track, 4 meters wide, has the dimensions shown in the following diagram. The ends of the track are semicircles with diameter 20 meters. What is the surface area of the track?

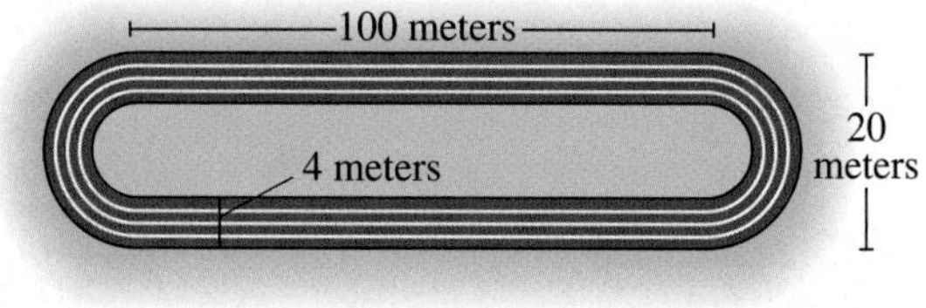

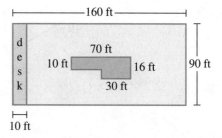

54. Measuring a running track. In Exercise 53, if the 100-meter dimension is increased to 120 meters, the 20-meter dimension is increased to 40 meters, and the width of the track is increased to 6 meters, what is the surface area of the track?

55. Tiling a hotel lobby. The lobby of The Golden Peacock hotel in Bangkok, Thailand, has dimensions 160 by 90 feet. The desk area and floating lily pool are shown in the diagram. (Assume all angles are right angles.) How many square feet of tile will be needed to cover the remaining area?

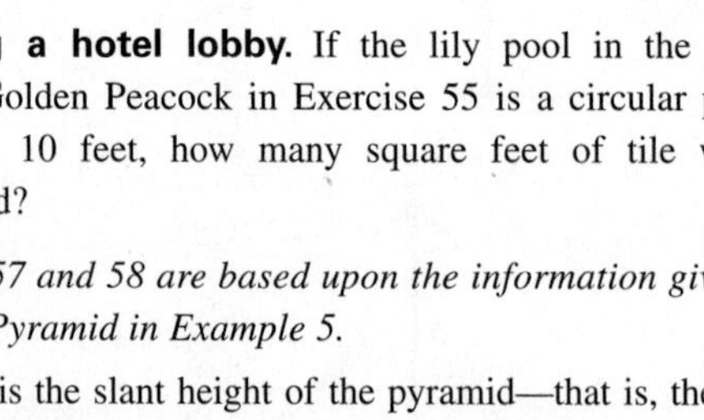

56. Tiling a hotel lobby. If the lily pool in the lobby of The Golden Peacock in Exercise 55 is a circular pool with radius 10 feet, how many square feet of tile would be needed?

Exercises 57 and 58 are based upon the information given about the Great Pyramid in Example 5.

57. What is the slant height of the pyramid—that is, the distance from its tip to the midpoint of one of its sides?

58. What is the area of one face of the pyramid?

59. Redo Example 6, but now assume that the sides of the basketball court are 18 feet long and the angle measures 72°.

60. Redo Example 6. However, now the tip of the court has been removed, as shown in the figure.

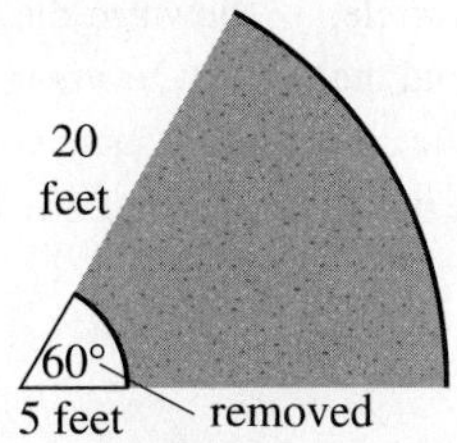

61. **Designing a Japanese garden.** A design for a rectangular Japanese stone garden has a circular pond in its center as shown. How many square yards will be covered with stones?

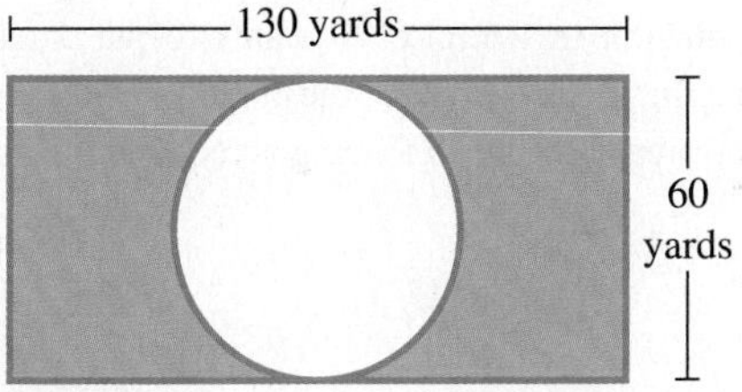

62. **A memorial garden.** A trapezoid-shaped flower bed in a memorial garden contains a statue with a round base having a diameter of 8 feet and centered inside the trapezoid as shown. The length of the top of the trapezoid is twice the diameter of the statue and the bottom is 4 feet longer than the top. What is the area outside the statue?

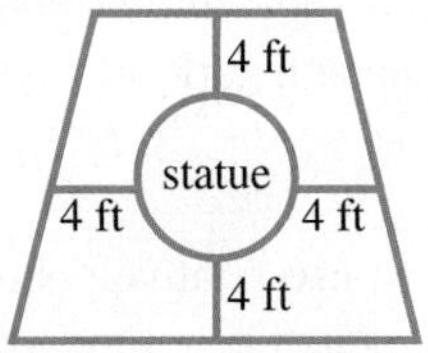

Communicating Mathematics

63. In this section, we derived formulas for the areas of the following figures. Arrange the terms in the order in which we did the derivation: triangle, trapezoid, rectangle, parallelogram.

In Exercises 64–66, state how you can remember the formula for the areas of each of the following.

64. a parallelogram

65. a triangle

66. a trapezoid

Between the Numbers

67. How many times larger is the area of a circle with a circumference of 6 inches than the area of a square with perimeter 6 inches?

68. Suppose you bend a 1-foot-long wire to make different regular polygons. What do you think happens to the area of the polygons as the number of sides increases? What shape do you think you could make with the wire to enclose the greatest area?

Challenge Yourself

69. A modern art museum has sides shaped like trapezoids with identical equilateral triangular windows with dimensions shown in the accompanying figure. What is the area of one side of the museum, excluding the area of the windows?

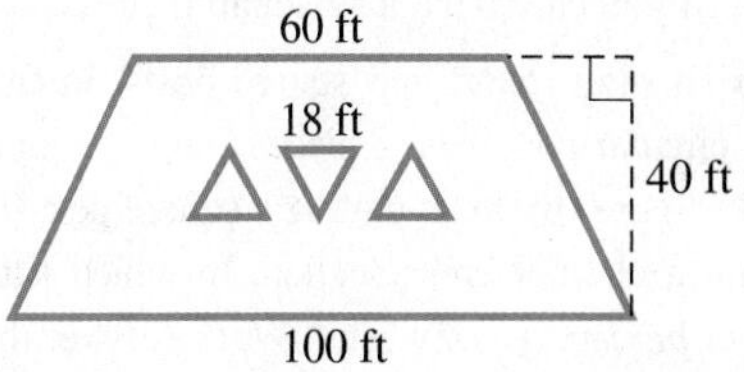

70. One side of an air traffic control tower, with dimensions given, is shown in the accompanying figure. Assume that all polygons are trapezoids. What is the area of the side, excluding the two congruent windows?

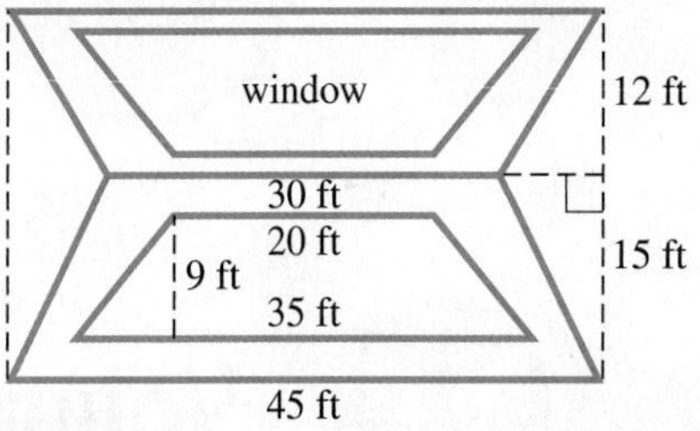

71. **True or false?** If we double the radius of a circle, the area also doubles.

72. **True or false?** If we double the radius of a circle, the circumference also doubles.

73. **Acres of pizza.** It is estimated that Americans eat 100 acres' worth of pizza a day. Let's assume that they are large pizzas with diameters of 14 inches.

a. Find the number of square inches in a large pizza.

b. An acre has 43,560 square feet and a square foot has 144 square inches. How many square inches are there in 100 acres?

c. How many large pizzas do Americans eat each day?

d. If a large pizza has eight slices, estimate how many slices per second Americans eat.

74. **Domino's delivers on Super Bowl Sunday.** If Domino's expects to deliver 9 million slices of pizza on Super Bowl Sunday, use the information in Exercise 73 to estimate how many acres' worth of pizza Domino's will deliver.

75. In the figure below, W, X, Y, and Z are the midpoints of the line segments on which they lie. How does the area of $WXYZ$ compare with the area of rectangle $ABCD$? Explain how you got your answer.

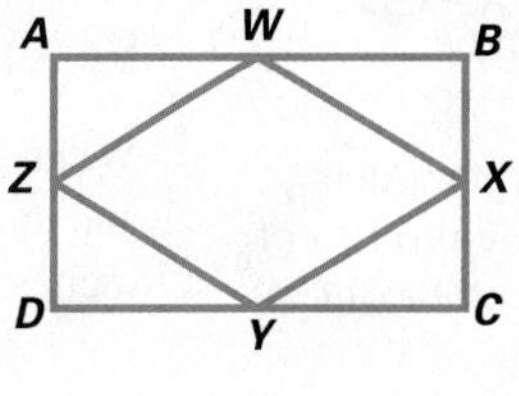

76. If W and Y were not the midpoints of the line segments on which they lie, would it change your answer in Exercise 75? Explain your answer.

77. You have 200 feet of fencing and want to enclose a rectangular area. What shape will enclose the most area? You might solve this problem by experimenting with different lengths and widths until you can make a reasonable guess.

78. Bernie is shaping round and square posts. In figure (a), he is cutting a circular post from a piece of wood with a square cross section. In figure (b), he is cutting a square post from a piece of wood with a circular cross section. In which situation is there the smaller *percentage* of waste? (*Hint:* Answer this question by making up numerical examples.)

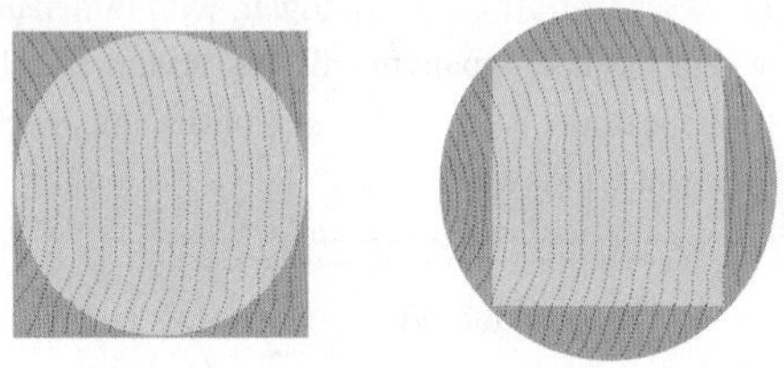

(a) (b)

79. *Extreme Makeover Home Edition* wants to create a housing development with lots clustered in circles and an open area in the center of the circle, as shown in the diagram. The side of a lot is 180 feet and the central common area is 11,304 square feet. Each lot in the cluster is the same size. What is the area of each lot? (We are using 3.14 for π.)

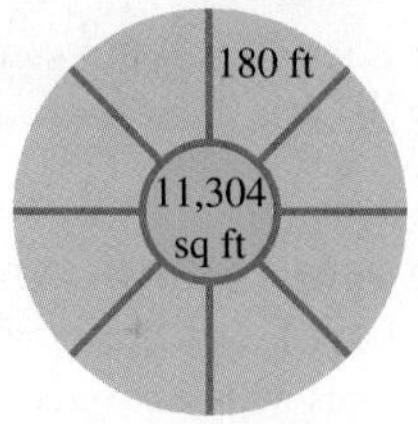

80. Ty Pennington in Exercise 79 wants to put a fence entirely around a single lot, except for the curved section away from the common area. How much fencing is required?

9.4 Volume and Surface Area

Objectives

1. Understand the idea of volume of basic three-dimensional objects.
2. Be able to apply the volume and surface area formulas for cylinders.
3. Understand the relationship between the volume and surface area formulas for cylinders, spheres, and cones.

Many times friends and relatives have asked me to use relatively simple mathematics to help them solve real problems in their lives.* For example, several years ago, a colleague in the psychology department asked me to help her determine if her oil deliveryman was cheating her. In order to solve her problem, I needed to know the basic formulas for computing volumes that you will learn in this section.

We will now extend the ideas of two-dimensional Euclidean geometry to three dimensions. Instead of squares and rectangles, we will discuss boxlike figures called *parallelepipeds*; instead of circles we will talk about cylinders, spheres, and cones; and instead of measuring perimeters and areas, we will calculate the volumes and surface areas of three-dimensional geometric figures.

Many applications require a knowledge of three-dimensional geometry. When designing a skyscraper, an architect needs to know the volume of concrete to pour for a foundation. A manufacturer must be able to compute the amount of materials required to fabricate a cylindrical water tank. A state highway department engineer may want to know how much road salt is contained in a conical storage shed.

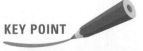

KEY POINT

The volume of a rectangular parallelepiped is the product of its length, width, and height.

Volume

When we measure length in one dimension, we use units such as inches, feet, centimeters, and meters. In measuring area in two dimensions, we use square units such as square inches and square centimeters. We measure the **volume** of a three-dimensional figure using *cubic units.* Even though we will study three-dimensional geometric objects, *we will still measure surface area in square units.*

*In Example 6 of Section 9.3, a friend asked me for help in laying out a strangely shaped basketball court for his son. The star problem in Example 3 of Section 9.2 was from a minister asking me to advise his youth group on how to construct a five-pointed star to place on their church steeple during Advent.

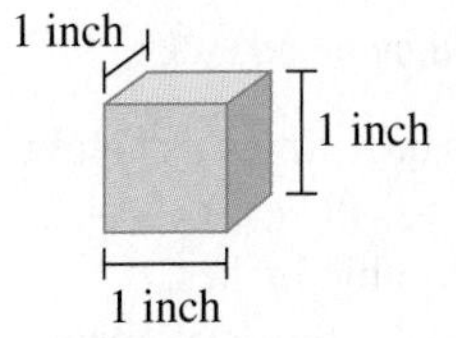

FIGURE 9.47 A cube whose volume is 1 cubic inch.

Figure 9.47 shows a cube whose edges are each 1 inch long, so the cube has a volume of 1 cubic inch. In determining the volume of a three-dimensional figure whose measurements are given in inches, we are really asking how many of these 1-inch cubes will fit inside the figure. Because it might be impossible to fit these cubes inside the figure exactly, we can imagine filling 1-cubic-inch containers with water and asking, "How many of these containers of water does it take to fill the figure exactly?"

It is easy to compute the volume of the box-shaped rectangular solid, called a *rectangular parallelepiped*,* shown in Figure 9.48. If we slice along the lines drawn at 1-inch intervals on the solid, we get four layers each containing 15 1-inch cubes. The volume of this solid is given by

$$\text{volume} = \text{length} \times \text{width} \times \text{height} = 5 \times 3 \times 4 = 60 \text{ cubic inches.}$$

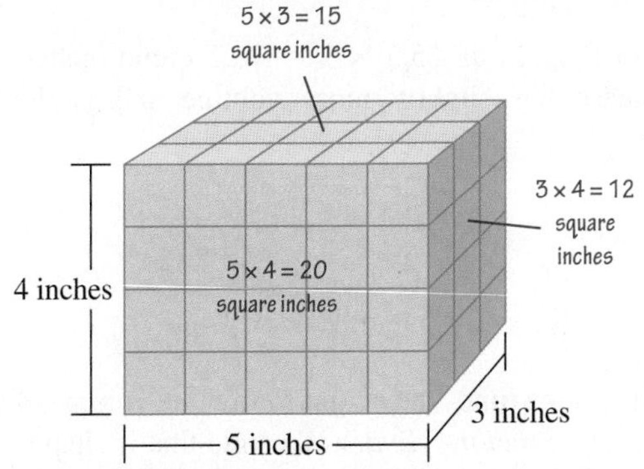

FIGURE 9.48 Rectangular solid sliced into 60 one-inch cubes.

It is also easy to find the surface area of the solid in Figure 9.48. Clearly the six sides of the solid have the following areas:

$$\text{areas of top and bottom} = \text{length} \times \text{width} = 5 \times 3 = 15,$$

$$\text{areas of front and back} = \text{length} \times \text{height} = 5 \times 4 = 20, \text{ and}$$

$$\text{areas of two sides} = \text{width} \times \text{height} = 3 \times 4 = 12.$$

The total surface area is therefore $(2 \times 15) + (2 \times 20) + (2 \times 12) = 94$ square inches. Figure 9.48 helps us remember these formulas for volume and surface area. 13

QUIZ YOURSELF 13

Find the volume and surface area of a rectangular solid with length 8 centimeters, width 6 centimeters, and height 3 centimeters.

VOLUME AND SURFACE AREA OF A RECTANGULAR SOLID (RECTANGULAR PARALLELEPIPED) If a rectangular solid has length l, width w, and height h, the volume of the solid is $V = lwh$ and the surface area of the solid is

$$S = 2lw + 2lh + 2wh.$$

Another way to look at the way we calculated the area for the rectangular solid above is that we multiplied the area of the base, which is lw, times the height, h. We can apply this approach to other figures.

KEY POINT

For many solids, volume equals base area times height.

VOLUME EQUALS AREA OF BASE TIMES HEIGHT If an object has a flat top and base and *sides perpendicular to the base*, as shown in Figure 9.49, then, if the area of the base is A and the height is h, the volume will be

$$V = A \cdot h.$$

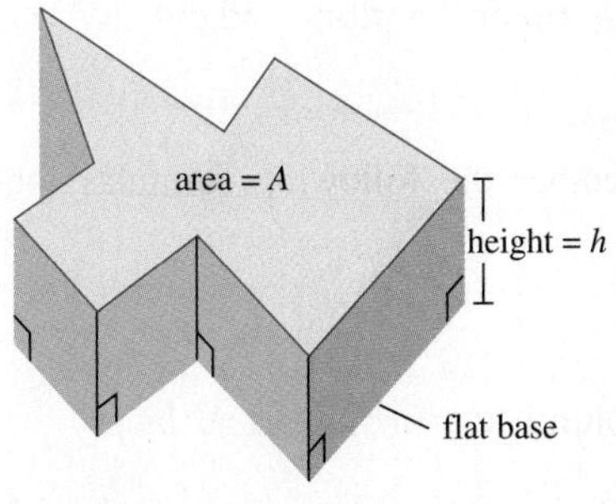

FIGURE 9.49 Volume = area of base times height.

EXAMPLE 1 Comparing Volumes

Figure 9.50 shows two blocks of cheese that are selling for the same price. Which block contains the greater volume?

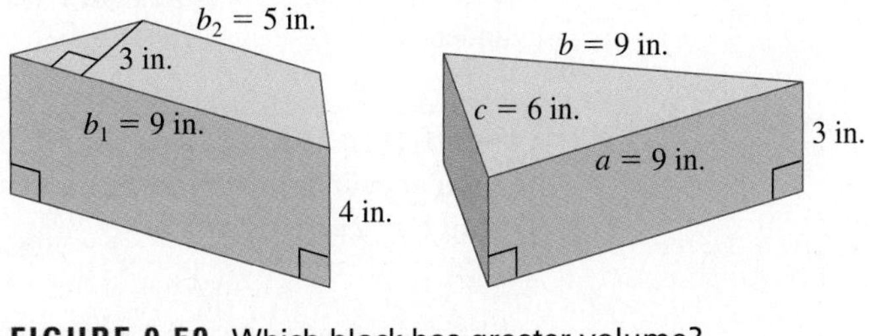

FIGURE 9.50 Which block has greater volume?

*A parallelepiped is a solid with six faces, each of which is a parallelogram. In this case, the parallelepiped is rectangular, which means the faces are rectangles.

SOLUTION: Review the Relate a New Problem to an Older One Strategy on page 9.

We calculate the volume of each block by multiplying the area of its base times its height. The area of the base of the trapezoidal block is $\frac{1}{2}(b_1 + b_2) \cdot h = \frac{1}{2}(9 + 5) \cdot 3 = 21$ square inches. Therefore, this block has a volume of $21 \times 4 = 84$ cubic inches.

We use Heron's formula to find the area of the base of the triangular block of cheese. We first must compute $s = \frac{1}{2} \times (9 + 9 + 6) = 12$. Then the area is

$$\sqrt{s(s-a)(s-b)(s-c)} = \sqrt{12(12-9)(12-9)(12-6)} = \sqrt{648} \approx 25.5.$$

Multiplying this area by the height of 3 gives us $25.5 \times 3 = 76.5$ cubic inches. Therefore, the trapezoidal block of cheese has slightly more volume and is the better deal.

Now try Exercises 9 to 14.

Cylinders

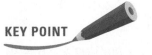

KEY POINT

Simple diagrams illustrate the formulas for finding the volume and surface area of a cylinder.

We can use this method of multiplying the base area times the height to find the volume of some common three-dimensional solids. A *right circular cylinder* is a solid that is shaped like a soup or tuna fish can (see Figure 9.51). We call these cylinders "right" because the sides are perpendicular to the base. The cylinder's base is a circle with area $A = \pi r^2$.* Thus,

$$\text{volume of cylinder} = \text{area of base} \times \text{height} = \pi r^2 \cdot h.$$

To find the surface area of the cylinder in Figure 9.51(a), imagine that we remove the top and bottom of the cylinder and then cut the side of the cylinder and open it up, as in Figure 9.51(b). If we flatten this curved surface, we get a rectangle with length $2\pi r$ and height h, shown in Figure 9.51(c). Therefore, the surface area of the side of the cylinder is $2\pi rh$. Adding to this the areas of the top and bottom of the cylinder, which are each πr^2, we get the total surface area of the cylinder.

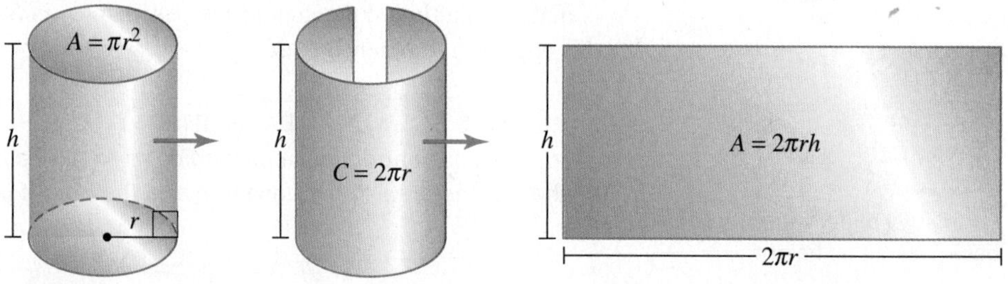

(a) circular cylinder **(b)** curved surface **(c)** flattened surface

FIGURE 9.51 (a) Right circular cylinder. (b) Remove top and bottom and cut side of cylinder. (c) Flatten side of cylinder.

The diagrams in Figure 9.51 will help you to remember the following formulas for cylinders.

> **VOLUME AND SURFACE AREA OF A RIGHT CIRCULAR CYLINDER** A right circular cylinder with radius r and height h has volume $V = \pi r^2 h$ and surface area $S = 2\pi rh + 2\pi r^2$.

We can use these formulas to compare purchases.

*Throughout this section, we again approximate π by 3.14.

EXAMPLE 2 Comparing Volumes of Commercial Products

A Pearl art supply store sells paint solvent in a cylindrical can that has a diameter of 4 inches and a height of 6 inches. The giant economy size can of the same solvent has a diameter twice as large (the height of the can is the same) and costs three times as much as the smaller can.

a) Which can is the better deal? b) Compare the surface areas of the two cans.

SOLUTION:

a) The radius of the smaller can is 2, so its volume is

$$V = \pi r^2 h \approx 3.14 \cdot (2)^2 \cdot 6 = 75.4 \text{ cubic inches.}$$

The radius of the larger can is 4, so its volume is

$$V = \pi r^2 h \approx 3.14 \cdot (4)^2 \cdot 6 = 301.4 \text{ cubic inches.}$$

Because the large can contains *four* times as much solvent but costs only *three* times as much, the larger can is the better deal.

b) The surface area of the smaller can is

$$2\pi rh + 2\pi r^2 = 2\pi(2)(6) + 2\pi(2)^2 \approx 2(3.14) \cdot 2 \cdot 6 + 2(3.14)(2)^2 = 100.5 \text{ square inches.}$$

The surface area of the larger can is

$$2\pi rh + 2\pi r^2 = 2\pi(4)(6) + 2\pi(4)^2 \approx 2(3.14) \cdot 4 \cdot 6 + 2(3.14)(4)^2 = 251.2 \text{ square inches.}$$

Now try Exercises 29 and 30. 14

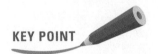

QUIZ YOURSELF 14

What is the volume and surface area of a right circular cylinder with a radius of 5 centimeters and height of 10 centimeters?

You may have been surprised in Example 2 that although the larger can contains four times the amount of solvent contained in the smaller can, it takes only two and a half times as much material to make the larger can. Manufacturers are interested in such relationships so they can minimize the amount of materials they need to package a product and, by doing so, reduce their cost. What is the most efficient shape of a container to contain the largest amount of a product? For example, what is the most efficient shape of a soup can?

KEY POINT

An efficient container has a small surface-area-to-volume ratio.

EXAMPLE 3 The Most Efficient Shape of a Can

What are the dimensions of a can that will contain 1 cubic foot of liquid and that will have the smallest amount of surface area?

SOLUTION:

We must find the radius, r, and the height, h, of the can. Our intuition tells us that if the radius is small, as in Figure 9.52(a), then the height must be large. On the other hand, if the radius is large, as in Figure 9.52(b), then the height must be small. Our problem then is to find the ideal radius that gives the minimum surface area (Figure 9.52(c)).

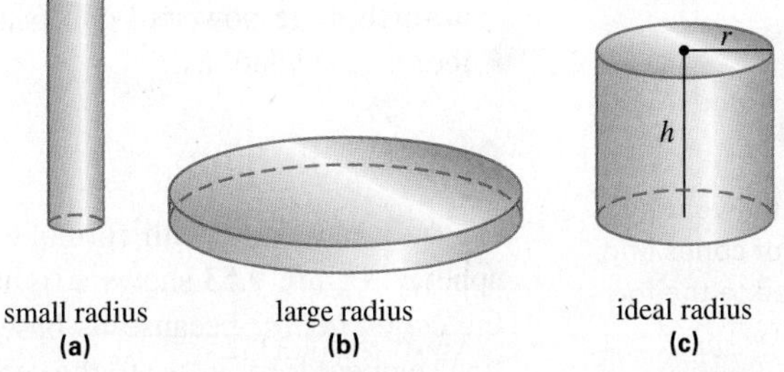

FIGURE 9.52 A can with a small radius requires a large height, whereas a can with a large radius requires a small height.

Because the volume of the can is to be 1, we can set $\pi r^2 h = 1$. Dividing both sides of this equation by πr^2, we get $h = \frac{1}{\pi r^2}$. Thus, as our intuition tells us, the height depends on the choice of radius.

Recall that the formula for finding the surface area of a cylinder with radius r and height h is

$$S = 2\pi rh + 2\pi r^2.$$

area of side (arrow to $2\pi rh$); area of top and bottom (arrow to $2\pi r^2$)

Substituting $\frac{1}{\pi r^2}$ for h, we get

$$S = 2\pi r\left(\frac{1}{\pi r^2}\right) + 2\pi r^2 = \frac{2\pi r}{\pi r^2} + 2\pi r^2 = \frac{2\not{\pi}\not{r}}{\not{\pi} r^{\not{2}}} + 2\pi r^2 = \frac{2}{r} + 2\pi r^2.$$

We can cancel π and r.

This simpler expression for the surface area S depends *only* on the radius r.

So we see that $S = \frac{2}{r} + 2\pi r^2$.

The neat thing about this equation is that now we have a formula for the surface area of our can that depends *only on the radius* r. So, as we vary r, we can calculate how the surface area changes. In Table 9.4, we begin with a very small radius, 0.1, which we expect would give us a large surface area. Then as the size of the radius increases, the surface area decreases. Eventually, as the radius becomes too large, the can is becoming too flat, and as Table 9.4 shows, the surface area starts to increase again.

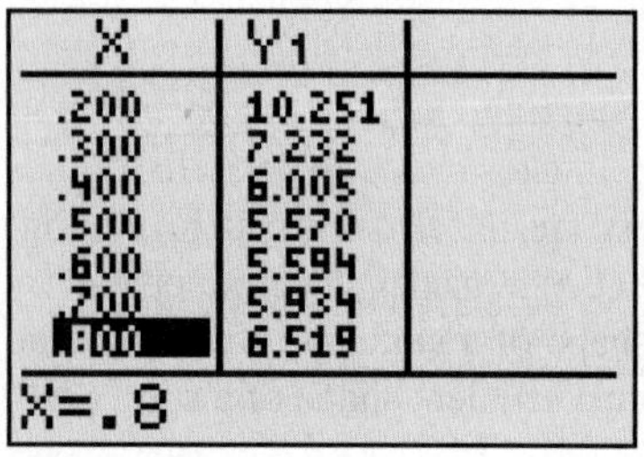

This graphing calculator screen confirms our calculations in Example 3.

Radius, r (feet)	Surface Area, $S = \frac{2}{r} + 2\pi r^2$ (square feet)
0.1	20.063
0.2	10.251 decreasing
0.3	7.232 decreasing
0.4	6.005 decreasing
0.5	5.570 decreasing
0.6	5.594 increasing
0.7	5.934 increasing
0.8	6.519 increasing

Ideal radius is between 0.5 and 0.6.

TABLE 9.4 Surface area first decreases, then increases as r increases.

You can see in Table 9.4 that the ideal radius is somewhere between 0.5 and 0.6. To get a better estimate of the ideal radius, we could use a graphing calculator to calculate the surface area for radii of sizes 0.51, 0.52, 0.53, and so on. If you were to do this, eventually you would find that $r \approx 0.542$ gives us the smallest surface area for a can with volume 1. (Actually, if you were ever to take a calculus course, you would learn a much more powerful and faster way to find this value for r without doing so many tedious calculations.)

KEY POINT

The formulas for the volume and surface area of cones and spheres are related to the formulas for cylinders.

Cones and Spheres

We will now work with formulas for the volume of two other solid figures: cones and spheres. Figure 9.53 shows a right circular cone with height h and base radius r. We call the cone *circular* because its base is a circle, and we use the adjective *right* because the line segment from its tip to the center of its base forms a right angle with the base.

VOLUME AND SURFACE AREA OF A RIGHT CIRCULAR CONE A right circular cone with height h and base radius r has volume $V = \frac{1}{3}\pi r^2 h$ and surface area $S = \pi r\sqrt{r^2 + h^2}$.

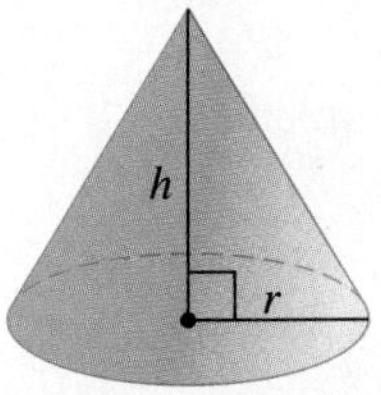

FIGURE 9.53 Right circular cone.

In the formula for the surface area of a cone, if we want to include the area of the base we must add πr^2.

PROBLEM SOLVING

Strategy: The Analogies Principle

We can use what we know about the volume of a cylinder to understand the formula for the volume of a cone. Figure 9.54 shows that a cone with height h and radius r easily fits inside a cylinder with height h and radius r. The cone does not appear to occupy even one-half of the volume of the cylinder, which is $V = \pi r^2 h$. Therefore, it is easy to remember that the volume of the cone is $V = \frac{1}{3}\pi r^2 h$.*

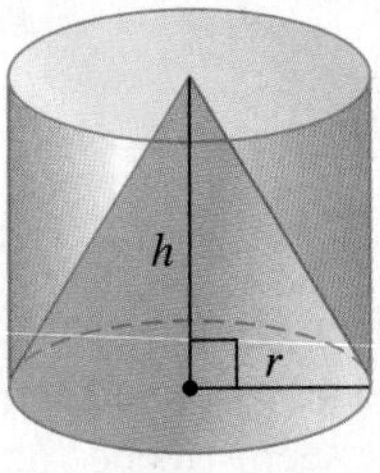

FIGURE 9.54 The cone has a volume $\frac{1}{3}$ that of the cylinder.

EXAMPLE 4 Finding the Volume and Surface Area of a Cone

What is the volume and surface area of a cone with a height of 10 meters and a base radius of 8 meters?

SOLUTION:

The volume is $V = \frac{1}{3}\pi r^2 h = \frac{1}{3}\pi(8)^2(10) = \frac{640\pi}{3} \approx 669.87$ cubic meters. The surface area is

$$S = \pi r\sqrt{r^2 + h^2} = \pi(8)\sqrt{8^2 + 10^2} = 8\sqrt{164}\pi \approx 321.69 \text{ square meters.}$$

Now try Exercises 3 and 6. 15

QUIZ YOURSELF 15

Find the volume and surface area of a right circular cone with a radius of 4 yards and a height of 5 yards.

EXAMPLE 5 The Volume of Conical Storage Buildings

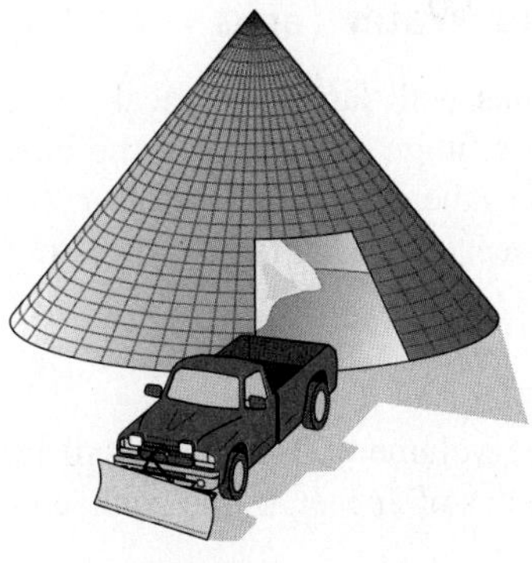

Many northern states stockpile salt to use for melting road ice. In one such state, the highway department builds sheds in the shape of right circular cones to store the salt. Current sheds have a diameter of 30 feet and a height of 10 feet. The department plans to build larger conical sheds to hold more salt. The larger shed will either have a 10-foot-longer diameter and the same height, or the same diameter and a 10-foot-greater height.

a) Calculate the volume for each of the proposed sheds.
b) Calculate the surface area for each of the proposed sheds.
c) Which design seems to be more economical?

SOLUTION:

a) *Keep height at 10 feet and increase diameter to 40 feet*: The radius of the cone will be $\frac{40}{2} = 20$ feet and the height is 10 feet, so the volume is

$$V = \frac{1}{3}\pi r^2 h = \frac{1}{3}\pi(\underbrace{20}_{\text{radius}})^2(\underbrace{10}_{\text{height}}) \approx 4{,}187 \text{ cubic feet.}$$

*Of course, we cannot tell just by looking at Figure 9.54 that the cone occupies exactly one-third of the cylinder; however, visualizing this diagram may help you remember the formula for the volume of a cone.

Keep diameter at 30 feet and increase height to 20 feet:

The radius of the cone is 15 feet and the height is now 20 feet, so the volume is

$$V = \frac{1}{3}\pi r^2 h = \frac{1}{3}\pi \overset{\text{radius}}{(15)^2}\overset{\text{height}}{(20)} \approx 4{,}710 \text{ cubic feet.}$$

b) The surface area for the shed with radius 20 and height 10 is

$$S = \pi(r)\sqrt{r^2 + h^2} = \pi(20)\sqrt{20^2 + 10^2} \approx 1{,}404 \text{ square feet.}$$

A shed with radius 15 and height 20 has a surface area of

$$S = \pi(r)\sqrt{r^2 + h^2} = \pi(15)\sqrt{15^2 + 20^2} \approx 1{,}178 \text{ square feet.}$$

c) Increasing the height by 10 feet and keeping the diameter at 30 feet gives us more volume and smaller surface area than if we increase the diameter to 40 feet and keep the height at 10 feet. Therefore, increasing the height to 20 feet seems to be the better design.

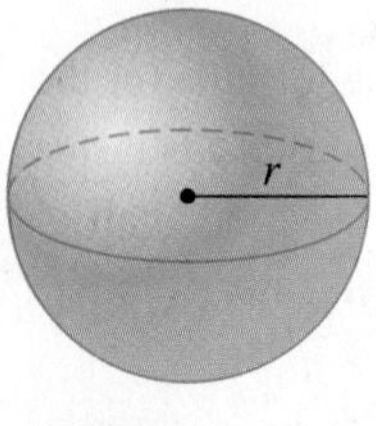

FIGURE 9.55 A sphere with radius r.

The last three-dimensional figure we consider is a sphere (Figure 9.55).

VOLUME AND SURFACE AREA OF A SPHERE A sphere with radius r has volume $V = \frac{4}{3}\pi r^3$ and surface area $S = 4\pi r^2$.

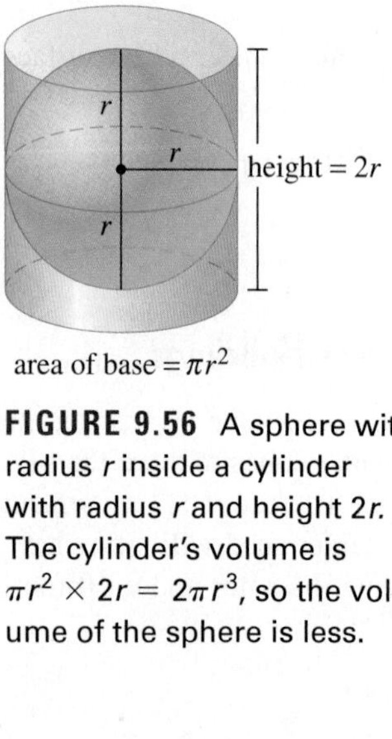

FIGURE 9.56 A sphere with radius r inside a cylinder with radius r and height $2r$. The cylinder's volume is $\pi r^2 \times 2r = 2\pi r^3$, so the volume of the sphere is less.

Figure 9.56* will help you remember the formula for the volume of a sphere. A sphere with radius r fits exactly inside a cylinder having radius r and height $2r$. The volume of this cylinder is $\pi r^2 \times 2r = 2\pi r^3$. Because the sphere does not completely fill the cylinder, you should remember that the volume of the sphere is *less than* $2\pi r^3$, or $\frac{4}{3}\pi r^3$.

Many containers, such as water tanks, are shaped like spheres.

EXAMPLE 6 Measuring the Capacity of Water Tanks

Metrodelphia is replacing existing spherical water tanks with larger spherical water tanks. The water commission insists that to provide for future expansion of the city, the new capacity of the tanks should be at least five times the capacity of the old water tanks. If the city purchases tanks that have a radius that is twice the radius of the old tanks, will these tanks satisfy the water commission?

SOLUTION:

Let the radius of the old water tanks be r feet, so the volume of each tank will be $V = \frac{4}{3}\pi r^3$ cubic feet. The new tanks will each have a radius of $2r$ feet, so the volume of the new tanks will be

$$V = \frac{4}{3}\pi \underbrace{(2r)^3}_{\text{New radius is twice old radius.}} = \frac{4}{3}\pi 8r^3 = 8\underbrace{\left(\frac{4}{3}\pi r^3\right)}_{\text{original volume}} \text{cubic feet.}$$

The new volume is eight times the volume of the original tanks. Therefore, the new tanks will exceed the specifications of the water commission.

Now try Exercises 4 and 7. 16

QUIZ YOURSELF 16

Find the volume of a sphere with a radius of 6 centimeters.

*Again, we can't tell by just looking at Figure 9.56 that the volume of the sphere is exactly $\frac{4}{3}\pi r^3$, but the diagram may help you remember that the sphere's volume is somewhat less than $2\pi r^3$.

EXERCISES 9.4

Estimate π by 3.14 in all exercises where needed.

Sharpening Your Skills

In Exercises 1–8, find a) the surface area and b) the volume of each figure. Approximate π by 3.14.

1.

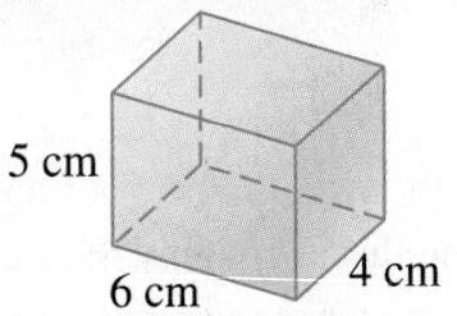

2.

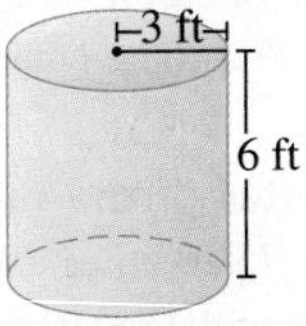

3.

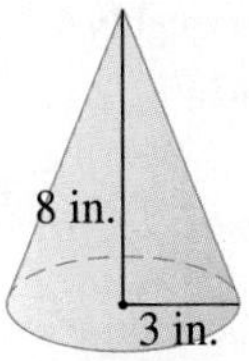

4.

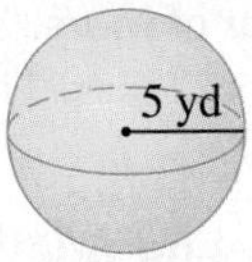

5.

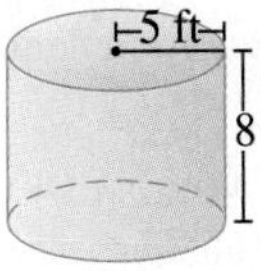

6.

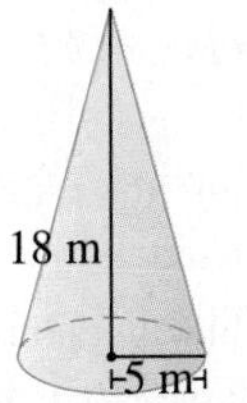

7.

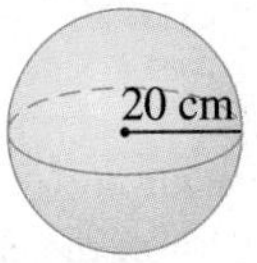

8.

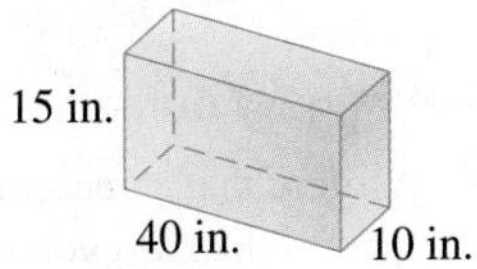

In Exercises 9–14, find the volume of each figure.

9.

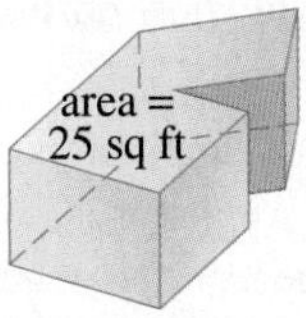

10.

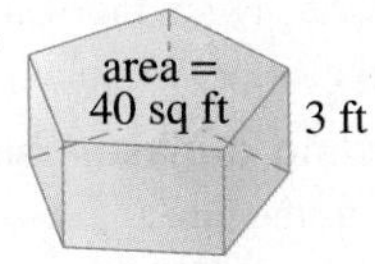

11.

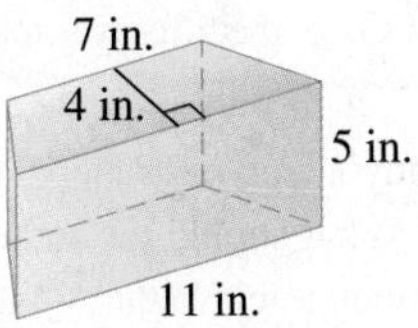

12.

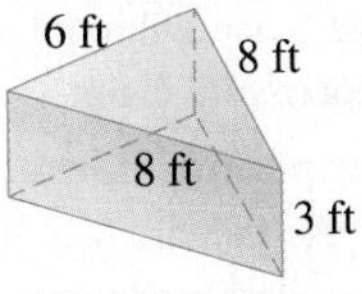

13.

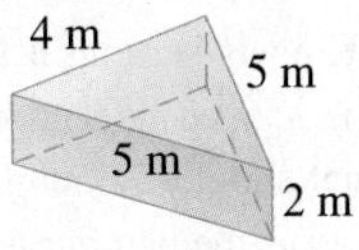

14.

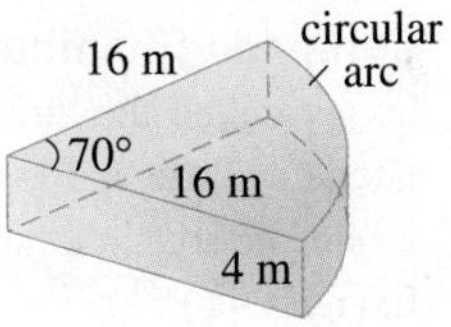

Applying What You've Learned

15. How many cubic inches are in a cubic foot?

16. How many cubic inches are in a cubic yard?

17. Punch bowl. A punch bowl is in the shape of a hemisphere (half of a sphere) with a radius of 9 inches. The cup part of a ladle is also in the shape of a hemisphere with a radius of 2 inches. If the bowl is full, how many full ladles of punch are there in the bowl?

18. Filling glasses. Assume that we are filling cylindrical glasses that have a diameter of 3 inches and a height of 4 inches from the punch bowl in Exercise 17. If the bowl is completely full, how many glasses can we fill?

19. Filling glasses. A cylindrical pitcher with radius 2.5 inches and 8 inches high is filled with a tropical drink. How many glasses shaped like an inverted cone with a height of 1.5 inches and radius 2 inches can be filled from this pitcher?

20. Filling juice containers. The Great American Orange Juice Factory is selling its Caribbean Orange Cocktail in collectible spherical containers. A standard half-gallon container of juice contains 115.5 cubic inches. How many of the spherical containers with diameters of 3 inches can be filled from a standard container?

21. Building a Japanese garden. Mariko needs about 16 wheelbarrows full of stone for her Japanese garden. Her wheelbarrow has vertical sides with the given shape and is $2\frac{1}{2}$ feet wide. How many cubic yards of stone should she order?

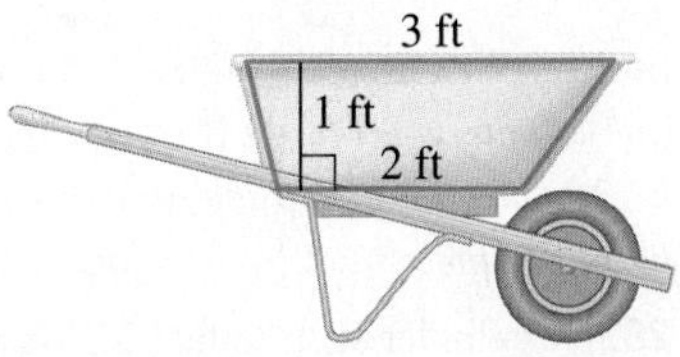

22. Juice barrel. A gallon contains 231 cubic inches. A cylindrical barrel full of juice concentrate has a diameter of 2 feet and is 3 feet high. How many gallons of concentrate are in the barrel?

23. Comparing cakes. One cake is rectangular with a length of 14 inches, a width of 9 inches, and a height of 3 inches. The other cake is a round cake with a radius of 5 inches and a height of 4 inches. Which cake has more volume?

24. Comparing ice cream scoops. Which has more volume: a single scoop of ice cream made with a 3-inch scoop, or two scoops of ice cream made with a $2\frac{1}{2}$-inch scoop? (Assume that both ice cream scoops form perfectly filled spheres of ice cream.)

Find the volume of the objects shown in Exercises 25–28.

25.

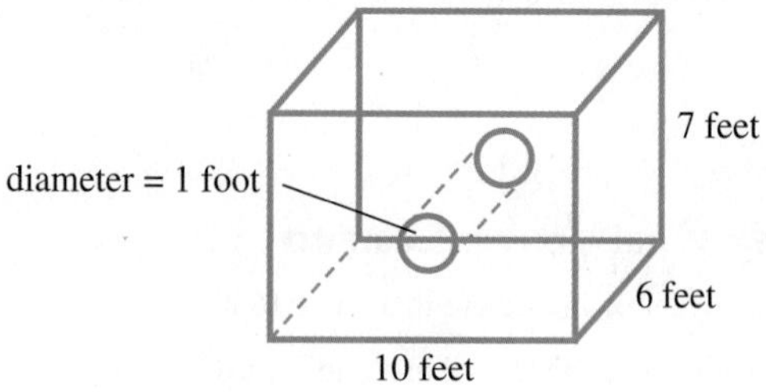

26.

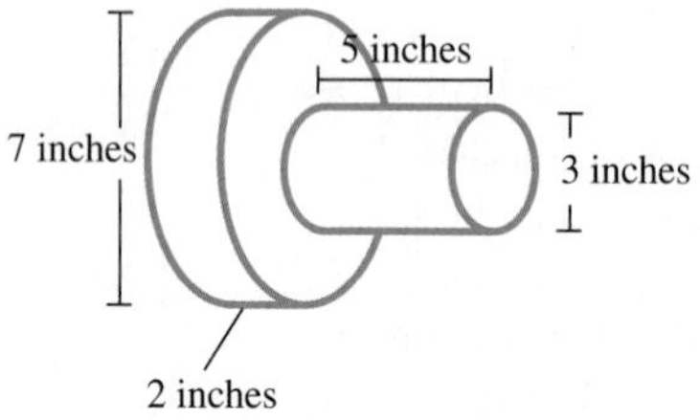

27.

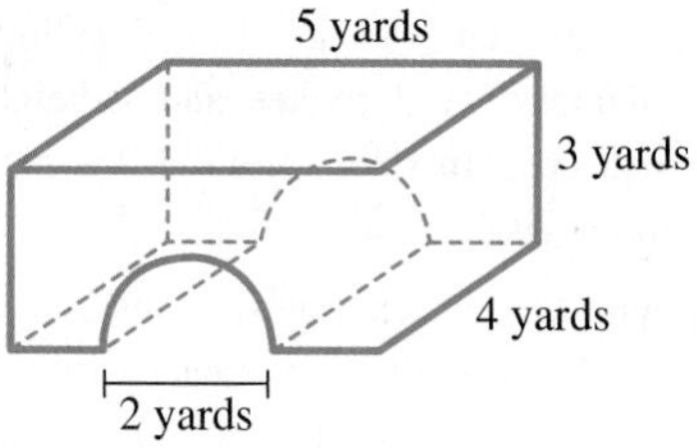

28.

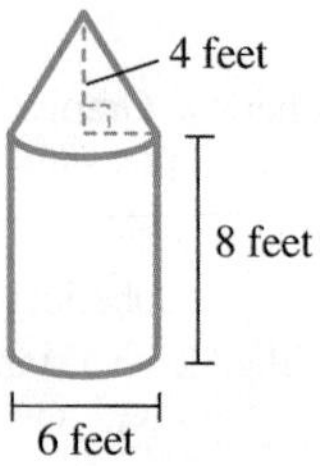

In Exercises 29–32, we describe a geometric object. Compute the volume you obtain if a) you increase the radius of the object by 2 inches and keep the height the same and b) you increase the height of the object by 2 inches and keep the radius the same.

29. A cylinder with radius 10 inches and height 5 inches. Which causes the volume to increase more: increasing its radius or height? Explain.

30. A cylinder with radius 4 inches and height 8 inches. Which causes the volume to increase more: increasing its radius or height? Explain.

31. A cone with radius 6 inches and height 3 inches. Which causes the volume to increase more: increasing its radius or height? Explain.

32. A cone with radius 5 inches and height 12 inches. Which causes the volume to increase more: increasing its radius or height? Explain.

33. Building an aviary. Luis plans to build a hummingbird aviary with a concrete floor shaped like a regular hexagon. Each side will measure 8 feet, as shown in the diagram, and the floor will be 6 inches deep. How many cubic feet of concrete are needed? Assume that Luis cannot buy a fraction of a cubic foot. (*Hint:* Divide the hexagon into equilateral triangles and use Heron's formula to find the area of the top surface of the aviary.)

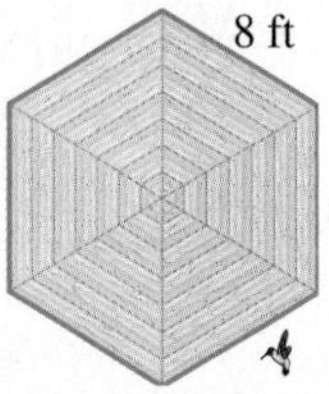

34. Constructing a patio. A homeowner is constructing a concrete patio that is 20 feet wide, 30 feet long, and 6 inches thick. Mixed concrete is sold by the cubic yard. How many cubic yards are needed for the project? Assume that the homeowner cannot buy a fraction of a cubic yard of mixed concrete. (*Hint:* You must convert cubic feet into cubic yards.)

35. Diameter of the moon. Earth has a diameter of approximately 7,920 miles and a volume that is roughly 49 times the volume of the moon. Find the diameter of the moon.

36. Diameter of Mars. The volume of Earth is roughly 6.7 times the volume of Mars. Find the diameter of Mars.

Communicating Mathematics

37. How does Figure 9.54 help you remember the formula for the volume of a right circular cone?

38. How does Figure 9.56 help you remember the formula for the volume of a sphere?

39. If we double the radius of a right circular cone, what effect does that have on the volume? Explain your answer.

40. If we double the radius of a sphere, what effect does that have on the volume? Explain your answer.

Challenge Yourself

41. Assume that three tennis balls fit in a can with no room to spare. Which do you think is larger—the height of the can or the circumference of the can? Explain your answer.

42. A spherical tank has a certain volume. If another spherical tank has twice the volume of the first tank, how do the two radii compare? Explain why this is so.

43. Cooling drinks. Assume that ice cools a liquid proportionally to the amount of surface area of the ice that is in contact with the liquid. Determine whether a large ice cube or a number of smaller ice cubes whose total volume is equal to the larger cube will cool a drink faster. Give specific examples to support your claim.

44. Comparing hamburgers. Wendy's serves square hamburgers that are $4 \times 4 \times \frac{1}{4}$ inches. What should the radius be of a round hamburger that is also $\frac{1}{4}$-inch thick so that it has the same volume?

45. Comparing hamburgers. Burger King is serving a hamburger called the "tri-burger," which is shaped like an equilateral triangle and is $\frac{1}{4}$-inch thick. What should the length of its side be so that it has the same volume as the hamburgers in Exercise 44?

46. If we cut off the top of a cone by making a horizontal slice, we get a figure called a *frustum* of a cone, as shown in the accompanying figure. Suppose that we cut off the top half of a cone that has radius r and height h. What is the volume of the remaining frustum? (*Hint:* The radius of the top of the frustum is $\frac{r}{2}$.)

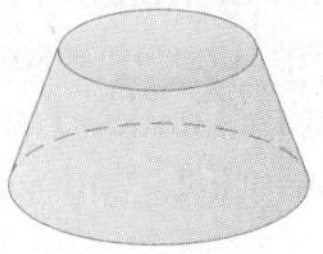

The given figure shows a pyramid with a square base; however, pyramids can have other polygons as their base. The line labeled h *indicates the height of the pyramid. A pyramid whose base has area* B *and that has height* h *has volume* $V = \frac{1}{3}Bh$. *Find the volume of the pyramids specified in Exercises 47–50.*

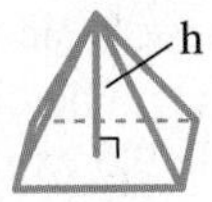

47. Volume of a pyramid. The base is a square with side 3 inches; height is 5 inches.

48. Volume of a pyramid. The base is an equilateral triangle with side 4 feet; height is 3 feet.

49. Volume of a pyramid. The base is a triangle with sides 5, 7, and 8 feet; height is 6 feet.

50. Volume of a pyramid. The base is a regular hexagon with side 4 yards; height is 5 yards.

51. Height of a pyramid. If a pyramid has a square base with side 3 feet and its volume is 10 cubic feet, what is the height of the pyramid?

52. Base of a pyramid. A pyramid with a square base has height 8 inches and volume 20 cubic inches. What is the length of the sides of the base?

9.5 The Metric System and Dimensional Analysis

Objectives

1. Understand the basic units of measurement in the metric system.
2. Make conversions between metric measurements.
3. Use dimensional analysis to make conversions between different measurement systems to solve applied problems.

If you were to travel to France or Poland, or even to Jamaica or India, you might be surprised to see a sign at a service station stating its price of gasoline as the equivalent of \$1.18 U.S. dollars. And you might be concerned about your safety as you travel down the highway and notice that the speed limit is 100. But it turns out that gas is not that cheap in France and speed limits are not as dangerous as it first seems in India. Your misunderstanding is due to the fact that most of the world measures things differently than we do in the United States. Here we use measurements such as inches, feet, gallons, and pounds, but as we will explain, the rest of the world uses meters, liters, and grams.

The Metric System

Unlike the United States, most of the world uses the **metric** system of measurement, which is officially known as the **Système International d'Unités** or the **SI system.** The U.S. system of measurement is called the **U.S. customary system.**

In order to work in the metric system, you need to understand the following.

1. The basic metric units that we use to measure length, weight, and volume.
2. The way these different measures of length, volume, and weight relate to each other. As you will see, the relationship between different metric measures is much simpler than the relationship between different customary measures.

In the metric system, length measurement is based on the **meter,** which is slightly more than a yard. If you were to meet a tall man on a train in Argentina, he would be about

2 meters tall. Volume is based on the **liter,** which is a little more than a quart. Instead of buying a half-gallon of iced apple-pear drink in Poland, you would carry home 2 liters from the market. The basic unit of weight in the metric system is the **gram,** which is about the weight of a large paper clip. One thousand grams is called a **kilogram,** which is about 2.2 pounds. If you were visiting Thailand and wanted to make about 2 pounds of chicken breasts with herb-lemongrass crust, you would ask your meat cutter for a kilogram of chicken.

KEY POINT

Units of measure in the metric system are based on powers of 10.

Although measurement in the metric system is based on meters, liters, and grams, we often use variations of these basic units to measure different quantities. For example, we might measure length in centimeters or kilometers, volume in milliliters, or weight in decigrams or kilograms. Table 9.5 explains the meaning of prefixes such as milli-, deka-, kilo-, centi-, and so on.

kilo-(k)	hecto-(h)	deka-(da)	base unit	deci-(d)	centi-(c)	milli-(m)
× 1,000	× 100	× 10		× 1/10 or × 0.1	× 1/100 or × 0.01	× 1/1,000 or × 0.001

TABLE 9.5 Some common metric prefixes.

For example, a kilometer is 1,000 meters, a centigram is $\frac{1}{100}$ of a gram or 0.01 gram, and a milliliter is $\frac{1}{1,000}$ of a liter, or 0.001 liter. Conversion of units in the metric system is quite simple. From Table 9.5, we see that 1,000 of something is always a "kilo" and $\frac{1}{100}$ of something is always a "centi." The same pattern of prefixes works for length, volume, and weight. In the U.S. customary system, this is certainly not the case. Three feet equal 1 yard, but 4 quarts equal 1 gallon. There are 16 ounces in a pound, but only 2 cups in a pint. There is no consistency or uniformity.

QUIZ YOURSELF 17

a) 1 kiloliter is equal to how many liters?
b) 1 gram is equal to how many centigrams?

We have listed the abbreviations for the metric prefixes in Table 9.5. Also, we use m for meter, g for gram, and L for liter. For example, to represent kilograms, we could use k for kilo and g for grams, thus we could write 10 kilograms as 10 kg. Similarly, we could write 25 milliliters as 25 mL. 17

PROBLEM SOLVING

Strategy: The Analogies Principle

It helps you to remember the meaning of some metric prefixes if you connect them with some common words. For example, "centi" reminds you of cent, which is one one-hundredth of a dollar. "Milli" might remind you of millennium, which is 1,000 years. "Deci" is found in the word *decimal* and a decimal system is based on powers of 10.

Highlight

How Many Yottabytes* Does Your iPhone 17 Have?

In the early days of personal computing, a computer costing over $2,000 might have about 64,000 = 64×10^3 bytes, or 64 **kilo**bytes, of memory. In 2012, the Apple iPhone 4S had 64 **giga**bytes, which is 64,000,000,000 = 64×10^9 bytes, or *one million times* the amount of memory of the early PCs. As the amount of memory increases, we need new names for larger powers of ten, as we show in the table.

It is already common when reading about powerful scientific supercomputers to see the prefixes "peta" and "exa" used, and we can only wonder what amazing technology we will carry in our pockets in the future.

10^3	10^6	10^9	10^{12}	10^{15}	10^{18}	10^{21}	10^{24}
kilo	mega	giga	tera	peta	exa	zetta	yotta

*By today's standards, a yottabyte is farfetched. Some estimate that the current cost of a yottabyte of memory would be around $100 trillion.

Metric Conversions

If you remember the meaning of the prefixes in Table 9.5, it is easy to convert from one unit to an other in the metric system. For example, in measuring length, a kilometer is 10 times as long as a hectometer and a meter is 100, or 10^2, times as long as a centimeter. A millimeter is $\frac{1}{1{,}000}$, or 10^{-3}, of a meter.

EXAMPLE 1 **Converting Units of Measurement in the Metric System**

Convert each of the following quantities to the unit of measurement that we specify.

a) 5 dekameters to centimeters

b) 2,300 milliliters to hectoliters

SOLUTION:

a) We can make this conversion as follows:

$$5 \text{ dekameters} = 5 \times (10 \text{ meters}) = 50 \text{ meters},$$
$$50 \text{ meters} = 50 \times (10 \text{ decimeters}) = 500 \text{ decimeters},$$
$$500 \text{ decimeters} = 500 \times (10 \text{ centimeters}) = 5{,}000 \text{ centimeters}.$$

Or, looking at Table 9.5 again, we see that every time we move one column to the right, we require 10 times as many objects. Note that because centimeters are much smaller than dekameters, we need many more centimeters than dekameters.

kilo-	hecto-	deka-	base unit	deci-	centi-	milli-
		1 of these	Equals 10 of these	Equals 100 of these	Equals 1,000 of these	

Moving three places to the right gives us 10^3 as many objects.

You can easily perform this calculation by moving the decimal point in 5.0 dekameters three places to the right to get 5,000 centimeters.

b) A quick way to solve this problem is to realize that because the prefix hecto- is five places to the left of milli- in Table 9.5, all we have to do to make the conversion is move the decimal point in 2,300 milliliters five places to the left to get

$$0\,2\,3\,0\,0\,.$$

or 0.023 hectoliter.

It is good to check if this makes sense to you. Remember that milliliters are much smaller than hectoliters, so therefore it does not require many hectoliters to represent many milliliters.

Now try Exercises 1 to 6. 18

QUIZ YOURSELF 18

a) 5.63 kiloliters is equal to how many deciliters?

b) 4,850 milligrams is equal to how many hectograms?

SOME GOOD ADVICE

Although it is tempting to simply memorize how to move the decimal point in doing conversions such as those we did in Example 1, it is important to *understand* why you are moving the decimal point and in which direction. If you want to make a number larger, you are multiplying by powers of 10, which means that you move the decimal point to the right. To make the number smaller, you move the decimal point to the left. In checking your answer, remember that it takes many small things to make one large thing, and vice versa.

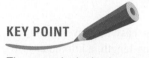

KEY POINT

The meter is the basic unit of length in the metric system.

In 1790, the French Academy of Science defined the **meter** to be one ten-millionth of the distance from the North Pole to the equator—a distance of about 39.37 inches. Since that time, scientists have redefined the meter several times, and currently the meter is the distance that light travels in a vacuum in $\frac{1}{299,792,458}$ of a second. A good way to visualize a meter is that it is slightly longer than a yardstick. Figure 9.57 shows the relationship between millimeters $\left(\frac{1}{1,000}\text{ of a meter}\right)$, centimeters $\left(\frac{1}{100}\text{ of a meter}\right)$, and inches.

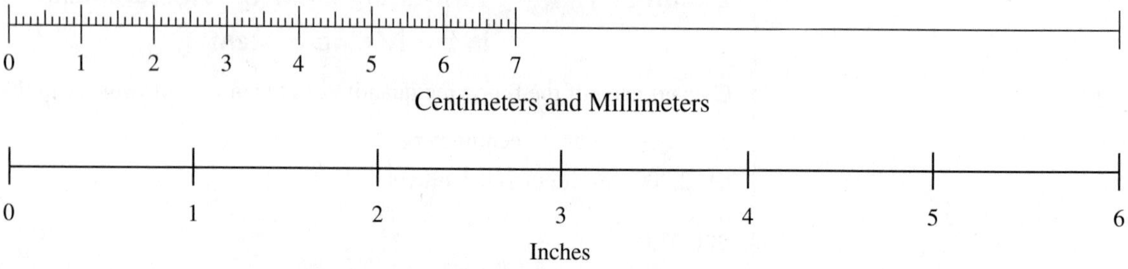

FIGURE 9.57 Comparing millimeters, centimeters, and inches.

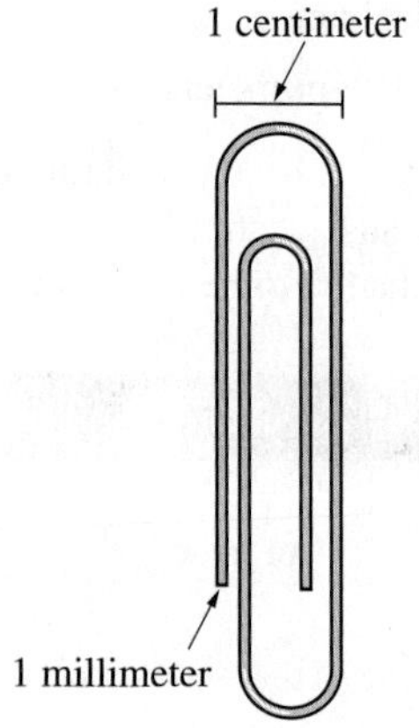

From Figure 9.57, you can see that a millimeter is about the thickness of a large paper clip, and the width of the paper clip is roughly one centimeter. One inch is about 2.5 centimeters, and a centimeter is roughly 0.4 inch. We often measure large distances in kilometers (1,000 meters). A kilometer is about 0.6 mile and a mile is approximately 1.6 kilometers.

EXAMPLE 2 Estimating Metric Lengths

Match each of the following items with one of these metric measurements: 2 m, 3 cm, 10 m, 16 km.

a) The length of a 30-foot yacht
b) The length of a 10-mile race
c) The length of your bed
d) The thickness of this book

SOLUTION:

a) Thirty feet equals 10 yards, which is about 10 m.
b) One mile is about 1.6 km, so a 10-mile race would be about 16 km long.
c) Your bed is probably a little more than 6 feet long, so it would be approximately 2 m long.
d) If you place the edge of your book on the diagram in Figure 9.57, you will see that the book is around 3 or 4 cm thick.

KEY POINT

We use dimensional analysis to convert from one system to the other.

1 foot = 12 inches
1 yard = 3 feet = 36 inches
1 mile = 5,280 feet

TABLE 9.6 Some relationships among units in the customary system.

Dimensional Analysis

Often we need to convert from customary units to metric units, or vice versa. Before we explain how to convert from one system to the other, let us examine how we convert one unit of length to another within our customary system. Recall the relationships among various units of length, which we give in Table 9.6.

To convert one quantity to another, we will use unit fractions. A *unit fraction* is a quotient such as $\frac{3\textit{ feet}}{1\textit{ yard}}$ or $\frac{1\textit{ foot}}{12\textit{ inches}}$ that has different units of measurement in the numerator and denominator and whose value is equal to 1. We will show you how to use unit fractions to make conversions in Example 3.

EXAMPLE 3 Converting Yards to Inches

Convert 5 yards to inches.

SOLUTION:

We want to have an answer that is in inches instead of yards, so we will multiply by the unit fraction $\frac{36\ inches}{1\ yard}$. Because we want the final answer in terms of inches, we put "inches" in the numerator. Thus,

$$5\ \cancel{yards} \times \frac{36\ inches}{1\ \cancel{yard}} = 5 \times 36\ inches = 180\ inches.$$

You can use dimensional analysis to make conversions between the customary and metric systems, but first you need to know some basic relationships between units of length in the two systems, which we give in Table 9.7.

1 inch	= 2.54 centimeters
1 foot	= 30.48 centimeters
1 yard	= 0.9144 meter
1 mile	= 1.6 kilometers

TABLE 9.7 Some basic relationships between customary and metric units of length.

From Table 9.7, we see that there are a number of unit fractions that we can use to make conversions such as $\frac{1\ inch}{2.54\ centimeters}$ and $\frac{0.9144\ meter}{1\ yard}$.

EXAMPLE 4 Converting Between the Metric and Customary Systems

a) In celebration of National Cookie Day, December 4, the residents of *Sesame Street* have baked a gigantic cookie, measuring 5 yards in circumference, for Cookie Monster. How many centimeters is the circumference of this cookie?

b) In recognition of Earth Day, students will be participating in a 0.3-kilometer, three-legged "Race to Save Mother Earth." How long is this race in feet?

SOLUTION: Review the Relate a New Problem to an Older One Strategy on page 9.

a) To make this conversion, we need a unit fraction that will convert yards to meters and then another to convert meters to centimeters.* Multiplying by the unit fraction $\frac{0.9144\ meter}{1\ yard}$ will convert yards to meters, and because 1 meter = 100 *centimeters*, we can use the unit fraction $\frac{100\ centimeters}{1\ meter}$ to finish the conversion. Thus,

converts yards to meters; converts meters to contimeters

$$5\ yards = 5\ \cancel{yards} \times \frac{0.9144\ \cancel{meter}}{1\ \cancel{yard}} \times \frac{100\ centimeters}{1\ \cancel{meter}}$$

$$= 5 \times 0.9144 \times 100\ centimeters = 457.2\ centimeters.$$

b) We will solve this problem similarly to the way we solved part a). We need one unit fraction, $\frac{1\ mile}{1.6\ kilometers}$, to convert kilometers to miles and then another, $\frac{5{,}280\ feet}{1\ mile}$, to convert miles to feet. Therefore,

converts kilometers to miles; converts miles to feet

$$0.3\ kilometer = 0.3\ \cancel{kilometer} \times \frac{1\ \cancel{mile}}{1.6\ \cancel{kilometers}} \times \frac{5{,}280\ feet}{1\ \cancel{mile}}$$

$$= \frac{0.3 \times 5{,}280}{1.6} = 990\ feet.$$

Now try the Exercises 23 to 36 that deal with length. 19

QUIZ YOURSELF 19

Convert 0.65 kilometer to feet.

The **liter** is defined to be 1 cubic decimeter (see Figure 9.58). That is, a liter is the volume of a cube that measures 1 decimeter (or 10 centimeters) on each side. From Figure 9.58, you can see that a liter is 10 × 10 × 10 = 1,000 cubic centimeters.

*Of course, we could convert meters to centimeters by moving the decimal point as we did in Example 1. Either approach will give you the same answer.

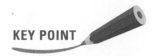

KEY POINT

The liter is the basic unit of volume in the metric system.

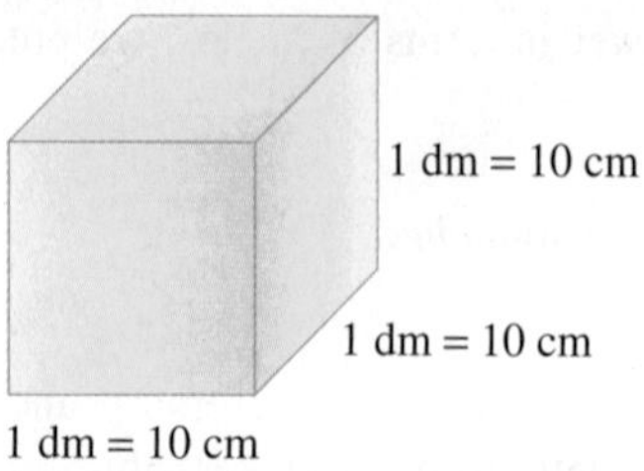

FIGURE 9.58 1 liter = 1 cubic decimeter = 1,000 cubic centimeters.

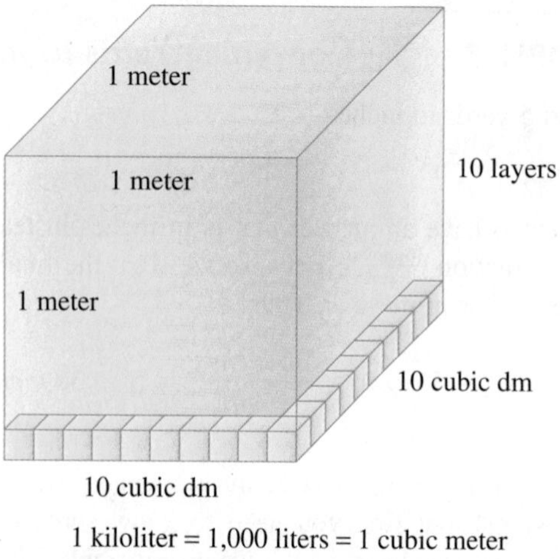

FIGURE 9.59 Representing a cubic meter in terms of liters.

As we mentioned earlier, a liter is slightly more than a quart. We can use the same prefixes (kilo-, hecto-, deka-, and so on) that we used in measuring length. We use a milliliter, $\frac{1}{1{,}000}$ of a liter, to measure small quantities, such as a dose of medicine. If you go to a clinic to get a flu shot, you might receive 2 milliliters of vaccine. We measure large volumes in kiloliters (1,000 liters). Figure 9.59 shows that a kiloliter is the same as a cubic meter. Each layer in Figure 9.59 contains $10 \times 10 = 100$ cubic decimeters, and because there are 10 layers, we have 1,000 cubic decimeters. In Canada, you might order 20 cubic meters of cement to build a patio floor. We also use cubic centimeters, *cc*, to measure volume.

EXAMPLE 5 Estimating Metric Volumes

Estimate the volume of each of the following items with one of these metric measurements: 3 mL (milliliters), 250 mL (milliliters), 100 L (liters), 500 m^3 (cubic meters), 1 dm^3 (cubic decimeter).*

a) The gas tank of a car
b) A large pile of gravel at a construction site
c) A glass of soda
d) A quart of orange juice

SOLUTION:

a) 100 liters is about 100 quarts, or 25 gallons, which is the size of the gas tank of a medium-size car.

b) 500 cubic meters is about 500 cubic yards, which is the size of a fairly good-size pile of gravel.

c) A glass of soda would be about $\frac{1}{4}$ of a quart. Because a quart is roughly equal to a liter and a liter is 1,000 milliliters, the glass of soda has a volume of about $\frac{1{,}000}{4} = 250$ milliliters.

d) A quart is roughly a liter and a liter is 1 cubic decimeter, so the volume of the orange juice is about 1 cubic decimeter.

We make conversions among units of volume in the metric system in exactly the same way we made conversions among units of length.

EXAMPLE 6 Estimating the Amount of Vaccinations

The East Side Community Health Clinic has on hand four bottles that each contain 0.35 L of flu vaccine. Each vaccination requires 2 mL of vaccine. How many patients can be treated with the amount of vaccine that the clinic has on hand?

SOLUTION:

The total amount of vaccine is $4 \times 0.35 = 1.4$ liters. From Table 9.5, we see that to convert liters to milliliters, we move the decimal point three places to the right.

1.4 L - - - → 14. dL - - - → 140. cL - - - → 1,400. mL

Therefore the clinic has 1,400 mL of vaccine on hand, or enough for $\frac{1{,}400}{2} = 700$ vaccinations.

*We will denote a liter by L to avoid confusion between lowercase "el" and the numeral 1.

Between the Numbers

But Officer—I Only Had…

In most states, you are considered legally drunk and unfit to drive if your blood alcohol content (BAC) is above 0.08. Now what exactly does that mean? The 0.08 stands for 0.08 gram per 100 milliliters of blood.

To put this in context, two beers contain about 30 grams of alcohol and an average-sized man has about 5 liters, or 5,000 milliliters, of blood. So dividing 30 by 5,000, we get $\frac{30}{5{,}000} = 0.006$ gram of alcohol per milliliter in your blood. So in 100 milliliters, there would be $100 \times 0.006 = 0.6$ gram of alcohol. Actually, if this amount of alcohol were absorbed immediately, it would kill you—fatalities often occur with a BAC between 0.4 and 0.5. But, because the alcohol is absorbed over a period of time and the liver also works to eliminate the alcohol, we would not expect two beers to be fatal.

You can find BAC calculators on the Internet that will estimate your blood alcohol content based on your weight, what you are drinking, and so on. When I did this, I was thinking about the custom that some college students have of taking 21 drinks on their 21st birthday. When I tried to calculate my BAC if I had 21 drinks in 6 hours, the calculator responded by saying, "This is not a reasonable number of drinks, enter another number." So I entered 10 drinks for a 6-hour period and the estimated BAC was 0.22, or about three times the legal limit. If we were to double this to 20 drinks, it is not unreasonable to expect a BAC above 0.4, which is in the potentially fatal range.

You can use dimensional analysis to make conversions between units of volume in the metric and customary systems, but first you need to know some basic relationships, which we give in Table 9.8.

2 cups = 1 pint	1 cup = 0.2366 liter
2 pints = 1 quart	1 quart = 0.9464 liter
32 fluid ounces = 1 quart	1 cubic foot = 0.03 cubic meter
4 quarts = 1 gallon	1 cubic yard = 0.765 cubic meter

TABLE 9.8 Relationships between units of volume in the metric and customary systems.

You can now use these relationships to define unit fractions to make conversions from one system to the other as we did earlier in this section.

EXAMPLE 7 Calculating the Volume of a Car's Gas Tank

Assume that an American car's gas tank has a capacity of 23 gallons. Supposing that you were driving this same car in France, how many liters of gasoline would you need to fill the tank?

SOLUTION: Review the Relate a New Problem to an Older One Strategy on page 9.

Looking at the shaded boxes in Table 9.8, we see that we can first use the unit fraction $\frac{4\ quarts}{1\ gallon}$ to convert gallons to quarts and then use the unit fraction $\frac{0.9464\ liter}{1\ quart}$ to convert quarts to liters. Thus,

converts gallons to quarts — converts quarts to liters

$$23\ gallons = 23\ \cancel{gallons} \times \frac{4\ \cancel{quarts}}{1\ \cancel{gallon}} \times \frac{0.9464\ liter}{1\ \cancel{quart}}$$

$$= 23 \times 4 \times 0.9464 = 87.07\ liters.$$

So you would need to purchase 87.07 L of gasoline.

Now try Exercises 23 to 36 that deal with volume.

KEY POINT

The gram is the basic unit of mass (weight) in the metric system.

The last type of measurement we will consider is mass, which, for simplicity's sake, you might think of as weight. Strictly speaking, mass and weight are not the same thing. The mass of an object depends on its molecular makeup and does not change. The weight of an object, however, depends on the gravitational pull on that object. Gravity is stronger on a large planet and weaker on a small planet. Because the moon's gravity is only $\frac{1}{6}$ of Earth's gravity, an elephant that weighs 2,400 pounds on Earth would weigh only 400 pounds on the moon. However, the mass of the elephant would be the same on both Earth and the moon.

The basic unit of mass in the metric system is the **gram,** which is defined to be the mass of 1 cubic centimeter (1 milliliter) of water at a certain specified temperature and pressure. A liter (roughly 1 quart) is 1,000 mL, so a liter of water has a mass of 1,000 g or 1 kg. It is common in the metric system to measure the mass of a large object in kilograms; 1 kilogram is approximately 2.2 pounds. So, if you were having a barbecue in Spain and wanted to broil about 8 or 9 quarter-pounders, you would go to the market and buy 1 kg of ground beef.

The pattern of prefixes, and abbreviations, and also the rules for making conversions are the same for mass as they were for length and volume.

EXAMPLE 8 Estimating Mass in the Metric System

Estimate the mass of each of the following items with one of these metric measurements: 1 g, 50 dag, 5 kg, 1,000 kg.

a) A decent-size steak
b) A large bag of sugar
c) A medium-size car
d) A large paper clip

SOLUTION:

a) The steak might weigh a pound or more. A kilogram is 2.2 pounds, so the steak would be about 500 grams, or 50 dag.

b) Typically, we buy sugar in 10-pound bags; 5 kg would be equal to $5 \times 2.2 = 11$ pounds, so the sugar would weigh a little less than 5 kg.

c) A car might weigh 2,200 pounds, which is 1,000 kg.

d) A paper clip is pretty light, so its weight might be about 1 g.

To make mass conversions between the metric and customary systems, you can use the information in Table 9.9.

16 ounces = 1 pound	1 ounce = 28 grams
2,000 pounds = 1 ton	2.2 pounds = 1 kilogram
	1 metric tonne = 1,000 kilograms
	1.1 tons = 1 metric tonne

TABLE 9.9 Relationships between units of mass in the metric and customary systems.

Again, we can use dimensional analysis to convert units in one system to units in the other.

EXAMPLE 9 Refurbishing a Church with Italian Marble

A contractor who is refurbishing a church has ordered slabs of Italian marble that weigh 1.8 metric tonnes each. If he wants to rent a crane to lift the slabs of marble, how much, in pounds, should the cranes be able to lift?

QUIZ YOURSELF 20

Convert 3.8 kg to pounds, then ounces.

SOLUTION:

We will convert metric tonnes to tons and then tons to pounds. The unit fraction $\frac{1.1\ tons}{1\ metric\ tonne}$ converts metric tonnes to tons and the unit fraction $\frac{2{,}000\ pounds}{1\ ton}$ will convert tons to pounds. Therefore,

$$1.8\ metric\ tonnes = 1.8\ \cancel{metric\ tonnes} \times \underbrace{\frac{1.1\ \cancel{tons}}{1\ \cancel{metric\ tonne}}}_{\text{converts metric tonnes to tons}} \times \underbrace{\frac{2{,}000\ pounds}{1\ \cancel{ton}}}_{\text{converts tons to pounds}}$$

$$= 1.8 \times 1.1 \times 2{,}000 = 3{,}960\ pounds$$

A crane that can lift 4,000 pounds would just barely do the job.

Now try Exercises 23 to 36 that deal with weight. **20**

1 meter = 1.0936 yards
1 mile = 1.609 kilometers
1 pound = 454 grams
1 metric tonne = 1.1 tons
1 liter = 1.0567 quarts

We will frequently use the equivalents between the metric and customary systems in the table to the left in doing the exercises. Also, your answers may sometimes differ slightly from ours due to roundoff discrepancies.

EXERCISES 9.5

Sharpening Your Skills

Use Table 9.5 to make the conversions in Exercises 1–6.

1. 2.4 kiloliters to deciliters

2. 240 centigrams to dekagrams

3. 28 decimeters to millimeters

4. 5.6 hectograms to centigrams

5. 3.5 dekaliters to deciliters

6. 7,600 centimeters to meters

In Exercises 7–16, match the italicized words in the left column with the metric measurement in the right column that corresponds to it. Before calculating your final answer, it would be wise to first convert the given measurement to basic metric units such as meters, grams, or liters.

7. A *quarter-pound* hamburger	**a.** 0.02159 dam
8. A *gallon* of milk	**b.** 946.3 mL
9. A *15-foot*-tall giraffe	**c.** 378.5 cl
10. *Five pounds* of potatoes	**d.** 1.77 dL
11. A *6-inch*-long ruler	**e.** 4,572 mm
12. A *quart* of motor oil	**f.** 0.0061 km
13. A book *eight and one-half inches* wide	**g.** 15.24 cm
14. A *12-ounce* gerbil	**h.** 1,135 dg
15. *Six ounces* of orange juice	**i.** 3.405 hg
16. A *20-foot*-long swimming pool	**j.** 2.27 kg

Pick the most appropriate measurement for each of the following items. Explain your answer.

17. The volume of a small juice glass

a. 125 mL **b.** 500 mL **c.** 2.5 L

18. The weight of books in your backpack

a. 500 g **b.** 80 hg **c.** 6 kg

19. The length of your nose

a. 4 dm **b.** 3 mm **c.** 5 cm

20. The volume of a bottle of wine

a. 0.25 L **b.** 0.2 kL **c.** 750 mL

21. The height of a Boston terrier

a. 100 mm **b.** 0.06 dam **c.** 0.6 hm

22. The weight of NFL lineman D'Brickashaw Ferguson

a. 136 kg **b.** 13,000 g **c.** 20 dag

Use dimensional analysis to make each of the following conversions. You may have to define several unit fractions to make the conversions. Also, because the constants that we give in the table are only approximations, depending on the method with which you do your computations, your answers may differ slightly from ours.

23. 18 meters to feet

24. 27 gallons to liters

25. 3 kilograms to ounces

26. 10,000 milliliters to quarts

27. 2.1 kiloliters to gallons

28. 47 pounds to kilograms

29. 10,000 deciliters to quarts

30. 507,820 milligrams to pounds

31. 176 centimeters to inches

32. 3 yards to millimeters

33. 45,000 kilograms to tons

34. 0.65 tonne to pounds

35. 2.6 feet to decimeters

36. 10 tons to tonnes

Applying What You've Learned

Rewrite each statement, replacing the metric measure by the corresponding customary measure.

37. Hold your ground. Don't give him 2.54 centimeters.

38. Tex is wearing a 40.16-liter hat.

39. It is first down and 9.14 meters to go.

40. 28 grams of prevention is worth 0.45 kilogram of cure.

41. Height of a mountain. The highest mountain in the world is Mount Everest, which is 8,850 meters high. How high is it in feet?

42. Height of a mountain. Mount Kilimanjaro, the highest mountain in Africa, is 19,340 feet. How high is that in meters?

43. Area of an oriental rug.

a. Find the number of square feet in a square meter.

b. If a small oriental rug measures 3 meters by 4 meters, how many square feet does the rug contain?

44. Area of a photograph.

a. What is the number of square inches in a square meter?

b. If the Jet Propulsion laboratory has a photograph of the Mars Rover measuring 1.5 meters by 2 meters, how many square inches does the photograph contain?

45. Volume of a Koi pond. If a rectangular Koi pond is 11 feet long, 8 feet wide, and 3 feet deep, how many liters does it contain?

46. Volume of a tank. A rectangular tank is 5 m wide, 8 m long, and 4 m deep.

a. What is the volume of the tank in cubic meters?

b. How many liters of water does the tank contain?

c. What is the weight of the water in kilograms?

47. Converting speed. In *The Amazing Race*, Rachel is driving a car that shows speed in both miles per hour and kilometers per hour. If she is traveling in Italy at 80 kilometers per hour, what is her speed in miles per hour?

48. Converting speed. TK is driving in Switzerland at a speed of 55 miles per hour. What is his speed in kilometers per hour?

49. Finding the volume of a swimming pool. Adrian's rectangular swimming pool is 20 ft wide, 40 ft long, and averages 6 ft in depth. How many kiloliters of water are needed to fill the pool?

50. Depth of a holding pond. If a holding pond at the fish hatchery is 6 meters long, 10 meters wide, and contains 240 kiloliters of water, how deep is the pond?

51. Buying canned food. Marco purchased a large can of chili that has a diameter of 10 cm and a height of 14 cm. Determine the volume of the chili in the can.

a. in liters. **b.** in ounces.

52. Radius of an oxygen tank. The volume of a cylindrical oxygen tank at Sacred Heart Hospital is 4 kiloliters. If its height is 5 meters, what is the radius of its base?

53. Purchasing fruit in Europe. While visiting Poland, Anthony purchased some red plums that cost $2.75 per kilogram. What is the cost of the plums per pound?

54. Measuring a medication. Justin is taking the anti-inflammatory drug Niamoxin and the instructions say that the patient must receive 10 mg for each 15 kg of body weight. If Justin weighs 275 lb, what dosage should he receive?

55. Buying gasoline. If gasoline costs $2.18 per liter in Gambia (Africa), what is its cost in dollars per gallon?

56. Buying gasoline. Europeans pay very heavy taxes on gasoline. When I was writing these exercises, gasoline in Germany cost $8.00 per gallon. What would the cost be in dollars per liter?

57. Fencing a dog pen. Serina wants to fence in a rectangular exercise pen for her golden labrador retriever. The pen is 35 ft wide and 62 ft long. The fencing is sold in whole meters. How much fencing should she buy?

58. In Exercise 57, what is the area of Serina's pen in square meters?

59. Calculating gas mileage. If a car averages 30 miles per gallon, how many kilometers per liter would that be?

60. Calculating gas mileage. If a car averages 15 kilometers per liter, how many miles per gallon would that be?

61. Buying flooring. If oak flooring costs $8.00 a square foot, how much is that cost per square meter?

62. Buying flooring. If vinyl flooring costs $98.00 a square meter, how much is that per square foot?

In the metric system, temperatures are measured on thermometers using the Celsius (also called centigrade) scale instead of the Fahrenheit scale that we commonly use. On the Celsius scale, water freezes at 0 degrees and boils at 100 degrees. To convert a temperature from one scale to a temperature on the other, you can use the following equation:

$$F = \frac{9}{5}C + 32,*$$

*This equation converts degrees Celsius to degrees Fahrenheit. You can also use it (plus algebra) to convert Fahrenheit to Celsius, which we recommend that you do. If you really want another equation to convert Fahrenheit to Celsius, you can use the equation $C = \frac{5}{9}(F - 32)$. However, we strongly recommend that you memorize only one equation and use algebra to do the second conversion.

where F is the Fahrenheit temperature and C is the Celsius temperature. Use this equation in Exercises 63–70 to convert each temperature to a corresponding measurement in the other system.

63. 149 degrees Fahrenheit

64. 95 degrees Fahrenheit

65. 60 degrees Celsius

66. 85 degrees Celsius

67. **Record temperature.** In 1922, the highest temperature ever recorded was 136° Fahrenheit in El Azizia, Libya. Convert this temperature to Celsius.

68. **Record temperature.** The highest temperature recorded in North America was 57° Celsius, which occurred in 1913 in Death Valley, California. What is this temperature in Fahrenheit?

69. **Extreme temperatures.** The highest temperature recorded in Antarctica was 59° Fahrenheit and the lowest was −129°. What is the difference between these two temperatures in Celsius degrees?

70. **Extreme temperatures.** The highest temperature recorded in Asia was 54° Celsius and the lowest was −68°. What is the difference between these two temperatures in Fahrenheit degrees?

In Exercises 71–74, use the fact that a hectare (pronounced "HEKtaire") is a square that measures 100 meters on each side.

71. What is the relationship between hectares and square meters?

72. What is the relationship between hectares and square kilometers?

73. Thiep purchased a rectangular piece of land that measures 0.75 km by 1.2 km. How many hectares of land is that?

74. If Ivanka purchased a rectangular piece of land that contains 40 hectares and is 0.65 km long, how long is the other dimension?

Communicating Mathematics

75. If you were converting hectometers to decimeters, would you have more hectometers or more decimeters? Why?

76. In converting milligrams to dekagrams, we would move the decimal point either four places to the right or to the left. Which is it? Explain how you would help a fellow classmate remember which to do.

77. If *a* kiloliters equals *b* dekaliters, which is larger, *a* or *b*? Explain how you arrived at your answer.

78. Explain the advantages that you see in using the metric system over the customary system. Be specific. Give concrete examples.

Between the Numbers

One frequently mentioned formula for calculating Blood Alcohol Content (BAC) is due to Dr. Erik Widmark and is given by*

$$\text{BAC} = \frac{A \times 5.14}{W \times r} - 0.015 \times H,$$

where A is the number of ounces of alcohol consumed, W is the weight of the person in pounds, r = 0.73 for men and 0.66 for women, and H is the number of hours since the person began drinking.

79. **Estimating blood alcohol content.** Assume that a regular beer contains 5% alcohol. Estimate the BAC for a 160-pound man who drinks four 12-ounce beers in two hours.

80. **Estimating blood alcohol content.** If a 120-pound woman drinks three 5-ounce glasses of wine (wine contains 12% alcohol) over a period of three hours, estimate her BAC.

Challenge Yourself

81. **Cost of gasoline in France.** Suppose that you are driving in France and see that the price of gasoline is 1.585 euros per liter. Use the fact that one U.S. dollar is worth 0.7564 euro to calculate the price of gasoline in dollars per gallon.

82. **Buying grapes in Mexico.** One U.S. dollar is worth 12.92 Mexican pesos. If, while visiting Mexico, you buy a kilogram of grapes for 30 pesos, how much do the grapes cost per pound in dollars?

83. **Buying dragon fruit in Thailand.** Assume that you buy 2 kilograms of dragon fruit at a market in Bangkok, Thailand, for 78 bahts (the exchange rate is 31 bahts for one U.S. dollar). What is the cost of the dragon fruit in dollars per pound?

84. **The price of coconut milk.** In India, a liter of coconut milk costs 328 rupees. How many U.S. dollars per quart does it cost? The exchange rate is 49 rupees for one U.S. dollar.

85. Research the history of the definitions of an inch and a yard. Report on what you find in your research.

86. Research the definitions of avoirdupois and troy weight. Use your research to explain why a pound of silver does not weigh the same as a pound of steak.

---

*This formula is an example of only *one* way that BAC can be calculated. There are different methods and many variables that affect the calculation of BAC, as well as controversy as to the dependability of these formulas, so you should not use this as a reliable way to calculate your own BAC.

9.6 Geometric Symmetry and Tessellations

Objectives

1. Understand the relationship between symmetry and rigid motions.
2. Recognize the symmetries of an object.
3. Determine when it is possible to tessellate a plane with polygons.

At what age do you think a child becomes aware of beauty? Two years? Four years? Wouldn't you think that a child has to live in a society for a while before *learning* what is considered to be beautiful?

Surprisingly, this is not so. Psychologist Judith Langlois of the University of Texas, Austin, has found that babies seem to have a "built-in" appreciation for beauty that agrees with adults' concept of beauty. In her experiments, she found that children as young as 3 months will stare longer at an "attractive" female face than at one that is not as "attractive." She repeated her experiments with Caucasian females and males, Afro-American females, and even faces of other babies, and always got the same results.*

So, just what *are* the babies recognizing? It's the topic of this section—symmetry.

Rigid Motions

Although you no doubt have an intuitive idea of what we mean by symmetry, it is not an easy concept to describe. In mathematics, we begin with an intuitive concept, such as symmetry, and define it precisely so that we can measure it and calculate with it. For example, in Figure 9.60, it is easy to see that the star has more symmetry than the arrowhead. But exactly what do we mean when we say that?

To understand symmetry, imagine that the arrowhead is made of a very stiff wire and is resting in shallow grooves cut into a piece of wood. Now suppose that you close your eyes and a friend picks up the arrowhead, flips it over, and returns it to its resting place. Figure 9.61(a) shows the arrowhead before the flip, and Figure 9.61(b) shows the arrowhead after the flip. Although many points on the arrowhead have been moved, the overall appearance of the arrowhead remains unchanged. When you open your eyes, you would not know whether the arrowhead was flipped. Other than this flip, there is really nothing else that we can do to the arrowhead that will move individual points yet keep the overall appearance of the object the same.

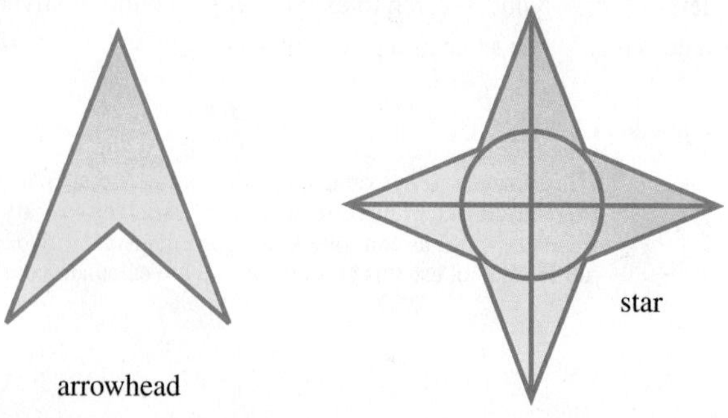

FIGURE 9.60 Objects possessing different amounts of symmetry.

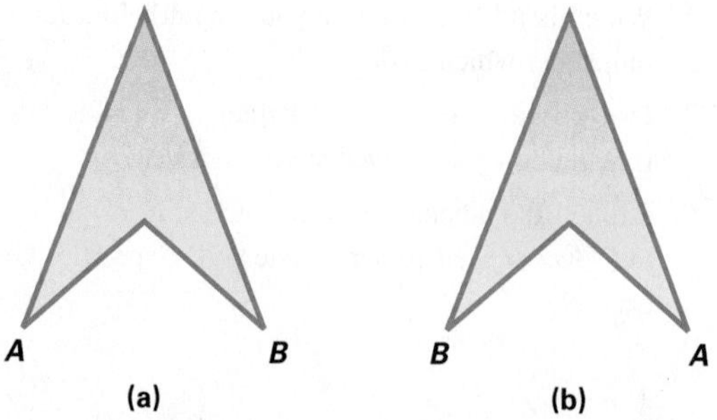

FIGURE 9.61 The arrowhead has been flipped over and returned to its resting place.

*http://www.psy.utexas.edu/psy/spotlights/pdfs/life_letters_092_langlois.pdf

On the other hand, there are many different ways to move the star that will change the position of individual points but leave the overall appearance of the star the same. Figure 9.62 shows two possibilities. In Figure 9.62(b), we have rotated the star in a counterclockwise direction so that *A* moves to the bottom of the star, *B* moves to what was *A*'s position, *C* moves to what was *B*'s position, and so on. In Figure 9.62(c), we have flipped the original star over, interchanging vertices *A* and *C*.

Figures 9.61 and 9.62 show examples of rigid motions.

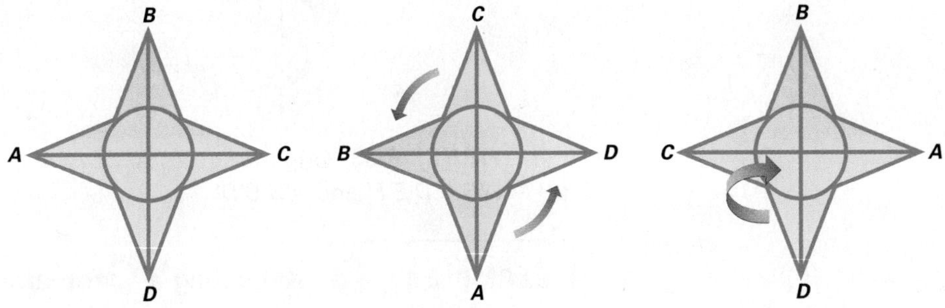

FIGURE 9.62 Two transformations of the star.

> **DEFINITION** A **rigid motion** is the action of taking a geometric object in the plane and moving it in some fashion to another position in the plane without changing its shape or size.

When we consider a rigid motion, we are interested only in the beginning and ending positions of the object. For example, suppose that in flipping the arrowhead, your friend drops it and it rolls across the floor. If your friend retrieves it and places it back in the grooves in the wood, most of what has happened to the arrowhead is irrelevant. All that matters mathematically is that from the beginning to the end of the process, the arrowhead has been flipped over.

You might think that there are many different ways that we could perform rigid motions; however, this is not the case.

KEY POINT

There are essentially only four possible rigid motions.

> Every rigid motion is essentially a reflection, a translation, a glide reflection, or a rotation.

By the word *essentially*, we mean that if we consider only the beginning and ending positions of the object, and ignore the intermediate motions, the rigid motion can be accomplished in only one of four ways.

We first discuss reflections.

> **DEFINITION** A **reflection** is a rigid motion in which we move an object so that the ending position is a mirror image of the object in its starting position.

In a reflection, there is a line, called the *axis of reflection*, that acts as a mirror to transform the figure from its original position to its final position. We say that the original figure has been *reflected about the axis of reflection* to produce the final figure. We show two reflections in Figure 9.63.

As in Figure 9.63, we label the vertices of the original polygons with *A, B, C,* and so on, and the vertices of the reflected polygons with *A′, B′, C′,* and so on.

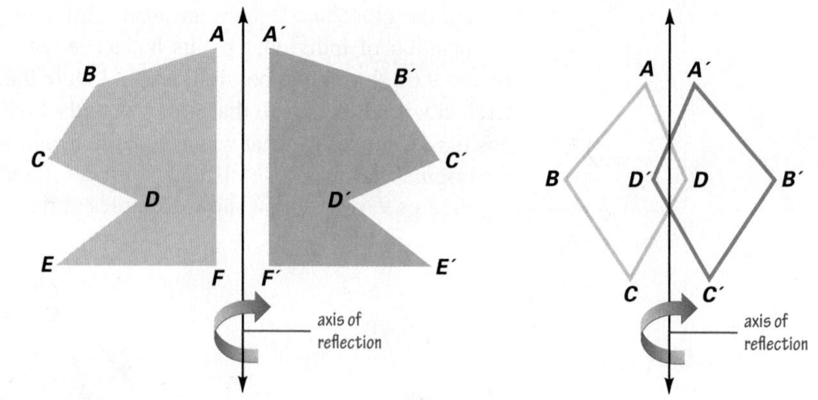

FIGURE 9.63 Original figures *ABCDEF* and *ABCD*; reflected figures *A′B′C′D′E′F′* and *A′B′C′D′*.

EXAMPLE 1 Reflecting a Geometric Object

Reflect polygon *ABCDE* about the axis of reflection *l* in Figure 9.64.

SOLUTION:

In order to reflect the polygon about line *l*, we must reflect each vertex *A*, *B*, *C*, *D*, and *E* about *l*.

1st: Reflect *A* about *l* by drawing a line segment from *A* to *A′* that is perpendicular* to *l* and also so that the distance from *A* to *l* is the same as the distance from *A′* to *l*.

2nd: Draw segments *BB′*, *CC′*, *DD′*, and *EE′* in a similar way, as shown in Figure 9.65.

3rd: Draw the reflected polygon by connecting vertices *A′*, *B′*, *C′*, *D′*, and *E′*, as shown in Figure 9.66.

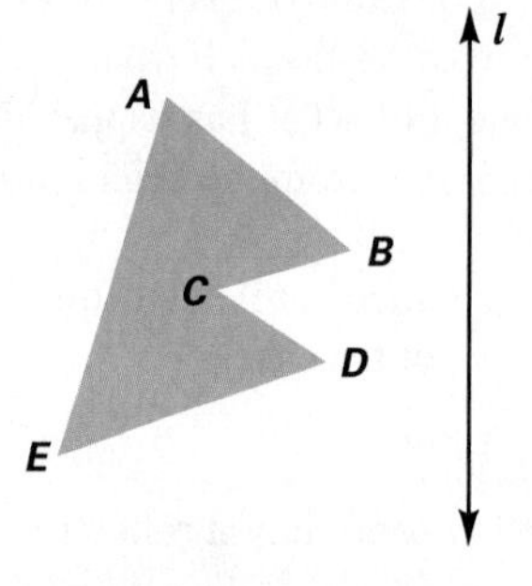

FIGURE 9.64 Polygon *ABCDE* is to be reflected about line *l*.

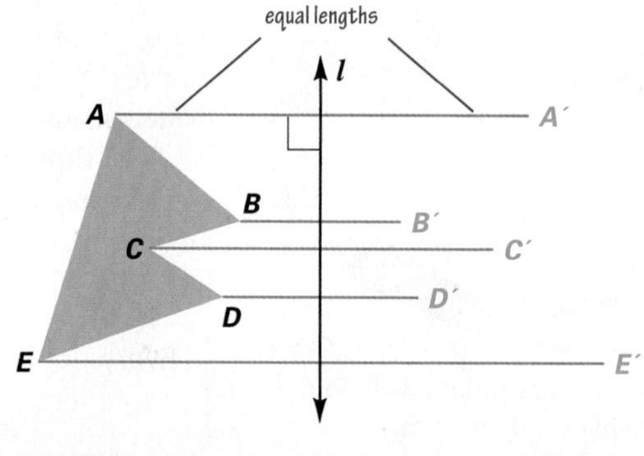

FIGURE 9.65 Reflecting points *A, B, C, D,* and *E* about *l*.

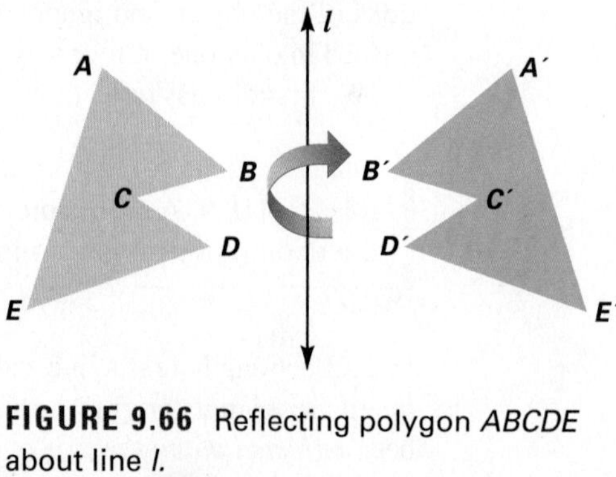

FIGURE 9.66 Reflecting polygon *ABCDE* about line *l*.

Now try Exercises 1 to 8. 21

QUIZ YOURSELF 21

Reflect the quadrilateral *ABCD* about line *l*.

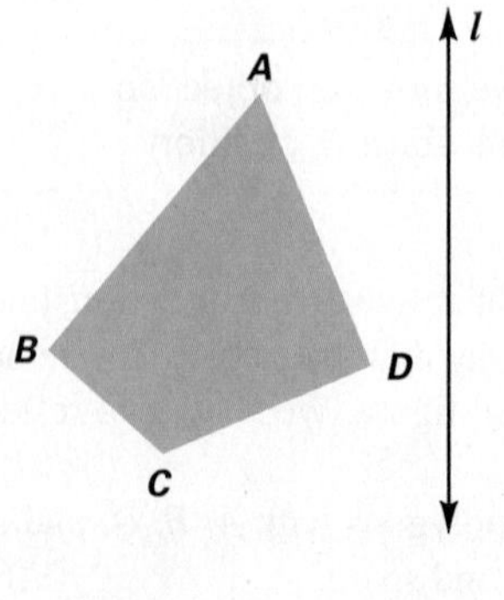

*Although we have not explained how to construct a line that is perpendicular to another line, you can draw a perpendicular freehand to get an idea of what the reflection looks like.

DEFINITION A **translation** is a rigid motion in which we move a geometric object by sliding it along a line segment in the plane. The direction and length of the line segment completely determine the translation. We represent the distance and direction of a translation by a line segment with an arrow on it, called the **translation vector.**

Figure 9.67 shows a translation.

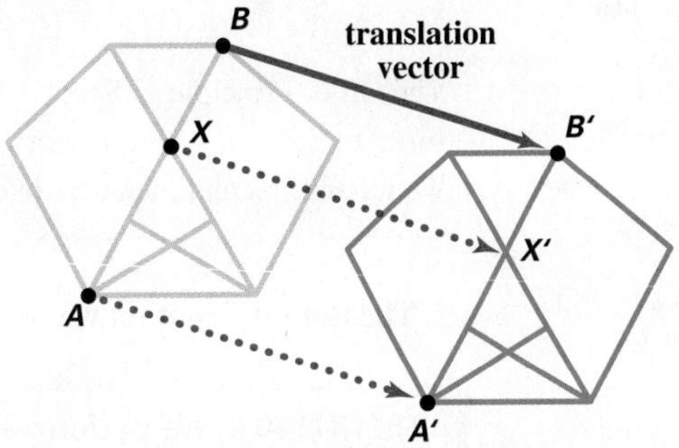

FIGURE 9.67 A translation vector determines a translation of an object.

Now try Exercises 11 to 14.

DEFINITION A **glide reflection** is a rigid motion formed by performing a translation (the glide) followed by a reflection.

EXAMPLE 2 Producing a Glide Reflection of a Geometric Object

Use the translation vector and the axis of reflection to produce a glide reflection of the object shown in Figure 9.68.

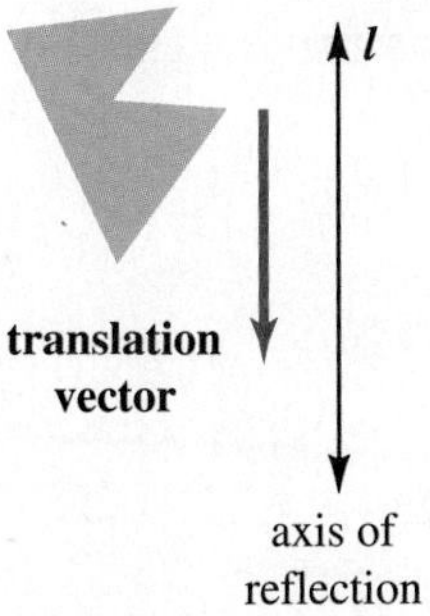

FIGURE 9.68 Forming a glide reflection of an object.

SOLUTION:

We will accomplish this glide reflection as follows:

1st: Place a copy of the translation vector at some point, say *Y*, on the object (Figure 9.69(a)).

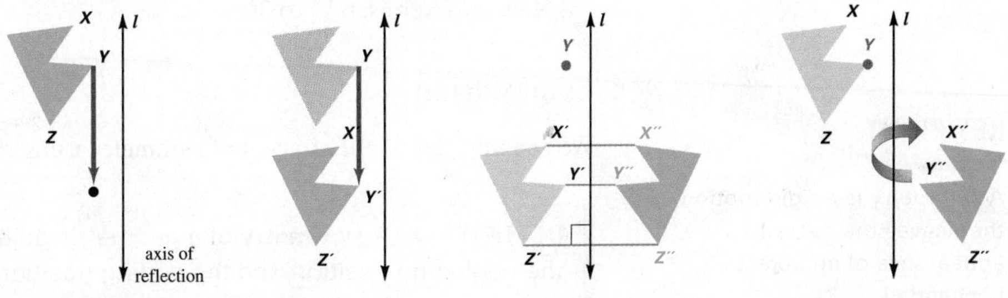

FIGURE 9.69 A glide reflection.

2nd: Slide the object along the translation vector so that the point *Y* coincides with the tip of the translation vector (Figure 9.69(b)).

QUIZ YOURSELF 22

Perform a glide reflection by first translating the triangle using the given translation vector, then reflecting the translated triangle about the axis of reflection.

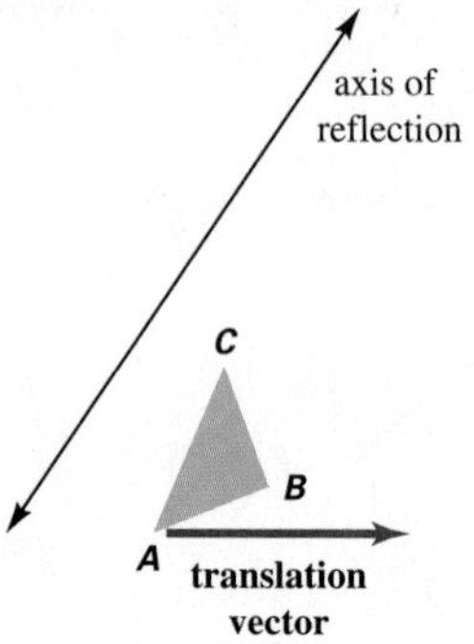

3rd: Reflect the object about the axis of reflection to get the final object (Figure 9.69(c)).

Figure 9.69(d) shows the final effect the glide reflection has in moving the original polygon to the polygon with vertices X'', Y'', and Z''.

Now try Exercises 15 and 16. 22

PROBLEM SOLVING

Strategy: The Order Principle

The Order Principle in Section 1.1 tells you to be careful about the order in which you perform a glide reflection. Performing a translation and then a reflection is not the same as if we reflect first and then translate.

The last rigid motion we will study is a rotation.

DEFINITION We perform a **rotation** by first selecting a point, called the *center of the rotation,* and then, while holding this point fixed, we rotate the plane about this point through an angle called the *angle of rotation.*

A good way to think of a rotation is to imagine that the plane is a piece of paper. If you stick a pin in the plane at the center of rotation and then rotate the plane, the plane will turn about the pin, and all points in the plane will move except the point where the pin is placed. We show a rotation in Figure 9.70.

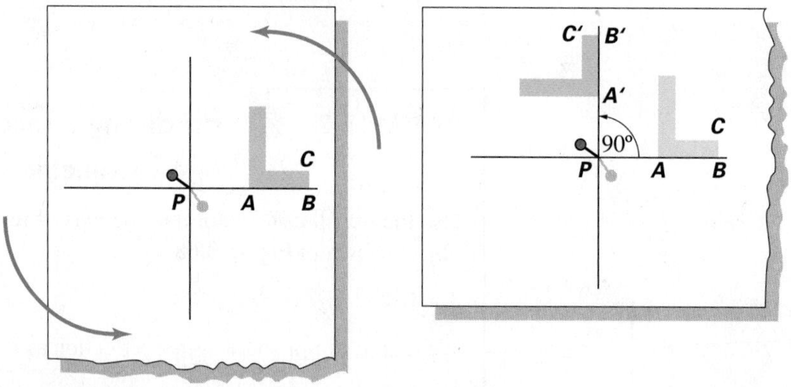

FIGURE 9.70 A rotation of 90° about point P.

Now try Exercises 17 to 20.

Symmetries

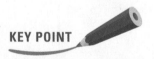

KEY POINT

A symmetry is a rigid motion that leaves the overall appearance of an object unchanged.

We can now define the concept of symmetry using the idea of a rigid motion.

DEFINITION A **symmetry** of a geometric object is a rigid motion such that the beginning position and the ending position of the object by the motion are exactly the same.

Looking back at Figure 9.62, we can say that the rotation and flip are two examples of symmetries of the star. The rotation is of course a rotation as we defined earlier, whereas the flip is a reflection.

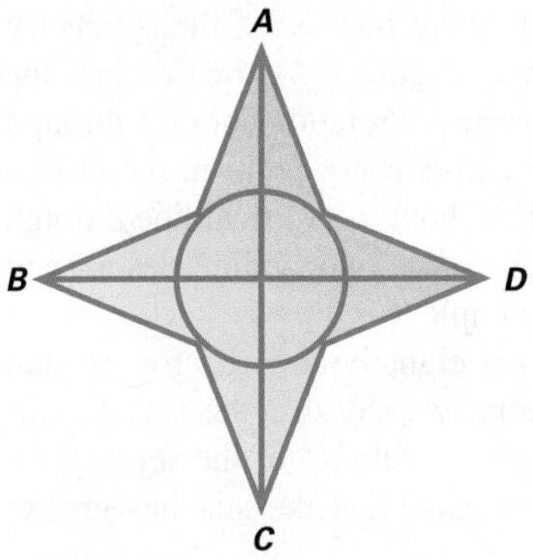

FIGURE 9.71 The star has many symmetries.

EXAMPLE 3 Symmetries of a Star

Find two symmetries (other than those shown in Figure 9.62) of the star in Figure 9.71.

SOLUTION:

There are clearly many ways to reflect and rotate the star to produce symmetries of the star. For example, in Figure 9.72, we reflect the star about the line l. We call l a *line of symmetry* for the star.

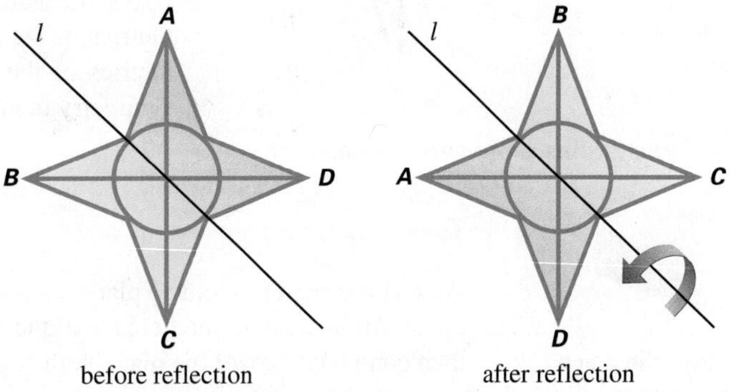

FIGURE 9.72 Reflection of the star about line *l*.

We can also rotate the star about its center, as shown in Figure 9.73.

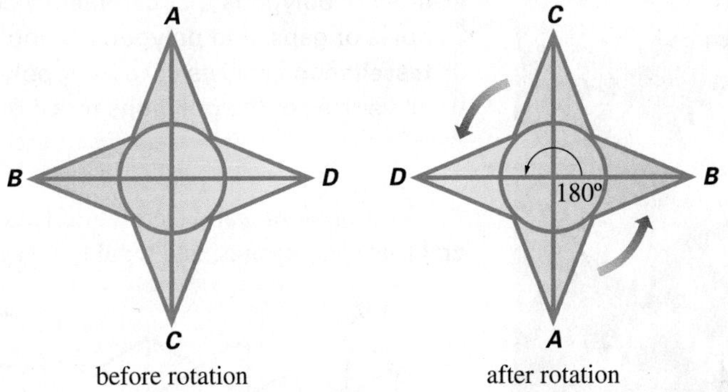

FIGURE 9.73 Rotation of the star about its center through an angle of 180°.

Now try Exercises 31 to 36.

Math in Your Life

What Makes a Bug Beautiful?

Surprisingly, what makes one insect attracted to another insect is the same thing we mentioned earlier: symmetry. Randy Thornhill and Steven Gangestad, professors at the University of New Mexico, have found that female scorpion flies are attracted to males with symmetrical wings. In another experiment, scientists discovered that by clipping the tail feathers of male swallows, thereby making them less symmetrical, the birds were less attractive to females and less likely to mate.

Biologists believe that animals with a high degree of symmetry have greater genetic diversity, which enables them to withstand environmental stress better and makes them more resistant to parasites. Lower symmetry goes hand in hand with lower survival rates and fewer offspring.

Professor Thornhill worked with Professor Karl Grammer of the University of Vienna to investigate whether symmetry is a factor in human attractiveness. They devised an index to measure facial symmetry with regard to placement of eyes, cheekbones, nose, and several other factors. Their research found that there is indeed a high correlation between facial symmetry and perceived attractiveness.

On the downside, other research suggests that women with asymmetrical breasts have a higher rate of breast cancer, and a study of West Indian men indicates that less symmetrical men are more susceptible to disease than their more symmetrical brothers.

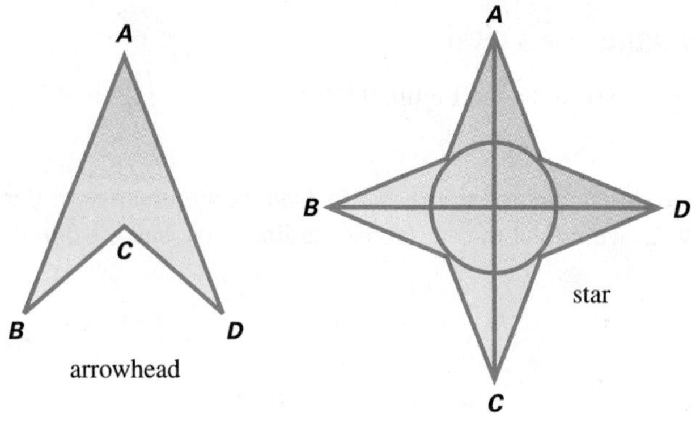

FIGURE 9.74 The star has more symmetries than the arrowhead.

Let's return to our question about measuring the symmetry in the arrowhead and the star. In Figure 9.74, we see that the arrowhead has only two symmetries: a rotation about A through an angle of 0° (this symmetry leaves every point in the arrowhead unchanged) and a reflection about the vertical line through points A and C. The star also has these two symmetries as well as many more, as we saw in Example 3.

It can be shown that there are eight symmetries for the star. Using the language of set theory, we say that the set of symmetries of the arrowhead is a proper subset* of the set of symmetries of the star. It is in this sense that the star has greater symmetry than the arrowhead.

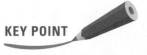

Tessellations

KEY POINT

Tessellations cover the plane with copies of polygons.

We can use rigid motions to place copies of a geometric figure at different positions in the plane. An interesting mathematical question is "Can we begin with a set of polygons and then completely cover the plane with copies of these polygons?"

DEFINITIONS A **tessellation** (or *tiling*) of the plane is a pattern made up entirely of polygons that completely covers the plane. The pattern must have no holes or gaps, and polygons cannot overlap except at their edges. A **regular tessellation** consists of regular polygons of the same size and shape such that all vertices of the polygons touch other polygons only at their vertices.

A designer of wallpaper, fabric patterns, or company logos should know that equilateral triangles, squares, and regular hexagons tessellate the plane, as we see in Figure 9.75.

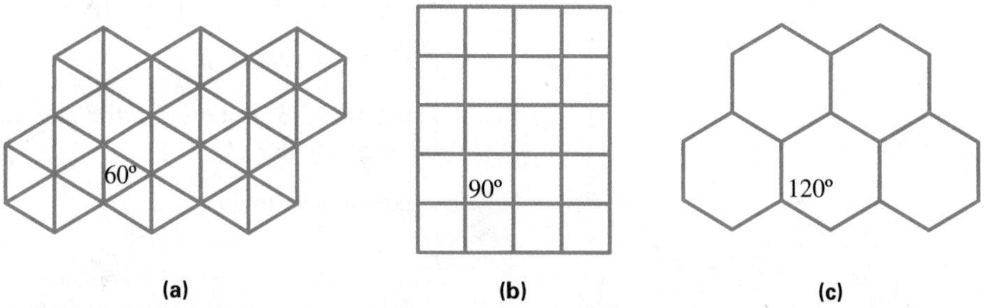

FIGURE 9.75 Regular tessellations of the plane using (a) triangles, (b) squares, and (c) hexagons.

It is natural to ask, "Are there any other regular polygons that tessellate the plane?" To answer this question, recall that in Section 9.2 we stated that for a regular n-sided polygon, each interior angle has measure $\frac{(n-2) \times 180°}{n}$. For example, the measure of each interior angle of a regular 12-sided polygon would be

$$\frac{(12 - 2) \times 180°}{12} = \frac{1{,}800°}{12} = 150°.$$

We will use this fact to determine which other regular polygons tessellate the plane.

*We discussed subsets in Section 2.2.

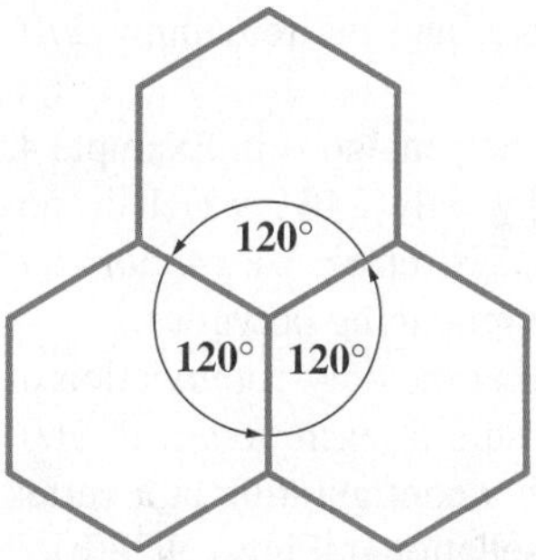

FIGURE 9.76 The angle sum around a vertex in a tessellation must add up to 360°.

EXAMPLE 4 Tessellating the Plane with Regular Polygons

Which regular polygons can tessellate the plane?

SOLUTION:

To understand the problem, let us look at a part of a tessellation using hexagons (Figure 9.76).

Notice that around a vertex of any regular tessellation, we must have an angle sum of 360°, and we also must have three or more polygons of the same shape and size. In Figure 9.75(a), we see that each vertex is surrounded by six 60° angles; in Figure 9.75(b), each vertex is surrounded by four 90° angles.

Let us now consider if it is possible to have a regular tessellation of the plane using pentagons. Recall that interior angles of a regular pentagon measure

$$\frac{(5-2)\times 180°}{5} = \frac{540°}{5} = 108°.$$

If we have three pentagons surrounding the vertex of a tessellation, as shown in Figure 9.77, the sum of the angles around the vertex is $3 \times 108° = 324° < 360°$. We do not have enough pentagons to completely surround the vertex; however, if we include a fourth pentagon, the sum of the angles around the vertex exceeds 360°, which causes the pentagons to overlap. Thus, there cannot be a regular tessellation using pentagons.

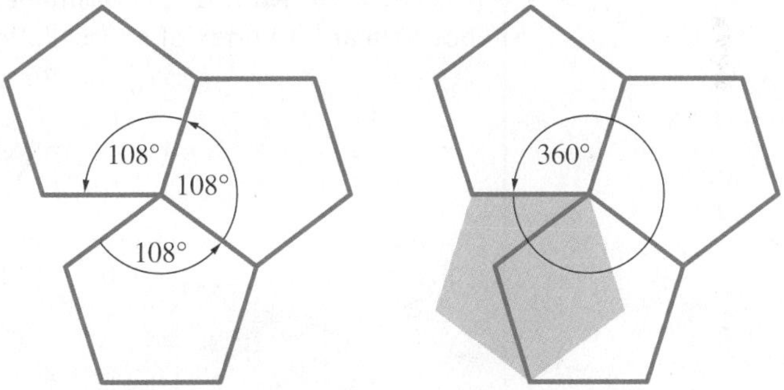

FIGURE 9.77 There is no regular tessellation of the plane using pentagons.

If a regular polygon has more than six sides, the measure of its interior angles exceeds 120°; therefore, we cannot have three such polygons surrounding a vertex of a tessellation.

Thus, we conclude that we can construct only three regular tessellations. They use equilateral triangles, squares, or regular hexagons.

Although there are only three regular tessellations, many tessellations are not regular. Is it possible to construct a tessellation using two different types of polygons that have edges of the same length?

EXAMPLE 5 Decorating with Nonregular Tessellations

The Spanish architect Antoni Gaudi specializes in producing decorative mosaics using ceramic tiles. At present, he has a number of tiles shaped like equilateral triangles and squares. All tiles have sides of the same length. Is it possible to produce a tessellation using a combination that contains both types of tiles?

SOLUTION: Review the Relate a New Problem to an Older One Strategy on page 9.

We can solve this problem by considering the different possibilities systematically. We will investigate tessellations that have one square at a vertex, two squares at a vertex, and so on.

We will begin by trying to construct a tessellation using only one square and the rest triangles; it would start out as shown in Figure 9.78.

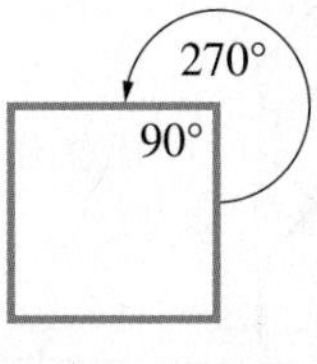

FIGURE 9.78 One square leaves 270° to be used for triangles.

Around the vertex 90° would be occupied by the square, and the remaining 270° would be occupied by triangles.

This gives us a problem similar to the one we had with the pentagons in Example 4. We need exactly 270 more degrees. Four triangles occupy only 240°, which is not enough, but five triangles occupy 300°, which is too much. Therefore, we *cannot* construct a tessellation with just one square and triangles as the remaining polygons.

Continuing this type of thinking, we see in Table 9.10 that the only combination of equilateral triangles and squares resulting in a total angle sum at each vertex of 360° occurs when we have two squares and three triangles. Such a configuration at a vertex is shown in Figure 9.79(a). We show a larger part of the tessellation in Figure 9.79(b).

Number of Squares at a Vertex	Amount of 360° Used by the Squares	Amount of 360° Remaining for the Triangles	Does 60° Divide This Number?	Is This Configuration Possible?
1	90°	270°	No	No
2	180°	180°	Yes	Yes
3	270°	90°	No	No
4	360°	0°	Yes	No (there will be no triangles)

TABLE 9.10 Number of possible equilateral triangle–square combinations possible at the vertex of a tessellation.

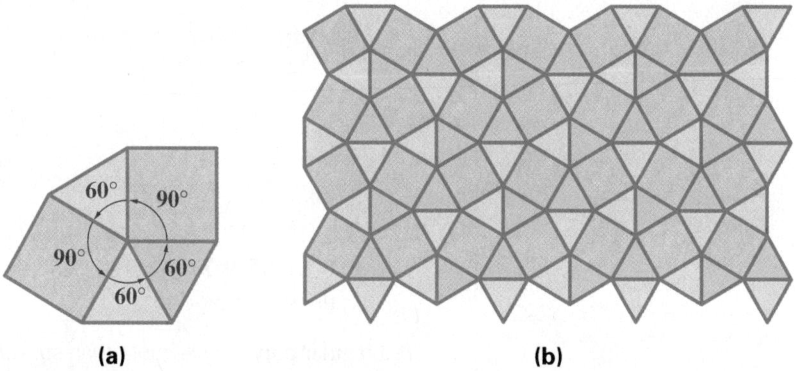

FIGURE 9.79 Two squares and three equilateral triangles around the vertex of a tessellation.

Now try Exercises 41 to 44. 23

QUIZ YOURSELF 23

Explain why the following tessellation is possible by considering the angles at each vertex in the tessellation.

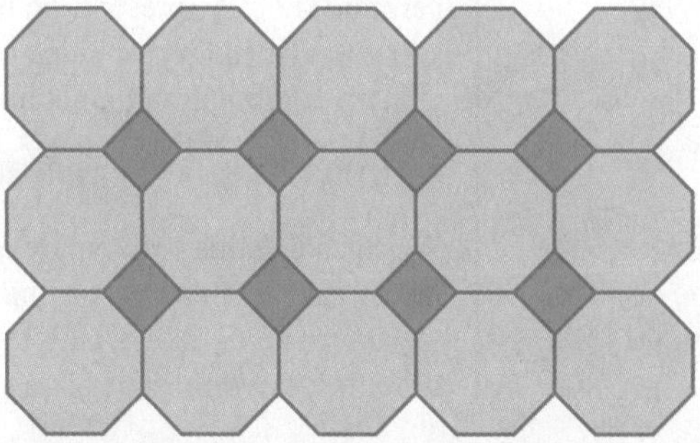

Highlight

Has All of Mathematics Been Discovered?

It may surprise you to know that mathematicians are still discovering new facts about the topics you studied in this section. You have seen that it is possible to tessellate the plane with certain types of regular polygons and nonregular polygons. For example, although we cannot tessellate the plane with regular pentagons, we can tessellate the plane with the following nonregular pentagon:

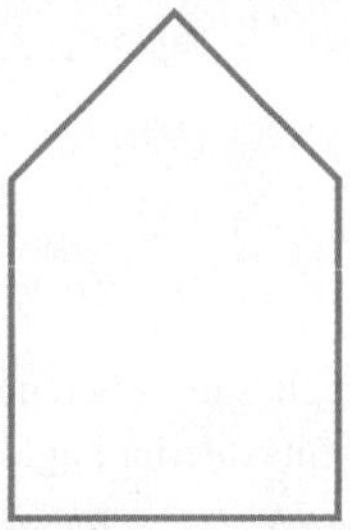

A simple question that a mathematician then asks is, "Exactly what types of pentagons can be used to tessellate the plane?" At one time, it was believed that there are only eight such pentagons. However, Marjorie Rice, a woman with only a high school background in mathematics, found a ninth pentagon that would tessellate the plane. By applying her new methods to this problem, she later went on to discover four more pentagonal tessellations. At this time, mathematicians do not know how many different types of convex pentagons will tile the plane. The example of Marjorie Rice is inspiring in that a person with little formal training in mathematics, but having interest and talent in mathematics, can make a significant contribution.

EXERCISES 9.6

Sharpening Your Skills

Use the following figure for Exercises 1–4. You may want to use graph paper to solve these exercises.

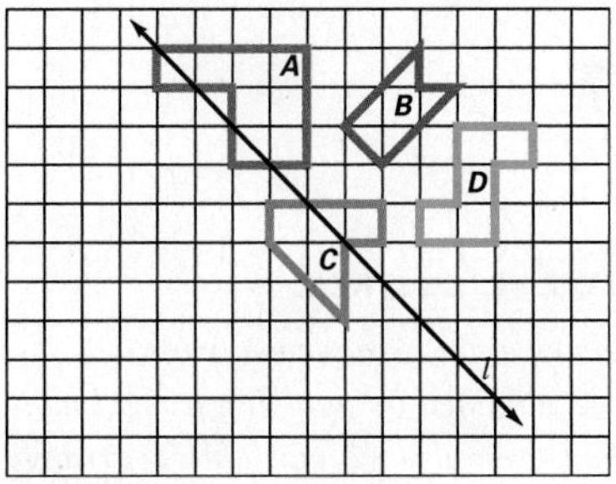

1. Reflect figure A about line l.
2. Reflect figure B about line l.
3. Reflect figure C about line l.
4. Reflect figure D about line l.

Use the following figure for Exercises 5 and 6.

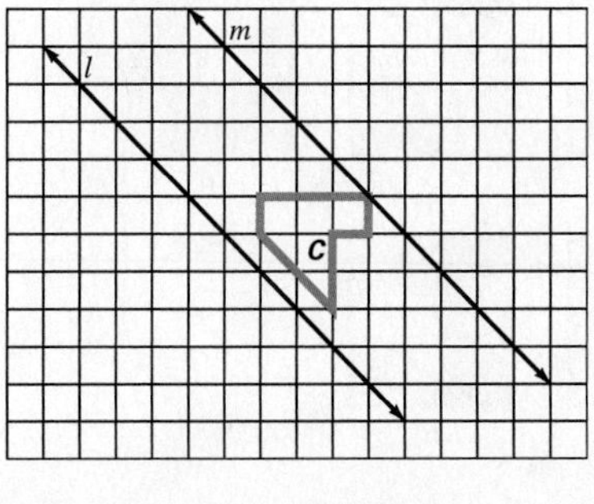

5. Reflect figure C about line l. Call the resulting figure C'. Then reflect figure C' about line m. Call the resulting figure C''.
6. Reflect figure C about line m. Call the resulting figure C'. Then reflect figure C' about line l. Call the resulting figure C''.

Use the following figure for Exercises 7 and 8.

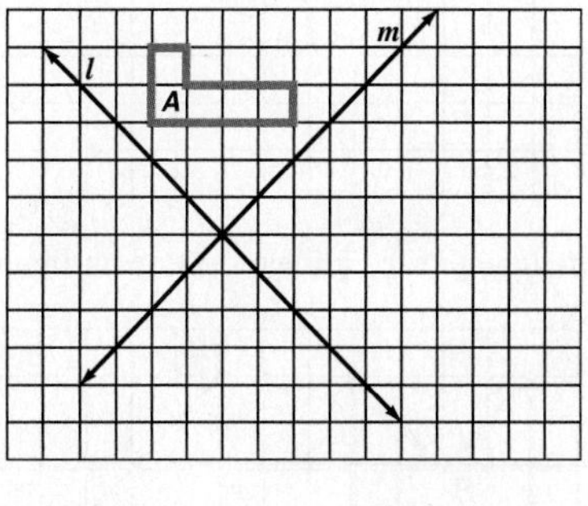

7. Reflect figure A about line l. Call the resulting figure A'. Then reflect figure A' about line m. Call the resulting figure A''.
8. Reflect figure A about line m. Call the resulting figure A'. Then reflect figure A' about line l. Call the resulting figure A''.
9. In Exercises 5 and 6, you were reflecting an object about one line and then another line that was parallel to the first line.
   a. Did the order in which you did the reflections make a difference? Explain.
   b. What is the effect of performing the two reflections on the object?

10. In Exercises 7 and 8, you were reflecting an object about one line and then another line that was perpendicular to the first line. Did the order in which you did the reflections make a difference? Explain.

Use the following figure for Exercises 11–14.

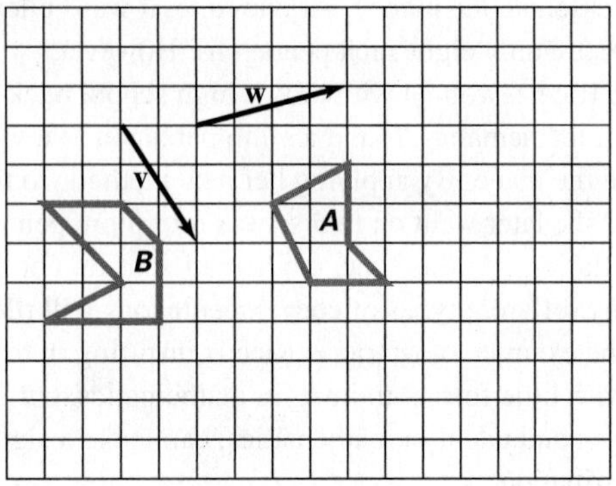

11. Translate figure A by vector **w**.

12. Translate figure B by vector **v**.

13. Translate figure A by vector **v**. Call the resulting figure A'. Then translate figure A' by vector **w**. Call the resulting figure A''.

14. If you reverse the order in which you do the translations in Exercise 13, does it make a difference in the resulting figure A''? Explain.

15. Perform the indicated glide reflection on figure A.

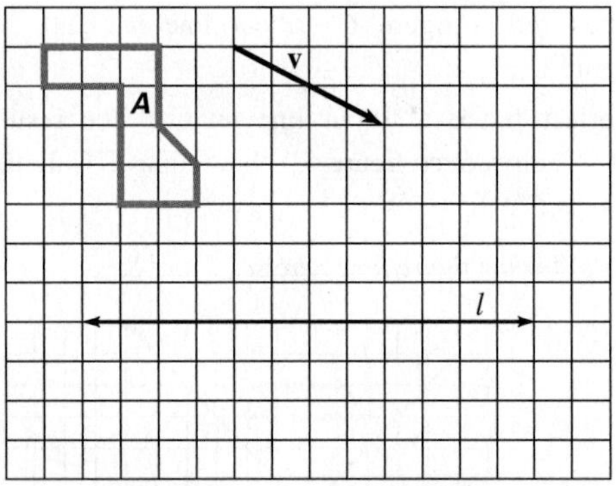

16. Perform the indicated glide reflection on figure B.

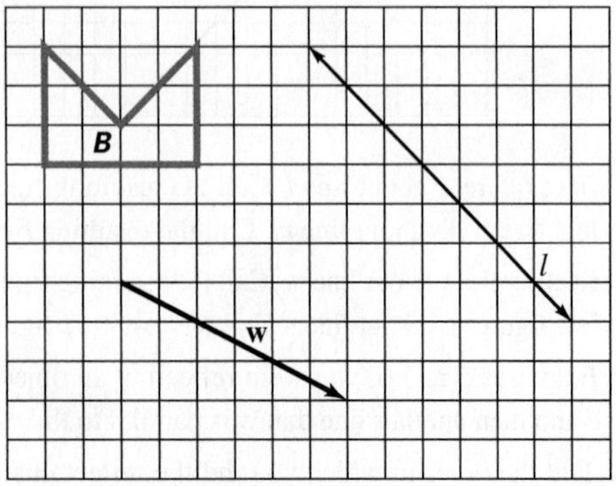

In Exercises 17–20, starting with the original position shown, rotate each shape about point P *for each of the following rotations:* 45°; 90°; 180°.

17.

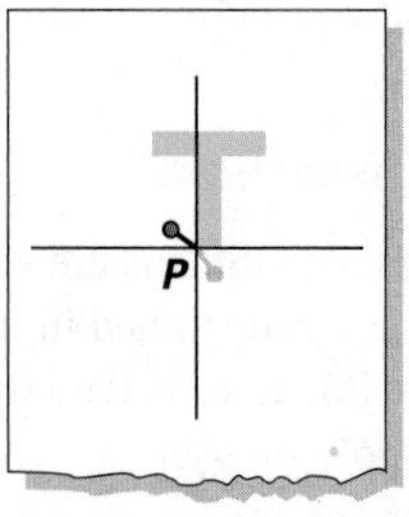

18.

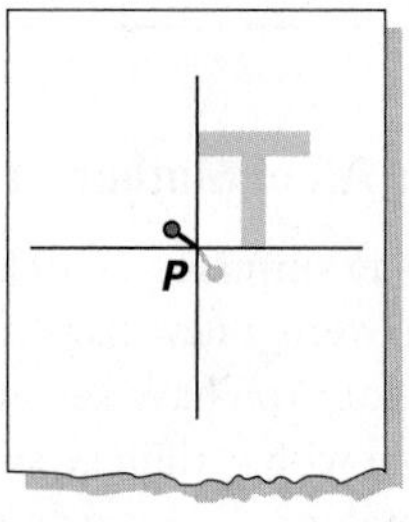

19.

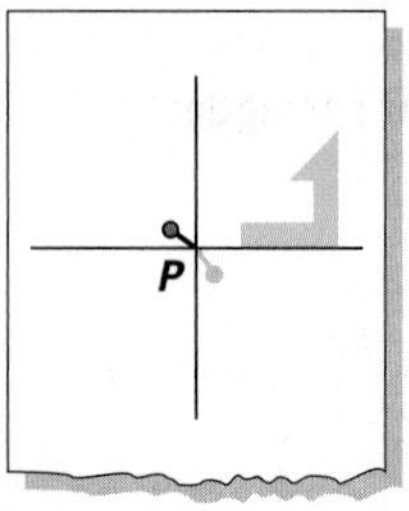

20.

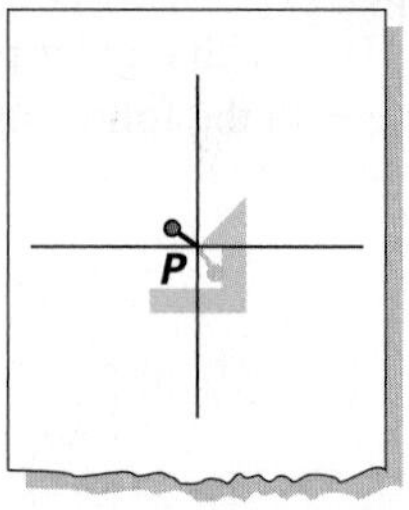

21. What is the interior angle sum of a regular 12-sided polygon? What is the measure of its interior angles?

22. What is the interior angle sum of a regular 15-sided polygon? What is the measure of its interior angles?

23. Explain why we can tessellate the plane with a regular hexagon, but not with a regular pentagon.

In Exercises 24–26, tesselate the plane with the given figure.

24.

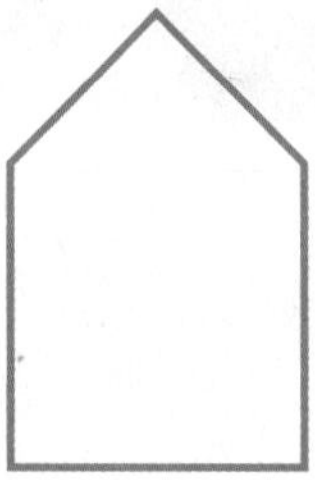

25.

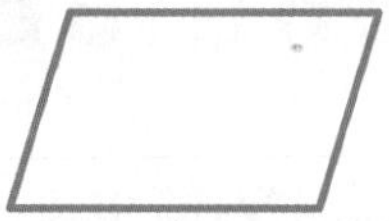

26.

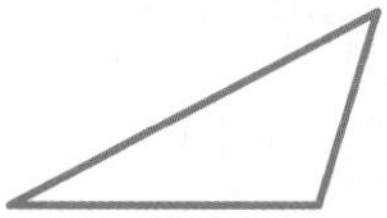

Applying What You've Learned

The following figure shows eight arrangements of tiles. Some of the arrangements can be obtained by applying a rigid motion to other arrangements. For example, we can obtain arrangement (e) by reflecting arrangement (a) about a vertical axis. Use these arrangements to solve Exercises 27–30.

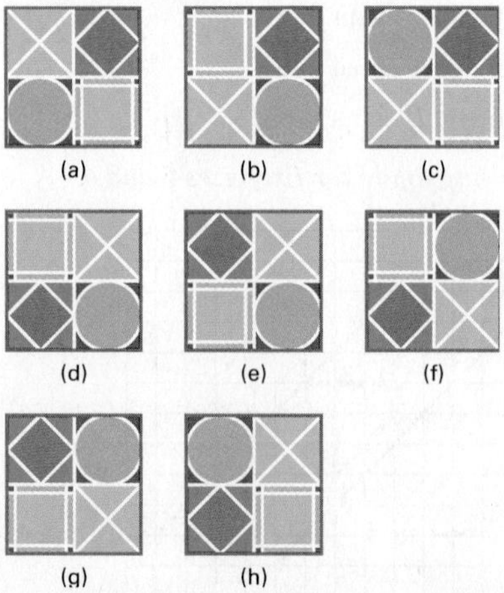

27. List all arrangements that we obtain by reflecting arrangement (b) about a line. (Don't forget reflections about diagonal lines.)

28. List all arrangements that we obtain by rotating arrangement (g) about its center.

29. List all arrangements that we obtain by applying a rigid motion to arrangement (f).

30. Explain why it is impossible to find a rigid motion that we can apply to arrangement (a) to obtain arrangement (b).

In Exercises 31–36, list all the reflectional symmetries for each object. Also find all rotational symmetries of the object using angles between 1° and 359°.

31.

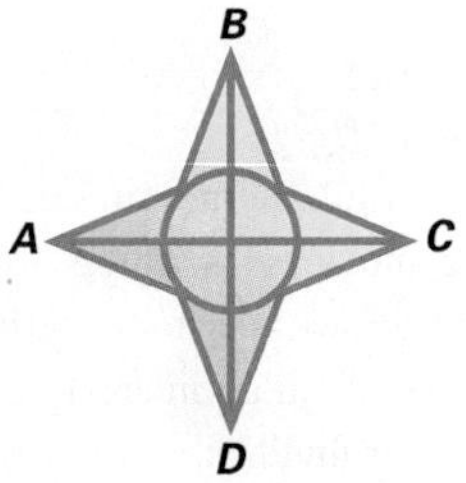

32.

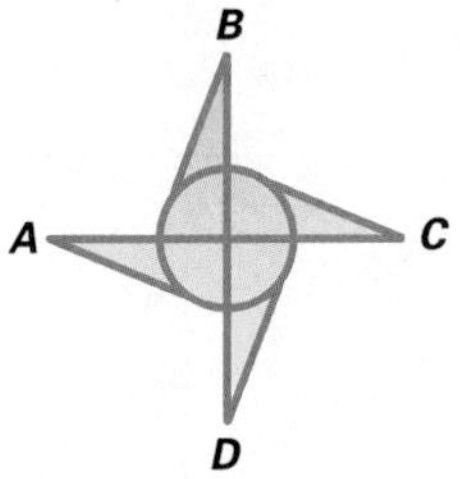

33.

(phone symbol)

34.

(hospital symbol)

35.

(Olympic symbol)

36.

(Chrysler symbol)

Communicating Mathematics

37. What are the four types of rigid motions?

38. If we tessellate the plane with polygons, what is the angle sum around each vertex? Why does this imply that we can tessellate the plane with a hexagon but not a pentagon?

39. What is the difference between a symmetry of an object and a rigid motion?

40. In the Problem Solving box on page 492, we stated that in doing a glide reflection, you must do the glide first and then the reflection because if you reverse these rigid motions, you do not get the same result. Yet, in Example 2, it appears that if we had reflected first and then translated, we would have gotten the same final figure. Is there a contradiction here?

Challenge Yourself

In Exercises 41–44, explain why each tessellation is possible by considering the angles at each vertex in the tessellation, as we did in Quiz Yourself 23.

41.

42.

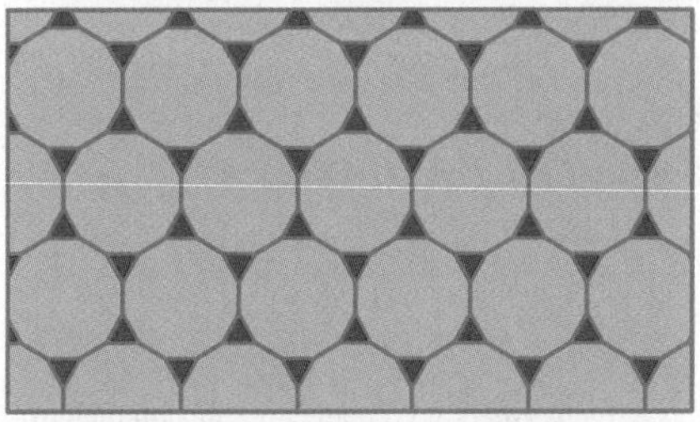

43.

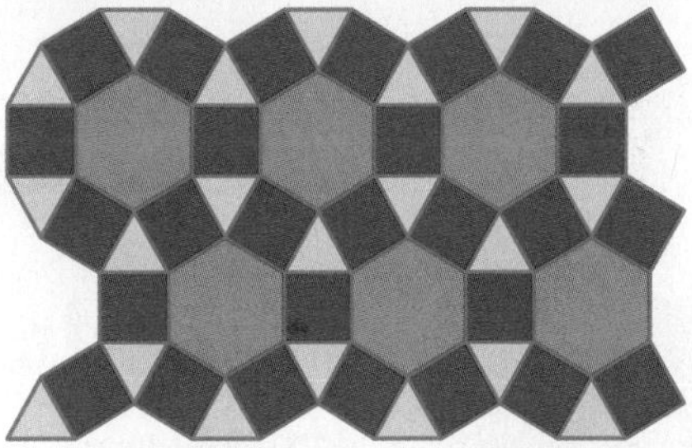

44.

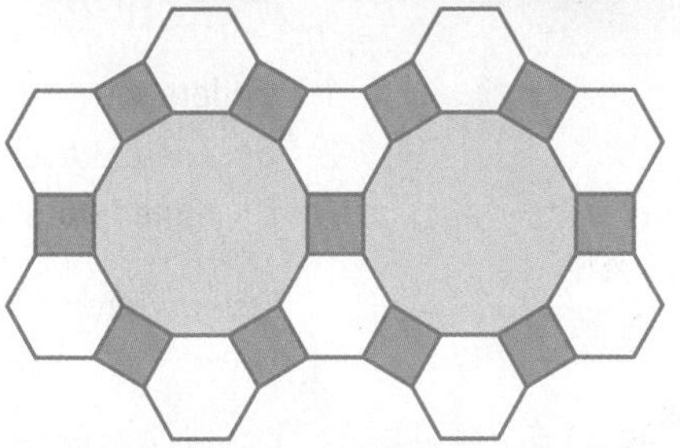

45. Use the given figure to explain why any convex quadrilateral will tessellate the plane.

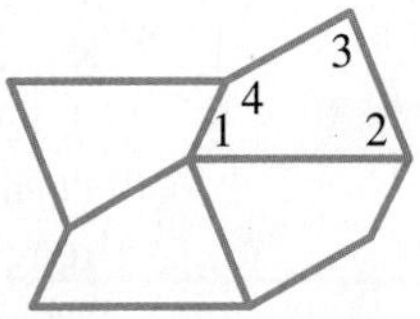

46. Use a drawing similar to the one in Exercise 45 to show that the given quadrilateral tessellates the plane.

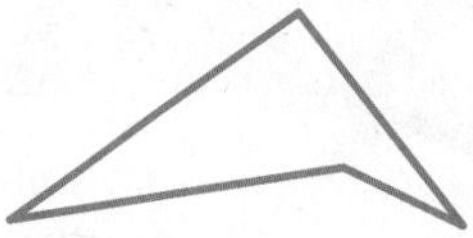

A perpendicular bisector of a chord of a circle passes through the center of a circle. Because the line* l *divides the segment* AB *into two equal parts and is also perpendicular to the segment* AB, *we know that the center of the circle lies somewhere on* l. *Use this information to estimate the center of rotation for the rotations shown in Exercises 47 and 48.*

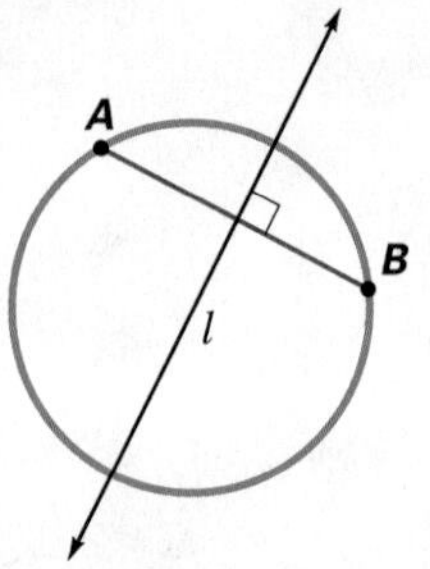

Figure for Exercises 47 and 48

*A *chord* of a circle is a line segment with both endpoints on the circle.

47.

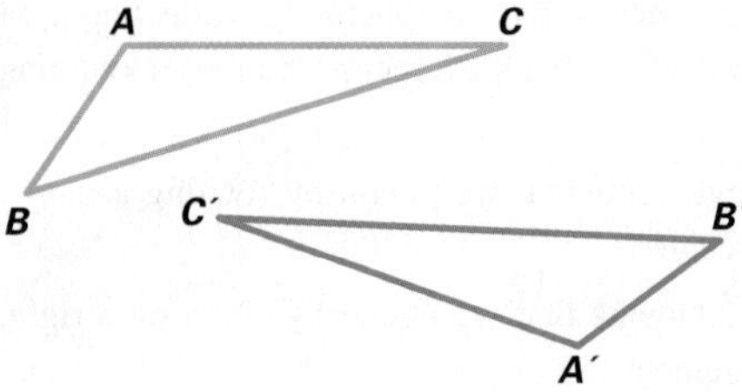

48.

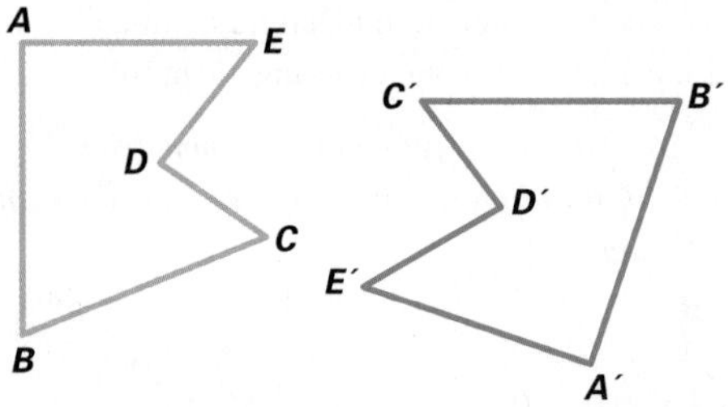

49. There are many interesting and beautiful Internet sites devoted to tessellations. Some sites have movies describing how to tessellate the plane, and others allow you to experiment with creating your own tessellations. Find an interesting site, experiment with it, and report on your findings.

Looking Deeper

9.7 Fractals

Objectives

1. Understand the self-similar nature of fractals.
2. Calculate the length and area of fractal objects.
3. Compute fractal dimension.

Everything in Nature can be viewed in terms of cones, cylinders, and spheres.

—Paul Cezanne (nineteenth-century impressionist artist)

Clouds are not spheres, mountains are not cones, coastlines are not circles, and bark is not smooth, nor does lightning travel in a straight line.

—Benoit Mandelbrot (former IBM research mathematician)

So, who's right? If you look around you, it is easy to see that Mandelbrot seems to be describing nature better than Cezanne. Mountain peaks, the ocean shore, and the clouds do not have nice smooth edges as drawn in a child's storybook, but, rather, they have rough, jagged edges that we cannot explain using traditional Euclidean methods. In this section, you will study a relatively new and different type of geometry, called **fractal geometry,** that describes real-life objects and patterns more accurately than we can by using Euclidean geometry and that has many important real-life applications.

Fractals

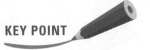

KEY POINT

Fractal objects are self-similar.

To understand how fractal geometry differs from Euclidean geometry, imagine a photograph of the edge of a large cloud taken from a weather satellite hundreds of miles away. The edge would not be a smooth curve, as clouds are often drawn in children's books; rather, it would be extremely jagged, as shown in Figure 9.80(a). If we enlarged a small portion of this edge, we would see a jagged curve that looks something like Figure 9.80(b). If we further enlarged a tiny portion of this smaller piece, we would still see an edge

containing the same type of jaggedness that was present in the original photograph. With a fractal object, no matter how much we magnify it, we still see patterns that are very similar to the patterns that were present in the original object. We say that an object with this property is **self-similar.**

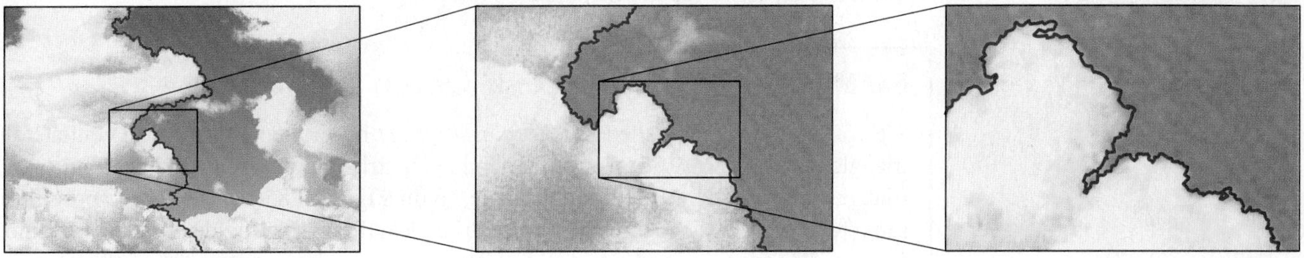

FIGURE 9.80 A cloud is self-similar.

Benoit Mandelbrot developed the theory of fractal geometry while working as a mathematician at IBM during the early 1960s. To get a better understanding of his geometry, we will construct a curve called the *Koch curve.*

EXAMPLE 1 The Koch Curve Is a Fractal

We begin the Koch curve by drawing a line segment* *AB* (step 0), which we divide into three equal parts. At segment *CD*, we construct an equilateral triangle ΔCED, and then remove segment *CD*, giving the object shown as step 1 in Figure 9.81.

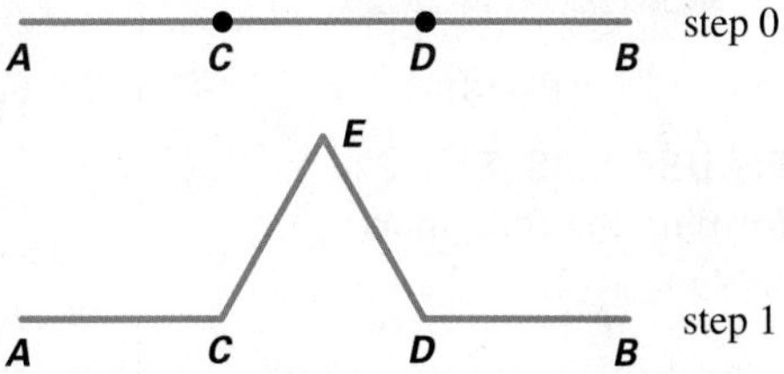

FIGURE 9.81 Beginning the Koch curve.

To continue the construction of the curve, we divide each of the four line segments in step 1 into three parts and replace the middle segment by a triangular bump, as we did in going from step 0 to step 1. The resulting curve is shown in step 2 of Figure 9.82. If we repeat this process again for the 16 line segments in step 2, we get the curve shown in step 3 of Figure 9.82.

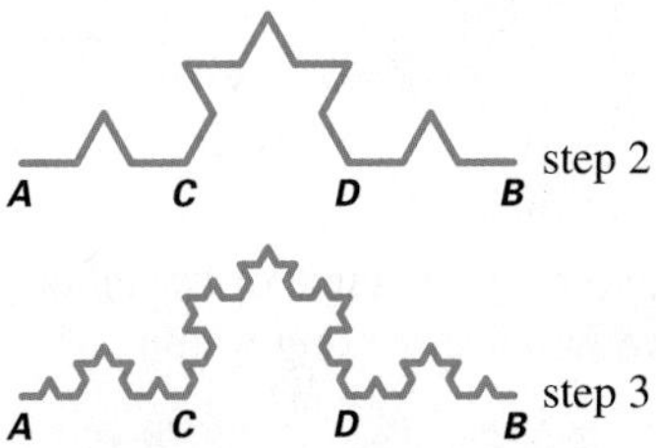

FIGURE 9.82 Steps 2 and 3 of the Koch curve.

To finish the Koch curve, we must repeat indefinitely the process of subdividing each line segment and replacing it with a line segment having a triangular bump.

In Figure 9.83, we see why the Koch curve is a fractal. If we enlarge a small portion of the curve, we see that the enlargement has the exact same structure as the original curve. No matter how many times we magnify this curve, we always see the same repeated pattern.

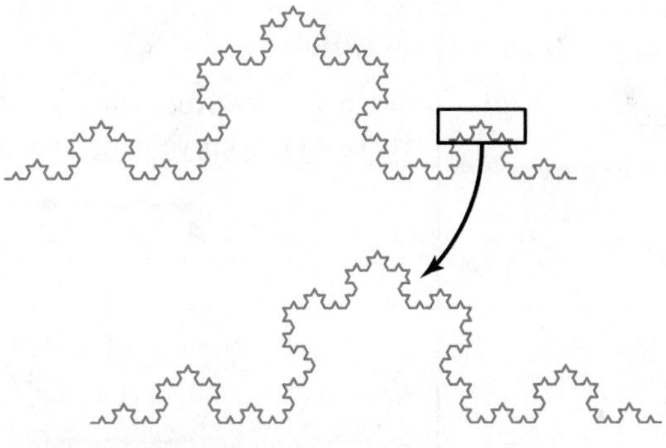

FIGURE 9.83 Small portions of the Koch curve repeat the same pattern as the original curve.

Now try Exercises 1 to 4. 24

QUIZ YOURSELF 24

How many line segments would there be in step 4 of the Koch curve?

*To avoid cluttering fractal drawings, we will not show the endpoints of line segments.

FIGURE 9.84 Fractal art.

Beautiful fractal art, such as in Figure 9.84, has the same self-similarity property as the Koch curve. If we were able to put the fractal in Figure 9.84 under a microscope, we would see the same beautiful patterns at every level of magnification.

We can begin with a two-dimensional object and, by applying some rule repeatedly, create a fractal, as we see in Example 2.

EXAMPLE 2 The Sierpinski Gasket Is a Fractal

We construct a fractal called the *Sierpinski gasket* by first constructing an equilateral triangle, as in Figure 9.85(a). We then divide this triangle into four smaller equilateral triangles and remove the middle triangle, as in Figure 9.85(b). Next, we apply this same rule to each of the three remaining triangles; that is, we divide each triangle into four smaller triangles and remove the center one. We show the results of this step in Figure 9.85(c).

(a) step 0 **(b)** step 1 **(c)** step 2

FIGURE 9.85 The first two steps in forming the Sierpinski gasket.

step 3

FIGURE 9.86 Step 3 in forming the Sierpinski gasket.

As with the Koch curve, we must continue this process of removing the center of each solid equilateral triangle indefinitely to complete this fractal. Figure 9.86 shows the next step in forming the Sierpinski gasket.

We cannot draw the entire gasket because to do this, we would have to draw smaller and smaller triangles that would eventually become so small that their size would be finer than the resolution of any printing device. 25

QUIZ YOURSELF 25

How many dark triangles would appear in step 5 of forming the Sierpinski gasket?

Length and Area

As in Euclidean geometry, we are interested in the length, area, and volume of fractal objects. When we investigate the length of the Koch curve, we find a surprising result.

EXAMPLE 3 The Length of the Koch Curve Is Infinite

Find the length of the Koch curve.

SOLUTION:

We began the Koch curve with a line segment that has length 1 (step 0). In step 1 we replaced this curve with a curve $\frac{4}{3}$ as long, as shown in Figure 9.87.

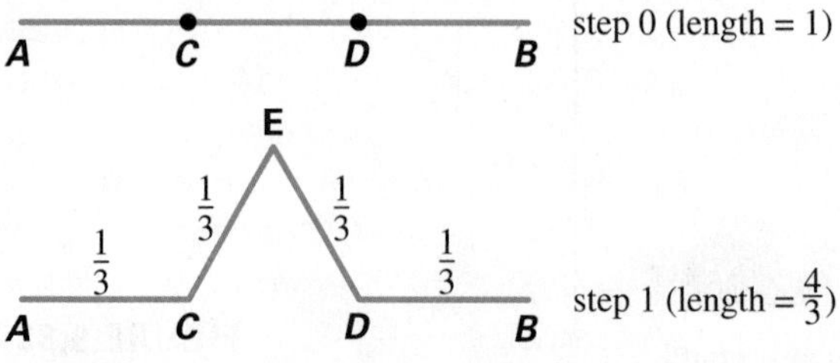

FIGURE 9.87 In step 1, the Koch curve is $\frac{4}{3}$ as long as the original line segment.

Figure 9.88 shows that in step 2, the curve consists of 16 small line segments with length $\frac{1}{9}$, so the length is now $\frac{16}{9}$.

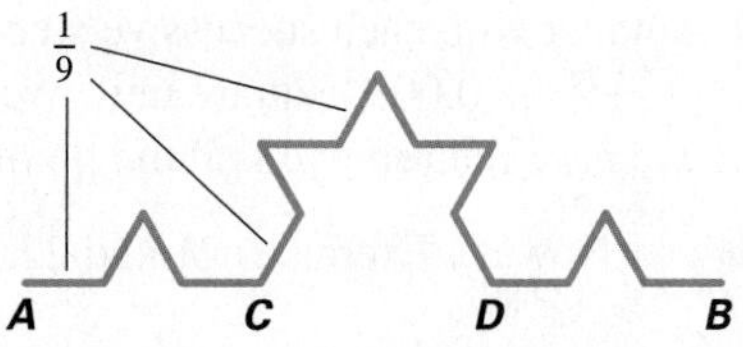

step 2 (length $= \frac{16}{9}$)

FIGURE 9.88 Each segment now has length $\frac{1}{9}$.

At each successive step, the curve is $\frac{4}{3}$ as long as the curve in the previous step. This means that the curve's length keeps growing larger and larger as we construct further steps of the curve. At step 40, we would find that the length of the Koch curve is $\left(\frac{4}{3}\right)^{40}$, which is slightly more than 99,437 units long. At stage 100, the curve is over 3 trillion units long! Of course, we do not stop at stage 100. Because we must go through an infinite number of stages to construct the whole curve, the total length of the Koch curve is therefore infinite.

QUIZ YOURSELF 26

What is the length of the Koch curve at step 3 of the construction process?

Now try Exercises 5 and 6. 26

EXAMPLE 4 The Area of the Sierpinski Gasket Is Zero

What is the area of the Sierpinski gasket?

SOLUTION:

To make the computations easier to follow, let us assume that we begin the Sierpinski gasket with an equilateral triangle with area 1 (see Figure 9.89). In step 1 of the construction, we have removed $\frac{1}{4}$ of the area, so the gasket now consists of three triangles, each with area $\frac{1}{4}$. The area of the dark triangles at step 1 is therefore $\frac{3}{4}$.

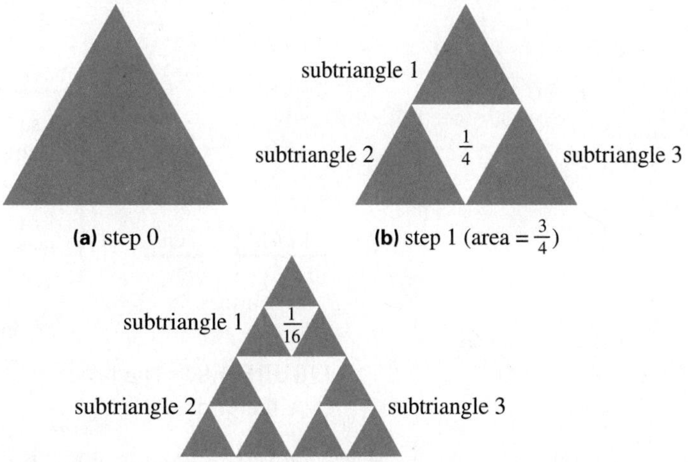

(c) step 2 (area $= \frac{9}{16}$)

FIGURE 9.89 Step 1 of the Sierpinski gasket has area $\frac{3}{4}$; step 2 has area $\frac{9}{16}$.

Now consider subtriangle 1. In step 2, we remove $\frac{1}{4}$ of its area, which is $\left(\frac{1}{4}\right)\left(\frac{1}{4}\right) = \frac{1}{16}$ of the original area, leaving three smaller triangles also with area $\frac{1}{16}$. The area that remains in subtriangle 1 is therefore $\frac{3}{16}$ of the original area. Likewise, $\frac{3}{16}$ of the original area remains in subtriangles 2 and 3 after we remove their centers. Therefore, in step 2 of constructing the gasket, the remaining area is

$$\frac{3}{16} + \frac{3}{16} + \frac{3}{16} = \frac{9}{16} = \left(\frac{3}{4}\right)\left(\frac{3}{4}\right).$$

We see that at each successive step in constructing the gasket, we get an area that is $\frac{3}{4}$ the area of the previous step. Therefore, the area of the gasket keeps getting smaller and

QUIZ YOURSELF 27

What is the area of the Sierpinski gasket at step 3 of the construction?

smaller with each successive step in the construction. For example, at step 20 the area is $\left(\frac{3}{4}\right)^{20} \approx 0.0032$ square unit. We conclude that the area of the gasket is 0, even though we have not removed all the points in the original triangle.

Now try Exercises 21 and 22. 27

Dimension

You may have a feeling that the Koch curve, because of all its "wiggling around," is somewhat thicker than the kinds of curves we draw in Euclidean geometry. Therefore, we might say that the Koch curve, in some sense, has a larger dimension than a smooth curve in Euclidean geometry. On the other hand, all the "wiggling" is not enough for the curve to fill entirely some region of the plane, which would make it a two-dimensional object. In order to make this concept of dimension more clear, consider the line segment, the square, and the cube in Figure 9.90.

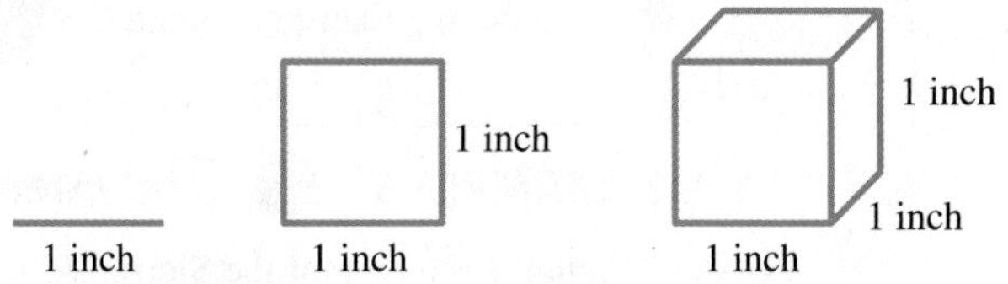

FIGURE 9.90 A unit of measurement in one, two, and three dimensions.

Imagine that we place the line, the square, and the cube into a three-dimensional copying machine. This copying machine will increase or decrease the size of any object that we place into it. If we set the size of our copies to two times the original, the copies of the line, square, and cube would come out looking as in Figure 9.91.

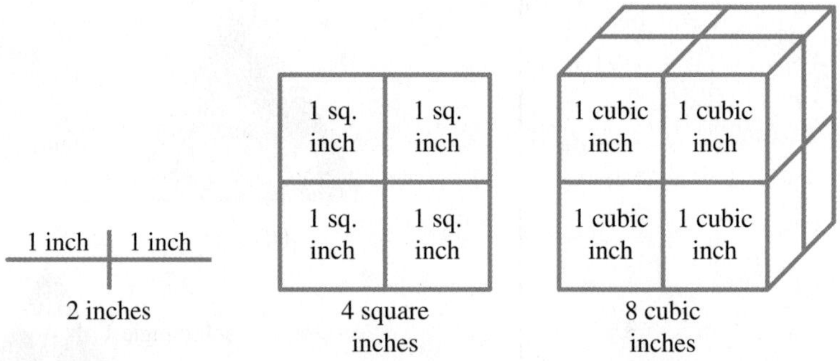

FIGURE 9.91 The line, square, and cube have been enlarged by a factor of 2.

A general way of looking at this is to say we used a scaling factor of s ($s = 2$ in this case). In the one-dimensional case, the copier returned a copy that had length equal to $s^1 = 2$ times the length of the original. In the two-dimensional case, the copier returned a copy containing an area equal to $s^2 = 4$ times the original area. For the cube, the copier returned a copy with $s^3 = 8$ times the volume of the original. It seems then that we can think of the dimension of an object as an exponent D that satisfies these types of equations. We now define the notion of dimension for fractals.

DEFINITION The **fractal dimension** of an object is a number D that satisfies the equation

$$n = s^D,$$

where s is a scaling factor and n is the amount by which the quantity we are measuring (length, area, volume) of the object changes when we apply the scaling factor to the object.

Highlight

Applications of Fractals

It is surprising that the fractal geometry that produces such strange and beautiful images can also describe many important natural phenomena. Scientists use fractal geometry to explain the similarities and differences between the way a tree branch divides repeatedly to form finer subbranches and the way the bronchi of the lungs subdivide to form a tree of airways inside our lungs. This same sort of branching occurs in a reverse fashion as small tributaries join together to form streams and eventually a river.

Geographers classify the roughness of coastlines according to their fractal dimension. The South African coast is relatively smooth, with a fractal dimension close to 1, whereas the extreme irregularity of Norway's coast has a fractal dimension of 1.52.

Understanding the fractal pattern seen in a lightning bolt also helps explain how electrical insulators break down when subjected to high voltages. This same pattern occurs at a microscopic level when crystals form.

Economists studying the stock market have found that if they graph the market over hundreds of days, then over hundreds of hours, and finally over hundreds of 30-second intervals, the graphs look remarkably similar. Cardiologists have learned that the healthy heart beats according to a fractal rhythm rather than a steady, regular rhythm, as we used to think.

Using percolation theory, mathematicians apply fractal geometry to describe the way coffee percolates in a coffeepot and the way groundwater seeps into the soil. It is surprising that this same area of fractal mathematics also explains how a fire "percolates" through a forest, how an epidemic "percolates" through a population, and how galaxies "percolate" throughout the universe.

Figure 9.92 shows an interesting fractal similarity between a satellite view of the rivers of Norway and the growth of ice crystals on a window pane.

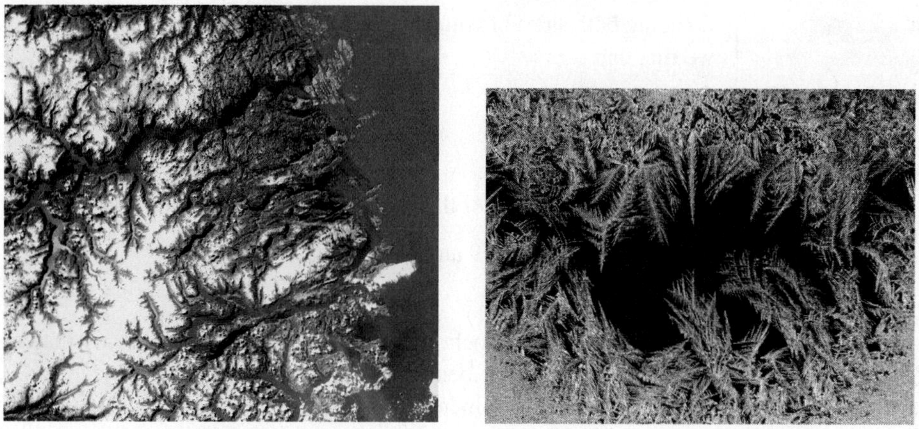

(a) (b)

FIGURE 9.92 (a) Rivers of Norway. (b) Ice crystals.

To understand the notion of fractal dimension, let's look again at the Koch curve.

EXAMPLE 5 The Fractal Dimension of the Koch Curve

What is the dimension of the Koch curve?

SOLUTION:

We need to understand what it means to magnify the curve by a factor. Consider Figure 9.93(a), where we show a picture of the *entire* Koch curve but because the picture is so small, we cannot see much detail. We next enlarge Figure 9.93(a) by a factor of 3 and display this enlargement in Figure 9.93(b).

We see that when each of the 16 tiny line segments in Figure 9.93(a) is enlarged, it appears as a line segment with a "bump" on it, as shown in Figure 9.94. The line segment with the bump is $\frac{4}{3}$ times as long as the line segment before enlargement.

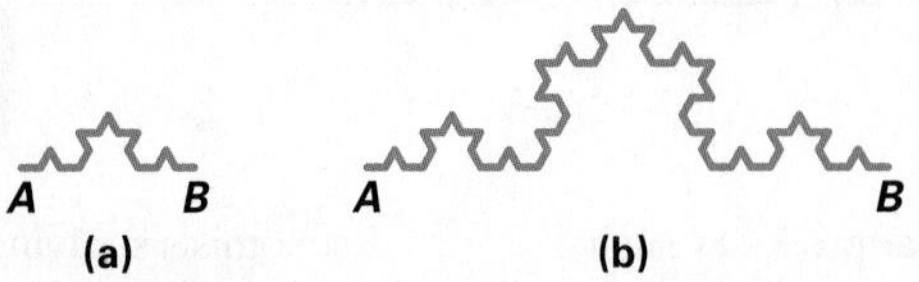

FIGURE 9.93 The Koch curve enlarged by a factor of 3.

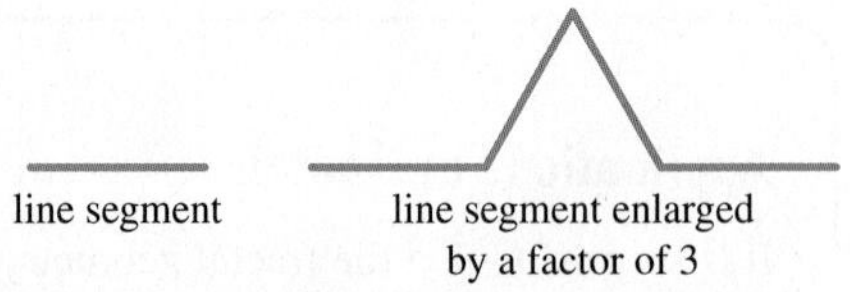

FIGURE 9.94 Each enlarged segment of the curve appears $\frac{4}{3}$ as long as the original.

What we are saying is that enlarging any portion of the curve by a scale factor of 3 appears to increase its length by 4. If we call the dimension of the curve D, this means that

$$3^D = 4. \qquad (1)$$

We now solve for D. To do this, we need to use the log key on our calculator. The log function has the property that $\log a^x = x \log a$; we will use this property to solve equation (1).

Taking the log of both sides of equation (1) gives us

$$\log 3^D = \log 4. \qquad (2)$$

Now using the property we just stated for log, we get

$$D \log 3 = \log 4. \qquad (3)$$

Dividing both sides of equation (3) by log 3 and using a calculator to evaluate the result, we find that

$$D = \frac{\log 4}{\log 3} \approx 1.26.$$

The fractal dimension of the Koch curve is therefore approximately 1.26.

Now try Exercises 9 and 10. 28

QUIZ YOURSELF 28

Assume that for a fractal curve, enlarging the curve by a factor of 4 increases its length by a factor of 8. What is the fractal dimension of the curve?

The idea that the Koch curve has dimension 1.26 means that in a certain sense it is thicker, or fills space better, than a one-dimensional object such as a line segment. However, because this dimension is less than 2, the Koch curve does not fill space as well as a two-dimensional object such as a solid square.

Artists use fractal geometry in movies to create beautiful mountains, clouds, and other natural-looking objects. We will show you a simple example of how to create a tree using fractals.

EXAMPLE 6 Drawing a Fractal Tree

Explain why the "tree" in Figure 9.95 is a fractal.

SOLUTION:

The basic pattern in the tree is determined by the line segments joining points A, B, C, and D. At the end of segment AB is a branching represented by segments BC and BD. Smaller versions of this Y-shaped motif occur repeatedly throughout the tree. The fractal pattern is clear. To add finer branches to the tree, we choose a branch such as DF and replace it with a small Y-shaped figure.

Now try Exercises 23 and 24.

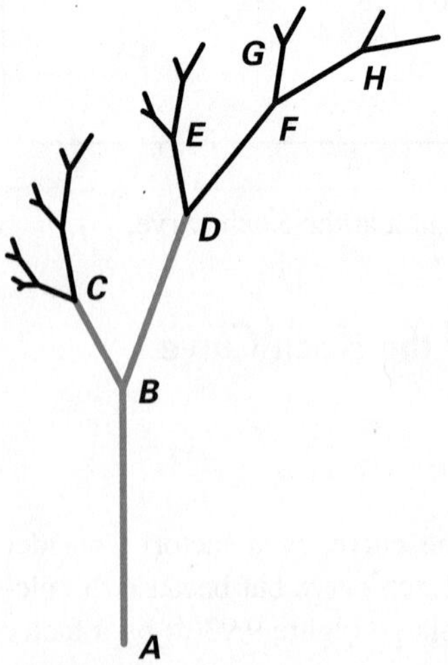

FIGURE 9.95 A fractal tree.

Natural-looking scenes such as those shown in Figure 9.96 are generated using techniques that are similar to the method we described in Example 6.

FIGURE 9.96 Fractal landscape.

EXERCISES 9.7

Sharpening Your Skills

Exercises 1–4 pertain to the Koch curve in Example 1.

1. How many line segments will the curve have in step 5?

2. What is the length of the curve at step 5?

3. What is the length of the curve at step 10?

4. True or false: Doubling the number of steps doubles the length of the curve.

In Exercises 5 and 6, you are given steps 0 and 1 for constructing a fractal.

a) Construct step 2 of the fractal.

b) Assume that the length of the line segment in step 0 is 1. Find the length of the curve in step 5.

5.

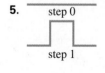

6.

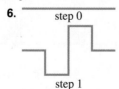

7. Solve $4^D = 6$.

8. Solve $4^D = 12$.

9. Find the fractal dimension of the curve described in Exercise 5.

10. Find the fractal dimension of the curve described in Exercise 6.

Applying What You've Learned

11. We can get a realistic "coastline" effect if we slightly vary the construction of the Koch curve. In the construction of the Koch curve in Example 1, whenever we added a bump to a line segment, we always added it above the curve. Now when we add a bump, we will use the following list of random numbers:

87127 03570 73103 16946 81852 94819
33108 72734 43411 31078

We will add bumps above the curve for even digits and below the curve for odd digits. The first random digit is even, so we will add the first bump above the curve; the second and third digits are odd, so we add the second and third bumps below the curve. The fourth digit is even, so we draw the fourth bump above the curve, and so on. We show steps 1 and 2 in the following diagram. Redraw steps 1 and 2 of this fractal; however, now use random digits beginning with 16946.

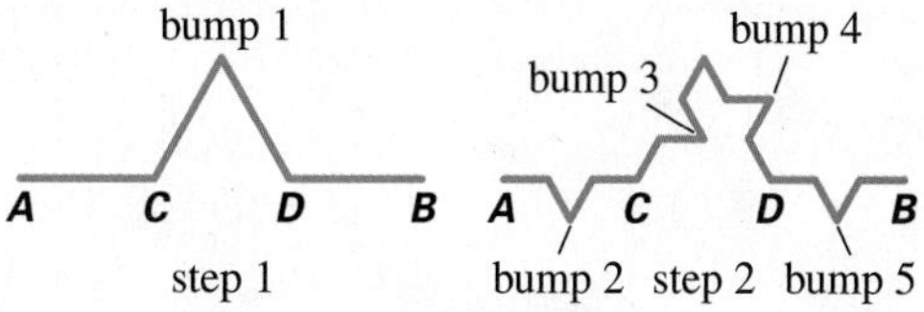

12. What is the dimension of the curve described in Exercise 11?

In Exercises 13 and 14, construct step 2 for each fractal.

13. The Sierpinski carpet

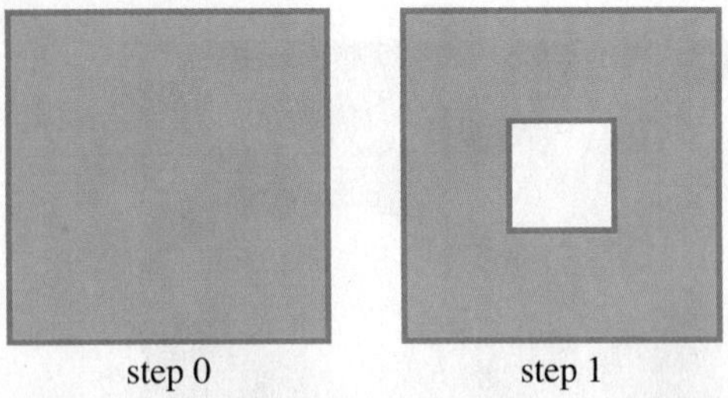

14.

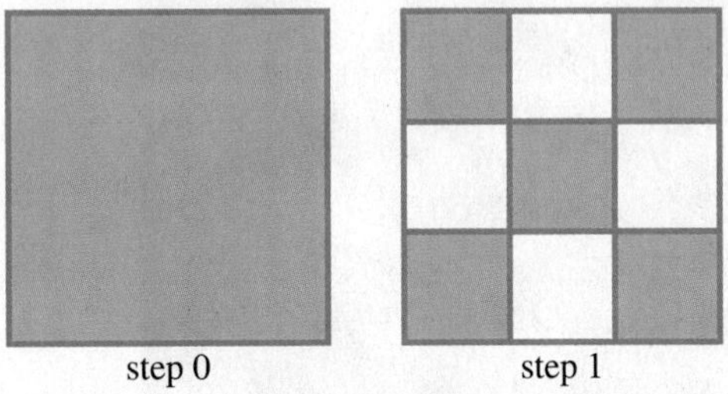

Communicating Mathematics

15. What do we mean when we say that a fractal is self-similar?

16. What is the intuitive meaning of fractal dimension?

Challenge Yourself

17. How did we argue that the length of the Koch curve is infinite?

18. How did we argue that the area of the Sierpinski gasket is 0?

19. Find a formula for the number of line segments at step n for the fractal in Exercise 5.

20. Find a formula for the number of line segments at step n for the fractal in Exercise 6.

21. Find a formula for the area of the Sierpinski gasket at step 10; at step n.

22. Find a formula for the area of the Sierpinski carpet in Exercise 13 at step 10; at step n.

23. Draw a fractal tree using the method of Example 6; however, vary the lengths and angles of the branches.

24. Draw a fractal tree using the method of Example 6; however, now use a motif with three branches.

25. The Mandelbrot set. There are numerous Web sites that have programs called "applets" that allow you to experiment with fractals. A particularly good site is Robert Devaney's site http://math.bu.edu/DYSYS/applets/. Several other sites illustrate a famous fractal called the **Mandelbrot set.** Search for "fractal zoom" to see a fascinating animation of this fractal.

26. Fractal applications. There are many Web sites that discuss real-life applications of fractals. For a starting point, you can go to http://library.thinkquest.org/26242/full/ for links to fractal applications in biology, chemistry, weather, human anatomy, art, music, and other areas. Choose an application that interests you and write a brief report on it.

CHAPTER SUMMARY

Section	Summary	Example
SECTION 9.1	A **ray** is a half line with its endpoint included. A piece of a line joining two points and including the points is called a **line segment. Parallel lines** are lines that lie on the same plane and have no points in common. Two lines that have a single point in common are called **intersecting lines.**	Discussion, p. 437
	An **angle** is formed by two rays that have a common endpoint. An **acute** angle has a measure between 0° and 90°. An angle with a measure of 90° is called a **right** angle. An **obtuse** angle has a measure between 90° and 180°. A **straight** angle has a measure of 180°. Two intersecting lines form two pairs of **vertical angles** having equal measures. Two angles are **complementary** if the sum of their measures is 90°. Two angles that have an angle sum of 180° are called **supplementary** angles. Two lines that intersect forming right angles are called **perpendicular** lines. Parallel lines cut by a **transversal** form several pairs of **equal** angles.	Discussion, pp. 438–440 Example 1, pp. 440–441
	A **circle** is the set of all points lying on a plane that are located at a fixed distance, called the **radius,** from a given point, called the **center.** A **diameter** of a circle is a line segment passing through the center, with both endpoints lying on the circle. The **circumference** is the distance around the circle.	Definitions, p. 441 Example 2, p. 441 Example 3, p. 442
SECTION 9.2	A plane figure is **closed** if we can draw it without lifting the pencil and the starting and ending points are the same. A plane figure is **simple** if we can draw it without lifting the pencil and in drawing it, we never pass through the same point twice, with the possible exception of the starting and ending points. A **polygon** is a simple, closed figure consisting only of line segments, called **edges,** such that no two consecutive edges lie on the same line. If all edges are the same length, the polygon is called **regular.** A polygon is **convex** if for any two points X and Y inside the polygon, the entire line segment XY also lies inside the polygon.	Definitions, p. 446 Definitions, p. 447 Definition, p. 447
	The **sum** of the measures of the interior angles of a convex polygon that has n sides is $(n - 2) \times 180°$.	Examples 1 and 2, p. 449
	Two polygons are **similar** if their corresponding sides are proportional and their corresponding angles are equal.	Definition, p. 451
	Non-Euclidean geometries have different properties than Euclidean geometry. The shortest distance between two points is not a straight line segment. Triangles do not have angle sums of 180°.	Highlight, p. 452
SECTION 9.3	If a rectangle has length l and width w, then the **perimeter** of the rectangle is $P = 2l + 2w$ and the **area** is $A = l \cdot w$.	Discussion, p. 457
	The **area of a parallelogram** with height h and base b is $A = h \cdot b$. The **area of a triangle** with height h and base b is $A = \frac{1}{2}h \cdot b$. **Heron's formula** states that for a triangle with sides of lengths a, b, and c, if we define the quantity $s = \frac{1}{2}(a + b + c)$, then the area of the triangle is $A = \sqrt{s(s - a)(s - b)(s - c)}$. A **trapezoid** with lower base b_1 and upper base b_2 and height h has **area** $A = \frac{1}{2}(b_1 + b_2) \times h$.	Discussion, pp. 457–458 Example 1, p. 458 Example 2, pp. 458–459 Example 3, p. 460
	The **Pythagorean theorem** states that in a right triangle with legs of lengths a and b and hypotenuse (the side opposite the right angle) of length c, then $a^2 + b^2 = c^2$.	Example 4, p. 461 Example 5, p. 461–462
	A circle with radius r has **circumference** $C = 2\pi r$ and **area** $A = \pi r^2$.	Example 6, p. 463
SECTION 9.4	If a **rectangular solid** has length l, width w, and height h, then the **volume** of the solid is $V = lwh$ and the **surface area** of the solid is $S = 2lw + 2lh + 2wh$. If an object has a flat top and base and *sides perpendicular to the base*, and if the area of the base is A and the height is h, the volume will be $V = A \cdot h$.	Discussion, p. 469 Example 1, pp. 469–470
	A **right circular cylinder** with radius r and height h has **volume** $V = \pi r^2 h$ and **surface area** $S = 2\pi rh + 2\pi r^2$.	Example 2, p. 471
	A **right circular cone** with base radius r and height h has **volume** $V = \frac{1}{3}\pi r^2 h$ and **surface area** $S = \pi r\sqrt{r^2 + h^2}$. A **sphere** with radius r has volume $V = \frac{4}{3}\pi r^3$ and **surface area** $S = 4\pi r^2$.	Examples 4 and 5, pp. 473–474 Example 6, p. 474

(Continued)

Section	Summary	Example
SECTION 9.5	In the **metric system,** length measurement is based on the **meter,** which is slightly more than a yard. Volume is based on the **liter,** which is a little more than a quart. The basic unit of weight is the **gram.** A pound is 454 grams.	Discussion, pp. 477–478
	The following table explains the **prefixes** that we use in the metric system:	Table 9.5, p. 478

kilo- (k)	hecto- (h)	deka- (da)	base unit	deci- (d)	centi- (c)	milli- (m)
$\times$ 1,000	$\times$ 100	$\times$ 10		$\times \frac{1}{10}$ or $\times$ 0.1	$\times \frac{1}{100}$ or $\times$ 0.01	$\times \frac{1}{1{,}000}$ or $\times$ 0.001

We use the following **equivalents** between the metric and customary systems in doing **dimensional analysis:** — Table, p. 485

1 meter = 1.0936 yards
1 mile = 1.609 kilometers
1 pound = 454 grams
1 metric tonne = 1.1 tons
1 liter = 1.0567 quarts

Section	Summary	Example
SECTION 9.6	A **rigid motion** is the action of taking a geometric object in the plane and moving it in some fashion to some other place in the plane without changing its shape or size. Every rigid motion is essentially a reflection, a translation, a glide reflection, or a rotation. A **reflection** moves an object so that the ending position is a mirror image of the object in its starting position. A **translation** slides an object along a line segment, called the **translation vector,** in the plane. A **glide reflection** is formed by performing a translation (the glide) followed by a reflection. We perform a **rotation** by first selecting a point, called the **center of rotation,** and then while holding this point fixed, rotating the plane about this point through an angle called the **angle of rotation.**	Definition, p. 489 Example 1, p. 490 Example 2, pp. 491–492 Definition, p. 492
	A **symmetry** of a geometric object is a rigid motion such that the beginning position and the ending position of the object are exactly the same.	Example 3, p. 493
	A **tessellation** (or tiling) of the plane is a pattern made up entirely of polygons that completely cover the plane with no gaps or overlapping polygons. The only regular polygons **that tessellate the plane** are triangles, squares, and hexagons.	Definition, p. 494 Example 4, p. 495
SECTION 9.7	A **fractal** object is **self-similar** in the sense that if we magnify it, we see the same patterns that were present in the original object. The **Koch curve** and the **Sierpinski gasket** are examples of fractals.	Discussion, pp. 500–501 Example 1, p. 501 Example 2, p. 502
	The Koch curve has infinite **length,** and the **area** of the Sierpinski gasket is zero.	Example 3, pp. 502–503
	The **fractal dimension** of an object is a number D that satisfies the equation $n = s^D$, where s is a scaling factor and n is the amount by which the quantity we are measuring changes when we apply the scaling factor to the object.	Definition, p. 504 Example 5, pp. 505–506

CHAPTER REVIEW EXERCISES

Section 9.1

1. In the given figure:
   a. Find a pair of acute, alternate exterior angles.
   b. Find a pair of obtuse, alternate interior angles.

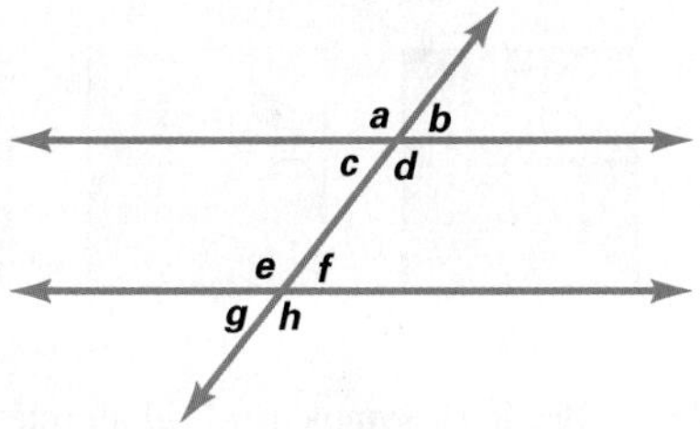

2. Find the measures of angles a, b, and c in the given diagram.

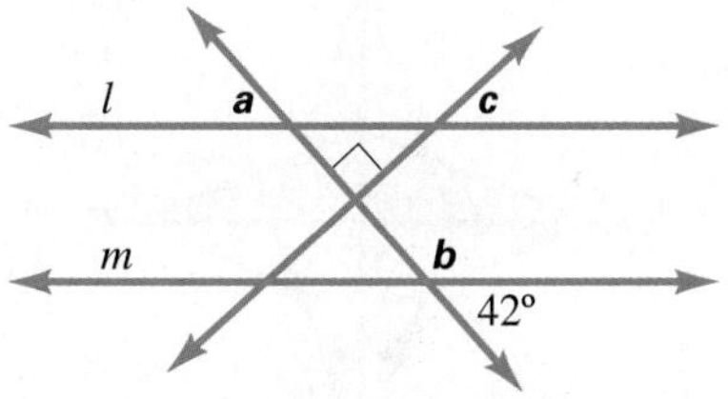

3. Assume that in the given diagram, the circumference is 24 inches and the length of arc DE is 3 inches. Find $m\angle BCD$.

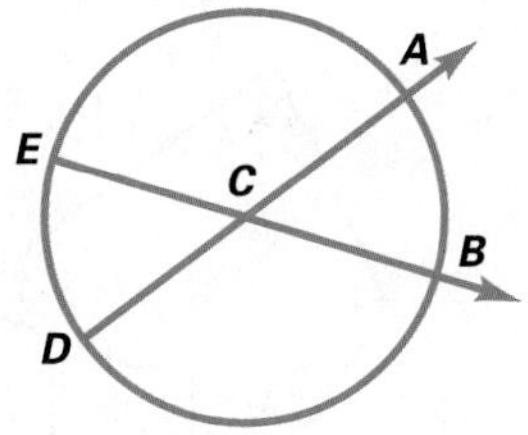

4. Solve for x in the given diagram.

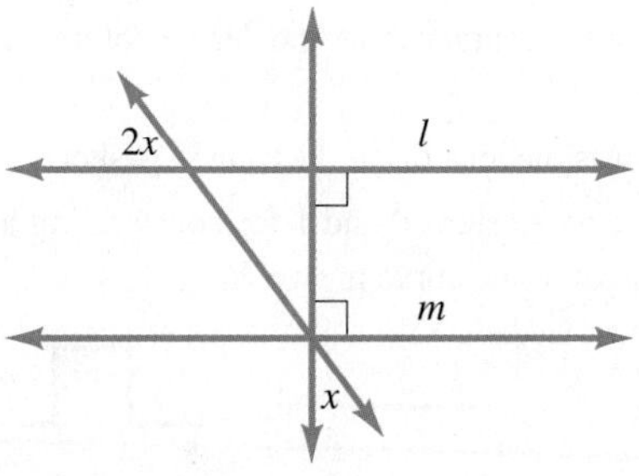

5. What is the shortest distance between two points on a sphere?

Section 9.2

6. What is the measure of an interior angle of a regular 18-sided polygon?
7. The given pair of figures in the next column are similar. What information do you know about the figure on the right?

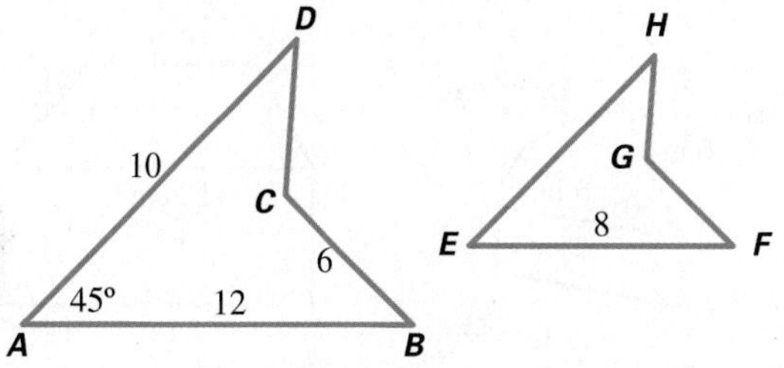

Figure for Exercise 7

8. Solve for x in the given diagram.

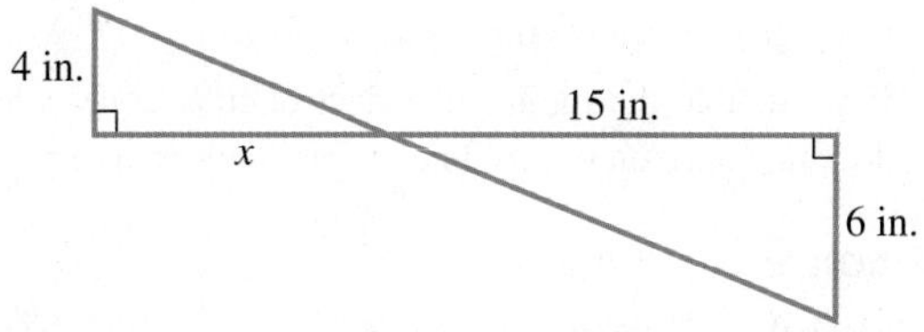

Section 9.3

9. Find the area of each figure.

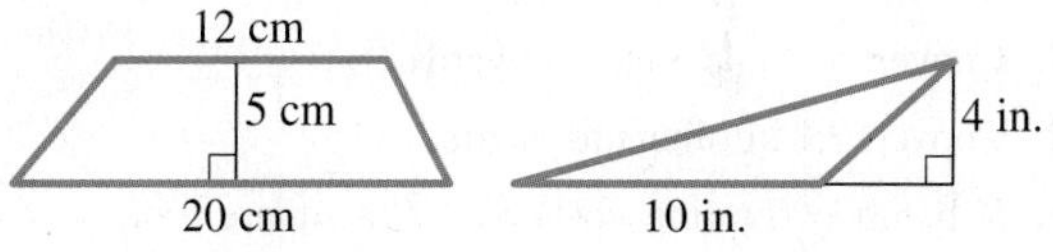

(a) (b)

10. Find the shaded area of each figure.

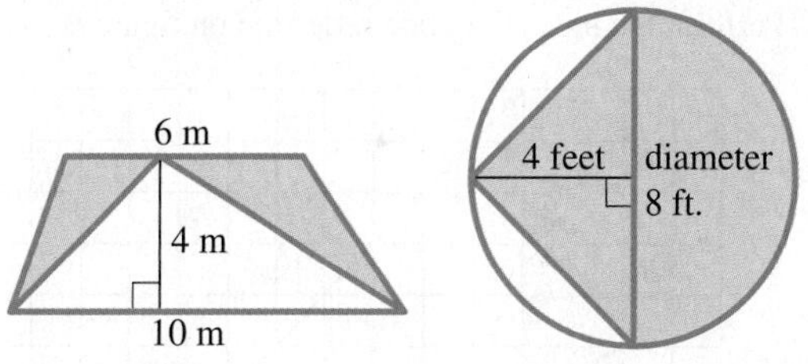

(a) (b)

11. a. Find the area of the triangle.
    b. Find the height h of the triangle.

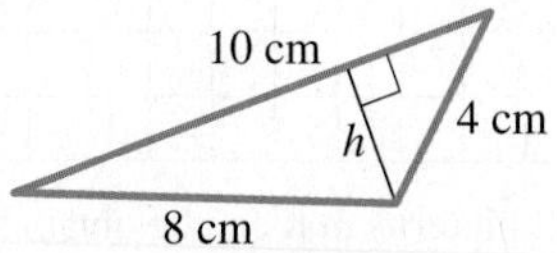

12. A running track, 5 meters wide, has the dimensions shown in the diagram. The ends of the track are semicircles with diameter 20 meters. What is the surface area of the track?

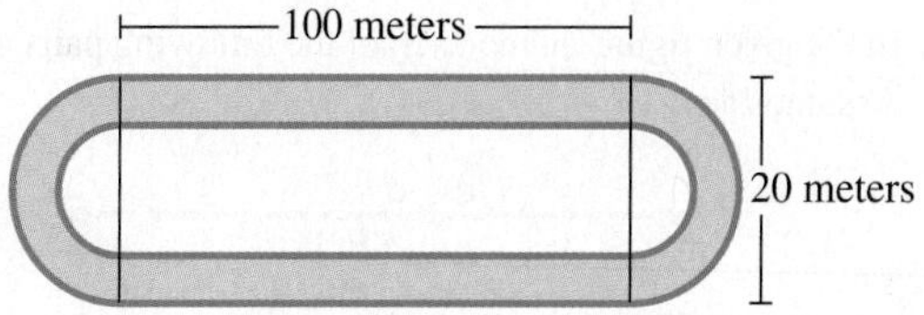

Section 9.4

13. Find the volume of each solid.

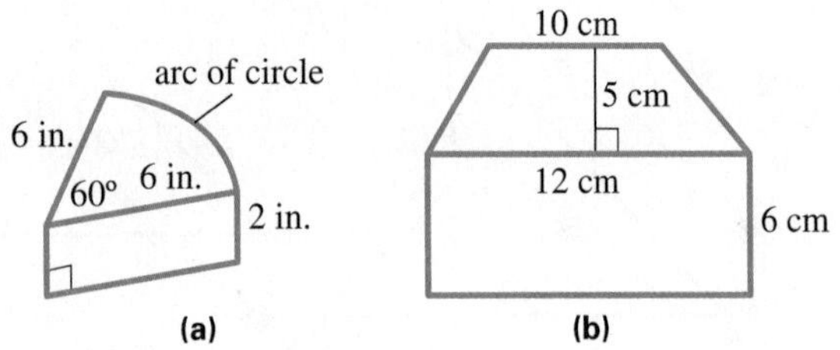

14. A punch bowl shaped like a hemisphere with a radius of 9 inches is full of punch. If we are filling cylindrical glasses that have a diameter of 3 inches and a height of 3 inches, how many glasses can we fill?

15. If we double the radius of a right circular cone, what effect does that have on the volume? Explain your answer.

Section 9.5

16. Make the following conversions:

a. 3,500 millimeters to meters

b. 4.315 hectograms to centigrams

c. 3.86 kiloliters to deciliters

17. Convert 514 decimeters to yards.

18. Convert 2.1 kiloliters to quarts.

19. If bamboo flooring costs \$11.00 a square foot, how much is that cost per square meter?

Section 9.6

20. Perform the indicated glide reflection on figure *B*.

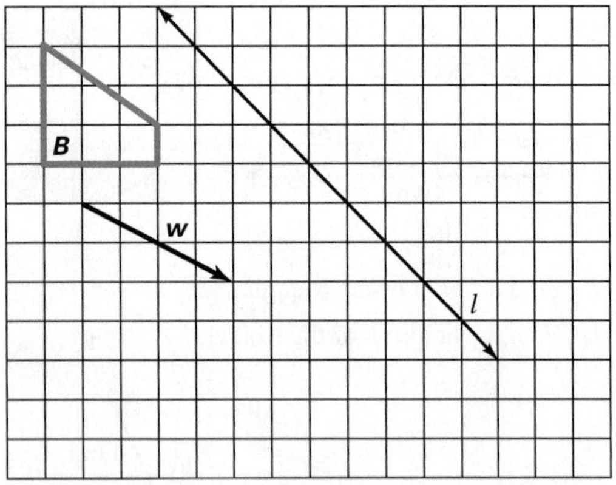

21. a. List all patterns that can be obtained by reflecting pattern (a) about a single line.

b. List all patterns that can be obtained by rotating pattern (a) about its center.

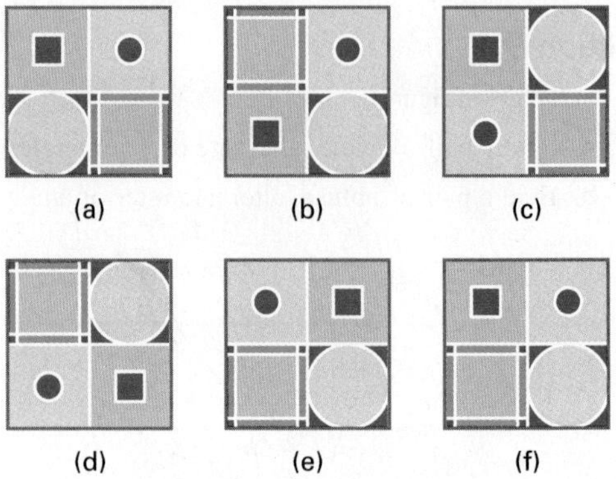

22. Find all reflectional symmetries and all rotational symmetries of the given object using angles between 1° and 359°.

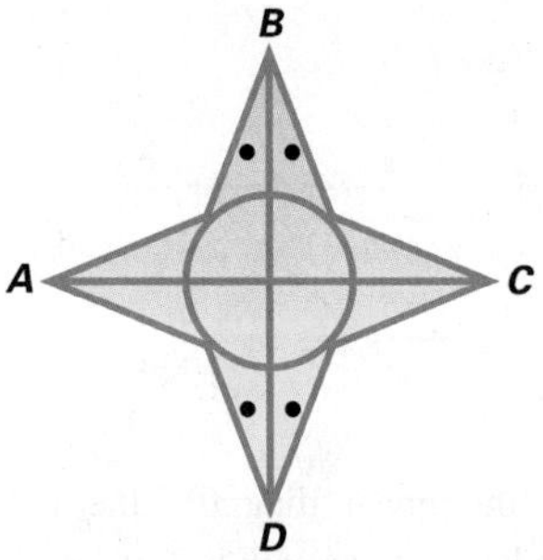

23. Tessellate the plane with the given quadrilateral.

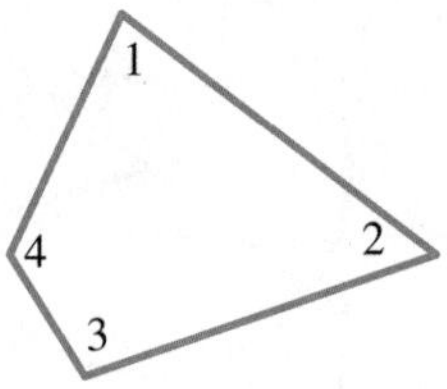

Section 9.7

24. How did we argue that the length of the Koch curve was infinite?

25. What was the area of the Sierpinski gasket?

26. You are given steps 0 and 1 for constructing a fractal. What is the length of the curve in step 8?

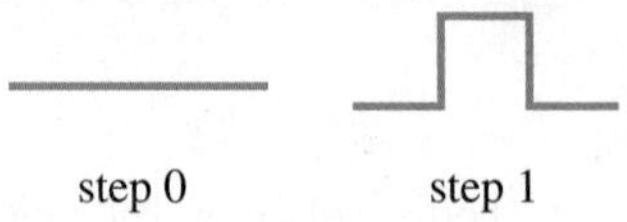

CHAPTER TEST

1. In the given figure, name each of the following pairs of angles. Assume $l \parallel m$.

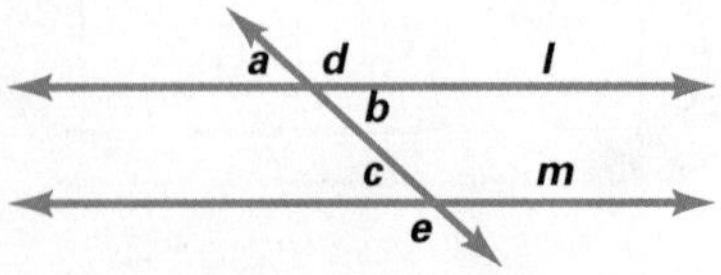

a. *a* and *b* **b.** *a* and *c* **c.** *d* and *e* **d.** *b* and *c*

2. What is the sum of the measure of the interior angles of a regular 12-sided polygon?

3. Find the measure of angles *a*, *b*, and *c* in the diagram on the next page. Assume $l \parallel m$.

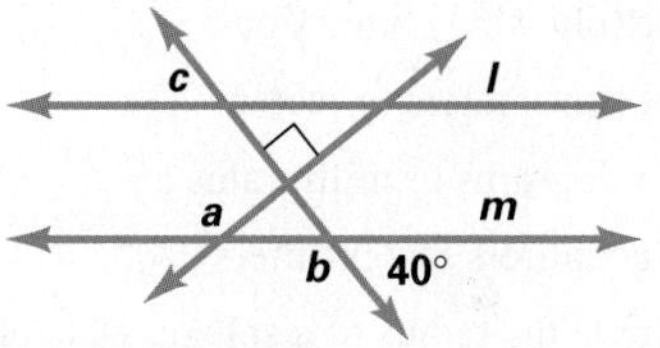

Figure for Exercise 3

4. A spherical water tank that has a radius of 15 feet is being replaced with a cylindrical tank that also has a radius of 15 feet. How high must the new tank be to contain the same amount of water as the old tank?

5. Find the volume of each solid.

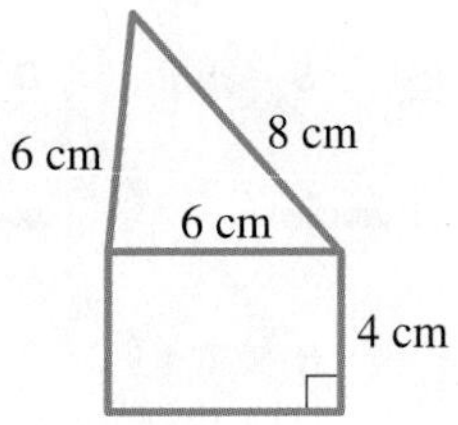

(a)

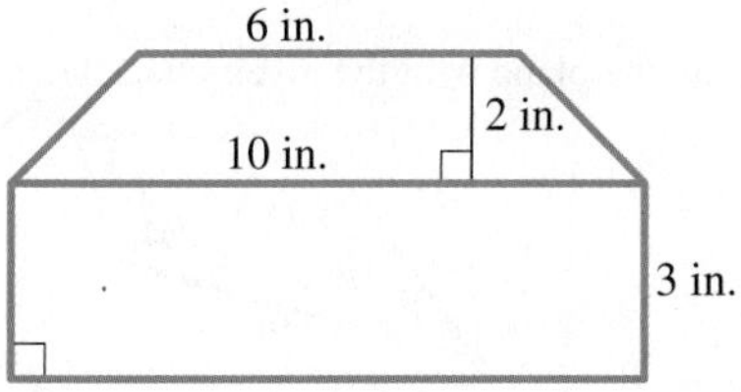

(b)

6. a. List all patterns that can be obtained by reflecting pattern (a) about a single line.

b. List all patterns that can be obtained by rotating pattern (a) about its center.

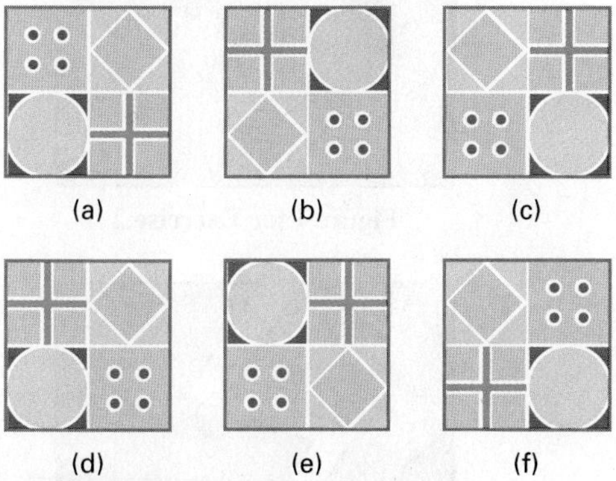

7. Assume that in the given diagram $m\angle BCD = 150°$ and the circumference is 36 inches. What is the length of arc DE?

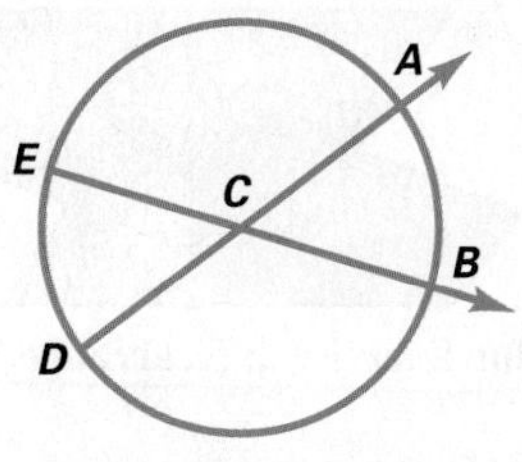

8. What is the length of the Koch curve after five steps?

9. Find the area of each figure.

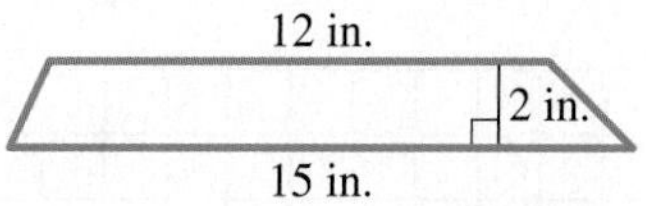

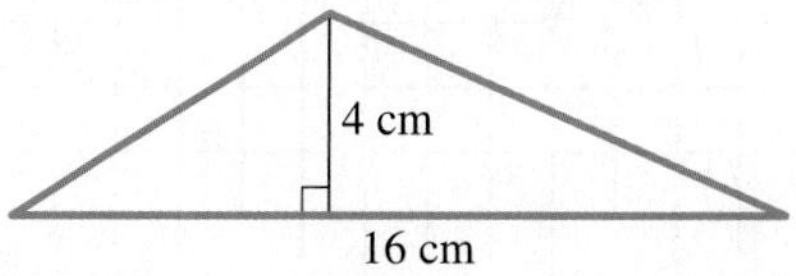

10. Find the shaded area of each figure.

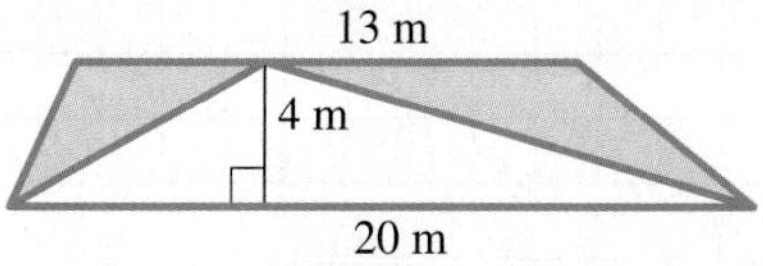

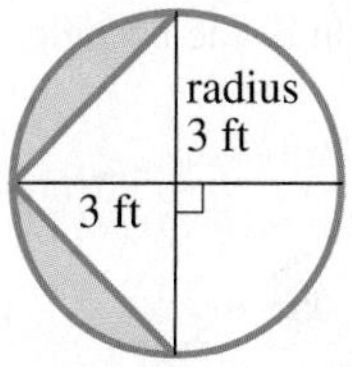

11. The given pair of figures are similar. What information do you know about the figure on the right?

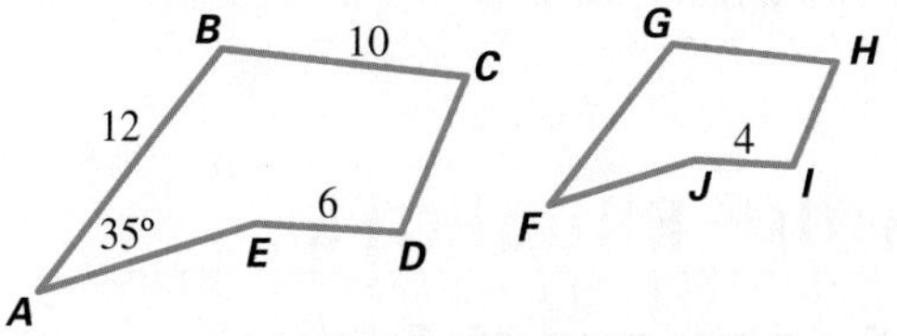

12. A pool is surrounded by a brick walkway as shown in the diagram. The pool is 3 feet deep and the walkway is 4 feet wide.

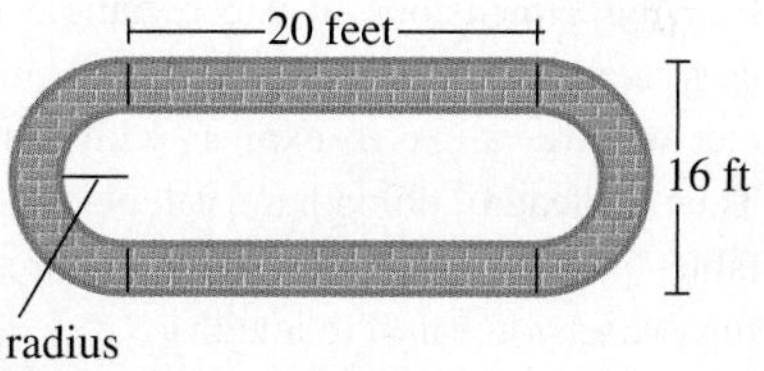

a. Find the surface area of the pool.

b. Find the volume of the pool.

c. Find the area of the walkway.

13. Solve for x in the following diagram:

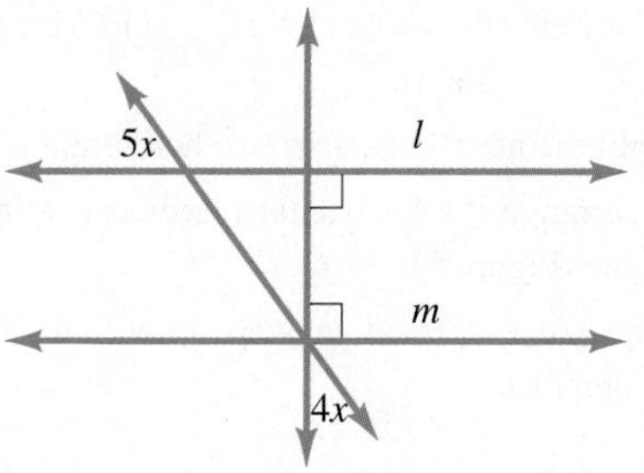

14. Reflect the given figure about the line $x = 1$, and then the line $y = 1$.

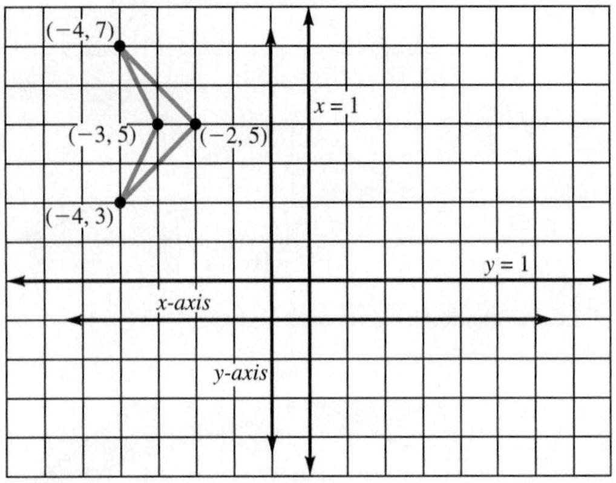

15. a. Find the area of the triangle.

b. Find the height of the triangle.

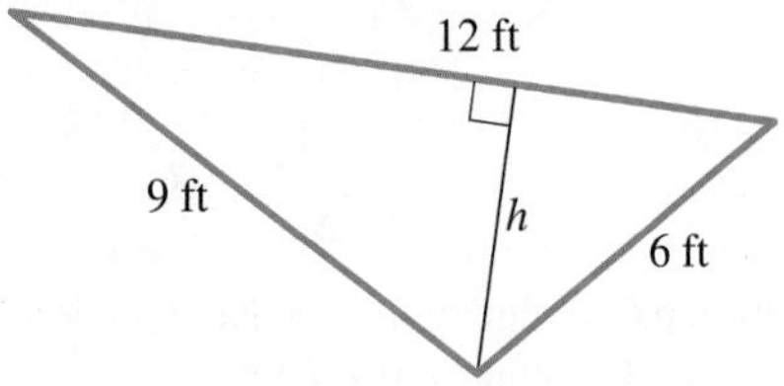

16. What was the area of the Sierpinski gasket?

17. Make the following conversions:

a. 2,400 centimeters to meters

b. 3.46 kilograms to milligrams

c. 2.14 dekaliters to centiliters

18. If we double the radius of a sphere, what effect does that have on the surface area?

19. Convert 18 yards to decimeters.

20. Convert 2,614.35 quarts to kiloliters.

21. Find all reflectional symmetries and all rotational symmetries of the given figure using angles between 1° and 359°.

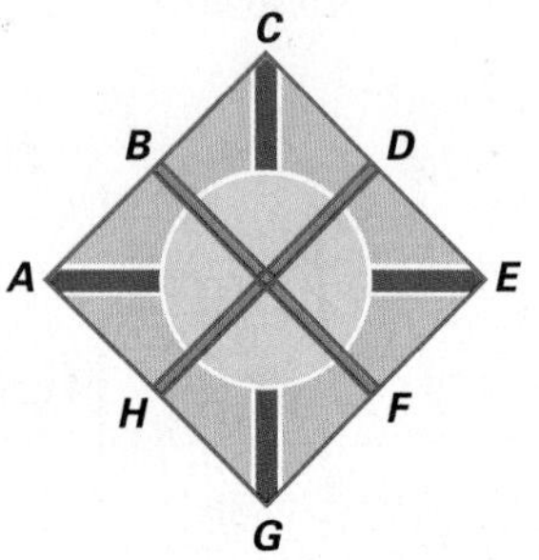

22. Tessellate the plane with the given quadrilateral.

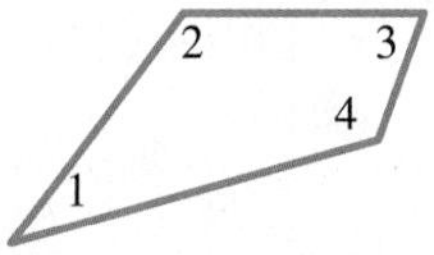

GROUP PROJECTS

1. Optimizing geometric figures.

a. Suppose that you have 24 inches of stiff wire to form a rectangle. Experiment by varying length and width to find the dimensions of the rectangle that contains the most area. (*Hint:* You can use the formula for the perimeter of a rectangle to express width in terms of length. Then, instead of doing hand calculations, you can use the table feature of a graphing calculator to calculate areas for you as you vary the length.)

b. Do the same for triangles.

2. Creating tessellations. Create a tessellation similar to the one shown as follows:

a. Start with a rectangle and split it vertically with a zigzag line or curve (Figures 1 and 2).

b. Rearrange the vertical pieces and join them to make a new figure (Figure 3).

c. Split the new figure into two horizontal pieces (Figure 4).

d. Rearrange the horizontal pieces and join them into one piece (Figure 5).

e. The piece formed in step d) will tessellate the plane (Figure 6).

Figure 1 for Exercise 2

Figure 2 for Exercise 2: Split Vertically

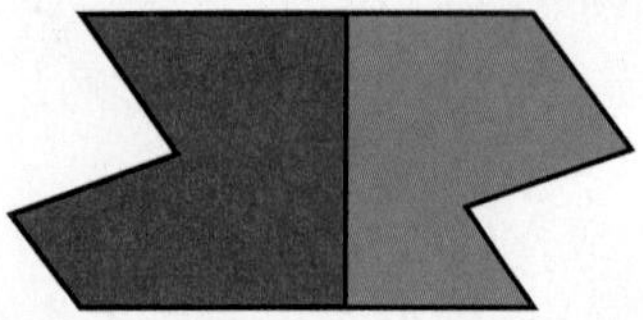

Figure 3 for Exercise 2: Rearrange Vertical Pieces

Figure 4 for Exercise 2: Split Horizontally

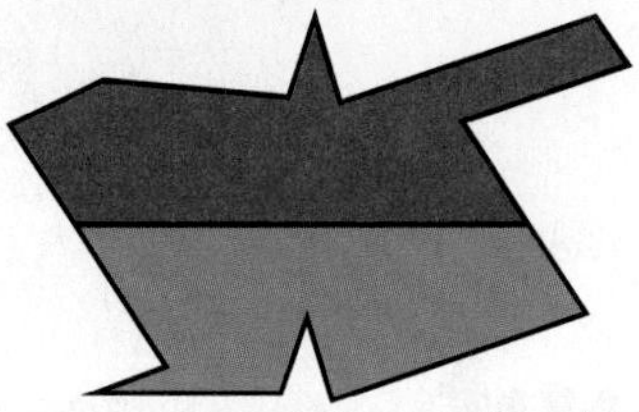

Figure 5 for Exercise 2: Rearrange Horizontal Pieces

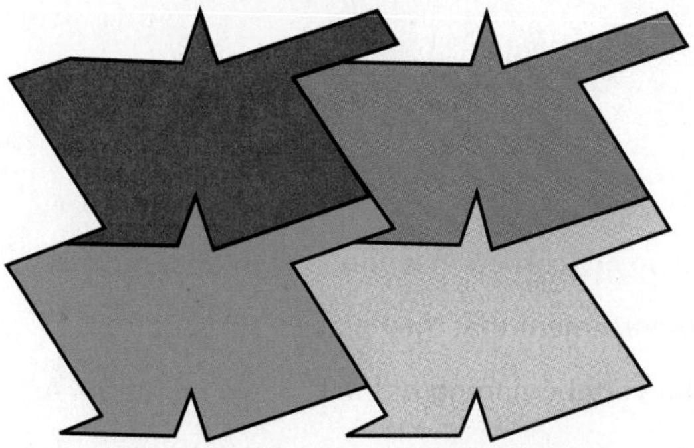

Figure 6 for Exercise 2: Tessellate Plane

3. **Escher drawings.** There are many sites on the Internet that discuss drawings similar to these two that were drawn by the Dutch artist Maurits Cornelis Escher. Escher formed these drawings by first creating a tessellation and then modifying it to create the image.

   a. Identify what tessellations Escher is using in Figures 1 and 2.

Figure 1 for Exercise 3

Figure 2 for Exercise 3

   b. Obtain other Escher drawings and analyze them as you did in part (a). A good place to start is www.mcescher.com.

   c. Research a Web site that explains how Escher made his drawings, and then try to make a simple one of your own.

USING TECHNOLOGY

1. **The best shape of a cone.** Determine the best shape for a cone having volume equal to 1 cubic unit, using a technique similar to the way we determined the best shape for the can in Example 3 in Section 9.4.

   a. Set the volume of the cone equal to 1 and solve for h in terms of r.

   b. Substitute the expression for h that you found in (a) in to the formula for the surface area of a cone. Don't forget to include the bottom of the cone.

   c. You now have an expression for the surface area of a cone that depends only on the radius r. Use either a graphing calculator or an Excel spreadsheet to find the radius r that will give the minimal surface area to enclose a volume of 1.

2. **Fractal applets.** Find a Web site that has applets to create fractals and use it to create several fractals of your own. Search for "fractals" + "applets".

10

Apportionment

How Do We Measure Fairness?

The founding fathers were so concerned that you and I would be fairly represented in the new government that they placed the requirements for state representation at the very beginning of the U.S. Constitution. Article I, Section 2, states:

> *Representatives and direct taxes shall be apportioned among the several states which may be included within the Union, according to their respective numbers.*

You might think this is a straightforward task, but apportioning representatives is not as easy as it might seem. The problem is that one state might deserve exactly 13.465 representatives, whereas another deserves 11.702. Just like dividing 10 party favors among four small children, the question is, "How do we decide who gets the extra whole representative?"

(continued)

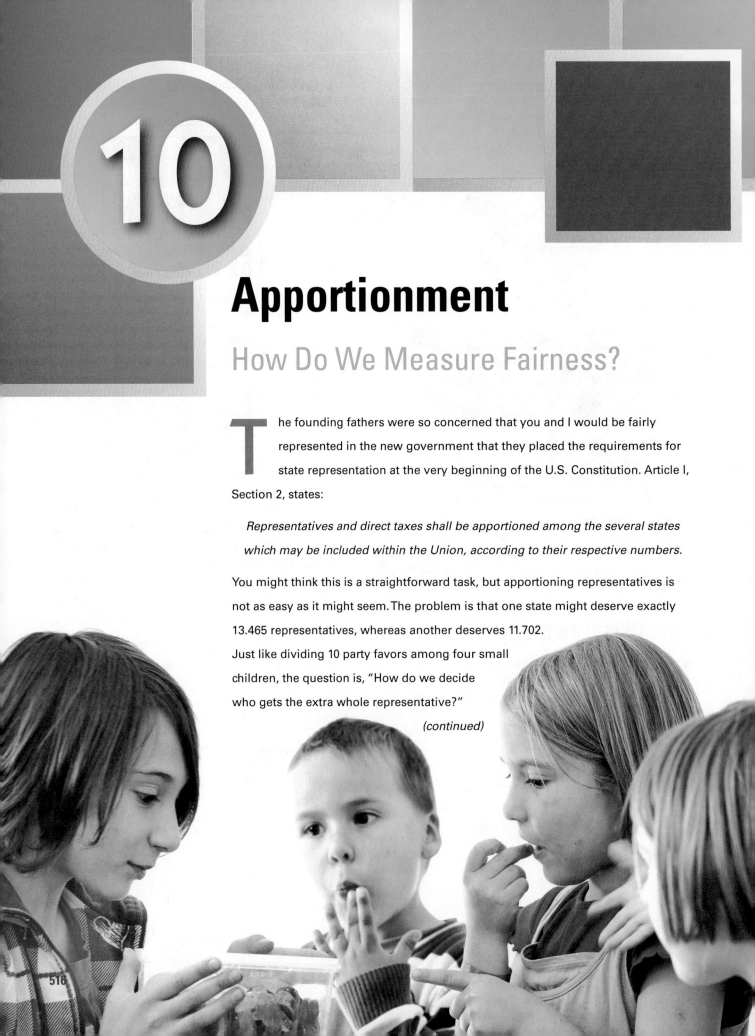

Daniel Webster believed that it was not possible to solve the apportionment problem perfectly. Speaking before the House of Representatives in 1832, he said,

> *The Constitution... must be understood, not as enjoining an absolute relative equality.... That which cannot be done perfectly must be done in a manner as near perfection as can be.**

As you will see shortly, it turns out that Webster's intuition about apportionment was correct, and by the end of this chapter, you will have a better understanding of the mathematics of this sticky problem.

In the first part of this chapter, we will introduce you to a variety of historical apportionment methods proposed by our founding fathers, and also examine their defects. Section 10.4 will present an interesting method for distributing a collection of items fairly among a group of people—a technique that you might someday find useful should you ever need to settle an estate or dissolve a business.

10.1 Understanding Apportionment

Objectives

1. Understand how the Hamilton method leads to the Alabama paradox.
2. Calculate the absolute and relative unfairness of an apportionment.

People often encourage you to use common sense when facing a problem, and this is often good advice; however, agreeing to a simple plan that seems to make sense now may lay a trap for you later. This is exactly what happened to our founding fathers when they approved a "common-sense" method, developed by Alexander Hamilton, to apportion the first U.S. Congress.

Years later, in 1881, Congress was surprised that, when using Hamilton's method, Alabama was entitled to 8 representatives in a House having 299 members but would receive only 7 representatives in a 300-member House. Under Hamilton's method, Alabama would receive fewer representatives in a larger House, *although no state had a change in population.* This strange situation, called the **Alabama paradox,** occurred again following the 1890 census, when Arkansas lost a representative as the House increased from 359 to 360 members.

After the 1890 census, as a different apportionment plan caused the number of Maine's representatives to fluctuate, Representative Littlefield remarked:

> *...God help the state of Maine when mathematics reach for her and undertake to strike her down in this manner in connection with her representation on this floor....*

In order to develop an apportionment method that avoids an Alabama paradox, we must first understand why the problem occurs. The following example will help explain it.

Naxxon, Aroco, and Eurobile have formed a consortium to develop an oil drilling platform off the coast of Africa. The companies have agreed to form a nine-member board with executives from the three companies to oversee the project. Each company will have at least one representative on the board, and additional board members will be assigned in proportion to the number of stockholders in each company. Naxxon has 4,700 stockholders, Aroco has 3,700 stockholders, and Eurobile has 1,600 stockholders.

Because the total number of stockholders in the three companies is $4{,}700 + 3{,}700 + 1{,}600 = 10{,}000$, this means that Naxxon is entitled to

$$\frac{4{,}700}{10{,}000} = 0.47 = 47\%$$

---

*Daniel Webster, *The Works of Daniel Webster*, Vol. III, 16th ed. (Boston: Little Brown & Company, 1872).

of the board's nine members. Naxxon is therefore entitled to exactly 47% × 9 = 0.47 × 9 = 4.23 members. We show similar calculations for the other partners in Table 10.1

Company	Percent of Stockholders	Board Members Deserved
Naxxon	$\frac{4{,}700}{10{,}000} = 0.47 = 47\%$	$0.47 \times 9 = 4.23$
Aroco	$\frac{3{,}700}{10{,}000} = 0.37 = 37\%$	$0.37 \times 9 = 3.33$
Eurobile	$\frac{1{,}600}{10{,}000} = 0.16 = 16\%$	$0.16 \times 9 = 1.44$
Total	100%	9.0

TABLE 10.1 Exact number of representatives allotted to each company.

A company cannot have a part of a board member. Therefore, because Naxxon is entitled to exactly 4.23 members, it will be given either 4 or 5. We will call 4 the *integer part* of 4.23 and 0.23 the *fractional part* of 4.23.

The Hamilton Apportionment Method

KEY POINT

The Hamilton method uses fractional parts to apportion representatives.

If we give each company its integer part as its number of board members, then Naxxon will have four members, Aroco will have three, and Eurobile will have one. Thus there will be only eight members on the board. To have the required nine, we must now decide which company gets the additional member. It seems reasonable to assign the member to Eurobile, which has the highest fractional part—namely, 0.44. In fact, this is exactly how the **Hamilton apportionment method** would allocate the last board member.

Highlight

Using Technology to Do Apportionment

Graphing calculators and spreadsheets can greatly reduce the tedium of doing the many repetitious computations required to do an apportionment. The Excel spreadsheet below shows one way to streamline your work.

Line 8 shows the formulas that we used to do the calculations on line 11. (We have omitted the equal signs on line 8 that would be necessary to make the formulas operational.) Once we defined the formulas on line 11, we dragged down the corner marker of the cells in each column to produce the computations on lines 12 and 13.

D11 f_x =B11/$B$14

	A	B	C	D	E	F	G
1							
2	This spreadsheet duplicates the calculations in Example 1. Line 8 shows the formulas						
3	we used to do the calculations. To make the formulas operational, you have to begin each						
4	formula with an equal sign as we show in cell D11.						
5							
6	Number of Board Members		10				
7							
8		Formulas:		B11/$B$14	$C$6*D11	FLOOR(E11,1)	E11-F11
9							
10		Number of Stockholders		Percent	No. of Reps.	Integer Part	Fractional Part
11		4700		0.47	4.7	4	0.7
12		3700		0.37	3.7	3	0.7
13		1600		0.16	1.6	1	0.6
14	Total	10000					
15							

HAMILTON APPORTIONMENT METHOD (APPLIED TO THE CONSORTIUM BOARD)

Follow these steps for each company.

1. Determine the exact number of board members due to the company by computing

$$\text{percent of stockholders} \times \text{size of board.}$$

2. Assign the integer part of the exact number of board members to each company.

If there are more members to be allocated, then go to step 3.

3. The first additional member goes to the company having the largest fractional part; the second additional member, if any, goes to the company with the second-largest fractional part. Continue in this manner until you have assigned all additional members.

To illustrate steps 2 and 3, we add three columns to Table 10.1 to get Table 10.2

Company	Percent of Stockholders	Step 1: Board Members Deserved	Step 2: Assign Integer Parts	Examine Fractional Parts	Step 3: Assign Additional Member
Naxxon	47	$0.47 \times 9 = 4.23$	4	0.23	4
Aroco	37	$0.37 \times 9 = 3.33$	3	0.33	3
Eurobile	16	$0.16 \times 9 = 1.44$	1	0.44	2
Total	100	9.0	8		9

TABLE 10.2 Using the Hamilton method to allocate nine board members.

We need to apportion one more.

The company with the largest fractional part gets the extra representative.

EXAMPLE 1 Using the Hamilton Apportionment Method

Suppose the consortium decides to increase the size of the board to 10 members. Use the Hamilton apportionment method to apportion the 10-member board.

SOLUTION: Review the Be Systematic Strategy on page 5.

We show the steps in making this apportionment in Table 10.3.

Company	Percent of Stockholders	Step 1: Board Members Deserved	Step 2: Assign Integer Parts	Examine Fractional Parts	Step 3: Assign Additional Members
Naxxon	47	$0.47 \times 10 = 4.7$	4	0.7	5
Aroco	37	$0.37 \times 10 = 3.7$	3	0.7	4
Eurobile	16	$0.16 \times 10 = 1.6$	1	0.6	1
Total	100	10.0	8		10

TABLE 10.3 Using the Hamilton method to allocate 10 board members.

We need to apportion two more representatives.

The companies with the two largest fractional parts get the extra representatives.

Eurobile loses one representative.

QUIZ YOURSELF 1

Reapportion the oil consortium board assuming that there are now 12 members on the board.

The last column of Table 10.3 shows the assignment of members according to the Hamilton method. We assigned the first 8 members using the integer parts of the exact amounts deserved by each company. Then we gave the 2 additional members to Naxxon and Aroco because they have the highest fractional parts.

Notice that increasing the size of the board to 10 members causes Eurobile to lose 1 board member.

Now try Exercises 1 to 6. 1

The reason that the Alabama paradox occurred in Example 1 was that when the board increased to 10, *we reassigned the 9 board members that had been apportioned earlier.* This allowed for the possibility of some companies losing board members.

In order to prevent this problem, we need a method so that once we assign seats, they are never reassigned at a later date. In order to make assignments on a "once-and-done" basis, we need a clear measure of who is most deserving of the next representative at each stage in the apportionment process.*

Measuring Fairness

KEY POINT

Average constituency measures the fairness of an apportionment.

To begin the development of such a measure, consider the representation of two hypothetical states A and B in the House of Representatives. Suppose state A has a population of 2,000,000 people and eight representatives and state B has 800,000 people and four representatives. Each of state A's representatives has an average of $\frac{2{,}000{,}000}{8} = 250{,}000$ constituents, whereas each representative from state B averages $\frac{800{,}000}{4} = 200{,}000$ constituents. Because a representative from state A serves more constituents than a representative from state B, it is fair to say that state A is more poorly represented in the House than state B. This leads to the following definitions.

DEFINITIONS The **average constituency** of a state is the quotient

$$\frac{\text{population of the state}}{\text{number of representatives from the state}}.$$

Comparing the representation of two states A and B, we say that state A is **more poorly represented** than state B if the average constituency of A is greater than the average constituency of B.

Math in Your Life

Just How Well Represented Are You?

It may surprise you that among the 165 countries that have a legislature, the United States is second only to India in the ratio of citizens to representatives in its legislature. This was not always the case. In 1792, Congress passed an apportionment act that set the size of the House at 103 representatives for a population of about 4 million, or a ratio of less than 40,000 people for each representative. This ratio has crept up until today, we have 435 representatives for a population of 307 million, or a ratio of about 706,000 people for each representative.

One way to look at this is that you are only $\frac{40}{706} \approx 5.7\%$ as well represented by your congressman as were U.S. citizens 200 years ago. To put this in perspective, the United Kingdom, with a population that is about one-fifth that of the United States, has a House of Commons with 659 members, or a ratio of about 1 representative for every 90,000 people. If the ratio of representatives to people were the same in the United States, we would have more than 3,000 representatives in our House of Representatives.

*As you will see, the method we develop has applications beyond allocating representatives to Congress.

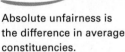

EXAMPLE 2 Determining Which State Is More Poorly Represented

Following the 2010 census, the U.S. House of Representatives was reapportioned and California, with a population of 37,341,989, was assigned 53 representatives. Florida, with a population of 18,900,773, was given 27. Calculate the average constituency for each state to determine which state is more poorly represented.

SOLUTION:

The average constituency of California is

$$\frac{\text{population of California}}{\text{number of representatives allocated to California}} = \frac{37{,}341{,}989}{53} \approx 704{,}565.83,$$

whereas the average constituency of Florida is

$$\frac{\text{population of Florida}}{\text{number of representatives allocated to Florida}} = \frac{18{,}900{,}773}{27} \approx 700{,}028.63.$$

Because California has a higher average constituency, it is more poorly represented.

QUIZ YOURSELF 2

a) If the 420-member electricians' union has three representatives on the United Labor Council, what is the average constituency of this group?
b) If the 440-member plumbers' union has four representatives on the council, are the electricians or the plumbers more poorly represented?

It would be ideal, of course, to have an apportionment in which the average constituencies are the same for all states because it would be consistent with the "one man, one vote"* concept. However, it is usually not possible to achieve this ideal when making an actual apportionment. If we cannot have equality, then we should try assigning the representatives to make the average constituencies as equal as possible. One measure of how close we come to this goal is called *absolute unfairness.*

DEFINITIONS Suppose that representatives are apportioned between two states A and B. We define the **absolute unfairness** of this apportionment as the difference between the larger average constituency and the smaller one. If state A has the larger average constituency, then the absolute unfairness is

(average constituency of state A) − (average constituency of state B).

If the two states have the same average constituencies, then we say that the two states are *equally well represented.*

KEY POINT

Absolute unfairness is the difference in average constituencies.

Notice that in computing the absolute unfairness, we subtract the smaller average constituency from the larger. Consequently, the absolute unfairness of an apportionment *cannot* be negative.

EXAMPLE 3 Finding Absolute Unfairness

Suppose the Weavers' Guild, with 1,542 members, has six delegates on the National Arts Commission and the Artists' Alliance, with 1,445 members, has five delegates. Calculate the absolute unfairness for this assignment of delegates.

SOLUTION:

The average constituency of the Weavers' Guild is $\frac{1{,}542}{6} = 257$ and the average constituency of the Artists' Alliance is $\frac{1{,}445}{5} = 289$. The Artists' Alliance is more poorly represented than the Weavers' Guild; the absolute unfairness of this apportionment is $289 - 257 = 32$.

QUIZ YOURSELF 3

Assume that state X has a population of 974,116 with four representatives and state Y has a population of 730,779 with three representatives. Compute the absolute unfairness for this apportionment.

*We have retained this terminology for historical reasons. See Between the Numbers in Section 10.3.

KEY POINT

Relative unfairness considers the size of constituencies in calculating absolute unfairness.

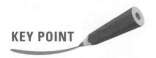

Although absolute unfairness measures the imbalance of an apportionment between two states, it is inadequate to compare the unfairness of two apportionments. You may think that a larger absolute unfairness indicates a greater imbalance.

This conclusion is wrong!

When measuring the unfairness of an apportionment, it is important to take into consideration the size of the states involved. The following examples will help you understand why we need a different measure of unfairness.

In the Republican 2012 Iowa presidential caucuses, Rick Santorum defeated Mitt Romney by a margin of 29,839 to 29,805. Because Santorum defeated Romney by only 34 votes, newspapers proclaimed:

SANTORUM AND ROMNEY FINISH IN DEAD HEAT

On the other hand, if Santorum and Romney were running for a seat on the school board of a small town and Santorum got 70 votes and Romney received 36, the local paper might read:

SANTORUM BURIES ROMNEY IN LANDSLIDE

Although the difference in votes is important, the significance of that difference depends on the number of votes received. A difference of 34 votes is considered minuscule when one candidate receives 29,805 votes, but a difference of 34 votes is considered large when one candidate receives only 36.

Similarly, the absolute unfairness of $704{,}565 - 700{,}028 = 4{,}537$ in Example 2 is *relatively* small due to the size of the constituencies for California and Florida. However, the absolute unfairness, 32, that we computed in Example 3 is *relatively* large when we compare it to the average constituencies of 289 and 257. From these examples, we see that we must consider the sizes of the average constituencies when measuring the unfairness of an apportionment.

DEFINITION When apportioning the representatives for two states, the **relative unfairness** of the apportionment is defined as

$$\frac{\text{the absolute unfairness of the apportionment}}{\text{the smaller average constituency of the two states}}.$$

EXAMPLE 4 Calculating Relative Unfairness

Compute the relative unfairness for the apportionment of representatives to California and Florida in Example 2 and the apportionment of delegates for the Weavers' Guild and the Artists' Alliance in Example 3.

SOLUTION: Review the Analogies Principle on page 13.

Recall that the absolute unfairness of the California-Florida apportionment was $704{,}565.83 - 700{,}028.63 = 4{,}537.2$. Because Florida had the smaller average constituency of 700,028, the relative unfairness of this apportionment is $\frac{4{,}537.2}{700{,}028.63} \approx 0.006$.

The absolute unfairness of the apportionment in Example 3 is $289 - 257 = 32$. Because the Weavers' Guild had the smaller average constituency, namely 257, the relative unfairness of this apportionment is $\frac{32}{257} \approx 0.125$.

The smaller relative unfairness for the California-Florida apportionment means that the apportionment is more fair than the apportionment for the Weavers' Guild and Artists' Alliance.

Now try Exercises 7 to 14. 4

QUIZ YOURSELF 4

If state A has a population of 11,710 and five representatives and state B has a population of 16,457 and seven representatives, calculate the relative unfairness of the apportionment.

We now have the necessary background concepts on unfairness to develop an apportionment criterion and principle that avoid an Alabama paradox. We will show this in Section 10.2.

EXERCISES 10.1

Sharpening Your Skills

In Exercises 1–6, use the Hamilton apportionment method to make the assignment.

1. **Apportioning a water authority.** Assume that California, Arizona, and Nevada are cooperating to build a dam to provide water to communities currently lacking adequate water supplies. Seats on the 11-member Southwest Water Authority, which governs the project, are assigned according to the number of customers in each state who use the water from the project. There are 56,000 customers in California, 52,000 in Arizona, and 41,000 in Nevada. Assign the seats on this authority.

2. **Apportioning a negotiations committee.** The employees of the Six Flags theme park are negotiating a new contract. There are 213 performers, 273 food workers, and 178 maintenance workers. The nine-person negotiations committee has members in proportion to the number of employees in each of the three groups. Assign members to the negotiations committee.
3. **Assigning booths at an art show.** The Civic Arts Guild is having a show. There is room for 31 booths and the guild has decided that the booths will be assigned in proportion to the type of members in the guild. The guild has 87 painters, 46 sculptors, and 53 weavers. Assign the booths to the three groups.
4. **Allocating a resident council.** A campus has three freshman dormitories. Building A has 78 units, building B has 36 units, and building C has 51 units. A 12-person resident council will set rules governing the complex. Membership in the council is to be proportional to the number of units in each building. Assign representatives to this council.
5. **Apportioning representatives.** In the 2010 census, Alabama's population in thousands was 4,803, Mississippi's was 2,978, and Louisiana's was 4,554. Apportion 17 members of the U.S. House of Representatives to these three states.
6. **Apportioning representatives.** In the 2010 census, Michigan's population in thousands was 9,912, Ohio's was 11,568, and Illinois' was 12,864. Apportion 48 members of the U.S. House of Representatives to these three states.
7. If the American Nurses Association has 177,408 members and three representatives on the National Health Board, what is the association's average constituency?
8. If the International Brotherhood of Electrical Workers has 819,000 members and 13 representatives on the Building Trades Council, what is its average constituency?
9. Which state is more poorly represented: state A with a population of 27,600 and 16 representatives, or state B with a population of 23,100 and 14 representatives? What is the absolute unfairness of this apportionment? What is the relative unfairness of this apportionment?
10. Which state is more poorly represented: state C with a population of 85,800 and 11 representatives, or state D with a population of 86,880 and 12 representatives? What is the absolute unfairness of this apportionment? What is the relative unfairness of this apportionment?
11. Recall that on a 10-member board, Naxxon, with 4,700 stockholders, received five members and Aroco, with 3,700 stockholders, received four members. Calculate the absolute and relative unfairness of this apportionment.
12. Redo Exercise 11 for Aroco and Eurobile. Recall that Eurobile had 1,600 stockholders and one board member.
13. **Apportioning representatives.** According to the 2010 census, Texas had a population of 25,268 thousand and Georgia had a population of 9,728 thousand. Texas was apportioned 36 members in the U.S. House and Georgia was given 14. Calculate the absolute and relative unfairness of this apportionment.
14. **Apportioning representatives.** According to the 2010 census, Washington had a population of 6,753 thousand and Oregon had a population of 3,849 thousand. Washington was apportioned 10 members in the U.S. House and Oregon was given 5. Calculate the absolute and relative unfairness of this apportionment.

Applying What You've Learned

15. Suppose an electronics company has three divisions: (D)igital, (C)omputers, and (B)usiness products. Division D has 140 employees, C has 85, and B has 30. Assume that a 12-member quality improvement council has membership on the council proportional to the number of employees in the three divisions.
    a. Apportion this council using the Hamilton method.
    b. Now increase the council's size to 13, and then to 14, and so on, redoing the apportionment each time until an Alabama paradox occurs.
    c. What is the first size above 12 when the paradox occurs, and what division loses a seat when the size of the council increases?

16. There are 47 local police, 13 federal agents, and 40 state police involved with drug enforcement in Metro City. A special nine-person task force will be formed to investigate a particular case. Officers will be assigned to this task force with membership proportional to the number of the three types of law enforcement officers.

a. Apportion this task force using the Hamilton method.

b. Now increase the task force's size to 10, and then to 11, and so on, redoing the apportionment each time until an Alabama paradox occurs.

c. What is the first size above 10 when the paradox occurs, and what group loses an officer when the size of the task force increases?

17. Urbandelphia General Hospital has three emergency clinics throughout the city. Statistics have been gathered regarding the number of patients seen on night shift for each clinic.

Clinic	Center City	South Street	West Side
Number of Patients	213	65	300

a. If there are 23 emergency physicians available, how should they be apportioned?

b. Increase the number of physicians until an Alabama paradox occurs. At what increase does the paradox occur and what clinic loses a doctor?

18. Center City Community College has three branch campuses with student enrollment in media shown below. Currently the college has 12 computer labs, but with its new electronic media concentration, the administration wants to increase the number of labs.

Campus	Midtown	Northeast	East River
Number of Students	40	180	240

a. How should the 12 labs be apportioned?

b. Increase the number of labs until an Alabama paradox occurs. At what increase does the paradox occur and what branch loses a lab?

Communicating Mathematics

19. What causes an Alabama paradox? How can an Alabama paradox be avoided?

20. If one apportionment has an absolute unfairness of 1,000 and another has an absolute unfairness of 50,000, must the first apportionment be a more fair apportionment than the second? Explain.

21. You have seen five examples of Alabama paradoxes in this section (Example 1 and Exercises 15–18). Can you describe a general pattern as to what "states" seem to be hurt by an Alabama paradox?

22. We discussed in Math in Your Life how you are "less represented" than citizens of the United States were 200 years ago. What did we mean by this?

Challenge Yourself

23. Assume that on the oil consortium board, Naxxon currently receives five representatives and Aroco receives four. Suppose one additional representative can be given to either Naxxon or Aroco.

a. Calculate the relative unfairness of the apportionment if the additional representative is given to Naxxon.

b. Calculate the relative unfairness of the apportionment if the additional representative is given to Aroco.

c. Based on your answers to parts (a) and (b), which company should get the additional representative? Why?

24. Assume that on the oil consortium board, Naxxon currently receives six representatives and Eurobile receives two. Suppose one additional representative can be given to either Naxxon or Eurobile.

a. Calculate the relative unfairness of the apportionment if the additional representative is given to Naxxon.

b. Calculate the relative unfairness of the apportionment if the additional representative is given to Eurobile.

c. Based on your answers to parts (a) and (b), which company should get the additional representative? Why?

25. Apportioning an additional representative. Currently, New York, with a population of 19,421 thousand, has 27 representatives and New Jersey, with a population of 8,808 thousand, has 12 representatives in the U.S. House of Representatives. Is New York or New Jersey more deserving of one additional representative? To make this decision first consider what the relative unfairness would be if the representative were given to New York. What would it be if the representative were given to New Jersey instead?

26. Apportioning an additional representative. Redo Exercise 25 for Tennessee, which has a population of 6,375 thousand and 9 representatives, and Maryland, which has a population of 5,790 thousand and 8 representatives.

27. Do a search on the Internet for "Hamilton apportionment applets." Use these programs to duplicate some of the calculations in this section and report on your findings.

10.2 The Huntington–Hill Apportionment Principle

Objectives

1. Understand the apportionment criterion.
2. Compute Huntington–Hill numbers.
3. Use the Huntington–Hill method to do apportionment.

You have seen that an Alabama paradox may occur if we *reapportion seats that have already been assigned*. We will now develop a method that avoids this paradox by *assigning only new seats* when the representative body increases in size.

The Apportionment Criterion

We want a method that tells us at each stage in the apportionment process who gets the next seat. We use relative unfairness and the following criterion in making the decision.

> **APPORTIONMENT CRITERION** When assigning a representative among several parties, make the assignment so as to give the smallest relative unfairness.

EXAMPLE 1 Using the Apportionment Criterion

In a general election, state A has a population of 13,680 and five representatives, and state B has a population of 6,180 and two representatives. Use the apportionment criterion to determine which state is more deserving of one additional representative.

SOLUTION:

We will show this solution in two stages. First, we will compute the relative unfairness of the apportionment if we give the additional representative to A. Then we will do the same computation if the representative is assigned to B.

Stage One: Assume that we give the representative to A instead of B. So A will have six representatives and B will have two. Their average constituencies are as follows:

$$\text{A's average constituency} = \frac{\text{A's population}}{\text{A's representatives}} = \frac{13{,}680}{6} = 2{,}280$$

smaller average constituency

$$\text{B's average constituency} = \frac{\text{B's population}}{\text{B's representatives}} = \frac{6{,}180}{2} = 3{,}090$$

Because state A has the smaller average constituency, the relative unfairness of this apportionment is

larger relative unfairness

$$\frac{\text{B's average constituency} - \text{A's average constituency}}{\text{A's average constituency}} = \frac{3{,}090 - 2{,}280}{2{,}280} = \frac{810}{2{,}280} \approx 0.355.$$

QUIZ YOURSELF 5

State A, with a population of 41,440, presently has seven representatives. State B, with a population of 25,200, has four representatives.

a) Determine the relative unfairness of the apportionment if we give an additional representative to state A.
b) Determine the relative unfairness of the apportionment if this representative is given instead to state B.
c) Use the apportionment criterion to decide which state should receive the additional representative.

Stage Two: Assume that we give the representative to B instead of A. So A will have five representatives and B will have three. Their average constituencies are as follows:

$$\text{A's average constituency} = \frac{\text{A's population}}{\text{A's representatives}} = \frac{13{,}680}{5} = 2{,}736$$

$$\text{B's average constituency} = \frac{\text{B's population}}{\text{B's representatives}} = \frac{6{,}180}{3} = 2{,}060$$

smaller average constituency

Now B has the smaller average constituency, so the relative unfairness of this apportionment is

$$\frac{\text{A's average constituency} - \text{B's average constituency}}{\text{B's average constituency}} = \frac{2{,}736 - 2{,}060}{2{,}060} = \frac{676}{2{,}060} \approx 0.328.$$

smaller relative unfairness

So we see that if we give the additional representative to state A, the relative unfairness would be 0.355 and if we instead assign the additional representative to B, the relative unfairness would be 0.328. Thus the additional representative should be assigned to B.

Now try Exercises 1 to 8. 5

Note that our solution in Example 1 violates the way the Hamilton method apportions eight representatives to states A and B. According to Hamilton's method, the exact number of representatives due state A is

$$\text{number of representatives} \times \frac{\text{population of A}}{\text{total population}} = 8 \times \frac{13{,}680}{13{,}680 + 6{,}180} = 8 \times \frac{13{,}680}{19{,}860} \approx 5.511,$$

whereas the exact number due state B is

$$\text{number of representatives} \times \frac{\text{population of B}}{\text{total population}} = 8 \times \frac{6{,}180}{13{,}680 + 6{,}180} = 8 \times \frac{6{,}180}{19{,}860} \approx 2.489.$$

Because A's fractional part (0.511) is larger than B's (0.489), Hamilton's method would assign the eighth representative to A.

The Huntington–Hill Method

Comparing the relative unfairness of two apportionments as we did in Example 1 is slow and tedious. There is an easier way to determine which state is more deserving of an additional representative. We will now describe the **Huntington–Hill apportionment principle** and explain its derivation at the end of this section.

THE HUNTINGTON–HILL APPORTIONMENT PRINCIPLE If states X and Y have already been allotted *x* and *y* representatives, respectively, then state X should be given an additional representative in preference to state Y provided that

$$\frac{(\text{population of Y})^2}{y \cdot (y+1)} < \frac{(\text{population of X})^2}{x \cdot (x+1)}$$

Otherwise, state Y should be given the additional representative. We will often refer to a number of the form $\frac{(\text{population of X})^2}{x \cdot (x+1)}$ as a **Huntington–Hill number.**

Historical Highlight

Is There a Perfect Apportionment Method?

The Huntington–Hill apportionment principle, developed by mathematicians Edward Huntington and Joseph Hill, was signed into law by Franklin D. Roosevelt in 1941 and is currently used to apportion the U.S. House of Representatives. Although this method avoids the Alabama paradox, it is not perfect.

You might wonder, "Does a perfect apportionment method exist?" To answer this question, consider two reasonable criteria that we expect an apportionment method to satisfy:

- The method should not be subject to the Alabama or other similar paradoxes that we discuss in Section 10.3
- An apportionment should satisfy the **quota rule.** That is, if the exact number of representatives due to a state is 31.46, then the number apportioned must be either 31 or 32. The state cannot receive 30 or 33 representatives.

In 1980, Michael Balinski and H. Peyton Young proved a surprising theorem, called the **Balinski and Young's impossibility theorem,** which states:

> *There is no apportionment method that avoids all paradoxes and at the same time satisfies the quota rule.*

Because any apportionment method must be flawed, politics often plays as large a role as mathematics when Congress discusses an apportionment method.

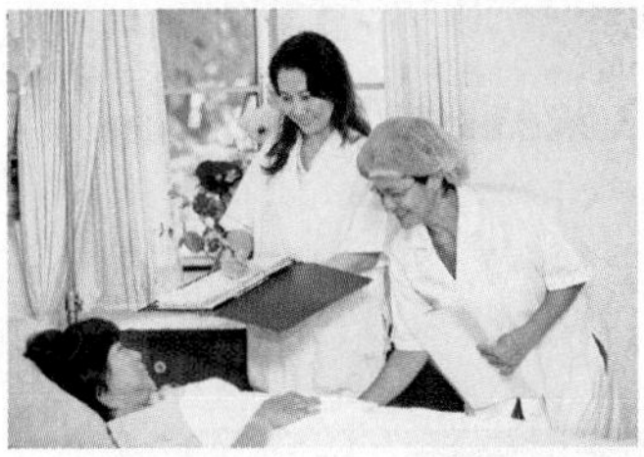

QUIZ YOURSELF 6

According to a recent census, Iowa had a population of approximately 3 million people and four representatives to the U.S. House of Representatives; and Nebraska had a population of 1.8 million people and three representatives. Use the Huntington–Hill apportionment principle to determine which state is more deserving of an additional representative.

EXAMPLE 2 Using the Huntington–Hill Apportionment Principle

There are 320 full-time nurses and 148 part-time nurses at Community General Hospital. The nursing supervisor has chosen 4 full-time nurses and 2 part-time nurses to serve on a committee to evaluate proposed nursing guidelines. Use the Huntington–Hill apportionment principle to decide whether the seventh nurse on the committee should be full-time or part-time.

SOLUTION:

We compute the Huntington–Hill numbers for the full-time nurses and for the part-time nurses:

larger Huntington–Hill number

$$\frac{(\text{number of full-time nurses})^2}{(\text{current representation}) \cdot (\text{current representation} + 1)} = \frac{(320)^2}{4 \cdot 5} = \mathbf{5{,}120}$$

$$\frac{(\text{number of part-time nurses})^2}{(\text{current representation}) \cdot (\text{current representation} + 1)} = \frac{(148)^2}{2 \cdot 3} \approx 3{,}651$$

Comparing these numbers, we find that the next nurse selected for the committee should be a full-time nurse.

Now try Exercises 9 to 16. 6

Apportioning the Oil Consortium Board

We will now use the Huntington–Hill apportionment principle to apportion the entire nine-member oil consortium board.

EXAMPLE 3 Using a Table of Huntington–Hill Numbers to Do Apportionment

Make a table of Huntington–Hill numbers to apportion the oil consortium board.

SOLUTION: Review the Be Systematic Strategy on page 5.

We begin by giving each company one seat, which is consistent with a provision in the U.S. Constitution that mandates each state must have at least one representative. So we have to apportion the remaining six representatives.

Recall Naxxon has 47 hundred stockholders, Aroco has 37 hundred, and Eurobile has 16 hundred. Huntington–Hill numbers have the form

$$\frac{(\text{number of stockholders in the company})^2}{(\text{current representation}) \cdot (\text{current representation} + 1)}.$$

(To keep our numbers small, we will ignore hundreds. This is acceptable as long as we are consistent in doing this for each company.)

Table 10.4 shows the first line of a table of Huntington–Hill numbers, where we are assuming that each company has one representative. Because Naxxon has the largest Huntington–Hill number in Table 10.4, it is given its second representative in preference to the other companies.

Naxxon	Aroco	Eurobile
$\frac{(47)^2}{1 \times 2} = 1{,}104.5$	$\frac{(37)^2}{1 \times 2} = 684.5$	$\frac{(16)^2}{1 \times 2} = 128$

Naxxon has largest Huntington–Hill number.

TABLE 10.4 Huntington–Hill numbers, assuming that each company has one representative.

Now that four representatives have been assigned, we will add more lines to Table 10.4, giving us Table 10.5, which we will use to allocate the remaining representatives. We cross off the first entry under Naxxon because we have already used that number. The symbol ④ indicates that Naxxon received the fourth representative.

Current Representation	Naxxon	Aroco	Eurobile
1	~~$\frac{(47)^2}{1 \times 2} = 1{,}104.5$~~ ④	$\frac{(37)^2}{1 \times 2} = 684.5$ ⑤	$\frac{(16)^2}{1 \times 2} = 128$
2	$\frac{(47)^2}{2 \times 3} = 368.2$ ⑥	$\frac{(37)^2}{2 \times 3} = 228.2$ ⑦	$\frac{(16)^2}{2 \times 3} = 42.7$
3	$\frac{(47)^2}{3 \times 4} = 184.1$ ⑧	$\frac{(37)^2}{3 \times 4} = 114.1$	$\frac{(16)^2}{3 \times 4} = 21.3$
4	$\frac{(47)^2}{4 \times 5} = 110.5$	$\frac{(37)^2}{4 \times 5} = 68.5$	*

TABLE 10.5 More Huntington–Hill numbers.

Aroco deserves fifth representative.

QUIZ YOURSELF 7

Use Table 10.5 to assign representative number 9 to the council.

The next four Huntington–Hill numbers in Table 10.5, in decreasing order, are

684.5, 368.2, 228.2, 184.1,

which tells us how to allocate board members 5 through 8. So we see that the first eight board members are assigned as follows:

1. Naxxon 2. Aroco 3. Eurobile 4. Naxxon (1,104.5)
5. Aroco (684.5) 6. Naxxon (368.2) 7. Aroco (228.2) 8. Naxxon (184.1)

We leave the assignment of board member 9 to Quiz Yourself 7.

Now try Exercises 17 to 26. 7

Notice that if we increase the size of the board to 10, the next largest Huntington–Hill number in Table 10.5 is 114.1, which means the tenth representative would be assigned to Aroco—avoiding an Alabama paradox.

If we represent this expression by the algebraic expression $\frac{s}{t} \times n$, we can then rewrite it as

$$\frac{s}{t} \times n = s \times \frac{n}{t} = \frac{s}{\frac{t}{n}}.$$

state's population

total population divided by number of representatives

That is, in calculating the exact number of representatives that a state deserves, we could have first calculated

$$\frac{\text{total population}}{\text{number of representatives being allocated}}$$

and then divided this number into the state's population. Although this may not seem as intuitively clear as our original approach, this idea of first computing a divisor and then dividing into the state's population will be the central theme of our calculations for the rest of this section.

DEFINITIONS In allocating a group of representatives among several states, we define the **standard divisor** by

$$\text{standard divisor} = \frac{\text{total population}}{\text{number of representatives being allocated}},$$

and a state's **standard quota** is defined to be

$$\text{standard quota} = \frac{\text{state's population}}{\text{standard divisor}}.$$

Intuitively, you can think of the standard divisor as follows:

$$\text{standard divisor} = \frac{\text{total population}}{\text{number of representatives}}$$
$$= \text{number each representative represents.}$$

We think of the standard quota as

$$\text{standard quota} = \frac{\text{state's population}}{\text{standard divisor}}$$
$$= \text{number of representatives state deserves.}$$

EXAMPLE 1 Finding the Standard Divisor and Standard Quota

a) Suppose we want to allocate eight representatives among several states that have a total population of 4,000,000. Find the standard divisor.

b) If state A has a population of 1,500,000, calculate its standard quota.

SOLUTION:

a) We calculate the standard divisor as follows:

$$\text{standard divisor} = \frac{\text{total population}}{\text{number of representatives}} = \frac{4{,}000{,}000}{8} = 500{,}000,$$

which equals the number of constituents that each representative represents.

QUIZ YOURSELF 8

Assume we are apportioning eight representatives among three states A, B, and C, which have populations of 3 million, 4 million, and 5 million, respectively.

a) Calculate the standard divisor for this apportionment.

b) Calculate each state's standard quota.

b) We compute the standard quota for state A as:

$$\text{standard quota for state A} = \frac{\text{state's population}}{\text{standard divisor}} = \frac{1{,}500{,}000}{500{,}000} = 3.$$

So state A deserves three representatives. 8

It is interesting to recalculate Naxxon's apportionment on the nine-member oil consortium board using this new approach. Recall that Naxxon had 4,700 stockholders of the 10,000 total. Thus the standard divisor is

$$\frac{\text{total number of stockholders}}{\text{number of board members}} = \frac{10{,}000}{9} = 1{,}111.11.$$

Naxxon's standard quota is $\frac{\text{number of Naxxon stockholders}}{\text{standard divisor}} = \frac{4{,}700}{1{,}111.11} = 4.23$, as we found before. Likewise, Aroco's standard quota is 3.33 and Eurobile's is 1.44.

SOME GOOD ADVICE

Keep in mind that the standard divisor is a single number that we calculate once and then use for the entire apportionment process. However, we must compute the standard quota individually for each state.

Recall that in using the Hamilton apportionment method, we always gave each state a number of representatives that was immediately below or immediately above its standard quota (the exact number of representatives that it was due). For example, if a state's standard quota was 4.375 representatives, we always gave it either 4 or 5 representatives. We will now make this notion more precise.

DEFINITIONS If we round the standard quota down, we call that number the **lower quota;** if we round the standard quota up, then we call that number the **upper quota.** If in making an apportionment, each state is allocated a number of representatives that is between its lower quota and upper quota, then we say the apportionment satisfies the **quota rule.**

EXAMPLE 2 Apportioning Representatives to Three States

Assume that we are apportioning eight representatives to three states: A, with population 5.4 million; B, with population 6.7 million; and C, with population 7.3 million. Find A's lower and upper quotas.

SOLUTION: Review the Analogies Principle on page 13.

To find A's upper and lower quotas, we must first find the standard divisor and then A's standard quota. The standard divisor for this apportionment is

$$\text{standard divisor} = \frac{\text{total population}}{\text{number of representatives}} = \frac{5.4 + 6.7 + 7.3}{8} = \frac{19.4}{8} = 2.425.$$

Dividing this into A's population, we get

$$\text{A's standard quota} = \frac{\text{A's population}}{\text{standard divisor}} = \frac{5.4}{2.425} \approx 2.227.$$

If we round 2.227 down, we get A's lower quota, which is 2. Rounding up, we get A's upper quota, which is 3. 9

QUIZ YOURSELF 9

In Example 2, find C's standard quota, lower quota, and upper quota.

We can use this new language to rephrase Hamilton's apportionment method that we discussed in Section 10.1. Although the terminology sounds different, this method gives exactly the same results as we found before.

HAMILTON'S APPORTIONMENT METHOD

a) Find the standard divisor for the apportionment (total population/total number of representatives).
b) Find the standard quota (state's population/standard divisor) for each state and round it down to its lower quota. Assign that number of representatives to each state.
c) If there are any representatives left over, assign them to states in order according to the size of the fractional parts of the states' standard quotas.

Perhaps the Hamilton method would be the apportionment method used today if its only problem was the Alabama paradox. However, there are other paradoxes that we will discuss now.

More Apportionment Paradoxes

KEY POINT

The Hamilton method can have the population paradox and the new states paradox.

In the early 1900s, it was discovered that Hamilton's method had another serious flaw in that a state with a faster-growing population could lose a representative to a state whose population was growing more slowly.

DEFINITION The **population paradox** occurs when state A's population is growing faster than state B's population, yet A loses a representative to state B. (We are assuming that the total number of representatives in the legislature is not changing.)

EXAMPLE 3 The Population Paradox Can Occur with the Hamilton Method

The graduate school at Great Eastern University is using the Hamilton method to apportion 15 graduate assistantships among the colleges of education, liberal arts, and business based on their undergraduate enrollments. The college of education has 940 undergraduate students, the college of liberal arts has 1,470, and the college of business has 1,600.

a) Use Hamilton's method to allocate the graduate assistantships to the three colleges.
b) Assume that after the allocation was made in part a), education gains 30 students, liberal arts gains 46, and the business enrollment stays the same. Reapportion the graduate assistantships again using Hamilton's method.
c) Explain how this illustrates the population paradox.

SOLUTION:

a) The standard divisor for this apportionment is

$$\frac{\text{total number of undergraduate students}}{\text{number of assistantships}} = \frac{4{,}010}{15} \approx 267.33.$$

In Table 10.8, we compute the standard quota and the number of assistantships to be allocated to each college.

College	Standard Quota (Exact Number Deserved)	Lower Quota (Integer Part)	Fractional Part	Assign 2 Additional Assistantships
Education	$\frac{940}{267.33} = 3.52$	3	0.52	4
Liberal arts	$\frac{1,470}{267.33} = 5.50$	5	0.50	5
Business	$\frac{1,600}{267.33} = 5.99$	5	0.99	6
Total		13		15

Divide enrollment of each college by the standard divisor.

Education and business have the largest fractional parts.

Colleges with largest fractional parts get extra assistantships.

TABLE 10.8 Apportioning the 15 graduate assistantships before the enrollments increase.

b) Table 10.9 shows how the apportionment would be done after the enrollments increase. The standard divisor is now

$$\frac{\text{new total number of undergraduate students}}{\text{number of assistantships}} = \frac{4,086}{15} \approx 272.4.$$

College	Number of Students	Standard Quota (Exact Number Deserved)	Lower Quota (Integer Part)	Fractional Part	Assign 2 Additional Assistantships
Education	940 + 30 = 970	$\frac{970}{272.4} = 3.56$	3	0.56	3
Liberal arts	1,470 + 46 = 1,516	$\frac{1,516}{272.4} = 5.57$	5	0.57	6
Business	1,600	$\frac{1,600}{272.4} = 5.87$	5	0.87	6
Total	4,086		13		15

Education loses one assistantship.

TABLE 10.9 Apportioning the 15 graduate assistantships after the enrollments increase.

c) Notice how the college of education lost an assistantship to the college of liberal arts. However, the percentage increase in the number of education students was

$$\frac{\text{increase in education students}}{\text{original number of education students}} = \frac{30}{940} = 0.0319 = 3.19\%.$$

But the increase in the number of liberal arts students was

$$\frac{\text{increase in liberal arts students}}{\text{original number of liberal arts students}} = \frac{46}{1,470} = 0.0313 = 3.13\%.$$

So, even though the college of liberal arts grew more slowly than the college of education, it was able to take one of education's assistantships away.

Now try Exercises 25 to 28.

Admittedly, it was a close call in Example 3, and to be truthful it can be a lot of work to find such a counterexample (a spreadsheet helps!), but the point is, Hamilton's method can allow the population paradox to occur.

When Oklahoma joined the union in 1907, the House of Representatives had to be reapportioned. Congress decided to increase the size of the House by five and give those

Between the Numbers

Subverting the Legislative Process

Until the 1960s, state legislatures had great power in apportioning legislative districts to favor small, rural districts over larger, urban districts. For example, in Texas, Speaker of the House Sam Rayburn presided over a rural district of 228,000 constituents, whereas Albert Thomas, of Houston, had over 800,000. In response to lawsuits, the Supreme Court ruled that representation must be fairly distributed among the people of one state—a ruling often called the "one man, one vote" rule.

Another way to subvert the democratic process is a technique called **gerrymandering,** which is the redrawing of boundaries of legislative districts to benefit one particular party. The given map shows a proposed restructuring of Florida's fifth legislative district (shown in blue) following the 2010 census.

In arguing in court to prevent such extremely irregularly shaped districts, critics often apply mathematical techniques such as comparing areas with boundaries to measure how "gerrymandered" a district is.

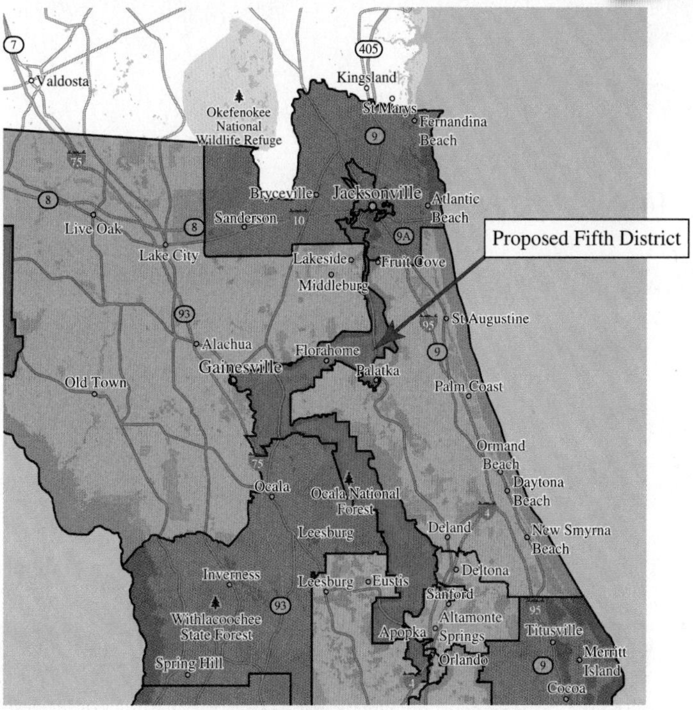

Source: Orlando Sentinel, May 17, 2012.

five representatives to Oklahoma. However, when the House was reapportioned, New York was required to give one of its seats to Maine. This is called the *new states paradox*.

DEFINITION The **new states paradox** occurs when a new state is added and its share of seats is added to the legislature, causing a change in the allocation of seats previously given to another state.

EXAMPLE 4 The New States Paradox Can Occur with the Hamilton Method

A small country, Namania, consists of three states A, B, and C with populations given in Table 10.10 on the following page. Namania's legislature has 37 representatives that are to be apportioned to these states using the Hamilton method.

a) Apportion these representatives using the Hamilton method.
b) Assume that Namania annexes the country Darelia, whose population is 3,000 (thousand). Give Darelia its current share of representatives using the current standard divisor and add that number to the total number of representatives of Namania. Reapportion Namania again using the Hamilton method.
c) Explain how the new states paradox has occurred.

SOLUTION:

a) The standard divisor for this apportionment is the total population, 12,140 (thousand), divided by the number of representatives, 37. So the standard divisor is $\frac{12{,}140}{37} = 328.11$. We will use this to find each state's standard quota in Table 10.10.

State	Population (thousands)	Standard Quota (Exact Number Deserved)	Lower Quota (Integer Part)	Fractional Part	Assign Additional Representative
A	2,750	$\frac{2{,}750}{328.11} = 8.38$	8	0.38	8
B	6,040	$\frac{6{,}040}{328.11} = 18.41$	18	0.41	19
C	3,350	$\frac{3{,}350}{328.11} = 10.21$	10	0.21	10
Total	12,140		36		37

State with highest fractional part gets extra representative.

TABLE 10.10 Apportioning representatives to states A, B, and C before the annexation of Darelia.

b) According to the apportionment in part a), using the standard divisor of 328.11, Darelia's standard quota is $\frac{3{,}000}{328.11} = 9.14$. This means that Darelia deserves nine representatives under the current apportionment, so we will add nine representatives to Namania's legislature and reapportion it. Now, because Namania's population has increased and we have added nine representatives, we must recalculate the standard divisor.

$$\text{standard divisor} = \frac{\text{population}}{\text{number of representatives}} = \frac{12{,}140 + 3{,}000}{37 + 9} = \frac{15{,}140}{46} \approx 329.13$$

We will use this to compute the new standard quotas in Table 10.11.

State	Population (thousands)	Standard Quota (Exact Number Deserved)	Lower Quota (Integer Part)	Fractional Part	Assign Additional Representative
A	2,750	$\frac{2{,}750}{329.13} = 8.36$	8	0.36	9
B	6,040	$\frac{6{,}040}{329.13} = 18.35$	18	0.35	18
C	3,350	$\frac{3{,}350}{329.13} = 10.18$	10	0.18	10
(D)arelia	3,000	$\frac{3{,}000}{329.13} = 9.11$	9	0.11	9
Total	15,140		45		46

B loses one representative.

TABLE 10.11 Apportioning representatives to states A, B, C, and D after the annexation of Darelia.

c) We saw in part b) that by adding Darelia, and increasing the size of the legislature according to the number of representatives that it deserved, B had to give one of its representatives to A. This illustrates the new states paradox.

Now try Exercises 29 to 32.

For the remainder of this section, we will discuss some alternative apportionment methods proposed by Jefferson, Adams, and Webster. These three methods are similar in that each uses a divisor that is different from the standard divisor, which is called a **modified divisor.** (As you will see, we find this modified divisor by trial and error.) When we divide a state's population by the modified divisor, we get the state's **modified quota.**

Once we find the modified quotas for each state, the question is what do we do with them? Jefferson says to round them down to get the required number of representatives, Adams says to round up, and Webster says to round in the usual way.

Other Apportionment Methods

KEY POINT

Jefferson's method rounds quotas down.

In Jefferson's method, we are going to round the modified quotas down, so we will need modified quotas that are *larger** than the standard quotas. This means that the modified divisor has to be *smaller* than the standard divisor to give us the larger quotas.

*Sometimes the standard divisor and quotas will work when using the Jefferson method.

JEFFERSON'S APPORTIONMENT METHOD

a) Use trial and error to find a modified divisor that is smaller than the standard divisor for the apportionment.
b) Calculate the **modified quota** (state's population/modified divisor) for each state and round it down. Assign that number of representatives to each state. (Keep varying the modified divisor until the sum of these assignments is equal to the total number being apportioned.)

We will now reapportion the oil consortium board using Jefferson's method.

EXAMPLE 5 Using Jefferson's Apportionment Method

Recall in Example 1 in Section 10.1, we were apportioning a nine-member board to oversee the oil consortium formed by Naxxon, Aroco, and Eurobile. We used Hamilton's method to apportion the board according to the number of stockholders of each company. Naxxon had 4,700 stockholders, Aroco had 3,700, and Eurobile had 1,600. Now use Jefferson's method to apportion the board.

SOLUTION:

In the discussion in Section 10.1, we found that the standard divisor for this apportionment was $\frac{10{,}000}{9} \approx 1{,}111.11$, and the standard quotas for the three companies were 4.23, 3.33, and 1.44. When we used Hamilton's method and rounded the standard quotas down, the total was eight and we had to assign the ninth representative.

In applying Jefferson's method, we want to find modified quotas so that when we round these quotas down, the total is nine. This means that the modified quotas must be larger than the standard quotas. In order to get larger modified quotas, we must have a modified divisor that is smaller than the standard divisor of 1,111.11.

Let's try a modified divisor of 1,050 to see what happens. We show the calculations in Table 10.12.

Modified Divisor = 1,050			
	Naxxon	**Aroco**	**Eurobile**
Number of stockholders	4,700	3,700	1,600
Standard quota	4.23	3.33	1.44
Modified quota	$\frac{4{,}700}{1{,}050} = 4.48$	$\frac{3{,}700}{1{,}050} = 3.52$	$\frac{1{,}600}{1{,}050} = 1.52$
Round modified quota down	4	3	1
			Total = 8

Too small—we need a smaller modified divisor.

TABLE 10.12 Using the Jefferson method to apportion the oil consortium board using a modified divisor of 1,050.

Modified Divisor = 935			
	Naxxon	**Aroco**	**Eurobile**
Number of stockholders	4,700	3,700	1,600
Standard quota	4.23	3.33	1.44
Modified quota	$\frac{4{,}700}{935} = 5.03$	$\frac{3{,}700}{935} = 3.96$	$\frac{1{,}600}{935} = 1.71$
Round modified quota down	5	3	1
			Total = 9

TABLE 10.13 Using the Jefferson method to apportion the oil consortium board using a modified divisor of 935.

Because the total number of board members assigned is too small, we need *larger* modified quotas. To do this, we must try *smaller* modified divisors. In making up this example, we used a spreadsheet and tried 1,000 and then 950, but neither of these modified divisors worked. Finally, when we used the modified divisor of 935, we found the results in Table 10.13. Of course, there are numbers other than 935 that would be acceptable modified divisors. We will ask you to investigate which other modified divisors would be acceptable in the exercises. 10

When we used the Hamilton method in Section 10.1 to assign representatives to the oil consortium board, Naxxon received 4, Aroco received 3, and Eurobile received 2. So we see that the Jefferson method gives a different apportionment than the Hamilton method does.

We will next illustrate how to use Adams's method to apportion the oil consortium board.

In Adams's method, we are going to round the modified quotas up, so we will need modified quotas that are *smaller* than the standard quotas. This means that the modified divisor has to be *larger* than the standard divisor to give us the smaller quotas.

QUIZ YOURSELF 10

Assume we are apportioning nine representatives among three groups A, B, and C, which have populations of 1,276, 1,427, and 2,697, respectively.

a) Calculate the standard divisor for this apportionment.
b) Use the modified divisor of 550 to calculate group C's modified quota.
c) Using Jefferson's method, how many representatives would C receive if we use the modified divisor 550?

ADAMS'S APPORTIONMENT METHOD

a) Use trial and error to find a modified divisor that is larger than the standard divisor for the apportionment.
b) Calculate the modified quota (state's population/modified divisor) for each state and round it up. Assign that number of representatives to each state. (Keep varying the modified divisor until the sum of these assignments is equal to the total number being apportioned.)

KEY POINT

Adams's method rounds quotas up.

Example 6 illustrates Adams's apportionment method.

EXAMPLE 6 Allocating iPads

The Springfield School District is distributing 124 of the newest iPads to computer labs in its three elementary schools. The iPads will be allocated proportionally to enrollments at the three schools. East Springfield has 470 students, West Ridge has 350, and North Valley has 280. Use the Adams method to allocate the iPads.

SOLUTION:

The total number of students is $470 + 350 + 280 = 1{,}100$, so the standard divisor is $\frac{1{,}100}{124} = 8.871$. The following table shows that if we begin by using 8.871 as a trial divisor, the total is too large, so we need a larger divisor. We then try the divisor 9.5, which gives a total that is too small, so we need a smaller trial divisor. Finally we try 9.0, which gives the desired total.

School	Divisor 8.871	Round Up	Divisor 9.5	Round Up	Divisor 9.0	Round Up
East	$\frac{470}{8.871} = 52.98$	53	$\frac{470}{9.5} = 49.47$	50	$\frac{470}{9.0} = 52.22$	53
West	$\frac{350}{8.871} = 39.45$	40	$\frac{350}{9.5} = 36.84$	37	$\frac{350}{9.0} = 38.89$	39
North	$\frac{280}{8.871} = 31.56$	32	$\frac{280}{9.5} = 29.47$	30	$\frac{280}{9.0} = 31.11$	32
Total		125		117		124

Total is too large, so we need a *larger* divisor than 8.871.

Total is too small, so we need a *smaller* divisor than 9.5.

Total is correct.

So, East Springfield will get 53 iPads, West Ridge will get 39, and North Valley will be given 32. 11

QUIZ YOURSELF 11

Assume we are apportioning 10 representatives among three groups A, B, and C, which have populations of 1,300, 950, and 2,550, respectively.

a) Calculate the standard divisor for this apportionment.
b) Use the modified divisor of 500 to calculate group A's modified quota.
c) Using Adams's method, how many representatives would A receive if we use the modified divisor 500?

As in the Jefferson method, there will be acceptable modified divisors other than 9.0 that we could have used in Example 6. We will pursue this question further in the exercises.

SOME GOOD ADVICE

In order to know how to adjust the modified divisors when using the Jefferson and Adams methods, think of this diagram, where the modified divisors and quotas can be thought of as being on a teeter-totter. When one goes up, the other goes down.

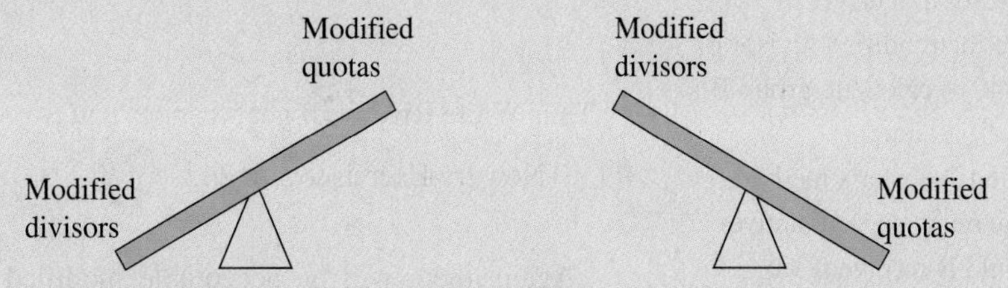

If you need more representatives, you need larger modified quotas, so you must use smaller modified divisors.

If you need fewer representatives, you need smaller modified quotas, so you must use larger modified divisors.

Finally, we will discuss Webster's apportionment method.

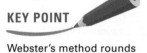

KEY POINT

Webster's method rounds quotas in the usual way.

In Webster's method, we round the modified quotas in the usual way. That is, if the quota has a decimal part of 0.5 or greater, we round up; otherwise, we round down. Again, as in the Jefferson and Adams methods, we find the modified divisor by trial and error.

WEBSTER'S APPORTIONMENT METHOD

a) Use trial and error to find a modified divisor.
b) Calculate the modified quota (state's population/modified divisor) for each state and round it in the usual way. Assign that number of representatives to each state. (Keep varying the modified divisor until the sum of these assignments is equal to the total number being apportioned.)

Example 7 illustrates how to use Webster's method to do apportionment.

EXAMPLE 7 Using Webster's Method to Allocate Canoes

Whitewater Adventures, Inc., has purchased 14 new canoes for its two locations and is going to allocate them according to the number of trips taken at each location last month. White River had 53 trips and Sweet Mountain had 84. Use Webster's method to allocate the canoes.

SOLUTION:

The total number of trips is $53 + 84 = 137$. It really doesn't matter how we pick our first trial divisor, so let's start with 8. In the table on the next page, we see that the trial divisor 8 gives us a total that is too large, so we need a larger divisor. We then try the divisor 10, which results in a total that is too small, so we need a smaller divisor. Finally we try 9.7, which gives the desired total of 14.

QUIZ YOURSELF 12

Assume we are apportioning eight representatives among three groups A, B, and C, which have populations of 2,700, 4,300, and 5,000, respectively.

a) Calculate the standard divisor for this apportionment.
b) Use the modified divisor of 1,540 to calculate group B's modified quota.
c) Using Webster's method, how many representatives would B receive if we use the modified divisor 1,540?

Location	Divisor 8	Round as Usual	Divisor 10	Round as Usual	Divisor 9.7	Round as Usual
White River	$\frac{53}{8} = 6.63$	7	$\frac{53}{10} = 5.3$	5	$\frac{53}{9.7} = 5.46$	5
Sweet Mountain	$\frac{84}{8} = 10.5$	11	$\frac{84}{10} = 8.4$	8	$\frac{84}{9.7} = 8.66$	9
Total		18		13		14

Divisor 8: Total is too large, so we need a *larger* divisor than 8.
Divisor 10: Total is too small, so we need a *smaller* divisor than 10.
Divisor 9.7: Total is correct.

Thus White River will get 5 canoes and Sweet Mountain will get 9.

Now try Exercises 5 to 24. 12

Again, there will be acceptable modified divisors other than 9.7. In the exercises, we will ask you to investigate what numbers would be acceptable modified divisors for Example 7.

PROBLEM SOLVING

Strategy: The Analogies Principle

In learning these apportionment methods, try to relate them to each other. In each case, first you must find a modified divisor and then round in one of three ways: Jefferson (down), Adams (up), or Webster (usual way).

The following diagram, although a little silly, will help you remember which way to round with each method. Think of the J in "Jefferson" as an arrow pointing down, and the A in "Adams" as the tip of an arrow pointing up. The W in "Webster" is pointing neither strictly up nor down.

Jefferson	Adams	Webster
Round down	Round up	Neither up … nor down

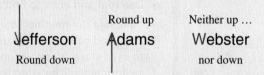

Don't bother memorizing how to select the modified divisor—you can always start with the standard divisor and remember that if your apportionment is too large, then you need smaller quotas and therefore you need a larger modified divisor. Likewise, if the apportionment is too small, you need a smaller modified divisor.

We summarize the four methods we have studied in Table 10.14. As you see from this table, the Jefferson, Adams, and Webster methods are not perfect because they all can fail to satisfy the quota rule. We will give examples of such violations in the exercises.

	Hamilton	Jefferson	Adams	Webster
Can have Alabama paradox	Yes	No	No	No
Can have population paradox	Yes	No	No	No
Can have new states paradox	Yes	No	No	No
Can violate quota rule	No	Yes	Yes	Yes

TABLE 10.14 Summary of four apportionment methods.

EXERCISES 10.3

Sharpening Your Skills

In Exercises 1–4, we give you a total population, state A's population, and the total number of representatives that are to be apportioned. Find:

a) The standard divisor for the apportionment and the standard quota for A.

b) The number of representatives that would be given to the state using Jefferson's method.

c) The number of representatives that would be given to the state using Adams's method.

d) The number of representatives that would be given to the state using Webster's method.

1. Total population 512,000; As population 84,160; number of representatives being apportioned 16
2. Total population 135,000; As population 48,000; number of representatives being apportioned 9
3. Total population 102,000; As population 19,975; number of representatives being apportioned 11
4. Total population 512,000; As population 84,160; number of representatives being apportioned 15

Applying What You've Learned

5. **Assigning seats on a water authority.** The 11-member Southwest Water Authority is planning a dam on the Gila River in Arizona. Seats on the authority will be assigned according to the number of customers in each state that will use the water from the project. There are 56,000 customers in California, 52,000 in Arizona, and 41,000 in Nevada. Find the standard divisor and each state's standard quota.
6. Use the Jefferson method to assign the seats on the Authority.
7. Use the Adams method to assign the seats on the Authority.
8. Use the Webster method to assign the seats on the Authority.
9. **Choosing representatives on a negotiations committee.** The employees of Six Flags theme park are negotiating a new contract. There are 213 performers, 273 food workers, and 178 maintenance workers. The 20-person negotiations committee has members in proportion to the number of employees in each of the three groups. Find the standard divisor and each group's standard quota.
10. Use the Jefferson method to apportion the members of the negotiations committee.
11. Use the Adams method to apportion the members of the negotiations committee.
12. Use the Webster method to apportion the members of the negotiations committee.
13. **Apportioning a labor council.** A labor council is being formed from the members of four unions. The electricians' union has 25 members, the plumbers have 18 members, the painters have 29, and the carpenters have 31 members. The council will have 20 representatives. Find the standard divisor and each union's standard quota.
14. Use the Jefferson method to assign the representatives to the council.
15. Use the Adams method to assign the representatives to the council.
16. Use the Webster method to assign the representatives to the council.
17. **Awarding assistantships.** Nineteen fellowships are to be apportioned to students in a writing program at a major university. The apportionment is to be based on the number of full-time graduate students in each area of the writing program. There are 30 fiction majors, 20 poetry majors, 17 majoring in technical writing, and 14 majoring in writing for the media. Find the standard divisor and each major's standard quota.
18. Use the Jefferson method to assign the fellowships.
19. Use the Adams method to assign the fellowships.
20. Use the Webster method to assign the fellowships.
21. **Scheduling fitness classes.** A health club instructor has a course load that allows her to teach six two-credit-hour classes. A preregistration survey indicates the following interests:

    56 want to take Pilates;

    29 want to take kickboxing;

    11 want to take yoga;

    4 want to take spinning.

    Find the standard divisor and the standard quotas for each interest area.
22. Use the Jefferson method to assign the number of classes to each area.
23. Use the Adams method to assign the number of classes to each area.
24. Use the Webster method to assign the number of classes to each area.

In Exercises 25–32, we use the Hamilton method to do the apportionment.

25. **The population paradox.** Ten representatives are apportioned among three states A, B, and C, which have the populations (in thousands) shown in the table. Ten years later, the population of A has increased by 20 thousand and the population of B has increased by 60 thousand, whereas

C's population has remained the same. The 10 representatives are reapportioned. Does this example illustrate the population paradox? Explain.

State	Population
A	570
B	2,557
C	6,873

26. The population paradox. The Metrodelphia Regional Transportation Authority has three lines, the Arrow, the Blue Line, and the City division. The numbers of passengers that use each line per day are given in the table below. Assume that 100 new cars are apportioned among the three lines according to the number of passengers using each line. One year later, the MRTA decides to reapportion the 100 cars among the three divisions. We show the change in ridership in the table. Does this example illustrate the population paradox? Explain.

Line	Passengers per Day (now)	Passengers per Day (one year later)
Arrow	23,530	23,930
Blue	5,550	5,650
City	70,920	71,100

27. The population paradox. A group of three regional airports has hired a company to increase security. The contract calls for 150 security personnel to be apportioned to the airports according to the number of passengers using each airport each week. Two years later, the number of passengers per week has increased at the three airports, so the 150 security personnel are reallocated. We show these data in the given table. Does this example illustrate the population paradox? Explain.

Airport	Passengers per Week (now)	Passengers per Week (two years later)
Allenport	120,920	121,420
Bakerstown	5,550	5,650
Columbia City	23,530	23,930

28. The population paradox. The current enrollments at a university's three campuses are shown below. The New State University's administration has decided to upgrade 18 computer labs according to the enrollments at the three campuses. One year later, the enrollments have increased at each campus and again it is decided to upgrade 18 labs. Does this example illustrate the population paradox? Explain.

Campus	Students (now)	Students (one year later)
Altus	7,200	7,480
Brightsfield	20,500	21,140
Center City	52,300	52,900

29. The new states paradox. Two states A and B have formed a 100-member commission to explore ways in which they can work jointly to improve the business climate in their states. The members will be apportioned to the commission according to the populations of the two states. After seeing the success of the commission, a third state asks to join and appoint members to the commission. States A and B agree and will allow state C to add as many members to the commission as they would deserve now using the current standard divisor. The populations of the three states are given below. Does this example illustrate the new states paradox? Explain.

State	Population (thousands)
A	2,000
B	7,770
C	600

30. The new states paradox. Redo Exercise 29, but now use the populations given below. Also assume that the original commission has 40 members.

State	Population (thousands)
A	12,550
B	4,450
C	8,250

31. The new states paradox. Redo Exercise 29, but now use the populations given below. Also assume that the original commission has 56 members. Assume that A, B, and C are the original states and D is the new state.

State	Population (thousands)
A	8,700
B	4,300
C	3,200
D	6,500

32. The new states paradox. Redo Exercise 31, but now use the populations given below. Also assume that the original commission has 70 members.

State	Population (thousands)
A	230
B	740
C	380
D	600

Exercises 33–36 illustrate that the Jefferson and Adams apportionment methods can violate the quota rule. In each case, determine which method illustrates the violation, and explain how the quota rule is not satisfied.

33. The quota rule. We are apportioning 200 representatives among the states A, B, C, D, E, and F. The populations of these states are given below.

State	Population
A	700
B	1,500
C	820
D	4,530
E	2,200
F	550

34. The quota rule. We are apportioning 100 representatives among the states A, B, C, D, E, and F. The populations of these states are given below.

State	Population
A	700
B	8,000
C	820
D	1,500
E	4,000
F	550

35. The quota rule. We are apportioning 500 representatives among the states A, B, C, D, E, and F. The populations of these states are given below.

State	Population
A	1,400
B	2,000
C	1,500
D	9,000
E	4,300
F	500

36. The quota rule. We are apportioning 400 representatives among the states A, B, C, D, E, and F. The populations of these states are given below.

State	Population
A	700
B	1,000
C	750
D	200
E	2,100
F	4,500

Communicating Mathematics

37. What is the intuitive meaning of *standard divisor* and *standard quota*? In doing an apportionment, how many times must we calculate the standard divisor? The standard quota?

38. How do the Jefferson, Adams, and Webster methods use the modified quotas differently?

39. Explain why the modified divisor in the Jefferson method is usually smaller than the standard divisor.

40. Explain why the modified divisor in the Adams method is usually greater than the standard divisor.

Between the Numbers

41. Investigate gerrymandering to find dramatic examples of "gerrymandered" districts. If possible, find examples in your state. If possible, find examples where a state's Supreme Court has overturned redistricting plans.

42. Go to the site www.mapcenter.org/region/gerrymeas.html or other sites to investigate how mathematics is used to study gerrymandering.

Challenge Yourself

43. In using the Adams method to apportion a legislature, Juanita got an assignment with too many representatives. What change must she make in her approach?

44. In using the Webster method to apportion a legislature, Joe got an assignment with too few representatives. What change must he make in his approach?

45. In Example 5, we found that a modified divisor of 935 worked in using Jefferson's apportionment method. Use trial and error to find the smallest and the largest integers that we could have used in this example as modified divisors.

46. In Example 6, we found that a modified divisor of 9.0 worked in using Adams's apportionment method. Use trial and error to find the smallest and the largest numbers to two decimal places that we could have used in this example as modified divisors.

47. In Example 7, we found that a modified divisor of 9.7 worked in using Webster's apportionment method. Use trial and error to find the smallest and the largest numbers to two decimal places that we could have used in this example as modified divisors.

48. Make up several hypothetical countries each consisting of several large and several small states. Experiment with different apportionments to see if the Jefferson method seems to favor large or small states.

49. Repeat Exercise 48 for the Adams method.

50. Repeat Exercise 48 for the Webster method.

Looking Deeper

10.4 Fair Division

Objectives

1. Understand the difference between discrete and continuous division of items.
2. Use the method of sealed bids to make a fair division of items.

How often do you agree to a deal only to find out after it is done that you will receive more than you expected? You will now learn how to divide any group of items among several interested parties that does exactly that. This method works in dividing an inheritance among the heirs of an estate, awarding property to a couple getting divorced, and dissolving a business among partners.

Fair Division

There are various ways to solve these problems, which are called **fair division** problems. We will now explain one method for handling *discrete* fair division problems as opposed to *continuous* fair division problems.

> DEFINITIONS **Discrete fair division** means that the items being distributed cannot be subdivided into fractional parts. **Continuous fair division** means that the items being distributed can be subdivided into parts and the division can be done as finely as we want.

QUIZ YOURSELF 13

Identify each situation as dealing with either discrete or continuous division of items.

a) Apportioning the seats in the U.S. House of Representatives.
b) Dividing a piece of land among three people.
c) Dividing a mixture of M&Ms, hard candy, and pieces of bubble gum among five children.
d) Sharing a can of soda among three people.

For example, we could not subdivide an antique car into parts to distribute it among heirs to an estate. On the other hand, we can subdivide a cake and distribute pieces of it among children at a birthday party. 13

Method of Sealed Bids

We will now discuss the *method of sealed bids*, which is based on work done by the Polish mathematician Hugo Steinhaus. We start by looking at the fair division of an estate consisting of one item being shared equally by several people. Our problem is to determine how each person, in his or her opinion, can get a fair share when the item cannot be physically divided. But first, we must explain what we mean by a person's *fair share* of an estate.

> DEFINITION A person's **fair share** of an estate is the person's estimate of the estate's value divided by the number of people sharing equally in the estate.

Wayne Malloy has died and left to three cousins—Dahlia, Cael, and DiDi—a rare painting that has been in the family for generations. The three cousins agree that the painting should not be sold, but each is entitled to an equal share of the bequest. Obviously, the painting cannot be divided, so they agree that one of them will get the painting and pay cash to the others.

Dahlia, Cael, and DiDi contact several appraisers to establish the painting's value. However, they receive such a wide range of estimates that it only causes confusion. Dahlia, who has taken a mathematics course in which she studied fair division, suggests that they proceed in the following way.

Each of them secretly writes down an estimate of the painting's value and places the bid in a sealed envelope. By doing this, the three of them will establish an opinion on the estate's value and, in turn, their fair share of the estate. For the process to work, each must promise to give an honest estimate and be content to receive a fair share. Next, they will open the bids and the person making the highest bid will get the painting.

Upon opening the envelopes, they find that Dahlia has valued the painting at \$84,000, Cael feels the painting is worth \$60,000, and DiDi thinks it is worth \$72,000. As agreed, Dahlia gets the painting because she had the highest bid. But now she must pay the estate an amount of money. The amount is her estimate of the estate's value less her fair share of the estimate.

Because Dahlia had valued the estate at \$84,000 and because the cousins are to share equally in the estate, Dahlia's estimate of her fair share is

$$\frac{1}{3} \times 84{,}000 = \$28{,}000.$$

Therefore, she pays the estate the difference:

value of painting — Dahlia's fair share

$$\$84{,}000 - \$28{,}000 = \$56{,}000.$$

— Dahlia owes estate

But should Cael and DiDi split that money and get \$28,000 apiece? It would be unfair if they did, because they each estimated the estate at a lower value. Furthermore, they agreed to be content with receiving their fair share. Because Cael bid \$60,000, he thinks that his fair share is

$$\frac{1}{3} \times 60{,}000 = \$20{,}000.$$

With DiDi's bid of \$72,000, she perceives that her fair share of the estate is

$$\frac{1}{3} \times 72{,}000 = \$24{,}000.$$

Giving Cael \$20,000 and DiDi \$24,000 from the estate leaves a balance of

Dahlia paid to estate — Cael and DiDi's fair shares

$$\$56{,}000 - \$20{,}000 - \$24{,}000 = \$12{,}000$$

— remaining balance

in the estate. Now what do we do with the balance? Because the three cousins have equal interests in the estate, we will divide this balance equally among them. Therefore, Cael and DiDi each get an additional \$4,000, and the remaining \$4,000 is returned to Dahlia.

Table 10.15 shows each person's estimate of the estate's value and his or her opinion of a fair share. Table 10.16 summarizes the accounting of cash based on our method of fair division. A plus sign indicates the person pays that amount to the estate,

	Dahlia $\left(\frac{1}{3}\right)$	Cael $\left(\frac{1}{3}\right)$	DiDi $\left(\frac{1}{3}\right)$
Bid on painting	\$84,000	\$60,000	\$72,000
Fair share of estate	$\frac{\$84{,}000}{3} = \$28{,}000$	$\frac{\$60{,}000}{3} = \$20{,}000$	$\frac{\$72{,}000}{3} = \$24{,}000$
Item obtained with high bid	Painting		

TABLE 10.15 Estimates of value and fair share.

Dahlia, Cael, and DiDi said they would be satisfied with these amounts.

	Dahlia ($\frac{1}{3}$)	Cael ($\frac{1}{3}$)	DiDi ($\frac{1}{3}$)
Pays to estate (+) or receives from estate (−)	\$56,000 (+)	\$20,000 (−)	\$24,000 (−)
Division of estate balance (\$12,000)	\$4,000 (−)	\$4,000 (−)	\$4,000 (−)
Summary of cash	Pays \$52,000	Receives \$24,000	Receives \$28,000

TABLE 10.16 Fair division accounting of estate cash.

whereas a minus sign means the person receives that amount from the estate. Observe in Table 10.16 that the amount paid to the estate must equal the sum of the amounts received from the estate.

It may appear at first that there is some unfairness in this division because DiDi receives more money than Cael. However, keep in mind that each party agreed to be content with his or her estimate of a fair share when using this method. But what is interesting is that each party ends up with more than expected. First, Dahlia gets the painting and pays \$4,000 less than her estimate of the others' fair share, namely, \$52,000 versus \$56,000. However, Cael and DiDi each receive \$4,000 more than their estimated fair share. This "win–win" outcome is due to Dahlia's higher assessment of the painting, which raised Cael and DiDi's share, whereas their smaller estimates lowered the amount that Dahlia had to pay to the estate. 14

QUIZ YOURSELF 14

Rosa, Juan, Carlos, and Luis have inherited a house from their parents. Because they have equal interests in the house and want to keep it in the family, they will use the method of sealed bids to decide who gets the house. That person will pay cash to the others. The table on the right shows the four bids on the house. Complete the table by showing each person's estimate of his or her fair share of the estate and who gets the house.

	Rosa ($\frac{1}{4}$)	Juan ($\frac{1}{4}$)	Carlos ($\frac{1}{4}$)	Luis ($\frac{1}{4}$)
Bid on house	\$250,000	\$240,000	\$260,000	\$200,000
Fair share of estate				
Item obtained with highest bid				

EXAMPLE 1 Dividing an Estate with Several Items

Matt, Annie, Connor, and Wilber have inherited equal portions of an estate left by McNamara and Troy. The estate consists of a house, an expensive car, and a boat. They decide to use the method of sealed bids in dividing the estate, and each of the four submits a sealed bid on each item in the estate. Their bids, and therefore their individual estimates of a fair share of the estate, are listed in Table 10.17. Determine how the estate should be divided using the method of sealed bids.

SOLUTION:

We see from the last line of Table 10.17 that Matt and Wilber will be satisfied if they each receive a combination of items and cash worth at least \$65,000, whereas Connor

	Matt ($\frac{1}{4}$)	Annie ($\frac{1}{4}$)	Connor ($\frac{1}{4}$)	Wilber ($\frac{1}{4}$)
Bid on house	\$165,000	\$180,000	\$160,000	\$190,000
Bid on car	\$55,000	\$50,000	\$35,000	\$40,000
Bid on boat	\$40,000	\$36,000	\$25,000	\$30,000
Total value	\$260,000	\$266,000	\$220,000	\$260,000
Estimate of fair share of estate	$\frac{1}{4} \cdot \$260{,}000 = \$65{,}000$	$\frac{1}{4} \cdot \$266{,}000 = \$66{,}500$	$\frac{1}{4} \cdot \$220{,}000 = \$55{,}000$	$\frac{1}{4} \cdot \$260{,}000 = \$65{,}000$

TABLE 10.17 Estimates of value and fair share.

would be happy with \$55,000, but Annie would not be satisfied with less than \$66,500. Remember that no one knows the others' estimates of the estate's value until the bids are opened. When the bids are opened, Wilber gets the house because he has the high bid of \$190,000. Because Matt bid the highest on both the car and the boat, he gets those items. Of course, Connor and Annie get none of the items but will receive cash from the estate. However, it is interesting to note that even though Annie did not submit the highest bid on any item, she had the largest estimate of the total value of the estate. Let us now do the final accounting of the cash transactions for the estate.

We start with those who receive items from the estate. Wilber gets the house, which is worth \$190,000 to him. But he thinks that his fair share of the total estate is worth \$65,000. Therefore, he receives considerably more than his fair share, namely the difference

$$\overset{\text{value of house}}{\$190{,}000} - \overset{\text{Wilber's fair share}}{\$65{,}000} = \overset{\text{Wilber pays estate}}{\$125{,}000}.$$

He must pay this difference to the estate. Matt receives the car and the boat from the estate. Based on his bids for these items, he receives a total value of

$$\$55{,}000 + 40{,}000 = \$95{,}000.$$

However, this is more than what he feels is his fair share of the estate. Therefore, he must also pay the estate the difference between the value of items received and his fair share, namely,

$$\overset{\text{value of boat and car}}{\$95{,}000} - \overset{\text{Matt's fair share}}{\$65{,}000} = \overset{\text{Matt pays estate}}{\$30{,}000}.$$

The combined cash paid to the estate by Wilber and Matt is

$$\$125{,}000 + \$30{,}000 = \$155{,}000.$$

Last, Connor and Annie receive their perceived fair shares from these payments. The cash accounting thus far is tabulated in Table 10.18.

	Matt $(\frac{1}{4})$	Annie $(\frac{1}{4})$	Connor $(\frac{1}{4})$	Wilber $(\frac{1}{4})$
Item obtained with high bid	Car, boat			House
Pays to estate (+) or receives from estate (−)	\$30,000 (+)	\$66,500 (−)	\$55,000 (−)	\$125,000 (+)

TABLE 10.18 Fair division accounting of estate.

The last line of Table 10.18 shows that there is a cash balance left in the estate, namely,

$$\underbrace{\$30{,}000 + \$125{,}000}_{\text{Matt and Wilber pay to estate}} - \underbrace{\$66{,}500 - \$55{,}000}_{\text{Annie's and Connor's fair shares}} = \$155{,}000 - \$121{,}000 = \underbrace{\$33{,}500}_{\text{balance}}.$$

Because the four heirs each have one-quarter interest in the estate, they each receive

$$\frac{1}{4} \cdot \$33{,}500 = \$8{,}375$$

of the balance. As with any fair division based on sealed bids, the four heirs get more than they expected. Both Matt and Wilber get items but pay less to the estate than they had expected. Although Annie and Connor do not receive items, they each get an amount

QUIZ YOURSELF 15

We continue the fair division problem discussed in Quiz Yourself 14. Because Carlos gets the house, he must pay an amount of cash to the estate, whereas the others will receive cash. Construct a table to show an accounting of the estate's cash transactions.

	Matt ($\frac{1}{4}$)	Annie ($\frac{1}{4}$)	Connor ($\frac{1}{4}$)	Wilber ($\frac{1}{4}$)
Item obtained with high bid	Car, boat			House
Pays to estate (+) or receives from estate (−)	\$30,000 (+)	\$66,500 (−)	\$55,000 (−)	\$125,000 (+)
Division of estate balance (\$33,500)	\$8,375 (−)	\$8,375 (−)	\$8,375 (−)	\$8,375 (−)
Summary of cash	Pays \$21,625	Receives \$74,875	Receives \$63,375	Pays \$116,625

TABLE 10.19 Fair division accounting of estate.

of cash that is more than they think is their respective fair share. Table 10.19 summarizes the fair division of the estate. As before, the total paid to the estate equals the total received from the estate and the estate is now closed with a zero balance. 15

Let us summarize fair division through the method of sealed bids. We explain the process in terms of dividing an estate; however, the method can be applied to other, similar situations.

THE METHOD OF SEALED BIDS (EQUAL INTERESTS) Each person having a stake in the fair division of the items makes a sealed bid on each item.

1. Each person's fair share is determined as follows:
   - Add the person's bids on the various items.
   - Divide this total by the number of people sharing in the estate.

   Because different people make different bids, what is a fair share for one person may not be the same as a fair share to another.
2. Open the bids. The person bidding the highest amount on an item gets that item. If several people have the same highest bid, then the decision of who gets the item can be based on random selection.
3. If the total value of the items awarded to a person is more than the person's fair share, then the person pays the difference in cash to the estate. If the total value is less than the person's fair share, then the person receives the difference in cash from the estate. If a person does not receive any items, then the person gets his or her fair share in cash from the estate.
4. If, after completing step 3, there is a cash balance in the estate, then it is divided equally among the heirs.

The method of sealed bids will work successfully only when

- The parties involved make an honest bid on each item;
- The person making the highest bid on an item is willing to accept the item and pay cash to the estate if the bid is larger than the fair share; and
- Most important, each person makes a bid without knowledge of the others' bids.

We can also use this method when the parties have different percentage interests in the estate. For example, three heirs to an estate may have, respectively, 50%, 30%, and 20% interests as stated in a will. In this case, a person's fair share is his or her estimate of the estate's value times his or her percentage interest in the estate. The exercises have problems of this type.

EXERCISES 10.4

Sharpening Your Skills

Identify each situation as dealing with either discrete or continuous division of items.

1. a. Dividing a collection of rare books among five people

b. Dividing a chocolate bar among several children

c. Apportioning the seats on a city council with representatives chosen from different neighborhoods of the city

2. a. Splitting candy collected at Halloween among three trick-or-treaters

b. Dividing an estate consisting of a house and a boat among five heirs

c. Dividing a piece of pie among four people

Use the method of sealed bids to complete the tables in Exercises 3–14.

3. Dividing an inheritance. Darnell and Joy want to determine who should inherit their rich aunt's antique car.

	Darnell $(\frac{1}{2})$	Joy $(\frac{1}{2})$
Bid on car	$110,000	$85,000
Fair share of estate	a.	b.
Item obtained with highest bid	c.	d.

4. Dividing an inheritance. Karl and Friedrich have inherited their father's copy of a rare book and want to decide who should receive it.

	Karl $(\frac{1}{2})$	Friedrich $(\frac{1}{2})$
Bid on book	$27,000	$35,000
Fair share of estate	a.	b.
Item obtained with highest bid	c.	d.

5. Dissolving a partnership. Dennis and Zadie have written a book together and now want to decide how to dissolve their partnership, with one of them obtaining sole rights to the copyright. Their current entitlement to the book is 40% belonging to Dennis and 60% belonging to Zadie.

	Dennis (40%)	Zadie (60%)
Bid on copyright	$20,000	$28,000
Fair share of copyright	a.	b.
Item obtained with highest bid	c.	d.

6. Dissolving a partnership. Matt and Tony own a small pizza stand at the shore, with Matt having 35% interest and Tony 65% interest. They want to dissolve their partnership, with one buying the other out.

	Matt (35%)	Tony (65%)
Bid on pizza stand	$120,000	$90,000
Fair share of pizza stand	a.	b.
Item obtained with highest bid	c.	d.

Applying What You've Learned

7. Dividing an inheritance. Three brothers—Ed, Al, and Jerry—want to decide who gets a valuable painting and statue that they have inherited from their grandfather.

	Ed $(\frac{1}{3})$	Al $(\frac{1}{3})$	Jerry $(\frac{1}{3})$
Bid on painting	$13,000	$8,000	$9,000
Bid on statue	$14,000	$13,000	$15,000
Total value	$27,000	a.	b.
Fair share of estate	c.	d.	e.

8. Dissolving a partnership. Franz, Ida, Bill, and Monica are members of an investment club. Their portfolio consists of three stocks—a computer manufacturer, an oil company, and a pharmaceutical company. They want to dissolve their group and determine who gets the stocks and who gets cash.

	Franz $(\frac{1}{4})$	Ida $(\frac{1}{4})$	Bill $(\frac{1}{4})$	Monica $(\frac{1}{4})$
Bid on computer stock	$75,000	$80,000	$70,000	$90,000
Bid on oil stock	$35,000	$40,000	$45,000	$40,000
Bid on pharmaceutical stock	$40,000	$30,000	$25,000	$35,000
Total value	a.	b.	c.	d.
Fair share of portfolio	e.	f.	g.	h.

9. Dividing an inheritance. Consider your answer to Exercise 3. Complete the settlement of the estate by completing the fair division accounting of the estate cash as indicated in the table.

	Darnell ($\frac{1}{2}$)	Joy ($\frac{1}{2}$)
Pays to estate (+) or receives from estate (−)		
Division of estate balance ($)		
Summary of cash		

10. Dividing an inheritance. Consider your answer to Exercise 4. Complete the settlement of the estate by completing the fair division accounting of the estate cash as indicated in the table.

	Karl ($\frac{1}{2}$)	Friedrich ($\frac{1}{2}$)
Pays to estate (+) or receives from estate (−)		
Division of estate balance ($)		
Summary of cash		

11. Dissolving a partnership. Consider your answer to Exercise 5. Complete the settlement of the copyright by completing the fair division accounting of the cash as indicated in the table.

	Dennis (40%)	Zadie (60%)
Pays to pool (+) or receives from pool (−)		
Division of pool balance ($)		
Summary of cash		

12. Dissolving a partnership. Consider your answer to Exercise 6. Complete the settlement of the pizza stand by completing the fair division accounting of the cash as indicated in the table.

	Matt (35%)	Tony (65%)
Pays to pool (+) or receives from pool (−)		
Division of pool balance ($)		
Summary of cash		

13. Dividing an inheritance. Consider your answer to Exercise 7. Complete the settlement of the estate by completing the fair division accounting of the estate cash as indicated in the table.

	Ed ($\frac{1}{3}$)	Al ($\frac{1}{3}$)	Jerry ($\frac{1}{3}$)
Items obtained with highest bid			
Pays to estate (+) or receives from estate (−)			
Division of estate balance ($)			
Summary of cash			

14. Dissolving a partnership. Consider your answer to Exercise 8. Complete the settlement of the portfolio by completing the fair division accounting of the cash as indicated in the table.

	Franz ($\frac{1}{4}$)	Ida ($\frac{1}{4}$)	Bill ($\frac{1}{4}$)	Monica ($\frac{1}{4}$)
Stock obtained with highest bid				
Pays to pool (+) or receives from pool (−)				
Division of pool balance ($)				
Summary of cash				

In Exercises 15 and 16, use the method of sealed bids to determine how the items and cash are distributed.

15. Dividing an inheritance. Betty and Dennis want to divide a diamond ring, an antique desk, and a collection of rare books that were bequeathed to them by their dear aunt who passed away. Use the method of sealed bids to divide the items and determine a fair division of the estate. Betty's and Dennis's estimates of these items are given in the table.

	Betty	Dennis
Ring	$16,000	$18,000
Desk	$4,500	$5,000
Books	$4,000	$3,000

16. Dissolving a partnership. Matt, Dani, and Christian have a partnership in which they own a chain of four fast-food restaurants (call them A, B, C, and D). They have decided to dissolve their partnership but keep ownership of the restaurants individually among themselves. Determine a fair division by using the method of sealed bids. The table shows the partners' estimates of the values of the individual restaurants.

	Matt	Dani	Christian
A	$170,000	$180,000	$160,000
B	$145,000	$150,000	$155,000
C	$200,000	$190,000	$210,000
D	$70,000	$50,000	$45,000

Table for Exercise 16

Communicating Mathematics

17. Why are the fair shares of an estate usually different?

18. In dividing an estate, why can some parties receive no items?

19. In using the method of sealed bids, why can some parties receive more in cash than they expected?

20. What factors must be in place for the method of sealed bids to be successful?

Challenge Yourself

In Exercises 21–24, we reconsider our opening example in which Dahlia, Cael, and DiDi are dividing Wayne Malloy's estate. In Exercises 25–28, we take another look at Example 1, in which Matt, Annie, Connor, and Wilber are dividing McNamara and Troy's estate. These questions are intended to be open ended, with no right or wrong answers. Just consider the implications of what might happen if the division is done when the rules of fairness are not being followed.

21. What can happen if Cael has prior knowledge of Dahlia's bid?

22. What can happen if Dahlia has prior knowledge of Cael's bid?

23. What can happen if Cael bids unfairly by bidding too low? Bidding too high?

24. What can happen if Dahlia bids unfairly by bidding too low? Bidding too high?

25. What can happen if Matt has prior knowledge of others' bids?

26. What can happen if Wilber has prior knowledge of others' bids?

27. What can happen if Annie bids unfairly by bidding too low? Bidding too high?

28. What can happen if Connor bids unfairly by bidding too low? Bidding too high?

CHAPTER SUMMARY

You will learn the items listed in this chapter summary more thoroughly if you keep in mind the following advice:

1. Focus on "remembering how to remember" the ideas. What pictures, word analogies, and examples help you remember these ideas?
2. Practice writing each item without looking at the book.
3. Make up 3 × 5 flash cards to break your dependence on the text. Use these cards to give yourself practice tests.

Section	Summary	Example
SECTION 10.1	The **Alabama paradox** occurred in 1881 when Alabama lost one representative when the size of the legislature increased. The **Hamilton** method apportions representatives by giving each state the integer part of the exact number of representatives due to it. Additional representatives are allocated to states that have the largest fractional part of their exact representation.	Discussion, pp. 517, 518 Example 1, pp. 519–520
	The **average constituency** of a state is defined as $\dfrac{\text{population of the state}}{\text{number of representatives of the state}}.$	Definitions, p. 520
	State A is **more poorly represented** than state B if A has a larger average constituency than B. If A is more poorly represented than B, the **absolute unfairness** is (average constituency of A) − (average constituency of B). The **relative unfairness** of an apportionment between two states is the absolute unfairness divided by the smaller average constituency.	Example 2, p. 521 Example 3, p. 521 Example 4, p. 522
SECTION 10.2	The **apportionment criterion** says that we assign a representative to a state to give the smallest relative unfairness.	Example 1, pp. 525–526
	The **Huntington–Hill apportionment principle** says that if states X and Y have been allocated x and y representatives, respectively, then X should be given the next representative instead of Y provided $\dfrac{(\text{population of Y})^2}{y \cdot (y+1)} < \dfrac{(\text{population of X})^2}{x \cdot (x+1)}.$ These quotients are called **Huntington–Hill numbers.**	Example 2, p. 527
	In **apportioning the oil consortium** board, each time we allocated a representative to a company, we then calculated Huntington–Hill numbers to decide who gets the next representative.	Example 3, pp. 527–528
	We can use the Huntington–Hill method in apportioning other objects where we do not want to reassign objects once they have already been assigned.	Example 4, p. 530
SECTION 10.3	In apportioning representatives, we defined the **standard divisor** by $\text{standard divisor} = \dfrac{\text{total population}}{\text{number of representatives being allocated}},$	Discussion, p. 535
	and a state's **standard quota** as $\text{standard quota} = \dfrac{\text{state's population}}{\text{standard divisor}}.$	Discussion, p. 535 Example 1, pp. 535–536 Example 2, p. 536
	If we round the standard quota down, we call that number the **lower quota;** if we round the standard quota up, we call that number the **upper quota.** The **quota rule** requires that each state's allocation of representatives be between its lower and upper quotas. **Hamilton's method** is as follows: a) Give each state its lower quota. b) Assign remaining representatives according to the size of the fractional parts of the states' standard quotas.	Definition, p. 536 Discussion, p. 537

Section	Summary	Example
	The **population paradox** occurs when A's population is growing faster than B's, yet A loses a representative to B. The **new states paradox** occurs when a new state is added and its share of seats is added to the legislature, causing a change in the allocation of seats previously given to another state.	Example 3, pp. 537–538 Example 4, pp. 539–540
	Jefferson's method is as follows: a) Find a modified divisor that is smaller than the standard divisor for the apportionment. b) Calculate the modified quota for each state and round it down to determine the number of representatives assigned to the state.	Example 5, p. 541
	Adams's method is as follows: a) Find a modified divisor that is larger than the standard divisor for the apportionment. b) Calculate the modified quota for each state and round it up to determine the number of representatives assigned.	Example 6, p. 542
	Webster's method is as follows: a) Use trial and error to find a modified divisor. b) Calculate the modified quota for each state and round it in the usual way to determine the number of representatives assigned.	Example 7, pp. 543–544
SECTION 10.4	**Discrete fair division** means that the items being distributed cannot be subdivided into fractional parts, as opposed to **continuous fair division,** where they can be subdivided.	Discussion, p. 548
	A person's **fair share** of an estate is that person's estimate of the estate's value divided by the number of people sharing equally in the estate.	Definition, p. 548
	In the **method of sealed bids (equal interests),** each person makes a sealed bid on each item, then: 1. Divide a person's estimate of the value of the estate by the number of people sharing in the estate to determine that person's fair share. 2. The person making the highest bid on an item must take it. 3. If the total value of items awarded to a person is greater than his or her fair share, then he or she must pay the estate; otherwise, that person receives cash from the estate. 4. If, after step 3, there remains a cash balance in the estate, then it is divided equally among the heirs.	Example 1, pp. 550–552

CHAPTER REVIEW EXERCISES

Section 10.1

1. What is the Alabama paradox?

2. The governing board of the American History Roundtable Society has 11 members. The seats are apportioned among the four divisions of the society—Revolutionary War (560 members), Civil War (524), World War I (431), and World War II (485)—according to the number of members in each division. Use the Hamilton method to apportion seats on the board.

3. Explain why an Alabama paradox can occur when using the Hamilton apportionment method.

4. Suppose state A has a population of 935,000 and five representatives, whereas state B has a population of 2,343,000 and 11 representatives.

a. Determine which state is more poorly represented, and calculate the absolute unfairness for this assignment of representatives.

b. Determine the relative unfairness for this apportionment.

Section 10.2

5. Suppose that Florida has a population of approximately 16.03 million and has 25 representatives and Texas has a population of 21.49 million and has 32 representatives.

a. Calculate the Huntington–Hill numbers for both Florida and Texas.

b. According to the Huntington–Hill apportionment principle, which of these states is more deserving to receive an additional representative?

6. Explain why the Huntington–Hill apportionment principle will avoid an Alabama paradox.

7. In the 2008 Summer Olympics, in addition to the host city of Beijing, several other cities—Qingdao, Tianjin, and Qinhuangdao—also hosted some events. You are to allocate 13 temporary train services to these cities. Each city is guaranteed at least one train service and you are to use the Huntington–Hill method and the given table of Huntington–Hill numbers to allocate the train services.

Current Representation	(B)eijing	(Q)ingdao	(T)ianjin	Qin(H)uangdao
1	21,012.5	13,448.0	3,362.0	27,378.0
2	7,004.2	4,482.7	1,120.7	9,126.0
3	3,502.1	2,241.3	560.3	4,563.0
4	2,101.3	1,344.8	336.2	2,737.8
5	1,400.8	896.5	224.1	1,825.2
6	1,000.6	640.4		1,303.7

8. An instructor at the Physical Fitness Institute can teach eight classes. A preregistration survey indicates the following interests:
   66 want to take tae-bo;
   39 want to take karate;
   18 want to take weight training;
   23 want to take tai chi.

   Assume that the instructor will teach at least one class for each area. Use the Huntington–Hill method to apportion the instructor's remaining four classes among the four areas.

Section 10.3

9. A health club instructor has a course load that allows her to teach eight classes. A preregistration survey indicates the following interests:
   8 want to take Pilates;
   64 want to take kickboxing;
   11 want to take yoga;
   31 want to take spinning.

   Apportion her classes using the Jefferson method.

10. Apportion the classes in Exercise 9 using the Adams method.

11. Apportion the classes in Exercise 9 using the Webster method.

12. A group of three regional airports has hired a company to increase security. The contract calls for 150 security personnel to be apportioned to the airports according to the number of passengers using each airport each week. One year later, the number of passengers per week has increased at the three airports, so the 150 security personnel are reallocated. We show these data in the given table. (We are using the Hamilton method.) Does this example illustrate the population paradox? Explain.

Airport	Passengers per Week (now)	Passengers per Week (one year later)
A	80,500	87,700
B	6,800	7,410
C	93,300	101,300

13. What do we mean by the new states paradox?

14. Which apportionment method(s) do not satisfy the quota rule?

Section 10.4

15. Three cousins—Tito, Omarosa, and Piers—are arguing over how to divide an inheritance of a rare antique rifle and a jewel-encrusted sword that belonged to their grandfather. Use the method of sealed bids to complete the table below and begin to solve their problem.

	Tito ($\frac{1}{3}$)	Omarosa ($\frac{1}{3}$)	Piers ($\frac{1}{3}$)
Bid on rifle	\$12,000	\$17,000	\$10,000
Bid on sword	\$15,000	\$13,000	\$11,000
Estimated total value of estate			
Fair share of estate			

16. Continuing with the situation in Exercise 15, complete the following table.

	Tito ($\frac{1}{3}$)	Omarosa ($\frac{1}{3}$)	Piers ($\frac{1}{3}$)
Items obtained with highest bid			
Pays to estate (+) or receives from estate (−)			
Division of estate balance (\$)			
Summary of cash			

CHAPTER TEST

1. What is the Alabama paradox?

2. Suppose state C has a population of 1,640,000 and eight representatives and state D has a population of 1,863,000 and nine representatives. Determine the absolute and relative unfairness of this assignment of representatives.

3. The Metropolitan Community College Arts Council will consist of eight members. The seats are to be apportioned according to student participation in the areas of art (47 students), music (111 students), and theater (39 students). Use the Hamilton method to apportion the council.

4. Explain why an Alabama paradox can occur with the Hamilton method.

5. Suppose that Arizona has a population of 5.23 million and has eight representatives and Oregon with a population of 3.61 million has five representatives.

   a. Calculate the Huntington–Hill number for each state.

   b. According to the Huntington–Hill method, which state is more deserving of an additional representative?

6. Explain why the Huntington–Hill apportionment method avoids an Alabama paradox.

7. You are to allocate 14 shuttle buses to four areas in a theme park: Mystery Mountain, Jungle Village, Great Frontier, and City Sidewalks. Each area is to have at least one bus. Use the given table to apportion the remaining buses to each area using the Huntington–Hill method.

Current Number of Buses	Mystery Mountain	Jungle Village	Great Frontier	City Sidewalks
1	6,612.5	3,612.5	2,112.5	4,050.0
2	2,204.2	1,204.2	704.2	1,350.0
3	1,102.1	602.1	352.1	675.0
4	661.3	361.3	211.3	405.0
5	440.8	240.8	140.8	270.0

8. Your college's community outreach program has 11 volunteers for an after-school tutoring program for elementary school students. Assume that any volunteer can tutor any subject. There are 16 requests for mathematics, 9 for reading, and 6 for study skills. Use the Huntington–Hill method to apportion the tutors to the three areas. Assume that each area will have at least one volunteer assigned.

9. Upscale Malls, Inc., has allocated 150 security personnel to its malls in Allenwood, Black Hills, and Colombia according to the weekly numbers of customers at each mall. Assume that Upscale builds a new mall in Devon, requiring new security personnel who are allocated to Devon according to the first apportionment. We show these data in the given table.

	Allenwood	Black Hills	Colombia	Devon
Number of customers per week	80,500	6,800	93,300	41,400

   Use the Hamilton method to do the apportionment. Does this example illustrate the new states paradox? Explain.

10. What is the population paradox?

11. The Relaxation Institute can offer nine sections of stress management courses per week. A preregistration survey shows the following interest:

    47 want to take massage;
    32 want to take aromatherapy;
    41 want to take yoga;
    21 want to take meditation.

    Assume that the institute will offer at least one section in each area. Apportion the remaining sections using the Webster method.

12. Apportion the sections in Exercise 11 using the Jefferson method.

13. Apportion the sections in Exercise 11 using the Adams method.

14. Which apportionment method(s) that you have studied satisfy the quota rule?

15. Three brothers—Larry, Moe, and Curly—are dissolving their company Stooge Investments, Inc., which consists of two branches, A and B. They want to keep the business in the family, so some of the brothers will buy the others out. Use the method of sealed bids to dissolve their partnership using the table below.

	Larry $(\frac{1}{3})$	Moe $(\frac{1}{3})$	Curly $(\frac{1}{3})$
Estimate of A's worth	\$13 million	\$9 million	\$10 million
Estimate of B's worth	\$17 million	\$18 million	\$14 million

GROUP PROJECTS

1. **Comparing apportionment methods.**

   a. Make up an example in which the Hamilton and the Huntington–Hill methods give you the same apportionment.

   b. Make up an example in which the Hamilton and the Huntington–Hill methods give you different apportionments.

2. **Apportionment paradoxes.** Make up new examples to illustrate the population paradox and new states paradox. Do this by trial and error by using spreadsheets or online applets that do apportionment.

3. **Comparing apportionment methods.** Try to find an example where the Jefferson, Adams, and Webster methods all give different apportionments.

4. **Creating fair division examples.**

   a. Make up an example of a method of sealed bids problem in which three people bid on three items and each person receives exactly one item.

   b. Make up an example of a method of sealed bids problem in which three people bid on three items and one person receives all three items.

USING TECHNOLOGY

The spreadsheet shows how to implement the Jefferson method. The highlighted cells show the formulas we used to do the calculations; however, you have to insert equal signs to make the formulas operational.

	A	B	C	D	E	F	G	H
1					The Jefferson Apportionment Method			
2								
3			This spreadsheet does the computations to illustrate the Jefferson					
4			apportionment method that we explained in Section 10.3					
5								
6								
7	**Jefferson**		Total No. of Stockholders		No. Board Members		Mod. Divisor	
8			10000		9		935	
9								
10			Population		Modified Quota		Mod. Q. (Rd. down)	
11	Naxxon		4700		C11/G8		TRUNC(E11)	
12	Aroco		3700		C12/G8		TRUNC(E12)	
13	Eurobile		1600		C13/G8		TRUNC(E13)	
14	Totals		10000				SUM(G11:G13)	
15								

1. Create this worksheet and then reproduce the computations in Example 5 of Section 10.3.

2. Modify this worksheet to implement the Adams method and then reproduce the computations in Example 6 of Section 10.3 (*Note:* You might want to use the CEILING function.)

3. Modify this worksheet to implement the Webster method and then reproduce Example 7 of Section 10.3. (*Note:* You might want to use the ROUND function.)

4. Below are four TI graphing calculator screens that show how to do the computations in Table 10.1 on page 518.

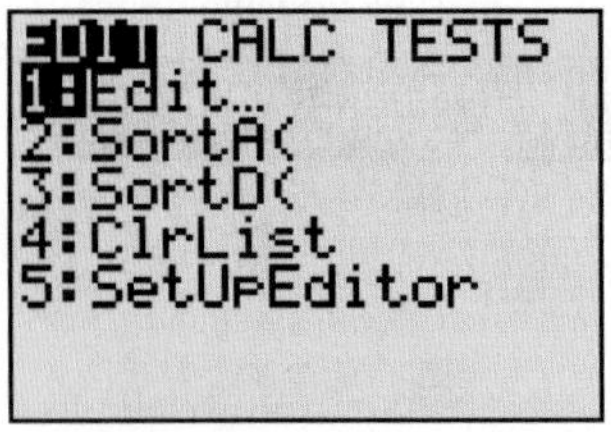

Screen 1

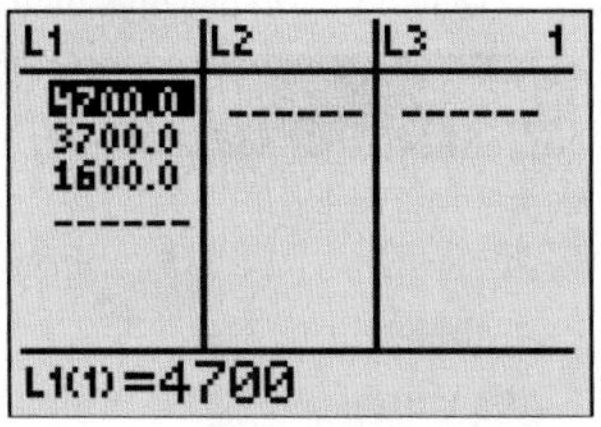

Screen 2

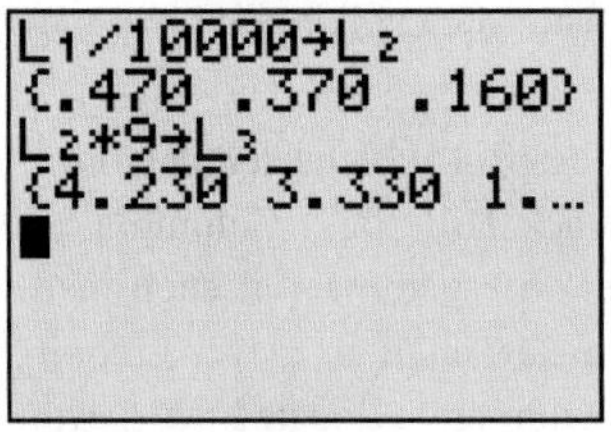

Screen 3

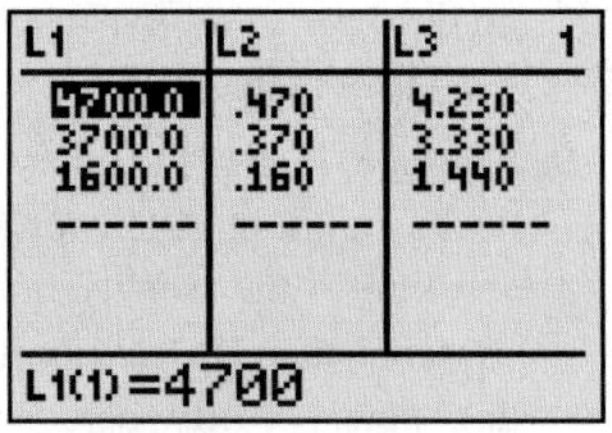

Screen 4

We begin by pressing STAT and then selecting EDIT, as shown in Screen 1. Screen 2 shows the data we entered in list L1. In Screen 3, we divide all entries in list L1 by 10,000 and store these in list L2, and then multiply all entries in L2 by 9, storing the results in L3. Screen 4 shows the contents of L1, L2, and L3 after performing these operations. Reproduce this example and then use this method to solve Exercises 1 and 3 in Section 10.1.

11

Voting

Using Mathematics to Make Choices

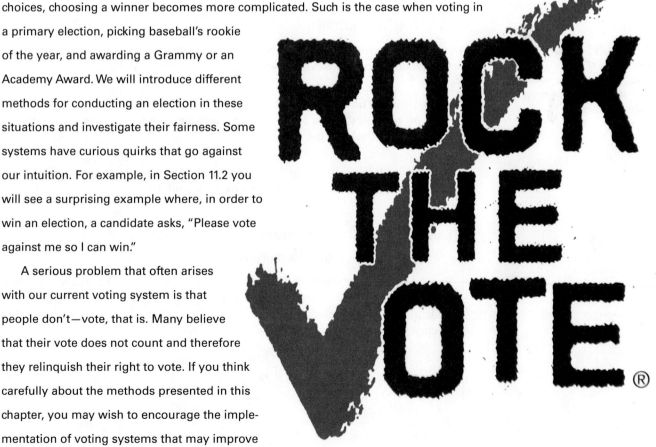

As a citizen of a free society, you have both a right and a duty to voice your opinion by voting. In addition to selecting political leaders, you will vote on many other issues: Are you in favor of raising taxes to improve your state's clean air and water standards? Should we eliminate the death penalty? Do you think we should use casino profits to improve education? At work, whom do you support to head your department?

With only two choices, the winner is the one receiving the most votes; however, when there are more than two choices, choosing a winner becomes more complicated. Such is the case when voting in a primary election, picking baseball's rookie of the year, and awarding a Grammy or an Academy Award. We will introduce different methods for conducting an election in these situations and investigate their fairness. Some systems have curious quirks that go against our intuition. For example, in Section 11.2 you will see a surprising example where, in order to win an election, a candidate asks, "Please vote against me so I can win."

A serious problem that often arises with our current voting system is that people don't—vote, that is. Many believe that their vote does not count and therefore they relinquish their right to vote. If you think carefully about the methods presented in this chapter, you may wish to encourage the implementation of voting systems that may improve society by addressing this problem.

11.1 Voting Methods

Objectives

1. Use the plurality method to determine the winner of an election.
2. Understand the Borda count voting method.
3. Use the plurality-with-elimination method to determine the winner of an election.
4. Determine the winner of an election using the pairwise comparison method.

Are you ever troubled by how our election system works? For example, was it fair that in the 2012 Republican primary in Florida, Mitt Romney was awarded *all 50* of the delegates to the Republican National Convention even though *more than half of the voters voted against him*? Newt Gingrich, who earned 32% of the votes, received none of the delegates.* Would it not have been more reasonable to split the delegates among the candidates according to the votes they received? Or was it fair that George W. Bush won the presidency in 2000 even though Al Gore defeated him in the popular vote?

Does it bother you (as it does me) that after a few small, early primaries, most candidates running for president cannot raise enough money and have to withdraw from the race before most of us get a chance to cast our votes? Is there not a better voting system that would keep candidates in the race until our state's primary rolls around?

As you will see, voting can be more complicated than it first seems and the winner of an election depends not only on the votes cast but also on how we agree to use those votes. In this section, we will introduce you to several different voting systems, and in Section 11.2 we will investigate their weaknesses.

The Plurality Method

KEY POINT

Using the plurality method, the person with the most votes wins an election.

The **plurality method** is the simplest way to determine the outcome of an election; the person earning the most votes is the winner.

DEFINITION The Plurality Method
Each person votes for his or her favorite candidate. The candidate receiving the most votes is declared the winner.

Many state and local elections use the plurality method because it is easy to determine the winner—all we need to do is tally the votes for each candidate.

EXAMPLE 1 Determining the Winner Using the Plurality Method

Imagine that student employees on your campus have decided to organize a union in order to improve their salaries and working conditions. As a first step, a group of 33 students, calling themselves the Undergraduate Labor Council (ULC), have just had an election to choose their president. The results of this election are as follows:

Ann	10
Ben	9
Carim	11
Doreen	3

*This was the case during the writing of this chapter, although there was some movement to have the delegates divided proportionally among the candidates running in the election.

Using the plurality method, who is the winner of this election?

SOLUTION:

Because Carim has the most votes, he is declared the winner.

Now try Exercises 1–3, 7, 11, and 15.

Note in Example 1 that using the plurality method, Carim won the election even though $\frac{22}{33} \approx 66.7\%$ of the ULC members voted against him.

The Borda Count Method

KEY POINT

Using the Borda count method, the person earning the most points wins an election.

Although it was too late once the election had been conducted, Ann realized that even though she was not the first choice of most voters, she was the second choice of a great many voters. Unfortunately for Ann, the ballot did not allow voters to state their second, third, and fourth preferences. A voting method called the **Borda count method** permits a voter to "fine-tune" his or her vote in the sense that the voter can designate not only a first choice but also a second choice, a third choice, and so on. In the ULC election, we could have used the Borda method by specifying that on a voter's ballot, the first choice would be given 4 points, the second choice 3 points, the third choice 2 points, and the fourth choice 1 point.

DEFINITION The Borda Count Method
If there are k candidates in an election, each voter ranks all candidates on the ballot. Then the first choice is given k points, the second choice is given $k-1$ points, the third choice $k-2$ points, and so on.* The candidate who receives the most total points wins the election.

To use the Borda count method in the ULC election in Example 1, voters must rank candidates on their ballots. For example, the given ballot would mean that a voter preferred Carim first, Ann second, Ben third, and Doreen fourth. Such a ballot is called a **preference ballot.**

1st	C	receives 4 points
2nd	A	receives 3 points
3rd	B	receives 2 points
4th	D	receives 1 point

With 33 voters, some preference ballots would be the same,† so in tallying the votes we group identical ballots together in a table called a **preference table.**

EXAMPLE 2 Determining the Winner Using the Borda Count Method

Assume that the ULC used the Borda count method to determine its president. Table 11.1 summarizes the preference ballots cast in the election. Who is the winner of the election?

*In some variations of the Borda count method, different numbers of points are awarded for first place, second place, third place, and so on. For example, we may give 5 points to first place, 3 to second, and 1 to third. We investigate these variations in the exercises.

†When we discuss permutations in Section 11.4, you will see that there are only 24 possible different preference ballots for A, B, C, and D.

Carim, Ann, Ben, and Doreen have the same number of first-place votes as before.

	Number of Ballots					
Preference	6	7	5	3	9	3
1st	C	A	C	A	B	D
2nd	A	C	D	D	A	A
3rd	B	B	B	B	D	C
4th	D	D	A	C	C	B

Ann has 18 second-place votes.

TABLE 11.1 Preference table for ULC election.

SOLUTION: Review the Be Systematic Strategy on page 5.

The numbers at the top of the columns in Table 11.1 show how many voters submitted the particular preference ballot in that column. For example, the 6 means that six voters had preference ballots choosing Carim first, Ann second, Ben third, and Doreen fourth. Notice, as before, Carim has 11 first-place votes, Ann has 10, Ben 9, and Doreen 3.

We can use Table 11.1 to calculate each student's point total, awarding 4 points for each first-place vote, 3 points for each second-place vote, and so on. For example, Ann's total is

$$10 \times 4 + 18 \times 3 + 0 \times 2 + 5 \times 1 = 40 + 54 + 0 + 5 = 99$$

ten first-place votes times 4 points

eighteen second-place votes times 3 points

five fourth-place votes times 1 point

Table 11.2 summarizes the election results.

	Number of Points				
Candidates	1st-Place Votes × 4 (Points)	2nd-Place Votes × 3 (Points)	3rd-Place Votes × 2 (Points)	4th-Place Votes × 1 (Points)	Total Points
A	$10 \times 4 = 40$	$18 \times 3 = 54$	$0 \times 2 = 0$	$5 \times 1 = 5$	99
B	$9 \times 4 = 36$	$0 \times 3 = 0$	$21 \times 2 = 42$	$3 \times 1 = 3$	81
C	$11 \times 4 = 44$	$7 \times 3 = 21$	$3 \times 2 = 6$	$12 \times 1 = 12$	83
D	$3 \times 4 = 12$	$8 \times 3 = 24$	$9 \times 2 = 18$	$13 \times 1 = 13$	67
				Total	330

TABLE 11.2 Points earned using the Borda method for the ULC election.

When we examine the voting, we see that Ann now wins the election.

Now try Exercises 4, 8, 12, and 16. 1

QUIZ YOURSELF 1

Use the preference table on the right to determine the winner of this election using the Borda count method. How many points does the winning candidate receive?

	Number of Ballots			
Preference	8	7	5	7
1st	C	D	C	A
2nd	A	A	B	D
3rd	B	B	D	B
4th	D	C	A	C

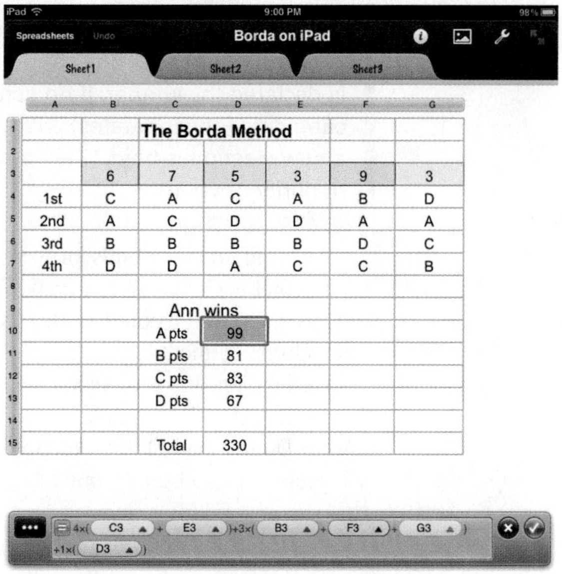

		The Borda Method				
	6	7	5	3	9	3
1st	C	A	C	A	B	D
2nd	A	C	D	D	A	A
3rd	B	B	B	B	D	C
4th	D	D	A	C	C	B
		Ann wins				
		A pts	99			
		B pts	81			
		C pts	83			
		D pts	67			
		Total	330			

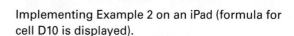

Implementing Example 2 on an iPad (formula for cell D10 is displayed).

PROBLEM SOLVING

Strategy: Verify Solutions

After completing a problem, double-check your answer for accuracy. In Example 2, we found that 330 points were awarded. This happened because each of the 33 voters was awarding 10 points.

Many polls use the Borda method to rank sports teams. For example, when this section was being written, there was no comprehensive playoff system to determine a national champion in college football, which often led to controversy as to which team is number one. One of several polls used in determining an overall ranking is conducted by the Associated Press. Each voter in this poll gives 25 points to his or her number-one choice, 24 to the second choice, and so on, to select the top 25 teams each week.

Because the outcomes of the plurality method and the Borda count method are different, you may wonder which is the "correct" method for determining the ULC president. We will not answer this question. Rather, knowing that these two methods yield different results, which do you believe is more appropriate?

The Plurality-with-Elimination Method

KEY POINT

The plurality-with-elimination method eliminates the weakest candidate before revoting.

When we used the plurality method in the ULC election in Example 1, we obtained the following results.

Ann	10
Ben	9
Carim	11
Doreen	3

One could argue that because Doreen is so far out of the running, perhaps we should eliminate her and conduct a new election with just Ann, Ben, and Carim as candidates. If no candidate earns a majority (more than 50%) in this second election, then we could eliminate the person receiving the smallest number of votes and then conduct a third election. This voting method is called the **plurality-with-elimination method.**

DEFINITION The Plurality-with-Elimination Method
Each voter votes for one candidate. A candidate receiving a majority of votes is declared the winner. If no candidate receives a majority of votes, then the candidate (or candidates) with the fewest votes is dropped from the ballot and a new election is held.* This process continues until a candidate receives a majority of votes.

To avoid many rounds of voting, we can simplify the elimination process. If we assume that each voter has a ranking of all the candidates, we can construct a preference table as in the Borda count method. Of course, we will use this table differently than we do in the Borda method.

Throughout this section we will assume that once a voter has ranked the candidates, *this ranking does not change during subsequent rounds of voting.* That is, if a voter prefers Ann to Doreen and Doreen to Ben in round one, if Doreen is eliminated, then the voter will prefer Ann to Ben in round two.

EXAMPLE 3 Determining the Winner Using the Plurality-with-Elimination Method

Determine the winner of the ULC election using the plurality-with-elimination method.

SOLUTION:

The preference table for this election is shown in Table 11.3.

Doreen has the fewest first-place votes, so she is eliminated.

	Number of Ballots					
Preference	6	7	5	3	9	3
1st	C	A	C	A	B	D̸
2nd	A	C	D̸	D̸	A	A
3rd	B	B	B	B	D̸	C
4th	D̸	D̸	A	C	C	B

Remaining candidates move up.

TABLE 11.3 Preference table for the ULC election.

From Table 11.3, we see that Doreen has only 3 first-place votes, so she is eliminated and a new election is held. Remember, *we are assuming that voters do not change their preferences from one round of voting to the next.* For example, because three voters had Doreen as their first choice, now that she is eliminated, those three voters now have Ann as their first choice. One way to think of this adjustment is to imagine that when we eliminate D from a column of the table, all candidates below D in that column move up one row. By eliminating Doreen, we get Table 11.4.

	Number of Ballots					
Preference	6	7	5	3	9	3
1st	C	A	C	A	B	A
2nd	A	C	B	B	A	C
3rd	B	B	A	C	C	B

Combine identical ballots.

TABLE 11.4 Preference table for the ULC election after Doreen is eliminated.

*To reduce the number of rounds of voting, some variations of this method may drop more than one candidate from the second round of voting. For example, it may be specified that only the top two candidates are allowed to proceed to the second round of voting.

We combine identical columns in Table 11.4 to get Table 11.5.

Ben now has the fewest first-place votes, so he is eliminated.

	Number of Ballots				
Preference	**6**	**10**	**5**	**3**	**9**
1st	C	A	C	A	B
2nd	A	C	B	B	A
3rd	B	B	A	C	C

TABLE 11.5 Preference table for the ULC election with three candidates.

We see now that Ann has 13 first-place votes, Ben has 9, and Carim has 11. Thus, Ben is eliminated and a third round of voting takes place. Table 11.6 summarizes the preferences at this point.

Ann has 22 first-place votes; Carim has 11.

	Number of Ballots				
Preference	**6**	**10**	**5**	**3**	**9**
1st	C	A	C	A	A
2nd	A	C	A	C	C

TABLE 11.6 Preference table for the ULC election after removing Ben.

In Table 11.6 we see that Ann has 22 first-place votes and Carim has 11, so Ann wins the ULC election.

Now try Exercises 5, 9, 13, and 17. 2

QUIZ YOURSELF 2

Use the preference table on the right to determine the winner of this election using the plurality-with-elimination method.

	Number of Ballots				
Preference	**8**	**9**	**5**	**4**	**2**
1st	C	E	B	A	A
2nd	A	D	C	D	C
3rd	B	B	E	B	B
4th	E	C	A	C	E
5th	D	A	D	E	D

The Pairwise Comparison Method

It is natural to expect that the winner of an election should be capable of beating each of the other candidates "head-to-head." We show a variation of this in the next method, the **pairwise comparison method.**

DEFINITION The Pairwise Comparison Method
Voters first rank all candidates. If A and B are a pair of candidates, we count how many voters prefer A to B and vice versa. Whichever candidate is preferred the most receives 1 point. If A and B are tied, then each receives $\frac{1}{2}$ point. Do this comparison, assigning points, for each pair of candidates. At the end, the candidate receiving the most points is the winner.

KEY POINT

In the pairwise comparison method, the candidate who can beat the most other candidates "head-to-head" is the winner.

As with the plurality-with-elimination method, we assume that if a voter prefers A to B and B to C, then the voter will prefer A to C. In Example 4, we see that we can use a voting method to choose among alternatives as well as decide elections.

EXAMPLE 4 Determining a Preference Using the Pairwise Comparison Method

In order to decide which new item to add to its menu, the Chipotle chain did a market survey in which customers were asked to rank their preferences for (T)acos, (N)achos, and (B)urritos. The chain will use the pairwise comparison method to decide which item to add to its menu. The results of counting the ballots are shown in Table 11.7(a). Which item should the chain select?

	Number of Ballots					
Preference	**2,108**	**864**	**1,156**	**1,461**	**1,587**	**1,080**
1st	T	T	N	N	B	B
2nd	N	B	T	B	T	N
3rd	B	N	B	T	N	T

TABLE 11.7(a) Preference table for voting on menu items.

SOLUTION: Review the Be Systematic Strategy on page 5.

We must compare a) T with N, b) T with B, and finally, c) N with B.

a) In comparing T with N, we will ignore all references to B in Table 11.7(b). You can see that 4,559 prefer T over N but only 3,697 prefer N over T, thus we award 1 point to option T.

1,156 + 1,461 + 1,080 = 3,697 prefer N over T.

	Number of Ballots					
Preference	**2,108**	**864**	**1,156**	**1,461**	**1,587**	**1,080**
1st	T	T	N	N	B	B
2nd	N	B	T	B	T	N
3rd	B	N	B	T	N	T

2,108 + 864 + 1,587 = 4,559 prefer T over N.

TABLE 11.7(b) Comparing T with N.

b) In comparing T with B, we now ignore all references to N, as we see in Table 11.7(c). In this comparison, you see that 4,128 prefer T over B and also 4,128 prefer B over T. So T and B each receive $\frac{1}{2}$ point.

1,461 + 1,587 + 1,080 = 4,128 prefer B over T.

	Number of Ballots					
Preference	**2,108**	**864**	**1,156**	**1,461**	**1,587**	**1,080**
1st	T	T	N	N	B	B
2nd	N	B	T	B	T	N
3rd	B	N	B	T	N	T

2,108 + 864 + 1,156 = 4,128 prefer T over B.

TABLE 11.7(c) Comparing T with B.

c) Finally we compare N with B. Table 11.7(a) tells us that 2,108 + 1,156 + 1,461 = 4,725 customers prefer N over B. Also, 864 + 1,587 + 1,080 = 3,531 customers prefer B over N. We award 1 point to N.

Because T has $1\frac{1}{2}$ points, N has 1 point, and B has $\frac{1}{2}$ point, option T is declared the winner.

Now try Exercises 6, 10, 14, and 18. 3

QUIZ YOURSELF 3

Redo Example 4 using the preference table on the right to determine which item to add to Chipotle's menu. We are again using the pairwise comparison method. Also list how many points each item gets.

	Number of Ballots					
Preference	**985**	**864**	**1,156**	**1,021**	**1,187**	**1,080**
1st	T	T	N	N	B	B
2nd	N	B	T	B	T	N
3rd	B	N	B	T	N	T

Table 11.8 summarizes the voting methods that we have discussed in this section.

Method	How the Winning Candidate Is Determined
Plurality	The candidate receiving the most votes wins.
Borda count	Voters rank all candidates by assigning a set number of points to first choice, second choice, third choice, and so on; the candidate with the most points wins.
Plurality-with-elimination	Successive rounds of elections are held, with the candidate receiving the fewest votes being dropped from the ballot each time, until one candidate receives a majority of votes.
Pairwise comparison	Candidates are compared in pairs, with a point being assigned the voters' preference in each pair. (In the case of a tie, each candidate gets a half point.) After all pairs of candidates have been compared, the candidate receiving the most points wins.

TABLE 11.8 Summary of voting methods.

Which of these voting methods is the best? Unfortunately, each method we have introduced has serious flaws, as you will see in Section 11.2.

Between the Numbers

Let the 12% Decide

In the 2009 primary election, 88% of the registered voters in Berks County, Pennsylvania, did not vote, thereby allowing the remaining 12% of us voters to decide who would represent them in our state and local governments. Voter apathy is not unique to Pennsylvania. Some say this apathy arises, in part, because of voters' frustration with plurality voting and the belief that their votes do not count.

Some suggest another method called **approval voting,** in which voters vote for as many candidates as they want. The winner is the person who earns the most votes of approval. Supporters, such as Stephen Brams, professor of politics at New York University, argue that approval voting tends to select the strongest candidate and reduces infighting among candidates. For an enlightening video by Professor Brams on approval voting, see http://bigthink.com/ideas/18726.

In recent years another method, called **instant runoff voting** (IRV),* has been gaining support and in fact was supported by both Barack Obama and John McCain prior to the 2008 election. With IRV, voters rank their candidates and, after the votes are counted, the weakest candidate is eliminated. However, if the eliminated candidate happens to be your first choice, your vote is not wasted because it then counts towards your second choice.

*For an entertaining explanation by the Muppets of how IRV works, go to http://www.fairvote.org/?page=1973. For more on IRV, see also http://www.fairvote.org/.

EXERCISES 11.1

Sharpening Your Skills

1. Four candidates running for a vacant seat on the town council receive votes as follows: Edelson, 2,156; Borowski, 1,462; Callow, 986; Lopez, 428.
   a. Is there a candidate who earns a majority?
   b. Who wins the election using the plurality method?
2. Five candidates running for mayor receive votes as follows: Kinison, 415; Sanchez, 462; Lichman, 986; Devlin, 428; Saito, 381.
   a. Is there a candidate who earns a majority?
   b. Who wins the election using the plurality method?

The university administration has asked a group of student leaders to vote on the aspects of college life to target for improvement over the next year. The choices were (D)ining facilities, (A)thletic facilities, (C)ampus security, and the (S)tudent union building. The votes are summarized in the following preference table. Use this information to answer Exercises 3–6.

	Number of Ballots					
Preference	**15**	**30**	**18**	**17**	**10**	**2**
1st	A	D	A	C	C	S
2nd	C	C	S	D	S	A
3rd	S	S	D	S	A	C
4th	D	A	C	A	D	D

3. **Voting on improving college life.** What option is selected using the plurality method?
4. **Voting on improving college life.** What option is selected using the Borda count method?
5. **Voting on improving college life.** What option is selected using the plurality-with-elimination method?
6. **Voting on improving college life.** What option is selected using the pairwise comparison method?

The drama society members are voting for the type of play they will perform next season. The choices are (D)rama, (C)omedy, (M)ystery, or (G)reek tragedy. The votes are summarized in the following preference table. Use this information to answer Exercises 7–10.

	Number of Ballots					
Preference	**10**	**15**	**13**	**12**	**5**	**7**
1st	C	D	C	M	M	G
2nd	M	M	G	D	G	C
3rd	G	G	D	G	C	M
4th	D	C	M	C	D	D

7. **Selecting a play.** What type of play is selected using the plurality method?

8. **Selecting a play.** What type of play is selected using the Borda count method?
9. **Selecting a play.** What type of play is selected using the plurality-with-elimination method?
10. **Selecting a play.** What type of play is selected using the pairwise comparison method?

Before a conference on "Trends in the Next Decade," a panel of participants voted on the topic for the keynote address. The choices were (T)echnology, (P)overty in Third World Countries, (E)conomy, (G)lobal Warming, and (F)oreign Policy. The votes are summarized in the following preference table. Use this information to answer Exercises 11–14.

	Number of Ballots				
Preference	**15**	**7**	**13**	**5**	**2**
1st	T	E	G	P	E
2nd	P	P	F	G	F
3rd	E	G	E	F	G
4th	F	F	P	E	P
5th	G	T	T	T	T

11. **Choosing a speaker.** What topic is selected using the plurality method?
12. **Choosing a speaker.** What topic is selected using the Borda count method?
13. **Choosing a speaker.** What topic is selected using the plurality-with-elimination method?
14. **Choosing a speaker.** What topic is selected using the pairwise comparison method?

A small employee-owned Internet company is voting on a merger with one of its competitors. The choices are (E)-Mall, (F)irestorm, (S)ecurenet, (T)echenium, and (L)ondram. The employees' votes are summarized in the following preference table. Use this information to answer Exercises 15–18.

	Number of Ballots				
Preference	**15**	**11**	**9**	**10**	**2**
1st	L	F	S	T	L
2nd	F	S	L	E	S
3rd	S	L	F	F	E
4th	E	E	T	L	F
5th	T	T	E	S	T

15. **Deciding on a merger.** What option is their first choice using the plurality method?

16. **Deciding on a merger.** What option is their first choice using the Borda count method?

17. **Deciding on a merger.** What option is their first choice using the plurality-with-elimination method?

18. **Deciding on a merger.** What option is their first choice using the pairwise comparison method?

Applying What You've Learned

Variations of the Borda count method are often used to determine winners of sports awards. In these methods, voters assign candidates a certain number of points for first place, fewer for second place, and so on. In Exercises 19 and 20, first determine the voting scheme being used and then complete the table.

19. **Awarding the Heisman Trophy.** Find a., b., and c. in the following table summarizing the voting for the 2011 Heisman Trophy, which was awarded to Robert Griffin III.

Player, Team	1st	2nd	3rd	Total Points
Robert Griffin III, Baylor	405	a.	136	1,687
Andrew Luck, Stanford	b.	250	166	1,407
Trent Richardson, Alabama	138	207	c.	978
Montee Ball, Wisconsin	22	83	116	348

20. **Awarding the Cy Young Award.** Find a., b., and c. in the following table summarizing the voting for the 2011 National League Cy Young Award, which was awarded to Clayton Kershaw.

Player, Team	1st	2nd	3rd	4th	5th	Total
Clayton Kershaw, Dodgers	27	a.	2			207
Roy Halladay, Phillies	b.	21	7			133
Cliff Lee, Phillies		5	17	9	1	90
Ian Kennedy, Diamondbacks	1	3	c.	18	3	76
Cole Hamels, Phillies				2	13	17

We can use each voting method described in this section to rank candidates in addition to determining the winner. The plurality method and the Borda count method can rank candidates in order of the most first-place votes and the points earned, respectively. Use this information to answer Exercises 21 and 22.

21. Devise a method for ranking candidates using the plurality-with-elimination method.

22. Devise a method for ranking candidates using the pairwise comparison method.

In Exercises 23–26, refer to the preference table that was given prior to Exercise 3 regarding choices for improvement of college life. Use that table to rank the choices using the specified method.

23. The plurality method

24. The Borda count method

25. The plurality-with-elimination method

26. The pairwise comparison method

In Exercises 27–30, refer to the preference table that was given prior to Exercise 11 regarding choices for the conference keynote address. Use that table to rank the choices using the specified method.

27. The plurality method

28. The Borda count method

29. The plurality-with-elimination method

30. The pairwise comparison method

31. What is the total number of points that all candidates can earn in an election using the Borda count method if there are four candidates and 20 voters?

32. What is the total number of points that all candidates can earn in an election using the Borda count method if there are five candidates and 18 voters?

33. What is the maximum number of points that a candidate can earn in an election using the Borda count method if there are five candidates and 22 voters?

34. What is the minimum number of points that a candidate can earn in an election using the Borda count method if there are six candidates and 20 voters?

35. What is the total number of points that all candidates can earn in an election using the pairwise comparison method if there are four candidates?

36. What is the total number of points that all candidates can earn in an election using the pairwise comparison method if there are five candidates?

37. What is the maximum number of points that a candidate can earn in an election using the pairwise comparison method if there are five candidates?

38. What is the minimum number of points that a candidate can earn in an election using the pairwise comparison method if there are five candidates?

Communicating Mathematics

39. What is a benefit of the Borda count method over the plurality method?

40. When is the election decided using the plurality-with-elimination method?

41. What is the philosophy behind the pairwise comparison method?

42. Which voting method do you prefer? Explain your reasons for choosing this method.

Between the Numbers

43. Approval voting. Go to the Web site suggested in Between the Numbers on page 569 (or some other Web site of your choosing) and then list three reasons in favor of approval voting. Find several examples where approval voting is used.

44. Instant runoff voting. Repeat Exercise 43 for instant runoff voting.

Challenge Yourself

In approval voting, a person can vote for more than one candidate. In Exercises 45–48, we will assume that voters can vote for their top three choices. The candidate who receives the most votes of approval wins the election. Use the preference tables specified to conduct the election.

45. Use the preference table given before Exercise 3.

46. Use the preference table given before Exercise 7.

47. Use the preference table given before Exercise 11.

48. Use the preference table given before Exercise 15.

In Exercises 49–51, assume that only two candidates are running in the election.

49. Explain why the winner using the Borda count method will have a majority of the votes.

50. Explain why the winner using the plurality-with-elimination method will have a majority of the votes.

51. Explain why the winner using the pairwise comparison method will have a majority of the votes.

52. You may find it interesting to research some of the controversies and suggestions for improving the election process in the United States. You might search for "proportional voting," "electronic voting," "Internet voting," "approval voting," and so on. Write a report on your findings.

11.2 Defects in Voting Methods

Objectives

1. Understand the majority criterion and how it is not satisfied by the Borda count method.
2. See how the plurality method violates Condorcet's criterion.
3. Recognize that the pairwise comparison method can fail to satisfy the independence-of-irrelevant-alternatives criterion.
4. See how the plurality-with-elimination method fails the monotonicity criterion.

At this point, you may prefer one of the voting systems that we discussed in Section 11.1 over the others. Are you wondering which one of them is the best? You may be surprised by our answer at the end of this section.

We will now begin our discussion with some common-sense conditions that we want any election to have. An election should satisfy the following *four fairness criteria*:

- the majority criterion
- Condorcet's criterion
- the independence-of-irrelevant-alternatives criterion
- the monotonicity criterion

We will discuss which methods satisfy these criteria.

The Majority Criterion

The first criterion that we will consider is the majority criterion.

KEY POINT

According to the majority criterion, a candidate with more than half the first-place votes wins an election.

DEFINITION The Majority Criterion
If a majority of the voters (more than half) rank a candidate as their first choice, then that candidate should win the election.

It is clear that the plurality method and the pairwise comparison method satisfy the majority criterion. (Verify this.) However, this is not the case with the Borda count method, as we show in Example 1.

EXAMPLE 1 The Borda Count Method Violates the Majority Criterion

Motor Tendency Magazine has narrowed down its choice for car of the year to the following four cars: the high-end luxury (A)udi R8, the sporty (B)MW Z4, the classic (C)orvette ZR1, and the massive (D)odge Viper. A three-person editorial panel, using the Borda count method, will make the final selection. Preference ballots for the panel members are as shown at left. Determine the winner using the Borda count method, and show that this outcome violates the majority criterion.

1st	C	C	D
2nd	D	D	A
3rd	A	B	B
4th	B	A	C

SOLUTION:

According to the Borda method, first-place votes are worth 4 points, second-place votes are worth 3 points, and so on. The following table summarizes the election results.

	Number of Points				
	1st-Place Votes × 4 (Points)	2nd-Place Votes × 3 (Points)	3rd-Place Votes × 2 (Points)	4th-Place Votes × 1 (Point)	Total
A	$0 \times 4 = 0$	$1 \times 3 = 3$	$1 \times 2 = 2$	$1 \times 1 = 1$	6
B	$0 \times 4 = 0$	$0 \times 3 = 0$	$2 \times 2 = 4$	$1 \times 1 = 1$	5
C	$2 \times 4 = 8$	$0 \times 3 = 0$	$0 \times 2 = 0$	$1 \times 1 = 1$	9
D	$1 \times 4 = 4$	$2 \times 3 = 6$	$0 \times 2 = 0$	$0 \times 1 = 0$	10

Although the Borda count winner is the Dodge Viper, the majority of first-place votes were cast for the Corvette ZR1.

Now try Exercises 1 and 2. 4

QUIZ YOURSELF 4

Use the Borda count method and the preference table on the right to determine whether the election satisfies the majority criterion.

	Number of Ballots			
Preference	8	10	5	2
1st	C	D	C	A
2nd	A	A	B	B
3rd	B	B	D	D
4th	D	C	A	C

KEY POINT

Condorcet's criterion says that a candidate who defeats everyone else head-to-head is the winner.

Condorcet's Criterion

The next condition, which was proposed by the Marquis de Condorcet, is another desirable property of a voting method.

Historical Highlight

The Marquis de Condorcet

Marie-Jean-Antoine-Nicolas Caritat, the Marquis de Condorcet, was an eighteenth-century French nobleman who was also a highly regarded philosopher and mathematician.

In 1785, Condorcet wrote a paper, *Essay on the Applications of Mathematics to the Theory of Decision Making*, in which he intended to prove that we could use mathematics to discover laws in the social sciences as precise as those in the physical sciences. He believed that scientists could employ mathematics to free society from its reliance on greedy capitalists and to develop insurance programs to protect the poor, and, once we discover appropriate moral laws, crime and war would disappear.

Although Condorcet's ideas may seem naive by twenty-first-century standards, he was a forward thinker whose views are still relevant today. He believed in abolishing slavery, opposed capital punishment, advocated free speech, and supported early feminists lobbying for equal rights. As far as Condorcet was concerned, human rights were derived from people's ability to use reason in forming moral concepts.

> *Women, having these same qualities, must necessarily possess equal rights. Either no individual of the human species has any true rights, or all have the same. And he who votes against the rights of another, of whatever religion, color, or sex, has thereby abjured his own.**

DEFINITION Condorcet's Criterion
If candidate X can defeat each of the other candidates in a head-to-head vote, then X is the winner of the election.

Clearly, a candidate with a majority of first-place votes can defeat every other candidate in a head-to-head election. However, Example 2 shows that the plurality method may violate Condorcet's criterion.

EXAMPLE 2 The Plurality Method Violates Condorcet's Criterion

A seven-member Olympic committee is voting using preference ballots to decide which one of (H)ungary, (I)ndia, and (T)aiwan will remain as a finalist country to host the 2016 Olympics. The partially completed preference ballots are as follows.

1st	H	H	H	I	I	T	T
2nd							
3rd							

Using the plurality voting method, Hungary is the winner. Complete the ballots so that in head-to-head voting, India defeats both Hungary and Taiwan.

SOLUTION:

Because Hungary defeats India in the first three ballots, we must force India to defeat Hungary on the remaining four ballots. We could do this as follows:

1st	H	H	H	I	I	T	T
2nd				H	H	I	I
3rd						H	H

At this point, India defeats Hungary by 4 to 3. We want India to also defeat Taiwan in head-to-head voting. Because Taiwan is currently leading India by 2 to 0, if we make Taiwan the last choice on the uncompleted ballots, we force India to defeat Taiwan. The final ballots are then as follows:

*This quote was taken from an article written by Condorcet in 1790, entitled "On Admission of Women to the Rights of Citizenship."

1st	H	H	H	I	I	T	T
2nd	I	I	I	H	H	I	I
3rd	T	T	T	T	T	H	H

We have shown that although Hungary wins the election using the plurality method, India defeats both Hungary and Taiwan in head-to-head competition, so Condorcet's criterion is not satisfied.

Now try Exercises 3, 4, 7, and 8. 5

QUIZ YOURSELF 5

Complete the following ballots so that not only does India defeat Hungary and Taiwan in head-to-head voting, but Taiwan also defeats Hungary in head-to-head voting.

1st	H	H	H	I	I	T	T
2nd						I	I
3rd						H	H

In Quiz Yourself 5, we see that although Hungary wins the election using the plurality method, most voters prefer both India and Taiwan over Hungary.

You must be careful to understand what we are saying in Example 2 when we state that the plurality method violates Condorcet's criterion. We are not claiming that *every* vote using the plurality method violates Condorcet's criterion. Example 2 simply provides a counterexample to the statement, "The plurality method satisfies Condorcet's criterion."

The Independence-of-Irrelevant-Alternatives Criterion

KEY POINT

The independence-of-irrelevant-alternatives criterion says that removing losers from the ballot does not affect the outcome of the election.

Suppose that after an election has been conducted using preference ballots, we discover that one of the losing candidates should never have been listed on the ballot. Rather than have a new election, we could decide to strike that candidate's name from each ballot and do a recount. It is a desirable condition of voting that the winner using the original count should also be the winner using the modified ballots.

DEFINITION Independence-of-Irrelevant-Alternatives Criterion
If candidate X wins an election, some nonwinners are removed from the ballot, and a recount is done, then X still wins the election.

EXAMPLE 3 The Plurality Method Violates the Independence-of-Irrelevant-Alternatives Criterion

The county board of supervisors is voting on methods to raise taxes to finance a new sports stadium. The options are levying a tax on (H)otel rooms, increasing the tax on (A)lcohol, and increasing the tax on (G)asoline. The board will use the plurality method to make its decision. The following preference table shows the results of the vote.

	Number of Ballots		
Preference	8	6	6
1st	A	H	G
2nd	G	G	A
3rd	H	A	H

A has most 1st-place votes.

Does this vote satisfy the independence-of-irrelevant-alternatives criterion?

SOLUTION:

What we are really asking is whether the removal of one of the losing options changes the outcome of the election. If, due to lobbying pressure, the board removes the tax on hotel rooms as an option, we get the following:

	Number of Ballots		
Preference	**8**	**6**	**6**
1st	A	G	G
2nd	G	A	A

— Now G has most 1st-place votes.

We see that the gasoline tax now wins by a vote of 12 to 8; thus, the plurality method does not satisfy the independence-of-irrelevant-alternatives criterion.

Example 4 shows that the pairwise comparison method also violates the independence-of-irrelevant-alternatives criterion.

EXAMPLE 4 The Pairwise Comparison Method Violates the Independence-of-Irrelevant-Alternatives Criterion

The 18-member board of the Universal Cheerleaders Association is using the pairwise comparison method to choose where to hold the next College Nationals competition. The places being considered are (A)tlanta, (B)oston, (C)leveland, and (D)isneyworld. The following table summarizes the preference ballots cast for A, B, C, and D:

	Number of Ballots			
Preference	**8**	**4**	**5**	**1**
1st	A	D	C	D
2nd	B	A	B	A
3rd	C	C	D	B
4th	D	B	A	C

Using the pairwise comparison method, who is the winner of this election? If any of the losing candidates are removed, does this change the results of the election?

SOLUTION: Review the Be Systematic Strategy on page 5.

For each pair of candidates, we must determine who is the winner in a head-to-head vote. The following table shows the results of these comparisons:

	Vote Results	Points Earned
A vs. B	A wins 13 to 5.	A gets 1 point.
A vs. C	A wins 13 to 5.	A gets 1 point.
A vs. D	D wins 10 to 8.	D gets 1 point.
B vs. C	Tie—each has 9.	B and C get $\frac{1}{2}$ point.
B vs. D	B wins 13 to 5.	B gets 1 point.
C vs. D	C wins 13 to 5.	C gets 1 point.

Thus A gets 2 points, B and C each get $1\frac{1}{2}$ points, and D gets 1 point, so Atlanta is the winner.

If we remove Boston and Cleveland, the preference table now looks like this:

	Number of Ballots			
Preference	8	4	5	1
1st	A	D	D	D
2nd	D	A	A	A

We see that Disneyworld now defeats Atlanta by 10 votes to 8. Because the removal of some original losing candidates changes the outcome of the election, this method does not satisfy the independence-of-irrelevant-alternatives criterion.

Now try Exercises 5 and 6. 6

QUIZ YOURSELF 6

Use the pairwise comparison method to determine the winner of the election summarized in the preference table on the right. Is the independence-of-irrelevant-alternatives criterion satisfied?

	Number of Ballots			
Preference	5	3	3	1
1st	W	Z	Y	Z
2nd	X	W	X	W
3rd	Y	Y	Z	X
4th	Z	X	W	Y

The Monotonicity Criterion

KEY POINT

The monotonicity criterion states that if a candidate wins an election and then gains more support, then that candidate will win a reelection.

Often before an election takes place, pollsters are hired to determine the preferences of the voters. If candidate X, the current front-runner, draws away support from one of his opponents, then it would seem likely that X has improved his chances of winning the election. This idea motivates our final voting criterion, the monotonicity criterion.

DEFINITION The Monotonicity Criterion
If X wins an election and in a reelection all voters who change their votes only change their votes to favor X, then X also wins the reelection.

If we are using the plurality method or the Borda count method, a candidate who wins an election and then gains more support will win any reelection.

Example 5 is the example that we described at the start of the chapter explaining why it may be an advantage for voters to vote against you in order for you to win an election.

EXAMPLE 5 The Plurality-with-Elimination Method Violates the Monotonicity Criterion

Table 11.9 summarizes the preference ballots cast for the president of the International Students' Organization (ISO), in which Michael Chang, Rukevwe Kwami, and Anna Woytek are candidates. The plurality-with-elimination method is being used to determine the winner. The day before the election, three supporters of Woytek, who had preferred Chang over her, tell her that because they know she is going to win the election, they are going to throw their support to her in the election tomorrow. Later that day, after talking with her uncle who is an expert in voting theory, Woytek calls the three new supporters and asks them to vote for Chang instead.

Highlight

Using Spreadsheets* to Find Counterexamples

You can use spreadsheets to avoid the tedium of generating preference tables by hand to find flaws in the voting methods. This Numbers spreadsheet shows that the Borda count method fails Condorcet's criterion.

By adjusting the numbers in the fourth line of the table in a trial-and-error fashion, we quickly found the desired counterexample.

Sheet1 | Sheet2 | Sheet3 | +

		Borda Violates Condorcet's Criterion				
	15	4	28	10	8	30
1st	C	D	C	A	A	B
2nd	A	A	B	D	B	C
3rd	B	B	D	B	C	A
4th	D	C	A	C	D	D
BORDA	A pts	217				
	B pts	286				
	C pts	292	C is Borda winner			
	D pts	155				
	Total	950				
	NVoters	95				
	Notice that although C won the election, B defeats every other					
	candidate in head-to-head competition as we see below.					
CONDORCET	A vs. B	A=	37	B=	58	B wins
	A VS C	A=	22	C=	73	
	A VS D	A=	63	D=	32	
	B VS C	B=	52	C=	43	B wins
	B VS D	B=	81	D=	14	B wins
	C VS D	C=	81	D=	14	

	Number of Ballots			
Preference	12	9	3	8
1st	W	K	C	C
2nd	C	W	W	K
3rd	K	C	K	W

K has fewest 1st-place votes and is eliminated.

TABLE 11.9 Preference table for the ISO president election.

If the three voters indicated in the highlighted column in Table 11.9 change their votes so that Woytek is first, Chang is second, and Kwami is third, why should this cause Woytek concern?

*I used Microsoft's Excel and Apple's Numbers to generate tables in this chapter. Numbers is available on mobile devices such as the iPhone, iPod touch, and the iPad; other spreadsheets are available for Android devices. Sample spreadsheets are available in MyMathLab.

SOLUTION:

Let us first determine the winner of the election using the ballots in Table 11.9. We see that Woytek gets 12 first-place votes, Chang 11, and Kwami 9. Thus, Kwami is eliminated and a runoff election is held. Because voters do not change their preferences in the runoff, we can simply eliminate all references to Kwami on the original ballots and recount the votes. The ballots will look like this.

	Number of Ballots			
Preference	**12**	**9**	**3**	**8**
1st	W		C	C
2nd	C	W	W	
3rd		C		W

Woytek wins 21 to 11.

With Kwami removed from the ballots, we see that Woytek defeats Chang by a vote of 21 to 11.

Now let us see what would have happened if the three voters who preferred Chang to Woytek to Kwami had switched their votes. The ballots before the first elimination would have been as follows:

	Number of Ballots			
Preference	**12**	**9**	**3**	**8**
1st	W	K	W	C
2nd	C	W	C	K
3rd	K	C	K	W

Chang is eliminated.

These three voters have changed their votes.

Because Chang has the fewest first-place votes, he is eliminated and the ballots now look like this:

	Number of Ballots			
Preference	**12**	**9**	**3**	**8**
1st	W	K	W	
2nd		W		K
3rd	K		K	W

Kwami wins 17 to 15.

It is now clear why Woytek did not want the votes changed; in the runoff election between Woytek and Kwami, she loses 17 to 15.

Now try Exercises 29, 31, and 32.

We have seen that each of the voting methods discussed in Section 11.1 violates an important voting criterion. Table 11.10 summarizes what we know about voting methods' defects at this point and indicates where we address this issue in the examples and exercises. In Table 11.10, a "Yes" entry means the criterion stated in the row is always satisfied by the method listed in the column. Otherwise, we reference a counterexample from this section showing that the method need not satisfy the stated criterion.

All the voting methods that you have studied so far have failed to satisfy at least one of the four fairness criteria, so you might be wondering, "Does any perfect voting method exist?" In 1951, an economist named Kenneth Arrow was studying decision making for a government think tank called the RAND Corporation. In his research, he discovered the following remarkable theorem that answers this question:

ARROW'S IMPOSSIBILITY THEOREM In any election involving more than two candidates, there is no voting method that will satisfy all of the four fairness criteria.

Thus, any voting method that we decide to use must have flaws.

	Plurality	Borda Count	Plurality with Elimination	Pairwise Comparison
Majority	Yes	No—Example 1	Yes	Yes
Condorcet's	No—Example 2	No—Exercise 3	No—Exercise 7	Yes
Independence-of-irrelevant-alternatives	No—Example 3	No—Exercise 5	No—Exercise 18	No—Example 4
Monotonicity	Yes	Yes	No—Example 5	Yes

TABLE 11.10 Flaws in voting methods.

EXERCISES 11.2

Sharpening Your Skills

Some of these exercises have no fixed solution method. Often, constructing a preference table and adjusting the entries by trial and error will eventually lead to a solution.

1. In the preference table, A has the majority of first-place votes. Who wins the election if we use the Borda count method?

	Number of Ballots			
Preference	**12**	**15**	**9**	**13**
1st	A	B	C	A
2nd	B	C	B	D
3rd	C	A	D	B
4th	D	D	A	C

2. In the preference table below, D has the majority of first-place votes. Who wins the election if we use the Borda count method?

	Number of Ballots			
Preference	**4**	**10**	**3**	**2**
1st	C	D	C	A
2nd	A	A	A	D
3rd	B	B	D	B
4th	D	C	B	C

3. Determining the legal drinking age. A state commission is voting on changing the legal drinking age. The options are A, lower the age to 18; B, lower the age to 19; C, lower the age to 20; and D, keep the age at 21. Use the preference table to determine the winner using the Borda count method. Show that Condorcet's criterion is not satisfied.

	Number of Ballots				
Preference	**8**	**10**	**14**	**3**	**10**
1st	C	D	C	A	B
2nd	A	A	B	D	C
3rd	B	B	D	B	A
4th	D	C	A	C	D

4. Voting for the president of a club. A chapter of the Sierra Club is voting for president. The candidates are (A)lvaro, (B)rown, (C)lark, and (D)ukevitch. Use the preference table to determine the winner using the Borda count method. Show that Condorcet's criterion is not satisfied.

	Number of Ballots				
Preference	**4**	**23**	**8**	**3**	**12**
1st	C	D	C	A	A
2nd	A	A	B	D	B
3rd	B	B	D	B	C
4th	D	C	A	C	D

5. Choosing a location for a research facility. Teach for America is considering (A)tlanta, (B)oston, (C)hicago, and (D)enver for a new research facility. A group of senior managers voted to determine where the facility will be located. Use the preference table to determine the city that was chosen using the Borda count method. Show that the independence-of-irrelevant-alternatives criterion is not satisfied.

	Number of Ballots					
Preference	**15**	**4**	**8**	**10**	**8**	**2**
1st	C	D	C	B	B	A
2nd	B	B	A	D	A	C
3rd	A	A	D	A	C	B
4th	D	C	B	C	D	D

6. Locating a new factory. The Land Mover Tractor Company is going to build a new factory in either (A)labama, (C)alifornia, (O)regon, or (T)exas. The vote of the board of directors is shown in the preference table. Determine the state that they chose using the Borda count method. Show that the independence-of-irrelevant-alternatives criterion is not satisfied.

	Number of Ballots					
Preference	**9**	**2**	**4**	**5**	**4**	**1**
1st	C	T	C	A	A	O
2nd	A	A	O	T	O	C
3rd	O	O	T	O	C	A
4th	T	C	A	C	T	T

7. Reducing a budget. Due to a decrease in state funding, a citizens' committee is recommending to the school board ways to reduce expenses. The options are A, reduce sports programs; B, reduce expenditures on art and music programs; C, increase class size; and D, defer maintenance on buildings. Use the preference table in the next column to determine the choice that the committee recommends using the plurality-with-elimination method. Show that Condorcet's criterion is not satisfied.

	Number of Ballots					
Preference	**9**	**12**	**4**	**5**	**4**	**1**
1st	C	D	C	A	A	B
2nd	A	A	B	D	B	C
3rd	B	B	D	B	C	A
4th	D	C	A	C	D	D

8. Voting on an award for best restaurant. A group of columnists is voting on the restaurant of the year. The choices are The (A)lamo, The (B)ar-B-Q, (C)hez Nous, and (D)anny's Place. Use the preference table to determine the winner using the plurality-with-elimination method. Show that Condorcet's criterion is not satisfied.

	Number of Ballots				
Preference	**8**	**11**	**3**	**4**	**3**
1st	B	D	B	C	C
2nd	C	C	A	D	A
3rd	A	A	D	A	B
4th	D	B	C	B	D

Use the following preference table for Exercises 9 and 10.

Preference	13	10	5
1st	A	B	C
2nd	B	C	B
3rd	C	A	A

9. Who wins this election using the pairwise comparison method? Why does this election not violate the majority criterion?

10. a. Who wins the election using the plurality-with-elimination method?

b. If the last five voters change their ballots to

C
A
B

who now wins the election using the plurality-with-elimination method? Is this a violation of the monotonicity criterion? Explain.

Applying What You've Learned

11. Complete the preference table so that the Borda count winner violates Condorcet's criterion.

Preference	a.	b.
1st	B	A
2nd	A	C
3rd	C	B

12. Complete the preference table so that A is the Borda count winner, but when we remove C, then B is the Borda count winner, thus violating the independence-of-irrelevant-alternatives criterion.

Preference	a.	b.	c.
1st	A	B	C
2nd	C	A	B
3rd	B	C	A

13. Make a preference table, similar to the one given in Example 2, with nine voters to choose among three choices, in which the plurality method violates Condorcet's criterion.

14. Make a preference table similar to the one given in Example 3, but with at least four different types of ballots and three candidates, in which the plurality method violates the independence-of-irrelevant-alternatives criterion.

15. Complete the preference table so that the plurality-with-elimination method violates Condorcet's criterion.

	Number of Ballots				
Preference	20	—	8	8	12
1st	C	D	A	A	B
2nd	A	A	D	B	C
3rd	B	B	B	C	A
4th	D	C	C	D	D

16. Does the plurality method satisfy the majority criterion?

17. Does the plurality-with-elimination method satisfy the majority criterion?

18. Voters are choosing among four choices. Make a preference table in which the plurality-with-elimination method violates the independence-of-irrelevant-alternatives criterion.

19. Presidential election. One of the several controversies of the 2000 presidential election was that Al Gore had more popular votes than George W. Bush, yet lost the election. A summary of the popular vote is given in the table below.

Candidate	Popular Vote
George W. Bush	50,456,002
Al Gore	50,999,897
Ralph Nader	2,882,955
Pat Buchanan	448,895
Harry Browne	384,431
Howard Phillips	98,020
John Hagelin	83,714
Other	51,186
Total	105,405,100

Assume that the election was held using the plurality-with-elimination method. Further assume:

i. All of Hagelin's, Phillips', and Buchanan's voters preferred Bush as their second choice.

ii. All Browne's voters preferred Nader second and Gore third.

iii. The "Other" voters had Bush and Gore split equally as their second choices.

iv. 70 percent of those whose first choice was Nader preferred Gore second and 30 percent preferred Bush second.

a. What are the vote totals after all are eliminated except Bush, Gore, and Nader?

b. What are the final vote totals?

c. Who wins and with what percent of the popular vote?

20. Presidential election. Refer to Exercise 19, keeping conditions i., ii., and iii. as before. What percent of Nader's voters would have to have Bush as their second choice in order for Bush to win the election?

21. A runoff election. The table shows the results of an election for the host of *Midnight Snack*, an all-night radio talk show. The candidates are (K)laus, (E)lijah, (C)aroline, and (D)amon.

Preference	3	7	5	3	9
1st	C	K	C	K	E
2nd	E	E	D	E	K
3rd	K	C	E	D	D
4th	D	D	K	C	C

The plurality voting method is being used; however, after the first round of voting, a runoff election between the top two candidates will be held to determine the winner. Who wins the election?

22. A runoff election. Repeat Exercise 21 using this table.

Preference	5	4	5	3	9
1st	K	E	K	E	C
2nd	C	C	D	C	E
3rd	E	K	C	D	D
4th	D	D	E	K	K

Communicating Mathematics

23. What are four criteria that we would like an election to satisfy?

24. What does Arrow's theorem tell us?

25. Explain why the pairwise comparison method satisfies the majority criterion.

26. Explain why the pairwise comparison method satisfies Condorcet's criterion.

27. Voters are choosing among three options. Make a preference table in which the Borda count winner violates the majority criterion.

28. Voters are choosing among five options. Make a preference table in which the Borda count winner violates the majority criterion.

Challenge Yourself

29. Make a preference table, similar to the one given in Example 5, with at least five different types of ballots and four choices that shows that the plurality-with-elimination method violates the monotonicity criterion.

30. Make a preference table, similar to the one given in Example 4, in which voters use the pairwise comparison method to vote on five choices and the vote violates the independence-of-irrelevant-alternatives criterion.

31. Voters are choosing among five choices. Make a preference table in which the plurality-with-elimination method violates the monotonicity criterion.

32. Voters are choosing among four choices. Make a preference table in which the plurality-with-elimination method violates the monotonicity criterion.

33. A famous example called *Condorcet's paradox* is illustrated in the preference table.

	Number of Ballots		
Preference	**1**	**1**	**1**
1st	A	B	C
2nd	B	C	A
3rd	C	A	B

Notice that two-thirds of the voters prefer A over B, two-thirds prefer B over C, and two-thirds prefer C over A. Make a preference table for five candidates A, B, C, D, and E such that 80% of the voters prefer A over B, 80% prefer B over C, 80% prefer C over D, 80% prefer D over E, and 80% prefer E over A.

34. In using the pairwise comparison method, as the number of candidates grows it is easy to see that the number of comparisons grows quite rapidly.

a. Complete the following table:

Candidates	Number of Comparisons
A, B	1
A, B, C	3
A, B, C, D	6
A, B, C, D, E	
A, B, C, D, E, F	

b. In Chapter 12, we will show that for k candidates, there are $k(k - 1)/2$ comparisons necessary. How many comparisons are necessary for 10 candidates? For 20?

35. One of the voting methods we have been discussing satisfies three of the four criteria in Exercise 23. Which method is it and what criterion does it fail to satisfy?

36. Research how the electoral college system is used to elect the president of the United States. How does this procedure conflict with the majority criterion?

11.3 Weighted Voting Systems

Objectives

1. Understand the numerical representation of a weighted voting system.
2. Find the winning coalitions in a weighted voting system.
3. Compute the Banzhaf power index of a voter in a weighted voting system.

When we vote, it would be nice if all of our votes carried the same weight, but this is sometimes not the case. If you bought stock in Facebook when it went public in 2012, it is unlikely that you and its CEO, Mark Zuckerberg, would have the same voting power. Often shareholders of a corporation have votes proportional to the number of shares of stock that they hold.

This inequality occurs in many other organizations. For example, not all members of the United Nations Security Council have the same voting power. The present council consists of 5 permanent members (Great Britain, France, the United States, China, and Russia) and 10 nonpermanent members. According to its rules, the council cannot pass a resolution unless all the permanent members vote "yes" and, in addition, four of the nonpermanent members must also vote "yes." There are other situations in which voting is done very differently from the methods we have discussed in this chapter. For example, when a jury votes in a criminal trial, a vote of 11 to 1 is not enough to convict the accused. In corporate decision making, a large stockholder may possess 40 or 50 times the voting power of a smaller stockholder. In this section, we will develop a method to measure the power of voters in a system in which not everyone has the same strength.

KEY POINT

We describe a weighted voting system by its quota and the weight of each voter.

Weighted Voting Systems

In order to understand the concept of a weighted voting system, consider the following situation. A corporation that owns the Phoenix Flames, a professional football team, has six stockholders, who own different amounts of stock. Let's say that Alicia Mendez and her son Ben each own 26% of the stock but Carl, Dante, Emily, and Felix each own 12%. (We represent these stockholders as A, B, C, D, E, and F.) Assume that each stockholder possesses as many votes as the percentage of stock owned. Voting in this corporation clearly does not reflect the "one person, one vote" principle. In fact, A and B possess a lot of power compared with C, D, E, and F. If a resolution requires a vote of more than 50% to pass, we see that A and B together can pass any resolution they want, whereas C, D, E, and F are considerably weaker in their ability to get their resolutions passed.

In this situation, we see a number of characteristics of weighted voting systems that are present in the examples we will study. First, there is a number of votes, 51, required for a resolution to pass. This number is called a *quota.* Second, there are voters, each of whom controls a number of votes. We call this number the voter's *weight.* We will make these ideas more precise.

DEFINITIONS A **weighted voting system** with n voters is described by a set of numbers that are listed in the following format:

[quota: weight of voter 1, weight of voter 2, . . . , weight of voter n]

The **quota** is the number of votes necessary in this system to get a resolution passed.

The numbers that follow, called **weights,** are the amount of votes controlled by voter 1, voter 2, etc.

EXAMPLE 1 Weighted Voting Systems

Explain each of the following weighted voting systems:

a) [51 : 26, 26, 12, 12, 12, 12]
b) [4 : 1, 1, 1, 1, 1, 1, 1]
c) [14 : 15, 2, 3, 3, 5]
d) [10 : 4, 3, 2, 1]
e) [12 : 1, 1, 1, 1, 1, 1, 1, 1, 1, 1, 1, 1]
f) [12 : 1, 2, 3, 1, 1, 2]
g) [39 : 7, 7, 7, 7, 7, 1, 1, 1, 1, 1, 1, 1, 1, 1, 1]

SOLUTION:

a) This is the stockholder situation that we described earlier. The following diagram explains how to interpret this system:

[51 : 26, 26, 12, 12, 12, 12]

Need 51 votes to pass a resolution.
A and B each have 26 votes.
C, D, E, and F have 12 votes each.

b) In this situation, there are seven voters having one vote each. Because the quota is four, a simple majority suffices to pass a resolution. This is an example of a "one person, one vote" situation.

[4 : 1, 1, 1, 1, 1, 1, 1]

Need 4 votes to pass a resolution.
A, B, C, D, E, F, G have 1 vote each.

c) The quota is 14; because voter 1 has 15 votes, he or she has total control. Because the other four voters have no power whatsoever in this system, we call voter 1 a **dictator.**

The quota is 14.
The dictator is the only person able to pass a resolution.

[14 : 15, 2, 3, 3, 5]

d) Notice that the sum of the votes is 10, which is also the quota. In this system, even though the first voter has greater weight than the others, in fact he or she has no more power, because the support of even the weakest voter is necessary to pass a resolution. A voter who can, by him- or herself, prevent a motion from passing has **veto power.**

The quota is 10.

$$[10 : \underbrace{4, 3, 2, 1}]$$

Every voter is needed to pass every resolution—all have the same power.

e) This describes our jury system for trying criminal cases. Because the quota is 12, every voter must vote for a resolution for it to pass. Each voter has veto power.

The quota is 12.

$$[12 : \underbrace{1, 1, 1, 1, 1, 1, 1, 1, 1, 1, 1, 1}]$$

Every voter is needed to pass every resolution—all have the same power.

f) In this system, the sum of all the possible votes is less than the quota, so no resolutions can be passed.

The quota is 12.

$$[12 : \underbrace{1, 2, 3, 1, 1, 2}]$$

not enough votes to pass any resolutions

g) This system describes the voting in the UN Security Council. Notice in the following diagram that the quota cannot be achieved unless all the first five voters vote for a resolution. In addition, 4 of the next 10 voters must also vote for it to pass a resolution.

$$[39 : \underbrace{7, 7, 7, 7, 7}, \underbrace{1, 1, 1, 1, 1, 1, 1, 1, 1, 1}]$$

need 39 votes to pass a resolution

Each of these voters has veto power.

Four of these votes are needed to pass a resolution.

Now try Exercises 1 to 12.

Coalitions

KEY POINT

Groups of voters form coalitions to pass resolutions.

The weighted voting system in Example 1(d) is interesting because even though it appears at first that voter 1 has more power than the other voters, this is not the case. The system [10 : 4, 3, 2, 1] behaves exactly the same as the system [4 : 1, 1, 1, 1]. Notice that the number of votes that a voter possesses is not the same thing as the power a voter has to pass resolutions. In order to describe the concept of power in a voting system, we need to introduce some more definitions. We first describe the subsets of voters that have the power to pass resolutions.

DEFINITIONS Any set of voters who vote the same way is called a **coalition.** The sum of the weights of the voters in a coalition is called the **weight of the coalition.** If a coalition has a weight that is greater than or equal to the quota, then that coalition is called a **winning coalition.**

In a weighted voting system [4 : 1, 1, 1, 1, 1, 1, 1], any coalition of four or more voters is a winning coalition.

PROBLEM SOLVING

Strategy: Be Systematic

Recall from Chapter 2 that a k-element set has 2^k subsets. Because one of these is the empty set, we see that a set of k voters can form $2^k - 1$ possible coalitions. For example, a set of five voters can form $2^5 - 1 = 32 - 1 = 31$ different coalitions. It is useful to list coalitions systematically. That is, consider all one-element sets, then all two-element sets, followed by the three-element sets, and so on.

EXAMPLE 2 Finding Winning Coalitions

A town has two large political parties, (R)epublican and (D)emocrat, and one small party, (I)ndependent. Membership on the town council is proportional to the size of the parties. We will assume that R has nine members on the council, D has eight, and I only three. Traditionally, each party votes as a single bloc, and resolutions are passed by a simple majority. List all possible coalitions and their weights, and identify the winning coalitions.

SOLUTION: Review the Be Systematic Strategy on page 5.

A coalition is a nonempty subset of the set of all parties {R, D, I}. We list these subsets and their weights in the following table. Because there are 20 members on the council, any coalition with a weight of 11 or more is a winning coalition.

Coalition	Weight	
{R}	9	
{D}	8	
{I}	3	
{R, D}	17	Winning
{R, I}	12	Winning
{D, I}	11	Winning
{R, D, I}	20	Winning

Now try Exercises 13 to 16. 7

QUIZ YOURSELF 7

Consider the weighted voting system [5 : 3, 2, 2, 1].

a) How many voters are there?
b) What is the quota?
c) List all winning coalitions.

It is interesting that even though party I has less representation on the council than either R or D, it still appears in as many winning coalitions as does R or D.

In Example 2, it seems that all three parties have the same amount of voting power. We will make this idea of power more precise shortly. The key to understanding a voter's power is knowing how many coalitions need the voter in order to win.

DEFINITION A voter in a winning coalition is called **critical** if it is the case that if he or she were to leave the coalition, then the coalition would no longer be winning.

EXAMPLE 3 Identifying Critical Voters in a Coalition

Identify the critical voters in the winning coalitions in the town council in Example 2.

SOLUTION:

Coalition	Weight		Critical Voters
{R}	9		
{D}	8		
{I}	3		
{R, D}	17	Winning	R, D
{R, I}	12	Winning	R, I
{D, I}	11	Winning	D, I
{R, D, I}	20	Winning	none

Remove any of these voters and the coalition no longer wins.

Now try Exercises 17 to 20.

The Banzhaf Power Index

KEY POINT

The Banzhaf power index measures voting power.

We now define how to measure voting power in a weighted voting system.

DEFINITION In a weighted voting system, a voter's **Banzhaf power index*** is defined as

$$\frac{\text{the number of times the voter is critical in winning coalitions}}{\text{the total number of times voters are critical in winning coalitions}}.$$

QUIZ YOURSELF 8

Reconsider the weighted voting system [5 : 3, 2, 2, 1] from Quiz Yourself 7, which had winning coalitions {A, B}, {A, C}, {A, B, C}, {A, B, D}, {A, C, D}, {B, C, D}, and {A, B, C, D}.

a) Determine the critical voters in each winning coalition.
b) Compute the Banzhaf power index for voters in this system.

EXAMPLE 4 Computing the Banzhaf Power Index

In Example 3, we saw that R, D, and I each were critical voters twice. Thus, R's Banzhaf power index is

$$\frac{\text{the number of times R is critical in winning coalitions}}{\text{the total number of times voters are critical in winning coalitions}} = \frac{2}{6} = \frac{1}{3}.$$

Similarly, D and I each have a Banzhaf power index of $\frac{1}{3}$.

Now try Exercises 29 to 34. 8

EXAMPLE 5 Calculating the Banzhaf Power Index Indirectly

In the law firm of Krook, Cheatum, and Associates, there are two senior partners (Krook and Cheatum) and four associates (W, X, Y, and Z). To change any major policy of the firm, a vote must be taken in which Krook, Cheatum, and at least two associates agree to the change. Calculate the Banzhaf power index for each member of this firm.

SOLUTION: Review the Be Systematic Strategy on page 5.

We can represent the members of the firm by {K, C, W, X, Y, Z}. At first it seems tempting to list all of the coalitions (subsets) of the firm and determine which are winning. However, this is a tedious job and is not necessary. Realize that every winning coalition includes {K, C}, so all we have to do is determine the subsets of {W, X, Y, Z} with two or more members and form the union of these with {K, C}.

*There are other, slightly different ways of defining this index, depending on which author you read.

2-Element Subsets of {W, X, Y, Z}	3-Element Subsets of {W, X, Y, Z}	4-Element Subsets of {W, X, Y, Z}
{W, X}, {W, Y}, {W, Z}, {X, Y}, {X, Z}, {Y, Z}	{W, X, Y}, {W, X, Z}, {W, Y, Z}, {X, Y, Z}	{W, X, Y, Z}

The winning coalitions of the firm and their critical members are therefore:

		Winning Coalitions	Critical Members
All voters are necessary to pass a resolution in these coalitions.	1	{K, C, W, X}	K, C, W, X
	2	{K, C, W, Y}	K, C, W, Y
	3	{K, C, W, Z}	K, C, W, Z
	4	{K, C, X, Y}	K, C, X, Y
	5	{K, C, X, Z}	K, C, X, Z
	6	{K, C, Y, Z}	K, C, Y, Z
Only K and C are critical in these coalitions.	7	{K, C, W, X, Y}	K, C
	8	{K, C, W, X, Z}	K, C
	9	{K, C, W, Y, Z}	K, C
	10	{K, C, X, Y, Z}	K, C
	11	{K, C, W, X, Y, Z}	K, C

From this table we see that K and C are critical members 11 times, whereas W, X, Y, and Z are each critical members only 3 times. The Banzhaf power indices for the members of this firm are therefore:

Members	Banzhaf Power Index
K, C	$\frac{11}{11 + 11 + 3 + 3 + 3 + 3} = \frac{11}{34}$
W, X, Y, Z	$\frac{3}{11 + 11 + 3 + 3 + 3 + 3} = \frac{3}{34}$

If we add the Banzhaf power indices for the members of the firm as follows:

$$\underset{K}{\frac{11}{34}} + \underset{C}{\frac{11}{34}} + \underset{W}{\frac{3}{34}} + \underset{X}{\frac{3}{34}} + \underset{Y}{\frac{3}{34}} + \underset{Z}{\frac{3}{34}} = \frac{34}{34} = 1$$

we see that the total is 1. This is always the case when computing Banzhaf power indices in a weighted voting system.

Now try Exercises 43 to 46.

Often the chair of a committee votes only in the case of breaking a tie. In fact, this is the way the U.S. Senate votes. The vice president of the United States presides over the 100-member U.S. Senate and votes only to break a tie. The mathematics to determine the Banzhaf power index for the vice president and the members of the Senate is too lengthy to include here. We will, however, analyze a much simpler example based on the same voting principle.

EXAMPLE 6 Finding the Banzhaf Power Index of a "Tie Breaker"

A five-person air safety review board is developing in-flight safety procedures to deal with skyjackings. The board is chaired by a federal aviation administrator (A) and consists of two senior pilots (S and T) and two flight attendants (F and G).

It was the intent when the board was established that the administrator have considerably less power than the pilots and flight attendants. Therefore, the administrator votes only in the case of a tie; otherwise, cases are decided by a simple majority. How much less power does the administrator have than the other members of the board?

SOLUTION:

To answer this question, we compute the Banzhaf power index for each member of the board. We first list the winning coalitions and the critical members of these coalitions. Clearly if three or four of {S, T, F, G} vote together, they form a winning coalition in the table below. We list these five coalitions first and then list the six ways ties can be formed, which then require the addition of the administrator to break them.

If we count all of the number of times members A, S, T, F, and G are critical members of some coalition, we get a total of 30. We see that any single board member (including the chair) is a critical member of exactly six coalitions. Therefore, each of the five board members has exactly the same Banzhaf power index, which is $\frac{6}{30}$. Even though it is not apparent at first, the chair of the board has exactly the same power as the other board members.

		Winning Coalitions	Critical Members
No tie	1	{S, T, F}	S, T, F
	2	{S, T, G}	S, T, G
	3	{S, F, G}	S, F, G
	4	{T, F, G}	T, F, G
	5	{S, T, F, G}	None
Tie broken by administrator	6	{A, S, T}	A, S, T
	7	{A, S, F}	A, S, F
	8	{A, S, G}	A, S, G
	9	{A, T, F}	A, T, F
	10	{A, T, G}	A, T, G
	11	{A, F, G}	A, F, G

In Example 6, the administrator and each member of the board were critical in exactly the same number of circumstances, by being one of a bare majority in favor of a motion. Similarly, in the U.S. Senate, the vice president and each senator have exactly the same number of opportunities to be critical by being on the winning side of a 51-to-50 vote. Thus, the vice president has the same power as each senator. In Chapter 12, you will learn principles of counting that will enable you to calculate the number of cases that we have to consider to prove this.

Historical Highlight

Blocking Coalitions, Banzhaf, and the Electoral College

Political scientists use a variation of the method you have studied in this section to compute the Banzhaf power index of the states voting in the electoral college that elects the president of the United States.

A state in the electoral college has as many votes as the total of its senators and representatives. For example, New York presently has 27 representatives and 2 senators and therefore has 29 votes. We can think of the electoral college as a weighted voting system consisting of the 50 states plus the District of Columbia. California has the most votes, 55, and there are 8 states such as Delaware having only 3.

Although the number of coalitions would be impractical to consider by hand, Banzhaf indices can be calculated using computer methods. The table below shows Banzhaf indices for several states from an earlier census.

State	Banzhaf Power Index (%)
California	11.44
Texas	6.20
New York	5.81
Florida	5.02
Pennsylvania	3.87
New Jersey	2.75
Arizona	1.83
New Mexico	0.91
Delaware	0.55

You can see in this table that California has almost three times the Banzhaf power of Pennsylvania and more than twenty times the power of Delaware. Keep in mind when you read these numbers that this is only one way that we can measure power in a weighted voting system. There are other methods and also much controversy about the way power is distributed in the electoral college. You can find many articles analyzing the electoral college on the Internet.

EXERCISES 11.3

Sharpening Your Skills

In Exercises 1–12, the weights represent voters A, B, C, and so on, in that order. Identify a) the quota, b) the number and weights of the voters, c) dictators, and d) those having veto power.

1. [5 : 1, 1, 1, 1, 1]
2. [15 : 5, 4, 3, 2, 1]
3. [11 : 10, 3, 4, 5]
4. [6 : 6, 1, 2, 2]
5. [15 : 1, 2, 3, 3, 4]
6. [11 : 1, 2, 3, 4]
7. [12 : 1, 3, 5, 7]
8. [16 : 1, 5, 7, 9]
9. [25 : 4, 4, 6, 7, 9]
10. [21 : 3, 5, 6, 8, 9]
11. [51 : 20, 20, 20, 10, 10]
12. [67 : 15, 15, 15, 15, 10, 10]

In Exercises 13–16, write out all winning coalitions in each voting system. Do not approach this randomly. Be systematic by considering coalitions from smallest to largest.

13. [12 : 1, 3, 5, 7]
14. [16 : 1, 5, 7, 9]
15. [25 : 4, 4, 6, 7, 9]
16. [23 : 3, 5, 6, 8, 9]

17. Find the critical voters in the winning coalitions that you found in Exercise 13.

18. Find the critical voters in the winning coalitions that you found in Exercise 14.

19. Find the critical voters in the winning coalitions that you found in Exercise 15.

20. Find the critical voters in the winning coalitions that you found in Exercise 16.

Applying What You've Learned

21. A theater guild. The Theater Guild consists of (P)erformers, (T)echnicians, and (S)upport staff. Representation on the guild is proportional to the number in each group, and the representatives of each group tend to vote in a bloc. A simple majority is required to pass a resolution. Assume there are five performers, four technicians, and two members of the support staff on the guild. List all the coalitions of {P, T, S} and their weights. Which are the winning coalitions?

22. A theater guild. Repeat Exercise 21, but now assume that there are eight performers, six technicians, and two members of the support staff on the guild.

23. **A state committee.** The college athletics procedures committee makes policy decisions affecting college athletics programs throughout the state. The committee consists of three (A)dministrators, four (C)oaches in athletics departments, three (T)eam captains, and two (N)onathlete students. Assume that each of these four groups votes in a bloc. To make a policy change, a vote of at least 8 is required. Find all the winning coalitions of {A, C, T, N}, and state their weights.

24. **A state committee.** Repeat Exercise 23, but now assume that there are four administrators, five coaches, four team captains, and three nonathletes. Also assume that a vote of 12 is required to make a policy change.

25. **A theater guild.** Determine the critical voters in the winning coalitions in Exercise 21.

26. **A theater guild.** Determine the critical voters in the winning coalitions in Exercise 22.

27. **A state committee.** Determine the critical voters in the winning coalitions in Exercise 23.

28. **A state committee.** Determine the critical voters in the winning coalitions in Exercise 24.

In Exercises 29–34, determine the Banzhaf power index for each voter in each weighted voting system.

29. The weighted voting system in Exercise 1
30. The weighted voting system in Exercise 3
31. The weighted voting system in Exercise 9
32. The weighted voting system in Exercise 11
33. The weighted voting system in Exercise 13
34. The weighted voting system in Exercise 15

35. The system [3 : 1, 1, 1, 1, 1] is an example of a "one person, one vote" situation.
    a. Calculate the Banzhaf power index for each person in this system.
    b. How does this conform with your intuition? Explain.

36. A 12-person jury corresponds to the weighted voting system

$$[12 : 1, 1, 1, 1, 1, 1, 1, 1, 1, 1, 1, 1].$$

    a. Calculate the Banzhaf power index for each person in this system.
    b. How does this conform with your intuition? Explain.

37. Consider the system [14 : 15, 2, 3, 3, 5] in which A is a dictator.
    a. Calculate the Banzhaf power index for each person in this system.
    b. How does this conform with your intuition? Explain.

38. Consider the system [12 : 1, 2, 3, 1, 1, 2] in which no resolution can be passed because the quota is too high.
    a. Explain why Banzhaf power indices cannot be calculated in this system.
    b. How does this conform with your intuition? Explain.

39. **Calculating power in the electoral college.** After the 2010 census, California was apportioned 53 representatives. Adding its 2 senators gave it 55 votes in the electoral college. Similarly Texas had 38 votes, and Florida had 29. Assume for the moment that these were the only three states in the United States and calculate the Banzhaf power index for each state. Assume that the quota in this weighted system is 62.

40. **Calculating power in the electoral college.** Repeat Exercise 39 for Illinois (20 votes), Michigan (16 votes), and Colorado (9 votes). Assume the quota is 23.

41. **Calculating power in the electoral college.** Repeat Exercise 39 for Pennsylvania (20 votes), North Carolina (15 votes), and New Mexico (5 votes). Assume the quota is 21.

42. **Calculating power in the electoral college.** Repeat Exercise 39 for Ohio (18 votes), Louisiana (8 votes), and Connecticut (7 votes). Assume the quota is 17.

In Example 5, we analyzed the voting power of the law partners in the firm of Krook, Cheatum, and Associates. Recall that to change a policy required the votes of Krook, Cheatum, and two associates. For each scenario described in Exercises 43–46, do the following:

a) Give an intuitive explanation as to how you think the voting power of Krook, Cheatum, and each associate is changed.

b) Calculate the Banzhaf power index for each member of the firm to see how this calculation corresponds to your intuition.

43. **A law firm.** Another associate is added to the original firm, so the firm now has two senior partners and five associates.

44. **A law firm.** Another senior partner, named Fair, is added to the original firm, so the firm now has three senior partners and four associates. Assume that two senior partners and two associates must vote to change a policy.

45. **A law firm.** One of the original associates, named Howe, is promoted to senior partner. The firm now has three senior partners and three associates. Assume that two senior partners and two associates must vote to change a policy.

46. **A law firm.** Krook resigns, leaving Cheatum as the only senior partner with four associates. Assume that Cheatum and two associates must vote to change a policy.

Communicating Mathematics

47. Is a voter's weight the same as the voter's Banzhaf index in a weighted voting system? Explain.

48. How do you compute the Banzhaf power index for a voter in a weighted voting system? What is the sum of the Banzhaf power indices in a weighted voting system?

49. True or false: If voter A has twice the weight of voter B in a weighted voting system, then A's Banzhaf index is twice that of B's. Explain.

50. True or false: In a weighted voting system, if all voters have the same power, then all voters must have the same weight. Explain.

Challenge Yourself

51. A **dummy** in a weighted voting system is a voter whose Banzhaf power index is zero. Determine two values for the quotient q in the following weighted voting system so that voter D is a dummy: [q : 16, 6, 6, 4].

52. Consider the weighted voting system $[14 : a, 3, 3, 2]$. Find the smallest value for a such that D will be a dummy.

53. We saw that in the voting system $[10 : 4, 3, 2, 1]$, all voters had the same power. What happens to voter A's power as we decrease the quotient q?

54. Continuing the thought in Exercise 53, what happens to voter D's power as we decrease the size of the quotient q?

In Exercises 55 and 56, devise a voting system that behaves with specifications that are similar to the UN Security Council's described in Example 1.

55. A committee has three standing members and six temporary members. The three standing members and two of the temporary members must vote for a resolution for it to be passed.

56. A committee has four standing members and eight temporary members. The four standing members and three of the temporary members must vote for a resolution for it to be passed.

Looking Deeper

11.4

The Shapley-Shubik Index

Objectives

1. Determine all the permutations of a set.
2. Find the pivotal voters in a coalition.
3. Calculate the Shapley-Shubik index for a voter in a weighted voting system.

Your local state representative is asking for your support in the upcoming election on the grounds that she recently voted to increase funding for your college. However, she omitted to mention that she initially opposed the bill and only threw in her support when it was clear that the bill already had enough votes to pass. In fact, her only motive for voting for the bill was to gain votes from the many constituents in her district who attend college.

From this example you can see that it may be important to know the order in which members join a coalition to make it a winner. In the discussion of the Banzhaf power index, we only considered the members of a coalition, specifically the critical members. The method we study here, which is based on the work of the mathematician Lloyd Shapley and the economist Martin Shubik, focuses not only on the makeup of winning coalitions but also on the order in which winning coalitions are formed.

Permutations

To understand the Shapley-Shubik index, we must make a clear distinction between sets whose elements have an order to them versus sets in which the order is unimportant. Recall from Chapter 2, when we use the notation {A, B, C}, order is not important; if we want, we could write this set as {B, C, A} or {C, A, B} instead. If we want to emphasize that the order of the elements in a set is important, then we must use a different notation.

DEFINITION An ordering of the elements of the set $\{x_1, x_2, x_3, \ldots, x_n\}$ in a straight line is called a **permutation*** of that set. We denote a permutation as follows:

$$(x_{i_1}, x_{i_2}, x_{i_3}, \ldots, x_{i_n}),$$

where x_{i_1} is the first element in the permutation, x_{i_2} is the second element in the permutation, and so on.

Using this definition we see that (A, B, C), (B, C, A), and (C, A, B) are all different permutations of the set {A, B, C} because the elements are listed in different orders.

*We will study permutations at much greater length in Chapter 12.

From now on, we will assume the following:

A permutation of voters specifies a coalition in which the voters were added one at a time.

PROBLEM SOLVING

Strategy: The Splitting-Hairs Principle

The Splitting-Hairs Principle in Section 1.1 stated that different notation usually means that you are dealing with different concepts. You can conclude that because {A, B, C} and (A, B, C) look different, they do not mean the same thing. Try to connect new notation with notation you have seen before. Recalling that the ordered pair of numbers (3, 4) does not mean the same thing as the ordered pair (4, 3) helps you understand the meaning of the notation (A, B, C).

It will be important to know how many different ways we can order the elements in a set. We will see how to generate different permutations of a set by looking at the problem graphically, as in the next example.

EXAMPLE 1 Finding Permutations of a Set

How many permutations are there of each set?

a) {A, B, C} b) {A, B, C, D}

SOLUTION: Review the Be Systematic Strategy on page 5.

a) If we think about forming the orderings by deciding which is the first element of the ordering, then which is the second, and finally which is the third element, we can visualize the permutations as shown in Figure 11.1, which is called a *tree diagram.*

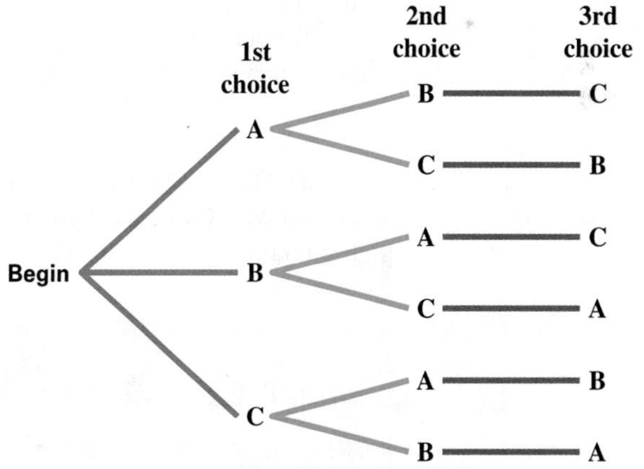

FIGURE 11.1 Tree diagram showing all permutations of {A, B, C}.

Following the six branches of this tree starting with "Begin," we generate the permutations (A, B, C), (A, C, B), (B, A, C), (B, C, A), (C, A, B), and (C, B, A). In this case, we found $3 \times 2 \times 1 = 6$ ways to order the three elements A, B, and C.

b) In this case, in Figure 11.2 we draw a tree beginning with four branches for our first choice, three branches at the next stage for our second choice, and so on.

From Figure 11.2, we see that there are $4 \times 3 \times 2 \times 1 = 24$ permutations of {A, B, C, D}. They are (A, B, C, D), (A, B, D, C), (A, C, B, D), . . . , (D, C, B, A).

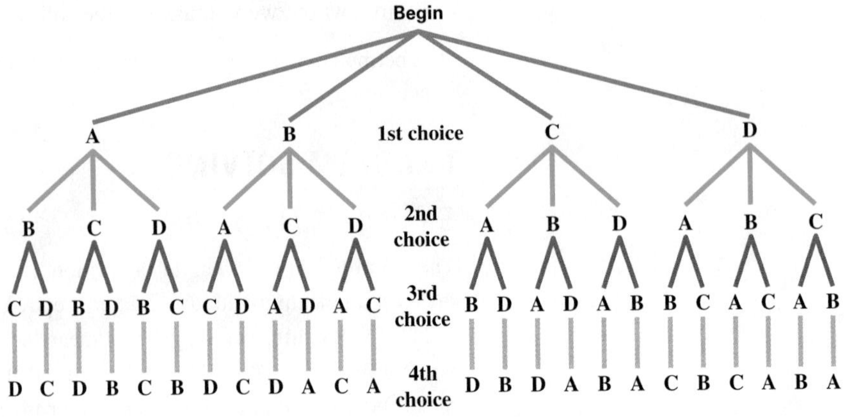

FIGURE 11.2 Tree diagram showing all permutations of {A, B, C, D}.

Now try Exercises 1 to 4. 9

QUIZ YOURSELF 9

How many permutations are there of a five-element set?

Example 1 leads us to state the following principle.

> **THE NUMBER OF PERMUTATIONS OF A SET** If we want to order a set with *n* elements, following the method in Example 1, there are *n* ways to choose the first element, followed by $n - 1$ ways to choose the second element, followed by $n - 2$ ways to choose the third element, and so on. Thus there are $n \times (n - 1) \times (n - 2) \times \cdots \times 2 \times 1$ permutations of these elements. This product is called *n* **factorial** and is written $n!$.

Pivotal Voters

In forming a coalition by adding voters one at a time, at some point we add a voter who makes the coalition a winning coalition. That voter is particularly important because he or she has changed a nonwinning coalition into a winning one.

> **DEFINITION** As we add voters to a coalition one at a time, the first voter added who makes the coalition a winning coalition is called a **pivotal voter** for the coalition.

EXAMPLE 2 Identifying Pivotal Voters in Committee Coalitions

A legislative committee has been formed to investigate the environmental effect of a controversial natural gas extraction method called hydraulic fracturing or "fracking." The members are (A)laimo, (B)rown, (C)abrera, and (D)illon. The committee members have different weights according to the size of their districts and can be modeled by the weighted voting system [5 : 3, 2, 2, 1]. Recall that A has three votes, B and C have two votes each, and D has one vote. Find the pivotal voter in each coalition.

a) (A, B, C, D) b) (A, D, C, B) c) (D, A, B, C) d) (D, C, A, B)

SOLUTION:

In each coalition, we are looking for the first voter who makes the coalition a winning one by giving it a voting weight of 5 or more.

QUIZ YOURSELF 10

Consider the weighted voting system [5 : 2, 2, 1, 1]. Find the pivotal voter in each coalition:

a) (A, C, B, D)
b) (C, D, B, A)

	Sum Weights of Coalition Members Until We Reach Quota of 5	Pivotal Voter
a)	(A, B, C, D) 3 + 2	B
b)	(A, D, C, B) 3 + 1 + 2	C
c)	(D, A, B, C) 1 + 3 + 2	B
d)	(D, C, A, B) 1 + 2 + 3	A

B is the first member to make the coalition winning.

10

The Shapley-Shubik Index

We are now in a position to define the Shapley-Shubik index for measuring voter power.

DEFINITION In a weighted voting system, the **Shapley-Shubik index** for a voter is

$$\frac{\text{the number of times the voter is pivotal in some permutation of the voters}}{\text{the total number of permutations of the voters}}.$$

EXAMPLE 3 Finding the Shapley-Shubik Index

Compute the Shapley-Shubik index for each voter in the weighted voting system [4 : 3, 2, 1].

SOLUTION: Review the Be Systematic Strategy on page 5.

Consider the voters to be A, B, and C in that order. We first list all permutations of these voters and determine in each permutation which voter is pivotal.

Sum Weights of Coalition Members Until We Reach Quota of 4	Pivotal Voter
(A, B, C) 3 + 2	B
(A, C, B) 3 + 1	C
(B, A, C) 2 + 3	A
(B, C, A) 2 + 1 + 3	A
(C, A, B) 1 + 3	A
(C, B, A) 1 + 2 + 3	A

B and C are pivotal only once.

A is pivotal four out of six times.

A's Shapley-Shubik index is

$$\frac{\text{the number of times A is pivotal in some permutation of the voters}}{\text{the total number of permutations of the voters}} = \frac{4}{6} = \frac{2}{3}.$$

Because B is pivotal one out of six times, B's index is $\frac{1}{6}$. Similarly, C's index is also $\frac{1}{6}$.

Now try Exercises 9 to 24.

Note in Example 3 that even though B has twice the votes of C, both B and C have the same voting power according to the Shapley-Shubik index. In general, the number of votes a person has is not the same as that person's voting power as measured according to the Shapley-Shubik index.

PROBLEM SOLVING

Strategy: Verify Solutions

As we recommended in Section 1.1, it is always good to check to see if your answers are reasonable. Notice when computing the Shapley-Shubik index, the sum of the indices of all the voters is always 1.

As you might expect, the Banzhaf and Shapley-Shubik indices do not measure power the same way.

EXAMPLE 4 The Banzhaf Power Index Differs from the Shapley-Shubik Index

Let us compute the Banzhaf power index for the voters in the system [4 : 3, 2, 1] in Example 3.

SOLUTION:

We summarize the computations in the following table:

Winning Coalition	Weight	Critical Voters
{A, B}	5	A, B
{A, C}	4	A, C
{A, B, C}	6	A

If we remove any voter, the coalition is no longer winning.

Because A is a critical voter three out of five times, her Banzhaf power index is $\frac{3}{5}$. Similarly, B and C have Banzhaf power indices of $\frac{1}{5}$ each.

Notice that both the Shapley-Shubik index and the Banzhaf power index rate A as the most powerful member of the weighted voting system in Examples 3 and 4. However, also note that the power indices are not identical for both methods.

Often, the chair of a committee has more power than the other members of the committee. It is possible to measure exactly how much greater this power is using the Shapley-Shubik index.

EXAMPLE 5 Comparing the Power of Committee Members

Assume a campus blog has an editorial board consisting of the managing editor and four other members. In order to begin to develop a story, the managing editor and two other board members must agree that the story is newsworthy. Use the Shapley-Shubik index to compare the power of the managing editor with that of the other board members.

SOLUTION:

Let us assume that the managing editor is A and the other board members are B, C, D, and E. We will solve this problem by first determining B's power. We can then deduce that C, D, and E also have this same power. From this information, we can easily determine the power that A has.

In forming permutations of {A, B, C, D, E}, we can think of filling slots in a list as follows:

___ ___ ___ ___ ___
1st 2nd 3rd 4th 5th

In order for B to be a pivotal voter, the managing editor A and one other board member must already have been selected for the 1st and 2nd slots and B must be in the 3rd slot. So, our slot diagram looks like this:

A is in slot 1 or 2.

___ ___ B ___ ___
1st 2nd 3rd 4th 5th

We will next consider two cases depending on whether the editor A is in slot 1 or slot 2.

Case 1: Assume that A is in slot 1, so the diagram now looks like this:

A ___ B ___ ___
1st 2nd 3rd 4th 5th

3 ways to fill this slot (2nd); 2 ways to fill this slot (4th); 1 way to fill this slot (5th)

This means that we can fill the remaining three slots in $3! = 6$ ways.

Case 2: Assume that A is in slot 2. The diagram now looks like this:

___ A B ___ ___
1st 2nd 3rd 4th 5th

Again, the remaining three slots can be filled in six ways.

From Cases 1 and 2, we see that there are exactly 12 ways to form a permutation in which B is pivotal. Thus, B's Shapley-Shubik index is

$$\frac{12}{5!} = \frac{12}{120} = \frac{1}{10}.$$

By a similar analysis, we would find that C, D, and E also have a Shapley-Shubik index of $\frac{1}{10}$. Because B, C, D, and E each are pivotal in 12 permutations, it follows that A is pivotal in the other $120 - 12 - 12 - 12 - 12 = 72$ permutations. Thus, A's index is

$$\frac{72}{5!} = \frac{72}{120} = \frac{6}{10}.$$

Therefore, A is six times as powerful as the other editorial board members.

EXERCISES 11.4

Sharpening Your Skills

In Exercises 1–4, use tree diagrams to find all the permutations of each set.

1. {X, Y, Z}

2. {P, Q, R}

3. {W, X, Y, Z}

4. {P, Q, R, S}

5. How many permutations are there of a six-element set?

6. How many permutations are there of an eight-element set?

7. How many permutations are there of a 12-element set?

8. How many permutations are there of a 13-element set?

9. For the weighted voting system [5 : 3, 2, 2], complete the following table, which is similar to the table given in Example 3:

Sum Weights of Coalition Members	Pivotal Voter
(A, B, C)	
(A, C, B)	
(B, A, C)	
(B, C, A)	
(C, A, B)	
(C, B, A)	

10. For the weighted voting system [7 : 4, 3, 2], complete the following table, which is similar to the table given in Example 3:

Sum Weights of Coalition Members	Pivotal Voter
(A, B, C)	
(A, C, B)	
(B, A, C)	
(B, C, A)	
(C, A, B)	
(C, B, A)	

In Exercises 11–16, determine the Shapley-Shubik index of each voter in each weighted voting system.

11. [6 : 3, 3, 2]
12. [8 : 4, 3, 3]
13. [8 : 3, 3, 2, 2]
14. [9 : 4, 3, 3, 1]
15. [4 : 2, 1, 1, 1]
16. [13 : 6, 6, 5, 3]

Applying What You've Learned

17. The system [3 : 1, 1, 1, 1, 1] is an example of a "one person, one vote" situation.

a. What is the Shapley-Shubik index for each person in this system?

b. Explain how you obtained your answer in part (a).

c. How does your answer in part (a) conform with your intuition?

18. Measuring power on a jury. We can consider a 12-person jury as the weighted system [12 : 1, 1, 1, 1, 1, 1, 1, 1, 1, 1, 1, 1].

a. What is the Shapley-Shubik index for each person in this system?

b. Explain how you obtained your answer in part (a).

c. How does your answer in part (a) conform with your intuition?

19. Consider the system [14 : 15, 2, 3, 3, 5], in which A is a dictator.

a. What is the Shapley-Shubik index for each person in this system?

b. Explain how you obtained your answer in part (a).

c. How does your answer in part (a) conform with your intuition?

20. Consider the system [12 : 1, 2, 3, 1, 1, 2], in which no resolution can be passed because the quota is too high.

a. What is the Shapley-Shubik index for each person in this system?

b. Explain how you obtained your answer in part (a).

c. How does your answer in part (a) conform with your intuition?

21. Measuring power on a committee. Committee {A, B, C, D} has chairperson A. For an item to be placed on the committee's agenda, the chairperson and at least one other committee member must agree to that item. What is the Shapley-Shubik index for each person in this system?

22. Measuring power on a theater guild. The Theater Guild consists of (P)erformers, (T)echnicians, and (S)upport staff. Representation on the guild is proportional to the number in each group, and the representatives of each group vote in a bloc. A simple majority is required to pass a resolution. There are five performers, four technicians, and two members of the support staff on the guild.

a. List all permutations of {P, T, S}.

b. Determine the pivotal voter in each permutation of {P, T, S}.

c. Find the Shapley-Shubik index for each voter in this system.

23. Measuring power on a state committee. The college athletics procedures committee makes policy decisions affecting college athletics programs throughout the state. The committee consists of three (A)dministrators, four (C)oaches in athletics departments, three (T)eam captains, and two (N)onathlete students. Assume that each of these four groups votes in a bloc. To make a policy change, a vote of at least 8 is required.

a. List all permutations of {A, C, T, N}.

b. Determine the pivotal voter in each permutation of {A, C, T, N}.

c. Find the Shapley-Shubik index for each voter in this system.

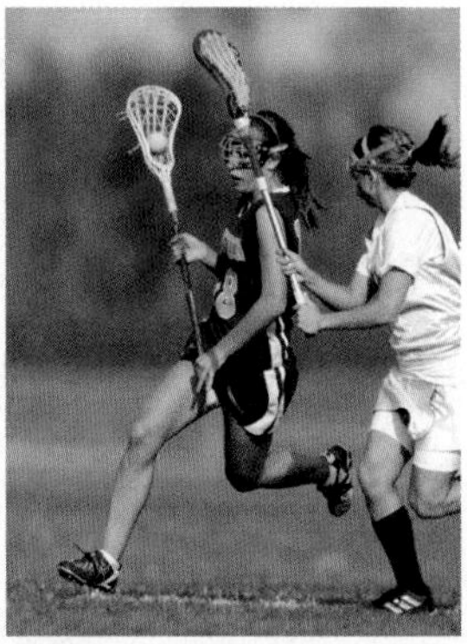

24. Measuring power in a law firm. In Example 5 of Section 11.3, we analyzed the voting power of the law partners in the firm of Krook, Cheatum, and Associates. Recall that to change a policy required the votes of Krook, Cheatum, and at least two of the associates: W, X, Y, and Z. Find the Shapley-Shubik index for each voter in this system.

A new social media company, Chirp, has an executive board consisting of the chief executive officer, (F)inch; the chief financial officer, (R)obin; the chief operating officer, (W)ren; and the treasurer, (C)rane. The council forms a weighted voting system of the form [6 : 4, 3, 2, 1]. Use this information in Exercises 25 and 26.

25. Measuring power on an executive board.

a. How many permutations of {F, R, W, C} are there?

b. List all permutations of {F, R, W, C}. Be systematic in doing this.

26. **Measuring power on an executive board.** Find the pivotal voter in each permutation of {F, R, W, C} and then calculate the Shapley-Shubik index for each member of the Chirp executive board.

27. **Measuring power among states.** After the 2010 census, Florida was apportioned 27 members in the U.S. House of Representatives; Illinois, 18; and New Jersey, 12. Assume that these 57 representatives constitute a commission investigating Medicare fraud. If representatives of each state vote as a bloc, then this is a weighted voting system of the form [29 : 27, 18, 12]. Determine the Shapley-Shubik index for each state.

28. **Measuring power among states.** Repeat Exercise 27 for New York (27 representatives), Pennsylvania (18), and Iowa (4) with quota 25.

Communicating Mathematics

29. Explain the difference between the Banzhaf index and the Shapley-Shubik index.

30. The terms *critical voter* and *pivotal voter* sound somewhat similar. What is the difference in their meanings?

31. In Example 5 we said that in order to be pivotal, B must be in the third slot. Explain why B could not be pivotal if B were in slots 1 or 2.

32. In Example 5, explain why B could not be pivotal if B were in slots 4 or 5.

Challenge Yourself

33. To compute the Shapley-Shubik index for 20 voters, we would have to consider 20! permutations. Assume that you have a computer that could find the pivotal voter in 1,000,000 permutations per second. Assuming a 365-day year, how many years would it take to find the pivotal voter in 20! permutations?

34. The electoral college consists of the 50 U.S. states plus the District of Columbia. If we were to consider these 51 parties as voters in a weighted voting system and wanted to calculate the Shapley-Shubik index for each voter, you might want to consider all of the permutations of these 51 objects. There will be 51! permutations of these 51 objects. Compute this with your calculator. Your answer will be in scientific notation (see Section 6.5). If you had a computer that could list 1,000,000 permutations per second, how many years would it take to list all 51! permutations. (Assume a year has 365 days.)

CHAPTER SUMMARY

You will learn the items listed in this chapter summary more thoroughly if you keep in mind the following advice:

1. Focus on "remembering how to remember" the ideas. What pictures, word analogies, and examples help you remember these ideas?
2. Practice writing each item without looking at the book.
3. Make up 3 × 5 flash cards to break your dependence on the text. Use these cards to give yourself practice tests.

Section	Summary	Example
SECTION 11.1	In the **plurality method,** each person votes for his or her favorite candidate and the candidate receiving the most votes is declared the winner.	Example 1, pp. 562–563
	With the **Borda count method,** each voter ranks k candidates. The first choice is given k points, the second choice is given $k - 1$ points, the third is given $k - 2$ points, and so on. The candidate receiving the most points wins the election. A voter uses a **preference ballot** to rank the candidates. A **preference table** summarizes the preference ballots.	Discussion, p. 563 Example 2, pp. 563–564
	In the **plurality-with-elimination method,** each voter votes for one candidate. A candidate who receives a majority of votes is the winner. If no candidate receives a majority of votes, then the candidates with the fewest votes are dropped from the ballot and a new election is held. This process continues until a candidate receives a majority of votes.	Example 3, pp. 566–567
	In the **pairwise comparison method,** voters rank all candidates. We then consider candidates two at a time. Whichever candidate, A or B, is preferred by most voters receives 1 point. If candidates A and B are tied, then each receives $\frac{1}{2}$ point. At the end, the candidate with the most points is the winner.	Example 4, p. 568
	Approval voting allows voters to vote for as many candidates on the ballot as they want. Candidates are not ranked and the candidate who receives the most votes of approval is the winner. In **instant runoff voting,** each voter ranks all candidates. If a voter's top candidate is eliminated, then that voter's vote is given to the second choice on the ballot.	Between the Numbers, p. 569
SECTION 11.2	The **majority criterion** states that if a majority of voters rank a candidate as their first choice, then that candidate wins the election.	Example 1, p. 573
	Condorcet's criterion says that if candidate X can defeat each of the other candidates in a head-to-head vote, then X is the winner of the election.	Example 2, pp. 574–575
	The **independence-of-irrelevant-alternatives criterion** states that if candidate X wins an election, and some nonwinners are removed from the ballot, and then a recount is done, X stills wins the election.	Example 3, pp. 575–576 Example 4, pp. 576–577
	The **monotonicity criterion** says that if X wins an election and in a reelection all voters who change their votes only change their votes to favor X, then X wins the election.	Example 5, pp. 577–579
	Arrow's impossibility theorem proves that there is no perfect voting method.	Statement, p. 580
SECTION 11.3	A **weighted voting system** with n voters is described by a set of numbers of the form [quota: weight of voter 1, weight of voter 2, . . . , weight of voter n]. The **quota** is the number of votes necessary to get a resolution passed. The **weights** are the amounts of votes controlled by voter 1, voter 2, and so on. A **dictator** is a voter such that resolutions can be passed only if—and even if only—the dictator votes for the resolution. A voter who can prevent a motion from passing has **veto power.**	Example 1, pp. 584–585
	A set of voters who vote the same way is called a **coalition.** The sum of the weights of the voters in a coalition is called the **weight of the coalition.** A coalition that has weight greater than or equal to the quota is called a **winning coalition.** A voter in a winning coalition is called **critical** when, if he or she were to leave the coalition, the coalition would no longer be winning.	Definition, p. 585 Example 2, p. 586 Example 3, p. 587

Section	Summary	Example
	In a weighted voting system, a voter's **Banzhaf power index** is defined as $$\frac{\text{the number of times the voter is critical in winning coalitions}}{\text{the total number of times voters are critical in winning coalitions}}.$$ A **blocking coalition** is a set of voters with enough votes to defeat a resolution.	Example 4, p. 587 Example 5, pp. 587–588 Example 6, p. 589 Historical Highlight, p. 590
SECTION 11.4	A **permutation** of the set $\{x_1, x_2, x_3, \ldots, x_n\}$ is an ordering of the elements in a straight line. We denote a permutation as $(x_{i_1}, x_{i_2}, x_{i_3}, \ldots, x_{i_n})$, where x_{i_1} is the first element in the permutation, x_{i_2} is the second element in the permutation, and so on. A set with n elements has $n \times (n-1) \times (n-2) \times \cdots \times 2 \times 1 = n!$ permutations.	Example 1, pp. 593–594 Discussion, p. 594
	As we add voters to a coalition one at a time, the first voter added who makes the coalition a winning coalition is called a **pivotal** voter.	Example 2, pp. 594–595
	In a weighted voting system, the **Shapley-Shubik index** for a voter is $$\frac{\text{the number of times the voter is pivotal in some permutation of the voters}}{\text{the total number of permutations of the voters}}.$$	Example 3, p. 595 Example 5, pp. 596–597

CHAPTER REVIEW EXERCISES

Section 11.1

1. Four candidates running for town council receive votes as follows: Myers, 2,156; Pulaski, 1,462; Harris, 986; Martinez, 428.

a. Is there a candidate who earns a majority?

b. Who wins the election using the plurality method?

2. Use the preference table to determine the winner of the election using the Borda count method.

	Number of Ballots					
Preference	**8**	**5**	**7**	**4**	**3**	**6**
1st	A	D	A	B	B	C
2nd	B	B	C	D	C	A
3rd	C	C	D	C	A	B
4th	D	A	B	A	D	D

3. Members of the chamber of commerce have been asked to vote on their preference for a topic for a speaker for their convention. The choices are (S)ocial justice, the (R)ole of government in a free society, (E)ducation in the future, and (G)lobal issues. Their preferences are summarized in the table. What option is their first choice using the plurality-with-elimination method?

	Number of Ballots				
Preference	**1,531**	**1,102**	**906**	**442**	**375**
1st	G	R	S	S	G
2nd	R	S	G	E	S
3rd	S	G	R	R	E
4th	E	E	E	G	R

4. Using the preference table, who wins the election using the pairwise comparison method?

	Number of Ballots			
Preference	**8**	**4**	**5**	**6**
1st	A	D	B	C
2nd	B	B	D	B
3rd	C	C	C	A
4th	D	A	A	D

Section 11.2

5. Consider the following three preference ballots:

Preference			
1st	R	R	D
2nd	D	D	P
3rd	P	Q	Q
4th	Q	P	R

Who is the winner of this election using the Borda count method? Does this election satisfy the majority criterion? Explain.

6. The Student Horror Organization of the College of Kokomo (SHOCK) is voting for its choice of a classic horror film to be featured as the theme of their spring horror festival. The finalists to be voted on by the members of SHOCK are *The (E)xorcist*, *(A)lien*, *The (N)ight of the Living Dead*, and *The (S)hining*. Use the preference table on the next page to determine the winner using the Borda count method. Is Condorcet's criterion satisfied in this election?

Preference	Number of Ballots				
	6	8	12	1	8
1st	E	S	E	A	N
2nd	A	A	N	S	E
3rd	N	N	S	N	A
4th	S	E	A	E	S

Table for Exercise 6

7. Use the preference table to determine the winner using the Borda count method. Is the independence-of-irrelevant-alternatives criterion satisfied? Explain.

Preference	Number of Ballots					
	11	3	6	9	4	3
1st	C	D	C	B	B	A
2nd	B	B	A	D	A	C
3rd	A	A	D	A	C	B
4th	D	C	B	C	D	D

8. The Wetherholds have narrowed the location of their family reunion down to: (D)ollywood, (C)leveland (to visit the Rock and Roll Hall of Fame), or (B)ridgeville, Delaware, to see the Punkin Chunkin World Championship. Use the plurality-with-elimination method to determine the winner of the election using the preference table. Is the independence-of-irrelevant-alternatives criterion satisfied? Explain.

Preference	Number of Ballots			
	10	7	2	4
1st	D	B	C	C
2nd	C	D	D	B
3rd	B	C	B	D

Section 11.3

9. Find the quota, find the weights of the voters, determine whether there is a dictator, and find those having veto power in the weighted voting system [17 : 1, 5, 7, 8].

10. Write out all the winning coalitions in the voting system [11 : 2, 3, 5, 7].

11. Determine the Banzhaf power index for each voter in the weighted voting system [11 : 2, 3, 5, 7].

12. Determine the Banzhaf power index for each voter in each weighted voting system. Explain why you would intuitively expect the calculations to come out as they did.

a. [10 : 1, 2, 3, 4] **b.** [10 : 11, 1, 3, 3, 2]

13. Committee {A, B, C, D} has chairperson A. For a resolution to be passed, the chairperson and at least one other member of the committee must support the resolution. What is the Banzhaf power index for each person on this committee?

Section 11.4

14. How many permutations are there in a seven-element set?

15. We have completed the first line in the table for the weighted voting system [6 : 4, 3, 2]. Complete the rest of the table.

Sum Weights of Coalition Members	Pivotal Voter
(A, B, C) 4 + 3	B
(A, C, B)	
(B, A, C)	
(B, C, A)	
(C, A, B)	
(C, B, A)	

16. Determine the Shapley-Shubik index of each voter in the voting system [6 : 4, 3, 2].

17. Consider the system [4 : 1, 1, 1, 1, 1, 1], which is an example of a "one person, one vote" situation. What is the Shapley-Shubik index for each person in this system?

18. The voting strength of a three-person committee in the U.S. House of Representatives corresponds to the number of representatives that each of their states has in the House: Minnesota (8), Wisconsin (7), and Delaware (1). Calculate the Shapley-Shubik index for each member of the committee using a quota of 9.

CHAPTER TEST

1. Four candidates running for town council receive votes as follows: Molina, 2,543; Sobieski, 1,532; Wilson, 892; Gambone, 473.

a. Is there a candidate who earns a majority?

b. Who wins the election using the plurality method?

2. The Alliance of Women Scientists took a survey of its membership regarding issues they want addressed. The choices were (R)esearch funding, (E)quality in the workplace, (A)ttracting more women to science, and (Q)uality of life. Their preferences are summarized in the table. What option is their first choice using the plurality-with-elimination method?

Preference	Number of Ballots				
	327	130	149	85	324
1st	E	R	A	E	R
2nd	R	E	Q	R	E
3rd	A	A	E	Q	Q
4th	Q	Q	R	A	A

3. Use the preference table to determine the winner using the Borda count method. Is Condorcet's criterion satisfied in this election? Explain.

	Number of Ballots				
Preference	**1,327**	**1,130**	**849**	**285**	**624**
1st	A	B	C	A	B
2nd	B	A	D	B	A
3rd	C	C	A	D	D
4th	D	D	B	C	C

4. Find the quota, find the weights of the voters, determine if there is a dictator, and find those having veto power in the weighted system [15 : 5, 3, 1, 3, 4, 2].

5. How many permutations are there in a six-element set?

6. Use the given preference table to determine the winner of the election using the Borda count method.

	Number of Ballots					
Preference	**3**	**5**	**8**	**2**	**6**	**5**
1st	A	B	C	A	B	D
2nd	B	A	D	B	A	C
3rd	C	D	A	C	D	B
4th	D	C	B	D	C	A

7. Consider the following three preference ballots. Who is the winner of this election using the Borda method? Does this election satisfy the majority criterion? Explain.

Preference			
1st	A	A	B
2nd	B	B	D
3rd	C	C	C
4th	D	D	A

8. Consider the system [5 : 1, 1, 1, 1, 1, 1, 1, 1], which is an example of a "one person, one vote" situation. What is the Shapley-Shubik index for each person in this system?

9. Use the plurality-with-elimination method to determine the winner of the election. Is the independence-of-irrelevant-alternatives criterion satisfied? Explain.

	Number of Ballots			
Preference	**35**	**71**	**36**	**14**
1st	A	B	C	D
2nd	B	A	D	A
3rd	C	C	A	C
4th	D	D	B	B

10. Determine the Banzhaf power index for each voter in each voting system. Explain why you would intuitively expect the calculations to come out as they did.

a. [15 : 2, 8, 3, 2] **b.** [13 : 15, 2, 4, 1, 3]

11. We have completed the first line in the table for the weighted voting system [7 : 5, 3, 3]. Complete the rest of the table.

Sum Weights of Coalition Members	Pivotal Voter
(A, B, C) 5 + 3	B
(A, C, B)	
(B, A, C)	
(B, C, A)	
(C, A, B)	
(C, B, A)	

12. Use the given preference table to determine who wins the election using the pairwise comparison method.

	Number of Ballots			
Preference	**23**	**47**	**83**	**21**
1st	A	B	D	C
2nd	B	A	C	B
3rd	C	C	A	A
4th	D	D	B	D

13. Use the preference table to determine the winner using the Borda count method. Is the independence-of-irrelevant-alternatives criterion satisfied? Explain.

	Number of Ballots					
Preference	**7**	**5**	**6**	**12**	**16**	**8**
1st	A	B	C	A	B	D
2nd	B	A	D	D	A	C
3rd	C	D	A	B	C	B
4th	D	C	B	C	D	A

14. Write out all the winning coalitions in the voting system [15 : 3, 4, 6, 8].

15. Determine the Shapley-Shubik index of each voter in the voting system [7 : 4, 4, 2].

16. Committee {A, B, C, D} has co-chairpersons A and B. For a resolution to be passed, the two co-chairpersons and at least one other member of the committee must support the resolution. What is the Shapley-Shubik index for each person on this committee?

GROUP PROJECTS

1. **Creating an interesting example.** Construct a preference table for an election with candidates A, B, C, and D such that by using the table
   A wins using the plurality method,
   B wins using the Borda count method,
   C wins using the plurality-with-elimination method,
   D wins using the pairwise comparison method.
2. **Conducting an election.** Choose a topic you think will be of interest to your class and have them use preference ballots to rank their preferences in choosing among four or five alternatives regarding that topic. Use the different voting methods we presented in this chapter to determine the winner. It would be good if, when you present the alternatives, there is not an obvious favorite, so that one alternative might win using one voting method but lose using another. Discuss why you think that one option wins using one method but loses using another.
3. **Voting power in the European Union.** The Council of Ministers of the European Union is an example of a weighted voting system. Make a presentation or write a paper regarding this system, and include the following information.
   a. List the current membership of the European Union
   b. Show the voting weight of each member of the Council of Ministers
   c. Calculate the Banzhaf or Shapley-Shubik measure of the voting power of the members of the council. (You should be able to find Shapley-Shubik calculators online.)
   d. Discuss the concerns expressed by various members of the union with regard to their power. Specifically mention the concerns of older members versus newer members and members from large countries versus members from small countries.

USING TECHNOLOGY

1. **Calculating the Banzhaf index.** Go to http://math.temple.edu/~cow/bpi.html to find a Banzhaf power calculator. (This calculator may calculate the Banzhaf index slightly differently than we do in this chapter.) Use this calculator to calculate the Banzhaf indices for Exercises 29 to 34 in Section 11.3.
2. **Implementing the Borda method.** Find an online calculator that implements the Borda method. Use that calculator to reproduce Example 2 in Section 11.1 and then use it to do Exercises 4 and 8 in Section 11.1.
3. **Using a spreadsheet to implement the Borda method.** Use Excel or Numbers to reproduce the spreadsheet on page 565. Modify the spreadsheet do Exercises 4 and 8 in Section 11.1.
4. **Constructing counterexamples with a spreadsheet.** Use Excel or Numbers to reproduce the spreadsheet on page 578. Modify it and use it to do Exercises 3 and 4 in Section 11.2.

12

Counting

Just How Many Are There?

If someone promised you a guaranteed share in the winnings of a multimillion-dollar lottery, and all it would cost would be $3,000, would you buy into it? Several years ago, I received just such an offer. A friend told me of an Australian-based syndicate that was raising $15 million to play small lotteries in which they would buy tickets to cover every possible combination of numbers. When offered this "chance of a lifetime," my first question was, "How many tickets must we buy to cover all possible ticket combinations?"

After watching a somewhat vague video about the syndicate founder's new *secret, nonconventional* mathematical theory that would guarantee our success, I chose not to invest. However, several months later I was stunned to read in my morning paper:

Australian Syndicate Wins Virginia Lottery

Did I make a mistake? After learning some basic principles of counting you will be better able to answer that question.

(continued)

On a more personal note, sometime, in a moment of desperation, you might be tempted to try to improve your financial position by playing some games of chance. After learning how these games are mathematically slanted against the player, you will understand why this can be a foolish move, and see how casino owners can become wealthy as gamblers lose their money.

12.1

Introduction to Counting Methods

Objectives

1. Count elements in a set systematically.
2. Use tree diagrams to represent counting situations graphically.
3. Use counting techniques to solve applied problems.

As a young child, when you first learned how to count, you probably wanted to count for everyone. I recall driving with my young children, and they would ask, "Do you want to hear me count to 100?" About two minutes later they would finish with, "..., 99, 100." Suppose instead that they offered to count to a familiar number that we hear in the news every day—1 billion. At a rate of one number per second, they would finish almost 32 years later. Obviously, when counting a large number of objects, we need to streamline the process.

In this section, you will begin to learn several techniques to count large sets of items systematically and effectively.

Systematic Counting

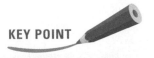

KEY POINT

We can count a set by listing its elements systematically.

One of the simplest ways to count a set is to list its elements.

EXAMPLE 1 Counting Sets by Listing

How many ways can we do each of the following?

a) Flip a coin.
b) Roll a single die (singular of *dice*).
c) Pick a card from a standard deck of cards (see Figure 12.1).
d) Choose a features editor from a five-person newspaper staff.

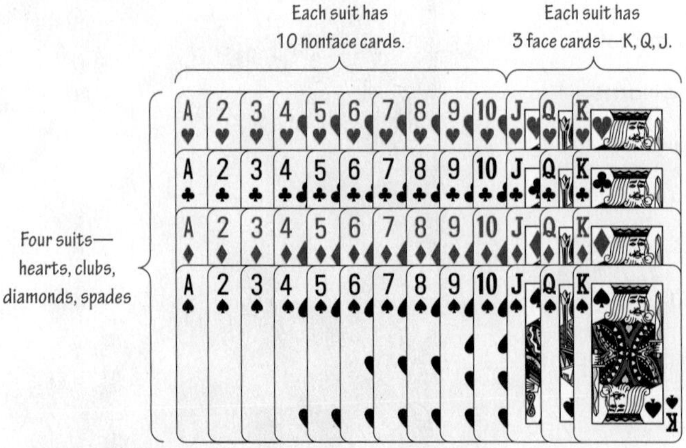

FIGURE 12.1 A standard deck of cards.

SOLUTION:

a) The coin can come up either heads or tails, so there are two ways to flip a coin.

b) The die has six faces numbered 1, 2, 3, 4, 5, and 6, so there are six ways the die can be rolled.

c) There are 52 different ways to choose a card from a standard deck (see Figure 12.1).

d) There are five ways to choose one of the five staff members to be editor.

For more complex sets, it is useful to list the elements systematically, as we suggested in Section 1.1 on problem solving.

EXAMPLE 2 Counting Animal Pairs

An environmental group plans to develop a fund-raising campaign featuring two endangered species. The list of candidates includes the (C)heetah, the (O)tter, the black-footed (F)erret, and the Bengal (T)iger. One animal will appear in a series of TV commercials and a different animal will appear in online advertisements. In how many ways can we choose the two animals for the campaign?

SOLUTION: Review the Be Systematic Strategy on page 5.

So as not to overlook any possibilities, we will list all pairs of animals systematically. We denote the animals by the letters C, O, F, and T. We begin by assuming that C is the animal selected for the TV commercials and then consider each of the other animals for the online campaign. That gives us

CO, CF, CT.

If we next list O as the animal used on TV with each of the other animals appearing online, we get

OC, OF, OT.

Continuing in this fashion, the complete list is

CO, CF, CT
OC, OF, OT
FC, FO, FT
TC, TO, TF.

Thus, there are 12 ways to select animals for the TV and online ads.

Now try Exercises 1 to 4. 1

QUIZ YOURSELF 1

A, B, C, D, E, F, and G are finalists in a downhill ski race. Medals will be awarded for first and second place. In how many different ways can we award these two medals? Do this either by listing all possible pairs or by reasoning abstractly, as we do in the discussion following the solution to Example 2.

Notice in Example 2 that if instead of 4 animals there were 10 animals under consideration, we could count without actually writing down the pairs. Clearly, if C were featured on TV, then there would be nine ways to complete the pair with an animal for the online ads. Similarly, if we used O on TV, there again would be nine ways to then choose an animal for the online campaign. It is easy to see that with each of the 10 choices for the TV animal, we could choose 9 animals to appear online. We would then have $10 \times 9 = 90$ ways to form the desired pairs.

KEY POINT

Tree diagrams help visualize counting situations that take place in stages.

Tree Diagrams

The Three-Way Principle in Section 1.1 recommended viewing situations graphically. It is good to keep this principle in mind when solving counting problems.

EXAMPLE 3 A Tree Diagram Shows How Three Coins Are Flipped

How many ways can three coins be flipped?

SOLUTION: Review the Three-Way Principle on page 13.

To emphasize that the three coins are different, let's assume we are flipping a penny, a nickel, and a dime. A **tree diagram** is a handy way to illustrate the possibilities. First, in Figure 12.2(a) we draw a tree with two branches to show the possibility of a head or a tail for the penny. Next, in Figure 12.2(b), we see that if the penny shows either a head or a tail, the nickel also can show a head or a tail.

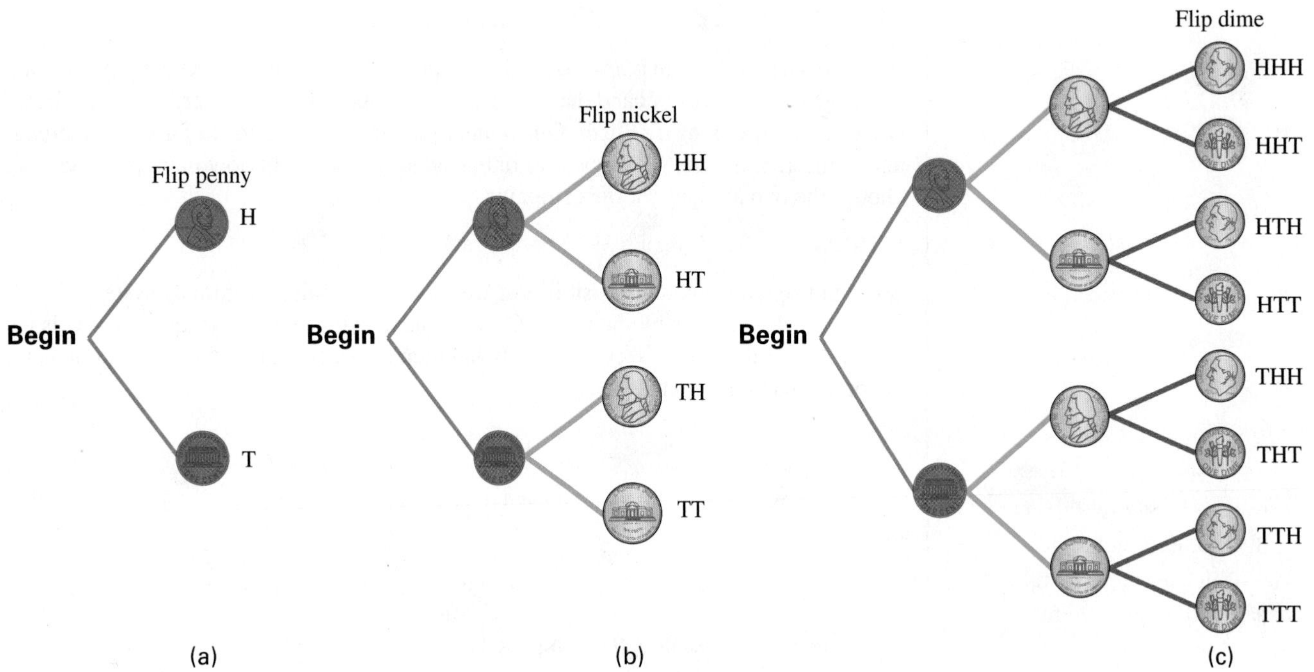

FIGURE 12.2 Tree diagrams representing all possible ways to flip (a) a penny; (b) a penny and then a nickel; (c) a penny, a nickel, and a dime.

QUIZ YOURSELF 2

How many ways are there to flip four coins?

So far there are four ways to flip the two coins. Finally, with each of these four possibilities, the dime can show a head or a tail, as we see in Figure 12.2(c).

Moving through the tree from left to right, we can trace eight branches corresponding to the ways the three coins can be flipped:

HHH, HHT, HTH, HTT, THH, THT, TTH, TTT 2

You will see the situation shown in Example 4 frequently in Chapter 13 when we discuss probability.

EXAMPLE 4 Rolling Two Dice

If we roll two dice, how many different pairs of numbers can appear on the upturned faces?

SOLUTION:

To emphasize that the dice are different, we assume one is red and the other green. Clearly, a red 2 and a green 3 is not the same as a red 3 and a green 2. We will use ordered pairs of the form (red number, green number) to represent the pairs showing on the dice. For example, (4, 5) represents a red 4 and a green 5. Figure 12.3 on page 609 illustrates this situation.

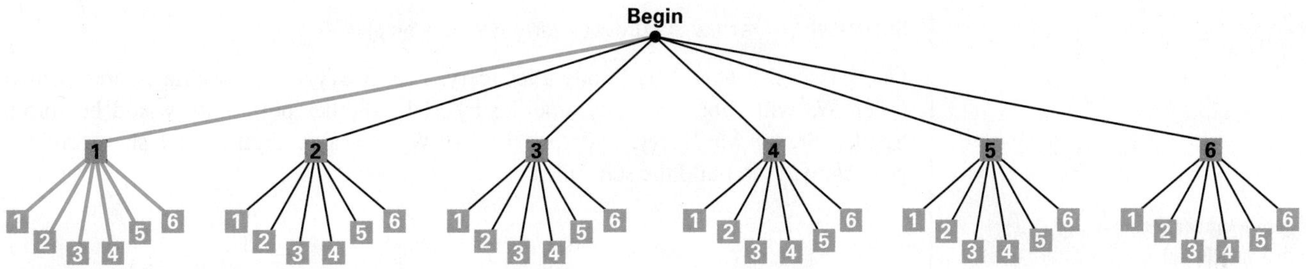

FIGURE 12.3 Tree diagram showing how many ways to roll two dice.

The leftmost set of six branches in Figure 12.3 shows pairs corresponding to 1 on the red die and either a 1, 2, 3, 4, 5, or 6 on the green. These branches correspond to the pairs (1, 1), (1, 2), (1, 3), (1, 4), (1, 5), and (1, 6). Similarly, the second set of six branches corresponds to the pairs (2, 1), (2, 2), (2, 3), (2, 4), (2, 5), and (2, 6). Continuing this pattern, we get the following 36 pairs:

(1, 1),	(1, 2),	(1, 3),	(1, 4),	(1, 5),	(1, 6),
(2, 1),	(2, 2),	(2, 3),	(2, 4),	(2, 5),	(2, 6),
(3, 1),	(3, 2),	(3, 3),	(3, 4),	(3, 5),	(3, 6),
(4, 1),	(4, 2),	(4, 3),	(4, 4),	(4, 5),	(4, 6),
(5, 1),	(5, 2),	(5, 3),	(5, 4),	(5, 5),	(5, 6),
(6, 1),	(6, 2),	(6, 3),	(6, 4),	(6, 5),	(6, 6),

Now try Exercises 5 to 8.

To be able to solve a wide class of counting problems, including the lottery question we posed at the beginning of the chapter, we need to develop some more counting techniques—but first we will introduce some more terminology. In some counting problems, objects can be repeated; in others, they cannot. To open a combination lock, we can use the same number for each turn of the dial. For example, 23-23-23 could be a valid combination. In other situations, such as choosing the pairs of animals in Example 2, we cannot use the same animal twice. If objects are allowed to be used more than once in a counting problem, we will use the phrase **with repetition.** If we do not want objects to be used more than once, we will say **without repetition.**

SOME GOOD ADVICE

A good mathematician often takes a complex situation and breaks it into simpler components to solve a problem. We used this approach in Examples 2, 3, and 4. Frequently in counting problems it is useful to think of a situation not as occurring all at once but rather in distinct stages. Think of a first thing happening, then a second, then a third, and so on. If we count the possibilities at each stage, we then can combine this information to arrive at a final answer.

EXAMPLE 5 Planning a Casino's Advertising Campaign

The programming director at the MGM Grand casino in Las Vegas is designing a *Sizzling Summer Spectacular*, featuring one of three winners of *American Idol*—(S)cotty McCreery, (L)ee DeWyze, or (J)ordin Sparks—each month. Assume the singers are being chosen without repetition. In how many ways can these singers be scheduled?

SOLUTION: Review the Draw Pictures Strategy on page 3.

One possibility would be Scotty McCreery, Lee DeWyze, and Jordin Sparks in that order. We will abbreviate this ordering by SLJ. Another possibility would be Jordin Sparks, Scotty McCreery, and then Lee DeWyze (JSL). Figure 12.4 shows all the possible ways to build the schedules.

QUIZ YOURSELF 3

If we allowed choices for singers to be repeated in Example 5, that is, where schedules such as SSJ and LLL would be possibilities, how many branches would there have been in the tree in Figure 12.4?

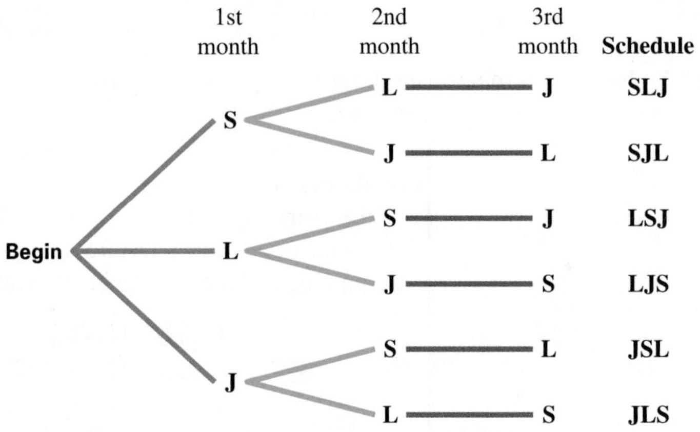

FIGURE 12.4 Tree diagram representing all different ways to schedule three singers (without repetition).

KEY POINT

We can solve counting problems by imagining a tree instead of drawing it.

We see that there are three singers the director can choose for the first month, then two possibilities for the second month, and finally only one singer remains for the third month. So the programming director can make $3 \times 2 \times 1 = 6$ different schedules.

Now try Exercises 9 to 12. 3

QUIZ YOURSELF 4

How many outfits could Amya have in Example 6 if she had four tops, two pairs of pants, and three jackets?

Visualizing Trees

Drawing large tree diagrams is tedious. After you have drawn several trees, you should be able to "see" the tree in your mind without actually drawing it.

EXAMPLE 6 Counting the Ways to Choose an Outfit

Amya has designed a new collection for the final competition in *Project Runway.* She has created five tops, four pairs of pants, and three jackets. If we consider an outfit to be a top, pants, and jacket, how many different outfits can Amya's models wear without repeating the exact same outfit?

SOLUTION:

We can imagine a tree diagram where the leftmost branches represent the choices of tops, the middle branches depict the choices of pants, and the rightmost branches show the choices for jackets.

We begin the tree with five branches, one for each top. Then to each branch we attach four branches representing the pants. So far we would have a tree with 20 branches. Finally, to each of these branches we attach three more representing the jackets, to give us a total of 60 different branches. Therefore, Amya's models can wear 60 different outfits.

Now try Exercises 19 and 20. 4

EXAMPLE 7 Counting in Geometric Figures

Consider the sequence of triangles in the figure, which shows the beginning stages of a geometric construction called Sierpinski's triangle.*

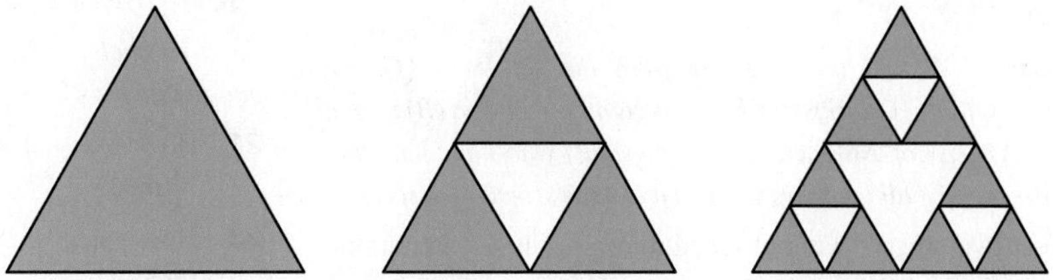

(a) Stage 0 (b) Stage 1 (c) Stage 2

Sierpinski's triangle

We began with an equilateral triangle (stage 0), divided it into four smaller, congruent equilateral triangles, and then removed the center triangle (stage 1). We then repeated this process on the smaller triangles to get stage 2. If we continued this process to stage 10, how many small triangles would we have?

SOLUTION: Review the Look for Patterns Strategy on page 6.

Notice in stage 0 we had one triangle, in stage 1 we had three triangles, and in stage 2 we had 9 triangles. We see that in going from one stage to the next, we increase the number of triangles by a factor of three.

We can continue this pattern in the following table:

Stage	0	1	2	3	4	5	6	7	8	9	10
Number of Triangles	1	3	9	27	81	243	729	2,187	6,561	19,683	59,049

So, we see that at stage 10, we would have 59,049 triangles.

Now try Exercises 35 and 36.

Math in Your Life

Can We Use Mathematics to Defeat Terrorism?

"It all started when I saw the movie *A Beautiful Mind*," says Jonathan Farley, a visiting professor at the Massachusetts Institute of Technology. He was struck by the line, "Mathematicians won the war," referring to how the research of John Nash was applied successfully to implement Cold War military strategy. Farley believes that we can use mathematics to reveal the organization of terrorist cells and identify unknown terrorists. Farley's company, called Phoenix Mathematical Systems Modeling, offers advice on the mathematics of terrorism and other applications of mathematics to national defense.

Gary Nelson, a senior researcher at the Homeland Security Institute, a quasigovernment institute, agrees that new mathematical methods are necessary to help security agencies deal with enormous amounts of unorganized data. At the University of Southern California, Jafar Adibi is developing computer programs, using a process called *data mining*, to sift through large databases of information. His programs apply mathematical techniques to search records regarding phone calls, places of worship, political affiliation, and family relations to find hidden links between known terrorists and their unknown confederates.

Jonathan Farley

*Sierpinski's triangle is an example of a fractal. For more on fractals, see Section 9.7. You can also find many interactive applets online that will show you the construction of the stages of Sierpinski's triangle.

EXERCISES 12.1

Sharpening Your Skills

In Exercises 1–4, you are selecting from the set W = {*Carrie (U)nderwood, Kelly (C)larkson, Chris (D)aughtry, Fantasia (B)arrino, and Clay (A)iken*} *of* American Idol *contestants who have had successful careers. List all the ways you can make the selection described.*

1. Select two singers without repetition; order is not important. For example, BB is not allowed and UC is the same selection as CU.

2. Select two singers with repetition; order is not important. For example, DD is allowed and BA is the same as AB.

3. Select two singers without repetition; order is important. For example, CD is not the same as DC.

4. Select two singers with repetition; order is important. For example, UU is allowed and DB is not the same as BD.

Draw a tree diagram that illustrates the different ways to flip a penny, nickel, dime, and quarter. Use this diagram to solve Exercises 5–8.

5. In how many ways can you get exactly one head?

6. In how many ways can you get no tails?

7. In how many ways can you get exactly two tails?

8. In how many ways can you get exactly three heads?

9. How many different two-digit numbers can you form using the digits 1, 2, 5, 7, 8, and 9 without repetition? For example, 55 is not allowed.

10. How many different two-digit numbers can you form using the digits 1, 2, 5, 7, 8, and 9 with repetition? For example, 55 is allowed.

11. How many different three-digit numbers can you form using the digits 1, 2, 5, 7, 8, and 9 without repetition?

12. How many different three-digit numbers can you form using the digits 1, 2, 5, 7, 8, and 9 with repetition?

In Exercises 13–18, assume you are rolling two dice: the first one is red and the second one is green. Use a systematic listing to determine the number of ways you can roll each of the following. For example, a total of 3 *can be rolled in two ways:* (1, 2) *and* (2, 1).

13. Roll a total of 5.

14. Roll a total of 7.

15. Roll both numbers the same.

16. Roll a 3 on the red die.

17. Roll a total less than 6.

18. Roll a total greater than 9.

Recall in Example 6 that Amya has designed different tops, pants, and jackets to create outfits for Project Runway. *How many different outfits can her models wear if she has designed the following:*

19. Counting outfits. Six tops, five pants, four jackets

20. Counting outfits. Seven tops, six pants, three jackets

Use the given diagram to solve Exercises 21 and 22.

Squares such as ABCD *and* EFGH *are called* 1×1 *squares. A square such as* CXGY *is called a* 2×2 *square.*

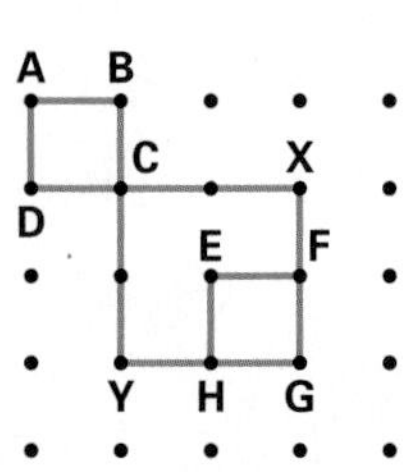

21. How many 2×2 squares can be formed?

22. How many 3×3 squares can be formed?

23. Answer Exercise 21 assuming the diagram has six rows with six dots in each row.

24. Answer Exercise 22 assuming the diagram has six rows with six dots in each row.

Applying What You've Learned

25. Assigning tasks. Angela's coworkers Pam, Phyllis, Jim, and Dwight have volunteered to help with the preparations for a party. How many ways can Angela assign someone to buy beverages, someone to arrange for food, and someone to send out invitations? Assume that no person does two jobs.

26. Making staff assignments. Suppose that the staff of a weekly newspaper, the *Southern California Sentinel*, consists of (A)drian, (B)rian, (C)armen, (D)avid, and (E)mily. The editor will choose a features editor and a sports editor from these five people. If the sports editor must be different from the features editor, in how many ways can the editor make this selection?

In Exercises 27 and 28, draw the tree diagram only if you must. Try to picture the tree mentally without actually putting it on paper.

27. If you drew a tree diagram showing how many ways five coins could be flipped, how many branches would it have?

28. If you drew a tree diagram showing how many ways six coins could be flipped, how many branches would it have?

29. The role-playing game *Dungeons and Dragons* uses a tetrahedral die that has four congruent triangular sides numbered 1, 2, 3, and 4. How many branches would a tree diagram have that shows the way two tetrahedral dice could be rolled?

30. *Dungeons and Dragons* also uses 12-sided dice. How many branches would a tree diagram have that shows the way two 12-sided dice could be rolled?

31. Counting license plates. An eyewitness to a crime said that the license plate of the getaway car began with the letters T, X, and L, but he could not remember the order. The rest of the plate had the numbers 8, 3, and 4, but again he did

not recall the order. How many license plates must the police investigate to find this car?

32. Counting license plates. In a small state, the license plate for a car begins with two letters, which may be repeated, and ends with three digits, which also may be repeated. How many license plates are possible in that state?

In Exercises 33 and 34, you are taking a true or false quiz.

33. Assume that you are taking a five-question true or false quiz.

a. How many different ways are there to answer all five questions?

b. How many ways are there to get no questions wrong? One question? Two questions?

c. If, not having studied, you simply guess at each answer, what are your chances of getting three or more correct?

34. Assume that you are taking a 10-question true or false quiz.

a. How many different ways are there to answer all 10 questions?

b. How many ways are there to get no questions wrong? One question? Two questions? Three questions?

c. If, not having studied, you simply guess at each answer, what are your chances of getting at least seven correct?

For Exercises 35 and 36, use the figures below. They show stages 0, 1, and 2 of the construction of a figure called Sierpinski's carpet (see Example 7). Note that to go from one stage to the next, we remove the center square from each square in the figure.

Stage 0 Stage 1 Stage 2

35. Sierpinski's carpet. How many shaded squares will there be in the carpet in stage 3?

36. Sierpinski's carpet. How many shaded squares will there be in the carpet in stage 5?

Prince William and Duchess Kate are attending the award-winning play War Horse *with their guests David Beckham and Elton John. They will be sitting in four adjacent seats. Use this information in Exercises 37 to 40. You may wish to draw tree diagrams or seating diagrams to better understand how to find your answers.*

37. Arranging seating in a theater. In how many ways can the four people be seated?

38. Arranging seating in a theater. In how many ways can the four be seated if the royal couple sits next to each other?

39. Arranging seating in a theater. In how many ways can the four people be seated if the royal couple sits in the center two seats?

40. Arranging seating in a theater. In how many ways can the four be seated if Elton John sits between William and Kate?

41. Stacking cans. In preparation for Thanksgiving Day, the Save-You-More Store has stacked cans of pumpkin pie filling in a triangular pyramid. Shown is a 3-by-3 row of cans. The top of the pyramid is a 1-by-1 row (only one can). The second row is a 2-by-2 row (three cans), the third row is a 3-by-3 row, and so on. If the pyramid has 12 rows, how many cans are there in the pyramid?

42. Stacking cans. Continuing Exercise 41, if the Save-You-More Store has 400 cans of cherry pie filling and wishes to build three triangular pyramids of equal height, how many rows will the pyramids have and how many cans will be left over?

In Exercises 43–46, you are buying a triple-deck ice cream cone with vanilla, strawberry, and chocolate as possible flavors. The flavors can be repeated or not, and we will consider two cones to be different if the flavors are the same but occur in different order.

43. How many different cones are possible?

44. How many cones have only vanilla and strawberry?

45. How many possible cones have only two different flavored scoops?

46. How many possible cones have all different flavors?

47. The base of a stack of oranges in a supermarket consists of five rows with five oranges in each row. The oranges in the next row will be placed as shown by the highlighted orange in the figure. How many oranges will there be in the stack?

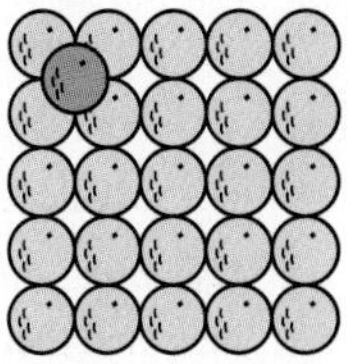

48. If the bottom row of oranges consisted of seven rows of seven oranges each, how many oranges would there be in the stack?

49. Finding patterns in a geometric figure. Consider the following diagram:

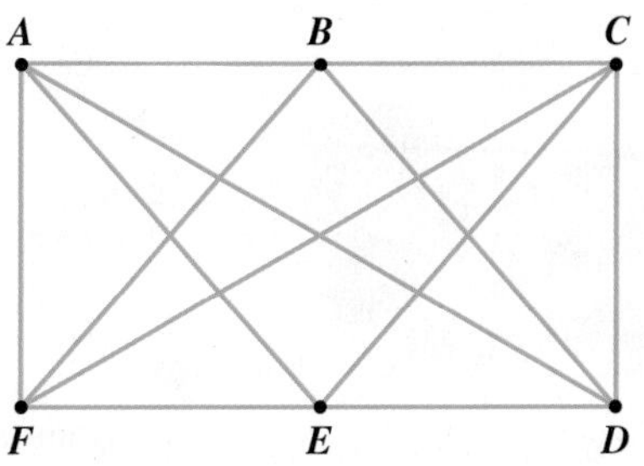

Using only the given lines joining the labeled points, how many triangles are there in this figure? (*Hint:* List all triples of the form *ABC*, *ABD*, *ABE*, and so on, and see which ones could be the vertices of triangles.)

50. Finding patterns in a geometric figure. Repeat Exercise 49 for the following figure. (*Hint:* If you start listing triples as you did in Exercise 49 and begin to see a pattern, you may not have to list all triples to find your answer.)

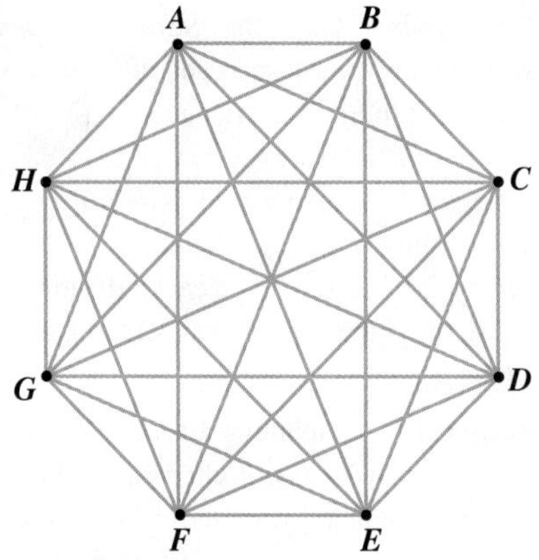

Communicating Mathematics

51. Mention three problem-solving strategies that are helpful in solving counting problems.

52. How are tree diagrams used in solving counting problems?

Challenge Yourself

53. Building class schedules. Anthony is building his class schedule for next semester. He has decided to take four classes: mathematics, English, sociology, and music. He intends to schedule classes only on Monday, Wednesday, and Friday mornings. The possible classes that he can take are

Mathematics:	MWF—9:00, 11:00, 12:00
English:	MWF—9:00, 10:00, 12:00
Sociology:	MWF—10:00, 11:00, 12:00
Music:	MWF—9:00, 10:00, 11:00

Use a tree diagram to determine the possible schedules he can have and then list all those schedules.

54. Building appointment schedules. A pharmaceutical salesperson wants to schedule appointments with three doctors to introduce a new line of antibiotics. The times that the doctors have available are as follows:

Dr. House: Monday at 10:00, Wednesday at 11:00, Thursday at 11:00

Dr. Cuddy: Monday at 9:00, Monday at 10:00

Dr. Wilson: Monday at 10:00, Wednesday at 11:00, Thursday at 11:00

Use a tree diagram to determine all possible schedules that would allow the salesperson to meet with each doctor. List those schedules.

Assume that you are a contestant on a game show called Wheel of Destiny. *The final round of this game, called the* Gem Round, *is played as follows. There are four treasure chests, each containing a gem. Two of the gems are red, one is blue, and one is green. The game show host will ask you to open the chests one at a time in any order that you choose. If the fourth chest you open contains* Your Final Gem, *then you win the grand prize.*

You must spin the Wheel of Destiny *to learn what your final gem will be. For example, when you spin the wheel, it may stop at "a green gem." This means that the round ends whenever you open a box containing the green gem.*

In Exercises 55–58, you are given the final gem. List the different orders in which you can select the colors red, blue, and green in order to win the grand prize. Drawing a tree will help you see the possibilities.

55. Enumerating game show possibilities. A green gem.

56. Enumerating game show possibilities. A blue gem.

57. Enumerating game show possibilities. A red gem.

58. If you were to draw tree diagrams to solve Exercises 55–57, how would they be different from the trees in Examples 3 and 4?

59. The figure in Exercise 50 was a regular octagon (eight-sided figure). If we had used instead a 10-sided figure with vertices labeled $A, B, C, \ldots, J$, how many triangles would have been formed? (One way to do this is to consider how many new triangles you could get if you added a ninth vertex, I, to the vertices in Exercise 50. Next find how many additional triangles would be formed if you added a tenth vertex, J, to the nine.)

60. In general, if the figures in Exercises 50 and 59 had n sides, verify for several cases that the number of triangles would be $\frac{n(n-1)(n-2)}{6}$.

12.2 The Fundamental Counting Principle

Objectives

1. Understand the fundamental counting principle.
2. Use slot diagrams to organize information in counting problems.
3. Know how to solve counting problems with special conditions.

While carrying a demanding course load, working at a part-time job, and staying up late to hang out with your friends, you have been neglecting good health habits. With no time for

exercise, snacking during late-night cram sessions, and substituting burgers and fries for Mom's home cooking, you have succumbed to the dreaded "Freshman Fifteen."* Maybe it's time to start an exercise program.

Assume that you have decided to work out your abs, arms, and cardiovascular system *in that order*, and your fitness center has six machines for the abs, four for arms, and eight for cardio. In how many different ways can you vary your workouts using different machines?

It is useful to relate a new problem such as this to problems that you have seen before. Whether flipping coins, rolling dice, or selecting clothes for a fashion show, you have seen the same basic pattern:

A first thing happens in *a* number of ways,
then a second thing happens in *b* number of ways,
then a third thing happens in *c* number of ways,
and so on.

We will solve this problem in Example 1.

EXAMPLE 1 Counting Workout Routines

If we have six ab machines, four arm machines, and eight cardio machines, how many different workouts can we have if we use one machine from each group?

SOLUTION:

You might start to draw a tree diagram in the following three steps:

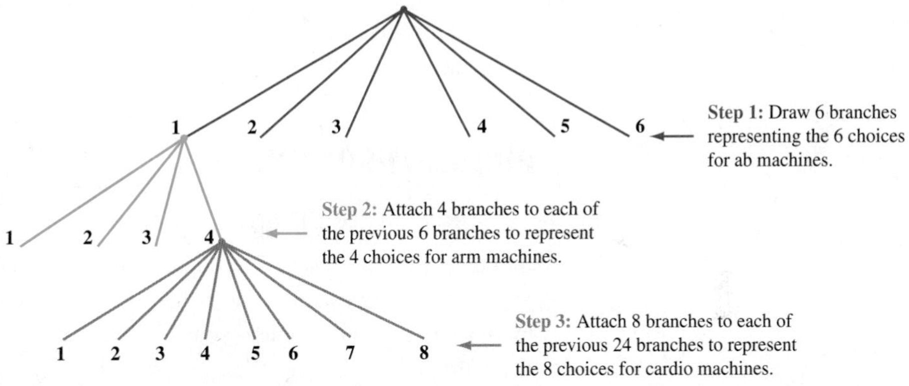
1
2
3
4
5
6
Step 1: Draw 6 branches representing the 6 choices for ab machines.
1
2
3
4
Step 2: Attach 4 branches to each of the previous 6 branches to represent the 4 choices for arm machines.
1
2
3
4
5
6
7
8
Step 3: Attach 8 branches to each of the previous 24 branches to represent the 8 choices for cardio machines.

We see that at first we have six red branches; then, after attaching four blue branches to each red branch, we have $6 \times 4 = 24$ blue branches. Finally, attaching 8 green branches to each of the 24 blue branches, we get $6 \times 4 \times 8 = 192$ total branches, representing the number of different possible workouts.

Notice it was not necessary to draw all 192 branches in Example 1. In fact, if you have a good imagination, you need not draw the tree at all, but only picture it in your mind.

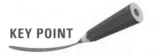

KEY POINT

The fundamental counting principle solves problems without listing elements or drawing tree diagrams.

The Fundamental Counting Principle

Continuing the above discussion, if you wanted to add a fourth or fifth type of exercise to your workout, we would add a fourth collection of branches and then a fifth collection of branches to the tree and so on. This thinking leads us to the following principle.

*Most associate the term "Freshman Fifteen" with the gain of 15 pounds by college freshmen during their first year at college.

THE FUNDAMENTAL COUNTING PRINCIPLE (FCP) If we want to perform a series of tasks and the first task can be done in *a* ways, the second can be done in *b* ways, the third can be done in *c* ways, and so on, then all the tasks can be done in $a \times b \times c \times \cdots$ ways.

We will revisit several situations from Section 12.1 to illustrate the power of the fundamental counting principle.

EXAMPLE 2 Using the FCP to Count Animal Pairs

Recall that an environmental group plans to launch an advertising campaign featuring two endangered species from among the following: the (C)heetah, the (O)tter, the black-footed (F)erret, and the Bengal (T)iger. One animal will appear in a series of TV commercials and a different animal will be featured in online advertisements. In how many ways can this be done?

SOLUTION:

We first have to choose the animal for the TV campaign, which can be done in four ways. The second task is to choose a *different* animal for the online ads. There are three ways to perform this task, so the total number of ways to choose both animals is

$$4 \times 3 = 12.$$

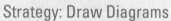

four ways to choose an animal for TV . . . . . . times three ways to choose a different animal for online ads

Now try Exercises 1 to 8.

PROBLEM SOLVING

Strategy: Draw Diagrams

If you get confused when using counting formulas as to when to multiply and when to add, you may find that a picture often helps you recall how we derived the formula. For example, by imagining a tree diagram, you will find the fundamental counting principle easier to remember.

EXAMPLE 3 Using the FCP in Coin and Dice Problems

a) How many ways can four coins be flipped?

b) How many ways can three dice (red, green, blue) be rolled?

SOLUTION:

a) The first task, flipping the first coin, can be done in two ways. The second, third, and fourth tasks also each consist of flipping a coin in one of two ways. Therefore, by the fundamental counting principle, the four coins can be flipped in

$$2 \times 2 \times 2 \times 2 = 16 \text{ ways.}$$

four tasks—each done in two ways

b) The first task is rolling the red die, which can be done in six ways. The second and third tasks also each can be done in six ways. Thus, the three dice can be rolled in $6 \times 6 \times 6 = 216$ ways.

Now try Exercises 9 to 12.

QUIZ YOURSELF 5

Assume that you are shopping for a new car. There are two different models to choose from, there are eight different colors available, each model has three different interior packages, and each model comes with either the plain or sport exterior trim package. How many different cars do you have to choose from?

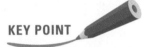

EXAMPLE 4 Counting Outfits Using the FCP

Matthew is working as a summer intern for a TV station and wants to vary his outfit by wearing different combinations of coats, pants, shirts, and ties. If he has three sports coats, five pairs of pants, seven shirts, and four ties, how many different ways can he select an outfit consisting of a coat, pants, shirt, and tie?

SOLUTION:

In selecting an outfit, we consider the following tasks:

Task	Number of Ways to Perform Task
Select coat	3
Select pants	5
Select shirt	7
Select tie	4

By the fundamental counting principle, Matthew has

$$3 \times 5 \times 7 \times 4 = 420 \text{ outfits.}$$

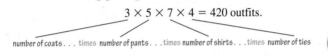

number of coats . . . times number of pants . . . times number of shirts . . . times number of ties

5

Slot Diagrams

Sometimes special conditions affect the number of ways we can perform the various tasks. A useful technique for solving problems such as these is to draw a series of blank spaces, as shown in Figure 12.5, to keep track of the number of ways to do each task. We will call such a figure a **slot diagram.**

KEY POINT

Slot diagrams help organize information before applying the fundamental counting principle.

1st task		2nd task		3rd task		4th task		5th task
Number of ways	×	Number of ways	×	Number of ways	×	Number of ways	×	Number of ways

FIGURE 12.5 A slot diagram helps organize counting problems.

EXAMPLE 5 Counting Keypad Patterns

To open your locker at the fitness center, you must enter five digits in order from the set 0, 1, 2, . . ., 9. How many different keypad patterns are possible if

a) any digits can be used in any position and repetition of digits is allowed?
b) the digit 0 cannot be used as the first digit, but otherwise any digit can be used in any position and repetition is allowed?
c) any digits can be used in any position, but repetition is not allowed?

SOLUTION: Review the Be Systematic Strategy on page 5.

a) We can use any of the 10 digits for each of the five tasks, as shown in the slot diagram in Figure 12.6. Thus, we see that there are $10 \times 10 \times 10 \times 10 \times 10 = 100{,}000$ possible keypad patterns.

1st task		2nd task		3rd task		4th task		5th task
10	×	10	×	10	×	10	×	10
Use any digit		Use any digit		Use any digit		Use any digit		Use any digit

FIGURE 12.6 The slot diagram shows five tasks, each of which can be done in 10 ways.

b) In this situation, we have only nine possible ways to select the first digit. The slot diagram is shown in Figure 12.7.

1st task		2nd task		3rd task		4th task		5th task
9	×	10	×	10	×	10	×	10
Can't use 0		Use any digit		Use any digit		Use any digit		Use any digit

FIGURE 12.7 The first task can be done in 9 ways; the remaining tasks each can be done in 10 ways.

Therefore, there are $9 \times 10 \times 10 \times 10 \times 10 = 90{,}000$ possible keypad patterns.

c) We can use any of the 10 digits for the first number. However, because repetitions are not allowed, we have only nine digits for the second number. After two numbers have been chosen, there are only eight possibilities for the third number. Similarly, there are seven for the fourth number and six for the fifth number, as we see in Figure 12.8.

1st task		2nd task		3rd task		4th task		5th task
10	×	9	×	8	×	7	×	6
Use any digit		Can't repeat digit		Can't repeat digits		Can't repeat digits		Can't repeat digits

FIGURE 12.8 Because repetition of digits is not allowed, there are fewer possibilities for each keypad combination.

So we see that there are $10 \times 9 \times 8 \times 7 \times 6 = 30{,}240$ possible keypad patterns.

Now try Exercises 13 to 16.

Handling Special Conditions

KEY POINT

In solving counting problems, consider special conditions first.

We will now look at a counting situation requiring a slightly more complex analysis than we needed in the previous examples.

EXAMPLE 6 Counting Seating Patterns with Special Conditions

Professor Bartlett teaches an advanced cognitive psychology class of 10 students. She has a visually challenged student, Louise, who must sit in the front row next to her tutor, who is also a member of the class. If there are six chairs in the first row of her classroom, how many different ways can Professor Bartlett assign students to sit in the first row?

SOLUTION:

In order to use the fundamental counting principle, we must identify the separate tasks in determining the seating.

We will first consider the special condition that Louise and her tutor must sit together.

Task 1: Decide which two seats Louise and her tutor will occupy.

Task 2: Decide where Louise and her tutor will sit in these two seats.

Task 3: Determine who sits in the remaining seats.

Task 1: As Figure 12.9 shows, there are five ways that Louise and her tutor can sit together.

<table>
<tr><th>Seat 1</th><th>Seat 2</th><th>Seat 3</th><th>Seat 4</th><th>Seat 5</th><th>Seat 6</th></tr>
<tr><td colspan="2">L and T</td><td>X</td><td>X</td><td>X</td><td>X</td></tr>
<tr><td>X</td><td colspan="2">L and T</td><td>X</td><td>X</td><td>X</td></tr>
<tr><td>X</td><td>X</td><td colspan="2">L and T</td><td>X</td><td>X</td></tr>
<tr><td>X</td><td>X</td><td>X</td><td colspan="2">L and T</td><td>X</td></tr>
<tr><td>X</td><td>X</td><td>X</td><td>X</td><td colspan="2">L and T</td></tr>
</table>

FIGURE 12.9 There are five ways to choose two seats for Louise and her tutor.

Task 2: Once we have determined which seats L and T occupy, there are two ways that L and T can sit within those seats—L sits either on the right or the left.

Task 3: After Louise and her tutor have been seated, we fill the remaining four seats from left to right. There are eight students left; thus, we have eight students for the first remaining seat, seven for the second seat, six for the third seat, and finally five for the last seat.

Therefore, the number of ways to seat the students in the first row is

$$5 \times 2 \times 8 \times 7 \times 6 \times 5 = 16{,}800.$$

five ways to choose the two seats for Louise and her tutor (5); two ways to seat Louise and the tutor in the two seats (2); number of ways to seat four of eight students in the remaining four seats ($8 \times 7 \times 6 \times 5$)

Now try Exercises 33 to 38. 6

QUIZ YOURSELF 6

Redo Example 6, but now assume that the class has 12 students and the first row has eight seats.

Math in Your Life

How Can Counting Make You Rich?

Most games of chance are based on mathematical principles that slightly favor casinos, allowing them to earn billions of dollars from customers willing to be exploited by playing casino games. Blackjack, however, is one of the casino table games that can be legally beaten by a skilled player. By a method called *card counting*, a person, or team of individuals, can recognize when the remaining cards to be played will favor the player, rather than the house.

In the 1990s, a team of students in the Boston area, called the MIT blackjack team,* who were highly skilled in mathematics and computer science, developed a devastatingly effective card-counting scheme that earned them over $10 million. As a result of their stunning success, these players were banned from playing blackjack at many casinos.

In order to recognize card counters and prevent them from playing, casinos have embraced cutting-edge techniques such as data mining, facial recognition software, and iris scan technology. James X. Dempsey, from the Center for Democracy and Technology, says that Las Vegas is the incubator for a host of surveillance technologies that are now used by the government to identify terrorists and by malls and amusement parks to enhance security.†

*You might find it interesting to research the MIT blackjack team on the Internet, read about their exploits in the book *Bringing Down the House* by Ben Mezrich (Free Press, 2002), or see the 2008 movie *21*.

†See *From Casinos to Counterterrorism*, at www.washingtonpost.com, October 22, 2007.

We will use the fundamental counting principle in Section 12.3 to develop further counting tools that will enable us to answer the lottery question we posed at the beginning of the chapter.

EXERCISES 12.2

Sharpening Your Skills

1. **Assigning responsibilities.** The board of an Internet start-up company has seven members. If one person is to be in charge of marketing and another in charge of research, in how many ways can these two positions be filled?
2. **Assigning positions.** If there are 12 members on the staff of the *Sandpiper*, a newspaper covering shore news, in how many ways can we choose an entertainment editor and a sales manager?
3. **Assigning officers.** The Equestrian Club has eight members. If the club wants to select a president, vice president, and treasurer (all of whom must be different), in how many ways can this be done?

4. **Assigning officers.** If the Chamber of Commerce has 20 members, in how many ways can they elect a different president, vice president, and treasurer?
5. **Building a home theater system.** Elaine is building a home theater system consisting of a tuner, an optical disc player, speakers, and a high-definition TV. If she can select from four tuners, eight speakers, three optical disc players, and five TVs, in how many ways can she configure her system?
6. **Counting schedules.** Jorge is using his educational benefits from the Navy to enroll in four courses to prepare him for a career as an architect: design, science, math, and social science. If there are seven design courses, five science courses, four math courses, and six social science courses that fit his schedule, in how many ways can he select his courses?
7. **Counting meal possibilities.** The early bird special at TGI Friday's features an appetizer, soup or salad, entrée, and dessert. If there are five appetizers, six choices for soup or salad, 13 entrées, and four desserts, how many different meals are possible? (We assume that you make a selection from each category.)
8. **Counting meal possibilities.** What would your answer to Exercise 7 be if you are allowed to skip some categories? For example, you may choose to skip the appetizer and dessert.

In games such as Dungeons and Dragons, *in addition to the common 6-sided dice, dice having 4, 8, 12, or 20 sides are also used, as shown in the figure.*

In Exercises 9–12, determine the number of possibilities for each situation.

9. The 8-sided die is rolled twice.
10. The 12-sided die is rolled twice.
11. The 8-sided die is rolled and then the 12-sided die is rolled.
12. The 20-sided die is rolled and then the 6-sided die is rolled.

In Exercises 13–16, using the digits 0, 1, 2, . . . , 8, 9, determine how many of each type of four-digit number can be constructed.

13. Zero cannot be used for the first digit; digits may be repeated.
14. The number must begin and end with an odd digit; digits may not be repeated.
15. The number must be odd and greater than 5,000; digits may be repeated.
16. The number must be between 5,001 and 8,000; digits may not be repeated.
17. **A true–false quiz.** In how many ways can you choose answers to an eight-question true–false quiz?
18. **A multiple-choice test.** In how many ways can you answer a ten-question multiple-choice test in which each question has four choices?
19. **Planning a trip.** Center City Community College's debate team is participating in a tournament in Albuquerque, New Mexico. They will fly from Boston to Chicago, then to Dallas, and finally to Albuquerque. If there are three possible flights to Chicago, four from Chicago to Dallas, and finally six from Dallas to Albuquerque, in how many ways can they schedule the trip?
20. **A piano competition.** In the Van Cliburn piano competition, the contestants must choose three pieces to perform: one of five from the classical period, one of seven from the romantic period, and one of eight from the modern period. In how many ways can a contestant choose pieces for his or her program?

Applying What You've Learned

21. Enumerating call letters. Radio stations in the United States that are east of the Mississippi River have call letters consisting of a W followed by three letters (W can be used again). How many different call letters of this type are possible?

22. Enumerating zip codes. Postal zip codes currently consist of five digits. For example, the zip code for Honolulu, Hawaii, is 96820. If we used letters instead of digits, how many letters would be required to make as many zip codes as five digits?

23. Counting musical patterns. In the early part of the twentieth century, the Austrian composers Arnold Schoenberg, Alban Berg, and Anton Webern created a system of music called *atonal music*, in which a note cannot be repeated in a composition until all other notes are used. In atonal music, a melody, called a *tone row*, consists of a sequence of 12 different notes. How many different tone rows are possible in this system?

24. Counting license plates. In a certain state, license plates currently consist of two letters followed by three digits. How many such license plates are possible? If the state department of transportation decides to change the plates to have three letters followed by two digits, how many plates will now be possible?

A pin tumbler lock has a series of pins, each of which must be positioned at the proper height to allow the lock to open. The security of the lock is determined by the number of pins in the lock, as well as the number of different possible lengths for the pins. Inserting the proper key in the lock aligns the pins properly and allows the interior of the lock to be turned.

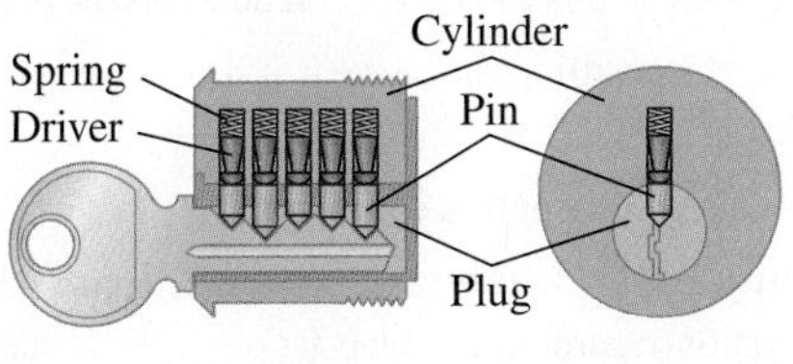

25. Counting keys. If a pin tumbler lock has five pins, each of three different lengths, how many different keys would you have to try in order to guarantee that you could open the lock?

26. Counting keys. If a pin tumbler lock has seven pins, each of four different lengths, how many different keys would you have to try to guarantee that you could open the lock?

27. Counting routes for an armored van. A Wells Fargo armored van must travel from the (D)iamond House, make a pickup at (J)avier's Jewelry, Inc, then stop at (E)mily's Emeralds, and finally go to the (B)ank. Use the following street map to answer the questions.

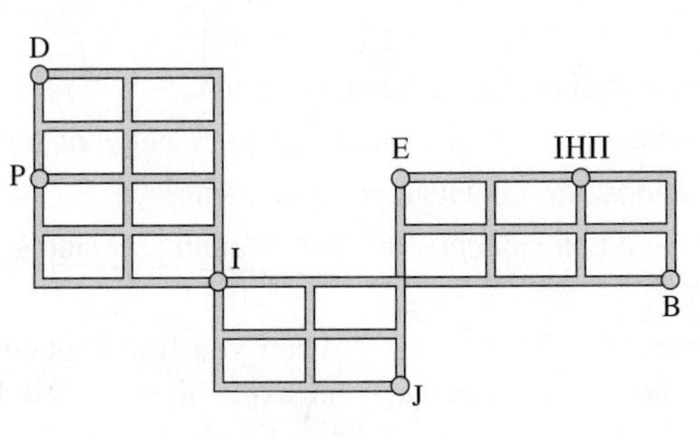

a. How many direct routes are there from D to I?

b. How many direct routes are there from I to J?

c. How many direct routes are there from J to E?

d. How many direct routes are there from E to B?

e. How many direct routes are there from D to B, passing through J and E?

28. Delivering pizzas. Refer to the map in Exercise 27. Carmen is driving from Papa's Pizza at location P, stopping at J, and then making a delivery at the IHP fraternity shown on the map. How many direct routes are possible?

29. Counting settings on an electronic device.

a. A computer interface for a Kawai digital studio piano has eight microswitches that can be set in either the "on" or "off" position. These switches must be set properly for the interface to work. In how many different ways can this group of switches be set?

b. If it takes 2 minutes to set the switches and test to see if the interface is working properly, what is the longest possible time that it would take to find the proper settings by trial and error?

30. Choosing singers. The college musical director has chosen to perform Faure's Après un Rêve as a vocal trio of a soprano, an alto, and a tenor. If she can choose from seven sopranos, three altos, and four tenors, in how many ways can she make her selection?

31. Counting pizza combinations. Mamma's Pizza advertises a special in which you can choose a thin crust, thick crust, or cheese crust pizza with any combination of different toppings. The ad says that there are almost 200 different ways that you can order the pizza. What is the smallest number of toppings available? Explain your thinking by referring to the fundamental counting principle.

32. Counting sandwich combinations. The Underground Railway Sandwich Shop has a sandwich special. You can choose whole wheat, rye, or an onion roll, with mayo or mustard, and lettuce or tomato, and a choice of one meat. What is the minimum number of meats available if there are at least 300 possible sandwich combinations? In your counting, don't forget the possibility of "both" or "neither" in the mayo–mustard choice and the lettuce–tomato choice. Explain your thinking by referring to the fundamental counting principle.

Exercises 33 and 34 are alternative versions of the question posed in Example 6 about the seating of Louise and her tutor in Professor Bartlett's class.

33. Counting seating arrangements. Assume that there are 12 students in the class, the front row has eight chairs, and the tutor, who is also a member of the class, must be seated next to Louise.

a. How many ways can we choose the seats for Louise and her tutor?

b. How many ways can we seat Louise and her tutor in those two seats?

c. How many ways can we fill the remaining six seats?

d. How many total ways can we seat students in the first row?

34. Counting seating arrangements. Assume that there are 10 students in the class, the front row has six chairs, and Louise will sit between her tutor and her friend Jonathan. It does not matter on which side of Louise the tutor sits.

Assume that we wish to seat three men and three women—Alex, Bonnie, Carl, Daria, Edith, and Frank—in a row of six chairs. Answer Exercises 35–38 using the fundamental counting principle. To apply this principle, first identify the separate tasks involved in making the seating arrangements.

35. Counting seating arrangements. In how many ways can we seat these people with no restrictions?

36. Counting seating arrangements. In how many ways can we seat these people if men and women must alternate?

37. Counting seating arrangements. In how many ways can we seat these people if all the men must sit together and all the women must sit together?

38. Counting seating arrangements. In how many ways can we seat these people if all the women must sit together but no woman sits on an end seat?

Communicating Mathematics

39. In solving counting problems, it is often useful to think of the situation as occurring in stages. Mention some examples in this section that illustrate that technique.

40. What is the relationship between trees, slot diagrams, and the fundamental counting principle?

41. Why would the number 29 be an unlikely answer for a problem that used the fundamental counting principle?

42. In Exercises 55–57 of Section 12.1, we described the *Gem Round* of the game show *Wheel of Destiny.* In this round, a contestant had to open a series of treasure chests containing a colored gem in a certain order to win the grand prize. There were two red gems, one blue gem, and one green gem. Explain why we cannot obtain the number of possible ways to select the gems in this round by using the fundamental counting principle.

43. Suppose that Yuki is selecting her schedule for next semester and has identified four English courses, three art courses, three psychology courses, and four history courses as possibilities to take. Assume that no two of the courses are offered at conflicting times.

a. Can the fundamental counting principle be used in solving this problem? Explain.

b. If any of the courses are offered at conflicting times, can you still use the fundamental counting principle? Explain.

44. Explain what must be present in a counting problem to make it possible to solve the problem using the fundamental counting principle.

Challenge Yourself

Sometimes in solving a counting problem it is not possible to use the FCP directly. However, if you divide the problem into smaller subcases, then the principle can be applied to the subcases and the results of the subcases added to get your answer. Use this approach in Exercises 45 and 46.

45. Dressing for an interview. Julio is dressing for an interview. He has a green, a brown, and a blue jacket. He has two pairs of tan slacks, one gray pair, one brown pair, and one blue pair. His shirts are white, gray, blue, and yellow. He has two red ties, two blue ties, and one brown tie. If he wears the blue jacket, he will not wear anything else that is blue or brown; otherwise there are no restrictions.

a. How many outfits can he select if he wears the blue jacket?

b. How many outfits can he wear if he does not wear the blue jacket?

c. How many total outfits can he wear?

46. Dressing for an interview. Repeat Exercise 45c, replacing the condition regarding the blue jacket with: If Julio wears the brown jacket, then he will wear nothing else that is blue or gray.

In Exercises 47–48, create a counting problem whose answer is the given number.

47. $6 \times 5 \times 4$ **48.** $9 \times 8 \times 7$

12.3 Permutations and Combinations

Objectives

1. Calculate the number of permutations of n objects taken r at a time.
2. Use factorial notation to represent the number of permutations of a set of objects.
3. Calculate the number of combinations of n objects taken r at a time.
4. Apply the theory of permutations and combinations to solve counting problems.

A student, in a moment of frustration, once said to me, "When you first explain a new idea, I understand what you are doing. Then you keep on explaining it until I don't

understand it at all!"* What I believe was frustrating this student is what you also may be dealing with in this chapter. At first we gave you simple, intuitive ways to count using systematic listing and drawing tree diagrams. Then, we went a little more deeply into the idea by introducing the fundamental counting principle. Now we are going to continue our line of thought by giving mathematical-sounding names and notation to the patterns that you have seen earlier.

For example, in Section 12.2, when we constructed keypad patterns in Example 5, you saw the product $10 \times 9 \times 8 \times 7 \times 6$. In Example 6, after seating Louise and her tutor, we used the pattern $8 \times 7 \times 6 \times 5$ to count the ways to fill the remaining four seats. We will now explore these patterns in more detail.

Permutations

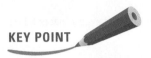

KEY POINT

Permutations are arrangements of objects in a straight line.

In both of these problems, we were selecting objects from a set and arranging them in order in a straight line. This notion occurs so often that we give it a name.

> **DEFINITION** A **permutation** is an ordering of distinct objects in a straight line. If we select r *different* objects from a set of n objects and arrange them in a straight line, this is called a *permutation of n objects taken r at a time.* The number of permutations of n objects taken r at a time is denoted by $P(n, r)$.[†]

The following diagram will help you to remember the meaning of the notation $P(n, r)$.

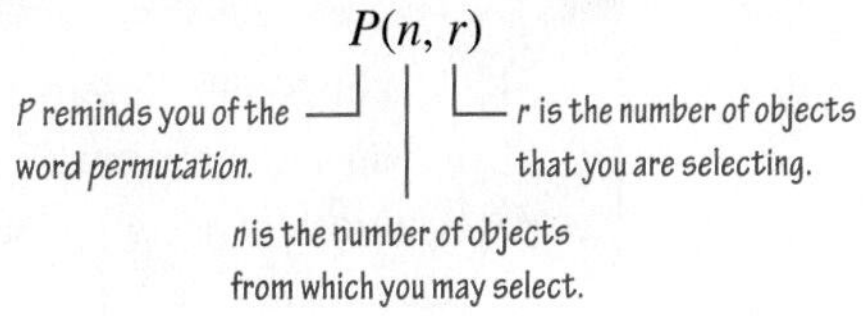

P(n, r)
P reminds you of the word permutation.
r is the number of objects that you are selecting.
n is the number of objects from which you may select.

For example $P(5, 3)$ indicates that you are counting permutations (straight-line arrangements) formed by selecting three different objects from a set of five available objects.

EXAMPLE 1 Counting Permutations

a) How many permutations are there of the letters a, b, c, and d? Write the answer using $P(n, r)$ notation.

b) How many permutations are there of the letters a, b, c, d, e, f, and g if we take the letters three at a time? Write the answer using $P(n, r)$ notation.

SOLUTION:

a) In this problem, we are arranging the letters a, b, c, and d in a straight line without repetition. For example, *abcd* would be one permutation, and *bacd* would be another. There are four letters that we can use for the first position, three for the second, and so on. Figure 12.10 is a slot diagram for this problem.

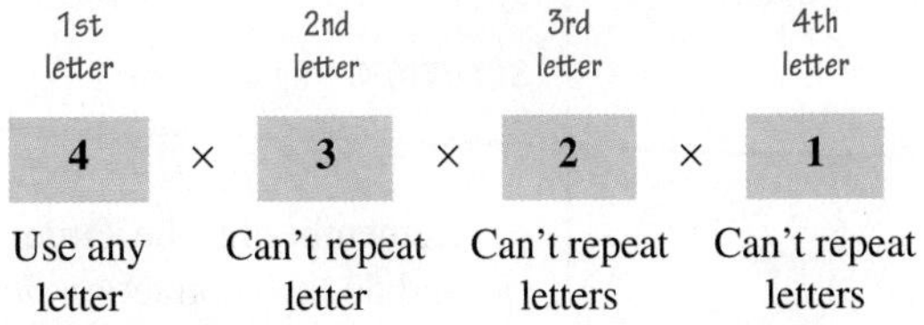

FIGURE 12.10 Slot diagram showing the number of ways to arrange four different objects in a straight line.

*I sincerely hope that this doesn't happen to you in this section.

[†]In Chapter 11 we treated permutations as ordered sets and used the notation (A, D, B, C). We will now use the less cumbersome notation ADBC instead when writing such permutations. Also, ${}_nP_r$ is another common notation for $P(n, r)$.

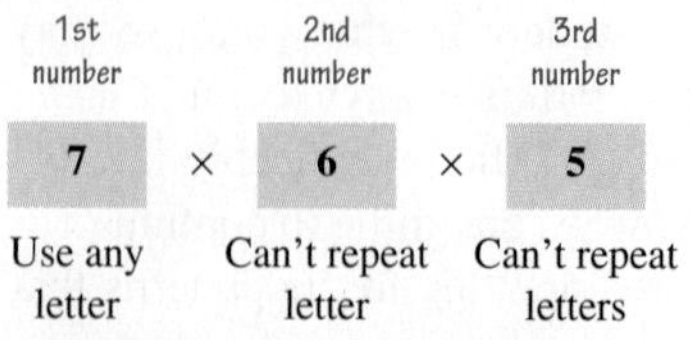

FIGURE 12.11 Slot diagram showing the number of ways to select three different objects from seven possible objects and arrange them in a straight line.

QUIZ YOURSELF 7

a) Explain in your own words the meaning of $P(8, 3)$.
b) Find $P(8, 3)$.

From Figure 12.10, we see that there are $4 \times 3 \times 2 \times 1 = 24$ permutations of these four objects. We can write this number more succinctly as

$$P(4, 4) = 24.$$

(number of permutations — P; of 4 objects — first 4; taken 4 at a time — second 4)

b) Figure 12.11 shows a slot diagram for this question. In this diagram, we see that because we have seven letters, we can use any one of the seven for the first position, one of the remaining six for the second position, and one of the remaining five for the third position.

We see that there are $7 \times 6 \times 5 = 210$ permutations of the seven objects taken three at a time. We write this as

$$P(7, 3) = 210.$$

(number of permutations — P; of 7 objects — 7; taken 3 at a time — 3)

7

SOME GOOD ADVICE

It is important to remember that an equation in mathematics is a symbolic form of an English sentence. In Example 1, when we write $P(7, 3) = 210$, we read this as "The number of permutations of seven objects taken three at a time is 210."

Factorial Notation

KEY POINT

We use factorial notation to express $P(n, r)$.

Suppose that we want to arrange 100 objects in a straight line. The number of ways to do this is $P(100, 100) = 100 \cdot 99 \cdot 98 \cdot \cdots \cdot 3 \cdot 2 \cdot 1$. Because such a product is very tedious to write out, we will introduce a notation to write such products more concisely.

DEFINITION If n is a counting number, the symbol $n!$, called *n factorial*, stands for the product $n \cdot (n - 1) \cdot (n - 2) \cdot (n - 3) \cdot \cdots \cdot 2 \cdot 1$. We define $0! = 1$.

EXAMPLE 2 Using Factorial Notation

Compute each of the following:

a) $6!$ b) $(8 - 3)!$ c) $\frac{9!}{5!}$ d) $\frac{8!}{5!3!}$

SOLUTION:

a) $6! = 6 \cdot 5 \cdot 4 \cdot 3 \cdot 2 \cdot 1 = 720.$

b) Remembering the Order Principle from Section 1.1, we work in parentheses first and do the subtraction *before* computing the factorial. Therefore,

$$(8 - 3)! = 5! = 5 \cdot 4 \cdot 3 \cdot 2 \cdot 1 = 120.$$

c) $$\frac{9!}{5!} = \frac{9 \cdot 8 \cdot 7 \cdot 6 \cdot \cancel{5 \cdot 4 \cdot 3 \cdot 2 \cdot 1}}{\cancel{5 \cdot 4 \cdot 3 \cdot 2 \cdot 1}} = 9 \cdot 8 \cdot 7 \cdot 6 = 3{,}024.$$

Cancel 5!.

Notice how canceling 5! from the numerator and denominator makes our computations simpler.

d) $\dfrac{8!}{5!3!} = \dfrac{8 \cdot 7 \cdot \cancel{6} \cdot \cancel{5} \cdot \cancel{4} \cdot \cancel{3} \cdot \cancel{2} \cdot \cancel{1}}{\cancel{5} \cdot \cancel{4} \cdot \cancel{3} \cdot \cancel{2} \cdot \cancel{1} \cdot \cancel{3} \cdot \cancel{2} \cdot \cancel{1}} = 8 \cdot 7 = 56.$

Cancel 5!. Cancel 3!, which equals 6.

Now try Exercises 1 to 8.

In Example 1(b), we first wrote $P(7, 3)$ in the form $7 \times 6 \times 5$. This product started as though we were going to compute 7!; however, the product $4 \cdot 3 \cdot 2 \cdot 1$ is missing. Another way of saying this is that $P(7, 3) = \frac{7!}{4!} = \frac{7!}{(7-3)!}$. We state this as a general rule.

FORMULA FOR COMPUTING $P(n, r)$

$$P(n, r) = \frac{n!}{(n-r)!}$$

We will use this formula in the computations in Example 3.

EXAMPLE 3 Counting Ways to Fill Positions at a Community Theater

The 12-person community theater group produces a yearly musical. Club members select one person from the group to direct the play, a second to supervise the music, and a third to handle publicity, tickets, and other administrative details. In how many ways can the group fill these positions?

SOLUTION:

This can be viewed as a permutation problem if we consider that we are selecting 3 people from 12 and then arranging those names in a straight line—director, music, administration. The answer to this question then is

$$P(12, 3) = \frac{12!}{(12-3)!} = \frac{12 \cdot 11 \cdot 10 \cdot \cancel{9} \cdot \cancel{8} \cdot \cancel{7} \cdot \cancel{6} \cdot \cancel{5} \cdot \cancel{4} \cdot \cancel{3} \cdot \cancel{2} \cdot \cancel{1}}{\cancel{9} \cdot \cancel{8} \cdot \cancel{7} \cdot \cancel{6} \cdot \cancel{5} \cdot \cancel{4} \cdot \cancel{3} \cdot \cancel{2} \cdot \cancel{1}} = 1{,}320$$

Cancellations such as this will save you from doing unnecessary computations.

ways to select 3 people from the 12 available for the three positions.

Now try Exercises 15 to 18.

SOME GOOD ADVICE

When talking with my students I often kid them by saying that to be a good mathematics student you have to be lazy—but in the right way. I don't mean that it is good to sleep through your math class or not do homework, but it is important to try to make your computations as easy as possible for yourself. For example, in Example 3 you had to compute the expression $P(12, 3)$. When doing this by hand, the weak student may compute this as

$$P(12, 3) = \frac{12!}{(12-3)!} = \frac{479{,}001{,}600}{362{,}880} = 1{,}320.$$

Working with such large numbers is tedious, time-consuming, and prone to error. However, if you simplify the computations as we did in Example 3, the computations go much more quickly. Some may argue that if you are using a calculator, then it doesn't make a difference as to how you do the computations. This is true provided that the numbers you are working with do not cause an overflow on your calculator. For example, try finding $\frac{500!}{498!}$ on your calculator without canceling before dividing.

KEY POINT

In forming combinations, order is not important.

Combinations

In order to introduce a new idea, let's change the conditions in Example 3 slightly. Suppose that instead of selecting three people to carry out three different responsibilities, we choose a three-person committee that will work jointly to see that the job gets done. In Example 3, if A, B, and C were selected to be in charge of directing, music, and administration, respectively, that would be different than if B directed, C supervised the music, and A handled administrative details. However, with the new plan, it makes no difference whether we say A, B, C or B, C, A. From Table 12.1, we see that all of the six different assignments of A, B, and C under the old plan are equivalent to a single committee under the new plan.

Directing	Music	Administration
A	B	C
A	C	B
B	A	C
B	C	A
C	A	B
C	B	A

TABLE 12.1 Six different assignments of responsibilities for the musical are now handled by one committee.

What we are saying for A, B, and C applies to any three people in the theater group. Thus, the answer 1,320 we found in Example 3 is too large, so we must reduce it by a factor of 6 under the new plan. Therefore, the number of three-person committees we could form would be $\frac{1,320}{6} = 220$. Remember that the factor of 6 we divided out is really 3! (the number of ways we can arrange three people in a straight line). This means that the number of three-person committees we can select can be written in the form

$$\frac{1,320}{6} = \frac{P(12, 3)}{3!}.$$

The reason this number is smaller is that now we are concerned only with *choosing a set* of people to produce the play, but the *order* of the people chosen *is not important.*

We can generalize what we just saw. If we are choosing r objects from a set of n objects and are not interested in the order of the objects, then to count the number of choices, we must divide $P(n, r)$* by $r!$. We now state this formally.

FORMULA FOR COMPUTING $C(n, r)$ If we choose r objects from a set of n objects, we say that we are forming a **combination** of n objects taken r at a time. The notation $C(n, r)$ denotes the number of such combinations.† Also,

$$C(n, r) = \frac{P(n, r)}{r!} = \frac{n!}{r! \cdot (n - r)!}.$$

We divide by $r!$ because we are not interested in the order of the objects.

SOME GOOD ADVICE

In working with permutations and combinations, we are choosing r different objects from a set of n objects. The big difference is whether the order of the objects is important. If *the order of the objects matters,* we are dealing with a permutation. If *the order does not matter,* then we are working with a combination.

Do not try to use the theory of permutations or combinations when that theory is not relevant. If a problem involves something other than simply choosing *different* objects (and maybe ordering them), then perhaps the fundamental counting principle may be more appropriate.

EXAMPLE 4 **Using the Combination Formula to Count Committees**

a) How many three-element sets can be chosen from a set of five objects?

b) How many four-person committees can be formed from a set of 10 people?

---

*Recall that $P(n, r) = \frac{n!}{(n - r)!}$.

† $_nC_r$ is another common notation for $C(n, r)$.

Historical Highlight

The Origins of Permutations and Combinations

As is the case with so many ideas in mathematics, the origins of some of the fundamental counting concepts go back to ancient times. The notion of permutations can be found in a Hebrew work called the *Book of Creation*, which may have been written as early as AD 200.

Even more surprising, there is evidence that Chinese mathematicians tried to solve permutations and combinations problems in the fourth century BC. We can trace other early counting formulas back to Euclid in the third century BC and the Hindu mathematician Brahmagupta in the seventh century.

SOLUTION:

a) Because order is not important in considering the elements of a set, it is clear that this is a combination problem rather than a permutation problem. The number of ways to choose three elements from a set of five is

$$C(5, 3) = \frac{5!}{3! \cdot (5-3)!} = \frac{5 \cdot 4 \cdot \cancel{3} \cdot \cancel{2} \cdot \cancel{1}}{\cancel{3} \cdot \cancel{2} \cdot \cancel{1} \cdot 2 \cdot 1} = \frac{20}{2} = 10.$$

b) The number of different ways to choose four elements from a set of 10 is

$$C(10, 4) = \frac{10!}{4! \cdot (10-4)!} = \frac{10 \cdot \overset{3}{\cancel{9}} \cdot \cancel{8} \cdot 7 \cdot \cancel{6} \cdot \cancel{5} \cdot \cancel{4} \cdot \cancel{3} \cdot \cancel{2} \cdot \cancel{1}}{\cancel{4} \cdot \cancel{3} \cdot \cancel{2} \cdot \cancel{1} \cdot \cancel{6} \cdot \cancel{5} \cdot \cancel{4} \cdot \cancel{3} \cdot \cancel{2} \cdot \cancel{1}} = 210.$$

Cancellation saves you work.

QUIZ YOURSELF 8

How many five-person committees can be formed from a set of 12 people?

Now try Exercises 19 to 22. 8

We can now address the question about the lottery syndicate that we posed at the beginning of this chapter.

EXAMPLE 5 How Many Tickets Are Necessary to Cover All Possibilities in a Lottery?

Recall that a syndicate intended to raise $15 million to buy all the tickets for certain lotteries. The plan was that if the prize was larger than the amount spent on the tickets, then the syndicate would be guaranteed a profit. To play the Virginia lottery, the player buys a ticket for $1 containing a combination of six numbers from 1 to 44. Assuming that the syndicate has raised the $15 million, does it have enough money to buy enough tickets to be guaranteed a winner?

Between the Numbers

How Not to Become a Millionaire

A survey sponsored by the Consumer Federation found that 27% of Americans think that they have a better chance of acquiring money for their retirement by playing the lottery than by systematically investing their money. Perhaps this is because many do not understand how unlikely it is that they will win a lottery.

For example, to win the Powerball game, you must first pick five numbers correctly from 59 possible and then also correctly pick the one Powerball number from 35 possible. So, according to our fundamental counting principle, the number of ways to make your Powerball selection is $C(59, 5) \times 35 = 175{,}223{,}510$. Even if you bought one million tickets, you would only have about a 0.0057 or $\frac{6}{10}$ of one percent chance that one of your tickets would win!

SOLUTION:

Because we are choosing a combination of 6 numbers from the 44 possible, the number of different tickets possible is $C(44, 6) = 7{,}059{,}052$. Thus, the syndicate has more than enough money to buy enough tickets to be guaranteed a winner.

In addition to wondering whether $15 million would cover all combinations in a lottery, there were some other practical concerns that caused me not to invest in the syndicate.

1. Suppose the syndicate did not raise all $15 million. What would be done with the money then? Would they play lotteries in which not all tickets can be covered? Doing this could lose all the money we invested.
2. What if there are multiple winners? In this case, the jackpot is shared and it is possible that the money won will not cover the cost of the tickets. This case would result in a loss to the syndicate.
3. Is it physically possible to actually purchase 7,059,052 tickets? Because there are $60 \times 60 \times 24 \times 7 = 604{,}800$ seconds in a week, it would take almost 12 weeks for one person buying one ticket every second to purchase enough tickets to cover all the combinations. In Virginia, the syndicate bought only about 5 million tickets, thus opening up the possibility that all the money could have been lost!
4. How much of the money raised will be used by the syndicate for its own administrative expenses?
5. Because the jackpot is to be shared with thousands of members of the syndicate and the prize money will be paid over a 20-year period, is the amount of return better than if the money had been placed in a high-interest-bearing account?

In light of these concerns, was I too cautious? What do you think? As a final note on this matter, since the Virginia lottery episode, some states have made it illegal to play a lottery by covering all ticket combinations.

As Example 6 shows, applications of counting appear in many gambling situations.*

EXAMPLE 6 Counting Card Hands

a) In the game of poker, five cards are drawn from a standard 52-card deck.† How many different poker hands are possible?

b) In the game of bridge, a hand consists of 13 cards drawn from a standard 52-card deck. How many different bridge hands are there?

SOLUTION:

a) $C(52, 5) = \dfrac{52!}{5!47!} = \dfrac{\overset{13}{\cancel{52}} \cdot \overset{17}{\cancel{51}} \cdot \overset{10}{\cancel{50}} \cdot 49 \cdot \overset{24}{\cancel{48}}}{\cancel{5} \cdot \cancel{4} \cdot \cancel{3} \cdot \cancel{2} \cdot 1} = 2{,}598{,}960.$

b) $C(52, 13) = \dfrac{52!}{13!39!} = 635{,}013{,}559{,}600.$

Combining Counting Methods

In some situations, it is necessary to combine several counting methods to arrive at an answer. We see this technique in the next example.

*We cover counting as it applies to gambling in Section 12.4 at the end of this chapter.

†Figure 12.1 shows that a standard 52-card deck contains 13 hearts, 13 spades, 13 clubs, and 13 diamonds. Each suit consists of A, K, Q, J, 10, 9, . . . , 3, 2. See Table 12.2, page 636, for the possible poker hands.

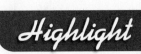

Highlight

Counting and Technology

Technology can save you work and ensure accuracy when doing calculations such as we saw in Example 6. Screens 1 and 2 illustrate how to use graphing calculators such as the TI-83 or 84 to compute $C(52, 5)$.

Screen 3 shows one of the many free apps that will do permutation and combination calculations quickly for you. Although I used an iPad, these apps are also widely available for smart phones and other mobile devices.

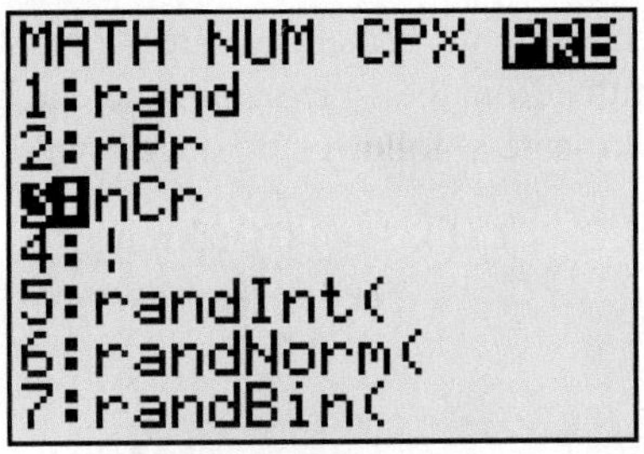

TI Screen 1

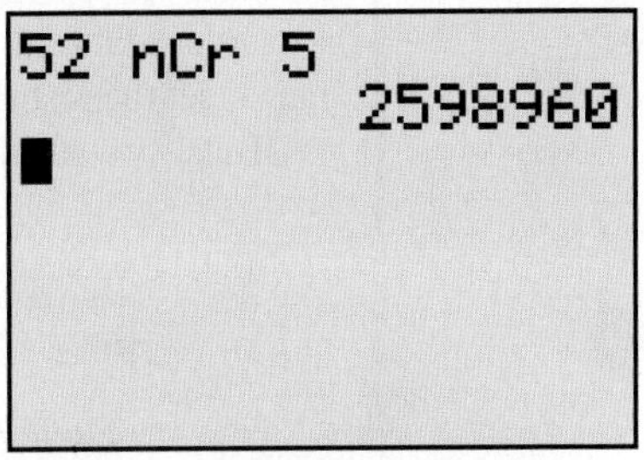

TI Screen 2

Screen 3: Free iPad App

EXAMPLE 7 Combining Counting Methods

The division of student services at your school is selecting two men and two women to attend a leadership conference in Honolulu, Hawaii. If 10 men and nine women are qualified for the conference, in how many different ways can management make its decision?

SOLUTION: Review the Be Systematic Strategy on page 5.

It is tempting to say that because we are selecting four people from a possible 19, the answer is $C(19, 4)$. But this is wrong. If we think about it, certain choices are unacceptable. For example, we could not choose all men or three women and one man.

We can use the fundamental counting principle once we recognize that the group can be formed in two stages.

Stage 1: Select the two women from the nine available. We can do this in $C(9, 2) = \frac{9!}{2!7!} = 36$ ways.

Stage 2: Select two men from the 10 available. We can make this choice in $C(10, 2) = \frac{10!}{2!8!} = 45$ ways.

Thus, choosing the women and then choosing the men can be done in $36 \cdot 45 = 1{,}620$ ways.

Example 8 shows another situation in which you have to use several counting techniques at the same time.

EXAMPLE 8 Forming a Governing Committee

Assume that you and 15 of your friends have formed a company called Net-Media, an Internet music and video provider. A committee consisting of a president, a vice president, and a three-member executive board will govern the company. In how many different ways can this committee be formed?

SOLUTION: Review the Be Systematic Strategy on page 5.

We can form the committee in two stages:

a) Choose the president and vice president.

b) Select the remaining three executive members.

Thus, we can apply the fundamental counting principle.

Because order is important in stage a), we recognize that we can choose the president and vice president in $P(16, 2)$ ways.

Order is not important in choosing the remaining three committee members, so we can select the rest of the committee from the 14 remaining people in $C(14, 3)$ ways.

Thus, you see that we can do stage a) followed by stage b) in

$$P(16, 2) \times C(14, 3) = 87{,}360 \text{ ways.}$$

Choose president and vice president from 16 people. — Choose 3 committee members from remaining 14 people.

Now try Exercises 55 to 66.

KEY POINT

The rows of Pascal's triangle count the subsets of a set.

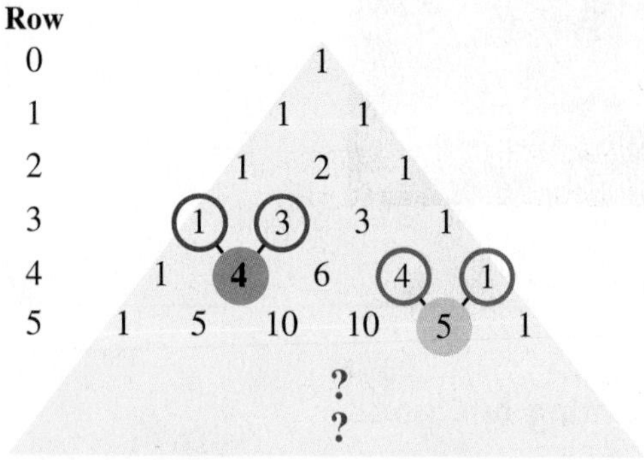

FIGURE 12.12 Pascal's triangle.

It is surprising how the same mathematical idea appears again and again in different cultures over a period of centuries. For example, Pascal's triangle in Figure 12.12 occurs in the writings of fourteenth-century Chinese mathematicians, which some historians believe are based on twelfth-century manuscripts. Pascal's triangle is also found in the work of eighteenth-century Japanese mathematicians.

In Pascal's triangle, we begin numbering the rows of the triangle with zero and also begin numbering the entries of each row with zero. For example, in Figure 12.12, the number 6 is the second entry in the fourth row. Each entry in this triangle after row 1 is the sum of the two numbers immediately above it, which are a little to the right and a little to the left.

Pascal's triangle has an interesting application in set theory. Suppose that we want to find all subsets of a particular set. Recall that the Be Systematic Strategy from Section 1.1 encourages us to list possibilities systematically to solve problems. We can find all subsets of $\{1, 2, 3, 4\}$ by first listing sets of size 0, then sets of size 1, 2, 3, and 4, as in the following list:

$\varnothing$ — 1 zero-element set

$\{1\}, \{2\}, \{3\}, \{4\}$ — 4 one-element sets

$\{1, 2\}, \{1, 3\}, \{1, 4\}, \{2, 3\}, \{2, 4\}, \{3, 4\}$ — 6 two-element sets

$\{1, 2, 3\}, \{1, 2, 4\}, \{1, 3, 4\}, \{2, 3, 4\}$ — 4 three-element sets

$\{1, 2, 3, 4\}$ — 1 four-element set

We see that there is one subset of size 0, four of size 1, six of size 2, four of size 3, and one of size 4. Note that this pattern,

$$1 \quad 4 \quad 6 \quad 4 \quad 1,$$

occurs as the fourth row in Pascal's triangle.

Similarly, the pattern

$$1 \quad 5 \quad 10 \quad 10 \quad 5 \quad 1,$$

which is the fifth row of Pascal's triangle, counts the subsets of size 0, 1, 2, and so on of a five-element set. In fact, the following result is true. 9

QUIZ YOURSELF 9

a) What is the seventh row of Pascal's triangle?

b) How many subsets of size 3 are there in a seven-element set?

PASCAL'S TRIANGLE COUNTS THE SUBSETS OF A SET The nth row of Pascal's triangle counts the subsets of various sizes of an n-element set.

Let us look at the subsets of the set $S = \{1, 2, 3, 4\}$ again. The sets $\{1\}, \{2\}, \{3\}$, and $\{4\}$ are the four different ways we can choose a one-element set from S. The sets

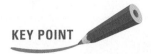

KEY POINT

The entries of Pascal's triangle are numbers of the form $C(n, r)$.

$\{1, 2\}, \{1, 3\}, \{1, 4\}, \{2, 3\}, \{2, 4\}$, and $\{3, 4\}$ are all the different ways that we can choose a two-element set from S. Because combinations are sets, this means that we can also use the entries in Pascal's triangle to count combinations.

> **ENTRIES OF PASCAL'S TRIANGLE AS $C(n, r)$** The rth entry of the nth row of Pascal's triangle is $C(n, r)$.

EXAMPLE 9 Relating Entries in Pascal's Triangle to Combinations

Interpret the numbers in the fourth row of Pascal's triangle as counting combinations.

SOLUTION:

The fourth row of Pascal's triangle is

$$1 \quad 4 \quad 6 \quad 4 \quad 1.$$

Because the leftmost 1 is the zeroth entry in the fourth row, we can write it as $C(4, 0)$. That is to say, $C(4, 0) = 1$. The first entry in the fourth row is 4, which means that $C(4, 1) = 4$. Similarly, $C(4, 2) = 6$, $C(4, 3) = 4$, and $C(4, 4) = 1$.

Now try Exercises 25 and 26. 10

QUIZ YOURSELF 10

Write the numbers in the third row of Pascal's triangle using $C(n, r)$ notation.

For counting problems that are not too large, it is quicker to use entries in Pascal's triangle to count combinations rather than to use the formula for $C(n, r)$ that we stated earlier, as you will see in Example 10.

EXAMPLE 10 Using Pascal's Triangle to Count Drug Combinations

Frequently, doctors experiment with combinations of new drugs to combat hard-to-treat illnesses such as AIDS and hepatitis. Assume that a pharmaceutical company has developed five antibiotics and four immune system stimulators. In how many ways can we choose a treatment program consisting of three antibiotics and two immune system stimulators to treat a disease? Use Pascal's triangle to speed your computations.

SOLUTION:

We can select the drugs in two stages: first select the antibiotics and then the immune system stimulators. Therefore, we can apply the fundamental counting principle, introduced in Section 12.2.

In selecting the antibiotics we are choosing three drugs from five, so this can be done in $C(5, 3)$ ways. Looking at the fifth row of Pascal's triangle, which is

1 5 10 10 5 1,

$C(5, 0)$ $C(5, 1)$ $C(5, 2)$ $C(5, 3)$

we see that the third entry (remember, we start counting row entries with 0) is 10. Thus, $C(5, 3) = 10$.

Choosing two immune system stimulators from four can be done in $C(4, 2)$ ways. The fourth row in Pascal's triangle is

1 4 6 4 1.

$C(4, 0)$ $C(4, 1)$ $C(4, 2)$

This means that $C(4, 2) = 6$. It follows that we can choose the antibiotics and then the immune system stimulators in $C(5, 3) \times C(4, 2) = 10 \times 6 = 60$ ways. The accompanying screen shows how to do the calculations with a graphing calculator.

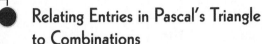

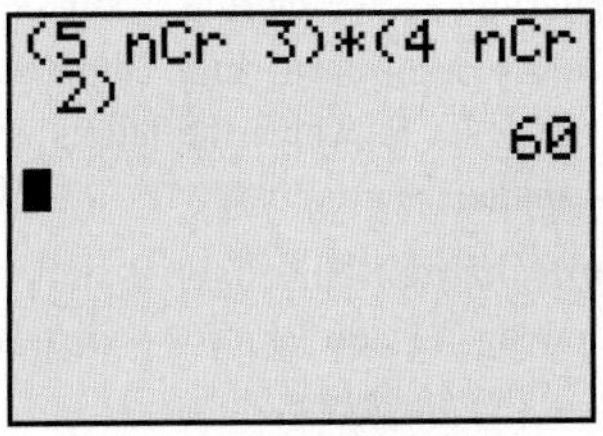

EXERCISES 12.3

Sharpening Your Skills

In Exercises 1–12, calculate each value.

1. 4!
2. 3!
3. $(8 - 5)!$
4. $(10 - 5)!$
5. $\frac{10!}{7!}$
6. $\frac{11!}{9!}$
7. $\frac{10!}{7!3!}$
8. $\frac{11!}{9!2!}$
9. $P(6, 2)$
10. $P(5, 3)$
11. $C(10, 3)$
12. $C(10, 4)$

Explain the meaning of each symbol in Exercises 13 and 14.

13. $P(10, 3)$
14. $C(6, 2)$

In Exercises 15–18, find the number of permutations.

15. Eight objects taken three at a time
16. Seven objects taken five at a time
17. Ten objects taken eight at a time
18. Nine objects taken six at a time

In Exercises 19–22, find the number of combinations.

19. Eight objects taken three at a time
20. Seven objects taken five at a time
21. Ten objects taken eight at a time
22. Nine objects taken six at a time
23. Find the eighth row in Pascal's triangle.
24. Find the tenth row in Pascal's triangle.

Use the seventh row of Pascal's triangle to answer Exercises 25 and 26.

25. How many two-element subsets can we choose from a seven-element set?
26. How many four-element subsets can we choose from a seven-element set?

In Exercises 27–30, describe where each number would be found in Pascal's triangle.

27. $C(18, 2)$
28. $C(19, 5)$
29. $C(20, 6)$
30. $C(21, 0)$

Applying What You've Learned

In Exercises 31–44, specify the number of ways to perform the task described. Give your answers using P(n, r) *or* C(n, r) *notation. The key in recognizing whether a problem involves permutations or combinations is deciding whether order is important.*

31. Quiz possibilities. On a biology quiz, a student must match eight terms with their definitions. Assume that the same term cannot be used twice.

32. Scheduling interviews. Producers wish to interview six actors on six different days to replace the actor who plays the phantom in Broadway's longest-running play, *The Phantom of the Opera.*

33. Choosing tasks. Gil Grissom on *CSI* wants to review the files on three unsolved cases from a list of 17. He does not care about the order in which he reads the files.

34. Selecting roommates. Annette has rented a summer house for next semester. She wants to select four roommates from a group of six friends.

35. Baseball lineups. Joe Torre has already chosen the first, second, fourth, and ninth batters in the batting lineup. He has selected five others to play and wants to write their names in the batting lineup.

36. Race results. There are seven boats that will finish the America's Cup yacht race.

37. Seeding players in a tournament. Novak Djokovic, Rafael Nadal, Roger Federer, and Andy Murray are invited to play in the Malaysian Open, in Kuala Lumpur. In how many ways can these players be seeded in the top four slots in the tournament?

38. Advertising tennis racquets. A tennis racquet manufacturer is planning a new advertising campaign. It will run ads in three different sports magazines featuring Victoria Azarenka, Maria Sharapova, and Petra Kvitova. If a different player is to be featured in each magazine, in how many ways can this be done?

39. Journalism awards. Ten magazines are competing for three identical awards for excellence in journalism. No magazine can receive more than one award.

40. Journalism awards. Redo Exercise 39, but now assume the awards are all different. No magazine can receive more than one award.

41. Lock patterns. In order to unlock a door, five different buttons must be pressed down on the panel shown. When the door opens, the depressed buttons pop back up. The order in which the buttons are pressed is not important.

42. Lock patterns. A bicycle lock has three rings with the letters A through K on each ring. To unlock the lock, a letter must be selected on each ring. Duplicate letters are not allowed, and the order in which the letters are selected on the rings does not matter.

43. Choosing articles for a magazine. The editorial staff of Oprah's *O Magazine* is reviewing 17 articles. They wish to select eight for the next issue of the magazine.

44. Choosing articles for a magazine. Redo Exercise 43, except that in addition to choosing the stories, the editorial staff must also decide in what order the stories will appear.

45. Reading program. Six players are to be selected from a 25-player Major League Baseball team to visit schools to support a summer reading program. In how many ways can this selection be made?

46. Reading program. If in Exercise 45 we not only want to select the players but also want to assign each player to visit one of six schools, how many ways can that assignment be done?

47. Writing articles. In Exercise 44, if it takes 1 minute to write a list of eight articles selected for the magazine, how many years would it take to write all possible lists of eight articles? Assume 60 minutes in an hour, 24 hours in a day, and 365 days in a year.

48. Writing assignments. In Exercise 46, if it takes 1 minute to write a list of six players with their assignments, how many years would it take to write all possible lists of assignments?

49. An incorrect answer. The editor must choose three articles for the front page of the campus newspaper from eight that have been suggested. Placement on the page is not important. Anna answered 336. Why was this answer marked wrong by her instructor?

50. Another incorrect answer. The disk jockey on the campus radio station will choose five songs from nine possibilities to open his show. In how many ways can he choose and then play these songs on the air? Noah, focusing on the word "choose," answered 126. Why is his answer incorrect?

A typical bingo card is shown in the figure. The numbers 1–15 are found under the letter B, 16–30 under the letter I, 31–45 under the letter N, 46–60 under the letter G, and 61–75 under the letter O. The center space on the card is labeled "FREE."

B	I	N	G	O
5	28	33	51	75
7	17	41	59	63
2	22	FREE	48	61
11	29	44	46	72
9	30	38	52	68

51. Bingo cards. How many different columns are possible under the letter B?

52. Bingo cards. How many different columns are possible under the letter N?

53. Bingo cards. How many different bingo cards are possible? (*Hint:* Use the fundamental counting principle.)

54. Bingo cards. Why is the answer to Exercise 53 not $P(75, 24)$?

In Exercises 55–68, use the fundamental counting principle.

55. Selecting astronauts. NASA wants to appoint two men and three women to send back to the moon in 2018. The finalists for these positions consist of six men and eight women. In how many ways can NASA make this selection?

56. Computer passwords. A password for a computer consists of three different letters of the alphabet followed by four different digits from 0 to 9. How many different passwords are possible?

57. Forming a public safety committee. Nicetown is forming a committee to investigate ways to improve public safety in the town. The committee will consist of three representatives from the seven-member town council, two members of a five-person citizens advisory board, and three of the 11 police officers on the force. How many ways can that committee be formed?

58. Choosing a team for a seminar. HazMat, Inc., is sending a group of eight people to attend a seminar on toxic waste disposal. They will send two of eight engineers and three of nine crew supervisors and the remainder will be chosen from the five senior managers. In how many ways can these choices be made?

59. Designing a fitness program. To lose weight and shape up, Sgt. Walden is considering doing two types of exercises to improve his cardiovascular fitness from running, bicycling, swimming, stair stepping, and Tae Bo. He is also going to take two nutritional supplements from AllFit, Energize, ProTime, and DynaBlend. In how many ways can he choose his cardiovascular exercises and nutritional supplements?

60. Counting assignment possibilities. Gina must write evaluation reports on three hospitals and two health clinics as part of her degree program in community health services. If there are six hospitals and five clinics in her vicinity, in how many ways can she complete her assignment?

61. Assigning passengers to vans. The 24-member college baseball team is traveling to a game in three vans, with eight in each van. In how many different ways can players be assigned to the vans? Where the players sit inside the vans is unimportant.

62. Reviewing video games. The editor of the *Video Gamer* is selecting four video games from ten possibilities to review in his online column. One will be the featured game and the other three will be covered in less detail and in no particular order. In how many ways can he make his selection?

63. Picking a team for a competition. The students in the 12-member advanced communications design class at Center City Community College are submitting a project to a national competition. They must select a four-member team to attend the competition. The team must have a team leader and a main presenter; the other two members have no particularly defined roles. In how many different ways can this team be formed?

64. Why is Exercise 63 neither a strictly permutations nor a strictly combinations problem?

65. Choosing an evaluation team. The academic computing committee at Sweet Valley College is in the process of evaluating different computer systems. The committee consists of five administrators, seven faculty, and four students. A five-person

subcommittee is to be formed. The chair and vice chair of the committee must be administrators; the remainder of the committee will consist of faculty and students. In how many ways can this subcommittee be formed?

66. Why is Exercise 65 neither a strictly permutations nor a strictly combinations problem?

An armored car must pick up receipts at a shopping mall and at a jewelry store and deliver them to a bank. For security purposes, the driver will vary the route that the car takes each day. Use the map to answer Exercises 67 and 68.

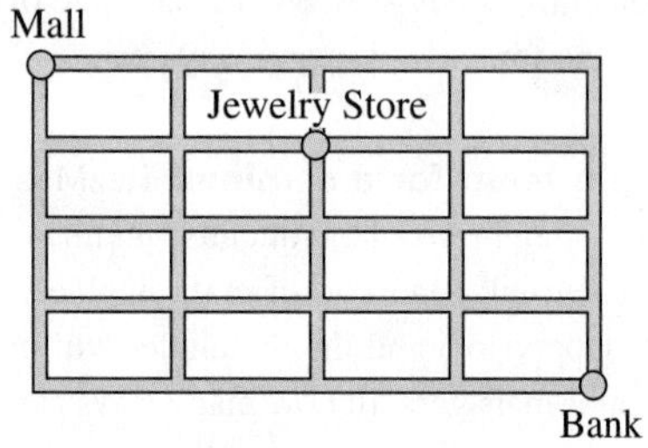

67. Armored car routes. How many different direct routes can the driver take from the shopping mall to the bank?

68. Armored car routes. How many direct routes can the driver take from the mall to the bank if he must also stop at the jewelry store?

69. Dividing volunteers into groups. Eighteen students have volunteered to be Campus Ambassadors to prospective new students and their families. In how many ways can the students be divided into three groups of six for training workshops?

70. A presidential advisory group. The president of your college has asked the president of student government to form an eight-person presidential advisory committee from among 17 student leaders. From the eight people selected, one will be chosen as chair, two others will be chosen as co-facilitators, and the other five on the committee will have no special roles. In how many ways can this committee be formed?

Communicating Mathematics

71. How do you distinguish a combination problem from a permutation problem?

72. In Example 7, we were choosing 4 people from a group of 19, yet our answer was not $C(19, 4)$. Explain this. How does the fundamental counting principle apply here?

73. Recall that combinations are subsets of a given set and that $C(n, r)$ is the number of subsets of size r in an n-element set. Give an intuitive explanation as to why you would expect each of the following equations to be true.

a. $C(5, 0) = 1$ **b.** $C(8, 7) = 8$

74. Explain why $C(n, r) = C(n, n - r)$. To understand this statement, recall the Three-Way Principle from Section 1.1, which states that making numerical examples is useful in gaining understanding of a situation.

Between the Numbers

75. Playing a lottery. Many states have a lottery called *The Daily Number*, where players pick a three-digit number such as 407, 556, or 333. Assume that it costs one dollar to play and the prize is \$500 if you pick the correct number. If on a given day you were to buy all numbers between 000 and 999 inclusive, how much would you win or lose?

76. Playing a lottery. Some lotteries ask you to choose five numbers from 1 to 39. If you were to buy 1,000 different tickets, what are your chances of winning?

77. Consider the question: How many four-digit numbers can be formed that are odd and greater than 5,000? Why is this not a permutation problem?

78. Consider the question: How many four-digit numbers can be formed using the digits 1, 3, 5, 7, and 9? Repetition of digits is not allowed. Why is this not a combination problem?

Challenge Yourself

79. Playing cards. How many five-card hands chosen from a standard deck contain two diamonds and three hearts?

80. Playing cards. How many five-card hands chosen from a standard deck contain two kings and three aces?

The following diagram shows a portion of Pascal's triangle found in a book called The Precious Mirror, *written in China in 1303. Use this diagram to answer Exercises 81–84.*

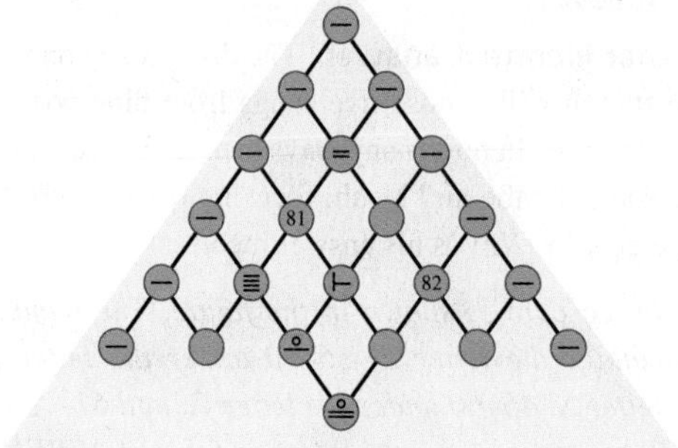

81. What symbol do you expect to find in the circle labeled 81?

82. What symbol do you expect to find in the circle labeled 82?

83. What number does the symbol  represent?

84. What number does the symbol  represent?

Consider rows n and n − 1 of Pascal's triangle, shown in the diagram to the right. Use this diagram to answer Exercises 85–89.

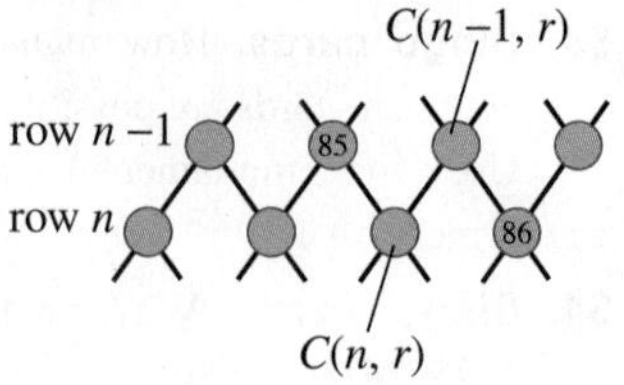

85. What expression will be in the circle labeled 85?

86. What expression will be in the circle labeled 86?

87. Complete the following equation: $\square + C(n - 1, r) = \square$

88. Complete the following equation: $C(n, r) + \square = \square$

89. Notice that the arrangement of numbers in each row of Pascal's triangle is symmetric. See how the 4s are balanced in the fourth row, the 5s are balanced in the fifth row, and so on. Can you give a set theory explanation to account for this?

Looking Deeper

12.4 Counting and Gambling

Objectives

1. Understand how to use the fundamental counting principle to analyze counting problems that occur in gambling.
2. Use the theory of permutations and combinations to compute the probability of drawing various card hands.

While visiting a casino, I saw a sign proclaiming:

$100,000 jackpot !!

The jackpot was formed by putting one nickel of every dollar gambled on slot machines into the jackpot. Periodically, a lucky player wins the jackpot. At that time, the jackpot is emptied and must start building all over again. If we look at this jackpot another way, the neon sign should perhaps proclaim:

$2,000,000 gambled on slot machines since the last jackpot !!

Operators of gambling establishments earn huge fortunes using the simple mathematical principle that the number of ways they can win is greater than the number of ways that you can win. In this section, we will use the theory of counting to explain why people win (and lose) at games such as slot machines and poker.

Applying the Fundamental Counting Principle to Gambling

Early mechanical slot machines had three wheels with 20 positions on each wheel containing symbols such as oranges, bells, lemons, cherries, and bars. A gambler put a coin in the machine and pulled a lever, causing the wheels to spin. When the wheels stopped, the machine displayed a symbol from each wheel in a window.* For certain arrangements of symbols, the player received a payoff. Arrangements of symbols that occur infrequently pay larger payoffs than those that occur more frequently. Figure 12.13 shows a typical arrangement of the symbols that might appear on each of the three wheels.

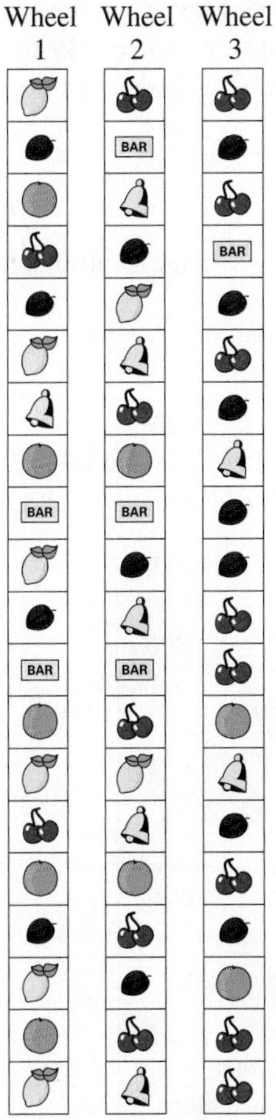

FIGURE 12.13 Three wheels of a typical mechanical slot machine.

EXAMPLE 1 Using the FCP to Analyze a Slot Machine

a) In how many different ways can the three wheels of the slot machine depicted in Figure 12.13 stop?

b) One payoff is for cherries on wheel 1 and no cherries on the other wheels. In how many different ways can this happen?

SOLUTION: Review the Be Systematic Strategy on page 5.

a) It is useful to think of the first wheel stopping, then the second, and finally the third. The solution becomes apparent if we draw a slot diagram. Using the fundamental counting principle, we see that the total number of different ways the three wheels can stop is $20 \times 20 \times 20 = 8{,}000$.

*Modern slot machines do not use levers and gear wheels as did early machines. Instead, they rely on computer chips that regulate the odds based on random number generators. Computers then control "virtual wheels" to determine the payoffs.

1st wheel		2nd wheel		3rd wheel	1st wheel		2nd wheel		3rd wheel
20	×	20	×	20	2	×	15	×	12
Number of ways first wheel can stop		Number of ways second wheel can stop		Number of ways third wheel can stop	Number of ways first wheel can stop with cherries		Number of ways second wheel can stop with no cherries		Number of ways third wheel can stop with no cherries

b) As before, think of the wheels stopping in stages. Again, a slot diagram helps with the counting. By the fundamental counting principle, we see that the total number of ways we can get cherries only on the first wheel is $2 \times 15 \times 12 = 360$.

EXAMPLE 2 **Counting the Number of Ways to Get Three Bars**

For the slot machine in Figure 12.13, the largest payoff is for three bars. In how many ways can three bars be obtained?

SOLUTION:

There are two ways a bar can come up on the first wheel, three ways a bar can come up on the second wheel, and only one way a bar can come up on the third wheel. By the fundamental counting principle, a bar on the first, second, and third wheels can occur in $2 \times 3 \times 1 = 6$ ways.

Now try Exercises 1 to 6.

From Examples 1 and 2 we would expect that the payoff for three bars to be much greater than the payoff for cherries on the first wheel only.

Counting and Poker

Counting principles also explain why one poker hand beats another (see Table 12.2 below). In poker, a hand that is less likely to occur beats a hand that occurs more commonly. We first determine the total number of possible poker hands.

Royal flush	10, J, Q, K, and A, all of the same suit
Straight flush	Five cards in sequence,* all of the same suit
Four of a kind	Four cards of one rank, plus another card
Full house	Three cards of one rank and two of another
Flush	All five cards are from the same suit
Straight	Any five cards in sequence
Three of a kind	Three cards of one rank and two other cards that are different from each other
Two pair	Two different pairs of cards and a fifth card that is different from the first four
Pair	Two cards of the same rank and three other cards, all of which are different from each other
Nothing	None of the above hands

TABLE 12.2 Poker hands in order of strength.

*In choosing a sequence of five cards, we allow the ace to be considered higher than the king and also as a one. For example, the sequence A, 2, 3, 4, 5 is allowable as well as the sequence 10, J, Q, K, A.

Historical Highlight

Blaise Pascal

While trying to answer questions about rolling dice and drawing cards, a group of seventeenth-century European mathematicians began to develop a theory of counting. The Frenchman Blaise Pascal, who was one of the central figures in this endeavor, wrote a paper on combinations in 1654.

Pascal was a child prodigy who became interested in Euclid's *Elements* at age 12. Within four years, he wrote a research paper of such quality that some of the leading mathematicians of the time refused to believe that a 16-year-old boy had written it.

Pascal temporarily abandoned mathematics to devote himself to philosophy and religion. However, while suffering from a toothache, he decided to take his mind off the pain by thinking about geometry. Surprisingly, the pain stopped. Pascal took this as a sign from heaven that he should return to mathematics. He resumed his research but soon became seriously ill with dyspepsia, a digestive disorder. Pascal spent the remaining years of his life in excruciating pain, doing little work until his death at age 39 in 1662.

EXAMPLE 3 Counting the Number of Poker Hands

How many ways can a five-card poker hand be drawn from a standard 52-card deck?

SOLUTION:

In drawing a five-card poker hand, we are choosing five cards from 52, which can be done in

$$C(52, 5) = \frac{52!}{5!47!} = \frac{52 \cdot 51 \cdot 50 \cdot 49 \cdot 48}{5 \cdot 4 \cdot 3 \cdot 2 \cdot 1} = 2{,}598{,}960 \text{ ways.}$$

As we will show in Examples 4 and 6, the reason four of a kind beats a full house is that there are fewer ways to obtain four of a kind than there are to obtain a full house. In these examples, we will make heavy use of the fundamental counting principle.

EXAMPLE 4 Drawing Four of a Kind

Determine the number of different ways we can obtain four of a kind when drawing five cards from a standard 52-card deck.

SOLUTION: Review the Be Systematic Strategy on page 5.

We can construct a four-of-a-kind hand in two stages.

Step 1: Pick the rank of the card of which we will have four of a kind. For example, the four-of-a-kind cards could be aces, kings, threes, and so on. This decision can be made in 13 ways.

Step 2: Now that we have chosen four cards, we select the fifth card. For example, if we have selected four kings, then we could pick a jack for the fifth card. We can choose any one of the remaining 48 cards for the fifth card.

Therefore, by the fundamental counting principle, we can build a four-of-a-kind hand in $13 \times 48 = 624$ ways.

Recall that a full house is a hand that has three cards of one rank and two of another rank. For example, three kings and two jacks make a full house. Before determining how many different ways we can draw a full house, we will attack a simpler question.

EXAMPLE 5 Drawing Three Cards of the Same Rank

Determine the number of different ways to draw three cards of the same rank when drawing three cards from a standard 52-card deck.

SOLUTION:

This problem can be divided into two stages. First, choose the rank of the three cards that will be the same. This decision can be made in 13 ways. Second, choose three cards from the four cards in the deck with this rank. This choice can be made in $C(4, 3) = 4$ ways. Using the fundamental counting principle, we see that the three cards of the same rank can be selected in $13 \times 4 = 52$ ways. 11

QUIZ YOURSELF 11

Determine the number of different ways to draw two cards of the same rank when drawing two cards from a standard 52-card deck.

EXAMPLE 6 Drawing a Full House

Determine the number of different ways we can obtain a full house when drawing five cards from a standard 52-card deck.

SOLUTION:

We can think of this problem as consisting of two simpler subproblems.

Subproblem A: Determine the number of ways we can choose the three cards having the same rank. We solved this subproblem in Example 5, where we found that this selection can be made in 52 ways.

Subproblem B: Determine the number of ways we can choose the remaining two cards having the same rank. It is tempting to use the answer found in Quiz Yourself 11, but this would be a mistake. Realize that at this stage, we are selecting the last two cards *after* we have already decided on the three-of-a-kind cards.

Because we have already used one rank for the three-of-a-kind cards, we have only 12 ranks remaining. For example, if we have chosen three kings, then kings cannot be used again for the remaining two cards. Thus, the rank of the last two cards can be chosen in only 12 ways. After we have decided the rank of the last two cards, we must select two of the four cards with this rank. This can be done in $C(4, 2) = 6$ ways. Using the fundamental counting principle, we see that the two-of-a-kind cards can be selected in $12 \times 6 = 72$ ways.

Now we are ready to answer the original question. In constructing a full house, we can first choose the three-of-a-kind cards in 52 ways. Then we can choose the two-of-a-kind cards in 72 ways. Using the fundamental counting principle a final time, we see that a full house can be constructed in $52 \times 72 = 3{,}744$ ways.

EXAMPLE 7 Drawing Three of a Kind

If we select five cards from a standard 52-card deck, in how many ways can we draw a hand with three of a kind?

SOLUTION:

We first must count the number of ways that we can select three cards of the same rank. In Example 5, we found the answer to this question to be 52.

Next we must select the remaining two cards. The fourth card can be chosen among the 12 remaining ranks in 48 ways, and then the fifth card (which must be a different value than the first four) can be chosen in 44 ways. It would seem then that the fourth and fifth cards could be selected in 48×44 ways.

However, there is a slight flaw in our thinking! Suppose, for example, we began with three kings and then for our fourth and fifth cards chose a queen and an eight *in that order.* We could then rearrange the queen and eight in our hand and the hand would be

unchanged. Thus, we recognize that in counting the ways to complete the hand fourth and fifth card, we have counted every possibility twice. Therefore, there are $\frac{48 \times 44}{2} = 1{,}056$ ways to select the last two cards.

Since the first three cards (of the same rank) can be selected in 52 ways and the last two cards can be selected in 1,056 ways, a hand with three of a kind can be chosen in $52 \times 1{,}056 = 54{,}912$ ways.

Now try Exercises 7 to 12.

EXERCISES 12.4

Applying What You've Learned

Exercises 1–6 are based on the slot machine shown in Figure 12.13.

1. Slot machines. In how many ways can we obtain cherries on only the first two wheels?

2. Slot machines. In how many ways can we obtain cherries on all three wheels?

3. Slot machines. In how many ways can we obtain oranges on all three wheels?

4. Slot machines. In how many ways can we obtain plums on all three wheels?

5. Slot machines. In how many ways can we obtain bells on all three wheels?

6. Slot machines. In how many ways can we obtain bars on all three wheels?

Exercises 7–12 deal with the poker hands described in Table 12.2. We assume that we are drawing five cards from a standard 52-card deck.

7. Playing poker. How many ways can we obtain a royal flush (10, J, Q, K, and A, all of the same suit)?

8. Playing poker. In constructing a straight flush, we first choose a suit and then choose a sequence of five cards within the suit.

a. How many ways can we choose the suit?

b. How many ways can we choose the sequence of five cards within the suit?

c. How many ways can we construct a straight flush?

9. Playing poker. To construct a flush, we first select a suit and then choose five cards without regard to order from that suit.

a. How many ways can we choose the suit?

b. How many ways can we choose the five cards?

c. How many ways can we construct a flush?

10. Playing poker. To construct a straight, we must choose five cards in sequence (not all of the same suit).

a. How many ways can we choose a sequence of five cards?

b. For each sequence in part (a) we must select a suit for the first card, a suit for the second card, and so on. In how many ways can we do this? (*Hint:* Draw a slot diagram for the five cards.)

c. Multiply the results from parts (a) and (b).

d. Subtract the number of royal flushes and straight flushes that are not royal, found in Exercises 7 and 8.

11. Playing poker. How many ways can we construct a five-card poker hand that contains two pairs?

12. Playing poker. How many ways can we construct a poker hand that contains only one pair and no other cards of any value?

Communicating Mathematics

13. Why is Example 5 not strictly a combinations problem?

14. Why are we using combinations rather than permutations in solving the examples in this section?

15. On the slot machine in Figure 12.13, would you expect the payoff for three cherries or three oranges to be larger? Explain.

16. On the slot machine in Figure 12.13, would you expect the payoff for three plums or three bells to be larger? Explain.

Challenge Yourself

17. Playing poker. How many poker hands are less in value than one pair? To solve this,

a. Calculate the number of hands of each type in Table 12.2.

b. Subtract the total of the hands found in part (a) from the total of all possible hands.

CHAPTER SUMMARY

You will learn the items listed in this chapter summary more thoroughly if you keep in mind the following advice:

1. Focus on "remembering how to remember" the ideas. What pictures, word analogies, and examples help you remember these ideas?
2. Practice writing each item without looking at the book.
3. Make 3×5 flash cards to break your dependence on the text. Use these cards to give yourself practice tests.

Section	Summary	Example
SECTION 12.1	A small set can be counted by **listing** its elements systematically.	Example 1, pp. 606–607 Example 2, p. 607
	Tree diagrams help visualize counting situations that happen in stages.	Example 3, p. 608 Example 4, pp. 608–609 Example 5, pp. 609–610
	It is often enough to **imagine** a tree without actually drawing it to solve a counting problem.	Example 6, p. 610
SECTION 12.2	The **fundamental counting principle** states that if we want to perform a series of tasks and the first task can be done in a ways, the second can be done in b ways, the third can be done in c ways, and so on, then all the tasks can be done in $a \times b \times c \times \cdots$ ways.	Examples 2–4, pp. 616–617
	Slot diagrams help organize information before applying the fundamental counting principle.	Example 5, pp. 617–618
	Any **special conditions** in a counting problem should be considered first.	Example 6, pp. 618–619
SECTION 12.3	A **permutation** is an ordering of objects in a straight line. We denote the **number of permutations of *n* objects taken *r* at a time** by the notation $P(n, r)$: $$P(n, r) = \frac{n!}{(n-r)!}$$	Example 1, pp. 623–624 Examples 2–3, pp. 624–625
	If we choose r objects from a set of n objects, we are forming a **combination of *n* objects taken *r* at a time.** The symbol $C(n, r)$ represents the number of such combinations.	Examples 4–5, pp. 626–628
	$$C(n, r) = \frac{n!}{r!\,(n-r)!}$$	Example 6, p. 628
	We can **combine counting methods** to solve applied counting problems.	Example 7, p. 629 Example 8, pp. 629–630
	The nth line of **Pascal's triangle** counts the subsets of various sizes of a set having n elements. The rth entry of the nth row of Pascal's triangle is $C(n, r)$.	Example 9, p. 631 Example 10, p. 631
SECTION 12.4	We can use the fundamental counting principle to solve gambling problems.	Examples 1–2, pp. 635–636
	We can use the theory of permutations and combinations to explain the ranking of poker hands.	Examples 3–7, pp. 637–638 pp. 638–639

CHAPTER REVIEW EXERCISES

Section 12.1

1. List all the ways you can select two different members from $S = \{P, Q, R, S\}$. The order in which you select the members is important. For example, PQ is not the same selection as QP.
2. The game *Dungeons and Dragons* uses a tetrahedral die that has four congruent triangular sides numbered 1, 2, 3, and 4. If you were to draw a tree diagram to show the number of ways that three tetrahedral dice could be rolled, how many branches would it have?
3. An outfit consists of a shirt, pants, tie, and jacket. If you have five shirts, four pairs of pants, six ties, and two jackets, how many different outfits can you make?

Section 12.2

4. The rock-climbing club has 14 members. If the club wants to select a president, vice president, and treasurer (all of whom must be different), in how many ways can this be done?
5. The early bird special at TGI Friday's consists of an appetizer, entrée, and dessert. If there are four appetizers, 12 entrées, and six desserts, how many different meals are possible? (Assume that you have to make a selection from each category.)
6. Hector is building a home theater system consisting of a tuner, an optical disc player, speakers, and a high-definition TV. If he can select from three tuners, six speakers, three optical disc players, and four TVs, in how many ways can he configure his system?
7. A campus bus must travel from a (D)ormitory to the (S)tudent center and then to the (F)itness center. How many direct routes are there from D to F, passing through S?

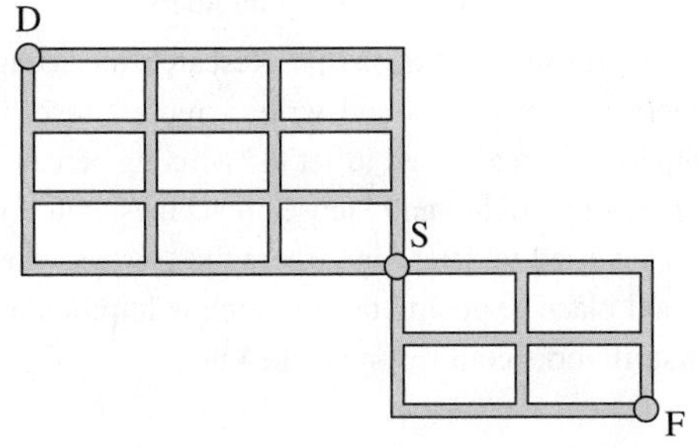

8. Assume that we want to seat Alex, Bonnie, Carl, Daria, Edith, and Frank in a row of six chairs. In how many ways can the seating be done if Alex and Bonnie must sit together?

Section 12.3

9. Explain the main difference between a permutation and a combination.
10. On a psychological test, a patient must match eight terms with their pictures. We are assuming that the same term cannot be used twice. In how many ways can this be done?
11. The research director of NASA must choose three experiments for the next space shuttle. If 17 experiments are proposed, in how many ways can this decision be made?
12. A password for a computer consists of three different letters of the alphabet followed by two different digits from 0 to 9. How many passwords are possible?
13. How is the formula for $P(n, r)$ related to the formula for $C(n, r)$?
14. Use the seventh row of Pascal's triangle to find the number of two-element subsets that can be chosen from a seven-element set.
15. Describe where $C(18, 2)$ would be found in Pascal's triangle.

Section 12.4

16. Using the slot machine in Figure 12.13, in how many ways can we obtain
    a. Plums on the first two wheels only?
    b. Plums on all three wheels?
17. In poker, a flush is a five-card hand all from the same suit. In how many different ways can we construct a flush? (*Hint:* Think of first selecting the suit and then selecting the five cards.)

CHAPTER TEST

1. List all the ways that you can select two different members from the set $S = \{A, B, C, D\}$. The order of the members is important. For example, AB is not the same as BA.
2. Some role-playing games use octahedral dice that have eight congruent triangular sides numbered 1, 2, . . . , 8. If you were to draw a tree diagram showing the number of ways that two octahedral dice could be rolled, how many branches would it have?
3. Allen, Burt, Consuella, Devaun, and Emily are attending a play. If Devaun and Emily want to sit together, in how many ways can we seat them in five seats?
4. Use the sixth row of Pascal's triangle to find the number of three-element subsets that can be chosen from a six-element set.
5. On the slot machine in Figure 12.13, how many ways are there to get bells on all three wheels?

6. Senior communication design students must present three projects for their senior seminar from a list of 12 possibilities. In how many ways can they make their decision?

7. The Social Action Club has 15 members. The club wants to elect a president, vice president, secretary, and public relations officer (all of whom must be different). In how many ways can this be done?

8. In poker, a straight flush is a sequence of consecutive cards all in the same suit. How many ways can you get a straight flush? (*Note:* AKQJ10 is one and 5432A is another.)

9. You are selecting your courses for next semester and you can choose one of three biology courses that will satisfy your science requirement. There are four courses that will satisfy your diversity requirement, and six courses that will fulfill your writing requirement. If you want to choose a biology course, a diversity course, and a writing course, in how many ways can you make your choice?

10. What is the main difference between a permutation and a combination?

11. Raphael is about to sign a sales agreement for his new car. He can choose one of 4 warranties, one of 3 exterior combos, one color from 11 available, and one of 6 interior packages. (Assume that he must choose a color and an interior but can decide not to take any of the other options.) In how many ways can he make his decision?

12. Six men and six women are playing in a mixed doubles tennis tournament. In how many ways can we match all six men with all six women as teams?

13. Write an equation that relates the expressions $P(n, r)$, $C(n, r)$, and $r!$.

14. Where would you find $C(12, 5)$ in Pascal's triangle?

15. In choosing your new cell phone, you must choose one of three purchase agreements, one of nine exterior designs, and one of two calling plans, and you have the option of buying extended service. In how many ways can you make your decision?

16. Viewers of the popular TV game show *Wheel of Fortune* are encouraged to "play at home." Home contestants submit their entries with a viewer ID that consists of three different letters of the alphabet followed by four different digits from 0 to 9. How many different viewer IDs are possible?

GROUP PROJECTS

1. State lotteries. Have each member of the group investigate the lotteries available in his or her home state or some other state of interest. Compute, or find in some other way, the number of ways to win the lottery. Discuss among the group which lottery appears to be the most favorable to the player.

2. Keno. The gambling game keno, which originated in China over 2,000 years ago, is popular in many casinos. In keno there are balls numbered from 1 to 80. The player chooses from 1 to 20 numbers, or "spots," as they are called, and marks them on a card. The casino then randomly picks 20 balls, called the *lucky* balls, from the 80. The remaining 60 balls are called the *unlucky* balls. The more lucky numbers the player has among his spots, the more he wins. With this in mind, calculate the following:

a. How many different ways can the casino choose the 20 lucky balls?

b. If a player plays five spots, how many different ways can the player have all lucky numbers?

c. If the player plays five spots, in how many different ways can the player have exactly three lucky numbers? (*Note:* Realize that to have three lucky numbers, the player must have also two unlucky numbers. This is similar to our analysis of counting the number of ways to get a full house in poker in Section 12.4.)

d. If the player plays 20 spots, in how many different ways can the player have 17 lucky numbers?

3. Slot machine technology. Research the technology of modern slot machines and write a report. Investigate how computer chips are used to set the winning percentage, what we mean by "tight" and "loose" machines, and how mathematics is used to determine where the "virtual wheels" stop. A good place to begin your research is http://entertainment.howstuffworks.com/slot-machine4.htm.

USING TECHNOLOGY

1. Comparing technologies. As a group, investigate various products available for various mobile devices, laptop and desktop computers, and graphing calculators that will help you in doing the calculations in this chapter. Share your research with the group. Then, as a group, vote on the technologies that you would, or would not, recommend to others.

2. Rating technologies. Choose several technologies that the group investigated in Exercise 1 and use them to reproduce the calculations in Examples 3 to 7 in Section 12.3. Explain which technology you *personally* find to be the most useful. Be specific with regard to which features you find to be most important. Also discuss which technologies you find to be least useful.

13

Probability

What Are the Chances?

Are you more likely to be killed by . . . a great white shark . . .
. . . or . . . a vending machine?

Where is the safest place to sit on an airplane in case of a crash?

Is there a greater chance of you winning the Powerball lottery or being struck by lightning?

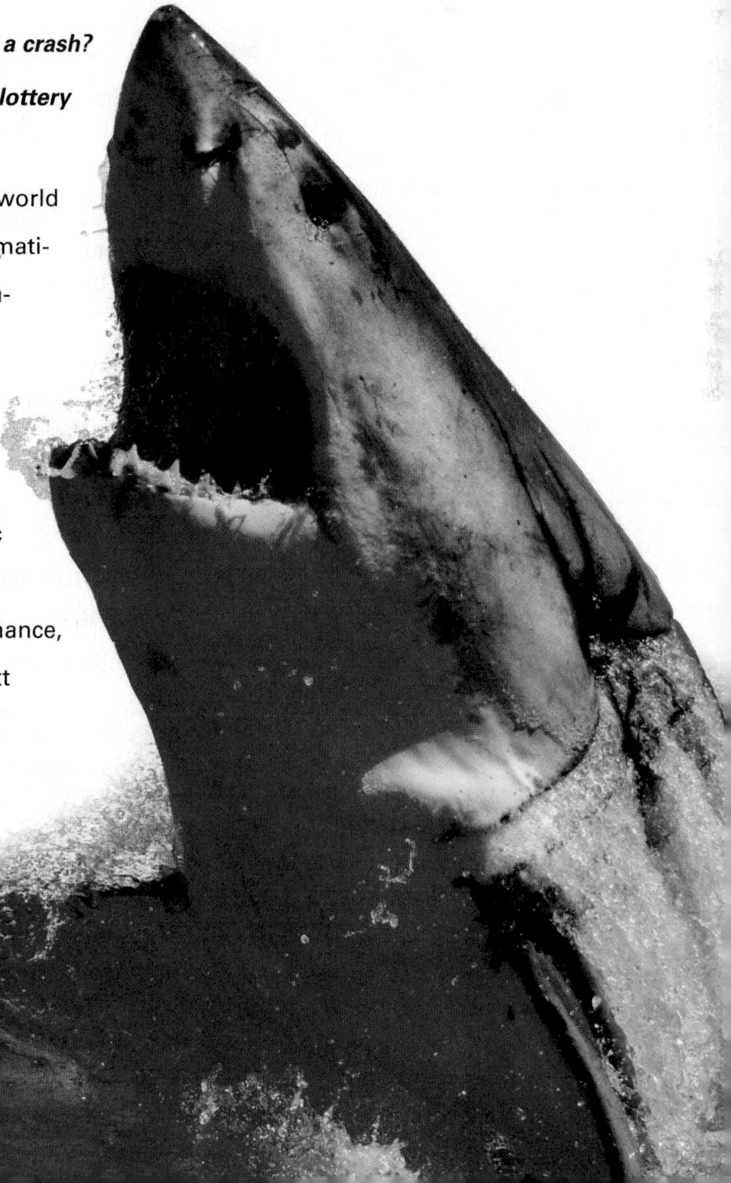

When we ask questions like these, we enter the world of probability that has been studied by mathematicians for hundreds of years. The rules of probability have made insurance companies and casino owners rich, helped farmers in impoverished countries grow more productive grains, and enabled pharmaceutical companies to develop drugs to cure our illnesses.

In this chapter we will introduce you to some of the basic concepts of probability and discuss the role it plays in your life—whether buying insurance policies, playing games of chance, investing in the stock market, or choosing a seat on your next flight to Disney World.

Later in the chapter, we show you a remarkable example of how a person who tests positive on a drug test can have a surprisingly high likelihood of being innocent.

13.1 The Basics of Probability Theory

Objectives

1. Calculate probabilities by counting outcomes in a sample space.
2. Use counting formulas to compute probabilities.
3. Understand how probability theory is used in genetics.
4. Understand the relationship between probability and odds.

Have you ever been camping in the rain? If not, try to imagine it. You are sitting in a tent with the rain flaps down, with nothing to do—the weather report was wrong and you wish that you had picked another weekend for this experience. As we all know, predicting weather is not an exact science. The weather is an example of a random phenomenon. **Random phenomena** are occurrences that vary from day-to-day and case-to-case. In addition to the weather, rolling dice in Monopoly, drilling for oil, and driving your car are all examples of random phenomena.

Although we never know exactly how a random phenomenon will turn out, we can often calculate a number called a **probability** that it will occur in a certain way. We will now begin to introduce some basic probability terminology.

Sample Spaces and Events

KEY POINT

Knowing the sample space helps us compute probabilities.

Our first step in calculating the probability of a random phenomenon is to determine the sample space of an experiment.

> DEFINITIONS An **experiment** is any observation of a random phenomenon. The different possible results of the experiment are called **outcomes.** The set of all possible outcomes for an experiment is called a **sample space.**

If we observe the results of flipping a single coin, we have an example of an experiment. The possible outcomes are head and tail, so a sample space for the experiment would be the set {head, tail}.

EXAMPLE 1 Finding Sample Spaces

Determine a sample space for each experiment.

a) We select an iPad from a production line and determine whether it is defective.
b) Three children are born to a family and we note the birth order with respect to gender.
c) We select one card from a standard 52-card deck (see Figure 12.1 on page 606), and then without returning the card to the deck, we select a second card. We will assume the order in which we select the cards is important.
d) We roll two dice and observe the pair of numbers showing on the top faces.

SOLUTION: Review the Be Systematic Strategy on page 5.

In each case, we find the sample space by collecting the outcomes of the experiment into a set.

a) This sample space is {defective, nondefective}.

b) In this experiment, we want to know not only how many boys and girls are born but also the birth order. For example, a boy followed by two girls is not the same as two girls followed by a boy. The tree diagram* in Figure 13.1 helps us find the sample space.

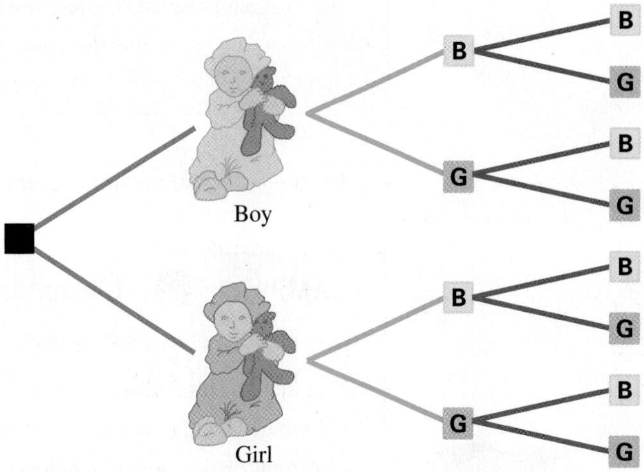

FIGURE 13.1 Tree diagram showing the genders of three children.

There are two ways that the first child can be born, followed by two ways for the second child, and finally two ways for the third. If we abbreviate "boy" by "b" and "girl" by "g," following the branches of the tree diagram gives us the following sample space:

$$\{\text{bbb, bbg, bgb, bgg, gbb, gbg, ggb, ggg}\}.$$

c) This sample space is too large to list; however, we can use the *fundamental counting principle* from Section 12.2 to count its members. Because we can choose the first card in 52 ways and the second card in 51 ways (we are not replacing the first card), we can select the two cards in $52 \times 51 = 2{,}652$ ways.

d) As we did in Example 4 of Section 12.1, we will think of the first die as being red and the second die as being green. As you can see, the pairs (1, 3) and (3, 1) are not the same. With this in mind, the sample space for this experiment consists of the following 36 pairs.

$$\begin{matrix} \{(1,1), & (1,2), & (1,3), & (1,4), & (1,5), & (1,6), \\ (2,1), & (2,2), & (2,3), & (2,4), & (2,5), & (2,6), \\ (3,1), & (3,2), & (3,3), & (3,4), & (3,5), & (3,6), \\ (4,1), & (4,2), & (4,3), & (4,4), & (4,5), & (4,6), \\ (5,1), & (5,2), & (5,3), & (5,4), & (5,5), & (5,6), \\ (6,1), & (6,2), & (6,3), & (6,4), & (6,5), & (6,6)\} \end{matrix}$$

KEY POINT

An event is a subset of a sample space.

Imagine that you are playing Monopoly and are on the verge of going bankrupt. If you roll a total of seven, you will lose the game and you want to know the probability that this will happen. At this point, you are only concerned with the *set* of pairs (1, 6), (2, 5), (3, 4), (4, 3), (5, 2), and (6, 1). This focus on particular *subsets* of the sample space is a recurring theme in probability theory.

DEFINITION In probability theory, an **event** is a subset of the sample space.

*We introduced tree diagrams in Chapter 12.

Keep in mind that *any* subset of the sample space is an event, including such extreme subsets as the empty set, one-element sets, and the whole sample space.

SOME GOOD ADVICE

Although we usually describe events verbally, you should remember that *an event is always a subset of the sample space.* You can use the verbal description to identify the set of outcomes that make up the event.

Example 2 illustrates some events from the sample spaces in Example 1.

EXAMPLE 2 Describing Events as Subsets

Write each event as a subset of the sample space.

a) A head occurs when we flip a single coin.
b) Two girls and one boy are born to a family.
c) A total of five occurs when we roll a pair of dice.

SOLUTION:

a) The set {head} is the event.
b) Noting that the boy can be the first, second, or third child, the event is {bgg, gbg, ggb}.
c) The following set shows how we can roll a total of five on two dice: {(1, 4), (2, 3), (3, 2), (4, 1)}.

Now try Exercises 1 to 8. **1**

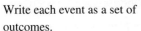

QUIZ YOURSELF 1

Write each event as a set of outcomes.

a) A total of six occurs when we roll two dice.
b) In a family with three children, there are more boys than girls.

We will use the notions of outcome, sample space, and event to compute probabilities. Intuitively, you may expect that when rolling a fair die, each number has the same chance, namely $\frac{1}{6}$, of showing. In predicting weather, a forecaster may state that there is a 30% chance of rain tomorrow. You may believe that you have a 50–50 chance (that is, a 50% chance) of getting a job offer in your field. In each of these examples, we have assigned a number between 0 and 1 to represent the likelihood that the outcome will occur. You can interpret probability intuitively as in the diagram below.

KEY POINT

The probability of an event is the sum of the probabilities of the outcomes in that event.

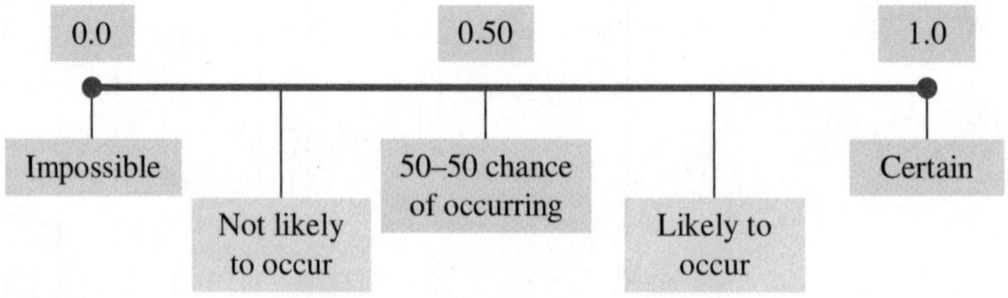

Intuitive Meaning of Probability

DEFINITIONS The **probability of an outcome** in a sample space is a number between 0 and 1 inclusive. The sum of the probabilities of all the outcomes in the sample space must be 1. The **probability of an event** E, written $P(E)$,* is defined as the sum of the probabilities of the outcomes that make up E.

*Do not confuse $P(E)$, which is the notation for the probability of an event, with $P(n, r)$, which is the notation for the number of permutations of n objects taken r at a time (see Section 12.3).

One way to determine probabilities is to use *empirical information.* That is, we make observations and assign probabilities based on those observations.

EMPIRICAL ASSIGNMENT OF PROBABILITIES If E is an event and we perform an experiment several times, then we estimate the probability of E as follows:

$$P(E) = \frac{\text{the number of times } E \text{ occurs}}{\text{the number of times the experiment is performed}}.$$

This ratio is sometimes called the *relative frequency* of E.

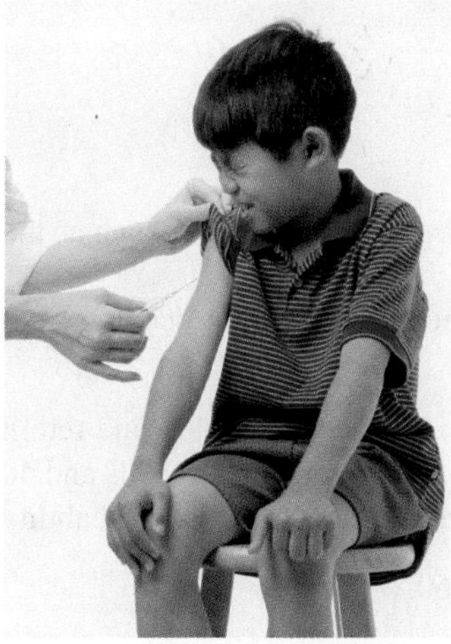

EXAMPLE 3 Using Empirical Information to Assign Probabilities

A pharmaceutical company is testing a new flu vaccine. The experiment is to inject a patient with the vaccine and observe the occurrence of side effects. Assume that we perform this experiment 100 times and obtain the information in Table 13.1.

Side Effects	None	Mild	Severe
Number of Times	72	25	3

TABLE 13.1 Summary of side effects of the flu vaccine.

Based on Table 13.1, if a physician injects a patient with this vaccine, what is the probability that the patient will develop severe side effects?

SOLUTION:

In this case, we base our probability assignment of the event that severe side effects occur on observations. We use the formula for the relative frequency of an event as follows:

$$P(\text{severe side effects}) = \frac{\text{the number of times severe side effects occurred}}{\text{the number of times the experiment was performed}}$$

$$= \frac{3}{100} = 0.03.$$

Thus, from this empirical information, we can expect a 3% chance that there will be severe side effects from this vaccine.

Now try Exercises 23 to 26. 2

QUIZ YOURSELF 2

Use Table 13.1 to find the probability that a patient receiving the flu vaccine will experience no side effects.

EXAMPLE 4 Investigating Marital Data

Table 13.2 summarizes the marital status of men and women in the United States between the ages of 18 and 34 in 2011. All numbers represent thousands. If we randomly pick a male, what is the probability that he is married but not separated?

	Now Married (except separated)	Widowed	Divorced or Separated	Never Married
Males	9,740	36	1,621	24,655
Females	12,279	104	2,328	20,449

TABLE 13.2 Marital status of men and women between the ages of 18 and 34 in the United States in 2011.

Source: U.S. Bureau of the Census.

SOLUTION:

It is important to recognize that only the line labeled "Males," which we highlighted in the table, is relevant to the question you were asked. Therefore, we consider the sample space, S, to be the

$$9{,}740 + 36 + 1{,}621 + 24{,}655 = 36{,}052$$

thousand males mentioned in Table 13.2.

The event, call it M, is the set of 9,740 thousand males that are married but not separated. Thus the probability that we would select a married male that is not separated is

$$P(M) = \frac{n(M)^*}{n(S)} = \frac{9{,}740}{36{,}052} \approx 0.27.$$

Now try Exercises 43 and 44, and 49 to 52. 3

QUIZ YOURSELF 3

Using Table 13.2, if we select a person who has never married, what is the probability that the person is a woman?

Counting and Probability

KEY POINT

We can use counting formulas to compute probabilities.

Another way we can determine probabilities is to use *theoretical information*, such as the counting formulas we discussed in Chapter 12.

In order to understand the difference between using theoretical and empirical information, compare these two experiments:

Experiment one—Without looking, draw a ball from a box, note its color, and return the ball to the box. If you repeat this experiment 100 times and get 60 red balls and 40 blue balls, based on this *empirical information*, you would expect that the probability of getting a red ball on your next draw to be $\frac{60}{100} = 0.60$.

Experiment two—Draw a 5-card hand from a standard 52-card deck. As you will soon see, you can use *theoretical information*, namely, combination formulas from Chapter 12, to calculate the probability that *all* cards will be hearts.

EXAMPLE 5 Using Counting Formulas to Calculate Probabilities

Assign probabilities to the outcomes in the following sample spaces.

a) We flip three fair coins. What is the probability of each outcome in this sample space?
b) We draw a 5-card hand randomly from a standard 52-card deck. What is the probability that we draw one particular hand?

SOLUTION: Review the Be Systematic Strategy on page 5.

a) This sample space has eight outcomes, as we show in Figure 13.2. Because the coins are fair, we expect that heads and tails are equally likely to occur. Therefore, it is reasonable to assign a probability of $\frac{1}{8}$ to each outcome in this sample space.

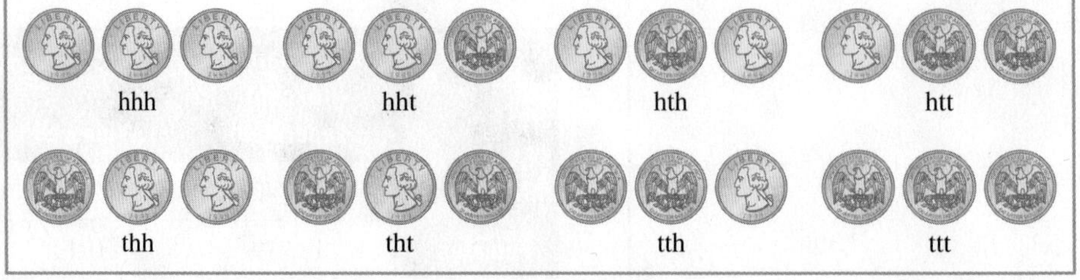

FIGURE 13.2 Eight theoretically possible outcomes for flipping three coins.

---

*Recall that $n(M)$ is the cardinal number of set M (see Section 2.1).

Math in Your Life*

Why Does It Always Happen to You?

When you wait at a toll booth, why does it seem that the line of cars next to you moves faster than your line? If you look into a sock drawer, do you notice how many unmatched socks there are? Why does the buttered side of a slice of bread almost always land face down if you drop the bread while making a sandwich? Do such annoyances affect only you? Or is there a mathematical explanation?

First let's consider the socks. Suppose you have 10 pairs of socks in a drawer and lose 1 sock, destroying a pair. Of the 19 remaining socks, there is only 1 unpaired sock. Therefore, if you lose a second sock, the probability it will be a paired sock is $\frac{18}{19}$. Now you have 2 unpaired socks and 16 paired socks. The probability that the third sock you lose will be part of a pair is still overwhelming. Continuing this line of thought, you see that it will be quite a while before probability theory predicts that you can expect to lose an unpaired sock.

The problem with the buttered bread is simple to explain. In fact you might conduct an experiment to simulate this situation with an object such as a computer mouse pad. If you slide the object off the edge of a table, as it begins to fall it rotates halfway around so that the top side is facing down. However, there is usually not enough time for the object to rotate back to the upward position before it hits the floor. You could conduct this experiment 100 times and determine the empirical probability that an object you slide off the table will land face down. So again, it's not bad luck—it's probability.

The slow-moving line at the toll booth is the easiest to explain. If we assume that delays in any line occur randomly, then one of the lines—yours, the one to the left, or the one to the right—will move fastest. Therefore, the probability that the fastest line will be yours is only $\frac{1}{3}$.

QUIZ YOURSELF 4

If we were to flip four coins, how many outcomes would there be in the sample space?

b) In Chapter 12, we found that there are $C(52, 5) = 2{,}598{,}960$ different ways to choose 5 cards from a deck of 52. Because we are drawing the cards randomly, each hand has the same chance of being drawn. Therefore, the probability of drawing any one hand is $\frac{1}{2{,}598{,}960}$. **4**

In a sample space with equally likely outcomes, it is easy to calculate the probability of any event by using the following formula.

CALCULATING PROBABILITY WHEN OUTCOMES ARE EQUALLY LIKELY If E is an event in a sample space S with all *equally likely outcomes,* then the probability of E is given by the formula:

$$P(E) = \frac{n(E)}{n(S)}.$$

EXAMPLE 6 Computing Probability of Events

a) What is the probability in a family with three children that two of the children are girls?

b) What is the probability that a total of four shows when we roll two fair dice?

†c) If we draw a 5-card hand from a standard 52-card deck, what is the probability that all 5 cards are hearts?

†d) Leonard, Sheldon, Penny, and Raj belong to *The Greens*, their school's 10-person environmental club. Two people from the club will be selected randomly to attend a conference in the Everglades. What is the probability that two of these four friends will be selected?

---

*This Math in Your Life is based on Robert Matthews, "Murphy's Law or Coincidence." *Reader's Digest*, March 1998, pp. 25–30.

†Questions c) and d) require counting formulas from Chapter 12.

SOLUTION:

In each situation, we assume that the outcomes are equally likely.

a) We saw in Example 1 that there are eight outcomes in this sample space. We denote the event that two of the children are girls by the set $G = \{bgg, gbg, ggb\}$. Thus,

$$P(G) = \frac{n(G)}{n(S)} = \frac{3}{8}.$$

b) The sample space for rolling two dice has 36 ordered pairs of numbers. We will represent the event "rolling a four" by F. Then $F = \{(1, 3), (2, 2), (3, 1)\}$. Thus,

$$P(F) = \frac{n(F)}{n(S)} = \frac{3}{36} = \frac{1}{12}.$$

c) From Example 5, we know that there are $C(52, 5)$ ways to select a 5-card hand from a 52-card deck. If we want to draw only hearts, then we are selecting 5 hearts from the 13 available, which can be done in $C(13, 5)$ ways. Thus, the probability that all five cards selected will be hearts is

$$\frac{C(13, 5)}{C(52, 5)} = \frac{1{,}287}{2{,}598{,}960} \approx 0.000495.$$

d) The sample space, S, consists of all the ways we can select two people from the 10 members in the club. As you know from Chapter 12, we can choose 2 people from 10 in $C(10, 2) = \frac{10!}{8! \cdot 2!} = \frac{10 \cdot 9}{2} = 45$ ways. The event, call it E, consists of the ways we can choose two of the four friends. This can be done in $C(4, 2) = 6$ ways. Because all elements in S (choices of two people) are equally likely, the probability of E is

$$P(E) = \frac{n(E)}{n(S)} = \frac{C(4, 2)}{C(10, 2)} = \frac{6}{45} = \frac{2}{15}.$$

Now try Exercises 9 to 14. 5

QUIZ YOURSELF 5

a) If we roll two fair dice, what is the probability of rolling a total of eight?
b) If we select 2 cards randomly from a standard 52-card deck, what is the probability that both are face cards?

SOME GOOD ADVICE

If the outcomes in a sample space are not equally likely, then you cannot use the formula $P(E) = \frac{n(E)}{n(S)}$. In that case, to find $P(E)$, you must add the probabilities of all the individual outcomes in E.

Suppose that we have a sample space with equally likely outcomes. An event, E, can contain none, some, or all of the outcomes in the sample space S, so we can say

$$0 \le n(E) \le n(S).$$

Dividing this inequality by the positive quantity $n(S)$, we get the inequality

$$\frac{0}{n(S)} \le \frac{n(E)}{n(S)} \le \frac{n(S)}{n(S)},$$

which simplifies to $0 \le \frac{n(E)}{n(S)} \le 1$. This gives us the first probability property listed below. The other properties are easy to see.

BASIC PROPERTIES OF PROBABILITY Assume that S is a sample space for some experiment and E is an event in S.

1. $0 \le P(E) \le 1$ 2. $P(\emptyset) = 0$ 3. $P(S) = 1$

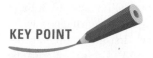

KEY POINT

Probability theory helps explain genetic theory.

Probability and Genetics*

In the nineteenth century, the Austrian monk Gregor Mendel noticed while cross-breeding plants that often a characteristic of the plants would disappear in the first-generation offspring but reappear in the second generation. He theorized that the first-generation plants contained a hidden factor (which we now call a *gene*) that was somehow transmitted to the second generation to enable the characteristic to reappear.

To check his theory, he selected a characteristic such as seed color—some peas had yellow seeds and some had green. Then when he was sure that he had bred plants that would produce only yellow seeds or green seeds, he was ready to begin his experiment. Mendel believed that one of the colors was *dominant* and the other was *recessive.* Which turned out to be the case, because when yellow-seeded plants were crossed with green-seeded plants, the offspring had yellow seeds. When Mendel crossed these offspring for a second generation, he found that 6,022 plants had yellow seeds and 2,001 had green seeds, which is almost exactly a ratio of 3 to 1. Because Mendel was skilled in mathematics as well as biology, he gave the following explanation for what he had observed.

We will represent the gene that produces the yellow seed by Y and the gene that produces the green seed by g. The uppercase Y indicates that yellow is dominant and the lowercase g indicates that green is recessive.†

Figure 13.3 shows the possible genetic makeup of the offspring from crossing a plant with pure yellow seeds and a plant with pure green seeds. Every one of these offspring has a Yg pair of genes. Because "yellow seed" is dominant over "green seed," every plant in the first generation will have yellow seeds.

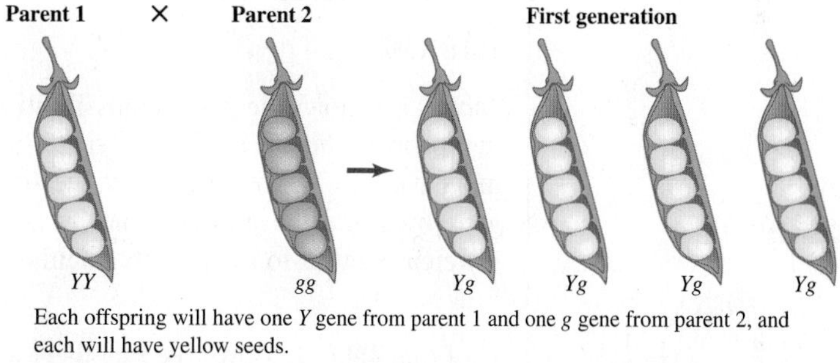
Parent 1
×
Parent 2
First generation
YY
gg
Yg
Yg
Yg
Yg
Each offspring will have one Y gene from parent 1 and one g gene from parent 2, and each will have yellow seeds.

FIGURE 13.3 The possible first-generation offspring we obtain by crossing pure yellow and pure green parents.

Figure 13.4 shows the possible genetic outcomes that can occur if we cross two first-generation pea plants. We summarize Figure 13.4 in Table 13.3, which is called a **Punnett square.**

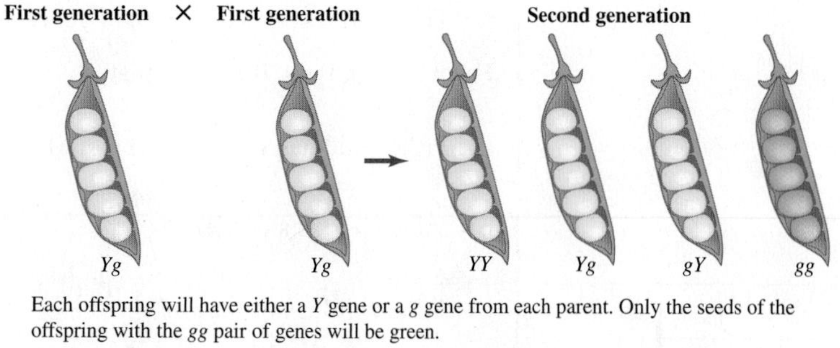
First generation
×
First generation
Second generation
Yg
Yg
YY
Yg
gY
gg
Each offspring will have either a Y gene or a g gene from each parent. Only the seeds of the offspring with the gg pair of genes will be green.

FIGURE 13.4 The possible second-generation offspring we obtain by crossing the first-generation offspring of pure yellow and pure green parents.

---

*Although Mendel's theory that we present in this section has been shown to be correct, you can find interesting articles that argue Mendel's results are "too perfect" and that Mendel faked some of his data.

†Biology books customarily use slightly different notation to indicate genes. For example, to indicate that yellow is dominant over green, you may see the notation *Yy*. The capital *Y* represents the yellow dominant gene and the lowercase *y* indicates the recessive green gene.

		First-Generation Plant	
		Y	*g*
First-Generation Plant	*Y*	*YY*	*Yg*
	g	*gY*	*gg*

TABLE 13.3 The genetic possibilities of crossing two plants that each have one yellow-seed and one green-seed gene.

As you can see in Table 13.3, there are four things that can happen when we cross two first-generation plants. We can get a *YY*, *Yg*, *gY*, or *gg* type of plant. Of these four possibilities, only the *gg* will result in green seeds, which explains why Mendel saw the recessive characteristic return in roughly one fourth of the second-generation plants.

EXAMPLE 7 Using Probability to Explain Genetic Diseases

Sickle-cell anemia is a serious inherited disease that is about 30 times more likely to occur in African American babies than in non–African American babies. A person with two sickle-cell genes will have the disease, but a person with only one sickle-cell gene will be a carrier of the disease.

If two parents who are carriers of sickle-cell anemia have a child, what is the probability of each of the following:

a) The child has sickle-cell anemia?

b) The child is a carrier?

c) The child is disease free?

SOLUTION:

Table 13.4 shows the genetic possibilities when two people who are carriers of sickle-cell anemia have a child. We will denote the sickle-cell gene by *s* and the normal gene by *n*. We use lowercase letters to indicate that neither *s* nor *n* is dominant.

		Second Parent	
		s	*n*
First Parent	*s*	*ss* has disease	*sn* carrier
	n	*ns* carrier	*nn* normal

TABLE 13.4 The genetic possibilities for children of two parents with sickle-cell trait.

From Table 13.4, we see that there are four equally likely outcomes for the child.

- The child receives two sickle-cell genes and therefore has the disease.
- The child receives a sickle-cell gene from the first parent and a normal gene from the second parent and therefore is a carrier.
- The child receives a normal gene from the first parent and a sickle-cell gene from the second parent and therefore is a carrier.
- The child receives two normal genes and therefore is disease free.

From this analysis, it is clear that

a) $P(\text{the child has sickle-cell anemia}) = \frac{1}{4}$.

b) $P(\text{the child is a carrier}) = \frac{1}{2}$.

c) $P(\text{the child is normal}) = \frac{1}{4}$.

Now try Exercises 31 to 38.

Odds

KEY POINT

In computing odds remember "against" versus "for."

We often use the word *odds* to express the notion of probability. When we do this, we usually state the odds *against* something happening. For example, before the 2011 Kentucky Derby, Las Vegas oddsmakers had set the odds against I'll Have Another, the eventual winner of the derby, to be 12 to 1.

When you calculate odds against an event, it is helpful to think of what is against the event as compared with what is in favor of the event. For example, if you roll two dice and want to find the odds against rolling a seven, you think that there are 30 pairs that will give you a nonseven and six that will give you a seven. Therefore, the odds against rolling a total of seven are 30 to 6. We write this as 30:6. Just like reducing a fraction, we can write these odds as 5:1. Sometimes you might see odds written as a fraction, such as 30/6, but, for the most part, we will avoid using this notation.

DEFINITION If the outcomes of a sample space are equally likely, then the **odds against an event** *E* are simply the number of outcomes that are against *E* compared with the number of outcomes in favor of *E* occurring. We would write these odds as $n(E'):n(E)$, where E' is the complement of event *E*. (We discussed set complements in Section 2.3.)

Recall that in Example 1, you saw there were eight ways for boys and girls to be born in order in a family. To find the odds against all children being of the same gender, you think of the six outcomes that are against this happening versus the two outcomes that are in favor, as in the accompanying diagram.

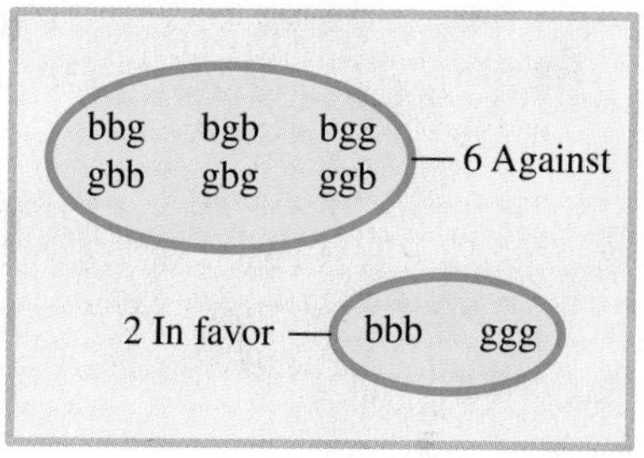

Therefore, the odds against all three children being of the same gender are 6:2, which we can reduce and rewrite as 3:1. We could also say that the *odds in favor* of all children being of the same gender are 1:3. It is important to understand that the odds in favor of an event are not the same as the probability of an event. In the example we are discussing, the probability of all children being the same gender is $\frac{2}{8} = \frac{1}{4}$.

PROBLEM SOLVING

Strategy: The Analogies Principle

Making an analogy with a real-life situation helps you to remember the meaning of mathematical terminology. For example, when General Custer fought Chief Sitting Bull at the battle of Little Big Horn, Custer had about 650 soldiers and Sitting Bull had 2,500 braves. If we think of how many were for Custer and how many were against him, we might say that the odds against Custer winning were 2,500 to 650. Similarly, the odds against Sitting Bull winning were 650 to 2,500.

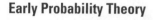

Historical Highlight

Early Probability Theory

We can trace probability theory back to prehistoric times. Archaeologists have found numerous small bones, called *astrogali*, that are believed to have been used in playing various games of chance. Similar bones were later ground into the shape of cubes and decorated in various ways. They eventually evolved into modern dice. Games of chance were played for thousands of years, but it was not until the sixteenth century that a serious attempt was made to study them using mathematics.

Among the first to study the mathematics of games of chance was the Italian Girolamo Cardano. In addition to his interest in probability, Cardano was an astrologer. This caused him frequent misfortune. In one instance, Cardano cast the horoscope of Jesus and for this blasphemy he was sent to prison. Shortly thereafter, he predicted the exact date of his own death and, when the day arrived, Cardano made his prediction come true by taking his own life.

EXAMPLE 8 Calculating Odds on a Roulette Wheel

A common type of roulette wheel has 38 equal-size compartments. Thirty-six of the compartments are numbered 1 to 36 with half of them colored red and the other half black. The remaining 2 compartments are green and numbered 0 and 00. A small ball is placed on the spinning wheel and when the wheel stops, the ball rests in one of the compartments. What are the odds against the ball landing on red?

SOLUTION: Review the Analogies Principle on page 13.

This is an experiment with 38 equally likely outcomes. Because 18 of these are in favor of the event "the ball lands on red" and 20 are against the event, the odds against red are 20 to 18. We can write this as 20:18, which we may reduce to 10:9.

Although we have defined odds in terms of counting, we also can think of odds in terms of probability. Notice in the next definition we are comparing "probability against" with "probability for."

PROBABILITY FORMULA FOR COMPUTING ODDS If E' is the complement of the event E, then the odds against E are

$$\frac{P(E')}{P(E)}.$$

You may have been surprised that we have not used the "$a:b$" notation in this definition of odds. However, we have a good reason for doing this. If the probability of an event E is 0.30, then it is awkward to say that the odds against E are 0.70 to 0.30. It is better to say

$$\text{odds against } E = \frac{\text{probability of } E'}{\text{probability of } E} = \frac{0.70}{0.30} = \frac{70}{30}.$$

We could then say that the odds against E are 70 to 30, or 7 to 3. We will use this approach in Example 9.

EXAMPLE 9 The Odds Against Surviving an Airplane Crash

The safest place to be seated in the event of an airplane crash is in an aisle seat above the wings. In this case, the probability of surviving the crash is 56%. What are the odds against you surviving?

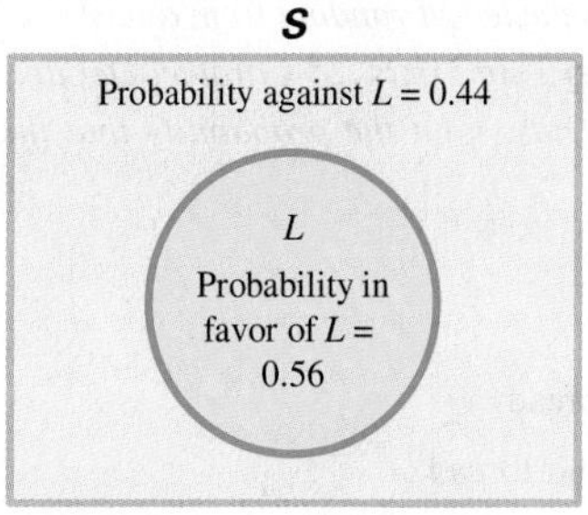

FIGURE 13.5 The odds against you surviving are the probability of you not surviving versus the probability of you surviving.

SOLUTION:

Call the event L (for live). Figure 13.5 illustrates this situation.

Thus the odds against you surviving are

$$\frac{P(L')}{P(L)} = \frac{0.44}{0.56} = \frac{0.44 \times 100}{0.56 \times 100} = \frac{44}{56} = \frac{11}{14}.$$

Thus we would say that if you sit over the wings, the odds against you surviving are 11 to 14.

Now try Exercises 15 to 18.

EXERCISES 13.1

Sharpening Your Skills

In Exercises 1–4, write each event as a set of outcomes. If the event is large, you may describe the event without writing it out.

1. When we roll two dice, the total showing is seven.

2. When we roll two dice, the total showing is five.

3. We flip three coins and obtain more (h)eads than (t)ails.

4. We select a red face card from a standard 52-card deck.

In Exercises 5–8, use the given spinner to write the event as a set of outcomes. Abbreviate "red" as "r," "blue" as "b," and "yellow" as "y."

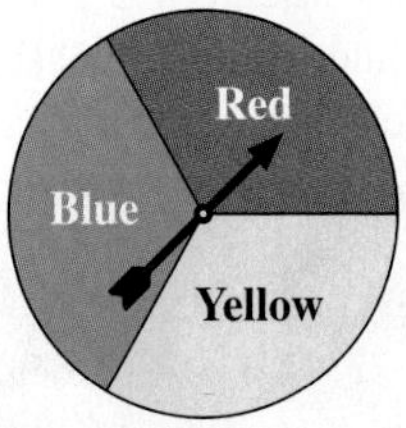

5. Red appears exactly once when we spin the given spinner two times.

6. Yellow appears at least once when we spin the given spinner two times.

7. Blue appears exactly twice in three spins of the spinner.

8. Red does not appear in three spins of the spinner.

9. We are rolling two four-sided dice having the numbers 1, 2, 3, and 4 on their faces. Outcomes in the sample space are pairs such as (1, 3) and (4, 4).

a. How many elements are in the sample space?

b. Express the event "the total showing is even" as a set.

c. What is the probability that the total showing is even?

d. What is the probability that the total showing is greater than six?

10. We are rolling two four-sided dice. One die has the numbers 1, 2, 3, and 4 on its faces. The other die has the numbers 3, 4, 5, and 6 on its faces.

a. How many elements are in the sample space?

b. Express the event "the total showing equals seven" as a set.

c. What is the probability that the total showing is equal to seven?

11. Singers (E)nrique, (K)aty, (R)ihanna, and (B)runo are performing a charity concert for world hunger and their order of appearance will be chosen randomly. We will represent outcomes in the sample space as strings of letters such as ERBK or KERB.

a. How many elements are in this sample space?

b. Express the event "Katy and Rihanna perform consecutively" as a set.

c. What is the probability that Katy and Rihanna perform consecutively?

12. We are flipping four coins. Outcomes in the sample space are represented by strings of Hs and Ts such as TTHT and HHTT.

a. How many elements are in this sample space?

b. Express the event "there are more heads than tails" as a set.

c. What is the probability that there are more heads than tails?

d. What is the probability that there are an equal number of heads and tails?

13. An experimenter testing for extrasensory perception has five cards with pictures of a (s)tar, a (c)ircle, (w)iggly lines, a (d)ollar sign, and a (h)eart. She selects two cards without replacement. Outcomes in the sample space are represented by pairs such as (s, d) and (h, c).

a. How many elements are in this sample space?

b. Express the event "a star appears on one of the cards" as a set.

c. What is the probability that a star appears on one of the cards?

d. What is the probability that a heart does not appear?

14. Amy and Louisa are going with three friends to see *Avatar II*. If they are seated randomly in a row in five consecutive seats, what is the probability that Amy and Louisa will sit together?

In Exercises 15–18, a) Find the probability of the given event. b) Find the odds against the given event.

15. A total of nine shows when we roll two fair dice.

16. A total of three shows when we roll two fair dice.

17. We draw a heart when we select 1 card randomly from a standard 52-card deck.

18. We draw a face card when we select 1 card randomly from a standard 52-card deck.

In Exercises 19–22, assume that we are drawing a 5-card hand from a standard 52-card deck.

19. What is the probability that all cards are diamonds?

20. What is the probability that all cards are face cards?

21. What is the probability that all cards are of the same suit?

22. What is the probability that all cards are red?

The residents of a small town and the surrounding area are divided over the proposed construction of a sprint car racetrack in the town. Use the following table to answer Exercises 23 and 24.

	Support Racetrack	Oppose Racetrack
Live in Town	1,512	2,268
Live in Surrounding Area	3,528	1,764

23. If a newspaper reporter randomly selects a person to interview from these people,

a. what is the probability that the person supports the racetrack?

b. what are the odds in favor of the person supporting the racetrack?

24. If a newspaper reporter randomly selects a person from town to interview,

a. what is the probability that the person supports the racetrack?

b. what are the odds in favor of the person supporting the racetrack?

Applying What You've Learned

25. In a given year, 2,048,861 males and 1,951,379 females were born in the United States. If a child is selected randomly from this group, what is the probability that it is a female?

26. Playing a carnival game. A fish pond at a carnival contains 84 goldfish, 30 tetras, 17 angel fish, and 29 cory cats. If Marcelo picks a fish at random from the pond, what is the probability that it is a goldfish?

In Exercises 27–30, a cookie is selected at random from a basket of Girl Scout Cookies containing 16 Thin Mints, 28 Caramel DeLites, 12 Shortbreads, and 24 Shout Outs. Find the probability that the specified cookie is selected.

27. a Thin Mint

28. not a Shout Out

29. a Caramel DeLite or Shortbread

30. neither a Thin Mint nor a Shortbread

31. The following table lists some of the empirical results that Mendel obtained in his experiments in cross-breeding pea plants.

Characteristics That Were Crossbred	First-Generation Plants	Second-Generation Plants
Tall vs. short	All tall	787 tall 277 short
Smooth seeds vs. wrinkled seeds	All smooth seeds	5,474 smooth 1,850 wrinkled

Assume that we are cross-breeding genetically pure tall plants with genetically pure short plants. Use this information to assign the probability that a second-generation plant will be short. How consistent is this with the theoretical results that Mendel derived?

32. Assume that we are cross-breeding genetically pure smooth-seed plants with genetically pure wrinkled-seed plants. Use the information provided in Exercise 31 to assign the probability that a second-generation plant will have smooth seeds. How consistent is this with the theoretical results that Mendel derived?

In Exercises 33 and 34, construct a Punnett square as we did in Table 13.4 to show the probabilities for the offspring.

33. One parent who has sickle-cell anemia and one parent who is a carrier have a child. Find the probability that the child is a carrier of sickle-cell anemia.

34. One parent who has sickle-cell anemia and one parent who is a carrier have a child. Find the probability that the child has sickle-cell anemia.

In cross-breeding snapdragons, Mendel found that flower color does not dominate as happens with peas. For example, a snapdragon with one red and one white gene will have pink flowers. In Exercises 35 and 36, analyze the cross-breeding experiment as we did in the discussion prior to Example 7.

35. a. Construct a Punnett square showing the results of crossing a purebred white snapdragon with a purebred red one.

b. What is the probability of getting red flowers in the first-generation plants? What is the probability of getting white? Of getting pink?

36. a. If we cross two pink snapdragons, draw a Punnett square that shows the results of crossing two of these first-generation plants.

b. What is the probability of getting red flowers in the second-generation plants? What about white? Pink?

37. Cystic fibrosis is a serious inherited lung disorder that often causes death in victims during early childhood. Because the gene for this disease is recessive, two apparently healthy adults, called *carriers*, can have a child with the disease. We will denote the normal gene by N and the cystic fibrosis gene by c to indicate its recessive nature.

a. Construct a Punnett square as we did in Example 7 to describe the genetic possibilities for a child whose two parents are carriers of cystic fibrosis.

b. What is the probability that this child will have the disease?

38. From the Punnett square in Exercise 37, what is the probability that a child of two carriers will

a. be normal?

b. be a carrier?

For Exercises 39–42, assume that a dart is thrown and hits somewhere in the diagram shown. Find the probability that the dart hits the specified area.

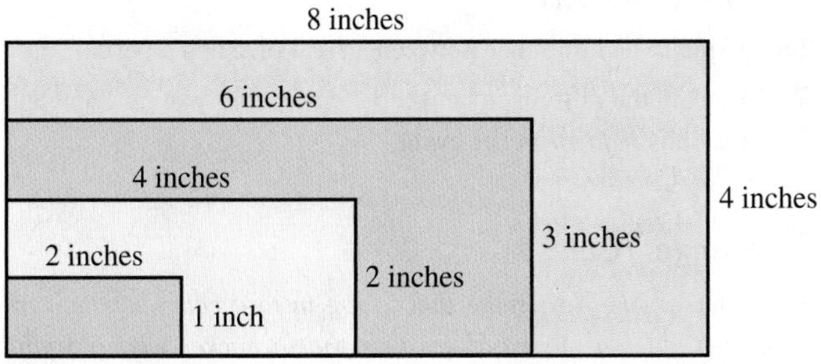

39. the yellow area

40. a green area

41. a yellow or blue area

42. a non-blue area

43. Assume that the following table summarizes a survey involving the relationship between living arrangements and grade point average for a group of students.

	On Campus	At Home	Apartment	Totals
Below 2.5	98	40	44	182
2.5 to 3.5	64	25	20	109
Over 3.5	17	4	8	29
Totals	179	69	72	320

If we select a student randomly from this group, what is the probability that the student has a grade point average of at least 2.5?

44. Using the data in Exercise 43, if a student is selected randomly, what is the probability that the student lives off campus?

Use this replica of the Monopoly game board to answer Exercises 45–48.

45. Assume that your game piece is on the Electric Company. If you land on either St. James Place, Tennessee Avenue, or New York Avenue, you will go bankrupt. What is the probability that you avoid these properties?

46. Assume that your game piece is on Pacific Avenue. If you land on either Park Place or Boardwalk, you will go bankrupt. What is the probability that you avoid these properties?

47. Your game piece is on Virginia Avenue. What is the probability that you will land on a railroad on your next move?

48. Your game piece is on Pennsylvania Avenue. What is the probability that you will have to pay a tax on your next move?

In Exercises 49–52 assume that we are randomly picking a person described in Table 13.2 from Example 4.

49. If we pick a divorced or separated person, what is the probability that the person is a woman?

50. If we pick a woman, what is the probability that she is married but not separated?

51. If we pick a never-married person, what is the probability that the person is a man?

52. If we pick a man, what is the probability that he is a widower?

Use spinners A, B, and C below to do Exercises 53 and 54.

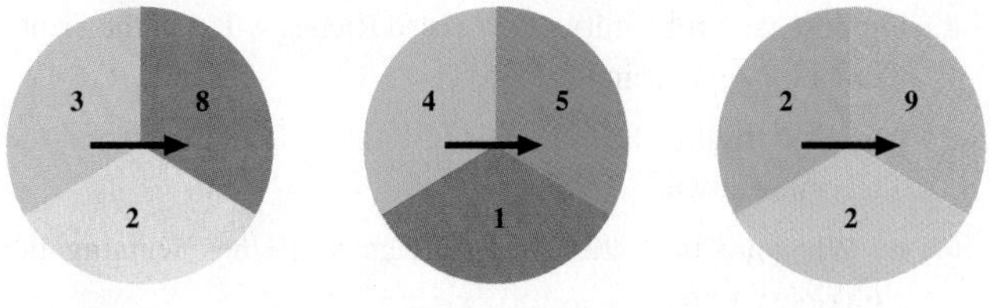

Spinner A Spinner B Spinner C

53. **Calculating the probability of winning a game.** You and I are going to play a game. You pick either spinner A or B and I pick the other. We each spin our spinner and whoever spins the higher number wins the game. (If a spinner lands on a boundary, that spinner is spun again.) Which spinner should you choose and what is the probability that you will win?

54. **Calculating the probability of winning a game.** Repeat Exercise 53 using spinners A and C. (If both spinners stop on a 2, we spin again.)

In horse racing, a trifecta is a race in which you must pick the first-, second-, and third-place winners in their proper order to win a payoff.

55. If eight horses are racing and you randomly select three as your bet in the trifecta, what is the probability that you will win? (Assume that all horses have the same chance to win.)

56. If 10 horses are racing and you randomly select 3 as your bet in the trifecta, what is the probability that you will win? (Assume that all horses have the same chance to win.)

57. If the odds against event E are 5 to 2, what is the probability of E?

58. If $P(E) = 0.45$, then what are the odds against E?

59. If the odds against the U.S. women's soccer team winning the World Cup are 7 to 5, what is the probability that they will win the World Cup?

60. If the odds against the U.S. men's volleyball team defeating China in the Summer Olympic Games were 9 to 3, what is the probability that the United States defeated China?

61. Suppose the probability that the Yankees will win the World Series is 0.30.

 a. What are the odds in favor of the Yankees winning the World Series?

 b. What are the odds against the Yankees winning the World Series?

62. Suppose the probability that Rags to Riches will win the Triple Crown in horse racing is 0.15.

 a. What are the odds in favor of Rags to Riches winning the Triple Crown?

 b. What are the odds against Rags to Riches winning the Triple Crown?

63. In the New York Lotto, the player must correctly pick six numbers from the numbers 1 to 59. What are the odds against winning this lottery?

64. Go to the National Safety Council's Web site at http://www.nsc.org/ to find the odds against you being killed by lightning in your lifetime. How many more times likely is it that you will be killed by lightning than win the New York Lotto?

We sometimes use terminology such as "the odds are 1 in 10" of an event. We will interpret this to mean the "the probability of the event is 1/10" or "the odds in favor of the event are 1 to 9."

65. The odds that a member of the U.S. Senate (selected randomly) has a law degree are 1 in 1.75. What is the probability that a member of the Senate (selected randomly) has a law degree?

66. If the manager of a major league baseball team is picked randomly, the odds that the manager has been with the team less than 4 years are 1 in 1.43. What is the probability that that manager has been with the team less than 4 years?

Communicating Mathematics

67. Arianna is solving the following probability problem on an exam that begins:

 We are selecting three digital picture frames from a production line to determine if they are defective or non-defective.

 She intends to use the formula $\frac{n(E)}{n(S)}$. Is this a valid approach? Explain.

68. If the odds against an event are a to b, what are the odds in favor of the event?

69. Explain the difference between an outcome and an event.

70. Explain the difference between the probability of an event and the odds *in favor* of the event.

Challenge Yourself

In Exercises 71–74, assume that to log in to a computer network you must enter a password. Assume that a hacker who is trying to break into the system randomly types one password every 10 seconds. If the hacker does not enter a valid password within 3 minutes, the system will not allow any further attempts to log in. What is the probability that the hacker will be successful in discovering a valid password for each type of password?

71. **Forming passwords.** The password consists of two letters followed by three digits. Case does not matter for the letters. Thus, Ca154 and CA154 would be considered the same password.

72. **Forming passwords.** Repeat Exercise 71, but now assume that passwords are case sensitive—that is, uppercase and lowercase letters are considered to be different.

73. **Forming passwords.** The password consists of any sequence of two letters and three digits. Case does not matter for the letters. Thus, B12q5 and b12Q5 would be considered the same password.

74. **Forming passwords.** Repeat Exercise 73, but now assume that passwords are case sensitive—that is, uppercase and lowercase letters are considered to be different.

75. Find some examples of advertising claims in the media. In what way are these claims probabilities?

76. **a.** Flip a coin 100 times. How do your empirical results compare with the theoretical probabilities for obtaining heads and tails?

b. Roll a pair of dice 100 times. How do your empirical results compare with the theoretical probabilities for rolling a total of two, three, four, and so on?

c. Toss an irregular object such as a thumbtack 1,000 times. After doing this, what probability would you assign to the thumbtack landing point up? What about point down? Does it matter what kind of thumbtack you use? Explain.

77. Investigate other genetic diseases, such as Tay-Sachs or Huntington's disease. Explain how the mathematics of the genetics of these diseases is similar or dissimilar to the examples we studied in this section.

13.2 Complements and Unions of Events

Objectives

1. Understand the relationship between the probability of an event and the probability of its complement.
2. Calculate the probability of the union of two events.
3. Use complement and union formulas to compute the probability of an event.

Suppose that you are attending a luncheon on your campus to develop a campaign to fight world hunger. Someone at your table might ask, "This room is pretty crowded. I wonder how many are here?" You look around and see that there are eight tables that seat 10 people each and notice that there are only six empty seats. So, without actually counting, you can quickly say that there are 74 people attending.

Similarly, you can often solve complex probability problems by applying a few simple, intuitive formulas. We will use some results from set theory in Chapter 2 to develop rules for computing the probability of the complement and the union of events.

Complements of Events

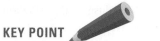

KEY POINT

We can compute the probability of an event by finding the probability of its complement.

You can frequently solve a mathematics problem by rephrasing it as an equivalent question that is easier to answer. For example, suppose that you want to calculate the probability of event E,* but find that E is too complicated to understand easily. If you remember that the total probability available in a sample space is 1, then it may be simpler to find the probability of E's complement and subtract that number from 1.

COMPUTING THE PROBABILITY OF THE COMPLEMENT OF AN EVENT
If E is an event, then $P(E') = 1 - P(E)$.

Of course we can state this result differently as $P(E) = 1 - P(E')$ or $P(E) + P(E') = 1$. We illustrate this formula in Figure 13.6.

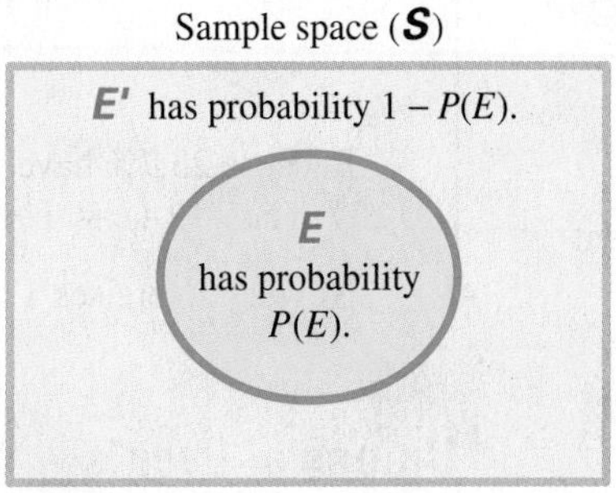
Sample space (S)
E′ has probability 1 – P(E).
E
has probability
P(E).

FIGURE 13.6 $P(E) + P(E') = 1$.

*Throughout the rest of this chapter, unless we state otherwise, E, F, and so on will represent events in the sample space S.

PROBLEM SOLVING

Strategy: The Analogies Principle

The Analogies Principle in Section 1.1 says that you can learn a new area of mathematics more easily if you connect it with an area that you already know. By drawing Venn diagrams, you can visualize many of the rules of probability theory that will help you remember them and use them properly.

EXAMPLE 1 Using the Complement Formula to Study Voter Affiliation

The accompanying graph shows how a group of first-time voters are classified according to their party affiliation. If we randomly select a person from this group, what is the probability that the person has a party affiliation?

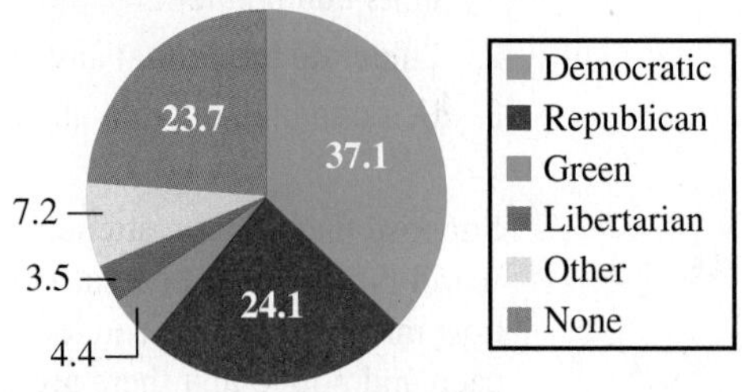

Percent of Voters According to Party Affiliation

SOLUTION:

Let A be the event that the person we select has some party affiliation. Rather than compute the probability of this event, it is simpler to calculate the probability of A', which is the event that the person we select has no party affiliation. We illustrate this situation in Figure 13.7.

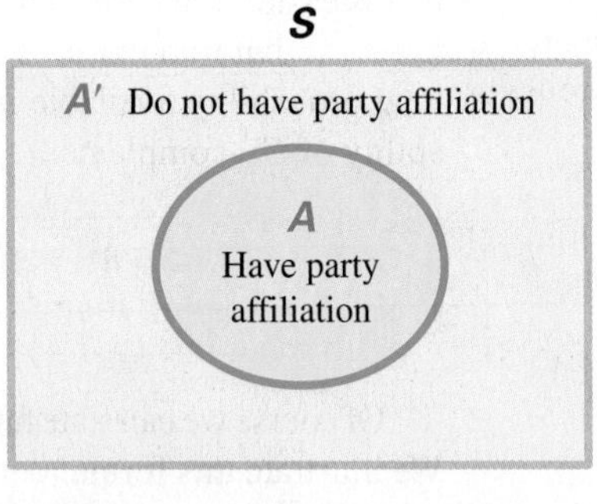

FIGURE 13.7 Calculate $P(A)$ by first finding $P(A')$.

Because 23.7% have no party affiliation, the probability of selecting such a person is 0.237. Thus, $P(A) = 1 - P(A') = 1 - 0.237 = 0.763$.

Now try Exercises 1 to 8. 6

QUIZ YOURSELF 6

If a pair of dice are rolled, what is the probability of rolling a total that is less than 11? (Use the complement rule.)

Unions of Events

KEY POINT

We often describe the union of events using the word *or.*

We often combine objects in a mathematical system to obtain other objects in the system. For example, in number systems, we add and subtract pairs of numbers to get other numbers. In probability, we join events using the set operations union and intersection, which

Between the Numbers

What Are Your Chances?

Once, while we were studying the probability of lotteries and casino games, one of my students, upon realizing his slim chances of winning a lottery, blurted out, "I'm never going to gamble again!"

If you did Exercises 63 and 64 from Section 13.1, you saw that the odds against your winning the New York Lotto are 45,057,473 to 1, which is much greater than being killed by lightning. To put in perspective what a long shot it is to win a big lottery, consider the odds on your dying from various events provided by the National Safety Council.*

Event (Dying From . . .)	Odds Against This Event	Number of Times More Likely Than Winning the New York Lotto
Lightning	79,113 to 1	570
Drowning	9,641 to 1	4,674
Assault by Firearms	308 to 1	146,290
Motor Vehicle Accident	87 to 1	517,902

we discussed in Section 2.3. We will investigate the union of events in this section and the intersection of events in Section 13.3. We often describe the union of events using the word *or.*

Figure 13.8 shows the union of events E and F. In computing $P(E \cup F)$, it is a *common mistake* to simply add $P(E)$ and $P(F)$. Because some outcomes may be in both E and F, you will count their probabilities twice. To compute $P(E \cup F)$ properly, you must subtract $P(E \cap F)$.

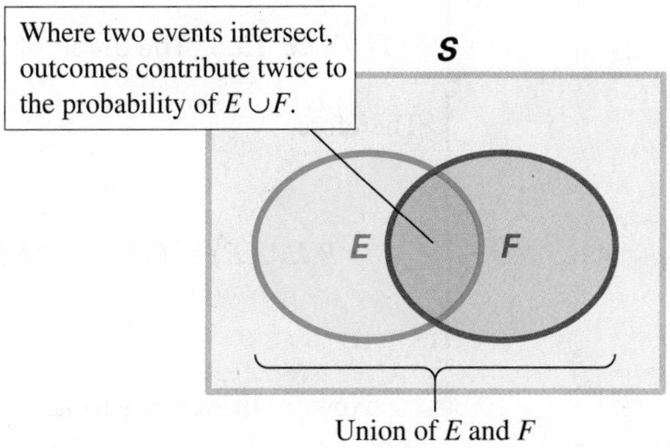
Where two events intersect, outcomes contribute twice to the probability of E ∪ F.
S
E
F
Union of E and F

FIGURE 13.8 The union of two events.

RULE FOR COMPUTING THE PROBABILITY OF A UNION OF TWO EVENTS If E and F are events, then

$$P(E \cup F) = P(E) + P(F) - P(E \cap F).$$

If E and F have no outcomes in common, they are called *mutually exclusive events.* In this case, because $E \cap F = \varnothing$, the preceding formula simplifies to

$$P(E \cup F) = P(E) + P(F).$$

*One reviewer rightly pointed out that these odds could be fine-tuned, so to speak. The odds against dying from lightning would no doubt be different if you lived in Oklahoma than if you lived in New York City. However, these odds for the country as a whole are as given by the National Safety Council.

EXAMPLE 2 Finding the Probability of the Union of Two Events

If we select a single card from a standard 52-card deck, what is the probability that we draw either a heart or a face card? (See Figure 12.1 for a picture of a standard deck of cards.)

SOLUTION: Review the Draw Pictures Strategy on page 3.

Let H be the event "draw a heart" and F be the event "draw a face card." We are looking for $P(H \cup F)$. There are thirteen hearts, twelve face cards, and three cards that are both hearts and face cards. Figure 13.9 helps us remember the formula to use.

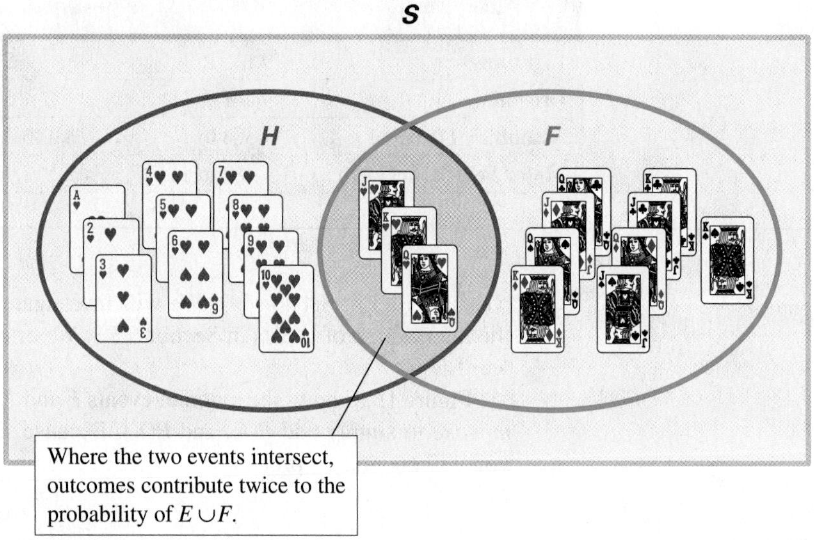
S
H
F
Where the two events intersect, outcomes contribute twice to the probability of E ∪ F.

FIGURE 13.9 The union of the events "draw a heart" and "draw a face card."

Therefore,

probability of a heart (13/52); probability of a face card (12/52)

$$P(H \cup F) = P(H) + P(F) - P(H \cap F) = \frac{13}{52} + \frac{12}{52} - \frac{3}{52} = \frac{22}{52} = \frac{11}{26}$$

probability of a heart that is a face card (3/52)

Now try Exercises 9 to 12.

SOME GOOD ADVICE

If you are given any three of the four quantities in the formula

$$P(E \cup F) = P(E) + P(F) - P(E \cap F),$$

you can use algebra to solve for the other. See Example 3.

EXAMPLE 3 Using Algebra to Find a Missing Probability

A magazine conducted a survey of readers age 18 to 25 regarding their health concerns. The editors will use this information to choose topics relevant to their readers. The survey found that 35% of the readers were concerned with improving their cardiovascular fitness and 55% wanted to lose weight. Also, the survey found that 70% are concerned with

either improving their cardiovascular fitness or losing weight. If the editors randomly select one of those surveyed to profile in a feature article, what is the probability that the person is concerned with both improving cardiovascular fitness *and* losing weight?

SOLUTION: Review the Choose Good Names for Unknowns Strategy on page 5.

If we let C be the event "the person wants to improve cardiovascular fitness" and W be the event "the person wishes to lose weight," then the word "and" tells us that we need to find $P(C \cap W)$.

We are given that 35% want to improve cardiovascular fitness, so $P(C) = 0.35$. Similarly, $P(W) = 0.55$. The event "a person wants to improve cardiovascular fitness or lose weight" is the event $C \cup W$, and we are told that $P(C \cup W) = 0.70$. Figure 13.10 will now help you to see what to do.

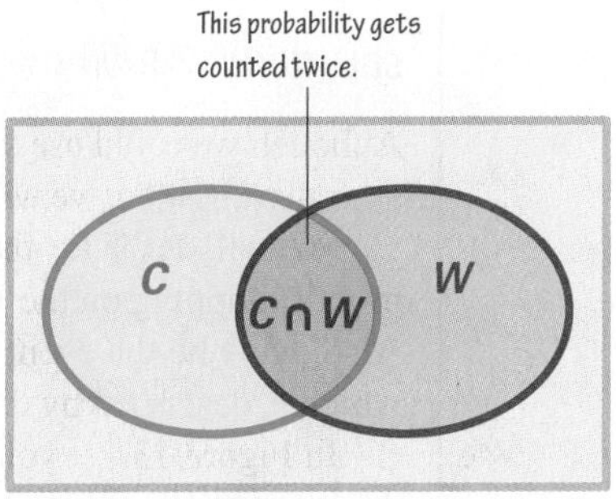

FIGURE 13.10 The union of events C and W.

From this diagram you can see that

$$P(C \cup W) = P(C) + P(W) - P(C \cap W).$$
$$\underset{0.70}{P(C \cup W)} \quad \underset{0.35}{P(C)} \quad \underset{0.55}{P(W)} \quad \underset{\text{unknown}}{P(C \cap W)}$$

This gives us the equation

$$0.70 = 0.35 + 0.55 - P(C \cap W).$$

Rewriting this equation as

$$P(C \cap W) = 0.35 + 0.55 - 0.70,$$

we find that $P(C \cap W) = 0.20$.

This means that there is a 20% chance that the person chosen will be interested in both improving cardiovascular fitness and losing weight.

Now try Exercises 13 to 16. 7

QUIZ YOURSELF 7

Suppose that A and B are events such that $P(A) = 0.35$, $P(A \cap B) = 0.15$, and $P(A \cup B) = 0.65$. Find $P(B)$.

We will not always draw diagrams to illustrate the rule for calculating the probability of the union of two events; however, we do encourage you to do so to aid you in your computations.

Combining Complement and Union Formulas

KEY POINT

We may use several formulas to calculate an event's probability.

In Example 4, we use both the complement formula and the union formula to compute an event's probability.

EXAMPLE 4 Finding the Probability of the Complement of the Union of Two Events

A survey of consumers comparing the amount of time they spend shopping on the Internet per month with their annual income produced the results in Table 13.5.

Annual Income	10 + Hours (T)	3–9 Hours	0–2 Hours	Totals
Above \$60,000 (A)	192	176	128	496
\$40,000–\$60,000	160	208	144	512
Below \$40,000	128	192	272	592
Totals	480	576	544	1,600

$n(T \cap A)$ (192); $n(T)$ (480); $n(A)$ (496); $n(S)$ (1,600)

TABLE 13.5 Survey results on Internet shopping.

Assume that these results are representative of all consumers. If we select a consumer randomly, what is the probability that the consumer neither shops on the Internet 10 or more hours per month nor has an annual income above \$60,000?

SOLUTION: Review the Draw Pictures Strategy on page 3.

Although we could use the probability techniques that we introduced in Section 13.1 to answer this question, we will use instead the formulas for the complement and union of events.

We will let T be the event "the consumer selected spends 10 or more hours per month shopping on the Internet." Event T corresponds to the first column in Table 13.5. Also, let A be the event "the consumer selected has an annual income above \$60,000," which is described by the first row in Table 13.5.

In Figure 13.11, you can see that the event "the consumer selected neither shops on the Internet 10 or more hours per month nor has an annual income above \$60,000" is the region outside of $T \cup A$, which is the complement of $T \cup A$.

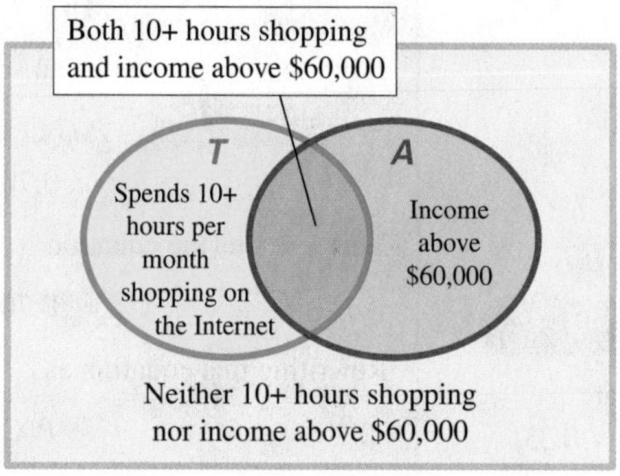

FIGURE 13.11 The event "neither T nor A" corresponds to the complement of $T \cup A$.

From Table 13.5, you see that the number of outcomes in T is

$$n(T) = 192 + 160 + 128 = 480.$$

The total number of outcomes in the sample space, S, is the total number of people surveyed, which is 1,600. Therefore,

$$P(T) = \frac{n(T)}{n(S)} = \frac{480}{1{,}600} = 0.30.$$

Similarly,

$$P(A) = \frac{n(A)}{n(S)} = \frac{496}{1{,}600} = 0.31.$$

Also, if you look at the intersection of the column labeled T and the row labeled A, you see that the number of people surveyed who shop on the Internet more than 10 hours per month and also earn above \$60,000 is 192. Therefore,

$$P(T \cap A) = \frac{n(T \cap A)}{n(S)} = \frac{192}{1{,}600} = 0.12.$$

Historical Highlight

Modern Probability Theory

Modern probability theory began in the seventeenth century when Blaise Pascal and his friend Pierre de Fermat began to study the mathematical principles of gambling. They were answering several questions posed to Pascal by Antoine Gombauld, the Chevalier de Mere:

> "How many times should a single die be thrown before we could reasonably expect two sixes?"
>
> "How should prize money in a contest be fairly divided in the case that the contest, for some reason, cannot be completed?"

In 1812, Pierre-Simon de Laplace presented classical probability theory in his work *Theorie Analytique des Probabilities.* Laplace boldly affirmed that all knowledge could be obtained by using the principles he set forth.

Physicists now use probability theory to study radiation and atomic physics, and biologists apply it to genetics and mathematical learning theory. Probability also is a theoretical basis for statistics, which is used in scientific, industrial, and social research.

We can now calculate the probability of the complement of $T \cup A$ as follows:

$$P((T \cup A)') = 1 - P(T \cup A) = 1 - [P(T) + P(A) - P(T \cap A)]$$
$$= 1 - [0.30 + 0.31 - 0.12] = 1 - 0.49 = 0.51.$$

This means that if we select a consumer randomly, there is a 51% chance that the consumer neither spends 10 or more hours per month shopping on the Internet nor has a yearly income above \$60,000.

Now try Exercises 19 to 22, 27, and 28.

EXERCISES 13.2

Sharpening Your Skills

In Exercises 1–8, use the complement formula to find the probability of each event.

1. If the probability that your DVD player breaks down before the extended warranty expires is 0.015, what is the probability that the player will not break down before the warranty expires?

2. If the probability that a vaccine you took will protect you from getting the flu is 0.965, what is the probability that you will get the flu?

3. If there is a 1 in 1,000 chance that you will pick the numbers correctly in tonight's lottery, what is the probability that you will not pick the numbers correctly?

4. If there is a 1 in 4 chance that it will rain for your Fourth of July barbecue, what is the probability that it won't rain?

5. If two dice are rolled, find the probability that neither die shows a five on it.

6. If two dice are rolled, find the probability that the total showing is less than 10.

7. If five coins are flipped, what is the probability of obtaining at least one head?

8. If five coins are flipped, what is the probability of obtaining at least one head and at least one tail?

9. If a single card is drawn from a standard 52-card deck, what is the probability that it is either a five or a red card?

10. If a single card is drawn from a standard 52-card deck, what is the probability that it is either a face card or a red card?

11. Probability and the weather. If the probability of rain is 0.6, the probability of fog is 0.4, and the probability of rain and fog is 0.15, what is the probability of rain or fog?

12. Probability and grades. Assume that the probability of you getting an A in math is 0.8, the probability of you making the dean's list is 0.72, and the probability of you getting an A in math and also making the dean's list is 0.6. What is the probability that you get an A in math or make the dean's list?

In Exercises 13–16, assume that A *and* B *are events.*

13. If $P(A \cup B) = 0.85$, $P(B) = 0.40$, and $P(A) = 0.55$, find $P(A \cap B)$.

14. If $P(A \cup B) = 0.75$, $P(B) = 0.45$, and $P(A) = 0.60$, find $P(A \cap B)$.

15. If $P(A \cup B) = 0.70$, $P(A) = 0.40$, and $P(A \cap B) = 0.25$, find $P(B)$.

16. If $P(A \cup B) = 0.60$, $P(B) = 0.45$, and $P(A \cap B) = 0.20$, find $P(A)$.

Applying What You've Learned

Use the following table from the U.S. Bureau of Labor Statistics, which shows the age distribution of those who earned less than the minimum wage in a recent year, to answer Exercises 17 and 18.

Age	Working Below Minimum Wage (thousands)
16–19	329
20–24	420
25–34	320
35–44	175
45–54	125
55–64	61
65 and older	53

17. If we select a worker randomly from those surveyed, what is the probability that the person is younger than 55?

18. If we select a worker randomly from those surveyed, what is the probability that the person is older than 19?

Use the following table that we presented in Example 4, relating the amount of time consumers spend shopping on the Internet per month with their annual income, to answer Exercises 19–22.

Annual Income	10 + Hours	3–9 Hours	0–2 Hours	Totals
Above $60,000	192	176	128	496
$40,000–$60,000	160	208	144	512
Below $40,000	128	192	272	592
Totals	480	576	544	1,600

19. What is the probability that a consumer we select randomly either spends 0–2 hours per month shopping on the Internet or has an annual income below $40,000?

20. What is the probability that a consumer we select randomly either spends 10 or more hours per month shopping on the Internet or has an annual income between $40,000 and $60,000?

21. What is the probability that a consumer we select randomly neither spends more than 2 hours per month shopping on the Internet nor has an annual income of $60,000 or less?

22. What is the probability that a consumer we select randomly neither spends more than 2 hours per month shopping on the Internet nor has an annual income below $40,000?

Probability of earning commissions. *Joanna earns both a salary and a monthly commission as a sales representative for an electronics store. The following table lists her estimates of the probabilities of earning various commissions next month. Use this table to calculate the probabilities in Exercises 23–26.*

Commission	Probability That This Will Happen
Less than $1,000	0.08
$1,000 to $1,249	0.11
$1,250 to $1,499	0.23
$1,500 to $1,749	0.30
$1,750 to $1,999	0.12
$2,000 to $2,249	0.05
$2,250 to $2,499	0.08
$2,500 or more	0.03

23. The probability that she will earn at least $1,000 in commissions

24. The probability that she will earn at least $1,500 in commissions

25. The probability that she will earn no more than $1,999 in commissions

26. The probability that she will earn less than $2,250 in commissions

27. If we draw a card from a standard 52-card deck, what is the probability that the card is neither a heart nor a queen? (*Hint:* Draw a picture of this situation before trying to calculate the probability.)

28. If we draw a card from a standard 52-card deck, what is the probability that the card is neither red nor a queen? (*Hint:* Draw a picture of this situation before trying to calculate the probability.)

29. Predicting a car repair. Your car's windshield wipers have not been working properly. The probability that you need a new motor is 0.55, the probability that you need a new switch is 0.4, and the probability that you need both is 0.15. What is the probability that you need neither a new motor nor a new switch?

30. Predicting final exam questions. From studying past World History exams, you think the probability of getting a question about Russia on your final is 0.75, the probability of getting a question about Poland is 0.6, and the probability of getting a question about both is 0.45. What is the probability that you will not get a question about Russia or Poland on your exam?

A college administration has conducted a study of 200 randomly selected students to determine the relationship between satisfaction with academic advisement and academic success. They obtained the following information: Of the 70 students on academic probation, 32 are not satisfied with advisement; however, only 20 of the students not on academic probation are dissatisfied with advisement. Use these data to answer Exercises 31–34. In each exercise, assume that we select a student at random.

31. What is the probability that the student is not on academic probation?

32. What is the probability that the student is satisfied with advisement?

33. What is the probability that the student is on academic probation and is satisfied with advisement?

34. What is the probability that the student is not on academic probation and is satisfied with advisement?

35. Selling defective cameras. A manufacturer has shipped 40 Flip cameras to a Great Buy store, of which six are defective. If Great Buy sold 18 cameras before discovering that some were defective, what is the probability that at least one defective camera was sold? (*Hint:* $C(34,18)/C(40,18)$ is the probability that no defective cameras were sold.)

36. Winning a raffle. The 35-member college ski club is holding a raffle. There will be three prizes awarded. If there are six freshmen in the club, what is that probability that at least one freshman will win a prize? (*Hint:* Consider the complement of this event.)

37. Serving spoiled food. The Sashimi restaurant has prepared 18 portions of seafood, two of which had been left out too long and spoiled. If 12 of the 18 portions are served randomly to customers, what is the probability that at least one customer will receive spoiled food? (*Hint:* Consider the complement of this event.)

38. Winning a prize. Eighteen students are being honored for their academic excellence at a banquet where various prizes will be awarded. Three of the students attending are international students. If four students are chosen randomly as finalists for a $500 gift certificate to the bookstore, what is the probability that at least one international student will be a finalist? (*Hint:* Consider the complement of this event.)

Communicating Mathematics

In Exercises 39 and 40, determine whether each statement is true or false for events A *and* B. *Explain your answer.*

39. $P(A) = P(A \cup B) - P(B)$

40. $P(A) + P(B) - P(A \cup B) = P(A \cap B)$

41. Many texts that discuss probability state that if events E and F are disjoint, then $P(E \cup F) = P(E) + P(F)$. Explain why it is really not necessary to state this formula.

42. If $P(E \cup F) = P(E) + P(F)$, what can you conclude about $P(E \cap F)$?

Between the Numbers

Many states have a lottery called The Daily Number in which a player wins by correctly selecting a three-digit number between 000 and 999 inclusive. The odds against winning are therefore 999 to 1. Use this information and the table on page 661 to do Exercises 43 and 44.

43. Comparing odds. How many times more likely is it to be killed by firearms than to win The Daily Number by playing a single time?

44. Comparing odds. How many times more likely is it to die in a motor vehicle accident than to win The Daily Number by playing a single time?

Challenge Yourself

45. If events A, B, and C are as pictured in this Venn diagram, write a formula for $P(A \cup B \cup C)$. Explain your answer.

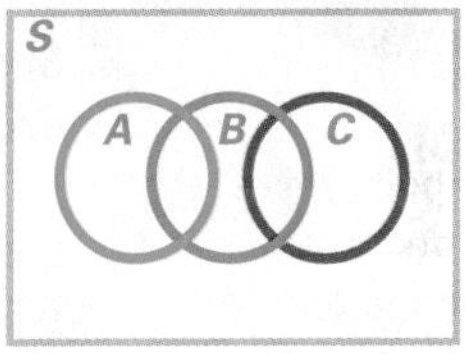

46. If events A, B, and C are as pictured in this Venn diagram, write a formula for $P(A \cup B \cup C)$. Explain your answer.

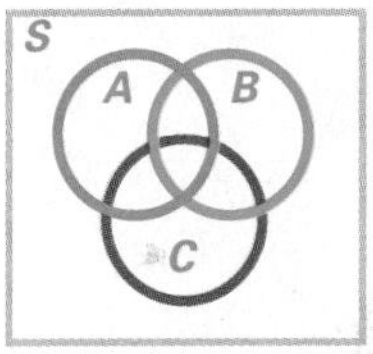

Use the given spinner to answer Exercises 47–50. Each green sector occupies 10% of the circle, each blue sector occupies 12%, each red sector occupies 9%, and the yellow sector occupies 4%. Assume that the spinner is spun once. What is the probability that

47. The spinner does not stop on yellow?

48. The spinner stops on an odd number or a green?

49. The spinner stops on neither an odd number nor a blue?

50. The spinner stops on neither green nor an even number?

13.3

Conditional Probability and Intersections of Events

Objectives

1. Understand how to compute conditional probability.
2. Calculate the probability of the intersection of two events.
3. Use probability trees to compute conditional probabilities.
4. Understand the difference between dependent and independent events.

You know how to compute the probability of complements and unions of events. We will now show you how to find the probability of intersections of events, but first you need to understand how the occurrence of one event can affect the probability of another event.

Conditional Probability

KEY POINT

Conditional probability takes into account that one event occurring may change the probability of a second event.

Suppose that you and your friend Marcus cannot agree as to which video to rent and you decide to settle the matter by rolling a pair of dice. You will each pick a number and then roll the dice. The person whose number comes up first as the total on the dice gets to pick the video. With your knowledge of probability, you know that the number you should pick is 7 because it has the highest probability of appearing, namely $\frac{1}{6}$.

To illustrate the idea of conditional probability, let's now change this situation slightly. Assume that your friend Janelle will roll the dice before you and Marcus pick your numbers. You are not allowed to see the dice, but Janelle will tell you something about the dice and then you will choose your number before looking at the dice.

Suppose that Janelle tells you that the total showing is an even number. Would you still choose a 7? Of course not, because once you know the *condition* that the total is even, you now know that the probability of having a 7 is 0. A good way to think of this is that once you know that the total is even, you must exclude all pairs from the sample space such as (1, 4), (5, 6), and (4, 3) that give odd totals.

In a similar way, suppose that you draw a card from a standard 52-card deck, put that card in your pocket, and then draw a second card. What is the probability that the second card is a king? How you answer this question depends on knowing what card is in your pocket. If the card in your pocket is a king, then there are three kings remaining in the 51 cards that are left, so the probability is $\frac{3}{51}$. If the card in your pocket is not a king, then the probability of the second card being a king is $\frac{4}{51}$. Why? This discussion leads us to the formal definition of conditional probability.

DEFINITION When we compute the probability of event F assuming that the event E has already occurred, we call this the **conditional probability** of F given E. We denote this probability as $P(F|E)$. We read $P(F|E)$ as "the probability of F given that E has occurred," or in a quicker way, "the probability of F given E."

Do not let this new notation intimidate you. The notation $P(F|E)$ simply means that you are going to compute a probability knowing that something else has already happened. For example, in our earlier discussion of Janelle, we said, "The probability of having a total of 7 knowing that the total is even is 0." We will restate this several times, each time increasing our use of symbols. So, we could say instead,

$$P(\text{having a total of 7 given that the total is even}) = 0;$$

or,

$$P(\text{having a total of 7} \mid \text{the total is even}) = 0.$$

If we now represent the event "total is 7" by F and "total is even" by E, then we could write our original statement as

$$\overbrace{P(F \mid E)}^{\text{Total is even.}} = 0.$$
$$\underbrace{}_{\text{Total is 7.}}$$

Similarly, let's return to the example of drawing two cards, and let A represent the event that we draw a king on our first card and put it in our pocket, and let B represent the event that we draw a king on our second card. Then we could write, "The probability that we draw a second king given that the first card was a king is $\frac{3}{51}$," as

First card was a king.
$$P(B \mid A) = \frac{3}{51}.$$
Second card is a king.

The Venn diagram in Figure 13.12 will help you remember how to compute conditional probability.

We drew E with a heavy line in Figure 13.12 to emphasize that when we assume that E has occurred, we can then think of the outcomes outside of E as being discarded from the discussion. In computing conditional probability, you will find it useful to consider the sample space to be E and the event as being $E \cap F$, rather than F. We will first state a special rule for computing conditional probability when the outcomes are equally likely (all have the same probability of occurring). We will state the more general conditional probability rule later.

We want the probability of the intersection outcomes, knowing that we're interested in only outcomes for set E.

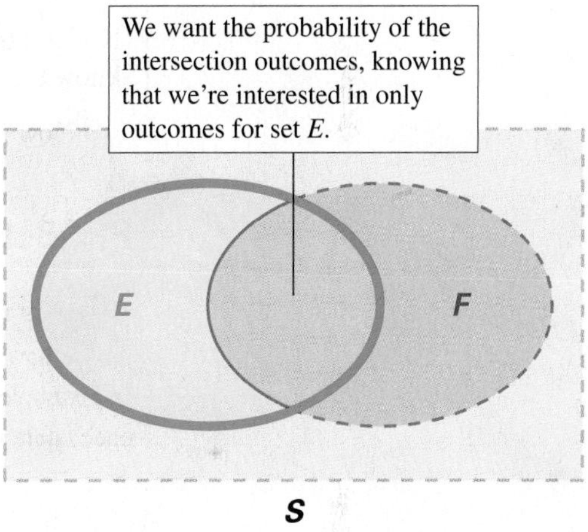

FIGURE 13.12 To compute the probability of F given E, we compare the outcomes in $E \cap F$ with the outcomes in E.

> **SPECIAL RULE FOR COMPUTING $P(F|E)$ BY COUNTING** If E and F are events in a sample space with equally likely outcomes, then $P(F|E) = \dfrac{n(E \cap F)}{n(E)}$.

EXAMPLE 1 ● Computing Conditional Probability by Counting

Assume that we roll two dice and the total showing is greater than nine. What is the probability that the total is odd?

SOLUTION: Review the Three-Way Principle on page 13.

This sample space has 36 equally likely outcomes. We will let G be the event "we roll a total greater than nine" and let O be the event "the total is odd." Therefore,

$$G = \{(4, 6), (5, 5), (5, 6), (6, 4), (6, 5), (6, 6)\}.$$

The set O consists of all pairs that give an odd total. Figure 13.13 shows you how to use the special rule for computing conditional probability to find $P(O|G)$.

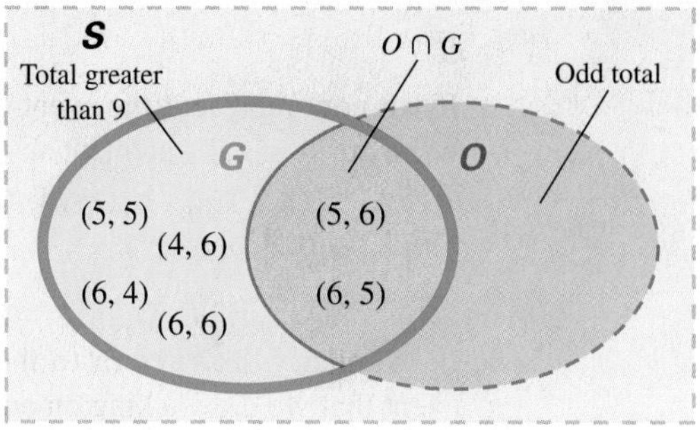

FIGURE 13.13 To compute $P(O \mid G)$, we compare the number of outcomes in $O \cap G$ with the number of outcomes in G.

Therefore,

$$P(O|G) = \frac{n(O \cap G)}{n(G)} = \frac{2}{6} = \frac{1}{3}.$$

Notice how the probability of rolling an odd total has changed from $\frac{1}{2}$ to $\frac{1}{3}$ when we know that the total showing is greater than nine.

Now try Exercises 1 to 14.

PROBLEM SOLVING

Strategy: The Order Principle

As we have emphasized often, the order in which we do things is important in mathematics. In Example 1, $P(O|G)$ does not mean the same thing as $P(G|O)$. To understand the difference, state what $P(G|O)$ represents and then compute $P(G|O)$.

It is important to remember that the special rule for computing conditional probability only works when the outcomes in the sample space are *equally likely*. Sometimes you may be solving a problem where that is not the case; or, you may have a situation where it is not possible to count the outcomes. In such cases, we need a rule for computing $P(F \mid E)$ that is based on probability rather than counting.

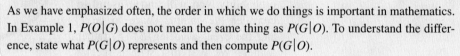

GENERAL RULE FOR COMPUTING $P(F|E)$ If E and F are events in a sample space, then $P(F \mid E) = \frac{P(E \cap F)}{P(E)}$.

We still can use Figure 13.12 to remember this rule; however, now instead of comparing the number of outcomes in $E \cap F$ with the number of outcomes in E, we compare the probability of $E \cap F$ with the probability of E.

EXAMPLE 2 Using the General Rule for Computing Conditional Probability

The state bureau of labor statistics conducted a survey of college graduates comparing starting salaries to majors. The survey results are listed in Table 13.6.

Major	$30,000 and Below	$30,001 to $35,000	$35,001 to $40,000	$40,001 to $45,000	Above $45,000	Totals (%)
Liberal arts	6*	10	9	1	1	27
Science	2	4	10	2	2	20
Social sciences	3	6	7	1	1	18
Health fields	1	1	8	3	1	14
Technology	0	2	7	8	4	21
Totals (%)	12	23	41	15	9	100

*These numbers are percentages.

TABLE 13.6 Survey comparing starting salaries to major in college.

If we select a graduate who was offered between $40,001 and $45,000, what is the probability that the student has a degree in the health fields?

SOLUTION: Review the Draw Pictures Strategy on page 3.

Each entry in Table 13.6 is the probability of an event. For example, the 8% that we highlighted is the probability of selecting a graduate in technology who earns between $40,001 and $45,000, inclusive. The 14% that we highlighted tells us the probability of selecting a graduate who majored in the health fields.

Let R be the event "the graduate received a starting salary between $40,001 and $45,000" and H be the event "the student has a degree in the health fields." *It is important in doing this problem that you identify clearly what you are given and what you must find.* We are given R and must find the probability of H, so we want to find $P(H|R)$, not $P(R|H)$.

Because we want the probability of H given R, we can, in effect, ignore all the outcomes that do not correspond to a starting salary of $40,001 to $45,000. We darken the columns we want to ignore in Table 13.7.

given

Major	$30,000 or Below	$30,001 to $35,000	$35,001 to $40,000	$40,001 to $45,000	Above $45,000	Totals (%)
Liberal arts	6	10	9	1	1	27
Science	2	4	10	2	2	20
Social sciences	3	6	7	1	1	18
Health fields	1	1	8	3	1	14
Technology	0	2	7	8	4	21
Totals (%)	12	23	41	15	9	100

$P(R)$ $P(H \cap R)$

TABLE 13.7 The columns we want to ignore are darkened.

In order to use the general rule for computing conditional probability, we first need to know $P(R)$ and $P(H \cap R)$, as you can see from Figure 13.14.

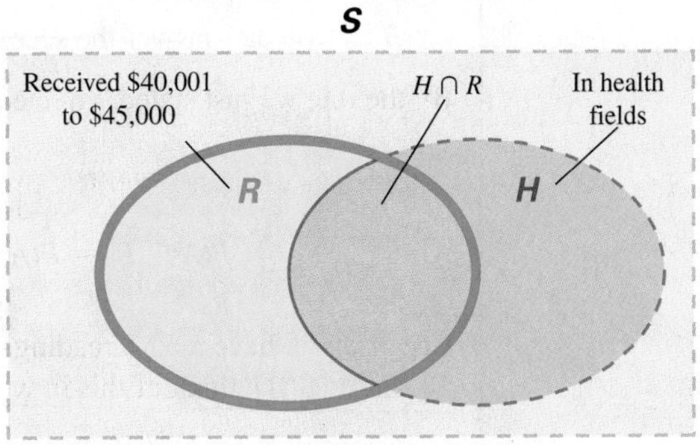

FIGURE 13.14 To compute $P(H|R)$, we compare $P(H \cap R)$ with $P(R)$.

QUIZ YOURSELF 8

Assume that we select a graduate who received more than \$45,000 as a starting salary. Use Table 13.7 to find the probability that the graduate has a degree in technology.

Therefore,

$$P(H|R) = \frac{P(H \cap R)}{P(R)} = \frac{0.03}{0.15} = 0.20.$$

If a graduate receives an offer between \$40,001 and \$45,000, then the probability that the person is in the health fields is 0.20, or 20%.

Now try Exercises 45 to 52. 8

The Intersection of Events

KEY POINT

We use conditional probability to find the probability of the intersection of two events.

We can now find a formula to compute the probability of the intersection of two events. The general rule for computing conditional probability states that

$$P(F \mid E) = \frac{P(E \cap F)}{P(E)}.$$

If we multiply both sides of this equation by the expression $P(E)$, we get

$$P(E) \cdot P(F \mid E) = P(E) \cdot \frac{P(E \cap F)}{P(E)}.$$

Canceling $P(E)$ from the numerator and denominator of the right side of this equation gives us the rule for computing the probability of the intersection of events E and F.

RULE FOR COMPUTING THE PROBABILITY OF THE INTERSECTION OF EVENTS If E and F are two events, then

$$P(E \cap F) = P(E) \cdot P(F \mid E).$$

This rule says that to find the probability of $P(E \cap F)$, you first find the probability of E and then multiply it by the probability of F, *assuming that E has occurred.*

EXAMPLE 3 Estimating Your Grade in a Class

Assume that for your literature final your professor has written questions on 10 assigned readings on cards and you are to randomly select two cards and write an essay on them. If you have read 8 of the 10 readings, what is your probability of getting two questions that you can answer? (We will assume that if you've done a reading, then you can answer the question about that reading; otherwise, you can't answer the question.)

SOLUTION:

We can think of this event as the intersection of two events A and B where

A is "you can answer the first question";

B is "you can answer the second question."

By the rule we just stated, you need to calculate

probability you can answer the first question; probability you can answer the second question, given that you answered the first question

$$P(A \cap B) = P(A) \cdot P(B \mid A).$$

Because you have read 8 readings from 10 that were assigned, we see that $P(A) = \frac{8}{10}$. To find $P(B|A)$, think of this in words, as we want to find

the probability that you can answer the second question, given that you have answered the first question.

There are now only seven questions that you can answer on the remaining nine cards so $P(B \mid A) = \frac{7}{9}$. Thus the probability that you get two questions on readings that you have done is

$$P(A \cap B) = P(A) \cdot P(B \mid A) = \frac{8}{10} \cdot \frac{7}{9} = \frac{56}{90} \approx 0.62.^*$$

Now try Exercises 15 to 30.

SOME GOOD ADVICE

A common mistake that you can make when computing conditional probability is to use the formula $P(A \cap B) = P(A) \cdot P(B)$. That is, you may forget to take into account that event A has occurred. Notice that if we had used this incorrect formula in Example 3, the probability that you can answer the second question would have been $\frac{8}{10}$, not $\frac{7}{9}$. In effect, we would have computed the second probability as though the first question had been returned to the cards.

We can extend the technique for calculating conditional probability to more than two events, as we show in Example 4.

EXAMPLE 4 Drawing Colored Balls from an Urn

After winning a carnival game, Evan will draw three colored balls consecutively (without replacement) from the urn in Figure 13.15 to determine his prize. If he draws three red balls, he can choose one of the most desirable prizes. What is the probability that he will draw a red ball, a red ball, and a green ball, in that order?

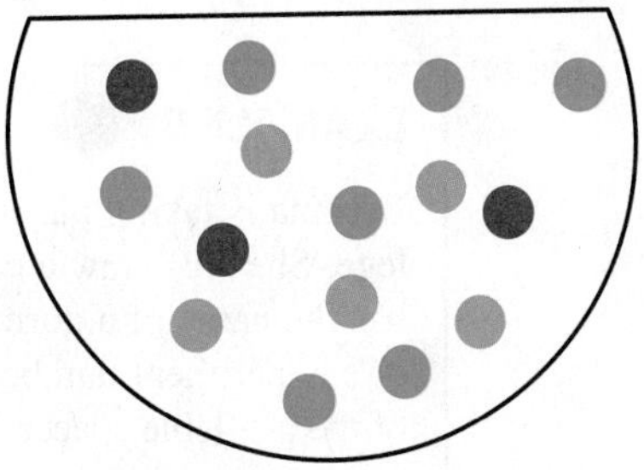

FIGURE 13.15 Drawing balls from an urn.

SOLUTION:

We can think of Evan drawing in stages as follows:

Stage 1: He first draws a red ball from the 15 balls in the urn.
Stage 2: He draws a second red ball, assuming the first ball is not replaced.
Stage 3: He then draws a green ball, assuming the first two red balls were not replaced.

*The perceptive student will recognize that we could have also solved this problem using the techniques of Section 13.2. That is, calculate $\frac{C(8, 2)}{C(10, 2)} = \frac{28}{45} \approx 0.62$. However, we chose our approach in this example to illustrate conditional probability.

Stage 1: The probability of Evan drawing a red ball first is $\frac{3}{15} = \frac{1}{5}$.

Stage 2: In drawing the second ball, we want the probability of drawing a second red ball given that the first ball drawn was red and *not replaced*. There are two red balls among the 14 remaining balls. So this probability is $\frac{2}{14} = \frac{1}{7}$.

Stage 3: Now we want the probability that a green ball is drawn, assuming the two red balls were not replaced. There are six green balls among the remaining 13, so this probability is $\frac{6}{13}$.

The probability of drawing two red balls followed by a green ball is

$$P(\text{red}) \cdot P(\text{red} \mid \text{red was drawn}) \cdot P(\text{green} \mid 2 \text{ reds were drawn}) = \frac{1}{5} \cdot \frac{1}{7} \cdot \frac{6}{13} \approx 0.013.$$

Now try Exercises 23 to 26. 9

QUIZ YOURSELF 9

In Example 4, what is the probability of Evan drawing (without replacement) a green ball, followed by a red ball, followed by a blue ball?

Probability Trees

KEY POINT

Trees help you visualize probability computations.

Recall that the Three-Way Principle in Section 1.1 tells you that drawing a diagram is a good problem-solving technique. You will find that it is often very helpful to draw a probability tree to help you understand a conditional probability problem.

USING TREES TO CALCULATE PROBABILITIES We can represent an experiment that happens in stages with a tree whose branches represent the outcomes of the experiment. We calculate the probability of an outcome by multiplying the probabilities found along the branch representing that outcome. We will call these trees **probability trees.**

Example 5 illustrates how useful trees can be in visualizing probability situations.

EXAMPLE 5 Selecting a Dormitory Room in a Lottery

Brianna is taking part in a lottery for a room in one of two new dormitories at her college. She will draw a card randomly to determine which room she will have. Each card has the name of a dormitory, (X)avier or (Y)oung, and also a two-person room number or an apartment number. Thirty percent of the available spaces are in X. Eighty percent of the available spaces in X are rooms and 40% of the spaces in Y are rooms.

a) Draw a probability tree to visualize this situation.

b) If Brianna selects a card for dormitory X, what is the probability that she will be assigned a room?

c) What is the probability that Brianna will be assigned an apartment in one of the two dormitories?

SOLUTION: Review the Draw Pictures Strategy on page 3.

a) In drawing the tree, we will think of Brianna's assignment happening in two stages.

Stage 1: She is assigned a dormitory.

Stage 2: She is assigned a room or an apartment.

We draw the tree in Figure 13.16, beginning with two branches corresponding to selecting dormitories X and Y. Both of these branches have two further branches representing the assignment of either a room or an apartment.

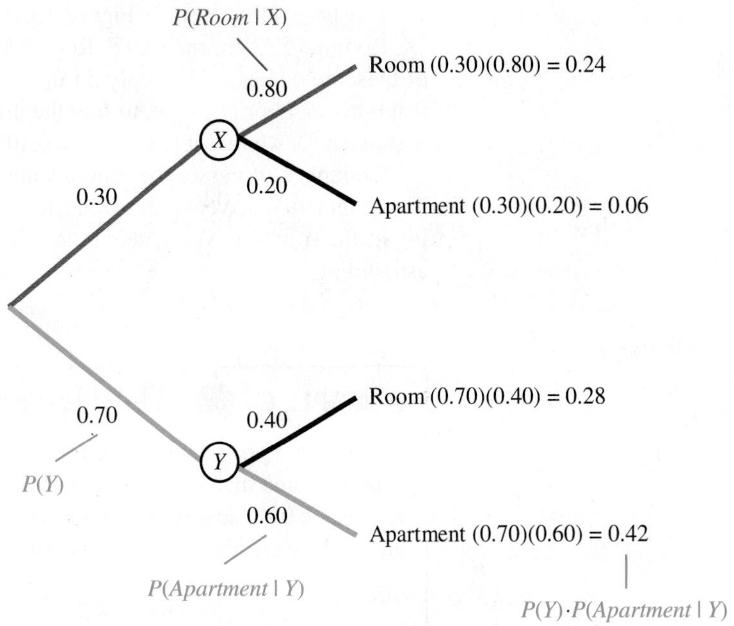

FIGURE 13.16 Probability tree for Brianna's room assignment.

We wrote various probabilities along the tree's branches in Figure 13.16. If you examine the lower branch (in green), you see that 0.70 is the probability that Brianna will be assigned to Y; thus, $P(Y) = 0.70$. The number 0.60 is a conditional probability. *Assuming the condition* that Brianna is assigned to Y, there is a 0.60 chance that she will then be assigned an apartment. Symbolically, this means that $P(Apartment|Y) = 0.60$.

The product $(0.70)(0.60) = 0.42$ is the probability that Brianna will be assigned to Y and also to an apartment.* We could have written this numeric equation more formally as

$$P(Y \cap Apartment) = P(Y) \cdot P(Apartment|Y),$$

which is the formula for computing the probability of the intersection of two events. Quiz Yourself 10 asks you to interpret some more of these probabilities.

b) The first branch of the tree—highlighted in red in Figure 13.16—tells us that once we know Brianna will be in X, the probability of her having a room is 80%, or 0.80.

c) Realize that there are two ways that Brianna can be given an apartment—either in dormitory X or in dormitory Y. Figure 13.17 shows how to find the probability that she is assigned an apartment.

In Figure 13.17, we see that the event "Brianna is assigned an apartment" is the union of two disjoint subevents with probabilities 0.06 and 0.42.

Therefore, the probability that Brianna is assigned an apartment is $0.06 + 0.42 = 0.48$.

Now try Exercises 57 and 58. 10

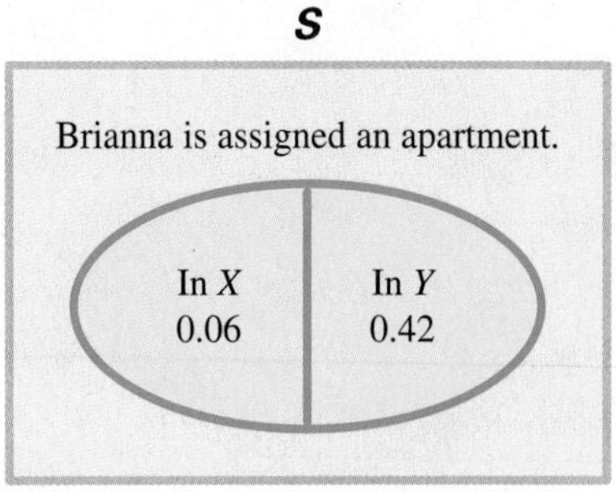

FIGURE 13.17 Probabilities for dormitories X and Y that Brianna is assigned an apartment.

QUIZ YOURSELF 10

a) In Example 5, if we know that Brianna has been assigned dormitory X, what is the probability that she will be assigned an apartment?

b) What is the probability that Brianna will be assigned a room?

*It may seem that we have gotten away from thinking of events as subsets of a sample space, but we have not. The "Y" represents the set of cards that have dormitory Y written on them, and "*Apartment*" represents the set of all cards that have the word "apartment" written on them.

The probability tree in Figure 13.16 describes four (non-equally likely) outcomes: (X, Room), (X, Apartment), (Y, Room), (Y, Apartment). We found the probability of each of these outcomes by multiplying the probabilities along the branch that corresponds to that outcome. For example, to find the probability of Brianna getting a room in dormitory Y (branch 3), we multiplied 0.70 by 0.40 to get 0.28.

Example 6 discusses the drug-testing example that we promised in the chapter opener. You might find it very interesting to estimate the answer to the problem before looking at the solution. Many people to whom I have shown this example found the results astounding.

EXAMPLE 6 Drug Testing

Assume that you are working for a company that has a mandatory drug-testing policy. It is estimated that 2% of the employees use a certain drug, and the company is giving a test that is 99% accurate in identifying users of this drug. What is the probability that if an employee is identified by this test as a drug user, the person is innocent?

SOLUTION:

Let D be the event "the person is a drug user" and let T be the event "the person tests positive for the drug." We are asking, then, "if we are given that the person tests positive, what is the probability that the person does not use the drug?" Realize that the complement of D, namely D', is the event "the person does not use the drug." So, we are asking for the conditional probability

$$P(D'|T).$$

Person does not use drug — given that person tests positive.

The probability tree in Figure 13.18 will give you some insight into this problem.

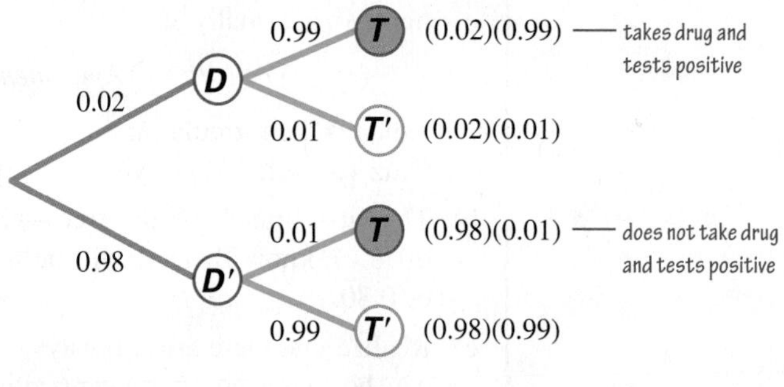

FIGURE 13.18 Tree showing probabilities for drug testing.

Recall that $P(D'|T) = \dfrac{P(D' \cap T)}{P(T)}$. Branches 1 and 3 in Figure 13.18 correspond to the drug test being positive. This means that $P(T) = (0.02)(0.99) + (0.98)(0.01)$. The event $D' \cap T$ corresponds to branch 3. Therefore, $P(D' \cap T) = (0.98)(0.01)$. From this, we see that the probability that an innocent person will test positive for the drug is

$$P(D'|T) = \frac{P(D' \cap T)}{P(T)} = \frac{(0.98)(0.01)}{(0.02)(0.99) + (0.98)(0.01)} = \frac{0.0098}{0.0296} \approx 0.331.$$

In other words, if a person tests positive, there is a roughly $\frac{1}{3}$ chance that the person is innocent!

Now try Exercises 69 and 70.

Math in Your Life

The Numbers Don't Lie—Or Do They?

- In an article in *Chance* magazine, Mary C. Meyer, a statistics professor at the University of Georgia, claims that a study shows the probability of your death in a automobile accident is slightly greater if your car has an airbag than if it doesn't.
- *The Ottowa Citizen* reports that Dennis Lindley, a professor of statistics at University College, London, has derived a formula based on probability that tells you the optimal age for you to settle down. His formula is $X + \frac{Y - X}{2.718}$, where X is the age at which you start looking for a spouse and Y is the age at which you expect to stop looking. According to Lindley's formula, if you start searching at age 16 and expect to stop looking at age 60, then the best age for you to choose a mate is at age $16 + \frac{60 - 16}{2.718} \approx 32$ years.
- A third article, from the Woods Hole Oceanographic Institute in Massachusetts, states that commercial fishing is one of the least safe occupations. If you are a commercial fisherman, you are 16 times more likely to die on the job than a firefighter or a police officer.

Truly, probability is all around us. Do you agree with the previous statements? Or do they go against your "common sense"? Do you feel that something is missing? As we have been emphasizing throughout this text, always remember that when someone builds a model, they decide what goes into the model and what is left out. Maybe the numbers don't lie—or, . . . maybe . . . sometimes they do.

Dependent and Independent Events

KEY POINT

Independent events have no effect on each other's probabilities.

As we have seen, knowing that a first event has occurred may affect the way we calculate the probability of a second event. Such was the case in Example 3 when we computed the probability of drawing a second card, given that a first card was drawn and not returned to the deck.

However, sometimes the occurrence of a first event has no effect whatsoever on the probability of the second event. This would be the case if you were drawing 2 cards with replacement from a standard 52-card deck. The probability of drawing a first king is $\frac{4}{52}$. If you return the king to the deck and draw again, the probability of the second king is also $\frac{4}{52}$. Therefore, drawing the first king would not affect the probability of drawing a second king. So you see that sometimes two events influence each other and other times they do not.

> **DEFINITIONS** Events E and F are **independent** events if
>
> $$P(F \mid E) = P(F).$$
>
> If $P(F \mid E) \neq P(F)$, then E and F are **dependent.**

This definition says that if E and F are independent, then knowing that E has occurred does not influence the way we compute the probability of F.

PROBLEM SOLVING

Strategy: The Analogies Principle

Recall that the Analogies Principle in Section 1.1 tells you that mathematical terminology, symbolism, and its equations are often based on real-life ideas. Although at times it may seem difficult, if you work hard to make the connection between the intuitive ideas and the mathematical formalism, you will be rewarded by your increased understanding of mathematics. If you understand how the everyday usage of the terms *independent* and *dependent* corresponds to their mathematical definitions, it will help you remember their meaning.

Highlight

Are Human Beings in Jeopardy?

Whether playing chess, making soup, practicing law, or playing *Jeopardy!*, what makes an expert different from the rest of us? Scientists studying this question have learned that organizing knowledge effectively makes a person an expert. Researchers in the area of artificial intelligence have developed complex computer programs called expert systems that rely heavily on conditional probability to draw conclusions from an organized body of information.

The "grandfather of expert systems" is an ingenious program called Mycin, developed at Stanford University, which diagnosis blood infections. Some other well-known expert systems are Hearsay, which understands spoken language; Prospector, which is able to predict the location of valuable mineral resources; and Internist, which can diagnose many internal diseases.

However, perhaps the most impressive display of a computer's ability to mimic human intelligence occurred in February of 2011, when IBM's "Watson" calmly defeated legendary *Jeopardy!* champions Ken Jennings and Brad Rutter in a challenge match of man versus machine.

For a fascinating video describing how Watson used probability to become a *Jeopardy!* champion, see http://www.youtube.com/watch?v=DywO4zksfXw.

EXAMPLE 7 Determining Whether Events Are Independent or Dependent

Assume we roll a red and a green die. Are the events F, "a five shows on the red die," and G, "the total showing on the dice is greater than 10," independent or dependent?

SOLUTION:

To answer this question, we must determine whether $P(G|F)$ and $P(G)$ are the same or different. There are three outcomes—(5, 6), (6, 5), and (6, 6)—that give a total greater than 10, so $P(G) = \frac{3}{36} = \frac{1}{12}$.

Now,

$$F = \{(5, 1), (5, 2), (5, 3), (5, 4), (5, 5), (5, 6)\}$$

and

$$G \cap F = \{(5, 6)\},$$

so

$$P(G|F) = \frac{P(G \cap F)}{P(F)} = \frac{1/36}{6/36} = \frac{1}{6}.$$

Because $P(G|F) \neq P(G)$, the events are dependent.

Now try Exercises 35 to 40. 11

QUIZ YOURSELF 11

The situation is the same as in Example 7. Are the events F, "a five shows on the red die," and O, "an odd total shows on the dice," dependent or independent?

EXERCISES 13.3

Sharpening Your Skills

In many of these exercises, you may find it helpful to draw a tree diagram before computing the probabilities.

In Exercises 1–4, assume that we are rolling two fair dice. First compute P(F) *and then* P(F | E)*. Explain why you would expect the probability of* F *to change as it did when we added the condition that* E *had occurred.*

1. E—an odd total shows on the dice.
F—the total is seven.

2. E—an even total shows on the dice.
F—the total is four.

3. E—a three shows on at least one of the dice.
F—the total is less than five.

4. E—a two shows on at least one of the dice.
F—the total is greater than five.

In Exercises 5–8, we are drawing a single card from a standard 52-card deck. Find each probability.

5. P(heart | red)
6. P(king | face card)
7. P(seven | nonface card)
8. P(even-numbered card | nonface card)

You are to randomly pick one disk from a bag that contains the disks shown below. Find each of the following probabilities. For example, P(*heart* | *yellow*) *means you are to find the probability of a heart being on the disk, given that the disk is yellow.*

9. P(heart | yellow)
10. P(pink | smiley face)
11. P(yellow | heart)
12. P(heart | blue)
13. P(heart | pink)
14. P(smiley face | blue)

Probability and drawing cards. *In Exercises 15–20, assume that we draw 2 cards from a standard 52-card deck. Find the desired probabilities.*

a) First assume that the cards are drawn without replacement.

b) Next assume that the cards are drawn with replacement.

15. The probability that we draw two jacks
16. The probability that we draw two hearts
17. The probability that we draw a face card followed by a non-face card
18. The probability that we draw a heart followed by a spade
19. The probability that we draw a jack and a king
20. The probability that we draw a heart and a spade
21. We are drawing 2 cards without replacement from a standard 52-card deck. Find the probability that we draw at least one face card. (*Hint:* Consider the complement.)
22. We are drawing 2 cards with replacement from a standard 52-card deck. Find the probability that we draw at least one heart. (*Hint:* Consider the complement.)

For Exercises 23 to 26 assume that you are drawing two balls without replacement from this urn.

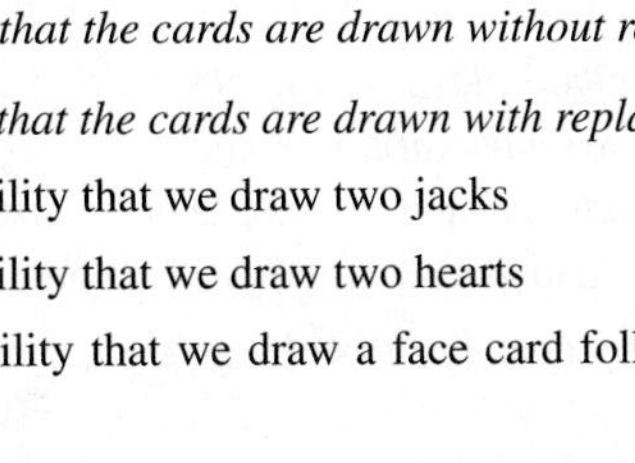

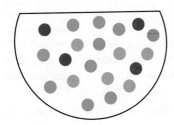

23. What is the probability that you will draw two red balls?
24. What is the probability that you will draw a green ball followed by a blue ball?
25. What is the probability that you will draw a red ball and a blue ball? (*Hint:* There are two ways that this can happen.)
26. What is the probability that neither ball is green?
27. We draw 3 cards without replacement from a standard 52-card deck. Find the probability of drawing exactly two hearts.
28. We draw 3 cards without replacement from a standard 52-card deck. Find the probability of drawing exactly two kings.
29. We roll a pair of dice three times. What is the probability of rolling an even total exactly once?
30. We roll a pair of dice three times. Find the probability that a total of five is rolled exactly twice.

The editors of Auto Web *have evaluated several (E)uropean and (J)apanese crossovers and given them either a (H)igh or (A)verage safety rating. The probability tree summarizes the results of the study. If a car from the study is selected randomly, find the probabilities indicated in Exercises 31–34.*

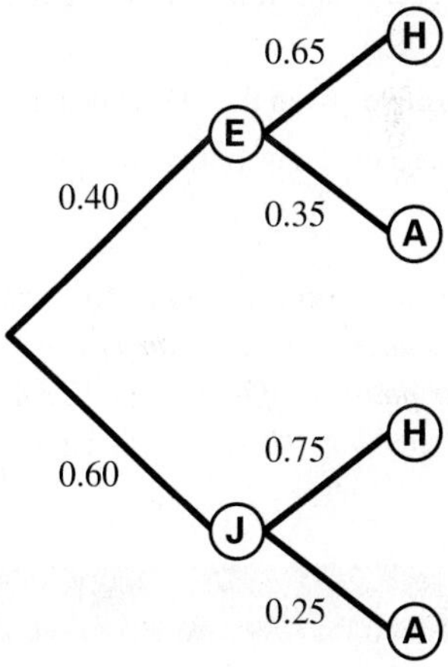

31. $P(A \mid J)$
32. $P(H \mid E)$
33. $P(E \cap A)$
34. $P(H)$

In Exercises 35–40, an experiment and two events are given. Determine if the events are independent or dependent.

35. A penny and a nickel are flipped; "head on the penny" and "tail on the nickel."
36. Two dice (one red, one green) are rolled; "less than three on the red" and "two on the green."
37. A card is drawn randomly from a standard 52-card deck; "the card is red" and "the card is a face card."
38. Two balls are selected randomly without replacement from the urn shown in Exercise 23; "the first ball is red" and "the second ball is green."
39. Two dice (one red, one green) are rolled; "the total is greater than nine" and "the total is even."
40. Two dice (one red, one green) are rolled; "a three shows on the red" and "the total is even."

Applying What You've Learned

According to U.S. government statistics, mononucleosis (mono) is four times more common among college students than the rest of the population. Blood tests for the disease are not 100% accurate. Assume that the following table of data was obtained regarding students who came to your school's health center complaining of tiredness, a sore throat, and slight fever. Use these data to answer Exercises 41–44.

	Has Mono	Doesn't Have Mono	Totals
Positive Blood Test	72	4	76
Negative Blood Test	8	56	64
Totals	80	60	140

If a student is selected from this group, what is the probability of each of the following?

41. The student has mono, given that the test is positive.

42. The student does not have mono, given that the test is positive.

43. The test is positive, given that the student has mono.

44. The test is negative, given that the student does not have mono.

Probability and political preferences. *The Political Action club has surveyed 240 students on your campus regarding the relationship between their political affiliation and their preference in the 2012 presidential election. The results are given in the following table.*

	Democrat	Republican	Independent	Totals
Preferred Obama	105	12	25	142
Preferred Romney	15	68	15	98
Totals	120	80	40	240

If a student is selected randomly from those surveyed, find the probability that the student

45. is a Democrat, given that the student preferred Romney.

46. preferred Obama, given that the student is an Independent.

47. preferred Romney, given that the student is a Republican.

48. is an Independent, given that the student preferred Obama.

Distracted driving. *The given graph summarizes National Highway Traffic Safety Administration statistics regarding distracted drivers below the age of 40 who were involved in fatal crashes in 2009. Assume in Exercises 49–52 that we are selecting a driver at random from the 2,829 tallied. For a more complete explanation of these statistics, see http://www.distraction.gov/research/PDF-Files/Distracted-Driving-2009.pdf.*

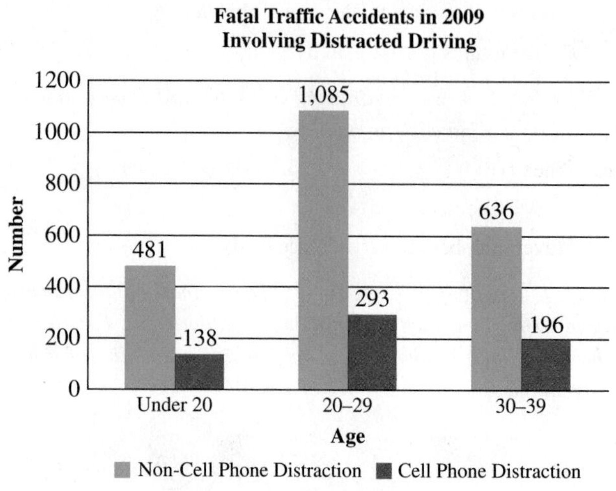

49. If a driver in this group is under 20, what is the probability that the accident was related to cell phone use?

50. If a driver in this group is between 20 and 29, what is the probability that the accident was not related to cell phone use?

51. If the accident was related to cell phone use, what is the probability that the driver was over 29?

52. If the accident was not related to cell phone use, what is the probability that the driver was between the ages of 20 and 29?

Assume that a softball player has a .300 batting average—to keep this simple we'll assume that this means the player has 0.30 probability of getting a hit in each at bat. Use this information to answer Exercises 53–56. Assume that the player bats four times.

53. What is the probability that she gets a hit only in her first at bat?

54. What is the probability that she gets exactly one hit? (Realize that there are four ways to do this.)

55. What is the probability that she gets exactly two hits?

56. What is the probability that she gets at least one hit? (*Hint*: Think complement.)

Selecting a dormitory room. *Exercises 57 and 58 refer to the tree diagram in Figure 13.16 of Example 5.*

57. If Brianna gets an apartment, what is the probability that she is in dorm X?

58. If Brianna gets a room, what is the probability that she is in dorm Y?

Testing a cold medication. *Imagine that you are taking part in a study to test a new cold medicine. Although you don't know exactly what drug you are taking, the probability that it is drug A is 10%, that it is drug B is 20%, and that it is drug C, 70%. From past clinical trials, the probabilities that these drugs will improve your condition are: A (30%), B (60%), and C (70%).*

59. Draw a tree to illustrate this drug trial situation.

60. What is the probability that you will improve given that you are taking drug B?

61. What is the probability that you will improve?

62. If you improve, what is the probability that you are taking drug B?

63. **Probability and exam questions.** Assume that either Professor Ansah or Professor Brunich has constructed the comprehensive exam that you must pass for graduation. Because each professor has extremely different views, it would be useful for you to know who has written the exam questions so that you can slant your answers accordingly. Assume that there is a 60% chance that Ansah wrote the exam. Ansah asks a question about international relations 30% of the time and Brunich asks a similar question 75% of the time. If there is a question on the exam regarding international relations, what is the probability that Ansah wrote the exam?

64. **Probability and exam questions.** Assume now that a third professor, Professor Ubaru, writes the exam 20% of the time, Brunich 30% of the time, and Ansah the rest. Ubaru asks a question about international relations 40% of the time, Brunich 35% of the time, and Ansah 25% of the time. If there is an international relations question on the exam, what is the probability that Brunich did not write the exam?

Product reliability. *You want to purchase a DVD drive for your laptop computer. Assume that 65% of the drives are made outside the United States. Of the U.S.-made drives, 4% are defective; of the foreign-made drives, 6% are defective. Determine each probability rounded to three decimal places.*

65. The probability that the drive you purchase is U.S. made and is not defective

66. The probability that the drive you purchase is foreign made and is defective

67. If your drive is defective, the probability that it is foreign made

68. If your drive is defective, the probability that it is made in the United States

Drug testing. *In Exercises 69 and 70, do computations similar to those in Example 6 using this revised information. Assume that 4% of the employees use the drug and that the test correctly identifies a drug user 98% of the time. Also assume that the test identifies a nonuser as a drug user 3% of the time.*

69. If an employee tests positive, what is the probability that the person is innocent?

70. If an employee tests negative, what is the probability that the person is a user?

Communicating Mathematics

71. If you know the conditional probability formula for $P(F \mid E)$, how do you find the probability formula for $P(E \cap F)$?

72. We say that events E and F are independent if $P(F \mid E) = P(F)$. Give an intuitive explanation of what this equation is saying.

73. Explain how the formal definition of dependent and independent events corresponds to your intuitive understanding of these words in English.

74. The formula $P(E \cap F) = P(E) \cdot P(F)$ is, in general, not true. When can we use this formula instead of the correct formula, $P(E \cap F) = P(E) \cdot P(F \mid E)$?

Challenge Yourself

The birthday problem. *A surprising result that appears in many elementary discussions on probability is called the birthday problem. The question simply stated is this: "If we poll a certain number of people, what is the probability that at least two of those people were born on the same day of the year?" For example, it may be that in the survey two people were both born on March 29.*

To solve this problem, we will use the formula for computing the probability of a complement of an event. It is clear that

$$P(\text{duplication of some birthdays}) = 1 - P(\text{no duplications})$$

To illustrate this, assume that we have three people. To have no duplications, the second person must have a birthday that is different from that of the first person, and the third person must have a birthday that is different from the first two. The probability that the second birthday is different from the first is $\frac{364}{365}$. The probability that the third person has a birthday different from the first two is $\frac{363}{365}$.

Therefore, for three people,

$$P(\text{duplication of birthdays}) = 1 - P(\text{no duplication of birthdays})$$

$$= 1 - \left(\frac{364}{365}\right)\left(\frac{363}{365}\right) = 0.0082.$$

In Exercises 75–79, we will look at several other cases.

75. Assume that we have 10 people. Find the probability that at least 2 of the 10 people were born on the same day of the year.

76. Repeat Exercise 75 for 20 people.

77. Find the smallest number of people such that the probability of two of them being born on the same day of the year is greater than 0.50.

78. Repeat the birthday problem for a room containing only people born in January (31 days). Find the smallest number of people who can be chosen before the probability that two have the same birthday is greater than 0.50.

79. Repeat Exercise 78 for a room containing people born in June, July, and August (92 days).

13.4 Expected Value

Objectives

1. Understand the meaning of expected value.
2. Calculate the expected value of lotteries and games of chance.
3. Use expected value to solve applied problems.

Life and Health Insurers' Profits Skyrocket 213% . . .*

How do insurance companies make so much money? When you buy car insurance, you are playing a sort of mathematical game with the insurance company. You are betting that you are going to have an accident—the insurance company is betting that you won't. Similarly, with health insurance, you are betting that you will be sick—the insurance company is betting that you will stay well. With life insurance, you are betting that, . . . well, . . . you get the idea.

Expected Value

Besides insurance companies, casinos also amass vast wealth using a mathematical theory called **expected value,** which we will now explain. Although we cannot predict exactly when you will have an accident or next win at a roulette wheel, we can accurately compute over the long term what will happen when millions of cases are considered.

To make this idea clear, assume that your school offers property insurance to cover personal items such as laptops, iPods, cell phones, and even books. Example 1 shows how an insurance company might go about setting your premium.

EXAMPLE 1 Evaluating an Insurance Policy

Suppose that you want to insure a laptop computer, an iPhone, a trail bike, and your textbooks. Table 13.8 lists the values of these items and the probabilities that these items will be stolen over the next year.

a) Predict what the insurance company can expect to pay in claims on your policy.
b) Is $100 a reasonable premium for this policy?

Item	Value	Probability of Being Stolen	Expected Payout by Insurance Company
Laptop	$2,000	0.02	0.02($2,000) = $40
iPhone	$400	0.03	0.03($400) = $12
Trail bike	$600	0.01	0.01($600) = $6
Textbooks	$800	0.04	0.04($800) = $32

TABLE 13.8 Value of personal items and the probability of their being stolen.

SOLUTION:

a) From Table 13.8 the company has a 2% chance of having to pay you $2,000, or, another way to look at this is the company expects to lose on average 0.02($2,000) = $40 by insuring your computer. Similarly, the expected loss on insuring your iPhone

*According to a report by Weiss Ratings, Inc., a provider of independent ratings of financial institutions.

is 0.03($400) = $12. To estimate, on average, what it would cost the company to insure all four items, we compute the following sum:

probability of being stolen

Computer iPhone Bike Books

$$0.02(\$2{,}000) + 0.03(\$400) + 0.01(\$600) + 0.04(\$800) = \$90$$

Computer iPhone Bike Books

cost

estimated cost to insurance company

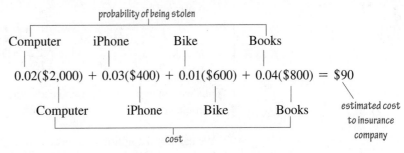

The $90 represents, *on average*, what the company can expect to pay out on a policy such as yours.

b) The $90 in part a) is telling us that if the insurance company were to write one million policies like this, it would expect to pay 1,000,000 × ($90) = $90,000,000 in claims. If the company is to make a profit, it must charge more than $90 as a premium, so it seems like a $100 premium is reasonable.

Now try Exercises 1 and 2. **12**

QUIZ YOURSELF 12

In Example 1, if you were to drop coverage on your iPhone and add coverage on your saxophone that cost $1,400, what would the insurance company now expect to pay out in claims if the probability of the saxophone being stolen is 4% and the probability of your books being stolen is reduced to 3%?

The amount of $90 we found in Example 1 is called the *expected value* of the claims paid by the insurance company. We will now give the formal definition of this notion.

DEFINITION Assume that an experiment has outcomes numbered 1 to n with probabilities $P_1, P_2, P_3, \ldots, P_n$. Assume that each outcome has a numerical value associated with it and these are labeled $V_1, V_2, V_3, \ldots, V_n$. The **expected value** of the experiment is

$$(P_1 \cdot V_1) + (P_2 \cdot V_2) + (P_3 \cdot V_3) + \cdots + (P_n \cdot V_n).$$

In Example 1, the probabilities were $P_1 = 0.02$, $P_2 = 0.03$, $P_3 = 0.01$, and $P_4 = 0.04$. The values were $V_1 = 2{,}000$, $V_2 = 400$, $V_3 = 600$, and $V_4 = 800$.

SOME GOOD ADVICE

Pay careful attention to what notation tells you to do in performing a calculation. In calculating expected value, you are told to *first* multiply the probability of each outcome by its value and *then* add these products together.

Expected Value of Games of Chance

EXAMPLE 2 Computing Expected Value When Flipping Coins

What is the number of heads we can expect when we flip four fair coins?

SOLUTION: Review the Be Systematic Strategy on page 5.

Recall that there are 16 ways to flip four coins. We will consider the outcomes for this experiment to be the different numbers of heads that could arise. Of course, these outcomes are not equally likely, as we indicate in Table 13.9. If you don't see this at first, you could draw a tree to show the 16 possible ways that four coins can be flipped. You would find that 1 of the 16 branches corresponds to no heads, 4 of the 16 branches would represent flipping exactly one head, and 6 of the 16 branches would represent flipping exactly two heads, and so on.

Number of Heads	Probability
0	$\frac{1}{16}$
1	$\frac{4}{16}$
2	$\frac{6}{16}$
3	$\frac{4}{16}$
4	$\frac{1}{16}$

TABLE 13.9 Probabilities of obtaining a number of heads when flipping four coins.

We calculate the expected number of heads by first multiplying each outcome by its probability and then adding these products, as follows:

$$\left(\underbrace{\frac{1}{16}}_{\text{probability of 0 heads}}\cdot 0\right)+\left(\underbrace{\frac{4}{16}}_{\text{probability of 1 head}}\cdot 1\right)+\left(\frac{6}{16}\cdot 2\right)+\left(\frac{4}{16}\cdot 3\right)+\left(\frac{1}{16}\cdot 4\right)=\frac{32}{16}=2.$$

Thus we can expect to flip two heads when we flip four coins, which corresponds to our intuition.

We can use the notion of expected value to predict the likelihood of winning (or more likely losing) at games of chance such as blackjack, roulette, and even lotteries.

EXAMPLE 3 The Expected Value of a Roulette Wheel

Although there are many ways to bet on the 38 numbers of a roulette wheel,* one simple betting scheme is to place a bet, let's say \$1, on a single number. In this case, the casino pays you \$35 (you also keep your \$1 bet) if your number comes up, and otherwise you lose the \$1. What is the expected value of this bet?

SOLUTION:

We can think of this betting scheme as an experiment with two outcomes:

1. Your number comes up and the value to you is +\$35.
2. Your number doesn't come up and the value to you is −\$1.

Because there are 38 equally likely numbers that can occur, the probability of the first outcome is $\frac{1}{38}$ and the probability of the second is $\frac{37}{38}$. The expected value of this bet is therefore

$$\left(\underbrace{\frac{1}{38}}_{\text{probability of winning}}\cdot \underbrace{35}_{\text{amount won}}\right)+\left(\underbrace{\frac{37}{38}}_{\text{probability of losing}}\cdot \underbrace{(-1)}_{\text{amount lost}}\right)=\frac{35-37}{38}=\frac{-2}{38}=-\frac{1}{19}=-0.0526.$$

This amount means that, on the average, the casino expects you to lose slightly more than 5 cents for every dollar you bet.

Now try Exercises 3 to 8.

The roulette wheel in Example 3 is an example of an unfair game.

DEFINITIONS If a game has an expected value of 0, then the game is called **fair.** A game in which the expected value is not 0 is called an **unfair game.**

Although it would seem that you would not want to play an unfair game, in order for a casino or a state lottery to make a profit, the game has to be favored against the player.

EXAMPLE 4 Determining the Fair Price of a Lottery Ticket

Assume that it costs \$1 to play a state's daily number. The player chooses a three-digit number between 000 and 999, inclusive, and if the number is selected that day, then the player wins \$500 (this means the player's profit is \$500 − \$1 = \$499).

*See Example 8 in Section 13.1 for a description of a roulette wheel.

Between the Numbers

A Fool and His Money Are Soon Parted*

While researching this topic, I came across numerous advertisements for books, pamphlets, charts, computer programs, and so forth claiming something like the following:

> *This book will make you a consistent lottery winner! Learn the scientific way to track numbers and take the mystery out of gambling. Make betting on the lottery an investment that earns sure-fire profits using my tested methods. [And so on, and so on. . . .]*

From time to time people have approached me with claims they have discovered a new system that defies the age-old counting and probability principles that you have been learning in this book—a system that will help them "beat the odds."

There is no such system! If you play the same numbers every day because "they are due to come up" or if you think you have found a pattern to the numbers, realize that the numbered balls bouncing around have no memory. The balls may or may not behave as they did last week, or last month, or last year.

Whether playing a carnival game, a casino game, or a state lottery, the game is usually slanted against you. That is, *your* expected value is negative and the more often you play, the more likely it is that whomever you are playing against will wind up with your money in their pocket.

a) What is the expected value of this game?

b) What should the price of a ticket be in order to make this game fair?

SOLUTION:

Outcome	Value	Probability
You win	\$499	$\frac{1}{1{,}000}$
You lose	$-\$1$	$\frac{999}{1{,}000}$

TABLE 13.10 Values and probabilities associated with playing the daily number.

a) There are 1,000 possible numbers that can be selected. One of these numbers is in your favor and the other 999 are against your winning. So, the probability of you winning is $\frac{1}{1{,}000}$ and the probability of you losing is $\frac{999}{1{,}000}$. We summarize the values for this game with their associated probabilities in Table 13.10. The expected value of this game is therefore

$$\Big(\underbrace{\frac{1}{1{,}000}}_{\text{probability of winning}}\cdot \underbrace{499}_{\text{amount won}}\Big) + \Big(\overbrace{\frac{999}{1{,}000}}^{\text{probability of losing}}\cdot \overbrace{(-1)}^{\text{amount lost}}\Big) = \frac{499-999}{1{,}000} = \frac{-500}{1{,}000} = -0.50. \quad (1)$$

This means that the player, on average, can expect to lose 50 cents per game. Notice that playing this lottery is 10 times as bad as playing a single number in roulette.

b) From part a) we see that if we pay a dollar to play this game, we can expect to lose \$0.50 per game on average. Therefore we should pay 50 cents less to make the game fair. We can verify this as follows:

Assume that you pay \$0.50 to play the game and as before, you are paid \$500 if you win and nothing if you lose. We recalculate equation (1) above as

$$\left(\frac{1}{1{,}000}\cdot(500-0.50)\right) + \left(\frac{999}{1{,}000}\cdot(-0.50)\right) = \frac{(500-0.50) - 999\cdot(0.50)}{1{,}000}$$

$$= \frac{499.50 - 499.50}{1{,}000} = 0.$$

Because the expected value is now zero, the game is fair.

Of course, no state would charge \$0.50 to play this game because by doing so, the state would not make any profit.

Now try Exercises 9 to 12.

*Thomas Tusser (1524–1580)

Historical Highlight

The History of Lotteries

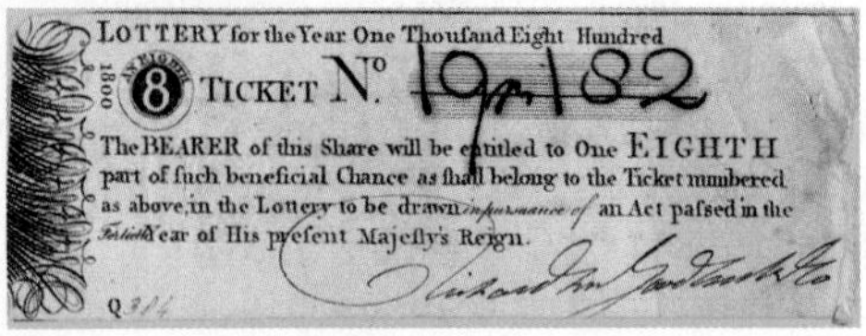

Lotteries have existed since ancient times. The Roman emperor Nero gave slaves or villas as door prizes to guests attending his banquets, and Augustus Caesar used public lotteries to raise funds to repair Rome.

The first public lottery paying money prizes began in Florence, Italy, in the early 1500s; when Italy became consolidated in 1870, this lottery evolved into the Italian National Lottery. In this lottery, five numbers are drawn from 1 to 90. A winner who guesses all five numbers is paid at a ratio of 1,000,000 to 1. The number of possible ways to choose these five numbers is $C(90, 5) = 43{,}949{,}268$. Thus, as with most lotteries, these odds make the lottery a very good bet for the state and a poor one for the ordinary citizen.

Lotteries also played an important role in the early history of the United States. In 1612, King James I used lotteries to finance the Virginia Company to send colonists to the New World. Benjamin Franklin obtained money to buy cannons to defend Philadelphia, and George Washington built roads through the Cumberland Mountains by raising money through lotteries they conducted. In fact, in 1776, the Continental Congress used a lottery to raise $10 million to finance the American Revolution.

Other Applications of Expected Value

Calculating expected value can help you decide what is the best strategy for answering questions on standardized tests such as the GMATs.

EXAMPLE 5 Expected Value and Standardized Tests

A student is taking a standardized test consisting of multiple-choice questions, each of which has five choices. The test taker earns 1 point for each correct answer; $\frac{1}{3}$ point is subtracted for each incorrect answer. Questions left blank neither receive nor lose points.

a) Find the expected value of randomly guessing an answer to a question. Interpret the meaning of this result for the student.

b) If you can eliminate one of the choices, is it wise to guess in this situation?

SOLUTION:

a) Because there are five choices, you have a probability of $\frac{1}{5}$ of guessing the correct result, and the value of this is $+1$ point. There is a $\frac{4}{5}$ probability of an incorrect guess, with an associated value of $-\frac{1}{3}$ point. The expected value is therefore

$$\left(\frac{1}{5}\cdot 1\right) + \left(\frac{4}{5}\cdot\left(-\frac{1}{3}\right)\right) = \frac{1}{5} + \frac{-4}{15} = \frac{3}{15} - \frac{4}{15} = -\frac{1}{15}.$$

Thus, you can expect to be penalized for guessing and should not do so.

b) If you eliminate one of the choices and choose randomly from the remaining four choices, the probability of being correct is $\frac{1}{4}$ with a value of $+1$ point; the probability of being incorrect is $\frac{3}{4}$ with a value of $-\frac{1}{3}$. The expected value is now

$$\left(\frac{1}{4}\cdot 1\right) + \left(\frac{3}{4}\cdot\left(-\frac{1}{3}\right)\right) = \frac{1}{4} + \frac{-1}{4} = 0.$$

You now neither benefit nor are penalized by guessing.

Now try Exercises 19 to 22. 13

QUIZ YOURSELF 13

Calculate the expected value as in Example 5(b), but now assume that the student can eliminate two of the choices. Interpret this result.

Businesses have to be careful when ordering inventory. If they order too much, they will be stuck with a surplus and might take a loss. On the other hand, if they do not order enough, then they will have to turn customers away, losing profits.

EXAMPLE 6 Using Expected Value in Business

Cher, the manager of the U2 Coffee Shoppe, is deciding on how many of Bono's Bagels to order for tomorrow. According to her records, for the past 10 days the demand has been as follows:

Demand for Bagels	40	30
Number of Days with These Sales	4	6

She buys bagels for \$1.45 each and sells them for \$1.85. Unsold bagels are discarded. Find her expected value for her profit or loss if she orders 40 bagels for tomorrow morning.

SOLUTION:

In calculating the expected profit (or loss) we have to consider:

1. The outcomes
2. The probabilities associated with the outcomes
3. The values (profit or loss) associated with each outcome.

1. The outcomes are: the demand for bagels is 30 or the demand for bagels is 40.
2. Based on past demand the probability of:

$$\text{the demand for 40 bagels is } \frac{4}{10} = 0.4\text{, and}$$

$$\text{the demand for 30 bagels is } \frac{6}{10} = 0.6.$$

3. If the demand is for 40 bagels, Cher will sell all of her bagels at a profit of

$$\$1.85 - \$1.45 = \$0.40 \text{ per bagel.}$$

If the demand is for 30 bagels, she will make

$$\underbrace{30(\$0.40)}_{\text{profit on 30 sold bagels}} - \underbrace{10(\$1.45)}_{\text{loss on 10 unsold bagels}} = \$12.00 - \$14.50 = -\$2.50.$$

That is, she will lose \$2.50. The following table summarizes our discussion.

Demand	Probability	Profit or Loss
40	0.4	\$16.00
30	0.6	−\$2.50

Thus, the expected value in profit or loss for ordering 40 bagels is

$$(0.40)(16) + (0.60)(-2.50) = +6.40 + (-1.50) = 4.90.$$

So she can expect a profit of \$4.90 if she orders 40 bagels.

Now try Exercises 37 and 38. 14

QUIZ YOURSELF 14

Redo Example 6 assuming that Cher orders 30 bagels.

EXERCISES 13.4

Sharpening Your Skills

In Exercises 1 and 2, we give the probabilities and values associated with the five outcomes of an experiment. Calculate the expected value for the experiment.

1.

Outcome	Probability	Value
A	0.1	4
B	0.3	6
C	0.4	−2
D	0.15	−4
E	0.05	8

Expected value: ________

2.

Outcome	Probability	Value
A	0.2	6
B	0.35	−4
C	0.1	−2
D	0.25	12
E	0.1	8

Expected value: ________

In Exercises 3 and 4, you are playing a game in which a single die is rolled. Calculate your expected value for each game. Is the game fair? (Assume that there is no cost to play the game.)

3. If an odd number comes up, you win the number of dollars showing on the die. If an even number comes up, you lose the number of dollars showing on the die.

4. You are playing a game in which a single die is rolled. If a four or five comes up, you win $2; otherwise, you lose $1.

In Exercises 5 and 6, you pay $1 to play a game in which a pair of fair dice are rolled. Calculate your expected value for the game. (Remember to subtract the cost of playing the game from your winnings.) Calculate the price of the game to make the game fair.

5. If a six, seven, or eight comes up, you win $5; if a two or 12 comes up, you win $3; otherwise, you lose the dollar you paid to play the game.

6. If a total lower than five comes up, you win $5; if a total greater than nine comes up, you win $2; otherwise, you lose the dollar you paid to play the game.

In Exercises 7 and 8, a card is drawn from a standard 52-card deck. Calculate your expected value for each game. You pay $5 to play the game, which must be subtracted from your winnings. Calculate the price of the game to make the game fair.

7. If a heart is drawn, you win $10; otherwise, you lose your $5.

8. If a face card is drawn, you win $20; otherwise, you lose your $5.

In Exercises 9–12, first calculate the expected value of the lottery. Determine whether the lottery is a fair game. If the game is not fair, determine a price for playing the game that would make it fair.

9. The Daily Number lottery costs $1 to play. You must pick three digits in order from 0 to 9 and duplicates are allowed. If you win, the prize is $600.

10. The Big Four lottery costs $1 to play. You must pick four digits in order from 0 to 9 and duplicates are allowed. If you win, the prize is $2,000.

11. Five hundred chances are sold at $5 apiece for a raffle. There is a grand prize of $500, two second prizes of $250, and five third prizes of $100.

12. One thousand chances are sold at $2 apiece for a raffle. There is a grand prize of $300, two second prizes of $100, and five third prizes of $25.

Applying What You've Learned

13. Evaluating a franchise's profits. Grace Adler is planning to buy a franchise from Home Deco to sell decorations for the home. The table below shows average weekly profits, rounded to the nearest hundred, for a number of the current franchises. If she were to buy a franchise, what would her expected weekly profit be?

Average Weekly Profit	Number Who Earned This
$100	4
$200	8
$300	13
$400	21
$500	3
$600	1

14. Preparing for a heat wave. For the past several years, the Metrodelphia Fire Department has been keeping track of the number of fire hydrants that have been opened illegally daily during heat waves. These data (rounded to the nearest ten) are given in the table below. Use this information to calculate how many hydrants the department should expect to be opened per day during the upcoming heat wave.

Hydrants Opened	Days
20	13
30	11
40	15
50	11
60	9
70	1

In Exercises 15–18, we describe several ways to bet on a roulette wheel. Calculate the expected value of each bet. We show a portion of a layout for betting on roulette in the diagram below. When we say that a bet pays "k to 1," we mean that if a player wins, the player wins k *dollars as well as keeping his or her bet. When the player loses, he or she loses* $1. *Recall that there are 38 numbers on a roulette wheel.*

15. Playing roulette. A player can "bet on a line" by placing a chip at location A in the figure. By placing the chip at A, the player is betting on 1, 2, 3, 0, and 00. This bet pays 6 to 1.

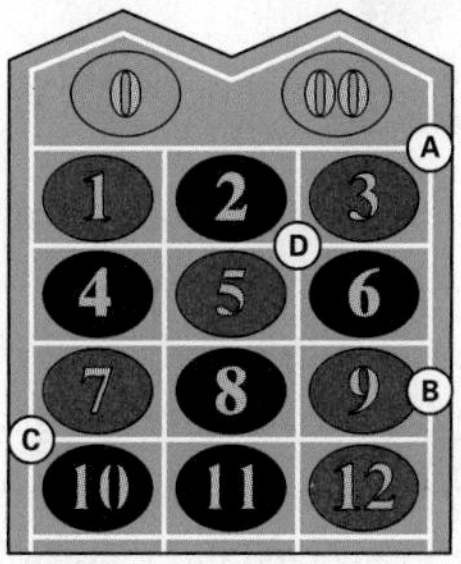

16. Playing roulette. A player can "bet on a square" by placing a chip at the intersection of two lines, as at location D. By placing a chip at D, the player is betting on 2, 3, 5, and 6. This bet pays 8 to 1.

17. Playing roulette. A player can "bet on a street" by placing a chip on the table, as at location B. The player is now betting on 7, 8, and 9. This bet pays 8 to 1.

18. Playing roulette. Another way to "bet on a street" is to place a chip as at location C. By placing a chip at location C, the player is betting on 7, 8, 9, 10, 11, and 12. This bet pays 5 to 1.

In Exercises 19–22, a student is taking the GRE, consisting of several multiple-choice questions. One point is awarded for each correct answer. Questions left blank neither receive nor lose points.

19. Expectation and standardized tests. If there are four options for each question and the student is penalized $\frac{1}{4}$ point for each wrong answer, is it in the student's best interest to guess? Explain.

20. Expectation and standardized tests. If there are three options for each question and the student is penalized $\frac{1}{3}$ point for each wrong answer, is it in the student's best interest to guess? Explain.

21. Expectation and standardized tests. If there are five options for each question and the student is penalized $\frac{1}{2}$ point for each wrong answer, how many options must the student be able to rule out before the expected value of guessing is zero?

22. Expectation and standardized tests. If there are four options for each question and the student is penalized $\frac{1}{2}$ point for each wrong answer, how many options must the student be able to rule out before the expected value of guessing is zero?

Assume that you have $10,000 to invest in stocks, bonds, or precious metals. Use this table, which shows the probabilities for gain or loss for these investments over the next year, in Exercises 23–26.

Stocks		Bonds		Metals	
Probability	Change	Probability	Change	Probability	Change
0.5	Gain 6%	0.6	Gain 4%	0.45	Gain 10%
0.3	No change	0.3	No change	0.15	No change
0.2	Lose 2%	0.1	Lose 3%	0.4	Lose 6%

23. Evaluating investments. What would be the expected gain or loss (in dollars) if you invested in stocks?

24. Evaluating investments. Repeat Exercise 23 for bonds.

25. Evaluating investments. Repeat Exercise 23 for precious metals.

26. Evaluating investments. What do you see as the advantages and disadvantages of each investment? Which investment do you personally feel the most comfortable with?

27. Assume that the probability of a 25-year-old male living to age 26, based on mortality tables, is 0.98. If a $1,000 one-year term life insurance policy on a 25-year-old male costs $27.50, what is its expected value?

28. Assume that the probability of a 22-year-old female living to age 23, based on mortality tables, is 0.995. If a $1,000 one-year term life insurance policy on a 22-year-old female costs $20.50, what is its expected value?

29. Your insurance company has a policy to insure personal property. Assume you have a laptop computer worth $2,200, and there is a 2% chance that the laptop will be lost or stolen during the next year. What would be a fair premium for the insurance? (We are assuming that the insurance company earns no profit.)

30. Assume that you have a used car worth $6,500 and you wish to insure it for full replacement value if it is stolen. If there is a 1% chance that the car will be stolen, what would be a fair premium for this insurance? (We are assuming that the insurance company earns no profit.)

31. A company estimates that it has a 60% chance of being successful in bidding on a $50,000 contract. If it costs $5,000 in consultant fees to prepare the bid, what is the expected gain or loss for the company if it decides to bid on this contract?

32. In Exercise 31, suppose that the company believes that it has a 40% chance to obtain a contract for $35,000. If it will cost $2,000 to prepare the bid, what is the expected gain or loss for the company if it decides to bid on this contract?

Communicating Mathematics

33. Explain in your own words the definition of expected value. If you are playing a game, such as roulette, one time, what does expected value tell you about what to expect?

34. What do we mean by a fair game? What do we mean by the fair price of a game? If you play a game such as a state lottery or a casino game, would you guess that the expected value of the game (for you) would be positive or negative? Why?

Between the Numbers

35. Investigate a lottery in your state. Gather as much detailed information as possible and make a report on your findings. If possible, calculate the expected value of that lottery.

36. Search online for advertisements for books, software, etc. that claim to show you how to win at playing lotteries. If you go to amazon.com, you can see specific information (table of contents, introductions, etc.) about such books. Report on the claims that you find in your research.

Challenge Yourself

37. Estimating daily profit. Nell's Bagels & Stuff, a local coffee shop, sells coffee, bagels, magazines, and newspapers. Nell has gathered information for the past 20 days regarding the demand for bagels. We list this information in the following table.

Demand for Bagels Sold	150	140	130	120
Number of Days with These Sales	3	6	5	6

Nell wants to use expected value to compute her best strategy for ordering bagels for the next week. She intends to order the same number each day and must order in multiples of 10; therefore, she will order either 120, 130, 140, or 150 bagels. She buys the bagels for 65 cents each and sells them for 90 cents.

a. What is Nell's expected daily profit if she orders 130 bagels per day? (*Hint:* First compute the profit Nell will earn if she can sell 150, 140, 130, and 120 bagels.)

b. What is Nell's expected daily profit if she orders 140 bagels per day?

38. Estimating daily profit. Mike sells the *Town Crier*, a local paper, at his newsstand. Over the past 2 weeks, he has sold the following number of copies.

Number of Copies Sold	90	85	80	75
Number of Days with These Sales	2	3	4	1

Each copy of the paper costs him 40 cents, and he sells it for 60 cents. Assume that these data will be consistent in the future.

a. What is Mike's expected daily profit if he orders 80 copies per day? (*Hint:* First compute the profit Mike will earn if he can sell 90, 85, 80, and 75 papers.)

b. What is Mike's expected daily profit if he orders 85 copies per day?

39. a. Calculate the expected total if you roll a pair of standard dice.

b. Unusual dice, called *Sicherman dice*, are numbered as follows.

Red: 1, 2, 2, 3, 3, 4 Green: 1, 3, 4, 5, 6, 8

Calculate the expected total if you roll a pair of these dice.

40. Suppose we have two pairs of dice (these are called *Efron's dice*) numbered as follows.

Pair One: Red: 2, 2, 2, 2, 6, 6 Green: 5, 5, 6, 6, 6, 6

Pair Two: Red: 1, 1, 1, 5, 5, 5 Green: 4, 4, 4, 4, 12, 12

If you were to play a game in which the highest total wins when you roll, what is the better pair to play with?

41. In playing a lottery, a person might buy several chances in order to improve the likelihood (probability) of winning. Does buying several chances change your expected value of the game? Explain.

42. A lottery in which you must choose 6 numbers correctly from 40 possible numbers is called a $\frac{40}{6}$ lottery. In general, an $\frac{m}{n}$ lottery is one in which you must correctly choose n numbers from m possible numbers. Investigate what kind of lotteries there are in your state. What is the probability of winning such a lottery?

Looking Deeper

13.5

Binomial Experiments

Objectives

1. Be able to compute binomial probabilities.
2. Use binomial probability to solve applied problems.

Have you ever taken a quiz for which you were not prepared? Perhaps it was a 10-question true–false quiz, or maybe a 5-question multiple-choice quiz. What would your chances of being successful on these types of tests be by purely guessing? Maybe you have bought a box of cereal containing a collectable figure. How many purchases must you make before you can reasonably expect to have the complete set of figures? In an apparently different situation, a pharmaceutical company makes a claim regarding the effectiveness of a new vaccine. How might we test the company's claim for its reliability?

Binomial Probability

KEY POINT

Binomial trials have only two outcomes.

All these questions are related to an important class of experiments in probability theory that deserve special consideration. These experiments are similar in that they all have two outcomes—one of the outcomes is referred to as "success" and the other as "failure." Such experiments are called *binomial trials.** We list several of these in Table 13.11. What we have decided to call a success or a failure is arbitrary. The point is that one outcome is considered a success and the other a failure.

Experiment	Success	Failure
Flip one coin	Head	Tail
Roll two dice	Roll a total of seven	Roll a total other than a seven
Test a computer chip	Chip functions properly	Chip is defective
Inoculate a person with a flu vaccine	Person does not get the flu	Person gets the flu
Buy a box of cereal	Obtain a new collectable figure	Obtain a collectable figure that you already have
Guess at a multiple-choice question	Guess correct answer	Guess incorrectly

TABLE 13.11 Examples of binomial trials.

PROPERTIES OF A BINOMIAL EXPERIMENT A sequence of binomial trials is called a **binomial experiment** and has the following properties.

1. The experiment is performed for a fixed number of trials.
2. The experiment has only two outcomes, "success" and "failure."
3. The probability of success is the same from trial to trial.
4. The trials are independent of each other.

*These are also often called Bernoulli trials, named after the brilliant seventeenth-century Swiss mathematician Jakob Bernoulli.

EXAMPLE 1 Experiments That Are Not Binomial Experiments

Explain why each experiment is not a binomial experiment.

a) Roll a single die until a six comes up.
b) A student takes a course and earns either an A, B, C, D, or F.
c) Select four cards without replacement and observe whether a heart is drawn.

SOLUTION:

a) The number of trials is not fixed. We do not know how many times the die will have to be rolled before a six comes up.
b) There are more than two outcomes.
c) Because the cards are not being replaced, the probability of a heart will change from draw to draw.

Now try Exercises 1 to 6.

In Example 2, we see a common pattern that occurs in computing binomial probabilities.

KEY POINT

The probabilities of binomial experiments have the same pattern.

EXAMPLE 2 A Binomial Experiment of Rolling a Pair of Dice Three Times

What is the probability of rolling a total of seven exactly once if we roll a pair of dice three times?

SOLUTION: Review the Three-Way Principle on page 13.

Because we can think of this experiment as occurring in three stages (roll dice first time, roll dice second time, roll dice third time), we can represent it by a tree diagram, as in Figure 13.19. We have placed the probability of each outcome of a stage of the experiment along the appropriate branch.

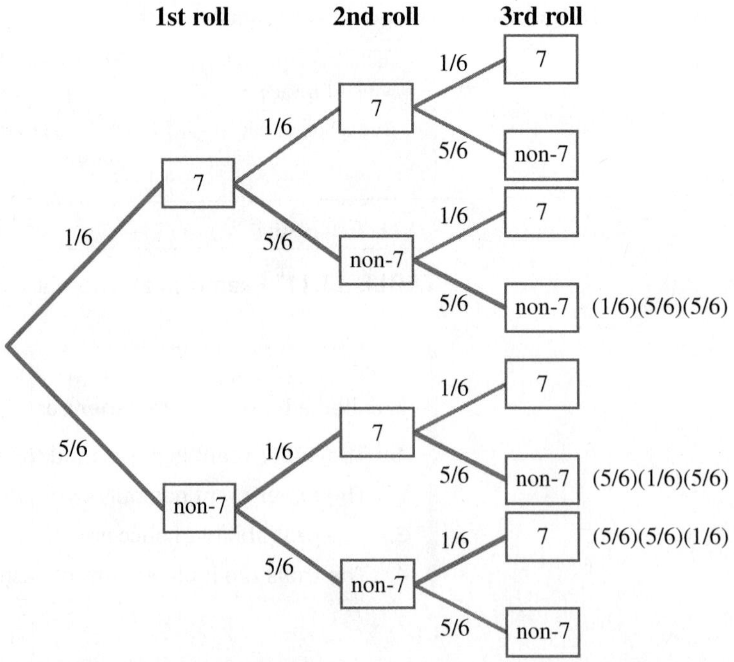

FIGURE 13.19 Rolling three dice can be thought of as a binomial experiment—in each roll we get a seven or a nonseven.

We have highlighted three branches in Figure 13.19 to show the three ways to roll exactly one seven in three rolls of the dice. If we consider rolling a seven a success and rolling a nonseven a failure, then we see that the probability of a success on the first roll and failure on rolls two and three is $\left(\frac{1}{6}\right)\left(\frac{5}{6}\right)\left(\frac{5}{6}\right)$. Similarly, we could get success on the second roll and failure on the other rolls with a probability of $\left(\frac{5}{6}\right)\left(\frac{1}{6}\right)\left(\frac{5}{6}\right)$. Finally, if the success occurs on the third roll, the probability is $\left(\frac{5}{6}\right)\left(\frac{5}{6}\right)\left(\frac{1}{6}\right)$. Thus, the probability of rolling exactly one seven on three rolls of the dice is $\left(\frac{1}{6}\right)\left(\frac{5}{6}\right)\left(\frac{5}{6}\right) + \left(\frac{5}{6}\right)\left(\frac{1}{6}\right)\left(\frac{5}{6}\right) + \left(\frac{5}{6}\right)\left(\frac{5}{6}\right)\left(\frac{1}{6}\right) = \frac{75}{216} \approx 0.3472.$

Now try Exercises 7 and 8.

In Example 2, we can write each one of the three probabilities in the form $\left(\frac{1}{6}\right)^1\left(\frac{5}{6}\right)^2$. We view this abstractly as

$$(\text{probability of success})^{(\text{number of successes})} \times (\text{probability of failure})^{(\text{number of failures})}$$

Suppose now that we want to compute the probability of obtaining exactly two sevens when a pair of dice is rolled five times. We will compute this probability in stages.

First, we have to choose which two of the five rolls correspond to sevens. From Chapter 12, recall that the number of ways to choose two objects from a set of five is $C(5, 2) = \frac{5!}{2!3!} = 10.$

Second, by doing computations similar to Example 2, we see that each of these 10 outcomes has a probability of $\left(\frac{1}{6}\right)^2\left(\frac{5}{6}\right)^3$. Thus the probability of rolling exactly two sevens in five rolls of a pair of dice is

probability of a seven — probability of a nonseven

$$C(5, 2)\left(\frac{1}{6}\right)^2\left(\frac{5}{6}\right)^3 = 10 \cdot \frac{1}{36} \cdot \frac{125}{216} = \frac{1,250}{7,776} \approx 0.1608.$$

number of ways to choose 2 of 5 rolls

We will now state this calculation in a general form.

FORMULA FOR COMPUTING BINOMIAL PROBABILITIES In a binomial experiment with n trials, if the probability of success in each trial is p, then the probability of exactly k successes is given by

$$C(n, k)(p)^k(1 - p)^{(n-k)}.$$

We will denote this probability by $B(n, k; p)$.

SOME GOOD ADVICE

If you connect new ideas with those you have seen before, it helps you understand mathematics as an integrated collection of concepts. Even though the binomial probability notation may seem complicated, the notation helps us remember its meaning. The B reminds us of the word *binomial.* We have seen the n, k pattern before in Chapter 12 in counting the number of combinations of n objects taken k at a time. Finally, the p reminds us of the word *probability*; in this case, the probability of a success in a binomial trial.

Application of Binomial Probability

Binomial probabilities arise in many real-life situations involving the testing of products.

EXAMPLE 3 Binomial Probability and Drug Testing

A pharmaceutical company claims that a new drug is effective 75% of the time. Assuming the company's claim is accurate, if we test the drug on 20 patients, what is the probability that exactly 15 people will benefit from the drug?

SOLUTION:

This is a binomial experiment with 20 trials in which success is that the drug benefits the patient. According to the company, the probability of success is 0.75 and the probability of failure is 0.25. The probability of exactly 15 successes is therefore $B(20, 15; 0.75)$.

Using the binomial probability formula,* we get

$$\begin{aligned} B(20, 15; 0.75) &= C(20, 15)(0.75)^{15}(1 - 0.75)^{(20-15)} \\ &= 15{,}504(0.75)^{15}(0.25)^{5} \approx 0.2023. \end{aligned}$$

15

QUIZ YOURSELF 15

If we roll a single die eight times, what is the probability of getting exactly two threes?

We will investigate further how to test the pharmaceutical company's claim in the exercises.

EXAMPLE 4 Binomial Probability and Guessing on Exams

Maria has fallen behind in her anthropology class and her instructor just announced a pop quiz consisting of 10 true–false questions. If she must get 6 or more correct to pass the quiz, what is the probability of passing the quiz by guessing?

SOLUTION:

To pass the quiz, Maria must get either 6, 7, 8, 9, or 10 correct. The probability of this is

probability of getting 6 correct

probability of getting 7 correct

$$B(10, 6; \tfrac{1}{2}) + B(10, 7; \tfrac{1}{2}) + B(10, 8; \tfrac{1}{2}) + B(10, 9; \tfrac{1}{2}) + B(10, 10; \tfrac{1}{2})$$
$$= 0.2051 + 0.1172 + 0.0439 + 0.0098 + 0.0010 = 0.377.$$

Thus, the probability that she will pass the quiz by randomly guessing is almost 40%.

Now try Exercises 13 to 22.

We conclude this section with another interesting application involving binomial trials. Manufacturers often package sports cards in small groups so that a collector will have to buy numerous packages in order to obtain a complete set. A collector might ask, "If I know the number of items in the complete set, how many purchases should I expect to make to get a complete set?" Before we answer the collector's question, we need to know more about binomial trials.

Suppose that we are making a sequence of binomial trials where success has probability p and that we repeat the trial until we obtain a success. How many trials do we expect to perform before we get a success? If the probability of success is small, then our intuition tells us that we can expect to perform many trials before we get a success. On the other hand, if the probability of success is large, then we should not have to perform many trials before obtaining a success. In fact, the following can be proved (we will not prove it here):

THE NUMBER OF BINOMIAL TRIALS WE CAN EXPECT BEFORE A SUCCESS If we repeat a binomial trial in which success has probability p, then the number of trials we can expect to perform before we get a success is $\frac{1}{p}$.

*Throughout this section, we will round these binomial probabilities to four decimal places. If you carry more decimal places, you will sometimes get answers that are slightly different from ours.

This is saying, for example, that if in a binomial trial the probability of success is $\frac{1}{100}$, then we should expect to perform the experiment 100 times before we get a success. We are now ready to answer a simplified version of the collector's question.

EXAMPLE 5 How Many Packages Should You Expect to Buy to Obtain a Complete Set of Collectibles?

Assume that a person is buying boxes of cereal, each of which contains a figure of one of five characters in a current movie. The packages are sealed, so the buyer does not know which figure is included with the cereal. How many purchases can this collector expect to make before obtaining the entire set of five figures?

SOLUTION: Review the Look for Patterns Strategy on page 6.

We will call the total number of purchases to obtain a complete set *T*. Then,

$$\begin{aligned} T = {} & \text{the number of purchases to obtain a first new figure } + \\ & \text{the number of purchases to obtain a second new figure } + \\ & \text{the number of purchases to obtain a third new figure } + \\ & \text{the number of purchases to obtain a fourth new figure } + \\ & \text{the number of purchases to obtain a fifth new figure.} \end{aligned}$$

The number of purchases required to obtain a first new figure is 1 because the very first purchase will give us a new figure.

At this point, we have one figure; before we open our next purchase, there are four possible new figures and one old figure. We can think of buying a box of cereal now as a binomial experiment in which the probability of getting a new figure (success) is $\frac{4}{5}$ and the probability of failure is $\frac{1}{5}$. The number of trials we can expect before we get a new figure is therefore $\frac{1}{4/5} = \frac{5}{4}$.

Once we have two different figures, we change our computations. Before we open the next box of cereal, there are possibly one of two old figures in the box or one of three new ones. Buying a box of cereal is now a binomial trial with probability of success $\frac{3}{5}$. Thus, we can expect to obtain the third new figure in $\frac{1}{3/5} = \frac{5}{3}$ trials.

Similarly, the expected number of purchases before we get the fourth figure is $\frac{5}{2}$ and the expected number of purchases before we get the fifth figure is $\frac{5}{1}$.

The total number of purchases *T* we can expect to make before we have a complete set is, therefore,

$$T = 1 + \frac{5}{4} + \frac{5}{3} + \frac{5}{2} + \frac{5}{1} = 11.42.$$

It is likely that the collector will obtain the complete set of five figures by the twelfth purchase.

Now try Exercises 25 and 26.

EXERCISES 13.5

Sharpening Your Skills

In Exercises 1–6, determine whether each experiment is a binomial experiment. If not, explain what property of a binomial experiment fails.

1. Select 4 cards with replacement from a standard 52-card deck and observe whether a king is drawn.

2. Roll a pair of dice five times and observe whether doubles are obtained.

3. Three coins are flipped until three heads are obtained.

4. A person buys five chances on the Daily Number lottery.

5. A student takes a 10-question true–false quiz.

6. Twenty people are injected with a new cold vaccine.

7. If we roll a pair of dice three times, what is the probability of rolling a total of five exactly once?

8. If we roll a pair of dice three times, what is the probability of rolling a total of eight exactly twice?

In Exercises 9 and 10, explain the meaning of each expression and compute its value.

9. $B\left(5, 3; \frac{1}{4}\right)$ **10.** $B\left(6, 2; \frac{1}{2}\right)$

In Exercises 11 and 12, explain the error in using the notation to indicate a binomial probability.

11. $B\left(3, 5; \frac{1}{3}\right)$ **12.** $B(4, 3; 2)$

Applying What You've Learned

13. If you are given a 12-question true–false quiz, what is the probability that you will answer *exactly* 9 of the questions correctly by guessing?

14. If you are given an eight-question multiple-choice quiz, what is the probability that you will answer *exactly* six of the questions correctly by guessing? Assume that each question has four options.

15. If you are given a 12-question true–false quiz, what is the probability that you will answer *at least* 9 of the questions correctly by guessing?

16. If you are given a six-question multiple-choice quiz, what is the probability that you will answer *at least* four of the questions correctly by guessing? Assume that each question has five options.

17. Dr. House has developed a new procedure that he believes can correct a life-threatening medical condition. If the success rate for this procedure is 80%, and the procedure is tried on 10 patients, what is the probability that at least 8 of them will show improvement?

18. You are planning a vacation at the beach and the weather forecast is that there is a 30% chance of rain for each of the next 3 days. What is the probability that you will have rain less than 2 days?

19. Chase Utley currently has a 0.250 batting average. If we ignore all other factors, what is the probability that if he comes to bat five times he gets exactly two hits in today's game?

20. Lisa Leslie, a WNBA basketball player, has a 50% foul shooting average. If she attempts eight foul shots today, what is the probability that she makes exactly four foul shots?

21. A hospital has an urgent need for three units of type A+ blood. Assume that approximately 30% of the population has this type of blood. If there are 12 people waiting to donate one unit of blood, what is the probability that the hospital will be able to meet its need? (*Hint:* Subtract the probability that fewer than three people have A+ blood from 1.)

22. A cellular phone communications network has redundancy built into it in the sense that if several of the components fail, the network may still be able to function properly. If a network has 15 components, each of which is 95% reliable, what is the probability that the network will fail if the network requires at least 12 of the components working in order to function properly? (*Hint:* Compute the probability that the network will not fail and subtract from 1.)

In Exercises 23 and 24, fill in the missing item to make the equation true. Give an intuitive reason why both sides of the equation must be equal.

23. $B\left(5, 3; \frac{1}{4}\right) = B(5, 2; ?)$

24. $B(10, 3; 0.4) = B(10, ?; 0.6)$

25. Assume that a child is buying a fast-food meal that also contains a (random) figure of one of six characters in a popular movie. How many purchases can she expect to make before she collects the entire set of six figures?

26. Assume that a child is buying packages of candy that also contain a (random) picture card of one of 10 players on the 2012 men's Olympic basketball team. How many purchases can he expect to make before he collects the entire set of 10 pictures?

Communicating Mathematics

27. What are the four properties of a binomial experiment?

28. If the probability of success in a binomial trial is $\frac{1}{5}$, how many trials can you expect to perform before you get a success?

Challenge Yourself

29. Assume that during the winter, 50% of the population will come down with the common cold. Suppose a company claims that it has developed a cold vaccine to lower the infection rate. Ten people are vaccinated and we agree that we will accept the claim of the company if four or fewer catch a cold. What is the probability that we are accepting a worthless claim?

30. Continuing the situation in Exercise 29, suppose we decide on a more strict requirement in order to accept the claim that the cold vaccine is effective. We agree that now we will accept the company's claim provided that 2 or fewer people of the 10 who were inoculated catch a cold. What is the probability that we reject a worthwhile vaccine?

CHAPTER SUMMARY

You will learn the items listed in this chapter summary more thoroughly if you keep in mind the following advice:

1. Focus on "remembering how to remember" the ideas. What pictures, word analogies, and examples help you remember these ideas?
2. Practice writing each item without looking at the book.
3. Make 3 × 5 flash cards to break your dependence on the text. Use these cards to give yourself practice tests.

Section	Summary	Example
SECTION 13.1	An observation of a **random phenomenon** is called an **experiment.** The different possible observations of the experiment are called **outcomes.** The set of all possible outcomes is called the **sample space** of the experiment. An **event** is a subset of the sample space. The **probability of an outcome** in a sample space is a number between 0 and 1, inclusive. The **probability of an event** is the sum of the probabilities of the outcomes in the event. We can assign the probability of an event **empirically** as $$P(E) = \frac{\text{the number of times } E \text{ occurs}}{\text{the number of times the experiment is performed}}.$$	Definitions, p. 644 Example 1, pp. 644–645 Example 2, p. 646
	This ratio is called the **relative frequency** of E. If E is an event in a sample space S with all equally likely outcomes, then $$P(E) = \frac{n(E)}{n(S)}.$$	Examples 3 and 4, pp. 647–648
	We can use **counting formulas** to compute probabilities. If E is an event in some sample space S, then 1. $0 \leq P(E) \leq 1$ 2. $P(\varnothing) = 0$ 3. $P(S) = 1$.	Example 5, pp. 648–649 Example 6, pp. 649–650 Discussion, p. 650
	Probability can help explain **genetic theory.**	Discussion, p. 651 Example 7, p. 652
	If the outcomes in the sample space are equally likely, the **odds against** event E are the number of outcomes against E occurring compared to the number of outcomes in favor of E occurring. That is, $\frac{n(E')}{n(E)}$. An alternate formula for computing odds using probability is $\frac{P(E')}{P(E)}$.	Definition, p. 653 Examples 8 and 9, pp. 654–655
SECTION 13.2	We can find the probability of event E by subtracting the probability of its **complement** from 1. That is, $P(E) = 1 - P(E')$.	Example 1, p. 660
	If E and F are events, we find the probability of their **union** as follows: $P(E \cup F) = P(E) + P(F) - P(E \cap F)$. If E and F are disjoint, this formula simplifies to $P(E \cup F) = P(E) + P(F)$.	Examples 2 and 3, pp. 662–663
	We sometimes combine the complement and union formulas to solve a problem.	Example 4, pp. 663–665
SECTION 13.3	If we compute the probability of an event F assuming that event E has already occurred, this is called the **conditional probability of *F* given *E*.** We denote this probability by $P(F \mid E)$. If E and F are events in a sample space with *equally likely outcomes*, then $P(F \mid E) = \frac{n(E \cap F)}{n(E)}$. In general, $P(F \mid E) = \frac{P(E \cap F)}{P(E)}$.	Definition, p. 668 Example 1, pp. 669–670 Example 2, pp. 670–672
	If E and F are events, then we calculate the probability of their **intersection** by using the following formula: $$P(F \cap E) = P(E) \cdot P(F \mid E).$$	Example 3, pp. 672–673 Example 4, pp. 673–674

(Continued)

Section	Summary	Example
	We use **probability trees** to visualize probability problems.	Example 5, pp. 674–675 Example 6, p. 676
	Events E and F are **independent** if $P(F \mid E) = P(F)$. Otherwise, they are **dependent.**	Definition, p. 677 Example 7, p. 678
SECTION 13.4	If an experiment has n outcomes with probabilities $P_1, P_2, P_3, \ldots, P_n$, and $V_1, V_2, V_3, \ldots, V_n$ are values associated with the outcomes, the **expected value** of the experiment is $(P_1 \cdot V_1) + (P_2 \cdot V_2) + (P_3 \cdot V_3) + \cdots + (P_n \cdot V_n)$.	Example 1, pp. 682–683 Definition, p. 683
	We use expected value to analyze games of chance.	Examples 2 and 3, pp. 683–684 Example 4, p. 684–685
	Expected value has many applications, such as evaluating test-taking strategies and business decision making.	Example 5, p. 686 Example 6, p. 687
SECTION 13.5	A **binomial trial** is an experiment that has two outcomes, labeled "success" and "failure." A binomial is a sequence of binomial trials that has the following properties: 1. The experiment is performed for a fixed number of trials. 2. Each trial has only two outcomes. 3. The probability of success is the same from trial to trial. 4. The trials are independent of each other.	Definitions, p. 691 Examples 1 and 2, pp. 692–693
	In a binomial experiment with n trials, if the probability of success in each trial is p, the probability of exactly k successes is given by $C(n, k)(p)^k(1 - p)^{(n-k)}$. We denote this probability by $B(n, k; p)$.	Discussion, p. 693
	Binomial probability can be used to solve wide-ranging applied problems.	Examples 3 and 4, p. 694 Example 5, p. 695

CHAPTER REVIEW EXERCISES

Section 13.1

1. Describe each event as a set of outcomes.

a. When three coins are flipped, we obtain exactly two heads.

b. When two dice are rolled, we obtain a total of eight.

2. If a single card is selected from a standard 52-card deck, what is the probability that a red face card is selected?

3. Explain the difference between empirical and theoretical probability. Give an example of each type of probability.

4. In cross-breeding pea plants, Mendel found that the characteristic "tall" dominated the characteristic "short." If we cross pure tall plants with pure short plants, what is the probability of tall plants in the second generation?

5. a. If the odds against the Dodgers winning the World Series are 17 to 2, what is the probability of them winning the Series?

b. If the probability of rain tomorrow is 0.55, what are the odds against rain?

Section 13.2

6. a. State the formula for computing the probability of the complement of an event.

b. Draw a diagram to explain this formula.

c. In what sort of situations would you be likely to use this formula?

7. If a single card is drawn from a standard 52-card deck, what is the probability that we obtain either a face card or a red card? Draw a diagram to illustrate this situation.

8. On the given spinner, each green sector occupies 10% of the circle, each blue sector occupies 12%, each red sector occupies 9%, and the yellow sector occupies 4%. Assume that the spinner is spun once. What is the probability that

a. The spinner does not stop on red?

b. The spinner stops on an even number or blue?

Section 13.3

9. Explain in your own words what we mean by *conditional probability*.

10. If a pair of fair dice is rolled, what is the probability that we roll a total of five if we are given that the total is less than nine?

11. Assume that 2 cards are drawn without replacement from a standard 52-card deck.

   a. What is the probability that two hearts are drawn?

   b. What is the probability that a queen and then an ace is drawn?

12. A pair of fair dice are rolled. Are events E and F independent?
   E—an odd total is obtained.
   F—the total showing is less than six.

13. Assume that the incidence of the HIV virus in a particular population is 4% and that the test correctly identifies the virus 90% of the time. Assume that false positives occur 6% of the time. If a person tests positive for the virus, what is the probability that the person actually has the virus?

Section 13.4

14. Based on mortality tables, the probability of a 20-year-old male living to age 21 is 0.99. What is the expected value of a \$1,000 one-year term life insurance policy on a 20-year-old male from the insurance company's point of view? Assume that the yearly premium is \$25.25.

15. A card is drawn from a standard 52-card deck. If a face card is drawn, you win \$15; otherwise, you lose \$4. Calculate the expected value for the game.

16. You are playing a game in which four fair coins are flipped. If all coins show the same (all heads or all tails), you win \$5. Calculate the price to play the game that would make the game fair.

Section 13.5

17. Calculate $B\left(8, 3; \frac{1}{2}\right)$.

18. If you are guessing on a 10-question true–false quiz, what is the probability that you will get 8 correct?

CHAPTER TEST

1. Describe each event as a set of outcomes.

   a. When two dice are rolled, we obtain a total greater than nine.

   b. When four coins are flipped, we get more heads than tails.

2. If we select a single card from a standard 52-card deck, what is the probability that we select a black king?

3. a. If the odds against the Dolphins winning the Super Bowl are 28 to 3, what is the probability of them winning the Super Bowl?

   b. If the probability of rain tomorrow is 0.15, what are the odds against rain?

4. If we draw a single card from a standard 52-card deck, what is the probability that we obtain either a heart or a king?

5. What is the difference in the meaning between $P(B \mid A)$ and $P(A \mid B)$?

6. We are rolling a pair of fair dice. Are events E and F independent?
   E—both dice show the same number.
   F—the total is greater than eight.

7. It costs \$1 to play a game in which two dice are rolled. If both dice show the same number, you win \$5; otherwise, you lose. What is the expected value of this game?

8. Calculate $B\left(10, 2; \frac{1}{4}\right)$.

9. If you are guessing on a five-question multiple-choice quiz where each question has four possible answers, what is the probability that you will get three correct?

10. a. Complete the following equation $P(E) +$ ____ $= 1$.

   b. Draw a diagram to illustrate this formula.

11. In pea plants, purple flower color dominates white. With snapdragons, however, a pure red flowering plant crossed with pure white produces a pink flowering plant. If we begin by crossing pure red and white snapdragons, what is the probability of pink flowers in the second-generation plants?

12. Assume that 2% of the Brazilian population had dengue fever and that a test correctly identified the fever 95% of the time. Assume that a false positive occurred 5% of the time. If a person tested positive for the fever, what was the probability that the person actually had the fever?

13. It costs \$2 to buy a raffle ticket. If there are 500 tickets sold, and there is one first prize of \$250, three second prizes of \$100, and five third prizes of \$50, what is the expected value of this raffle?

14. If a pair of dice is rolled, what is the probability of rolling an even total given that the total is less than five?

15. Assume that 2 cards are drawn without replacement from a standard 52-card deck.

   a. What is the probability that two face cards are drawn?

   b. What is the probability that a king and an ace are drawn? (*Hint:* There are two cases.)

GROUP PROJECTS

1. **Comparing empirical and theoretical probabilities.** Have a group of people flip coins 1,000 times, and then determine the ratio of heads obtained to the total flips. How close is the ratio $\frac{\text{Number of heads obtained}}{\text{Total number of flips}}$ to the theoretical probability of 0.5?
2. **Comparing empirical and theoretical probabilities.** Repeat Exercise 1, but have the group roll a single die instead. How close is the ratio $\frac{\text{Number of 5s obtained}}{\text{Total number of rolls}}$ to the theoretical probability of $\frac{1}{6} \approx 0.17$?
3. **Comparing empirical and theoretical probabilities.** Repeat Exercise 1, but now have the group roll two dice and record the totals shown on the two dice. Compute the ratios $\frac{\text{Number of 2s obtained}}{\text{Total number of rolls}}$, $\frac{\text{Number of 3s obtained}}{\text{Total number of rolls}}$, etc., with their theoretical probabilities of $\frac{1}{36} \approx 0.028$, $\frac{2}{36} \approx 0.056$, etc.
4. **Simulating Mendel's experiments.** Recall that in Mendel's pea plant experiment that we discussed in Section 13.1, pea plants could have yellow seeds or green seeds and if a plant had both the yellow and green seed genes, the seed color would be yellow. That is, yellow was dominant over green. Mendel predicted that if we crossed parents who had one yellow and one green gene, then we could expect the following results for the offspring.

		Second Parent	
		Y	*g*
First Parent	*Y*	*YY*	*Yg*
	g	*gY*	*gg*

That is, the offspring would be one-quarter *YY*, one-quarter *Yg*, one-quarter *gY*, and one-quarter *gg*. Only the *gg*-type offspring would have green seeds. You can simulate this by flipping a penny and a nickel. The penny represents the first parent and the nickel represents the second parent. A head means the child inherited the yellow gene and a tail represents the green gene. So HH corresponds to *YY*, HT corresponds to *Yg*, etc. Flip the coins 1,000 times and see how closely your simulation corresponds to the predicted probabilities.

USING TECHNOLOGY

You can replicate experiments very quickly using technology. Repeat Group Projects Exercises 1 to 4 using the technology of your choice. For example, you can use the following technology:

Graphing calculators. Graphing calculators have random number generators built into them. For example, on the TI-83 and TI-84, if you press MATH, then move the cursor to the right to PRB, then press ENTER twice, you will get a number like .5489861799. If you generate a list of such numbers, you can use the digits in the list to simulate flipping a coin. For example, if you think of an odd digit as heads and an even digit as tails, then the number above corresponds to HTTHTTHHHH. Generate 1,000 digits to see how close your simulated coin-flipping probability of a head corresponds to the theoretical probability of $\frac{1}{2}$.

Excel spreadsheets. Use the RANDBETWEEN function to generate random numbers. For example you can simulate flipping coins by using RANDBETWEEN(1,2) to generate a series of 1s and 2s representing heads and tails. RANDBETWEEN(1,6) randomly generates integers from 1 to 6 that simulates rolling dice. This figure shows how to simulate rolling two dice in Excel.

A2 fx =RANDBETWEEN(1,6)

	A	B	C	D	E	F
1	Die 1	Die 2	Total			
2	5	5	10			
3	3	6	9			
4	3	1	4			
5	5	3	8			
6	4	4	8			

Mobile applications. Here is an image of a dice simulation done with a free app that generates random numbers on an iPad or a smart phone.

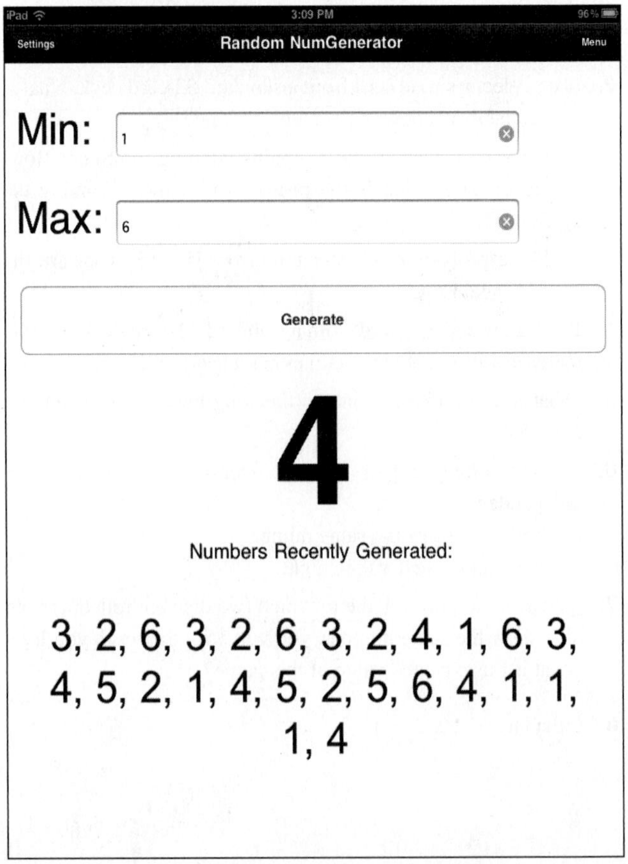

14

Descriptive Statistics

What a Data Set Tells Us

If you logged on to your favorite home page today, checked the news on your smart phone or the Kindle Fire Newsstand, or listened to the radio as you drove to class, you quite likely encountered statistics regarding our satisfaction with Congress, the job outlook for new graduates, the increase (or decrease) in consumer confidence, the state of marriage among 20- to 29-year-olds, and so on, and so on.

Statistics are truly all around us, but we must be careful. Just because a study states something to be so, that does not mean that we can trust the information presented. As an educated person, it is important for you to be aware of how some statistical studies can be flawed and how others can be used to manipulate you.

On the positive side, researchers analyze statistical data to design safer automobiles and to improve the quality of the products that we use every day. Traffic engineers use statistical simulations not only to improve the traffic flow on the roads you drive to school, but also to reduce the amount of time that you spend waiting to ride The Twilight Zone Tower of Terror at Disney World.

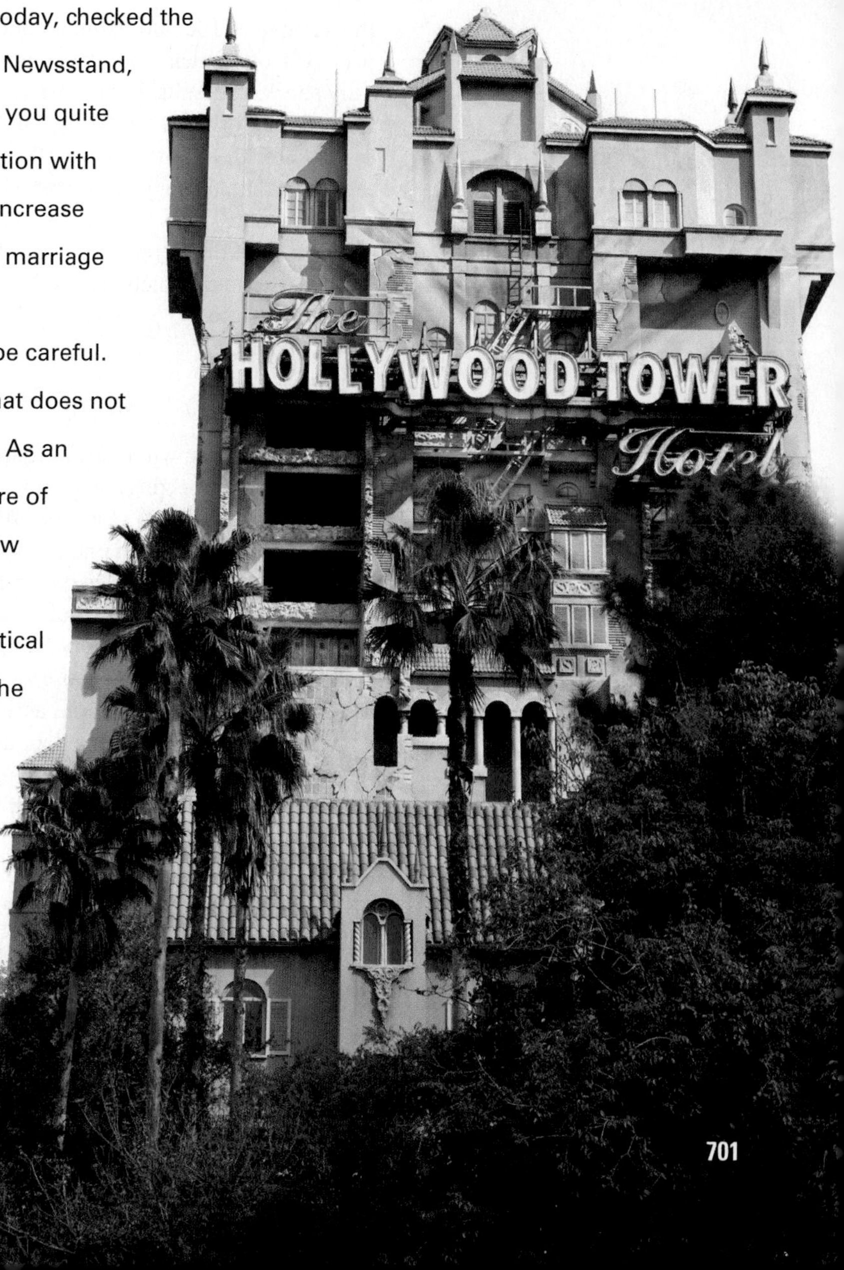

(continued)

In this chapter, you will learn how to organize, summarize, and visualize data so that you can better recognize what the data are telling you and understand the relationships among them.

Section 14.4 gives an interesting explanation of how manufacturers can seem to be able to predict when your Blu-ray player will break down, and we will show how *you* can write an effective warranty.

14.1 Organizing and Visualizing Data*

Objectives

1. Understand the difference between a sample and a population.
2. Organize data in a frequency table.
3. Use a variety of methods to represent data visually.
4. Use stem-and-leaf displays to compare data.

In a survey of 100,000 women conducted by *Cosmopolitan Magazine*, it was found that over 70% of women who were married for more than 5 years had had an affair. These are truly shocking results, but before you swear off marriage completely, there are some questions that you should be asking, such as who conducted the study? How were the women selected for the study? Of the 100,000 women, how many responded to the questions in the study? Did those being surveyed have any particular biases?

It may reassure you to know that a larger survey of 200,000 women found that only 15% of the women reported that they had been unfaithful. These contradictory results make us wonder, which survey is correct? As you will learn shortly, perhaps we should trust neither survey.

Populations and Samples

In this chapter you will study **statistics,** an area of mathematics in which we are interested in gathering, organizing, analyzing, and making predictions from numerical information called **data.** In the two marriage surveys, it would have been ideal if the researchers had been able to contact each one of the millions of married women in the United States. This set of all married women is called the **population.** Of course, this is impractical, so the researchers contacted only a subset of the population, called a **sample.** It is very important that the sample is typical of the population as a whole. In fact, in the two studies just mentioned, both samples were chosen very poorly, and we should trust neither survey.

We will describe a sample as **biased** if it does not accurately reflect the population as a whole with regard to the data that we are gathering. Bias often occurs if we use poor sampling techniques. There are many ways in which bias can creep into a sample. It could occur because of the way in which we decide how to choose the people to participate in the survey. This is called **selection bias.** As an example, if we were to do a phone survey in the middle of a weekday afternoon, we would probably get an overrepresentation of retirees and stay-at-home parents in our sample. Call-in surveys conducted by local news shows and radio stations are also prone to selection bias.

It might seem we would get better, more reliable information if we were to walk around a town and ask people "randomly" to take part in our survey. We put the word

*Many of the calculations that we do in this chapter can be done easily using computers and graphing calculators.

randomly in quotes because studies have found that selection bias can occur using this method since interviewers tend to choose people who are better dressed and who look cooperative, thus skewing the sample.

Another issue that can affect the reliability of a survey is the way we ask the questions, which is called **leading-question bias.** For example, in a Roper poll conducted for the American Jewish Committee on the Holocaust, people were asked, "Does it seem possible or does it seem impossible to you that the Nazi extermination of the Jews never happened?" The use of double negatives in this question caused confusion in the way people responded to the survey. When the question was worded this way, 22% of those surveyed said that it was possible that the Holocaust did not occur. A new survey was conducted in which the question was rephrased, "Does it seem possible to you that the Nazi extermination of the Jews never happened, or do you feel certain that it happened?" In the new survey, only 1% of those surveyed stated that it was possible that the Holocaust never occurred.

We have only touched on the notion of how important it is that statistical conclusions are based on reliable, nonbiased data. There are whole books written on sampling theory. For now, we just want you, as an educated consumer of technical information, to be aware that just because a study says something, as the song says, "It ain't necessarily so!"

We will now turn our focus on what to do once we have obtained reliable data.

Frequency Tables

When we gather information about a population, we often end up with a large collection of numbers. Unless we can organize the data in a meaningful way, it is nearly impossible to interpret these facts. For example, if you glance at the financial section of a newspaper, you will find several pages containing thousands of numerical facts about the daily performance of various stocks in the stock market. This amount of detail about the market's activity seems overwhelming; it is difficult to understand the general pattern of changes in stock prices from those lists of numbers. However, if you were to tune in to the evening news, the commentator might summarize this set of data by saying, "The Dow Jones lost 99.59 points today, closing at 12,251.7. Losers outnumbered winners by three to one." For any large amount of data to be comprehensible, we must organize it and present it so that we can see patterns, trends, and relationships.

> **DEFINITIONS** We refer to a collection of numerical information as **data** or a **distribution.** A set of data listed with their frequencies is called a **frequency distribution.**

KEY POINT

A frequency table is one method used to organize data.

Sometimes we want to show the percent of the time that each item occurs in a frequency distribution. In this case, we call the distribution a **relative frequency distribution.** We often present a frequency distribution as a **frequency table.** In a frequency table, we list the values in one column and the frequencies of the values in another column, as we show in Example 1. We can also present a relative frequency distribution in table form.

EXAMPLE 1 Using Tables to Summarize TV Program Evaluations

Assume that 25 viewers were surveyed to evaluate a preview of an episode of the CBS drama *CSI*. The possible evaluations are

(E)xcellent, (A)bove average, a(V)erage, (B)elow average, (P)oor.

After the show, the 25 evaluations were as follows:

A, V, V, B, P, E, A, E, V, V, A, E, P, B, V, V, A, A, A, E, B, V, A, B, V

Construct a frequency table and a relative frequency table for this list of evaluations.

Historical Highlight

Presidential Polls

In 1936, pollster George Gallup boldly declared that the *Literary Digest* would incorrectly predict Alfred Landon to defeat Franklin Roosevelt for reelection as president of the United States. Gallup's claim seemed far-fetched because he used a sample of only 50,000, whereas the *Digest* intended to survey 10 million.

When Roosevelt won, it was clear that Gallup's superior sampling methods were more important than having a large sample. Because the *Digest* used telephone directories, magazine subscription lists, and membership lists of clubs and organizations, it had sampled people in the higher economic classes. Thus, its survey suffered from extreme selection bias and greatly overrepresented Republicans.

After this fiasco, pollsters developed *quota sampling*, in which samples would reflect the makeup of the population. The idea was to have the same percentage of men, women, Catholics, Jews, blacks, whites, and so on in a sample as there were in the population. However, when pollsters used quota sampling to forecast the 1948 presidential election, disaster struck again. All the major polling organizations wrongly predicted that Thomas Dewey would defeat the incumbent Harry S. Truman.

One major reason that the 1948 polls failed was that the polling stopped too soon, missing a late trend toward Truman. Even though quotas were met, selection bias crept in because interviewers had too much freedom in choosing whom to interview within those categories.

Evaluation	Frequency
E	4
A	7
V	8
B	4
P	2
Total	25

TABLE 14.1 Frequency table summarizing viewer evaluations of a police drama.

SOLUTION:

If we count the number of Es, As, and so on in the list, we get the results shown in Table 14.1.

By organizing the data in this table, we can see the distribution of favorable and unfavorable evaluations more quickly. Notice that the sum of the frequencies in Table 14.1 is 25, which is the number of viewers asked to evaluate the program.

We construct a relative frequency distribution for these data by dividing each frequency in Table 14.1 by 25. For example, because there are 4 Es, the relative frequency of the score E is $\frac{4}{25} = 0.16$. Table 14.2 shows the relative frequency distribution for the set of evaluations.

The sum of the relative frequencies in Table 14.2 is 1; however, in other examples, the sum of the relative frequencies may not be exactly 1 due to rounding. 1

Evaluation	Relative Frequency
E	$\frac{4}{25} = 0.16$
A	$\frac{7}{25} = 0.28$
V	$\frac{8}{25} = 0.32$
B	$\frac{4}{25} = 0.16$
P	$\frac{2}{25} = 0.08$
Total	1.00

TABLE 14.2 Relative frequency table summarizing viewer evaluations of a police drama.

QUIZ YOURSELF 1

Construct a frequency table and a relative frequency table for the following distribution:

1, 2, 7, 2, 6, 5, 2, 7, 8, 8,
1, 3, 10, 7, 9, 1, 7, 3, 5, 2

If there are many different values in a data set, we may group the data values into classes to make the information more understandable. Although there is no hard-and-fast rule, generally using 8 to 12 classes will give a good presentation of the data. We illustrate how to group data in Example 2.

EXAMPLE 2 Grouping Data Values into Classes

Suppose that 40 school counselors take a test to evaluate their skill at identifying signs of bullying and earn the following scores:

79, 62, 87, 84, 53, 76, 67, 73, 82, 68,
82, 79, 61, 51, 66, 77, 78, 66, 86, 70,
76, 64, 87, 82, 61, 59, 77, 88, 80, 58,
56, 64, 83, 71, 74, 79, 67, 79, 84, 68

Construct a frequency table and a relative frequency table for these data.

SOLUTION:

Because there are so many different scores in this list, the frequency of each score will be very small; constructing a frequency table as we did in Example 1 would not give us any useful information.

We must decide how to group the scores before making a table. The smallest score is 51 and the largest is 88. The difference, $88 - 51 = 37$, suggests that if we take a range of 40 and divide it into equal parts, we might get a reasonable grouping of the data. We will group the data into classes, each containing five values. The first class contains numbers from 50 to 54, the second contains numbers from 55 to 59, and so on. Counting the number of scores that fall into each class gives us the frequencies in the second column of Table 14.3.

Range of Scores on Bullying Awareness Test	Frequency	Relative Frequency
50–54	2	0.05
55–59	3	$\frac{3}{40} = 0.075$
60–64	5	0.125
65–69	6	0.15
70–74	4	0.10
75–79	9	0.225
80–84	7	0.175
85–89	4	0.10
Total	40	1.00

TABLE 14.3 Frequency table and relative frequency table for scores on bullying awareness test.

To find the relative frequencies, we divide each count in the second column by 40, which is the total number of scores. For example, in the row labeled 55–59, we divide 3 by 40 to get 0.075 in the third column.

Table 14.3 helps us see patterns in the data. For example, since a large number of the counselors have test scores below 70, this might mean that some counselors need extra training regarding their sensitivity to signs of bullying.

Representing Data Visually

KEY POINT

We use bar graphs to represent frequency distributions graphically.

The saying "A picture is worth a thousand words" certainly applies when working with large sets of data. By presenting data graphically, we can observe patterns more easily. A **bar graph** is one way to visualize a frequency distribution. In drawing a bar graph, we specify the classes on the horizontal axis and the frequencies on the vertical axis. If we are graphing a relative frequency distribution, then the heights of the bars correspond to the size of the relative frequencies, as we show in Example 3.

EXAMPLE 3 Drawing a Bar Graph of the Viewer Evaluation Data

a) Draw a bar graph of the frequency distribution of *CSI* viewers' responses summarized in Table 14.1 in Example 1.

b) Draw a bar graph of the relative frequency distribution of the *CSI* viewers' responses summarized in Table 14.2 in Example 1.

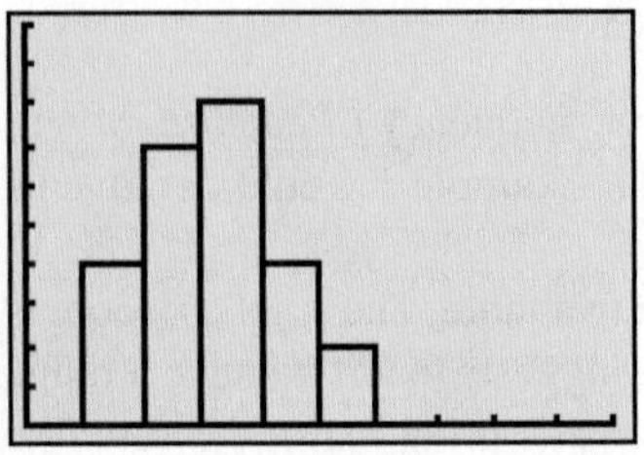

Graphing calculator representation of Figure 14.1(a).*

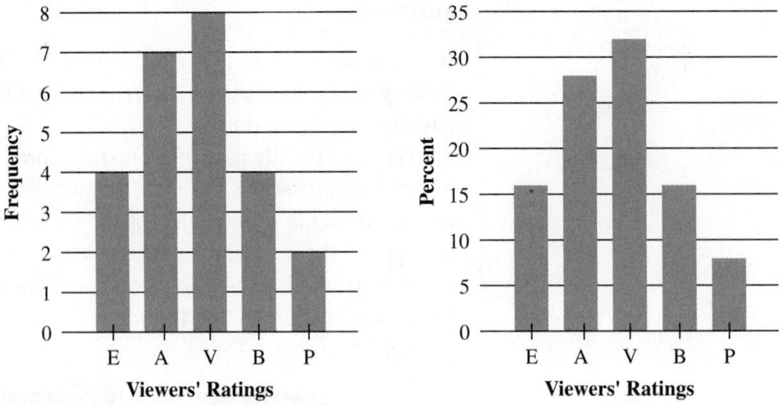

(a) (b)

FIGURE 14.1 (a) Bar graph of frequency distribution of viewers' ratings. (b) Bar graph of relative frequency distribution of viewers' ratings.

SOLUTION:

a) Because the largest frequency is 8, we labeled the vertical axis from 0 to 8. Next we drew five bars of heights 4, 7, 8, 4, and 2 to indicate the frequencies of the evaluations E, A, V, B, and P, as is shown in Figure 14.1(a).

b) In Figure 14.1(b), we labeled the vertical axis from 0 to 35 because the largest relative frequency was 0.32, or 32%. Although both bar graphs have the same shape, it is usually better to draw a bar graph of relative frequency distributions when comparing two different data sets.

QUIZ YOURSELF 2

Draw a bar graph representing the relative frequencies that you found in Quiz Yourself 1.

Now try Exercises 1 to 6. 2

If we are comparing two data sets of different sizes, graphing the relative frequencies, rather than the actual values in the data sets, allows us to compare the distributions. In this case, instead of drawing two separate bar graphs, we could show both distributions on a single graph, using, say, red for the bars in the first distribution and green for the bars in the second.

Until now, the data we have been organizing and graphing could not take on fractional values. By this we mean that in Example 1, a viewer could evaluate *CSI* as above average or excellent but could not give a rating between those two. Similarly, in Example 2, a score on the bullying awareness test could be 78 or 79, but a score between these two numbers, such as 78.56, was not possible. A variable quantity that cannot take on arbitrary values is called *discrete.* Other quantities, called *continuous* variables, can take on arbitrary values. Weight is an example of a continuous variable. We may say that a person weighs 150 pounds; however, with a more accurate scale, we may find that the person actually weighs 150.3 or perhaps 150.314 pounds.

We use a special type of bar graph called a **histogram** to graph a frequency distribution when we are dealing with a continuous variable quantity. We also may use a histogram when the variable quantity is not continuous, but has a very large number of different possible values. Money is an example of such a quantity.

As with a bar graph, we specify classes for a histogram. With a histogram, however, we do not allow any spaces between the bars above each class. If a data value falls on the boundary between two data classes, then you must make it clear as to whether you are counting that value in the class to the right or to the left of the data value. We show how to draw a histogram for a frequency distribution in Example 4. As with bar graphs, we can also draw a histogram for a relative frequency distribution.

*Note that in this graphing calculator screen, there are no spaces between the bars as in the graph in Figure 14.1(a).

Pounds Lost	Frequency
0 to 10	14
10+ to 20	23
20+ to 30	17
30+ to 40	8
40+ to 50	3
Total	65

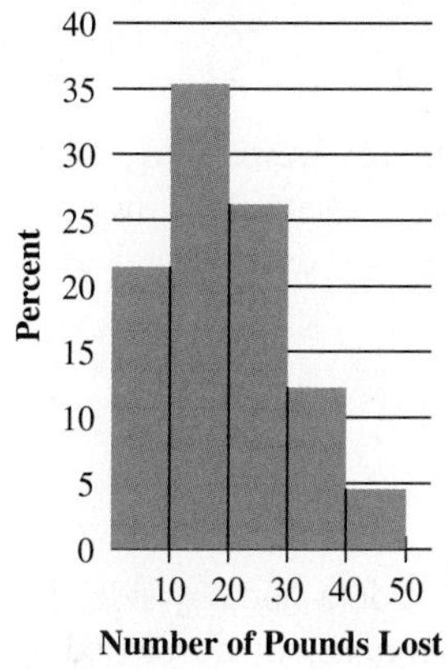

FIGURE 14.2 Histogram of weight loss at the New You Clinic.

EXAMPLE 4 Drawing a Histogram to Represent Weight-Loss Data

The New You Clinic has the following data regarding the weight lost by its clients over the past 6 months. Draw a histogram for the relative frequency distribution for these data.

SOLUTION:

We first must find the relative frequency distribution. Because there are 65 data values, we divide each frequency by 65 to obtain the corresponding relative frequency distribution in the third column of Table 14.4.

Pounds Lost	Frequency	Relative Frequency
0 to 10	14	0.215
10+ to 20	23	0.354
20+ to 30	17	0.262
30+ to 40	8	0.123
40+ to 50	3	0.046
Total	65	1.00

TABLE 14.4 Frequency and relative frequency distributions of weight loss at the NewYou Clinic.

We now draw this histogram exactly like a bar graph, as shown in Figure 14.2, except that we do not allow spaces between the bars. Also, we label the endpoints of the class intervals on the horizontal axis.

Now try Exercises 7, 8, 11, and 12.

When we look at the histogram in Figure 14.2, we can see that the majority of the clients lost between 10 and 30 pounds. There is really no strict rule as to how to construct a histogram. It is up to you to decide whether to group the data and how large each data class should be; however, it is customary to have data classes all of the same size.

EXAMPLE 5 Determining Information from a Graph

Figure 14.3 shows the number of Atlantic hurricanes over a period of years (Source: Colorado State Tropical Prediction Center). Use this bar graph to answer the following questions.

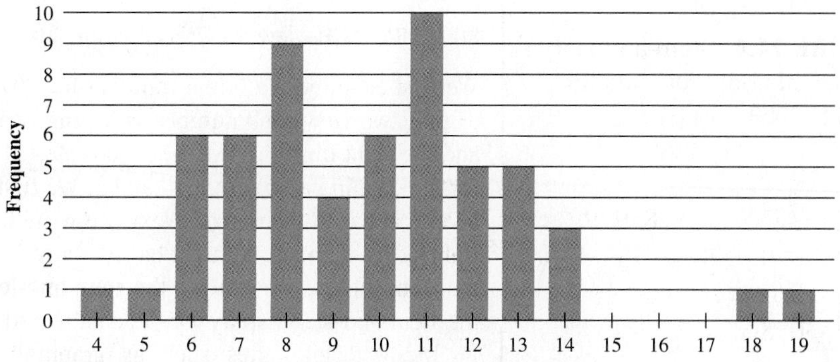

FIGURE 14.3 Number of hurricanes per year.

a) What was the smallest number of hurricanes in a year during this period? What was the largest?
b) What number of hurricanes per year occurred most frequently?

c) How many years were the hurricanes counted?

d) In what percentage of the years were there more than 10 hurricanes?

SOLUTION:

a) The smallest number of hurricanes in any year during this time period was 4. The largest number was 19.

b) The number of hurricanes per year that occurred most frequently corresponds to the tallest bar in Figure 14.3, which appears over the number 11. Therefore, 11 hurricanes occurred in 10 different years.

c) To find the total number of years for which these data were gathered, we add the heights of all of the bars to get

$$1 + 1 + 6 + 6 + 9 + 4 + 6 + 10 + 5 + 5 + 3 + 1 + 1 = 58 \text{ years.}$$

d) First, we count the number of years in which there were more than 10 hurricanes. If we add the heights of the bars above these values, we get $10 + 5 + 5 + 3 + 1 + 1 = 25$. Because there are 58 years of data, we calculate $\frac{25}{58} = 0.431$, which is approximately 43%.

Now try Exercises 13 to 18.

Stem-and-Leaf Displays

KEY POINT

A stem-and-leaf display is another way to display data.

A **stem-and-leaf display** is an effective way to present two sets of data "side by side" for analysis. This technique is used in an area called *exploratory data analysis*, developed by John Tukey, a mathematician who worked at Princeton University and Bell Labs.

Some sports fans believe that home run records have become meaningless in recent years because of the use of steroids by baseball players. In Example 6, we use a stem-and-leaf display to investigate whether there has been an increase in home run production in the National League recently.

EXAMPLE 6 Using Stem-and-Leaf Home Run Records from Two Eras

The following are the number of home runs hit by the home run champions in the National League for the years 1975 to 1989 and for 1996 to 2010.

a) 1975–1989: 38, 38, 52, 40, 48, 48, 31, 37, 40, 36, 37, 37, 49, 39, 47

b) 1996–2010: 47, 49, 70, 65, 50, 73, 49, 47, 48, 51, 58, 50, 48, 47, 42

Compare these home run records using a stem-and-leaf display.

SOLUTION: Review the Draw Pictures Strategy on page 3.

We first examine the home run data for 1975 to 1989. In constructing a stem-and-leaf display, we view each number as having two parts. The left digit is considered the stem and the right digit the leaf. For example, 38 has a stem of 3 and a leaf of 8. The stems for the data in part a) are 3, 4, and 5. We first list the stems in numerical order and draw a vertical bar to their right. Next, we write the leaves corresponding to each stem to the right of its stem and vertical bar. We also list the leaves in increasing order away from the stem. Figure 14.4 shows the stem-and-leaf display for the data in part a). We show the stem-and-leaf display for the data in part b) in Figure 14.5.

Stems	Leaves
3	1 6 7 7 7 8 8 9
4	0 0 7 8 8 9
5	2

FIGURE 14.4 Stem-and-leaf display of home run data for 1975 to 1989.

4	2 7 7 7 8 8 9 9
5	0 0 1 8
6	5
7	0 3

FIGURE 14.5 Stem-and-leaf display of home run data for 1996 to 2010.

Many technologies such as graphing calculators, Microsoft Word and Excel, Numbers for iPads, and various apps for tablet computers allow you to enter a frequency table and a graph will then be drawn. Here is a graph of the paired stem-and-leaf data from Example 6 as shown in Numbers on an iPad.*

*Excel spreadsheets, which can also be used in the Apple spreadsheet Numbers, are available in Pirnot's MyMathLab.

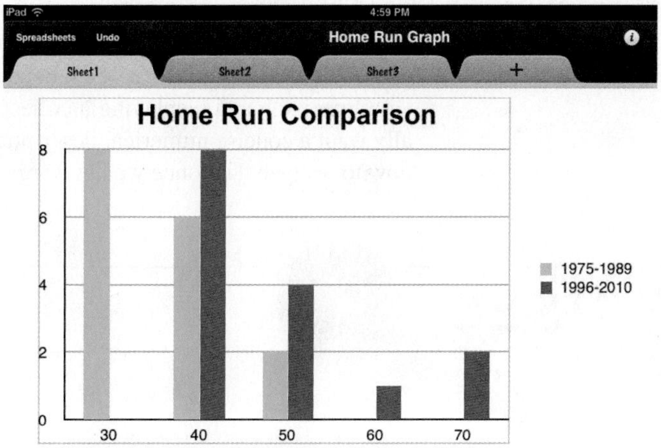

We can compare these data by placing these two displays side by side as we show in Figure 14.6. Some call this display a *back-to-back stem-and-leaf display*.

1975–1989		1996–2010
9 8 8 7 7 7 6 1	3	
9 8 8 7 0 0	4	2 7 7 7 8 8 9 9
2	5	0 0 1 8
	6	5
	7	0 3

FIGURE 14.6 Combined stem-and-leaf display of home run data.

From Figure 14.6, we can clearly see the pattern that National League home run champions hit significantly more home runs from 1996 to 2010 than from 1975 to 1989.

Now try Exercises 9 and 10. 3

QUIZ YOURSELF 3

Use a stem-and-leaf display to represent the following collection of scores:

92, 68, 77, 98, 88,
75, 82, 62, 84, 67,
62, 91, 82, 73, 66,
81, 63, 90, 83, 71

Historical Highlight

Florence Nightingale*

It may surprise you to see a reference to the legendary nurse Florence Nightingale in a discussion of statistics. Although she is best known for her compassion toward the sick and her work in improving hospital sanitation, she was also trained in mathematics. As a young woman, she had the good fortune to study with James Sylvester, one of the most eminent British mathematicians of the nineteenth century.

While serving as a nurse at a military hospital during the Crimean War, she was disturbed by the high mortality rate of her patients. Using statistical methods that she invented, she persuaded her superiors to carry out hospital reforms. As a result of improved sanitation, deaths decreased, and some credit her with saving the British Army during the Crimean War.

In recognition for her revolutionary work in developing a statistical approach to medicine, Nightingale was elected to the Statistical Society of England. She also was an advisor on military health during the American Civil War, and in 1874, she became an honorary member of the American Statistical Association. The renowned statistician Karl Pearson described Florence Nightingale as a prophetess in the development of applied statistics.

*This note is based on the biography of Florence Nightingale by Cynthia Audain, located at the Biographies of Women Mathematicians Web site (www.agnesscott.edu/lriddle/women/women.htm) at Agnes Scott College, Atlanta, Georgia. Also, we used "Mathematical Education in the Life of Florence Nightingale," by Sally Lipsey, *The Newsletter of the Association for Women in Mathematics*, Vol. 23, No. 4 (1993), 11–12.

Because the numbers in Example 6 were two-digit numbers, we used single digits for the stems. If we had to represent a number such as 325, we could use either a stem of 32 and a leaf of 5 or a stem of 3 and a leaf of 25, depending on which way presented the data more clearly.

Organizing and displaying a collection of data is generally not our final goal; we usually want a concise numerical description of a set of data. In Section 14.2, we will discuss how to analyze data once we have organized it.

EXERCISES 14.1

Sharpening Your Skills

In Exercises 1 and 2, construct a frequency table, a relative frequency table, and a bar graph for the data given.

1. The number of hours of flight time for 20 amateur pilots last month:

7, 8, 6, 5, 7, 10, 2, 7, 9, 5, 8, 8, 10, 9, 6, 5, 10, 7, 9, 8

2. The number of passengers per car on an Amtrak Acela train:

38, 39, 38, 37, 40, 38, 38, 37, 40, 38, 38, 39, 38, 37, 40, 37, 40, 38, 38, 37, 40, 38, 39, 38, 37, 40, 38, 38, 39, 38

In Exercises 3 and 4, construct a bar graph for the data given in each frequency table.

3. This table contains the EPA mileage ratings for 38 domestic cars.

MPG	19	20	21	22	23	24	25	26	27	28	29	30
Frequency	2	5	3	2	1	8	1	2	3	2	7	2

4. This table contains the weight loss by 30 participants in the *Biggest Loser* show.

Weight Loss (Pounds)	0	1	2	3	4	5	6	7	8	9	10
Frequency	2	3	3	0	1	6	1	4	3	2	5

In Exercises 5 and 6, construct a bar graph for the relative frequency table for the data given.

5. The following are the ages of 60 people who volunteered to work for the Katrina relief effort:

21, 23, 27, 22, 23, 29, 28, 24, 24, 25, 27, 22, 26, 26, 23, 23, 26, 28, 25, 24, 28, 27, 24, 23, 29, 28, 24, 22, 27, 26, 22, 24, 26, 21, 24, 28, 24, 25, 22, 25, 27, 21, 23, 26, 23, 23, 27, 27, 23, 21, 22, 27, 26, 23, 25, 29, 24, 27, 27, 26

6. The following are the ages of 40 ExecuCorps volunteers who are advising young entrepreneurs:

51, 56, 57, 52, 53, 59, 58, 52, 54, 55, 53, 56, 58, 55, 54, 58, 57, 52, 53, 59, 52, 54, 53, 51, 54, 58, 54, 55, 52, 55, 52, 57, 57, 53, 51, 52, 57, 56, 53, 55

In Exercises 7 and 8, group the data as indicated and construct a histogram.

7. The following are the heights (in inches) of the players in four Eastern Conference teams of the WNBA for a recent season:

Chicago Sky:	69, 71, 74, 74, 78, 68, 68, 71, 75, 69, 76, 68, 69, 74, 65, 73
NY Liberty:	72, 67, 72, 73, 75, 77, 73, 69, 67, 75, 76, 76, 73, 74, 69, 70
Washington Mystics:	68, 72, 75, 68, 70, 69, 68, 74, 74, 69, 73, 76, 76, 80, 73, 78
Atlanta Dream:	72, 77, 71, 80, 69, 73, 69, 75, 66, 68, 77, 74, 76, 73, 72, 75

Use classes of width 2, starting at 64.5.
(*Source:* www.wnba.com)

8. The following are the scores on a 100-point language aptitude test given to 60 people who are applying for the Peace Corps:

83, 71, 92, 87, 56, 64, 41, 95, 88, 91, 78, 73, 81, 79, 59, 73, 81, 93, 84, 66, 74, 51, 85, 78, 81, 98, 63, 91, 89, 64, 74, 61, 92, 77, 86, 79, 63, 91, 86, 91, 58, 83, 81, 77, 89, 83, 61, 83, 94, 76, 78, 61, 84, 88, 87, 68, 83, 71, 85, 64

Use classes of width 10, starting at 40.5.

In Exercises 9 and 10, represent the two sets of data on a single stem-and-leaf display.

9. A: 29, 32, 34, 43, 47, 43, 22, 38, 42, 39, 37, 33, 42, 18, 22, 39, 21, 26, 18, 43

B: 32, 38, 22, 39, 21, 26, 28, 16, 13, 20, 21, 29, 22, 24, 33, 47, 23, 22, 18, 33

10. X: 29, 42, 34, 44, 47, 43, 22, 38, 42, 59, 41, 16, 47, 43, 42, 18, 22, 49, 21, 26, 18, 45, 24, 40

Y: 32, 48, 22, 59, 21, 26, 28, 16, 14, 20, 17, 45, 21, 29, 22, 24, 34, 47, 23, 22, 18, 45, 21, 16

Applying What You've Learned

In Exercises 11 and 12, group the data as indicated and construct a histogram.

11. Gasoline prices. The following table contains the average price of one gallon of unleaded regular gasoline for 2009 to 2011. Use classes of width 50 cents, starting at $1.75. If a number is on a class boundary, count it in the higher class.

Year	Jan.	Feb.	Mar.	Apr.	May	Jun.	Jul.	Aug.	Sep.	Oct.	Nov.	Dec.
2009	1.79	1.93	1.95	2.06	2.27	2.63	2.54	2.63	2.57	2.56	2.66	2.62
2010	2.73	2.66	2.78	2.86	2.87	2.74	2.74	2.75	2.70	2.80	2.85	2.99
2011	3.09	3.17	3.55	3.82	3.93	3.70	3.65	3.63	3.61	3.47	3.42	3.28

Source: U.S. Bureau of Labor Statistics

12. Mishandled baggage. The following data are the number of reports of mishandled baggage per 1,000 passengers for 10 U.S. airlines during 3 months of 2010. Use classes of width 1.0, starting at 1.5. If a number is on a class boundary, count it in the higher class.

Month	Amer.	Am. Eagle	Comair	Cont.	Delta	Frontier	Jet Blue	SW	UA	US
Oct.	2.8	5.8	3.9	2.1	2.8	2.2	2.1	2.9	2.4	2.2
Nov.	2.9	6.0	3.8	2.3	2.7	2.2	1.9	3.1	2.4	1.9
Dec.	4.4	8.9	7.6	4.4	5.0	3.3	2.9	4.8	4.1	3.5

Source: U.S. Department of Transportation

13. The manager at the local Starbucks counted the customers once each hour over a busy weekend. We summarize the results she obtained in the following bar graph; use it to answer the following questions.

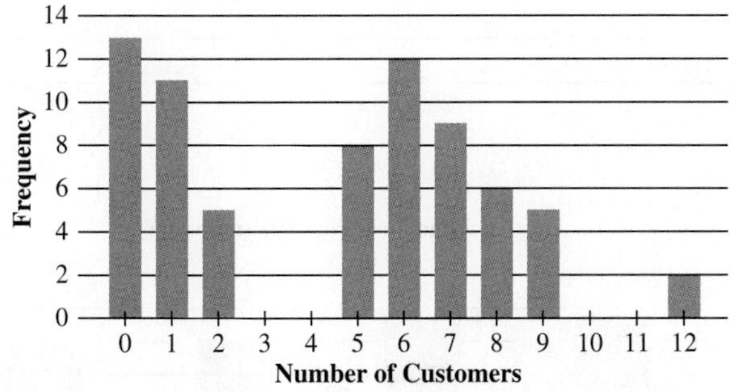

a. What was the largest number of customers in the bar, and how often did it occur?

b. What was the most frequently occurring nonzero customer count?

c. For how many hours were the customers counted?

d. For what fractional part of the total number of hours were there more than six customers in the bar?

14. Scheduling a rec center. Because of budget cutbacks, the campus rec center surveyed the number of students using the center hourly between 9 PM and midnight for a semester to decide if it should reduce its hours. Use the results in the given graph to answer the following questions.

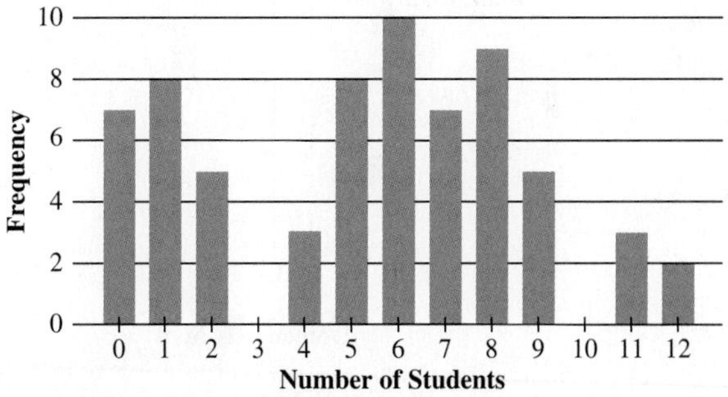

a. What was the smallest number of students in the rec center and how many times did it occur?

b. What was the smallest student count between 5 and 8 inclusive? How many times did that occur?

c. For how many hours was the survey taken?

d. For what fractional part of the total number of hours were there less than five students present?

The Consumer Price Index (CPI) measures inflation. The base for the CPI is the period 1982–84, which is considered to be 100. In 2011, the CPI was 224.94, which means, on average, goods and services that cost \$100 in 1982–84 cost \$224.94 in 2011. Use the graph below, which shows the percent of change in the CPI each year from 2001 to 2011, to do Exercises 15–18.

CPI Change (%) 2001 to 2011

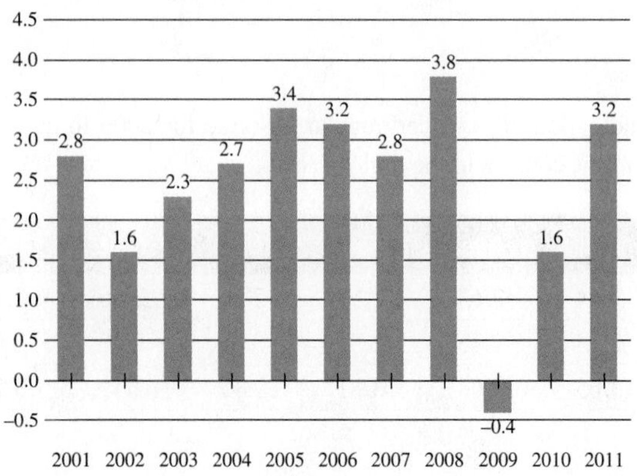

15. What year had the largest percent of increase in the CPI?

16. What year had the smallest *positive* percent of increase in the CPI?

17. During which three-year period did the CPI grow the fastest?

18. Some might say that the CPI decreased from 2005 to 2007. Is this correct? Explain.

Comparing wage data. *The following bar graphs compare women's and men's hourly wages in a recent year. Use these graphs to answer Exercises 19–24. Because you are estimating your answers by looking at the graphs, your answers may not agree exactly with ours.*

Percentage Earning Less than \$7.51 per Hour

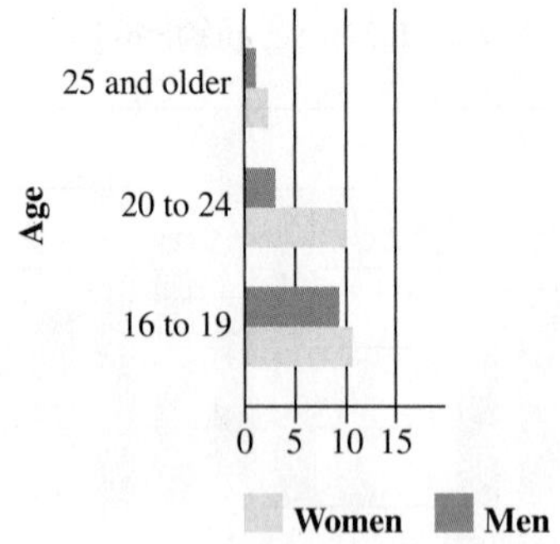

Percentage Earning \$7.51 to \$9.99 per Hour

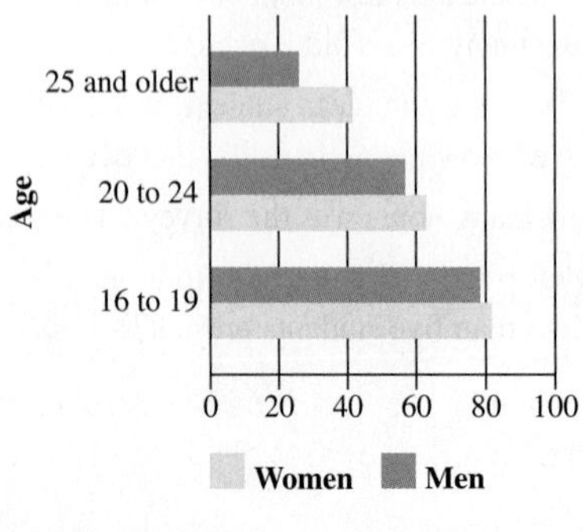

Percentage Earning at Least \$10.00 per Hour

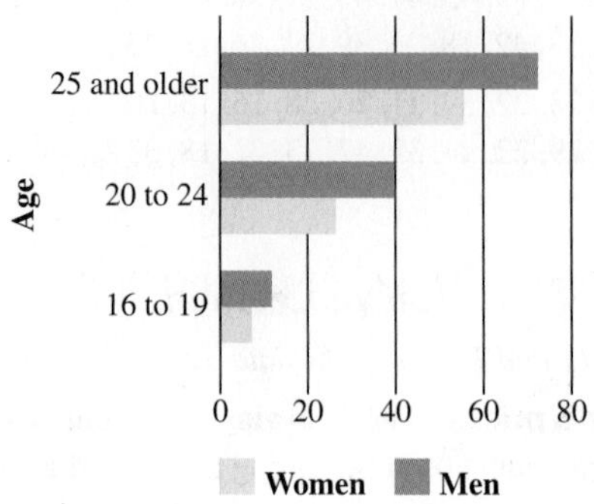

19. What percent of women ages 20 to 24 earned \$7.50 or less per hour?

20. What percent of men age 25 and older earned between \$7.51 and \$9.99 per hour?

21. In what age and wage categories do men seem to have the biggest advantage (in terms of a percentage difference) over women?

22. In what age and wage categories do women seem to have the biggest advantage (in terms of a percentage difference) over men?

23. In what age and wage categories do women and men seem to be most equal?

24. What general conclusions can you draw from these data?

For Exercises 25–28, use the four graphs of the same (hypothetical) data regarding the number of college students involved with volunteerism.

(A) Student Volunteers (millions)

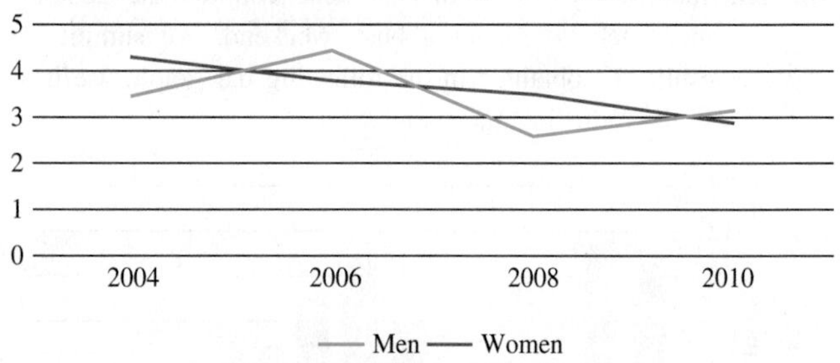

(B) Student Volunteers (millions)

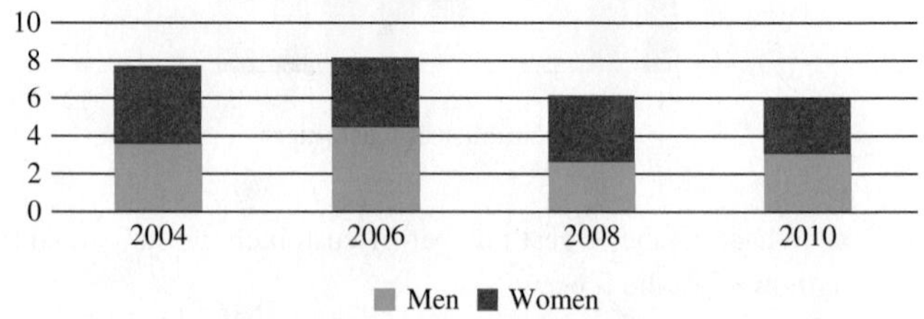

(C) Student Volunteers (millions)

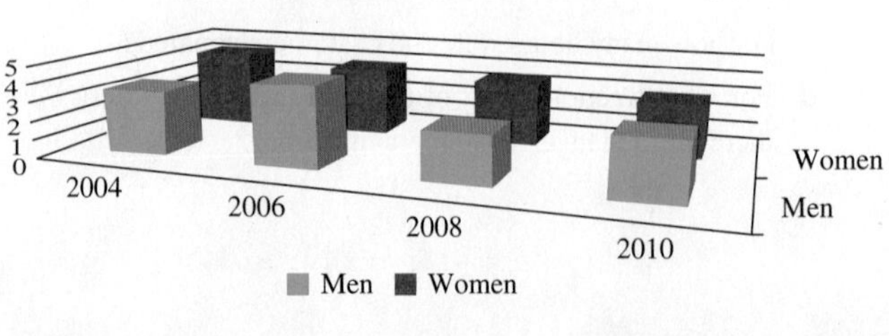

(D) Student Volunteers (millions)

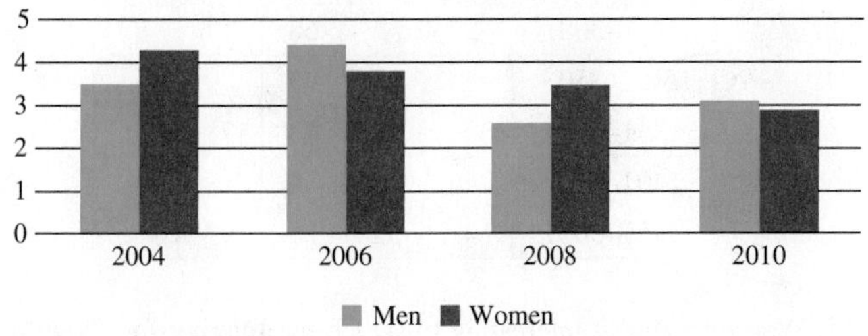

25. Which graph(s) do you think makes the information most understandable?

26. Which graph(s) do you find hardest to understand?

27. Which graph(s) makes it easiest to see what years there were more women than men volunteering?

28. What is the purpose of graph B?

29. **Comparing training programs.** Mary Kay gives sales training to its newly hired employees. To determine how effective the training is, the firm compared the monthly sales of a group that has completed the training with a group that has not. The following numbers indicate thousands of dollars of sales last month. Represent the two sets of data on a single stem-and-leaf display. Does the orientation program seem to be succeeding?

No Training: 19, 22, 34, 23, 27, 43, 42, 28, 32, 29, 41, 26, 28, 26, 43, 40

With Training: 29, 21, 39, 44, 41, 36, 37, 29, 43, 45, 28, 32, 28, 33, 36, 32

30. **Comparing weight-loss programs.** A hospital is testing two weight-loss programs to determine which program is more effective. The following data represent the amount of weight lost by comparable clients for a year in each program. Represent the two given sets of data on a single stem-and-leaf display. Which program seems to be more effective?

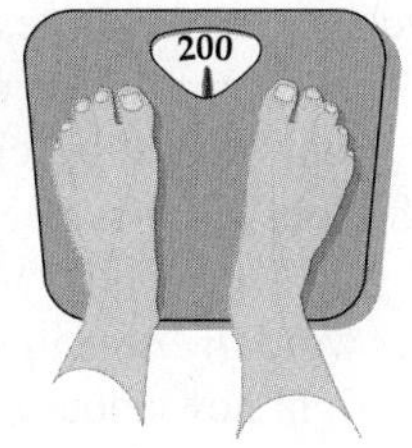

Program A: 19, 32, 27, 34, 33, 36, 47, 32, 25, 52, 29, 26, 37, 28, 26, 43, 31, 40

Program B: 29, 21, 39, 44, 41, 36, 37, 26, 26, 43, 45, 28, 32, 28, 33, 36, 53, 39

31. **Super Bowl scores.** Given are the points scored by the NFC and the AFC in the Super Bowl in a recent 20-year period. Make a back-to-back stem-and-leaf plot of these data. Does either conference seem to have an advantage in scoring?

NFC: 31, 21, 31, 23, 17, 17, 10, 21, 29, 48, 17, 7, 23, 19, 24, 35, 27, 49, 30, 52

AFC: 25, 17, 17, 27, 14, 29, 21, 24, 32, 21, 20, 34, 16, 34, 31, 21, 17, 26, 13, 17

32. **Law School Aptitude Test scores.** The following are two lists of 40 LSAT scores. Group A studied in a traditional way and group B used a new online tutoring system to prepare for the exam. Make a back-to-back stem-and leaf plot of these data. Does either group seem to score better?

A: 140, 154, 140, 123, 121, 158, 174, 155, 157, 127, 155, 154, 122, 164, 160, 142, 159, 154, 163, 141, 141, 129, 172, 142, 152, 146, 137, 173, 154, 149, 155, 144, 152, 149, 170, 151, 146, 149, 140, 146

B: 155, 144, 151, 136, 149, 165, 150, 153, 135, 142, 158, 173, 155, 151, 127, 142, 173, 140, 162, 147, 141, 137, 147, 161, 155, 165, 161, 177, 143, 126, 149, 164, 161, 144, 149, 164, 157, 173, 128, 147

33. Graph the data in Exercise 31, using a graph similar to graph D in the instructions to Exercises 25–28 in the first column.

34. Graph the data in Exercise 32, using a graph similar to graph D in the first column.

35. **Ages of Academy Award winners.** These lists are the ages of actors (M) and actresses (F) at the time they won an Academy Award from 1980 to 2012. Draw a graph similar to graph D in the first column and see if you notice any patterns to the data.

F: 31, 74, 33, 49, 38, 61, 21, 41, 26, 80, 42, 29, 33, 35, 45, 49, 39, 34, 26, 25, 33, 35, 35, 28, 30, 30, 29, 61, 32, 33, 45, 29, 62

M: 37, 76, 39, 52, 45, 35, 61, 43, 51, 32, 42, 54, 52, 37, 38, 31, 45, 60, 46, 40, 36, 47, 29, 43, 37, 37, 38, 45, 50, 48, 60, 50, 39

36. **Exercise and academic performance.** After reading a report from The American Society of Exercise Physiologists, your school has implemented a test program to see if there is a relationship between regular exercise and academic achievement. List E is the set of 40 test scores of those who exercised; N is the set of 40 scores of those who did not. Draw a graph similar to graph D in the first column and see if you observe anything about the two sets of data.

E: 84, 77, 77, 86, 72, 61, 74, 76, 85, 60, 66, 91, 80, 78, 89, 80, 88, 64, 69, 88, 77, 94, 76, 84, 77, 74, 62, 68, 85, 60, 70, 95, 69, 77, 81, 88, 74, 63, 87, 90

N: 73, 67, 88, 77, 71, 88, 69, 78, 93, 73, 74, 82, 64, 69, 78, 73, 75, 69, 68, 63, 67, 76, 65, 77, 80, 80, 71, 79, 69, 73, 73, 86, 85, 75, 70, 64, 61, 85, 73, 70

Communicating Mathematics

37. What is the difference between a population and a sample? Mention several things that can result in choosing a poor sample of the population.

38. What type of graph is useful in comparing two different sets of data?

39. What is the difference between discrete and continuous variables? What type of graph do we use to represent a frequency distribution of a continuous variable?

40. What do you see as an advantage in grouping data? A disadvantage?

Challenge Yourself

41. **Graphing movie data.** According to http://boxofficemojo.com/alltime/adjusted.htm at the time I was revising this edition, the top three grossing movies of all time, when adjusted for ticket price inflation, were *Gone with the Wind*, *Star Wars*,

and *The Sound of Music*. Go to this site (or a similar one) and find the top 25 grossing movies of all time and construct a histogram of the data you find. Explain how you decided on the class width and the first class. Mention any problems that the data presented in making your histogram.

42. Graphing television viewing data. Go to http://www.nielsen.com/us/en/insights/top10s/television.html (or some other similar site) to obtain data regarding the most watched shows for a recent week. Construct a histogram of these data and discuss how you decided on the class width, the first class, and any problems that you had in making your histogram.

43. How might you present three sets of data in the same graph?

44. The following table is an example of a double-stem display. Without having this table explained to you, try to interpret its meaning and list the data items in the table.

(201–250)	2	34	45	49	
(251–300)	2	57	68	77	82
(301–350)	3	23	45		
(351–400)	3	62	73	78	
(401–450)	4	12	34		
(451–500)	4	82	89	93	

45. You may find it interesting to search the Internet for "how to lie with statistics." In addition to many references to Darrell Huff's classic book *How to Lie With Statistics*, you will find other sites that explain how to use statistics to mislead your audience. Find a site that interests you and report on your findings.

14.2 Measures of Central Tendency

Objectives

1. Compute the mean, median, and mode of distributions.
2. Find the five-number summary of a distribution.
3. Apply measures of central tendency to compare data.

On *American Idol*, Randy Jackson tells a contestant, "You won't win because your voice is just *average*." Your child psych professor claims the *average* child laughs several hundred times a day, and in your sociology class you learn that in 1900, the *average* worker died while still on the job. A news blog claims the *average* Facebook user has 130 friends.

As you can see, we use the word *average* in many different ways. As practical applications, you might be concerned right now about your grade point average, and in the future, you may want to know the average starting salary in your chosen career or the average school taxes in the neighborhood where you buy a new home.

Statisticians are also interested in describing the average of a set of data and measure it in several different ways called **measures of central tendency.** The four measures of central tendency that we will discuss are the mean, median, mode, and quartiles. Except for the mode, each number will give you some sense of the center of a set of data.

The Mean and the Median

KEY POINT

The mean is the most common notion of average.

When considering changing your cell phone company, you may want to know the average number of minutes that you have been calling over the past 6 months. To calculate this average, you would add the number of minutes for each of the 6 months and divide by 6. When we calculate an "average" this way, we are calculating the mean.

To calculate the mean and other numbers in statistics, we have to add lists of numbers. We use the Greek letter Σ (capital sigma) to indicate a sum.* If we have n data values, $x_1, x_2, x_3, \ldots, x_n$, then we will represent the sum of these data values by Σx. For example, we will write the sum of the data values 7, 2, 9, 4, 10 by $\Sigma x = 7 + 2 + 9 + 4 + 10 = 32$.

*Although we could avoid some of this formal notation in our discussion, it is good for you to become comfortable with it if you are to ever study statistics more deeply.

> **COMPUTING THE MEAN** If a data set contains n data values, the **mean** $\bar{x}$ of the data set is
>
> $$\bar{x} = \frac{\sum x}{n}$$

We represent the mean of a *sample* of a population by $\bar{x}$ (read as "x bar"), and we will use the Greek letter μ (lowercase mu) to represent the mean of the whole *population.* Unless we state otherwise, we will assume that data sets in this chapter are samples rather than populations.

EXAMPLE 1 Finding the Mean Number of Accidents

Union leaders criticizing NAFTA's outsourcing of jobs to other countries claim that safety in factories in other countries is not as good as it is in the United States. As a result, The National Motor Corporation has been studying its safety record at its Mexican factory and found that the number of accidents over the past 5 years was 25, 23, 27, 22, and 26. Find the mean annual number of accidents for this 5-year period.

SOLUTION:

To calculate the mean, we add the number of accidents and divide by 5, as follows:

Add data values.

$$\bar{x} = \frac{\sum x}{n} = \frac{25 + 23 + 27 + 22 + 26}{5} = \frac{123}{5} = 24.6$$

number of data values

This tells us that over the past 5 years, the factory in Mexico has averaged between 24 and 25 accidents per year. 4

QUIZ YOURSELF 4

Find each of the following for the data set 9, 12, 22, 6, 5, 15, 12, 25.

a) Σx b) n c) $\bar{x}$

As you saw in Example 1, the five data values balance on either side of the mean, as shown in Figure 14.7. Notice that just like children on a seesaw, a few numbers further from the mean (the balancing point) will balance with more numbers that are close to the mean.

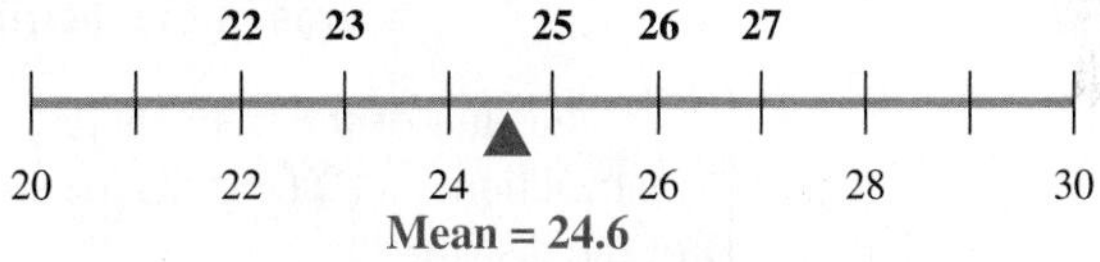

FIGURE 14.7 A set of data values balances about its mean.

We often use the mean to compare data to see trends. In Example 1, we can ask whether the 5-year mean of 24.6 accidents per year is good or bad. We do not know unless we compare it with some other data. For example, suppose that National Motor also has a Seattle, Washington, factory of roughly the same size and workload, but that factory had a mean of 31.8 accidents per year over the past 5 years. This difference in means shows that perhaps a problem exists at the Seattle factory instead, and management must take steps to correct it.

In Example 1, each data value occurred once. Often, however, some values in a data set occur several times, in which case we use a frequency table to compute the mean.

EXAMPLE 2 Computing the Mean of a Frequency Distribution of Water Temperatures

The Environmental Protection Agency (EPA) suspects that hot water discharged from a nuclear power plant is responsible for a recent fish kill. To investigate this problem, the agency has recorded the water temperature at a point downstream from the plant for the

last 30 days. The given graph summarizes the information obtained. What is the mean temperature for this distribution?

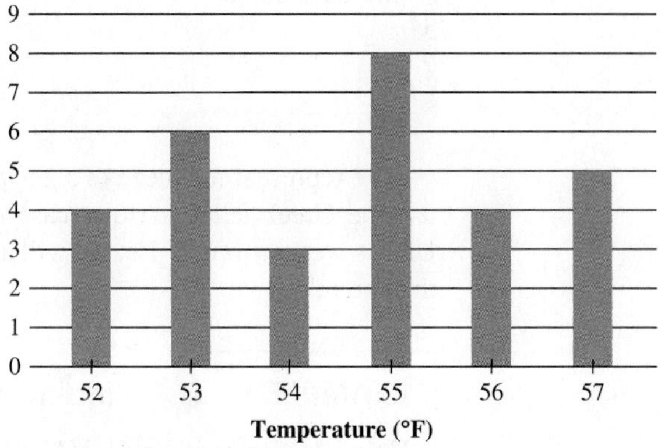

SOLUTION:

From this graph we see that 52 occurs four times in the data, so instead of summing 52 + 52 + 52 + 52, we can write $52 \cdot 4$ instead. Table 14.5 lists the products of the raw scores and their frequencies.

Temperature (°F), x	Frequency, f	Product, $x \cdot f$
52	4	$52 \cdot 4 = 208$
53	6	$53 \cdot 6 = 318$
54	3	$54 \cdot 3 = 162$
55	8	$55 \cdot 8 = 440$
56	4	$56 \cdot 4 = 224$
57	5	$57 \cdot 5 = 285$
Totals	$\Sigma f = 30$	$\Sigma(x \cdot f) = 1{,}637$

sum of frequencies — sum of products

TABLE 14.5 The number of scores and the sum of the scores in the distribution of water temperatures.

The sum of the frequencies, Σf, is the same as n, the total number of scores in the distribution, and $\Sigma(x \cdot f)$ is the sum of all scores in the distribution. Therefore, the mean of the distribution is

$$\frac{\sum(x \cdot f)}{\sum f} = \frac{\text{sum of scores}}{\text{number of scores}} = \frac{1{,}637}{30} \approx 54.6°\text{F}.$$

With this information, the EPA may conclude that the water is not too hot and look for another reason for the fish kill.

As you saw in Example 2, when you compute the mean of a frequency distribution, *you must multiply the data values by their frequencies before adding them.*

COMPUTING THE MEAN OF A FREQUENCY DISTRIBUTION We use a frequency table to compute the mean of a data set as follows:

1. Write all products $x \cdot f$ of the scores times their frequencies in a new column of the table.
2. Represent the sum of the products you calculated in step 1 by $\Sigma(x \cdot f)$.
3. Denote the sum of the frequencies by Σf.
4. The mean is then $\dfrac{\sum(x \cdot f)}{\sum f}$.

SOME GOOD ADVICE

A common mistake students sometimes make when computing the mean of a frequency distribution is to divide by the number of entries in the first column of the frequency table. This is the number of different data values, but it does not count repeated values. You must instead be sure to divide by the sum of the frequencies. 5

QUIZ YOURSELF 5

A paramedic service kept track of the number of calls per day that it received over a 2-week period. Use the information in the following frequency table to find the mean number of daily calls over this period.

Number of Calls, x	4	5	6	7	8	9	10
Frequency, f	1	4	1	2	3	2	1

One of the drawbacks in using the mean to represent the average value in a data set is that one or two extreme scores can have a strong influence on the mean.

EXAMPLE 3 The Effect of Extreme Scores on the Mean

Table 14.6 lists the yearly earnings of some celebrities.

Rank	Celebrity	Earnings (2011) (millions of dollars)
1	Oprah Winfrey	290
2	U2	195
3	Tyler Perry	130
4	Bon Jovi	125
5	Jerry Bruckheimer	113
6	Steven Spielberg	107
7	Elton John	100
8	Lady Gaga	90
9	Simon Cowell	90
10	James Patterson	84

TABLE 14.6 Earnings of some top celebrities.

a) What is the mean of the earnings of the celebrities on this list?

b) Is this mean an accurate measure of the "average" earnings for these celebrities?

SOLUTION:

a) Summing the earnings and dividing by 10 gives us

$$\bar{x} = \frac{290 + 195 + 130 + 125 + 113 + 107 + 100 + 90 + 90 + 84}{10}$$
$$= \$132.4 \text{ million.}$$

b) To answer our second question, notice that eight of the celebrities have earnings below the mean, whereas only two have earnings above the mean. Therefore, the mean in this example does not give an accurate sense of what is "average" in this set of data because it was unduly influenced by Oprah Winfrey's and U2's earnings. 6

QUIZ YOURSELF 6

To see the effect that Oprah's and U2's earnings have on the mean in Example 3, recalculate the mean with these two extreme scores removed. (Note that the number of scores will now be 8, not 10.)

Highlight

Using a Graphing Calculator in Statistics*

Doing statistical calculations by hand when you have a large amount of data can be tedious and prone to errors. This is why people who do statistical computations in real applications use technology to make the work more manageable.

Graphing calculators such as the TI-83 and TI-84 have very powerful built-in statistical capabilities. To use these capabilities, we first store our data in lists. Figure 14.8(a) shows some of the data from Example 3 stored in list L1. Then Figure 14.8(b) displays a menu of choices to process the data, and finally in Figure 14.8(c), we see the results of the calculations. In addition to other results, which are not relevant now, you can see the number of data, $n = 10$; the sum of the data, $\sum x = 1324$; and the mean, $\bar{x} = 132.4$.

To calculate a different mean, we could replace the data in L1 or store new data in L2 and redo the computations. When using 1-Var Stats, you can specify the list containing the data as follows: 1-Var Stats L1 or 1-Var Stats L2.

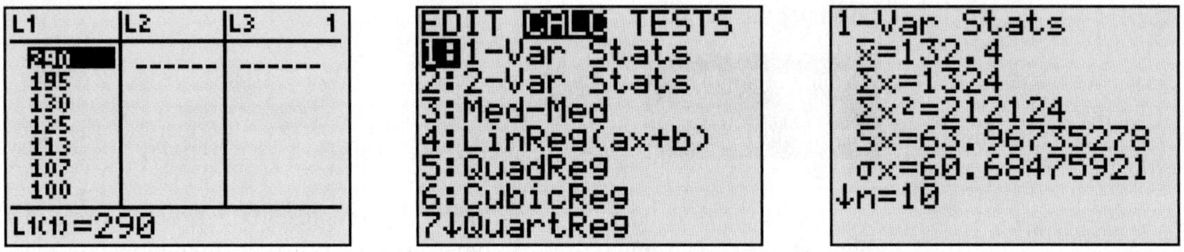

FIGURE 14.8 (a) Data. (b) Menu of choices. (c) Results of calculations.

Extreme scores in a data set, such as Oprah Winfrey's and U2's earnings, are called *outliers*, and you must decide what to do with them when you analyze data. In Example 3, you may believe that it is best to discard these two extreme values and calculate the mean based on the remaining eight earnings values. Or you may decide that it is better to describe the data using some other measure than the mean.

A measure that describes the middle of a data set is called the *median.*

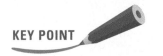

KEY POINT

The median tells us the middle of a set of data.

COMPUTING THE MEDIAN If we arrange a set of data values in increasing (or decreasing) order, the **median** is the middle value in the list of values. However, there are two cases to consider.

1. If there is odd number of values, then the median is the value in the middle position.
2. If there is an even number of values, then the median is the average of the two middle values.

One reason that we often use the median to describe the middle of a set of data is that the median is not affected by a few outliers.

SOME GOOD ADVICE

A common error made when finding the median is forgetting to arrange the values in increasing numerical order.

When Barack Obama was campaigning for president, some political observers thought that he was too young and inexperienced to be president. In Example 4, we investigate how Obama's age compares with that of earlier presidents.

*We will show you how to use other technologies to do statistical calculations in the Using Technology exercises at the end of this chapter.

T. Roosevelt	42
Taft	51
Wilson	56
Harding	55
Coolidge	51
Hoover	54
F. D. Roosevelt	51
Truman	60
Eisenhower	61
Kennedy	43
L. Johnson	55
Nixon	56
Ford	61
Carter	52
Reagan	69
G. H. W. Bush	64
Clinton	46
G. W. Bush	54
Obama	47

TABLE 14.7 Ages of U.S. presidents at inauguration.

EXAMPLE 4 Finding the Median of Presidents' Ages

Table 14.7 lists the ages at inauguration of the presidents who assumed office between 1901 and 2008. Find the median age for this distribution to see how Barack Obama's age at inauguration compares with that of other presidents.

SOLUTION:

The distribution of ages, when arranged in increasing order, is

Median is middle score.

42, 43, 46, 47, 51, 51, 51, 52, 54, 54, 55, 55, 56, 56, 60, 61, 61, 64, 69.

Because there are 19 scores in this distribution, the middle score is the tenth score, which is 54. Because Obama's age when he took office is well below the median, we agree with the opinion that he was young when he took office.

PROBLEM SOLVING

Strategy: The Analogies Principle

The Analogies Principle in Section 1.1 says that making associations with everyday words helps you understand the meaning of mathematical ideas. Recalling that the strip between a divided highway is called the *median* will help you remember that the median is the middle number in a distribution.

Often, consumer protection agencies test products to see whether they match the weight or volume stated on the package. In Example 5, we use the median to evaluate the accuracy of packaging information.

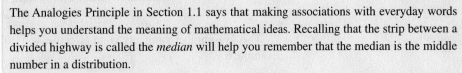

EXAMPLE 5 Using a Frequency Table to Find the Median

Acting on a tip from a dissatisfied customer, an agent of the Consumer Protection Agency purchased 50 quarts of a particular brand of milk at various supermarkets to see whether they contained 32 ounces of milk. The results of this survey are reported in Table 14.8. What is the median for this distribution?

Number of Ounces, x	Frequency, f
27	8
28	5
29	12
30	16
31	9
	$\Sigma f = 50$

Median is between the 25th and 26th scores.

25 scores here (27–29)

25 scores here (30–31)

TABLE 14.8 Number of ounces of milk in 50 quart cartons.

SOLUTION:

Notice that because the 50 scores in Table 14.8 are in increasing order, the two middle scores are in positions 25 and 26. Counting the frequencies, we see that 29 ounces is in position 25 and 30 ounces is in position 26. Therefore, the median for this distribution is $\frac{29+30}{2} = 29.5$. Because the median is far below the advertised 32 ounces, the variation is probably not due to randomness in filling the containers and perhaps the company packaging this milk should be fined.

Now try Exercises 1 to 8 and 13 to 16 (we discuss the mode later in this section). 7

QUIZ YOURSELF 7

Find the median for the following frequency distribution.

x	72	86	91	95	100
Frequency, f	2	4	3	7	3

The Five-Number Summary

KEY POINT

The five-number summary describes the position of data items in a data set.

DEFINITION The Five-Number Summary
The median divides a data set into two halves. The set of numbers below the median is called the **lower half** and the set of numbers above the median is called the **upper half.** The median of the lower half is called the **first quartile** and is indicated by Q_1; the **third quartile,** denoted by Q_3, is the median of the upper half. The **five-number summary** of a set of data values consists of the following:

minimum value, Q_1, median, Q_3, maximum value

The five-number summary is an effective way to describe a set of data, as we see in Example 6.

EXAMPLE 6 Finding the Five-Number Summary of Presidents' Ages

Consider the list of ages of the presidents from Example 4:

42, 43, 46, 47, 51, 51, 51, 52, 54, 54, 55, 55, 56, 56, 60, 61, 61, 64, 69

Find the following for this data set:

a) the lower and upper halves
b) the first and third quartiles
c) the five-number summary

SOLUTION: Review the Three-Way Principle on page 13.

It helps to consider the following diagram.

Median
42, 43, 46, 47, (51), 51, 51, 52, 54, **54**, 55, 55, 56, 56, (60), 61, 61, 64, 69
Lower half — Upper half

a) Because the median is in the tenth position in the list, the lower half is 42, 43, 46, 47, 51, 51, 51, 52, 54 and the upper half is 55, 55, 56, 56, 60, 61, 61, 64, 69.

b) The first quartile is the median of the lower half, which you can see from the diagram is 51 (circled). So, $Q_1 = 51$.

The third quartile is the median of the upper half, which is 60 (circled). Thus $Q_3 = 60$.

c) The five-number summary for this data set is 42, 51, 54, 60, 69.

Now try Exercises 17 to 20. 8

QUIZ YOURSELF 8

a) Find the five-number summary of the entertainers' salaries listed in Example 3.
b) Represent the five-number summary in part a) by a box-and-whisker plot.

We represent the five-number summary by a graph called a **box-and-whisker plot.** The five-number summary for the president's ages in Example 6 is 42, 51, 54, 60, 69. We show the box-and-whisker plot for this summary in Figure 14.9.

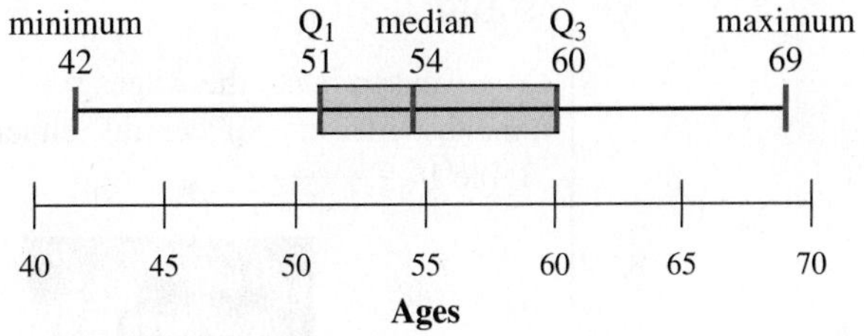

FIGURE 14.9 Box-and-whisker plot of ages of some U.S. presidents at inauguration.

The horizontal axis in Figure 14.9 represents the ages of the presidents and includes the smallest and largest ages. Next, we draw a box above the axis from the first quartile to the third quartile to show where the ages between Q_1 and Q_3 lie. We draw a vertical line through the box to show the median. Last, the lines extending to the left and right of the box—the "whiskers"—show the extreme values in the distribution. From this graph we can see quickly where the extreme values, the median, and the middle 50% of the data lie.

We often want to know which data value occurs the most frequently in a data set. For example, a fashion designer might want to know the most common dress size, or an automobile manufacturer might be interested in the most common height of U.S. drivers. The mode is a measure that is easy to find to describe the most prevalent value in a data set.

KEY POINT

The mode is the most frequent value in a data set.

COMPUTING THE MODE The **mode** of a set of data is the data value that occurs most frequently. If two values occur most frequently, then each is a mode. If more than two values occur most frequently, we will say that there is no mode.

EXAMPLE 7 Finding the Mode of a Data Set

Find the mode for each data set.

a) 5, 5, 68, 69, 70 b) 3, 3, 3, 2, 1, 4, 4, 9, 9, 9

c) 98, 99, 100, 101, 102 d) 2, 3, 4, 2, 3, 4, 5

SOLUTION:

a) The mode is 5 because that is the most frequently occurring value.

b) There are two modes, namely 3 and 9.

In parts c) and d) there is no mode because more than two values occur most often.

Comparing Measures of Central Tendency

For a set of data, the mean, median, and mode are usually not the same. Therefore, when presenting a summary of data, you may want to emphasize one measure over another.

EXAMPLE 8 Which Measure of Central Tendency Is the Best?

Assume that you are negotiating the contract for your union at Magnum Industrial Corporation. To prepare for the next negotiation session, you have gathered annual wage data and found that three workers earn \$30,000, five workers earn \$32,000, three workers earn \$44,000, and one worker earns \$50,000. In your negotiations, which measure of central tendency should you emphasize?

SOLUTION:

You want to make the wages seem as low as possible; therefore, you should choose the smallest measure of central tendency. We list the frequency distribution of wages in Table 14.9.

Salary (thousands \$), *x*	30	32	44	50
Frequency, *f*	3	5	3	1

TABLE 14.9 Wage distribution at Magnum Industrial Corporation.

We see in Table 14.9 that the mode is \$32,000. Because there are 12 salaries, we look at positions 6 and 7. Both contain \$32,000; therefore, the median salary is also \$32,000.

We use Table 14.9 to compute the mean as follows:

$$\bar{x} = \frac{\Sigma(x \cdot f)}{\Sigma f} = \frac{30 \cdot 3 + 32 \cdot 5 + 44 \cdot 3 + 50 \cdot 1}{12} = \frac{432}{12} = 36$$

The mean is therefore \$36,000.

It is to the union's benefit to present the median (or the mode) of \$32,000. No doubt management will claim that the mean of \$36,000 is the average salary.

Now try Exercises 21 to 28.

In describing data, which measure—mean, median, or mode—is most appropriate? This is not an easy question to answer. The three measures are often different, and it is really up to you to decide how you want to summarize the data.

If you are considering the mean, remember that one or two extreme values in a distribution have an undue influence on the mean. This is why, for example, in scoring ice

Between the Numbers

You Can Observe a Lot by Just Watching (Yogi Berra)

Several years ago, I attended a town meeting held by a local state representative. As the presentation unfolded, I realized that the slide presentation was carefully crafted to appeal to the large number of senior citizens in the audience. Pointing to a graph (similar to the one in this note*), he remarked that to keep taxes down, he was keeping an eye on the increases for higher education that had occurred over the past few years.

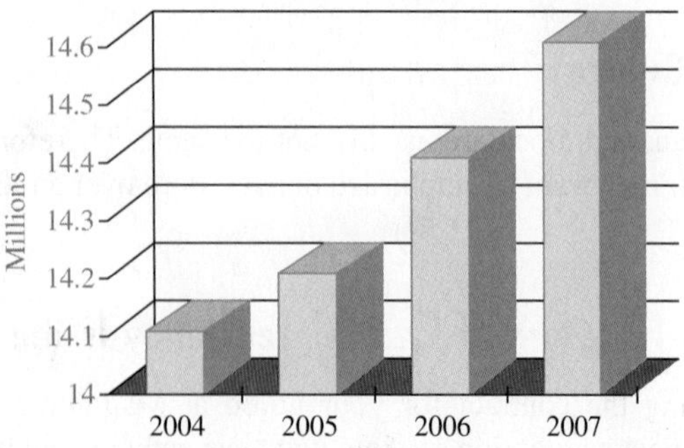

Does anything bother you about this graph? It should. Notice that this particular increase for higher education was \$0.5 million over 3 years. That amounts to an average increase of less than 1.2% per year. Also, the labels on the vertical axis are in light gray and begin with \$14 million, which makes the data hard to see and causes the graph to appear to be rapidly increasing. How would you present these same data to make the case to your state legislature for greater appropriations for higher education? As an intelligent consumer of information, always be aware that people can mislead you by the way they present their data.

*To be honest, I did not write down the exact numbers from the presentation, but the data and the spirit of the talk were essentially as I describe it.

skating at the Olympics the highest and lowest scores are discarded. If a data set has a few outliers, you may decide that it is best to discard them before computing the mean.

If the scores in a distribution are symmetrically balanced on either side of the mean, then the mean, median, and mode (assuming there is only one mode) will all be the same. However, some distributions are highly asymmetric, so there may be many small scores on one side of the mean and a few large scores on the other. In such a case, the mean is not the best measure of the center of the data. For this reason, when we read about the average family income or the average cost of a new home, the median is often used rather than the mean.

The mode emphasizes what scores occur most frequently. For example, if a shoe store sells mostly size 9s and size 11s, then it would not make sense in a report to say that the average size of the shoes sold was size 10.

In addition to describing the center of a data set, it is also important to understand how data values are spread out. In the next section, we will compute numbers to measure the spread of data that gives us a better understanding of a distribution.

EXERCISES 14.2

Sharpening Your Skills

Find the mean, median, and mode for the following distributions.

1. 4, 6, 8, 3, 9, 11, 4, 7, 5

2. 8, 9, 4, 2, 10, 5, 5, 3, 3

3. 4, 6, 4, 6, 7, 9, 3, 9, 10, 11

4. 12, 11, 7, 9, 8, 6, 4, 5, 10, 1

5. 7, 3, 1, 5, 8, 6, 2, 5, 9, 4

6. 12, 4, 4, 8, 4, 7, 9, 8, 7, 7

7. 7, 8, 6, 5, 7, 10, 2, 7, 9, 5, 8, 8, 10, 9, 6, 5, 10, 7, 9, 8

8. 8, 8, 6, 6, 7, 10, 4, 7, 9, 5, 9, 8, 10, 10, 6, 5, 10, 8, 9, 8

In Exercises 9 to 12, use the given graph to find the mean, median, and mode of the distribution.

9.

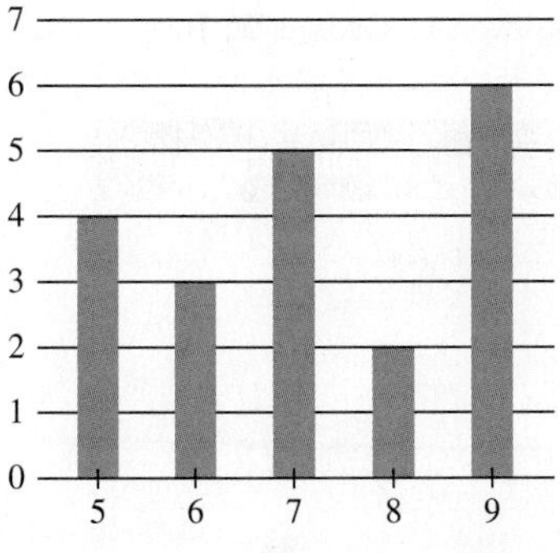

10.

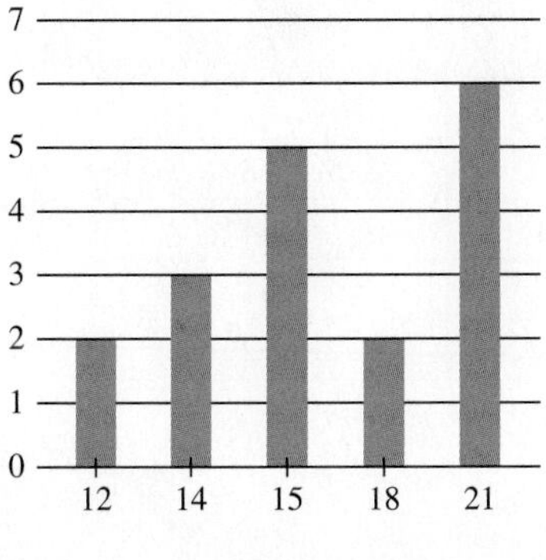

11.

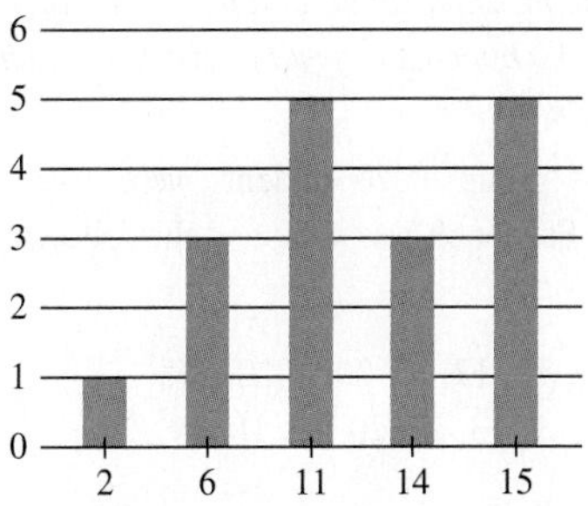

12.

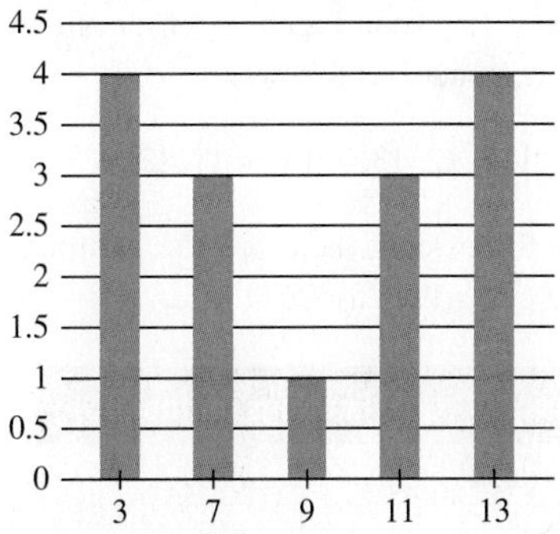

Find the mean, median, and mode for the following frequency distributions.

13.

x	*f*
3	2
4	5
6	3
7	1
8	4
10	3
11	2

14.

x	*f*
2	3
5	2
7	3
8	2
9	1
10	5
11	2

15.

x	f
5	1
8	3
11	6
12	2
14	5
15	4
18	3

16.

x	f
3	5
6	2
8	8
11	3
15	6
17	2
23	1

In Exercises 17–20, a) give the five-number summary for the distribution, and b) draw a box-and-whisker plot.

17. 11, 23, 25, 17, 26, 31, 45, 18, 41, 26, 31, 33, 48, 44, 53

18. 21, 24, 15, 45, 18, 31, 26, 41, 23, 18, 44, 27, 36, 21, 43

19. 31, 25, 41, 33, 28, 34, 37, 41, 33, 29, 49, 32, 38, 45, 30

20. 13, 24, 27, 45, 32, 29, 28, 39, 25, 21, 37, 36, 42, 34, 49

Applying What You've Learned

In Exercises 21–28, find the mean, median, and mode for each data set. In each case, explain which measure you believe best describes a typical value in the distribution.

21. Facebook usage. A group of 20 students were asked how many times they visit Facebook per week and the following results were obtained:

6, 20, 9, 18, 17, 25, 20, 4, 22, 25,
13, 5, 6, 9, 13, 23, 20, 13, 18, 14

22. Real friends. After one semester, a group of 15 freshmen were asked how many new (non-Facebook) friends they had made at college and responded as follows:

11, 9, 6, 9, 11, 7, 12, 13, 7, 14, 6, 11, 13, 9, 7

23. According to www.forbes.com, here are the earnings of a number of deceased celebrities for 2011.

Celebrity	Earnings (millions $)
Michael Jackson	170
Elvis Presley	55
Marilyn Monroe	27
Charles Schulz	25
John Lennon	12
Elizabeth Taylor	12
Albert Einstein	10
Theodor Geisel (Dr. Seuss)	9
Jimi Hendrix	7
Stieg Larsson	7

24. Major League Baseball payrolls. Here are the top 10 payrolls for Major League Baseball teams in 2011.

Team	Payroll (millions $)
Boston Red Sox	162
Chicago Cubs	125
Chicago White Sox	128
Detroit Tigers	106
Los Angeles Angels	139
Minnesota Twins	113
New York Mets	119
New York Yankees	203
Philadelphia Phillies	173
San Francisco Giants	118

25. Carbon dioxide emissions. Many believe that excess carbon dioxide in the atmosphere contributes to global warming. Below are the concentrations of carbon dioxide in the atmosphere from 2000 to 2009.

Year	CO_2 (parts per million)
2000	367
2001	369
2002	371
2003	373
2004	375
2005	377
2006	379
2007	381
2008	383
2009	385

Source: U.S. Department of Energy

26. Presidential vetoes. The following is a summary of the number of vetoes (including pocket vetoes) for U.S. presidents from Franklin D. Roosevelt to George W. Bush.

President	Number of Vetoes
F. D. Roosevelt	635
Truman	250
Eisenhower	181
Kennedy	21
L. Johnson	30
Nixon	43
Ford	66
Carter	31
Reagan	78
G. H. W. Bush	44
Clinton	37
G. W. Bush	12

Source: The World Almanac, 2008

27. Health club use. A local health club surveyed the number of days that 28 of its members visited the club during the past month. The results are given below.

4, 8, 7, 9, 18, 6, 7, 5, 5, 15, 8, 11, 12, 8, 17,
7, 14, 5, 4, 16, 10, 13, 17, 7, 18, 17, 8, 6

28. Ages of Supreme Court justices. At the start of the 2011–2012 term of the Supreme Court, the nine justices and their years of birth were Scalia (1936), Kennedy (1936), Thomas (1948), Ginsburg (1933), Breyer (1938), Roberts (1955), Alito (1950), Sotomayor (1954), Kagan (1960). (*Source: World Almanac and Book of Facts 2011*) For data, use the ages of the justices on January 1, 2011. (None was born on New Year's Day.)

Many colleges assign numerical points to grades as follows: A – 4, B – 3, C – 2, D – 1, and F – 0. Then your grade point average (GPA) is computed by multiplying the number of credits for each course by its numerical grade, adding these products, and dividing by the number of credits. For example, if you get an A in a 3-credit history course and a D in a 2-credit personal fitness course, then your grade point average would be $\frac{3 \times 4 + 2 \times 1}{3 + 2} = \frac{14}{5} = 2.8.$ *In Exercises 29 and 30, use this method to compute the GPA for each semester grade report.*

29.

Course	Credits	Grade
English	3	A
Speech	2	B
Mathematics	3	A
History	3	F
Physics	4	B

30.

Course	Credits	Grade
Philosophy	3	C
Health	2	A
Mathematics	3	B
Psychology	3	D
Music	1	B

31. Exam scores. Izzy had an 84 and an 86 on his first two tests and believed that he did well enough on his final exam to keep his B average. However, when he got his grade he received a D. He checked with his instructor and learned that the instructor had made a mistake by transposing the digits when he recorded the final exam grade. If Izzy's incorrect average (mean) for the class was 69, what was his correct final exam score?

32. Hours worked. The mean number of hours that Raphael worked over the past four weeks is 38.75. If he works 42 hours this week, what will the mean number of hours be that he will have worked over the 5-week period?

33. Mileage ratings. The EPA has determined the number of miles per gallon (MPG) for 58 foreign cars, as given in the following frequency table. Find the mean, median, and mode for these data.

Table for Exercises 33 and 34

MPG	19	20	21	22	23	24	25	26	27	28	29	30
Frequency	2	5	3	5	4	8	8	6	3	5	7	2

34. Mileage ratings. The Rolls-Royce and Jaguar XJ12 have ratings of 11 MPG. Suppose this new score with its frequency of 2 is included in the table in Exercise 33. What are the new mean, median, and mode of the distribution?

35. Exam scores. Assume that in your History of Film class you have earned test scores of 78, 82, 56, and 72, and only one test remains.

a. If you need a mean score of 70 to earn a C, then what must you obtain on the final test?

b. If you need a mean score of 80 to earn a B, then what must you obtain on the final test?

36. Exam scores. Assume that in your Abnormal Psychology class you have earned test scores of 74, 81, 56, and 70, and only one test remains.

a. If you need a mean score of 70 to earn a C, then what must you obtain on the final test?

b. If you need a mean score of 80 to earn a B, then what must you obtain on the final test?

Credit card companies often compute the average daily balance on your account as follows: If you start on the first of a 31-day month with a balance due of $100, and then on the 5th you charge another $50 and on the 27th you charge another $20, they would say that on days 1, 2, 3, and 4 you owed $100; on days 5, 6, 7, . . . , 25, and 26 you owed $150; and on days 27, 28, 29, 30, and 31 you owed $170. Therefore, to calculate the average daily balance, they would compute*

$$\frac{4 \cdot 100 + 22 \cdot 150 + 5 \cdot 170}{31} = \frac{4{,}550}{31} = \$146.77.$$

Use this method to answer Exercises 37 and 38.

37. Credit card balance. Assume that in a 31-day month you begin with a $50 balance due on your credit card, then charge an item for $75 on the 10th of the month and a $120 item on the 25th of the month. What is your average daily balance on your credit card for this month?

38. Credit card balance. Assume that in a 31-day month you begin with an $80 balance due on your credit card, then charge an item for $60 on the 5th of the month and a $100 item on the 20th of the month. What is your average daily balance on your credit card for this month?

39. Draw box-and-whisker plots for the distributions of the number of home runs hit for the two given time periods in Example 6 in Section 14.1.

40. In considering your answer to Exercise 39, which do you feel conveys the home run information better—the stem-and-leaf displays or the box-and-whisker plots?

*The average daily balance method is covered thoroughly in Section 8.3.

Draw a box-and-whisker plot of the data presented in the graphs in Exercises 41 and 42.

41. **Top 10 Parent Internet Companies**

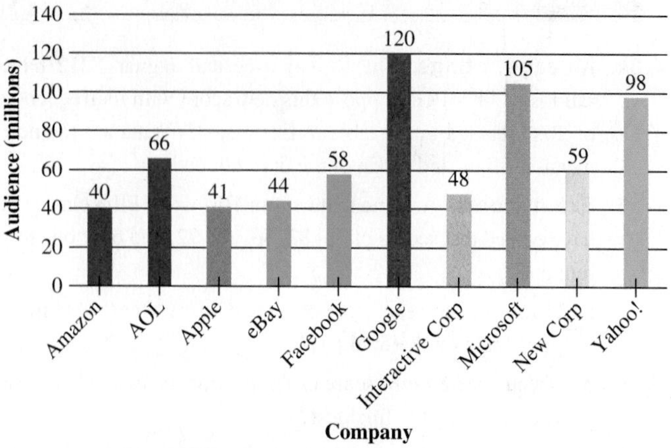

Source: New York Times Almanac 2011

42. **Top 15 Recording Artists in 2011**

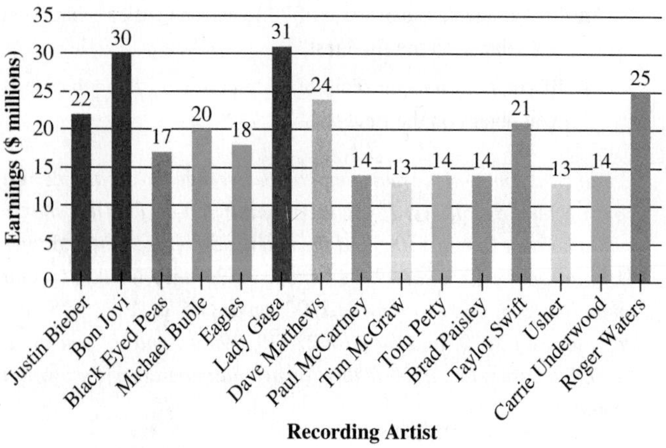

Source: Billboard.com

A college placement office has made a comparative study of the starting salaries for graduates in various majors. Use the following box-and-whisker plots, which describe the starting salaries for business, education, engineering, and liberal arts, to answer Exercises 43–48.

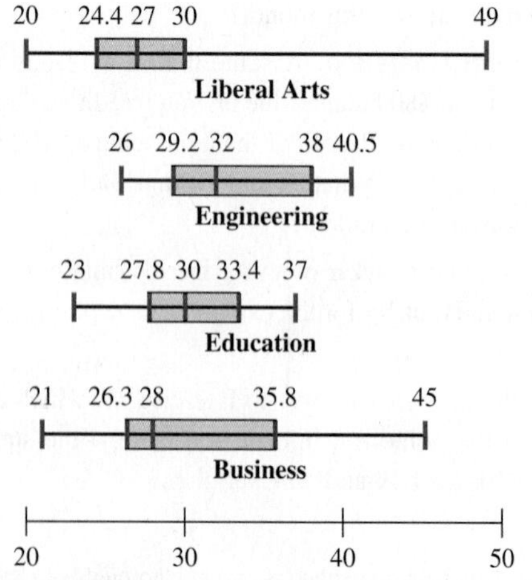

43. Salaries for college graduates. Which major, in general, receives the worst salary offers?

44. Salaries for college graduates. If a student wants to receive $32,000 or more for a starting salary, which major offers the best chance of achieving that?

45. Salaries for college graduates. Does education or business have the higher median salary? Does this imply that salaries in education are generally higher than those in business? Discuss.

46. Salaries for college graduates. What is the significance of the long whisker in the liberal arts plot?

47. Salaries for college graduates. What is the significance of the short whisker in the engineering plot?

48. Salaries for college graduates. Try to describe specific situations that might lead a student to a misinterpretation of the information contained in these plots.

Communicating Mathematics

49. What is the difference between Σx and $\Sigma x \cdot f$?

50. What are the five numbers in a box-and-whisker plot? In a box-and-whisker plot, what does the box represent? The whiskers?

51. Give three separate real-life examples when the mean, median, or mode would be the appropriate measure of the average in a distribution. Explain why you think that each measure would be appropriate in the given situation.

52. Which measure of central tendency do you think is being used in each of the following statements?

"I read that the Golden Bears averaged 7 yards per carry in Saturday's game."

"I saw all of our team's basketball games and I would say that Gunner averages 21 points a game."

"The average cost of a new house is $123,000."

"The typical new car is in the $14,000–$18,000 range."

Between the Numbers

53. Manipulating data. Choose some set of data involving a topic that may be politically charged such as inflation, unemployment, taxes, etc. (If you can't find some data that you like, make up some reasonable data.) Present the data in two different graphs—one in which you minimize the trend in the data and another in which you exaggerate the trend.

54. Analyzing misleading graphs. Locate a graph in the media or online that appears to be drawn to mislead you. Mention specifically which features of the graph might be misleading.

Challenge Yourself

55. The Super Bowl. Many have criticized the Super Bowl as uninteresting because the games have not been close contests. Use the methods that you have learned in this chapter to make an argument for or against this statement using as data the margins of victory for each Super Bowl since 1967.

56. Those who play fantasy baseball often are interested in comparing the statistics of various players that they are interested in picking for their team. For example, you might be interested in comparing the Phillies' Roy Halladay with the Giants' Tim Lincecum. Choose a pair of players and make a comparison using the methods of this section. You can find statistics about players at www.baseball-reference.com.

57. Suppose there are nine scores from 1 to 10 in a distribution. If possible, give an example of each of the following. If you think that it is not possible to give the requested distribution, try to explain why you believe this.

a. a distribution in which the mean, median, and mode are each 5

b. a distribution in which the median is 3 and the mean is 5

c. a distribution in which the median is 5 and the mean is less than 5

58. Suppose there are nine scores from 1 to 10 in a distribution. If possible, give an example of each of the following. If you think that it is not possible to give the requested distribution, try to explain why you believe this.

a. a distribution in which the mean is 5, the median is 4, and the mode is 2

b. a distribution in which the mean is 3 and the median is 5

c. a distribution in which the mean is 5 and the median is less than 5

If possible, give examples of data sets (sets of numbers) that have the properties specified in Exercises 59–62. It is very helpful to use spreadsheets such as Excel or Numbers to make up different examples very quickly.

59. The mean, median, and mode are all the same.

60. The mean, median, and mode are all different.

61. The mean is larger than the median.

62. The median is larger than the mean.

63. Suppose that distributions X and Y have means $\bar{x}$ and $\bar{y}$, respectively. If we combine X and Y into a single distribution, is the mean of this new distribution $\bar{x} + \bar{y}$? Explain.

64. Suppose that distributions X and Y have medians a and b, respectively. If we combine X and Y into a single distribution, is the median of this new distribution $a + b$? Explain.

65. Give an example of two distributions that have the same mean, median, and mode. In one distribution, the data values should be tightly bunched together; in the other, the data values should be spread apart.

14.3 Measures of Dispersion

Objectives

1. Compute the range of a data set.
2. Understand how the standard deviation measures the spread of a distribution.
3. Use the coefficient of variation to compare the standard deviations of different distributions.

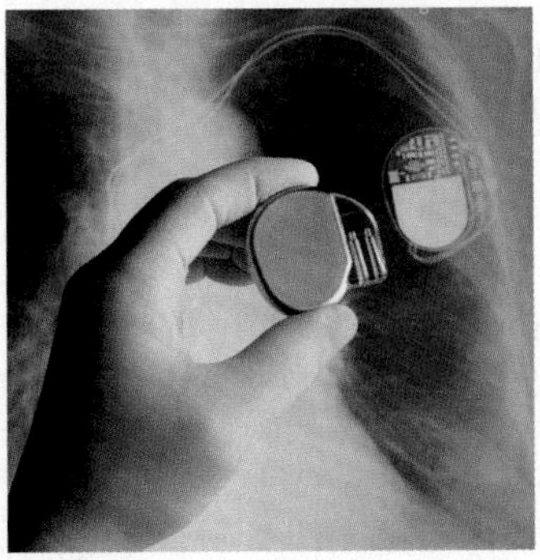

Let's imagine that you are a cardiac surgeon faced with deciding which of two heart pacemaker batteries you should choose. Because you must operate to replace a failing battery, it is important that you choose the one that will last the longest. Assume that battery A lasts for a mean time of 45,000 hours (slightly more than 5 years) and battery B has a mean time of 46,000 hours.

On the surface, it would seem that you should choose battery B. However, suppose that we also tell you that in testing the batteries, we found that all of A's times were within 500 hours of the mean, but B's times varied widely. In fact, some of B's times were as much as 2,500 hours below the mean. So B's times could be as low as $46{,}000 - 2{,}500 = 43{,}500$ hours, whereas A's times were never below 44,500. Based on this information, it appears that battery A is the better choice.

The point of our story is that although the mean and median tell you something about a distribution, they do not tell the whole story. For example, the following two distributions both have a mean and median of 25, but Y's values are much more spread out than X's:

$$X: 24, 25, 25, 25, 25, 26 \qquad Y: 1, 2, 3, 47, 48, 49$$

From these examples, it is clear that we need to develop some method for measuring the spread of a distribution.

KEY POINT

The range is a crude measure of the spread of a data set.

The Range of a Data Set

It is clear from the discussion of pacemaker batteries that a numerical measure of the dispersion, or spread, of a data set can provide useful information. One simple way to describe the spread of a set of data is to subtract the smallest data value from the largest.

DEFINITION The **range** of a data set is the difference between the largest and smallest data values in the set.

EXAMPLE 1 Comparing Heights

Find the range of the heights of the people listed in the accompanying table.

Person	Height	Height in Inches
Robert Wadlow (World's Tallest Person)	8 feet, 11 inches	107 inches
Tim Tebow	6 feet, 3 inches	75 inches
Madge Bester (World's Shortest Woman)	2 feet, 2 inches	26 inches
LeBron James	6 feet, 8 inches	80 inches

SOLUTION:

The range of these data is

$$\text{range} = \text{largest data value} - \text{smallest data value} = 107 - 26 = 81 \text{ inches}$$
$$= 6 \text{ feet, } 9 \text{ inches.}$$

The range is generally not useful to measure the spread of a distribution because it can be influenced by a single outlier. For example, the range of the distribution 2, 2, 2, 2, 3, 4, 100 is $100 - 2 = 98$.

Historical Highlight

The Origins of Statistics

Ring a ring of roses,
A pocket full of posies,
*Asha Asha, We all fall down.**

It may have been King Henry VII's fear of the dreaded Black Plague that prompted him to begin publishing weekly Bills of Mortality in 1532. John Graunt, a merchant, noticed patterns in these reports regarding deaths due to accident, suicide, and disease and concluded that social phenomena do not occur randomly. He published his observations in a paper titled, "*Natural and Political Observations... Made upon the Bills of Mortality.*" King Charles II, impressed by Graunt's work, nominated him to the Royal Society of London even though he had few academic credentials.

Edmund Halley, the noted English astronomer, continued the work of Graunt in a paper titled "*An Estimate of the Degrees of the Mortality of Mankind, Drawn from Curious Tables of the Births and Funerals at the City of Breslaw; with an Attempt to Ascertain the Price of Annuities upon Lives.*" By studying the mathematics of life expectancy, Halley helped lay the foundations for actuarial science, which insurance companies use today to determine their premiums.

*Some see this rhyme as describing the rosy rash, the pockets full of medicinal herbs, the sneezing and eventual death associated with the Black Plague. If so, it owes its origin, like statistics, to that horrible time in European history.

Standard Deviation

KEY POINT

The standard deviation is a reliable measure of dispersion.

A better way to find the spread of data is to use the standard deviation, a measure based on calculating the distance of each data value from the mean.

> **DEFINITION** If x is a data value in a set whose mean is $\bar{x}$, then $x - \bar{x}$ is called *x's deviation from the mean.*

To illustrate the deviation from the mean, consider distribution *A*: 16, 14, 12, 21, 22, whose mean is 17. We list the deviation of each data value from the mean in Table 14.10.

Data Value, x	Deviation from Mean, $x - \bar{x}$
16	−1
14	−3
12	−5
21	4
22	5
Total	0

TABLE 14.10 Deviations from the mean for data values in distribution *A*.

It might seem reasonable to measure the spread in *A* by averaging the deviations from the mean in Table 14.10. However, this will not work. As you see in Figure 14.10, for scores above the mean that have positive deviations from the mean, there must be scores below the mean that have negative deviations from the mean. When we add the positive and negative deviations, they cancel each other, giving a total of zero.

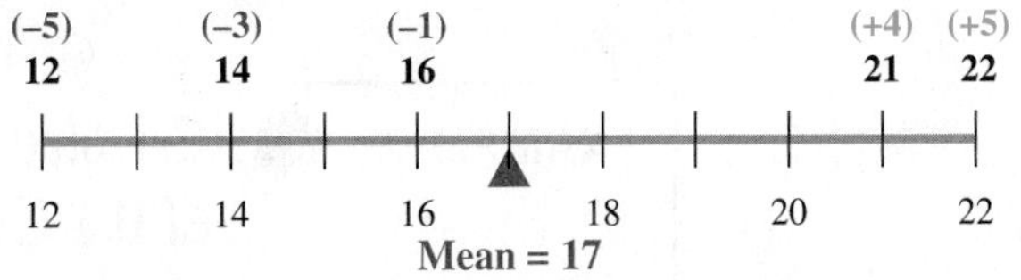

FIGURE 14.10 Values having positive and negative deviations from the mean must balance about the mean.

To avoid this cancellation, we square each of these deviations, as shown in Table 14.11.

Data Value, x	Deviation from Mean, $x - \bar{x}$	Deviation Squared, $(x - \bar{x})^2$
16	−1	1
14	−3	9
12	−5	25
21	4	16
22	5	25

TABLE 14.11 The squares of deviations from the mean for distribution *A*.

If we now average* these squared deviations, we get

$$\frac{1 + 9 + 25 + 16 + 25}{\underbrace{5 - 1}_{n-1}} = \frac{76}{4} = 19.$$

Clearly, this number is too large to represent the spread of scores from the mean. To compensate for the fact that we squared the deviations in doing our calculations, we take the square root of this number to get $\sqrt{19} \approx 4.36$. This number is a more reasonable measurement of the way scores vary from the mean. The quantity that we just calculated is called the *standard deviation.*

> **DEFINITION** We denote the **standard deviation** of a *sample* of n data values by s,† which is defined as follows:
>
> $$s = \sqrt{\frac{\sum (x - \bar{x})^2}{n - 1}}$$

*When doing this calculation for a *sample* (as opposed to a *population*), statisticians divide by $n - 1$ instead of n. There are technical reasons for doing this that are beyond the scope of this text. In doing this calculation for a population, we would divide by n rather than $n - 1$.

†We represent the standard deviation of a population by σ (instead of s), which we calculate using the formula

$$\sigma = \sqrt{\frac{\sum (x - \mu)^2}{n}}.$$

We compute the standard deviation in four steps.

COMPUTING THE STANDARD DEVIATION To compute the standard deviation for a sample consisting of n data values, do the following:

1. Compute the mean of the data set; call it $\bar{x}$.
2. Find $(x - \bar{x})^2$ for each score x in the data set.
3. Add the squares found in step 2 and divide this sum by $n - 1$; that is, find

$$\frac{\sum (x - \bar{x})^2}{n - 1},$$

which is called the **variance**.

4. Compute the square root of the number found in step 3.

EXAMPLE 2 Calculating the Standard Deviation of the Cost of Plane Flights

While planning for spring break, suppose you went to Orbitz.com to find the cheapest flight from Philadelphia to Orlando and found the following prices:

$195, $213, $208, $219, $210, $215

Find the standard deviation of this data set.

SOLUTION:

Step 1: We find the mean, which is

$$\frac{195 + 213 + 208 + 219 + 210 + 215}{6} = \frac{1{,}260}{6} = 210.$$

Step 2: We next calculate the squares of the deviations of the data values from the mean, as shown in the following table:

Price, x	Deviation from Mean, $x - \bar{x}$	Deviation Squared, $(x - \bar{x})^2$
195	$195 - 210 = -15$	$(-15)^2 = 225$
213	$213 - 210 = 3$	$3^2 = 9$
208	$208 - 210 = -2$	$(-2)^2 = 4$
219	$219 - 210 = 9$	$9^2 = 81$
210	$210 - 210 = 0$	$0^2 = 0$
215	$215 - 210 = 5$	$5^2 = 25$
		$\sum (x - \bar{x})^2 = 344$

Step 3: We average these squared deviations to get

$$\frac{225 + 9 + 4 + 81 + 0 + 25}{\underbrace{6 - 1}_{n-1}} = \frac{344}{5} = 68.8.$$

Step 4: Taking the square root of the result in step 3, we get the standard deviation $s = \sqrt{68.8} \approx 8.29$.

Now try Exercises 1 to 10. 9

QUIZ YOURSELF 9

Find the standard deviation of the following sample of data values:

3, 4, 5, 6, 4, 2, 0, 8, 4

In Example 3, we calculate the standard deviation of a distribution using a frequency table.

EXAMPLE 3 Using a Frequency Table to Compute the Standard Deviation

Anya is considering investing in APPeal, a software company that specializes in smart phone applications. She intends to compute the standard deviation of its recent prices to measure how steady its stock price has been recently. The following are the closing prices for the stock for the past 20 trading sessions:

37, 39, 39, 40, 40, 38, 38, 39, 40, 41,

41, 39, 41, 42, 42, 44, 39, 40, 40, 41

What is the standard deviation for this data set?

SOLUTION: Review the Be Systematic Strategy on page 5.

In Table 14.12, we see that the sum of the 20 closing prices is 800, so the mean is $\frac{800}{20} = 40$. We have added columns to the table for the deviations from the mean, the squares of these deviations, and so on.

Closing Price, x	Frequency, f	Product, $x \cdot f$	Deviation, $(x - 40)$	Deviation Squared, $(x - 40)^2$	Product, $(x - 40)^2 \cdot f$
37	1	37	−3	9	9
38	2	76	−2	4	8
39	5	195	−1	1	5
40	5	200	0	0	0
41	4	164	1	1	4
42	2	84	2	4	8
44	1	44	4	16	16
	$\Sigma f = 20$	$\Sigma(x \cdot f) = 800$			$\Sigma(x - 40)^2 \cdot f = 50$

TABLE 14.12 Computations necessary to find the standard deviation of APPeal closing prices.

When calculating the standard deviation, we must multiply the squared deviation of each price from the mean by its frequency. We list these products in the column labeled "Product, $(x - 40)^2 \cdot f$," and show the sum of these products at the bottom of the column.

We can now calculate the standard deviation. Remember that Σf is the number of data values that we have been representing by n. The standard deviation is therefore

$$s = \sqrt{\frac{\sum (x - 40)^2 \cdot f}{n - 1}} = \sqrt{\frac{50}{19}} \approx 1.62.$$

This relatively small standard deviation indicates that the closing prices for the APPeal stock have not been varying very much lately. If Anya is a cautious investor, she will find APPeal's price stability attractive.

Now try Exercises 11 to 16.

We summarize the method we use for finding the standard deviation in the formula on the next page.

FORMULA FOR COMPUTING THE SAMPLE STANDARD DEVIATION FOR A FREQUENCY DISTRIBUTION We calculate the standard deviation, s, of a sample that is given as a frequency distribution as follows:

$$s = \sqrt{\frac{\sum(x - \overline{x})^2 \cdot f}{n - 1}}$$

where $\overline{x}$ is the mean of the distribution, f is the frequency of data value x, and $n = \Sigma f$ is the number of data values in the distribution. 10

QUIZ YOURSELF 10

The investor in Example 3 also gathered the closing stock prices for WebRanger, a Internet software company. We list these stock prices in Table 14.13. The mean closing price is 42. Find the standard deviation for this distribution.

Closing Price, x	36	37	38	39	40	41	42	43	44	45
Frequency, f	2	0	0	3	1	3	0	1	5	5

TABLE 14.13 WebRanger closing stock prices for 20 days.

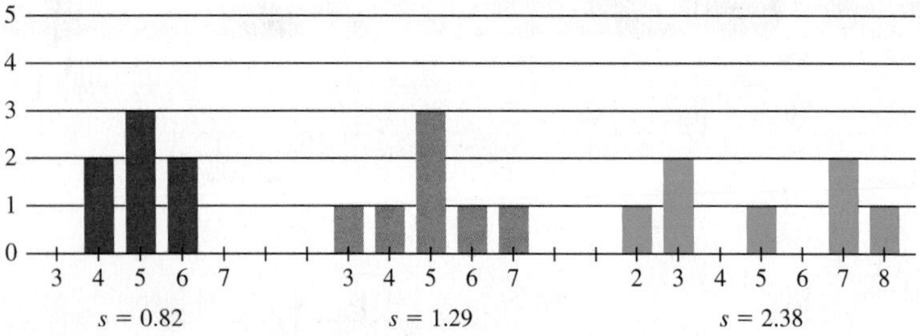

FIGURE 14.11 As the spread of the distribution increases, so does the standard deviation.

Comparing the standard deviation of 1.62 for APPeal with the standard deviation of 3.01 for WebRanger in Quiz Yourself 10, we see that WebRanger's stock is much more volatile than APPeal's stock.

We have said that the standard deviation describes the spread of a distribution. In Figure 14.11, all three distributions have a mean and median of 5; however, as the spread of the distribution increases, so does the standard deviation.

Highlight

Using a Graphing Calculator to Find the Standard Deviation*

Because calculating the standard deviation can become tedious, we often use a calculator to save work, as we show in the accompanying TI calculator screen. We have stored the data from Example 3 in list L1, and then used the calculator to find the standard deviation, as we see in the screen below.

With such powerful technology available, it is very important that you think critically to set up problems correctly rather than just focusing on doing computations.

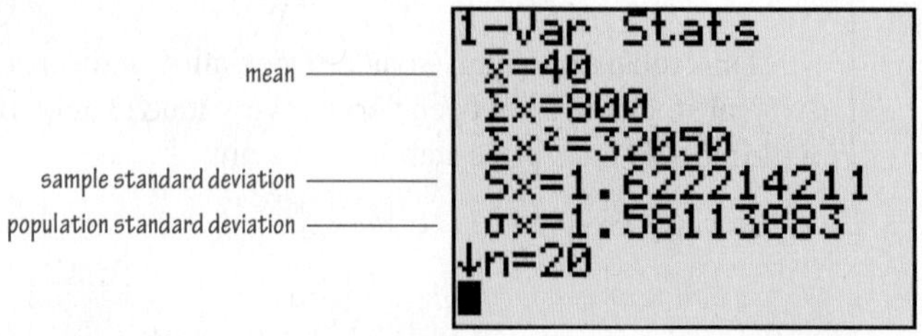

*Technology resources are available through Pirnot's MyMathLab.

The Coefficient of Variation

KEY POINT

We use the coefficient of variation to compare the standard deviations of different data sets.

In Example 3, we used the standard deviation with one set of data. If you use the standard deviation to compare two sets of data, the data must be similar and have comparable means. The following situation shows why.

Suppose we are comparing the body weights of two groups of people and find the standard deviation to be 3 pounds for the first group and 10 pounds for the second group. Can we state that there is more uniformity in the first group than the second? Before answering this question, here is more information. The first group consists of preschool children, whereas the second is a group of National Football League linemen. Clearly, the standard deviation of 3 is more significant, *relatively speaking*, for a group of preschoolers with a mean weight of 30 pounds than the standard deviation of 10 for a group of football players with a mean weight of 300 pounds. In order to use the standard deviation effectively to compare different sets of data, we must first make the numbers comparable. We do this by finding the coefficient of variation.

DEFINITION For a set of data with mean $\overline{x}$ and standard deviation s, we define the **coefficient of variation,** denoted by CV, as

$$CV = \frac{s}{\overline{x}} \cdot 100\%.$$

Note that the coefficient of variation compares the standard deviation to the mean and is expressed as a percentage. The coefficient of variation for the group of preschool children is

$$CV = \frac{3}{30} \cdot 100\% = 10\%.$$

(3: standard deviation; 30: mean)

In contrast, the coefficient of variation for the group of NFL football linemen is

$$CV = \frac{10}{300} \cdot 100\% = 3.3\%.$$

(10: standard deviation; 300: mean)

Because the coefficient of variation is larger for the group of preschoolers, we conclude that there is more variation, *relatively speaking*, in their weights than in the weights of the football players.

EXAMPLE 4 Using the Coefficient of Variation to Compare Data

Use the coefficient of variation to determine whether the women's 100-meter race (328 feet) or the men's marathon (26 miles) has had more consistent times over the five Olympics listed in Table 14.14.

	Women's 100 Meters	Men's Marathon
2008	10.78 sec	2 h, 6 m, 32 sec (7,592 sec)
2004	10.93 sec	2 h, 10 m, 55 sec (7,855 sec)
2000	10.75 sec	2 h, 10 m, 11 sec (7,811 sec)
1996	10.94 sec	2 h, 12 m, 36 sec (7,956 sec)
1992	10.82 sec	2 h, 13 m, 23 sec (8,003 sec)

TABLE 14.14 Olympic times in women's 100-meter race and men's marathon.

SOLUTION:

Using a calculator, we found the mean time for the 100-meter race is 10.844 and the standard deviation is approximately 0.087. For the marathon, the mean is 7,843.4 and the standard deviation is approximately 160.11.

We find the coefficient of variation for the women's 100-meter race is

$$\frac{0.087}{10.844} \cdot 100\% = 0.80\%.$$

For the men's marathon, the coefficient of variation is

$$\frac{160.11}{7{,}843.4} \cdot 100\% = 2.04\%.$$

Thus, using the coefficient of variation as a measure, there is more variation for the marathon than for the 100-meter race.

Now try Exercises 31 to 36.

EXERCISES 14.3

You may want to use a calculator or computer software, such as a spreadsheet, to solve the exercises in this section. Your answers may differ slightly from ours due to differences in the way we round off our calculations. Assume that we are computing the sample standard deviation unless we specify otherwise.*

Sharpening Your Skills

Find the range, mean, and standard deviation for the following data sets.

1. 19, 18, 20, 19, 21, 20, 22, 18, 18, 17
2. 89, 72, 100, 87, 65, 98, 77, 92
3. 5, 7, 9, 4, 6, 8, 7, 10
4. 8, 4, 7, 6, 5, 5, 4, 9
5. 2, 12, 3, 11, 14, 5, 8, 9
6. 21, 3, 5, 11, 7, 1, 9, 7
7. 18, 3, 8, 7, 7, 9, 13, 7
8. 22, 18, 15, 21, 21, 15, 19, 13
9. 3, 3, 3, 3, 3, 3, 3
10. 4, 6, 4, 6, 4, 6, 4, 6
11. The following frequency table shows the number of fires per month in a city. Complete the table entries to find the mean and standard deviation for this distribution.

Number, x	Frequency, f	Product, $x \cdot f$	Deviation, $(x - \bar{x})$	Deviation², $(x - \bar{x})^2$	Product, $(x - \bar{x})^2 \cdot f$
2	1				
3	1				
4	0				
5	2				
6	3				
7	4				
8	4				
9	3				
10	2				
	$\Sigma f =$	$\Sigma x \cdot f =$			$\Sigma(x - \bar{x})^2 \cdot f =$

*Technology resources are available in Pirnot's MyMathLab.

12. The following frequency table summarizes the number of job offers made to graduates of a Microsoft network administrator certification program. Complete the table entries to find the mean and standard deviation for this distribution.

Number, x	Frequency, f	Product, $x \cdot f$	Deviation, $(x - \bar{x})$	Deviation^2, $(x - \bar{x})^2$	Product, $(x - \bar{x})^2 \cdot f$
4	2				
5	0				
6	3				
7	5				
8	4				
9	5				
10	1				
	$\Sigma f =$	$\Sigma x \cdot f$			$\Sigma(x - \bar{x})^2 \cdot f =$

Find the mean and standard deviation for the following frequency distributions.

13.

x	f
2	2
3	4
4	2
5	0
6	4

14.

x	f
8	3
9	2
10	1
11	2
12	3

15.

x	f
3	1
4	1
5	0
6	2
7	5
8	3
9	3

16.

x	f
12	3
13	0
14	5
15	0
16	1
17	2
18	3

Find the mean and standard deviation of the distributions given by the graphs in Exercises 17 and 18.

17.

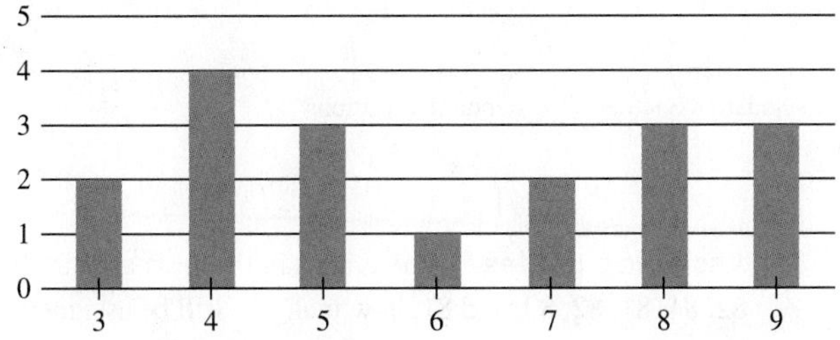

18.

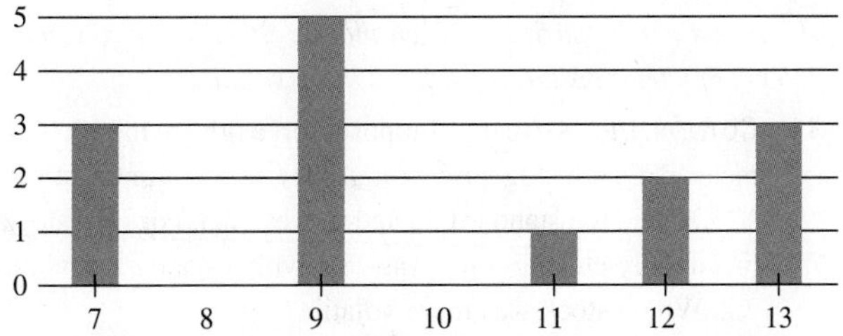

Applying What You've Learned

19. Summarizing test score data. The following are the scores of 20 people who took a paramedics licensing test. Find the mean and standard deviation for these data.

Score	72	73	78	84	86	93
Frequency	7	1	3	6	2	1

20. Summarizing age data. The following are the ages of 16 veterans of the Iraq War who are serving on a panel discussing post-traumatic stress disorder. Find the mean and standard deviation of these data.

Age	30	29	28	26	25	23
Frequency	6	3	2	2	2	1

21. Summarizing tax data. The following table lists the state income tax rate for a single person earning $50,000 per year for several states.

State	Income Tax (%)
Arizona	4.24
California	9.55
Hawaii	8.25
Georgia	6.00
New Jersey	5.53
New Mexico	4.90
Pennsylvania	3.07
Michigan	4.35

Source: World Almanac and Book of Facts 2011

a. Find the find the mean and standard deviation for these data.

b. Does the mean you found in part a) represent the average state income tax paid by the people of these states? Explain.

22. Summarizing vital statistics. The following is a partial roster of the 2011 San Antonio Silver Stars of the WNBA. Find the mean and standard deviation of these data.

Player	Weight (lbs.)
Danielle Adams	239
Jayne Appel	206
Shameka Christon	174
Becky Hammon	174
Loree Moore	165
Ziomara Morrison	183
Porsha Phillips	173
Sophia Young	165

Source: www.wnba.com

23. Summarizing customer data. The manager at the local Starbucks counted the customers once each hour over a busy weekend. We summarize the results she obtained in the bar graph below. Find the mean and standard deviation of these data.

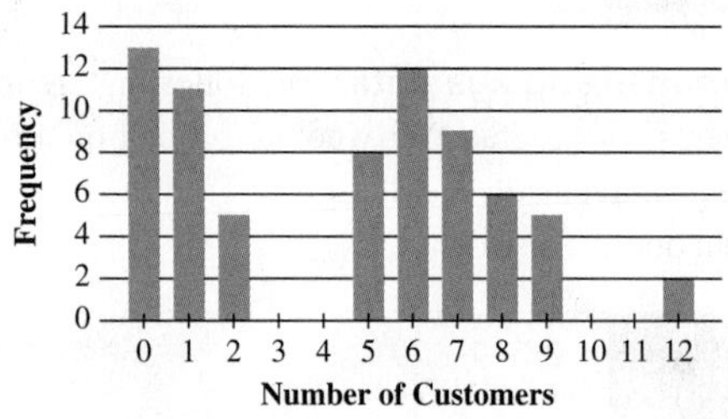

24. Measuring service times. In order to improve service, the local Subway recorded the waiting time (in minutes) for customers at a local franchise and obtained the following graph. Find the mean and standard deviation for this graph.

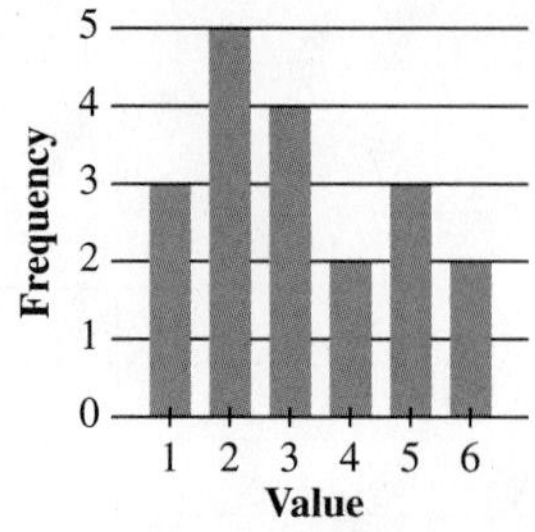

25. Family incomes. The following table gives the annual incomes for eight families, in thousands of dollars. Find the number of standard deviations family H's income is from the mean.

Family	A	B	C	D	E	F	G	H
Annual Income (in \$ thousands)	47	48	50	49	51	47	49	51

26. Family incomes. The following table gives the annual incomes for eight families, in thousands of dollars. Find the number of standard deviations family A's income is from the mean.

Family	A	B	C	D	E	F	G	H
Annual Income (in \$ thousands)	49	51	52	51	51	50	52	52

Professor Jones assigns grades in her modern poetry class according to where a student's average is located in the following diagram. (Averages on a border are assigned the higher grade.) Use this diagram in Exercises 27–30.

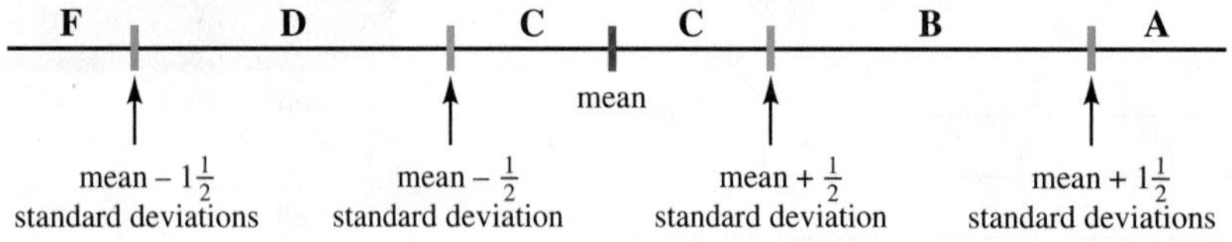

27. Assigning grades. The averages in the class are 80, 76, 81, 84, 79, 80, 90, 75, 75, and 80. What grade does the person earning the 76 get?

28. Assigning grades. The grades in the class are 72, 71, 73, 70, 71, 79, 65, 73, 74, and 72. What grade does the person earning the 74 get?

29. Assigning grades. If the grades in the class are 80, 75, 81, 83, 80, 78, 84, 77, and 76, how many As will be assigned?

30. Assigning grades. If the grades in the class are 80, 81, 81, 82, 81, 81, 82, 80, and 81, how many Fs will be assigned?

In Exercises 31–34, we present information on the performance of various stocks during a given month. Stocks with greater coefficients of variation are considered more volatile.

31. Comparing stocks. Suppose for a given month that the mean daily closing price for Apple Inc. common stock was 123.76 and the standard deviation was 12.3. For Dell stock, the mean daily closing price was 78.6 with a standard deviation of 7.2. Which stock was more volatile?

32. Comparing stocks. Suppose for a given month that the mean daily closing price for Netflix was 37.4 and the standard deviation was 4.1. For Blockbuster, the mean daily closing price was 18.6 with a standard deviation of 3.2. Which stock was more volatile?

33. Comparing stocks. The Dow Jones Industrial Average (DJIA) measures the stock prices of a group of 30 stocks. If, for a given week, the DJIA had a mean daily closing price of 11,261.12 with a standard deviation of 72.17 and WebMaster stock had a mean closing price of 37.6 with a standard deviation of 1.7, did WebMaster have more or less volatility than the DJIA during that week?

34. Comparing stocks. If, for a given week, the DJIA had a mean daily closing price of 10,834.8 with a standard deviation of 144.5 and WebRanger stock had a mean closing price of 123.6 with a standard deviation of 1.2, did WebRanger have more or less volatility than the DJIA during that week?

35. Price stability. In the following table, we list the monthly price (in dollars) of a pound of coffee and a gallon of unleaded gasoline for 2011. Find the *population* standard deviation of each set of data.

2011	Coffee Prices \$/lb	Gasoline Prices \$/gal
Jan.	4.42	3.09
Feb.	4.22	3.17
Mar.	4.64	3.55
Apr.	5.10	3.82
May	5.13	3.93
Jun.	5.23	3.70
Jul.	5.55	3.65
Aug.	5.77	3.63
Sep.	5.65	3.61
Oct.	5.51	3.47
Nov.	5.64	3.42
Dec.	5.44	3.28

Source: Bureau of Labor Statistics, U.S. Department of Energy

36. Comparing price stability. Consider the table given in Exercise 35. Use the coefficient of variation to determine whether coffee prices or gasoline prices were more stable in 2011.

37. Comparing family incomes. Consider the tables given in Exercises 25 and 26, respectively. Suppose that the eight families are the same in both tables and the incomes are for two different years that are 2 years apart. Which family had the greatest improvement in annual income over the 2-year period with respect to the other families? Explain how you arrived at your conclusion.

38. Comparing family incomes. Suppose that the mean family income in the United States was \$48,000 with a standard deviation of \$1,000 in year X. Also, the mean family income was \$51,000 with a standard deviation of \$2,250 three years later, in year $X + 3$. If a family earned \$50,000 during year X and \$54,000 three years later, then in which of these 2 years did the family have a better annual income in relation to the rest of the population? Explain how you arrived at your answer.

39. Laptop batteries. A particular brand of laptop was sampled with regard to the time it can be used before it requires recharging. The mean time was calculated to be 7.8 hours with a standard deviation of 1.78 hours. Calculate the coefficient of variation for this example.

40. Study times. A group of students were sampled with regard to how much time they expect to spend studying for the final exam in their math course. The mean time was 6.4 hours with a standard deviation of 2.1 hours. Calculate the coefficient of variation for this example.

Communicating Mathematics

Determine whether the statements in Exercises 41–43 are true or false. Explain your answer in words, or give appropriate counterexamples to support your answers.

41. a. The standard deviation of the set of numbers $-2, 2, -2, 2, -2, 2, -2, 2$ is zero.

b. If the standard deviation of a set of data is zero, then all the numbers in the data set are the same.

42. The more numbers there are in a distribution, the larger the standard deviation.

43. If we consider two distributions A and B, if A has the greater mean, then A will also have the greater standard deviation.

Use the following graphs for Exercise 44.

44. a. Which data set has the smallest standard deviation?

b. The largest?

c. Rank the data sets in order from smallest standard deviation to largest. You can do this just by looking at the graphs and without doing any calculations.

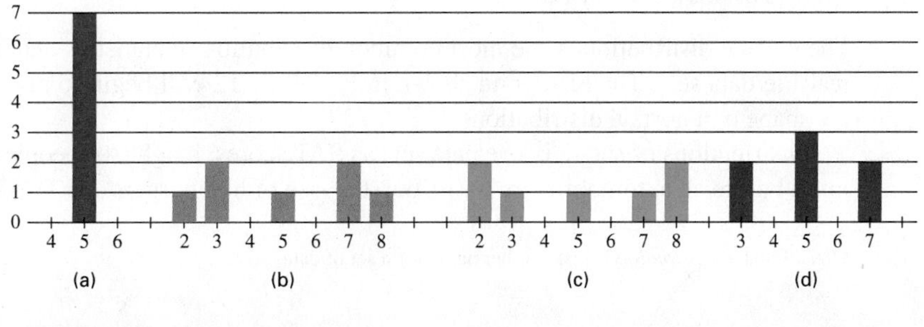

Challenge Yourself

45. **a.** Pick a data set consisting of any five numbers. Compute the mean and the standard deviation of this data set. (Consider the data set to be a sample.)

b. Add 20 to each number in the data set you created in part (a), and compute the mean and standard deviation for this new data set.

c. Subtract 5 from each number in the data set you created in part (a), and compute the mean and standard deviation for this new data set.

d. What conclusion can you make about changes in the mean and the standard deviation when the same number is added to or subtracted from each score in a data set?

e. Use the conclusion reached in part (d) to simplify the calculation of the mean and standard deviation for the distribution 598, 597, 599, 596, 600, 601, 602, 603.

46. **a.** Pick any five numbers and compute the mean and standard deviation for this data set.

b. Multiply each number in the data set you created in part (a) by 4, and compute the mean and standard deviation for this new data set.

c. Multiply each number in the data set you created in part (a) by 9, and compute the mean and standard deviation for this new data set.

d. What conclusion can you make about changes in the mean and standard deviation when each score in a data set is multiplied by the same number?

e. The mean is 4 and the standard deviation is 2.2 for the data set 3, 4, 7, 1, 5. Consider your conclusion from part (d), and compute the mean and standard deviation for the data set 15, 20, 35, 5, 25.

47. Consider the two distributions given by the bar graphs in Exercises 23 and 24. Can you simply look at the graphs to decide which of the two distributions has the greater standard deviation? If so, describe what you look for in the graphs on which to base your decision.

48. Consider the grading method used in Exercises 27–30.

a. Construct a distribution of course averages (not all the same) in which no one gets an A or an F.

b. Construct a distribution of course averages in which no one gets a C.

14.4 The Normal Distribution

Objectives

1. Understand the basic properties of the normal curve.
2. Relate the area under a normal curve to *z*-scores.
3. Make conversions between raw scores and *z*-scores.
4. Use the normal distribution to solve applied problems.

If the shoe fits, wear it.
—*Well-known proverb*

In addition to shoes, what about the fit of movie seats, men's ties, and headroom in cars? How do manufacturers decide how wide the seats at your favorite movie theater should be? Or, how long must a tie be so that it is neither too long nor too short for most men? How do we design a car so that most people will not bump their heads on its roof? Or, how high should a computer table be so that it does not put an unnecessary strain on your wrists and arms? In the relatively new science of **ergonomics,** scientists gather data to answer questions such as these and to ensure most people fit comfortably into their environments. Much of the work of these scientists is based on a curve that you will study in this section called the **normal curve,** or **normal distribution.***

The Normal Distribution

KEY POINT

Many different types of data sets are normal distributions.

The normal distribution is the most common distribution in statistics and describes many real-life data sets. The histogram shown in Figure 14.12 will begin to give you an idea of the shape of a normal distribution.

Distributions of such diverse data sets as SAT scores, heights of people, the number of miles before an automobile tire wears out, the size of hamburgers at a fast-food restaurant,

*Recall that a *distribution* is just another name for a set of data.

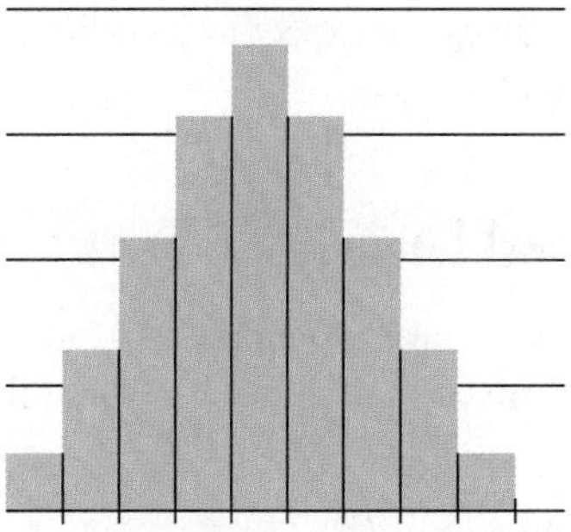

FIGURE 14.12 A normal distribution.

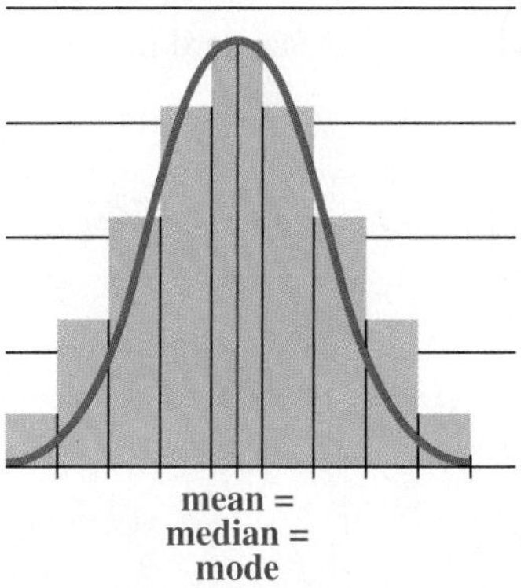

FIGURE 14.13 Smoothing a histogram of a normal distribution into a normal curve.

and the number of hours of use before a certain brand of DVD player breaks down are all examples of normal distributions.

There are several patterns in Figure 14.12 that are common to all normal distributions. First, if we were to take a wire and attach it to the tops of the bars in the histogram and then smooth out the wire to make a curve, the curve would have a bell-shaped appearance, as shown in Figure 14.13. This is why a normal distribution is often called a **bell-shaped curve.**

Second, the mean, median, and mode of a normal distribution are the same. Third, the curve is symmetric with respect to the mean. This means that if you see some pattern in the graph on one side of the mean, then the mean acts as a mirror to give a reflection of that same pattern on the other side of the mean. Fourth, the area under a normal curve equals 1.

In Figure 14.14, we have marked the mean and two points on the normal curve called inflection points. An **inflection point** is a point on the curve where the curve changes from being curved upward to being curved downward, or vice versa. For a normal curve, inflection points are located 1 standard deviation from the mean. Also, because a normal curve is symmetric with respect to the mean, one-half, or 50%, of the area under the curve is located on each side of the mean.

In normal distributions, approximately 68% of the data values occur within 1 standard deviation of the mean, 95% of the data values lie within 2 standard deviations of the mean, and 99.7% of the data values lie within 3 standard deviations of the mean. We refer to these facts as the **68-95-99.7 rule.** Figure 14.14 illustrates the 68-95-99.7 rule. In discussing normal distributions, we usually assume that we are dealing with an *entire population* rather than a *sample*, so in Figure 14.14 we represent the mean by μ and the standard deviation by σ (rather than $\bar{x}$ and s).

We summarize the properties of a normal distribution.

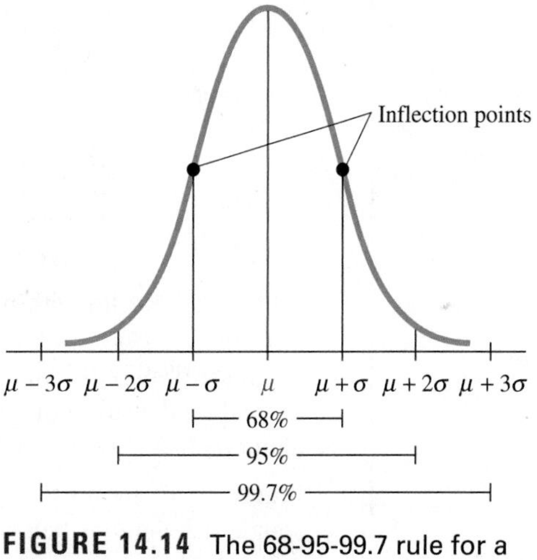

FIGURE 14.14 The 68-95-99.7 rule for a normal distribution.

PROPERTIES OF A NORMAL DISTRIBUTION

1. A normal curve is bell shaped.
2. The highest point on the curve is at the mean of the distribution.
3. The mean, median, and mode of the distribution are the same.
4. The curve is symmetric with respect to its mean.
5. The total area under the curve is 1.
6. Roughly 68% of the data values are within 1 standard deviation from the mean, 95% of the data values are within 2 standard deviations from the mean, and 99.7% of the data values are within 3 standard deviations from the mean.*

*To keep this discussion simple, we are approximating these percentages. Shortly, we will use a table to slightly improve our estimate of the percentage of scores within 1 and 2 standard deviations from the mean.

We can use the 68-95-99.7 rule to estimate how many values we expect to fall within 1, 2, or 3 standard deviations of the mean of a normal distribution.

EXAMPLE 1 The Normal Distribution and Intelligence Tests

Suppose that the distribution of scores of 1,000 students who take a standardized intelligence test is a normal distribution. If the distribution's mean is 450 and its standard deviation is 25,

a) how many scores do we expect to fall between 425 and 475?

b) how many scores do we expect to fall above 500?

SOLUTION: Review the Draw Pictures Strategy on page 3.

a) As you can see in Figure 14.15, the scores 425 and 475 are 1 standard deviation below and above the mean, respectively.

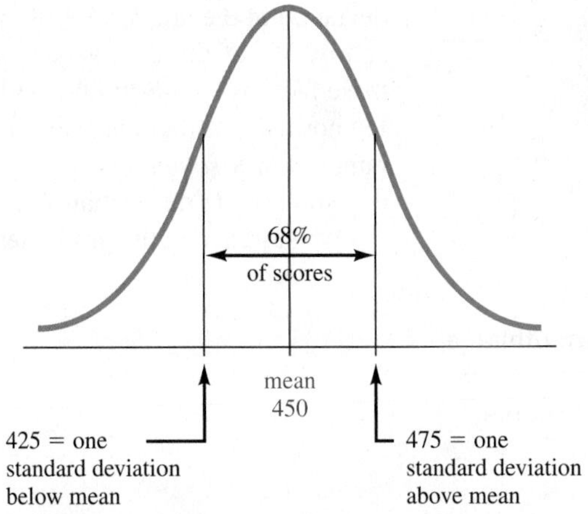

FIGURE 14.15 Sixty-eight percent of the scores lie within 1 standard deviation of the mean.

From the 68-95-99.7 rule, we know that 68%, or about 0.68 of the scores, lie within 1 standard deviation of the mean. Because we have 1,000 scores, we can expect that about $0.68 \times 1{,}000 = 680$ scores are in the range 425 to 475.

b) Figure 14.16 shows that 95% of the scores in a normal distribution fall between 2 standard deviations below the mean and 2 standard deviations above the mean.

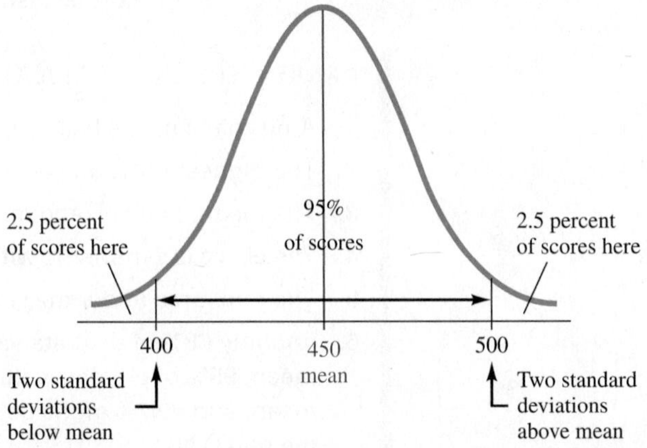

FIGURE 14.16 Five percent of the scores lie more than 2 standard deviations from the mean.

QUIZ YOURSELF 11

Use the information given in Example 1 to determine the following.

a) How many scores do we expect to fall between 400 and 500?

b) How many scores do we expect to fall below 400?

KEY POINT

Areas under a normal curve represent percentages of values in the distribution.

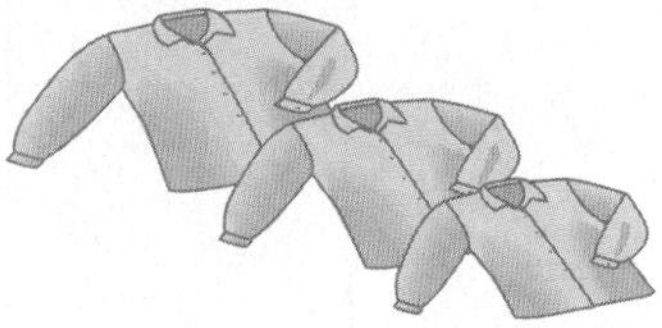

This means that 5%, or 0.05, of the scores lie more than 2 standard deviations above or below the mean. Thus, we can expect to have $0.05 \div 2 = 0.025$ of the scores to be above 500. Again multiplying by 1,000, we can expect $0.025 \times 1{,}000 = 25$ scores to be above 500.

Now try Exercises 1 to 12. 11

z-Scores

In Example 1, we estimated how many values were within 1 standard deviation of the mean. It is natural to ask whether we can predict how many values lie other distances from the mean. For example, knowing that the weights of women between ages 18 and 25 form a normal distribution, a clothing manufacturer may want to know what percentage of the female population is between 1.3 and 2.5 standard deviations above the mean weight for women. It is possible to find this by using Table 14.15, which is the table of areas under the standard normal curve. The standard normal distribution has a mean of 0 and a standard deviation of 1.

Table 14.15 gives the area under this curve between the mean and a number called a *z*-score. A ***z*-score** represents the number of standard deviations a data value is from the mean. In Example 1, we found that for a normal distribution with a mean of 450 and a standard deviation of 25, the value 500 was 2 standard deviations above the mean. Another way of saying this is that the value 500 corresponds to a *z*-score of 2.

Notice that Table 14.15 gives areas only for positive *z*-scores—that is, for data that lie above the mean. We use the symmetry of a normal curve to find areas corresponding to negative *z*-scores.

Example 2 shows how to use Table 14.15. Because the total area under a normal curve is 1, we can interpret areas under the standard normal curve as percentages of the data values in the distribution. Also note that because the mean is 0 and the standard deviation for the standard normal curve is 1, the value of a *z*-score is also the same as the number of standard deviations that *z*-score is from the mean. After you have practiced working with the standard normal distribution, we will show you how to use the techniques that you have learned to solve problems involving real-life data.

Math in Your Life

How Worried Are You About Your Next Test?*

People often argue about how much of your physical, intellectual, and psychological makeup is due to your heredity and how much is due to your environment. Scientists have identified genes that affect characteristics such as height, body fat, blood pressure, and IQ. Surprisingly, even human anxiety has been associated with genetic makeup.

It has been estimated that there are as many as 15 genes that can affect your anxiety level. If you have all these genes, then you have a greater tendency† to be anxious, and if you have none of them, then you tend to be less anxious. If we were to plot a histogram of the number of anxiety genes present in a large number of people, we would find very few people have almost none of these genes, many people have a moderate number of these genes, and very few people have almost all of them. If you were to plot the distribution of anxiety genes present in the population, it would look very much like the normal curves that you are studying in this section.

*This note is based on a series of PowerPoint slides by Dr. Lee Bardwell, professor of genetics at the University of California at Irvine.

†It is believed that 45% of the variance in human anxiety is due to genetic factors. The other 55% is due to other factors.

z	A	z	A	z	A	z	A	z	A	z	A
.00	.000	.56	.212	1.12	.369	1.68	.454	2.24	.488	2.80	.497
.01	.004	.57	.216	1.13	.371	1.69	.455	2.25	.488	2.81	.498
.02	.008	.58	.219	1.14	.373	1.70	.455	2.26	.488	2.82	.498
.03	.012	.59	.222	1.15	.375	1.71	.456	2.27	.488	2.83	.498
.04	.016	.60	.226	1.16	.377	1.72	.457	2.28	.489	2.84	.498
.05	.020	.61	.229	1.17	.379	1.73	.458	2.29	.489	2.85	.498
.06	.024	.62	.232	1.18	.381	1.74	.459	2.30	.489	2.86	.498
.07	.028	.63	.236	1.19	.383	1.75	.460	2.31	.490	2.87	.498
.08	.032	.64	.239	1.20	.385	1.76	.461	2.32	.490	2.88	.498
.09	.036	.65	.242	1.21	.387	1.77	.462	2.33	.490	2.89	.498
.10	.040	.66	.245	1.22	.389	1.78	.463	2.34	.490	2.90	.498
.11	.044	.67	.249	1.23	.391	1.79	.463	2.35	.491	2.91	.498
.12	.048	.68	.252	1.24	.393	1.80	.464	2.36	.491	2.92	.498
.13	.052	.69	.255	1.25	.394	1.81	.465	2.37	.491	2.93	.498
.14	.056	.70	.258	1.26	.396	1.82	.466	2.38	.491	2.94	.498
.15	.060	.71	.261	1.27	.398	1.83	.466	2.39	.492	2.95	.498
.16	.064	.72	.264	1.28	.400	1.84	.467	2.40	.492	2.96	.499
.17	.068	.73	.267	1.29	.402	1.85	.468	2.41	.492	2.97	.499
.18	.071	.74	.270	1.30	.403	1.86	.469	2.42	.492	2.98	.499
.19	.075	.75	.273	1.31	.405	1.87	.469	2.43	.493	2.99	.499
.20	.079	.76	.276	1.32	.407	1.88	.470	2.44	.493	3.00	.499
.21	.083	.77	.279	1.33	.408	1.89	.471	2.45	.493	3.01	.499
.22	.087	.78	.282	1.34	.410	1.90	.471	2.46	.493	3.02	.499
.23	.091	.79	.285	1.35	.412	1.91	.472	2.47	.493	3.03	.499
.24	.095	.80	.288	1.36	.413	1.92	.473	2.48	.493	3.04	.499
.25	.099	.81	.291	1.37	.415	1.93	.473	2.49	.494	3.05	.499
.26	.103	.82	.294	1.38	.416	1.94	.474	2.50	.494	3.06	.499
.27	.106	.83	.297	1.39	.418	1.95	.474	2.51	.494	3.07	.499
.28	.110	.84	.300	1.40	.419	1.96	.475	2.52	.494	3.08	.499
.29	.114	.85	.302	1.41	.421	1.97	.476	2.53	.494	3.09	.499
.30	.118	.86	.305	1.42	.422	1.98	.476	2.54	.495	3.10	.499
.31	.122	.87	.308	1.43	.424	1.99	.477	2.55	.495	3.11	.499
.32	.126	.88	.311	1.44	.425	2.00	.477	2.56	.495	3.12	.499
.33	.129	.89	.313	1.45	.427	2.01	.478	2.57	.495	3.13	.499
.34	.133	.90	.316	1.46	.428	2.02	.478	2.58	.495	3.14	.499
.35	.137	.91	.319	1.47	.429	2.03	.479	2.59	.495	3.15	.499
.36	.141	.92	.321	1.48	.431	2.04	.479	2.60	.495	3.16	.499
.37	.144	.93	.324	1.49	.432	2.05	.480	2.61	.496	3.17	.499
.38	.148	.94	.326	1.50	.433	2.06	.480	2.62	.496	3.18	.499
.39	.152	.95	.329	1.51	.435	2.07	.481	2.63	.496	3.19	.499
.40	.155	.96	.332	1.52	.436	2.08	.481	2.64	.496	3.20	.499
.41	.159	.97	.334	1.53	.437	2.09	.482	2.65	.496	3.21	.499
.42	.163	.98	.337	1.54	.438	2.10	.482	2.66	.496	3.22	.499
.43	.166	.99	.339	1.55	.439	2.11	.483	2.67	.496	3.23	.499
.44	.170	1.00	.341	1.56	.441	2.12	.483	2.68	.496	3.24	.499
.45	.174	1.01	.344	1.57	.442	2.13	.483	2.69	.496	3.25	.499
.46	.177	1.02	.346	1.58	.443	2.14	.484	2.70	.497	3.26	.499
.47	.181	1.03	.349	1.59	.444	2.15	.484	2.71	.497	3.27	.500
.48	.184	1.04	.351	1.60	.445	2.16	.485	2.72	.497	3.28	.500
.49	.188	1.05	.353	1.61	.446	2.17	.485	2.73	.497	3.29	.500
.50	.192	1.06	.355	1.62	.447	2.18	.485	2.74	.497	3.30	.500
.51	.195	1.07	.358	1.63	.449	2.19	.486	2.75	.497	3.31	.500
.52	.199	1.08	.360	1.64	.450	2.20	.486	2.76	.497	3.32	.500
.53	.202	1.09	.362	1.65	.451	2.21	.487	2.77	.497	3.33	.500
.54	.205	1.10	.364	1.66	.452	2.22	.487	2.78	.497		
.55	.209	1.11	.367	1.67	.453	2.23	.487	2.79	.497		

TABLE 14.15 Standard normal distribution.

PROBLEM SOLVING

Strategy: The Three-Way Principle

The Three-Way Principle in Section 1.1 suggests that you can often understand a problem graphically. This is certainly true in solving problems involving the normal curve. If you draw a good picture of the situation, it often becomes more clear to you as to what to do to set up and solve the problem.

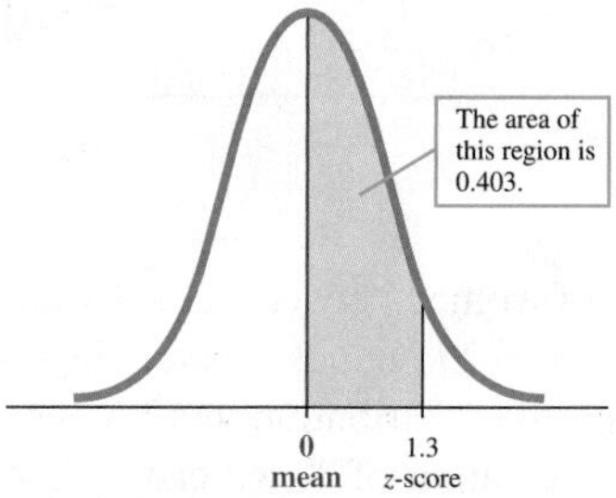

FIGURE 14.17 Area under the standard normal curve between $z = 0$ and $z = 1.3$.

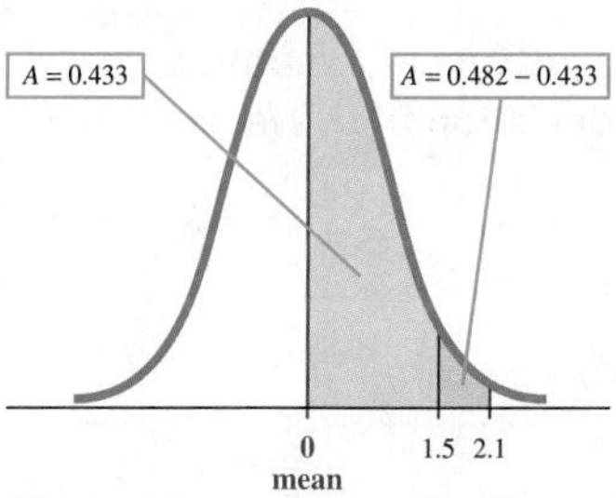

FIGURE 14.18 Finding the area under the standard normal curve between $z = 1.5$ and $z = 2.1$.

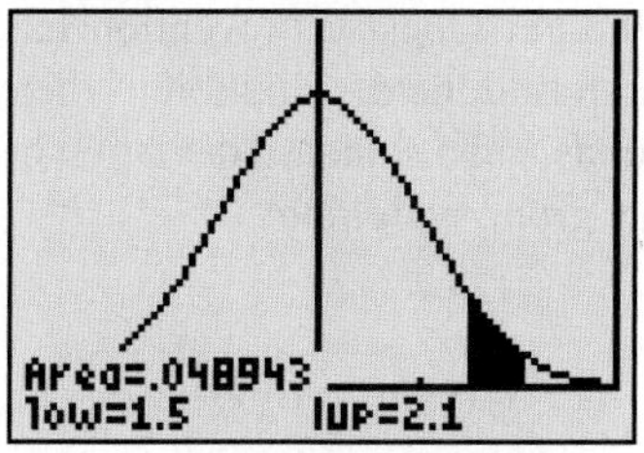

TI screen showing area under standard normal curve between $z = 1.5$ and $z = 2.1$.

EXAMPLE 2 Finding Areas under the Standard Normal Curve

Use Table 14.15 to find the percentage of the data (area under the curve) that lies in the following regions for a standard normal distribution:

a) between $z = 0$ and $z = 1.3$

b) between $z = 1.5$ and $z = 2.1$

c) between $z = 0$ and $z = -1.83$

SOLUTION: Review the Draw Pictures Strategy on page 3.

a) The area under the curve between $z = 0$ and $z = 1.3$ is shown in Figure 14.17. We find this area by looking in Table 14.15 for the z-score 1.30. We see that A is 0.403 when $z = 1.30$. Thus, we expect 0.403, or 40.3%, of the data to fall between 0 and 1.3 standard deviations above the mean.

b) Figure 14.18 shows the area we want. Areas in Table 14.15 correspond to regions from $z = 0$ to the given z-score. Therefore, to find the area under the curve between $z = 1.5$ and $z = 2.1$, we must first find the area from $z = 0$ to $z = 2.1$ and then subtract the area from $z = 0$ to $z = 1.5$. From Table 14.15, we see that when $z = 2.1$, $A = 0.482$, and when $z = 1.5$, $A = 0.433$. To finish this problem, we can set up our calculations as follows:

larger area from $z = 0$ to $z = 2.1$	0.482	subtract areas, not z-scores
− smaller area from $z = 0$ to $z = 1.5$	−0.433	
	0.049	— area we want

This means that in the standard normal distribution, the area under the curve between $z = 1.5$ and $z = 2.1$ is 0.049, or 4.9%.

c) Due to the symmetry of the normal distribution, the area between $z = 0$ and $z = -1.83$ is the same as the area between $z = 0$ and $z = 1.83$ (Figure 14.19). From Table 14.15 we see that when $z = 1.83$, $A = 0.466$. Therefore, 46.6% of the data values lie between 0 and −1.83.

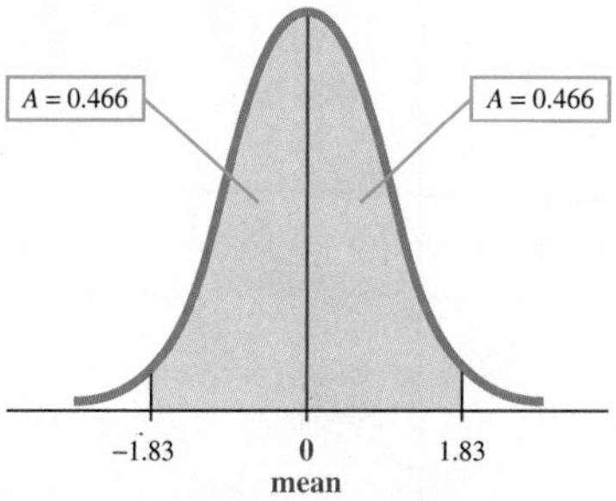

FIGURE 14.19 Finding the area under the standard normal curve between $z = -1.83$ and $z = 0$.

Now try Exercises 15 to 34. 12

QUIZ YOURSELF 12

Use Table 14.15 to find the following areas under the standard normal curve:

a) between $z = 0$ and $z = 1.45$

b) between $z = 1.23$ and $z = 1.85$

c) between $z = 0$ and $z = -1.35$

SOME GOOD ADVICE

A common mistake in solving a problem such as Example 2(b) is to subtract 1.5 from 2.1 to get 0.6 and then wrongly use this for the z-score in Table 14.15. If you think about the graph of the standard normal curve, clearly the area between $z = 0$ and $z = 0.6$ is not the same as the area between $z = 1.5$ and $z = 2.1$.

Converting Raw Scores to z-Scores

KEY POINT

We convert values in a nonstandard normal distribution to z-scores.

A real-life normal distribution, such as the set of all weights of women between ages 18 and 25, may have a mean of 120 pounds and a standard deviation of 25 pounds. Such a distribution will have the properties that we stated earlier for a normal distribution, but because the distribution does not have a mean of 0 and a standard deviation of 1, we cannot use Table 14.15 directly, as we did in Example 2. We *can* use Table 14.15, however, if we first convert nonstandard values, called *raw scores*, to z-scores. The following formula shows the desired relationship between values in a nonstandard normal distribution and z-scores.

FORMULA FOR CONVERTING RAW SCORES TO z-SCORES Assume a normal distribution has a mean of μ and a standard deviation of σ. We use the equation

$$z = \frac{x - \mu}{\sigma}$$

to convert a value x in the nonstandard distribution to a z-score.

You can also use this same formula to convert z-scores into raw scores as in Exercises 41 to 46. Simply substitute for z, μ, and σ and then solve for x.

EXAMPLE 3 Converting Raw Scores to z-scores

In an effort to combat dog obesity, The Association for Pet Obesity Prevention has weighed a group of Basenjis (African barkless dogs) and found the distribution of their weights to be normally distributed with a mean of 20 pounds and a standard deviation of 3 pounds. Find the corresponding z-score for a dog in this group weighing:

a) 25 pounds b) 16 pounds.

SOLUTION:

a) Figure 14.20 gives us a picture of this situation. We will use the conversion formula $z = \frac{x - \mu}{\sigma}$, with the raw score $x = 25$, the mean $\mu = 20$, and the standard deviation

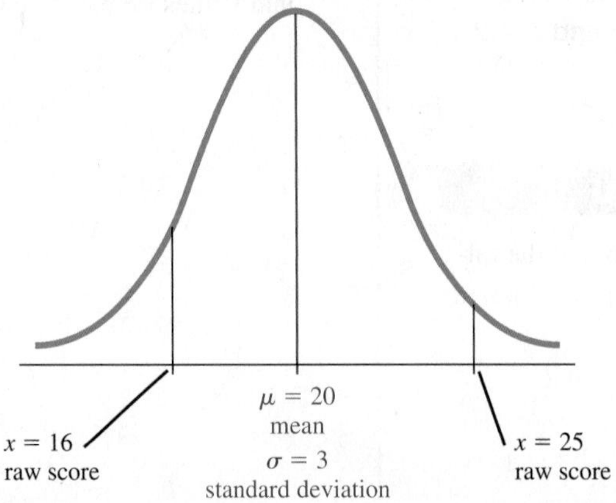

FIGURE 14.20 Normal distribution with mean of 20 and standard deviation of 3.

QUIZ YOURSELF 13

Suppose a normal distribution has a mean of 50 and a standard deviation of 7. Convert each raw score to a z-score.

a) 59 b) 50 c) 38

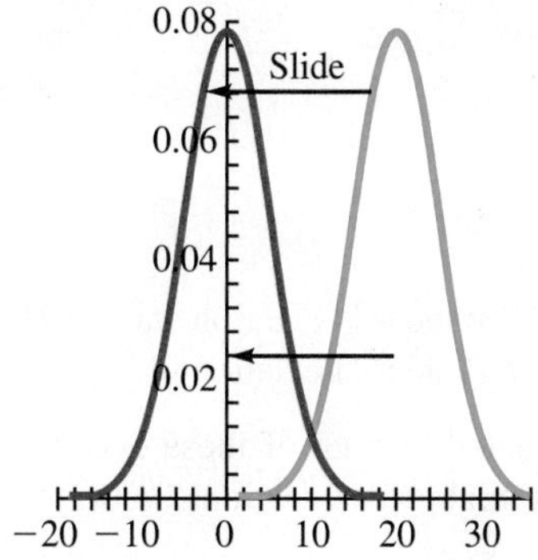

FIGURE 14.21 (a) First slide the distribution 20 units to the left to get a mean of 0.

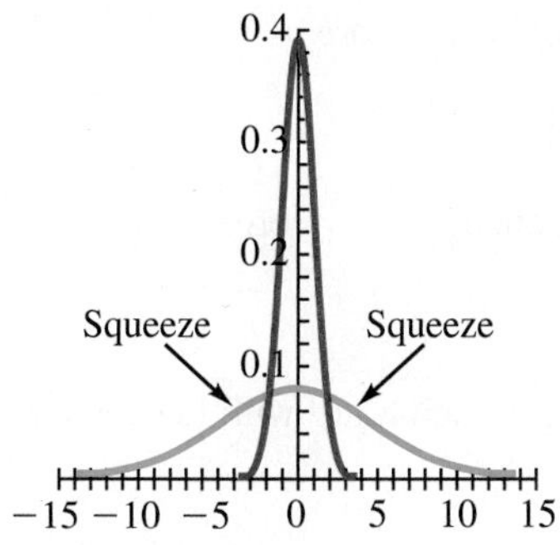

FIGURE 14.21 (b) Then squeeze the distribution to get a standard deviation of 1.

$\sigma = 3$. Making the substitutions, we have, $z = \frac{25 - 20}{3} = \frac{5}{3} = 1.67$. We can interpret this result as telling us that in this distribution, 25 is 1.67 standard deviations above the mean.

b) We will use Figure 14.20 and the conversion formula again, but this time with $x = 16$, $\mu = 20$, and $\sigma = 3$. Therefore, the corresponding z-score is $z = \frac{16 - 20}{3} = \frac{-4}{3} = -1.33$. This tells us that 16 is 1.33 standard deviations below the mean.

Now try Exercises 35 to 40. 13

The Three-Way Principle in Section 1.1 suggests that you can remember the formula for converting raw scores to z-scores by interpreting the formula geometrically. In Example 3, we had a normal distribution with a mean of 20 and standard deviation 3. In order to make the mean of 20 in the nonstandard distribution correspond to the mean of 0 in the standard normal distribution, you can think of "sliding" the distribution to the left 20 units, as we show in Figure 14.21(a). That is why we compute an expression of the form $x - 20$ in the numerator of the z-score formula. Because the standard deviation was 3, we divided $x - 20$ by 3 to "narrow" the distribution so that it would have a standard deviation of 1, as you can see in Figure 14.21(b), and therefore allow us to use Table 14.15.

Applications

Normal distributions have many practical applications.

EXAMPLE 4 Interpreting the Significance of an Exam Score

Suppose that to qualify for a management training program offered by your employer you must score in the top 10% of those employees who take a standardized test. Assume that the distribution of scores is normal and you received a score of 72 on the test, which had a mean of 65 and a standard deviation of 4. What percentage of those who took this test had a score below yours?

SOLUTION:

We will find the number of standard deviations your score is above the mean and then determine what percentage of scores fall below this number. First, we calculate the z-score, which corresponds to your 72:

$$z = \frac{72 - 65}{4} = \frac{7}{4} = 1.75$$

Now, looking in Table 14.15 for $z = 1.75$, we find that $A = 0.460$. Therefore, 46% of the scores fall between the mean and your score (Figure 14.22).

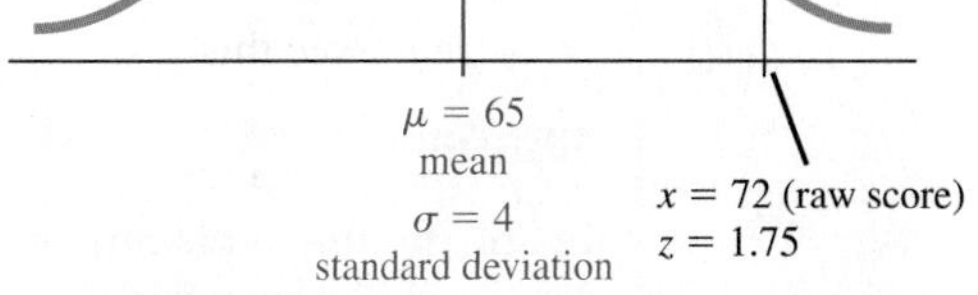

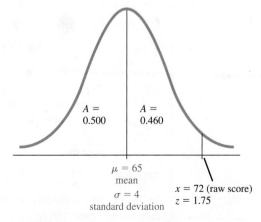

FIGURE 14.22 Normal distribution with mean of 65 and standard deviation of 4.

However, we must not forget the scores that fall below the mean. Because a normal curve is symmetric, another 50% of the scores fall below the mean. So, there are 50% + 46% = 96% of the scores below yours. Congratulations! You qualify for the program.

Baseball fans often argue about who the best players of all time were. Our next example shows one way we can use statistics to compare outstanding players from different eras.

EXAMPLE 5 Using *z*-Scores to Compare Data

Consider the following information:

Ty Cobb hit .420 in 1911.
Ted Williams hit .406 in 1941.

In the 1910s, the mean batting average was .266 and the standard deviation was .0371.
In the 1940s, the mean batting average was .267 and the standard deviation was .0326.

Assume that the batting averages were normally distributed in both of these decades. Use *z*-scores to determine which batter was ranked higher in relationship to his contemporaries.

SOLUTION:

We will convert each of the two batting averages to a corresponding *z*-score for the distribution of batting averages for the decade in which the athlete played.

In the 1910s, Ty Cobb's average of .420 corresponded to a *z*-score of

$$\frac{.420 - .266}{.0371} = \frac{.154}{.0371} = 4.1509.$$

In the 1940s, Ted Williams's average of .406 corresponded to a *z*-score of

$$\frac{.406 - .267}{.0326} = \frac{.139}{.0326} = 4.2638.$$

We see that in using the standard deviation to compare each hitter with his contemporaries, Ted Williams was ranked as the better hitter.

Now try Exercises 69 and 70.

We conclude our discussion of normal distributions with the example we promised in the chapter opener, that is, explaining how manufacturers can use statistics to write effective warranties for their products.

EXAMPLE 6 Using the Normal Distribution to Write Warranties

To increase sales, Panasonic plans to offer a warranty on a new Viera Smart TV. In testing the TV, quality control engineers found that it has a mean time to failure of 3,000 hours with a standard deviation of 500 hours. Assume that the typical purchaser will use the TV for 6 hours per day. If the manufacturer does not want more than 5% to be returned as defective within the warranty period, how long should the warranty period be to guarantee this?

SOLUTION:

In solving this problem, we first work with the standard normal distribution and then convert the answer to fit our nonstandard normal distribution. Figure 14.23 gives a picture of the situation for the standard normal curve.

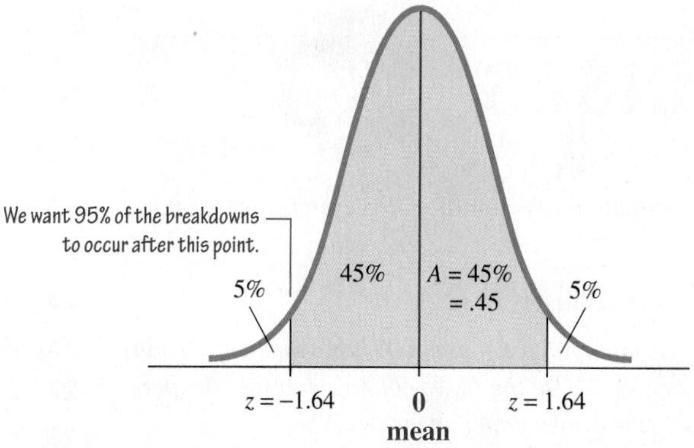

FIGURE 14.23 Finding the area under the standard normal curve for a negative *z*-score.

In Figure 14.23, we see that we need to find a z-score such that at least 95% of the entire area is *beyond* this point. Observe that this score is to the left of the mean and is negative. Therefore, we use symmetry and find the z-score such that 95% of the entire area is *below* this score. However, Table 14.15 gives the areas only for the upper half of the distribution. This is not a problem, though, because we know that 50% of the entire area lies below the mean. Therefore, our problem reduces to finding a z-score greater than 0 such that 45% of the area lies between the mean and that z-score.

In Table 14.15, we see that if $A = 0.450$, the corresponding z-score is 1.64. This means that 95% of the area underneath the standard normal curve falls below $z = 1.64$. By symmetry, we also can say that 95% of the values lie above -1.64. We must now interpret this z-score in terms of the original normal distribution of raw scores.

Recall the equation that relates values in a distribution to z-scores, namely,

$$z = \frac{x - \mu}{\sigma}. \tag{1}$$

Previously we used this equation to convert raw scores to z-scores. We now use the equation in answering the reverse question. Specifically, what is x when $\mu = 3{,}000$, $\sigma = 500$, and $z = -1.64$? Substituting these numbers in equation (1), we get

$$-1.64 = \frac{x - 3{,}000}{500}. \tag{2}$$

We solve for x in this equation by multiplying both sides by 500:

$$-1.64 \cdot (500) = \frac{x - 3{,}000}{500} \cdot (500)$$

QUIZ YOURSELF 14

Redo Example 6, but now assume that the mean time to failure is 4,200 hours, with a standard deviation of 600, and the manufacturer wants no more than 2% to be returned as defective within the warranty period.

Simplifying the equation, we get

$$-820 = x - 3{,}000.$$

Last, we add 3,000 to both sides to obtain $x = 2{,}180$ hours. We expect 95% of the breakdowns to occur beyond 2,180 hours.

Because we know that owners will use the TV about 6 hours per day, we divide 2,180 by 6 to get $\frac{2{,}180}{6} = 363.3$ days. Therefore, if the manufacturer wants to have no more than 5% of the breakdowns occur within the warranty period, the warranty should be for one year. 14

EXERCISES 14.4

Assume that all distributions mentioned in this exercise set are normal distributions.

Sharpening Your Skills

Assume that the distribution in Exercises 1–6 has a mean of 10 and a standard deviation of 2. Use the 68-95-99.7 rule to find the percentage of values in the distribution that we describe.

1. Between 10 and 12
2. Between 12 and 14
3. Above 14
4. Below 8
5. Above 12
6. Below 10

Assume that the distribution in Exercises 7–12 has a mean of 12 and a standard deviation of 3. Use the 68-95-99.7 rule to find the percentage of values in the distribution that we describe.

7. Below 9
8. Between 12 and 15
9. Above 6
10. Below 12
11. Between 15 and 18
12. Above 18

Use the following graph of the standard normal distribution and Table 14.15 to do Exercises 13 and 14.

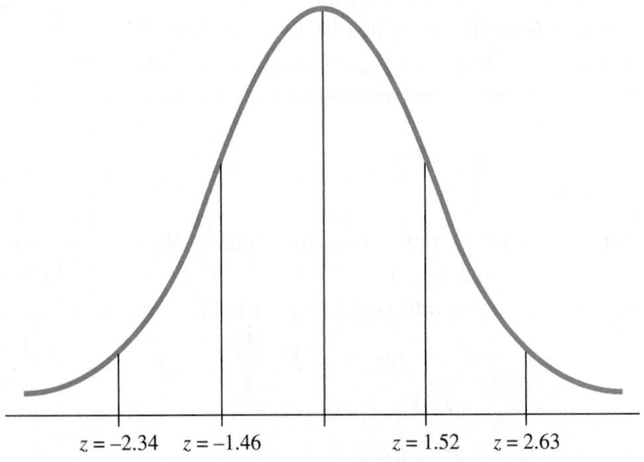

13. Find the areas indicated by the given z-scores:

a. between 0 and 1.52
b. between 0 and 2.63
c. between 1.52 and 2.63
d. above 2.63

14. Find the areas indicated by the given z-scores:

a. between 0 and -1.46
b. between 0 and -2.34
c. between -2.34 and -1.46
d. below -2.34

Use Table 14.15 to find the percentage of area under the standard normal curve that is specified.

15. Between $z = 0$ and $z = 1.23$
16. Between $z = 0$ and $z = 2.06$
17. Between $z = 0$ and $z = -0.75$
18. Between $z = 0$ and $z = -1.35$
19. Between $z = 1.25$ and $z = 1.95$
20. Between $z = 0.37$ and $z = 1.23$
21. Between $z = -0.38$ and $z = -0.76$
22. Between $z = -1.55$ and $z = -2.13$
23. Above $z = 1.45$
24. Below $z = -0.64$
25. Below $z = -1.40$
26. Above $z = 0.78$
27. Below $z = 1.33$
28. Above $z = -0.46$

29. Find a z-score such that 10% of the area under the standard normal curve is above that score.

30. Find a z-score such that 20% of the area under the standard normal curve is above that score.

31. Find a z-score such that 12% of the area under the standard normal curve is below that score.

32. Find a z-score such that 24% of the area under the standard normal curve is below that score.

33. Find a z-score such that 60% of the area is below that score.

34. Find a z-score such that 90% of the area is below that score.

In Exercises 35–40, we give you a mean, a standard deviation, and a raw score. Find the corresponding z-score.

35. Mean 80, standard deviation 5, $x = 87$
36. Mean 100, standard deviation 15, $x = 117$
37. Mean 21, standard deviation 4, $x = 14$
38. Mean 52, standard deviation 7.5, $x = 61$
39. Mean 38, standard deviation 10.3, $x = 48$
40. Mean 8, standard deviation 2.4, $x = 6.2$

In Exercises 41–46, we give you a mean, a standard deviation, and a z-score. Find the corresponding raw score.

41. Mean 60, standard deviation 5, $z = 0.84$
42. Mean 20, standard deviation 6, $z = 1.32$
43. Mean 35, standard deviation 3, $z = -0.45$
44. Mean 62, standard deviation 7.5, $z = -1.40$
45. Mean 28, standard deviation 2.25, $z = 1.64$
46. Mean 8, standard deviation 3.5, $z = -1.25$

Applying What You've Learned

Due to random variations in the operation of an automatic coffee machine, not every cup is filled with the same amount of coffee. Assume that the mean amount of coffee dispensed is 8 ounces and the standard deviation is 0.5 ounce. Use Figure 14.14 to solve Exercises 47 and 48.

47. Analyzing vending machines.

a. What percentage of the cups should have at least 8 ounces of coffee?

b. What percentage of the cups should have less than 7.5 ounces of coffee?

48. Analyzing vending machines.

a. What percentage of the cups should have at least 8.5 ounces of coffee?

b. What percentage of the cups should have less than 8 ounces of coffee?

A machine fills bags of candy. Due to slight irregularities in the operation of the machine, not every bag gets exactly the same number of pieces. Assume that the number of pieces per bag has a mean of 200 with a standard deviation of 2. Use Figure 14.14 to solve Exercises 49 and 50.

49. Analyzing vending machines.

a. How many of the bags do we expect to have between 200 and 202 pieces in them?

b. How many of the bags do we expect to have at least 202 pieces in them?

50. Analyzing vending machines.

a. How many of the bags do we expect to have between 198 and 200 pieces in them?

b. How many of the bags do we expect to have at least 196 pieces in them?

51. Product reliability. Suppose that for a particular brand of LCD television set, the distribution of failures has a mean of 120 months with a standard deviation of 12 months. With respect to this distribution, 140 months corresponds to what z-score?

52. Product reliability. Redo Exercise 51 using 130 months instead of 140.

53. Distribution of heights. Assume that in the NBA the distribution of heights has a mean of 6 feet, 8 inches, and a standard deviation of 3 inches. If there are 324 players in the league, how many players do you expect to be over 7 feet tall?

54. Distribution of heights. Redo Exercise 53, but now we want to know how many players you would expect to be over 6 feet, 6 inches.

55. Distribution of heart rates. Assume that the distribution of 21-year-old women's heart rates at rest is 68 beats per minute with a standard deviation of 4 beats per minute. If 200 women are examined, how many would you expect to have a heart rate of less than 70?

56. Distribution of heart rates. Redo Exercise 55, but now we ask how many you would expect to have a heart rate less than 75.

In Exercises 57 and 58, we assume that the mean length of a human pregnancy is 268 days.

57. Birth statistics. If 95% of all human pregnancies last between 250 and 286 days, what is the standard deviation of the distribution of lengths of human pregnancies? What percentage of human pregnancies do we expect to last at least 275 days?

58. Birth statistics. Babies born before the 37th week (that is, before day 252 of a pregnancy) are considered premature. What percentage of births do we expect to be premature? (See Exercise 57.)

59. Analyzing Internet use. Comcast has analyzed the amount of time that its customers are online per session. It found that the distribution of connect times has a mean of 37 minutes and a standard deviation of 11 minutes. For this distribution, find the raw score that corresponds to a z-score of 1.5.

60. Analyzing customer service. The distribution of times that customers spend waiting in line in a supermarket has a mean of 3.6 minutes and a standard deviation of 1.2 minutes. For this distribution, find the raw score that corresponds to a z-score of -1.3.

61. Assigning grades. Assume that final grades in your math class have a mean of 72 and a standard deviation of 8. If your professor plans to give an A to the top 10% of the class, what is the cutoff for an A?

62. Assigning grades. Continuing the situation in Exercise 61, if the bottom 15% of class will be given an F, what is the cutoff for an F?

63. Cross country racing. If the distribution of race times for a cross country race has a mean of 85 minutes and a standard deviation of 9 minutes, what is the cutoff time for a runner to finish in the top 20% of the runners?

64. Weight lifting. In a power lifting competition the distribution of total weight lifted has a mean of 1,100 pounds with a standard deviation of 20 pounds. What is the cutoff of total weight lifted for a competitor to finish in the bottom 30% of the competition?

65. Strength of cables. A certain type of cable has a mean breaking point of 150 pounds with a standard deviation of 8 pounds. What weight should we specify so that we expect 95% of the cables not to break supporting that weight?

66. Strength of cables. Repeat Exercise 65 if the mean is 180 pounds, the standard deviation is 11 pounds, and we expect 90% of the cables not to break.

67. Rainiest city. According to MSNBC.com, the rainiest city in the United States is Mobile, Alabama, with a mean annual rainfall of 67 inches. Assume that the standard deviation for the distribution of annual rainfall is 10 inches. In a 20-year period, how many years would we expect to have a rainfall in Mobile over 72 inches?

68. Rainiest city. Continuing the situation in Exercise 67, in a 20-year period, how many years would we expect the rainfall in Mobile to be less than 61 inches?

In Exercises 69 and 70, use the information and methods of Example 5 to make the indicated comparisons.

69. Comparing athletes. In 1949, Jackie Robinson hit .342 for the Brooklyn Dodgers; in 1973, Rod Carew hit .350 for the Minnesota Twins. In the 1970s the mean batting average was .261 and the standard deviation was .0317. Determine which batting average was more impressive.

70. Comparing athletes. In 1940, Joe DiMaggio hit .352 for the New York Yankees; in 1975, Bill Madlock hit .354 for the Pittsburgh Pirates. In the 1970s the mean batting average was .261 and the standard deviation was .0317. Determine which batting average was more impressive.

71. Writing a warranty. A manufacturer plans to provide a warranty on Wii Fit. In testing, the manufacturer has found that the set of failure times for Wii Fit is normally distributed with a mean time of 2,000 hours and a standard deviation of 800 hours. Assume that the typical purchaser will use Wii Fit for 2 hours per day. If the manufacturer does not want more than 4% returned as defective within the warranty period, how long should the warranty period be to guarantee this? (Assume that there are 31 days in each month.)

72. Writing a warranty. Redo Exercise 71, but now assume that the mean of the distribution of failure times is 2,500 hours, the standard deviation is 600 hours, and the manufacturer wants no more than 2.5% returned during the warranty period.

73. Analyzing investments. A certain mutual fund over the last 15 years has had a mean yearly return of 7.8% with a standard deviation of 1.3%.

a. If you had invested in this fund over the past 15 years, in how many of these years would you expect to have earned at least 9% on your investment?

b. In how many years would you expect to have earned less than 6%?

74. Analyzing investments. A certain bond fund over the last 12 years has had a mean yearly return of 5.9% with a standard deviation of 1.8%.

a. If you had invested in this fund over the past 12 years, in how many of these years would you expect to have earned at least 8% on your investment?

b. In how many years would you expect to have earned less than 4%?

75. Analyzing the SATs. Assume that the math SAT scores are normally distributed with a mean of 500 and a standard deviation of 100. If you score 480 on this exam, what percentage of those taking the test scored below you?

76. Analyzing the SATs. Assume that the verbal SAT scores are normally distributed with a mean of 500 and a standard deviation of 100. If you score 520 on this exam, what percentage of those taking the test scored below you?

Communicating Mathematics

77. a. If a raw score corresponds to a z-score of 1.75, what does that tell you about that score in relationship to the mean of the distribution?

b. What if the raw score corresponds to a z-score of -0.85?

78. Without looking at Table 14.15, can you determine whether the area under the standard normal curve between $z = 0.5$ and $z = 1.0$ is greater than or less than the area between $z = 1.5$ and $z = 2.0$? Explain.

79. Explain why, if you want to compute the area under a normal curve between $z = 1.3$ and $z = 2.0$, it is not correct to subtract $2.0 - 1.3 = 0.7$ and then use the z-score of 0.7 to find the area.

80. Explain how you would estimate the mean and the standard deviation of the two normal distributions in the given figure.

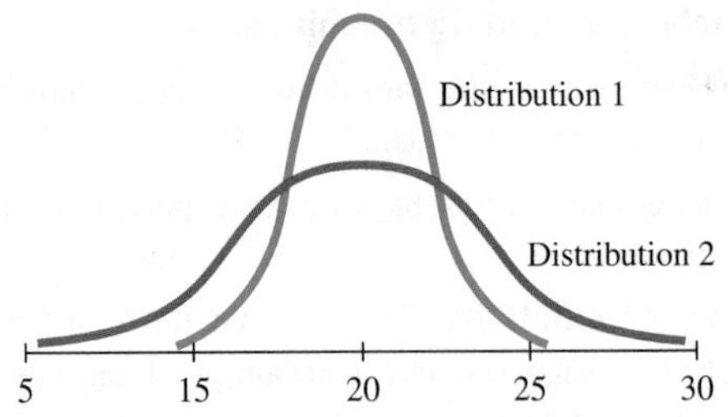

Challenge Yourself

81. If you had a large collection of data, how might you go about determining whether the distribution was normal?

82. Discuss whether you feel that the method we used to compare Ty Cobb and Ted Williams would be a valid way to compare the great African American athletes who played in the Negro leagues with the great white athletes of the same era.

The n*th percentile in a distribution is a number* x *such that* n*% of the values in that distribution fall below* x. *For example, 20% of the values in the distribution fall below the 20th percentile.*

83. What z-score in the standard normal distribution is the 80th percentile?

84. What z-score in the standard normal distribution is the 30th percentile?

85. If a distribution has a mean of 40 and a standard deviation of 4, what is the 75th percentile?

86. If a distribution has a mean of 80 and a standard deviation of 6, what is the 35th percentile?

87. Finding percentiles. If you score a 75 on a commercial pilot's exam that has a mean score of 68 and a standard deviation of 4, to what percentile does your score correspond?

88. Finding percentiles. If you know that the mean salary for your profession is \$53,000 with a standard deviation of \$2,500, to what percentile does your salary of \$57,000 correspond?

Looking Deeper

14.5 Linear Correlation

Objectives

1. Be able to construct a scatterplot to show the relationship between two variables.
2. Understand the properties of the linear correlation coefficient.
3. Use linear regression to find the line of best fit for a set of data points.

How can we tell whether two sets of data are related? For example, is there a relationship between

- the amount of dietary fat eaten and the number of pounds a person is overweight?
- time spent at tutoring sessions and grades?
- the amount of cell phone use and cancer?

In each case, we seek some connection between the first quantity and the second. We say there is a **correlation** between two variables if one of them is related to the other in some way—for example, "Is there a correlation between the amount of dietary fat and the number of pounds a person is overweight?" Although there are many different types of correlation, later in this section we will discuss one particular type called *linear correlation.*

Scatterplots

In order to determine whether there is a correlation between two variables, we obtain pairs of data, called *data points*, relating the first variable to the second. For example, we might examine 100 patients and relate their dietary fat to the number of pounds they are overweight. To understand such data, we plot data points in a graph called a **scatterplot.**

EXAMPLE 1 Constructing a Scatterplot

An instructor wants to know whether there is a correlation between the number of times students attended tutoring sessions during the semester and their grades on a 50-point examination. The instructor has collected data for 10 students, as shown in Table 14.16. Represent these data points by a scatterplot and interpret the graph.

Number of Tutoring Sessions, x	18	6	16	14	0	4	0	10	12	18
Exam Grade, y	42	31	46	41	25	38	28	39	42	44

TABLE 14.16 Tutoring sessions attended and exam grades.

SOLUTION:

We plot the points (18, 42), (6, 31), (16, 46), and so on in Figure 14.24. The scatterplot shows that as the number of tutoring sessions increases, the grades also generally increase. Based on this scatterplot, the instructor might conclude that there is some relationship between the number of study sessions students attend and their grades. 15

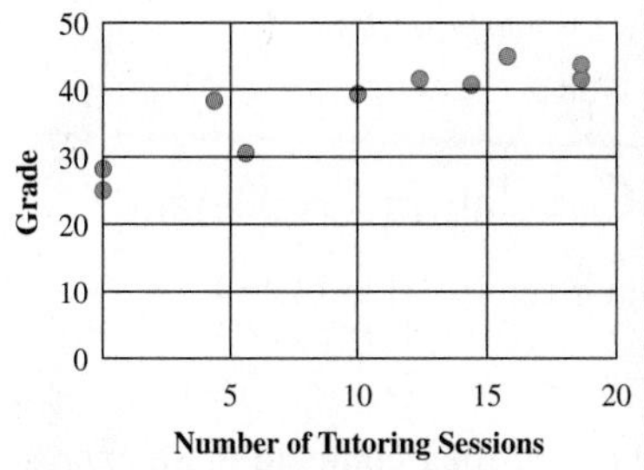

FIGURE 14.24 Scatterplot of data in Table 14.16.

QUIZ YOURSELF 15

Draw a scatterplot of the data shown in Table 14.17 and interpret the graph.

x	5	16	14	0	10	20	3	2
y	39	20	29	41	30	15	42	44

TABLE 14.17 Paired data for variables x and y.

In Example 1, we must be careful not to conclude that there is a cause-and-effect relationship between the number of tutoring sessions attended and the students' grades. There are probably many factors that affect the students' grades, such as class attendance, motivation, and the amount of time spent doing homework. Statisticians often say, "Correlation does not imply causation." Based on the scatterplot, there seems to be some kind of relationship between the two variables; however, at this point we cannot say more than this.

Linear Correlation

We will now discuss one type of correlation called *linear correlation.* **Linear correlation** exists between two variables if, when we graph them in a scatterplot, the points in the graph tend to lie in a straight line. Although this definition may seem vague, there is a number called the *linear correlation coefficient* that allows us to compute to what degree the points of a scatterplot lie along a straight line. The derivation of this number is beyond the scope of this text, but we can explain its meaning and show you how to use it.

FORMULA FOR COMPUTING THE LINEAR CORRELATION COEFFICIENT If we have pairs of data for two variables x and y, the linear correlation coefficient* for these data, denoted by r, is given by the formula

$$r = \frac{n\sum xy - (\sum x)(\sum y)}{\sqrt{n(\sum x^2) - (\sum x)^2}\sqrt{n(\sum y^2) - (\sum y)^2}} \qquad (1)$$

The number n is the number of pairs of data we have for x and y.

We will first show you how to compute the linear correlation coefficient and we will then explain how to interpret its meaning.

EXAMPLE 2 Computing the Linear Correlation Coefficient

Compute the linear correlation coefficient for the data given in Table 14.16.

SOLUTION:

We will first describe each expression in equation (1) and then compute these expressions in Table 14.18.

*In this section, we will assume that we are dealing with samples. In this case, the linear correlation coefficient is sometimes called the *sample* linear correlation coefficient.

Σx is the sum of the first coordinates of the data pairs.

Σy is the sum of the second coordinates of the data pairs.

Σx^2 is the sum of the squares of the first coordinates of the data pairs.

Σy^2 is the sum of the squares of the second coordinates of the data pairs.

Σxy is the sum of the products of the first and second coordinates of the data pairs.

x	y	x^2	y^2	xy
18	42	324	1,764	756
6	31	36	961	186
16	46	256	2,116	736
14	41	196	1,681	574
0	25	0	625	0
4	38	16	1,444	152
0	28	0	784	0
10	39	100	1,521	390
12	42	144	1,764	504
18	44	324	1,936	792
$\Sigma x = 98$	$\Sigma y = 376$	$\Sigma x^2 = 1{,}396$	$\Sigma y^2 = 14{,}596$	$\Sigma xy = 4{,}090$

TABLE 14.18 Computations to calculate the correlation coefficient.

The linear correlation coefficient is therefore

$$r = \frac{n\sum xy - (\sum x)(\sum y)}{\sqrt{n(\sum x^2) - (\sum x)^2}\sqrt{n(\sum y^2) - (\sum y)^2}}$$

$$= \frac{10 \cdot 4{,}090 - (98)(376)}{\sqrt{10 \cdot (1{,}396) - 98^2}\sqrt{10 \cdot (14{,}596) - 376^2}}$$

$$= \frac{4{,}052}{\sqrt{4{,}356}\sqrt{4{,}584}} \approx 0.9068.$$

Now try Exercises 3 to 6. 16

QUIZ YOURSELF 16

Compute the linear correlation coefficient for the data shown in Table 14.17.

We will now investigate the meaning of the correlation coefficient. There is a *positive correlation* between the variables x and y if whenever x increases or decreases, then y changes in the same way. We will say that there is a *negative correlation* between the variables x and y if whenever x increases or decreases, then y changes in the opposite way. The correlation coefficient, r, allows us to measure the linear correlation between two variables and has the following properties.

PROPERTIES OF THE LINEAR CORRELATION COEFFICIENT

1. r is a number between -1 and 1, inclusive. If $r = 1$ or -1, then the scatterplot lies on a straight line.
2. If r is positive, there is positive correlation between the variables. If r is negative, there is negative correlation between the variables.
3. If r is close to 1, then there is a significant positive linear correlation between the variables and the scatterplot is close to lying along a straight line that rises from left to right.
4. If r is close to -1, then there is a significant negative linear correlation between the variables and the scatterplot is close to lying along a straight line that falls from left to right.
5. If r is close to 0, then there is little linear correlation between the variables.

We show the relationship between scatterplots and the linear correlation coefficient in Figure 14.25.

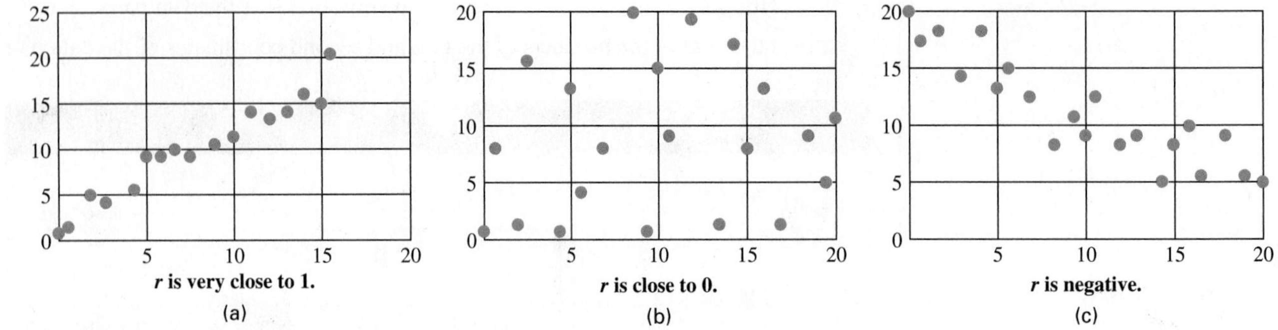

FIGURE 14.25 The value of r is related to the shape of the scatterplot.

n	$\alpha = .05$	$\alpha = .01$
4	.950	.999
5	.878	.959
6	.811	.917
7	.754	.875
8	.707	.834
9	.666	.798
10	.632	.765
11	.602	.735
12	.576	.708
13	.553	.684
14	.532	.661
15	.514	.641
16	.497	.623
17	.482	.606
18	.468	.590
19	.456	.575
20	.444	.561

TABLE 14.19 Critical values for the linear correlation coefficient.

If r is close to 1 or -1, then there is significant positive or negative linear correlation. But how close to 1 or -1 does r have to be for us to conclude that there is significant linear correlation? Table 14.19 contains a list of critical values that we can use to decide whether there is significant linear correlation between two variables. We use this table as follows:

1. Compute the linear correlation coefficient r for n pairs of data.
2. Go to line n in Table 14.19.
3. If the absolute value* of r exceeds the number in the column labeled $\alpha = .05$ on line n, then there is less than a .05 (5%) chance that the variables do not have significant linear correlation. That is, we can be 95% confident that there is significant linear correlation between the variables.
4. If the absolute value of r exceeds the number in the column labeled $\alpha = .01$ on line n, then there is less than a .01 (1%) chance that the variables do not have significant linear correlation. In other words, we can be 99% confident that there is significant linear correlation between the variables.

We will now use the linear correlation coefficient to determine how confident we are that there is a linear relationship between two variables.

EXAMPLE 3 Determining Correlation Between Car Weight and Mileage

Let us suppose that you are considering buying a used car. After talking to several car dealers, you are convinced that there is a relationship between the weight of the car and gas mileage. To check this, you gather data regarding gas mileage of cars with various weights, summarized in Table 14.20, to see whether there is a linear correlation between weight and gas mileage for these data.

Weight (hundreds of pounds)	29	28	31	24	25	30	24	28	32	26
City MPG	21	22	22	23	23	21	24	21	20	22

TABLE 14.20 Car weight versus gas mileage.

*The absolute value of a nonnegative number is the number itself. The absolute value of a negative number is its opposite. For example, the absolute value of 5 is 5, and the absolute value of -13 is 13. The absolute value of 0 is 0.

Find the correlation coefficient for these data and determine whether there is linear correlation at the 5% or 1% level.

SOLUTION: Review the Order Principle on page 12.

We will represent the weight by x and the gas mileage by y and by doing calculations similar to those in Example 2, we find that $\Sigma x = 277$, $\Sigma y = 219$, $\Sigma x^2 = 7{,}747$, $\Sigma y^2 = 4{,}809$, and $\Sigma xy = 6{,}040$. Substituting these values in the formula for the correlation coefficient, we get $r = -0.8507$. This number is close to -1, so we expect that there is significant negative linear correlation for this sample of data.

Because there are 10 pairs of data, we will use line 10 of Table 14.19 to determine how confident we can be that there is a significant linear correlation. The absolute value of r is 0.85, which exceeds .765 in line 10 of Table 14.19. Therefore, we can be 99% confident that there is significant negative linear correlation between the variables of car weight and gas mileage.

Now try Exercises 7 to 10.

Although the computations in Example 3 show that we can be very sure that there is a significant linear correlation between the variables of car weight and gas mileage, remember that you cannot assume that there is a cause-and-effect relationship between these variables.

The Line of Best Fit

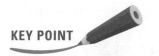

KEY POINT

We use linear regression to find the line of best fit.

In Example 2, you saw that there was a strong positive linear correlation between the number of times that students attended tutoring sessions and their scores on an examination. We will now show you how to find the line that best models the data that we used in Example 2. This line is called **the line of best fit.*** Although the line of best fit usually does not pass through many of the data points, it is the line that minimizes the sum of the squares of all the vertical distances from the data points to the line, as we show in Figure 14.26.

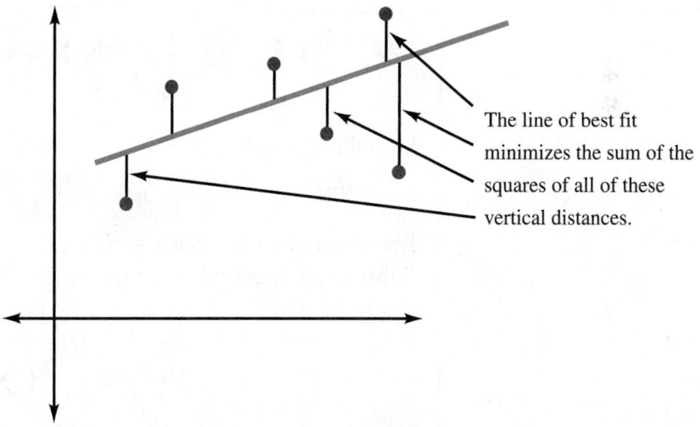
The line of best fit
minimizes the sum of the
squares of all of these
vertical distances.

FIGURE 14.26 The line of best fit is the best linear approximation for a set of data points.

Recall that (except for vertical lines) we can always write a linear equation in the form $y = mx + b$, where m is the slope of the line and b is the y-intercept.† Notice in the next definition that the calculations we do to find the slope and the y-intercept of the line of best fit are similar to the calculations that you did in finding the linear regression coefficient.

*The line of best fit is also called the *regression line* or the *least squares line.*

†If you are a little rusty on this, see Section 7.1.

Highlight

Using Technology to Compute the Linear Correlation Coefficient*

When you are working with a large set of data, a statistical calculator will speed your work. The following screens are taken from a graphing calculator. Screen (a) shows the data for the car weights and gas mileage from Example 3. Screen (b) shows the command (we chose 4) we enter to instruct the calculator to compute the linear correlation coefficient. Screen (c) shows the value of r, which is the same as we computed in Example 3.

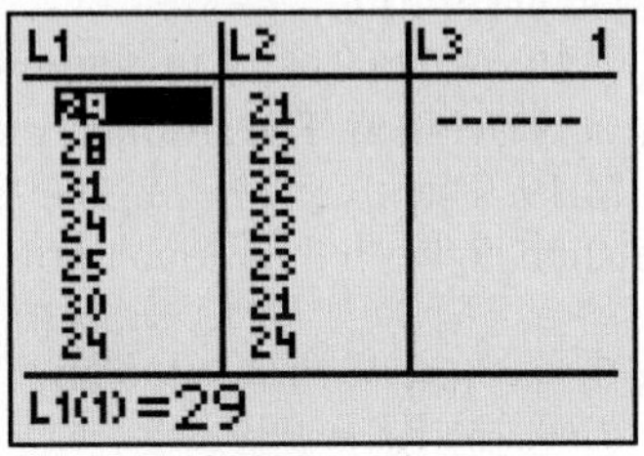

(a) Data have been entered.

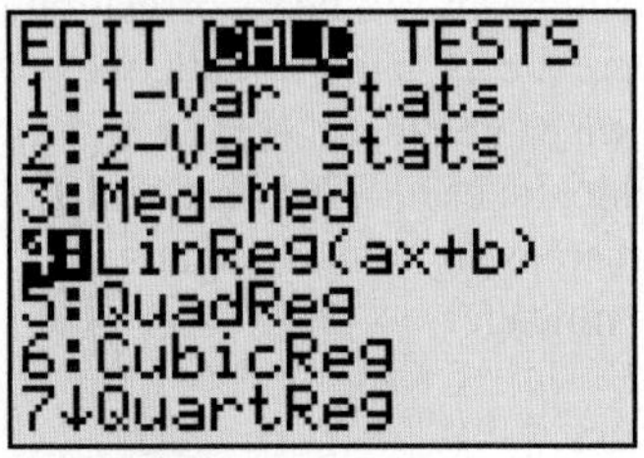

(b) Select command.

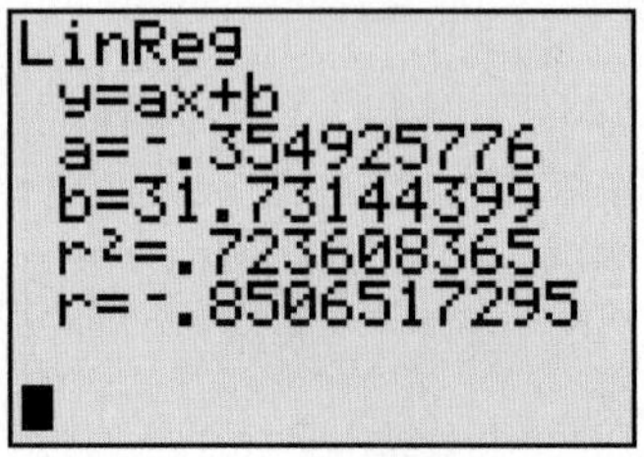

(c) Calculator computes linear correlation coefficient.

DEFINITION The **line of best fit** for a set of data points of the form (x, y) is of the form $y = mx + b$, where

$$m = \frac{n\sum xy - (\sum x)(\sum y)}{n(\sum x^2) - (\sum x)^2} \quad \text{and} \quad b = \frac{\sum y - m(\sum x)}{n}.$$

We will use this definition to find the line of best fit for the data points in Example 2.

EXAMPLE 4 Finding the Line of Best Fit for a Set of Data Points

Find the line of best fit for the set of data points we used in Example 2.

SOLUTION: Review the Order Principle on page 12.

Recall in Example 2 that $n = 10$, $\Sigma xy = 4{,}090$, $\Sigma x = 98$, $\Sigma y = 376$, and $\Sigma x^2 = 1{,}396$. Substituting these values in the formula for the slope of the line, we get

$$m = \frac{n\sum xy - (\sum x)(\sum y)}{n(\sum x^2) - (\sum x)^2} = \frac{10 \cdot 4{,}090 - 98 \cdot 376}{10 \cdot 1{,}396 - (98)^2}$$

$$= \frac{40{,}900 - 36{,}848}{13{,}960 - 9{,}604} = \frac{4{,}052}{4{,}356} \approx 0.9302.$$

The y-intercept is given by

$$b = \frac{\sum y - m(\sum x)}{n} = \frac{376 - (0.9302)98}{10} = \frac{376 - 91.1596}{10} \approx \frac{284.84}{10} = 28.48.$$

Thus, the line of best fit for the data in Example 2 is

$$y = 0.93x + 28.48.$$

*Technology resources are available at Pirnot's MyMathLab.

Highlight

Health Scares

If you eat apples, drink coffee, or use a cell phone, you can breathe a little easier. According to the American Council on Science and Health, these are three of the greatest unfounded health scares of the last 50 years.

In one study, Alar, a chemical used for ripening apples, was thought to cause childhood cancer. Later studies showed that Alar produced cancer only at a dosage of over 100,000 times what a child would consume by eating apples or drinking apple juice. Nonetheless, this scare cost the apple industry about \$375 million.

In another study, Harvard researchers found that drinking two cups of coffee a day doubled the risk of pancreatic cancer. When other researchers were unable to reproduce these results, the council stated, "This brief scare illustrates the danger of putting too much credence in a single study without analyzing any possible biases or confounding factors."

More recently, several cancer patients claimed that electromagnetic fields from their cell phones caused their illness. The cell phone industry sponsored studies that determined there is no evidence that cell phones are a cause of cancer.*

These examples point out that you must look at reported studies very carefully to decide whether they are reliable.

In Example 2, we found that the linear correlation coefficient was close to 1, which meant that it was reasonable for us to model the data by a line. The best line to do that is the line that we just found.

Now try Exercises 11 to 14.

As you can see, doing calculations to find the regression coefficient and the line of best fit can be both lengthy and tedious. For that reason, most people who do these computations use a graphing calculator or a computer program. The Highlight that we showed earlier to do the correlation computations in Example 3 regarding the gas mileage data also shows the slope and the y-intercept of the line of best fit. However, the TI-83 calls the slope a instead of m. As you can see from the Highlight, the line of best fit for the car mileage data is

$$y = -0.355x + 31.73.$$

Notice that the slope of the line is negative, which is consistent with the fact that we found negative linear correlation in Example 3.

EXERCISES 14.5

Sharpening Your Skills

In Exercises 1 and 2, state what kind of correlation, if any, each scatterplot indicates.

1.

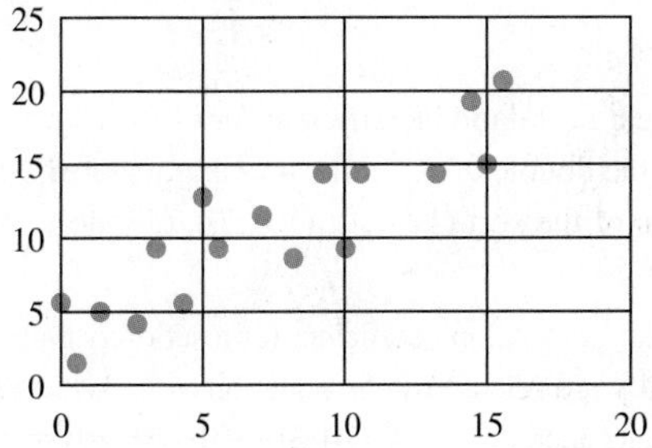

2.

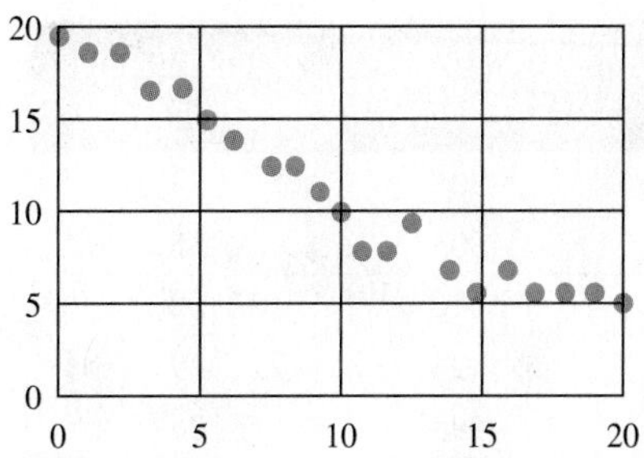

*However, more recent studies show that there may be a correlation between the use of cell phones and certain types of cancer.

For Exercises 3–6, do the following.

a) Plot the xy *pairs in a scatterplot.*

b) Estimate the linear correlation coefficient between the two data sets.

c) Calculate the linear correlation coefficient to see how accurate your guess was.

3. (3, 5), (7, 8), (4, 6), (6, 7)

4. (4, 5), (6, 8), (5, 3), (5, 6)

5. (11, 5), (15, 8), (12, 3), (12, 6)

6. (3, 12), (7, 10), (4, 8), (6, 0)

In Exercises 7–10, use Table 14.19 and the linear correlation coefficient you found in the given exercise to determine whether we can be 95% or 99% confident that there is significant linear correlation between x *and* y.

7. Exercise 3 **8.** Exercise 4

9. Exercise 5 **10.** Exercise 6

In Exercises 11–14, find the line of best fit for the data in the specified exercise.

11. Exercise 3 **12.** Exercise 4

13. Exercise 5 **14.** Exercise 6

Applying What You've Learned

In Exercises 15–18, determine the linear correlation coefficient for the data and also determine whether we can be 95% or 99% confident that there is significant linear correlation between the two variables.

15. The following table lists the number of years of education past high school and the annual income of five people.

Years of Education Past High School	Annual Income (in $ thousands)
0	23
1	22
2	27
2	28
5	35

16. The accompanying table lists the average SAT score and teachers' salaries for certain states.

State	SAT Score	Teacher Salary (in $ thousands)
Connecticut	901	44
Delaware	903	35
Kansas	1,040	30
Kentucky	994	29
Louisiana	993	26

Source: Statistical Abstract of the United States

17. As a car begins to accelerate, the gas mileage is poor. As the speed increases, the gas mileage begins to increase. The gas mileage increases for a while, then as the speed increases further, the mileage begins to decrease. The accompanying table of data illustrates this.

Speed in Miles per Hour	Mileage in Miles per Gallon
30	26
40	31
50	33
60	31
70	26

18. The accompanying table is based on data from the National Center for Health Statistics (www.cdc.gov/nchs).

Years of Education	Percentage Who Are Overweight
8	62.7
10	61.1
12	65.5
16	57.6
18	53.7

In Exercises 19–22, find the line of best fit for the data in the specified exercise.

19. Exercise 15 **20.** Exercise 16

21. Exercise 17 **22.** Exercise 18

Communicating Mathematics

23. In Example 3, we found the absolute value of the linear correlation coefficient r to be 0.85, which was greater than 0.765 in the line labeled 10 in Table 14.19. What did that tell us?

24. What is the purpose of finding the line of best fit?

25. a. Describe a situation in which you would expect to have a significant positive correlation between two data sets.

b. Describe a situation in which you would expect to have a significant negative correlation between two data sets.

c. Describe a situation in which you would expect to have no correlation between two data sets.

Challenge Yourself

26. What is the linear correlation coefficient for a set of xy pairs if each y score is the double of the corresponding x score? We are looking at pairs of the sort (12, 24), (10, 20), (23, 46), and so on.

27. What is the linear correlation coefficient for a set of xy pairs if in each pair, x and y are related by the equation $y = 3x + 5$? We are looking at pairs such as (4, 17), (10, 35), (21, 68), and so on.

CHAPTER SUMMARY

You will learn the items listed in this chapter summary more thoroughly if you keep in mind the following advice:

1. Focus on "remembering how to remember" the ideas. What pictures, word analogies, and examples help you remember these ideas?
2. Practice writing each item without looking at the book.
3. Make 3 × 5 flash cards to break your dependence on the text. Use these cards to give yourself practice tests.

Section	Summary	Example
SECTION 14.1	We refer to a collection of numerical information as **data.** The entire set of data that we are studying is called the **population.** A subset of the population is called a **sample.** When selecting a sample for analysis, it is important that the sample is not **biased** and fairly represents the whole population.	Discussion, p. 702
	A set of data listed with their frequencies is called a **frequency distribution.** We often present a frequency distribution in a **frequency table.** A **relative frequency distribution** shows the percentage of time that each value occurs in a frequency distribution.	Example 1, pp. 703–704 Example 2, pp. 704–705
	In drawing a **bar graph,** we specify the classes on the horizontal axis and frequencies on the vertical axis. We draw bars (rectangles) above each class having a height equal to the frequency of the class. A **histogram** is a type of bar graph used to graph a frequency distribution when we are dealing with a continuous distribution.	Example 3, pp. 705–706 Example 4, p. 707
	A **stem-and-leaf display** is an effective way to present two sets of data "side by side" for analysis.	Example 6, pp. 708–709
SECTION 14.2	The **mean** of a data set is $\frac{\Sigma x}{n}$, where Σx is the sum of the data values and n is the number of values. We represent the mean of a sample by $\bar{x}$ and the mean of a population by μ. We compute the mean of a frequency distribution using the formula $\frac{\Sigma(x \cdot f)}{\Sigma f}$, where $x \cdot f$ represents the product of a data value x and its frequency f, and Σf is the sum of the frequencies of the data values in the distribution.	Example 1, p. 715 Example 2, pp. 715–716
	To find the **median,** first arrange the data in order. If there is an odd number of values, the median is the value in the middle. If there is an even number of values, the median is the mean of the two middle values. The **mode** is the most frequently occurring value. If two values are most frequent, then we say that there are two modes. We do not allow more than two modes.	Examples 4 and 5, p. 719 Example 7, p. 721
	If a data set is arranged in increasing order, the set of values below the median is called the **lower half.** The set of values above the median is called the **upper half.** The **first quartile,** Q_1, is the median of the lower half. The **third quartile,** Q_3, is the median of the upper half. The **five-number summary** of a data set is the minimum value, the first quartile, the median, the third quartile, and the maximum value. We draw a **box-and-whisker plot** as follows: First draw a horizontal axis with a scale that includes the extreme values of the data. Then draw a box from Q_1 to Q_3 with a vertical line in the box to indicate the median. Finally, draw whiskers (lines) from Q_3 to the largest value and from Q_1 to the smallest value.	Example 6, p. 720 Discussion, pp. 720–721
	We can summarize data in various ways depending on which measure of central tendency we choose to use.	Example 8, pp. 721–722
SECTION 14.3	The **range** is the difference between the largest and smallest values in a data set.	Example 1, p. 728
	The **standard deviation** of a sample, indicated by s, is defined as $$s = \sqrt{\frac{\Sigma(x - \bar{x})^2}{n - 1}},$$	Definition, p. 729 Example 2, p. 730

(Continued)

Section	Summary	Example
	where $\bar{x}$ is the mean of the sample, and n is the number of data values. For a population, we use the formula $\sigma = \sqrt{\frac{\sum (x - \mu)^2}{n}}$.	
	If f represents the frequency of data value x in the distribution, then the standard deviation is given by the formula $s = \sqrt{\frac{\sum (x - \bar{x})^2 \cdot f}{n - 1}}$. The number $n = \Sigma f$ is the number of data values in the distribution.	Example 3, p. 731
	The **coefficient of variation** expresses the standard deviation of a sample as a percentage of the mean and is calculated as follows: $CV = \frac{s}{\bar{x}} \cdot 100\%$	Example 4, pp. 733–734
SECTION 14.4	A **normal curve** has the following properties: 1. It is bell shaped. 2. The highest point on the curve is at the mean of the distribution. 3. The mean, median, and mode of the distribution are the same. 4. The curve is symmetric with respect to its mean. 5. The total area under the curve is 1.	Discussion, p. 739
	6. Roughly 68% of the data values are within 1 standard deviation from the mean, 95% of the values are within 2 standard deviations from the mean, and 99.7% of the values are within 3 standard deviations from the mean.	Example 1, pp. 740–741
	A **z-score** represents the number of standard deviations a data value is from the mean of a data set. We can use a table to relate positive z-scores to the amount of area under the standard normal curve between the mean and the given z-scores.	Example 2, p. 743
	If a normal distribution has a mean of μ and a standard deviation of σ, the equation $z = \frac{x - \mu}{\sigma}$ converts a data value to a z-score.	Example 3, pp. 744–745
	The normal distribution has many useful applications.	Example 4, pp. 745–746 Example 5, p. 746 Example 6, pp. 746–747
SECTION 14.5	A **scatterplot** is a graph of a set of data points.	Example 1, p. 751
	The **linear correlation coefficient** measures to what degree points of a scatterplot lie along a straight line.	Example 2, pp. 752–753 Example 3, pp. 754–755
	The **line of best fit** is the line that best models a set of data points.	Example 4, pp. 756–757

CHAPTER REVIEW EXERCISES

Section 14.1

The following data set represents the number of automobile accidents that have occurred in a city each day over the past month.

8, 8, 6, 6, 7, 10, 4, 6, 5, 10, 8, 9, 8, 7, 8, 8,
6, 9, 5, 9, 8, 10, 8, 8, 6, 10, 6, 5, 10, 8, 9

1. Construct a frequency table for these data.

2. Construct a relative frequency table for these data.

3. Display the relative frequency table in Exercise 2 as a bar graph.

4. An operator of Six Flags' Superman roller coaster counted the number of people on the ride at 10-minute intervals for a certain period of time. The results he obtained are summarized in the accompanying bar graph. Use the graph to answer the following questions.

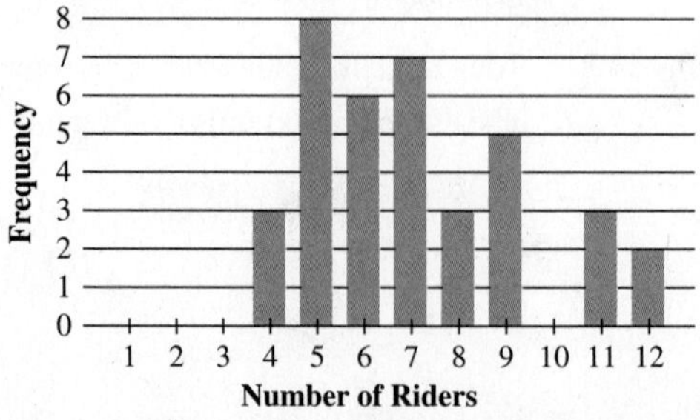

a. What was the smallest number of people on the ride and how often did it occur?

b. What was the most frequently occurring rider count?

c. For how many 10-minute intervals were the riders counted?

5. The lists represent the ages of actors (M) and actresses (F) at the time they won an Academy Award from 1980 to 2008.

F: 31, 74, 33, 49, 38, 61, 21, 41, 26, 80, 42, 29, 33, 35, 45, 49, 39, 34, 26, 25, 33, 35, 35, 28, 30, 30, 29, 61, 32

M: 37, 76, 39, 52, 45, 35, 61, 43, 51, 32, 42, 54, 52, 37, 38, 31, 45, 60, 46, 40, 36, 47, 29, 43, 37, 37, 38, 45, 50

a. Represent these two sets of data on a single stem-and-leaf plot.

b. Do you notice any patterns in comparing the two plots?

Section 14.2

6. The following is the number of regular-season wins by the Detroit Pistons over 12 seasons ending in 2008: 59, 53, 64, 54, 54, 50, 50, 32, 42, 29, 37, 54. Find the mean, median, and mode for this distribution.

7. Briefly describe what information is conveyed by the mean, median, and mode.

8. Calculate the five-number summary for the following distribution:

10, 8, 6, 12, 6, 3, 11, 7, 6, 17, 4, 9, 13, 20, 7

9. Construct a box-and-whisker plot for the distribution given in Exercise 8.

Section 14.3

10. Calculate the sample standard deviation for the following distribution:

4, 6, 7, 3, 5, 6, 4, 5

11. Explain what the standard deviation tells you about a distribution.

Section 14.4

12. State several basic properties of the normal distribution.

13. If a distribution has a mean of 80 and a standard deviation of 7, to what z-score does the raw score 85 correspond?

14. If a distribution has a mean of 60 and a standard deviation of 5, the z-score 1.35 corresponds to what raw score?

15. Assume that 1,000 values are normally distributed with a mean of 72 and a standard deviation of 4. How many values can we expect to be between 75 and 82?

16. Suppose you earned an 82 on a history exam that had a mean score of 78 and a standard deviation of 3, and you also earned an 84 on an anthropology exam that had a mean of 79 and a standard deviation of 4. On which exam did you do better in relationship to the rest of the class?

17. Redo Example 6 in Section 14.4 regarding setting a warranty on a TV. Now assume that the mean time to failure is 2,500 hours with a standard deviation of 500 and we want no more than 4% to be returned before the warranty expires.

Section 14.5

18. State what kind of correlation, if any, is indicated by the following scatterplots.

a.

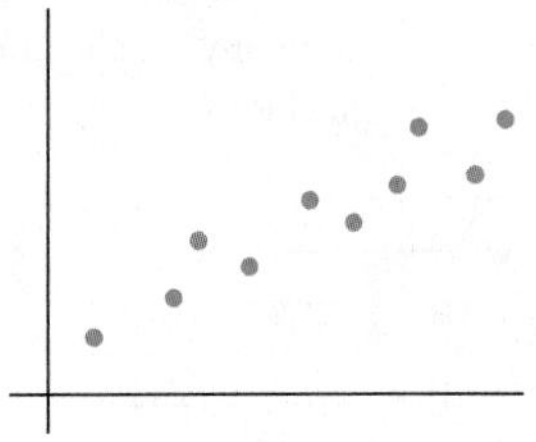

b.

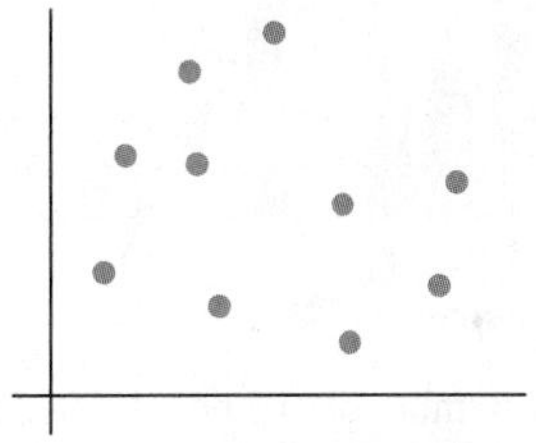

19. **a.** Calculate the correlation coefficient for the following set of data points:

(3, 5), (4, 9), (5, 10), (6, 13)

b. Find the line of best fit for these points.

CHAPTER TEST

The following data set represents the number of visits to the health center that have occurred at your school over the past month.

7, 7, 8, 8, 6, 11, 5, 7, 6, 12, 4, 9, 8, 6, 9, 8, 9, 7, 6, 7, 7, 8, 6, 10, 9, 11, 8, 8, 7, 9, 10, 6, 7, 5, 5, 6, 6, 7, 9, 5

1. Construct a frequency table for these data.

2. Construct a relative frequency table for these data.

3. Display the relative frequency table in Exercise 2 as a bar graph.

4. Explain what the standard deviation tells you about a distribution.

5. Calculate the standard deviation for the following distribution:

3, 4, 5, 6, 5, 6, 8, 9, 8

6. State several basic properties of the normal distribution.

7. We have listed the number of home runs that Babe Ruth hit during the seasons he played for the Yankees, and the number of home runs that Hank Aaron hit during his best 15 years. Represent the two sets of data on a single stem-and-leaf display.

Ruth: 54, 59, 35, 41, 46, 25, 47, 60, 54, 46, 49, 46, 41, 34, 22

Aaron: 44, 30, 39, 40, 34, 45, 44, 32, 44, 39, 44, 38, 47, 34, 40

8. The following is a list of wins by the Boston Red Sox from 1998 to 2007:

92, 94, 85, 82, 93, 95, 98, 95, 86, 96

a. Calculate the five-number summary for this distribution.

b. Construct a box-and-whisker plot for this distribution.

9. Suppose that you earned an 85 on a statistics exam that had a mean of 78 and a standard deviation of 4, and an 88 on a sociology exam that had a mean of 80 and a standard deviation of 5. On which exam did you do better in relationship to the rest of the class?

10. The webmaster at Match.com recorded the number of hits at 1-minute intervals for a certain period of time. The results she obtained are summarized in the accompanying bar graph. Use the graph to answer the following questions.

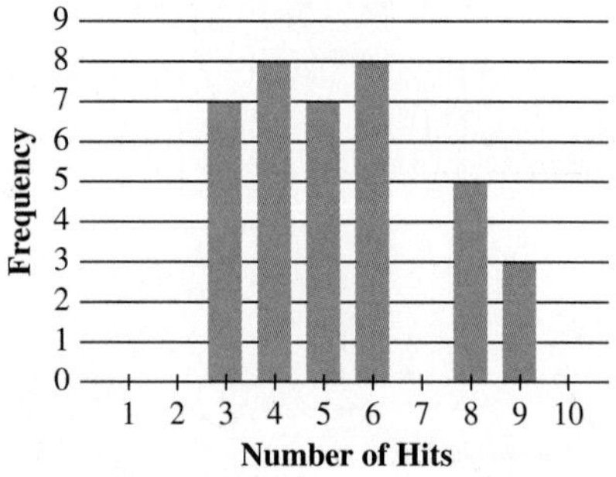

a. What was the smallest number of hits per minute, and how often did that occur?

b. What was the most frequent number of hits?

c. For how many 1-minute intervals were the number of hits counted?

11. State what kind of correlation, if any, is indicated by the following scatterplots:

a.

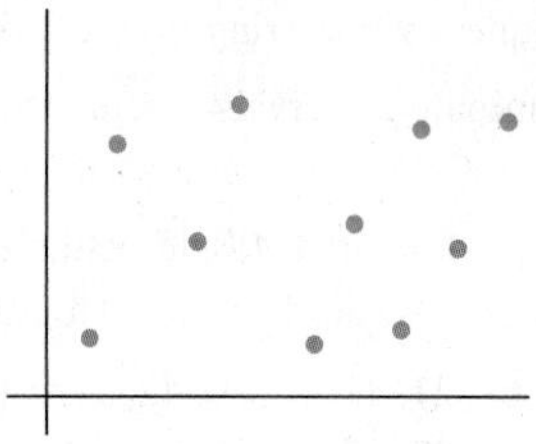

b.

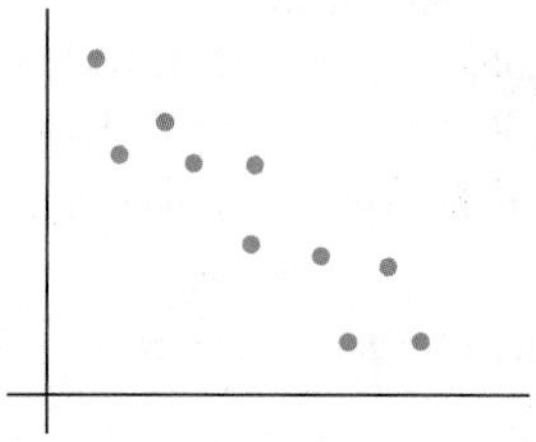

12. Briefly describe what information is conveyed by the mean, median, and mode.

13. Find the mean, median, and mode of the following distribution:

9, 8, 6, 11, 5, 9, 7, 6, 3, 10, 11, 9, 9, 6, 10, 7, 8, 9, 5, 9

14. If a distribution has a mean of 50 and a standard deviation of 6, the z-score 1.83 corresponds to what raw score?

15. If a distribution has a mean of 75 and a standard deviation of 5, to what z-score does the raw score 82 correspond?

16. Assume that 1,000 values are normally distributed with a mean of 54 and a standard deviation of 5. How many values can we expect to be above 58?

17. Redo Example 6 in Section 14.4 regarding setting a warranty on a TV. Now assume that the mean time to failure is 2,800 hours with a standard deviation of 400 and we want no more than 6% to be returned before the warranty expires.

18. a. Calculate the correlation coefficient for the following set of data points:

(3, 6), (4, 8), (5, 7), (6, 10)

b. Find the line of best fit for these points.

GROUP PROJECTS

1. Presenting data with a stem-and-leaf display. Find some data that interest you and present them in a stem-and-leaf display similar to the way we presented the age data for men and women on winning Academy Awards in the Chapter Review Exercises. Write a brief report on some conclusions that you draw from your presentation.

2. Finding correlation between sets of data. Find some data relating two quantities similar to our example regarding the weights of cars and gas mileage in Example 3 of Section 14.5. Find the correlation coefficient for these data points and also find the line of best fit. You might consider the following types of data pairs: (weight of current car, current gas mileage of same car), (price of airfare between two cities one day before flight, price of airfare between same two cities one month before flight), (average cost of tuition for a given year, consumer price index for that year), (height of winner in presidential election, height of runner-up), (weight of a football running back, average yards gained per carry by that running back).

3. Simpson's paradox. A famous paradox in statistics, called **Simpson's paradox,** can be illustrated as follows: In each of the years 1995, 1996, and 1997, David Justice had a higher batting average than Derek Jeter, yet if we look at their batting

averages over the three-year period 1995–97, Jeter's average is higher than Justice's.

a. Research this paradox and give details of the computations that illustrate the problem.

b. Search the Internet for interesting examples of Simpson's paradox, such as voting on the Civil Rights Act of 1964, the University of Berkeley Gender Bias Case, the Low Birth Weight Paradox, or some other of your choosing. Give a detailed report on your findings.

USING TECHNOLOGY

You can use technology to do many of the calculations in this chapter easily.

1. **Graphing calculators.** You can follow the Highlights on pages 718, 732, and 756 to do statistical calculations on the TI graphing calculators. To perform calculations with normal distributions, press [2nd] [DISTR] and then select option 2 to choose "normalcdf". In part (a) of the figure below, we have calculated the area under the standard normal curve between $z = 1.5$ and $z = 2.1$.

   The screens shown in part (b) of the figure illustrate how to graph the area under the standard normal curve between $z = 1.23$ and $z = 1.85$. Before graphing, we used the ClrDraw command from the DRAW menu to clear all previous graphs.

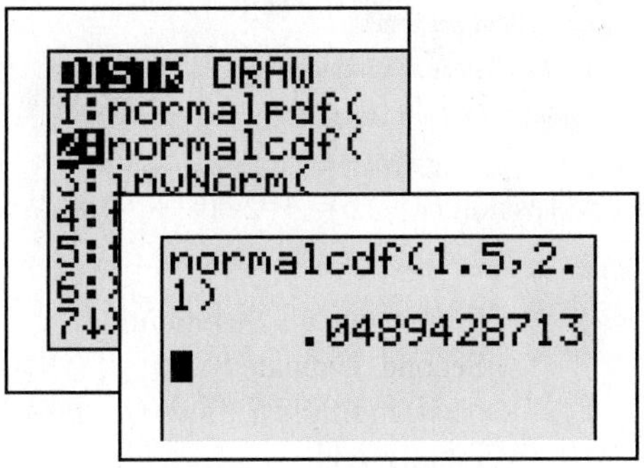

(a)

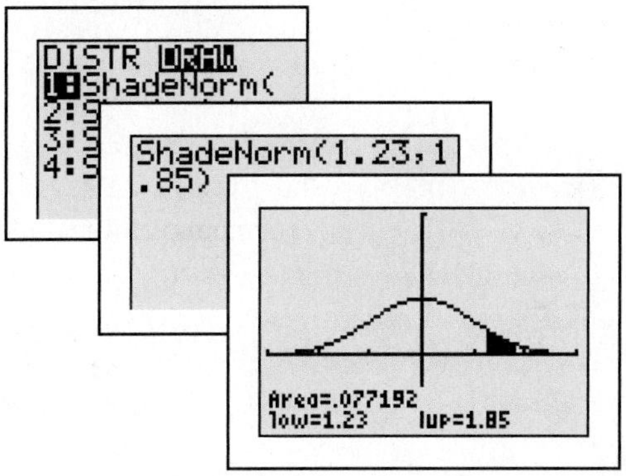

(b)

Use a calculator's statistical functions to do Exercises 9, 17, and 22 in Section 14.3, and Exercises 13, 19, 27, and 28 in Section 14.4.

2. **Excel spreadsheets.** To use the Analysis ToolPak in Excel to do statistical calculations, do the following: Open Excel and enter your data. Next, on the File tab, select Options and select Add Ins from the menu that appears. Then from the Add Ins menu, select Analysis ToolPak and click on OK.

   To get the analysis shown, we selected the data tab and clicked on the Data Analysis option. In the new window that opened, we chose Descriptive Statistics and specified the input range of cells and where we wanted the output to be placed. Use the ToolPak to do Exercises 9 to 12 in Section 14.2, and Exercises 7 and 23 in Section 14.3.

	A	B	C	D
1	37		*Column1*	
2	38			
3	38		Mean	40.00
4	39		Standard Error	0.36
5	39		Median	40.00
6	39		Mode	39.00
7	39		Standard Deviation	1.62
8	39		Sample Variance	2.63
9	40		Kurtosis	0.74
10	40		Skewness	0.49
11	40		Range	7.00
12	40		Minimum	37.00
13	40		Maximum	44.00
14	41		Sum	800.00
15	41		Count	20.00

3. **Minitab.** Download a trial version of Minitab, a popular, easy-to-use statistical program. If you enter a data set in column C1, click on the stat tab, choose Basic Statistics, and then select Display Descriptive Statistics, Minitab gives the output below. Use Minitab to do Exercises 7, 19, and 23 in Section 14.3.

Descriptive Statistics: C1

Variable	N	N*	Mean	SE Mean	StDev	Minimum	Q1	Median	Q3	Maximum
C1	6	0	210.00	3.39	8.29	195.00	204.75	211.50	216.00	219.00

Appendix A

Basic Mathematics Review

Order of Operations

To do arithmetic and algebraic calculations correctly, you must understand the order in which we perform operations. Many people use the memory device that we explain in the following diagram:

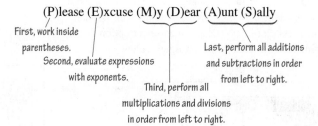

(P)lease (E)xcuse (M)y (D)ear (A)unt (S)ally

First, work inside parentheses.

Second, evaluate expressions with exponents.

Third, perform all multiplications and divisions in order from left to right.

Last, perform all additions and subtractions in order from left to right.

For example, we would evaluate the expression $(3 + 5)^2 + 2 \cdot (7 + 4) + 9$ as follows:

$(3 + 5)^2 + 2 \cdot (7 + 4) + 9$	
$8^2 + 2 \cdot 11 + 9$	First, work inside parentheses.
$64 + 2 \cdot 11 + 9$	Second, evaluate 8^2.
$64 + 22 + 9$	Next, multiply 2 times 11.
95	Finally, add.

Solving Equations

To solve equations, first, if you can, simplify by collecting like terms on each side of the equation. Then you can add or subtract the same quantity from both sides of the equation. You can also multiply or divide both sides of the equation by the same quantity, provided you do not divide by zero. We could solve the equation $3x - 17 + 7x = 15 + 6x + 8$ as follows:

like terms (3x, 7x) like terms (15, 8)

$3x - 17 + 7x = 15 + 6x + 8$	
$10x - 17 = 23 + 6x$	Collect like terms.
$10x - 17 + 17 = 23 + 6x + 17$	Add 17 to both sides.
$10x = 40 + 6x$	Simplify.
$10x - 6x = 40 + 6x - 6x$	Subtract $6x$ from both sides.
$4x = 40$	Simplify.
$x = 10$	Divide both sides by 4.

If an equation has decimal or fractional coefficients, you can make the equation easier to solve if you simplify the coefficients by multiplying both sides of the equation by a suitable quantity before solving.

For example, you could multiply both sides of the equation $2.4x + 4.12 = 1.3x + 5$ by 100 to eliminate decimals and then solve the resulting equation $240x + 412 = 130x + 500$.

To solve the equation $\frac{1}{3}x + \frac{1}{2} = x - \frac{5}{4}$, you could multiply both sides of the equation by 12 to get the equivalent equation $(12)\frac{1}{3}x + (12)\frac{1}{2} = (12)x - (12)\frac{5}{4}$, which simplifies to $4x + 6 = 12x - 15$.

Working with Fractions (Rational Numbers)

For a review of the properties and operations on rational numbers, see the following examples in the text:

Equality of Rational Numbers: Section 6.3, Example 1
Reducing Rational Numbers: Section 6.3, Example 2
Addition and Subtraction: Section 6.3, Example 3
Multiplication: Section 6.3, Example 4
Division: Section 6.3, Example 5
Improper Fractions and Mixed Numbers: Section 6.3, Examples 7 and 8

Working with Decimals

To add or subtract decimals, line up the decimal points as in the following example:

$$\begin{array}{r} 27.36 \\ 142.405 \\ +\ 3.2 \\ \hline 172.965 \end{array}$$

Line up decimal points.

To multiply decimals, we add the decimal places to obtain the final answer as follows:

$$\begin{array}{rl} 5.304 & \text{——— 3 decimal places} \\ \times 6.27 & \text{——— 2 decimal places} \\ \hline 33.25608 & \text{——— 3 + 2 decimal places} \end{array}$$

To divide decimals, move the decimal point in the divisor and dividend before dividing as in the following example:

$$1.27\overline{)93.472}\quad \text{quotient } 73.6$$

Move decimal point two places to the right.

Working with Percents

For a review of the basic principles in working with percents, see the following examples:

Meaning of Percent: Introduction to Section 8.1
Conversions between decimals and percents: Section 8.1, Example 1
Conversions between fractions and percents: Section 8.1, Example 2
Solving equations involving percents: Section 8.1, Example 6

Answers to Quiz Yourself Problems

Chapter 1

1.

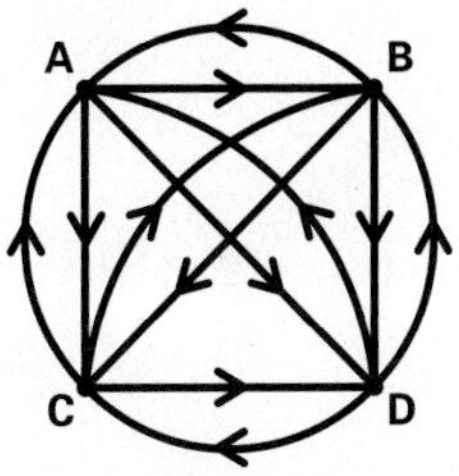

2. students taking both Zumba and Pilates; taking Pilates but not Zumba

3.

Line 4:	No	No	Yes
Line 7:	No	Yes	Yes

4. **a.** 1 6 15 20 15 6 1

   1 7 21 35 35 21 7 1

   b. 1 and 100

5. 132

6. 4, 18, 28 tons

7. Math (9), English (12), Sociology (11), Art History (10)

   Math (12), English (9), Sociology (10), Art History (11)

8. Any numbers will work.

9. Almost any choice of numbers for a, b, and c will work. For example, let

   $a = 3, b = 4$, and $c = 5$. Then $\frac{a+b}{a+c} = \frac{3+4}{3+5} = \frac{7}{8}$.

   However, $\frac{b}{c} = \frac{4}{5}$.

10. **a.** We are first adding two numbers and then taking the square root of the sum.

    b. We are first taking the square roots of two numbers and then forming the sum of these square roots.

    c. To "first add and then take the square root" is not the same as to "first take the square roots and then add."

11. **a.** There is an extra line in the symbol on the right.

    b. The symbol on the left has a pair of braces around the ∅ symbol.

    c. The symbol on the left is rounded; the symbol on the right is pointed.

    d. The symbol on the right is the number 0; the symbol on the left is the number 0 with braces around it.

12. **a.** intersection of streets

    b. happening at the same time

13. **a.** 31 + 7

    b. 5 + 41

14. No, there are numerous counterexamples.

15. 479,001,600

16. 400,000,000 × 0.15 = 60,000,000

Chapter 2

1. **a.** {Monday, Tuesday, . . . , Sunday}

   b. $\{x : x$ is a natural number less than or equal to $60\}$

2. **a.** well defined **b.** not well defined

3. Yes, both sets have no elements and represent the empty set.

4. **a.** true **b.** true **c.** false

5. **a.** 10 **b.** 2 **c.** 50

6. **a.** true **b.** false

7. **a.** true **b.** false

8. $\varnothing, \{1\}, \{2\}, \{3\}, \{1, 2\}, \{1, 3\}, \{2, 3\}, \{1, 2, 3\}$

9. **a.** $2^5 = 32$ **b.** $2^{26} = 67{,}108{,}864$

10. **a.** r_1; volunteers who will neither cook nor serve food

    b. r_1 and r_4; the volunteers who will not cook

11. $M \cup B = \{m, a, t, h, e, i, c, s, b, u, y\}$, $M \cap B = \{a, t, e\}$

12. **a.** $C' = \{c, i, k, o, q\}$

    b. $B - A = \{2, 7, 8\}; A - B = \{1, 5\}$

13. **a.** $\{1, 3, 4, 6, 8, 9, 10\}$ **b.** $\{3, 4, 6, 10\}$

    c. $\{1, 4, 8, 9, 10\}$ **d.** $\{1, 3, 4, 6, 8, 9, 10\}$

14. **a.** $A' \cap B' \cap C$

    b. $(A \cap B \cap C') \cup (A' \cap B \cap C')$

15. Region x is B^+ and region y is A^-.

16. 57

17. There are several correct answers; one would be

Carreras	Domingo	Pavarotti
↕	↕	↕
Shakira	Beyonce	Gaga

18. There are many correct answers for this, one is

1	2	3	⋯	n	⋯
↕	↕	↕		↕	
2	3	4	⋯	$n + 1$	⋯

19. **a.** $6, \frac{5}{2}, \frac{4}{3}$ **b.** $\frac{2}{6}$

Chapter 3

1. **a.** simple **b.** compound **c.** compound

2. **a.** $\sim d \vee \sim i$ **b.** $\sim d \wedge i$

3. **a.** $\sim f \rightarrow \sim q$ **b.** $f \leftrightarrow q$

4. **a.** At least one professor is not friendly.

   b. No dogs bite.

5. **a.** true **b.** false **c.** false **d.** true

6.

			2	1	5	3	4	
p	q	$(p$	$\wedge$	$\sim q)$	$\vee$	$(\sim p$	$\vee$	$q)$
T	T	T	F	F	T	F	T	T
T	F	T	T	T	T	F	F	F
F	T	F	F	F	T	T	T	T
F	F	F	F	T	T	T	T	F

7.

		2		1	
p	q	$\sim$	$(p$	$\vee$	$q)$
T	T	F	T	T	T
T	F	F	T	T	F
F	T	F	F	T	T
F	F	T	F	F	F

1	3	2
$(\sim p)$	$\wedge$	$(\sim q)$
F	F	F
F	F	T
T	F	F
T	T	T

The statements have identical truth tables and so are logically equivalent.

8. It is not the case that: the car is more than five years old or has been driven over 50,000 miles.

9.

			2	1	6	3	5	4
p	q	$(p$	$\wedge$	$\sim q)$	$\rightarrow$	$(\sim p$	$\vee$	$q)$
T	T	T	F	F	T	F	T	T
T	F	T	T	T	F	F	F	F
F	T	F	F	F	T	T	T	T
F	F	F	F	T	T	T	T	F

10. Inverse: If the price of downloading videos does not increase, then people will not copy them illegally.

Contrapositive: If people will not copy them illegally, then the price of downloading videos does not increase.

11. **a.** If you travel on the Amazon, then you have updated your immunization.

b. If you increase your cardiovascular fitness, then you exercise three times a week.

12. On the third line of the truth table, the premises are true, but the conclusion is false.

13. fallacy of the inverse; invalid

14.

invalid

15. valid

16. Answers will vary; however, here are some possibilities:

a. Some Bs are not As.

b. Some As are not Cs.

c. Some Bs are not Cs.

17. **d.** The truth value of "*Gone with the Wind* is not a great movie" is 0.27.

e. "You find mathematics uninteresting" has a truth value of 0.3.

18. **a.** 0.45 **b.** 0.82

19. **a.** 0.45 **b.** 0.55

20. **a.** 0.80 **b.** 0.55 **c.** 0.80

21. **a.** $(\sim i) \vee c = 0.95$ **b.** $(\sim i) \vee c = 0.75$

Chapter 4

1. The graph is connected. A, B, D, E, and F are even. C and G are odd. G has degree 3. Edge CG is a bridge.

2. Graph (a) can be traced; C, B, A, C, D, B, A, D.

Graph (b) cannot be traced; it has four odd vertices.

3. **a.** CABCFEGDFIH

b. No. Because C and H are both odd vertices, we must begin at one and end at the other.

4. A, B, C, and D are the odd vertices in the graph. If we duplicate edge CD and also all edges along the indicated path from A to B, the new graph will be Eulerian.

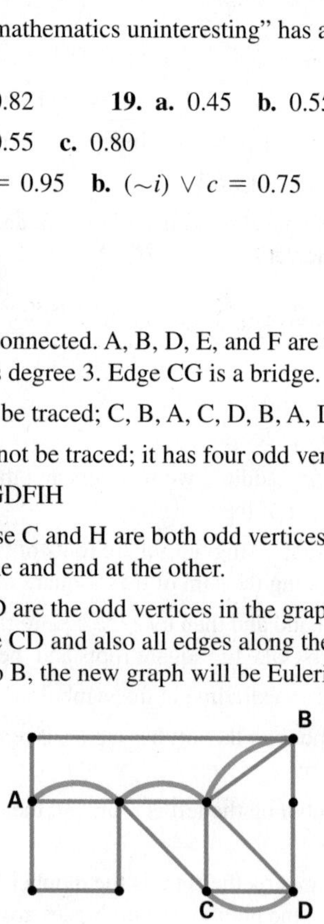

5.

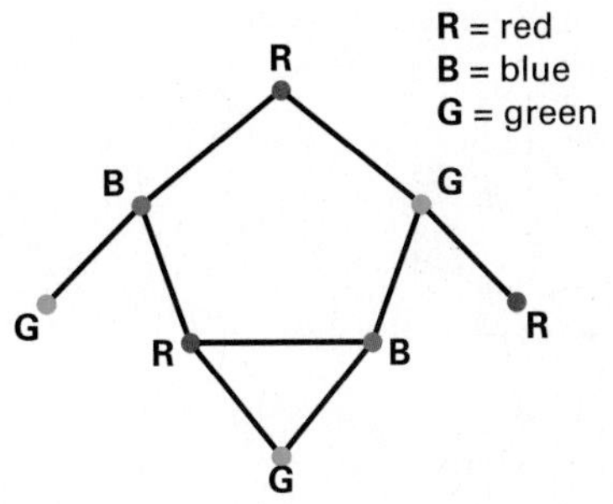

6. **a.** BDACB **b.** ADBCA

7. 8! = 40,320

8. ABDCA has the minimal weight of 11. (Of course, the reverse circuit ACDBA also has the same weight.)

9. MAPNCM; $1,200

10. Manny

11. **a.** yes; ABC and AFBC **b.** no **c.** 4

12.

From \ To	A	B	C	D
A	0	1	0	0
B	0	0	0	0
C	1	2	0	0
D	1	2	1	0

13.

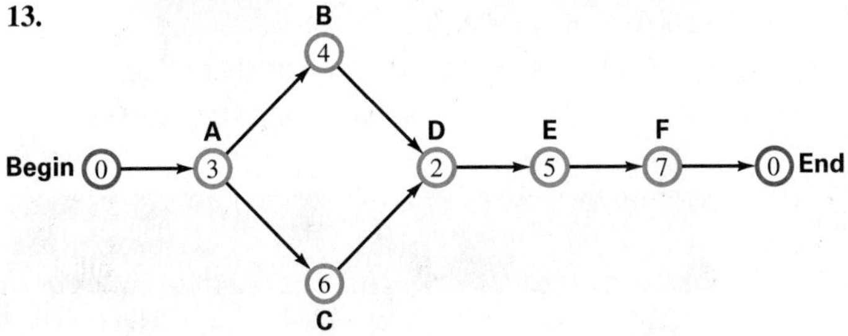

14. **a.** Begin, B, D, E, G

b. Begin, B, D, E, G, End

c. To begin on the tenth day of the project

d. 15 days

Chapter 5

1. **a.** 21,346

b. 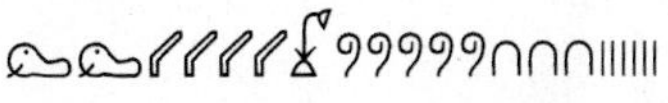

2. **a.**  **b.** 

3. **a.** $1 + 8 + 16$ **b.** $43 + 344 + 688 = 1{,}075$

4. **a.** We cannot subtract L; we can only subtract I from V and X.

b. 548 **c.** $9 \times 100 = 900$; $5 \times 1{,}000 = 5{,}000$

5. **a.** $12 \times 60^2 + 23 \times 60 + 32 = 44{,}612$

b. 

6. 868,682

7.

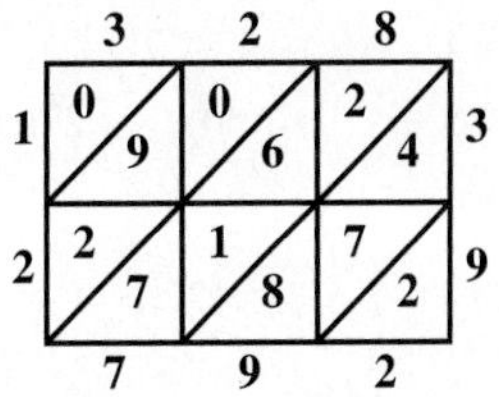

Product is 12,792.

8. 1, 2, 10_3, 11_3, 12_3, 20_3, 21_3, 22_3, 100_3, 101_3

9. 142

10. 423_5

11. 11_5 and 10_5

12. **a.** $1{,}243_6$ **b.** 154_6

13. 13_5 and 31_5

14. **a.** $3{,}133_5$ **b.** $3{,}402_8$

15. **a.** 3 **b.** 6

16. **a.** false **b.** true

17. **a.** 2 **b.** 7 **c.** 2

18. **a.** 5 **b.** 0, 3, 6

19. 33

Chapter 6

1. **a.** prime **b.** composite **c.** composite **d.** prime

2. **a.** yes **b.** yes **c.** no **d.** yes

e. yes **f.** no **g.** yes **h.** yes

3. **a.**

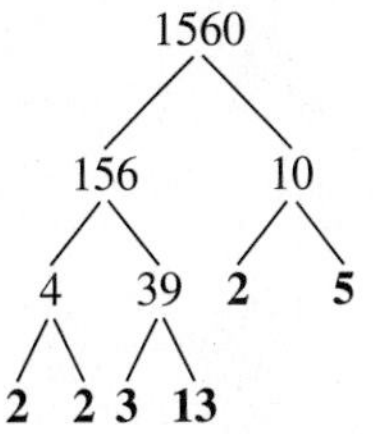

b. $1{,}560 = 2 \cdot 2 \cdot 2 \cdot 3 \cdot 13 \cdot 5 = 2^3 \cdot 3 \cdot 5 \cdot 13$

4. 42

5. **a.** +10 **b.** −5 **c.** −10 **d.** +22

6. **a.** $(+5) - (+12) = (+5) + (-12) = -7$

b. $(-3) + (+9) = +6$

7. **a.** +24 **b.** −56

8. −24

9. **a.** equal **b.** not equal

10. $\frac{3}{4}$

11. **a.** $\frac{29}{24}$ and $\frac{17}{48}$ **b.** $\frac{1}{2} + \frac{1}{2} \neq \frac{1+1}{2+2}$

12. **a.** $\frac{14}{5}$ **b.** $\frac{10}{9}$

13. **a.** $\frac{5}{22}$ **b.** $\frac{12}{7}$

14. **a.** $5\frac{3}{4}$ **b.** $\frac{13}{5}$

15. $0.\overline{63}$

16. **a.** $\frac{2{,}548}{10{,}000} = \frac{637}{2{,}500}$ **b.** $\frac{6}{11}$

17. **a.** $3\sqrt{5}$ **b.** $\frac{\sqrt{5}}{4}$

18. $\frac{\sqrt{15}}{5}$

19. **a.** $11\sqrt{5}$ **b.** 0

20. **a.** $3 + (\sqrt{2} + 7) = (3 + \sqrt{2}) + 7$ **b.** $5 \times 6 = 6 \times 5$

c. $(-2) \times (-3) = 6$, which is not a negative integer.

21. $3^6 = 729$ **22. a.** 5^{12} **b.** $(-2)^8 = 256$

23. a. $\frac{1}{36}$ **b.** 64 **c.** 1

24. a. $3^8 \cdot 3^{-6} = 3^{8+(-6)} = 3^2 = 9$

b. $(2^{-2})^{-1} = 2^{(-2)\cdot(-1)} = 2^2 = 4$

25. a. 5.372841×10^4 **b.** 0.00245 **26.** 50 **27.** 440

28. 354,294 **29.** 144 and 233

Chapter 7

1. a. $x = 2 - \frac{2}{3}y$ **b.** $W = \frac{P - 2L}{2}$

2. x-intercept, $(4\frac{1}{2}, 0)$; y-intercept, $(0, 6)$

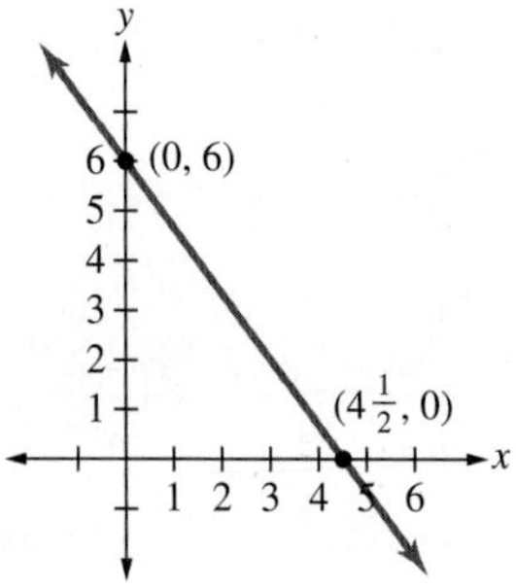

3. $\frac{5}{4}$

4. $y = -4x + 5$ **5.** $y = -\frac{5}{9}x + \frac{10}{3}$

6. $x = 4$ and $x = -11$

7.

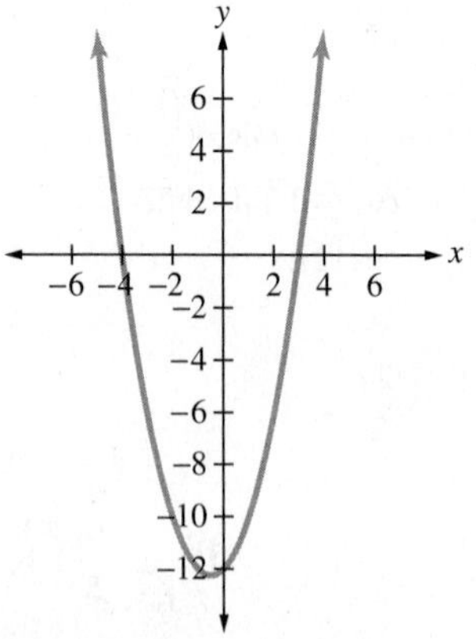

8. a. \$1,451.61 **b.** \$8,611.05 **9.** slightly over 121.1 million

10. 2045 **11. a.** $y = 180(1.04)^x$ **b.** \$256.20

12. 0.6096, or 60.96% **13.** 4.5 hours

14. 104,000 **15.** 133 **16.** 12 minutes

17. (3, 1) **18.** (−4, 5)

19. a. There are no solutions.

b. The system represents a pair of distinct parallel lines.

20. h) and j)

21.

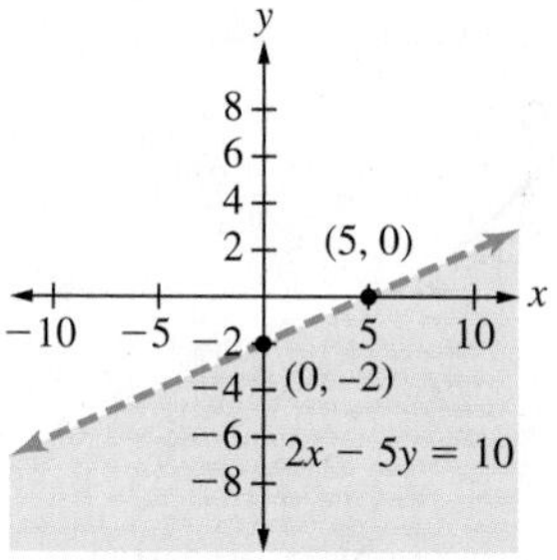

22.

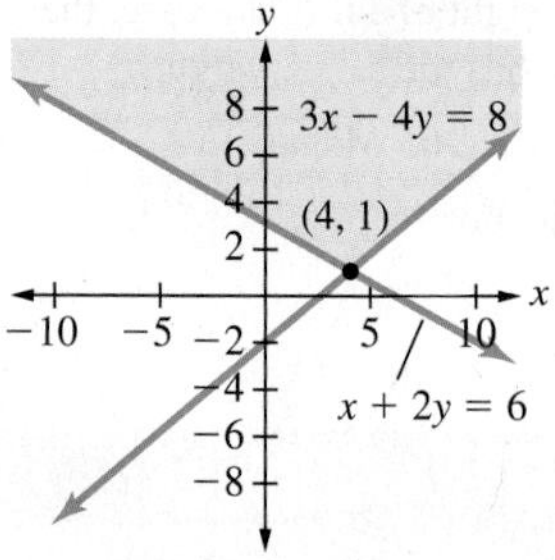

23. 496 mg **24. a.** $\frac{1}{2}$; unstable **b.** −6; stable

Chapter 8

1. a. 0.1745 **b.** 0.0005 **c.** 245% **d.** 2.5% **2.** 24%

3. a. 9 **b.** 75 **c.** 30% **4.** \$18,271.50

5. \$2,000 **6.** \$2,928.20 **7.** \$4,440.73

8. \$1,126.49 **9.** 1.511391594 **10.** 17.67 years

11. \$36.30 **12.** \$289.25; \$5.06 **13.** \$175; \$125; \$9,875

14.

Day	Balance	Number of Days × Balance
1, 2	\$240	2 × 240 = 480
3, 4, 5, 6, 7, 8, 9, 10	\$264	8 × 264 = 2,112
11, 12, 13, 14, 15, 16, 17, 18, 19, 20, 21, 22	\$204	12 × 204 = 2,448
23, 24, 25, 26, 27, 28, 29, 30	\$216	8 × 216 = 1,728

$$\text{average daily balance} = \frac{6{,}768}{30} = \$225.60.$$

15. $\frac{x^4 - 1}{x - 1}$ **16.** \$1,907.40 **17.** \$95.47

18. 240 months, or 20 years **19.** \$253.94

20.

4	\$1,199.10	\$997	\$202.10	\$199,197.60

21. \$9,493.49 **22. a.** \$840 **b.** \$709.99 **c.** \$31,202.40

23. a. \$17.76 **b.** about 11% **24.** 13% **25.** 18.81%

Chapter 9

1. **a.** a line **b.** line segment *CD*
 c. a ray **d.** a half line
2. **a.** vertical angles **b.** a right angle
 c. supplementary angles **d.** an obtuse angle
 e. an acute angle
3. **a.** 7 **b.** 8 **c.** 6 **d.** 4
4. **a.** 104° **b.** 76° **5.** 5 in. **6.** 360°
7. 135° **8.** 126 square units **9.** 17.32 sq in.
10. 18 sq ft **11.** approximately 17.3 in.
12. circumference $\approx$ 50.24; area $\approx$ 200.96
13. volume = 144 cu cm; surface area = 180 sq cm
14. volume $\approx$ 785 cu cm; surface area $\approx$ 471 sq cm
15. volume $\approx$ 83.73 cu yd; surface area $\approx$ 80.42 sq yd
16. $\frac{4}{3}\pi(6)^3 \approx 904$ cu cm **17. a.** 1,000 L **b.** 100 cg
18. **a.** 56,300 dL **b.** 0.04850 hg
19. 2,145 ft **20.** 8.36 lb.; 133.76
21.

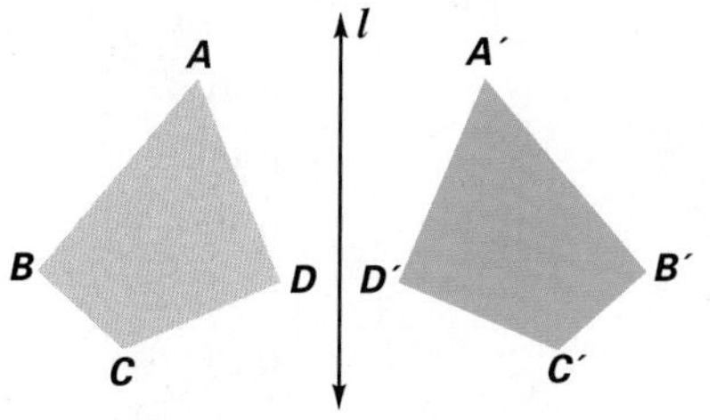

22.

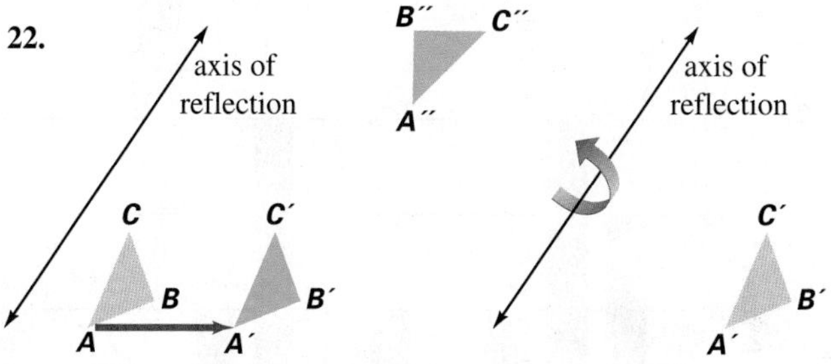

23. Each vertex is the vertex of two octagons and one square. The interior angle of each octagon is 135° and the interior angle of the square is 90°. The tessellation is possible because 135° + 135° + 90° = 360°.
24. 256 **25.** 243 **26.** $\frac{64}{27}$ **27.** $\frac{27}{64}$ **28.** 1.5

Chapter 10

1. On a 12-member board, Naxxon gets 6 representatives, Aroco 4, and Eurobile 2.
2. **a.** The average constituency of the electricians is 140.
 b. The electricians are more poorly represented.
3. 64 **4.** $\frac{2{,}351 - 2{,}342}{2{,}342} = \frac{9}{2{,}342} \approx 0.004$
5. **a.** 0.216 **b.** 0.175
 c. State B should get the additional representative.
6. For Iowa, the Huntington–Hill number is $\frac{3^2}{4 \times 5} = 0.45$; for Nebraska, the Huntington–Hill number is $\frac{(1.8)^2}{3 \times 4} = 0.27$. Therefore, Iowa should get the additional representative.
7. The Huntington–Hill numbers for the three companies are as follows: Naxxon, 110.5; Aroco, 114.1; Eurobile, 128. Because 128 is the largest, Eurobile gets the ninth representative.
8. **a.** 1,500,000 **b.** A: 2 B: 2.67 C: 3.33
9. 3.01; 3; 4 **10. a.** 600 **b.** 4.9 **c.** 4
11. **a.** 480 **b.** 2.6 **c.** 3
12. **a.** 1,500 **b.** 2.79 **c.** 3
13. **a.** discrete **b.** continuous **c.** discrete **d.** continuous
14.

	Rosa ($\frac{1}{4}$)	Juan ($\frac{1}{4}$)	Carlos ($\frac{1}{4}$)	Luis ($\frac{1}{4}$)
Bid on house	$250,000	$240,000	$260,000	$200,000
Fair share of estate	$62,500	$60,000	$65,000	$50,000
Item obtained with highest bid			House	

15.

	Rosa ($\frac{1}{4}$)	Juan ($\frac{1}{4}$)	Carlos ($\frac{1}{4}$)	Luis ($\frac{1}{4}$)
Pays to estate (+) or receives from estate (−)	$62,500 (−)	$60,000 (−)	$195,000 (+)	$50,000 (−)
Division of estate balance ($22,500)	$5,625 (−)	$5,625 (−)	$5,625 (−)	$5,625 (−)
Summary of cash	Receives $68,125	Receives $65,625	Pays $189,375	Receives $55,625

Chapter 11

1. A wins with 78 points. **2.** C wins.
3. N wins with 2 points. B gets 1 point, and T gets none.
4. The winner is A with 67 points. Because C has the majority of first-place votes, the majority criterion is not satisfied.
5.

1st	H	H	H	I	I	T	T
2nd	I	I	I	T	T	I	I
3rd	T	T	T	H	H	H	H

6. Using the pairwise comparison method, the winner is W. However, if X and Y are removed, then Z wins the election. The independence-of-irrelevant-alternatives criterion is not satisfied.
7. **a.** 4 (Let's call them A, B, C, and D.) **b.** 5
   c. {A, B}, {A, C}, {A, B, C}, {A, B, D}, {A, C, D}, {B, C, D}, {A, B, C, D}
8. **a.**

	Winning Coalitions	Critical Members
1	{A, B}	A, B
2	{A, C}	A, C
3	{A, B, C}	A
4	{A, B, D}	A, B
5	{A, C, D}	A, C
6	{B, C, D}	B, C, D
7	{A, B, C, D}	None

   b. A, $\frac{5}{12}$; B, $\frac{3}{12}$; C, $\frac{3}{12}$; D, $\frac{1}{12}$
9. $5! = 120$ **10. a.** B **b.** A

Chapter 12

1. $7 \times 6 = 42$ **2.** 16 **3.** $3 \times 3 \times 3 = 27$ **4.** 24
5. $2 \times 8 \times 3 \times 2 = 96$ different cars **6.** 2,116,800
7. **a.** $P(8, 3)$ is the number of ways we can select three different objects from a set of eight and arrange them in a straight line.
   b. 336
8. 792 **9. a.** 1 7 21 35 35 21 7 1 **b.** 35
10. $C(3, 0)$ $C(3, 1)$ $C(3, 2)$ $C(3, 3)$
11. $13 \times C(4, 2) = 13 \times 6 = 78$

Chapter 13

1. **a.** {(1, 5), (2, 4), (3, 3), (4, 2), (5, 1)}
   b. {bbb, bbg, bgb, gbb}
2. 0.72 **3.** 0.453 **4.** 16
5. **a.** $\frac{5}{36}$ **b.** $\frac{C(12, 2)}{C(52, 2)} = \frac{66}{1,326} = \frac{11}{221}$
6. $1 - \frac{3}{36} = \frac{33}{36} = \frac{11}{12}$ **7.** 0.45 **8.** $\frac{0.04}{0.09} = 0.44$
9. $\frac{7}{15} \cdot \frac{3}{14} \cdot \frac{5}{13} \approx 0.038$ **10. a.** 0.20 **b.** 0.52
11. Because $P(O) = \frac{1}{2} = P(O \mid F)$, the events are independent.
12. \$126 **13.** $+\frac{1}{9}$; it is now to the student's advantage to guess.
14. \$12.00 **15.** $B\left(8, 2; \frac{1}{6}\right) = 0.2605$

Chapter 14

1.

Score	Frequency	Relative Frequency
1	3	0.15
2	4	0.20
3	2	0.10
4	0	0.00
5	2	0.10
6	1	0.05
7	4	0.20
8	2	0.10
9	1	0.05
10	1	0.05

2.

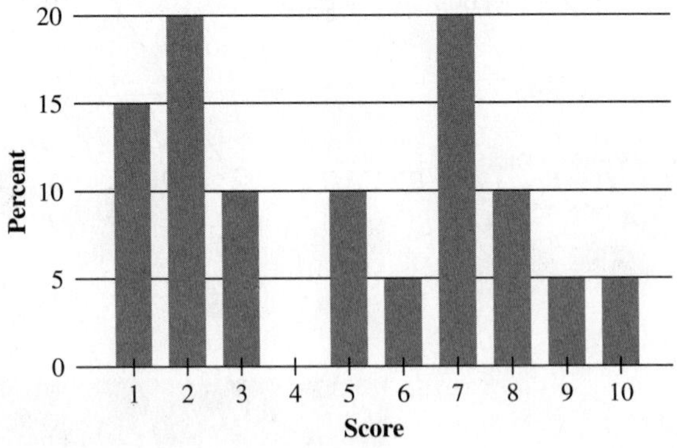

3.

6	2 2 3 6 7 8
7	1 3 5 7
8	1 2 2 3 4 8
9	0 1 2 8

4. **a.** $\Sigma x = 106$ **b.** $n = 8$ **c.** $\bar{x} = 13.25$

5. The mean number of calls is $\dfrac{\Sigma(x \cdot f)}{\Sigma f} = \dfrac{96}{14} \approx 6.86$.

6. \$104.88 million **7.** 95

8. **a.** 84, 90, 110, 130, 290

b.

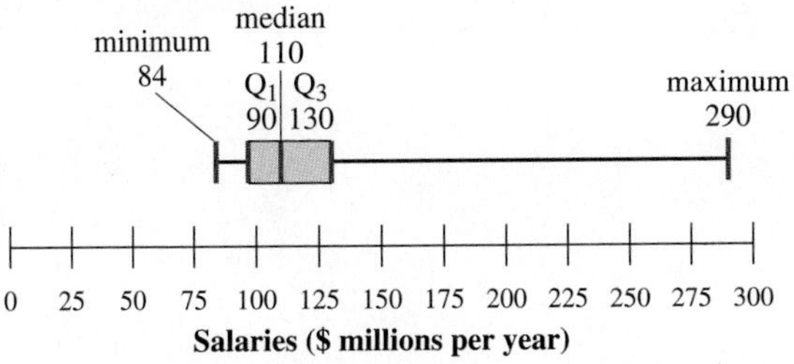

9. 2.29 **10.** $s = \sqrt{\dfrac{\sum(x - 42)^2 \cdot f}{n - 1}} = \sqrt{\dfrac{172}{19}} \approx 3.01$

11. **a.** 950 **b.** 25

12. **a.** 0.427 **b.** $0.468 - 0.391 = 0.077$ **c.** 0.412

13. **a.** 1.29 **b.** 0 **c.** -1.71

14. 16 months

15.

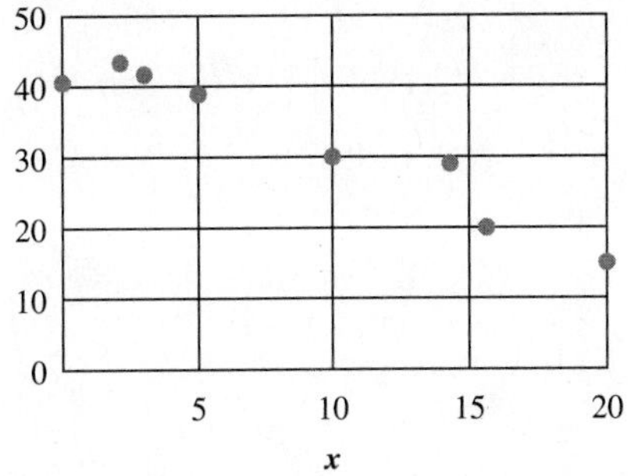

We see in this graph that as x increases, y generally decreases.

16. -0.9718

Answers to Exercises

Chapter 1

Section 1.1

1.

3.
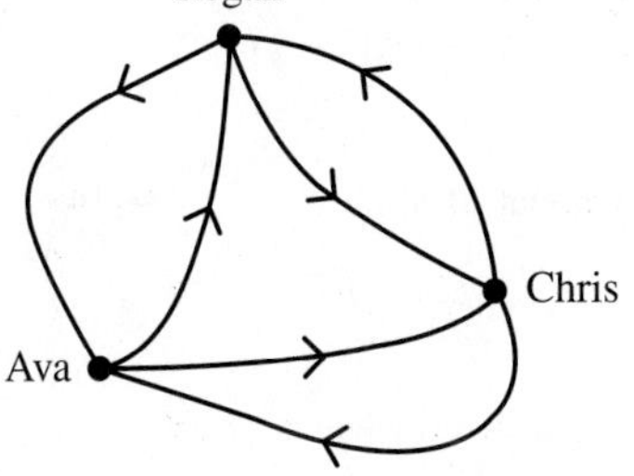

5. H, W, S

7. *s* for amount invested in stocks; *b* for amount invested in bonds

9. HH, HT, TH, TT

11. 32

13. BR, BU, BE, BL, RU, RE, RL, UE, UL, EL

15. (1, 1), (1, 2), (1, 3), (1, 4), (2, 1), (2, 2), (2, 3), (2, 4), (3, 1), (3, 2), (3, 3), (3, 4), (4, 1), (4, 2), (4, 3), (4, 4)

17. r_3 is the set of people who are good singers and appeared on *American Idol*. r_4 is the set of people who appeared on *American Idol* and are not good singers.

19. 35, 42, 49, 56, 63

21. *bf, cd, ce, cf, cg*

23. 21, 34, 55, 89, 144

25. Possible answer: Three people A, B, C can be lined up in 6 ways; four people can be lined up in 24 ways.

27. Possible answer: With three letters there are 9 ways; with five letters there are 25 ways.

29. Possible answer: For only air conditioning (*A*) and CD player (*C*) there are four possibilities: neither, *A*, *C*, *AC*. With three options there are eight possibilities.

31. False; September has 30 days.

33. False; almost any numbers give a counterexample.

35. False; A is the grandfather of C.

37. false **39.** not the same **41.** the same

43. 5 is a number; {5} is a number with braces around it.

45. One is uppercase, the other is lowercase.

47. The order of the numbers is different.

49. Answer will vary. **51.** Answer will vary.

53. 107, 214 **55.** 9 in the store and 6 giving lessons

57. 36 **59.** \$3,500 at 8% and \$4,500 at 6% **61.** 3 students

63. LCHPL, LCPHL, LHCPL, LHPCL, LPCHL, LPHCL

65.

Math	English	Sociology	Art History
9	11	12	10
9	12	10	11
10	9	12	11
10	11	12	9
12	9	10	11
12	11	10	9

67.

Math	English	Art History
9	11	10
9	12	10
9	12	11
10	9	11
10	11	9
10	12	9
10	12	11
12	9	10
12	9	11
12	11	9
12	11	10

69. 79

71. Yes; if the 8% raise is done first, you will pay more over the two years than if the 5% raise is done first.

73. Answers will vary. **75.** Answers will vary.

77. 216 **79.** (41, 43), (59, 61) **81.** 55 **83.** 252

Section 1.2

1. inductive **3.** deductive **5.** inductive

7. deductive **9.** inductive

11. 16 **13.** 96 **15.** 21

17.
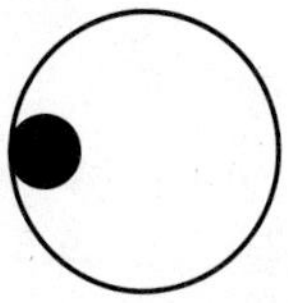

19.
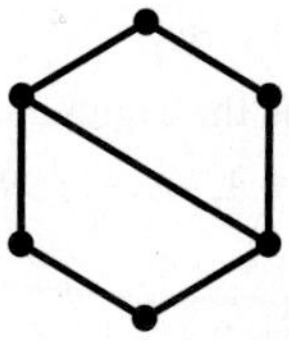

21. *Hint:* Think of prime numbers.

	X	X		X		X				X		X	

23. 5 + 11 **25.** 7 + 13

27. $1 + 2 + 3 + 4 + 5 = \frac{5 \times 6}{2}, 1 + 2$

29. $1 + 3 + 5 + 7 + 9 = 25, 1 +$

31. 30

33. The total of all the numbers in the square is $1 + 2 + 3 + \cdots + 16 = 136$, so the numbers in each of the four rows is $\frac{136}{4} = 34$. The same is true for the columns and diagonals. From this you can deduce the missing numbers.

7	6	12	9
10	11	5	8
13	16	2	3
4	1	15	14

35. The algebraic interpretation of the steps in the trick are as follows:

a. Call the number n. **b.** $3n$ **c.** $3n + 9$

d. $\frac{3n + 9}{3} = \frac{3n}{3} + \frac{9}{3} = n + 3$ **e.** $n + 3 - n = 3$

In this trick, you will always get the number 3.

37. The algebraic interpretation of the steps in the trick are as follows:

a. Call the number n. **b.** $8n$ **c.** $8n + 12$

d. $\frac{8n + 12}{4} = \frac{8n}{4} + \frac{12}{4} = 2n + 3$ **e.** $2n + 3 - 3 = 2n$

In this trick, you will always get a result that is twice the number that you started with.

39. Adriana (political issues), Caleb (solar power), Ethan (water conservation), and Julia (recycling)

41. 36644633 **43.** 986763

45. 720 for six cities; 5,040 for seven cities

47.

49. By looking at examples, inductive reasoning leads us to make conjectures that we then try to prove.

51. Answers will vary.

53. Answers will vary.

55. 53, 107, 213 (previous term plus twice the term before the previous term)

57. 47, 76, 123 (each term is the sum of previous two terms)

59. 20 **61. a.** 60 **b.** 210 **63.** 50

65. Answers will vary.

67. If you expand the expression as follows

$$\begin{aligned}(2n + 5)50 + 1{,}763 - 1{,}995 &= 100n + 250 + 1{,}763 - 1{,}995\\ &= 100n + 2{,}013 - 1{,}995\\ &= 100n + 18\end{aligned}$$

you see that $1{,}763 + 250$ gives a multiple of 100 plus $2{,}013 - 1{,}995$, which is your age. If you have already had your birthday, then you need to add the extra year, which is why we would then add 1,764.

on 1.3

$40 + 190 + 40 = 290$ **3.** $35 - 15 = 20$

$= 80$ **7.** $\frac{18}{3} = 6$ **9.** $0.1 \times 800 = 80$

11. $9\% \times 1{,}000 = 0.09 \times 1{,}000 = 90$

13. $4 \times 5 \times 6 = 120$

15. $\frac{325}{50} = 6.5$ more hours, 7:30 PM

17. roughly \$18.00

19. $3 \times \$3 + 4 \times \$1.50 + \$3 = \$18$

21. It seems safe. Alicia probably weighs less than 200 pounds, so that leaves $2{,}300 - 200 = 2{,}100$ pounds for the 21 students. They probably do not weigh 100 pounds each.

23. $\$40{,}000 \times 4\% = \$40{,}000 \times 0.04 = \$1{,}600$

25. roughly 1,000

27. Her total expenses are about \$100/month; $\frac{100}{7} \approx 14$,

$12 \times 14 = \$168$

29. \$42,000; \$41,100

31. college graduates; those with associate degrees

33. category 2

35. response 2

37. 17.3 million **39.** 24.7 million

41. \$875 billion **43.** \$156 billion

45. 213,019 **47.** 18.5%

49. Answers will vary. **51.** Answers will vary.

53. Answers will vary.

55. If you divide the grassy area into rectangles, you get 10,354 square feet. They need slightly over two bags of fertilizer.

57. Answers will vary.

Chapter Review Exercises

1. Understand the problem; devise a plan; carry out your plan; check your answer.

3. 10

5. 8 hours as a stockperson; 12 hours as a ski instructor

9. a. deductive **b.** inductive

11.

		X
		X

13. a. n **b.** $8n$ **c.** $8n + 12$

15. a. $210 - 60 = 150$ **b.** $6 \times 15 = 90$

17. a. \$21,000; \$8,000 **b.** ii **c.** from 1980–81 to 1990–91

Chapter Test

1. Answers will vary. **2.** Part b) is false. $\frac{3}{4 + 5} \neq \frac{3}{4} + \frac{3}{5}$

3. 31 million **4. a.** 4,320 **b.** 2,280

5. a. 36,000 **b.** 36,500

8. **a.** inductive **b.** deductive **10.** $3,200 **11.** 141

12. cde, cdf, cdg

13.

	X	X	
		X	X

14. $23 + 37 = 60$

15. False; Suppose the laptop cost $1,000. With a 10% discount, the computer would cost $900. If the price is increased by 10%, the computer would cost $990, not $1,000.

16. **a.** Call the number you choose n.

b. We get $4n$.

c. Next we get $4n + 40$.

d. Dividing by 2 gives us $2n + 20$.

e. Subtracting gives us $2n$, or twice the original number.

Chapter 2

Section 2.1

1. {10, 11, 12, 13, 14, 15}

3. {17, 18, 19, 20, 21, 22, 23, 24, 25}

5. {4, 8, 12, 16, 20, 24, 28}

7. {Sunday, Monday, Tuesday, Wednesday, Thursday, Friday, Saturday}

9. $\emptyset$ **11.** $\emptyset$

13. $\{x : x$ is a multiple of 3 between 3 and 12 inclusive$\}$

15. {28, 29, 30, 31}

17. {January, February, March, . . . ,December}

19. {101, 102, 103, . . .}

21. $\{x : x$ is an even natural number between 1 and 101$\}$

23. well defined **25.** not well defined

27. not well defined **29.** well defined

31. $\notin$ **33.** $\in$ **35.** $\in$ **37.** $\in$ **39.** $\notin$

41. $\in$ **43.** 6 **45.** 0 **47.** 4 **49.** 2 **51.** 1

53. finite **55.** infinite **57.** 4.5 **59.** Sony

61. Angela Merkel **63.** Sunday

65. $\{x : x$ is a humanities elective$\}$

67. {History012, History223, Geography115, Anthropology111}

69. {AZ, FL, GA, LA, NJ, NM, TX, VA}

71. $\{x : x$ had an average price of gasoline above 380$\}$

73. {Amazon, Apple, eBay, Facebook, Interactive Corp, News Corp}

75. $\{x : x$ had an audience of less than 50 million$\}$

77. Answers will vary.

79. $\{\emptyset\}$ is not empty; it contains one element, $\emptyset$.

81. Precise definitions are important, not only in mathematics but in everyday life.

83. Answers will vary.

Section 2.2

1. equal **3.** not equal **5.** equal **7.** equal

9. true **11.** false **13.** true **15.** equivalent

17. equivalent **19.** not equivalent **21.** equivalent

23. not equivalent **25.** {1, 2}, {1, 3}, {2, 3}

27. {1, 2, 3}, {1, 2, 4}, {1, 3, 4}, {2, 3, 4}

29. 32; 31 **31.** T **33.** L **35.** 64 **37.** 16

39. $2^7 = 128$ **41.** $2^8 = 256$

43. 7 **45.** {5P, 10P, 25D}

47. {5P, 10P, 25S} or {5P, 10P, 25D}

49. **a.** {1} is a subset, not an element of {1, 2, 3}.

d. 3 is an element, not a subset of {1, 2, 3}.

51. At the first branching in the tree, the "yes" or "no" indicates that in forming a subset of {1, 2}, we will either take the 1 or omit it. The second branchings in the tree indicate whether we are going to take the 2 as a member of the subset that we are forming. The tree shows all possible ways that we can decide to take or not to take 1 and 2 in forming a subset of {1, 2}. The top branch corresponds to the subset {1, 2}, the second branch corresponds to the subset {1}, and so on.

53. **a.** 25 is not a power of 2

b. He confused 5^2 with 2^5 **c.** 32

55. Answers will vary. **57.** 31

59. The fifth line counts the number of subsets of sizes 0, 1, 2, 3, 4, and 5 of a five-element set.

61. 84 **63.** The total in the nth row is 2^n. **65.** 16

67. If $A \subseteq B$ and $B \subseteq C$, then $A \subseteq C$. **69.** n!

Section 2.3

1. {1, 3, 5} **3.** {1, 2, 3, 4, 5, 6, 7, 8} **5.** {1, 3, 5, 7, 9}

7. {1, 2, 3, 4, 5, 6, 7, 8, 9, 10} **9.** {1, 3, 5, 7} **11.** {9}

13. {potato chip, bread, pizza} **15.** {apple, fish, banana}

17. {fish} **19.** 7 **21.** 5

23.

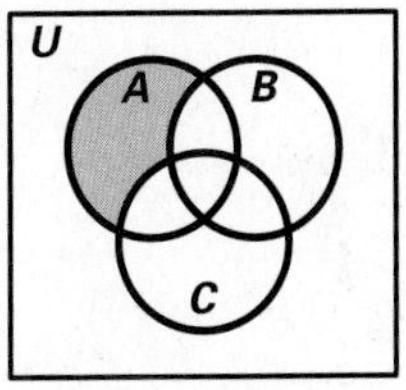

25.

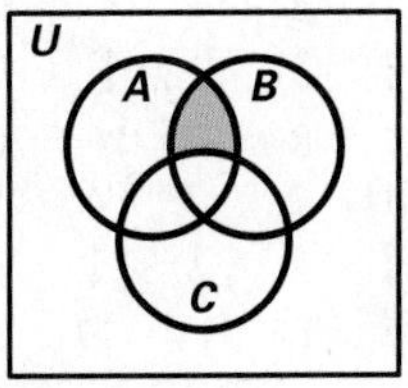

27. 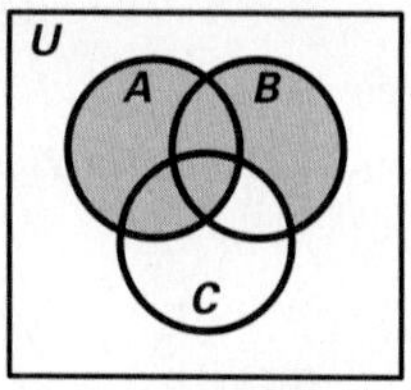

29. 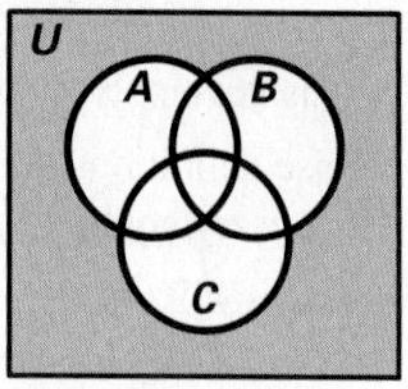

31. $B - A$ **33.** $(A \cup B)'$ **35.** $A \cap B \cap C$

37. $(A \cup C) - B$ **39.** equal **41.** 30 **43.** 28

45. 20 **47.** 27 **49.** $\{d, f, g\}$ **51.** $\{c, d, f, g\}$

53. $\{b, d, f, g, h\}$ **55.** $\{d, f\}$ **57.** $\{m, mc, hc\}$

59. $\{m, mc, bc, c, hc\}$ **61.** {FL, NJ, TX} **63.** {CA, NY}

65. $\{a, c, e\}$ **67.** $\{f\}$ **69.** $\{a, c\}$

71. *Union* implies "joining together," as in a labor union; *Intersection* implies "overlapping," as in streets.

73. Answers will vary.

77. false **79.** true **81.** A **83.** B

85. **a.** $A \cap B = B \cap A$ is true.

b. $(A \cup B) \cup C = A \cup (B \cup C)$ is true.

Section 2.4

1. 2, 3 **3.** 2, 3, 4 **5.** 2, 3, 5, 6 **7.** 4, 7 **9.** 7

11. 25 **13.** 18 **15.** 16 **17.** 20 **19.** 19

21. $n(A) = 7, n(B) = 3, n(C) = 12$

23. $n(A) = 18, n(B) = 15, n(C) = 14$

25. $n(A) = 5, n(B) = 14, n(C) = 9$

27. 59 **29.** 28 **31.** 30, 49

33. **a.** 76 **b.** 16 **c.** 21 **35.** **a.** 74 **b.** 45 **c.** 8

37. If none use both the bus and the train, then $68 + 59 = 127$ use either the bus or the train. If we add the 44 who use only the subway and the 83 that use none of the three, the total exceeds 200.

39. 158 **41.** 20 **43.** A⁻ **45.** A⁻, B⁻, or AB⁻

47. A ∩ B ∩ Rh **49.** B⁻, O⁻

51. The region r_2 is the part of A that is outside B. It does not take into account elements in $A \cap B$.

53. Answers will vary. **55.** 8 **57.** Answers will vary.

Section 2.5

1.
1 2 3 4 5 … n …
$\updownarrow$ $\updownarrow$ $\updownarrow$ $\updownarrow$ $\updownarrow$ $\updownarrow$
4 8 12 16 20 … $4n$ …

3.
1 2 3 4 5 … n …
$\updownarrow$ $\updownarrow$ $\updownarrow$ $\updownarrow$ $\updownarrow$ $\updownarrow$
8 11 14 17 20 … $3n + 5$ …

5.
1 2 3 4 5 … n …
$\updownarrow$ $\updownarrow$ $\updownarrow$ $\updownarrow$ $\updownarrow$ $\updownarrow$
2 4 8 16 32 … 2^n …

7.
1 2 3 4 5 … n …
$\updownarrow$ $\updownarrow$ $\updownarrow$ $\updownarrow$ $\updownarrow$ $\updownarrow$
1 1/2 1/3 1/4 1/5 … $1/n$ …

9.
1 2 3 4 5 … n …
$\updownarrow$ $\updownarrow$ $\updownarrow$ $\updownarrow$ $\updownarrow$ $\updownarrow$
3 6 9 12 15 … $3n$ …

11.
1 2 3 4 5 … n …
$\updownarrow$ $\updownarrow$ $\updownarrow$ $\updownarrow$ $\updownarrow$ $\updownarrow$
1 4 7 10 13 … $3n - 2$ …

13. Match {2, 4, 6, 8, 10, . . .} with {4, 6, 8, 10, 12, . . .}; in general, match $2n$ with $2n + 2$.

15. Match {7, 10, 13, 16, 19, . . .} with {10, 13, 16, 19, 22, . . .}; in general, match $3n + 4$ with $3n + 7$.

17. Match {2, 4, 8, 16, 32, . . .} with {4, 8, 16, 32, 64, . . .}; in general, match 2^n with 2^{n+1}.

19. Match {1, 1/2, 1/3, 1/4, 1/5, . . .} with {1/2, 1/3, 1/4, 1/5, 1/6, . . .}; in general, match $\frac{1}{n}$ with $\frac{1}{n+1}$.

21. Match {1/2, 1/4, 1/6, 1/8, 1/10, . . .} with {1/4, 1/6, 1/8, 1/10, 1/12, . . .}; in general, match $\frac{1}{2n}$ with $\frac{1}{2n+2}$.

23. 6 **25.** 25

27. We showed that the set of positive rational numbers is in a one-to-one correspondence with the natural numbers.

29. It is impossible to put {1, 2, 3, 4, 5} in a one-to-one correspondence with any of its 31 proper subsets.

31. They already had been matched as 1, 1/2, 1, and 2.

33. 6

35. If we take the union of {1}, which has cardinal number 1, and {2, 3, 4, . . .}, which has cardinal number $\aleph_0$, we get {1, 2, 3, 4, . . .}, which has cardinal number $\aleph_0$.

37. From this figure we see that every point on the semicircle matches with exactly one point on the line and vice versa.

Chapter Review Exercises

1. **a.** $\{x : x$ is an even natural number between 1 and 19$\}$

b. $\{x : x$ is a month in the year$\}$

c. {New Hampshire, New Jersey, New Mexico, New York}

d. $\varnothing$

3. 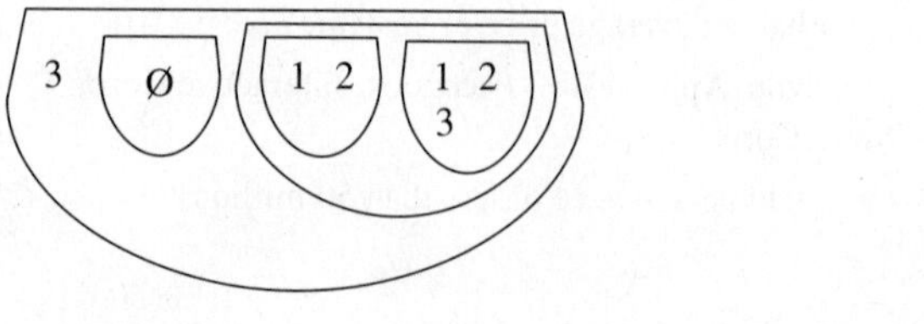

5. **a.** Yes; order of elements does not matter.

b. Yes; duplication of elements does not matter.

c. No; the first set has elements such as 1,002 that are not present in the second set

7. **a.** not equivalent **b.** equivalent **c.** equivalent **d.** equivalent

9. **a.** $\{5, 7, 9\}$ **b.** $\{2, 3, 4, 5, 7, 8, 9, 10\}$ **c.** $\{1, 4, 6, 7, 10\}$ **d.** $\{7\}$

11. **a.**

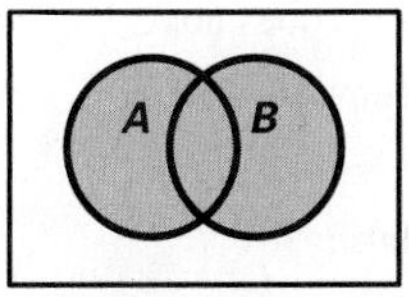

b.

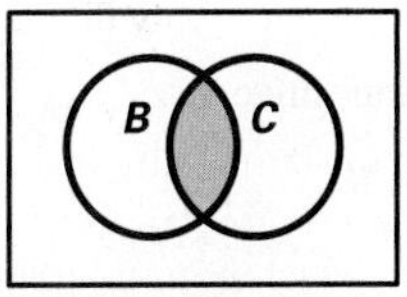

c.

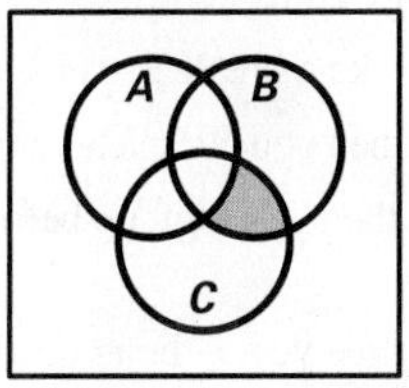

d.

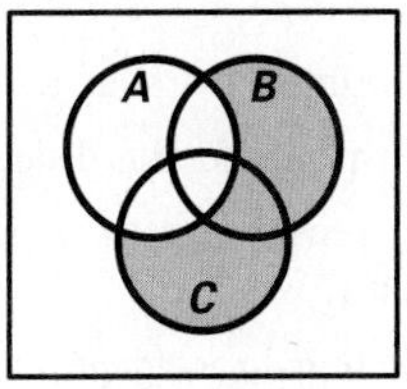

13. **a.** closure, commutativity, associativity, identity

b. closure, commutativity, associativity, identity

c. union distributes over intersection; intersection distributes over union

15. **a.** 10 **b.** 24 **17.** **a.** B^- **b.** O^-

19. $\{1, 2, 3, 4, \ldots\}$ can be put in a one-to-one correspondence with $\{2, 4, 6, 8, \ldots\}$.

21. We chose a digit that was different from the third decimal place of the third number in the list.

Chapter Test

1. **a.** $\{x : x \text{ is a natural number greater than } 100\}$

b. {January, February, March, . . . , December} **c.** $\varnothing$

2. **a.** Yes; the order of elements does not matter.

b. No; the set $\{1\}$ is not the same as the number 1.

c. Yes; the sets contain the same elements.

3. **a.** not equivalent **b.** equivalent **c.** equivalent

4. **a.** $\{1, 3, 5, 7, 9\}$ **b.** $\{3, 5\}$ **c.** $\{1, 3, 4, 6, 7, 8, 9, 10\}$ **d.** $\{2, 4, 6, 7, 8, 10\}$

5. $\varnothing$ contains no elements but $\{\varnothing\}$ contains one element, namely $\varnothing$.

6. **a.** 8 **b.** 1

7. 256

8.

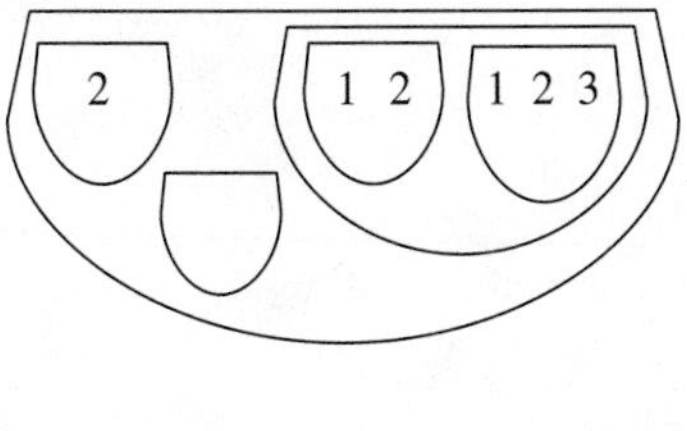

9. **a.** True; every element of the first set is a member of the second.

b. True; every element of the first set is a member of the second.

c. False; $\varnothing$ is not one of the numbers 1, 2, 3.

10. $A' \cup B'$

11. **a.**

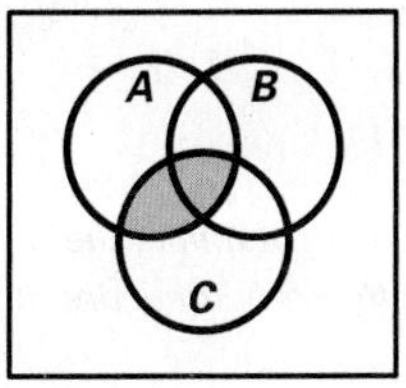

b.

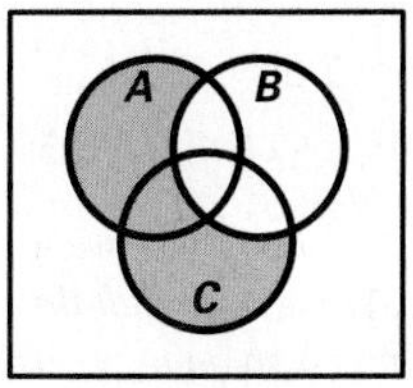

12. **a.** 10 **b.** 64

13. 89, 64

14. An infinite set can be placed in one-to-one correspondence with a proper subset of itself.

15. We can put the set $\{1, 2, 3, \ldots\}$ in a one-to-one correspondence with $\{2, 4, 6, \ldots\}$.

16. 1/4

17. We made the fifth decimal place of x to be different from the fifth decimal place of the fifth number in the list.

18. **a.** $A \cap B \cap Rh'$ **b.** $A' \cap B' \cap Rh$

Chapter 3

Section 3.1

1. statement **3.** not a statement

5. not a statement **7.** statement

9. not a statement **11.** compound; *if … then, and, or*

13. compound; *or* **15.** simple

17. compound; *if … then, or* **19.** compound; *if … then*

21. $g \vee \sim c$ **23.** $(\sim c) \rightarrow (g \vee a)$

25. The radial tires are included or the sunroof is not extra.

27. If the radial tires are included, then the sunroof is extra or the power windows are not optional.

29. There exists at least one snake that is not poisonous.

31. All personal items are covered by your renter's insurance.

33. All scientists believe that it is not true that an asteroid collision led to the extinction of the dinosaurs.

35. false (Nadia) **37.** false (all sophomores are athletes)

39. false (Lennox) **41.** true

43. false; All green happy faces are not in yellow boxes. (No green happy faces are in yellow boxes.)

45–53. We have not provided answers for these exercises. These statements are complex and sophisticated. You will probably find that you, your classmates, and your instructor do not always agree as to which connectives are present. Nevertheless, it is interesting to try to determine the form of these statements.

55. $p \wedge (q \vee r)$ 57. $(p \vee q) \wedge (p \vee r)$

59. true 61. false

63. a. "Some are not." b. "None are."

Section 3.2

1. 3 2 1 3. 4 3 2 1 5. T 7. T

9. F 11. T 13. yes; 6 15. no

In Exercises 17–25, we will give the final column of the truth table. We assume that the columns in the truth table are labeled p, q, r, *and so on in the usual way.*

17. FTFF 19. FFTF

21. FTTT 23. TFTTFFFF

25. FFFFTFFF 27. exclusive *or*

29. inclusive *or* 31. Bill is not tall or Bill is not thin.

33. Christian will not apply for a loan and he will not apply for work study.

35. Ken does not qualify for a rebate and he does not qualify for a reduced interest rate.

37. The number x is equal to 5 or s is odd.

39. yes 41. no 43. yes 45. yes

47. true 49. false 51. true

53. It is not true that: the earned income tax did reduce the tax you owe and gave you a refund.

55. It is not true that: you are single or the head of a household.

57. no 59. yes 61. false 63. false

65. $p \vee (q \vee r)$ 67. $p \vee \sim q$

69. We use the inclusive *or.* When filling in the tables in the standard fashion, the first line of the inclusive *or* table would have a *true* whereas the first line of the exclusive *or* table would have a *false.*

71. Let the subsets of $\{p, q, r\}$ indicate which of the variables should be true. For example, the subset $\{p, r\}$ specifies that p and r should be true and that q is false. This corresponds to the line labeled TFT in the truth table.

73. The stroke connective is logically equivalent to $\sim(p \wedge q)$.

75.

p	q	$(p\|p)$	$\|$	$(q\|q)$
T	T	F	T	F
T	F	F	T	T
F	T	T	T	F
F	F	T	F	T

Section 3.3

In exercises 9–19, we assume that the columns of the truth tables are labeled p, q, r, *and so on in the usual way.*

1. true 3. true 5. true 7. false

9. FTTT 11. FTFF 13. FFTTFTFT

15. TTTTTTTT 17. TTTTTFFT 19. TTTT

21. If it pours, then it rains.

23. If you do not buy the all-weather radial tires, then they will not last for 80,000 miles.

25. If its sides are not all equal in length, then a geometric figure is not an equilateral triangle.

27. If x does not evenly divide 6, then x does not evenly divide 9.

29. contrapositive 31. converse

33. $q \rightarrow \sim p$ 35. $q \rightarrow \sim p$

37. false 39. false

41. equivalent 43. not equivalent

45. If I finish my workout, then I'll take a break.

47. If you qualify for this deduction, then you complete Form 3093.

49. If you receive a free cell phone, then you sign up before March 1.

51. If you remain accident free for three years, then you get a reduction on your auto insurance.

53. true 55. true 57. true

59. false 61. true 63. false

65. If you can be claimed by someone else as a dependent, then your gross income is not more than \$2,250.

67. If you decrease the amount being withheld from your pay, then the amount you overpaid is large.

75. Jamie has not been a member for 10 years.

79. $r \rightarrow (p \wedge q)$

81. $p \vee (r \wedge (p \wedge q))$ is logically equivalent to p.

83.

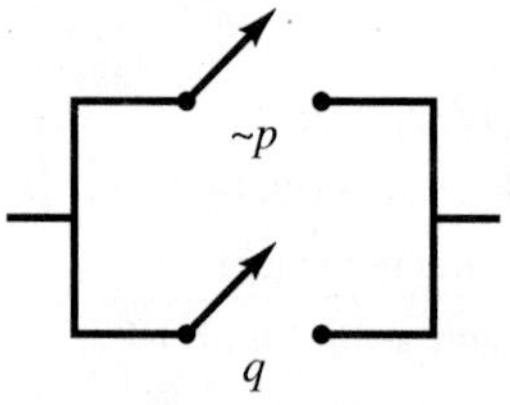

85.

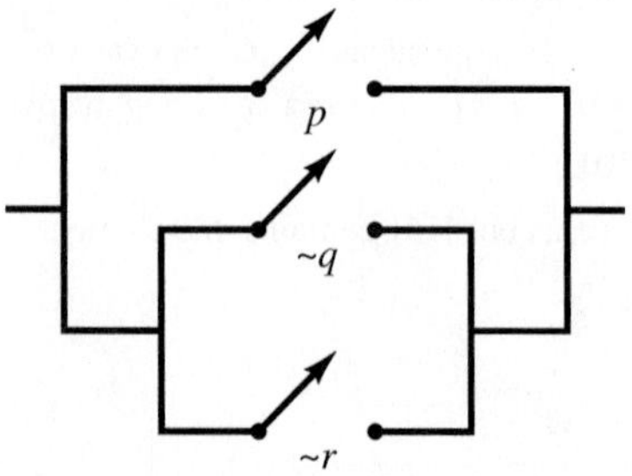

Section 3.4

1. law of detachment; valid

3. fallacy of the converse; invalid

5. disjunctive syllogism; valid
7. fallacy of inverse; invalid
9. law of syllogism; valid
11. law of contraposition; valid
13. fallacy of the inverse; invalid
15. law of syllogism; valid
17. valid 19. invalid 21. invalid
23. invalid 25. valid 27. invalid
29. Malik does not have the most expensive Dish TV package.
31. Minxia will attend school in Hawaii.
33. invalid 35. valid 37. valid 39. invalid
41. With the hypotheses $p \rightarrow q$ and p, we "detach" the p from the first conditional to get the conclusion, q; "disjunctive" reminds us of the word "disjunction," which contains an "or."
47. Voldemort is a knave and Dumbledore is a knight.
49. Rubeus is a knave and Bellatrix is a knight.
51. If $a \wedge b$ is true, then b is true. If b is true and $b \rightarrow c$ is true, by the law of detachment, then c is true. The contrapositive of $d \rightarrow \sim c$ is $c \rightarrow \sim d$. If both c and $c \rightarrow \sim d$ are true, then $\sim d$ is true.

Section 3.5

1. valid 3. invalid 5. invalid 7. invalid
9. valid 11. invalid 13. invalid 15. valid
17. Some taxes should be abolished.
19. Some teams that wear red uniforms do not play in a domed stadium.
21. Some opera singers have dogs.
23. No ballet dancers are firefighters.
25. There are many possible diagrams.
27. There are many possible diagrams.

Section 3.6

9. 0.05 11. 0.25 13. 0.85 15. 0.15
17. 0.27 19. 0.64 21. 0.29 23. 0.71
25. sales trainee (0.50) 27. attend Good Old State (0.60)
29. If we only allowed the values 1 (for true) and 0 (for false), then the rules for computing truth values in fuzzy logic are exactly the same as the rules for computing truth tables.

Chapter Review Exercises

1. **a.** not a statement **b.** statement **c.** not a statement
3. **a.** It is not true that: Antonio is fluent in Spanish and he has not lived in Spain for a semester.
   b. Antonio is not fluent in Spanish or he has not lived in Spain for a semester.
5. **a.** true **b.** false **c.** false
7. **a.** FFTF **b.** FFFFFTFF
9. **a.** logically equivalent b. not logically equivalent
11. **a.** TTTF **b.** TFTFFFTT
13. **a.** If the Heat get to the finals, then they beat the Lakers.
    b. If you are an astronaut, then you have a pilot's license.
15. invalid 17. invalid 19. **a.** 0.18 **b.** 0.53 **c.** 0.53

Chapter Test

1. **a.** statement **b.** not a statement
2. **a.** There is at least one rock star who is not a fine musician.
   b. No dogs are aggressive.
3. **a.** $p \vee (\sim f)$ **b.** $\sim((\sim p) \wedge f)$
4. **a.** It is not true that: the Tigers will win the series or Verlander will not win the Cy Young Award.
   b. The Tigers will not win the series and Verlander will not win the Cy Young Award.
5. 64 6. **a.** true **b.** false **c.** false
7. **a.** 0.25 **b.** 0.62 8. **a.** TFTT **b.** FFTFTFTF
9. **a.** If you go to Wikipedia, then you will get enough sources for your research term paper.
   b. If Ticketmaster will mail the concert tickets, then you will pay a fee.
10. **a.** You cannot take the final exam and you cannot write a term paper.
    b. I will finish the painting and I will show it at the gallery.
11. **a.** logically equivalent **b.** not logically equivalent
12. Converse: If it is gold, then it glitters.
    Inverse: If it does not glitter, then it is not gold.
    Contrapositive: If it is not gold, then it does not glitter.
13. **a.** true **b.** true **c.** true 14. **a.** FTTT **b.** FTTFTFTF
15. valid
16. **a.** fallacy of the inverse **b.** disjunction syllogism
17. $(\sim p \vee q)$ 18. invalid 19. invalid

Chapter 4

Section 4.1

1. connected; odd: A, B; even: C, D
3. connected; odd: A, B, E, F; even: C, D
5. not connected; all vertices are even
7. connected; odd: A, B, E, F; even: C, D
9. yes
11. no; four odd vertices
13. no; not connected
15. no; four odd vertices

In Exercises 17–19, we put spaces in the sequences of vertices to make them more readable.

17. ABFHI GCBEF GECDA

19. duplicate edges: AJ, IJ, DE, EF; ABCDE FEDHF GHICA JIJA

21. 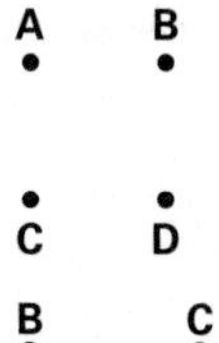

23. impossible

25. 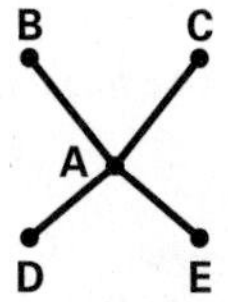

27. 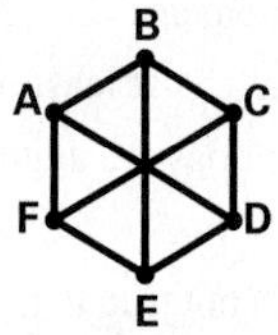

29. Remove AB

31. Remove AB

33. No; if we represent this map by a graph, it has four odd vertices.

In Exercise 35, after duplicating edges as indicated, follow edges in numerical order.

35. 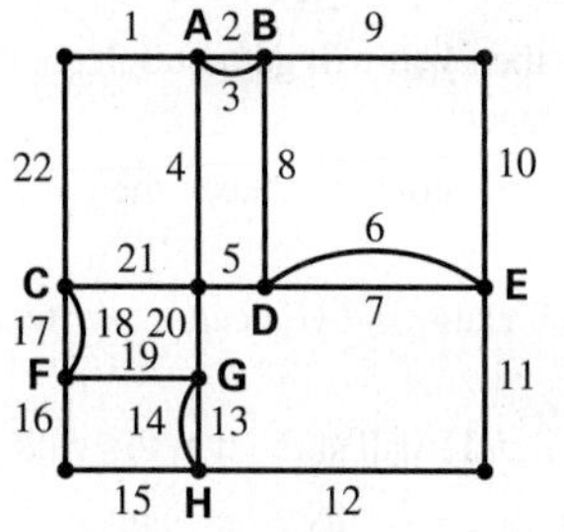

37. 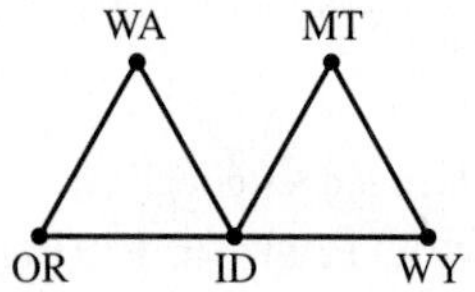

39. 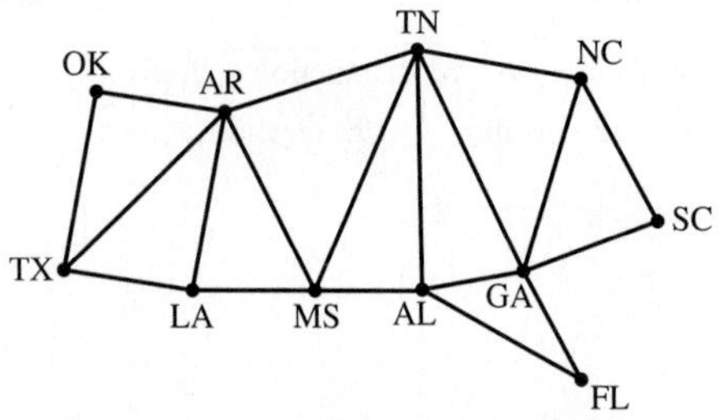

41. 3

43. 3

45. The trip is possible because the graph in Exercise 37 has no odd vertices.

47. The trip is not possible because the graph in Exercise 39 has more than two odd vertices.

49. Not possible. In the graph model, B will be an odd vertex, which is neither a beginning nor ending vertex in the attempt to trace the graph.

51. The graph below relates the people who do not get along with each other. Because this graph can be colored with three colors, three tables will be required. One possible seating: Table 1—Peter, Chris, Glenn, and Dianne. Table 2—Carter, Lois, Tom, and Cleveland; Table 3—Barbara, Steve, and Meg.

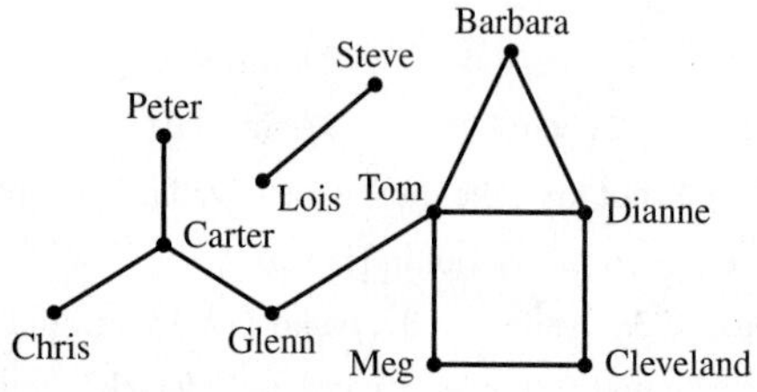

53. Model the committees with vertices. Two vertices are joined if they have common members. Color the graph and then committees that are colored the same can meet at the same time. This graph requires three colors, so three meeting times are needed.

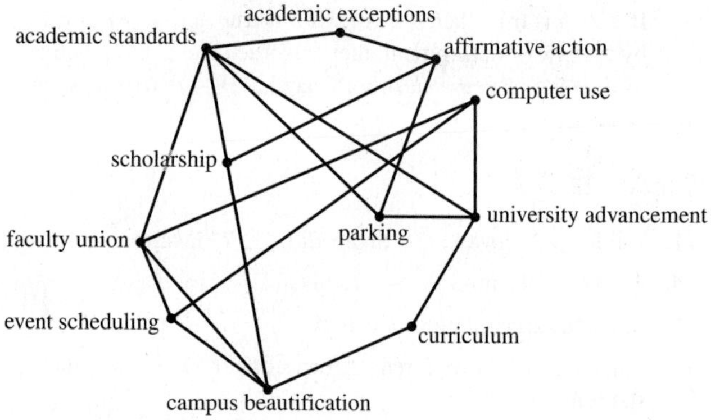

55. Every time we go into a vertex we must also leave it. The only possible vertices that we don't both enter and leave are the beginning and ending vertices.

57. No. Since there is a circuit in an Eulerian graph, removing one edge will not disconnect it.

61. 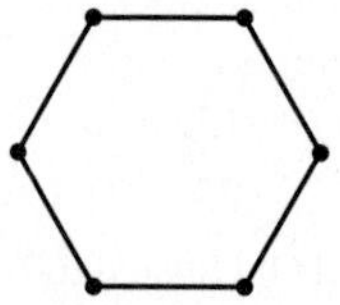

63. A graph containing the following cannot be colored with three colors.

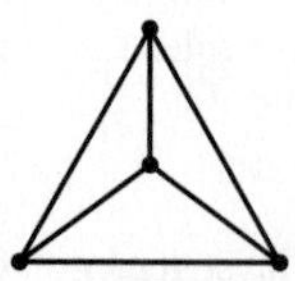

65. One sequence is C, F, E, C#, D, A, E♭, B.

67. One possible schedule: Day one—Anth215, Phy212; Day two—Bio325, Mat311; Day three—Chem264, Fin323, ComD265

Section 4.2

1. a. ADBCEA **b.** EBCDAE

3. a. ADCBFEA **b.** EADCBFE

5. ABCDEA, ABCEDA, ABDCEA, ABECDA

7. 6! = 720

9.

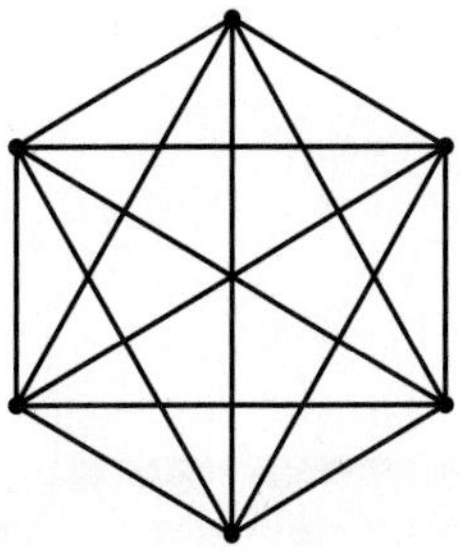

11. a. Yes **b.** Yes **c.** Yes **13. a.** No **b.** No **c.** No

15. a. 20 **b.** 23 **17.** ACBDA has weight 14.

19. ACBDA has weight 99. **21.** ABDECA

23. ADCFEBA **25.** ABEDCA **27.** ADBECFA

29. 120 **31.** 40,320

33. a. (We won't count reversals of circuits.) 720/2 = 360

b. 6 hours **c.** PNBRAMCP **d.** PNBCMRAP

35. a. ABMDHEFA **b.** ABMDHEFA

37. a. XFEBACDPX

39. A Hamilton circuit goes through every vertex but does not travel over every edge as an Euler circuit does.

41. a. After starting with vertex A, we have 9 choices for the first vertex, 8 choices for the second, 7 for the third, and so on. The total number of ways to choose all the vertices is $9 \times 8 \times 7 \times 6 \times 5 \times 4 \times 3 \times 2 \times 1 = 9!$.

43. Answers will vary.

45. 362,880/1,000 = 362.88 seconds, or roughly 6 minutes

47. A, C, E, F, B, D **49.** ABEDCA has weight 19

Section 4.3

1. a. ABCE, ABCDE **b.** ABC **c.** ABCE **d.** not possible **e.** ABCEBA

3. a. BCE **b.** ABCFG **c.** ACFGCE **d.** not possible **e.** ABCFGC

5.

From \ To	A	B	C	D	E
A	1	1	1	0	0
B	1	1	1	1	1
C	1	1	0	2	2
D	1	2	0	1	1
E	2	1	1	1	1

7.

From \ To	A	B	C	D	E	F	G
A	0	1	2	0	1	1	0
B	0	0	1	0	1	1	0
C	0	0	0	0	1	1	1
D	0	0	1	0	2	1	0
E	0	0	0	0	0	0	0
F	0	0	1	0	0	0	1
G	0	0	1	0	1	1	0

9. a. Kevin **b.** One way: Reverse the direction on the edge KD.

11. chief financial officer, sales manager, production manager, president, marketing director

13.

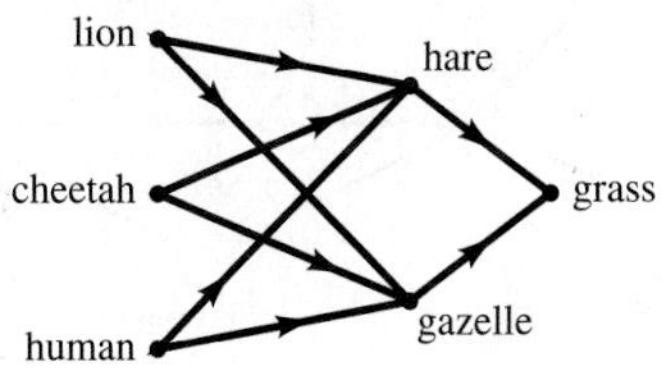

15. Florida (6), Mississippi (5), Alabama and LSU (3), Arkansas (1), Georgia (0)

17. Feinstein (5), Murciano (4), Johnson (3), Cordaro (1), Lee (0)

19. Fuji (8), Dasani (7), Propel (6), Vitaminwater (3), Aquafina (2)

21. $\begin{vmatrix} 0 & 1 & 0 & 0 \\ 0 & 0 & 1 & 0 \\ 1 & 0 & 0 & 1 \\ 1 & 0 & 0 & 0 \end{vmatrix}$

23.

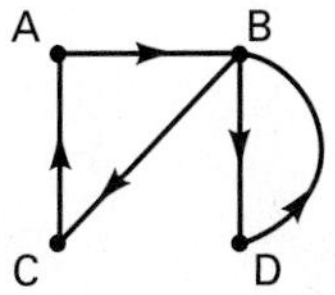

25.

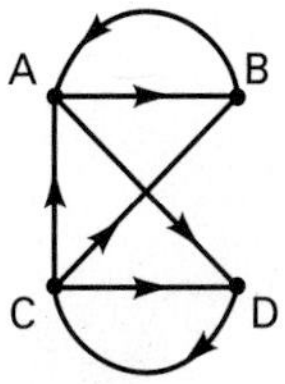

27. when the relationship goes in only one direction

Section 4.4

1. a. Begin, A, B, E, I **b.** Begin, A, B, E **c.** day 14 **d.** day 16 **e.** 17 days **f.** Begin, A, B, E, I, J, End

3. a. Begin, B, E, F, G **b.** Begin, B, E, F, H **c.** Begin, B, E, F, G, I, End **d.** to begin on day 6 **e.** to begin on day 9 **f.** 14 days

5.

Task	Day Task Can Begin
A	1
B	1
C	3
D	3
E	6
F	8
G	8
H	12

7.

Task	Day Task Can Begin
A	1
B	1
C	1
D	3
E	4
F	5
G	7
H	7
I	11
J	11

9.

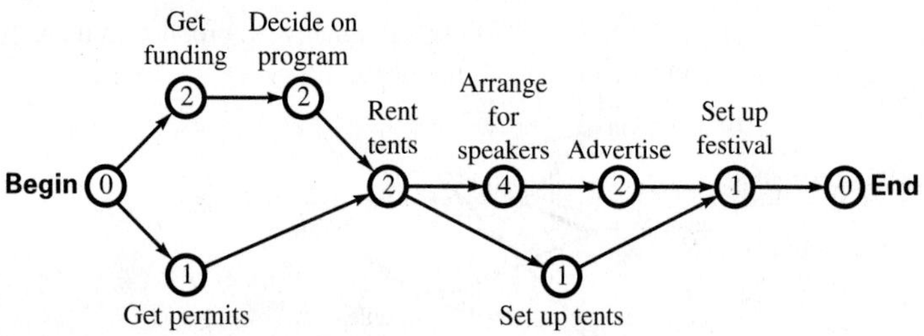

Task	Week Task Can Begin
Get funding	1
Get permits	1
Decide on program	3
Rent tents	5
Arrange for speakers, etc.	7
Advertise	11
Set up tents, booths, etc.	7
Set up festival	13

11.

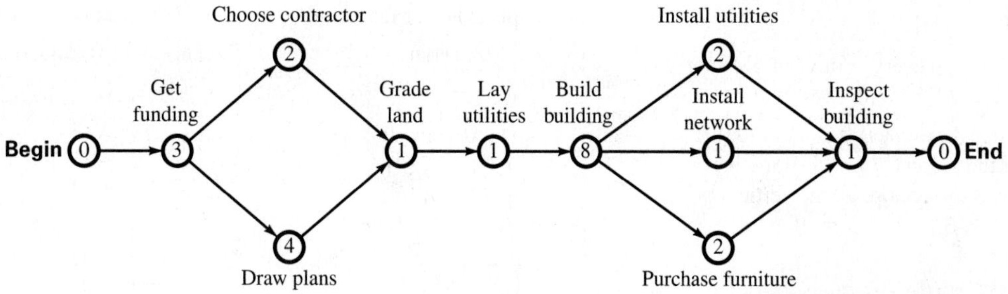

Task	Month Task Can Begin
Get funding	1
Choose contractor	4
Draw plans	4
Grade land	8
Lay utilities	9
Build building	10
Install utilities	18
Install network	18
Purchase furniture	18
Inspect building	20

13.

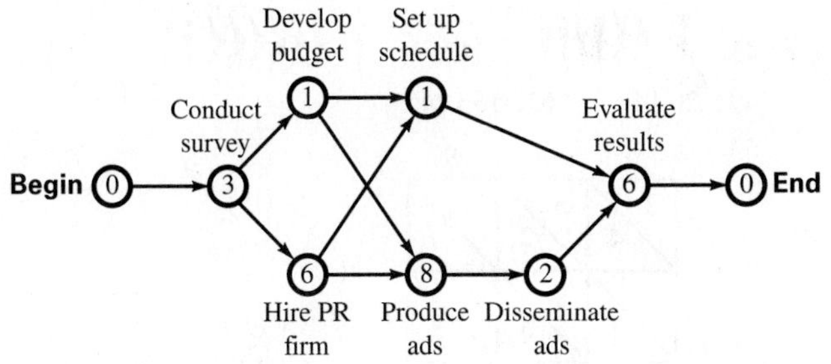

Task	Month Task Can Begin
Conduct survey	1
Develop budget	4
Hire PR firm	4
Set up production schedule	10
Produce ads	10
Disseminate ads	18
Evaluate results	20

Chapter Review Exercises

1. **a.** 12 edges.
   b. F, G, H, and I are odd; the rest are even.
   c. yes
   d. FH, GH, and HI are bridges.
3. Graph (a) can be traced because it has only two odd vertices. Graph (b) cannot be traced because it has more than two odd vertices.
5. Duplicate edges BC, EI, HL, and NO. The resulting graph will have all even edges, which can be traced using Fleury's algorithm. This tracing will give the indicated route.
7. Four canoes; those who can travel together: A, K, and N; B, C, and M; D and E; J and L
9. ABDECA has weight 13. **11.** ABDECA
13. Directed graphs are used when the relationship modeled may apply in one direction. For example, object X is related to Y, but Y may not be related to X.
15. **a.** Begin, A, C, F, H **b.** Begin, A, C, F, I, End **c.** day 10 **d.** 18 days

Chapter Test

1. **a.** 9 **b.** odd—E, F; even—A, B, C, D, G, H **c.** yes **d.** yes, EF
2. **a.** no—more than two odd vertices **b.** yes
3. (one way) ABDACEGIHGFEDFA
4. ADBCEA, ADBECA, ADCBEA, ADCEBA, ADEBCA, ADECBA
5. 6. Two colors suffice.

7. ABDECA and ACEBDA have weight 19.
8. ACEBDA **9.** ACEBDA
10. **a.** BAEF **b.** none exists
11. **a.** Begin, X, P, B, Z **b.** Begin, X, P, B, Z, End **c.** 16th day **d.** 18 days
12. D

Chapter 5

Section 5.1

1. 324 **3.** 21, 324 **5.** 2,121,210

7. 𓍢𓍢𓎆𓎆𓎆𓏺𓏺𓏺𓏺𓏺 **9.** 𓆼𓆼𓆼𓍢𓍢𓎆𓎆𓎆𓏺𓏺𓏺𓏺𓏺

11. 𓆐𓆐𓂭𓂭𓂭𓂭𓆼𓆼𓆼𓆼𓆼𓍢𓍢𓍢𓎆

13. 𓆼𓍢𓎆𓎆𓎆𓎆𓎆𓏺𓏺𓏺 **15.** 𓁨𓁨𓆐𓆐𓂭𓆼𓍢𓍢𓍢𓎆

17. 𓍢𓍢𓎆𓎆𓎆𓎆𓎆𓎆𓎆𓏺𓏺𓏺𓏺 **19.** 𓆼𓆼𓆼𓆼𓆼𓆼𓍢𓎆𓎆𓎆𓎆𓏺𓏺𓏺𓏺𓏺𓏺

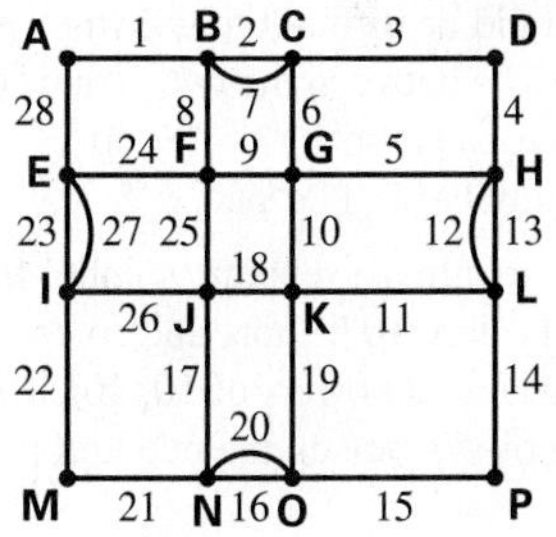

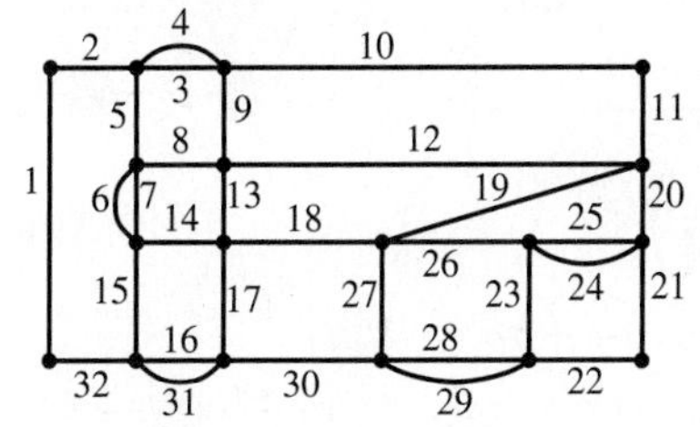

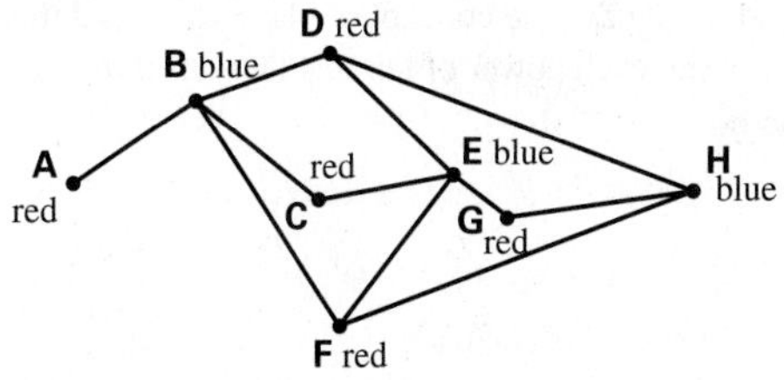

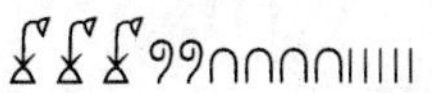

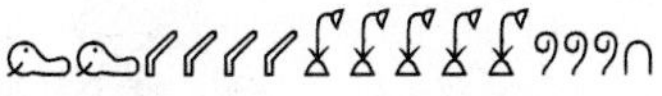

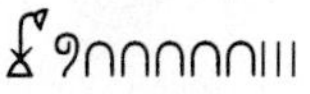

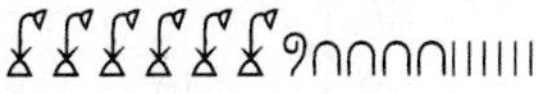

21. 602 **23.** 2,646 **25.** 564 **27.** 1963

29. 7,544 **31.** 50,262 **33.** 501,420 **35.** CCLXXVIII

37. CDXLIV **39.** $\overline{\text{IV}}$DCCXCV **41.** $\overline{\text{LXXXIX}}$CDXXIII

43. 436 **45.** 5,067 **47.** 9,999 **49.** 四百九十五

51. 二千八百五 **53.** 九千八百四十六 **55.** 2,650

57. 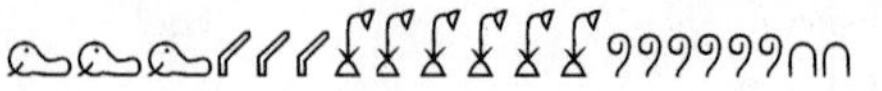

59. 36 **61.** 54 **63.** MCMXXXIX **65.** MCMXCIV

67. 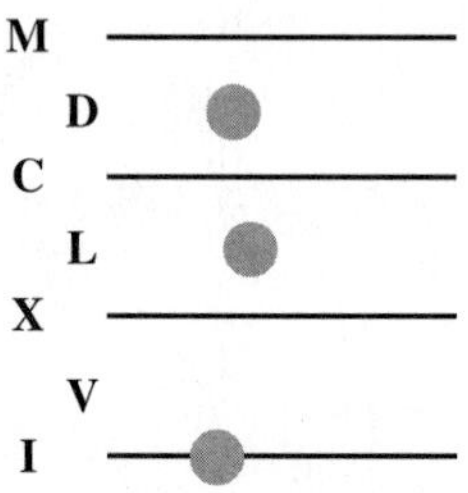

DLI

69. 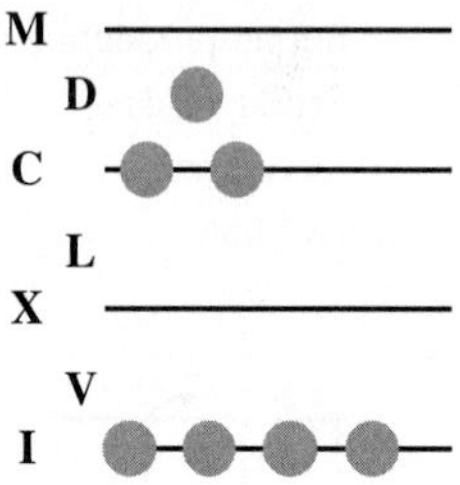

DCCIIII

71. 一千五百二十三；一千二十七

73. A number tells us "how many." A numeral is a symbol for writing a number.

75. They did not have the concept of place value and therefore had to write each power of ten that they used in their numeral.

77. 9,999,999

79. **a.** ∝∝∝∇ ⊗ ⊗ ⊗ ⊗ √√; ∝∝∝∝∝∝∝∇∇√√√

b. ≈∇∇∇⊗⊗⊗⊗ √√√√√

c. ∞∞∞∞ ⊗⊗⊗⊗⊗⊗√

81. $\frac{1}{2} + \frac{1}{6}$ **83.** $\frac{1}{4} + \frac{1}{28}$

Section 5.2

1. 731 **3.** 8,187

5. 

7. 

9. 

11. 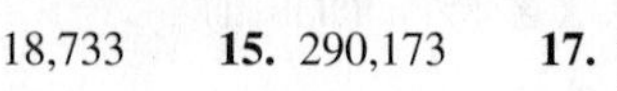

13. 18,733 **15.** 290,173 **17.** 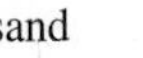 **19.** 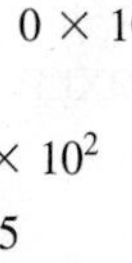

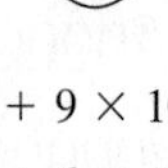

21. five hundred **23.** five thousand

25. $2 \times 10^4 + 5 \times 10^3 + 3 \times 10^2 + 8 \times 10^1 + 9 \times 10^0$

27. $2 \times 10^5 + 7 \times 10^4 + 8 \times 10^3 + 0 \times 10^2 + 6 \times 10^1 + 3 \times 10^0$

29. $1 \times 10^6 + 2 \times 10^5 + 0 \times 10^4 + 0 \times 10^3 + 0 \times 10^2 + 4 \times 10^1 + 5 \times 10^0$

31. 5,368 **33.** 370,082 **35.** $1 \times 10^3 + 7 \times 10^2$

37. 3,288 **39.** 6,238 **41.** 142 **43.** 3,655

45.  **47.** 

49. 20,148 **51.** 136,245

53. 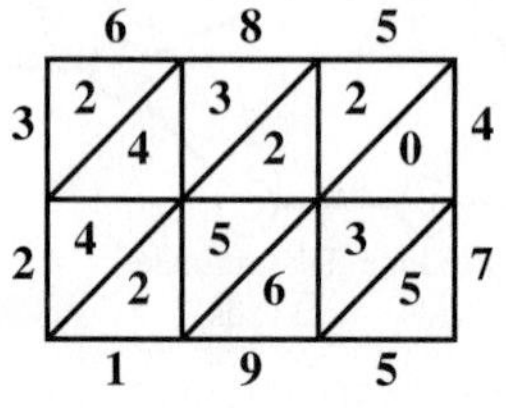

$685 \times 47 = 32{,}195$

55. 3,936 **57.** 5,544 **59.** 199,260

61. one mina and 22 shekels

63. We can write numerals more efficiently, and we do not need different symbols for different powers of the base.

65. In the galley method, we compute all partial products separately before we add them to get the final product. In our method, we combine several smaller partial products into a single partial product before adding.

67. In writing , it could be difficult to tell whether there is one space or two spaces between the two groups of symbols. If there is one space, the number is $1 \times 60 + 5$. If there are two spaces, the number is $1 \times 60^2 + 5$.

69. The ancient Chinese system requires a separate symbol for each power of 10, e.g., one billion, 10 billion, and so on. The Hindu-Arabic system uses different powers of 10, 10^9, 10^{10}, and so on. The positioning of symbols in the numeral indicates these powers of 10.

71. 四三六 **73.** 五〇六七 **75.** 九〇〇九

77. 3×10^{-1}; 7×10^{-2}; 5×10^{-3}

79. $2 \times 10^2 + 5 \times 10^0 + 6 \times 10^{-1} + 3 \times 10^{-4}$

Section 5.3

1. 23_5, 30_5 **3.** 1010_2, 1100_2 **5.** EE_{16}, $F0_{16}$

7. 117 **9.** 184 **11.** 39 **13.** 117 **15.** 183

17. 452 **19.** 756 **21.** 3,336 **23.** $2{,}314_5$

25. $12{,}302_6$ **27.** $1{,}100{,}111_2$ **29.** $1{,}011{,}110_2$

31. $6{,}513_8$ **33.** $AE8_{16}$ **35.** DEA_{16} **37.** $4{,}143_5$

39. 6082_9 **41.** 11154_7 **43.** $2E53_{16}$

45. $110{,}000_2$ **47.** $2{,}041_5$ **49.** 1524_7 **51.** 648_{16}

53. $100{,}110_2$ **55.** $2{,}043_5$ **57.** $24{,}041_5$

59. 1145 $R11_5$ **61.** 445 $R24_5$ **63.** $1{,}355_8$; $2ED_{16}$

65. $1{,}751_8$; $3E9_{16}$ **67.** $10{,}100{,}110_2$

69. $101{,}000{,}111{,}110_2$ **71.** 754_{16} **73.** $1{,}654_8$

75. CANDY **77.** LOVE

79. 21 quarters, 1 dime, 1 nickel, and 3 pennies

81. *b* **83.** 6

85. 64; 256. In base 8, we combine the eight symbols, 0 through 7, with themselves; in a base-16 system, we combine the sixteen symbols 0, 1,... 9, A, B,..., F.

87. One; we can count from 000 to 111 (decimal 0 to 7) in three binary places, which is the same as we can count in one octal place.

89. 4123_5 **91.** 3774_9

93. We will use the letters P, G, B, and Y to represent the pink, green, blue, and yellow faces, respectively. G, B, Y, GP, GG, GB, GY, BP, BG, BB, BY, YP, YG, YB, YY, GPP, GPG, GPB

95. GYP **97.** $x = 4, y = 5$ **99.** $x = A, y = 5$

Section 5.4

1. 7 **3.** 4 **5.** true **7.** true **9.** false **11.** 2

13. 7 **15.** 0 **17.** 6 **19.** 7 **21.** 6 **23.** 6

25. 3 **27.** 4 **29.** 3, 8 **31.** 24 **33.** 42

35. 3 **37.** 28 **39.** 23

41. In 2010 or 2011, 1982; after 2011, 1994.

43. 9 **45.** 6 **47.** 6 **49.** 3

51. Perform the operation as usual and then count off the result on an *m*-hour clock.

53. **a.** Check digits help ensure that numbers have no errors.

b. Modular arithmetic is often used in computing check digits.

55. 15 **57.** 10 **59.** 4

Chapter Review Exercises

1. 1,232,210 **3.** 1,961 **5.** 7,593

7. The Roman system had the subtraction principle, which allowed people to write certain numbers more efficiently. They also had a multiplication principle.

9.  **11.** 4,738

13. **a.** 375 **b.** 1,440 **15.** $6{,}513_8$ **17.** 1345 $R20_5$

19. $1{,}431_5$ **21.** **a.** true **b.** false

23. **a.** 3, 7, 11 **b.** no solutions

Chapter Test

1. MMMDCLXXXV **2.** It has place value and zero.

3. 144; 2622 **4.** 2,123,420

5. **a.** 379 **b.** 1,320 **6.** **a.** false **b.** true

7.

8. 5,348 **9.** 3 **10.** 10, X, ∩ **11.** $1{,}100{,}010_2$

12. 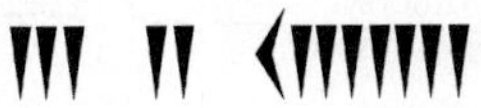 **13.** 15,946 **14.** $4{,}401_8$

15. **a.** 8 **b.** 8 **c.** 2 **16.** 2,264

17. They had to use more symbols to write their numerals.

18. **a.** 2, 5, 8, and 11 **b.** no solutions **19.** 110_5, $R2_5$

20. 1,534 **21.** $6{,}536_8$; $D5E_{16}$ **22.** $3{,}133_5$

Chapter 6

Section 6.1

1. true **3.** false **5.** false **7.** true

9. 2, 3, 5, 6, 10 **11.** 3, 5, 9 **13.** yes

15. no **17.** 4 **19.** 20 **21.** 9 **23.** 12

25. 53, 59, 61, 67, 71, 73, 79, 83, 89, 97 **27.** $3 \times 7 \times 11$

29. prime **31.** prime **33.** 7×17 **35.** $2^2 \times 5 \times 7^2$

37. $3^3 \times 23$ **39.** 11×29 **41.** 4; 120 **43.** 14; 280

45. 72; 864 **47.** 21; 3,969 **49.** 28 **51.** 45

53. 108 **55.** 360 **57.** 6 and 2,520

59. 4 and 11,200 **61.** 8; 2, 4, 8 **63.** 15,000

65. 2231 **67.** 140 **69.** 6 **71.** 3 feet by 3 feet

73. 432 **75.** 17

77. Three will always divide multiples of 9 that is, $5 \cdot 999 + 7 \cdot 99 + 1 \cdot 99$; so, all we have to check is to see if 3 divides the sum of the digits $5 + 7 + 1 + 2$, which it does.

79. not prime **81.** prime

83. **a.** 3, 7, 31, 127, 8191. **b.** 11; $2^{11} - 1 = 2{,}047 = 23 \times 89$

85. 59 and 61; 71 and 73; 101 and 103

87. No. Both 4 and 6 divide 12, but $4 \times 6 = 24$, which does not divide 12.

89. 15 will divide a number if both 3 and 5 divide the number.

Section 6.2

1. +3 **3.** +9 **5.** +19 **7.** +23 **9.** −13

11. −65 **13.** +50 **15.** +35 **17.** −56 **19.** +48

21. −38 **23.** +72 **25.** −3 **27.** +15 **29.** +5

31. −6 **33.** +6 **35.** −26 **37.** −12 **39.** −48

41. −6 **43.** −2 **45.** 0 **47.** not possible **49.** true

51. false: $(-5) + (+8) = +3$ **53.** false; $\frac{-6}{-3} = +2$

55. true **57.** 57,260 feet **59.** 7,201 feet

61. 263 degrees **63.** 39 degrees **65.** 66 **67.** 247 years

69. 1,492 years **71.** 37 degrees **73.** 12,350 **75.** Tuesday

77. We use the notion of adding signed numbers to define subtraction.

79. When multiplying or dividing signed numbers, if both numbers have the same sign, then the result is positive; if the signs are different, then the result is negative.

81. If $\frac{8}{0} = x$, then $8 = 0 \cdot x$, which is not possible.

83. For the next 4 days you spend \$6, so in 4 days, your net change in finances will be $-\$24$. Thus, $(+4)(-6) = -24$.

85. For the past 3 days you have gained \$7, so 3 days ago, you had \$21 less. Thus, $(-3)(+7) = -21$.

87. $a = 2, b = -3, c = -7, d = -8, e = -6$

89. The sum of the nine entries is 9×37.

91. The reasoning is similar to Exercise 90, except now there is no center number. So, put a point, C, in the middle of the 4-by-4 square, and notice that any two numbers that are symmetric with respect to C have the same total. There are eight such pairs in a 4-by-4 box; therefore, add two numbers in the box that are symmetric with respect to the center, C, and multiply this total by 8 to get the total of all 16 numbers.

Section 6.3

1. equal **3.** not equal **5.** not equal **7.** equal

9. $\frac{3}{7}$ **11.** $-\frac{1}{3}$ **13.** $\frac{9}{14}$ **15.** $\frac{13}{14}$ **17.** $\frac{7}{6}$

19. $-\frac{1}{3}$ **21.** $\frac{5}{48}$ **23.** $\frac{59}{24}$ **25.** $-\frac{1}{24}$ **27.** $\frac{1}{3}$

29. $\frac{1}{3}$ **31.** $-\frac{21}{5}$ **33.** $-\frac{3}{8}$ **35.** $-\frac{33}{4}$ **37.** 2

39. $\frac{62}{81}$ **41.** $-\frac{49}{10}$ **43.** $\frac{79}{5} = 15\frac{4}{5}$ **45.** $\frac{29}{3} = 9\frac{2}{3}$

47. $6\frac{3}{4}$ **49.** $8\frac{1}{15}$ **51.** $\frac{11}{4}$ **53.** $\frac{55}{6}$ **55.** 0.75

57. 0.1875 **59.** $0.\overline{81}$ **61.** $0.\overline{307692}$ **63.** $\frac{16}{25}$

65. $\frac{209}{250}$ **67.** $\frac{61}{5}$ **69.** $\frac{4}{9}$ **71.** $\frac{7}{37}$ **73.** $\frac{7}{22}$

75. $\frac{1}{4}$ **77.** $\frac{5}{48}$ **79.** $\frac{27}{44}$ **81.** 24; no

83. $9\frac{23}{24}$ miles **85.** $102\frac{1}{3}$ feet **87.** $3\frac{3}{4}$ tablespoons

89. The small tube costs 49.5 cents per ounce; the large tube costs 50.5 cents per ounce; so the smaller tube is the better buy.

91. $\frac{9}{40}$

93. He can get $3\frac{5}{9}$ strips from a 12-foot strip and $4\frac{4}{9}$ strips from a 15-foot strip. Thus, there is more waste per strip from the 12-foot strips.

101. $\frac{5}{4}$ **103.** $2\frac{23}{32}$ feet **105.** $29\frac{1}{4}$ inches **107.** 10 inch

Section 6.4

1. rational **3.** irrational **5.** rational **7.** irrational

9. $\sqrt{9} = 3$ **11.** $3\sqrt{2}$ **13.** $5\sqrt{3}$ **15.** $3\sqrt{21}$

17. not possible **19.** $8\sqrt{5}$ **21.** not possible **23.** 6

25. $6\sqrt{5}$ **27.** $14\sqrt{3}$ **29.** 2 **31.** $\frac{4}{3}$ **33.** $\frac{2\sqrt{3}}{3}$

35. $\frac{3\sqrt{5}}{5}$ **37.** $2\sqrt{6}$ **39.** $\frac{5\sqrt{22}}{11}$ **41.** $\frac{\sqrt{5}}{2}$

43. 0.435; 0.43512112111211112 . . .

45. 0.45785; 0.4578512112111211112 . . .

47. 0.6; 0.612112111211112 . . . **49.** a, d, b, c **51.** a, c, d, b

53. false; $\sqrt{2} \times \sqrt{2} = 2$, which is rational

55. true

57. false; we cannot divide by the rational number 0

59. distributive **61.** commutative (addition)

63. associative (addition) **65.** identity element for addition

67. commutative (addition) **69.** $4\sqrt{6} \approx 9.8$ miles

71. $20\sqrt{6} \approx 49.0$ mph **73.** 4.97 seconds

75. 15 feet **77.** 200 pounds **79.** 10 seconds

81. **a.** 12.85 seconds **b.** 411 feet per second **c.** over 280 mph

83. **a.** 1.414213562 is a terminating decimal and hence a rational number; however, $\sqrt{2}$ is irrational.

b. π is an irrational number and thus cannot be equal to the rational number $\frac{22}{7}$.

85. No, because a number that is rational and irrational would have to have both a repeating and a nonrepeating decimal expansion.

87. "I like green eggs and ham."

93. **a.** We can cancel common factors from the numerator and denominator.

b. Square both sides of a).

c. Multiply through equation by b^2.

d. 2 divides the left side of c), so 2 divides the right side of c).

e. If a were odd, then a^2 would be odd, but it is not.

f. a is even.

g. Substitute $2k$ for a in c).

h. Square $2k$.

i. Divide both sides of h) by 2.

j. 2 divides b^2.

k. If b were odd, then b^2 would be odd.

l. We assumed that $\frac{a}{b}$ was reduced, so a and b cannot both be even.

95. 2 **97.** 5, 12, 13 plus many others.

Section 6.5

1. 32 **3.** −16 **5.** −9 **7.** 9 **9.** 1 **11.** 729

13. 117,649 **15.** $\frac{1}{25}$ **17.** $\frac{1}{729}$ **19.** −3 **21.** 25

23. 36 **25.** 1 **27.** 8 **29.** 4.356×10^6

31. 7.83×10^2 **33.** 2.4×10^{-3} **35.** 8.0×10^{-3}

37. 32,500 **39.** 0.00178 **41.** 63 **43.** 0.00000045

45. 2.381×10^7 **47.** 8.4×10^2 **49.** 6.0×10^{11}

51. 3.6×10^2 **53.** 4.0×10^{-5} **55.** 3.36×10^{-1}

57. 2.6311×10^{-6} **59.** 8.076×10^{10} **61.** 1.564×10^{13}

63. 4.0×10^{-7} **65.** 1.0×10^{19} **67.** 3.72×10^{12}

69. 1.0×10^{-7} **71.** 1.4×10^{-14} **73.** $\$1.3 \times 10^4$

75. 1.05×10^8 **77.** 2.224×10^1

79. **a.** 5.902×10^4 **b.** 5.902×10^7

c. The 130-pound person weighs as much as 59 million mosquitoes.

81. 8.46×10^1 **83.** 1.459×10^3 **85.** 3.17×10^1

87. 320 million **89.** 1.12×10^5 hours, or almost 13 years

91. 3.24×10^1 seconds **93.** 5.87×10^{12} miles

95. **a.** 125 is too large; in scientific notation, *a* must be between 0 and 10; 1.25436×10^5.

b. 0 is too small; 5.37×10^7.

97. $(2 + 3)^2 \neq 2^2 + 3^2$. On the left side, we add first and then square; on the right side, we square first and then add.

99. 4.45×10^{11} **101.** no **103.** 96,906.57 miles

105. 2.0×10^6

Section 6.6

1. arithmetic; 17; 20 **3.** geometric; 648; 1,944

5. geometric; $\frac{1}{16}$; $\frac{1}{32}$ **7.** arithmetic; −10; −15

9. geometric; 32; 64 **11.** arithmetic; 3.5; 4.0

13. 35; 220 **15.** 86; 660 **17.** 10.5; 115 **19.** 436

21. 59,049 **23.** $\frac{1}{64}$ **25.** 0.00002 **27.** 144

29. 377 **31.** \$36,810 **33.** \$1,475.11 **35.** 4.10 feet

37. 220 million **38.** 1,399 million **39.** Answers will vary.

41. Answers will vary. **43.** $\frac{(n - k + 1)(a_k + a_n)}{2}$

45. $F_n^2 + F_{n+1}^2 = F_{2n+1}$ **47.** $335 = 21 \times 3 + 34 \times 8$

49. 4.919 **51.** 33.988

Chapter Review Exercises

1. 71, 73, 79, 83, 89 **3.** 191 is prime; $441 = 3^2 7^2$.

5. LCM = 7,920; GCD = 264 **7.** yes; yes; no

9. **a.** 9 **b.** 11 **c.** 72 **d.** −16

11. If $\frac{-24}{-8} = c$, then $-24 = -8 \cdot c$. Thus, $c = +3$.

13. If $\frac{5}{0} = c$, then $5 = 0 \cdot c$, which is impossible. **15.** \$67

17. **a.** $\frac{5}{27}$ **b.** $\frac{18}{37}$ **19.** $14\frac{1}{2}$ **21.** $\frac{1}{8}$ pound

23. Answers will vary. **25.** $6\sqrt{3}$ **27.** $\frac{8}{\frac{4}{2}}$ does not equal $\frac{\frac{8}{4}}{2}$.

29. In evaluating -2^4, we first raise 2 to the 4th power and then take its opposite to get −16. In evaluating $(-2)^4$, we raise −2 to the 4th power to get 16.

31. 1,325,000; 0.0000863 **33.** 6.06×10^1

35. **a.** 91 **b.** 1,425 **37.** 55 and 610

Chapter Test

1. 101, 103, 107, 109, 113 **2.** 14

3. 241 is prime; $539 = 7^2 \cdot 11$

4. The number is divisible by 3, 4, 6, and 8.

5. 156; 10,296

6. $GCD = 2^3 \cdot 3^6 \cdot 5^2 \cdot 7^2$; $LCM = 2^8 \cdot 3^9 \cdot 5^3 \cdot 7^9$

7. **a.** −6 **b.** +3 **c.** −54 **d.** +8

8. **a.** arithmetic; 27, 30 **b.** geometric; 486, 1458

9. If you spend \$5 for the next 7 days, you will have \$35 less.

10. Solve $-21 = (\ ) \times 3$; the solution is −7. **11.** 69 degrees

12. To divide 8 by 0, we would have to solve the equation $8 = (\ ) \times 0$, which is impossible.

13. 377 and 6,765

14. No; when we cross multiply, we get different results.

15. **a.** $\frac{1}{60}$ **b.** $\frac{16}{3}$ **16.** $15\frac{2}{3}$ **17.** $\frac{4}{3}$

18. **a.** $\frac{573}{1{,}000}$ **b.** $\frac{19}{33}$ **19.** **a.** 182 **b.** 1,930

20. $23\frac{5}{8}$ inches by $33\frac{1}{4}$ inches

22. Rational numbers have repeating expansions, and irrational numbers have nonrepeating expansions.

23. **a.** $6\sqrt{5}$ **b.** $\frac{\sqrt{15}}{5}$

25. **a.** 8 **b.** $\frac{1}{81}$ **c.** 125 **d.** 27 **26.** 39,366

27. In evaluating $(-3)^2$, we square −3 to get 9; in evaluating -3^2, we first square 3 and then take its opposite to get −9.

28. 1.546×10^7; 6.23×10^{-9}

29. 4.35

Chapter 7

Section 7.1

1. 5 **3.** $-\frac{4}{5}$ **5.** 30 **7.** 25 **9.** $w = \frac{P - 2l}{2}$

11. $\mu = x - \sigma z$ **13.** $r = \frac{A - P}{Pt}$ **15.** $x = \frac{-3y + 6}{2}$

17. $l = \frac{V}{wh}$ **19.** $h = \frac{S - 2\pi r^2}{2\pi r}$

21. intercepts: (4, 0), (0, 6) **23.**

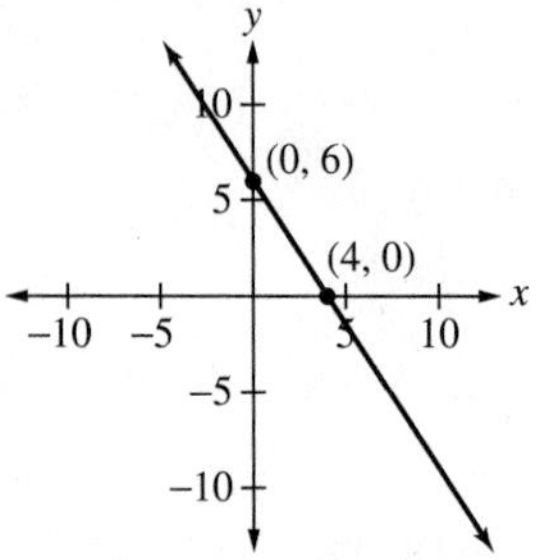

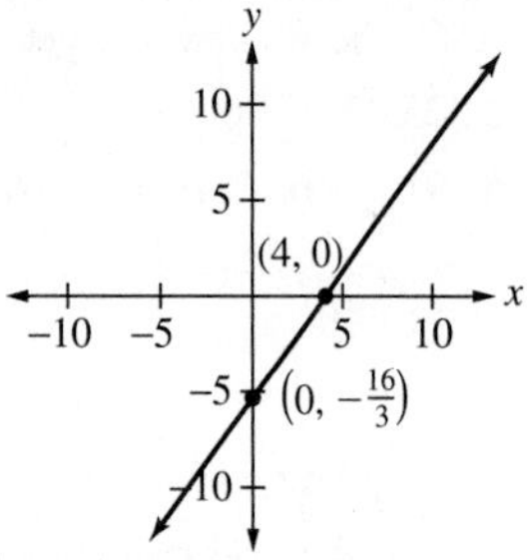

25. intercepts: (9, 0), (0, 6) **27.**

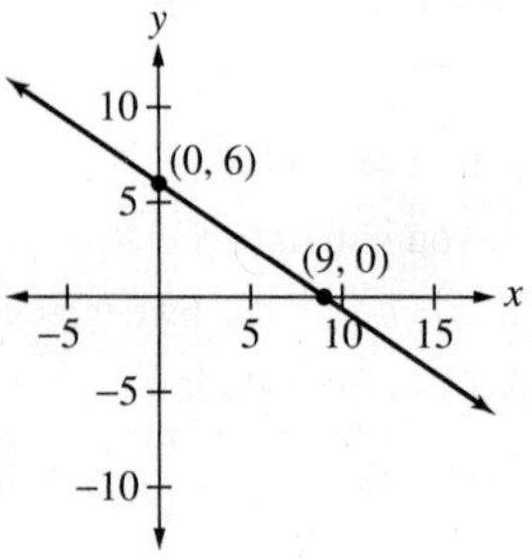

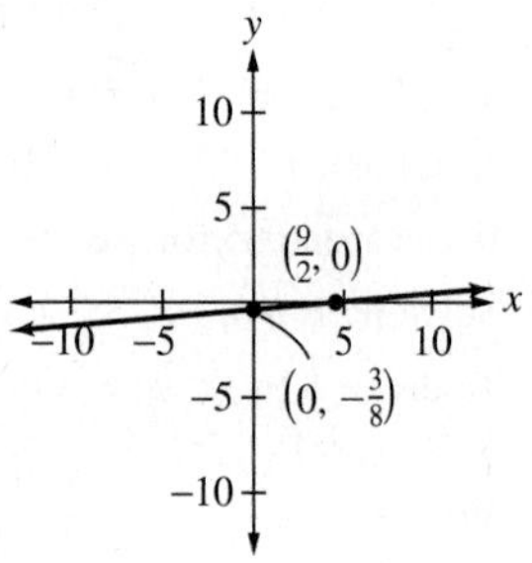

29. intercepts: (8, 0), (0, −0.4) **31.**

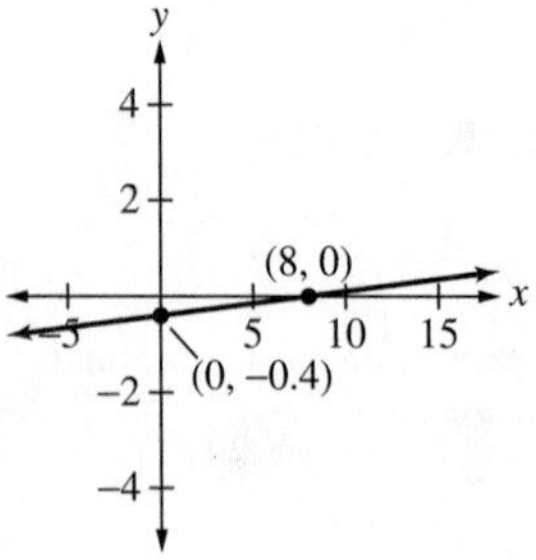

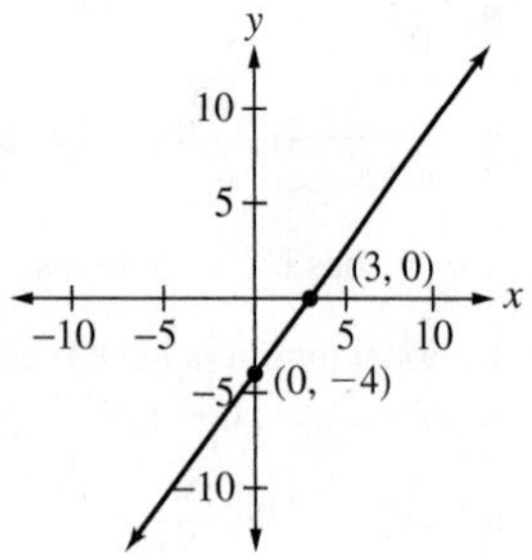

33. *a* **35.** *c* **37.** $\frac{3}{4}$ **39.** $-\frac{4}{5}$ **41.** $\frac{5}{2}$ **43.** *a, d*

45. *f* **47.** The rise is zero.

49. y-intercept = −3; slope = 4

51. y-intercept = −3; slope = −5

53. 168 **55.** 32 **57.** $d = 0.74t + 80$ **59.** 94.8

61. **a.** 7.98 **b.** $y = 7.98t + 229.4$ **63.** by 2012

65. **a.** $r = 0.47t + 2.17$ **b.** $11.57 billion

67. $d = 1{,}832t + 22{,}900$; $33,832

69. $5.60d + 7.35p = 133$ **71.** $35t + 55c = 14{,}500$

73. $9e + 15c = 342$ **75.** after 60 tokens

77. Answers will vary.

81. 7; If Chuck sells more than seven systems, Buy More is better for him.

83. 3.268% **85.** **a.** $y = 105d$ **b.** $d = \frac{y}{105}$

87. $p = 3{,}500e$

89. The slope is the same regardless of which two points you choose.

91. $y = 3x - 5$ **93.** $y = 5x - 7$ **95.** 17.75 feet

Section 7.2

1. $y = 3x - 5$ **3.** $y = 4x + 14$ **5.** $y = -2x + 11$

7. $y = -5x - 31$ **9.** $y = 2x - 1$ **11.** $y = \frac{1}{4}x + \frac{31}{4}$

13. $y = -\frac{6}{9}x - \frac{10}{19}$ **15.** $y = \frac{7}{2}x + 13$

17. $y = 0.1t + 80.6$; 82.7 years

19. **a.** $y = 0.5t + 7.8$ **b.** 16.8 million **c.** by 2022

21. $y = 750t + 15{,}000$; $23,250

23. **a.** −17.3 **b.** $v = -17.3t + 481$ **c.** 239

25. **a.** $y = 0.058t + 2.62$ **b.** 3.606 million **c.** 2032

27. $y = -0.0154t + 8.5$; 2452

29. **a.** 1.77 million **b.** $l = 1.77t + 146$
c. 156.62 million **d.** 2034

31. the slope of the line

35.

	2003 (year 0)	2004 (year 1)	2005 (year 2)	2006 (year 3)	2009 (year 6)
Actual Data	146	147	149.2	151.3	154.1
Values predicted by your model	146	147.77	149.54	151.31	156.62
Values predicted by the line of best fit	145.66	147.47	149.28	151.09	156.52

Section 7.3

1. 2, 8 **3.** $1, \frac{3}{2}$ **5.** $-3, \frac{2}{3}$ **7.** $-\frac{3}{5}, 4$

9. opening down; vertex (3, 1); (2, 0), (4, 0); (0, −8)

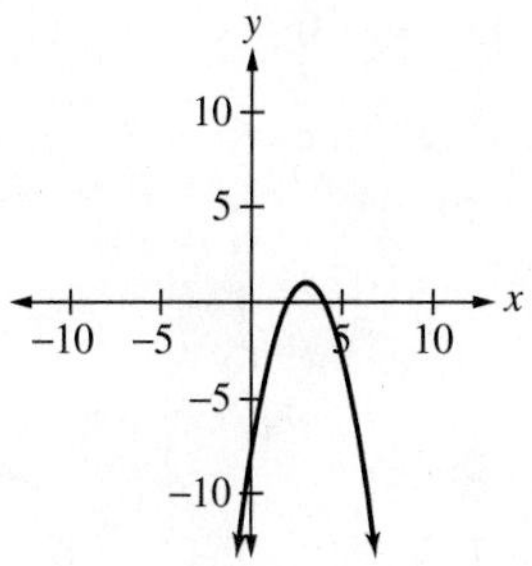

11. opening down; vertex (1, 9); $\left(-\frac{1}{2}, 0\right)$, $\left(\frac{5}{2}, 0\right)$; (0, 5)

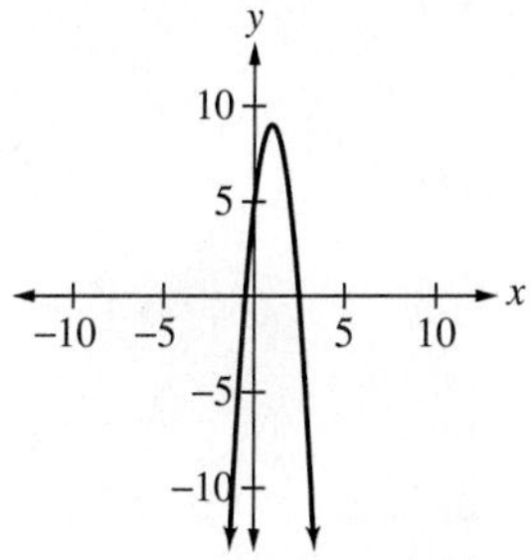

13. opening up; vertex $\left(\frac{1}{2}, -3\right)$; $\left(\frac{1 + \sqrt{3}}{2}, 0\right)$, $\left(\frac{1 - \sqrt{3}}{2}, 0\right)$; (0, −2)

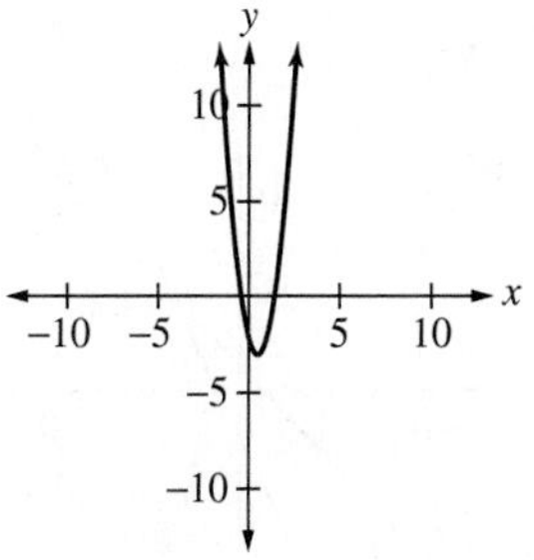

15. opening up; vertex $\left(-\frac{7}{6}, -\frac{121}{12}\right)$; (−3, 0), $\left(\frac{2}{3}, 0\right)$; (0, −6)

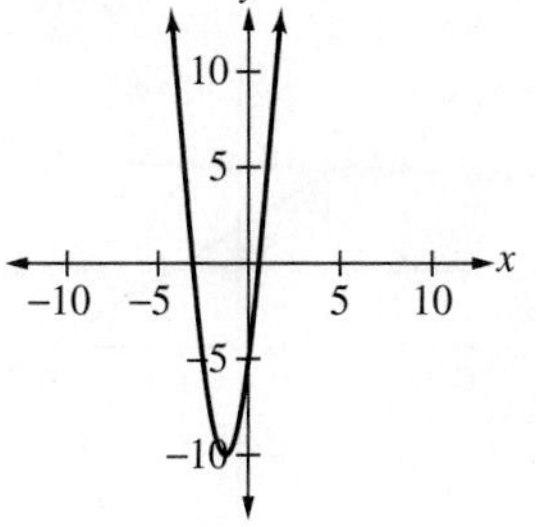

17. opening down; vertex $\left(\frac{7}{2}, \frac{1}{4}\right)$; (3, 0), (4, 0); (0, −12)

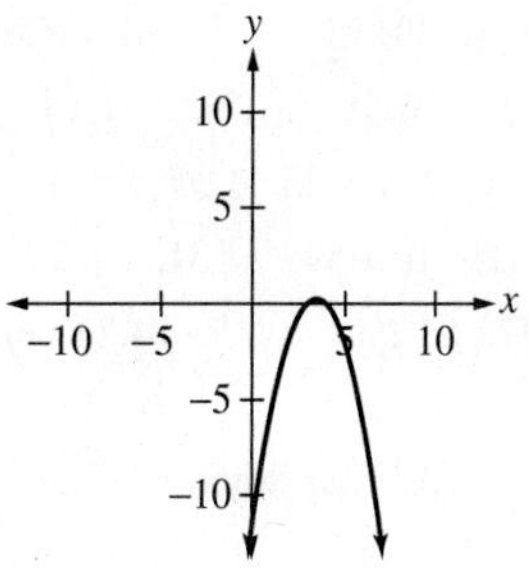

19. **b.** 8.125

21. **a.** week 2 **b.** 4 million **c.** week 3

23. **a.**

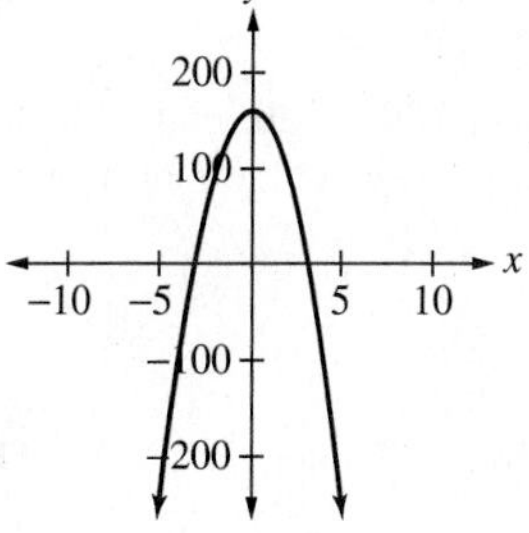

b. t (time) cannot be negative. **c.** 3.16 seconds

25. **a.** the parabola **b.** the line **c.** the parabola

27. Linear regression gives the line of best fit for a set of data; quadratic regression gives the parabola of best fit.

29. The crate begins falling at a point above the ground and picks up speed as it falls. A linear equation would not be appropriate, because the rate of change of the distance of the crate above the ground is not constant.

31. **a.**

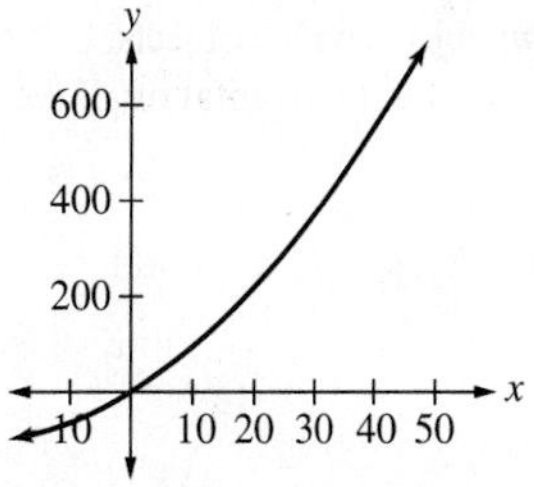

b. t cannot be negative

c. The runner is running more slowly when t is closer to 0.

d. 10.45 seconds

35. **a.**

	2002 (year 0)	2003 (year 1)	2004 (year 2)	2005 (year 3)	2009 (year 7)
Values predicted by parabola of best fit	81.01	82.87	89.43	100.69	192.73

b. The parabola gives better approximations until at least 2005. In 2009, the approximation is quite bad.

Section 7.4

1. \$1,050 **3.** \$4,100 **5.** \$6,381.40 **7.** \$4,665.60

9. 3.24% **11.** 3.33% **13.** 176.51 **15.** 124.14

17. 291.29 **19.** 1.86 **21.** 0.51 **23.** 1.57

25. 1.32% **27.** −0.25% **29.** 0.3669 **31.** 0.7291

33. 35.2% **35.** 30.9% **37.** 63.2% **39.** 14.21 years

41. 2111 **43.** 61 years **45.** 27 years **47.** 6.63 million

49. 198 years **51.** 307 mg **53.** after 8 hours

55. **a.** $y = 120(1.0299)^x$ **b.** \$161.11

57. 80.55% **59.** 4.5%

61. when the rate of growth of a quantity is proportional to the amount present

63. \$150 **65.** double the interest rate

67. The population will exceed 300 by the end of the fourth year; fishing can begin toward the end of the fourth year.

69. choose the one cent

Section 7.5

1. 4 **3.** 62.5 **5.** 36 **7.** 65 **9.** 1 **11.** 100

13. $\frac{1}{16}$ **15.** 25 **17.** 14 **19.** $\frac{4}{3}$ **21.** 192 **23.** 2.6 mg

25. 35 miles **27.** 3.25 hours **29.** 12,000 **31.** 5 or 6

33. 9.23 minutes **35.** 126 pounds **37.** 12 feet

39. 386.49 euros **41.** 18,000 gal **43.** 400 feet

45. 172 feet **47.** 588 gal **49.** 6.4 pounds per square inch

51. *y* increases. **53.** inversely **55.** Strength doubles.

57. Strength triples. **59.** 132 + 171 = 303 feet

61. In *m*:*n*, we are comparing *m* objects with *n* objects. In *n*:(*m* + *n*), we are comparing the *n* objects with the total of all the *m* + *n* objects.

63. 7:9

Section 7.6

1.

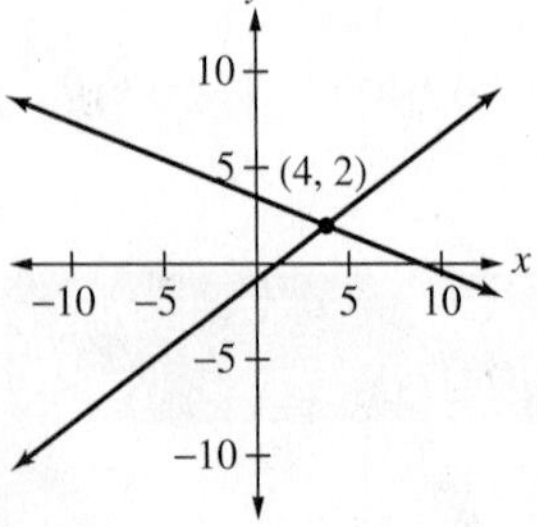

3.

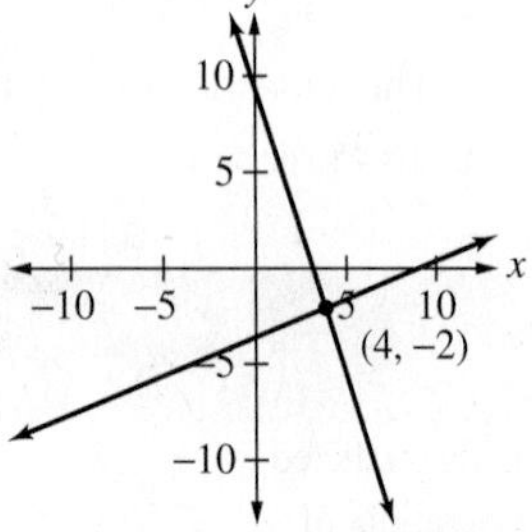

5. (1, 4) **7.** (−6, −2) **9.** no solution

11. infinite number of solutions **13.** no solution

15. $\left(\frac{1}{4}, \frac{1}{2}\right)$ **17.** $\left(\frac{1}{5}, \frac{2}{5}\right)$ **19.** (−1, 3)

21. (−1, 4) **23.** no solution **25.** a, d

27.

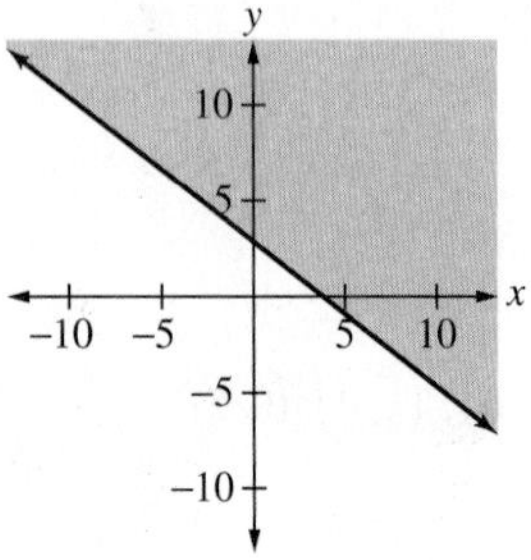

29.

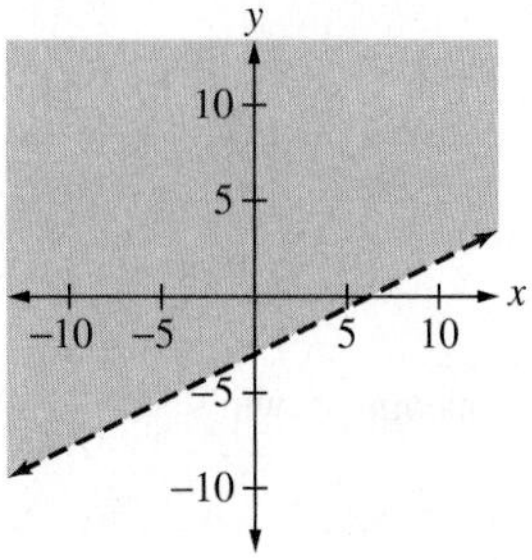

31.

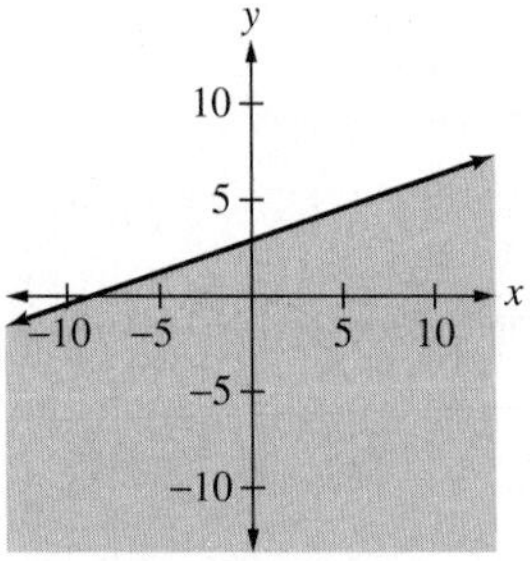

33.

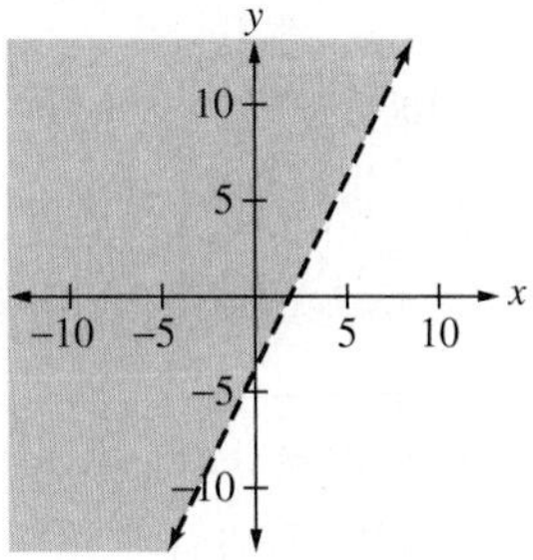

35.

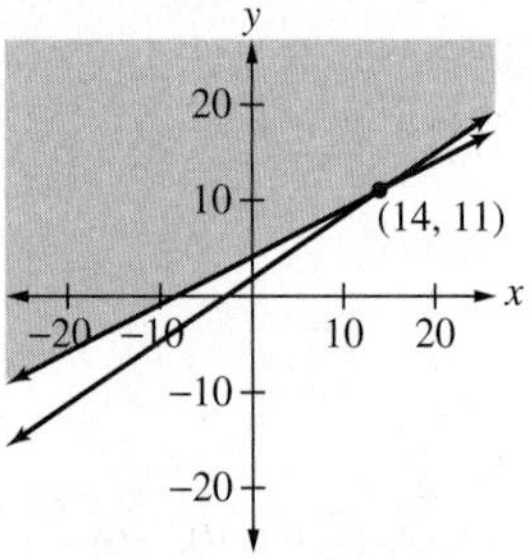

37.

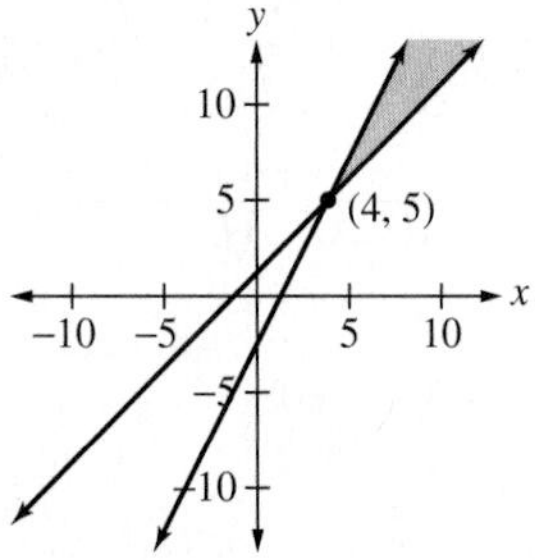

39.

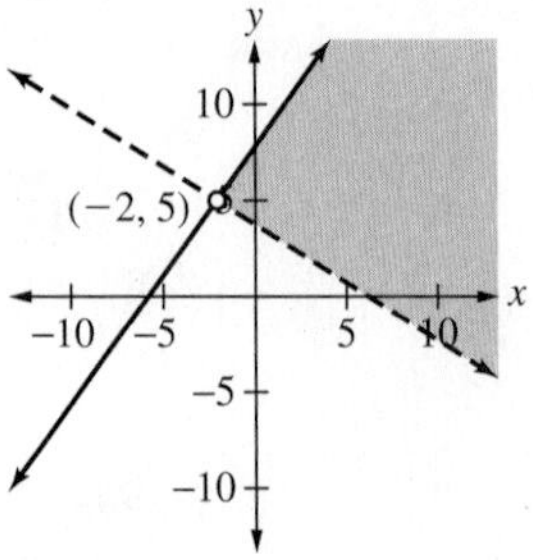

41.

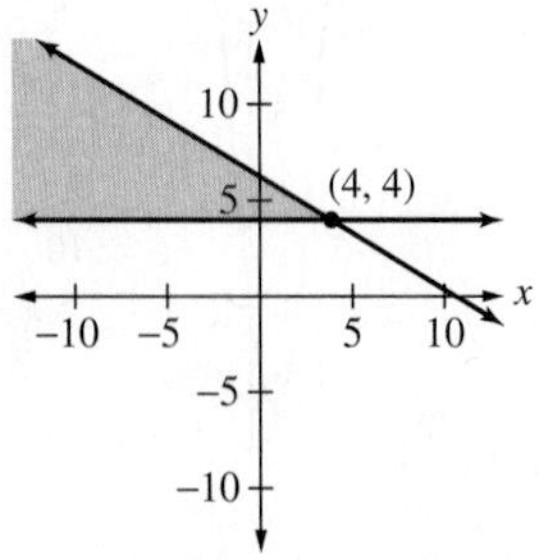

43.

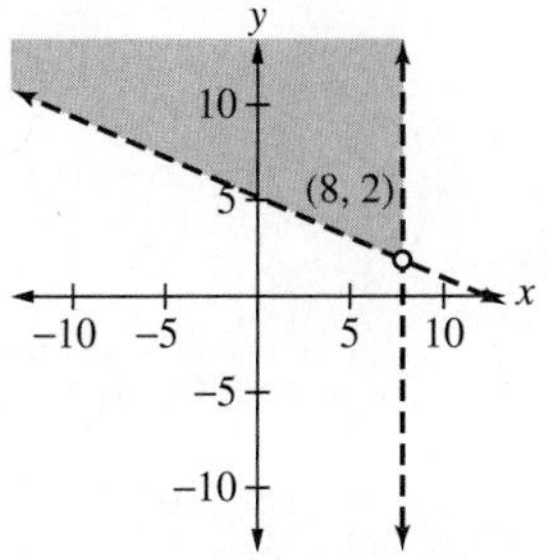

45. Tennessee 73; Georgia 43

47. 4 bagels, 5 ounces of cream cheese

49. After 11 months, WorldCom is the better deal.

51. Atlanta 54 million; Beijing 44 million

53. U.S. $y = -0.5x + 13$; China $y = 1.6x + 2$

55. \$11 per hour   57. \$34   59. 4,500 pairs; \$180,000

61.

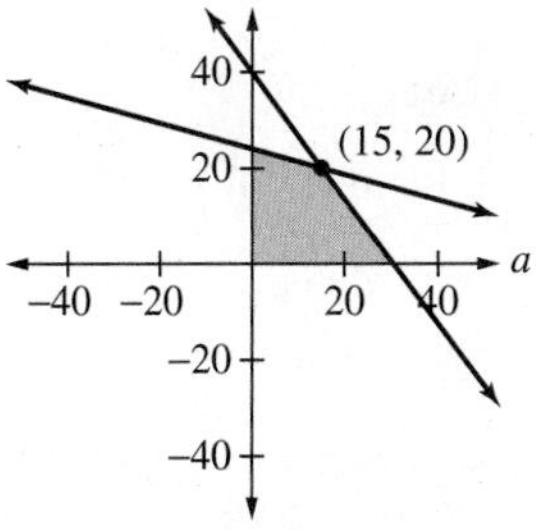

63.

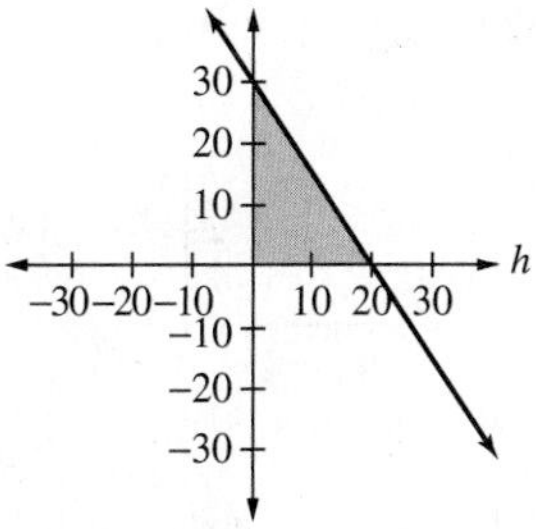

65.

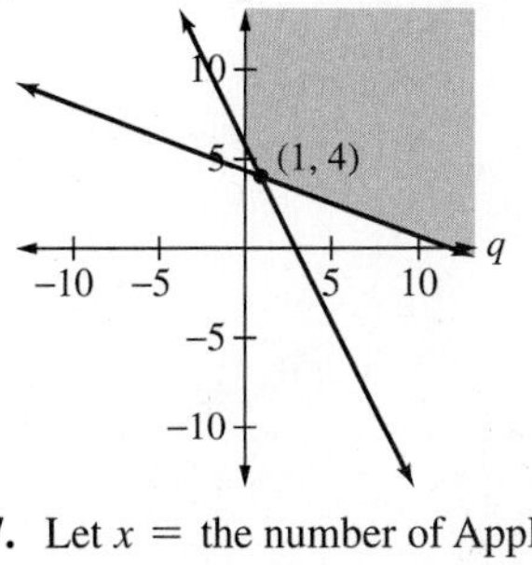

67. Let x = the number of Apple phones and y = the number of Android phones.

$$x + y \geq 30$$
$$x + y \leq 60$$
$$x \geq 2y$$
$$x \geq 0, y \geq 0$$

69. **a.** There can be one solution, an infinite number of solutions, or no solution.

b. If in solving we get a unique value for either x or y, then the lines intersect in a single point and the system has one solution. If in solving we get a false statement such as $2 + 2 = 0$, then the lines are parallel and the system has no solution. If in solving we get a true statement that does not restrict either x or y, then both lines are the same, and we have an infinite number of solutions.

71. The solution will be a half-plane that can be determined by using the one-point test.

73. Answers will vary.

75.

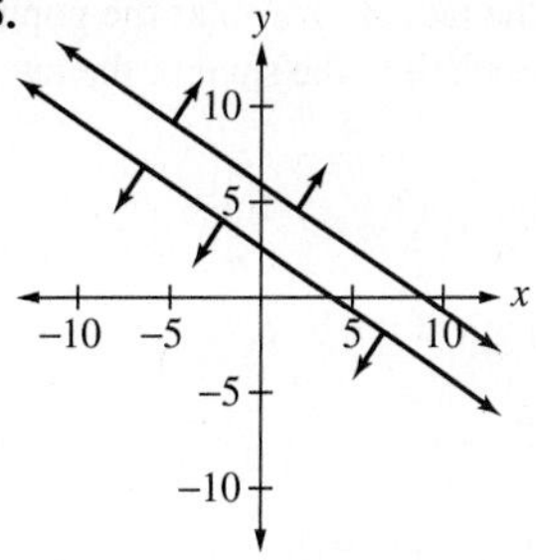

77.

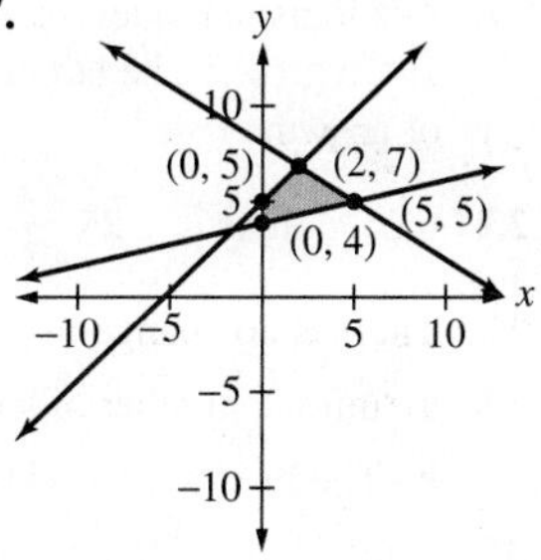

79.

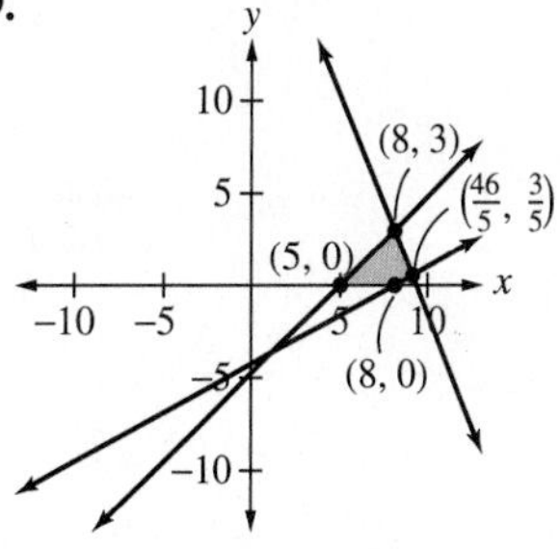

81.
$$3y \leq 43 - 5x$$
$$11y \geq 27 - 2x$$
$$3y \geq 1 + 2x$$

83.
$$3y \leq 43 - 5x$$
$$11y \geq 27 - 2x$$
$$3y \leq 1 + 2x$$

Section 7.7

1. 5, 9   3. 82   5. 18.2464

7. $a = -3$; unstable   9. $a = \frac{16}{3}$; stable

11. $A_{n+1} = 1.05A_n$, $n = 0, 1, 2, \ldots$, where $A_0 = 1{,}000$; \$1,102.50

13. $P_{n+1} = [1 + (0.08)(1 - P_n)]P_n$, $n = 0, 1, 2, \ldots$, where $P_0 = 0.30$; 33.4%

15. $D_{n+1} = 0.6D_n + 250$, $n = 0, 1, 2, \ldots$, where $D_0 = 0$; 490 mg

17. **a.** $D_{n+1} = 1.05(D_n + 3000)$, $D_0 = 0$ **b.** \$13,576.89

19. There are equations relating A_1 to A_0, A_2 to A_1, A_3 to A_2, and so on.

21. If $-1 < m < 1$, then the equilibrium value will be stable.

23. 900 years   25. 11.8 pounds   27. 333

Chapter Review Exercises

1. **a.** 4 **b.** $-\frac{1}{2}$ 3. **a.** $s = 200 + 10(h - 40)$ **b.** 260

5. $\frac{3}{4}$ 7. after 10 months   9. $y = \frac{5}{3}x - 1$

11. It is the linear model that best approximates a set of data points.

13. $\frac{1}{2}, -4$

15. a technique by which we find a quadratic equation that best models a set of data

17. \$12,641.72 **19.** 0.5075

21. In a logistic model we adjust the rate of growth as the population grows. The larger the population, the smaller the rate of growth.

23. 6.25 gallons **25.** $\frac{7}{3}$

27. There is no change in the strength of the beam.

29. **a.** infinite number of solutions

b. The lines are the same line.

31. United States 70; opponent 53 **33.** \$18

35.

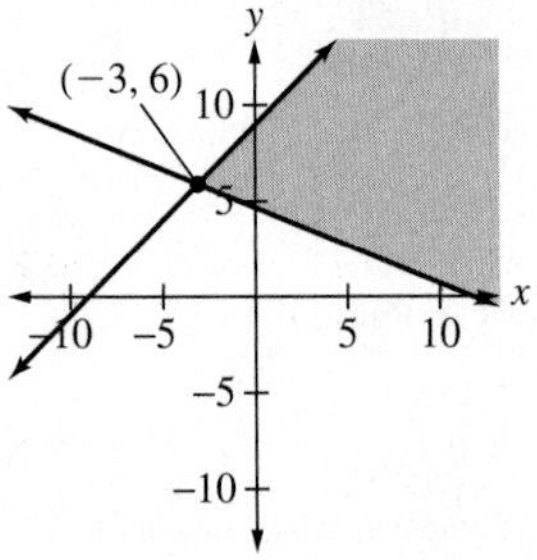

37. 27, 111, 447 **39.** 1,400 mg

Chapter Test

1. **a.** −12 **b.** 3.36 **2.** $b = \frac{X}{a} - 1$ **3.** $\frac{7}{5}$

4.

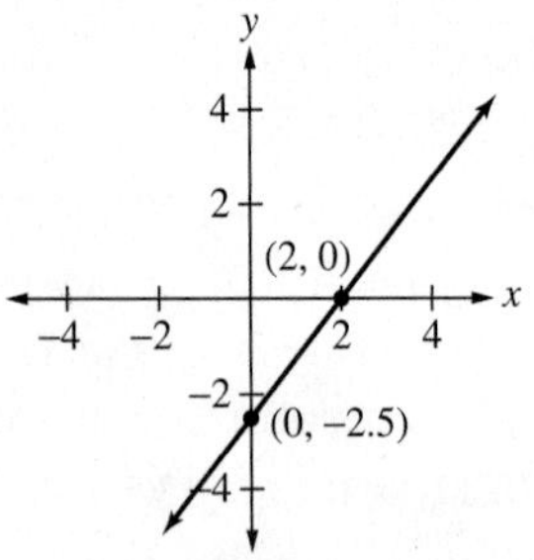

5. **a.** $c = 30 + 0.045(m - 1200)$ **b.** \$44.40

6. $y = 16(1.024)^x$; \$21.78 **7.** $1 - b, 2 - d, 3 - a, 4 - c$

8. after 36 rentals **9.** when the rate of change is constant

10. $y = \frac{3}{4}x + \frac{7}{2}$ **11.** $6.30d + 8.25l = 137.70$

12. a process by which we find the line of best fit for a set of data points

13. $n = 0.3t + 34.7$; 40.1 million

14. 22, 90, 362 **15.** $-\frac{2}{3}$, unstable

16. **a.** down **b.** $\left(\frac{11}{2}, \frac{25}{4}\right)$ **c.** (3, 0), (8, 0) **d.** (0, −24)

e.

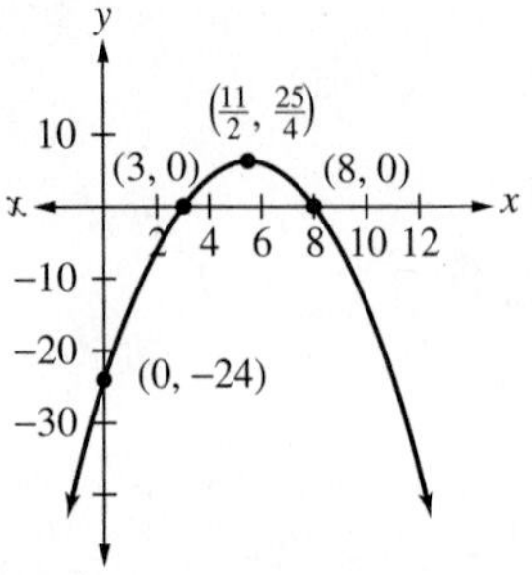

17. \$6,044.63 **18.** slightly over 29 years

19. quadratic regression **20.** during the fourth week

21. **a.** no solutions **b.** The lines are parallel.

22. **a.** infinite number of solutions **b.** The lines are the same.

23. **a.** (2, −3) **b.** The lines intersect in a single point.

24. 606.15 mg **25.** 0.4048 **26.** 1,800

27. 12 hours **28.** 23 boxes

29.

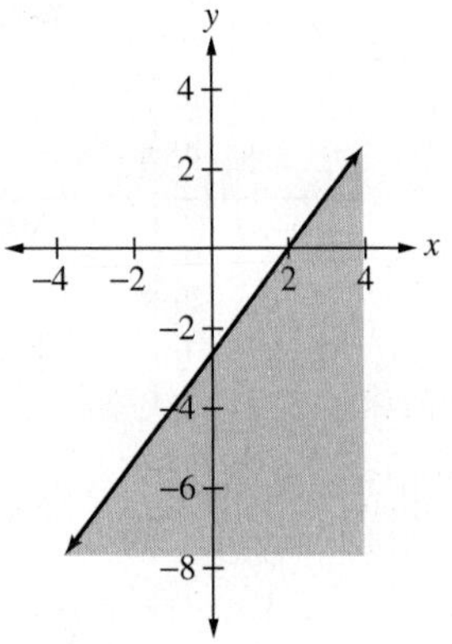

30. 10 or 11

31.

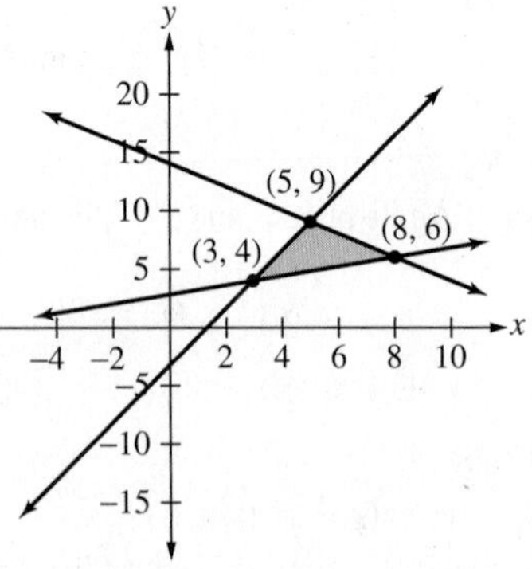

32. It would double. **33.** $\frac{4}{3}$ **34.** 20

35.

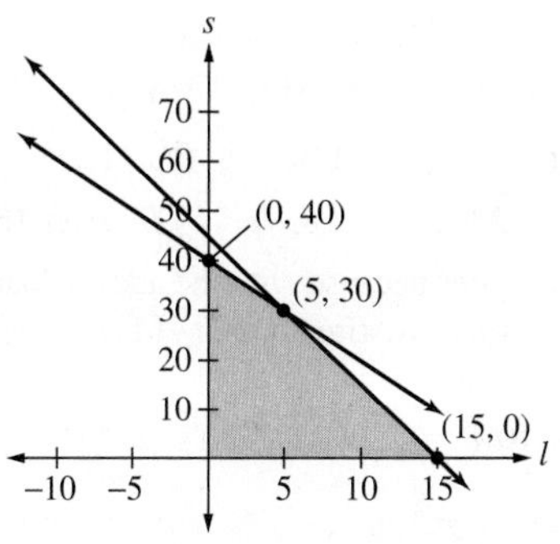

Chapter 8

Section 8.1

1. 0.78 **3.** 0.08 **5.** 0.2735 **7.** 0.0035 **9.** 43%

11. 36.5% **13.** 145% **15.** 0.2% **17.** 75%

19. 31.25% **21.** 250% **23.** 15% **25.** 350

27. 19.6 **29.** 17.5% **31.** 128 **33.** 9.4%

35. $207 thousand **37.** 806 **39.** 31,942% **41.** 31%

43. 42.9% **45.** 1,269 **47.** 9.3% **49.** 76.4 million

51. a. 6% **b.** She divided by 21,065 instead of 19,875.

53. 11.5% **55.** 24.5% **57.** 22.9% **59.** $37,800

61. $680 **63.** $12,800 **65.** $35,771.50

67. $8,507.50 **69.** 19.6% **71.** 167.8

77. For each $1,000 make $20.10 more over three years with 3%, 1%, then 2%.

79. $10,794.52

81. No. The price after the reduction will be less than the original price.

83. The half-gallon is 7.48 cents an ounce and the $1\frac{1}{2}$ quarts is 8.1 cents an ounce, so the price increase is 8.3%.

Section 8.2

1. $240 **3.** 5% **5.** $3,100 **7.** $1,500 **9.** 6%

11. $6,381.41 **13.** $4,686.64 **15.** $23,457.76

17. $4,885.48 **19.** 7.76% **21.** 6.14%

23. 4.95% compounded quarterly **25.** 2.096 **27.** 14.207

29. a. 20.15 **b.** 20 **31. a.** 23.45 **b.** 23.3

33. 4.56% **35.** 10.34 **37. a.** $4,896 **b.** $1,296

39. $66.67 **41.** 12.5% **43.** 75% **45.** $12,278.88

47. a. 462.1% **b.** $185.50 **49. a.** 164.7% **b.** $6,422.36

51. $1,712.15 **53. a.** 4.74% **b.** 4.67% **55.** 10.07 years

57. $146,973.08 **59.** $4,196,126,573 **61.** 68 cents

63. In computing compound interest, interest is paid on the principal plus previously earned interest.

65. when $m = 1$

67. between 1% and 3% of $350, or between $3.50 and $10.50

69. 10.5170863% **71.** 10.5170918%

73. $48,717.95 **75.** $1,442,015.79 per minute

Section 8.3

1. $46.50 **3.** $37.40 **5.** $243.20; $63.47

7. $3,019.68; $197.91 **9.** 288% **11.** 384%

13. a. $313.50 **b.** down to $256.50 **c.** $3,982.46

15. a. $536.67 **b.** down to $429.34 **c.** $6,614.91

17. $5,866.67 **19.** $13,373.34 **21.** $4.38 **23.** $7.96

25. $9.67 **27. a.** $2,880 **b.** $4,080 **29.** $5.37

31. $4.95 **33.** $6.77 **35.** $5.03

37. add-on method, $87.50; credit card, $82.50 **39.** $61.44

41. The simple interest for the length of the loan is added to the principal and the total is divided by the number of months of the loan to determine the payments.

43. $34,447.92 **45.** Answers will vary.

47. Answers will vary. **49.** Answers will vary.

Section 8.4

1. $\dfrac{x^8 - 1}{x - 1}$ **3.** $710.59 **5.** $21,669.48

7. $23,008.28 **9.** $85,785.11 **11.** $12,148.68

13. $7,463.67 **15.** $162.14 **17.** $193.75 **19.** 2.7268

21. 1.9527 **23.** 42.62 **25.** 30.91 **27.** 22.43

29. $2,435.99 **31.** $6,286.36 **33.** $121,417.91

35. 79 years **37.** 56 years **39.** $102.78 **41.** $98.31

43. a. $301,354.51; $226,015.88

b. $193,354.51; $145,015.88

c. $247,110.70 ; $199,913.02; tax-deferred earns $47,197.68 more

45. a. $146,709.85; $102,696.90

b. $50,709.85; $35,496.90

c. $102,696.90; $92,047.83; tax-deferred earns $10,649.07 more

47. a. $402,627.32; $301,970.49

b. $192,627.32; $144,470.49

c. $281,839.12; $258,629.34; tax-deferred earns $23,209.78 more

49. 61 months **51.** 47 months

53. Julio: $133,685.32; Max: $109,729.02 **55.** $2,436.65

57. To find the future value we solve for A; to find the payments we solve for R.

61. Think of this as making investments rather than payments. This is the same problem as investing $10,000 for 5 years or making ordinary annuity payments for 5 years. At 8%, you pay less by making payments. At 3%, paying cash is better.

63. a. Think of an annuity due as an ordinary annuity that earns one more month of interest. So if A is the future value of an ordinary annuity, then $A\left(1 + \frac{r}{m}\right)$ is the value of the corresponding annuity due.

b. ordinary annuity $1,966.81; annuity due $1,976.64

Section 8.5

1. $126.81 **3.** $97.06 **5.** $719.46

7. $97.64 **9.** $251 **11.** $1,435

	Payment	Interest Paid	Paid on Principal	Balance
13.	$126.82	$35.06	$91.76	$4,115.81
15.	$246.01	$30.54	$215.47	$4,147.02

17. a. $600

b.

Payment Number	Payment	Interest Paid	Paid on Principal	Balance
1	$600.00	$500.00	$100.00	$99,900.00
2	$600.00	$499.50	$100.50	$99,799.50
3	$600.00	$499.00	$101.00	$99,698.50

c.

Payment Number	Payment	Interest Paid	Paid on Principal	Balance
1	$700.00	$500.00	$200.00	$99,800.00
2	$700.00	$499.00	$201.00	$99,599.00
3	$700.00	$498.00	$202.00	$99,397.00

19. $302.80; $3,034.40 **21.** $278.96; $2,390.08

23. a. $954 **b.** $1,468

25. a. $1,181.40 **b.** $1,614.80

27. The present value of the lottery winnings is $425,678.19; this is slightly better than the lump sum of $425,000.

29. $13,593.02

31. The present value of the retirement plan is $38,900.73; this is slightly worse than the lump sum of $40,000.

33. $492.80

35. $880.80; $753.36; $30,585.60

37. $456.48; $431.69; $892.44

39. a. $70.24 **b.** $25,286.40

41. a. $119.25 **b.** $42,930

43. You must pay interest on whatever you borrow now. If you borrow $12,000, your monthly payments would exceed $200.

45. If there are fees to refinance that exceed the amount that you will save.

47. The initial lower rate makes the mortgage affordable at first; however, as the rate increases, the payments may become excessive.

49. A: $4,500 + $2,800 + $270,648 = $277,948;

B: $2,500 + $1,400 + $302,400 = $306,300

Section 8.6

1. 12% **3.** $15 **5.** $13 **7.** 13% **9.** 14%

11. 12.45% **13.** 13.67% **15.** 15% **17.** 11%

19. 15% **21.** 15% **23.** A **25.** B **27.** over 16%

29. The amortized loan is better because with the add-on loan you are paying interest on the entire amount of the loan for the whole length of the loan.

31. 10%; probably 12 × 10% = 120%; $313.84; 314%

Chapter Review Exercises

1. 12.45% **3.** 68.75% **5.** $1,806 million

7. $160.46 **9.** 10% **11.** $5,496.33 **13.** 3.13%

15. $326.70; $45.75 **17.** $4.57 **19.** $50.85

21. 30 months **23.** $126.82

25. The present value is $490,907.37; the lump sum is the better option.

27. $248.75 **29.** 15.2%

Chapter Test

1. 36.24% **2.** 0.2345 **3.** 43.75% **4.** $18.40

5. $3,655 **6.** 35% **7.** 11.77% **8.** 12% **9.** 15%

10. $4,618.54 **11.** 174 months **12.** $951.43; $1,801.93

13. 8.3 years **14.** $7,052.42 **15.** $327.60, $78.65

16. 4.55% **17.** 2.8% **18.** $6.74 **19.** $23,697.27

20. 2.19 **21.** 15.2% **22.** $47.15 **23.** $293.01

24. about 20 months **25.** $59.26 **26.** $8,707.50

27. The annuity is the better choice because it has a present value of $811,089.58.

28. a. $1,127.84

b.

Payment Number	Amount of Payment	Interest Payment	Applied to Principal	Balance
				$140,000.00
1	$1,127.84	$875.00	$252.84	$139,747.16
2	$1,127.84	$873.42	$254.42	$139,492.74

29. $531.89 **30.** $347.34

Chapter 9

Section 9.1

1. 7 and 9 **3.** 7 and 3 **5.** 10 **7.** 9 and 10

9. false **11.** true **13.** false **15.** *e*, *d*

17. *c*, *g* **19.** 60°, 150°

21. No complementary angle; supplementary angle is 60°.

23. 38.8°, 128.8°

25. $m\angle a = 144°, m\angle b = 36°, m\angle c = 144°$

27. $m\angle a = 45°, m\angle b = 135°, m\angle c = 45°$

29. $m\angle a = 52°, m\angle b = 90°, m\angle c = 128°$

31. Arc AB has length 6 ft. 33. $m\angle ACB = 120°$

35. circumference = 1,200 mm 37. 6 ft 39. 144°

41. 60° 43. 18° 45. 4 in. 47. 60°

49. 12 51. 14.4° 53. 40°

55. The sum of supplementary angles is 180°; the sum of complementary angles is 90°.

57. "Alternate" means that they are on opposite sides of the transversal. "Exterior angles" means that they are outside the parallel lines.

59. Yes; they could all be right angles.

61. No; the angle opposite would have to be obtuse also.

63. It equals 90°. 65. 120°

67.

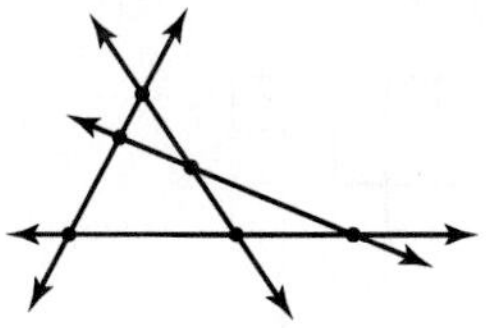

69. 180

Section 9.2

1. false 3. true 5. false 7. polygon

9. not a polygon; not made of line segments

11. 22.5° 13. 60° 15. 4; 720° 17. 162° 19. 18

21. 5 in. 23. $m\angle E = 40°$, length of EH is 7

25. $m\angle I = 35°$, length of HI is 84 27. 48° 29. 450 ft

31. 75 ft 33. 192.5 feet 35. 132° 37. 54° 39. 20.88 ft

41. scalene (no sides equal), isosceles (two sides equal), equilateral (three sides equal)

43. Divide the hexagon into four triangles so the interior angle sum is $4 \times 180 = 720°$. Relate a new problem to an older one.

45. possible

47. Not possible; the angle sum would be greater than 180°.

51. The large angle measures 132°; the two smaller angles measure 24°.

53. 25°

55. The measure of the interior angles gets larger. For very large n, the interior angles have measures close to 180°.

57. $\frac{(n-2)90}{n}$ degrees

Section 9.3

1. 160 ft^2 3. 140 $in.^2$ 5. 108 cm^2 7. 72 yd^2

9. 78.5 cm^2 11. 50.24 m^2 13. 72 yd^2 15. 6.88 m^2

17. 0.86 m^2 19. **a.** 6 ft **b.** 18 sq ft 21. 42.15 cm^2

23. 59.81 m^2 25. 6.62 cm^2 27. 13 m 29. 6.93 yd

31. 69 m^2 33. 7 $in.^2$ 35. 3.5 $in.^2$ 37. area

39. perimeter 41. 127.28 ft 43. 2 m 45. 24 $in.^2$

47. 12 yd^2 49. 86.13 ft^2

51. The large pizza is the better buy; it costs 4.5 cents per square inch versus 5.3 cents per square inch for the medium.

53. 1,000.96 m^2 55. 12,620 ft^2 57. 186.38 m

59. **a.** 22.61 ft **b.** 203.47 ft^2 61. 4,974 yd^2

63. rectangle, parallelogram, triangle, trapezoid

65. A triangle is one-half of a parallelogram.

67. 1.27 times as large 69. 2,779.1 ft^2 71. false

73. **a.** 153.86 $in.^2$ **b.** 627,264,000 $in.^2$
c. 4,076,849 pizzas **d.** 377 slices

75. The area of $WXYZ$ is one-half of the area of $ABCD$.

Section 9.4

1. **a.** 148 cm^2 **b.** 120 cm^3

3. **a.** 80.48 $in.^2$ **b.** 75.36 $in.^3$

5. **a.** 408.2 ft^2 **b.** 628 ft^3

7. **a.** 5,024 cm^2 **b.** 33,493.33 cm^3

9. 75 ft^3 11. 180 $in.^3$ 13. 18.33 m^3 15. 1,728

17. 91 19. 25 21. about 3.7 yd

23. the rectangular cake 25. 415.29 ft^3 27. 53.72 yd^3

29. 2,198; 2,260.8; the radius is being squared, so increasing it contributes greater to the increase.

31. 188.4; 200.96; the radius is being squared, so increasing it contributes greater to the increase.

33. 84 ft^3 35. 2,164 mi

37. We see that the cone occupies less than half of the cylinder, so its volume is less than half of $\pi r^2 h$.

39. The volume is four times as large, because in the formula for the volume of a cone, the radius is squared.

41. Let d be the diameter of the can. The height of the can is $3d$; the circumference of the can is $\pi d \approx 3.14d$, which is larger.

43. The smaller cubes have more surface area. For example, compare the surface area of one 3-in. cube with the surface area of twenty-seven 1-in. cubes.

45. 6.08 in. 47. 15 $in.^3$ 49. 34.64 ft^3 51. 3.33 ft

Section 9.5

1. 24,000 dL 3. 2,800 mm 5. 350 dL 7. 1,135 dg

9. 4,572 mm 11. 15.24 cm 13. 0.02159 dam

15. 1.77 dL

17. a; a juice glass is about one-eighth of a quart or about one-eighth of a liter.

19. c; an inch is about 2.5 cm, so 5 cm would correspond to a 2-in. nose.

21. b; the dog might be about 2 ft tall, or about $\frac{2}{3}$ of a meter.

23. 59.05 ft **25.** 105.73 oz **27.** 554.77 gal

29. 1,056.7 qt **31.** 69.29 in. **33.** 49.56 tons

35. 7.92 dm **37.** Don't give him an inch.

39. It is first down and 10 yards to go. **41.** 29,035.08 ft

43. **a.** 10.76 ft^2 **b.** 129.12 ft^2 **45.** 7,475.65 L

47. 49.72 **49.** 135.93 kL **51. a.** 1.099 L **b.** 37.16 oz

53. $1.25 per pound **55.** $8.25 **57.** 60 m

59. 12.75 km per liter **61.** $86.11 **63.** 65°C

65. 140°F **67.** 58°C **69.** 104.4°C

71. 1 hectare = 10,000 sq m **73.** 90

75. More decimeters because decimeters are smaller than hectometers.

77. Kiloliters are larger than dekaliters. Therefore, we need fewer kiloliters, so *b* is larger than *a*.

79. 0.08 **81.** $7.93 per gallon **83.** $0.57 per pound

Section 9.6

1–3.

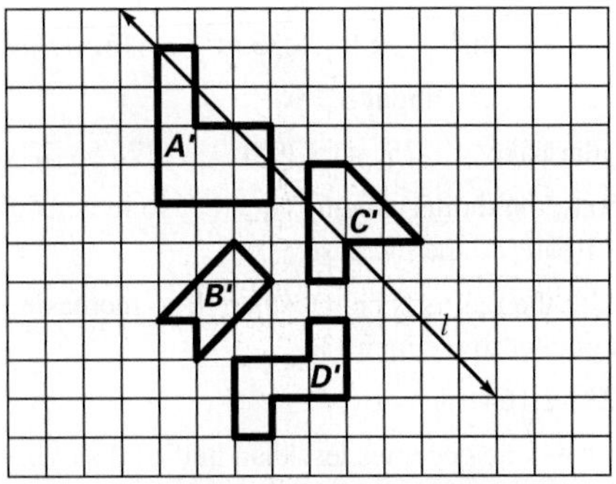

5.

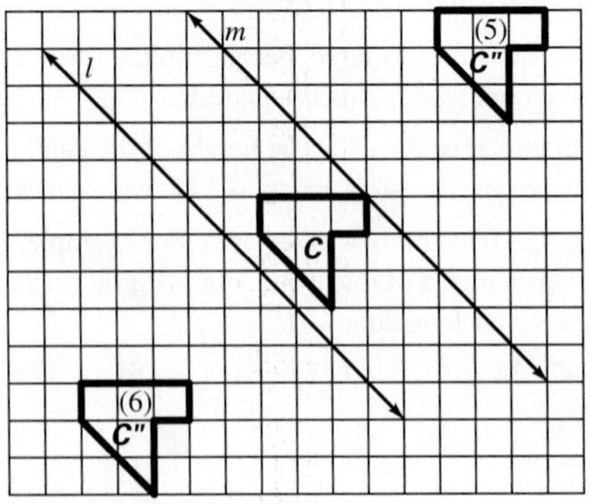

7.

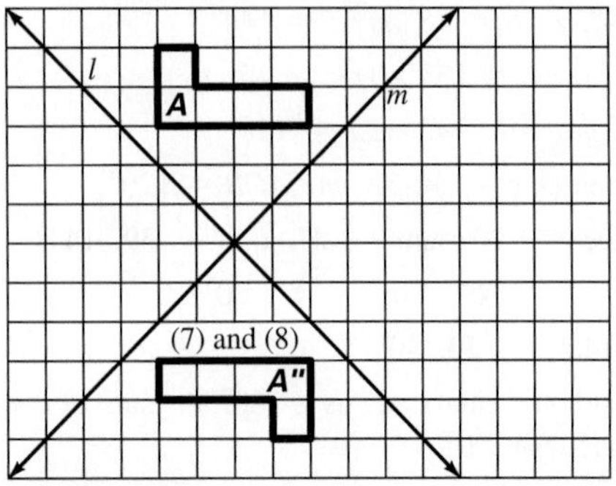

9. Yes; the effect is the same as if we performed a translation.

11.

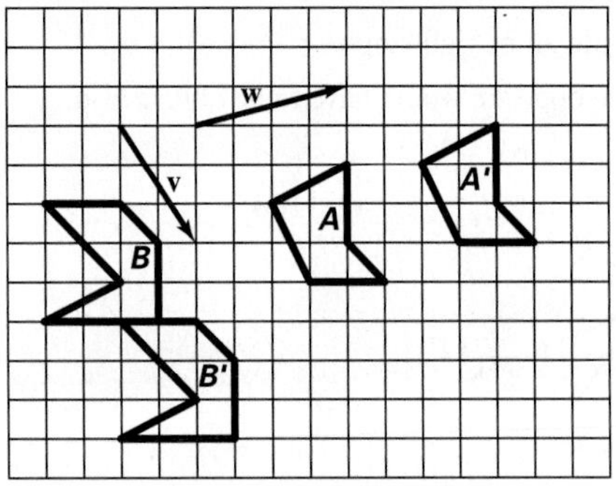

13.

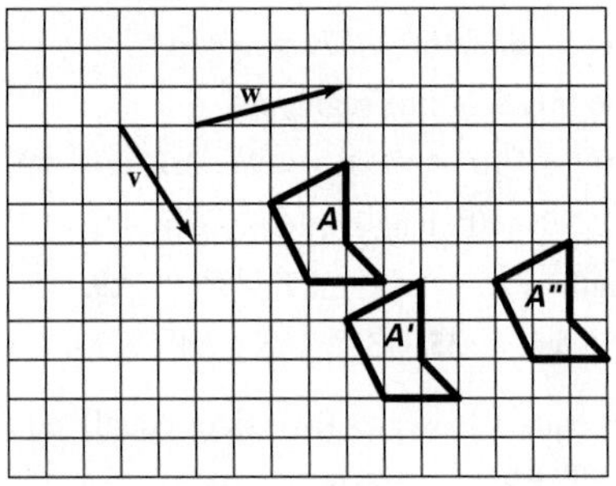

15.

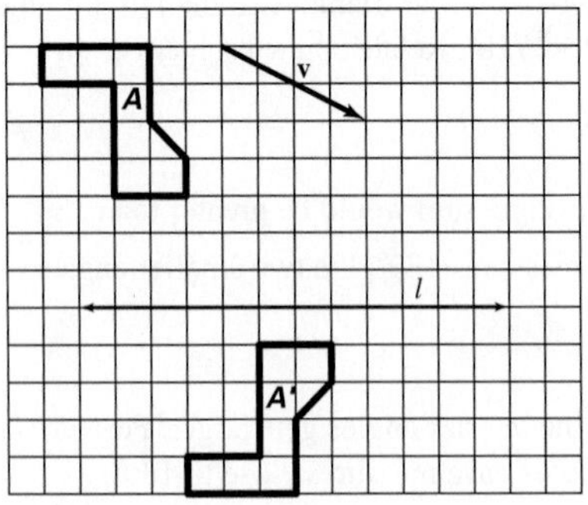

17.

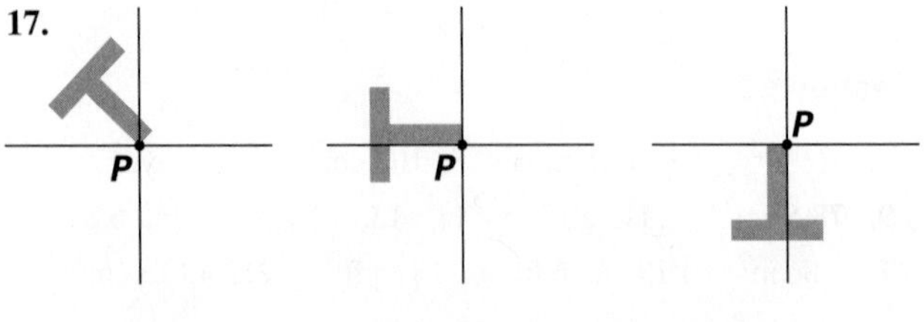

19. 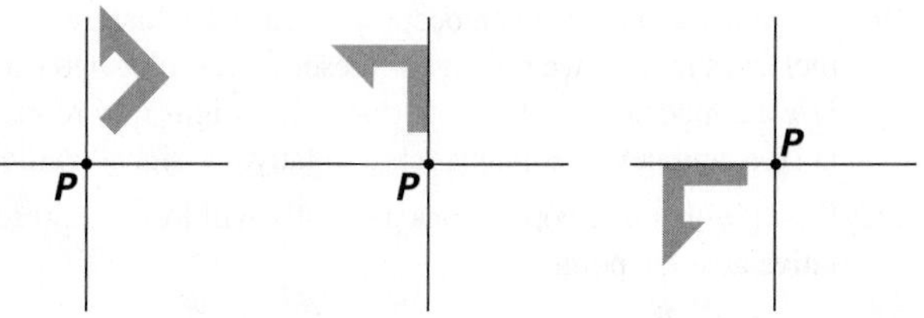

21. 1,800°; 150°

23. Using the interior angles of a regular pentagon, we cannot obtain an angle sum of 360° around a point.

25. 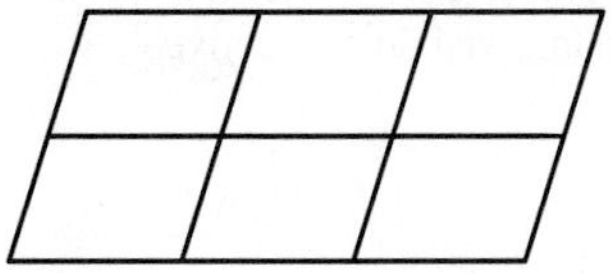

27. (c), (d) **29.** (a), (e)

31. reflectional symmetries: about a vertical line, a horizontal line, and two diagonal lines; rotational symmetries: 90°, 180°, 270°

33. reflectional symmetries: about a vertical line; no rotational symmetries

35. reflectional symmetries: about a vertical line; no rotational symmetries

37. reflection, translation, glide reflection, rotation

39. With a symmetry, the beginning and ending positions of the object are the same; with a rigid motion, this does not have to be the case.

41. At the points where the vertices of the equilateral triangles and regular hexagons meet, the angle sum is $120° + 60° + 120° + 60° = 360°$.

43. At the points where the vertices of the equilateral triangle, the squares, and the regular hexagon meet, the angle sum is $60° + 90° + 120° + 90° = 360°$.

45. Because the sum of the interior angles of a convex quadrilateral is 360°, you can always arrange four copies of the quadrilateral around a point to make an angle sum of 360°.

47. In the figure, we constructed the perpendicular bisectors of segments *AA'* and *CC'*. The point where these perpendicular bisectors meet is the center of rotation.

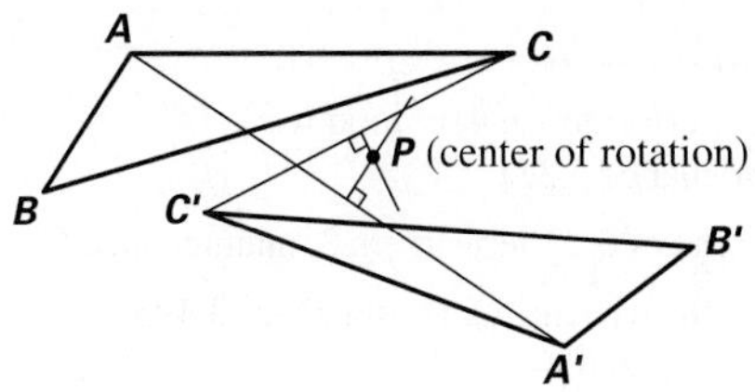

Section 9.7

1. $4^5 = 1{,}024$

3. $\left(\frac{4}{3}\right)^{10} \approx 17.76$

5. **a.** 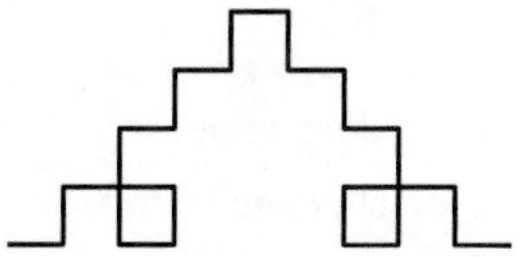 **b.** $\left(\frac{5}{3}\right)^5$

7. 1.29 **9.** 1.46

11. 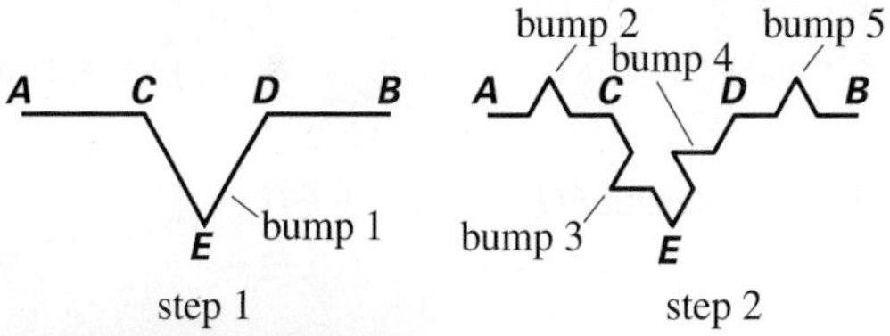

13. 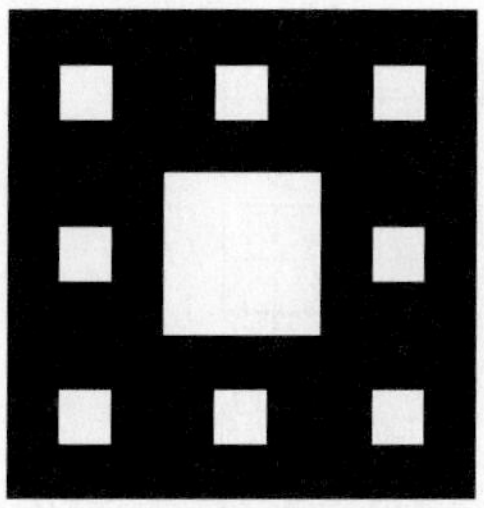

step 2

15. No matter how much we magnify the object, we see patterns that are similar to the patterns seen in the original object.

17. We saw that the length of the curve at step $n + 1$ was $\frac{4}{3}$ the length at step n, so as n increased, the length of the curve would keep getting larger and larger and would exceed any fixed number.

19. 5^n **21.** $\left(\frac{3}{4}\right)^{10} \approx 0.06$; $\left(\frac{3}{4}\right)^n$

Chapter Review Exercises

1. **a.** *b*, *g* **b.** *e*, *d* **3.** 135°

5. an arc of a great circle

7. $m\angle E = 45°$, length of $EH = \frac{20}{3}$, length of $GF = 4$

9. **a.** 80 cm^2 **b.** 20 $in.^2$

11. **a.** 15.2 cm^2 **b.** 3.04 cm

13. **a.** 37.68 $in.^3$ **b.** 330 cm^3

15. The volume will be four times as large because in the formula for the volume of a cone, we square the radius.

17. 56.21 yd **19.** $118.40 **21. a.** *c*, *e* **b.** *d*

23. 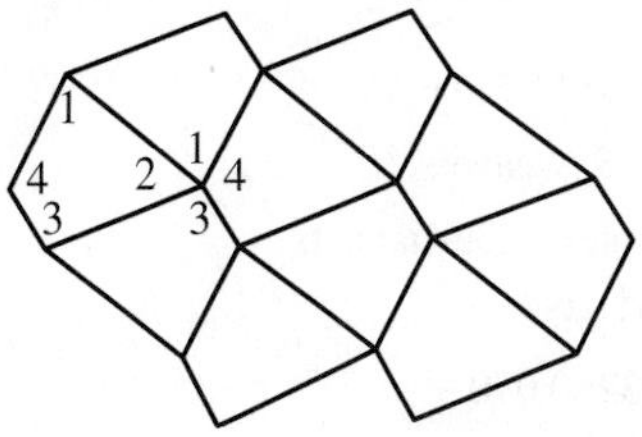

25. zero

Chapter Test

1. **a.** vertical angles **b.** corresponding angles
   c. alternate exterior angles **d.** alternate interior angles
2. 1,800° **3.** $m\angle a = 130°, m\angle b = 140°, m\angle c = 40°$
4. 20 ft **5. a.** 71.55 cm^3 **b.** 48 $in.^3$
6. **a.** *d, e,* and *f* **b.** *b, c* **7.** 3 in. **8.** $\left(\frac{4}{3}\right)^5 = 4.21$ units
9. 27 $in.^2$; 32 cm^2 **10.** 26 m^2; 5.13 ft^2
11. $m\angle F = 35°$, length $FG = 8$, length $GH = \frac{20}{3}$
12. **a.** 210.24 ft^2 **b.** 630.72 ft^2 **c.** 310.72 ft^2
13. 10°
14.

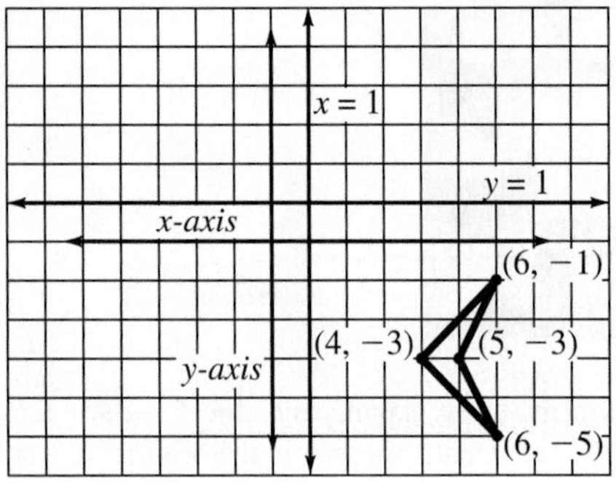

15. **a.** 26.14 ft^2 **b.** 4.4 ft **16.** zero
17. **a.** 24 m **b.** 3,460,000 mg **c.** 2,140 cL
18. The surface area is four times as large.
19. 164.59 dm **20.** 2.47 kL
21. reflectional symmetries about lines through *AE, BF, CG,* and *DH*; rotational symmetries through angles of 90°, 180°, and 270°.
22.

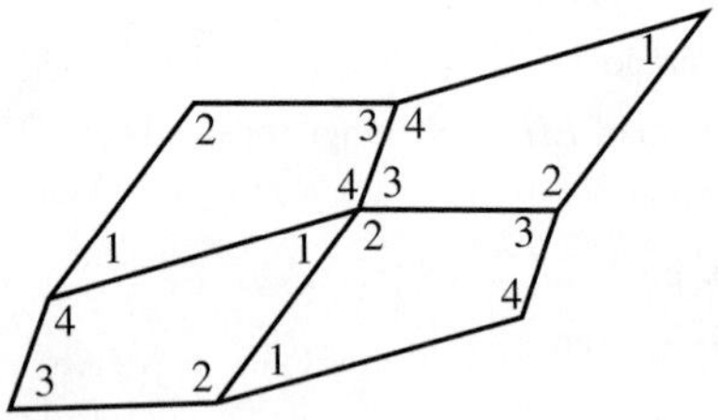

Chapter 10

Section 10.1

1. California: 4; Arizona: 4; Nevada: 3
3. painters: 14; sculptors: 8; weavers: 9
5. Alabama: 7; Mississippi: 4; Louisiana: 6
7. 59,136 **9.** A; 75; 0.045
11. 15; 0.016 **13.** 7.032; 0.010
15. **c.** Paradox occurs when assigning fourteenth seat; business.
17. **a.** Center City, 8; South Street, 3; West Side, 12
    b. In going from 31 to 32, South Street loses a physician.
19. An Alabama paradox can occur when, as a legislature increases in size, we reassign representatives allocated earlier. Use an apportionment system that only assigns new representatives, and never reassigns representatives assigned earlier.
21. States with small populations generally will lose a representative at some point.
23. **a.** 0.181 **b.** 0.270
    c. Giving the extra representative to Naxxon results in a smaller unfairness.
25. If given to New York: 0.058; if given to New Jersey: 0.062; New York deserves the additional representative.

Section 10.2

1. **a.** 0.162 **b.** 0.434 **c.** musicians
3. a. 0.012 b. 0.254 c. red line
5. a. 0.018 b. 0.075 c. New York
7. **a.** 0.074 **b.** 0.100 **c.** New Jersey
9. 41,223.2
11. 1,470.9
13. 0.469
15. 0.327
17. E, P, C, C, E, P, C
19. T, M, D, M, T, M, T, M, T, D, M
21. **a.** Bronx, 3; Manhattan, 3; Queens, 4
    b. B, M, Q, Q, M, B, Q, M, Q, B
23. U, I, O, O, U, O, U, I, O, O, U
25. T, M, F, T, F, T, F, T, M, T, F, T
27. Center 1, 4; Center 2, 1; Center 3, 2
29. It minimizes relative unfairness.
31. 5.7 million

Section 10.3

1. **a.** 32,000; 2.63 **b.** 2 **c.** 3 **d.** 3
3. **a.** 9,272.73; 2.15 **b.** 2 **c.** 3 **d.** 2
5. 13,545.45; CA 4.134, NV 3.027, AZ 3.839
7. CA 4, NV 3, AZ 4
9. 33.2; performers 6.416, food workers 8.223, maintenance 5.361
11. performers 6, food workers 8, maintenance 6
13. 5.15; electricians 4.854, plumbers 3.495, painters 5.631, carpenters 6.019
15. electricians 5, plumbers 4, painters 5, carpenters 6
17. 4.263; fiction 7.037, poetry 4.692, technical 3.988, media 3.284
19. fiction 7, poetry 5, technical 4, media 3
21. 16.667; Pilates 3.36, kickboxing 1.74, yoga 0.66, spinning 0.24
23. Pilates 2, kickboxing 2, yoga 1, spinning 1

25. yes, because A's population has increased faster than B's, but A loses a representative to B in the reapportionment

27. yes, because the number of passengers per week at Bakerstown has increased faster than at Columbia City, but Bakerstown loses a security guard to Columbia City

29. Yes; when C's representatives are added to the commission, B loses a representative to A.

31. No; when D's representatives are added to the commission, there is no change in A's, B's, or C's appointment.

33. D's standard quota is 87.961; however, using the Jefferson method, D would receive 89 representatives, which is greater than its upper quota.

35. D's standard quota is 240.642; however, using the Adams method, D would receive 239 representatives, which is less than its lower quota.

37. The standard divisor is the number of constituents that each representative has. The standard quota is the exact number of representatives due a state. We calculate the standard divisor once and calculate the standard quota for each state.

39. Because the Jefferson method rounds the modified quotas down, we often need larger modified quotas and hence smaller modified divisors.

43. She needs to try a larger modified divisor.

45. smallest 926; largest 940 **47.** smallest 9.64; largest 9.88

Section 10.4

1. **a.** discrete **b.** continuous **c.** discrete

3. **a.** \$55,000 **b.** \$42,500 **c.** car

5. **a.** \$8,000 **b.** \$16,800 **d.** copyright

7. **a.** \$21,000 **b.** \$24,000 **c.** \$9,000 **d.** \$7,000 **e.** \$8,000

9.

	Darnell $(\frac{1}{2})$	Joy $(\frac{1}{2})$
Pays to estate (+) or receives from estate (−)	\$55,000 (+)	\$42,500 (−)
Division of estate balance (\$12,500)	\$6,250 (−)	\$6,250 (−)
Summary of cash	Pays \$48,750	Receives \$48,750

11.

	Dennis (40%)	Zadie (60%)
Pays to pool (+) or receives from pool (−)	\$8,000 (−)	\$11,200 (+)
Division of pool balance (\$3,200)	\$1,280 (−)	\$1,920 (−)
Summary of cash	Receives \$9,280	Pays \$9,280

13.

	Ed $(\frac{1}{3})$	Al $(\frac{1}{3})$	Jerry $(\frac{1}{3})$
Items obtained with highest bid	Painting		Statue
Pays to estate (+) or receives from estate (−)	\$4,000 (+)	\$7,000 (−)	\$7,000 (+)
Division of estate balance (\$4,000)	\$1,333.33 (−)	\$1,333.33 (−)	\$1,333.33 (−)
Summary of cash	Pays \$2,666.67	Receives \$8,333.33	Pays \$5,666.67

15. Betty receives the books, Dennis receives the ring and desk; Dennis pays \$9,125 to Betty.

17. Each person has a different estimate of the value of the estate and does not know the others' estimates.

19. Those who think that items in the estate are worth more receive the items and then must contribute cash to the estate, which is then (partially) distributed to those who made low estimates of the items in the estate.

Chapter Review Exercises

1. When adding a member to the legislature and without changing populations, a state loses a representative.

3. If the number of representatives increases, we reassign representatives that have been assigned earlier.

5. **a.** Florida, 0.395; Texas, 0.437 **b.** Texas

7. BQTHHBQHBHQBT

9. Pilates, 0; kickboxing, 5; yoga, 1; spinning, 2

11. Pilates, 1; kickboxing, 4; yoga, 1; spinning, 2

13. The new states paradox occurs when a new state is added, and its share of seats is added to the legislature, causing a change in the allocation of seats previously given to another state.

15.

	Tito $(\frac{1}{3})$	Omarosa $(\frac{1}{3})$	Piers $(\frac{1}{3})$
Bid on rifle	\$12,000	\$17,000	\$10,000
Bid on sword	\$15,000	\$13,000	\$11,000
Estimated total value of estate	\$27,000	\$30,000	\$21,000
Fair share of estate	\$9,000	\$10,000	\$7,000

Chapter Test

1. When adding a member to the legislature and without changing populations, a state loses a representative.

2. 2,000; 0.0098

3. art 2, music 4, theater 2

4. If the number of representatives increases, we reassign representatives that have been assigned earlier.

5. **a.** Arizona, 0.380; Oregon, 0.434 **b.** Oregon
6. Once a representative is assigned, it is not reassigned at a later date.
7. Mystery Mountain 4, Jungle Village 3, Great Frontier 3, City Sidewalks 4
8. mathematics 6, reading 3, study skills 2
9. No, because when Devon is added, no mall has a change in the number of personnel assigned to it.
10. The population paradox occurs when state A grows faster than state B but, with no change in the number of representatives, A loses a representative to B.
11. massage 3, aromatherapy 2, yoga 3, meditation 1
12. massage 3, aromatherapy 2, yoga 3, meditation 1
13. massage 3, aromatherapy 2, yoga 2, meditation 2
14. only Hamilton's
15. Larry gets Branch A and pays $1,666,666.67; Moe gets Branch B and pays $7,666,666.67; Curly gets $9,333,333.33

Chapter 11

Section 11.1

1. **a.** no **b.** Edelson **3.** A (33 votes)
5. D **7.** C (23 votes) **9.** M **11.** T (15 votes)
13. G **15.** L (17 votes) **17.** L

19. The Borda count method is being used with 3 points for first, 2 for second, and 1 for third.
21. The first eliminated is in last place, the second eliminated is in second-to-last place, and so on.

23. A, D, C, S **25.** D, A, C, S **27.** T, G, E, P, F
29. G, T, E, P, F **31.** 200 **33.** 110 **35.** 6 **37.** 4

39. Voters can state their second, third, and fourth preferences.
41. A candidate who wins an election should be able to beat the other candidates head-to-head.

45. S (92 votes) **47.** E (37 votes)

49. With two candidates, two points will be awarded for first place and one for second. Whoever has the higher point total will have more first-place votes.
51. There is only one comparison. The candidate with the majority of first-place votes wins that comparison.

Section 11.2

1. B
3. C wins; B defeats everyone else head-to-head
5. B wins; if we remove A, then C defeats B 105 to 104.
7. C wins; however, A defeats all other options head-to-head
9. B wins; no one has a majority of first-place votes
11. There are many correct answers; one is (a) 5; (b) 4.
13. Answers will vary.
15. There are many correct answers; one is 30.
17. yes
19. **a.** Bush: 51,112,224; Gore: 51,025,490; Nader: 3,267,386
b. Bush: 51,977,110; Gore: 53,427,989 **c.** Gore: 50.69%
21. E wins 17 to 10
23. majority, Condorcet's, independence-of-irrelevant alternatives, monotonicity
27. Answers will vary.
29. Answers will vary.
35. pairwise comparison; independence-of-irrelevant alternatives

Section 11.3

1. **a.** 5 **b.** Each voter has 1 vote.
c. no dictator **d.** Each voter has veto power.
3. **a.** 11 **b.** A, 10 votes; B, 3; C, 4; D, 5
c. no dictator **d.** No voter has veto power.
5. **a.** 15 **b.** A, 1 vote; B, 2; C, 3; D, 3; E, 4
c. no dictator
d. No voter has veto power (no resolutions can be passed).
7. **a.** 12 **b.** A, 1 vote; B, 3; C, 5; D, 7
c. no dictator **d.** C and D have veto power.
9. **a.** 25 **b.** A, 4 votes; B, 4; C, 6; D, 7; E, 9
c. no dictator **d.** C, D, and E have veto power.
11. **a.** 51 **b.** A, 20 votes; B, 20; C, 20; D, 10; E, 10
c. no dictator **d.** No voter has veto power.
13. {C, D}, {A, C, D}, {B, C, D}, {A, B, C, D}
15. {A, C, D, E}, {B, C, D, E}, {A, B, C, D, E}
17. C and D are critical in all coalitions.
19. All are critical in {A, C, D, E} and {B, C, D, E}; C, D, and E are critical in {A, B, C, D, E}.
21.

Coalition	Weight	
{P}	5	
{T}	4	
{S}	2	
{P, T}	9	Winning
{P, S}	7	Winning
{T, S}	6	Winning
{P, T, S}	11	Winning

23.

Coalition	Weight	
{A}	3	
{C}	4	
{T}	3	
{N}	2	
{A,C}	7	
{A, T}	6	
{A, N}	5	
{C, T}	7	
{C, N}	6	
{T, N}	5	
{A, C, T}	10	Winning
{A, C, N}	9	Winning
{A, T, N}	8	Winning
{C, T, N}	9	Winning
{A, C, T, N}	12	Winning

25. All voters are critical in {P, T}, {P, S}, and {T, S}; none are critical in {P, T, S}.

27. All voters are critical in all coalitions, with the exception of {A, C, T, N}, which has no critical voters.

29. All have index $\frac{1}{5}$.

31. A and B have index $\frac{1}{11}$; all others have index $\frac{3}{11}$.

33. A and B have index 0; C and D have index $\frac{1}{2}$.

35. **a.** All have index $\frac{1}{5}$.

37. **a.** A has index 1; others have index 0.

39. Each has index $\frac{1}{3}$.

41. Pennsylvania has index $\frac{3}{5}$; North Carolina and New Mexico each has $\frac{1}{5}$.

43. K and C have index $\frac{13}{36}$; all others have index $\frac{1}{18}$.

45. All have index $\frac{1}{6}$.

47. No; for example, in the system [10 : 4, 3, 2, 1] voters have different weights but all have the same index, $\frac{1}{4}$.

49. False; see answer to 47.

51. 22 and 28

53. It is generally larger.

Section 11.4

1. (X, Y, Z), (X, Z, Y), (Y, X, Z), (Y, Z, X), (Z, X, Y), (Z, Y, X)

3. There are 24 permutations: (W, X, Y, Z), (W, X, Z, Y), (W, Y, X, Z), . . . , (Z, Y, X, W).

5. 720

7. 479,001,600

9.

Sum Weights of Coalition Members Until We Reach Quota of 5	Pivotal Voter
(A, B, C) 3 + 2	B
(A, C, B) 3 + 2	C
(B, A, C) 2 + 3	A
(B, C, A) 2 + 2 + 3	A
(C, A, B) 2 + 3	A
(C, B, A) 2 + 2 + 3	A

11. A and B have index $\frac{1}{2}$; C has index 0.

13. A and B have index $\frac{5}{12}$; C and D have index $\frac{1}{12}$.

15. A has index $\frac{1}{2}$; the others have index $\frac{1}{6}$.

17. **a.** Each has index $\frac{1}{5}$.

19. A has index 1; the others have index 0.

21. A has index $\frac{3}{4}$; the others have index $\frac{1}{12}$.

23. **c.** Each has index $\frac{1}{4}$.

25. **a.** 24

b. (F, R, W, C), (F, R, C, W), (F, C, R, W), . . . , (C, W, R, F)

27. All have index $\frac{1}{3}$.

29. The Banzhaf index for voter A is the number of times that A is a critical member of a coalition divided by the total number of times all voters are critical members of some coalition. The Shapley-Shubik index is the number of times A is pivotal in some coalition divided by the total number of permutations of all voters.

31. Before B can be pivotal, the managing editor and one other member of the board must already have been chosen.

33. almost 77,147 years

Chapter Review Exercises

1. **a.** no **b.** Myers **3.** social justice

5. D wins; no; R has the majority of first-place votes

7. B wins; no; if A is removed, then C wins with 80 points.

9. Quota, 17; A, 1 vote; B, 5; C, 7; D, 8; no dictator; B, C, and D have veto power.

11. A and B have index $\frac{1}{10}$, C has index $\frac{3}{10}$, and D has index $\frac{1}{2}$.

13. A, $\frac{7}{10}$; B, $\frac{1}{10}$; C, $\frac{1}{10}$; D, $\frac{1}{10}$

15.

Sum Weights of Coalition Members Until We Reach Quota of 6	Pivotal Voter
(A, B, C) 4 + 3	B
(A, C, B) 4 + 2	C
(B, A, C) 3 + 4	A
(B, C, A) 3 + 2 + 4	A
(C, A, B) 2 + 4	A
(C, B, A) 2 + 3 + 4	A

17. $\frac{1}{6}$

Chapter Test

1. a. no **b.** Molina **2.** E

3. Yes; A wins the election and can defeat B, C, and D head-to-head.

4. Quota, 15; A has 5 votes, B has 3, C has 1, D has 3, E has 4, F has 2; A and E have veto power.

5. $6! = 720$ **6.** B

7. B wins; however, A has majority of first-place votes. Majority criterion is not satisfied.

8. $\frac{1}{8}$

9. No; A wins the election; however, if we remove C, then B wins.

10. a. Each voter has index $\frac{1}{4}$; each voter is necessary to pass a resolution.

b. A has index 1 because A is a dictator. The rest have index 0.

11.

Sum Weights of Coalition Members Until We Reach Quota of 7	Pivotal Voter
(A, B, C) 5 + 3	B
(A, C, B) 5 + 3	C
(B, A, C) 3 + 5	A
(B, C, A) 3 + 3 + 5	A
(C, A, B) 3 + 5	A
(C, B, A) 3 + 3 + 5	A

12. C

13. A is winner; however, if we remove D, then B wins. Independence-of-irrelevant-alternatives criterion is not met.

14. {A, B, D}, {A, C, D}, {B, C, D}, {A, B, C, D}

15. A, $\frac{1}{2}$; B, $\frac{1}{2}$; C, 0 **16.** A, $\frac{5}{12}$; B, $\frac{5}{12}$; C, $\frac{1}{12}$; D, $\frac{1}{12}$

Chapter 12

Section 12.1

1. UC, UD, UB, UA, CD, CB, CA, DB, DA, BA

3. UC, UD, UB, UA, CU, CD, CB, CA, DU, DC, DB, DA, BU, BC, BD, BA, AU, AC, AD, AB

5. 4 **7.** 6 **9.** 30 **11.** 120

13. (1, 4), (2, 3), (3, 2), (4, 1)

15. (1, 1), (2, 2), (3, 3), (4, 4), (5, 5), (6, 6)

17. (1, 1), (1, 2), (2, 1), (1, 3), (2, 2), (3, 1), (1, 4), (2, 3), (3, 2), (4, 1)

19. 120 **21.** 9 **23.** 16 **25.** 24 **27.** 32 **29.** 16

31. 36 **33. a.** $2^5 = 32$ **b.** 1; 5; 10 **c.** 16 out of 32

35. 512 **37.** 24 **39.** 4 **41.** 364

43. 27 **45.** 18 **47.** 55 **49.** 14

51. Drawing pictures, being systematic, looking for patterns

53. Follow the branches in the tree diagram to make the schedules.

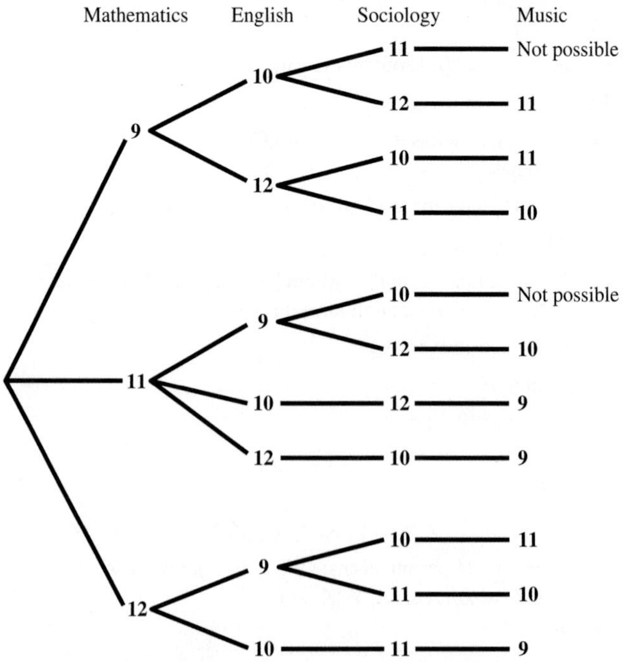

55. RRBG, RBRG, BRRG

57. RGBR, RBGR, GRBR, GBRR, BRGR, BGRR

59. 120

Section 12.2

1. 42 **3.** 336 **5.** 480 **7.** 1,560 **9.** 64

11. 96 **13.** 9,000 **15.** 2,500 **17.** 256

19. 72 **21.** 17,576 **23.** 479,001,600 **25.** 243

27. a. 15 **b.** 6 **c.** 1 **d.** 10 **e.** 900

29. a. $2^8 = 256$ **b.** over $8\frac{1}{2}$ hours **31.** 6

33. a. 7 **b.** 2 **c.** 151,200 **d.** 2,116,800

35. 720 **37.** 72 **39.** All examples use this technique.

41. When we use the fundamental counting principle, we usually get an answer that is the product of smaller numbers. The number 29 is prime, so it is not the product of smaller numbers.

45. a. 18 **b.** 200 **c.** 218

Section 12.3

1. 24 **3.** 6 **5.** 720 **7.** 120 **9.** 30 **11.** 120

15. 336 **17.** 1,814,400 **19.** 56 **21.** 45

23. 1 8 28 56 70 56 28 8 1 **25.** 21

27. the second entry in the 18th row

29. the sixth entry in the 20th row

31. $P(8, 8)$ **33.** $C(17, 3)$ **35.** $P(5, 5)$

37. $P(4, 4)$ **39.** $C(10, 3)$ **41.** $C(9, 5)$

43. $C(17, 8)$ **45.** 177,100 **47.** 1,865

49. Order is not important, but Anna used $P(8, 3)$ instead of $C(8, 3)$.

51. $P(15, 5) = 360{,}360$

53. $P(15, 5) \times P(15, 5) \times P(15, 4) \times P(15, 5) \times P(15, 5) =$ a VERY large number

55. $C(6, 2) \times C(8, 3) = 15 \times 56 = 840$

57. 57,750 **59.** 60

61. $C(24, 8) \times C(16, 8) = 9{,}465{,}511{,}770$

63. 5,940 **65.** 3,300 **67.** 70

69. $C(18, 6) \times C(12, 6) = 17{,}153{,}136$

71. In a permutation problem, we are choosing and arranging objects in order; in a combination problem, we are choosing, but not arranging, the objects.

73. a. There is only one way to choose no objects from a set of five objects.

b. There are eight ways to omit one object from a set of eight objects.

75. You lose $500.

77. Because of the special conditions on the numbers, we cannot use all digits at each position in the number.

79. $C(13, 2) \times C(13, 3) = 78 \times 286 = 22{,}308$

81. 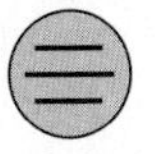 **83.** 6 **85.** $C(n - 1, r - 1)$

87. $C(n - 1, r - 1) + C(n - 1, r) = C(n, r)$

89. The number of ways that we can choose an r-element set from n elements is the same as the number of ways that we can choose $n - r$ elements from n elements; for example, the number of ways that we can choose two elements from five elements is the same as the number of ways that we can choose three elements from five elements.

Section 12.4

1. 120 **3.** 20 **5.** 10 **7.** 4

9. a. 4 **b.** 1,287 **c.** $5{,}148 - 40 = 5{,}108$ **11.** 123,552

13. We are not simply choosing three cards from a possible 52.

15. three oranges **17.** 1,302,540

Chapter Review Exercises

1. *PQ, PR, PS, QP, QR, QS, RP, RQ, RS, SP, SQ, SR*

3. 240 **5.** 288 **7.** 120

9. In a permutation, order is important; in a combination, it is not.

11. 680 **13.** $C(n, r) = \dfrac{P(n, r)}{r!}$

15. the second entry in row 18 **17.** 5,108

Chapter Test

1. *AB, AC, AD, BA, BC, BD, CA, CB, CD, DA, DB, DC*

2. 64 **3.** 48 **4.** 20 **5.** 10

6. $C(12,3) = 220$ **7.** $P(15, 4) = 32{,}760$

8. 40; excluding royal flushes, $40 - 4 = 36$ **9.** 72

10. In a permutation, order is important; in a combination, it is not.

11. 1,320 **12.** 720 **13.** $C(n, r) = \dfrac{P(n, r)}{r!}$

14. the fifth entry of the twelfth row **15.** 108

16. $P(26, 3) \times P(10, 4) = 78{,}624{,}000$

Chapter 13

Section 13.1

1. {(1, 6), (2, 5), (3, 4), (4, 3), (5, 2), (6, 1)}

3. {HHH, HHT, HTH, THH}

5. {(r, b), (r, y) (b, r), (y, r)}

7. {(b, b, r), (b, r, b), (r, b, b), (b, b, y), (b, y, b), (y, b, b)}

9. a. 16

b. {(1, 1), (1, 3), (2, 2), (2, 4), (3, 1), (3, 3), (4, 2), (4, 4)}

c. $\frac{1}{2}$ **d.** $\frac{3}{16}$

11. a. 24

b. {KREB, KRBE, RKEB, RKBE, EKRB, BKRE, ERKB, BRKE, EBKR, BEKR, EBRK, BERK}

c. 0.5

13. a. 20

b. {(s, c), (s, w), (s, d), (s, h), (c, s), (w, s), (d, s), (h, s)}

c. $\frac{2}{5}$ **d.** $\frac{3}{5}$

15. a. $\frac{1}{9}$ **b.** 8 to 1 **17. a.** $\frac{1}{4}$ **b.** 3 to 1

19. 0.000495 **21.** 0.002 **23. a.** $\frac{5}{9}$ **b.** 5 to 4

25. 0.49 **27.** 0.2 **29.** 0.5 **31.** 0.26

33. The probability that the child is a carrier is $\frac{1}{2}$.

		Second Parent	
		s	*n*
First	*s*	*ss*	*sn*
Parent	*s*	*ss*	*sn*

35. a.

		First-Generation Plant	
		r	*r*
First-Generation	*w*	*wr*	*wr*
Plant	*w*	*wr*	*wr*

b. All flowers will be pink. $P(\text{pink}) = 1$; all other probabilities are 0.

37. a.

		Second Parent	
		N	*c*
First	*N*	*NN*	*Nc*
Parent	*c*	*cN*	*cc*

b. $P(\text{disease}) = \frac{1}{4}$

39. 0.1875 **41.** 0.625 **43.** 0.43 **45.** $\frac{22}{36} = \frac{11}{18}$

47. $\frac{2}{36} = \frac{1}{18}$ **49.** 0.59 **51.** 0.55

53. Pick A, which has a $\frac{5}{9}$ probability of winning to B's $\frac{4}{9}$.

55. $\frac{1}{336}$ **57.** $\frac{2}{7}$ **59.** $\frac{5}{12}$ **61. a.** 3 to 7 **b.** 7 to 3

63. 45,057,473 to 1 **65.** 0.571

67. No. The likelihood of a frame being defective or non-defective is probably not equal. The outcomes are not equally likely, so the formula does not apply.

69. An outcome is an element in a sample space; an event is a subset of a sample space.

71. 0.000027 **73.** 0.0000027

Section 13.2

1. 0.985 **3.** $\frac{999}{1{,}000}$ **5.** $\frac{25}{36}$ **7.** $\frac{31}{32}$ **9.** $\frac{7}{13}$ **11.** 0.85

13. 0.10 **15.** 0.55 **17.** 0.923 **19.** 0.54 **21.** 0.08

23. 0.92 **25.** 0.84 **27.** $\frac{9}{13}$ **29.** 0.2 **31.** 0.65

33. 0.19 **35.** 0.981 **37.** 0.902 **39.** false

41. If E and F are disjoint, then $P(E \cap F) = 0$. **43.** 3.24 times

45. $P(A) + P(B) + P(C) - P(A \cap B) - P(B \cap C)$

47. 0.96 **49.** 0.14

Section 13.3

1. $\frac{1}{6}, \frac{1}{3}$ **3.** $\frac{1}{6}, \frac{2}{11}$ **5.** $\frac{1}{2}$ **7.** $\frac{1}{10}$ **9.** $\frac{2}{4}$ **11.** $\frac{2}{3}$

13. 0 **15. a.** $\frac{1}{221}$ **b.** $\frac{1}{169}$ **17. a.** $\frac{40}{221}$ **b.** $\frac{30}{169}$

19. a. $\frac{8}{663}$ **b.** $\frac{2}{169}$ **21.** $\frac{7}{17}$ **23.** 0.032 **25.** 0.126

27. $\frac{117}{850}$ **29.** 0.375 **31.** 0.25 **33.** 0.14

35. independent **37.** independent **39.** dependent

41. 0.947 **43.** 0.90 **45.** 0.153 **47.** 0.85 **49.** 0.223

51. 0.313 **53.** 0.1029 **55.** 0.2646 **57.** 0.125

59.

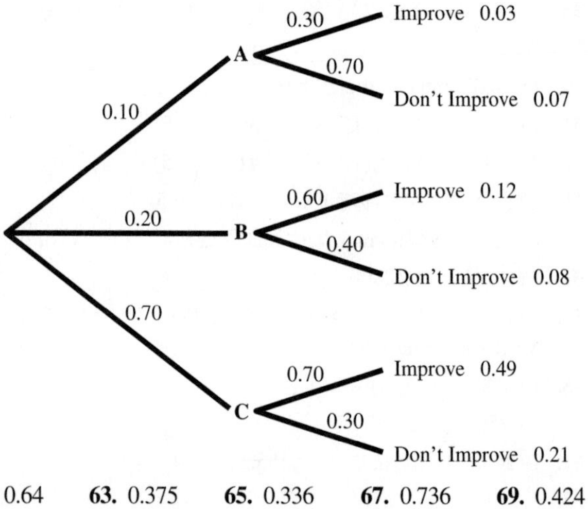

61. 0.64 **63.** 0.375 **65.** 0.336 **67.** 0.736 **69.** 0.424

71. Multiply both sides of the equation by $P(E)$ to get $P(E \cap F) = P(F|E) \cdot P(E)$.

73. Answer will vary. **75.** 0.117 **77.** 23 **79.** 12 people

Section 13.4

1. 1.2 **3.** −$0.50; no **5.** $1.39; $2.39

7. −$2.50; $2.50 **9.** −$0.40; $0.60 **11.** −$2.00; $3.00

13. $328 **15.** −$0.08 **17.** −$0.29

19. Student should guess; expected value is $\frac{1}{16}$. **21.** two options

23. $260 **25.** $210 **27.** −$7.50 **29.** $44 **31.** $25,000

33. Expected value predicts what we can expect to happen long term when an experiment is repeated many times. Expected value does not predict what will happen for any one repetition of an experiment.

37. a. $29.80 **b.** $27.35 **39. a.** 7 **b.** 7 **41.** no

Section 13.5

1. yes **3.** No; the number of trials is not fixed.

5. yes **7.** 0.2634 **9.** 0.0879 **11.** k is larger than n.

13. 0.0537 **15.** 0.0730 **17.** 0.6778 **19.** 0.2637

21. 0.7472 **23.** $\frac{3}{4}$ **25.** 14.7

27. The experiment is performed a fixed number of trials. The experiment has two outcomes: "success" and "failure." The probability of success is the same from trial to trial. The trials are independent of each other.

29. 0.3770

Chapter Review Exercises

1. **a.** {HHT, HTH, THH}
b. {(2, 6), (3, 5), (4, 4), (5, 3), (6, 2)}

3. To calculate the empirical probability of an event E, we divide the number of times the event occurs by the number of times the experiment is performed. To calculate theoretical probability, we use known mathematical techniques such as counting formulas.

5. **a.** $\frac{2}{19}$ **b.** 45 to 55 or 9 to 11 **7.** $\frac{8}{13}$

9. Answers will vary. **11.** **a.** $\frac{1}{17}$ **b.** $\frac{4}{663}$

13. 0.385 **15.** \$0.38 **17.** 0.219

Chapter Test

1. **a.** {(4, 6), (5, 5), (6, 4), (5, 6), (6, 5), (6, 6)}
b. {HHHT, HHTH, HTHH, THHH, HHHH}

2. $\frac{1}{26}$ **3.** **a.** $\frac{3}{31}$ **b.** 17 to 3 **4.** $\frac{4}{13}$

5. When calculating $P(B|A)$, we are calculating the probability of B assuming that event A has occurred. In calculating $P(A|B)$, we are assuming that B has occurred and then calculating $P(A)$.

6. no **7.** $-\frac{1}{6}$ **8.** 0.2816 **9.** 0.0879

10. **a.** $P(E')$ **b.** See Figure 13.6. **11.** $\frac{1}{2}$ **12.** 0.279

13. −\$0.40 **14.** $\frac{2}{3}$ **15.** **a.** $\frac{11}{221}$ **b.** $\frac{8}{663}$

Chapter 14

Section 14.1

1.

x	Frequency	Relative Frequency
2	1	0.05
3	0	0.00
4	0	0.00
5	3	0.15
6	2	0.10
7	4	0.20
8	4	0.20
9	3	0.15
10	3	0.15

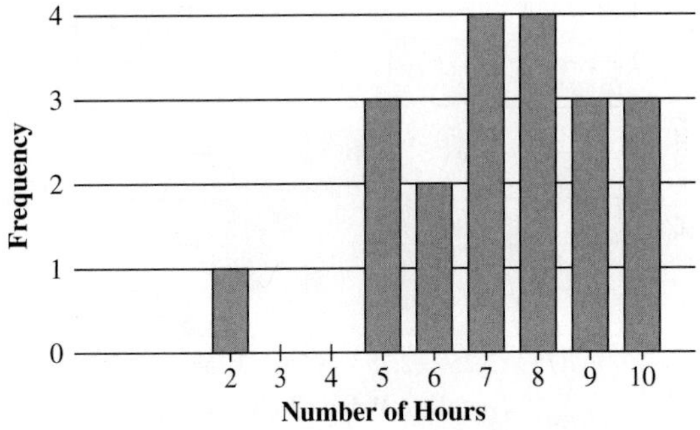

3.

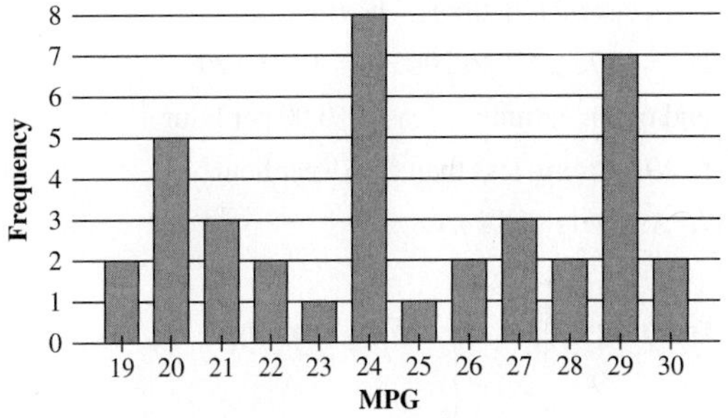

5.

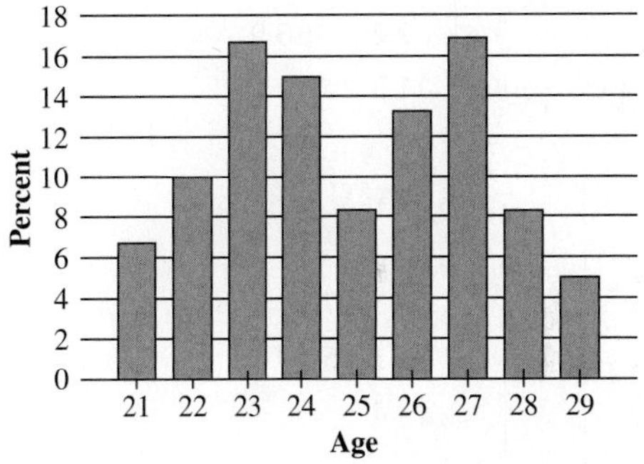

7.

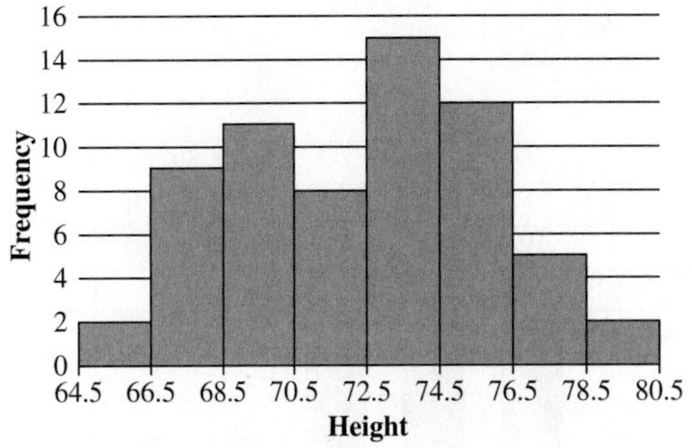

9.

A		B
8 8	1	3 6 8
9 6 2 2 1	2	0 1 1 2 2 2 3 4 6 8 9
9 9 8 7 4 3 2	3	2 3 3 8 9
7 3 3 3 2 2	4	7

11.

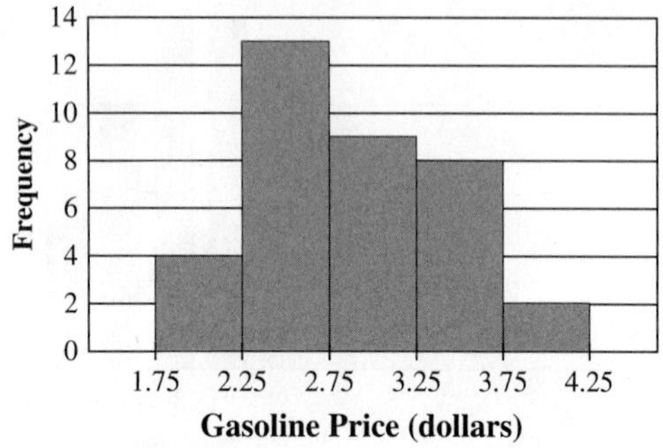

13. a. 12 occurred two times **b.** 6 **c.** 71 **d.** $\frac{22}{71}$

15. 2008 **17.** 2003 to 2005 **19.** 10.2%

21. 25 and older, earning at least $10.00 per hour

23. 16 to 19, earning less than $7.51 per hour

25. to 27. Answers will vary.

29.

No Training		With Training
9	1	
9 8 8 7 6 6 3 2	2	1 8 8 9 9
4 2	3	2 2 3 6 6 7 9
3 3 2 1 0	4	1 3 4 5

31.

NFC		AFC
7		
9 7 7 7 0	1	3 4 6 7 7 7 7
9 7 4 3 3 1 1	2	0 1 1 1 4 5 6 7 9
5 1 1 0	3	1 2 4 4
9 8	4	
2	5	

33.

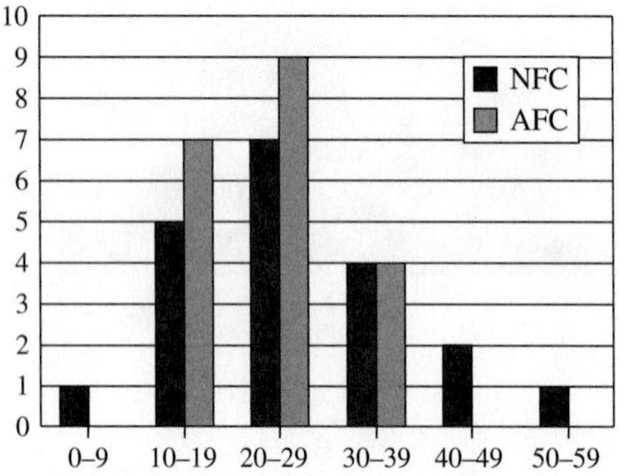

35.

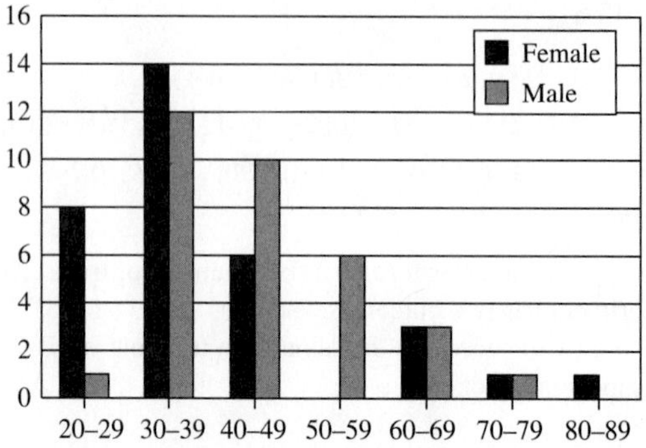

37. The population is the set of all objects being studied; a sample is a subset of the population; selection bias, leading-question bias

39. A continuous variable can take on arbitrary values; a discrete variable cannot; a histogram

Section 14.2

1. mean, 6.3; median, 6; mode, 4

3. mean, 6.9; median, 6.5; no mode

5. mean, 5; median, 5; mode, 5

7. mean, 7.3; median, 7.5; modes, 7, 8

9. mean, 7.15; median, 7; mode, 9

11. mean, 11.29; median, 11; modes, 11 and 15

13. mean, 6.75; median, 6.5; mode, 4

15. mean, 12.6; median, 13; mode, 11

17. a. min, 11; Q_1, 23; median, 31; Q_3, 44; max, 53

b.

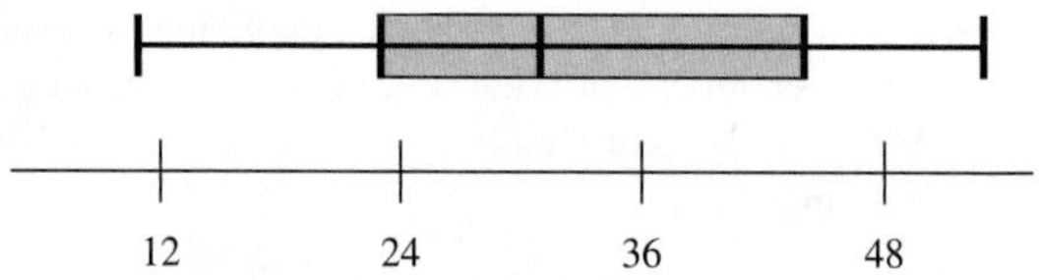

19. a. min, 25; Q_1, 30; median, 33; Q_3, 41; max, 49

b.

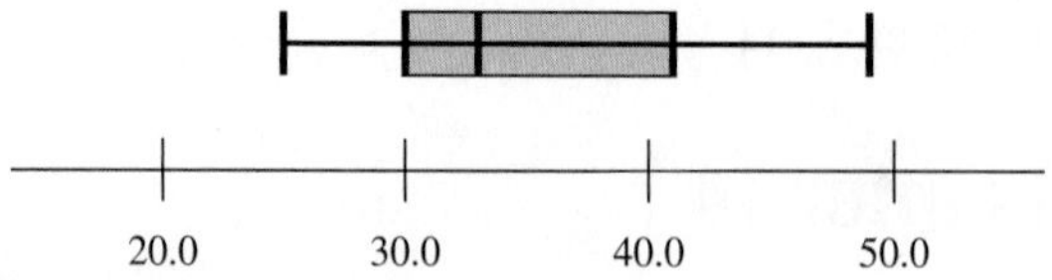

21. mean, 15; median, 15.5; modes, 13 and 20

23. mean, 33.4; median, 12; modes, 7 and 12

25. mean, 376; median, 376; no mode

27. mean, 10.07; median, 8; modes, 7 and 8

29. 2.8 **31.** 73

33. mean, 24.74; median, 25; modes, 24, 25

35. a. 62 **b.** 112

37. $130.32

39. a. b.

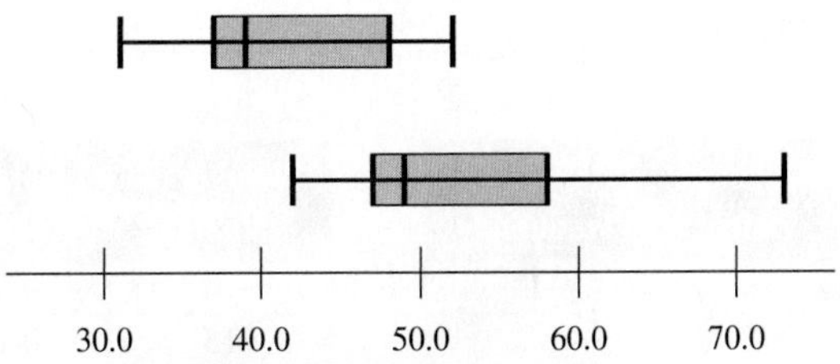

41.

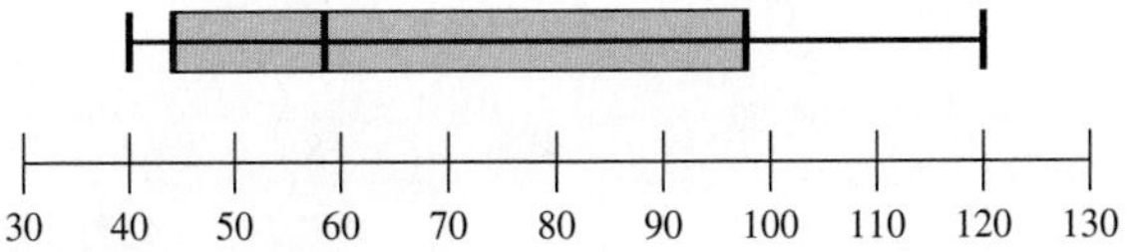

43. liberal arts **45.** education; no

47. The maximum and minimum salaries are not too far from the middle 50% of the salaries.

49. With Σx, we are adding the individual scores in the distribution; with $\Sigma x \cdot f$, we are multiplying each score x by its frequency f before adding.

Section 14.3

1. range, 5; mean, 19.2; standard deviation, 1.55

3. range, 6; mean, 7; standard deviation, 2

5. range, 12; mean, 8; standard deviation, 4.34

7. range, 15; mean, 9; standard deviation, 4.57

9. range, 0; mean, 3; standard deviation, 0

11. 7; 2.13

13. mean, 4; standard deviation, 1.60

15. mean, 7; standard deviation, 1.73

17. 6; 2.14 **19.** 79; 6.59

21. **a.** 5.74; 2.17

b. No; The states have different-size populations. The mean only represents the average of the tax rates in effect for these states.

23. 4.41; 3.38

25. Family H's income is 1.25 standard deviations above the mean.

27. D **29.** 1

31. *CV* for Apple Inc. is 9.9%; *CV* for Dell is 9.2%; Apple was more volatile.

33. DJIA has a *CV* of 0.6%; WebMaster has a *CV* of 4.5%; WebMaster was more volatile.

35. coffee 0.49; gasoline 0.24

37. Family G went from the mean income to 0.94 standard deviations above the mean.

39. 22.82% **41.** **a.** false **b.** true **43.** false

Section 14.4

1. 34% **3.** 2.5% **5.** 16% **7.** 16% **9.** 97.5%

11. 13.5% **13.** **a.** 0.436 **b.** 0.496 **c.** 0.060 **d.** 0.004

15. 39.1% **17.** 27.3% **19.** 8% **21.** 12.8%

23. 7.3% **25.** 8.1% **27.** 90.8% **29.** 1.28

31. −1.175 **33.** 0.2525 **35.** 1.4 **37.** −1.75

39. 0.97 **41.** 64.2 **43.** 33.65 **45.** 31.69

47. **a.** 50% **b.** 16% **49.** **a.** 34% **b.** 16%

51. 1.67 **53.** $0.092 \times 324 = 29.808$ (we expect about 29 or 30)

55. $69.2\% \times 200 = 138.4$ **57.** 9 days; 21.8% **59.** 53.5

61. 83 **63.** 92.56 min **65.** 163.12 **67.** 6

69. Robinson was 2.3 standard deviations above the mean batting average for the 1940s, and Carew was 2.81 standard deviations above the mean for the 1970s. Carew was more dominant.

71. about 10 months

73. **a.** 2.685 (2 or 3 years) **b.** 1.26 (1 or 2 years)

75. 42.1%

77. **a.** The score is 1.75 standard deviations above the mean.

b. The score is 0.85 standard deviations below the mean.

79. By looking at the graph of the normal curve, we see that there is clearly more area under the curve between 0.0 and 0.7 than there is between 1.3 .and 2.0.

83. 0.84 **85.** 42.692 **87.** 96th

Section 14.5

1. positive correlation

3. $r = 0.99$

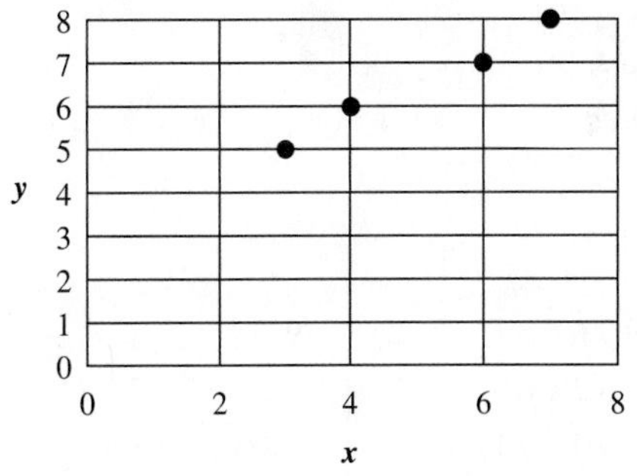

5. $r = 0.74$

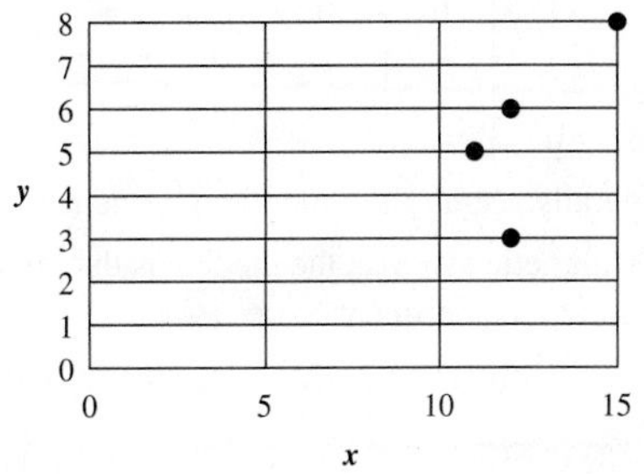

7. We can be 95% confident, but not 99% confident.

9. neither **11.** $y = 0.7x + 3$ **13.** $y = 0.89x - 5.61$

15. 0.96; we can be 99% confident that there is positive linear correlation.

17. 0.00; neither **19.** $y = 2.64x + 21.71$

21. $y = 0 \cdot x + 29.4$

23. We can be 99% confident that there is significant negative linear correlation between the variables of car weight and gas mileage.

27. 1

Chapter Review Exercises

1.

Number of Accidents	Frequency
4	1
5	3
6	6
7	2
8	10
9	4
10	5

3.

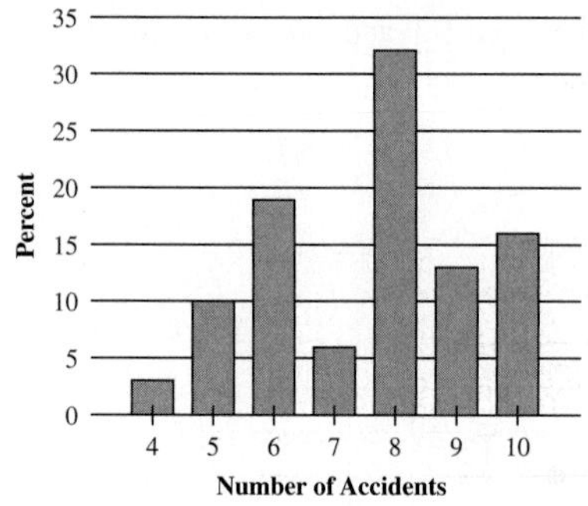

5. a.

F		*M*
9 9 8 6 6 5 1	2	9
9 8 5 5 5 4 3 3 3 2 1 0 0	3	1 2 5 6 7 7 7 7 8 8 9
9 9 5 2 1	4	0 2 3 3 5 5 5 6 7
	5	0 1 2 2 4
1 1	6	0 1
4	7	6
0	8	

b. The actors generally seem to be older than the actresses.

7. The mean is the arithmetic average; the median is the middle score; the mode is the most frequent score.

9.

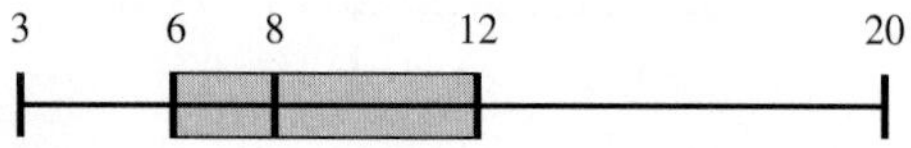

11. the spread of the distribution **13.** 0.71

15. 221 **17.** 8.9 months **19. a.** 0.977 **b.** $y = 2.5x - 2$

Chapter Test

1.

Number of Visits	Frequency
4	1
5	4
6	8
7	9
8	7
9	6
10	2
11	2
12	1

2.

Number of Visits	Relative Frequency
4	0.025
5	0.100
6	0.200
7	0.225
8	0.175
9	0.150
10	0.050
11	0.050
12	0.025

3.

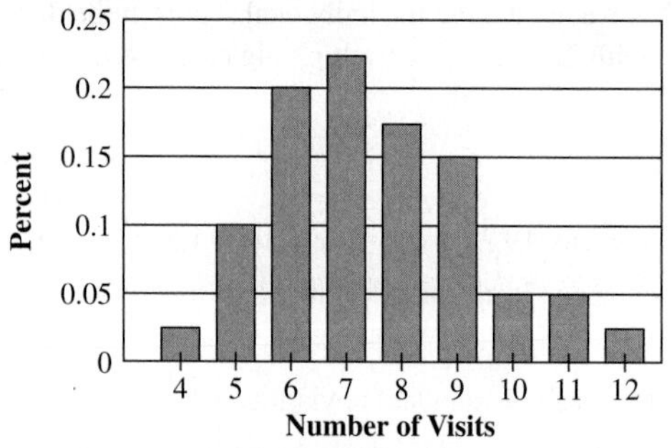

4. the spread of the distribution

5. 2 **6.** See page 739

7.

Ruth		Aaron
5 2	2	
5 4	3	0 2 4 4 8 9 9
9 7 6 6 6 1 1	4	0 0 4 4 4 4 5 7
9 4 4	5	
0	6	

8. a. 82, 86, 93.5, 95, 98

b.

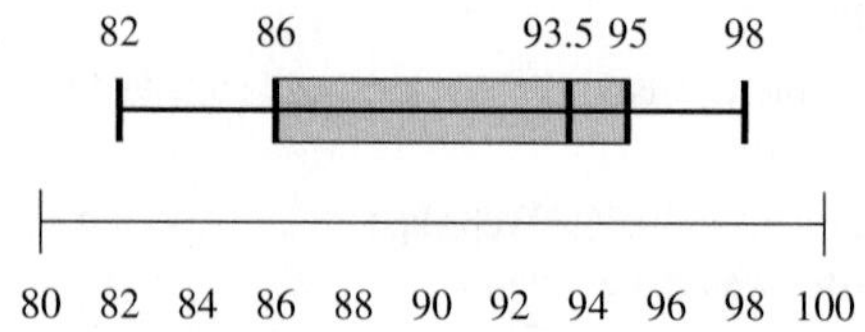

9. the statistics exam

10. a. three hits; seven times **b.** four and six **c.** 38

11. a. none **b.** negative

12. The mean is the arithmetic average; the median is the middle score; the mode is the most frequent score.

13. mean, 7.85; median, 8.5; mode, 9

14. 60.98 **15.** 1.40 **16.** 212

17. 18 months **18. a.** 0.832 **b.** $y = 1.1x + 2.8$

Credits

Photo Credits

p. 1: Bruce Rolff/Shutterstock; p. 4: Yuri Arcurs/Shutterstock; p. 15: Blend Images/Shutterstock; p. 25: Joe Rimkus Jr./MCT/Newscom; p. 30: Radius Images/Glow Images; p. 33: JupiterImages; p. 42: Exactostock/SuperStock; p. 45: Jonathan Alcorn/Zuma Press/Newscom; p. 48: Library of Congress Prints and Photographs Division [LC-USZ62-60242]; p. 53: Liu Yu/Xinhua/Photoshot/Newscom; p. 59: Nick Stubbs/Dreamstime; p. 69: Nemeziyaa/Dreamstime; p. 74: Mark Eaton/Dreamstime; p. 85: Getty Images; p. 87: Anweber/Shutterstock; p. 94: UK History/Alamy; p. 96: Jennifer Graylock/AP Images; p. 118: Rud/Dreamstime; p. 120: Vereshchagin Dmitry/Shutterstock; p. 122: Juan Silva/The Image Bank/Getty Images; p. 129: Tihis/Shutterstock; p. 131: Vlorzor/Fotolia; p. 135: NASA; p. 142: MNStudio/Dreamstime; p. 144: Irina Zavyalova/Dreamstime; p. 149: Msivyi/Dreamstime; p. 152: Imageegami/Shutterstock; p. 156: Shots Studio/Shutterstock; p. 163: SweetCrisis/Shutterstock; p. 168: Jarenwicklund/Dreamstime; p. 171: John S. Peterson/Icon SMI AYA/Newscom; p. 182: Menna/Shutterstock; p. 190: 20th Century Fox/Photoshot; p. 191: Peter Endig/DPA/Picture-Alliance/Newscom; p. 194: Javarman/Shutterstock; p. 199: World History Archive/Alamy; p. 201: Interfoto/Alamy; p. 203: Ivan Vdovin/Alamy; p. 208: Science Source/Photo Researchers, Inc.; p. 213: Purestock/Alamy; p. 219: Album/Oronoz/Newscom, AKG Images/Newscom, Science and Society/SuperStock; p. 226: Rachel Youdelman/Pearson Education, Inc.; p. 227: Flyinglife/Shutterstock; p. 232: JPL/Caltech/NASA; p. 243: SuthinSoonthorn/Shutterstock; p. 251: Daniel Prudek/Shutterstock; p. 258: Rudiuk/Shutterstock; p. 265: Lalo85/Dreamstime, Interfoto/Alamy; p. 282: NASA; p. 284: Condor 36/Shutterstock; p. 292: Cloki/Shutterstock, David H. Seymour/Shutterstock, Christopher Walker/Shutterstock, Ozerov Alexander/Shutterstock; p. 293: Mircea Bezergheanu/Shutterstock; p. 300: Lexus, a division of Toyota Motor Sales, U.S.A., Inc.; p. 314: Picture Contact BV/Alamy; p. 328: Caprasilana/Fotolia; p. 338: Tom Ang/Fotolia; p. 339: Kris Positief/Fotolia; p. 353: Shutterstock; p. 363: Tihis/Shutterstock; p. 370: Mark R. Higgins/Fotolia; p. 378: Gleam/Fotolia; p. 390: Jonathan Noden-Wilkinson/Shutterstock; p. 393: Glenda M. Powers/Shutterstock; p. 397: Bryant Jayme/Shutterstock; p. 406: Gelpi/Shutterstock; p. 414: Steffen Foerster/Shutterstock; p. 423: Kurhan/Shutterstock, Forgiss/Fotolia; p. 430: Tatyana Pavlova/Fotolia; p. 436: The Photos/Fotolia; p. 438: Brian Erickson/Shutterstock; p. 451: Handout/KRT/Newscom; p. 454: Losevsky Pavel/Shutterstock; p. 461: Andre Klaassen/Shutterstock; p. 465: Richard Coencas/Shutterstock; p. 488: Stuart Monk/Shutterstock; p. 494, Invisible/Shutterstock; p. 502: JupiterImages; p. 505: Geoff Hardy/Shutterstock, World History Archive/Newscom; p. 507: Courtesy of Professor Anne Burns; p. 515: Mark Grenier/Shutterstock, Jennifer Gottschalk/Shutterstock; p. 516: S. Kobold/Fotolia; p. 522: Win McNamee/Getty Images; p. 523: James M Phelps, Jr/Shutterstock; p. 527: Apple's Eyes Studio/Shutterstock; p. 532: Roman Gorielov/Fotolia; p. 534: Corbis; p. 561: S. Bukley/Newscom; p. 571: ZUMA Press/Alamy; p. 575: Aberenyi/Fotolia; p. 581: JoLin/Shutterstock; p. 583: Paul Sakuma/AP Images; p. 586: Gary Hathaway/Shutterstock; p. 592: Monkey Business Images/Shutterstock; p. 598: Shawn Pecor/Shutterstock; p. 605: Katharine Androitis/Alamy; p. 606: Corbis Flirt/Alamy; p. 610: Graylock/ABACAUSA.com/Newscom; p. 611: Charles Krupa/AP Images; p. 616: Neelsky/Shutterstock; p. 620: Margo Harrison/Shutterstock; p. 621: Noah Strycker/Shutterstock; p. 625: Gino Santa Maria/Fotolia; p. 628: Paul J. Richards/AFP/Getty Images/Newscom; p. 632: Beth Anderson/Pearson Education, Inc.; p. 633: NASA; p. 635: Akhilesh Sharma/Fotolia; p. 643: Puddingpie/Fotolia; p. 647: GeoM/Shutterstock; p. 651: Interfoto/Alamy; p. 653: Classic Image/Alamy; p. 654: Katya252728/Fotolia; p. 657: MONOPOLY ® & © 2012 Hasbro, Inc. Used with

permission; p. 658: Cao Zichen/Imaginechina/AP Images; p. 662: Galyna Andrushko/Shutterstock; p. 666: KairoCosta/Fotolia; p. 671: ClickPop/Fotolia; p. 681: Roger Weber/Digital Vision/Getty Images; p. 684: PhotoDisc/Getty Images; p. 686: Mary Evans Picture Library/Alamy; p. 701: Ron Buskirk/Alamy; p. 707: National Oceanic and Atmospheric Administration (NOAA); p. 709: David Cole/Alamy; p. 710: Caro/Alamy; p. 717: Allstar Picture Library/Alamy; p. 727: PhotoDisc/Getty Images; p. 746: Curtis Management Group/ZUMA Press/Newscom; p. 749: Ollirg/Fotolia; Cover: Rubik's Cube ® used by permission of Seven Towns Ltd. www.rubiks.com/Photo by Pearson Education, Inc.; Apps Icon: Bannosuke/Fotolia; Pencil Icon: Prakapenka/Shutterstock Film Icon: Borat/Shutterstock; Gear Icon: Kuttly/Shutterstock; Newspaper Icon: Vectorlib-com/Shutterstock

Text Credits

p. 30: Data from James M. Henslin, *Sociology: A Down-to-Earth Approach*, Ninth Edition. Copyright © 2007 by Allyn & Bacon; p. 36: Data from U.S. Department of Commerce, The Statistical Abstract of the United States, 2011–2012, 130th Edition, (Table 473); p. 49: Based on data from www.arizonoagasprices.com; p. 84: Microsoft Excel screenshot. Copyright © Microsoft Corporation. All Rights Reserved. Reprinted with permission of Microsoft Corporation; p. 89: From A History of Formal Logic by Innocentius M. Bochenski, © 1961 Notre Dame Press; p. 105: Data from the Human Society of the United States website (compiled from the American Pet Products Association 2011–2012 National Pet Owners Survey.) Copyright © 2011 by the Humane Society of the United States. Reprinted with permission; p. 104: Based on data from *Essentials of Sociology*, Eighth Edition, by James Henslin © 2008 by Allyn & Bacon; p. 115: Excerpt from "The Argument Clinic" sketch from *Monty Python's The Flying Circus*. Copyright © 2007 by Python (Monty) Pictures. Reprinted with permission; p. 141: Microsoft Excel screenshot. Copyright © Microsoft Corporation. All Rights Reserved. Reprinted with permission of Microsoft Corporation; p. 161: Based on material from The History of Mathematics: An Introduction by David M. Burton; p. 193: Based on The History of Mathematics: An Introduction by David M. Burton (1999); p. 231: Microsoft Excel screenshot. Copyright © Microsoft Corporation. All Rights Reserved. Reprinted with permission of Microsoft Corporation; p. 236: Based on Mathematicians Explore Cicada's Mysterious Link with Primes by Michael Stroh, from *The Baltimore Sun*, May 10, 2004; p. 246: Based on Mathematics: From the Birth of Numbers by Jan Gullberg, 1997; p. 269: Based on The History of Mathematics by David Burton; p. 284: Source: United States Government; p. 299: Microsoft Excel screenshot. Copyright © Microsoft Corporation. All Rights Reserved. Reprinted with permission of Microsoft Corporation; pp. 318, 319: Data from the World Almanac and Book of Facts 2011; p. 319: Based on World Almanac and Book of Facts 2011; p. 319: Source Statistical Abstract of the U.S. 2011–2012, from http://www.washingtonpost.com; p. 343: Based on The Census Count by Lynne Billard (in PS: Political Science and Politics, December 2000), according to USA Today; p. 387: Source: New York Times Almanac 2011; p. 402: Source is eBizMBA.com, as reported in USA Today; p. 403: Based on "Generation Broke: The Growth of Debt Among Younger Americans" at www.demos-usa.org/pub295.cfm; p. 404: Based on A History of Interest Rate by S. Homer and R. Sylla (1991); p. 414a: Source: based on data from NY Times Almanac 2011; p. 437: From the Great Circle Mapper website, www.gcmap.com. Copyright © 2012 by the Great Circle Mapper. Reprinted with permission; p. 500: Conversations with Cezanne, edited by Michael Doran; p. 500: The Fractal Geometry of Nature by Benoit B. Mandelbrot; p. 516: United States Constitution, Article 1, Section 2; p. 539: Map from "Redrawing Florida's Political Maps," by Wes Meltzer from THE ORLANDO SENTINEL. Copyright © 2012 by The Orlando Sentinel. Reprinted with permission; p. 565: Screen shot reprinted with permission from Apple Inc.; p. 578: Microsoft Excel screenshot. Copyright © Microsoft Corporation. All Rights Reserved. Reprinted with permission of Microsoft Corporation; p. 582: From the Federal Election Commission. www.fec.com; p. 629: Screen shot reprinted with permission from

Apple, Inc.; p. 676: Based on The Probability of Penalizing the Innocent Due to Bad Test Results by Tom Rogers Source, data from the Colorado State Tropical Prediction Center; p. 700: Microsoft Excel screenshot. Copyright © Microsoft Corporation. All Rights Reserved. Reprinted with permission of Microsoft Corporation; p. 700: Screen shot reprinted with permission from Apple Inc.; p. 706: Screen shot printed with permission of Minitab Inc. All such material remains the exclusive property and copyright of Minitab Inc. All rights reserved; p. 708: Screen shot reprinted with permission from Apple Inc.; p. 763: Microsoft Excel screenshot. Copyright © Microsoft Corporation. All Rights Reserved. Reprinted with permission of Microsoft Corporation.

Index

Historical Index

People

Abel, Niels Henrick, 325
Ahmes, 193
al-Khowarizmi, 203, 325
Appel, Kenneth, 151
Aristotle, 89, 123, 130
Augustus Caesar, 686
Babbage, Charles, 219
Bolyai, Janos, 443
Boole, George, 63, 95
Brahmagupta, 246
Cantor, Georg, 73, 77, 79
Cardano, Girolamo, 246, 325, 653
Caritat, Marie-Jean-Antoine, 574
Champollion, Jean Francois, 193
Cohen, Paul, 79
Condorcet, Marquis de, 574
DeMorgan, Augustus, 63, 151
Descartes, René, 246, 303, 309
Diophantus, 245
Eckert, J. Presper, 219
Eratosthenes, 31, 234, 238
Euclid, 237
Euler, Leonhard, 143–145, 266
Fermat, Pierre de, 665
Fibonacci (Leonardo of Pisa), 203, 291
Fourier, Jean-Baptiste, 193
Franklin, Benjamin, 408, 686
Gallup, George, 704
Galois, Evariste, 325
Gauss, Carl Friedrich, 221, 237, 443
Germain, Sophie, 259
Gödel, Kurt, 101
Graunt, John, 728
Guthrie, Francis, 151
Haken, Wolfgang, 151
Halley, Edmund, 728
Hamilton, Alexander, 517
Hamilton, William Rowan, 151, 161
Hilbert, David, 73, 79
Hill, Joseph, 527
Huntington, Edward, 527
Hypatia, 460
James I, King of England, 686
Lagrange, Joseph, 246
Laplace, Pierre-Simon de, 665
Leibniz, Gottfried, 89, 101, 219
Lobachevsky, Nicholai, 443
Lukasiewicz, Jan, 124
Mauchly, John, 219
Mendel, Gregor, 651
Napier, John, 206
Napoleon, 193
Nero, 686
Nightingale, Florence, 709
Pappus of Alexandria, 456
Pascal, Blaise, 219, 246, 665
Pearson, Karl, 709
Poincaré, Henri, 73
Polya, George, 14
Pythagoras of Samos, 265, 269
Rhind, A. Henry, 193
Riemann, Bernhard, 443
Russell, Bertrand, 101
Saint Thomas Aquinas, 89
Sun Tzu, 221
Tarski, Alfred, 124
Tukey, John, 708
Washington, George, 686
Whitehead, Alfred North, 101

Topics

Abacists vs. algorists, 195
Abacus, 195, 218–219
Analytical geometry, 303, 304
Apportionment, 517–519, 527
Banzhaf indices, 590
Boolean algebra, 63
Cardinal numbers, 79
Continuum hypothesis, 79
Credit cards, 404
Cubic equations, 325
Estimation of Earth's
circumference, 31
Exploratory data analysis, 708
Female mathematicians, 259
Fibonacci sequence, 291–293
Genetics, 651–652
Graph theory, 145
Hindu-Arabic numerals, 195
Human rights, 574
IQ tests, 19
Logic, 89, 101, 124
Lotteries, 686
Mayan mathematics and astronomy, 203
Negative numbers, 246
Non-Euclidean geometry, 443, 452
Number theory, 238
Polls, 704
Probability theory, 653, 665
Problem solving, 14
Pythagoreans, 269
Pythagorean theorem, 265
Quadratic equations, 325
Quartic equations, 325
Quintic equations, 325
Random walk problem, 14
Rhind papyrus, 193
Rosetta stone, 193
Set theory, 73, 79
Sophie Germain prime, 241
Statistics, 709, 728

Applications Index

Astronomy
Astronauts, 633
Astronomical distances, 285
Circumference of earth, 442, 445
Diameter of moon and Mars, 476
Earth to sun distance, 284
Light-year calculation, 285
Planets in solar system, 47
Space flight, 284–285
Space shuttle, 641

Automotive
Accidents, 73–74, 760
Air bags, 127
Braking distance of car, 59, 347
Car of the year, 573
Car repair, 666
Car value, 423
Car warranty, 102
Commuting costs, 312
Depreciation on new car, 388
Distracted drivers, 680
EPA mileage, 710
Foreign cars, 132–133
Gas mileage, 387, 486, 758
Gas prices, 35, 49, 486, 711
Gas tank capacity, 483
Mileage ratings, 725
Motor vehicle production, 362
Payments for new car, 417, 421, 423, 427
Purchasing a new car, 15, 53, 66, 83, 397, 406
Purchasing a used car, 754–755
Purchasing decision, 134–135
Safety ratings, 679
Sales, 381, 387
Servicing a car, 243
Speed of vehicle, 273, 344, 346, 486, 758
Tire warranty estimation, 35

Biology and Life Sciences
Atoms in person's body, 282–283
Cicada emergence patterns, 243
Cross-breeding plants, 656–657, 698
Dinosaur weight comparisons, 9
Food chain, 173
Human weight vs. mosquito weight, 284
Leatherback turtles, 342
Marine biologists, 373
Population growth, 373
Wildlife growth, 291, 369
Wildlife management, 370
Wildlife population estimation, 346–347

Business
Advertising, 183, 263, 311, 363, 632
Candy packaging, 227
Claims adjuster, 387
Coffee mug promotional, 220
Commissions, 546, 666
Committee formation, 630–631
Compounding raises, 388
Contract negotiation, 388, 689–690, 721–722
Cupcake production, 315–316
Daily profit, 690
Drug testing, 676
Electronic company apportionment, 523
Executives board, 598–599
Factory/facility locations, 581
Health club charges, 311
Interviews, 622, 632
iPad covers, 243
ISBNs, 228
Job comparison, 312
Job offers, 735
Law firms, 57, 587–588, 591, 598
Loan payments, 423
Logo design, 450–451
Magazines, 632–633
Merchandise display, 243
Merchandise markups, 387
Mergers, 570–571
Newspaper, 612, 620, 633
Paperwork flow, 173
Partnership dissolving, 553–555
Product ordering, 687
Profit evaluation, 688
Profits, 328
Résumés, 15
Safety record, 715
Salaries, 387, 750
Small businesses, 363
Standardized testing, 745–746
Textbook publishing, 265
Training programs, 713
Wage data, 712

Chemistry
Carbon dating, 369–370
Gas pressure, 347
Radioactive decay, 339

Construction
Bookshelf construction, 264
Building preservation, 16
Cabinet building, 264
Church refurbishment, 484–485
Creek dredging, 263
Egyptian computation, 194
Experimental house, 183
Furniture manufacturing, 311, 354–355
Gazebo supports, 455
Japanese furniture makers, 454–456
Laptop batteries, 737
Paint job estimation, 264
Paint purchase estimation, 37
Road resurfacing, 263
Scaffolding, 456
Shopping center, 169–171
Student union building, 182
Tessellation of tiles, 495–496
Tile floor, 243, 466
Tile wall, 17–18, 263
Triangles to build bridge, 451–452
Waste minimization, 264

Consumer Information
Air bags, 127
Apartment expense, 35
Art supply purchasing, 263
Car warranty, 102
Cell phone redundancy, 696
Computer purchase, 35
Credit card policy, 114
Customer service, 749
Electronic device sales, 36
Fast food, 56, 66, 568
Gasoline prices, 35, 49
Grocery bill estimation, 29
Jet-ski rental plans, 308–309
Milk packaging, 719
Paint buying, 265
Plant purchase, 35
Product reliability, 681, 749
Refund policy, 108
Rent-to-own agreement, 428, 430
Supply purchasing estimation, 35
Tip estimation, 35
Tire warranty estimation, 35
Unit pricing, 264
Vending machine analysis, 748–749
Waiting times, 736
Warranties, 746–747, 750, 761

Economics. *See also* Finance
Bubbles, 22, 27
Budget reduction, 581
Coin mints, 56
CPI, 385, 398, 712
Currency conversion, 312
Defense spending, 380
Inflation, 335, 339–340, 387, 388, 398
Money conversion, 347
National debt calculations, 282, 284, 399
Price stability, 737
Stock market, 248, 252, 311, 387
Stocks, 56, 297, 731, 736–737
Supply and demand, 353–354, 362
Taxes, 8–9, 35, 93, 102, 104, 114, 384, 387, 397, 433, 575–576, 735

Education
Academic advisement vs. academic success, 667
Academic services survey, 74
Activities puzzle, 24–25
Athletics committee, 591
Borrowing for education, 397
Class schedules, 9–10, 16, 614, 620, 622
College choice decision, 27
College choice survey, 82
College fund, 369, 398
College life improvements, 570, 571
Competitions, 633
Computer lab apportionment, 524
Cost of college education, 132, 319
Counselors on bullying, 704–705
Dean's list, 347–348
Dormitory board, 532
Dormitory rooms, 674–675, 680
Enrollment, 319, 546
Exam questions, 666, 681, 694, 696
Exam scores, 710, 713, 725, 735, 761
Exercise vs. academic performance, 713
Fellowships, 533, 545
Finals schedules, 157
General education electives, 49
GPA computation, 725

Grading, 311, 672–673, 736, 749
Graduate school, 537–538
Helicopter parents, 113–114
Internships, 104
iPad allocation, 542
ISO ballots, 577–579
Learning studies, 46
Living arrangements vs. grade point average, 67
Perceptions of students, 74
Planning committee, 16
Presidential advisory group, 634
Prize winning, 667
Quiz taking, 613, 632
Reading programs, 633
Rec center, 711
Resident council allocation, 523
Roommate selection, 632
Salaries vs. major in college, 670–672, 726
SATs, 750, 758
Scholarships, 74
Seating arrangements, 621–622
Secondary school teacher demand, 320
Senior group project, 182
Standardized tests, 686, 689, 740–741
Student data, 55–56, 93
Student loan debt, 311, 383–384, 402
Student services, 629
Study times, 737
Tuition, 17, 385, 393–394, 423, 428
Tutoring services, 362, 751
Unions, 562–563
Visually challenged student, 618–619
Volunteers, 634, 712–713
Writing assignments, 633
Years of education vs. annual income, 758
Years of education vs. obesity, 758

Engineering

Beam replacement/strength, 344, 345, 347, 348, 373
Electrical circuits, 94, 105

Environment

Alternative energy, 14
Carbon dioxide emissions, 724
Creek dredging, 263
Endangered species, 607, 616
Energy usage, 285
Environmental club, 649–650
Experimental house construction, 183
Global warming, 15, 69–70
Land use, 284
Mountain height, 486
Pollution, 15, 243–244
Rain forest destruction protest, 74
Speeches, 27
Toxic waste disposal seminar, 633
Water temperatures, 715–716
Water usage, 347

Finance. *See also* Economics

Account values, 390
Amortization, 417, 418–419, 421, 423, 430, 434
Annuities, 410–416, 423, 433–434
Apple Inc. revenue, 329
APR calculations, 426–430, 434
Bank account growth, 294, 365
Bank communication links, 7–8
Borrowing, 397
Business profits, 328
Carpet cost, 346
Celebrity yearly earnings, 717, 724
Charge computations, 311
College faculty salaries, 311
College fund, 369
College tuition, 393–394
Compounding raises, 388
Compound interest, 331, 339, 369, 391, 397, 399
Cookies, 381, 386
Credit cards, 403, 725
Direct satellite receiver sales, 320
DVDs, 325–327, 395
Earning calculations, 311
Family incomes, 736
Federal government revenue estimation, 36
Finance charges, 401, 404, 405–407
Gender pay gap estimation, 35
Health care spending, 284
Home financing, 418–419, 422, 423
Homeless shelter, 374
Interest, 389, 392–393, 397, 407, 433
Interest compounded, 373
Interest loan, 400
Interest rate, 340
Investments, 16, 136, 312, 362, 397, 398, 689, 750
Job hour calculations, 16
Living expenses, 263, 304–305
Loans, 406, 407, 422, 423, 433
Lottery winnings, 290, 423, 627–628
Manhattan purchase, 398
Manufacturing, 362
Medicare spending, 285
Money devaluation, 398
Money growth, 340
Mortgage payments, 423, 424
Mutual funds, 16
Pizza sales, 386
Price calculation, 388
Price estimation in foreign country, 341
Rental payments, 311
Retirement savings, 398
Revenue information, 311
Route planning, 166
Salary calculation, 288, 294
SAT scores vs. teachers' salaries, 758
Scholarship allocation, 263
State funding estimation, 37
Student loan debt, 311, 383–384, 402
Tuition increase, 17
Years of education vs. annual income, 758

General Interest

Advertising campaign, 183
Alphabet code problem, 15
Art show booth assignment, 523
Autographs, 386
Bagels, 362
Bail bondsman, 397
Barbecue grills, 387
Birthday problem, 681
Birth order, 644–646, 649–650
Birth times, 251
Books, 362
Cable strength, 749
Camera defections, 667
Candy sales, 433
Canned food, 486, 613
Chain letters, 294
Cheesecake division, 297
Child gender problem, 15
Chinese zodiac, 227
Classified information, 173
Committee meetings, 152–153, 156
Community service, 57–58
Conflict avoidance, 156
Contestants, 57
Cookies, 381, 386, 481, 656
Decorating project, 265
Dog weights, 744–745
Drink preferences, 174
Elevator capacity estimation, 35
Elevator travel, 251
Emergency communications network, 174
Fair trade items, 363
Family reunion, 602
Fashion, 610, 612, 617, 641
Fast food, 56, 66, 374, 387, 568
Fencing, 468, 486
Fertilizer estimation, 346
Fires in city, 734
Flooring purchases, 486
Food plans, 312
Geography and extreme distances, 251
Grass mowing, 346
Handshake counting problem, 3–4
Heating oil calculation, 347
Heat waves, 688
Heights of people, 45, 728
Holding pond, 486
Horizon viewing distance, 273
House buying decision, 136
Housework, 30–31
Housing for low-income renters, 362
Ice cream, 613
Influence measurement, 169–171, 174, 186
Inheritance division, 550–555, 558
Insurance policies, 682–683, 689, 699
Internet shopping, 663–666
Jeans display, 225–226
Jewelry, 353–354, 386
Keynote address topics, 570, 571
Koi pond, 486
License plates, 612–613, 621
Life expectancy, 319
Locks, 617–618, 621, 632
Magic square, 26
Mail costs, 386
Marijuana legalization, 109
Marital status, 104, 647–648
Mason-Dixon line, 455
Mirror hanging, 264
M&M estimation, 33
Names, 68
News media survey, 75
Noncitizens living in U.S., 47
Number trick, 23, 26–28
Oriental rug measurement, 486
Packages containing random collectibles, 695, 696
Party preparations, 612
Pendulum swing, 273–274

Pet ownership, 105
Photographs, 486
Photography lighting, 347
Pizza sales, 386
Product display, 294
Psychological test, 641
Racial-ethnic makeup, 264
Rainiest city, 749
Recipe modification, 263, 297
Research facility location, 581
Retirement planning, 423
Roman art dealer, 196–197
Rumors, 168–169
Sandwich length, 16
Seating arrangements, 168, 641
Sleep data, 320
Smoking rate, 314
Spanish-speaking radio stations, 386
Spoiled food, 667
Swimming pool, 486
Temperature changes, 251, 252, 487
Unsolved case files, 632
Vacation activities survey, 74
Vacation home refinancing, 423
Volunteer's ages, 710
Water tank, 486
Weather/extreme weather, 696, 707–708, 749
Weaver's Guild, 521, 522
Wedding tasks, 186
Wheelchair accessibility, 312
Women in armed services, 263
World hunger, 15
World issues survey, 74–75
Zip codes, 621

Geom etry

Accessibility ramp, 454
Acres of pizza, 467
Angle measure of star, 450–451
Arc of circle, 441
Area of fenced enclosure, 468
Area of home plate, 465
Area of housing development lots, 468
Area of playground, 458
Area of pyramid, 466
Area of running track, 466
Area of side of air traffic control tower, 467
Area of stained glass window, 466
Area of tile floor, 466
Area of trapezoid in base of statue, 460
Area of wall in art museum, 467
Area of wire shapes, 458–459, 467
Basketball court layout, 463, 466–467
Camping cot, 454
Circle division, 22, 27
Circumference of earth, 442, 445
Diameter of spheres, 476
Dimensions of most efficient shape of can, 471–472
Distances on baseball diamond, 465
Earth's circumference estimation, 37
Egyptian method, 194
Escher drawings, 515
Fertilizer purchase estimation, 37
Fractal tree, 506–507
Gazebo supports, 455
Graphics program, 258–259
Height of basketball player, 454
Height of pyramid, 461–462, 466
Irregular area estimation, 33–34
Japanese furniture makers, 454–456
Japanese garden design, 467
Koch curve, 501, 502–503, 505–507
Length of dollars end-to-end, 285
Length of pond, 454
Length of statue's shadow, 454
Lottery winnings, 290
Mandelbrot set, 508
Mason-Dixon line, 455
Memorial garden design, 467
Metric conversions, 481–487
Radius of flower bed, 466
Scaffolding, 456
Shape of posts, 468
Sierpinski gasket, 502, 503–504, 508
Squaring base of shed, 274
Surveillance camera position, 454
Symmetry with tiles, 498–499
Tessellation of tiles, 495–496
Triangles to build bridge, 451–452
Volume and surface area of cone, 473–474
Volume of various containers, 471, 474, 475–477
Width of river, 454
Windmill support, 454

Government and Politics

Arms race, 368–369
Caseworkers, 532
Chamber of commerce, 601, 620
Elections, 27, 525–526, 582
Electoral college, 591
Gallup poll, 113
Governing board apportionment, 557
Internships, 104
Juries, 598
Labor council, 532, 533, 545
Legal drinking age, 580
Legislative committee, 594–595
Oil consortium board apportionment, 524, 527–528, 541
Police officers, 524, 530, 534
Political parties, 586, 587
Political preferences, 680
Presidential popularity ratings, 328
Presidential vetoes, 724
President of club, 580
Presidents' ages, 719, 720
Public safety committee, 633
Reelections, 16
Representatives, 521–524, 532–536, 539–540, 545–547, 557, 599, 602
State committees, 598
Student action committee, 186
Supreme Court justices, 725
Unions, 562–563
Voter affiliation, 660
Voting, 563–564, 566–567, 571
Water authority, 523, 545

Labor

Carpenters, 532
Earning calculations, 311
Electrical worker's constituency, 523
Hours worked, 106–107, 725
Job hour calculations, 16
Job interview mistakes, 113
Job offers, 97, 135–136, 361–362
Labor council, 532, 533, 545
Labor force increase, 320, 321
Lawn service scheduling, 243
Overtime computation, 311, 372
Paint job estimation, 264

Medical

AIDS growth, 335–336
Antibiotic levels, 369
Antibodies in blood, 365
Appointment schedules, 614
Blood alcohol content, 487
Blood donations, 696
Blood types, 64, 70, 75, 82
Colds, 680–681, 696
Cystic fibrosis, 657
Disease spreading, 172, 173
Drug combinations, 631
Drug concentration, 339
Drug dose calculation, 346
Drug testing, 681, 694
Evaluation reports, 633
Exercise vs. academic performance, 713
Flu vaccination, 647
Health care spending, 284
Health concerns, 662
Health program organization, 183
Heart rates, 749
HIV incidence, 699
Hospital emergency clinic apportionments, 524
Medical personnel, 533
Medical procedure, 696
Medicine measurements, 486
Mononucleosis, 680
Nurses, 243, 320, 523, 527
Nutritional requirements, 360, 361, 363
Obesity, 758
Oxygen tank radius, 486
Post-traumatic stress disorder, 735
Sickle-cell anemia, 652, 656
Smoking rate, 314
Supply storage, 243
Vaccination amounts, 482
Weight loss, 707, 710, 713

Physics

Bouncing ball, 329
Bullet speed, 274
Emergency food supply drop, 328
Height of bouncing ball, 294
Model rockets, 328
MP3 player free fall, 274
Skydiving, 347
Spring strength, 347

Probability

Academic advisement vs. academic success, 667
Age distribution vs. minimum wage, 666
Airplane crash survival, 654–655
Batting averages, 696
Birthday problem, 681
Birth order, 644–646, 649–650
Blood donations, 696
Business profits, 688

Camera defects, 667
Card drawing, 644–645, 648–650, 662, 679, 688, 698–699
Carnival game, 656
Car repair, 666
Car safety ratings, 679
Cell phone redundancy, 696
Coin flipping, 4, 14, 608, 612, 616, 646, 648, 683–684, 698
Cold medication, 680–681
Cold vaccination, 696
Commission earnings, 666
Contract bids, 689–690
Cookie selection, 656
Cross-breeding plants, 656–657, 698
Cystic fibrosis, 657
Daily profit of businesses, 690
Dart game, 657
Dice rolling, 17, 608–609, 612, 616, 644–646, 649–650, 669–670, 678–679, 688, 690, 692–693, 699
Distracted drivers, 680
Dormitory rooms, 674–675, 680
Drawing balls from urn, 673–674, 679
Drug testing, 676, 681, 694
Environmental club, 649–650
Exam questions, 681, 694, 696
Final exam question, 666
Flu vaccination, 647
Gender selection, 656
Grade estimation, 672–673
Health concerns, 662
Heat waves, 688
HIV incidence, 699
Horse racing, 658
Insurance policies, 682–683, 689, 699
Internet shopping, 663–666
Investments, 689
iPad selection, 644–645
Living arrangements vs. grade point average, 67
Lotteries, 684–685, 688, 690
Marital status, 647–648
Medical procedure, 696
Mononucleosis, 680
Monopoly board game, 657
Odds terminology usage, 658, 667, 698
Packages containing random collectibles, 695, 696
Password formation, 658
Political preferences, 680
Prize winning, 667
Product ordering, 687
Product reliability, 681
Raffle winning, 667
Random person exercise, 657
Roulette wheel, 654, 684, 689
Shooting average, 696
Sickle-cell anemia, 652, 656
Softball, 680
Spinners, 657–658, 667, 698
Spoiled food, 667
Standardized tests, 686, 689
Starting salaries vs. major in college, 670–672
Voter affiliation, 660
Weather, 696
Winning a game, 658

Sports and Entertainment

Academy Award winners, 45, 713, 761
Amusement parks, 149–150, 347, 523, 545, 760–761
Arts alliance, 533
Audition scheduling, 161
Awards, 571, 632
Band shell decorating, 263
Baseball, 16, 26, 465, 632, 696, 708–709, 724, 727, 746, 750
Baseball stacking, 28
Basketball, 361, 373, 395–396, 463, 466–467, 696, 710, 749, 761
Bass fishing, 340
Bingo, 633
Boat race, 632
Canoe allocation, 543–544
Carnival game, 656
Casinos, 610–611
Celebrity yearly earnings, 717
Cheerleader competition, 576–577
Coffee shop, 711, 736
College athletics committee, 591
Concert organization, 180
Contestants, 15, 57
Cross country, 35, 749
Dart game, 657
Digital music sales, 316–317
Disney World hotel, 373
Drama society, 570
DVDs, 325–328, 395
Earth Day festival, 182
Emmy Award winners, 52
Entertainment book, 100–101
Entertainment system, 397
Equestrian club, 620
Fitness activities, 5, 59, 61, 74, 533, 545, 558, 615, 633, 725
Football, 16, 56, 361, 726
Gambling, 263, 294, 628, 635–639, 641, 642, 654, 684, 689
Game shows, 614, 622
Half-marathon training, 243
Home theater system, 426–427, 620, 641
Horse racing, 658
Jet-ski rental plans, 308–309
Lotteries, 290, 423, 627–628, 634, 684–685, 688, 690
Marching band arrangement, 227
Media user survey, 72
Miniature golf course, 445
Monopoly board game, 657
Movie club charges, 311
Movies, 66–67, 244, 328, 601–602
Musical patterns, 621
Music downloads, 387
Musicians, 532, 612, 621
Music video sales, 387
Nutritional requirements, 360, 363
Olympics, 53, 558, 574–575, 733–734
Online music survey, 75
Piano competition, 620
Pilots, 710, 750
Pizza, 466, 467, 621
Playing cards, 60, 634
Radio, 621, 633
Raffle winning, 667
Ranking teams, 174
Recording artists, 14, 76, 726
Restaurants, 423, 568, 581, 620, 621, 641
Rock climbing club, 641
Role-playing games, 14–15, 612, 620, 641
Route planning, 166
Running a race, 329, 481
Running shoes cost, 433
Running track, 466
Scuba equipment payments, 423
Softball, 680
Sports cards, 243
Sports stadium, 575
Tennis racquet advertising, 632
Theater, 590, 591, 598, 613, 625
Tournament seeding, 632
Trumpet valves, 157
TV viewer surveys, 70–71, 703–706
Video games, 633
Weight lifting, 749
Winning a game probability, 658
World's Strongest Man competition, 274

Statistics and Demographics

Age distribution vs. minimum wage, 666
Birth statistics, 749
Crime statistics, 319
Distracted drivers, 680
Gallup poll, 113
Immigration estimation, 36
Labor force, 320, 321, 329
Legal immigrants in U.S., 66
Mononucleosis, 680
Population, 30, 35, 284, 294, 306–307, 311, 333, 334, 336–339, 342, 387
Poverty data, 320
Price of new home, 386
Prison population, 328
Sports statistics, 382–383
Starting salaries vs. major in college, 670–672
Vital statistics, 736

Technology

ASCII coding system, 220
Blu-ray disks, 30
Cell phone comparison, 312
Computer circuits, 94, 105
Computer evaluation team, 633–634
Computer-generated animations, 285
Computer memory, 285
Computer passwords, 633, 641
Computer purchase, 35, 406
Computer virus, 125–126
Digital music sales, 316–317
Direct satellite receiver sales, 320
Electronic device sales estimation, 36
Electronic device settings, 621
Graphics program, 258–259
Internet companies, 49–50, 620
Internet searches, 43
Internet usage, 749
iPad allocation, 542
iPad selection, 644–645
iPhone options, 5–6
Laptop batteries, 737
Media user survey, 72
Music downloads, 387
News media survey, 75
Parent Internet companies, 726
Password formation, 658

Product reliability, 681
Satellite TV systems, 361, 373
Smart phones, 363
Social media, 60, 74, 596–597, 724
Space colony tasks, 176–179
Spreadsheets, 84
Warranties, 746–747
Web browser survey, 71–72
Web site visitors, 386
Word processing charges, 311
YouTube viewing habit estimation, 35–36

Travel

Airplane crash survival, 654–655
Airplane flight costs, 730
Airports, 362, 546, 558
Air safety review board, 589
Audition scheduling, 161
Boat position, 446
Borrowing for trip, 397
Bullet trains, 241–242
Coconut milk prices, 487
Commuting costs, 312
Flight scheduling, 243
Foreign, 320–321, 329, 341, 373
Fruit purchases, 486, 487
Gasoline prices, 487
Map coloring, 151–152, 155
Map distance estimations, 32
Map routes, 17
Mass transit survey, 75
Mishandled baggage, 711
Passengers in van, 633
Route efficiency/planning, 150–151, 154, 155, 164, 166–167, 621, 634, 641
Sales routes, 16, 20–21, 27
Speed of vehicle, 346
Time based on 7-hour clock, 227
Tourist travel, 362
Trains, 532, 546, 558, 710
Transit authority, 533
Travel time estimation, 35
Trip planning, 620
Vacation activities survey, 74

Subject Index

Abacists, 195
Abacus, 195, 218–219
Abel, Niels Henrick, 325
Absolute unfairness, 521–522
Acute angles, 439
Adam's apportionment method, 542, 544
Addition
  associative property of, 271, 272
  in base-5 system, 213–214
  closure property for, 271
  commutative property of, 271
  of decimals, 765
  of integers, 245–247, 249–250
  in modulo-*m* system, 224
  of radicals, 270
  of rational numbers, 255–256
Addition algorithm, 204
Additive inverse, 271
Add-on interest method, 400–401
Adibi, Jafar, 611
Ahmes, 193
Alabama paradox, 517, 520, 525, 527
Algebra
  Boolean, 63
  to find missing probability, 662–663
Algorists, 195
Algorithms
  addition, 204
  best edge, 164–165
  brute force, 162–163
  explanation of, 149, 162
  Fleury's, 149–150
  nearest neighbor, 163–164
  subtraction, 205
al-Khowarizmi, 203, 325
Always principle, 10–11, 20, 271
Amortization
  explanation of, 416–417
  loan refinancing and, 421–422
  payment calculation and, 417
  present value of annuity and, 419–421
Amortization schedules, 418–419
Analogies principle, 13, 52, 61, 212, 250, 281, 378, 380, 440, 473, 544, 653, 677, 719
Analytical geometry, 303, 304
Ancient Egyptian numeration system. *See* Egyptian numeration system *and,* 88, 97
Angle of rotation, 492
Angles
  acute, 439
  central, 441
  complementary, 439
  explanation of, 438–439
  formed by cutting parallel lines with transversal, 440
  initial side of, 438
  measurement of, 438, 440–442
  obtuse, 439
  polygons and, 449–451
  right, 439
  straight, 439
  supplementary, 439
  terminal side of, 438
  vertex of, 438
  vertical, 439
Annual percentage rate (APR)
  estimation of, 428
  explanation of, 425–426
  on graphing calculator, 429
  tables of, 426–427
Annuities
  explanation of, 408–409
  future value of, 408, 410–411, 413
  ordinary, 408, 410–411
  present value of, 419–421
  simplifying computations for, 409–410
  sinking funds as, 411–412
Appel, Kenneth, 151
Apportionment
  Adam's method of, 542, 544
  applications for, 529–530
  on calculator, 518, 560
  criterion, 525
  fair division and, 548–552
  Hamilton's method of, 518–520, 536–540, 544
  historical background of, 517–518
  Huntington-Hill principle of, 525–531
  Jefferson's method of, 540–542, 544
  measuring fairness of, 520–522
  use of, 525–526
  Webster's method of, 543–544
Approval voting methods, 569
Approximations
  of annual percentage rate, 428
  of e, φ, and π, 266–267
Arad, Michael, 451
Area
  estimation of irregular, 33–34
  inner, 34
  outer, 34
  of parallelograms, 457
  under standard normal curve, 741–743
  of trapezoids, 459–460
  of triangles, 457–459
Arguments
  explanation of, 115
  forms of, 118–119
  invalid, 116–118
  valid, 115–117, 119–120
Aristotle, 89, 123, 130
Arithmetic sequences, 286–288, 290
ARMs, 419
Arrow's impossibility theorem, 580
Artificial intelligence, 111
Associative properties, 271–272
Average constituency, 520, 521
Average daily balance method, 402–404
Axis of reflection, 489, 490
Axis of symmetry, 324

Babbage, Charles, 219
Babylonian numeration system, 200–202
Back-substitution, 350, 354
Balinski and Young's impossibility theorem, 527
Banzhaf power index
  calculation of, 587–588, 590
  explanation of, 587
  Shapley-Shubik index vs., 596
  of tie breaker, 588
Bar graphs, 705–708
Base, 276
Base-2 system, 210. *See also* Binary system
Base-5 system
  arithmetic in, 213–217
  converting from decimal to, 212
  explanation of, 209–210
Base-8 system, 210
Base-16 system, 210
Bell-shaped curve, 739
Best edge algorithms, 164–165
Be systematic strategy, 5, 586
Bias, 702–703
Biased samples, 123, 702
Biconditional statements
  explanation of, 89, 111
  truth table for, 112
Binary digit, 218
Binary system
  explanation of, 210
  use of, 217–218
Binomial experiments
  examples of, 692–693
  explanation of, 691
Binomial probabilities
  applications of, 694–695
  explanation of, 691–693
  formula to compute, 693
Binomial trials, 691, 693–695
Bits, 218
Boole, George, 63, 86
Boolean algebra, 63
Borda count method
  determining winner in, 563–565
  explanation of, 563
  majority criterion and, 573
  summary of flaws in, 580
Box-and-whisker plot, 720–721
Boylai, Janos, 443
Brahmagupta, 246
Brams, Stephen, 569
Bridge, in graphs, 145
Brute force algorithm, 162–163
Butterfly effect, 368

Calculators/graphing calculators
  annual percentage rate on, 429
  apportionment on, 518, 560
  drawing graphs on, 305
  interest on, 393
  linear correlation coefficient on, 756
  line of best fit on, 318
  log function on, 333
  permutations and combinations on, 629
  rational numbers on, 254
  scientific notation on, 280
  standard deviation of frequency distribution on, 732
  statistical calculations on, 718
Cantor, Georg, 73, 77, 79
Capture-recapture method, 342
Cardano, Girolamo, 246, 325, 653
Card counting, 619
Cardinal numbers, of set, 47, 77–79
Centimeters, 480

Central angle, of circle, 441
Central tendency. *See* Measures of central tendency
Champollion, Jean Francois, 193
Change, percent of, 380–381
Chinese numeration system, 197–198
Circles
  applications of, 441, 442
  area of, 462–463
  circumference of, 441, 462–463
  explanation of, 441
  great, 443
Circular reasoning, 123
Circumference, 441
Closed-ended credit agreements. *See* Installment loans
Closed plane figures, 446, 447
Closure property, 271
Coalitions
  critical voters in, 586, 587
  explanation of, 585–586
  pivotal voters in, 594–595
  weight of, 585
  winning voters in, 585, 586
Coefficient of variation, 733–734
Cohen, Paul, 79
Combinations
  explanation of, 626
  formula for, 626–627
  origin of, 627
  relating entries in Pascal's triangle to, 631
Combined variation, 345
Common logarithmic function, 394
Common ratio, 288–289
Commutative properties, 271–272
Compatible numbers, 29–30
Complement, of sets, 60–62
Complementary angles, 439
Complement formula, 659, 660
Complete graphs, 159
Composite numbers, 234
Compound interest
  explanation of, 330–331, 390
  method to calculate, 391–392
Compound interest formula
  explanation of, 331–332, 392
  solving for unknowns with, 392–396
Compound statements
  explanation of, 87, 90
  truth tables for, 97–100
Conclusion
  to statement, 115, 116
  of syllogisms, 123
Conditional probability. *See also* Probability
  counting to compute, 669–670
  dependent and independent events and, 677–678
  explanation of, 668–669
  general rule to compute, 670–672
  intersection of events and, 672–674
  probability trees and, 674–676
Conditional statements
  alternate wording of, 110–111
  derived forms of, 108–110
  explanation of, 89, 106–117
  truth tables for, 107–108
Condorcet's criterion
  explanation of, 573–574
  plurality method and, 574–575
Cones, volume and surface area of, 472–474
Congruencies, in modulo-*m* systems, 222, 223, 225–226
Conjunction
  explanation of, 88
  truth values of, 96, 132
Connected graphs, 145
Connectives
  explanation of, 87–90
  fuzzy, 131–133
  usage of, 95
Constant of proportionality, 343
Constant of variation, 343
Consumer loans
  add-on interest method for, 400–401
  average daily balance method for, 402–404
  comparing financing methods for, 404–405
  explanation of, 400
  historical background of, 399
  unpaid balance method for, 401–402
Consumer Price Index (CPI), 385
Continuous fair division, 548
Continuous variables, 706
Continuum hypothesis, 79
Contrapositive, of statements, 108–110
Converse, of statements, 108–110
Convex polygons, 447
Corner point, 359
Correlation, 751. *See also* Linear correlation
Correlation coefficient. *See* Linear correlation coefficient
Countable sets, 77–78
Counterexample principle, 11, 20
Counterexamples, method to find, 272
Counting methods
  combinations and, 626–628
  conditional probability and, 669–670
  factorial notation and, 624–625
  fundamental counting principle and, 615–617
  gambling and, 635–639
  method to combine, 628–631
  permutations and, 623–624
  probability and, 648–650
  slot diagrams and, 617–618
  for special conditions, 618–619
  systematic, 606–607
  tree diagrams to illustrate, 607–611
Counting numbers. *See* Natural numbers
Critical path, 178
Critical voters, 586, 587
Cross-multiplication principle, 341
Cross multiplying, 253, 254
Cross products, 253, 254, 341
Cubic equations, 325
Cylinders, volume and surface area of, 470–472

Data. *See also* Descriptive statistics; Statistics
  explanation of, 702, 703
  organization of, 702–705
  visual representation of, 705–710
Data analysis, exploratory, 708
Data mining, 611
Data points
  building linear model with two, 314–316
  explanation of, 751
  line of best fit for, 316–319, 755–756
Decimal notation, 211
Decimals
  addition and subtraction of, 764
  converted to scientific notation, 280
  division of, 765
  multiplication of, 765
  repeating, 260–262
  writing percent as, 379
Decision making, fuzzy, 133–135
Deductive reasoning, 23–25
Degree, of vertex, 145
Demand equation, 354
DeMorgan, Augustus, 63, 151
DeMorgan's laws, for logic, 101–102
Dempsey, James X., 619
Denominators
  explanation of, 253
  rationalizing, 269
Dependent events, 677–678
Dependent systems, 352
Descartes, René, 246, 303, 309
Descriptive statistics
  linear correlation and, 751-757
  measures of central tendency and, 714–723
  measures of dispersion and, 727–734
  normal distribution and, 738–747
  organizing and visualizing data and, 702–710
Deviation. *See* Standard deviation
Diameter, 441
Difference, of sets, 61
Dimensional analysis, 480–485
Diophantus, 245
Directed edges, 168
Directed graphs
  applications for, 169–172
  explanation of, 168–169
Directed path, 169
Direct variation, 343–344
Discrete fair division, 548
Discrete variables, 706
Discriminant, of equation, 323
Disjoint sets, 59
Disjunction
  explanation of, 88
  truth values of, 132
Disjunctive syllogism, 118
Dispersion. *See* Measures of dispersion
Distributions, 703. *See also* Data; Frequency distributions; Measures of dispersion; Normal distribution
Divisibility
  explanation of, 233
  tests for, 235–237
Division
  in base-5 system, 217
  of decimals, 765
  of integers, 248–250
  of radicals, 267–268
  of rational numbers, 257–259
Divisors
  explanation of, 233
  greatest common, 238–239
  modified, 540, 541
  standard, 534–537
Doubling time, 395
Draw diagram strategy, 616

Dynamic systems
explanation of, 364–365
modeling with, 365–366
stability and, 366–369

e, 266–267
Eckert, J. Presper, 219
Edges
directed, 168
of graph, 143, 144
of polygons, 447
Effective interest rate, 397
Egyptian hieroglyphics, 193
Egyptian numeration system, 191–194
Electoral college, 590
Elements, of sets, 43, 46, 47
Elimination method, 350–352
Empirical information, 647, 648
Empty sets, 45
Equality, of rational numbers, 253–254
Equally likely outcomes, 649–650
Equal sets, 51
Equations
cubic, 325
discriminant of, 323
equivalent, 301
exponential, 332–334
linear, 301–309, 313–319 (*See also* Linear equations)
method to solve, 764
percent, 382–384
quadratic, 321–327
quartic, 325
quintic, 325
recursive, 337
Equilateral triangles, 448
Equilibrium point, 353, 354
Equilibrium values, for dynamic system, 366–368
Equivalent equations, 301
Equivalent sets, 54–55
Eratosthenes, 31, 233, 238
Ergonomics, 738
Estimation. *See also* Rounding
of graphical data, 30–34
rounding for, 28–29
use of compatible numbers for, 29–30
Euclid, 237
Euclidean algorithm, 239
Euclidean geometry, 437, 452
Euclid's fifth postulate, 443
Euler, Leonhard, 143–145, 266
Euler circuits
explanation of, 148
Fleury's algorithm to find, 149–150
Euler diagrams
explanation of, 91
for syllogisms, 123–128
Eulerian graphs, 148
Eulerizing graphs, 150–151
Euler path, 148
Euler's theorem, 146–147
Events
complements of, 659–660
computing probability of, 649–650
dependent and independent, 677–678
intersection of, 672–674
in probability theory, 645
as subsets, 645, 646
union of, 660–662
Even vertex, 145
Excel, modeling data with, 376–377
Exclusive *or*, 96
Existential quantifiers, 90
Expected value
applications for, 686–687
explanation of, 682–683
games of change and, 683–686
Experiments, 644
Exploratory data analysis, 708
Exponential equations
compound interest formula as, 332
explanation of, 332
log function to solve, 334
modeling with, 333–334
Exponential growth, 330–331
Exponent property of log function, 333, 394–395
Exponents
evaluating expressions with, 276
explanation of, 275–276
power rule for, 277
product rule for, 276–277
quotient rule for, 278
use of rules for, 279
zero, 279

Factorial notation, 624–625
Factoring
to find least common multiple and greatest common divisors, 240
natural numbers, 237–238
Factors, 233
Factor tree, 237
Fair division
continuous, 548
discrete, 548
explanation of, 548
sealed bids method of, 548–552
Fair game, 684, 685
Fairness measures, 520–522
Fair share, 548–549
Fallacy of the converse, 118
Fallacy of the inverse, 118
False analogy, 123
Farley, Jonathan, 611
Fermat, Pierre de, 665
Fibonacci (Leonardo of Pisa), 203, 291
Fibonacci sequence, 291–293
Finance charge, 400, 426
Finite sets, 47
Five-number summary, 720–721
Fleury's algorithm, 149–150
Four-color problem, 151–152
Fourier, Jean-Baptiste, 193
Fractal dimension
explanation of, 504–505
of Koch curve, 505–506
Fractals
applications of, 505
dimension of, 504–506
explanation of, 500501
Koch curve as, 501–503, 505–506
Sierpinski gasket as, 502–504
Fractal tree, 506–507
Fractions
converted to percent, 380
improper, 259–260
unit, 480–481
Frequency distributions
computing mean of, 715–717
explanation of, 703
standard deviation of, 730–731
Frequency tables
to compute standard deviation, 731
explanation of, 703–705
finding median with, 719
Fundamental counting principle (FCP)
applied to gambling, 635–636
examples using, 616–617
explanation of, 615–616
slot diagrams and, 617–618
for special conditions, 617–618
Fundamental theorem of arithmetic, 237, 238
Future value
of annuity, 408, 410–411, 413
explanation of, 389
simple interest to compute, 389–390
Fuzzy logic
connectives in, 131–133
decision making with, 133–135
explanation of, 130–131

Galley method, for multiplication, 205–206
Gallup, George, 704
Galois, Evariste, 325
Gambling
counting and poker and, 636–639
fundamental counting principle applied to, 635–636
Games of change, 683–685
Gauss, Carl Friedrich, 221, 237, 443
Genetics, probability and, 651–652
Geometric sequences, 288–290
Geometry
analytical, 303, 304
angles, 438–441
area, 457–460
circles, 441–442, 462–463
cones and spheres, 472–474
cylinders, 470–472
Euclidean, 437, 452
fractal, 500–507
metric system measurements, 477–485
non-Euclidean, 443, 452
perimeter, 456–457
points, lines and planes, 437, 438
polygons, 446–452
Pythagorean theorem, 460–462
Riemannian, 452
rigid motions, 488–492
spherical, 443
symmetries, 492–494
tessellations, 494–496
volume, 468–470
Germain, Sophie, 259
Gerrymandering, 539
Glide reflection, 491–492
Gödel, Kurt, 101
Goldbach's conjecture, 19
Golden ratio, 292. *See also* φ
Golden rectangles, 292, 293
Grammer, Karl, 493

Grams, 378, 484
Graphing calculators. *See* Calculators/ graphing calculators
Graphs
bar, 705–708
bridge in, 145
complete, 159
connected, 145
directed, 168–172
edges of, 143, 144
Eulerian, 148
Eulerizing, 150–151
Euler's theorem and, 146, 147
explanation of, 143–144
Fleury's algorithm and, 149–150
of linear equations, 303–309
making estimates of information in, 30–34
map coloring and, 151–153
path in, 148
PERT diagrams and, 176–180
of quadratic equations, 322, 324, 326
stem-and-leaf, 708–710
supply and demand, 353–354
of systems of inequalities, 356–359
of systems of linear equations, 349, 351
tracing, 144–147
traveling salesperson problem and, 157–165
vertex of, 143–145
vertices of, 143–145
Graunt, John, 728
Great circle, 443
Greatest common divisors (GCD)
applications of, 241–242
explanation of, 238–239
to find least common multiple, 241
Guthrie, Francis, 151

Haken, Wolfgang, 151
Half line, 437, 438
Half-plane, 356
Halley, Edmund, 728
Hamilton, Alexander, 517
Hamilton, William Rowan, 151, 161
Hamilton apportionment method
explanation of, 518–520, 536, 537
new states paradox and, 539–540
population paradox and, 537–539
Hamilton circuits
explanation of, 158, 159
method to find, 159–161
Hamilton paths, 158–159
Heron's formula for area of triangle, 458–459
Hexadecimal system, 210, 218
Hilbert, David, 73, 79
Hill, Joseph, 527
Hindu-Arabic numeration system
algorists vs. abacists and, 195
converting between Babylonian and, 201–202
converting between Chinese and, 197
converting between Egyptian and, 192
converting between Roman numerals and, 194–196
explanation of, 203–205
as place value system, 203–205
Histograms, 706–707
Horner's method, 211
Huntington, Edward, 527
Huntington-Hill apportionment principle
derivation of, 530–531
explanation of, 526–531
Huntington-Hill numbers
explanation of, 526
spreadsheet to compute, 529
table of, 527–528
Hypatia, 460

if and only if, 88–89
if . . . then, 88, 108, 111
Improper fractions, 259–260. *See also* Fractions
Inches, 480, 481
Incidence matrix, 175
Inclusion-Exclusion Principle, 64
Inclusive *or,* 96
Inconsistent systems, 351
Independence-of-irrelevant-alternatives criterion, 575–577
Independent events, 677–678
Inductive reasoning
examples of, 19–21
explanation of, 19
Goldbach's conjecture and, 19
incorrect, 22
Inequalities
linear, 355–358
systems of, 358–360
Infinite sets
examples of, 77–78
explanation of, 47, 76–79
set of natural numbers as, 77
Inflation, 385, 394
Inflection point, 739
Information
empirical, 647, 648
reliability of, 702–703
theoretical, 648
Initial side, of angle, 438
Inner area, 34
Installment loans, 400
Installments, 400
Instant runoff voting method, 569
Integers. *See also* Numbers; Real numbers
adding and subtracting, 245–247, 249–250
explanation of, 245
multiplying and dividing, 247–250
Intercepts, 304–305
Interest. *See also* Consumer loans
on add-on interest loans, 400–401
average daily balance method to compute, 402–404
compound, 330–331, 390–396
explanation of, 388
simple, 389–390
unpaid balance method to compute, 401–402, 404
Interest rate, 389, 397. *See also* Annual percentage rate (APR)
Intersecting lines, 437, 438
Intersection
of events, 672–674
of sets, 59–60, 63–64
Invalid arguments, 116–118
Invalid syllogisms, 123, 125–127
Inverse, of statements, 108–110
Inverse variation, 343, 344
Irrational numbers
explanation of, 266–267
origin of, 265
Isosceles triangles, 448

Jefferson's apportionment method, 540–542, 544, 560
Joint variation, 344

Kilograms, 378
Koch curve
dimension and, 504
as fractal, 501
fractal dimension of, 505–506
length of, 502–503
Koenigsberg bridge problem, 144, 147

Lagrange, Joseph, 259
Landerstein, Karl, 64
Laplace, Pierre-Simon de, 665
Law of contraposition, 117, 118
Law of demand, 353
Law of detachment, 116, 118
Law of supply, 353
Law of syllogism, 118
Leading-question bias, 703
Least common multiple (LCM)
applications of, 241–242
explanation of, 239
prime factorization to find, 240–241
use of greatest common divisors to find, 241
Leibniz, Gottfried, 89, 101, 219
Length
of directed path, 169
estimation of, 480
of path, 148
Lindley, Dennis, 677
Linear correlation
explanation of, 751, 752
line of best fit and, 755–757
method to compute, 752–753
scatterplots and, 751–752
Linear correlation coefficient
application of, 754–755
on calculator, 756
critical values for, 754
explanation of, 318–319, 752
formula to compute, 752–753
properties of, 753
Linear equations. *See also* Equations; Systems of linear equations
equivalent, 301, 302
explanation of, 301
graphs of, 303–309
modeling with, 313–319
solutions to, 301–303
in standard form, 301
Linear inequalities
methods to solve, 356–358
in two variables, 355–358
Linear models
explanation of, 313
point and slope to build, 313–314
with real data, 316–319
with two points, 314–316
Linear regression, 318

Line of best fit
  explanation of, 318, 755–756
  on graphing calculator, 318
  method to find, 756–757
Lines
  explanation of, 437
  intersecting, 437, 438
  parallel, 437, 438
  perpendicular, 439
  slope-intercept form of, 307–309
  slope of, 305–307
Line segments, 437, 438
Liters, 378
Loans
  amortized, 416, 417
  consumer, 399–405 (*See also* Consumer loans)
  installment, 400
  refinancing, 421–422
Lobachevsky, Nicholai, 443
Log function
  explanation of, 333
  exponent property of, 333, 394–395
  to solve exponential equations, 334
Logic
  argument validity and, 115–120
  biconditional and, 111–112
  conditional and, 106–111
  connectives in, 87–90, 95
  Euler diagrams to verify syllogisms and, 123–128
  fuzzy, 130–135
  historical background of, 86, 89, 101, 124
  negation and, 87, 95–96
  quantifiers in, 90–92
  statements in, 86–92
  truth tables and, 95–103 (*See also* Truth tables)
Logically equivalent statements
  derived forms of conditionals and, 109–110
  explanation of, 100–103
Logistic growth model, 336–337
Logistic models, 336–338
Lower quota, 536
Lowest terms, 254
Lukasiewicz, Jan, 124

Majority criterion, 572–573
Markup, method to compute, 381
Mass, units of, 484
Mathematics review, 764–765
Matrices, incidence, 175
Mauchly, John, 219
Mayan numeration system, 202–203
Mean
  decision to use, 722–723
  deviation of, 729
  effect of extreme scores on, 717–718
  explanation of, 714–715
  of frequency distribution, 715–717
  limitations of, 727
  population, 715
Measures of central tendency
  comparison of, 721–723
  explanation of, 714
  five-number summary and, 720–721
  mean and, 714–718
  median and, 718–721
Measures of dispersion
  coefficient of variation and, 733–734
  range of data set and, 728
  standard deviation and, 729–732
Median
  explanation of, 718
  five-number summary and, 720–721
  limitations of, 727
  methods to find, 719–720
  use of, 723
Members. *See* Elements
Mendel, Gregor, 651
Mersenne prime, 241
Meters, 477–478, 480
Metric system
  common prefixes used in, 478
  converting between customary systems and, 481–485
  converting units of measurement in, 479–480
  dimensional analysis and, 480–485
  estimating mass in, 484
  explanation of, 477–478
  relationships between units of volume in, 483–484
Meyer, Mary C., 677
Millimeters, 480
Mixed numbers, conversions between improper fractions and, 259–260
Mode, 721, 723
Models/modeling
  directed graphs in, 169–172
  with exponential equations, 333–335
  guidelines for creating, 171, 309, 313
  with linear equations, 313–319
  logistic, 335–338
  with quadratic equations, 325–327
  with systems of inequalities, 360
  with systems of linear equations, 352–355
  use of Excel for, 376–377
Modified divisors, 540, 541
Modified quotas, 540, 541
Modular arithmetic, 226
Modulo-*m* systems
  arithmetic operations in, 224–225
  calculations in, 223
  congruencies in, 222, 223, 225–226
  explanation of, 221–222
Modulus, 221
Monotonicity criterion, 577–579
Mortgages, ARMs, 419
Multiplication
  associative property of, 271, 272
  in base-5 system, 214–216
  closure property for, 271
  commutative property of, 271, 272
  of decimals, 765
  galley method for, 205–206
  of integers, 247–250
  in modulo-*m* system, 225
  of radicals, 267–268
  of rational numbers, 256–259
Multiplicative inverse, 271
Multiplicative system, 197

Napier, John, 206
Napier's rods, 206–207
Napoleon, 193
Nash, John, 611
Natural numbers
  explanation of, 233–234
  prime factorization of, 237–238
Nearest neighbor algorithm, 163–164
Negative quantifiers, 91–92
Negative statements
  explanation of, 87
  with quantifiers, 91–92
  truth value of, 95–96, 131–132
Nelson, Gary, 611
Networks, 143. *See also* Graphs
New states paradox, 539–540
n factorial, 624
Nightingale, Florence, 709
No, in syllogisms, 125
Nominal rate, 397
Non-base-10 systems
  arithmetic in, 212–217
  types of, 209–212
Non-Euclidean geometry, 443, 452
Normal curve. *See* Area under standard normal curve; Normal distribution
Normal distribution
  applications for, 745–747
  explanation of, 738–739
  properties of, 739
  standard deviations of mean of, 740–741
  z-scores and, 741, 742
*n*th term
  of arithmetic sequence, 287
  of geometric sequence, 289
Null sets. *See* Empty sets
Number line, addition of integers as movement on, 245–246
Numbers. *See also* Integers; Real number system
  cardinal, 47, 77–79
  composite, 234
  explanation of, 191
  irrational, 265–267
  natural, 233–234, 237–238
  negative, 246
  opposite, 247
  prime, 19, 234–235, 237
  rational, 78, 253–262
  whole, 245
Number theory, 233
Numerals, 191
Numeration systems
  Babylonian, 200–202
  binary, 217–218
  Chinese, 197–198
  Egyptian, 191–194
  galley method and, 205–206
  hexadecimal, 218
  Hindu-Arabic, 192, 194–197, 203–205
  Mayan, 202–203
  modular, 221–226
  multiplicative, 197
  Napier's rods and, 206–207
  non-base-10, 209–217
  octal, 218
  place value, 200–207
  Roman, 194–197
  simple grouping, 191–192
Numerators, 253

Obtuse angles, 439
Octal system, 218
Odd vertices, 145, 146
One-point test, 357–358
One-to-one correspondence, 54, 76, 77. *See also* Infinite sets
Opposites, 247
or, 88, 96, 97
Ordered pairs
  as solution to linear equations, 301
  as solution to linear inequalities, 356
  as solution to quadratic equations, 322
  as solution to systems of linear equations, 349
Order of operations, 12, 764
Order principle, 12, 88, 492, 670
Ordinary annuity. *See also* Annuities
  explanation of, 408
  future value of, 410–411
Outcomes
  equally likely, 649–650, 653
  explanation of, 644
  probability of, 646
Outer area, 34
Outliers, 718

Pairwise comparison method
  determining winner with, 568–569
  explanation of, 567
  independence-of-irrelevant-alternatives criterion and, 576–577
  summary of flaws in, 580
Pappus of Alexandria, 456
Parabolas
  of best fit, 327
  explanation of, 322, 324, 325
Paradoxes
  Alabama, 517, 520, 525, 527
  new states, 539–540
  population, 537–539
  standard divisors and quotas and, 534–537
Parallelepipeds, 468
Parallel lines, 437, 438
Parallelograms, 448, 457
Pascal, Blaise, 219, 246, 637, 665
Pascal's triangle
  counting problems and, 630–631
  explanation of, 6, 7
Path
  directed, 169
  in graphs, 148
Patterns, 6–7
Pearson, Karl, 709
Pentagons, 449, 450
Percent
  of change, 380–381
  converting fraction to, 380
  explanation of, 379
  written as decimals, 379
Percent equation
  applications of, 382–385
  explanation of, 382
Perimeter, of polygons, 456–457
Permutations
  on calculator, 629
  explanation of, 592–593, 623
  method to count, 623–624
  origin of, 627
  in set, 593–594
Perpendicular lines, 439
PERT diagrams, 176–180
φ, 266–267, 292
Pivotal voters, 594–595
π, 266–267, 462
Place value systems
  Babylonian, 201–202
  explanation of, 200–201
  galley method and, 205–206
  Hindu-Arabic, 203–205 (*See also* Hindu-Arabic numeration system)
  Mayan, 202–203
  Napier's rods and, 206–207
Plane, 437
Plane figures, 446, 447
Plurality method
  Condorcet's criterion and, 574–575
  determining winner using, 562–563
  explanation of, 562
  independence-of-irrelevant-alternatives criterion and, 575–576
  summary of flaws in, 580
Plurality-with-elimination method
  determining winner with, 566–567
  explanation of, 565–566
  monotonicity criterion and, 577–579
  summary of flaws in, 580
Poincaré, Henri, 73, 142
Points
  corner, 359
  explanation of, 437
  used to find linear equations, 315, 316
Poker, 636–639
Polls, 704
Polya, George, 14
Polygons
  angle sum of, 449–450
  area of, 456–457
  classification of, 447, 448
  convex, 447
  explanation of, 446–448
  perimeter of, 456
  similar, 451–452
  tessellating plane with regular, 495
  vertex of, 447, 489
Population, 702
Population mean, 715
Population paradox, 537–539
Positional systems. *See* Place value systems
Power rule for exponents, 277
Preference tables, 563–565
Premises
  explanation of, 115, 116, 119
  of syllogisms, 123
Present value
  of annuity, 419–421
  explanation of, 389
  method to find, 390
Prime factorization, to find greatest common divisor, 239
Prime numbers
  explanation of, 19, 234, 237
  method to find, 234–235
  very large, 241
Principal, 389
Probability
  basic properties of, 650
  binomial, 691–695
  complements of events and, 659–660, 663–665
  conditional, 668–678 (*See also* Conditional probability)
  counting and, 648–650
  empirical assignment of, 647
  of events, 646
  expected value and, 682–687
  explanation of, 644
  formula to compute odds, 654
  genetics and, 651–652
  odds and, 652–655
  of outcome, 646
  sample spaces and events and, 644–648
  unions of events and, 660–665
  use of algebra to find missing, 662–663
Probability trees, 674–676
Problem-solving strategies
  analogies principle as, 13
  counterexample principle as, 11
  drawing pictures as, 3–4
  good organization as, 5–6
  guessing as, 8–9
  looking for patterns as, 6–7
  naming unknowns as, 5
  order principle as, 12
  Polya's, 2–3
  relating new problems to old problems as, 9–10
  simplifying problems as, 7–8
  splitting-hairs principle as, 12–13
  for survey problems, 67–72
  three-way principle as, 13
  use of "always" principle as, 10–11
  work-place applications for, 3
Product rule for exponents, 276–277
Proper subsets, 52
Proportions, 341, 342
Protractors, 439
Pseudosphere, 443, 452
Pythagorean theorem
  applications of, 461–462
  explanation of, 265, 460–461

Quadratic equations
  explanation of, 321–322
  graphs of, 322, 324, 326
  methods to solve, 325
  modeling with, 325–327
  quadratic formula to solve, 323
  verifying solutions for, 322
  vertex of graph of, 322
Quadratic formula, 323
Quadratic regression, 327
Quantifiers
  existential, 90
  explanation of, 90
  negative, 91–92
  in syllogisms, 123
  universal, 90
Quartic equations, 325
Quintic equations, 325
Quota
  modified, 540, 541
  standard, 534–537
  in weighted voting system, 584
Quota rule, 527
Quotient rule for exponents, 277–278

Quotients
as repeating decimals, 261–262
standard, 534–537

Radicals
adding expressions containing, 270
explanation of, 267
method to simplify, 268
multiplying and dividing, 267–268
rationalizing the denominator and, 269
Radical sign, 267
Radicand, 267
Radius, 441
Random phenomena, 644
Ratio, 341
Rationalizing the denominator, 269
Rational numbers
adding and subtracting, 255–256
decimal expansions of, 261
equality of, 253–254
explanation of, 253
improper fractions as, 259–262
method to reduce, 254–255
multiplying and dividing, 256–259
reducing, 254–255
rule to simplify, 254
set of, 78, 253
Raw scores, 744–745
Rays, 437, 438
Real numbers
properties of, 270–272
set of, 270
Real number system
computing with radicals and, 267–271
irrational numbers and, 266–267
Reasoning
deductive, 23–25
inductive, 19–22
Rectangles
area of, 457
explanation of, 448
perimeter of, 457
Rectangular solids, volume and surface area of, 469
Recursive equations, 337
Refinancing, loan, 421–422
Reflection
axis of, 489, 490
explanation of, 489
of geometric object, 490–491, 493
glide, 491–492
Regression
linear, 318
quadratic, 327
Regression coefficient, 757
Regular tessellations, 494
Relate new problem to older one strategy, 9, 315, 449
Relative frequency distributions, 703–704
Relative unfairness, 522
Repeating decimals, 260–262
Resource table, 355
Rhind, A. Henry, 193
Rhind paprus, 193
Rhombus, 448
Rice, Marjorie, 497
Riemann, Bernhard, 443
Riemannian Geometry, 452
Right angles, 439
Right circular cones, 472–473
Right circular cylinders, 470
Rigid motions
explanation of, 488–489
glide reflection as, 491–492
reflection as, 489, 490
rotation as, 492
translation as, 491
Roman numeration system, 194–197
Rosetta stone, 193
Rotation, 492, 493
Rounding, 28–29. *See also* Estimation
Rule of 70, 395
Russell, Bertrand, 101

Saint Thomas Aquinas, 89
Samples
biased, 123, 702
explanation of, 702–703
mean of, 714–718
standard deviation of, 729–732
Sample space
with equally likely outcomes, 649–650, 653
explanation of, 644
method to find, 644–645
Scalene triangles, 448
Scatterplots, 751–752, 754
Scientific notation
applications of, 282–283
conversions between standard notation and, 281
converting decimals to, 280
explanation of, 279–280
Sealed bids method of fair division, 548–552
Selection bias, 702–703
Self-similar objects, 501
Sequences
arithmetic, 286–288, 290
common difference of, 286
explanation of, 286
Fibonacci, 291–293
geometric, 288–290
terms of, 286
Set-builder notation, 44
Set operations
complement as, 60–62
difference as, 61
explanation of, 57–58
intersection as, 59–60, 63–64
order of, 61–64
union as, 58–59, 62–64
Sets
cardinal number of, 47, 77–79
countable, 77–78
disjoint, 59
elements of, 43, 46, 47
empty, 45
equal, 51
equivalent, 54–55
explanation of, 43
finding permutations in, 593–594
finite, 47
historical background of, 73
infinite, 47, 76–79
of integers, 245, 247
method to represent, 43–44
problem-solving techniques for, 52–54
of rational numbers, 78, 253
of real numbers, 270
subsets of, 51–54, 630–631
universal, 46
Venn diagrams of, 52
well-defined, 44–46
Shapley-Shubik index, 595–597
Sierpenski gasket
area of, 503–504
as fractal, 502
Sieve of Eratosthenes, 234, 235
Signed numbers, 249
Similar polygons, 451–452
Simple grouping system, 191–192
Simple interest, 389–390
Simple interest formula, 389
Simple plane figures, 446
Simple statements, 87
Sinking funds, 411–412
68-95-99.7 rule
application of, 740–741
explanation of, 739
Slippery slope, 123
Slope
explanation of, 305–306
finding linear equation from, 313–314
method to find, 306–307
Slope-intercept form, 307–309, 313
Slot diagrams, 617–618
Some, in syllogisms, 126–127
Sophie Germain prine, 241
Spheres, volume and surface area of, 474
Spherical geometry, 443
Splitting-hairs principle, 12–13, 158, 222, 593
Spreadsheets
to compute Huntington-Hill numbers, 529
to compute Jefferson's apportionment method, 560
to find counterexamples, 578
Squares, 448
Standard deviation
calculation of, 730–731
explanation of, 729
frequency table to compute, 731–732
graphing calculator to find, 732
Standard divisors
apportionment and, 534–537
explanation of, 535
Standard form, of linear equations, 301
Standard notation, 281
Standard quota
apportionment and, 534–537
explanation of, 535
Statements
biconditional, 89, 111–112
compound, 87, 90
conditional, 89, 106–111
conjunction in, 88
connectives in, 87–90
disjunction in, 88
explanation of, 86–87
fuzzy, 131
logically equivalent, 100–103
negation of, 87, 91–92
quantifiers in, 90–92
simple, 87

Statistics. *See also* Descriptive statistics
  applied, 709
  explanation of, 702–703
  frequency tables and, 703–705
  on graphing calculator, 718
  origins of, 728
  overview of, 701
  populations and samples and, 702–703
  stem-and-leaf displays and, 708–710
  visual representation of data and, 705–708
Stem-and-leaf displays, 708–710
Straight angles, 439
Subsets
  explanation of, 51–52
  identification of, 52, 53
  method to find, 53–54
  proper, 52
  of sets, 51–54, 630–631
Subtraction
  in base-5 system, 214
  of decimals, 765
  of integers, 247, 249–250
  in modulo-*m* system, 224
  of rational numbers, 256
Subtraction algorithm, 205
Subtraction principle, 194–195
Sun Tzu, 219
Supplementary angles, 439
Supply and demand graphs, 353–354
Supply equation, 354
Surface area
  of cones and spheres, 472–474
  of cylinders, 470–472
  explanation of, 469
  of rectangular solid, 469
Survey problems
  contradictions in, 71–72
  examples of, 69–71
  explanation of, 69
  Venn diagrams and, 68–69
Syllogisms
  disjunctive, 118
  explanation of, 123
  invalid, 123, 125–127
  valid, 123–124, 127–128
Symbolic logic, 86
Symmetry
  explanation of, 488, 492
  measurement of, 494
  method to find, 493
Systematic counting, 606–607
Système International d'Unités
    (SI system), 477
Systems of inequalities
  explanation of, 358
  method to solve, 358–359
  modeling with, 360
Systems of linear equations
  dependent, 352
  elimination method to solve, 350–352
  explanation of, 348–349
  inconsistent, 351
  modeling with, 352–355
  solutions to, 349–350

Tarski, Alfred, 124
Tautology, 99
Terminal side, of angle, 438
Terms, of sequence, 286
Tessellations
  explanation of, 494
  nonregular, 495–496
  regular, 494
  with regular polygons, 495
Theoretical information, 648
Thornhill, Randy, 493
Three-way principle, 13, 92, 239, 255, 277,
    607, 745
Tractrix, 221, 443
Translation, 491
Translation vector, 491
Tranversal, 440
Trapezoids
  area of, 459–460
  explanation of, 448
Traveling Salesperson Problem (TSP)
  best edge algorithm and, 164–165
  brute force and, 162–163
  explanation of, 20, 157–159
  Hamilton circuits and, 158–161
  Hamilton paths and, 158–159
  nearest neighbor algorithm and, 163–164
Tree diagrams
  explanation of, 4, 10, 608
  to illustration counting problems, 608–610
Trial and error method, 225
Triangles
  area of, 457–459
  classification of, 448
  sum of angles of, 449
Truth tables. *See also* Logic
  alternative method to construct, 102–103
  for biconditionals, 112
  for complex argument with three
    variables, 119–120
  for compound statements, 97–100
  for conditional statements, 107–108
  conjunctions and, 96–97
  DeMorgan's laws and, 101, 102
  explanation of, 95–96
  for invalid arguments, 117–118
  for logically equivalent statements,
    100–101
  for negation, 95–96
  number of lines in, 99
  for statement with three variables, 99–100
  for valid arguments, 115–117
Truth value
  of conditional, 133
  of conjunction, 132
  of conjunctions, 96
  of disjunction, 132
  of fuzzy logic statements, 131
  of negations, 95–96, 131–132
Tukey, John, 708

Unfair game, 684
Union
  of events, 660–662
  of sets, 58–59, 62–64
Union formula, 661
Unit fractions, 480–481
Universal quantifiers, 90
Universal sets, 46
Unpaid balance method, 401–402, 404
Upper quota, 536
U.S. Customary system
  converting between metric units
    and, 481–482
  explanation of, 477

Valid arguments, 115–117, 119–120
Valid syllogisms, 123–124, 127–128
Variables, 706
Variance, 730
Variation
  coefficient of, 733–734
  combined, 345
  constant of, 343
  direct, 343–344
  explanation of, 343
  inverse, 343, 344
  joint, 344
Venn diagrams
  conditional probability and, 669
  explanation of, 52
  naming, 68–69
Verify your answer strategy, 391, 565
Vertex
  of angle, 438
  of graphs, 143–145
  of parabolas, 322
  of polygons, 447, 489
Vertical angles, 439
Volume
  of cones and spheres, 472–474
  of cylinders, 470–472
  explanation of, 468–470
  method to compare, 469–470
  units of, 482, 483
Voters
  Banzhaf power index of, 587–590
  critical, 586, 587
  pivotal, 594–595
  winning, 585, 586
Voting methods
  approval, 569
  Arrow's impossibility theorem and,
    579–580
  Banzhaf power index and, 590
  Borda count, 563–565, 573, 580
  Condorcet's criterion for, 573–575
  defects in, 572–580
  independence-of-irrelevant-alternatives
    criterion for, 575–577
  instant runoff, 569
  majority criterion for, 572–573
  monotonicity criterion for, 577–579
  pairwise comparison, 567–569, 576–577
  permutations and, 592–594
  pivotal voters and, 594–595
  plurality, 562–563, 574–576, 580
  plurality-with elimination, 565–567,
    577–580
  Shapley-Shubik index and, 595–597
  weighted, 583–589

Walker, Peter, 451
Webster's apportionment method, 543–544
Weighted voting systems
  background of, 583
  Banzhaf power index and, 587–590
  coalitions and, 585–587
  explanation of, 584–585

Weights, 584
Well-defined sets, 44–46
Whitehead, Alfred North, 101
Winning coalitions, 585, 586
Word problems. *See also* Problem-solving strategies
  guessing to solve, 8–9
  visualizing conditions in, 3–4

x-intercept, 304

Yards, 481
y-intercept, 304

Zero
  as identity element for addition, 271
  origin of, 245, 246
Zero exponent, 279
Z-scores
  comparing data with, 746
  converting raw scores to, 744–745
  explanation of, 741
  table listing, 742